# INTRODUCTORY ELECTRIC CIRCUITS

## Electron Flow Version

# INTRODUCTORY ELECTRIC CIRCUITS

## Electron Flow Version

Robert T. Paynter

*With Technical Contributions by Toby Boydell*

Prentice Hall
Upper Saddle River, New Jersey ■ Columbus, Ohio

**Library of Congress Cataloging-in-Publication Data**
Paynter, Robert T.
Introductory electric circuits: electron flow version/Robert T. Paynter.
p. cm.
Includes index.
ISBN 0-02-392500-0 (alk. paper)
1. Electric circuits. I. Title.
TK454.P29 1999 98-7299
621.319′2—dc21 CIP

Cover Photo: Copyright Orion/FPG International
Editor: Linda Ludewig
Developmental Editor: Carol Robison
Production Editor: Rex Davidson
Design Coordinator: Karrie M. Converse
Text Designer: Anne D. Flanagan
Cover Designer: Ceri Fitzgerald
Production Manager: Patricia A. Tonneman
Illustrations: The Clarinda Company
Marketing Manager: Ben Leonard

This book was set in Times Roman and Helvetica by The Clarinda Company and was printed and bound by R. R. Donnelley & Sons Company. The cover was printed by Phoenix Color Corp.

Photo credits: Anne Murphy

Printed in the United States of America

10 9 8 7 6 5 4 3 2 1

ISBN: 0-02-392500-0

Prentice-Hall International (UK) Limited, *London*
Prentice-Hall of Australia Pty. Limited, *Sydney*
Prentice-Hall of Canada, Inc., *Toronto*
Prentice-Hall Hisanoamericana, S. A., *Mexico*
Prentice-Hall of India Private Limited, *New Delhi*
Prentice-Hall of Japan, Inc., *Tokyo*
Simon & Schuster Asia Pte. Ltd., *Singapore*
Editora Prentice-Hall do Brasil, Ltda., *Rio de Janeiro*

# Preface: To the Instructor

When I wrote the second edition of *Introductory Electronic Devices and Circuits* (1990), I incorporated a series of tools for learning that no one had used, to my knowledge, prior to that time. These tools for learning included **objective identifiers** that tie the text material to the chapter objectives, **summary illustrations,** and **marginal notes** designed to supplement the body of the text. In the fourth edition of the devices text (1997), the list of learning aids was expanded to include **highlighted lab references** that tie the text material to specific lab exercises. The response to that text has been overwhelming. As a result, it has served as a design model for *Introductory Electric Circuits.*

As always, my goal has been to produce an introductory textbook that students can really use in their studies. For this reason, the most useful learning aids from the *Devices* textbook have been incorporated into *Introductory Electric Circuits* (see pages vi and vii):

*Note:* The learning aids listed are identified (by number) on the following pages.

① **Performance-based objectives** provide a handy overview of the chapter organization and a road map to student learning.

② **Objective identifiers** in the margins cross-reference the objectives with the chapter material. These identifiers help the students to locate quickly the material that enables them to fulfill an objective.

③ **Marginal notes** include a running glossary of new terms, notes that highlight the differences between theory and practice, and brief reminders of principles covered in earlier sections and chapters.

④ **In-chapter practice problems** are included in the examples to provide your students with an immediate opportunity to apply the principles and procedures demonstrated in the examples.

⑤ **Summary illustrations** provide a convenient summary of circuit operating principles and applications. Many provide comparisons between related circuits.

The following items have also been incorporated into *Introductory Electric Circuits* to help reinforce student learning:

- **Section review questions:** Each section of the text ends with a series of questions that students can use to check their learning.
- **Chapter summaries:** Each chapter ends with a detailed summary of the major points covered in the chapter, including an **equation summary** and a **key terms list.**
- **Practice problems:** An extensive set of practice problems appears at the end of each chapter. In addition to standard practice problems, most of the problem sets include **troubleshooting practice problems** and **the Brain Drain** (challengers).

## FEATURES

In addition to the learning aids listed above, *Introductory Electric Circuits* contains several features designed to help your students learn the principles covered in the textbook (and more). Some of those features are described on the following pages.

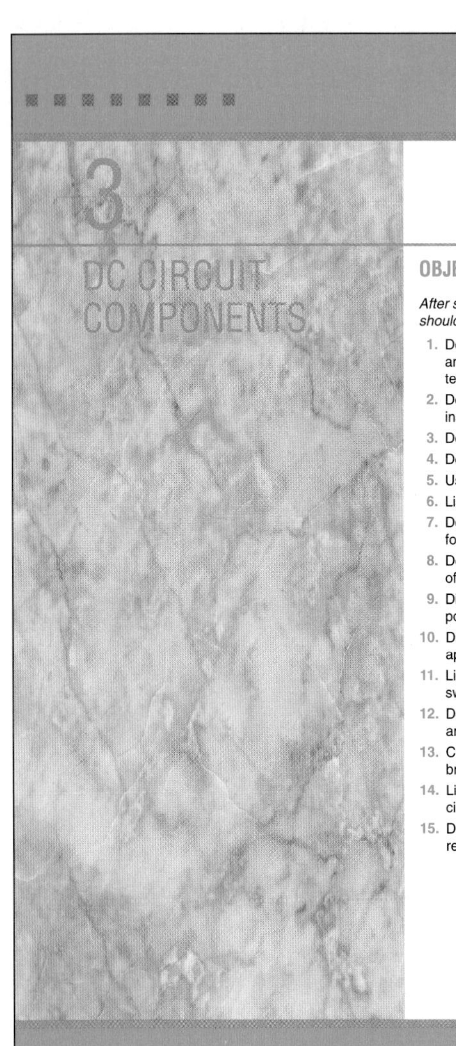

## OBJECTIVES

*After studying the material in this chapter, you should be able to:*

1. Describe the commonly used types of wire and the American Wire Gauge (AWG) system of wire sizing.
2. Describe the ratings used for conductors and insulators.
3. Describe the various types of resistors.
4. Describe the various resistor ratings.
5. Use the standard resistor color code.
6. List the guidelines for replacing resistors.
7. Describe the operation of, and applications for, commonly used potentiometers.
8. Describe the construction and characteristics of commonly used batteries.
9. Discuss the operation and use of a basic dc power supply.
10. Discuss the concept of voltage polarity as it applies to dc power supplies.
11. List and describe the various types of switches.
12. Describe the construction, operation, ratings, and replacement procedures for fuses.
13. Compare and contrast fuses with circuit breakers.
14. List and disc circuits on b
15. Describe the rent, voltage

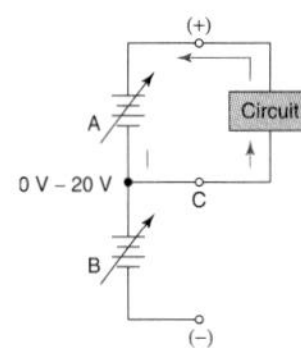


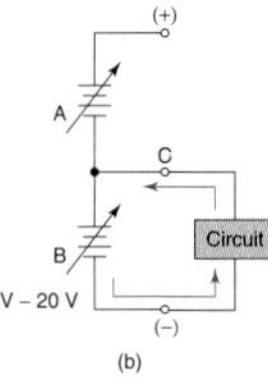


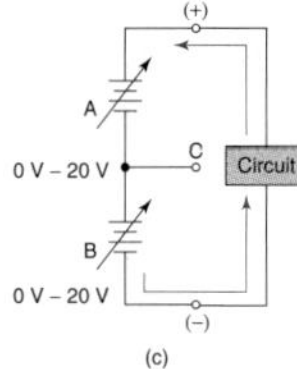


**FIGURE 3.35 Power supply connections.**

### *Positive Voltage Versus Negative Voltage*

*OBJECTIVE 10* ▶

Voltages are normally identified as having a specific polarity; that is, as being either positive or negative. For example, consider the positive and negative terminals of the voltage source shown in Figure 3.36. We could describe this source in either of two ways. We could say that:

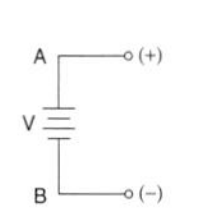


**FIGURE 3.36**

1. Side A is positive with respect to side B, meaning that side A of the source is more positive than side B.
2. Side B is negative with respect to side A, meaning that side B of the source is more negative than side A.

While both of these statements are correct, they do not assign a specific polarity to the source; that is, we have not agreed on whether it is a positive voltage source or a negative voltage source.

Assigning voltage polarities.

To assign a polarity to any voltage, we must agree on a reference point. For example, if we agree that side A of the voltage source in Figure 3.36 is the reference point, then we would say that the output from the source is a negative dc voltage because side B is more negative than our reference point. At the same time, if we agree that side B of the source is our reference point, then we would say that the output from the source is a positive dc voltage because side A of the source is more positive than our reference point.

**Common terminal**
The terminal of a dc power supply that is common to two (or more) voltage sources and thus is used as the reference point.

The **common terminal** of the dc power supply in Figure 3.33 is common to both voltage sources and therefore is used as the reference point; that is, the output polarity of either source is described in terms of its relationship to this point. For example, refer back to Figure 3.34. The common is connected to the negative terminal of source A. Since the top side of A is positive with respect to the common (reference) point, we say that the output from A is a positive dc voltage. By the same token, the bottom side of source B is negative with respect to the common, so we say that the output from source B is a negative dc voltage. In each case, *the output voltage polarity describes its relationship to the common terminal*. This is why the control for source A is labeled *+ Volts* and the control for source B is labeled *− Volts*. In both cases, the label indicates the output polarity with respect to the common terminal.

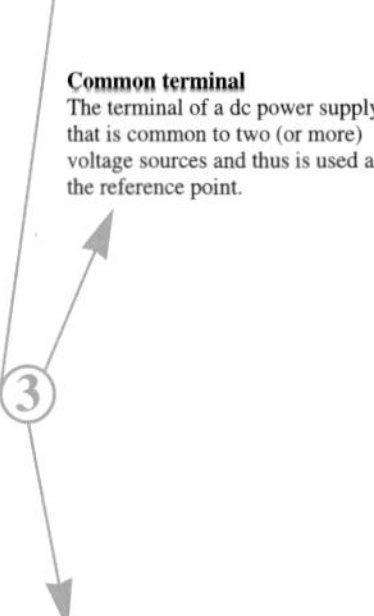


### *Using a dc Power Supply*

When you are in the process of constructing a circuit, the circuit is built and inspected for errors before the dc power supply is turned on. Once you are sure that you have built the circuit correctly:

Connecting a dc power supply to a properly constructed circuit.

1. Adjust the voltage controls on the dc power supply to their 0 $V_{dc}$ settings.
2. Connect the dc power supply to the circuit.

82 CHAPTER 3 / DC CIRCUIT COMPONENTS

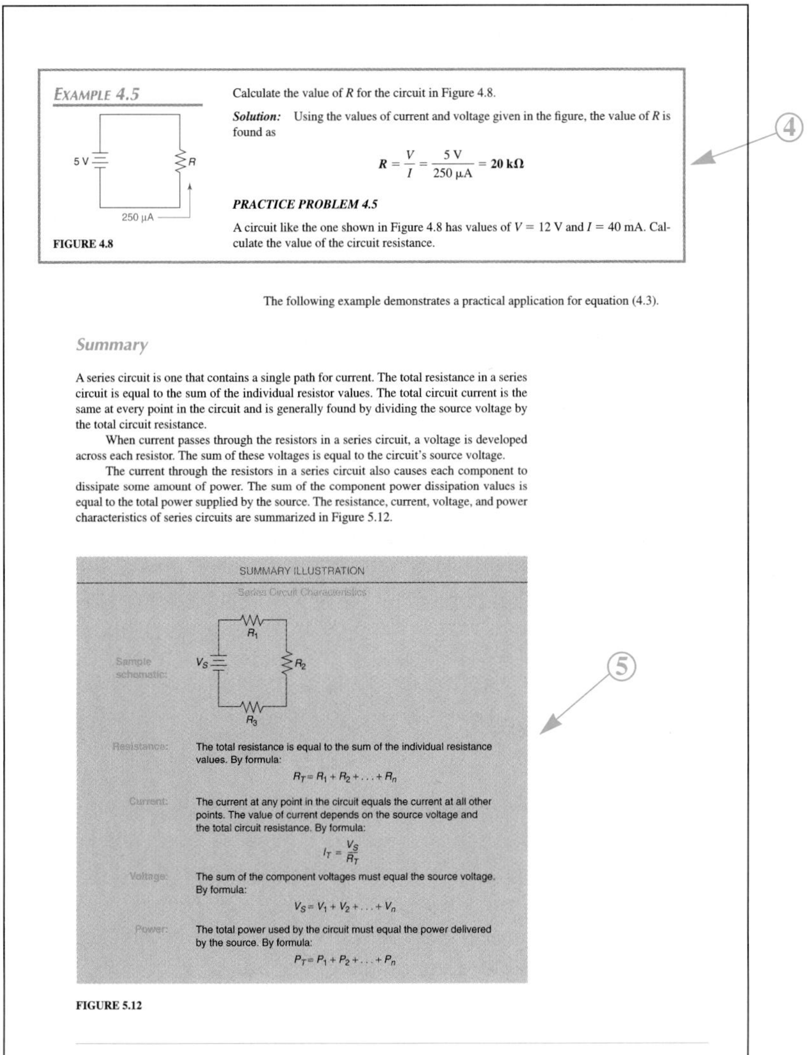

*EXAMPLE 4.5*

Calculate the value of $R$ for the circuit in Figure 4.8.

***Solution:*** Using the values of current and voltage given in the figure, the value of $R$ is found as

$$R = \frac{V}{I} = \frac{5\text{ V}}{250\ \mu\text{A}} = 20\text{ k}\Omega$$

***PRACTICE PROBLEM 4.5***

A circuit like the one shown in Figure 4.8 has values of $V = 12$ V and $I = 40$ mA. Calculate the value of the circuit resistance.



**FIGURE 4.8**

The following example demonstrates a practical application for equation (4.3).

*Summary*

A series circuit is one that contains a single path for current. The total resistance in a series circuit is equal to the sum of the individual resistor values. The total circuit current is the same at every point in the circuit and is generally found by dividing the source voltage by the total circuit resistance.

When current passes through the resistors in a series circuit, a voltage is developed across each resistor. The sum of these voltages is equal to the circuit's source voltage.

The current through the resistors in a series circuit also causes each component to dissipate some amount of power. The sum of the component power dissipation values is equal to the total power supplied by the source. The resistance, current, voltage, and power characteristics of series circuits are summarized in Figure 5.12.

SUMMARY ILLUSTRATION

Series Circuit Characteristics

Sample schematic:



Resistance: The total resistance is equal to the sum of the individual resistance values. By formula:

$$R_T = R_1 + R_2 + \ldots + R_n$$

Current: The current at any point in the circuit equals the current at all other points. The value of current depends on the source voltage and the total circuit resistance. By formula:

$$I_T = \frac{V_S}{R_T}$$

Voltage: The sum of the component voltages must equal the source voltage. By formula:

$$V_S = V_1 + V_2 + \ldots + V_n$$

Power: The total power used by the circuit must equal the power delivered by the source. By formula:

$$P_T = P_1 + P_2 + \ldots + P_n$$

**FIGURE 5.12**

## *Electronics Workbench® (EWB) Applications Problems*

In response to reviewer input, **Electronics Workbench® (EWB)** software has been fully integrated into this text. This simulation software has been incorporated so that instructors can choose (on an individual basis) whether or not to include it in their curricula. The EWB CD-ROM packaged with the text contains exercises that were developed and written by **George Shaiffer** (Pike's Peak Community College, Colorado Springs, CO). These exercises relate directly to various figures throughout the text. An EWB logo is used to identify those figures with an EWB file. A list of **EWB Applications Problems** at the end of each chapter provides the file numbers for the appropriate figures. (The directions for accessing the individual files are included with the CD.) In addition, the CD-ROM contains a tutorial that instructs students how to operate EWB and how to simulate circuits. The CD-ROM also includes a locked version of Electronics Workbench® Student Version 5.0 that can be unlocked by calling Interactive Image Technologies. Instructions for unlocking the software are included on the CD-ROM.

*A personal thought:* The use of simulation software in the classroom is still a matter of great debate. Many instructors see it as an invaluable learning tool, while others believe that its use should be limited to solving minor design problems encountered by more advanced students. (I count myself among the latter.) I believe that the method used to incorporate EWB into the textbook and lab manual will make it valuable for those who wish to use it while keeping it unobtrusive for those who don't.

### Functional Use of Color

In my *Devices* textbook, I opted for a *functional* use of color. The same approach to a full-color format has been used in this text. Color is used to help your students distinguish between the parts and values that make up the various circuits and graphs. It is also used to draw their attention to certain figures and marginal notes.

The use of color in a textbook can be very helpful or simply distracting. I believe that the goal of limiting the use of color to enhance your students' learning experience has been accomplished in this textbook.

### Text Content

It became clear early that a choice would have to be made regarding the content of this book. I could do my best to cram every conceivable topic into a limited number of pages or to explain thoroughly a more limited range of topics. I chose the latter option. It is my belief that an in-depth approach to a narrower range of topics provides a more solid foundation for growth.

### Lab Manual

The *Laboratory Manual to Accompany Introductory Electric Circuits* was written by **William Muckler,** a colleague (and good friend) of mine. Like the textbook, the lab manual uses EWB to supplement (not replace) the exercises. The EWB exercises for the lab manual were developed and written by **John Reeder** (Merced College, Merced, CA). Like the text, the lab manual EWB exercises are incorporated so that instructors can choose (on an individual basis) whether or not to include them in their curricula.

### Ancillaries

- Laboratory Manual
- Solutions Manual to the Lab Manual
- Instructor's Resource Manual
- Windows PH Custom Test
- PowerPoint Transparencies
- Web Site: www.prenhall.com/paynter

## ACKNOWLEDGMENTS

A project of this size could not have been completed without help from many capable and concerned individuals. First and foremost, I would like to acknowledge the efforts of **Toby Boydell,** Seva Electronics, Ontario, Canada. Toby has played a major role in all my work of recent years, and this text was certainly no exception. He provided quality content, editorial, and technical assistance during every phase of the project. He also provided the solutions for the *Instructor's Resource Manual* for this text, the test item file, and the answers in Appendix G. I would also like to acknowledge the efforts of **Scott Bisland** (Sematech, Inc.) and **Richard Parrett** (ITT Technical Institute, St. Louis) for the quality input they provided from start to finish.

A special thanks goes out to the following professionals for their reviews of selected chapters:

**Byron Paul** (Bismark State College)

**Phillip D. Anderson** (Muskegon Community College)

**Jeff Bigelow** (Oklahoma Christian University)

**Keith Bunting** (Randolph Community College)

**Thomas Elliott** (The Victoria College)

**David Longobardi** (Antelope Valley College)

**J. W. Roberts** (West Georgia Tech)

**Neal Willison** (Oklahoma State University, Oklahoma City)

**Larry J. Wheeler** (Senior Engineer, Lockheed Martin Aerospace)

**L'Houcine Zerrouki** (ITT Technical Institute, Seattle)

**Ted Rodriguez** (Skagit Valley College)

**Mauro Caputi** (Hofstra University)

**Frank Chandler** (Randolph Community College)

**John Reeder** (Merced College)

**David Juhola** (Applications Engineer, Canon, USA)

**Peter Kerckhoff** (DeVry, Kansas City)

**John Dyckman** (Pennsylvania Institute of Technology)

**Fred Evangelisti** (American River College)

**Dan Landiss** (St. Louis Community College)

**James A. Lookadoo** (Pittsburgh State University)

**Saeed Shaikh** (Miami Dade Community College)

**Michael Howard** (Ivy Tech State College, Evansville)

The photographs that appear in the textbook and lab manual were produced by **Anne Murphy** (Anne Murphy Studio, Webster Groves, MO). The materials used in the photos were supplied by St. Louis Community College at Forest Park and **Tom Gowan** (Yorkshire TV and VCR Repair, Webster Groves, MO).

I would like to thank the staff at Prentice-Hall and others for their "behind-the-scenes" efforts on this textbook. The following people deserve special recognition:

**Carol Robison,** my developmental editor, for overseeing the review process and the initial stages of production. Although she has since moved on, Carol played a major role in the development of this text. (Good luck, Carol!)

**Marianne L'Abbate,** my copy editor, who has provided (once again) the highest quality copy editing possible.

**Rex Davidson,** my production editor, for helping me throughout the production process and helping to keep things on track.

**Linda Ludewig,** editor, for having enough faith in the project to go out on many limbs, providing for the use of a four-color format in a first edition text.

and

**Dave Garza,** editor-in-chief, for believing in the project and providing the support needed to make it a reality.

Finally, a special thanks goes out to my family and friends (especially Wayne Newcomb, Dick Arnoldy, and Rich Reeves) for their constant support and patience.

*Bob Paynter*

# Preface: To the Student

## "WHY AM I LEARNING THIS?"

Have you ever found yourself asking this question? If you have, then take a moment to read this preface.

I have always believed that any subject is easier to learn if you understand *why* it is being taught and how it relates to your long-term goals. For this reason, we're going to take a moment to discuss:

1. How the material in this book relates to a career in electronics technology
2. How you can get the most out of this course

Few developments have affected our lives over the past thirty years as profoundly as those made in electronics technology. Most of the electronic "gizmos" we take for granted, such as cellular phones, laptop computers, home theaters, pagers, and personal audio systems, have been developed during that time. These items, and many others, have been made possible by advances in production technology. As a result of these advances, many electronic systems that once filled an entire room can now be held in the palm of your hand. Even so, these systems are extremely complex devices that contain a wide variety of components. And, each of these components operates according to one or more fundamental principles. These components and their operating principles are the subject of this book.

Learning how to work on various electronic systems begins with learning the components and principles that are common to all of them. While these principles may not always have a direct bearing on *how* to repair a specific electronic system, they must be learned if you are to understand *why* things work the way they do. Learning why things work the way they do allows you to grow beyond the scope of any textbook (or course).

The material in this book forms the foundation for the courses that are to follow. This means that learning the material is critical if your knowledge is to advance beyond the point where it is now. The next question is:

## "HOW CAN I GET THE MOST OUT OF THIS COURSE?"

There are several steps you can take to ensure that you will successfully complete this course and advance to the next. The first is to accept the fact that *learning electronics requires active participation on your part.* If you are going to learn the material in this book, you must take an active role in your education. It's like learning to play a musical instrument. If you want to learn how to play a musical instrument, you have to practice on a regular basis. You can't learn how to play a musical instrument simply by "reading the book." The same can be said about learning electronics. You must be actively involved in the learning process.

How do you get involved in the learning process? Here are some habits that will take you a long way toward successfully completing your course of study:

1. *Attend class on a regular basis.*
2. *Take part in classroom problem-solving sessions.* This means getting out your calculator and solving the problems along with the rest of the class.
3. *Do all the assigned homework.* Circuit analysis is a skill. As with any skill, you gain competency only through practice.
4. *Take part in classroom discussions.* More often than not, classroom discussions can serve to clarify points that may be confusing otherwise.
5. *Read the material before it is discussed in class.* When you know what is going to be discussed in class, read the related material *before* the discussion. That way, you'll know which parts (if any) are causing you problems before the class begins.
6. *Become an active reader.*

Being an *active reader* means that you must do more than simply "read the book." When you are studying new material, there are several things that you should do:

1. *Learn the terminology.* You are taught new terms because you need to know what they mean and how and when to use them. When you come across a new term in the book, take time to commit that term to memory. How do you know when a new term is being introduced? Throughout this text, new terms are identified in the margins. When you see a new term and its definition in the margin, stop and learn the term before going on to the next section.
2. *Use your calculator to work through the examples.* When you come across an example, get your calculator out and try the example for yourself. When you do this, you develop the skill necessary to solve the problems on your own.
3. *Solve the example practice problems.* Most of the examples in this book end with a practice problem that is identical in nature to the example. When you see these problems, solve them. Then you can check your answer(s) by looking them up at the end of the chapter.
4. *Use the chapter objectives to measure your learning.* Each chapter begins with an extensive list of performance-based objectives. *These objectives tell you what you should be able to do as a result of learning the material.*

Throughout this text, *objective identifiers* are included in the margins. For example, if you look on page 8, you'll see "Objective 4" printed in the margin. This identifier tells you that this is the point where you are taught the skill mentioned in Objective 4 on the opening page of the chapter. These identifiers can be used to help you with your studies. If you don't know how to perform the action called for in a specific objective, just flip through the chapter until you see the appropriate identifier. At that point in the chapter, you'll find the information you need to successfully meet the objective.

## ONE FINAL NOTE

A lot of work is involved in being an active learner. However, the extra effort will pay off in the end. Your understanding of electronics will be better as a result of your efforts. I wish you the best of success.

*Bob Paynter*

# Contents


## 1 INTRODUCTION 1

1.1 Electronics as a Career 2
1.2 Electronic Systems 4
1.3 Electronic Components and Units 7
1.4 Scientific Notation 14
1.5 Engineering Notation 21
Chapter Summary 25
Key Terms/Practice Problems/Answers to the Example Practice Problems/EWB Applications Problems

## 2 PRINCIPLES OF ELECTRICITY 29

2.1 The Starting Point: Elements, Atoms, and Charge 30
2.2 Current 37
2.3 Voltage 41
2.4 Resistance and Conductance 43
2.5 Conductors, Insulators,and Semiconductors 45
Chapter Summary 50
Equation Summary/Key Terms/Practice Problems/The Brain Drain/Answers to the Example Practice Problems/EWB Applications Problems

## 3 DC CIRCUIT COMPONENTS 53

3.1 Conductors and Insulators 54
3.2 Resistors 58
3.3 Potentiometers 69

3.4 Batteries 75
3.5 dc Power Supplies 80
3.6 Switches and Circuit Protectors 83
3.7 Circuit Construction 90
3.8 Measuring Current, Voltage, and Resistance 97
Chapter Summary 100
Equation Summary/Key Terms/Practice Problems/The Brain Drain/Answers to the Example Practice Problems/EWB Applications Problems

4
CIRCUIT FUNDAMENTALS 105
4.1 Ohm's Law 106
4.2 Power 113
4.3 Miscellaneous Topics 119
Chapter Summary 125
Equation Summary/Key Terms/Practice Problems/The Brain Drain/Answers to the Example Practice Problems/EWB Applications Problems

5
SERIES CIRCUITS 132
5.1 Series Circuit Characteristics 133
5.2 Voltage Relationships: Kirchhoff's Voltage Law, Voltage References, and the Voltage Divider 140
5.3 Series Circuit Analysis, Fault Symptoms, and Troubleshooting 146
5.4 Related Topics 155
Chapter Summary 165
Equation Summary/Key Terms/Practice Problems/Troubleshooting Practice Problems/The Brain Drain/Answers to the Example Practice Problems/EWB Applications Problems

6
PARALLEL CIRCUITS 178
6.1 Parallel Circuit Characteristics 179
6.2 Current Relationships: Kirchhoff's Current Law, Current Sources, and the Current Divider 187

6.3 Parallel Circuit Analysis, Fault Symptoms, and Troubleshooting 193
Chapter Summary 199
Equation Summary/Key Terms/Practice Problems/Troubleshooting Practice Problems/The Brain Drain/Answers to the Example Practice Problems/EWB Applications Problems

# 7
## SERIES-PARALLEL CIRCUITS 207

7.1 An Introduction to Series-Parallel Circuits 208
7.2 Analyzing Series-Parallel Circuits 216
7.3 Circuit Loading 224
7.4 The Wheatstone Bridge 233
7.5 Variable Voltage Dividers, Fault Symptoms, and Troubleshooting 238
Chapter Summary 245
Equation Summary/Key Terms/Practice Problems/Troubleshooting Practice Problems/The Brain Drain/Answers to the Example Practice Problems/Answers to the Section 7.1 Review Problems/EWB Applications Problems

# 8
## CIRCUIT ANALYSIS: THEOREMS AND CONVERSIONS 255

8.1 Superposition 256
8.2 Voltage and Current Sources 263
8.3 Thevenin's Theorem 268
8.4 Applications of Thevenin's Theorem 279
8.5 Norton's Theorem 289
8.6 Other Network Theorems and Conversions 297
Chapter Summary 305
Equation Summary/Key Terms/Practice Problems/The Brain Drain/Answers to the Example Practice Problems/EWB Applications Problems

# 9
## COMPLEX ANALYSIS TECHNIQUES 317

9.1 Solving Simultaneous Equations 318
9.2 Branch Current Analysis 330

9.3 Mesh Current Analysis 336
9.4 Node Voltage Analysis 343
Chapter Summary 346
Key Terms/Practice Problems/The Brain Drain/Answers to the Example Practice Problems/EWB Applications Problems

## 10 AN INTRODUCTION TO MAGNETISM 352

10.1 Magnetism: An Overview 353
10.2 Magnetic Characteristics of Materials 358
10.3 Electromagnetism 363
10.4 Related Topics 370
Chapter Summary 373
Equation Summary/Key Terms/Practice Problems/Answers to the Example Practice Problems

## 11 AC CHARACTERISTICS 378

11.1 Alternating Current (ac): Overview and Time Measurements 373
11.2 Magnitude Values and Measurements 389
11.3 Sine Waves: Phase Measurements and Instantaneous Values 404
11.4 Related Topics 423
Chapter Summary 435
Equation Summary/Key Terms/Practice Problems/The Brain Drain/Answers to the Example Practice Problems/EWB Applications Problems

## 12 POLAR AND RECTANGULAR NOTATION 453

12.1 Vectors 454
12.2 Polar Notation 456
12.3 Rectangular Notation 459
12.4 Vector Notation Conversions 468
Chapter Summary 474
Equation Summary/Key Terms/Practice Problems/The Brain Drain/Answers to the Example Practice Problems

## 13 INDUCTORS 480

13.1 Introduction to Inductance 481
13.2 The Phase Relationship Between Inductor Current and Voltage 489
13.3 Mutual Inductance 491
13.4 Inductive Reactance ($X_L$) 501
13.5 Related Topics 506
Chapter Summary 513
Equation Summary/Key Terms/Practice Problems/The Brain Drain/Open to Discussion/Answers to the Example Practice Problems/EWB Applications Problems

## 14 RESISTIVE-INDUCTIVE (*RL*) CIRCUITS 522

14.1 Series *RL* Circuits 523
14.2 Power Characteristics and Calculations 539
14.3 Parallel *RL* Circuits 546
14.4 Series-Parallel Circuit Analysis 560
14.5 Square Wave Response: *RL* Time Constants 567
Chapter Summary 584
Equation Summary/Key Terms/Practice Problems/The Brain Drain/Open for Discussion/Answers to the Example Practice Problems/EWB Applications Problems

## 15 TRANSFORMERS 596

15.1 Transformers: An Overview 597
15.2 Transformer Voltage, Current, and Power 601
15.3 Load Effects 610
15.4 Related Topics 617
Chapter Summary 625
Equation Summary/Key Terms/Practice Problems/Troubleshooting Practice Problems/The Brain Drain/Answers to the Example Practice Problems/EWB Applications Problems

## 16 CAPACITORS 632

16.1 Capacitors and Capacitance: An Overview 633
16.2 Alternating Voltage and Current Characteristics 638
16.3 Series and Parallel Capacitors 643
16.4 Capacitive Reactance ($X_C$) 649
16.5 Related Topics 653
Chapter Summary 663
Equation Summary/Key Terms/Practice Problems/The Brain Drain/Answers to the Example Practice Problems/EWB Applications Problems

## 17 RESISTIVE-CAPACITIVE (*RC*) CIRCUITS 671

17.1 Series *RC* Circuits 672
17.2 Power Characteristics and Calculations 687
17.3 Parallel *RC* Circuits 693
17.4 Series-Parallel *RC* Circuit Analysis 703
17.5 Square Wave Response: *RC* Time Constants 710
Chapter Summary 723
Equation Summary/Key Terms/Practice Problems/The Brain Drain/Answers to the Example Practice Problems/EWB Applications Problems

## 18 *RLC* CIRCUITS 734

18.1 Series *LC* Circuits 735
18.2 Parallel *LC* Circuits 742
18.3 Resonance 749
18.4 Series and Parallel *RLC* Circuits 758
18.5 Series-Parallel *RLC* Circuit Analysis 771
Chapter Summary 781
Equation Summary/Key Terms/Practice Problems/Answers to the Example Practice Problems/EWB Applications Problems

## 19 FREQUENCY RESPONSE AND PASSIVE FILTERS 789

19.1 Frequency Response: Curves and Measurements 790
19.2 Amplitude Measurements: dB Power Gain 806
19.3 Amplitude Measurements: dB Voltage and Current Gain 814
19.4 Low-Pass Filters 818
19.5 High-Pass Filters 835
19.6 Bandpass and Notch Filters 842
Chapter Summary 859
Equation Summary/Key Terms/Practice Problems/Answers to the Example Practice Problems/EWB Applications Problems

Appendix A Conversions and Units 874
Appendix B Component Standard Values and Color Codes 879
Appendix C Common Schematic Symbols 882
Appendix D Glossary 885
Appendix E Selected Equation Derivations 893
Appendix F Admittance Analysis of Parallel *RL* and *RC* Circuits 903
Appendix G Answers to Selected Odd-Numbered Problems 909
Index 911

1

# INTRODUCTION

## OBJECTIVES

*After studying the material in this chapter, you should be able to:*

1. Compare and contrast the duties and working environments of electronics engineers, electronics technicians, and electronics engineering technicians.
2. Describe the purposes served by communications, computer, industrial, and biomedical systems.
3. Identify and describe the most commonly used pieces of electronics equipment.
4. Describe the function, unit of measure, and schematic symbols for resistors.
5. Describe the function, unit of measure, and schematic symbols for capacitors.
6. Describe the function, unit of measure, and schematic symbols for inductors.
7. Compare and contrast vacuum tubes and semiconductors.
8. Discuss the difference between passive components and active components.
9. Work with values that are in scientific notation.
10. Convert numbers back and forth between standard form and scientific notation, with or without the use of a calculator.
11. Identify the commonly used engineering notation prefixes, along with their symbols and ranges.
12. Work with values written in engineering notation.

The fact that you are reading this book says something about you: that you have chosen a career in electronics technology or a related field. Good choice! The Bureau of Labor Statistics (a division of the U.S. Department of Labor) predicts that the need for highly trained professionals will continue to grow well into the twenty-first century. As a result of industry growth and personnel changes, thousands of new positions continue to open every year. By enrolling in this course, you have taken the first step toward filling one of those positions.

Before we get into the nuts and bolts of electricity and electronics, we will take a brief look at the technical career positions within this field. We will also look at the most common electronic systems, components, and measurements, along with the methods that engineers and technicians commonly use to represent extremely large and extremely small numbers.

## 1.1 ELECTRONICS AS A CAREER

**Electronics**
The field that deals with the design, manufacturing, installation, maintenance, and repair of electronic equipment.

**Electronics** is the field that deals with the design, manufacturing, installation, maintenance, and repair of electronic circuits and systems. Television receivers, audio systems, word processors, radar systems, and industrial control systems are just a few examples of electronic systems.

While a variety of trained professionals (such as salespersons, technical writers, equipment assemblers, and customer service representatives) are employed in the electronics industry, the primary technical duties are performed by electronics engineers, electronics engineering technicians, and electronics technicians. In this section, we will discuss the similarities and differences between engineers and technicians.

*OBJECTIVE 1* ▶

### Electronics Engineers

The *Dictionary of Occupational Titles* (a publication of the U.S. Department of Labor) describes an electronics engineer as someone who:

What an electronics engineer does.

1. Designs electronic components, circuits, and systems.
2. Develops new applications for existing electronics technologies.
3. Assists field technicians in solving unusual problems.
4. Supervises the construction and initial testing of system *prototypes* (initial working models).

As you can see, the electronics engineer is generally involved in the design and manufacturing aspects of electronics. In most cases, engineers are graduates of four-year college or university degree programs. In some cases, they are former electronics engineering technicians who have demonstrated an in-depth understanding of circuit and system design principles.

### Electronics Technicians

The *Dictionary of Occupational Titles* describes an electronics technician as someone who:

What an electronics technician does.

1. Locates and repairs faults in electronic systems and circuits using electronic test equipment.
2. Performs periodic maintenance (aligning, calibrating, etc.) on electronic systems.
3. Maintains records of system faults and maintenance.
4. Installs systems purchased by customers.

In contrast to the engineer, the electronics technician generally deals with the maintenance and repair aspects of electronics. An electronics technician usually works as either an in-house technician or as a field-service representative (FSR). While in-house technicians and FSRs are both involved in maintaining and repairing electronic systems, there are several significant differences between these two positions:

*In-house* and *field-service* positions.

1. The in-house technician generally works at a single location, while the FSR travels between various customer locations.
2. The in-house technician usually has little or no contact with the customers, while the FSR maintains close contact with customers. (It is often said that 90% of the FSR's job is to maintain and repair the *customer.*)
3. The in-house technician generally performs more complex maintenance and testing procedures than the FSR. The FSR usually performs only the maintenance and repair work needed to get the customer's system up and running as soon as possible.

Electronics technicians generally receive their training in the military, trade schools, or colleges with applied science programs.

## Electronics Engineering Technicians

An electronics engineering technician (EET) performs some of the duties of both the engineer and the technician. Basically, an EET is a technician who also:

What an engineering technician does.

1. Builds and tests prototype systems as directed by an engineer.
2. Maintains records on the performance of prototype systems.
3. Recommends circuit and system modifications based on the testing of prototypes.
4. Takes part in developing installation, maintenance, and testing procedures for newly designed systems.

As you can see, the EET is a technician who is involved in the design and manufacturing aspects of electronics. Most companies require a technician to have a significant amount of field experience before being considered for an EET position in a research and development group. However, graduates of accredited EET programs have been known to go straight from school into this type of working environment.

## Summary

Electronics is the field that deals with the design, manufacturing, installation, maintenance, and repair of electronic systems. Electronics engineers are involved primarily in the design and manufacturing aspects of this field, while the *electronics* technician is involved primarily in the maintenance and repair aspects. At a point somewhere between the two is the electronics engineering technician (EET). The EET is a technician who is involved, to an extent, in the design, construction, testing, and modification of system prototypes.

After gaining significant experience, it is not uncommon for a technician to be promoted to an EET position. After gaining additional experience, it is possible (though very difficult) to be promoted to an engineering position.

### Section Review

1. What is electronics?
2. What are the job functions of an electronics engineer?
3. What are the job functions of an electronics technician?
4. Discuss the similarities and differences between in-house technicians and field-service representatives.
5. What are the job functions of an engineering technician?

OBJECTIVE 2 ▸

## 1.2 ELECTRONIC SYSTEMS

*A Practical Consideration:* The classifications listed here are extremely broad. Each can be broken down into several areas of specialization.

In the broadest sense, most common electronic systems can be classified as communications, computer, industrial, or biomedical systems. In this section, we will look at each of these classifications. We will also take a brief look at some common types of test equipment.

### Communications Systems

**Communications systems**
Systems designed to transmit and/or receive information.

**Telecommunications**
The area that deals with the transmission of data between two or more locations.

**Communications systems** are designed to transmit and/or receive information. For example, radio systems transmit audio information from the radio station to your AM or FM receiver. Television systems transmit both audio and video information from the television station to your television.

**Telecommunications** is the area of communications that deals with the transmission of data between two or more locations. The Internet is perhaps the best known example of a telecommunications system. Telecommunications terminals usually communicate with each other via telephone lines, satellite links, or cable (wire or optical).

### Computer Systems

**Computer systems**
Systems designed to store and process information.

**Computer systems** are designed to store and process information. Personal computers, word processors, printers, and mainframes are a few examples of computer systems.

It should be noted that many electronic systems can be classified in more than one category. For example, a facsimile (fax) machine transmits documents from one point to another over the telephone lines. While the fax is a computer system (it processes and stores the information in the document), it is also classified as a telecommunications system. Thus, the line between one type of system and another is not always clearly defined.

### Industrial Systems

**Industrial systems**
Systems designed for use in manufacturing environments.

**Industrial systems** are those designed for use in industrial (manufacturing) environments. Control circuits for industrial welders, robotics systems, power distribution systems, and motor control systems (such as automotive electronic systems) are some examples of industrial systems.

### Biomedical Systems

**Biomedical systems**
Systems designed for use in diagnosing, monitoring, and treating medical problems.

**Biomedical systems** are systems designed for use in diagnosing, monitoring, and treating medical problems. Electrocardiographs, x-ray machines, diagnostic tread mills, and surgical laser systems are all examples of biomedical systems. Since biomedical systems are used in the diagnosis and treatment of medical problems, biomedical technicians are generally required to have some practical knowledge of anatomy and physiology.

### An Important Point

You may be wondering what you will need to learn in school to work on a specific type of system. While certain electronic principles apply more to some types of systems than to others, you will not encounter any of these principles until later in your education. *In this course, you will be taught the principles of electronics that apply equally to all the electronic systems mentioned in this section.* As you study, keep in mind that the material you are learning *does* apply to your area of interest, even though the connection is not always easy to see at first.

## Electronic Test Equipment

◄ *OBJECTIVE 3*

Many electronic system failures are not visible to the naked eye. For this reason, engineers and technicians must use a wide variety of test equipment to measure circuit values, and thus analyze and diagnose circuit failures. Many pieces of electronic test equipment are almost universal; that is, they are used by most engineers and technicians to take the most commonly needed measurements.

The **digital multimeter (DMM),** shown in Figure 1.1, is used to measure the most basic electrical quantities: *voltage, current,* and *resistance.* In addition to measuring these quantities, many DMMs are capable of making a variety of more complex tests and measurements.

The basic DMM is a descendant of the **volt-ohm-milliammeter (VOM).** The VOM (shown in Figure 1.2) is used to measure voltage, current, and resistance, just like the DMM. However, the DMM is preferred over the VOM because it is easier to read and usually provides more accurate readings. At the same time, there are some circuit tests that are easier to perform using a VOM.

The **frequency counter** (shown in Figure 1.3a) is used to measure the number of times that certain events occur every second. For example, the **function generator** (Figure 1.3b) can be used to generate a variety of waveforms. The frequency counter can be used to determine how many waveforms are produced each second by the function generator.

The **oscilloscope** (shown in Figure 1.4) is used to take a variety of voltage and time measurements. The oscilloscope may look a bit intimidating, but it is relatively easy to use with proper training. It is also one of the most versatile pieces of test equipment.

**Digital multimeter (DMM)**
A meter that measures voltage, current, and resistance. Many DMMs are capable of making additional measurements, depending on the model.

**Volt-ohm-milliammeter (VOM)**
The forerunner of the DMM. The VOM uses an analog scale (rather than a digital display) for a readout.

**Frequency counter**
A piece of test equipment that measures the number of times certain events occur every second.

**Oscilloscope**
A piece of test equipment that provides a visual display for a variety of voltage and time measurements.

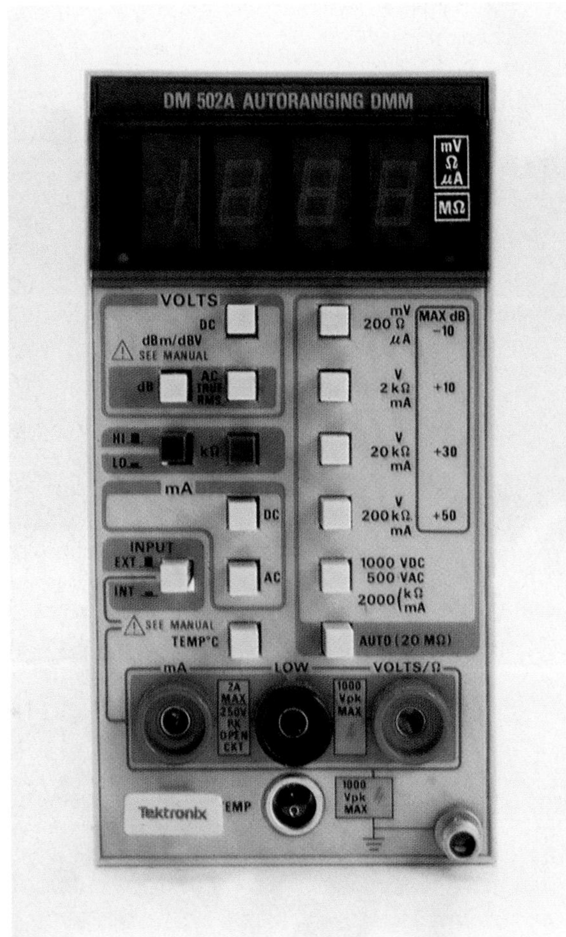

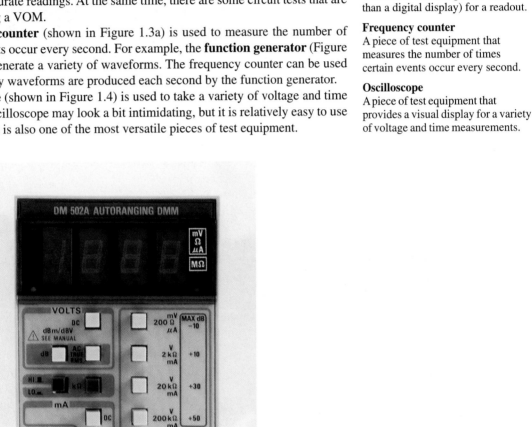

**FIGURE 1.1 A digital multimeter (DMM).**

EWB

FIGURE 1.2 An analog volt-ohm-milliammeter (VOM).

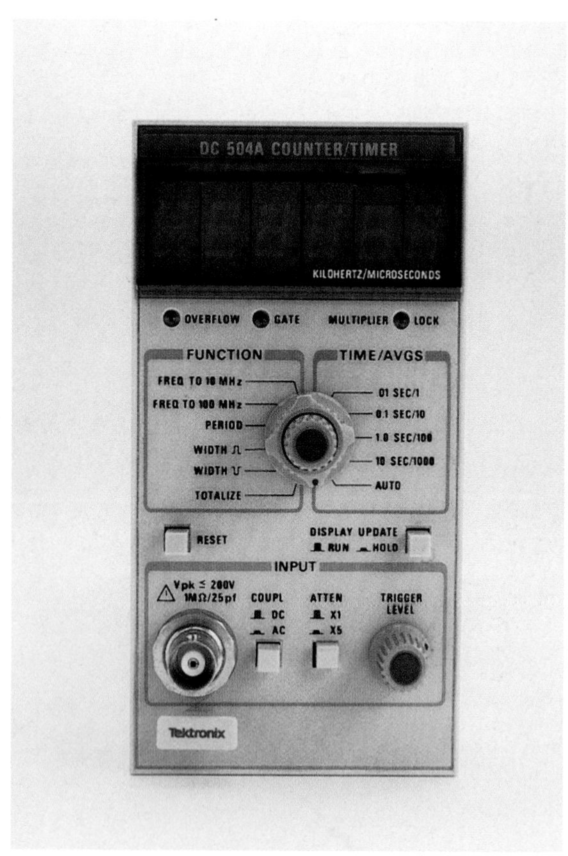


(a)

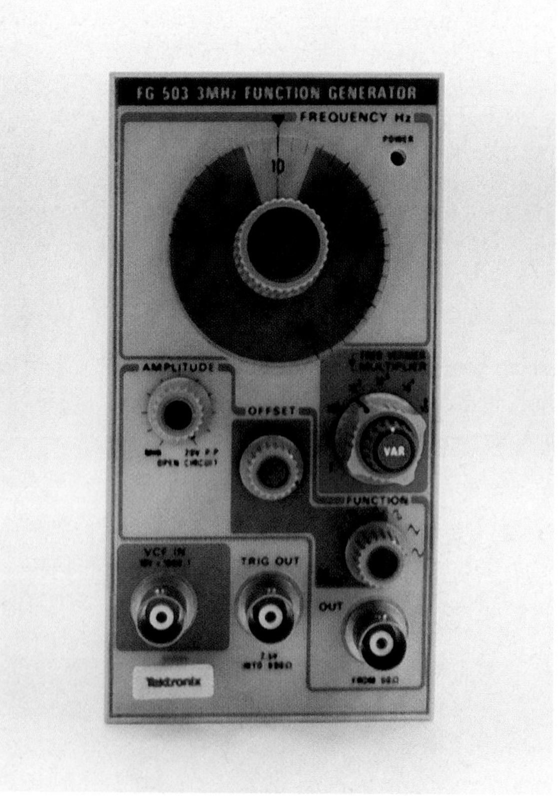


(b)

FIGURE 1.3 Frequency counter and function generator.

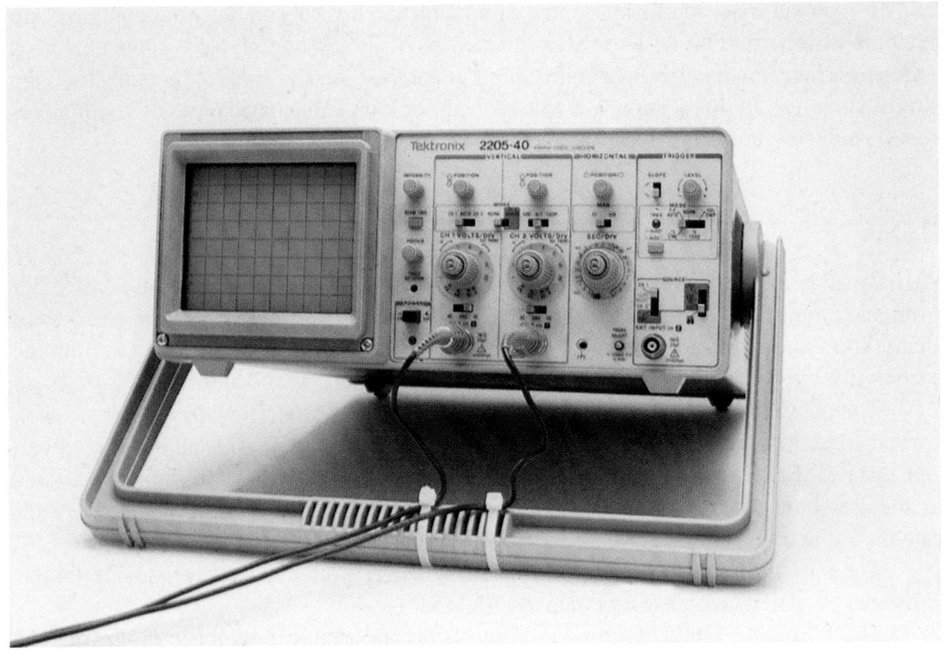


EWB

**FIGURE 1.4 An oscilloscope.**

## An Important Point

When used and maintained properly, any piece of test equipment will provide you with accurate information, even after it has been used for many years. However, if a piece of test equipment is not used and maintained properly, it will provide faulty information and may even be damaged or destroyed. For this reason, you must make sure that you understand *how* to use a given piece of test equipment before attempting to use it. When it comes to using any piece of test equipment, the rule of thumb is simple: *if you aren't sure, ask before acting!* Learn the proper technique, then use your test equipment to take the measurements you need. In the long run, you'll be glad you did.

### Section Review

1. What are communications systems?
2. What are computer systems?
3. What are industrial systems?
4. What are biomedical systems?
5. What is a DMM? What is it generally used to measure?
6. Why are DMMs usually preferred over VOMs?
7. What is a frequency counter?
8. What is an oscilloscope?
9. What precaution should be taken before using any piece of test equipment?

# 1.3 ELECTRONIC COMPONENTS AND UNITS

All electronic systems contain circuits. Loosely defined, a **circuit** is a group of components that performs a specific function. The function performed by a given circuit depends on the components used and how they are connected together.

**Circuit**
A group of components that performs a specific function.

It may surprise you to know that almost all circuits, no matter how complex, are made up of the same basic types of components. While the number of components used and their arrangement varies from one circuit to another, the *types* of components used are almost universal. In this section, we will take a brief look at the basic types of components and the units used to express their values.

*OBJECTIVE 4* ►

## Resistors

**Resistor**
A component used to limit current in a circuit.

We have all heard the term *current* as it applies to a river or stream: It is the flow of water from one point to another. In electronics, the term *current* is used to describe the flow of charge from one point in a circuit to another. The **resistor** is a component used to limit the current in a circuit, just as a valve can be used to limit the flow of water.

The photo in Figure 1.5a shows several types of resistors. The *wire-wound resistor, carbon composition resistor, metal film resistor,* and *integrated resistor* all have fixed values; that is, their values are determined by their physical construction and cannot be altered by the user. Note that the color bands on the carbon composition resistor are used to indicate the value of the component.

**Potentiometer**
A resistor whose value can be varied (within limits) by the user.

**Schematic diagram**
A diagram representing the components and connections within a circuit or system; the electronic equivalent of a blueprint.

The **potentiometer** is a *variable* resistor. It differs from the other resistors in that its value can be varied (within a set range) by the user.

The *schematic symbols* shown in Figure 1.5b are used in **schematic diagrams,** the electronics equivalents of blueprints. The schematic diagram of a given circuit shows you how the components are interconnected. Note that the schematic symbols for resistors do not indicate the composition of the components, only whether they are fixed in value or variable.

Resistance (*R*) is measured in ohms (Ω).

Resistance is represented using the letter *R*. The unit of measure for resistance is the ohm, which is represented using the Greek letter omega (Ω). Thus, if we wanted to describe a 10 ohm resistor in an equation, we would write:

$$R = 10\ \Omega$$

Note that the higher the ohmic value of a resistor, the more it restricts the current in a given circuit.

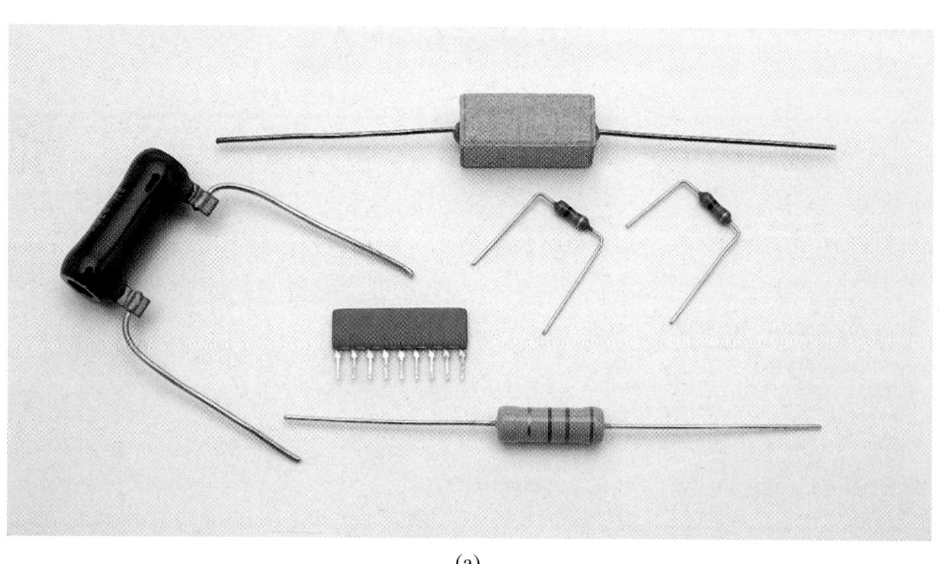

(a)

Fixed:

Variable:

or

(b) Schematic symbols

EWB

**FIGURE 1.5 Resistors.**

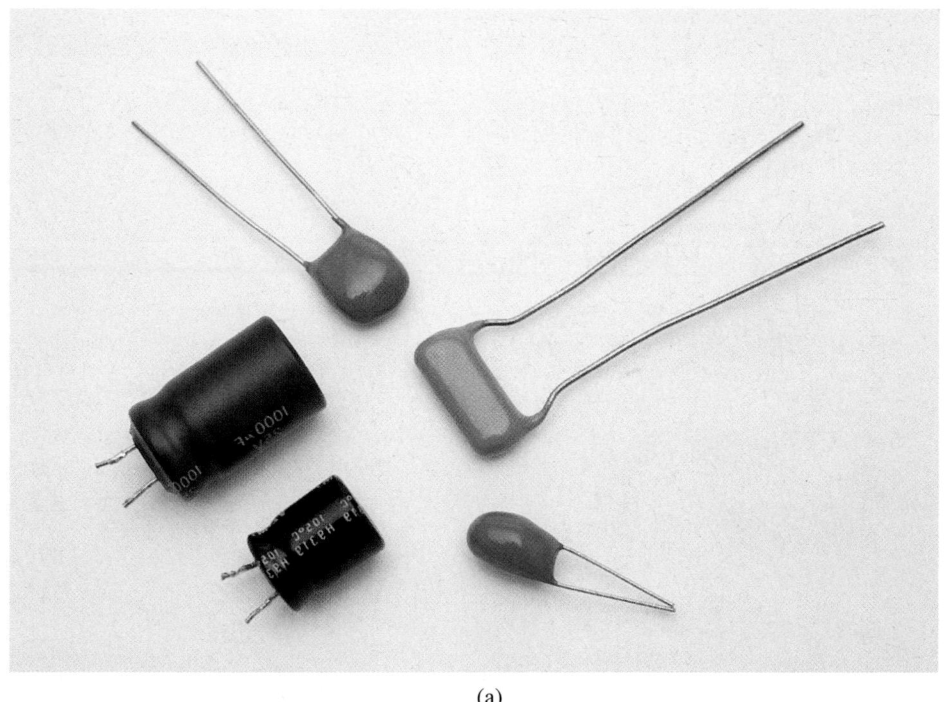

(a)

Fixed:

or

Variable:

(b) Schematic symbols

**FIGURE 1.6 Capacitors.**

## Capacitors

◄ *OBJECTIVE 5*

A **capacitor** is a component that opposes a change in voltage. While the applications for such a component are too complex to discuss at this time, rest assured that capacitors are used for a variety of purposes in all kinds of circuits. Several types of capacitors and the most common capacitor schematic symbols are shown in Figure 1.6.

**Capacitor**
A component that opposes a change in voltage.

Capacitance is represented using the letter *C*. The unit of measure for capacitance is the farad (F). Thus, if we wanted to describe a 2.2 farad capacitor in an equation, we would write:

$$C = 2.2 \text{ F}$$

Capacitance *(C)* is measured in farads (F).

Capacitors are often referred to as *condensers* in older electronics manuals.

## Inductors

◄ *OBJECTIVE 6*

An **inductor** is a component that opposes a change in current. While this seems similar (in purpose) to a capacitor, inductors and capacitors are nearly opposites in terms of their electrical characteristics. Several types of inductors and the basic inductor schematic symbols are shown in Figure 1.7.

**Inductor**
A component that opposes a change in current.

Inductance is represented using the letter *L*. The unit of measure for inductance is the henry (H). Thus, if we wanted to describe a 3.3 henry inductor in an equation, we would write

$$L = 3.3 \text{ H}$$

Inductance *(L)* is measured in henries (H).

It should be noted that inductors are often referred to as *coils* or *chokes*.

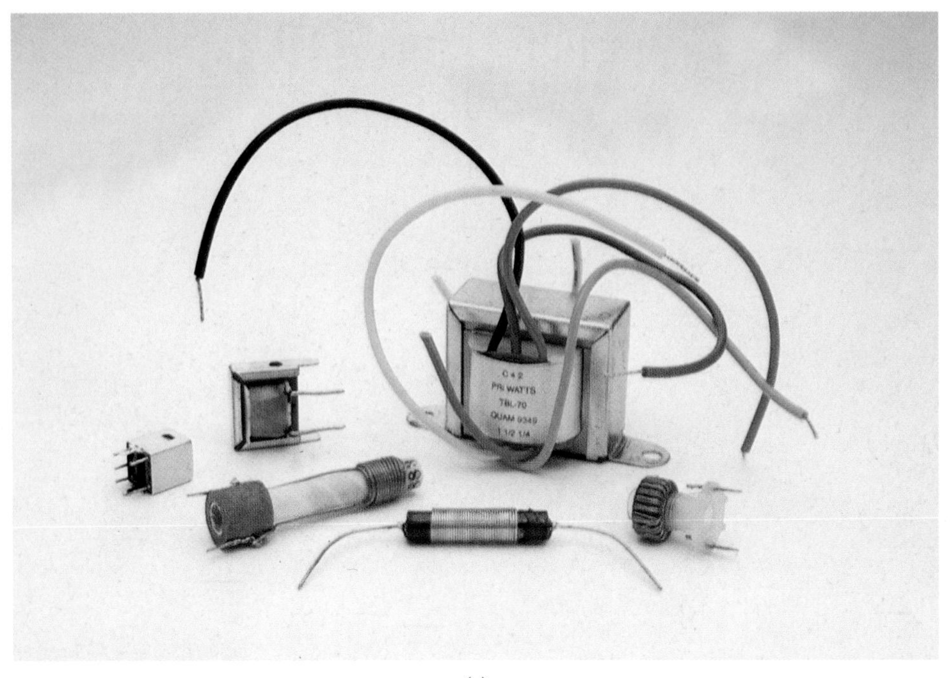

(a)

Fixed:

Variable:

Transformer:

(b) Schematic symbols

**FIGURE 1.7 Inductors.**

*OBJECTIVE 7* ▶

## *Vacuum Tubes*

**Vacuum tube**
A component whose operation is based on the flow of charge through a vacuum.

The term **vacuum tube** is used to describe an entire group of components that share one characteristic: their operation is based on the flow of charge through a vacuum. Several vacuum tubes and vacuum tube schematic symbols are shown in Figure 1.8.

Unlike the operation of resistors, capacitors, and inductors, vacuum tube operation varies from one type of vacuum tube to another. Also, there is no single unit of measure for vacuum tubes. Note that vacuum tubes like those shown in Figure 1.8 are used primarily in high-power systems, such as radio and television transmitters. They are also used extensively in high-fidelity audio applications (such as musical instrument amplifiers).

**Cathode-ray tube (CRT)**
A vacuum tube used as the video display for many televisions, video games, etc.

Another type of vacuum tube used extensively is the **cathode-ray tube (CRT).** The CRT (Figure 1.9) is used as the video display for many televisions, video games, computer terminals, and oscilloscopes.

## *Semiconductors*

**Semiconductors**
A group of components that began to replace vacuum tubes in the early 1950s. Semiconductors are smaller, cheaper, more efficient, and more rugged than their vacuum tube counterparts.

At one time, vacuum tubes were used in almost every type of electronic system. However, a group of components called **semiconductors** started to emerge in the late 1940s. These components are capable of performing most of the functions of vacuum tubes, even though they are smaller, cheaper, far more efficient, and more rugged. Some common semiconductors and their schematic symbols are shown in Figure 1.10.

Because of the advantages that semiconductors offer, they have replaced vacuum tubes in the vast majority of applications. However, semiconductors do not (yet) have the power-handling capabilities of vacuum tubes, which is why vacuum tubes are still used in many high-power applications.

(a)

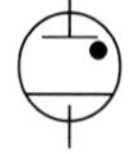

(b) Schematic symbols

**FIGURE 1.8 Vacuum tubes.**

**FIGURE 1.9 A cathode-ray tube (CRT).**

## *Integrated Circuits (ICs)*

**Integrated circuits,** or **ICs,** are semiconductor components that contain entire groups of components or circuits housed in a single package. Several types of ICs are shown in Figure 1.11.

**Integrated circuits (ICs)** Semiconductors that contain groups of components or circuits housed in a single package.

The development of ICs led to the availability of home computers and many of the other consumer electronic systems that we now take for granted. Prior to the 1970s, the cost of producing a computer (as well as the size required) kept computers out of the home market. However, when ICs made it possible to construct entire circuits in a relatively small space, home computers (and other consumer systems) became a reality.

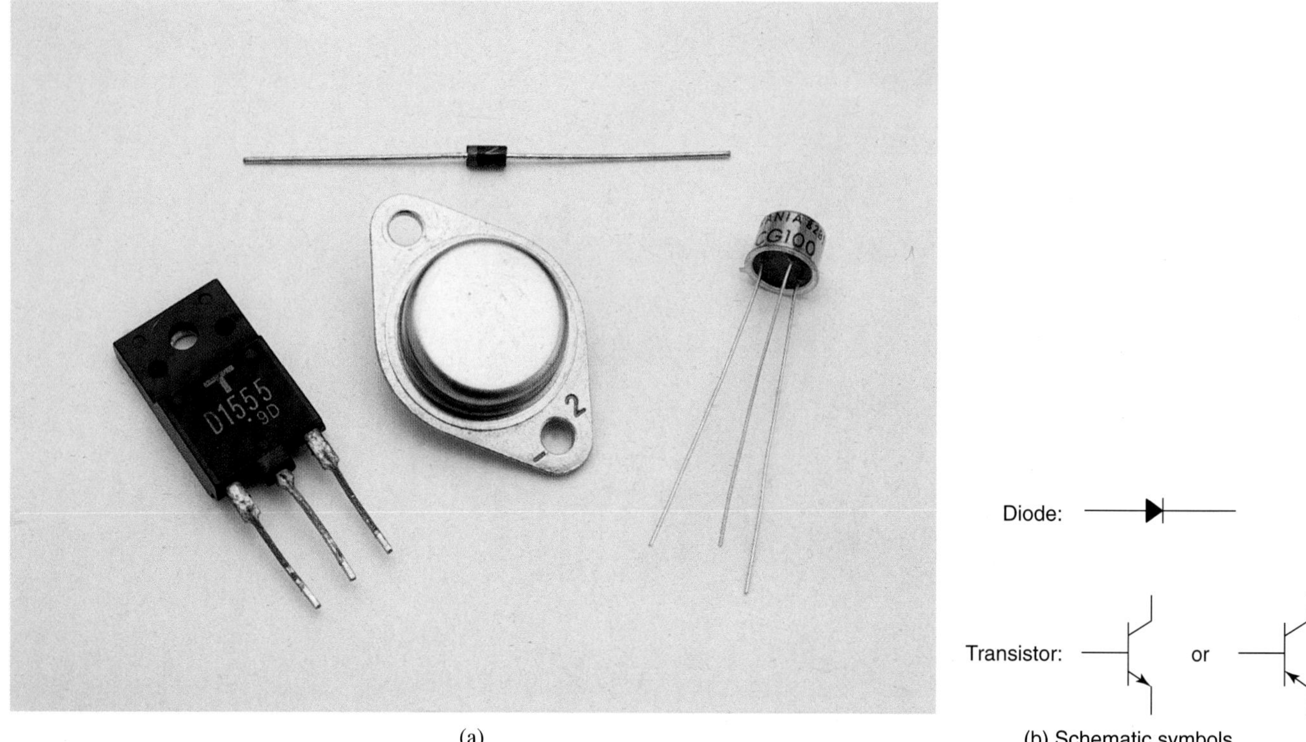


**FIGURE 1.10 Semiconductors.**

**FIGURE 1.11 Integrated circuits.**

*OBJECTIVE 8* ▸

## Putting It All Together

**Passive components** Components with values that are not controlled by an external power source.

**Active components** Components with values that are controlled by an external power source.

Resistors, inductors, and capacitors are all classified as **passive components** because their values are not controlled by an external power source. In contrast, vacuum tubes and most semiconductors are classified as **active components** because their values are controlled by an external power source; that is, their operating characteristics can be varied (within limits) by an external power source.

Most electronic circuits contain both active and passive components. In many cases, the passive components are used to help control the operation of the active components in one fashion or another. This point will become clear to you as your study of electronics advances.

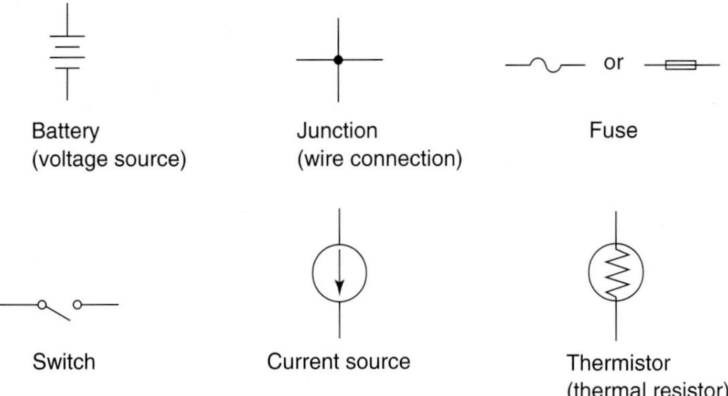


**FIGURE 1.12 More schematic symbols.**

## Other Schematic Symbols

Figure 1.12 shows many of the common schematic symbols that we will use in this text. A more complete set of schematic symbols can be found in Appendix C.

## Other Electrical Quantities and Units

The difference between quantities and units.

You have been shown that resistance *(R)* is measured in ohms (Ω), capacitance *(C)* is measured in farads (F), and inductance *(L)* is measured in henries (H). There are other electrical quantities that have their own unit of measure. Some of the more common quantities and their units of measure are listed in Table 1.1. As you can see, the table heading reads *"Quantity . . .* is measured in . . . *Unit."* While the difference between quantities and units may not be obvious at first, consider the following relationships that you already know:

*Temperature* is measured in *degrees.*

*Age* is measured in *years.*

In the first relationship, temperature is the quantity and degrees is the unit of measure. In the second, age is the quantity and years is the unit of measure.

## One Final Note

Some of the quantities listed in this section tend to have extremely high values (such as a frequency of 10,000,000 Hz), while others tend to have extremely low values (such as a capacitance of 0.000000001 F). Since many electronic values tend to be extremely high or

**TABLE 1.1 Common Quantities and Their Units**

| *Quantity* | . . . is measured in . . . | *Unit* |
|---|---|---|
| Charge *(Q)* | | coulombs (C) |
| Conductance *(G)* | | siemens (S)[a] |
| Current *(I)* | | amperes (A) |
| Frequency *(f)* | | hertz (Hz) |
| Impedance *(Z)* | | ohms (Ω) |
| Power *(P)* | | watts (W) |
| Reactance *(X)* | | ohms (Ω) |
| Time *(t)* | | seconds (s) |
| Voltage *(V)* | | volts (V) |

[a]Conductance is sometimes given in *mhos,* represented by an upside-down omega.

low in value, they are usually represented using either scientific notation or engineering notation. Scientific notation and engineering notation allow us to represent easily very large and very small numbers, as you will see in the following sections.

### Section Review

1. What is a resistor?
2. How does a potentiometer differ from other types of resistors?
3. Draw the resistor schematic symbols.
4. What is a schematic diagram?
5. What is the unit of measure for resistance? What symbol is used to represent this unit of measure?
6. What is a capacitor?
7. Draw the capacitor schematic symbols.
8. What is the unit of measure for capacitance?
9. What is an inductor?
10. Draw the inductor schematic symbol.
11. What is the unit of measure for inductance?
12. What are vacuum tubes? How do they differ from semiconductors?
13. What distinguishes passive and active components?
14. Describe the relationship between quantities and units.

## 1.4 SCIENTIFIC NOTATION

One of the most basic principles of electronics, called Ohm's law, can be stated mathematically as

$$V = IR$$

(which reads, "voltage equals current times resistance"). Now, consider the following typical values of current *(I)* and resistance *(R):*

$$I = 0.000045 \text{ A} \qquad R = 1{,}500{,}000 \ \Omega$$

The odds of making a mistake when entering these values into your calculator are better than average. However, the odds of making a mistake are greatly reduced when we use scientific notation. In this section, we will discuss scientific notation and the process for converting a given number into this form. We will also work through some practice problems involving scientific notation.

*OBJECTIVE 9* ▸

### Scientific Notation

**Scientific notation**
A system that represents a value as the product of a number and a whole-number power of ten.

In **scientific notation,** a given value is represented as the product of a number and a whole-number power of ten. For example, look at the following values and their scientific notation equivalents:

The number portion (shown in **bold type**) is always written as a number between 1 and 9.999. (The most significant digit must fall in the units position.)

| Value | Scientific Notation Equivalent |
|---|---|
| 2300 | $\mathbf{2.3} \times 10^{3}$ |
| 47,500 | $\mathbf{4.75} \times 10^{4}$ |
| 0.0683 | $\mathbf{6.83} \times 10^{-2}$ |
| 0.000000355 | $\mathbf{3.55} \times 10^{-7}$ |

**TABLE 1.2**

| | |
|---|---|
| $10{,}000{,}000 = 1 \times 10^{7}$ | |
| $1{,}000{,}000 = 1 \times 10^{6}$ | |
| $100{,}000 = 1 \times 10^{5}$ | Positive powers of ten |
| $10{,}000 = 1 \times 10^{4}$ | |
| $1{,}000 = 1 \times 10^{3}$ | |
| $100 = 1 \times 10^{2}$ | |
| $10 = 1 \times 10^{1}$ | |
| $1 = 1 \times 10^{0}$ | |
| $0.1 = 1 \times 10^{-1}$ | |
| $0.01 = 1 \times 10^{-2}$ | |
| $0.001 = 1 \times 10^{-3}$ | Negative powers of ten |
| $0.0001 = 1 \times 10^{-4}$ | |
| $0.00001 = 1 \times 10^{-5}$ | |
| $0.000001 = 1 \times 10^{-6}$ | |
| $0.0000001 = 1 \times 10^{-7}$ | |

In each case, the value is represented as a number (shown in bold type) multiplied by a whole-number power of ten. Note that the number is written as a value between 1 and $9.\overline{999}$. *This is always the case when a given number is written in scientific notation.*

Table 1.2 lists the most commonly used positive and negative powers of ten. Several observations can be made with the help of this list. First, note that all values greater than (or equal to) ten have a *positive* power of ten, while all the values less than one have a *negative* power of ten. The power of ten is *zero* for all values between 1 and $9.\overline{999}$ (zero is not considered to be either positive or negative). Second, note that the power of ten in each case equals the number of places that the decimal point must be shifted to end up on the immediate *right* of the one (1). For example, the first line in the table reads:

$$10{,}000{,}000 = 1 \times 10^{7}$$

*A Practical Consideration:* When a number does not contain a decimal point, its assumed position is at the end of the number (just as a period falls at the end of a sentence).

Imagine moving the decimal point from its implied position of 10,000,000.0 to the left so that it falls to the immediate *right* of the one (1). It would have to be moved seven places, and the power of ten is seven. A close look at the table shows that this principle holds true for all the numbers shown.

*OBJECTIVE 10*

At this point, we can establish the rules for converting a number from standard form to scientific notation:

Standard notation to scientific notation conversion.

1. Count the number of positions that the decimal point must be shifted so that it falls to the immediate right of the most significant digit in the number.
2. If the decimal point must be shifted to the *left,* the power of ten is *positive.* If the decimal point must be shifted to the *right,* the power of ten is *negative.* If the decimal point does not need to be shifted (i.e., it already falls to the immediate right of the most significant digit), the power of ten is *zero.*

The following series of examples illustrates the process for converting a number from standard form to scientific notation.

### Example 1.1

Convert the number 465,000 to scientific notation.

***Solution:*** To place the decimal point to the immediate right of the most significant digit (the 4), it must be shifted five places to the left. Thus, the power of ten is positive five, and the number is written as

$$465{,}000 = 4.65 \times 10^5$$

***PRACTICE PROBLEM 1.1***

Convert the number 1,238,000 to scientific notation.

### Example 1.2

Convert the number 0.00794 from standard form to scientific notation.

***Solution:*** To place the decimal point at the immediate right of the most significant digit (the 7), it must be shifted three places to the right. Thus, the power of ten is negative three, and the number is written as

$$0.00794 = 7.94 \times 10^{-3}$$

***PRACTICE PROBLEM 1.2***

Convert the number 0.000000000993 to scientific notation.

### Example 1.3

Convert the number 5.98 from standard form to scientific notation.

***Solution:*** Since the decimal point already falls to the immediate right of the most significant digit (the 5), it does not need to be shifted and the power of ten is zero. Thus,

$$5.98 = 5.98 \times 10^0$$

***PRACTICE PROBLEM 1.3***

Convert the following numbers from standard form to scientific notation:

6,800 0.00358 9.5 10 0.02

## Using Your Calculator

*A Practical Consideration:* Some calculators use labels other than EE or EXP to denote the exponent key. If you cannot locate the exponent key on your calculator, refer to its instruction booklet.

A value in scientific notation is entered into your calculator using the exponent key, which is most commonly labeled as EE or EXP. When the exponent key is pressed, it tells the calculator that the next value entered is a power of ten. For example, the number $5.28 \times 10^2$ would be entered into your calculator as follows:

5.28 EE 2

Once the value is entered into your calculator, you should get a display that is similar to the one shown in Figure 1.13. Note that the value (5.28) is shown on the left of the display, while the power of ten (02) is shown on the right.

Several important points should be made at this time. First, the exponent key takes the place of the "× 10" portion of a value written in scientific notation. In other words, you would not enter $3.4 \times 10^8$ using

3.4 × 10 EE 8 (incorrect)

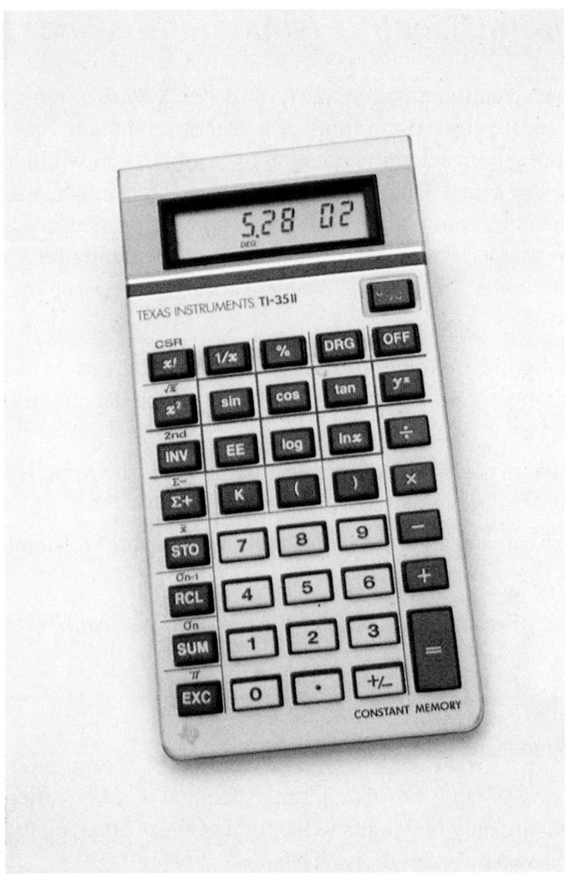


**FIGURE 1.13 A calculator display.**

The calculator assumes the "× 10" portion of the value when the EE key is pressed. Thus, the value would simply be entered as

3.4 EE 8 (correct)

Note that using the incorrect key sequence causes an incorrect value to be entered into your calculator. Second, negative powers of ten are entered into the calculator using the sign (+/−) key. When the power of ten is negative, the value is entered as before. Then, after pressing the EE key, the sign key is pressed. For example, the number $5.5 \times 10^{-2}$ would be entered into the calculator as follows:

5.5 EE +/− 2

Note that the calculator will allow you to press the sign key either before or after entering the number 2. However, for the exponent to be negative (rather than the value of 5.5), the sign key must be pressed *after* the EE key.

## Calculator Modes

Most scientific calculators are capable of operating in the scientific notation mode. When this mode is used, all values entered in standard form are converted automatically to scientific notation, and all values displayed are given in proper scientific notation form. Your calculator instruction booklet describes the procedure for switching your calculator to this operating mode.

## *Converting from Scientific Notation to Standard Form*

Even though we use scientific notation, most of us don't *think* in terms of scientific notation. For example, most people don't think of a year as containing $3.65 \times 10^2$ days. After you have been working with scientific notation for a while, you will find yourself forming a mental image of the actual value of any number written in scientific notation. In the meantime, you'll need to know how to convert a number written in scientific notation to standard form if the results of a problem are to have any meaning for you. A number written in scientific notation is converted to standard form by reversing the process described on page 15, as follows:

Scientific notation to standard notation conversion.

1. Shift the decimal point a number of positions equal to the power of ten. (For example, to convert the number $1.48 \times 10^4$ to standard form, the decimal point will need to be shifted four places.)
2. If the power of ten is positive, shift the decimal point to the right. If the power of ten is negative, shift the decimal point to the left.
3. If necessary, zeros are added to the value to allow the required number of decimal point shifts.

The following series of examples illustrates the use of this conversion technique.

### EXAMPLE 1.4

Convert $2.34 \times 10^3$ to standard form.

***Solution:*** The power of ten is positive 3, so the decimal point is shifted three places to the right. Since there are only two digits to the right of the decimal point, a zero is added to provide the required number of digits. Thus,

$$2.34 \times 10^3 = 2340.0$$

As you can see, the decimal point has been shifted three places to the right.

***PRACTICE PROBLEM 1.4***

Convert the number $6.22 \times 10^5$ to standard form.

### EXAMPLE 1.5

Convert the number $7.92 \times 10^{-5}$ to standard form.

***Solution:*** Since the power of ten is negative 5, the decimal point must be shifted five places to the left. There is only one digit to the left of the decimal point, so four zeros must be added to the front of the number to allow the required number of shifts. Thus,

$$7.92 \times 10^{-5} = .0000792$$

As you can see, the decimal point has been shifted five places to the left.

***PRACTICE PROBLEM 1.5***

Convert the number $6.17 \times 10^{-2}$ to standard form.

When dealing with standard form values that are less than one (1), many technicians and engineers add another zero to the left of the decimal point to firmly establish its position. Thus, the value of .0000792 from Example 1.5 may be written as 0.0000792. The addition of the zero before the decimal point does not change the value of the number.

Now that you have been shown how to convert numbers back and forth between scientific notation and standard form, we will look at solving problems that are written in scientific notation.

## Solving Scientific Notation Problems

When using your calculator to solve problems written in scientific notation, the values are entered using the key sequence given earlier in this section. For example, to solve the problem

$$(2.34 \times 10^4) + (6.2 \times 10^3) = \_\_\_\_\_\_$$

you would use the following key sequence:

2.34 EE 4 + 6.2 EE 3 =

As you can see, the numbers and operations are entered in the same order that they are shown in the problem. This is demonstrated as the following series of examples illustrates.

### EXAMPLE 1.6

Determine the key sequence needed to solve the following problem:

$$\frac{3.45 \times 10^{-4}}{2.22 \times 10^2} = \_\_\_\_\_\_$$

***Solution:*** The problem is entered exactly as it is shown. The needed key sequence is:

3.45 EE +/− 4 ÷ 2.22 EE 2 =

When this key sequence is used, your calculator gives you a result of $1.55 \times 10^{-6}$.

***PRACTICE PROBLEM 1.6***

Use your calculator to solve the following problem:

$$\frac{5.6 \times 10^4}{3.3 \times 10^{-2}} = \_\_\_\_\_\_$$

### EXAMPLE 1.7

Determine the key sequence needed to solve the following problem:

$$(6.1 \times 10^3)(3.9 \times 10^{-2}) + 628 = \_\_\_\_\_\_$$

***Solution:*** Again, we simply enter the problem the way it is written, as follows:

6.1 EE 3 × 3.9 EE +/− 2 + 628 =

After pressing the equals key, the calculator gives you a result of $8.659 \times 10^2$.

***PRACTICE PROBLEM 1.7***

Use your calculator to solve the following problem:

$$(5.5 \times 10^5) + \frac{3.3 \times 10^2}{1.2 \times 10^{-2}} = \_\_\_\_\_\_$$

You may have noticed that the value *628* in Example 1.7 was not entered into the calculator in scientific notation. It was not necessary to do so because the calculator is capable of solving problems that contain numbers in both forms.

## *Order of Operations*

Calculators are designed to solve mathematical operations in a predictable order. This order of operations is as follows:

1. Single-variable operations, such as radicals, square roots, and reciprocals.
2. Multiplication/division (in the order entered)
3. Addition/subtraction (in the order entered)

It is important to know the order of operations for your calculator because it affects the way you enter a given equation. For example, let's say that you want your calculator to solve the following problem:

$$\frac{10}{2+3} = \_\_\_\_\_\_$$

If you enter this problem into your calculator using

$$10 \quad \div \quad 2 \quad + \quad 3 \quad =$$

the result given by your calculator will be 8, which is clearly wrong. The problem in this case is the order of operations: the calculator performed the division operation *before* it performed the addition operation. To solve the equation correctly, you must indicate that the values in the denominator (2 and 3) are to be combined first. This is accomplished by using the parentheses keys, as follows:

$$10 \quad \div \quad ( \quad 2 \quad + \quad 3 \quad ) \quad =$$

When entered as shown above, the calculator provides the correct result: 2.

Avoiding order-of-operation errors.

There is no fail-safe rule that prevents order-of-operations errors. However, you can reduce your chances of making a mistake if you use the following guidelines:

1. When an equation contains any parentheses, enter them into your calculator as shown. This ensures that any groupings given in the equation are followed by the calculator.
2. When dealing with fractions, solve any unknowns in the numerator and the denominator before solving the fraction. For example, in the problem:

   $$\frac{24+62}{98} = \_\_\_\_\_\_$$

   the addition problem in the numerator must be solved before the division. This can be accomplished by adding your own parentheses, as follows:

   $$( \quad 24 \quad + \quad 62 \quad ) \quad \div \quad 98 \quad =$$

   When the parentheses are used, the calculator solves the addition problem before the division problem is entered. You could also use another equals sign (rather than the parentheses), as follows:

   $$24 \quad + \quad 62 \quad = \quad \div \quad 98 \quad =$$

   Again, the addition problem is solved before the division problem is entered into the calculator.

You will find that most order-of-operation errors can be avoided when you follow these simple guidelines.

## One Final Note

You have been shown how to convert numbers back and forth between scientific notation and standard form, how to enter numbers written in scientific notation into your calculator, and how to solve basic problems involving numbers written in scientific notation. After a while, you will find that working with scientific notation is actually easier than working with numbers written in standard form.

In the next section, you will be introduced to *engineering notation,* another method commonly used to represent extremely large and extremely small values. As you will see, engineering notation is nothing more than a slightly altered form of scientific notation.

### Section Review

1. Describe the scientific notation method of representing values.
2. List the steps for converting a number from standard form to scientific notation.
3. Describe the function of the EE button on your calculator.
4. List the steps for converting a number from scientific notation to standard form.

# 1.5 ENGINEERING NOTATION

◄ *OBJECTIVE 11*

You have probably heard people refer to 1000 meters as 1 kilometer. And you have probably heard people use the term *megabucks* to refer to millions of dollars.

Kilo- and mega- are only two of the prefixes used in **engineering notation,** a form of scientific notation where standard prefixes are used to designate specific ranges of values. Most of the prefixes used and the ranges they represent are shown in Table 1.3.

**Engineering notation**
A form of scientific notation where standard prefixes are used to designate specific ranges of values.

When a number is written in engineering notation, the appropriate prefix is used in place of the power of ten. For example, here are some values and their engineering notation equivalents:

| Value | Engineering Notation Equivalent |
|---|---|
| $12.3 \times 10^3$ ohms | 12.3 kΩ (kilo-ohms) |
| $284 \times 10^6$ hertz | 284 MHz (mega-hertz) |
| $4.7 \times 10^{-3}$ seconds | 4.7 ms (milli-seconds) |
| $51 \times 10^{-12}$ farads | 51 pF (pico-farads) |

In the first value, the $\times\ 10^3$ has been replaced by the prefix kilo (k). In the second, the $\times\ 10^6$ has been replaced by the prefix mega (M), and so on.

**TABLE 1.3 Engineering Notation Ranges, Prefixes, and Symbols**

| *Range* | *Prefix* | *Symbol* |
|---|---|---|
| $(1 \text{ to } 999) \times 10^{12}$ | tera- | T |
| $(1 \text{ to } 999) \times 10^{9}$ | giga- | G |
| $(1 \text{ to } 999) \times 10^{6}$ | mega- | M |
| $(1 \text{ to } 999) \times 10^{3}$ | kilo- | k |
| $(1 \text{ to } 999) \times 10^{0}$ | (none) | (none) |
| $(1 \text{ to } 999) \times 10^{-3}$ | milli- | m |
| $(1 \text{ to } 999) \times 10^{-6}$ | micro- | μ |
| $(1 \text{ to } 999) \times 10^{-9}$ | nano- | n |
| $(1 \text{ to } 999) \times 10^{-12}$ | pico- | p |

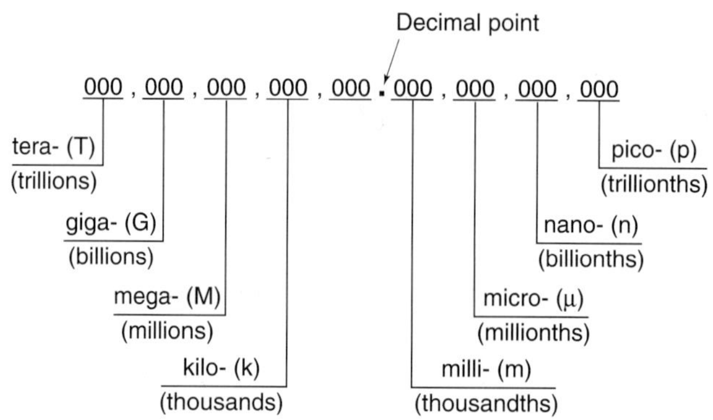


**FIGURE 1.14 Engineering notation groups.**

Why engineering notation prefixes are used.

The advantage of using engineering notation is that it makes it much easier to convey large and small values to others. For example, it is much easier to say, "I need a 510 k-ohm resistor," than it is to say, "I need a $5.1 \times 10^5$ ohm resistor." Writing values in engineering notation is also much faster than writing them in standard form or scientific notation.

After a while, you'll find that working with engineering notation is easy. In the meantime, you'll need to be shown how to convert numbers back and forth between engineering notation and the other numeric forms.

*OBJECTIVE 12* ▶

## Converting Numbers to Engineering Notation

Converting values from standard form to engineering notation is simple when you keep in mind the fact that engineering notation prefixes are used to group digits in the same fashion that you do. For example, consider the number 1,000,000,000.000,000,000. The commas separate the digits into groups of three, as do the prefixes used in engineering notation. In fact, the prefixes separate the numbers at the same positions as the commas, as shown in Figure 1.14. As you can see, kilo- is just another way of saying thousands, giga- is just another way of saying billions, and so on. By the same token, milli- is just another way of saying thousandths, nano- is just another way of saying billionths, and so on.

Standard notation to engineering notation conversion.

*A Practical Consideration:* Most calculators will perform this conversion automatically. However, learning to perform conversions mentally decreases your dependence on your calculator.

When you want to convert a number from standard form to engineering notation, use the following procedure:

1. Determine the appropriate prefix by noting the position of the most significant digit in the number.
2. Reading left to right, replace the first comma with a decimal point.

The following series of examples illustrates the use of this procedure.

### EXAMPLE 1.8

Convert 3,420 Hz to engineering notation.

***Solution:*** Since the most significant digit (the 3) is in the thousands, the prefix we will use is kilo- (k). Replace the comma with a decimal point:

$$3{,}420 \text{ Hz} = 3.42 \text{ kHz}$$

***PRACTICE PROBLEM 1.8***

Convert 357,800 Hz to engineering notation.

**EXAMPLE 1.9**

Convert 54,340,000,000 W to engineering notation.

***Solution:*** Since the most significant digit (the 5) is in the billions, the prefix we will use is giga- (G). Replacing the first comma with a decimal point, we get the following result:

$$54{,}340{,}000{,}000 \text{ W} = 54.34 \text{ GW}$$

Note that it is unnecessary to keep the remaining zeros in the original number because the prefix giga- indicates that the value is in billions.

***PRACTICE PROBLEM 1.9***

Convert 82,982,000 W to engineering notation.

**EXAMPLE 1.10**

Convert 0.00346 A to engineering notation.

***Solution:*** Since the most significant digit (the 3) is in the thousandths, the prefix we will use is milli- (m). If we were to use a comma to form groups of three digits (starting at the decimal point), the number would be written as 0.003,46. Replacing this comma with a decimal point, we obtain the following result:

$$0.00346 \text{ A} = 3.46 \text{ mA}$$

***PRACTICE PROBLEM 1.10***

Convert 0.0000492 A to engineering notation.

Converting from scientific notation to engineering notation starts by solving a simple problem:

Scientific notation to engineering notation conversion.

1. Determine the highest whole-number multiple of three that is less than (or equal to) the power of ten in the number to be converted.
2. Determine the difference between the value found in Step 1 and the power of ten in the number.

For example, let's say that you want to convert $6.8 \times 10^7$ to engineering notation:

1. Six (6) is the highest whole-number multiple of three that is less than 7.
2. The difference between 6 and 7 is one (1).

Why solve this problem? The six indicates the power of ten for the prefix to be used, and the one indicates the number of places that the decimal point must be shifted to the right. Therefore,

$$6.8 \times 10^7 = 68 \times 10^6 = 68 \text{ M}$$

Note that mega- corresponds to $10^6$ and that the decimal point in the number has been shifted one place to the right. Examples 1.11 and 1.12 further illustrate the use of this simple conversion technique.

There is a point that needs to be made at this time. When converting numbers that have negative powers of ten, you need to remember that more negative numbers are considered to be lower in value. That's why $-9$ is used in Example 1.12 as the highest whole-number multiple of three that is less than $-8$.

### Example 1.11

Convert $3.82 \times 10^{11}$ Hz to engineering notation.

***Solution:*** The highest whole-number multiple of three that is less than 11 is 9, and the difference between 9 and 11 is two. Since $10^9$ corresponds to the prefix giga-, this is the prefix we will use. Also, we need to shift the decimal point two places to the right. Therefore,

$$3.82 \times 10^{11} \text{ Hz} = 382 \times 10^9 \text{ Hz} = 382 \text{ GHz}$$

***PRACTICE PROBLEM 1.11***

Convert $4.55 \times 10^4$ Hz to engineering notation.

### Example 1.12

Convert $3.22 \times 10^{-8}$ A to engineering notation.

***Solution:*** The highest whole-number multiple of three that is less than $-8$ is $-9$, which corresponds to the engineering prefix nano- (n). The difference between $-8$ and $-9$ is 1, so the decimal point in the value must be shifted one place to the right. Therefore,

$$3.22 \times 10^{-8} \text{ A} = 32.2 \times 10^{-9} \text{ A} = 32.2 \text{ nA}$$

***PRACTICE PROBLEM 1.12***

Convert $7.62 \times 10^{-1}$ A to engineering notation.

As you can see, converting from scientific notation to engineering notation is relatively simple. In fact, it won't be long before you'll be doing it in your head (if you aren't already).

## Other Engineering Notation Prefixes

In this section, you have been introduced to the most commonly used engineering notation prefixes and their value ranges. Some less frequently used prefixes and their value ranges are as follows:

exa- (E) is used for (1 to 999) $\times 10^{18}$

peta- (P) is used for (1 to 999) $\times 10^{15}$

fempto- (f) is used for (1 to 999) $\times 10^{-15}$

atto- (a) is used for (1 to 999) $\times 10^{-18}$

## One Final Note

The prefixes listed above (and in Table 1.3) all show a value range of 1 to 999. There are cases, however, when values outside this range are used to keep the notation constant. A classic example of using constant notation can be found on the AM dial of any radio. The standard AM dial is labeled using the following sequence:

530 kHz 600 kHz 700 kHz 800 kHz 1000 kHz 1200 kHz 1400 kHz 1700 kHz

The last four values shown could be written as 1 MHz, 1.2 MHz, 1.4 MHz, and 1.7 MHz. However, they are written as shown in the sequence so that the kHz notation remains constant. By the same token, consider the following sequence of values:

100 V  10 V  1 V  ½V

In some references, the last value might be written as 0.5 V (rather than 500 mV) to maintain the constant notation. Other references may list it as 500 mV to stay within the value limits normally placed on the engineering notation prefixes.

### Section Review

1. List the commonly used engineering prefixes, along with their symbols and ranges.
2. What purpose is served by using engineering notation?
3. List the steps for converting a number from standard form to engineering notation.
4. How do you convert a number from scientific notation to engineering notation?

## Chapter Summary

Here is a summary of the major points made in this chapter:

1. Electronics is the field that deals with the design, manufacturing, installation, maintenance, and repair of electronic systems.
2. Electronics engineers are generally involved in the design and manufacturing aspects of electronics.
3. Electronics technicians are generally involved in the maintenance and repair aspects of electronics. Technicians generally work as in-house technicians or field-service representatives.
4. Engineering technicians perform some of the duties of an engineer and some of the duties of a technician.
5. Communications systems are designed to transmit and/or receive information.
6. Computer systems are designed to process and store information.
7. Industrial systems are designed for use in industrial (manufacturing) environments.
8. Biomedical systems are designed for use in diagnosing and treating medical problems.
9. Digital multimeters (DMMs) are used to measure voltage, current, and resistance.
   - **a.** Many DMMs are capable of performing a variety of additional circuit and component tests.
   - **b.** The DMM is a descendant of the volt-ohm-milliammeter (VOM).
10. The oscilloscope is used to make a variety of voltage and time measurements.
11. Always make sure that you have learned the proper technique for using a piece of test equipment before attempting to use it.
12. The resistor is a component designed to provide a specific amount of resistance.
13. The potentiometer is a variable resistor.
14. Schematic diagrams are the electronic equivalents of blueprints. The schematic diagram of a circuit shows you how the components are connected.
15. Resistance is measured in ohms, represented by the Greek letter omega (Ω).
16. The capacitor is a component that opposes a change in voltage.
17. Capacitance is measured in farads (F).

18. The inductor is a component that opposes a change in current. (The capacitor and inductor are nearly opposites in terms of their characteristics.)
19. Inductance is measured in henries (H).
20. Vacuum tube operation is based on the flow of charge through a vacuum.
    a. Unlike resistors, capacitors, and inductors, vacuum tubes have no single unit of measure.
    b. The most commonly used vacuum tube is the cathode-ray tube (CRT). The CRT is used as a video display.
21. Passive components require no external power source to operate.
22. Active components require an external power source to operate.
23. Using scientific notation greatly reduces the chances of making a mistake when entering extremely large and extremely small values into your calculator. In scientific notation, a value is represented as the product of a number between 1 and 10 and a whole-number power of ten.
24. The exponent (EE or EXP) key on your calculator is used to enter powers of ten.
25. Engineering notation is a form of scientific notation where standard prefixes are used to designate specific value ranges. The most commonly used prefixes are listed in Table 1.3.

## Key Terms

The following terms were introduced and defined in this chapter:

| | | |
|---|---|---|
| active component | electronics engineering technician | oscilloscope |
| biomedical systems | electronics technician | passive component |
| capacitor | engineering notation | potentiometer |
| choke | farad (F) | resistor |
| circuit | field-service representative | schematic diagram |
| coil | frequency counter | scientific notation |
| communications systems | henry (H) | semiconductor |
| computer systems | inductor | vacuum tube |
| condenser | industrial systems | volt-ohm-milliammeter (VOM) |
| digital multimeter (DMM) | in-house technician | |
| electronics | ohm | |
| electronics engineer | | |

## Practice Problems

1. Convert the following numbers to scientific notation without the use of your calculator.

| | |
|---|---|
| **a.** 3,492 | **f.** 1,220,000 |
| **b.** 922 | **g.** −82,000 |
| **c.** 23,800,000 | **h.** 1 |
| **d.** −476,000 | **i.** 3,970,000,000 |
| **e.** 22,900 | **j.** −3,800 |

2. Convert the following numbers to scientific notation without the use of your calculator.

| | |
|---|---|
| **a.** 0.143 | **f.** 0.0039 |
| **b.** 0.000427 | **g.** 0.07018 |
| **c.** −0.0000401 | **h.** −0.629 |
| **d.** 0.000000023 | **i.** 0.0000097 |
| **e.** −0.0381 | **j.** 0.0000000000702 |

3. Write the key sequence you would use to enter each of the following values into your calculator.

a. $1.83 \times 10^{23}$
b. $-5.6 \times 10^{4}$
c. $4.2 \times 10^{-9}$
d. $-9.22 \times 10^{-7}$
e. $3.3 \times 10^{-12}$
f. $10^{6}$

4. Convert each of the following values to standard form without the use of your calculator.

a. $3.2 \times 10^{4}$
b. $6.6 \times 10^{8}$
c. $9.24 \times 10^{0}$
d. $5.7 \times 10^{12}$
e. $-1.1 \times 10^{5}$
f. $-4.98 \times 10^{2}$
g. $8.77 \times 10^{6}$
h. $-1.26 \times 10^{1}$
i. $2.84 \times 10^{3}$
j. $4.45 \times 10^{7}$

5. Convert each of the following values to standard form without the use of your calculator.

a. $4.43 \times 10^{-2}$
b. $-2.8 \times 10^{-5}$
c. $7.62 \times 10^{-4}$
d. $3.38 \times 10^{-10}$
e. $-5.8 \times 10^{-8}$
f. $6.34 \times 10^{-1}$
g. $1.11 \times 10^{-9}$
h. $9.02 \times 10^{-3}$
i. $-8.3 \times 10^{-6}$
j. $4.44 \times 10^{-7}$

6. Write the key sequence you would use to solve each of the following problems with your calculator.

a. $(2.43 \times 10^{4})(-3.9 \times 10^{-4}) =$ _______
b. $\dfrac{-4 \times 10^{6}}{2.8 \times 10^{-2}} + 30 =$ _______
c. $(4.5 \times 10^{-5}) + (-1.1 \times 10^{-4}) =$ _______
d. $(8 \times 10^{-2})(5.1 \times 10^{3}) - (9 \times 10^{2}) =$ _______

7. Use your calculator to solve each of the following problems.

a. $(1 \times 10^{4})(2.2 \times 10^{-2}) + 96 =$ _______
b. $\dfrac{2.4 \times 10^{6}}{3.3 \times 10^{5}} - (3 \times 10^{2}) =$ _______
c. $(9 \times 10^{0}) + (9 \times 10^{-1}) - (-4 \times 10^{-2}) =$ _______
d. $\dfrac{(-6.2 \times 10^{-4})(-3 \times 10^{3})}{6.8 \times 10^{-3}} =$ _______

8. Use your calculator to solve each of the following problems. Write your answers in standard form.

a. $(8.3 \times 10^{-3})(3.3 \times 10^{6}) - (1.1 \times 10^{2}) =$ _______
b. $(4.22 \times 10^{0}) - (1.2 \times 10^{1})(3.3 \times 10^{-2}) =$ _______
c. $(-5 \times 10^{3}) + \dfrac{4 \times 10^{2}}{6.28 \times 10^{-1}} =$ _______
d. $\dfrac{4.82 \times 10^{4}}{(2.28 \times 10^{-2})(1.8 \times 10^{-1})} =$ _______

9. Write each of the following values in proper engineering notation.

| | |
|---|---|
| **a.** 38,400 m | **f.** 1,800 Ω |
| **b.** 234,000 W | **g.** 3.2 A |
| **c.** 44,320,000 Hz | **h.** 2,500,000 Ω |
| **d.** 175,000 V | **i.** 4,870,000,000 Hz |
| **e.** 60 W | **j.** 22,000 Ω |

10. Write each of the following values in proper engineering notation.

| | |
|---|---|
| **a.** 0.022 H | **f.** 0.288 A |
| **b.** 0.00047 F | **g.** 0.00000000244 s |
| **c.** 0.00566 m | **h.** 0.000033 H |
| **d.** 0.000055 A | **i.** 0.000000000051 F |
| **e.** 0.935 s | **j.** 0.00355 C |

11. Convert each of the following values to engineering notation form.

| | |
|---|---|
| **a.** $2.2 \times 10^{7}$ Ω | **f.** $4.7 \times 10^{-5}$ F |
| **b.** $3.3 \times 10^{3}$ W | **g.** $1 \times 10^{-1}$ H |
| **c.** $7 \times 10^{11}$ Hz | **h.** $6.6 \times 10^{-4}$ s |
| **d.** $5.1 \times 10^{5}$ Ω | **i.** $4 \times 10^{-8}$ S |
| **e.** $8.5 \times 10^{4}$ W | **j.** $2.2 \times 10^{-10}$ F |

## Answers to the Example Practice Problems

| | | | |
|---|---|---|---|
| **1.1** | $1.238 \times 10^{6}$ | **1.5** | 0.0617 |
| **1.2** | $9.93 \times 10^{-10}$ | **1.6** | $1.7 \times 10^{6}$ |
| **1.3** | $6.8 \times 10^{3}$ | **1.7** | $5.8 \times 10^{5}$ |
| | $3.58 \times 10^{-3}$ | **1.8** | 357.8 kHz |
| | $9.5 \times 10^{0}$ | **1.9** | 82.982 MW |
| | $1 \times 10^{1}$ | **1.10** | 49.2 μA |
| | $2 \times 10^{-2}$ | **1.11** | 45.5 kHz |
| **1.4** | 622,000 | **1.12** | 762 mA |

## EWB Applications Problems

| Figure | EWB File Reference |
|---|---|
| 1.1 | EWBA1_1.ewb |
| 1.2 | EWBA1_2.ewb |
| 1.4 | EWBA1_4.ewb |
| 1.5 | EWBA1_5.ewb |

# 2

# PRINCIPLES OF ELECTRICITY

## OBJECTIVES

*After studying the material in this chapter, you should be able to:*

1. Describe the relationship among atoms, elements, and matter.
2. Describe the structure of the atom.
3. Explain the concept of charge as it relates to the atom.
4. State the relationship among current, charge, and time.
5. Discuss the electron flow and conventional current approaches to defining current.
6. Compare direct current (dc) and alternating current (ac).
7. Define voltage and describe its relationship to energy and charge.
8. Define resistance and conductance, and describe the relationship between the two.
9. Compare and contrast conductors, insulators, and semiconductors.
10. Discuss the relationship between resistance and each of the following: resistivity, material length, material cross-sectional area, and temperature.

When you mention the term *electricity,* most people form a mental image of lightning, a power transmission station, or some other evidence of the *presence* of electricity. Others form a mental image of a computer, television, or some other electronic system that operates through the *use* of electricity. Very few, however, actually understand what electricity *is.*

In this chapter, we will discuss electricity and its sources, its properties, and the relationships between them. As you will see, the study of electricity starts with a discussion of a basic building block of matter, the atom.

## 2.1 THE STARTING POINT: ELEMENTS, ATOMS, AND CHARGE

A battery has two terminals: a negative terminal and a positive terminal. While the terms *negative* and *positive* are commonly associated with electricity, most people don't know that they describe charges found in the atom. In this section, we will look at matter, the structure of the atom, and the nature of charge.

*OBJECTIVE 1* ▶

### Matter

**Matter**
Anything that has weight and occupies space.

**Element**
A substance that cannot be broken down into a combination of simpler substances.

**Matter** can be defined as anything that has weight and occupies space. You, and everything around you, are classified as matter.

All matter is made up of individual elements or some combination of elements. An **element** is a substance that cannot be broken down into a combination of simpler substances. For example, copper (an element) cannot be broken down into a combination of other elements.

As far as we know, all matter in the universe is made up of various combinations of the approximately 108 known elements. While these elements each have their own physical and chemical properties, they all share one very important trait: they are all made up of atoms.

*OBJECTIVE 2* ▶

### Atomic Structure

**Atom**
The smallest particle of matter that retains the physical characteristics of an element.

The **atom** is the smallest particle of matter that retains the physical characteristics of an element. The simplest model of the atom, called the Bohr model, is represented in Figure 2.1. The model shown in Figure 2.1 represents the helium atom, which is one of the simplest atoms. As you can see, the atom contains a central core (called the nucleus) that is orbited by two particles, called electrons. The electrons revolve around the nucleus in a fashion that is similar to the orbit of the planets around the sun. The nucleus of the atom is

**FIGURE 2.1 Helium atom.**

Nucleus (core) contains protons and neutrons

Electrons orbit the nucleus of the atom

The atom is made up of protons and neutrons, which form the nucleus, and orbiting electrons.

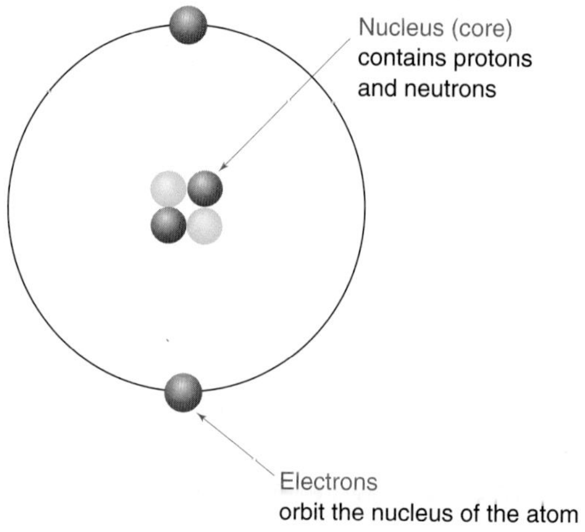

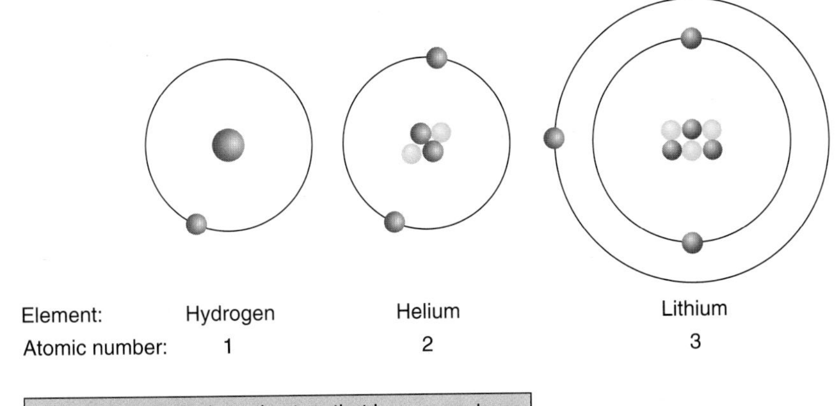


**FIGURE 2.2**

made up of two other types of particles: protons and neutrons. The neutron is the largest of the three particles. (Each neutron weighs approximately 1800 times as much as an electron.) The atom itself is so small that its total mass is approximately $6.2 \times 10^{-24}$ grams.

The number of electrons and protons contained in the atom varies from one element to another. This point is illustrated in Figure 2.2. As you can see, the hydrogen atom has one proton and one electron, the helium atom has two of each, and the lithium atom has three of each. Note that the number of protons in each atom is used as the **atomic number** for that element. Thus, the elements represented in Figure 2.2 have the atomic numbers shown. Figure 2.2 also serves to illustrate a very important point: in its natural state, a given atom contains an equal number of protons and electrons. The importance of this point will be discussed in the next section.

**Atomic number**
The number of protons in the nucleus of an atom. Elements can be identified by their atomic number.

Electrons travel around the nucleus of the atom in orbital paths that are referred to as **shells.** For example, look at the model of the copper atom shown in Figure 2.3. As you can see, the copper atom contains 29 electrons that travel around the nucleus in four different orbital shells. The innermost shell, called the K shell, contains two electrons, the L shell contains eight electrons, the M shell contains eighteen electrons, and the N shell contains a single electron.

**Shell**
A term used to describe the orbital paths of electrons.

Each orbital shell has a restriction on the number of electrons it can hold. For a given orbital shell, the maximum possible number of electrons can be found as $2n^2$, where $n$ is the number of the shell. For example, the third orbital shell (or M shell) can hold up to $2(3^2) = 18$ electrons.

**FIGURE 2.3 Copper atom.**

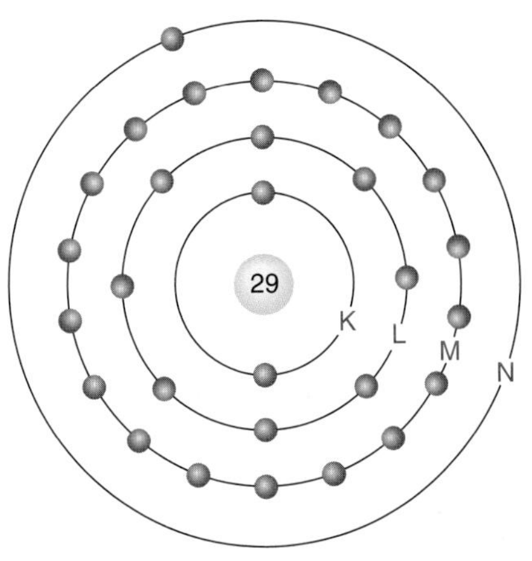

**FIGURE 2.4 Potassium atom.**

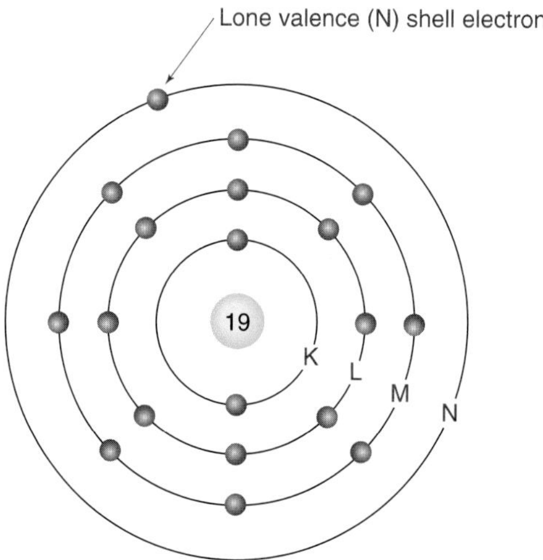


**Valence shell**
For a given atom, the outermost orbital shell containing electrons.

Another restriction applies only to the outermost shell, called the **valence shell.** The valence shell of an atom cannot contain more than eight electrons. For example, if the M shell is the valence (outermost) shell of an atom, it can hold no more than eight electrons, even though it is capable of holding up to eighteen electrons otherwise. This point can be seen by comparing the atoms shown in Figures 2.3 and 2.4. The copper atom in Figure 2.3 demonstrates the ability of the M shell to hold eighteen electrons. Now, consider the potassium atom in Figure 2.4. This atom has eight electrons in the M shell and one electron in its valence shell (the N shell). It would seem that this electron would orbit in the M shell because that shell is capable of holding eighteen electrons. However, the valence shell cannot hold more than eight electrons, so the added electron in potassium is forced to travel in the N shell.

There are several points that should be made regarding the valence shell of a given atom:

**Valence electron(s)**
The electron(s) orbiting in the valence shell of an atom. The electron(s) farthest from the nucleus of a given atom.

1. The electrons that occupy the valence shell of an atom are referred to as **valence electrons.**
2. If the valence shell of an atom contains eight electrons, the shell is said to be complete. A valence shell that contains fewer than eight electrons (like the valence shell of the copper atom) is said to be incomplete.

Note that the valence shell of an atom is very important because it determines the electrical characteristics of the element. This point will be examined in the next section.

*OBJECTIVE 3* ▶

## Charge

A planet remains in a fixed orbit around the sun because of a balance of forces: the centrifugal (outward) force balances with the gravitational (inward) force, keeping the planet in its orbit. This principle of balanced forces can also be applied to electrons orbiting the nucleus of an atom.

Electrons orbiting the nucleus of an atom are held in their orbits by a balance of forces. Like planets, electrons are acted upon by centrifugal (outward) forces. However, the inward force that acts on electrons is not gravity. It is an electrical force called *charge.*

**Charge**
A force that causes two particles to be attracted to, or repelled from, each other.

**Charge** is a force that causes two particles to be attracted to, or repelled from, each other. It was discovered in the early seventeenth century that there are two types of electrical charge. We refer to these charges as positive charge and negative charge. Note that the terms *positive* and *negative* are used to signify that the charges are opposites.

**FIGURE 2.5 Atomic charges.**

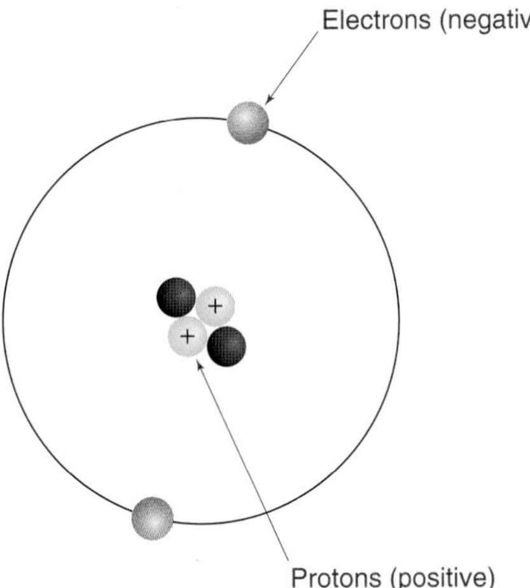


Since positive and negative charges are opposites, their effects cancel each other. For example, if an atom contains two positive charges and two negative charges, these charges effectively cancel each other, and the net charge on the atom is *zero*.

Protons and electrons are the source of charge in the atom. The protons each have a positive charge and the electrons each have a negative charge. This point is illustrated in Figure 2.5. The atom shown has a balance of positive and negative charges, and thus, a net charge of zero. The same situation is common for all atoms in their natural states. Note that neutrons are electrically neutral (which is how they got their name).

Proton → positive
Electron → negative
Neutron → neutral

## Attraction and Repulsion

One of the most fundamental laws of charges is that like charges repel each other and opposite charges attract each other. This law is illustrated in Figure 2.6. As you can see, the two positive charges are repelling each other, as are the two negative charges. However, a negative charge is attracted to a positive charge, and vice versa.

The force that two charges exert on each other is described by **Coulomb's law.** Coulomb's law states that the magnitude of force between two charges is directly proportional to the product of the charges and inversely proportional to the square of the distance between them. Coulomb's law is illustrated in Figure 2.7.

**Coulomb's law**
The magnitude of force between two charges is directly proportional to the product of the charges and inversely proportional to the square of the distance between them.

The force of attraction between the negatively charged electron and the positively charged proton is the inward force that offsets the centrifugal force of an orbiting electron. When these forces are balanced, the electron remains in its orbit.

*An interesting point:* You have been told that electron orbits are similar in nature to those of planets. Like electrons, planets are held in their orbits by a balance of outward and inward forces. In this case, gravity is the inward force. Newton's universal law of gravitation, which defines the gravitational force between two bodies, is given as

$$F_G = G\frac{m_1 m_2}{r^2}$$

If you compare this relationship to Coulomb's law (Figure 2.7), the similarities speak for themselves. (Both define the force of attraction that offsets centrifugal force, keeping the bodies in orbit.)

## Ions

When an outside force is applied to an atom, the balance between the forces acting on an electron can be disturbed, causing the electron to leave its orbit. This point is illustrated in Figure 2.8.

The forces associated with charges are commonly represented as outward and inward forces:

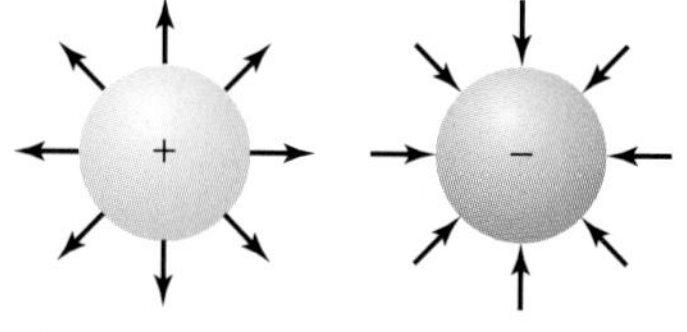


Two outward forces repel:

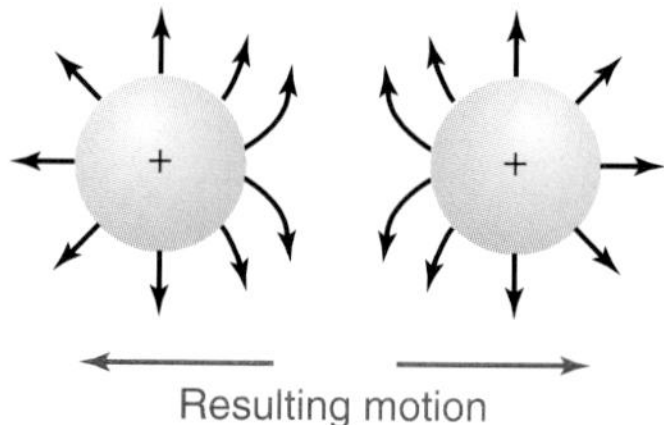


Two inward forces repel:

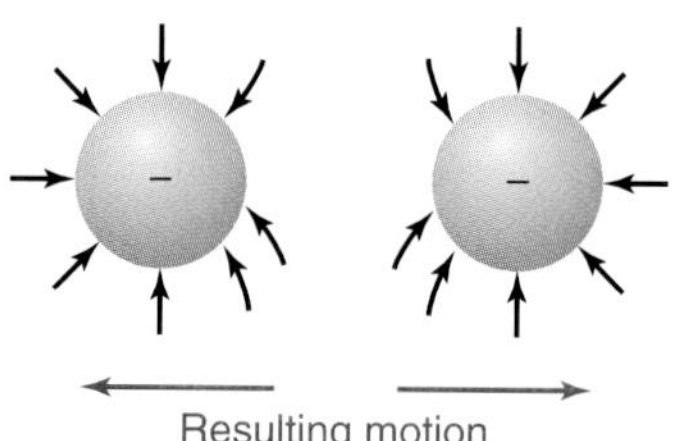


An outward force attracts an inward force:

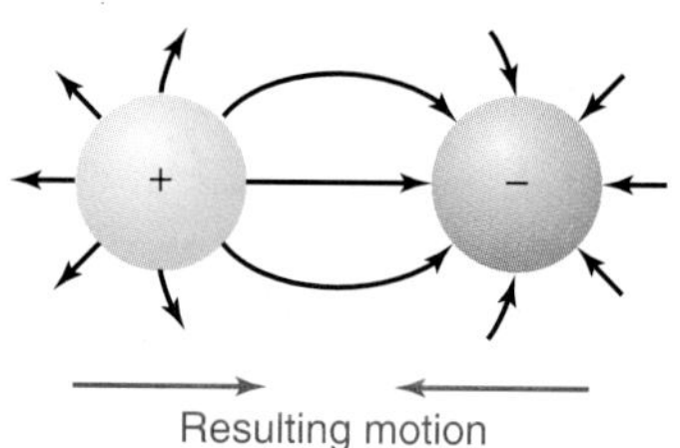


**FIGURE 2.6 Attraction and repulsion of charges.**

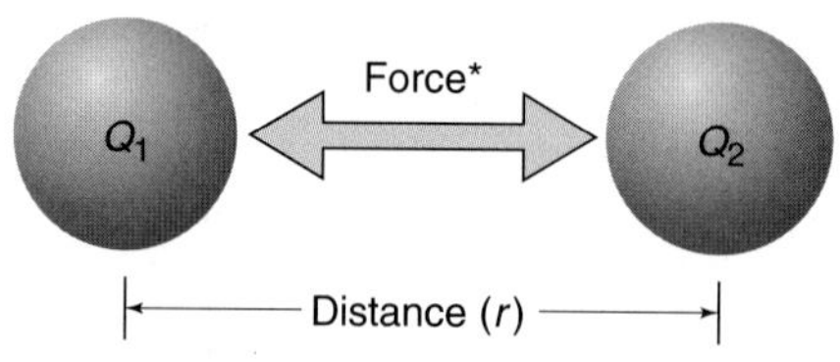


*Coulomb's Law*:

$$F = k\frac{Q_1 Q_2}{r^2}$$

$k = 9.0 \times 10^9 \frac{\text{N} \cdot \text{m}^2}{\text{C}^2}$ (the "constant of proportionality")
$Q_1 =$ the charge on the first body
$Q_2 =$ the charge on the second body
$r =$ the distance between the bodies

*Will cause unlike charges to be attracted and like charges to be repelled.

**FIGURE 2.7 Coulomb's law.**

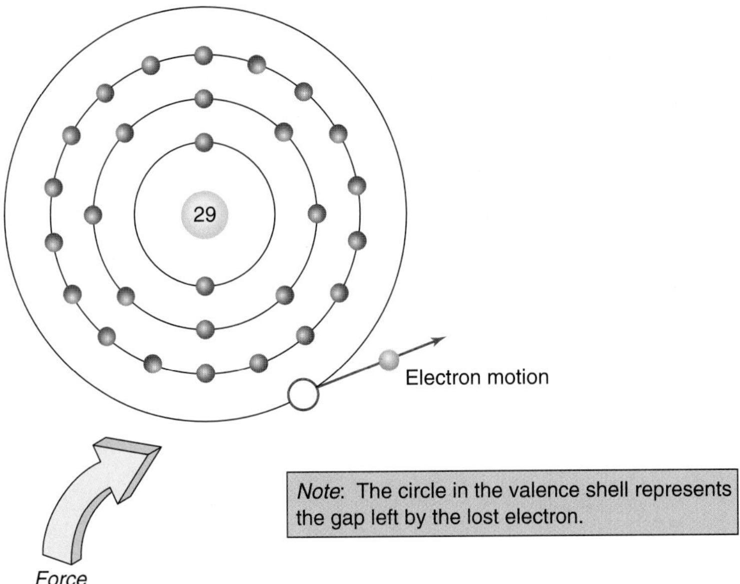
29
Electron motion
Note: The circle in the valence shell represents the gap left by the lost electron.
Force

**FIGURE 2.8 Generating a positive ion and a free electron.**

When the outside force causes the electron to leave its orbit, the atom is left with fewer electrons than protons, and thus has a net charge that is positive. For example, consider the copper atom shown in Figure 2.8. When the valence electron is forced to leave its orbit, the atom contains only twenty-eight electrons (negative charges). At the same time, it still contains twenty-nine protons (positive charges). Thus, the atom contains more positive charges than negative charges, and has a positive net charge. An atom with a positive net charge is referred to as a **positive ion.**

**Positive ion**
An atom that has fewer electrons than protons and thus a net charge that is positive.

Just as an outside force can cause an atom to lose an electron, it can also cause an atom to effectively gain an electron. When this occurs, the atom has more electrons than protons, and thus, a net charge that is negative. In this case, the atom is referred to as a **negative ion.**

**Negative ion**
An atom that has fewer protons than electrons and thus a net charge that is negative.

It is important to note that ions are formed by electron movement. Protons are fixed in the nucleus of the atom, and thus do not move from one atom to another. This is why the atomic number of an element is based on the number of protons (which is fixed) rather than on the number of electrons (which can vary).

## Free Electrons

When an outside force causes an electron to break free from its parent atom (as shown in Figure 2.8), that electron is referred to as a **free electron.** The term *free electron* means that the electron is not bound to any particular atom, and thus is free to drift from one atom to another.

**Free electron**
An electron that is not bound to any particular atom and is free to drift from one atom to another.

If a free electron drifts into the vicinity of a positive ion, the electron will neutralize the ion. For example, consider the free electron and the atom shown in Figure 2.9. When the electron nears the atom, it is attracted by the overall positive charge on the ion. As a result, the electron is pulled into orbit around the nucleus of the ion. When this happens, the atom once again has an equal number of negative and positive charges, and thus, a net charge of zero.

Figure 2.9 and its related discussion serve to illustrate an important point about positive and negative charges: positive and negative charges always seek to neutralize each other. As you will see in the next section, this is the basis of one of the primary electrical properties: current.

**FIGURE 2.9 Neutralizing a positive ion.**

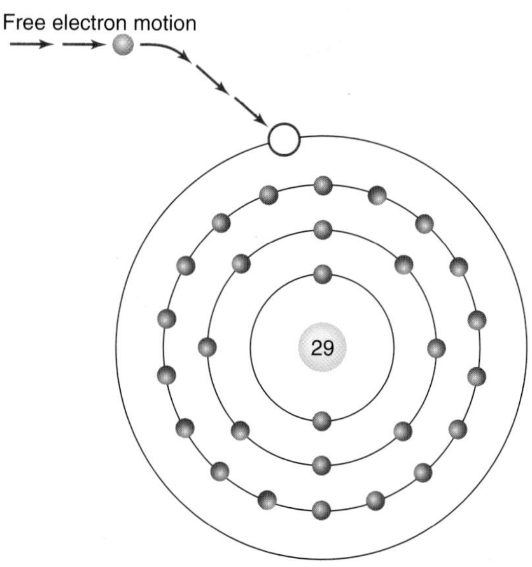


## The Bottom Line

The atom consists of a central core, called the nucleus, and one or more electrons that orbit the nucleus. The nucleus contains neutrons and protons. Electrons and protons exhibit a property referred to as charge. Neutrons (the largest of the three particles) are electrically neutral.

Charge is a force that causes two bodies to be attracted to, or repelled from, each other. There are two types of charge, called positive charge and negative charge.

A positive charge is always attracted to a negative charge, and vice versa. Two like charges (i.e., two positive charges or two negative charges) are repelled from each other.

Protons are positively charged particles and electrons are negatively charged particles. Since they have opposite charges, protons and electrons are attracted to each other. This attraction offsets the centrifugal (outward) force on the orbiting electrons, causing them to remain in their orbits.

An atom in its natural state contains an equal number of protons and electrons, and thus has a net charge of zero. If an outside force causes an atom to lose an electron, the atom is left with a positive net charge and is referred to as a positive ion. If an outside force causes an atom to gain an electron, the atom is left with a negative net charge and is referred to as a negative ion.

Positive and negative charges always seek to neutralize each other. If a free electron (one that has broken free of its parent atom) drifts into the vicinity of a positive ion, the electron falls into orbit around the atom. When this occurs, the net charge on the atom returns to zero.

## Section Review

1. What is matter?
2. What is an element? What trait is common to all elements?
3. What is an atom?
4. Describe the physical makeup of the atom.
5. How do atoms vary from one element to another?
6. What is the relationship between the number of protons and the number of electrons in an atom that is in its natural state?
7. What is an orbital shell?
8. What is the valence shell of an atom?

9. What is the restriction on the number of electrons that can occupy the valence shell of an atom?
10. What is a complete valence shell? What is an incomplete valence shell?
11. What are the primary forces that act on an electron that is orbiting the nucleus of an atom?
12. What is charge?
13. What are the two types of charges? How do they relate to each other?
14. What is the net charge on an atom in its natural state? Explain.
15. Identify the sources of charge in the atom.
16. What is the fundamental law of charges?
17. What are positive ions and negative ions? How is each created?
18. What is a free electron?
19. Describe what happens when a free electron drifts into the vicinity of a positive ion.

## 2.2 CURRENT

Current, voltage, and resistance are three of the most basic electrical properties. Stated simply, **current** is the directed flow of charge through a conductor, **voltage** is the force that causes the directed flow of charge (current), and **resistance** is the opposition to current provided by the conductor. In this section, we will take a close look at the first of these properties, current.

**Current**
The directed flow of charge through a conductor.

**Voltage**
The force that causes the current.

**Resistance**
An opposition to current.

As you know, an outside force can cause the valence electron in a copper atom to become a free electron. In copper (and many other elements), very little external force is needed to cause the generation of free electrons. In fact, the thermal energy (heat) present at room temperature (25°C) is sufficient to cause free electrons to be generated. The number of electrons generated varies directly with temperature.

In copper, the motion of the free electrons is random when no directing force is applied; that is, the free electrons move in every direction, as shown in Figure 2.10. Since the free electrons are moving in every direction, the net flow of electrons in any one direction is zero.

Now, assume that we apply an external force to the copper that causes all the electrons to move in the same direction, as shown in Figure 2.11. In this case, a negative potential is applied to one end of the copper and a positive potential is applied to the other. The free electrons move away from the negative potential (because like charges repel). At

**FIGURE 2.10 Random electron motion in copper.**

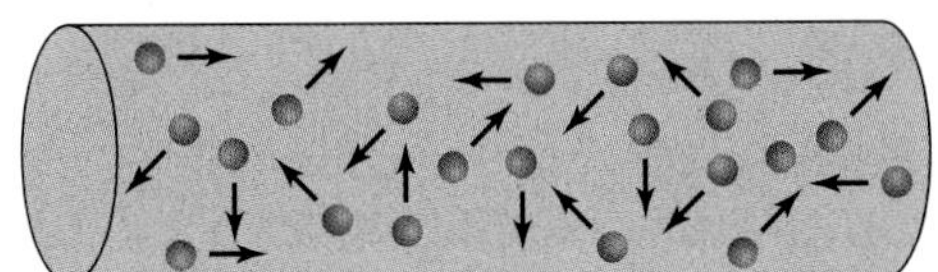

**FIGURE 2.11 Directed electron motion in copper.**

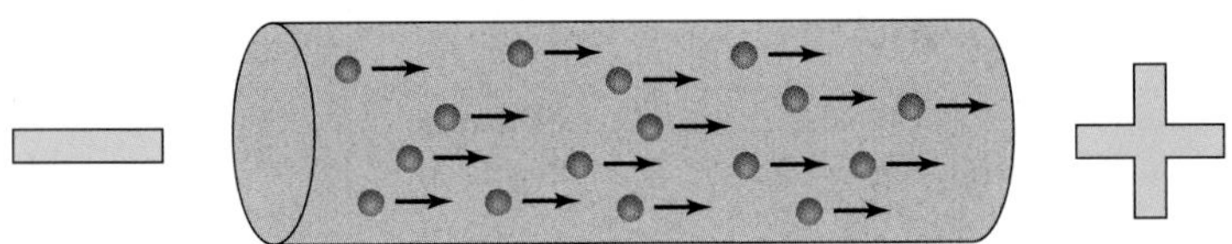

the same time, they are drawn toward the positive potential (because opposite charges attract). Thus, the free electrons all move from negative to positive, and we can say that we have a directed flow of charge (electrons). This directed flow of charge is referred to as *current.*

*OBJECTIVE 4* ▶

Current is represented by the letter $I$, which stands for *intensity.* The intensity of current depends on the amount of charge passing a given point and the amount of time required to do so. In other words, the intensity of current is measured in *charge per unit time.* Stated mathematically,

$$I = \frac{Q}{t} \tag{2.1}$$

where $I$ = the intensity of the current
$Q$ = the amount of charge
$t$ = the time (in seconds) required for the charge $(Q)$ to pass

**Coulomb (C)**
The basic unit of charge; equal to the combined charge of approximately $6.25 \times 10^{18}$ electrons.

**Ampere (A)**
The basic unit of current. A rate of charge flow equal to 1 coulomb per second.

The charge of a single electron is too small to provide a practical unit of charge. Therefore, the unit of charge used is the **coulomb (C),** which represents the combined charge of approximately $6.25 \times 10^{18}$ electrons. When one coulomb of charge passes a point in one second, we say that we have one **ampere (A)** of current. Note that the ampere (A) is the basic unit of measure for current $(I)$, given as

$$1 \text{ ampere} = 1 \text{ coulomb/second}$$

The total current passing a point (in amperes) can be found by dividing the total charge (in coulombs) by the time (in seconds). The calculation of current is demonstrated in the following examples.

## EXAMPLE 2.1

Three coulombs of charge pass through a point in a copper wire every two seconds. Calculate the total current through the wire.

***Solution:*** Using equation (2.1), the current is found as

$$\boldsymbol{I} = \frac{Q}{t} = \frac{3\text{ C}}{2\text{ s}} = 1.5\text{ C/s} = \mathbf{1.5\text{ A}}$$

***PRACTICE PROBLEM 2.1***

Ten coulombs of charge pass through a point in a copper wire every four seconds. Calculate the total current through the wire.

## EXAMPLE 2.2

Five coulombs of charge pass through a point in a copper wire every 20 seconds. Calculate the total current through the wire.

***Solution:*** Using equation (2.1), the total current through the wire is found as

$$\boldsymbol{I} = \frac{Q}{t} = \frac{5\text{ C}}{20\text{ s}} = 0.25\text{ C/s} = 250 \times 10^{-3}\text{ A} = \mathbf{250\text{ mA}}$$

***PRACTICE PROBLEM 2.2***

One coulomb of charge passes through a point in a wire every 100 seconds. Calculate the total current through the wire.

It should be noted that, in practice, current is not commonly calculated using equation (2.1) because it is neither easy nor practical to measure coulombs of charge directly. You are shown how to calculate current in this manner because it helps you to better understand the ampere as a measure of charge per unit time. As you will be shown in Chapter 4, there are far more practical ways of calculating the total current through a given point in a circuit.

## *Electron Flow Versus Conventional Current*

◄ *OBJECTIVE 5*

There are actually two ways to view current, and both are used in practice. The *electron flow* view defines current as the flow of charge from *negative to positive.* The *conventional current* view defines current as the flow of charge from *positive to negative.* Which viewpoint is correct? Both are.

If we assume that "the flow of charge" means the flow of electrons (as shown earlier), then we are taking the electron flow view of current. If we assume that it means the flow of positive ions, then we are taking the conventional current view. Both approaches to defining current are illustrated in Figure 2.12.

Assume that atom 1 (labeled A1) in Figure 2.12 is a positive ion, and that the potentials (+ and −) are applied as shown. In the first line, the valence electron from A2 is shown jumping to A1. The result is that A1 has been neutralized by the added electron and A2 (which lost an electron) has become a positive ion. In the second line, the same action causes A2 to be neutralized and A3 to become a positive ion, and so on. The result is that electrons are flowing from negative to positive, and a positive ion is *effectively* flowing from positive to negative.

The electron flow view of current would describe the action in Figure 2.12 as the flow of electrons from negative to positive. The conventional current approach would describe the same action as the flow of positive ions from positive to negative. As you can see, both approaches are valid.

Conventional current is sometimes referred to as *hole flow.*

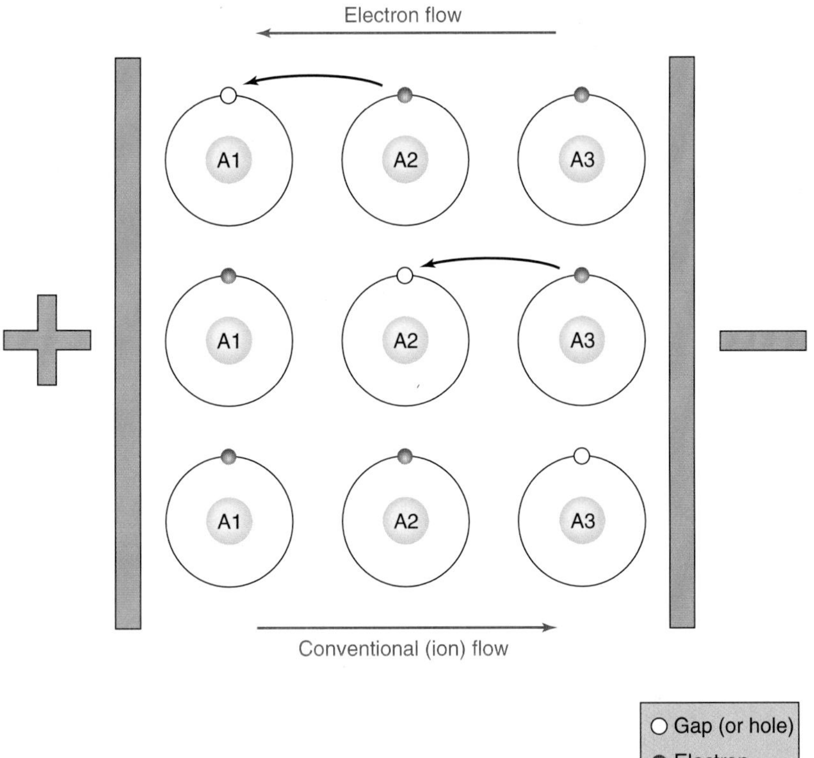


**FIGURE 2.12 Conventional and electron flow.**

The approach taken to current does not affect any circuit calculations or measurements. Even so, you should take the time to get comfortable with both approaches because both are used by many engineers, technicians, and technical publications. In this text, we will take the electron flow approach to current; that is, we will view current as the flow of electrons from negative to positive.

OBJECTIVE 6 ▶

## Direct Current Versus Alternating Current

Current is generally classified as being either direct current (dc) or alternating current (ac). The differences between direct current and alternating current are illustrated in Figure 2.13.

**Direct current (dc)**
A term used to describe current that is unidirectional; i.e., charge flows in one direction only.

**Direct current (dc)** is unidirectional; that is, the flow of charge is always in the same direction. The term *direct current* usually implies that the current has a fixed value. For example, the graph in Figure 2.13a shows that the current has a constant value of 1 A. While a fixed value is implied, direct current *can* change in value (as demonstrated in Figure 2.13b). However, the *direction* of the current does not change.

**Alternating current (ac)**
A term used to describe current that is bidirectional; i.e., the flow of charge changes direction periodically.

**Alternating current (ac)** is bidirectional; that is, the direction of current changes from one point in time to another. For example, in Figure 2.13c, the graph shows that the current builds to a peak value in one direction, and then builds to a peak value in the other direction. Note that the alternating current represented in Figure 2.13c not only changes direction but is constantly changing in value.

Through the first ten chapters of this text, we will deal exclusively with direct current and dc circuits. Then, our emphasis will shift to alternating current and ac circuits. As your electronics education continues, you will find that most electronic systems contain both dc and ac circuits.

EWB

**FIGURE 2.13 Direct current (dc) versus alternating current (ac).**

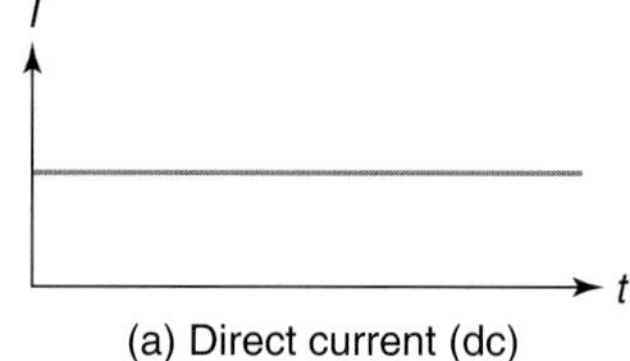

(a) Direct current (dc)

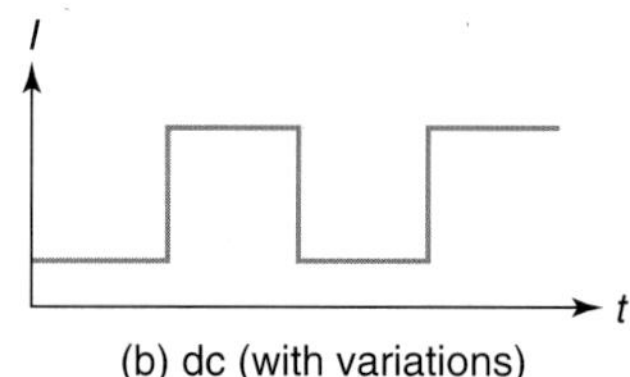

(b) dc (with variations)

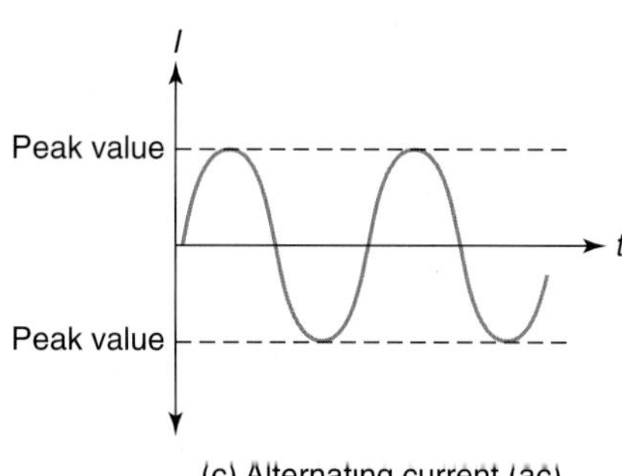

(c) Alternating current (ac)

## The Bottom Line

Free electrons are generated in copper at room temperature. When undirected, the motion of these free electrons is random, and the net flow of electrons in any one direction is zero.

When directed by an outside force, free electrons are forced to move in a uniform direction. This directed flow of charge is referred to as current.

Current is represented by the letter *I*, which stands for *intensity*. The intensity of current depends on the amount of charge moved and the time required to move it.

Current is measured in amperes (A). When one coulomb of charge passes a point every second, you have one ampere of current.

There are two approaches to defining current. The electron flow approach views current as the flow of charge (electrons) from negative to positive. The conventional current approach views it as the flow of charge (positive ions) from positive to negative. Both approaches are widely used and are technically correct. The approach taken does not affect the outcome of any circuit calculations or measurements.

Most electronic systems contain both direct current (dc) and alternating current (ac) circuits. In dc circuits, the flow of charge is always in the same direction. In ac circuits, the flow of charge changes direction periodically.

### Section Review

1. How are free electrons generated in a conductor at room temperature?
2. When undirected, the net flow of free electrons in any single direction is zero. Why?
3. What is current? What factors affect the intensity of current?
4. What is a coulomb?
5. What is the basic unit of current? How is it defined?
6. Why isn't current normally calculated in terms of charge and time?
7. Discuss the electron flow approach to describing current.
8. Discuss the conventional current approach to describing current.
9. Describe the difference between direct current (dc) and alternating current (ac).

## 2.3 VOLTAGE

◄ *OBJECTIVE 7*

**Voltage** can be described as the potential that causes the directed flow of charge (current) in a circuit. In this section, we'll take a closer look at voltage and the means by which it produces current.

**Voltage**
The potential that causes the directed flow of charge (current) in a circuit. Voltage is often referred to as *difference of potential.*

As you know, a positive ion is an atom that has lost an electron and has a net charge that is positive. A negative ion is an atom that has gained an electron and has a net charge that is negative.

The battery shown in Figure 2.14 has two terminals. One of the terminals is shown to have an excess of positive ions and is described as having a *positive potential.* The other is shown to have an excess of negative ions and is described as having a negative potential. Between the two terminals, we have a difference of potential, or voltage (*V*).

If we connect the two terminals of the battery with a copper wire (as shown in Figure 2.15), a current is produced in the wire as the negative and positive charges seek to neutralize each other. In other words, electrons flow from the negative terminal to the positive terminal until all the atoms in both terminals have a net charge of zero (i.e., the battery is dead).

*A Practical Consideration:* The terminal connection shown in Figure 2.15 is used to illustrate a point. *In practice, a wire is never connected across the terminals of any voltage source.* The reason for this will be made clear in Chapter 3.

There are several important points that should be made at this time:

1. Voltage is a force that moves electrons. For this reason, it is often referred to as *electromotive force,* or *emf.*

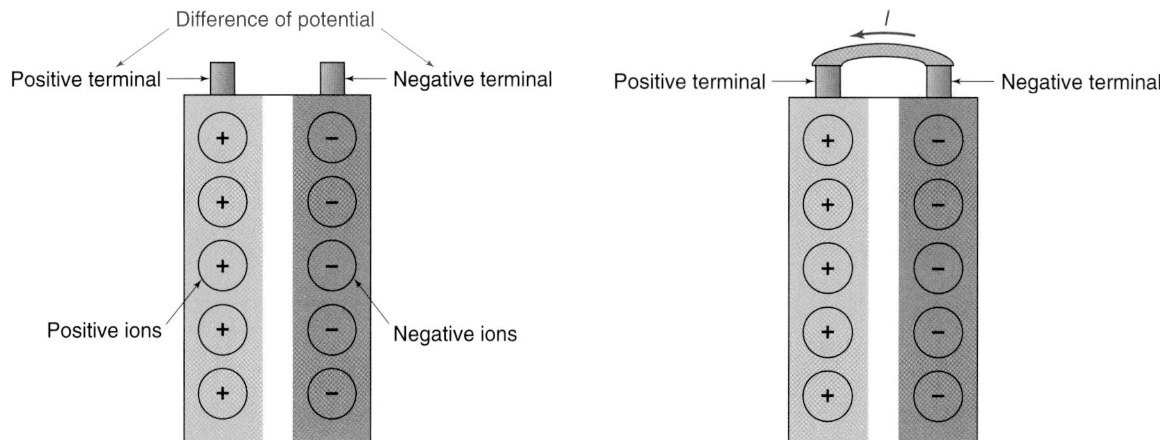


FIGURE 2.14

FIGURE 2.15

2. Current and voltage are not the same thing. Current is the flow of charge between the two terminals. The voltage is the difference of potential between the two terminals that generates the current. In other words, current occurs as a result of an applied voltage (difference of potential).

**Voltage**
The energy required to move a coulomb of charge.

Technically defined, **voltage** ***(V)*** is the amount of energy required to move a coulomb of charge. By formula,

$$V = \frac{E}{Q} \tag{2.2}$$

where $V$ = the applied voltage
$E$ = the energy used (in joules)
$Q$ = the charge moved (in coulombs)

A one-volt difference of potential uses one joule of energy to move one coulomb of charge. Thus,

$$1 \text{ volt} = 1 \text{ joule/coulomb}$$

**Volt (V)**
The unit of voltage (difference of potential). One volt is equal to 1 joule per coulomb.

Note that the **volt (V)** is the basic unit of voltage. The calculation of voltage using values of energy and charge is demonstrated in the following example.

## EXAMPLE 2.3

*A Practical Consideration:* Neither joules nor coulombs are easily measured, so equation (2.2) is rarely used. However, the example is included to demonstrate the relationship among joules, coulombs, and volts.

Ten joules are used by a battery to move 2.5 coulombs of charge. Calculate the voltage rating of the battery.

***Solution:*** Using equation (2.2), the rating of the battery is found as

$$V = \frac{E}{Q} = \frac{10 \text{ j}}{2.5 \text{ C}} = 4 \text{ j/C} = \mathbf{4\ V}$$

***PRACTICE PROBLEM 2.3***

A battery uses 2 j of energy to move 40 mC of charge. Calculate the voltage rating of the battery.

1. What is voltage?
2. How does voltage produce current through a wire?
3. What is a joule?
4. What is a volt?

**Section Review**

## 2.4 RESISTANCE AND CONDUCTANCE

◄ *OBJECTIVE 8*

Resistance and conductance are basically opposites. While resistance is the opposition to current in a circuit, **conductance** is a measure of the ability of a circuit to conduct (pass current).

**Conductance**
A measure of the ability of a circuit to conduct (pass current).

### Resistance

All elements provide some amount of opposition to current. This opposition to current is called *resistance.* The higher the resistance of an element, component, or circuit, the lower the current produced by a given voltage.

Resistance (R) is measured in *ohms.* Ohms are represented using the Greek letter *omega* (Ω). Technically defined, one **ohm** is the amount of resistance that limits current to 1 ampere when a 1 volt difference of potential is applied. This definition is illustrated in Figure 2.16.

**Ohm**
The unit of measure for resistance, represented using the Greek letter *omega* (Ω). One ohm is the amount of resistance that limits current to 1 A when a 1 V difference of potential is applied.

The schematic diagram shown in Figure 2.16 contains a battery and a resistor (a component designed to provide a specific amount of resistance). As shown in the figure, a resistance of 1 Ω limits the current to 1 A when 1 V is applied. Note that the long end-bar on the battery schematic symbol represents the positive terminal, and the short end-bar represents the negative terminal.

### Conductance

The opposite of resistance is conductance. While resistance is a measure of opposition to current, conductance is a measure of the ease with which current will pass through a component or circuit. Conductance *(G)* is found as the reciprocal of resistance. By formula,

$$G = \frac{1}{R} \qquad \textbf{(2.3)}$$

where $G$ = the conductance of the circuit or component
$R$ = the resistance of the circuit or component

Conductance is most commonly measured in siemens (S). However, some publications use the *mho* unit, which is symbolized by an upside-down omega. The following example illustrates the calculation of conductance when resistance is known.

Conductance is measured in Siemens (S). (Some publications still use *mhos* as the unit of measure, symbolized by an upside-down omega.)

**FIGURE 2.16**

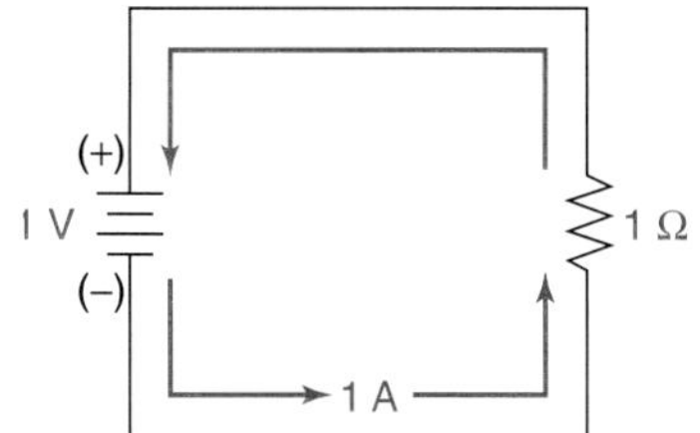


EWB

### EXAMPLE 2.4

Calculate the conductance of a 10 kΩ resistor.

***Solution:*** Using equation (2.3), the conductance of the resistor is found as

$$G = \frac{1}{R} = \frac{1}{10\ \text{k}\Omega} = 1 \times 10^{-4}\ \text{S} = \mathbf{100\ \mu S}$$

***PRACTICE PROBLEM 2.4***

Calculate the conductance of a 33 kΩ resistor.

Just as resistance can be used to calculate conductance, conductance (when known) can be used to calculate resistance, as follows:

$$R = \frac{1}{G} \tag{2.4}$$

### EXAMPLE 2.5

A circuit has a conductance of 25 mS. Calculate the resistance of the circuit.

***Solution:*** Using equation (2.4), the circuit resistance is found as

$$\mathbf{R} = \frac{1}{G} = \frac{1}{25\ \text{mS}} = \mathbf{40\ \Omega}$$

***PRACTICE PROBLEM 2.5***

A circuit has a conductance of 10 mS. Calculate the resistance of the circuit.

## Putting It All Together

We have now defined charge, current, voltage, resistance, and conductance. For convenience, these electrical properties are summarized in Table 2.1.

Many of the properties listed in Table 2.1 can be defined in terms of the others. For example, in our discussion on resistance, we said that one ohm is the amount of resistance that limits current to 1 A when a 1 V difference of potential is applied. By the same token, we can redefine the ampere and the volt as follows:

1. One ampere is the amount of current generated when a 1 V difference of potential is applied across 1 Ω of resistance.

**TABLE 2.1 Basic Electrical Properties**

| *Property* | *Unit* | *Description* |
|---|---|---|
| Charge *(Q)* | Coulomb (C) | The combined charge of $6.25 \times 10^{18}$ electrons. |
| Current *(I)* | Ampere (A) | The flow of charge; equal to 1 coulomb per second. |
| Voltage *(V)* | Volt (V) | Difference of potential. The energy required to move a given amount of charge; equal to 1 joule per coulomb. |
| Resistance *(R)* | Ohm (Ω) | Opposition to current. One ohm of resistance limits current to 1 A when 1 V is applied. |
| Conductance *(G)* | Siemens (S) | The reciprocal of resistance. A measure of the relative ease with which current will pass through a component or circuit. |

2. One volt is the difference of potential required to generate 1 A of current through 1 Ω of resistance.

As you will see in Chapter 4, these definitions relate closely to the most basic law of electronics, *Ohm's law.*

**Section Review**

1. What is resistance?
2. What is the basic unit of resistance? How is it defined?
3. What is conductance? What is its unit of measure?

# 2.5 CONDUCTORS, INSULATORS, AND SEMICONDUCTORS

In Section 2.4, you were told that conductance is the reciprocal of resistance. This means that a material with low resistance has high conductance and will conduct (pass current) easily. At the same time, a material with high resistance has low conductance and will not conduct easily. All materials are classified according to their ability (or inability) to conduct, as shown in Table 2.2.

A good example of a conductor is copper. Copper wire (which is used to interconnect components in a circuit) passes current with little opposition. A good example of an insulator is rubber. Rubber is used to coat the handles of many tools that are used in electrical work (such as pliers, screwdrivers, etc.) It takes an extremely high voltage per unit volume to force rubber to conduct. A good example of a semiconductor is graphite (a form of carbon), which is used to make many common resistors. Graphite limits the amount of current that can be generated by a given amount of voltage. In this section, we will take a look at some of the characteristics of conductors, insulators, and semiconductors.

## Conductors

Conductors are materials that easily give up (or accept) free electrons. The less energy needed to cause the flow of free electrons, the better the conductor.

In Section 2.1, you were told that the valence shell of an atom determines its electrical characteristics. The conductivity of a given element is actually determined by:

1. *Its number of valence shell electrons.* The more valence shell electrons that an atom contains, the more difficult it is to force the atom to give up (or accept) free electrons. The best conductors contain one valence electron per atom.
2. *The number of atoms per unit volume.* With more atoms per unit volume, more free electrons can be generated by a given amount of voltage. The best conductors contain a high number of atoms per unit volume.

*A Practical Consideration:* The best conductors (like copper) contain one valence electron, but many metal conductors contain two or three. For example, aluminum (a conductor) has three valence electrons.

**TABLE 2.2 Material Classifications**

◄ *OBJECTIVE 9*

| *Material* | *Characteristics* |
|---|---|
| Conductor | Extremely low resistance (high conductance). Allows current to pass with very little voltage applied. |
| Insulator | Extremely high resistance (low conductance). Allows current to pass only when an extremely high voltage per unit volume is applied. |
| Semiconductor | Resistance that falls about midway between that of a conductor and an insulator. Limits the current at a given voltage. |

**FIGURE 2.17**

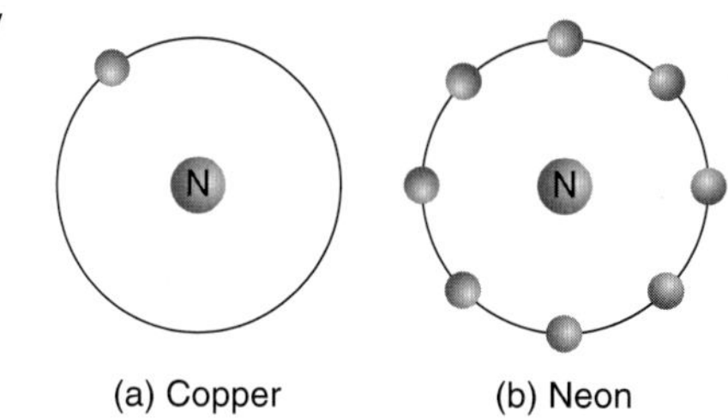


(a) Copper (b) Neon

The first of these principles is easy to understand when you consider the fact that an atom will give up electrons only as easily as it will accept them. For example, consider the atoms represented in Figure 2.17. (For simplicity, only the nucleus and valence shell of each atom are shown.) As you know, the valence shell of an atom can hold up to eight electrons. The copper atom represented in Figure 2.17a contains only one valence electron, so it can easily accept more. Since the copper atom will easily accept an electron, it will easily give up an electron. In contrast, the neon atom represented in Figure 2.17b has a complete valence shell, and therefore will not easily accept an electron. Since the neon atom will not easily accept an electron, it will not easily give up an electron. Thus, copper is a much better conductor than neon.

*A Practical Consideration:* The concept of "atoms per unit volume" applies only to gases. For solids, the number of atoms per unit volume generally has little effect on conductivity.

The second point should need little explanation. If one element has twice as many atoms per unit volume as another, it will generate twice as many free electrons at a given value of applied voltage.

## Insulators

Insulators are materials that normally block current. For example, the insulation that surrounds a power cord prevents current from reaching you when it is touched. Some elements (like neon) are natural insulators. Most insulators, however, are compounds such as rubber, Teflon, and mica (among others).

As you have probably guessed, conductors and insulators have opposite characteristics. In general, insulators are materials that have:

1. Complete valence shells.
2. Very few atoms per unit volume.

With this combination of characteristics, an extremely high voltage is required to force an insulator into conduction.

## Semiconductors

Semiconductors are materials that are neither good conductors nor good insulators. For example, graphite (a form of carbon) does not conduct well enough to be considered a conductor. At the same time, it does not block current well enough to be considered an insulator. Some other examples of semiconductors are silicon, germanium, and gallium arsenide.

In general, semiconductors have the following characteristics:

1. Valence shells that are half complete (i.e., contain four valence electrons).
2. A relatively high number of atoms per unit volume.

These characteristics combine to form a material that is neither a good conductor nor a good insulator.

## Other Factors That Affect Resistance

The resistance of a conductor, insulator, or semiconductor at a given temperature depends on three factors:

◄ *OBJECTIVE 10*

1. The *resistivity* of the material used.
2. The length of the material.
3. The cross-sectional area of the material.

The **resistivity** of an element or compound is the resistance of a specified volume of the element or compound. Resistivity is commonly rated in one of two ways:

**Resistivity**
The resistance of a specified volume of an element or compound.

1. Ohms per circular-mil foot.
2. Ohms per cubic centimeter.

The volumes used in these two ratings are illustrated in Figure 2.18.

The **mil** is a unit of length equal to one-thousandth of an inch (0.001 in.). The **circular-mil** is a unit of area found by squaring the diameter (in mils) of the conductor. One *circular-mil foot* of material has a diameter of one mil and a length of one foot (as shown in Figure 2.18a). When this volume is used to measure the resistivity of a given material, the rating is given in *circular mil-ohms per foot* (CM-Ω/ft).

**Mil**
A unit of length equal to one-thousandth of an inch.

**Circular-mil**
A unit of area found by squaring the diameter (in mils) of a conductor.

The second unit of volume used for rating resistivity is the cubic centimeter. A cubic centimeter of material is represented in Figure 2.18b. When this volume is used to measure the resistivity of a given material, the rating is given in *ohm-centimeters* (Ω-cm).

Table 2.3 shows the resistivity ratings of some elements that are commonly used in electronics. Note that resistivity is given in either circular mil-ohms per foot (CM-Ω/ft) or ohm-centimeters (Ω-cm), depending on the volume used in the measurement. The resistivity rating of copper, for example, is 10.37 CM-Ω/ft or $1.723 \times 10^{-6}$ Ω-cm.

*A Practical Consideration:* Resistivity is most commonly given in ohm-centimeters. The circular mil-ohms per foot rating is used for lengths of cable.

The resistivity ratings of semiconductors and insulators are typically much higher than those of conductors. For example, carbon (as shown in the table) has a rating of $2.1 \times 10^4$ CM-Ω/ft. The typical range of values for insulators is between $10^{11}$ and $10^{16}$ CM-Ω/ft!

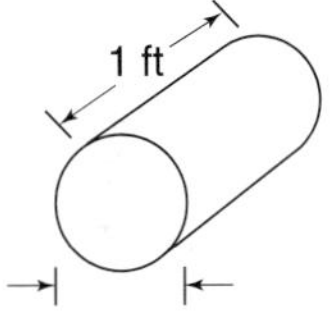

(a) Circular-mil-foot

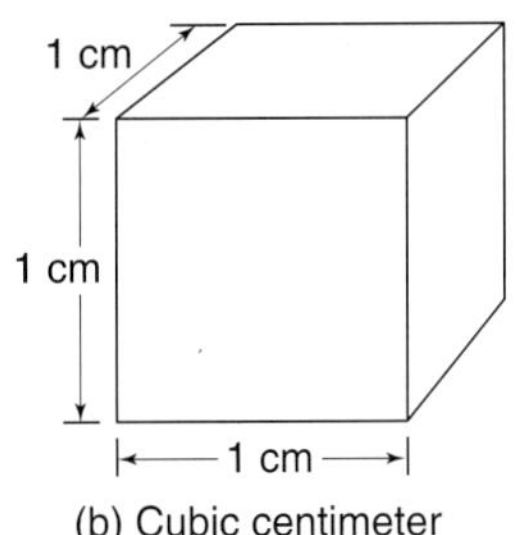

(b) Cubic centimeter

**FIGURE 2.18 Units of volume for resistivity measurements.**

## Calculating the Resistance of a Conductor

When the length and cross-sectional area of a conductor are known, the resistance of the conductor can be found as

$$R = \rho \frac{l}{A} \quad \textbf{(2.5)}$$

where
$\rho$ = the resistivity of the material
$l$ = the length of the conductor
$A$ = the cross-sectional area of the conductor

Note that the Greek letter rho *($\rho$)* is commonly used to represent resistivity. The following examples illustrate the calculation of the resistance of a conductor.

**TABLE 2.3 Resistivity Ratings of Some Common Elements**

| | *Resistivity* | |
|---|---|---|
| *Element* | *CM-Ω/ft* | *Ω-cm* |
| Silver | 9.9 | $1.645 \times 10^{-6}$ |
| Copper | 10.37 | $1.723 \times 10^{-6}$ |
| Gold | 14.7 | $2.443 \times 10^{-6}$ |
| Aluminum | 17.0 | $2.825 \times 10^{-6}$ |
| Iron | 74.0 | $12.299 \times 10^{-6}$ |
| Carbon | $2.1 \times 10^4$ | $3.50 \times 10^{-3}$ |

## EXAMPLE 2.6

Calculate the resistance of a piece of copper that is six inches long and has a diameter of 0.005 inch.

***Solution:*** Copper has a resistivity rating of 10.37 Ω/CM-ft (as shown in Table 2.3). Six inches is half a foot (0.5 ft) and 0.005 inch is 5 mils. With a diameter of 5 mils, the cross-sectional area is found as

$$\text{Area} = (5\text{ mils})^2 = \mathbf{25\ CM}$$

Now, using equation (2.5), the resistance of the copper is found as

$$R = \rho\frac{l}{A} = (10.37\text{ CM-}\Omega/\text{ft})\frac{0.5\text{ ft}}{25\text{ CM}} = \mathbf{207\ m\Omega}$$

With this kind of resistance, it is easy to see that copper provides very little opposition to current and therefore is considered to be an excellent conductor.

***PRACTICE PROBLEM 2.6***

Calculate the resistance of an 18 inch length of copper that has a diameter of 0.005 inch.

## EXAMPLE 2.7

Calculate the resistance of a 25 cm length of copper that has a cross-sectional area of 0.04 cm$^2$.

***Solution:*** Copper has a resistivity rating of $1.723 \times 10^{-6}$ Ω-cm (as shown in Table 2.3). Using this value and equation (2.5), the resistance of the copper is found as

$$R = \rho\frac{l}{A} = (1.723 \times 10^{-6}\ \Omega\text{-cm})\frac{25\text{ cm}}{0.04\text{ cm}^2} = \mathbf{1.08\ m\Omega}$$

***PRACTICE PROBLEM 2.7***

Calculate the resistance of a 25 cm length of copper that has a cross-sectional area of 0.1 cm$^2$.

## What Equation (2.5) Shows Us

Every equation serves two purposes. First, it shows us how to calculate a value of interest. Second (and probably most important), it shows us the relationship between the value of interest and its variables. For example, equation (2.5) shows us that

1. The resistance of a material is directly proportional to its resistivity; that is, an increase (or decrease) in resistivity causes similar change in resistance.
2. The resistance of a material is directly proportional to its length.
3. The resistance of a material is inversely proportional to its cross-sectional area; that is, an increase in cross-sectional area causes a decrease in resistance, and vice versa.

The first statement can be demonstrated by replacing the copper in Example 2.7 with aluminum, which has a resistivity of $2.825 \times 10^{-6}$ Ω-cm. If we do this, we get a total resistance of 1.77 mΩ, which is higher than the value obtained in the example.

The second statement is demonstrated by comparing the results of Example 2.6 and Practice Problem 2.6. The third statement is demonstrated by comparing the results of Example 2.7 and Practice Problem 2.7. Since the practice problems are included for you to solve, the comparison of the results is left up to you. Note, however, that the diameters are equal. Only the lengths have changed.

## The Effects of Temperature on Resistance

For some materials, an increase in temperature results in an increase in resistance. These materials are said to have a **positive temperature coefficient.** For others, the opposite is true; that is, an increase in temperature results in a decrease in resistance (and vice versa). These materials are said to have a **negative temperature coefficient.**

Conductors have positive temperature coefficients. Semiconductors and insulators generally have negative temperature coefficients. Thus, the resistance of a conductor will vary directly with temperature, while the resistance of a semiconductor or insulator will vary inversely with temperature.

**Positive temperature coefficient** A rating which indicates that resistance varies directly with temperature.

**Negative temperature coefficient** A rating which indicates that resistance varies inversely with temperature.

## Relative Conductivity

Just as conductance is the reciprocal of resistance, conductivity is the reciprocal of resistivity. **Relative conductivity** is the conductivity of a material as compared to that of copper. As shown in Table 2.4, relative conductivity is given as a percentage.

Since copper is the most commonly used conductor, its relative conductivity is used as a reference for the other materials and is therefore given as 100%. Silver is rated at 105%, which is 5% higher than copper and indicates that silver is a better conductor than copper. The other elements listed have ratings lower than that of copper and therefore do not conduct as easily as copper does. Note the value associated with carbon. This value indicates that copper is approximately 2,000 times as conductive as carbon.

**Relative conductivity** The conductivity of a material as compared to the conductivity of copper, expressed as a percentage.

**TABLE 2.4 Relative Conductivity Ratings of Some Common Elements**

| *Material* | *Relative Conductivity (%)* |
|---|---|
| Silver | 105.0 |
| Copper | 100.0 |
| Gold | 70.5 |
| Aluminum | 61.0 |
| Iron | 14.0 |
| Carbon (graphite) | 0.05 |

### Section Review

1. What is a conductor? An insulator? A semiconductor?
2. List the factors that determine the conductivity of an element.
3. List the factors that determine the resistance of a conductor, semiconductor, or insulator.
4. What is resistivity?
5. Describe the units of measure commonly used to express resistivity.
6. What is meant by the term *positive temperature coefficient?*
7. What type of material has a positive temperature coefficient?
8. What is meant by the term *negative temperature coefficient?*
9. What types of materials have negative temperature coefficients?
10. What is relative conductivity?

## Chapter Summary

Here is a summary of the major points made in this chapter:

1. The atom is made up of neutrons, protons, and electrons. Neutrons and protons make up the nucleus (core) of the atom. Electrons orbit the nucleus.
2. In its natural state, an atom contains an equal number of electrons and protons.
3. Electrons travel in orbital paths called shells.
4. The valence (outermost) shell of an atom cannot contain more than eight electrons.
5. Charge is a force that causes two particles to be attracted to, or repelled from, each other. The two types of charge are referred to as positive charge and negative charge.
6. Protons have a positive charge and electrons have a negative charge.
7. When an atom has an equal number of protons and electrons, the net charge of the atom is zero.
8. Like charges repel each other. Unlike charges attract each other.
9. When an atom loses an electron, it becomes a positive ion. When an atom gains an excess electron, it becomes a negative ion.
10. A free electron is one that has broken away from its parent atom and is free to drift to another atom.
11. Positive and negative charges always seek to neutralize each other.
12. *Current* is the directed flow of charge, measured in amperes (or "amps").
13. Electrons flow from negative to positive.
14. The coulomb is the combined charge of approximately $6.25 \times 10^{18}$ electrons.
15. A charge flow rate of one coulomb per second is equal to one ampere.
16. The *electron flow* approach defines current as the flow of electrons from negative to positive. The *conventional current* approach defines current as the flow of positive ions from positive to negative.
17. Direct current (dc) is unidirectional. Alternating current (ac) is bidirectional.
18. Voltage is a difference of potential that causes current. It is also referred to as electromotive force, or emf.
19. The unit of measure for voltage is the volt: equal to one joule per coulomb. The joule is the basic unit of energy.
20. Resistance is the opposition to current, measured in ohms ($\Omega$). One ohm is the amount of resistance that limits current to one ampere when a one-volt difference of potential exists.
21. Conductance is the reciprocal of resistance, measured in siemens (S) or mhos.
22. A conductor has extremely low resistance and easily passes current.
23. An insulator has extremely high resistance and allows current only when an extremely high voltage is present.
24. A semiconductor is neither a good conductor nor a good insulator.
25. The conductivity of a material is determined by its number of valence shell electrons (per atom) and the number of atoms per unit volume.
26. Conductors have few valence electrons per atom and a large number of atoms per unit volume.
27. Insulators have complete valence shells and few atoms per unit volume.
28. Semiconductors have valence shells that are half complete, a relatively high number of atoms per unit volume, and atoms spaced relatively far apart.
29. The resistance of a conductor, semiconductor, or insulator is determined by its resistivity, length, and cross-sectional area.
30. Resistivity is the resistance of one circular-mil-foot or one cubic centimeter of a material.

31. Conductors have low resistivity ratings, while insulators have extremely high resistivity ratings.
32. Conductors have positive temperature coefficients, which means that their resistance varies directly with temperature.
33. Semiconductors and insulators have negative temperature coefficients, which means that their resistance varies inversely with temperature.
34. The relative conductivity of an element is its conductivity as compared to copper, given as a percentage.

## Equation Summary

| Equation Number | Equation | Section Number |
|---|---|---|
| **(2.1)** | $I = \frac{Q}{t}$ | 2.2 |
| **(2.2)** | $V = \frac{E}{Q}$ | 2.3 |
| **(2.3)** | $G = \frac{1}{R}$ | 2.4 |
| **(2.4)** | $R = \frac{1}{G}$ | 2.4 |
| **(2.5)** | $R = \rho\frac{l}{A}$ | 2.5 |

## Key Terms

The following terms were introduced and defined in this chapter:

alternating current (ac)
ampere (A)
atom
Bohr model
charge
circular-mil
circular-mil-foot
circular-mil-ohms per foot (CM-Ω/ft)
conductance
conductor
coulomb
current
direct current (dc)
electromotive force (emf)
element
free electron
insulator
ion
matter
mil
negative ion
negative temperature coefficient
nucleus
ohm (Ω)
ohm-centimeter (Ω-cm)
positive ion
positive temperature coefficient
relative conductivity
resistance
resistivity
semiconductor
shell
thermal energy
valence shell
voltage

## Practice Problems

1. One coulomb of charge passes a point every 20 seconds. Calculate the value of the current through the point.
2. A total charge of $2.5 \times 10^{-3}$ C passes a point every 40 seconds. Calculate the value of the current through the point.
3. A total charge of $4.0 \times 10^{-6}$ C passes a point every second. Calculate the value of the current through the point.
4. A total charge of $50 \times 10^{-3}$ C passes a point every 2.5 seconds. Calculate the value of the current through the point.

5. Five joules of energy are required to move a total charge of $100 \times 10^{-3}$ C. Calculate the value of the applied voltage.
6. Ten joules of energy are required to move a total charge of 5 C. Calculate the value of the applied voltage.
7. One joule of energy is required to move a total charge of 10 C. Calculate the value of the applied voltage.
8. Two joules of energy are required to move a total charge of 80 C. Calculate the value of the applied voltage.
9. Calculate the conductance of a 4.7 Ω resistor.
10. Calculate the conductance of a 1.5 MΩ resistor.
11. Calculate the conductance of a 510 kΩ resistor.
12. Calculate the conductance of a 22 kΩ resistor.
13. A circuit has 50 μS of conductance. Calculate the circuit resistance.
14. A circuit has 4.25 mS of conductance. Calculate the circuit resistance.
15. A circuit has 250 nS of conductance. Calculate the circuit resistance.
16. A circuit has 375 μS of conductance. Calculate the circuit resistance.
17. Calculate the resistance of a 4 in. length of copper that has a cross-sectional area of 50 CM.
18. Calculate the resistance of a 30 cm length of copper that has a cross-sectional area of 0.08 $cm^2$.

## The Brain Drain

19. A 10 V source uses two joules of energy per second. Calculate the current generated by this voltage source.
20. The current through a wire is equal to 100 mA. The energy used by the voltage source is $2.5 \times 10^{-3}$ j. What is the value of the voltage source?
21. If one value increases at the same rate that another decreases (or vice versa), the two values are said to be inversely proportional. Prove that resistance and conductance are inversely proportional.
22. If two values increase (or decrease) at the same rate, the two values are said to be directly proportional. Prove that current is directly proportional to charge and inversely proportional to time.

## Answers to the Example Practice Problems

**2.1** 2.5 A
**2.2** 10 mA
**2.3** 50 V
**2.4** 30.3 μS
**2.5** 100 Ω
**2.6** 622 mΩ
**2.7** 431 μΩ

## EWB Applications Problems

| Figure | EWB File Reference |
|---|---|
| 2.13 | EWBA2_13.ewb |
| 2.16 | EWBA2_16.ewb |

# 3 DC CIRCUIT COMPONENTS

## OBJECTIVES

*After studying the material in this chapter, you should be able to:*

1. Describe the commonly used types of wire and the American Wire Gauge (AWG) system of wire sizing.
2. Describe the ratings used for conductors and insulators.
3. Describe the various types of resistors.
4. Describe the various resistor ratings.
5. Use the standard resistor color code.
6. List the guidelines for replacing resistors.
7. Describe the operation of, and applications for, commonly used potentiometers.
8. Describe the construction and characteristics of commonly used batteries.
9. Discuss the operation and use of a basic dc power supply.
10. Discuss the concept of voltage polarity as it applies to dc power supplies.
11. List and describe the various types of switches.
12. Describe the construction, operation, ratings, and replacement procedures for fuses.
13. Compare and contrast fuses with circuit breakers.
14. List and discuss the guidelines for building circuits on breadboards.
15. Describe the procedures for measuring current, voltage, and resistance.

There are two aspects to every science: *theory* and *practice.* Theory deals with the concepts of the science, while practice deals with using those concepts to accomplish something. In Chapter 2, you took the first step toward understanding the theoretical side of electricity and electronics. Now it is time to take the first step toward understanding its practical side.

In this chapter, you will be introduced to the most basic circuit components, along with their practical ratings. You will also be shown how to measure current, voltage, and resistance. Then, in Chapter 4, we will take the first step in combining this practical information with the concepts covered in Chapter 2.

## 3.1 CONDUCTORS AND INSULATORS

In Chapter 2, we discussed the structure and characteristics of conductors and insulators. In this section, you will be shown some of the various types of conductors and insulators, their ratings, and their applications.

### *Wires*

As you have probably figured out by now, copper is the most commonly used conductor. Copper wires come in a variety of types and sizes, some of which are shown in Figure 3.1.

All wires are either solid or stranded, as shown in Figure 3.1a. The solid wire contains a solid core conductor. The stranded wire contains a group of very thin wires wrapped together to form a larger diameter wire. While solid wires are easier to manufacture, stranded wires are more flexible. Both solid and stranded wires are usually coated with an insulating layer.

Wires come in a wide range of diameters. Approximately 42 sizes fall between the extremes shown in Figure 3.1b. Wire sizes are discussed in detail later in this section.

### *Why Copper Is Used*

You may recall that the resistance of a conductor is directly proportional to the resistivity of the element used; that is, the lower the resistivity of an element, the lower its resistance per unit length. Of all the readily available elements, only silver has a lower resistivity than copper and therefore lower resistance per unit length. At the same time, silver is much more expen-

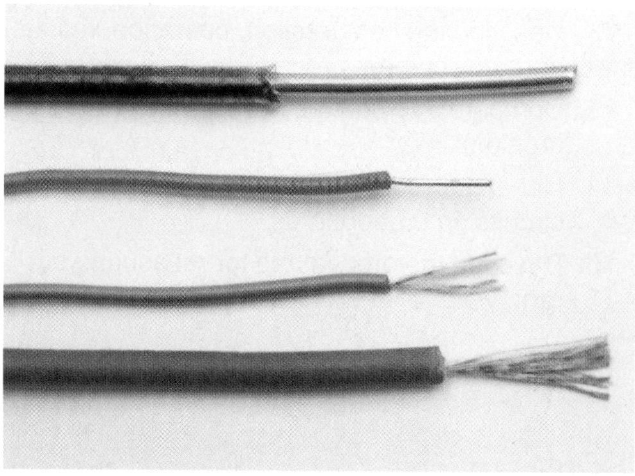

(a)

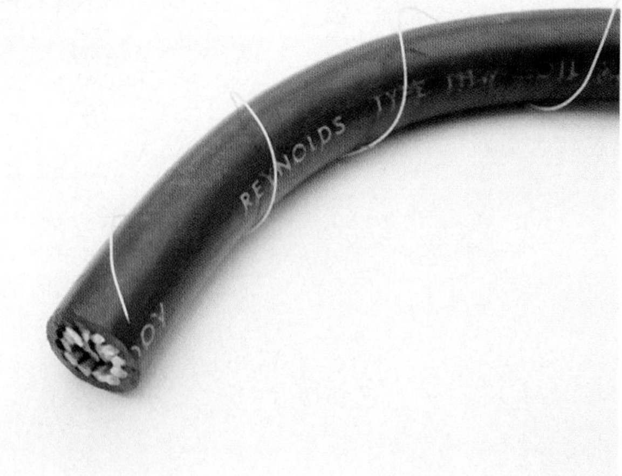


(b)

**FIGURE 3.1 Several sizes of solid and stranded copper wire.**

sive than copper. Thus, copper is used because it provides the best balance between resistance and cost per unit length. (Cost is *always* a consideration in circuit and systems engineering.)

Because of its low cost, aluminum would seem to be an alternative to copper for use in electrical and electronic circuits. However, aluminum has a characteristic drawback that limits its use: it has very low elasticity. When changes in temperature cause aluminum to expand and contract, it does not return to its original shape. This can cause problems when aluminum is connected to a copper connector (which has high elasticity).

## Wire Sizes

◄ *OBJECTIVE 1*

Wires are produced in standard sizes. The size of a wire is determined by its cross-sectional area. The **American Wire Gauge (AWG)** system uses numbers to identify these standard wire sizes, as shown in Table 3.1. For example, 24 gauge wire has a cross-sectional area of 404.01 CM and a resistance of 25.67 Ω per 1000 ft.

**American Wire Gauge (AWG)** A system that uses numbers to identify standard wire sizes (cross-sectional areas).

**TABLE 3.1 Standard AWG Sizes and Resistances**

| *AWG Number* | *Area (CM)* | *Ω/1000 ft* |
|---|---|---|
| 0000 | 211,600 | 0.0490 |
| 000 | 167,810 | 0.0618 |
| 00 | 133,080 | 0.0780 |
| 0 | 105,530 | 0.0983 |
| 1 | 83,694 | 0.1240 |
| 2 | 66,373 | 0.1563 |
| 3 | 52,634 | 0.1970 |
| 4 | 41,742 | 0.2485 |
| 5 | 33,102 | 0.3133 |
| 6 | 26,250 | 0.3951 |
| 7 | 20,816 | 0.4982 |
| 8 | 16,509 | 0.6282 |
| 9 | 13,094 | 0.7921 |
| 10 | 10,381 | 0.9989 |
| 11 | 8,234.0 | 1.260 |
| 12 | 6,529.0 | 1.588 |
| 13 | 5,178.4 | 2.003 |
| 14 | 4,106.8 | 2.525 |
| 15 | 3,256.7 | 3.184 |
| 16 | 2,582.9 | 4.016 |
| 17 | 2,048.2 | 5.064 |
| 18 | 1,624.3 | 6.358 |
| 19 | 1,228.1 | 8.051 |
| 20 | 1,021.5 | 10.15 |
| 21 | 810.10 | 12.80 |
| 22 | 642.40 | 16.14 |
| 23 | 509.45 | 20.36 |
| 24 | 404.01 | 25.67 |
| 25 | 320.40 | 32.37 |
| 26 | 254.10 | 40.81 |
| 27 | 201.50 | 51.47 |
| 28 | 159.79 | 64.90 |
| 29 | 126.72 | 81.83 |
| 30 | 100.50 | 103.2 |
| 31 | 79.70 | 130.1 |
| 32 | 63.21 | 164.1 |
| 33 | 50.13 | 206.9 |
| 34 | 39.75 | 260.9 |
| 35 | 31.52 | 329.0 |
| 36 | 25.00 | 414.8 |
| 37 | 19.83 | 523.1 |
| 38 | 15.72 | 659.6 |
| 39 | 12.47 | 831.8 |
| 40 | 9.89 | 1,049.0 |

At first glance, there is no apparent pattern to the area and resistance values given in Table 3.1. However, a closer look reveals the following:

1. If you go up three sizes from any gauge, you'll find a wire with approximately half the area and twice the resistance.
2. If you go up ten sizes from any gauge, you'll find a wire with approximately one-tenth the area and ten times the resistance.

For example, compare the ratings shown in Table 3.1 for 30 gauge wire to those shown for 33 gauge wire and 40 gauge wire. You will see that both of the above relationships hold true.

Here's an interesting side point: the wires shown in Figure 3.1b are the 0000 gauge and 40 gauge wires. The difference between their sizes is even easier to appreciate when you consider the following:

1. 1000 ft (304.8 m) of 40 gauge copper wire weighs approximately 0.03 lb (13.6 g).
2. 1000 ft (304.8 m) of 0000 gauge copper wire weighs approximately 641 lb (291 kg).

In other words, a length of 0000 gauge wire is approximately 21,400 times as heavy as an equal length of 40 gauge wire!

*OBJECTIVE 2* ▶

## Current Capacity

As shown in Table 3.1, the lower the gauge of a wire, the lower its resistance. The lower the resistance of a wire, the higher its current capacity at a specified voltage. In other words, lower gauge wires have less resistance (opposition to current), so they can pass more current at a given voltage than higher gauge wires.

**Ampacity**
From *amp*ere cap*acity,* the rated limit on current for a given wire gauge.

A term often used to describe the current capacity of a wire is **ampacity,** which stands for *ampere capacity.* The higher the ampacity of a wire, the more current it can carry. Note that:

1. Ampacity varies inversely with wire gauge.
2. Ampacity is rated in amperes.
3. Only wire gauges 0000 through 14 have actual ampacity ratings. The higher gauge wires are generally used in relatively low-current applications, making ampacity ratings unnecessary.

When replacing residential and commercial electrical wiring, the ampacity rating of the wire used is an important consideration. However, when replacing a wire in an electronic circuit, other factors (such as the diameter of a wire) are generally considered. This point will be illustrated later in this chapter when we cover circuit construction.

## PC Board Traces

Most electronic circuits are built on printed circuit boards, or *PC boards.* A PC board is shown in Figure 3.2. Usually, components are mounted on one side of the board, although double-sided or multi-layered PC boards have components mounted on both sides. The conductors that connect the components are commonly referred to as **traces.** Traces are most commonly made of copper.

**Traces**
The conductors that connect components on a printed circuit board.

Like wires, traces have a limit on the amount of current they can carry. Under normal circumstances, a trace is designed to carry a lot more current than is required by the circuit.

(a) Top

(b) Bottom

**FIGURE 3.2 A printed circuit (PC) board.**

## Insulator Ratings

Any insulator can be forced to conduct if a high enough voltage is applied. For this reason, an insulator is rated by its **average breakdown voltage.** The average breakdown rating of an insulator is the voltage that will cause the insulator to conduct, in kilo-volts per centimeter (kV/cm). Some average breakdown voltage ratings are listed in Table 3.2.

**Average breakdown voltage**
The voltage that will cause an insulator to conduct; rated in kV/cm.

An important point should be made at this time: there are basically two types of ratings. The first type describes the physical and/or electrical characteristics of a component. For example, Table 3.1 describes a 24 gauge wire as one that has an area of 404.01 CM and a resistance of 25.67 Ω per 1000 ft. These ratings merely describe the physical and electrical characteristics of 24 gauge wire.

Another type of rating, called a **parameter,** is a limit. This type of rating is a warning as much as anything else. It gives you a value and implies that the component will not function as designed if the value is exceeded.

**Parameter**
A limit.

The values given in Table 3.2 are all parameters; that is, they are limits that cannot be exceeded if the insulator is to block the flow of charge. For example, mica will block current as long as the applied potential is less than 2000 kV/cm. However, if this value is exceeded, mica may conduct.

You will find that many components have both types of ratings. In fact, the resistor (as you will see in the next section) has two commonly used ratings: one that describes its electrical characteristics and one that is a parameter.

**TABLE 3.2 Typical Insulator Breakdown Voltage Ratings**

| *Material* | *Average Breakdown Voltage (kV/cm)* |
|---|---|
| Air | 30 |
| Rubber | 270 |
| Paper | 500 |
| Teflon | 600 |
| Glass | 900 |
| Mica | 2,000 |

## The Bottom Line

Conductors and insulators are integral parts of any circuit. Conductors are used to pass current from one point to another, while insulators are used to prevent current from passing between points.

Wires are rated according to their cross-sectional areas. The American Wire Gauge (AWG) standard uses numbers to identify wires with specific cross-sectional areas. The lower the AWG number of a wire, the greater its cross-sectional area, the lower its resistance, and the greater its current-carrying capability (ampacity).

Insulators are designed to block current, and are rated according to their average breakdown voltages. The average breakdown voltage of an insulator is the voltage at which it will break down and conduct, given in kV/cm. The average breakdown voltage rating is a parameter, or limit.

**Section Review**

1. Contrast solid and stranded wires.
2. Why is copper the preferred conductor in most applications?
3. What is the relationship between a wire and one that is three wire gauges higher?
4. What is the relationship between a wire and one that is ten wire gauges higher?
5. What is ampacity?
6. What is the relationship between ampacity and wire gauge?
7. Define average breakdown voltage.
8. What is a parameter? Why are parameters important?

*OBJECTIVE 3* ▶

## 3.2 RESISTORS

**Resistor**
A component designed to provide a specific amount of resistance.

**Potentiometer**
A variable resistor.

As you know, a **resistor** is a component designed to provide a specific amount of resistance. Resistors are classified as being either fixed or variable. A fixed resistor is one that has a specific value that cannot be adjusted by the user. A variable resistor, or **potentiometer,** is a resistor that can be adjusted within a specified range.

In this section, we will focus on fixed resistors and their ratings. Potentiometers and their ratings are covered in Section 3.3.

### *Carbon-Composition Resistors*

**Carbon-composition resistor**
A component that uses carbon to provide a desired value of resistance.

What determines the value of a carbon-composition resistor.

The most commonly used resistor is the **carbon-composition resistor.** The construction of the carbon-composition resistor is illustrated in Figure 3.3. As you can see, the resistor has two metal leads (conductors) that are separated by carbon. The relatively high resistivity of the carbon is the source of the resistor's opposition to current.

The value of a given carbon-composition resistor is determined by several factors. The first is the purity of the carbon. By adding impurities (other elements) to carbon during the manufacturing process, the resistivity of the carbon can be increased or decreased, depending on the amount and type of impurity used.

The value of a given carbon-composition resistor is also determined by the spacing between its leads. The greater the spacing between the leads, the higher the value of a given resistor. This point is illustrated in Figure 3.4. As you can see, the current through the resistor in Figure 3.4a is encountering a given amount of carbon. Let's assume that the current through the resistor in Figure 3.4b is passing through twice as much carbon. Since the resistance of a material is directly proportional to its length [as given in equation (2.5)], the current in Figure 3.4b is encountering twice the resistance as the current in Figure 3.4a. Therefore, the resistor in Figure 3.4b has twice the value of the resistor in Figure 3.4a.

**FIGURE 3.3 Carbon-composition resistor.**

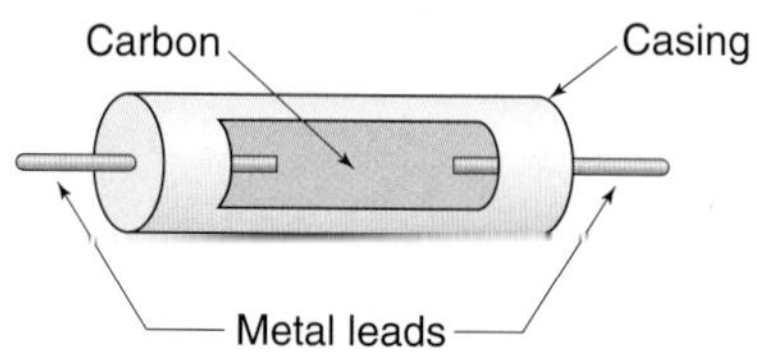

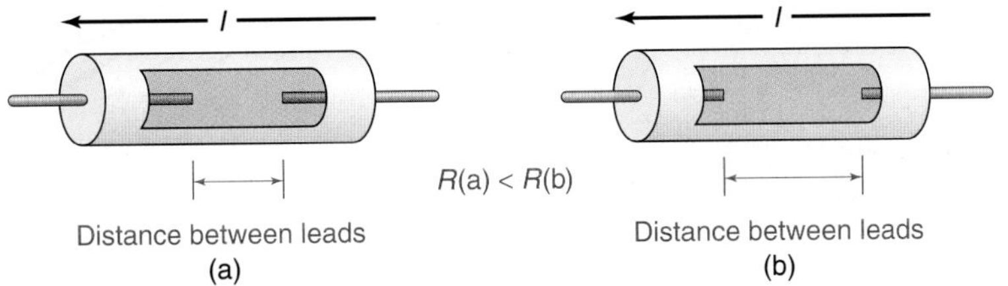


**FIGURE 3.4 Lead spacing.**

By varying the impurities used and the length between resistor leads, it is possible to produce carbon-composition resistors with a wide variety of values. Later in this section, you will be provided with a list of standard resistor values.

## Other Types of Resistors

A **wire-wound resistor** uses the resistivity of a length of wire to produce the desired resistance. A wire-wound resistor is shown in Figure 3.5. Note that the value of the component is determined by the length of the wire (which is wrapped in a series of loops) between the leads.

**Wire-wound resistor**
A resistor that uses the resistivity of a length of wire to produce a desired value of resistance.

Wire-wound resistors are used primarily in high-power applications; that is, applications where the components must be able to *dissipate* (throw off) a relatively high amount of heat. A wire-wound resistor has more surface area than a comparable carbon-composition resistor (because it is hollow) and therefore can dissipate more heat.

The **metal-film resistor** shown in Figure 3.6 is a component that has a very low temperature coefficient. This means that its resistance doesn't vary much when temperature increases or decreases.

**Metal-film resistor**
A resistor with a low temperature coefficient.

**FIGURE 3.5 A wire-wound resistor.**

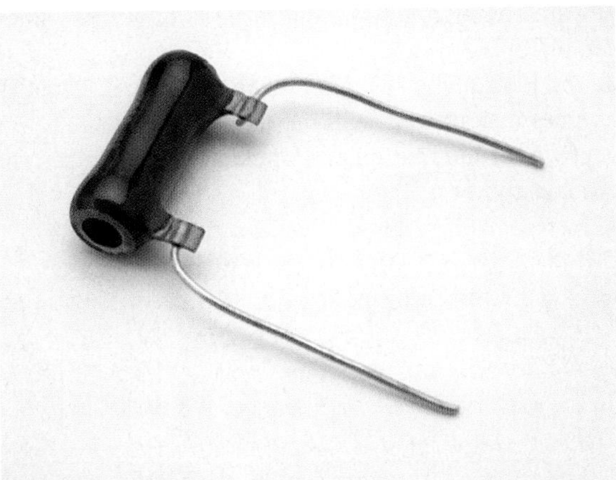

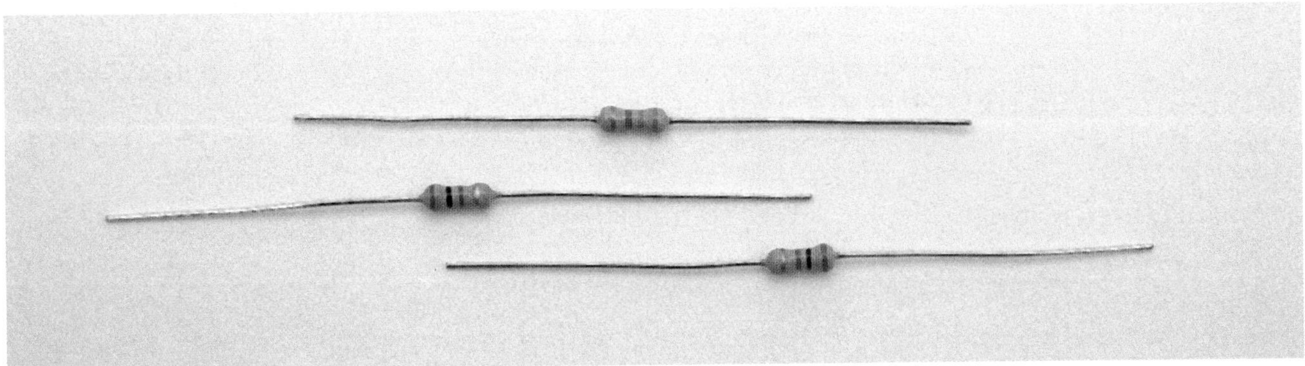

**FIGURE 3.6 Several metal-film resistors.**

*Note:* The SIP resistor pack (photo) is shown larger than actual size. The component is about the same size as a small paper clip (20 mm × 5 mm).

(a)

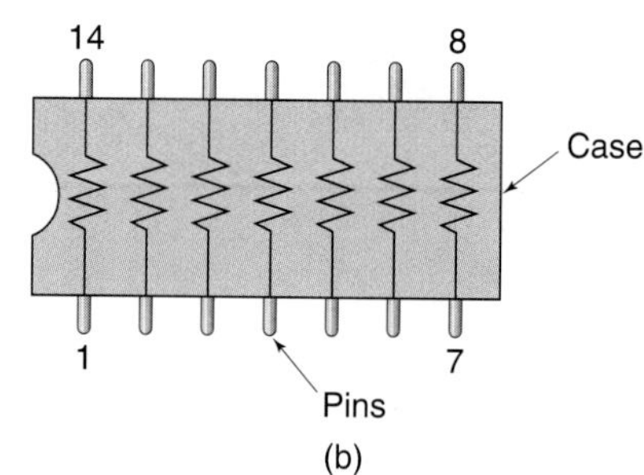


(b)

**FIGURE 3.7 Integrated resistors.**

**Integrated resistor**
A micro-miniature resistor made using semiconductor materials other than carbon. Because they are so small, many of them can be housed in a single casing.

**Integrated resistors** are micro-miniature components made using semiconductors other than carbon. These resistors are so small that several of them can be packaged in a single casing. For example, the resistor pack in Figure 3.7a is a nine-pin SIP (single in-line package). It contains eight resistors, each with its own input pin, which are connected to a common output pin. The resistor pack represented in Figure 3.7b is a 14-pin DIP (dual in-line package) that contains the components shown. Note that the resistor packs shown have the advantage of small size, but are limited to low-power applications.

*OBJECTIVE 4* ▶

## Standard Resistor Values

Resistors are commercially produced in a variety of values. The standard resistor values are listed in Table 3.3.

If you look at any row of numbers in Table 3.3, you'll see that the row contains a series of values that all start with the same two digits. The only difference between the values is a power-of-ten multiplier. For example, the values shown in the last row could have been given as follows:

$$91 \times 10^{-2}$$
$$91 \times 10^{-1}$$
$$91 \times 10^{0}$$
$$91 \times 10^{1}$$
$$91 \times 10^{2}$$
$$91 \times 10^{3}$$
$$91 \times 10^{4}$$
$$91 \times 10^{5}$$

Thus, we could designate the value of any resistor using only the first two digits and the power of ten. As you will see later in this section, this is the basis of the resistor color code (a method used to designate the value of a given resistor).

## Resistor Tolerance

**Tolerance**
The range of values for a resistor, given as a percentage of its nominal (rated) value.

Any commercially produced resistor is guaranteed to have a measured value that falls within a specified range. This range of values is called the **tolerance** of the component and is given as a percentage of the rated value. The most common resistors have 2%, 5%, or 10% tolerances, which means that their measured values are guaranteed to fall within ±2%, ±5%, or ±10% of their rated values.

**TABLE 3.3 Standard Resistor Values (2% tolerance and higher)**

| Ω | | | | kΩ | | | MΩ | |
|---|---|---|---|---|---|---|---|---|
| 0.10 | 1.0 | 10 | 100 | 1.0 | 10 | 100 | 1.0 | 10 |
| 0.11 | 1.1 | 11 | 110 | 1.1 | 11 | 110 | 1.1 | 11 |
| 0.12 | 1.2 | 12 | 120 | 1.2 | 12 | 120 | 1.2 | 12 |
| 0.13 | 1.3 | 13 | 130 | 1.3 | 13 | 130 | 1.3 | 13 |
| 0.15 | 1.5 | 15 | 150 | 1.5 | 15 | 150 | 1.5 | 15 |
| 0.16 | 1.6 | 16 | 160 | 1.6 | 16 | 160 | 1.6 | 16 |
| 0.18 | 1.8 | 18 | 180 | 1.8 | 18 | 180 | 1.8 | 18 |
| 0.20 | 2.0 | 20 | 200 | 2.0 | 20 | 200 | 2.0 | 20 |
| 0.22 | 2.2 | 22 | 220 | 2.2 | 22 | 220 | 2.2 | 22 |
| 0.24 | 2.4 | 24 | 240 | 2.4 | 24 | 240 | 2.4 | |
| 0.27 | 2.7 | 27 | 270 | 2.7 | 27 | 270 | 2.7 | |
| 0.30 | 3.0 | 30 | 300 | 3.0 | 30 | 300 | 3.0 | |
| 0.33 | 3.3 | 33 | 330 | 3.3 | 33 | 330 | 3.3 | |
| 0.36 | 3.6 | 36 | 360 | 3.6 | 36 | 360 | 3.6 | |
| 0.39 | 3.9 | 39 | 390 | 3.9 | 39 | 390 | 3.9 | |
| 0.43 | 4.3 | 43 | 430 | 4.3 | 43 | 430 | 4.3 | |
| 0.47 | 4.7 | 47 | 470 | 4.7 | 47 | 470 | 4.7 | |
| 0.51 | 5.1 | 51 | 510 | 5.1 | 51 | 510 | 5.1 | |
| 0.56 | 5.6 | 56 | 560 | 5.6 | 56 | 560 | 5.6 | |
| 0.62 | 6.2 | 62 | 620 | 6.2 | 62 | 620 | 6.2 | |
| 0.68 | 6.8 | 68 | 680 | 6.8 | 68 | 680 | 6.8 | |
| 0.75 | 7.5 | 75 | 750 | 7.5 | 75 | 750 | 7.5 | |
| 0.82 | 8.2 | 82 | 820 | 8.2 | 82 | 820 | 8.2 | |
| 0.91 | 9.1 | 91 | 910 | 9.1 | 91 | 910 | 9.1 | |

*Reference:* Precision resistors (1% tolerance and lower) are available in many values not listed here. A list of precision resistor standard values is provided in Appendix B.

To determine the guaranteed range of measured values for a resistor:

1. Multiply the rated value of the resistor by its tolerance to get the maximum variation in resistance.
2. Add the maximum variation to the rated value of the component to find the upper limit.
3. Subtract the maximum variation from the rated value of the component to find the lower limit.

The use of this procedure is demonstrated in the following example.

### EXAMPLE 3.1

Determine the guaranteed range of measured values for a 47 kΩ resistor that has a 2% tolerance.

***Solution:*** First, the maximum variation in resistance is found as

$$(47\ \text{k}\Omega) \times 2\% = 940\ \Omega\ (0.94\ \text{k}\Omega)$$

This maximum variation is added to the rated value of the component, as follows:

$$47\ \text{k}\Omega + 0.94\ \text{k}\Omega = 47.94\ \text{k}\Omega \quad \text{(maximum)}$$

The maximum variation is now subtracted from the rated value of the component, as follows:

$$47\ \text{k}\Omega - 0.94\ \text{k}\Omega = 46.06\ \text{k}\Omega \quad \text{(minimum)}$$

The guaranteed range of measured values for this resistor is 46.06 kΩ to 47.94 kΩ.

***PRACTICE PROBLEM 3.1***

Determine the guaranteed range of values for a 68 kΩ resistor that has a 2% tolerance.

The higher the tolerance of a resistor, the wider its possible range of values. This point is illustrated in the following example.

### *Example 3.2*

Determine the ranges of guaranteed values for a 33 kΩ resistor with a 2% tolerance and a 33 kΩ resistor with a 5% tolerance.

***Solution:*** For the 2% tolerance resistor:

$$(33\text{ k}\Omega) \times 2\% = 660\ \Omega\ (0.66\text{ k}\Omega)$$
$$33\text{ k}\Omega + 0.66\text{ k}\Omega = 33.66\text{ k}\Omega \quad \text{(maximum)}$$
$$33\text{ k}\Omega - 0.66\text{ k}\Omega = 32.34\text{ k}\Omega \quad \text{(minimum)}$$

For the 5% tolerance resistor:

$$(33\text{ k}\Omega) \times 5\% = 1.65\text{ k}\Omega$$
$$33\text{ k}\Omega + 1.65\text{ k}\Omega = 34.65\text{ k}\Omega \quad \text{(maximum)}$$
$$33\text{ k}\Omega - 1.65\text{ k}\Omega = 31.35\text{ k}\Omega \quad \text{(minimum)}$$

As you can see, the 5% tolerance component has a wider range of guaranteed values.

***PRACTICE PROBLEM 3.2***

Determine the guaranteed range of values for a 910 kΩ resistor with a 5% tolerance.

Several points should be made with regard to the tolerance of resistors:

1. Lower tolerance components are considered to be higher quality components because their measured values are guaranteed to fall within a narrower range, and thus closer to the design value. In fact, the ideal (perfect) resistor, if it could be produced, would have a tolerance rating of 0%. As a result, its rated and measured values would always be equal.
2. Even though their production has decreased, you will see resistors with tolerances of 10% and even 20% in some circuits and systems. (Later in this section, you'll be shown how the tolerance of a resistor is indicated.)
3. Higher tolerance resistors cost less to produce than low-tolerance components. The choice of components (for circuit design) is generally made on the basis of required quality versus acceptable cost. (Cost is *always* a consideration.)

*OBJECTIVE 5* ▶

## *The Resistor Color Code*

In most cases, the value of a resistor is indicated by a series of color bands on the component. For example, look at the resistors shown in Figure 3.8. Each of the resistors shown has four color bands. These bands are numbered as shown in the figure. Note that the band closest to the end of the component is the first color band. The rest are then numbered in order.

In our discussion on standard resistor values, it was stated that the value of a resistor could be designated using the first two digits of the value and a power-of-ten multiplier. The first three bands on a resistor designate its rated value in this fashion, as follows:

*A Practical Consideration:* Many precision resistors are identified using printed values rather than color bands. When color bands are used on precision resistors, they are interpreted in a slightly different fashion. The band designations for precision resistors are given in Appendix B.

| | |
|---|---|
| Band 1: | The color of this band designates the first digit in the resistor value. |
| Band 2: | The color of this band designates the second digit in the resistor value. |
| Band 3: | The color of this band designates the power-of-ten multiplier for the first two digits. (In most cases, this is simply the number of zeros that follow the first two digits.) |

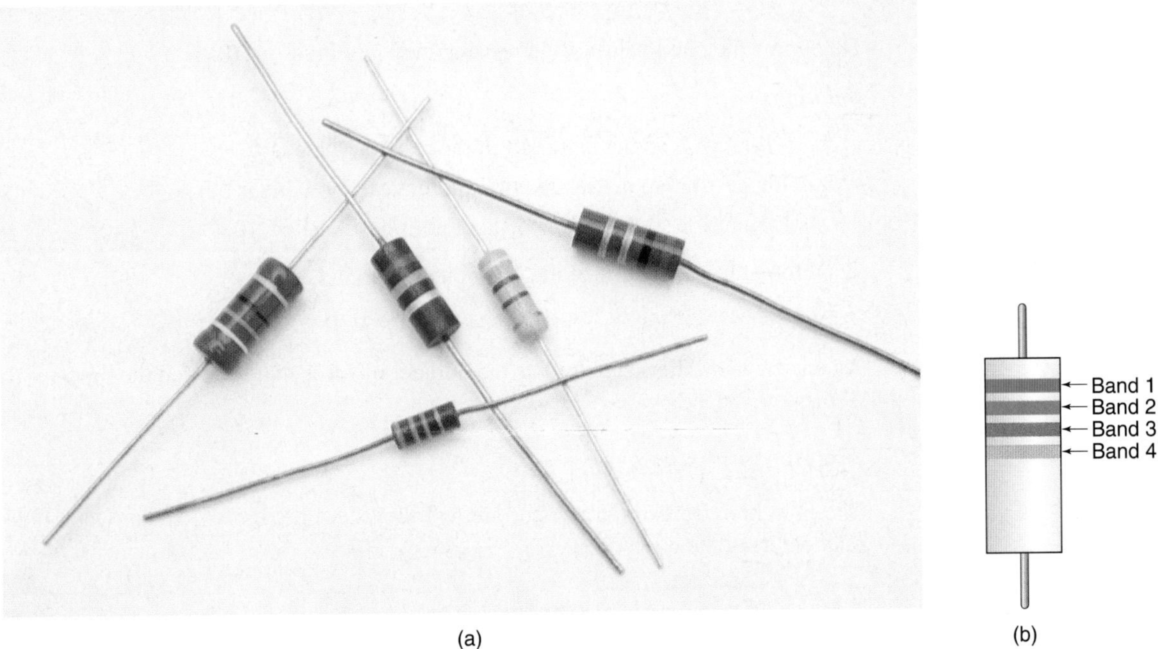


**FIGURE 3.8 Resistor color bands.**

The colors that appear in these three bands are coded as shown in Table 3.4. The following series of examples shows how the first three color bands are used to determine the value of a given resistor.

**TABLE 3.4 Color Band Coding**

| *Color* | *Value* | *Color* | *Value* |
|---|---|---|---|
| Black | 0 | Green | 5 |
| Brown | 1 | Blue | 6 |
| Red | 2 | Violet | 7 |
| Orange | 3 | Gray | 8 |
| Yellow | 4 | White | 9 |

### EXAMPLE 3.3

Determine the rated value of the resistor shown in Figure 3.9.

***Solution:***

Brown = 1, so the first digit in the resistor value is 1.

Green = 5, so the second digit in the resistor value is 5.

Red = 2, so the power-of-ten multiplier is $10^2 = 100$.

Therefore, the rated value of the resistor is found as

$$(15 \times 100)\ \Omega = 1500\ \Omega = 1.5\ \text{k}\Omega$$

Note that the multiplier band value (2) was equal to the number of zeros in the value of the resistor.

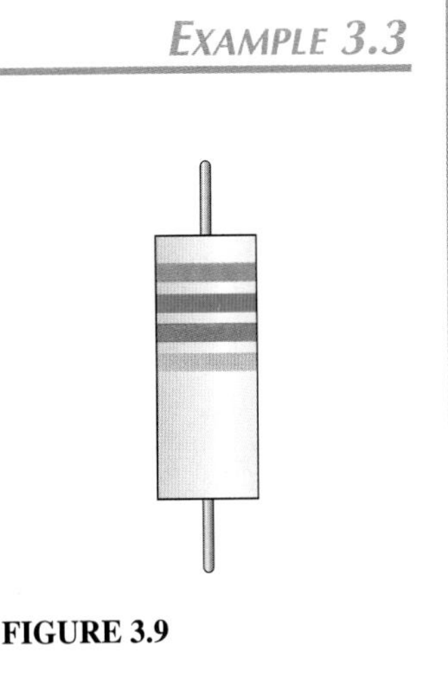

**FIGURE 3.9**

***PRACTICE PROBLEM 3.3***

The first three bands on a resistor are as follows: yellow, violet, orange. Determine the rated value of the component.

## EXAMPLE 3.4

Determine the rated value of the resistor shown in Figure 3.10.

***Solution:***

Yellow = 4, so the first digit in the resistor value is 4.

Orange = 3, so the second digit in the resistor value is 3.

Brown = 1, so the power-of-ten multiplier is $10^1 = 10$.

Therefore, the rated value of the component is found as

$$(43 \times 10)\ \Omega = 430\ \Omega$$

Again, the multiplier band told us the number of zeros that followed the first two digits of the component's value.

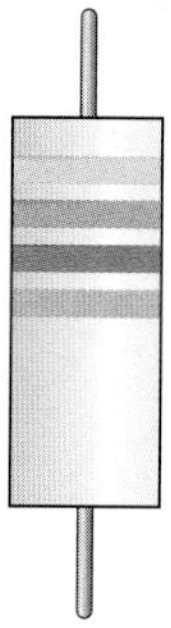

**FIGURE 3.10**

***PRACTICE PROBLEM 3.4***

The first three bands on a resistor are as follows: white, brown, green. Determine the rated value of the component.

Black multiplier bands.

When the multiplier band on a resistor is black, you have to be careful not to make a common mistake. Beginning technicians often see a black multiplier band, see that black corresponds to zero, and assume that there is a zero after the first two digits. A black multiplier band indicates that there are no zeros following the first two digits. This point is illustrated in the following example.

## EXAMPLE 3.5

Determine the rated value of the resistor shown in Figure 3.11.

***Solution:***

Green = 5, so the first digit in the value is 5.

Blue = 6, so the second digit in the value is 6.

Black = 0, so the multiplier is $10^0 = 1$.

Therefore, the rated value of the component is found as

$$(56 \times 1)\ \Omega = 56\ \Omega$$

Note that there are no zeros following the first two digits in the resistor's value.

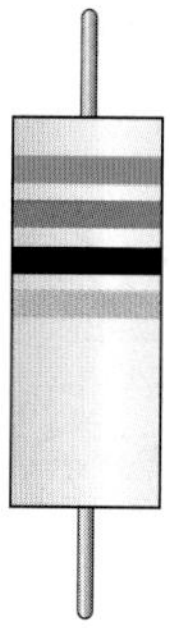

**FIGURE 3.11**

***PRACTICE PROBLEM 3.5***

The first three bands on a resistor are as follows: blue, gray, black. Determine the rated value of the component.

Gold and silver multiplier bands.

There are two other colors that may appear in the multiplier band. These colors and their multiplier values are as follows:

Gold = −1

Silver = −2

When the multiplier band is gold, a decimal point falls between the first two digits in the value. This is illustrated in the following example.

### EXAMPLE 3.6

Determine the rated value of the resistor shown in Figure 3.12.

***Solution:***

Red = 2, so the first digit in the component value is 2.

Red = 2, so the second digit in the component value is also 2.

Gold = −1, so the multiplier is $10^{-1} = 0.1$.

Therefore, the rated value of the component is found as

$$(22 \times 0.1)\ \Omega = 2.2\ \Omega$$

Note that the gold multiplier band placed the decimal point between the first two digits in the resistor's value.

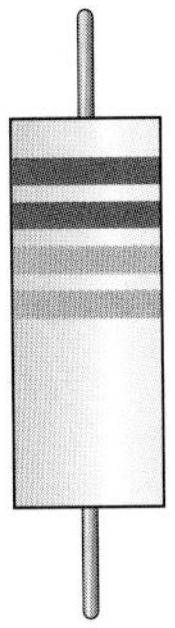

FIGURE 3.12

***PRACTICE PROBLEM 3.6***

The first three bands on a resistor are as follows: violet, green, gold. Determine the rated value of the component.

When the multiplier band is silver, the decimal point falls in front of the two digits. This is illustrated in the following example.

### EXAMPLE 3.7

Determine the rated value of the resistor shown in Figure 3.13.

***Solution:***

Gray = 8, so the first digit in the component value is 8.

Red = 2, so the second digit in the component value is 2.

Silver = −2, so the multiplier is $10^{-2} = 0.01$.

Therefore, the rated value of the component is found as

$$(82 \times 0.01)\ \Omega = 0.82\ \Omega$$

Note that the silver multiplier band placed the decimal point in front of the two digits in the resistor's value.

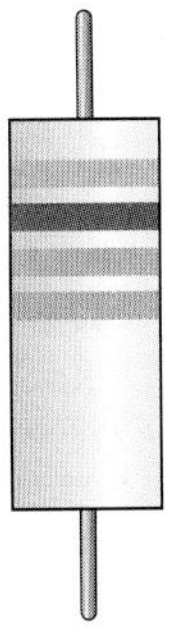

FIGURE 3.13

***PRACTICE PROBLEM 3.7***

The first three bands on a resistor are as follows: brown, gray, silver. Determine the rated value of the component.

As often as not, you will need to be able to determine the color code for a specified resistance so that you can locate the needed component. When this is the case, take a moment to write the component value in standard form. The first two colors are determined by the first two digits. The color of the multiplier band is determined (in most cases) by the number of zeros that follow the first two digits. This is illustrated in the following example.

## EXAMPLE 3.8

You need to locate a 360 kΩ resistor. Determine the colors of the first three bands on the component.

***Solution:*** Written in standard form, 360 kΩ = 360,000 Ω. Therefore, the first three bands are coded for 3, 6, and 4 (which is the number of zeros in the value). The colors that correspond to these numbers are as follows:

3 = orange

6 = blue

4 = yellow

These are the colors (in order) of the first three bands.

***PRACTICE PROBLEM 3.8***

Determine the colors of the first three bands of a 24 kΩ resistor.

*A Practical Consideration:* Most manufacturers have significantly reduced production of 10% and 20% tolerance resistors. However, you need to know how these tolerances are represented because you will see them in many systems.

The fourth band on a resistor designates its tolerance. The colors used in the tolerance band are as follows:

Red = 2%

Gold = 5%

Silver = 10%

When there is no fourth band, the resistor has a 20% tolerance. The following example shows how the first four bands on a resistor are used to determine the guaranteed range of values for the component.

## EXAMPLE 3.9

Determine the guaranteed range of values for the resistor in Figure 3.14.

***Solution:***

Green = 5, so the first digit in the component value is 5.

Brown = 1, so the second digit in the component value is 1.

Red = 2, so the multiplier is $10^2 = 100$.

Gold = 5% (the tolerance of the component).

The rated value of the component is 5100 Ω, or 5.1 kΩ, and

$$(5.1\ \text{k}\Omega) \times 5\% = 255\ \Omega$$
$$5.1\ \text{k}\Omega + 255\ \Omega = 5.355\ \text{k}\Omega \quad \text{(maximum)}$$
$$5.1\ \text{k}\Omega - 255\ \Omega = 4.845\ \text{k}\Omega \quad \text{(minimum)}$$

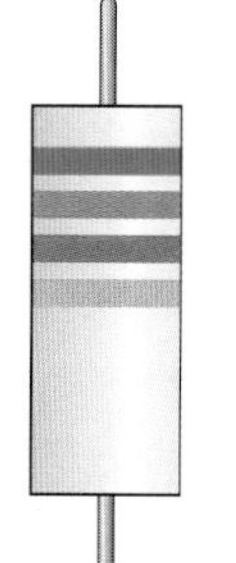

**FIGURE 3.14**

***PRACTICE PROBLEM 3.9***

The first four bands on a resistor are as follows: brown, blue, green, silver. Determine the guaranteed range of measured values for the component.

*Reference:* The fifth band on a precision resistor is interpreted differently. The band designations for precision resistors are given in Appendix B.

A fifth band on a non-precision resistor designates the reliability of the component; that is, it tells you the percentage of resistors that will fail to fall within tolerance after 1000 hours of use. For example, if the reliability rating of a resistor is 1%, it means that no more than 1% (one out of every 100) resistors will fall out of tolerance after 1000 hours of use. The colors used to indicate reliability are coded as follows:

Brown = 1% (1 out of every 100)

Red = 0.1% (1 out of every 1000)

Orange = 0.01% (1 out of every 10,000)

Yellow = 0.001% (1 out of every 100,000)

If a fifth band is not present (which is usually the case), the component does not have a reliability rating.

*A Practical Consideration:* Reliability bands are rarely seen on standard resistors used in commercial equipment or systems. (They were developed primarily for use in military equipment and systems.)

## Color Code Summary

The color code used for carbon-composition resistors is summarized for you in Table 3.5.

**TABLE 3.5 Summary of the Standard Resistor Color Code for Carbon-Composition Resistors**

| *Band* | *Designates* | *Color Values* | |
|---|---|---|---|
| 1<br>2<br>3 | First value digit<br>Second value digit<br>Multiplier equals number of zeros after the first two digits. | Black = 0<br>Brown = 1<br>Red = 2<br>Orange = 3<br>Yellow = 4 | Green = 5<br>Blue = 6<br>Violet = 7<br>Gray = 8<br>White = 9 |
| 3 | Multiplier—values that are less than zero | Gold = −1<br>Silver = −2 | ($10^{-1} = 0.1$)<br>($10^{-2} = 0.01$) |
| 4 | Tolerance | Red= 2%<br>Gold = 5% | Silver = 10%<br>(no band) = 20% |
| 5 | Reliability | Brown = 1%<br>Red = 0.1%<br>Orange = 0.01%<br>Yellow = 0.001% | |

## Precision Resistors

Advances in resistor technology have led to the production of precision resistors: resistors with tolerance ratings lower than 2%. Precision resistors with tolerances as low as 0.001% are now available commercially. As a result of their lower tolerances, precision resistors have a unique list of standard values and a unique color code. A list of precision resistor values can be found (with a diagram of the color code) in Appendix B.

## Resistor Power Ratings

In the last section, you were introduced to a type of component rating called a parameter. As you may recall, a parameter is a limit that should never be exceeded.

In Chapter 4, we will discuss the concept of power in depth. For now, we will simply say that the **power rating** of a component is a measure of its ability to dissipate heat, measured in watts (W).

**Power rating** A measure of a component's ability to dissipate heat, measured in watts (W).

All resistors have power ratings. If the power rating of a resistor is exceeded, the resistor will probably be destroyed as a result of excessive heat.

*Note:* From left to right, the resistors shown have power ratings of 1 W, ½ W, ¼ W, and ⅛ W. The resistors shown are approximately 50% larger than actual size.

**FIGURE 3.15 Carbon-composition resistors and power ratings.**

Resistor power rating indicators

The power rating of a carbon-composition resistor is indicated by the size of the component. As Figure 3.15 shows, the larger a resistor, the higher its power rating. The higher the power rating of a resistor, the more heat it can withstand. Note that the other types of resistors (such as wire-wound resistors) usually have their power ratings printed on them.

*OBJECTIVE 6* ►

## Resistor Substitution

When replacing a resistor, there are three rules to follow:

*A Practical Consideration:* The diameter of the leads on a resistor varies directly with the power rating; that is, the higher the power rating of a resistor, the wider the diameter of its leads. This may limit your ability to substitute a higher power resistor for a low-power resistor on a PC board.

1. Make sure that the substitute component has the same rated value as the original component.
2. Make sure that the tolerance of the substitute component is equal to (or lower than) that of the original component.
3. Make sure that the power rating of the substitute component is equal to (or greater than) that of the original component.

As long as these guidelines are followed, the new component will have the proper rated value, will be within circuit tolerance, and will not be damaged from excessive heat.

## Section Review

1. Describe the physical construction of the carbon-composition resistor.
2. How is the value of a carbon-composition resistor indicated?
3. How is the value of a wire-wound resistor indicated?
4. In what applications are wire-wound resistors primarily used?
5. What are metal-film resistors?
6. What are integrated resistors?
7. Refer to Table 3.3. What is the pattern for each row of the table? What does this pattern form the basis for?
8. What is the tolerance of a resistor?
9. How do you determine the guaranteed range of measured values for a resistor?
10. What is indicated by the color of each of the first three bands on a resistor?

11. List the colors commonly found in the first three bands of a resistor and the value represented by each.
12. What is indicated by a black multiplier band on a resistor?
13. What is indicated by a gold multiplier band on a resistor?
14. How do you determine the color code for a specific standard resistor value?
15. What does the fourth color band on a resistor indicate?
16. List the colors commonly found in the fourth band and give the value indicated by each of them.
17. What does the fifth color band on a resistor indicate? Explain this rating.
18. What does the power rating of a resistor indicate?
19. Why is the power rating of a resistor considered to be a parameter?
20. What are the guidelines for replacing one resistor with another?

## 3.3 POTENTIOMETERS

◄ *OBJECTIVE 7*

**Potentiometer**
A three-terminal resistor whose value can be adjusted (within limits) by the user.

The potentiometer is the most commonly used variable resistor. The **potentiometer** is a three-terminal resistor whose value can be adjusted (within set limits) by the user. To help in our discussion, the terminals of a typical potentiometer are identified as shown in Figure 3.16. The potentiometer, or *pot,* is designed so that the resistance between the middle terminal (b) and each of the outer terminals (a and c) changes when the control shaft is turned. Turning the control shaft in one direction reduces the resistance between terminals a and b (designated $R_{ab}$), and increases the resistance between terminals b and c (designated $R_{bc}$). Turning the control shaft in the opposite direction has the opposite effect. By setting the control shaft to a specific position, $R_{ab}$ and/or $R_{bc}$ can be set to any desired value within the limits of the component.

(a)

$R_{ac}$

a c

$R_{ab}$ b $R_{bc}$

(b)

**FIGURE 3.16 The potentiometer.**

## Potentiometer Construction and Operation

**Wiper arm**
The potentiometer terminal connected to the sliding contact; the control shaft of a potentiometer.

The construction of a typical pot is illustrated in Figure 3.17. The a and c terminals are connected to the ends of a length of carbon. The b terminal, which is referred to as the **wiper arm,** is connected to a sliding contact. By turning the control shaft, the position of the sliding contact is moved along the surface of the carbon. Figure 3.18 shows how turning the control shaft affects the component resistance values.

When the sliding contact is in the center position (Figure 3.18a), equal amounts of carbon are between the wiper arm and each of the outer terminals. Therefore, $R_{ab}$ is equal to $R_{bc}$. The resistance between the outer terminals is equal to the sum of these two resistances. By formula,

$$R_{ac} = R_{ab} + R_{bc} \quad \textbf{(3.1)}$$

Thus, if $R_{ac}$ for the potentiometer in Figure 3.18 is 10 kΩ, placing the sliding contact at the center position will provide values of $R_{ab} = R_{bc} = 5$ kΩ. The sum of these two values is, of course, 10 kΩ.

If the control shaft is turned counterclockwise, the sliding contact moves as shown in Figure 3.18b. The amount of carbon between the wiper arm and terminal a decreases, so $R_{ab}$ decreases. At the same time, the amount of carbon between the wiper arm and terminal c increases, so $R_{bc}$ increases. Note that the sum of the two resistances is still equal to $R_{ac}$, as shown in the figure.

**FIGURE 3.17 Potentiometer construction.**

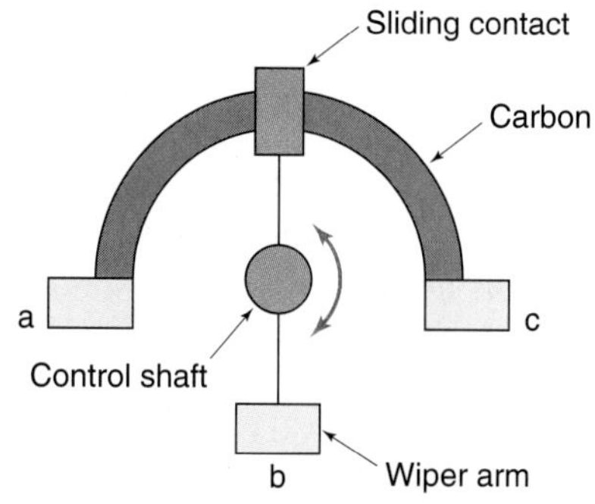


*Note:* Most potentiometers are constructed so that the moving contact can rotate approximately 350°.

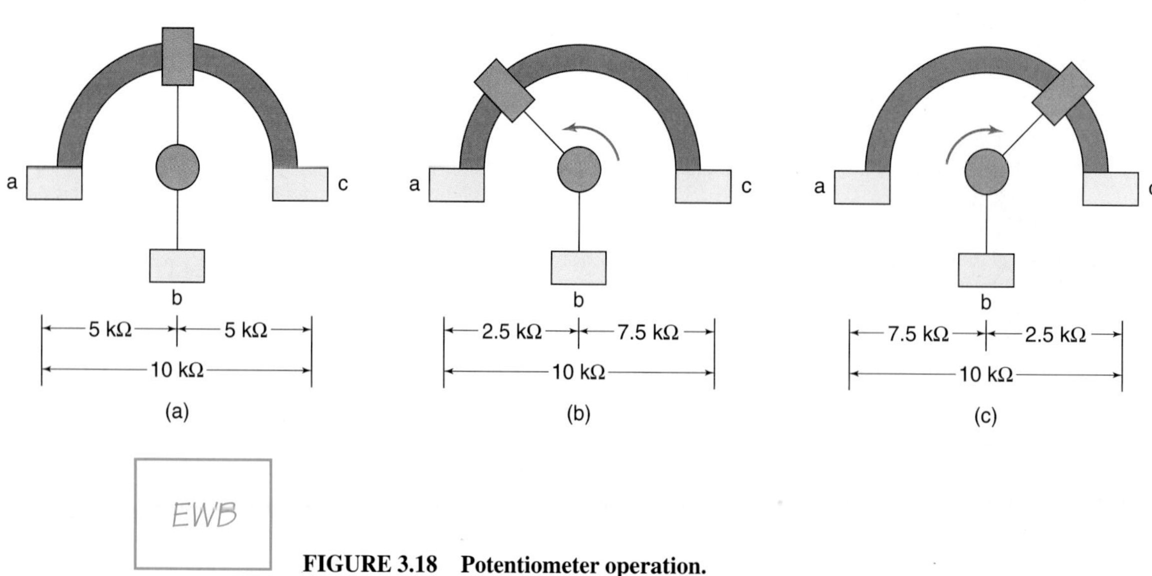


EWB

**FIGURE 3.18 Potentiometer operation.**

As Figure 3.18c shows, turning the control shaft clockwise causes the sliding contact to move toward terminal c. As a result, $R_{bc}$ decreases and $R_{ab}$ increases. As always, the sum of the two equals $R_{ac}$. If the control shaft is turned fully clockwise, $R_{bc}$ drops to 0 Ω and $R_{ab}$ increases to 10 kΩ (the full value of $R_{ac}$).

## Potentiometer Taper

The **taper** of a potentiometer is a measure of how its resistance changes as the control shaft is rotated between its extremes. There are two types of taper: linear and nonlinear.

**Taper**
A measure of how the resistance of a potentiometer changes as the control shaft is rotated between its extremes. Tapers are designated as either linear or nonlinear.

The resistance of a linear-taper pot changes at a linear rate as the shaft is turned; that is, it changes at a constant rate. At one-quarter turn, the two resistances change by 25%, at one-half turn, the two change by 50%, and so on.

The resistance of a nonlinear-taper pot changes at a nonlinear rate as the shaft is turned; that is, it changes at varying rates. For example, if you turn the control shaft from its left extreme to its right extreme, the change in resistance is initially low (per degree of rotation), and increases as you continue to turn the shaft.

Nonlinear-taper pots are commonly used in audio applications, such as the volume control on your television. Because of this, the component is often referred to as an *audio-taper* pot.

## Potentiometer Ratings

A potentiometer is rated in terms of its resistance and power dissipation capability, just like a resistor. Specifically:

Potentiometer ratings.

1. The resistance rating of a pot is the resistance between its outer terminals. Since the resistance between the wiper arm and either outer terminal can be adjusted to equal $R_{ac}$, the resistance rating of a pot is the maximum possible resistance setting for the component.
2. The power dissipation rating is a measure of the maximum amount of heat that the component can dissipate.

The resistance rating of a pot is normally printed on the component. The resistance may be written as a straightforward value (such as 5 kΩ, 10 kΩ, etc.) or in a code similar to the resistor color code. For example, a pot may be marked as shown in Figure 3.19. In this case, the 5 is the first digit in the value, the 0 is the second digit, and the 2 is the number of zeros that follow the first two digits. Thus, the pot shown in the figure would have a maximum value of 5 kΩ.

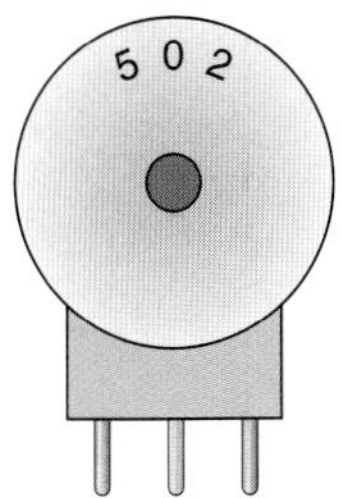


**FIGURE 3.19**

The power rating of a pot is not normally printed on the component because manufacturers tend to produce entire groups of potentiometers that all have the same power ratings. For example, the Clarostat series 380 potentiometers are all 2 W components; that is, the company has produced a line of potentiometers that have different resistance ratings but equal power ratings. In most cases, you'll need to consult the manufacturer's catalog if you want to determine the maximum power rating of a given potentiometer.

## Carbon-Composition Versus Wire-Wound Pots

Most potentiometers are classified as either carbon-composition or wire-wound. The wire-wound potentiometer (shown in Figure 3.20) works in the same basic fashion as the carbon-composition pot. However, it uses a wire-wound resistive element, much like the wire-wound resistor. Note that wire-wound pots tend to have much higher power dissipation ratings than the carbon-composition type.

FIGURE 3.20 **A wire-wound potentiometer.**

## Trimmer Potentiometers

**Trimmer potentiometer**
A carbon-composition pot designed for low-power applications. Trimmers are produced as stand-up and lay-down components.

The **trimmer potentiometer** is used in low-power applications. Some trimmer pots are shown in Figure 3.21. The trimmer pot on the right is a lay-down type and is designed to lay flat on a PC board. The one on the left is a stand-up type and is designed to stand upright on a PC board.

## Multi-Turn Potentiometers

The control shaft on most potentiometers can be turned nearly one full rotation (350°) or less. However, some pots have control shafts that are geared to allow rotations that are greater than 350°. One such component is the ten-turn potentiometer shown in Figure 3.22. The control shaft on this pot can be rotated through ten full turns between its extremes.

*Note:* The pots in Figure 3.21 are shown at approximately twice their actual size.

**FIGURE 3.21 Trimmer potentiometers.**

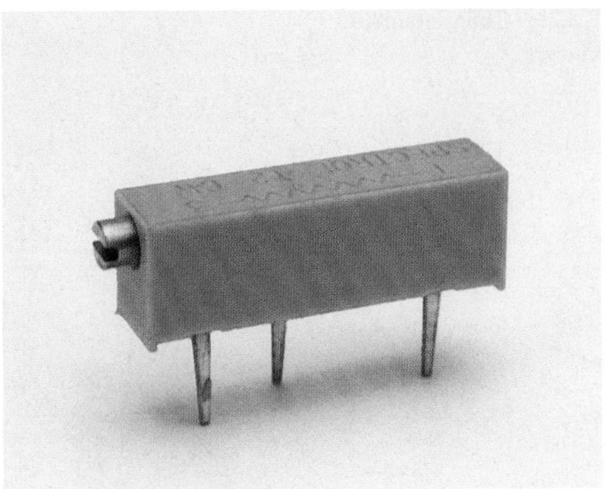

**FIGURE 3.22 A ten-turn potentiometer.**

*Note:* The pot in Figure 3.22 is shown at approximately twice its actual size.

The advantage of multi-turn potentiometers is that they have better **resolution** than single-turn pots; that is, one degree of rotation causes a much smaller change in resistance. For example, let's say that we have two linear 10 kΩ potentiometers. One of them is a 300° pot (a common value) and the other is a ten-turn pot. As the control shaft on the 300° pot is turned, the resistance of the component changes by

$$\frac{10\ \text{k}\Omega}{300°} = 33.33\ \Omega/\text{degree}$$

As the control shaft on the ten-turn pot is turned, the resistance of the component changes by

$$\frac{10\ \text{k}\Omega}{3600°} = 2.78\ \Omega/\text{degree}$$

where 3600° is the total number of degrees in ten 360° rotations. To change the resistance of the ten-turn pot by 1 kΩ would take one full rotation. To change the resistance of the 300° pot by the same amount would take approximately $\frac{1}{10}$ of one rotation. Thus, it is much easier to set the ten-turn pot to a specific value. At the same time, the ten-turn pot is far more expensive, and therefore it is used only when this high degree of resolution is required.

**Resolution**
For a potentiometer, the change in resistance per degree of control shaft rotation. High-resolution pots (those with low resistance per degree ratings) allow more exact control over the adjusted value of the potentiometer.

## Gang-Mounted Potentiometers

In some cases, potentiometers are gang-mounted. Two gang-mounted pots are shown in Figure 3.23. The pots on the left share a common control shaft. Therefore, when the value of one pot is varied, the value of the other is varied by the same amount. This type of pot is commonly used as a stereo volume control, where the volume of the left and right channels must be changed at the same rate.

The gang-mounted pots on the right actually have separate control shafts. The outer shaft controls one of the pots while the inner shaft controls the other. Normally, a knob is connected to the control. When turned, the knob rotates the inner control shaft. When pushed in and turned, the knob rotates the outer control shaft. This type of pot is used on many car stereos, where a single control knob is used to vary both volume and tone.

## Decade Boxes

**Decade box**
A device that contains a series of potentiometers, each with different resistance ratings. Adjusting the individual pots allows the user to set the overall (total) resistance to specific values over a wide range.

A **decade box** is a device that consists of a series of potentiometers, each with different resistance ratings. A typical decade box is shown in Figure 3.24. The decade box in Figure 3.24 has six pots, an input terminal, and an output terminal. By setting the various

FIGURE 3.23 Gang-mounted potentiometers.

pots, the resistance between the terminals can be set at any value between 1 Ω and approximately 1 MΩ. Decade boxes are used almost exclusively in lab situations where a specific resistance value is needed for an experiment or circuit design test.

## The Bottom Line

Potentiometers are three-terminal components whose resistance can be varied by the user. When the control shaft on a pot is turned, the resistance between the middle terminal (called the wiper arm) and each of the outer terminals is varied. As $R_{ab}$ increases, $R_{bc}$ decreases, and vice versa. The sum of the two resistances always equals the resistance between the two outer terminals ($R_{ac}$).

The resistance of a linear-taper pot varies at a constant rate as the control shaft is rotated. The resistance of a nonlinear-taper pot varies at a changing rate as the control shaft is rotated. Nonlinear-taper potentiometers are often referred to as audio-taper pots.

FIGURE 3.24 A decade box.

Pots are normally rated in terms of their outer terminal resistance ($R_{ac}$) and maximum power dissipation. The resistance rating of a pot is usually printed on the component. The manufacturer's catalog must be consulted to determine a pot's power dissipation rating. Carbon-composition pots tend to be low-power components, while wire-wound pots tend to be high-power components.

### Section Review

1. What is a potentiometer?
2. Briefly describe the operation of a potentiometer.
3. Describe the construction of a typical potentiometer.
4. Describe the relationship among the three terminal resistances of a potentiometer.
5. List and describe the various types of potentiometer taper.
6. List and describe the various potentiometer ratings.
7. In terms of application, what is the difference between carbon-composition pots and wire-wound pots?
8. What is a trimmer potentiometer?
9. What is a multi-turn potentiometer? What advantage does a multi-turn pot have over a standard potentiometer?
10. Describe the two types of gang-mounted potentiometers.
11. What is a decade box? What are decade boxes used for?

## 3.4 BATTERIES

◄ *OBJECTIVE 8*

In simpler dc circuits, voltage is obtained from a battery. A **battery** is a component that produces a difference of potential (or emf). Most common batteries produce a difference of potential as a result of a continuous chemical reaction. In this section, we will discuss a variety of batteries and their characteristics.

**Battery**
A component that produces a difference of potential (emf). Batteries convert chemical, thermal, or light energy into electrical energy.

### Cells

A battery is made up of one or more cells. A **cell** is a single unit designed to produce electrical energy through thermal (heat), chemical, or optical (light) means. In a typical battery, cells produce a voltage through chemical means. A basic cell is represented in Figure 3.25.

The cell shown in Figure 3.25 has three parts: two terminals called **electrodes** and a chemical called the **electrolyte.** When the copper electrode comes in contact with the elec-

**Cell**
A single unit designed to produce electrical energy through thermal (chemical) or optical (light) means.

**Electrodes**
The terminals of a battery. Typically, the negative electrode is made of copper and the positive electrode is made of zinc.

**Electrolyte**
The chemical in a cell that interacts with the terminals, producing a difference of potential between them.

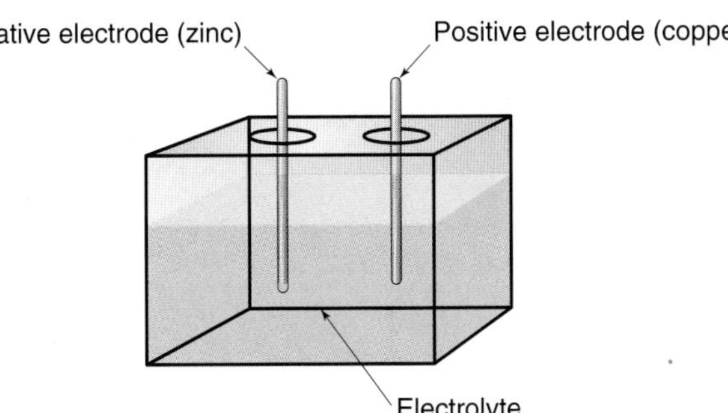


**FIGURE 3.25 A chemical cell.**

FIGURE 3.26

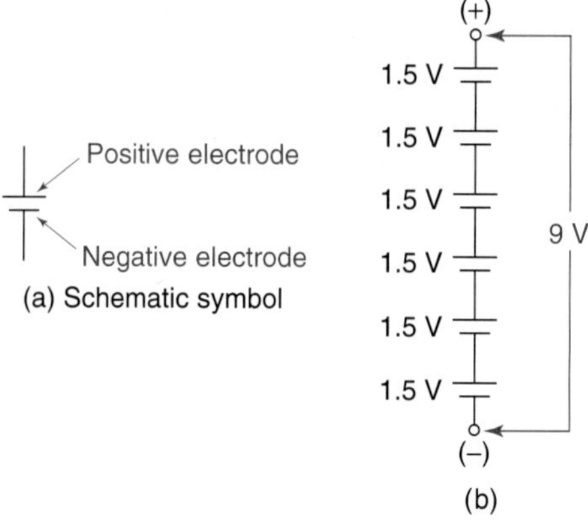


trolyte, the chemical reaction causes the copper to gain electrons. By the same token, when the zinc electrode comes in contact with the electrolyte, the chemical reaction causes the zinc to lose electrons. As a result, we have an excess of electrons at the copper electrode and a deficiency of electrons at the zinc electrode, and a difference of potential (or voltage) is created between the electrodes.

The schematic symbol for a single cell is shown in Figure 3.26a. A battery may contain a single cell or a series of cells (as shown in Figure 3.26b). Note that the voltage across the terminals of the battery is equal to the sum of the cell voltages. For example, the six 1.5 V cells in Figure 3.26b combine to form a 9 V battery.

## Primary Versus Secondary Cells

**Primary cell**
A cell that cannot be recharged. Often called dry cells because they contain dry electrolytes.

**Secondary cell**
A cell that can be recharged. Often called wet cells because they contain liquid electrolytes.

In some cells, the chemical action between the electrolyte and the electrodes causes the electrodes to be destroyed in a relatively short time. Once the electrodes have been destroyed, the chemical action stops and the battery is dead. This type of cell, called a **primary cell,** cannot be recharged. Note that primary cells are often referred to as *dry cells* because they contain dry electrolytes.

A **secondary cell** is a type of cell that can be recharged because the electrodes are not destroyed by the chemical action of the battery. Once a secondary cell has lost its charge, it can be recharged from an external current source. Note that secondary cells are often referred to as *wet cells* because they contain liquid electrolytes.

## Battery Capacity

**Capacity**
A measure of how long a battery will last at a given output current, measured in ampere-hours (Ah).

Batteries are generally rated in terms of their output voltage and capacity. The **capacity** rating of a battery is a measure of how long the battery will last at a given output current.

The unit of measure of capacity is the ampere-hour (Ah), and it is found as the product of discharge current (the current supplied by the battery) and time. For example, a battery with a capacity of 1 Ah will last for:

1. 1 hour at a discharge rate of 1 A.
2. 0.5 hour at a discharge rate of 2 A.
3. 2 hours at a discharge rate of 500 mA.
4. 10 hours at a discharge rate of 100 mA.

Note that, in each of the above cases, the product of discharge current and time (in hours) is equal to one. Also note that the battery lasts longer at lower discharge current values.

FIGURE 3.27 Carbon-zinc batteries.

## Types of Batteries

One of the most common types of batteries is the carbon-zinc battery. Two carbon-zinc batteries are shown in Figure 3.27. The battery on the left contains a single 1.5 V primary cell. The 9 V battery on the right contains six of these cells. Most carbon-zinc batteries have a shelf life of about one year, meaning that a given battery will still work one year after it has been produced (assuming that it hasn't been used).

*A Practical Consideration:* The name of a battery identifies the electrolyte used.

The alkaline battery looks just like the carbon-zinc battery. However, the alkaline battery has a much higher capacity rating and a much longer shelf life. The alkaline cell is a 1.55 V primary cell.

The mercury battery contains primary cells rated at 1.35 V. This type of battery is most commonly used in watches, cameras, hearing aids, etc. Mercury batteries are nearly identical to the batteries shown in Figure 3.28.

*Note:* The batteries in Figure 3.28 are shown at approximately twice their actual size.

FIGURE 3.28 Lithium-iodine batteries.

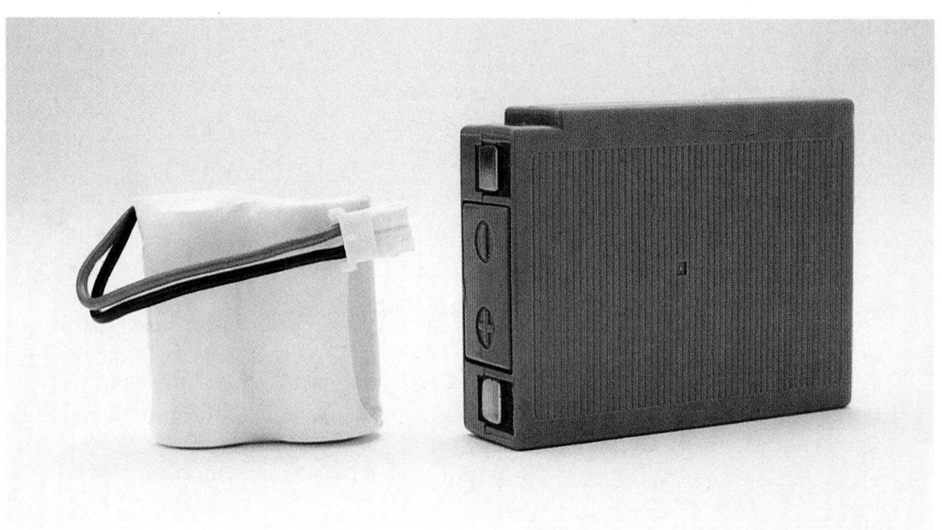

**FIGURE 3.29 A lead-acid battery.**

The lithium-iodine batteries shown in Figure 3.28 contain primary cells rated at 2.8 V. The average capacity for these cells is approximately 1.5 Ah.

One of the most common rechargeable batteries is the lead-acid battery shown in Figure 3.29. Each secondary cell is rated at 2 V. These types of batteries can have extremely high capacity ratings, some (like car batteries) in the range of 100 to 300 Ah.

Another common rechargeable battery is the nickel-cadmium, or Ni-Cd, battery. Several Ni-Cd batteries are shown in Figure 3.30. Each of the secondary cells is rated at approximately 1.2 V. At equal capacity ratings, a Ni-Cd battery is capable of higher output currents than a lead-acid battery.

**FIGURE 3.30 Ni-Cd batteries.**

## *Connecting Batteries*

In many applications, more than one battery is required to supply needed voltage or current. Figure 3.31 shows one of the two methods used to connect several batteries. When two (or more) batteries are connected as shown in Figure 3.31, the batteries are said to be connected in *series*. Note that the negative terminal of one battery is connected to the positive terminal of the other. When two or more batteries are connected in this fashion, the results are as follows:

1. The total voltage provided by the batteries is equal to the sum of their individual voltages.
2. The maximum current that can be supplied by the two does not increase. In other words, the maximum possible output current for the two is equal to the maximum possible for either battery alone.

Thus, the batteries in Figure 3.31 would supply a total voltage of 3 V at a maximum current of 80 mA.

The other method of connecting two (or more) batteries is shown in Figure 3.32. When two or more batteries are connected in this fashion, the batteries are said to be connected in *parallel*. Note that the two batteries are connected negative-to-negative and positive-to-positive. The results of connecting two or more batteries in this fashion are as follows:

1. The total voltage is equal to the individual battery voltages. In other words, the total battery voltage does not increase as a result of the connection.
2. The maximum possible output current is equal to the sum of the battery currents.

Thus, the connection shown in Figure 3.32 would have an output voltage of 1.5 V with a maximum output current of 160 mA.

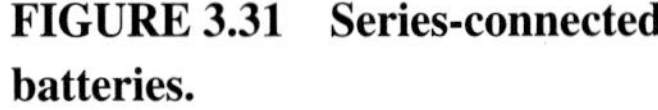

**FIGURE 3.31 Series-connected batteries.**

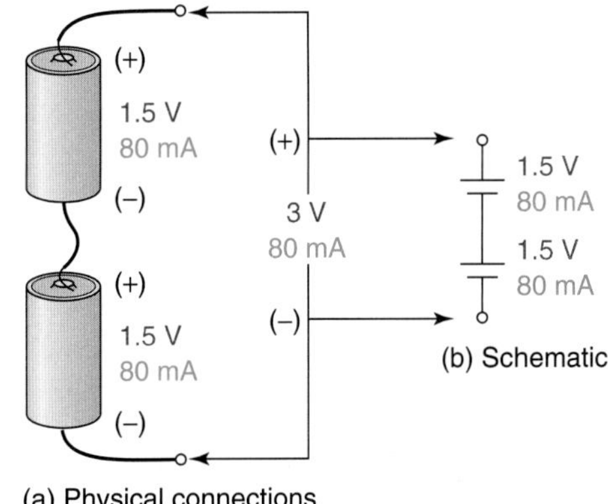


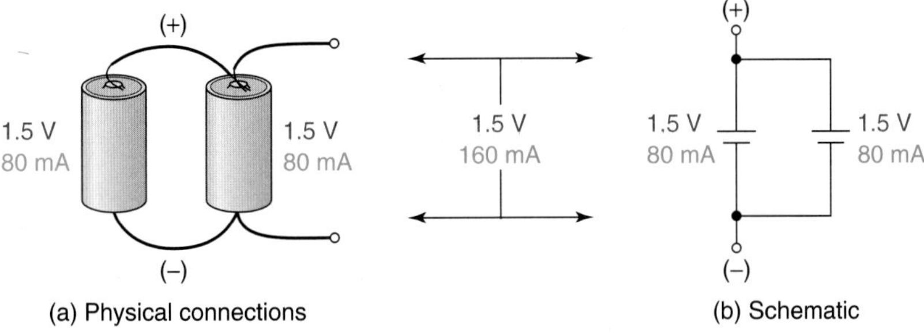


**FIGURE 3.32 Parallel-connected batteries.**

The bases for the results of series and parallel battery connections are discussed in future chapters. However, the information here has been presented to help you understand why many devices (such as flashlights) contain more than one battery. If the batteries are connected in series, the purpose is usually to increase the total voltage. If they are connected in parallel, the purpose is usually to increase the maximum possible current.

### Section Review

1. What is a battery?
2. What is a cell?
3. Briefly describe the operation of a cell.
4. How are cells connected to form a battery?
5. What are the differences between primary and secondary cells?
6. What is battery capacity?
7. What is the unit of measure for capacity?
8. What is the result of connecting batteries in series? In parallel?

## 3.5 dc POWER SUPPLIES

*OBJECTIVE 9* ▶ Even though batteries can provide dc voltages, their use in many circuits is impractical. For example, let's say we have a circuit that needs a 15 $V_{dc}$ input. To obtain this value with common D cells, we would have to connect ten of them in series. This is hardly practical.

**dc power supply**
A piece of equipment with dc outputs that can be adjusted to any value within its design limits.

One method of obtaining a variety of dc voltages is to use a piece of equipment called a **dc power supply.** A dc power supply has variable dc outputs that can be adjusted to any value within its design limits. The front panel of a typical dc power supply is represented in Figure 3.33. It should be noted that dc power supplies are produced by a variety of manufacturers, and that not all of them have the same features. However, the controls shown in Figure 3.33 are typical of those found on most dc power supplies.

### dc Outputs

As you can see, there are three dc outputs on the control panel represented in Figure 3.33. A simplified representation of these outputs is shown in Figure 3.34. The power supply contains two internal voltage sources with a common connection point between the two. By turning the +*Volts* control, the output from voltage source A is varied. By turning the −*Volts* control, the output from voltage source B is varied. (The meanings of the + and − signs will be explained in a moment.)

Normally, the dc power supply is connected to a circuit in one of three fashions. The first is shown in Figure 3.35a. When connected as shown, the power supply output current takes the path indicated by the arrows. In this case, voltage source B is not a part of the circuit, and the maximum power supply output voltage is 20 $V_{dc}$.

When connected as shown in Figure 3.35b, the power supply output current takes the path indicated by the arrows. In this case, voltage source A is not part of the circuit, and the maximum power supply output voltage is (once again) 20 $V_{dc}$.

When connected as shown in Figure 3.35c, the power supply output current takes the path indicated. In this case, both voltage sources are part of the circuit. Since the voltage sources are connected in series, the maximum power supply output voltage is equal to the sum of the maximum source voltages, 40 $V_{dc}$.

In most cases, a dc power supply is connected as shown in Figure 3.35a or b. The connection shown in Figure 3.35c is used only when the required voltage exceeds the maximum possible output from either of the voltage sources.

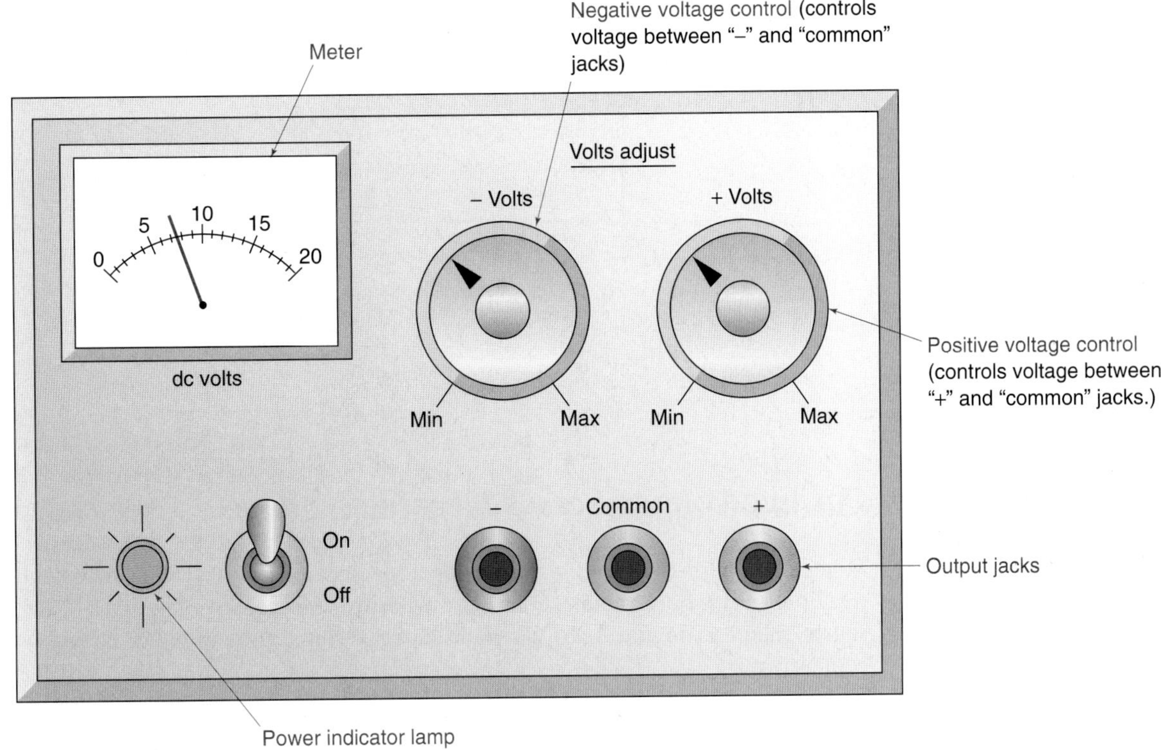


*Note*: This is a representative dc power supply panel. Many different panel designs are produced by various manufacturers.

**FIGURE 3.33 A dc power supply panel.**

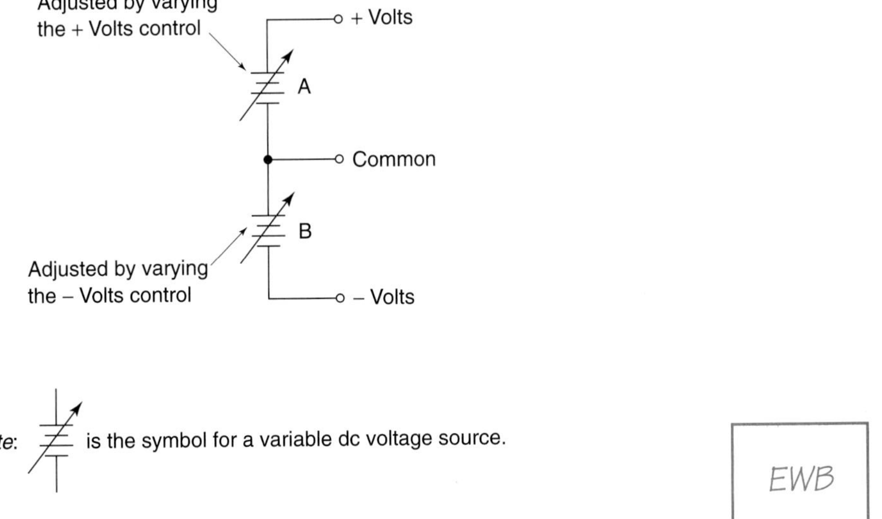


*Note*: is the symbol for a variable dc voltage source.

EWB

**FIGURE 3.34 A simplified representation of a dc power supply.**

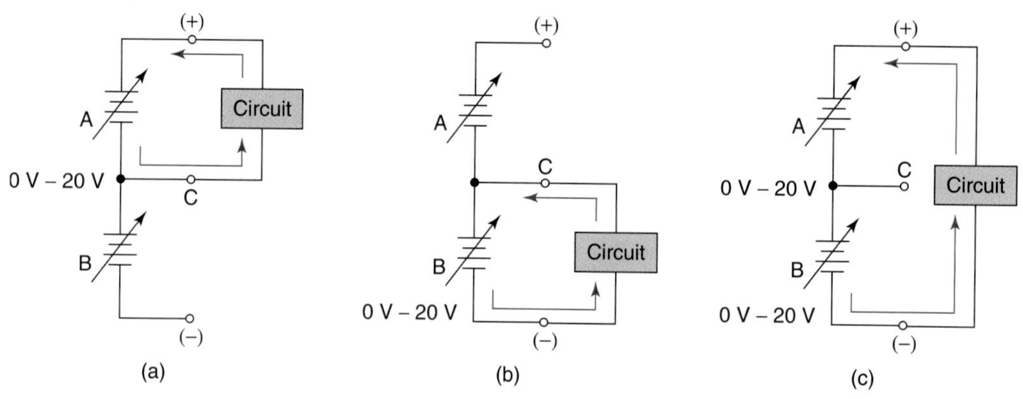


**FIGURE 3.35 Power supply connections.**

## Positive Voltage Versus Negative Voltage

*OBJECTIVE 10* ▶ Voltages are normally identified as having a specific polarity; that is, as being either positive or negative. For example, consider the positive and negative terminals of the voltage source shown in Figure 3.36. We could describe this source in either of two ways. We could say that:

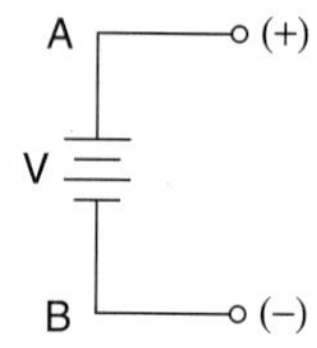


**FIGURE 3.36**

1. Side A is positive with respect to side B, meaning that side A of the source is more positive than side B.
2. Side B is negative with respect to side A, meaning that side B of the source is more negative than side A.

While both of these statements are correct, they do not assign a specific polarity to the source; that is, we have not agreed on whether it is a positive voltage source or a negative voltage source.

Assigning voltage polarities.

To assign a polarity to any voltage, we must agree on a reference point. For example, if we agree that side A of the voltage source in Figure 3.36 is the reference point, then we would say that the output from the source is a negative dc voltage because side B is more negative than our reference point. At the same time, if we agree that side B of the source is our reference point, then we would say that the output from the source is a positive dc voltage because side A of the source is more positive than our reference point.

**Common terminal**
The terminal of a dc power supply that is common to two (or more) voltage sources and thus is used as the reference point.

The **common terminal** of the dc power supply in Figure 3.33 is common to both voltage sources and therefore is used as the reference point; that is, the output polarity of either source is described in terms of its relationship to this point. For example, refer back to Figure 3.34. The common is connected to the negative terminal of source A. Since the top side of A is positive with respect to the common (reference) point, we say that the output from A is a positive dc voltage. By the same token, the bottom side of source B is negative with respect to the common, so we say that the output from source B is a negative dc voltage. In each case, *the output voltage polarity describes its relationship to the common terminal.* This is why the control for source A is labeled + *Volts* and the control for source B is labeled − *Volts.* In both cases, the label indicates the output polarity with respect to the common terminal.

## Using a dc Power Supply

When you are in the process of constructing a circuit, the circuit is built and inspected for errors before the dc power supply is turned on. Once you are sure that you have built the circuit correctly:

Connecting a dc power supply to a properly constructed circuit.

1. Adjust the voltage controls on the dc power supply to their 0 $V_{dc}$ settings.
2. Connect the dc power supply to the circuit.

3. Turn on the dc power supply.
4. Slowly increase the output from the power supply to the desired value.

### One Final Note

We have now touched on some of the basics of working with dc power supplies. As with all electronic equipment, you should take the time to familiarize yourself thoroughly with your dc power supply before attempting to use it.

**Section Review**

1. What is a dc power supply?
2. What determines the polarity of a voltage?
3. What is the common terminal of a power supply?
4. List the steps that you should take when connecting a dc power supply to a circuit.

## 3.6 SWITCHES AND CIRCUIT PROTECTORS

In any circuit, one of the requirements for current is that there must be a complete path between the terminals of the voltage source. Without a complete path, charge cannot flow between the terminals of the source.

Switches and circuit protectors (such as fuses and circuit breakers) are devices used to make or break the current path in a circuit under certain circumstances. Switches allow you to physically make or break a current path. For example, a light switch allows you to turn a light on or off by making or breaking the circuit. Circuit protectors automatically break a current path under certain circumstances. For example, fuses and circuit breakers are designed to protect equipment and circuits from excessive current.

### Switches

◄ *OBJECTIVE 11*

A **switch** is a device that allows you to make or break the connection between two or more points in a circuit. A switch is connected in a circuit as shown in Figure 3.37. Note that a path for current is made when the switch is closed. The current path is broken when the switch is opened.

A switch is made up of one or two moving contacts, called **poles.** A single pole switch, like the one represented in Figure 3.37, has only one moving contact. A double pole switch, like the one represented in Figure 3.38, has two moving contacts.

The switches represented in Figures 3.37 and 3.38 can be closed in only one position; that is, they can make or break a connection with only one nonmoving contact. The nonmoving contact is referred to as a **throw,** and the switches shown are referred to as single-throw switches. In contrast, several double-throw switches are shown in Figure 3.39.

**Switch**
A device that allows you to make or break the connection between two or more points in a circuit.

**Pole**
The moving contact in a mechanical switch.

**Throw**
The nonmoving contact in a mechanical switch.

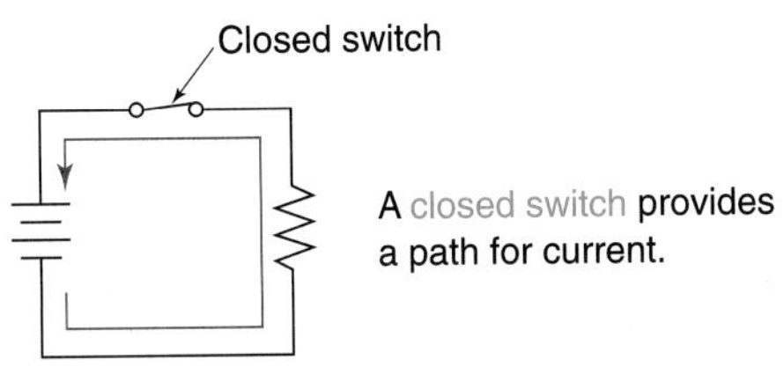


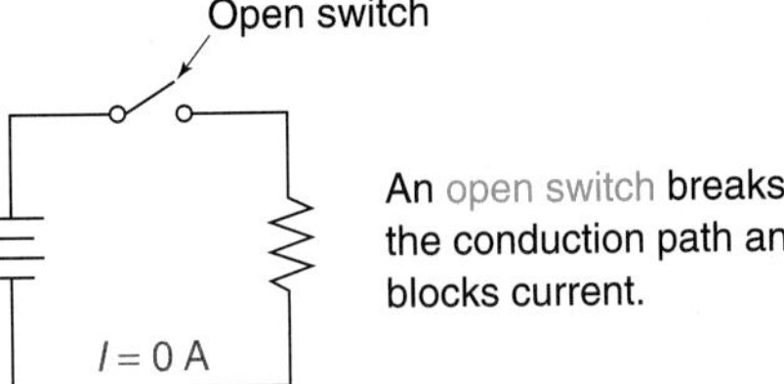

**FIGURE 3.37 Closed and open switches.**

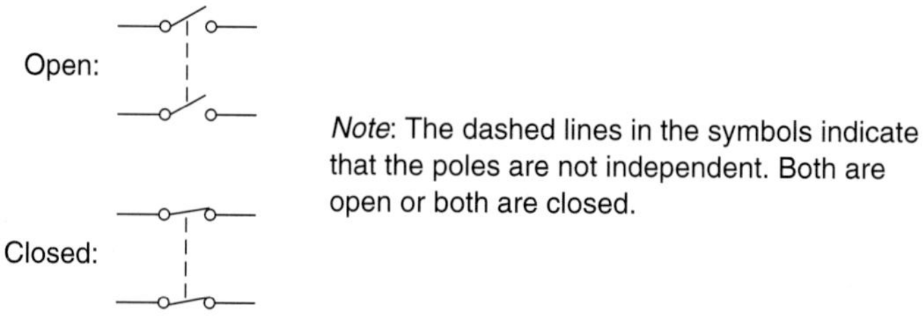

Open:

Closed:

*Note*: The dashed lines in the symbols indicate that the poles are not independent. Both are open or both are closed.

**FIGURE 3.38 Double-pole switches.**

EWB

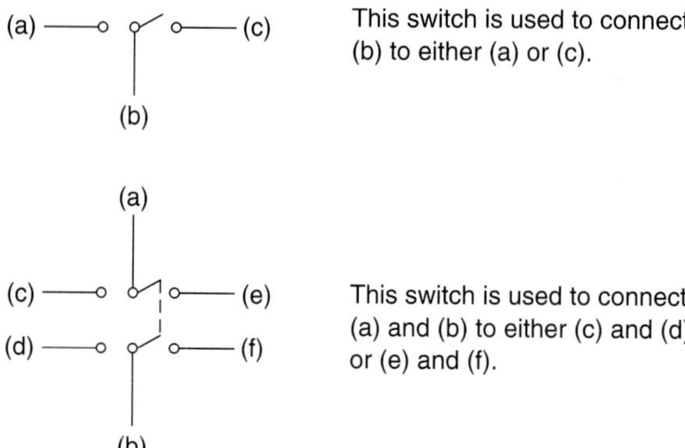

(a)
(b)
(c)
This switch is used to connect (b) to either (a) or (c).
(a)
(c)
(e)
(d)
(f)
(b)
This switch is used to connect (a) and (b) to either (c) and (d) or (e) and (f).

**FIGURE 3.39 Double-throw switches.**

Both of these switches have a moving contact that can be made or broken with either of two nonmoving contacts.

Switches are generally described in terms of the number of poles and throws they contain. For example, the switch represented in Figure 3.37 is described as a single-pole, single-throw (SPST) switch, meaning that it has a single moving contact that can be made to close in only one position. The switch in Figure 3.38 is described as a double-pole, single-throw (DPST) switch, meaning that it has two moving contacts that can be made to close in only one position. The switches shown in Figure 3.39 are the single-pole, double-throw (SPDT) switch and the double-pole, double throw (DPDT) switch.

Any of the switches shown up to this point can easily be identified by the number of terminals on its case, as follows:

1. An SPST switch has two terminals.
2. An SPDT switch has three terminals.
3. A DPST switch has four terminals.
4. A DPDT switch has six terminals.

The number of terminals for a given switch can be seen by taking another look at the switches shown in Figures 3.37 through 3.39.

## Normally Open and Normally Closed Switches

**Normally closed (NC) switch** A pushbutton switch that normally makes a connection between its terminals. Pushing the button breaks the connection.

**Normally open (NO) switch** A pushbutton switch that normally has a broken path between its terminals. Pushing the button makes the connection.

Some switches are designed so that they normally make or break a connection. For example, the **normally closed (NC) switch** is a pushbutton switch that normally makes a connection between its terminals. When the button is pushed, the switch breaks the connection. The schematic symbol for an NC switch is shown in Figure 3.40a.

The **normally open (NO) switch** is the opposite of the NC switch; that is, the current path between its terminals is normally open. When the button is pushed, the switch makes the connection. The schematic symbol for this type of switch is shown in Figure 3.40b.

**FIGURE 3.40 Normally closed and normally open switches.**

**Momentary switch**
A switch that makes or breaks the connection between its terminals only while the button is pressed.

**Rotary switch**
A switch with one or more poles and a number of throws.

Some NC and NO switches are classified as **momentary switches;** that is, the connection is made or broken only as long as the button is pushed. Others maintain the made or broken connection until the button is pressed again.

## Rotary Switches

The simplest explanation of a **rotary switch** is that it is a switch with one or more poles and a number of throws. The schematic symbol for a simple rotary switch is shown in Figure 3.41. As you can see, this particular switch has one pole and three throws. By turning the control shaft, a connection is made between the pole and one of the throws. An example of a rotary switch is the channel switch on older television sets.

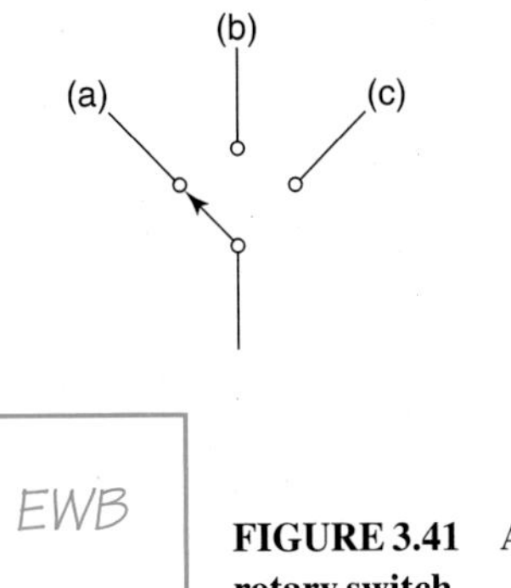


EWB

**FIGURE 3.41 A rotary switch.**

## DIP Switches

**Dual in-line package (DIP) switches** are SPST switches grouped in a single case. A group of DIP switches and the schematic symbol are shown in Figure 3.42. Note that DIP switches are the most commonly used switches in newer electronic circuits and systems.

**Dual in-line package (DIP) switches**
SPST switches grouped in a single case.

## Fuses

In some ways, a fuse is like a normally closed switch; that is, under normal circumstances, a fuse allows current to pass. However, unlike a normally closed switch, a **fuse** is designed to open automatically if its current exceeds a specified value. The simplest type of fuse and its schematic symbol are shown in Figure 3.43.

◄ *OBJECTIVE 12*

**Fuse**
A device designed to open automatically if its current exceeds a specified value.

(a)

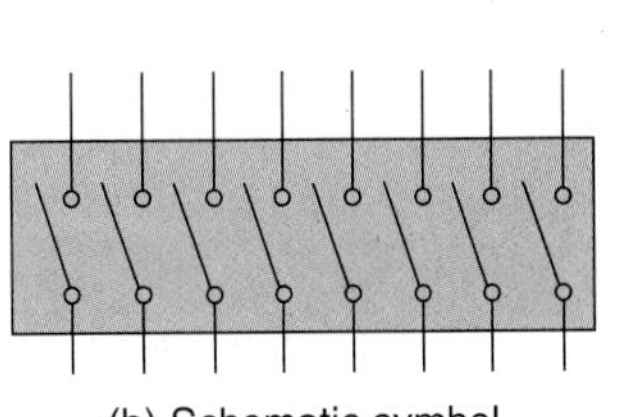

(b) Schematic symbol

**FIGURE 3.42 DIP switches.**

*Note:* The switch in Figure 3.42a is shown at approximately twice its actual size.

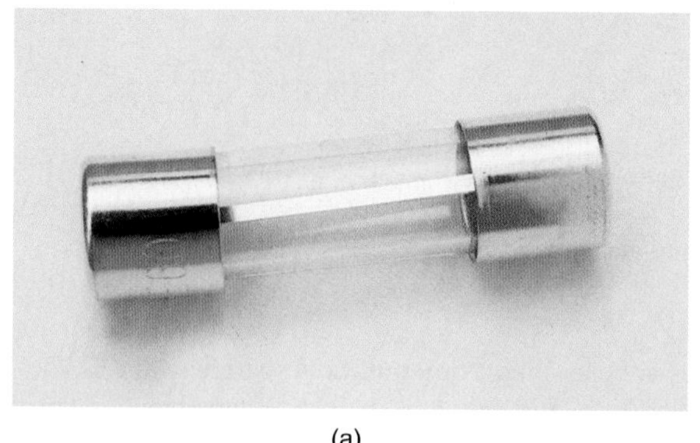

(a)

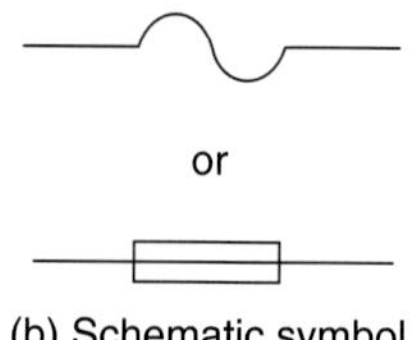


(b) Schematic symbol

**FIGURE 3.43 A fuse and its schematic symbol.**

As current passes through a conductor, heat is produced. The amount of heat varies directly with the value of current; that is, as current increases, so does the amount of heat produced.

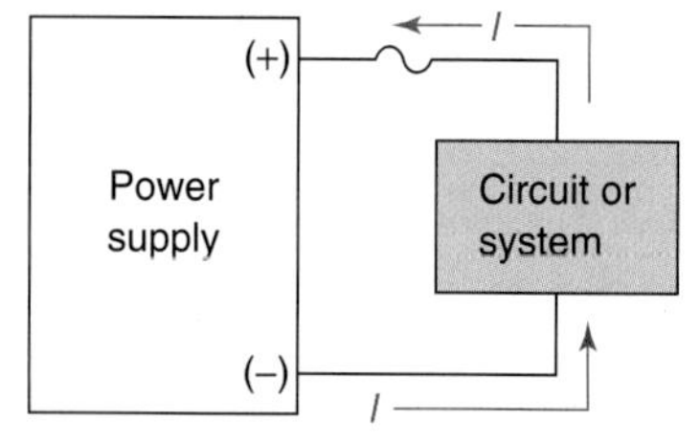


**FIGURE 3.44**

The fuse in Figure 3.43 contains a thin conductor designed to melt at relatively low temperatures. If the current through the component reaches a specified level, the conductor heats to its melting point. When it melts, the connection between the ends of the fuse is broken, the fuse is destroyed, and it acts as an open switch.

The purpose of a fuse is to protect a circuit or system from excessive current. For this reason, the fuse is normally placed between the power supply and the circuit or system, as shown in Figure 3.44. As you can see, the total circuit current in Figure 3.44 originates at the power supply and passes through the fuse. As long as the fuse is good, the current path in the circuit is complete. However, if a problem develops that causes the circuit current to exceed the current rating of the fuse, the component opens and the current path is broken. This protects the circuit from excessive current.

## *Fuse Ratings*

Fuse current ratings.

Fuses have a current rating and a voltage rating. The current rating of a fuse is the maximum allowable fuse current. If the current rating of a fuse is exceeded, the fuse will "blow" (open).

To understand the voltage rating of a fuse, we must establish one of the characteristics of an open circuit. In an open circuit, the applied voltage is felt across the open. For example, consider the circuit shown in Figure 3.45a. The applied voltage is shown to be 50 V. A difference of potential exists between points A and B. This difference of potential is equal to the applied voltage, 50 V. (The basis of this principle is covered in Chapter 4. For now, we are interested only in what happens when a circuit opens. Later, you will be shown *why* it happens.)

Fuse voltage ratings.

The voltage rating of a fuse is the maximum amount of voltage that an open fuse can block. The explanation of this rating starts with the circuit in Figure 3.45b. As you can see,

EWB

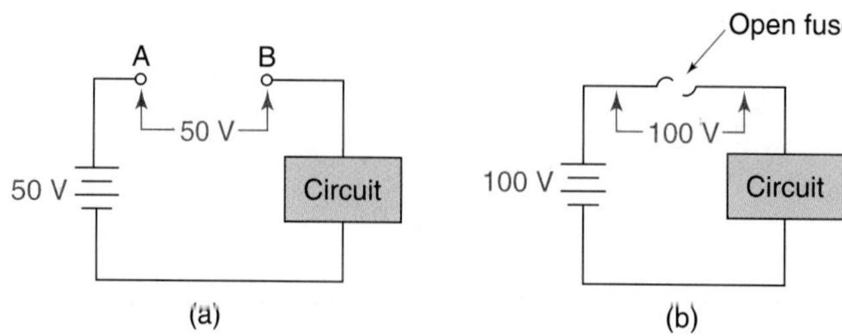


**FIGURE 3.45 The applied voltage is felt across an open fuse.**

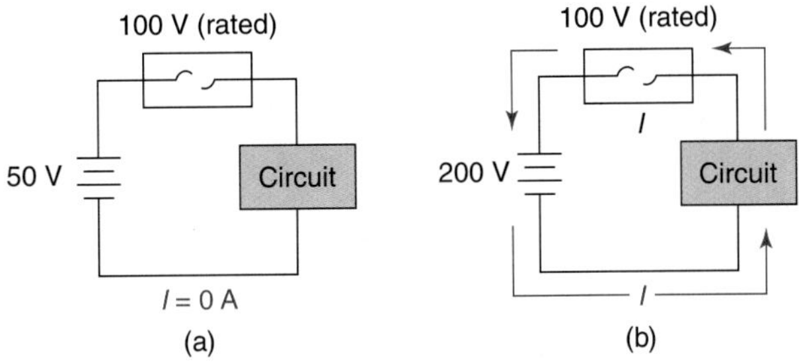


**FIGURE 3.46**

the fuse in the circuit has opened. Once it has opened, the applied voltage (100 V) is felt across the fuse.

You may recall that the breakdown voltage rating of an insulator is the difference of potential that will cause the insulator to break down and conduct. The voltage rating of a fuse is basically the same thing. If the voltage across an open fuse exceeds its voltage rating, the air in the fuse may ionize (charge), causing the open fuse to conduct again. For example, consider the two circuits in Figure 3.46. In Figure 3.46a, the applied volt-age is less than the voltage rating of the fuse. In this case, the open fuse does not allow conduction, and the circuit is protected. However, the applied voltage in Figure 3.46b is greater than the voltage rating of its fuse. Because of this, the open fuse ionizes and conducts. The result is that the circuit is no longer protected.

## Replacing a Fuse

When a fuse is blown, it must be replaced. When replacing a fuse, these guidelines should be followed:

1. Make sure that all power is removed from the system. For example, if you're replacing the fuse in a stereo system, make sure that the system is unplugged.
2. Use an exact replacement; that is, use one that is the same size and has the same current and voltage ratings as the original fuse.
3. Never replace a fuse with one that has a higher current rating!

The first of these guidelines is designed to protect both you and the circuit. As often as not, a fuse is connected directly to the power cord of an electronic system. When this is the case, replacing a fuse with the system plugged in can result in a surge of current. Not only could this surge hurt you, it could also severely damage the system.

The second guideline needs no explaining. Whenever possible, any component should be replaced with an identical component.

The third guideline is critical. If you replace a fuse with one that has a higher current rating, you run the risk of damaging the circuit and possibly starting a fire. As you know, current produces heat. If you replace a fuse with one that has a higher current rating, a problem could cause more current to be drawn through the circuit than it was designed to handle. The resulting heat could cause the circuit to be destroyed and could even result in a fire.

Once a fuse has been replaced, one of two things will happen:

1. The circuit operation will return to normal.
2. The new fuse will blow when power is restored.

If the circuit operation returns to normal, then the problem was probably nothing more than an old or defective fuse. If the new fuse blows, then there is a problem that is causing excessive current to be drawn from the power supply. In this case, the problem must be diagnosed and repaired before attempting to replace the fuse again.

## *Types of Fuses*

There are basically three types of fuses: high-speed instantaneous, normal instantaneous, and time delay. As you have probably guessed, these categories rate fuses in terms of the time required for them to blow.

High-speed instantaneous fuses react faster to an overload (excessive current) condition than any other type of fuse. The time lags of these fuses are rated as follows:

High-speed instantaneous.

1. They can carry 110% of their current ratings for at least four hours.
2. At 150% of their current ratings, they will blow within ten seconds (maximum).
3. At 200% of their current ratings, they will blow within five seconds (maximum).

High-speed instantaneous fuses are also known as *fast-acting* or *quick-acting* fuses.

Normal instantaneous fuses are somewhat slower to respond to an overload condition. The time lags of these fuses are rated as follows:

Normal instantaneous.

1. They can carry 110% of their current ratings for at least four hours.
2. At 135% of their current ratings, they will blow within one hour.
3. At 200% of their current ratings, they will blow within five seconds (maximum).

Normal instantaneous fuses are also known as *instantaneous* or *normal-blow* fuses.

Time delay fuses have the same time lag ratings as normal-blow fuses, with two exceptions:

Time delay.

1. At 200% of their current ratings, they require a minimum of five seconds to blow.
2. At 400% of their current ratings, they will blow within two seconds (maximum).

These ratings indicate that time delay fuses can handle short-duration overloads without blowing. The value of this feature can be seen by looking at the graph shown in Figure 3.47. Many appliances, such as clothes dryers, produce a current surge when they are turned on. The graph in Figure 3.47 represents such a surge. As you can see, at the turn-on time, the current surges to a relatively high value. Then it settles down very quickly to an average value. If a normal-blow fuse was used in the circuit of such an appliance, the current surge would most likely blow the fuse. However, a time delay fuse can handle such a surge without reacting. Note that time delay fuses are also known as *lag* or *slow-blow* fuses.

Fuses come in a variety of shapes and sizes, as shown in Figure 3.48. All of the fuses shown fall into one of the categories listed. The ratings for each fuse are given on the fuse itself. The type of fuse can be determined by looking up the part number in the manufacturer's catalog.

**FIGURE 3.47 A current surge.**

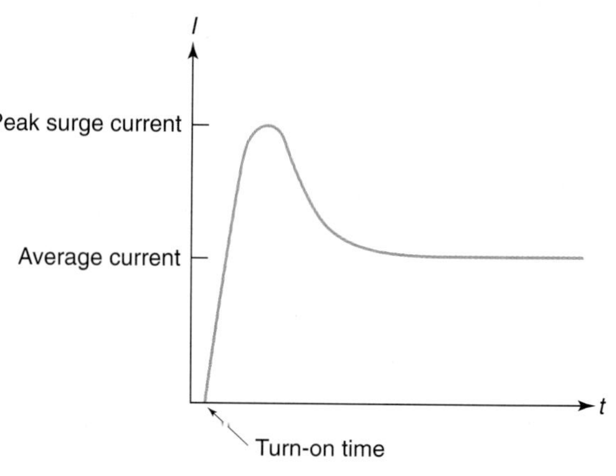

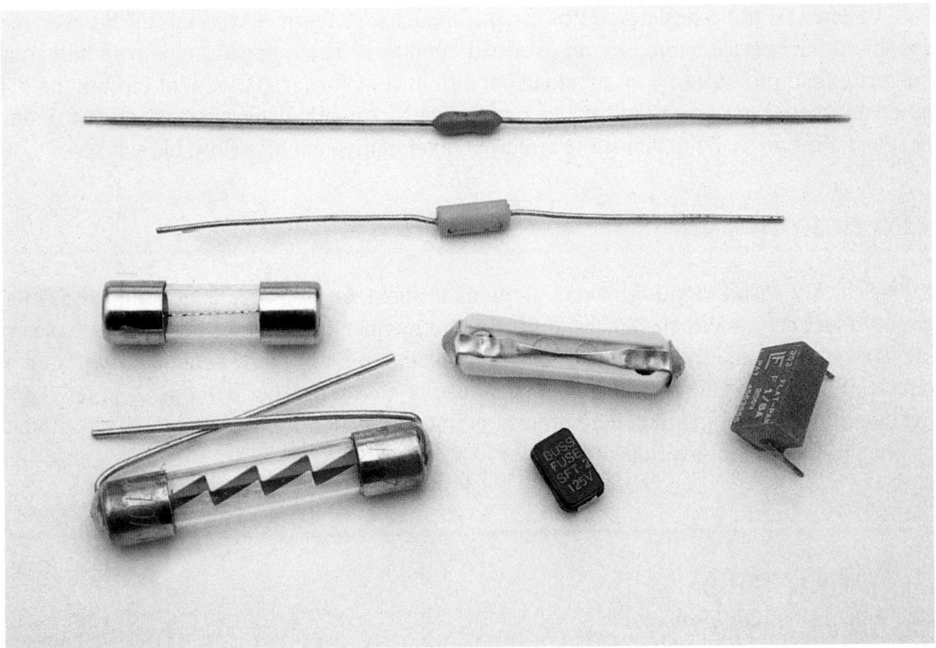


**FIGURE 3.48 Fuses of various shapes and sizes.**

## *Circuit Breakers*

Circuit breakers are designed to protect circuits from overload conditions, just like fuses. However, unlike fuses, circuit breakers are not destroyed when activated. A typical circuit breaker and its schematic symbol are shown in Figure 3.49. ◄ *OBJECTIVE 13*

In essence, a circuit breaker acts as a normally closed switch. When an overload condition develops, the circuit breaker opens and the circuit is protected. Pressing the reset switch on the circuit breaker returns it to its normally closed position, and the device can be used again.

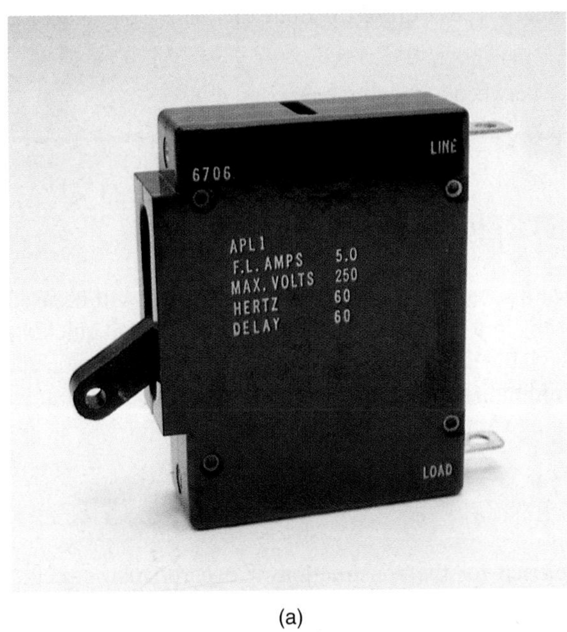


(a)

(b) Schematic symbol

**FIGURE 3.49 A circuit breaker and its schematic symbol.**

Types of circuit breakers.

There are three basic types of circuit breakers. A thermal-type circuit breaker responds to the heat generated by an overload condition. The magnetic-type responds to a magnetic field produced by an overload condition. (In Chapter 10, we will discuss the relationship between current and magnetism.) The thermomagnetic-type responds to both heat and a magnetic field, and is the circuit breaker equivalent of a slow-blow fuse.

### The Bottom Line

Switches, fuses, and circuit breakers allow us to make or break the conduction path in a circuit under certain circumstances. Switches are manual devices that allow us to make or break a circuit manually. Fuses and circuit breakers normally allow conduction through a circuit. If an overload condition develops, a fuse or circuit breaker will open and interrupt the circuit path. In doing this, they protect a circuit or system from the excessive heat that can be produced in an overload condition.

### Section Review

1. What is a switch?
2. What is a circuit protector?
3. What is a pole? What is a throw?
4. List and describe the four basic types of switches. State how each can easily be identified.
5. What is a normally open switch? A normally closed switch?
6. What is a momentary switch?
7. Describe rotary switches.
8. Describe DIP switches.
9. What is a fuse?
10. Describe the operation of a fuse.
11. List and describe the common fuse ratings.
12. List the guidelines for changing a fuse.
13. List and describe the three primary types of fuses. Include their alternate names.
14. What is a circuit breaker?
15. What is the primary difference between a circuit breaker and a fuse?

## 3.7 CIRCUIT CONSTRUCTION

When working in lab, you will build a variety of circuits. These circuits will be used to demonstrate many of the concepts covered in the text. If you can build a circuit quickly and correctly, the time you spend in the lab will be very productive. The more time you spend analyzing a circuit (rather than building it), the more you will learn. In this section, we'll discuss some of the basics of building circuits in the lab.

### Breadboards

**Breadboard**
A physical base used for the construction of experimental circuits.

**Prototype**
The first working model of a circuit or system.

A **breadboard** is a physical base used for the construction of experimental circuits. In other words, it provides a physical foundation that allows you to build temporary circuits rapidly. A typical breadboard is shown in Figure 3.50. Note that breadboards are often referred to as *protoboards* because they are frequently used to build **prototype** circuits (initial working models).

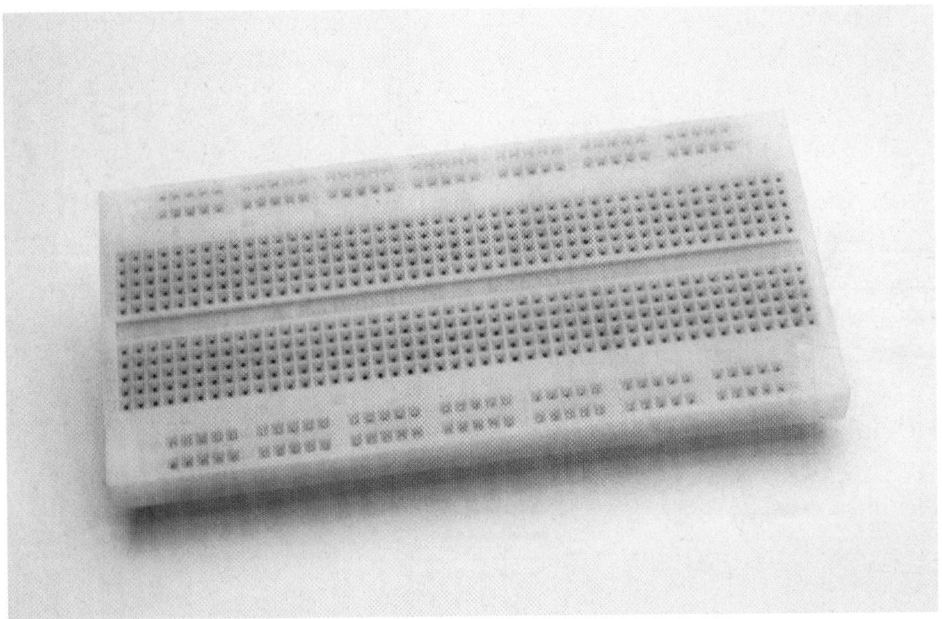

FIGURE 3.50 **A typical breadboard.**

Components are mounted in the small holes in the breadboard. In some cases, the connections between components are made with wires. However, the breadboard contains copper conductors that are also used for this purpose. For example, Figure 3.51 shows a close-up of a breadboard and its internal connections. The circles in the figure represent the holes on the board, and the lines connecting the circles represent the breadboard's internal conductors.

As Figure 3.51 illustrates, there are two rows of interconnected holes along the top of the breadboard. These two rows are commonly used for the dc power supply connections, as shown in Figure 3.52. When connected as shown, the holes in the top row are all connected to the +Volts (or −Volts) output of the power supply and the holes in the bottom row are all connected to the common terminal of the power supply. Note the wires that are added to bridge the gaps in the rows.

Components are normally mounted in the vertical groups of holes (columns). When two components are mounted in the same column, they are connected by the internal breadboard conductors, as illustrated in Figure 3.53.

**FIGURE 3.51 Typical breadboard internal connections.**

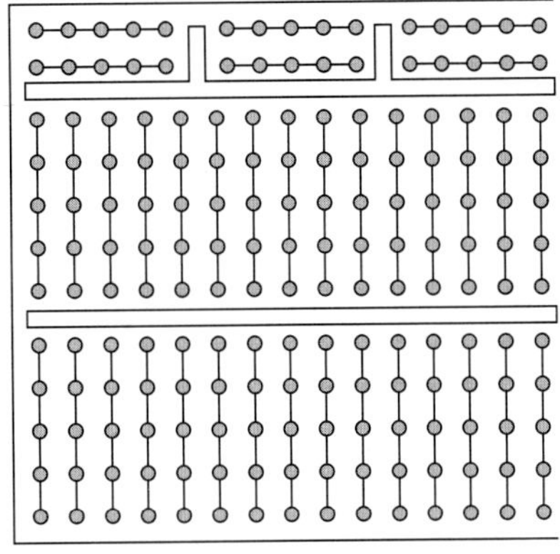

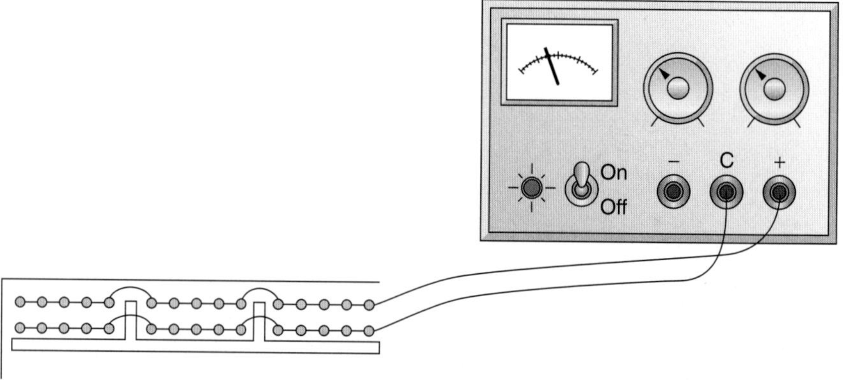


**FIGURE 3.52 Breadboard power supply connections.**

**FIGURE 3.53**

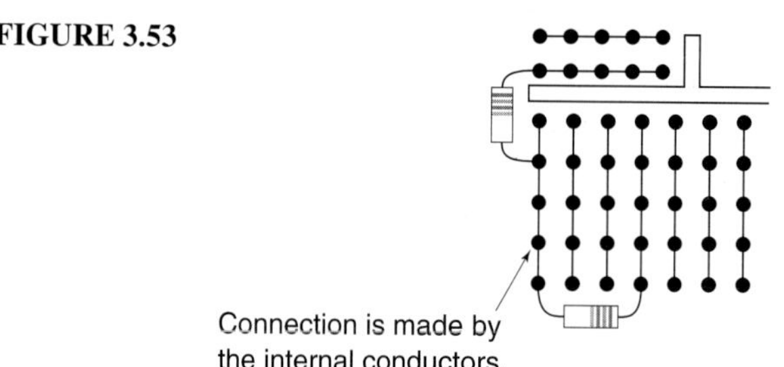


## *Circuit Construction*

*OBJECTIVE 14* ▶ Several guidelines should be followed when building a circuit on a breadboard:

1. Whenever possible, build the circuit so that it looks exactly like the schematic.
2. When using wires to connect components:
   a. Use 22 or 24 gauge solid wire.
   b. Keep the wires as short as possible (within reason).
   c. Lay the wires down flat on the breadboard.
3. Use ¼ W and/or ½ W resistors, making sure that the resistor leads are not too large for the holes on the protoboard.

When you follow the first of these guidelines, you will always be able to identify any component in the circuit. For example, look at the schematic and circuit shown in Figure 3.54. The resistors in the circuit are laid out in the same position as they are shown in the schematic. Thus, when going from the schematic to the circuit, you will always be able to easily identify the components. In contrast, look at the circuit construction in Figure 3.55. This is the same circuit as shown in Figure 3.54. However, because of the wiring, there is a good chance that one resistor will be mistaken for another.

The second guideline identifies the wire gauges that should be used and how they should be laid out. The wire gauges listed (22 and 24) are the best to use because they provide the best mechanical connection to the breadboard. At the same time, lower gauge wires (which have larger diameters) may cause the copper connectors in the board to be damaged or destroyed over time. The wires used should be kept short and laid flat on the board to keep them out of the way.

The third guideline also relates to the copper connectors in the breadboard. The leads on a ⅛ W resistor may not be large enough to provide a good mechanical connection with the breadboard. At the same time, the leads on a 1 W (or higher) resistor will almost always damage the holes on the breadboard (because of their diameters).

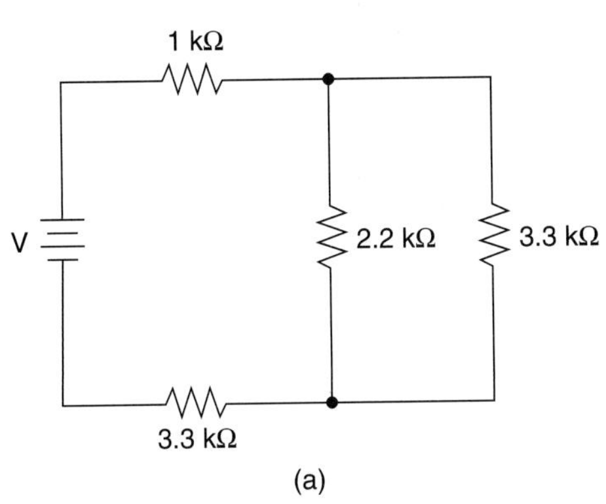


(a)

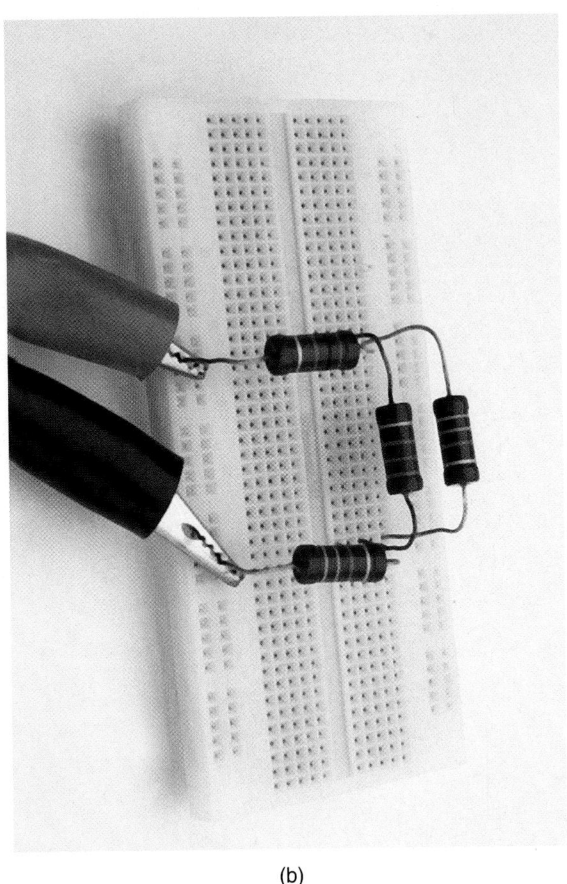

(b)

**FIGURE 3.54**

**FIGURE 3.55 A poorly constructed circuit.**

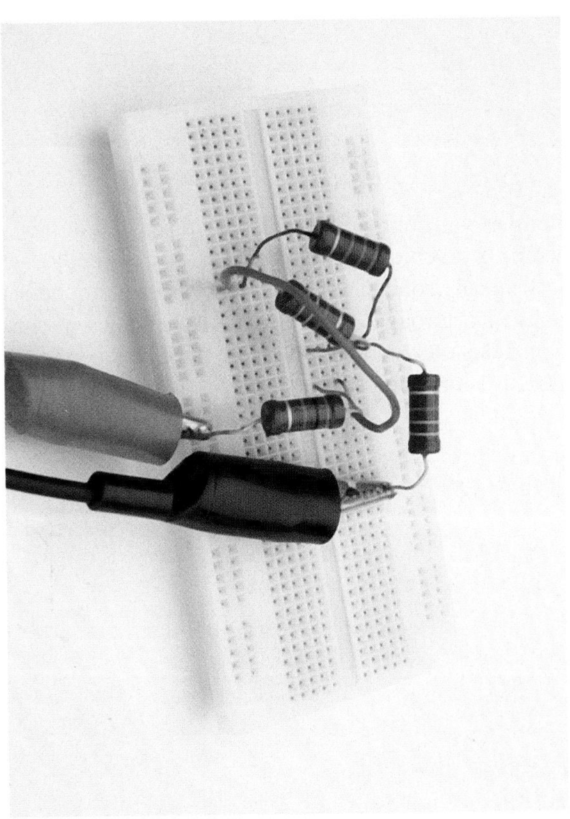

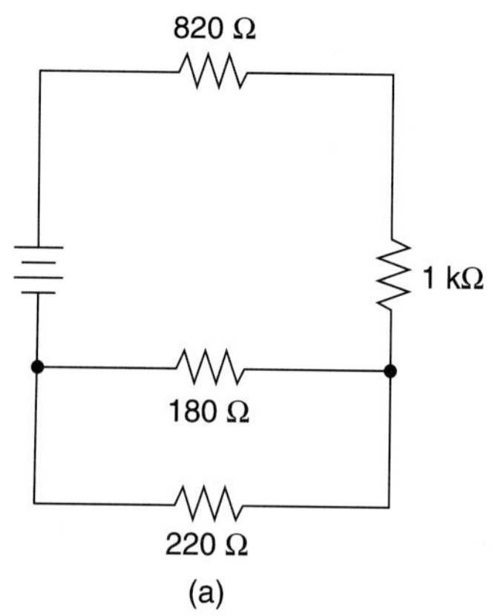

(a)

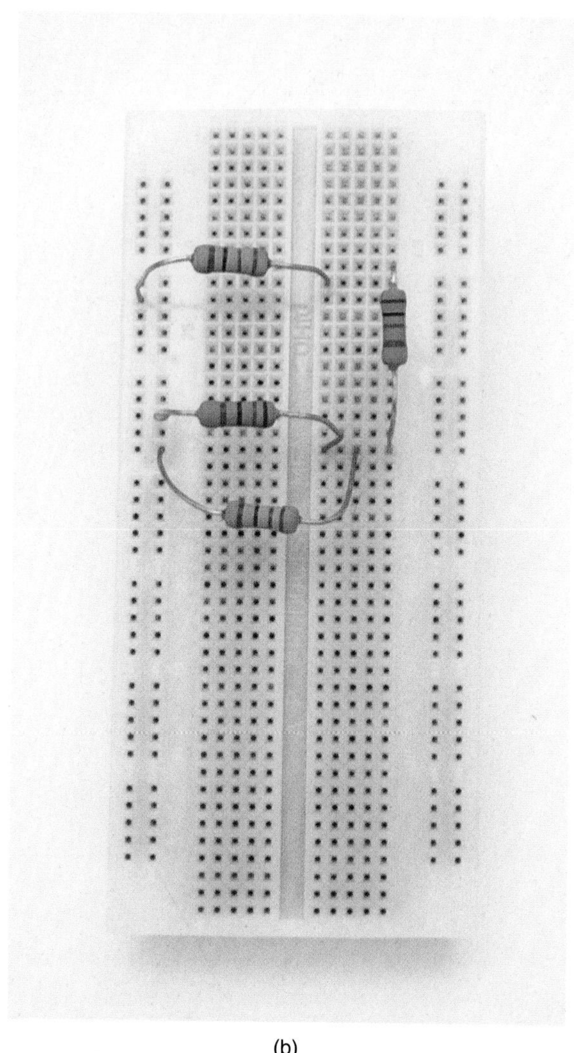
(b)

**FIGURE 3.56**

As long as you follow the guidelines discussed in this section, you will always be able to identify a given component rapidly (and correctly), your component connections will be mechanically solid, and your breadboard will not be damaged.

Figures 3.56 through 3.58 show the schematics of some circuits and their proper construction. Take a few moments to trace through the schematics and their circuits to see how the various connections were made.

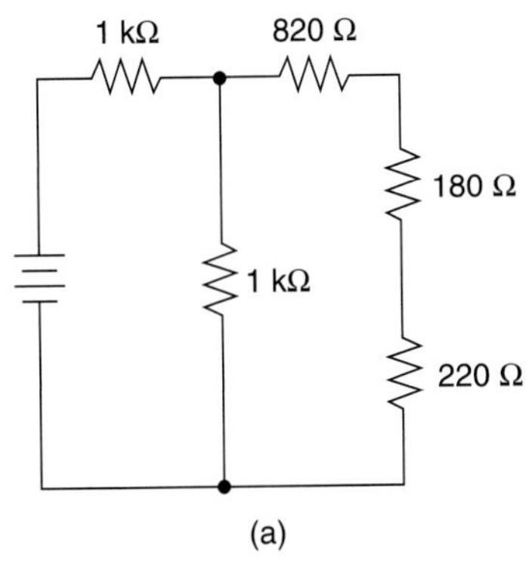

(a)

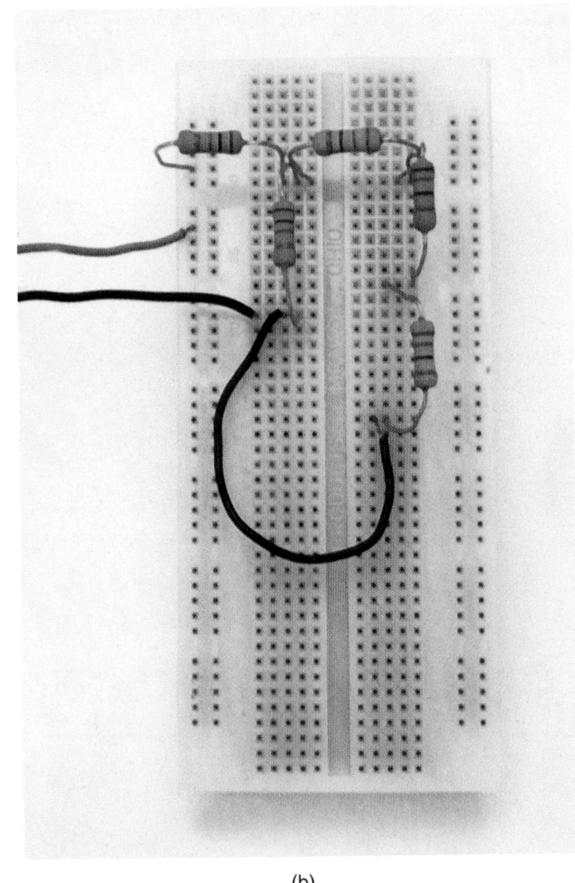
(b)

FIGURE 3.57

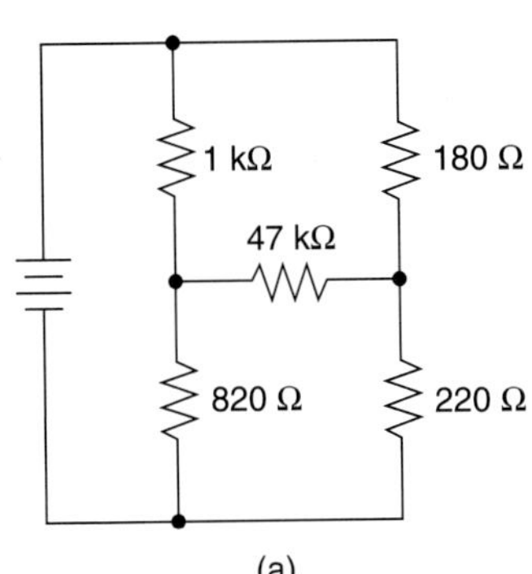

(a)

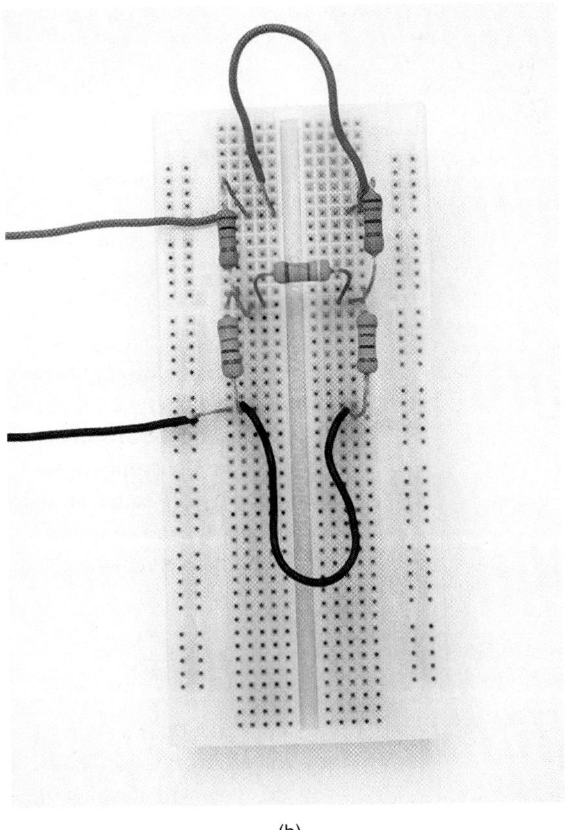
(b)

FIGURE 3.58

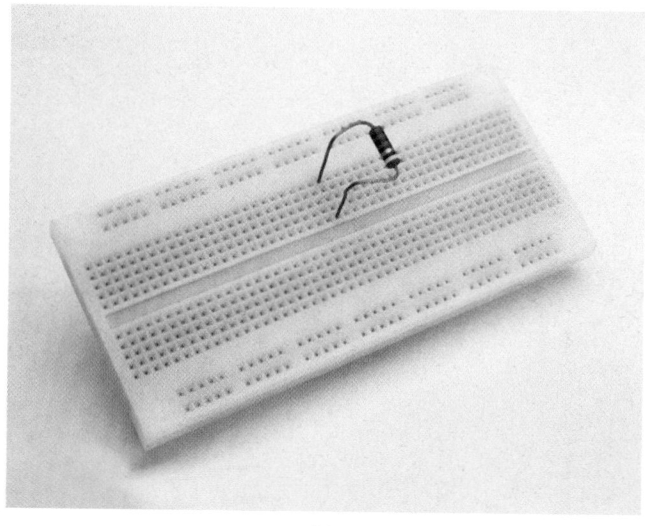

(a)

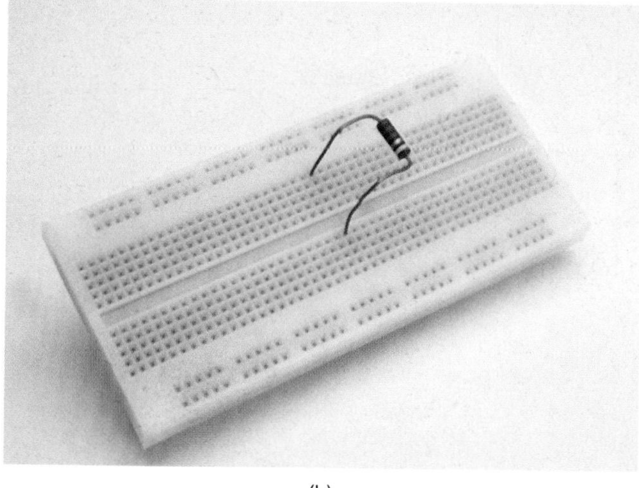

(b)

**FIGURE 3.59 Component connections on a breadboard.**

A word of caution: you should never mount both ends of a given component in the same column, as shown in Figure 3.59a. If you mount a component as shown, the component is shorted by the breadboard conductor; that is, the conductor in the breadboard effectively bypasses the component. As a result, any current in the circuit will bypass the component, which may cause an overload condition for the power supply. In contrast, when connected as shown in Figure 3.59b, the current in the circuit will pass through the resistance (since there is no alternate route).

## One Final Note

Circuit construction is a skill. As with any skill, competence is developed through practice. When you are working in the lab, take the time to build your circuits carefully and neatly. If you do, you will develop the skill needed to build even the most complex circuits quickly and correctly.

## Section Review

1. What is a breadboard? By what other name is it commonly known?
2. Describe the construction of a typical breadboard.
3. List the guidelines for building a circuit on a breadboard. What purpose is served by each of these guidelines?
4. What precaution should you take when mounting the two ends of a component on a breadboard. Why?

# 3.8 MEASURING CURRENT, VOLTAGE, AND RESISTANCE

In the lab manual that accompanies this text, there is a series of three exercises dealing with measuring current, voltage, and resistance. This section serves as an introduction to that series of exercises.

## Multimeters

Many types of meters are used to measure various electrical properties. For example, the *ammeter* is used to measure current, the *voltmeter* is used to measure voltage, and the *ohmmeter* is used to measure resistance. As you know, a digital-multimeter (DMM) is used to measure

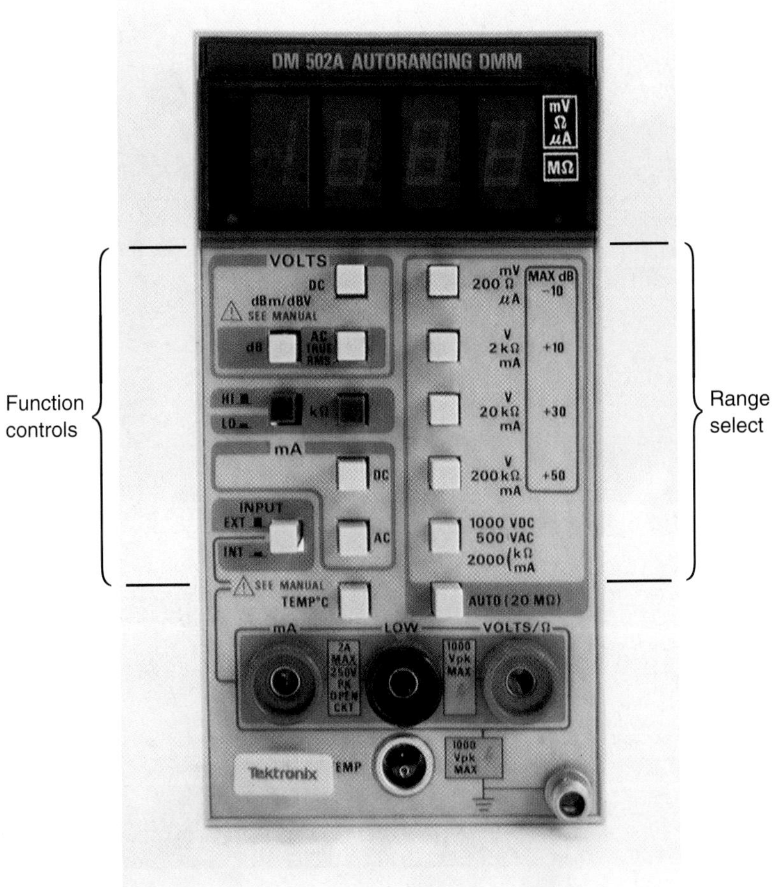


**FIGURE 3.60 A typical DMM.**

current, voltage, and resistance. The function controls on the DMM shown in Figure 3.60 are used to select the type and range of a given measurement. In the lab, you will be shown how to set the controls on your particular DMM. In this section, you will be shown how to connect the meter to a circuit or component to take the desired measurement.

*OBJECTIVE 15* ▶

## Measuring Current

Current is measured by inserting the meter in the current path. For example, let's say that we want to measure the current in the circuit shown in Figure 3.61a. The way to measure this current is to insert the meter so that the current must pass through the meter. This is accomplished by breaking the current path and inserting the meter in its place, as shown in Figure 3.61b. The current in this circuit must go through the meter because there is no other path for it to take. The photo in Figure 3.62 shows the actual connection of the DMM for measuring current in a circuit. Note that the meter leads are used to bridge the break between the resistors.

**FIGURE 3.61 Ammeter connection.**

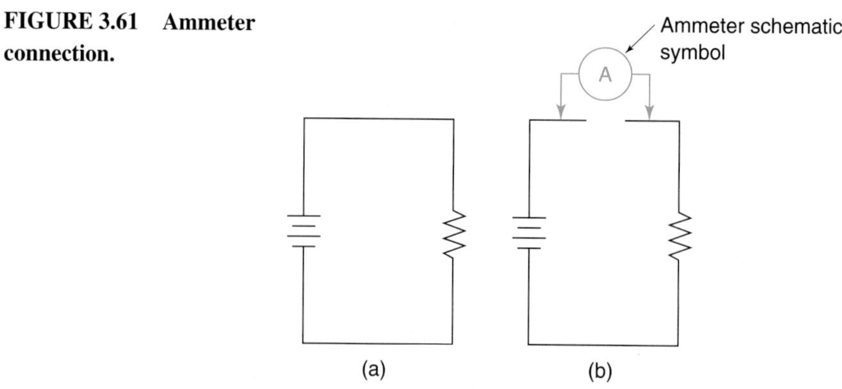


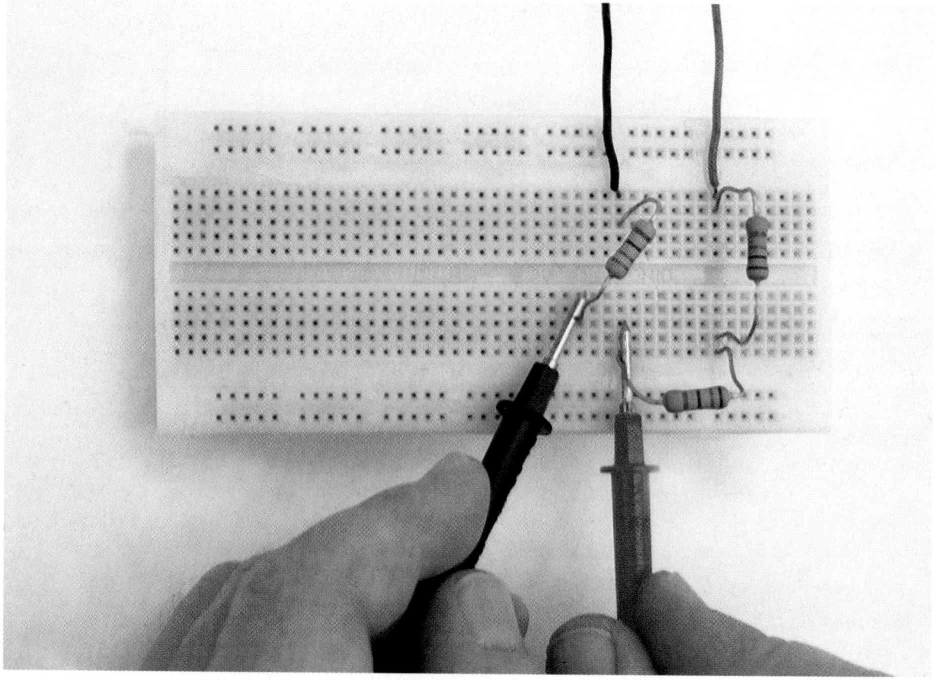

**FIGURE 3.62 Proper DMM connections for measuring current.**

**FIGURE 3.63 Voltmeter connection.**

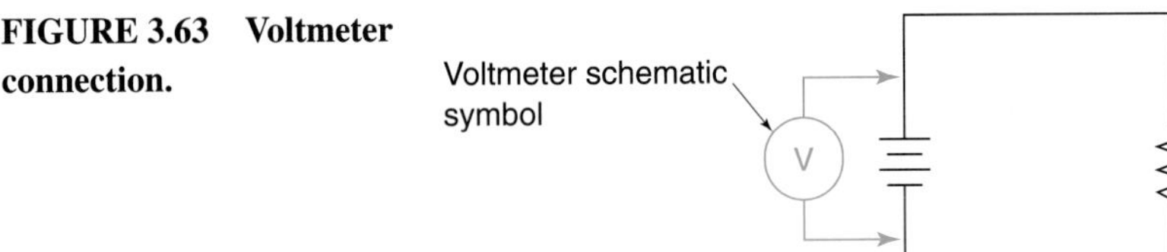


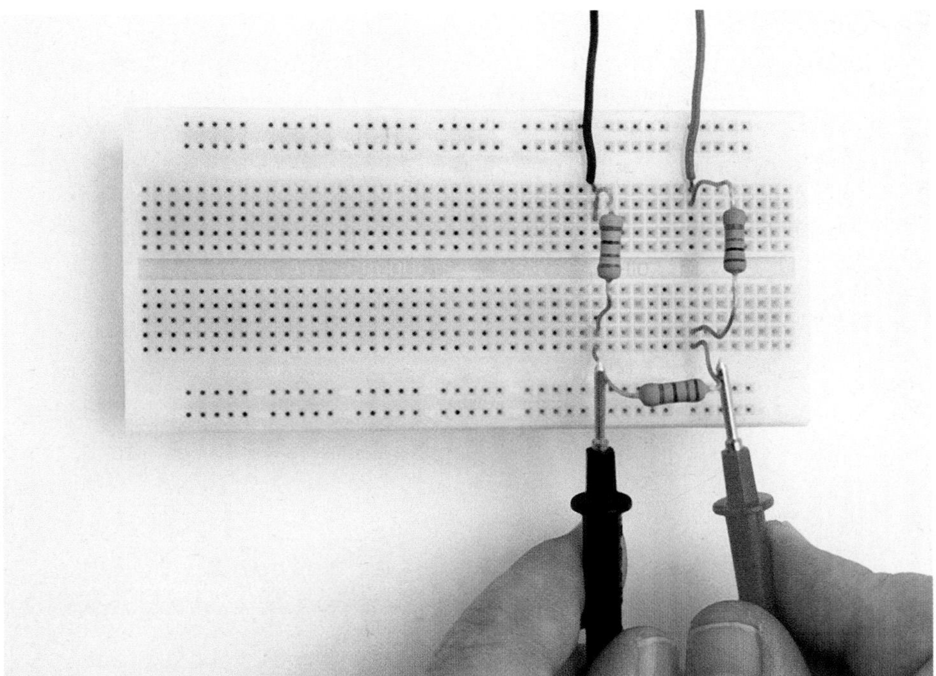

**FIGURE 3.64 Proper DMM connections for measuring voltage.**

## Measuring Voltage

Voltage is measured by connecting the voltmeter "across" a component: one of the voltmeter leads is connected to one side of the component, while the other lead is connected to the other side of the component, as shown in Figure 3.63. As you can see, the circuit current path is not broken when measuring voltage. The two meter leads are simply connected to the two sides of the component. The actual connection of a DMM for measuring the voltage across a resistor is shown in Figure 3.64.

## Measuring Resistance

Measuring resistance can get a little tricky because *resistance cannot be measured with power applied to the circuit.* In most cases, the safest and most reliable method of measuring resistance is to simply remove the resistor from the circuit. Once the resistor is removed from the circuit, the meter is connected across the component, just like a voltmeter. This point is illustrated in Figure 3.65.

An important point needs to be made at this time: if you hold both leads of a resistor while measuring its resistance, you'll get a faulty reading. When you hold both ends of a resistor, the resistance of your skin is placed in parallel with the component. As you will learn in Chapter 7, placing one resistance in parallel with another affects total resistance. To avoid this problem, you should hold only one resistor lead, as shown in Figure 3.66.

**FIGURE 3.65 Ohmmeter connection.**

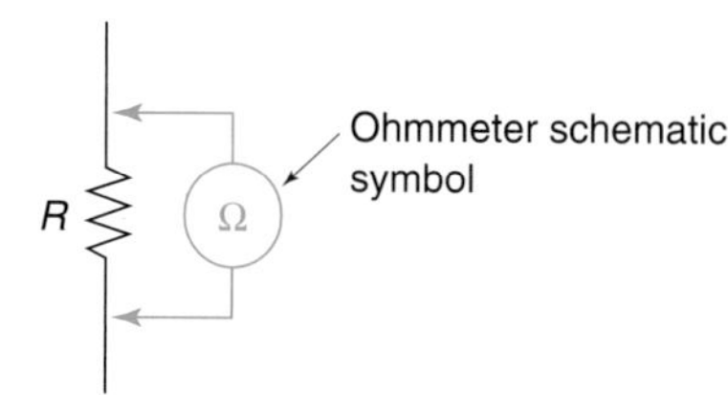


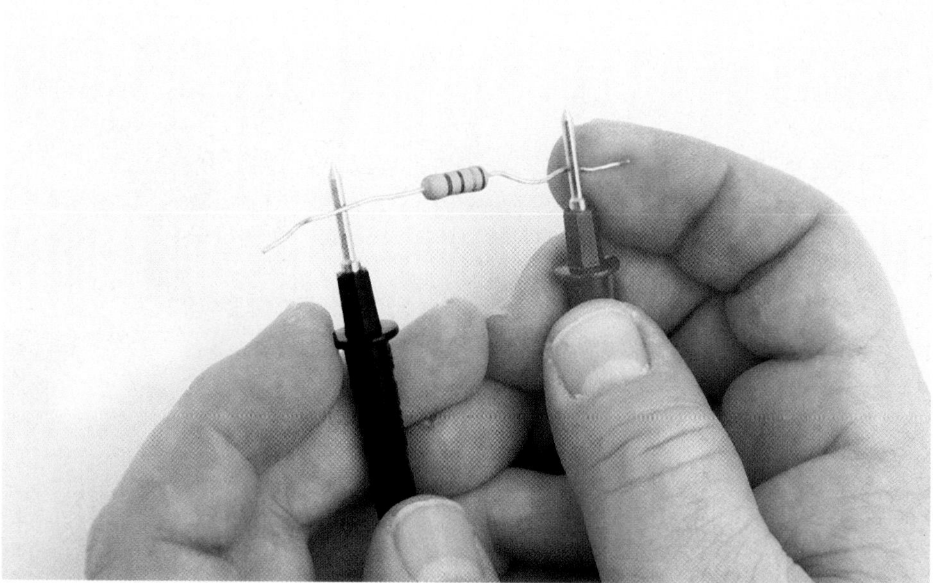

**FIGURE 3.66 Proper DMM connections for measuring resistance.**

## One Final Note

There is a lot to know about the proper use of test equipment. When you are in the lab, take the time to become thoroughly familiar with the meters before attempting to use them. This will help you to avoid obtaining faulty readings and possibly damaging the meters.

### Section Review

1. What does an ammeter measure? How is it connected in a circuit?
2. Why is an ammeter connected as described in Question 1?
3. How is a voltmeter connected in a circuit?
4. How is an ohmmeter connected to a component?
5. What precautions must be taken when measuring resistance?

## Chapter Summary

Here is a summary of the major points made in this chapter:

1. Theory deals with the concepts of a science. Practice deals with using those concepts to accomplish something.

2. Copper is the most commonly used conductor because it provides the best balance between low resistance and cost.
3. Solid wire contains a solid conductor. Stranded wire contains a group of thin wires wrapped together to form a larger diameter wire. Solid wire is cheaper to produce. Stranded wire is more flexible.
4. The American Wire Gauge (AWG) system sizes wires by cross-sectional area.
5. If you go up three wire gauges from any size wire, you'll find a wire with half the diameter and twice the resistance (per unit length).
6. If you go up ten wire gauges from any size wire, you'll find a wire with one-tenth the diameter and ten times the resistance (per unit length).
7. Ampacity is the maximum amount of current that a wire can carry, measured in amperes.
8. Circuits are usually mounted on printed circuit boards (PC boards). The copper conductors that interconnect the components are called traces.
9. The average breakdown voltage rating of an insulator is the voltage that will cause it to break down and conduct, given in kV/cm.
10. A resistor is a component designed to provide a specific amount of resistance.
    a. The value of a fixed resistor cannot be varied by the user.
    b. The value of a variable resistor can be adjusted by the user.
11. The most common type of resistor is the carbon-composition resistor. The value of a carbon-composition resistor is determined by the amount and purity of carbon between its leads.
12. Wire-wound resistors use the resistivity of a length of wire to produce the desired resistance. These resistors are used primarily in high-power applications.
13. Metal-film resistors are precision components with low temperature coefficients.
14. Integrated resistors are extremely small components. Several of these resistors are usually housed in a single case.
15. The value of any standard-value resistor can be given using the first two digits of its value and a power of ten. (Table 3.3 lists the resistor standard values.)
16. The tolerance of a resistor is the guaranteed range of measured values for the component, given as a percentage of the rated value.
17. The resistor color code is used to indicate the rated value, tolerance, and reliability of a resistor (as shown in Table 3.5).
18. The reliability rating of a resistor indicates the percentage of resistors that will fail to fall within tolerance after 1000 hours of use.
19. Resistors also have power ratings; that is, they are rated in terms of the amount of heat they can dissipate. The size of a carbon-composition resistor indicates its power rating.
20. When replacing a resistor, make sure that:
    a. The substitute component has the same resistance rating as the original.
    b. The tolerance of the substitute component is equal to (or less than) that of the original.
    c. The power capability of the substitute component is equal to (or greater than) that of the original.
21. A potentiometer is a three-terminal resistor whose value can be adjusted (within set limits) by the user. A potentiometer is usually referred to as a pot.
22. The center terminal of a pot is called the wiper arm. By adjusting the control shaft, the resistance between the wiper arm and each of the outer terminals is varied.
23. The sum of the two wiper arm resistances ($R_{ab}$ and $R_{bc}$) is equal to the resistance between the two outer terminals ($R_{ac}$).

24. The taper of a potentiometer is a measure of how its resistance varies as the control shaft is turned.
    a. The resistance of a linear taper pot varies at a constant rate.
    b. The resistance of a nonlinear-taper pot varies at a changing rate.
25. Pots are rated in terms of their maximum outer terminal resistance and power dissipation capability.
26. Standard pots are either carbon-composition or wire-wound.
27. Trimmer pots are small potentiometers used in low-power applications.
28. Multi-turn pots are high-resolution pots.
29. A decade box is a piece of lab equipment that contains a series of pots. The total resistance of the box can be varied over a wide range.
30. A battery is a component that produces a difference of potential from a continuous chemical reaction.
31. Batteries are made up of one or more cells. A cell is a single unit designed to produce electrical energy through thermal, chemical, or optical means.
32. Primary cells (or dry cells) are not rechargeable. Secondary cells (or wet cells) can be recharged.
33. The capacity of a battery is a measure of how long it will last at a given output current, measured in ampere-hours (Ah).
34. Batteries are connected in series to increase their total output voltage. They are connected in parallel to increase their total output current.
35. A dc power supply is a piece of lab equipment that provides variable dc voltages.
36. Voltages are generally assigned a polarity; that is, a label of being positive or negative.
37. The common terminal on a dc power supply is the reference to which polarity is assigned.
38. To have current in a circuit, there must be a complete path between the terminals of the voltage source.
39. A switch allows you to make or break the current path in a circuit manually.
40. A switch contains:
    a. One or more movable contacts, called poles.
    b. One or more nonmoving contacts, called throws.
41. Some switches are classified as normally open or normally closed.
42. A momentary switch is one that opens or closes only for as long as the button is pressed.
43. A fuse is a circuit protector designed to break a current path when its current exceeds a specified value.
44. The current rating of a fuse is the maximum allowable fuse current.
45. The voltage rating of a fuse is the maximum amount of voltage that an open fuse is designed to block.
46. Never replace a fuse with one that has a higher current rating.
47. Fuses are classified according to the amount of time required for them to blow (open).
48. Circuit breakers are normally closed switches designed to open under overload conditions. Pushing the reset button on an open circuit breaker restores it to normal operation.
49. A breadboard (or protoboard) is a physical base for constructing circuits. They contain internal conductors that allow you to connect two or more components.
50. An ammeter is used to measure current (as shown in Figure 3.61).
51. A voltmeter is used to measure voltage (as shown in Figure 3.63).
52. An ohmmeter is used to measure resistance (as shown in Figure 3.65).

## Equation Summary

| Equation Number | Equation | Section Number |
|---|---|---|
| **(3.1)** | $R_{ac} = R_{ab} + R_{bc}$ | 3.3 |

## Key Terms

The following terms were introduced and defined in this chapter:

ammeter
ampacity
audio taper
average breakdown voltage
breadboard
capacity
carbon-composition resistor
cell
circuit breaker
common
dc power supply
decade box
dry cell
electrode
electrolyte
fixed resistor
fuse
gang-mounted potentiometers
high-speed instantaneous fuse
integrated resistor
ionize
linear taper
metal-film resistors
momentary fuse
multi-turn potentiometer
negative voltage
nonlinear taper
normal-blow fuse
normal-instantaneous fuse
normally closed (NC) switch
normally open (NO) switch
ohmmeter
parameter
polarity
positive voltage
potentiometer
power rating
primary cell
printed circuit board
protoboard
reliability
resistor color code
resolution
rotary switch
secondary cell
slow-blow fuse
solid wire
stranded wire
taper
throw
time delay fuse
tolerance
trace
trimmer potentiometer
variable resistor
voltmeter
wet cell
wiper arm
wire-wound resistor

## Practice Problems

**1.** Determine the guaranteed range of values for a 110 kΩ resistor with a 10% tolerance.

**2.** Determine the guaranteed range of values for a 1.5 MΩ resistor with a 20% tolerance.

**3.** Determine the guaranteed range of values for a 360 kΩ resistor with a 5% tolerance.

**4.** Determine the guaranteed range of values for a 6.8 Ω resistor with a 10% tolerance.

**5.** For each of the color codes listed, determine the rated value of the component.

| Band 1 | Band 2 | Band 3 | Value |
|---|---|---|---|
| a. Brown | Blue | Brown | ____ |
| b. Red | Violet | Green | ____ |
| c. Orange | White | Yellow | ____ |
| d. Green | Brown | Black | ____ |
| e. Yellow | Violet | Gold | ____ |
| f. Green | Blue | Orange | ____ |

**6.** For each of the color codes listed, determine the rated value of the component.

| Band 1 | Band 2 | Band 3 | Value |
|---|---|---|---|
| a. Red | Black | Brown | ____ |
| b. Orange | Black | Silver | ____ |
| c. Green | Blue | Black | ____ |
| d. White | Brown | Green | ____ |
| e. Brown | Blue | Blue | ____ |
| f. Blue | Gray | Silver | ____ |

7. Determine the colors of the first three bands for each of the following resistor values.

| | |
|---|---|
| **a.** 33 kΩ | **e.** 120 kΩ |
| **b.** 15 Ω | **f.** 10 MΩ |
| **c.** 910 kΩ | **g.** 1.8 Ω |
| **d.** 2.2 kΩ | **h.** 0.24 Ω |

8. Determine the colors of the first three bands for each of the following resistor values.

| | |
|---|---|
| **a.** 11 Ω | **e.** 160 kΩ |
| **b.** 20 MΩ | **f.** 0.13 Ω |
| **c.** 680 Ω | **g.** 9.1 MΩ |
| **d.** 3.3 Ω | **h.** 36 Ω |

9. For each of the color codes listed, determine the rated component value and the guaranteed range of measured values.
   - **a.** Red, yellow, yellow, gold
   - **b.** Orange, blue, gold, red
   - **c.** Gray, red, black
   - **d.** Blue, gray, silver, silver
10. For each of the color codes listed, determine the rated component value and the guaranteed range of measured values.
   - **a.** Yellow, orange, silver, gold
   - **b.** Brown, gray, yellow, red
   - **c.** Yellow, violet, orange, silver
   - **d.** Brown, green, silver, gold

## The Brain Drain

11. What is the resistance of a 12.8 cm length of 40 gauge wire?
12. What length of 14 gauge wire (in cm) has a resistance of 2 mΩ?

## Answers to the Example Practice Problems

| | | | |
|---|---|---|---|
| **3.1** | 66.64 kΩ to 69.36 kΩ | **3.6** | 7.5 Ω |
| **3.2** | 864.5 kΩ to 955.5 kΩ | **3.7** | 0.18 Ω |
| **3.3** | 47 kΩ | **3.8** | Red, yellow, orange |
| **3.4** | 9.1 MΩ | **3.9** | 1.44 MΩ to 1.76 MΩ |
| **3.5** | 68 Ω | | |

## EWB Applications Problems

| Figure | EWB File Reference |
|---|---|
| 3.18 | EWBA3_18.ewb |
| 3.34 | EWBA3_34.ewb |
| 3.37 | EWBA3_37.ewb |
| 3.39 | EWBA3_39.ewb |
| 3.41 | EWBA3_41.ewb |
| 3.45 | EWBA3_45.ewb |

# 4

# CIRCUIT FUNDAMENTALS

## OBJECTIVES

*After studying the material in this chapter, you should be able to:*

1. Describe the relationship among voltage, current, and resistance.
2. Predict how a change in either voltage or resistance will affect circuit current.
3. Use Ohm's law to calculate any of the following values, given the other two: current, voltage, resistance.
4. List the primary applications of Ohm's law.
5. Discuss power and its relationship to current and voltage.
6. Calculate power given any two of the following: current, voltage, resistance.
7. Describe the relationship between power and heat.
8. Determine the minimum acceptable power rating for a resistor in any circuit.
9. Define and calculate efficiency and kilowatt-hours (kWh).
10. Calculate any two of the following values given the other two: current, voltage, resistance, power.
11. Calculate the range of current values for a given circuit.
12. Define the terms *load* and *full load.*
13. List the characteristics of an open circuit.
14. List the characteristics of a short circuit.

In Chapter 2, you were shown that current is the flow of charge in a circuit, voltage is the difference of potential that generates current, and resistance is the opposition to current provided by the circuit. You were also shown that the units of current, voltage, and resistance can be defined in terms of each other, as follows:

1. One ampere is the amount of current that is generated when a 1 V difference of potential is applied across 1 Ω of resistance.
2. One volt is the difference of potential required to generate 1 A of current through 1 Ω of resistance.
3. One ohm is the amount of resistance that limits current to 1 A when a 1 V difference of potential is applied.

These statements clearly indicate that there is a relationship among current, voltage, and resistance. This relationship is defined by what is called Ohm's law. Ohm's law is one of the most powerful tools that a technician has. As you will see, it can be used to:

1. Predict how the current in a circuit will respond to a change in voltage or resistance.
2. Calculate the value of any of the basic circuit properties (current, voltage, or resistance) when the other two values are known.
3. Determine the possible causes of a fault in a circuit.

In this chapter, we will discuss Ohm's law and its applications. We will also discuss another basic circuit property, power, and its relationship to current, voltage, and resistance. Finally, we will take a look at some miscellaneous topics that apply to basic circuits.

## 4.1 OHM'S LAW

*OBJECTIVE 1* ▶

In the early nineteenth century, Georg Simon Ohm, a German physicist, defined several cause-and-effect relationships between the values of current, voltage, and resistance in a circuit. Specifically, he found that current is directly proportional to voltage and inversely proportional to resistance. This statement, which has come to be known as **Ohm's law,** tells us that:

**Ohm's law**
A law stating that current is directly proportional to voltage and inversely proportional to resistance.

1. The current through a fixed resistance:
   a. Increases when the applied voltage increases.
   b. Decreases when the applied voltage decreases.
2. The current generated by a fixed voltage:
   a. Decreases when the resistance increases.
   b. Increases when the resistance decreases.

We'll start our discussion of Ohm's law by taking a closer look at these two statements.

### The Relationship Between Current and Voltage

As shown earlier, voltage is the difference of potential that causes the flow of charge (current) in a circuit. According to Ohm's law, the greater the difference of potential, the greater the resulting current. This relationship is illustrated in Figure 4.1.

*A Practical Consideration:* Current varies in direct proportion to voltage only if resistance is fixed (does not change). If resistance changes, current and voltage will not be directly proportional.

In Figure 4.1a, the voltmeter shows the applied voltage to be at half its maximum possible value. At the same time, the ammeter shows the current to be at half its maximum value. If the applied voltage is doubled as shown in Figure 4.1b, the circuit current also doubles. In other words, the current has increased proportionally to an increase in voltage. By the same token, a decrease in the applied voltage causes a proportional decrease in current.

In Figure 4.1, the current is changing as a result of a change in voltage. In other words, the change in voltage is the *cause* and the change in current is the *effect.*

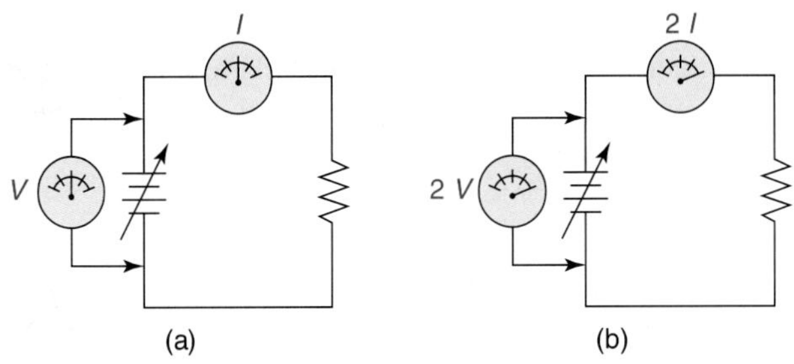


**FIGURE 4.1 The relationship between current and voltage.**

## The Relationship Between Current and Resistance

As you know, resistance is an opposition to current in a circuit. According to Ohm's law, the greater the opposition, the lower the current at a given voltage. This relationship is illustrated in Figure 4.2.

In Figure 4.2a, the applied voltage and circuit current are both at their maximum values, as indicated by the voltmeter and the ammeter. If the circuit resistance is doubled as shown in Figure 4.2b, the circuit current drops to half its original value, even though the applied voltage has not changed. In other words, the current has decreased proportionally to an increase in resistance. By the same token, a decrease in the circuit resistance causes the circuit current to increase proportionally.

*A Practical Consideration:* Current varies in inverse proportion to resistance only if voltage is fixed (does not change). If voltage changes, current and resistance will not be inversely proportional.

In this case, the current is changing as a result of a change in resistance. In other words, the change in resistance is the cause, and the change in current is the effect.

## Predicting Circuit Behavior

◄ *OBJECTIVE 2*

The cause-and-effect angle to Ohm's law can be used to predict the response that current will have to a change in voltage or resistance. For example, consider the circuit in Figure 4.3. This circuit contains a dc power supply and a potentiometer that is wired as a variable resistor, or *rheostat.* Note that the circuit current enters through one of the outer terminals of the pot and exits through the wiper arm. By turning the control shaft, we can change the total circuit resistance.

Ohm's law allows us to predict the effect that adjusting the pot will have on the circuit current. If we increase the circuit resistance by adjusting the pot, Ohm's law tells us that the circuit current will decrease. If we decrease the circuit resistance by adjusting the pot, Ohm's law tells us that the circuit current will increase.

Ohm's law also allows us to predict the effect that adjusting the dc power supply will have on the circuit current. If we increase the output voltage from the power supply, Ohm's law tells us that the circuit current will increase. If we decrease the output voltage from the power supply, Ohm's law tells us that the circuit current will decrease.

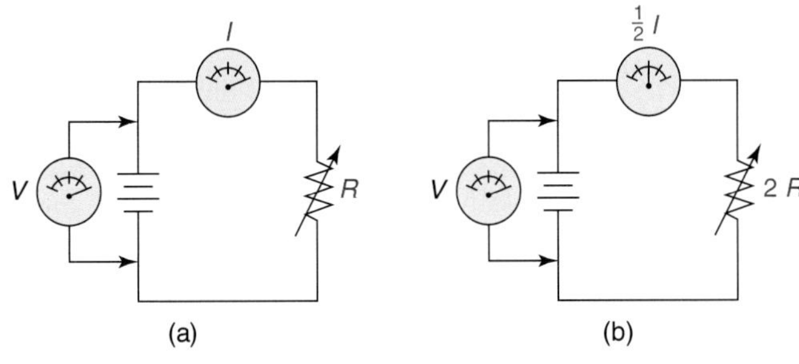


**FIGURE 4.2 The relationship between current and resistance.**

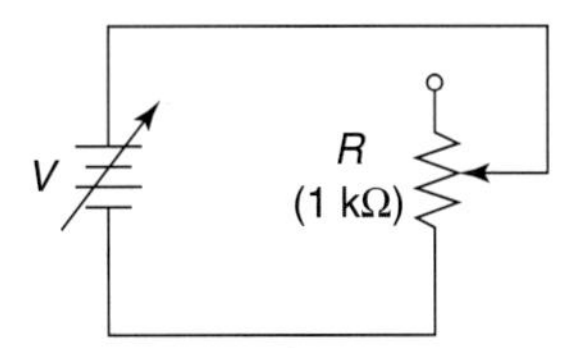


**FIGURE 4.3**

As you can see, Ohm's law can be used to predict how a change in circuit voltage or resistance will affect circuit current. At the same time, there is a subtle message here: *when you adjust the value of any circuit component, there will always be some type of response to the change.* Before you adjust any circuit component, you should always take a moment to consider the possible impact of doing so.

## Using Ohm's Law to Calculate Circuit Values

*OBJECTIVE 3* ▶ Ohm's law also provides a specific mathematical relationship among the values of current, voltage, and resistance in a circuit. This relationship is given as

$$I = \frac{V}{R} \tag{4.1}$$

where $I$ = the circuit current, in amperes

$V$ = the applied voltage, in volts

$R$ = the circuit resistance, in ohms

Equation (4.1) allows us to calculate the value of current in a circuit when the values of the applied voltage and circuit resistance are known. The following example demonstrates the use of this equation.

### EXAMPLE 4.1

Calculate the value of current for the circuit in Figure 4.4.

***Solution:*** Using the values of $V$ and $R$ given in the figure, the value of the circuit current is found as

$$\boldsymbol{I} = \frac{V}{R} = \frac{10\text{ V}}{2\ \Omega} = \mathbf{5\ A}$$

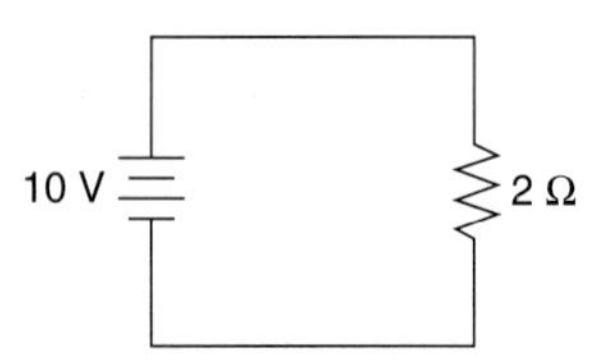


**FIGURE 4.4**

***PRACTICE PROBLEM 4.1***

A circuit like the one in Figure 4.4 has values of $V = 18$ V and $R = 100\ \Omega$. Calculate the value of the current in the circuit.

Ohm's law can be applied to a variety of practical situations. The following example demonstrates one practical application for equation (4.1).

### EXAMPLE 4.2

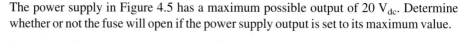

The power supply in Figure 4.5 has a maximum possible output of 20 $V_{dc}$. Determine whether or not the fuse will open if the power supply output is set to its maximum value.

***Solution:*** Using the values of $V = 20$ V and $R = 33\ \Omega$, the maximum circuit current is found as

$$\boldsymbol{I} = \frac{V}{R} = \frac{20\text{ V}}{33\ \Omega} = \mathbf{606\ mA}$$

Since the fuse is rated at ½ A, or 500 mA, it will open if the power supply is set to its maximum limit.

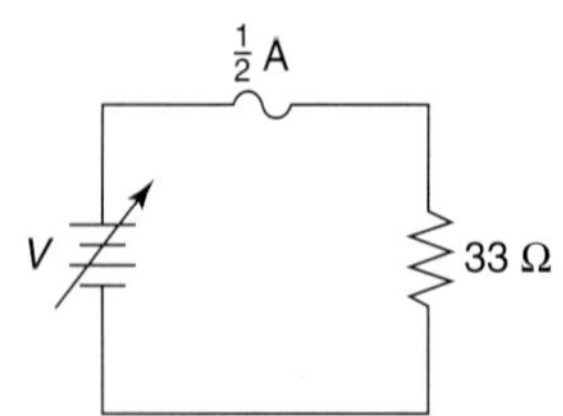


**FIGURE 4.5**

***PRACTICE PROBLEM 4.2***

A circuit like the one in Figure 4.5 has a 1 A fuse, an 18 Ω resistor, and a maximum power supply output of 20 V. Determine whether or not the fuse will open if the power supply is set to its maximum output value.

If we multiply both sides of equation (4.1) by $R$, we get

$$V = IR \tag{4.2}$$

Equation (4.2) allows us to calculate the value of applied voltage when the values of current and resistance are known. The use of this equation is demonstrated in the following example.

### EXAMPLE 4.3

Calculate the value of the applied voltage in the circuit shown in Figure 4.6.

***Solution:*** Using the values of $I$ and $R$ given in the figure, the value of the applied voltage is found as

$$V = IR = (1\text{mA})(10\ \text{k}\Omega) = \mathbf{10\ V}$$

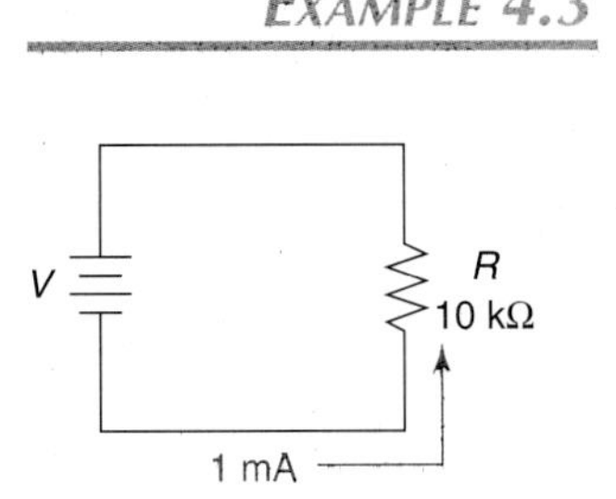


FIGURE 4.6

***PRACTICE PROBLEM 4.3***

A circuit like the one in Figure 4.6 has values of $I = 500\ \mu\text{A}$ and $R = 3.3\ \text{k}\Omega$. Calculate the value of the applied voltage for the circuit.

The following example demonstrates a practical application for equation (4.2).

### EXAMPLE 4.4

We have constructed the circuit in Figure 4.7 on a breadboard. Determine the power supply setting that will provide a circuit current of 25 mA.

***Solution:*** The needed input voltage is found using the value of the circuit resistance and the desired current, as follows:

$$V = IR = (25\ \text{mA})(300\ \Omega) = \mathbf{7.5\ V}$$

Thus, if we set the power supply for an output of 7.5 V, we will obtain the desired value of 25 mA for the circuit current.

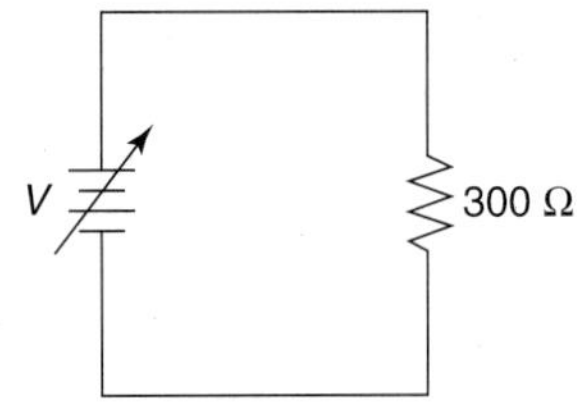

FIGURE 4.7

***PRACTICE PROBLEM 4.4***

A circuit like the one in Figure 4.7 has a 3.3 kΩ resistor. Determine the power supply setting that will provide a circuit current of 4 mA.

If we divide both sides of equation (4.2) by $I$, we get

$$R = \frac{V}{I} \tag{4.3}$$

This equation can be used to calculate the value of circuit resistance when the current and voltage values are known. This calculation is demonstrated in the following example.

## EXAMPLE 4.5

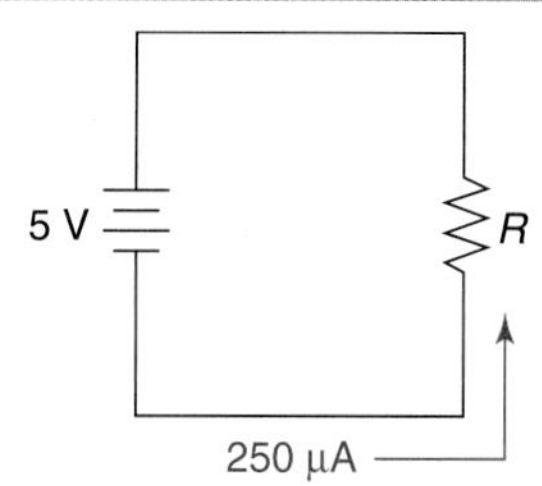


FIGURE 4.8

Calculate the value of $R$ for the circuit in Figure 4.8.

***Solution:*** Using the values of current and voltage given in the figure, the value of $R$ is found as

$$\boldsymbol{R} = \frac{V}{I} = \frac{5\text{ V}}{250\ \mu\text{A}} = \mathbf{20\ k\Omega}$$

***PRACTICE PROBLEM 4.5***

A circuit like the one shown in Figure 4.8 has values of $V = 12$ V and $I = 40$ mA. Calculate the value of the circuit resistance.

The following example demonstrates a practical application for equation (4.3).

## EXAMPLE 4.6

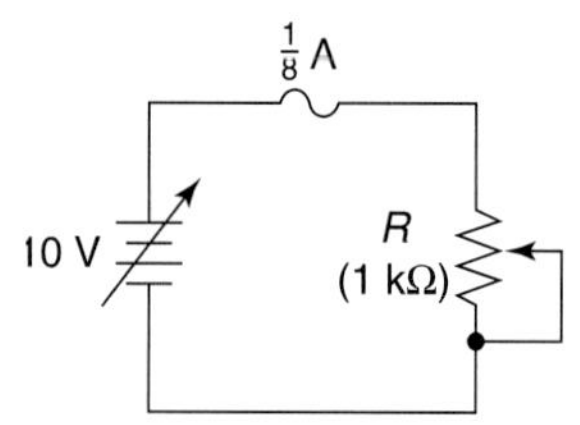


FIGURE 4.9

Determine the minimum allowable setting for the potentiometer in Figure 4.9.

***Solution:*** The circuit has a ⅛ A (125 mA) fuse, so the maximum allowable current is 125 mA. Using $I = 125$ mA and $V = 10$ V, the minimum allowable setting for the pot is found as

$$\boldsymbol{R} = \frac{V}{I} = \frac{10\text{ V}}{125\text{ mA}} = 80\ \Omega$$

If the pot is set to any value less than 80 Ω, the current will exceed 125 mA and the fuse will open. (Try using equation (4.1) to prove this to yourself.)

***PRACTICE PROBLEM 4.6***

A circuit like the one in Figure 4.8 has a 15 V source and a ¼ A fuse. Determine the minimum allowable setting for the potentiometer in the circuit.

## Verifying the Current Relationships

*Remember:* Voltage and current are directly proportional only when resistance does not change.

It was stated earlier that the current through a fixed resistance is directly proportional to the applied voltage. This relationship is verified in the following example.

## EXAMPLE 4.7

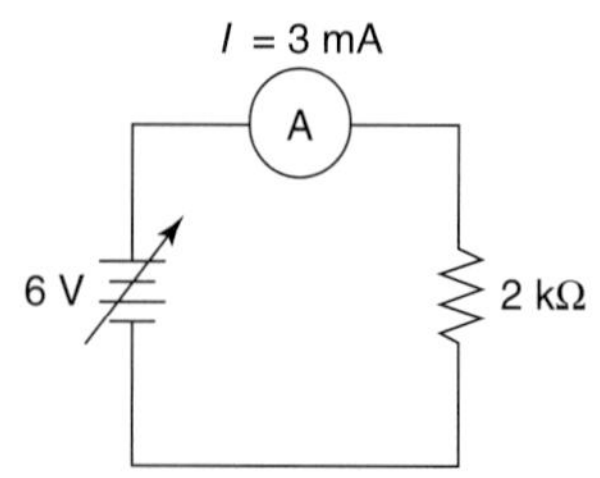


FIGURE 4.10

The circuit in Figure 4.10 has the values shown. If the applied voltage doubles, how does the circuit current respond?

***Solution:*** Using $V = 12$ V, the new value of $I$ is found as

$$\boldsymbol{I} = \frac{V}{R} = \frac{12\text{ V}}{2\text{ k}\Omega} = \mathbf{6\ mA}$$

Thus, when the applied voltage doubles, the circuit current also doubles.

***PRACTICE PROBLEM 4.7***

Verify that the circuit current in Figure 4.10 will drop to half its value if the applied voltage is cut in half.

You have been told that the current generated by a fixed voltage is inversely proportional to the circuit resistance. This relationship is verified in the following example.

*Remember:* Current and resistance are inversely proportional only when voltage does not change.

## EXAMPLE 4.8

The circuit shown in Figure 4.11 has the values shown. If the circuit resistance doubles, how does the circuit current respond?

***Solution:*** Using $R = 20\text{ k}\Omega$, the new value of $I$ is found as

$$I = \frac{V}{R} = \frac{10\text{ V}}{20\text{ k}\Omega} = \mathbf{500\ \mu A}$$

As you can see, doubling the circuit resistance causes the current to drop to half its original value.

***PRACTICE PROBLEM 4.8***

Verify that the current in Figure 4.11 will double if the circuit resistance is cut in half.

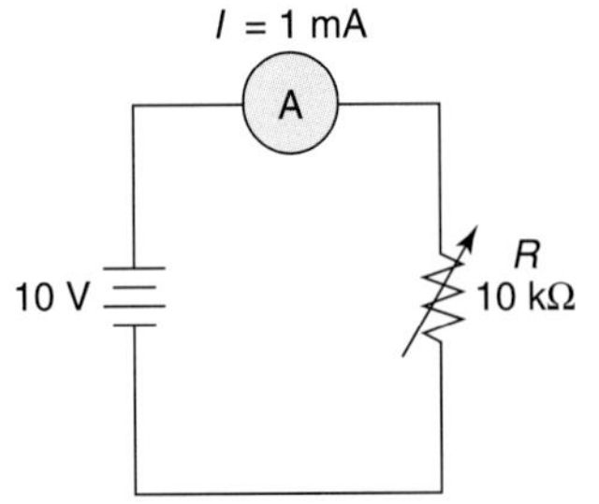

$I$ = 1 mA

A

10 V

$R$ 10 kΩ

**FIGURE 4.11**

## Troubleshooting with Ohm's Law

**Troubleshooting** is the process of locating faults in a circuit or system. When a circuit or system fails to operate as expected, a technician is called to locate the source of the problem.

**Troubleshooting**
The process of locating faults in a circuit or system.

When a circuit doesn't operate as expected, Ohm's law is one of the best tools you have to help diagnose the problem. For example, Ohm's law tells you that:

**1.** Current *increases* as a result of:
   **a.** An increase in the applied voltage.
   **b.** A decrease in the circuit resistance.

**2.** Current *decreases* as a result of:
   **a.** A decrease in the applied voltage.
   **b.** An increase in the circuit resistance.

The first of these statements can be used to determine the possible causes of the problem shown in Figure 4.12a. According to Ohm's law, the measured current in the circuit should be approximately 1 mA. However, the ammeter shows a reading of 10 mA. This high value of current indicates (according to statement 1) that either the applied voltage is too high or the circuit resistance is too low.

The second statement can be used to determine the possible causes of the problem shown in Figure 4.12b. Ohm's law tells us that the measured current in the circuit should be approximately 10 mA. However, the ammeter shows a reading of 1.5 mA. This low value of current indicates (according to statement 2) that either the applied voltage is too low or the circuit resistance is too high.

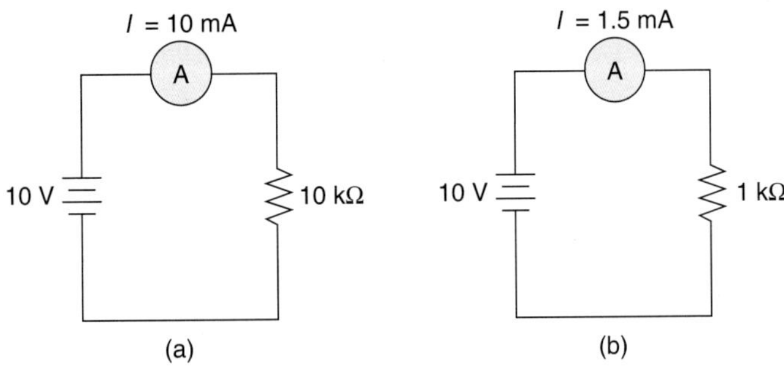


**FIGURE 4.12 Current readings indicating circuit faults.**

In each of these cases, Ohm's law has served a very valuable purpose: *it has told you the types of problems you are looking for.* Once you know the types of problems you are looking for, a few simple measurements with a multimeter will tell you which of those problems is actually present.

## One Final Note

OBJECTIVE 4 ▶ In this section, you have been shown how Ohm's law can be used to:

1. Predict the response of a circuit to a change in voltage or resistance.
2. Calculate the value of current, voltage, or resistance in a circuit.
3. Determine the possible causes of a problem in a simple circuit.

As you progress through the rest of this chapter (and the chapters to follow), you will see that Ohm's law is the most frequently used electronics principle. For this reason, you should take the time to become thoroughly familiar with this law and all that it can do for you.

## Summary

Ohm's law states that current is directly proportional to voltage and inversely proportional to resistance. These cause-and-effect relationships are summarized in Table 4.1.

The relationships given in Table 4.1 can be used to predict the response that the current in a circuit will have to a change in voltage or resistance. They can also be used to help in circuit troubleshooting: the process of finding a fault in a circuit or system.

Ohm's law also gives us the means of calculating any of the basic circuit properties (current, voltage, or resistance) when the values of the other two are known. The three mathematical forms of Ohm's law are:

$$I = \frac{V}{R} \qquad V = IR \qquad R = \frac{V}{I}$$

*Remember:* Each relationship given in Table 4.1 is true only if the third variable ($V$ or $R$) is fixed.

**TABLE 4.1 A Summary of Ohm's Law Relationships**

| *Cause* | *. . . Results in . . .* | *Effect* |
|---|---|---|
| Increased voltage | | Increased current |
| Decreased voltage | | Decreased current |
| Increased resistance | | Decreased current |
| Decreased resistance | | Increased current |

## Section Review

1. What is Ohm's law?
2. Describe the relationship between voltage and current.
3. Describe the relationship between resistance and current.
4. What happens to the current through a rheostat when you adjust its value? Explain your answer.
5. List the three Ohm's law equations.
6. What is troubleshooting?
7. List the possible causes of high circuit current.
8. List the possible causes of low circuit current.
9. What purpose does Ohm's law serve for troubleshooting?
10. List the three most common applications of Ohm's law.

# 4.2 POWER

In Chapter 3, you were told that the power rating of a component is a measure of its ability to dissipate (throw off) heat. In this section, we will take a closer look at what power is, how its value is calculated, and how it relates to heat.

## Power

◄ *OBJECTIVE 5*

Technically defined, **power *(P)*** a measure of energy used per unit time. When a voltage source uses one joule of energy per second to generate current, we say that it has used one watt (W) of power. By formula,

**Power (*P*)**
A measure of energy used per unit time. One watt (W) of power is equal to a rate of one joule per second.

$$1 \text{ watt} = 1 \text{ joule/second}$$

This relationship is illustrated in Figure 4.13. As you can see, we have a 1 V source that is generating 1 A of current through 1 Ω of resistance. Now, consider the following relationships that were introduced in Chapter 2:

$$1 \text{ A} = 1 \text{ coulomb/second}$$

$$1 \text{ V} = 1 \text{ joule/coulomb}$$

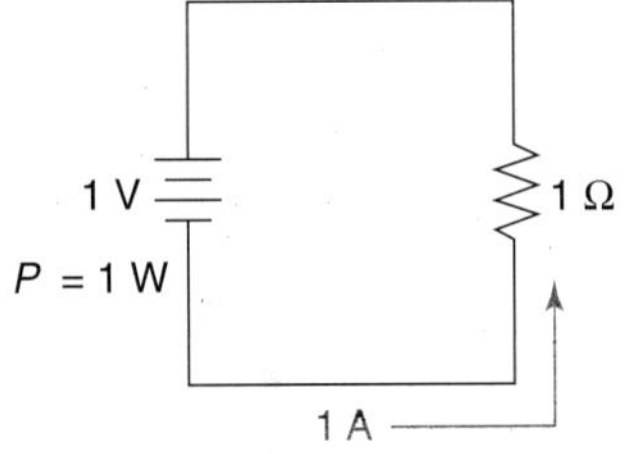


FIGURE 4.13

The power supply in Figure 4.13 is using one joule of energy to move one coulomb of charge per second. Therefore, it is using one joule of energy per second. This rate of energy use is equal to one watt (W) of power. The relationship among power, voltage, and current can be seen by comparing the circuit in Figure 4.13 with those in Figure 4.14.

In Figure 4.14a, 1 V is used to generate 2 A of current. In this case, the power supply is using one joule of energy per second to generate each of the two coulombs of charge. Therefore, the power supply is using 2 joules/second, or 2 W.

In Figure 4.14b, 2 V is used to generate 1 A of current. In this case, the power supply is using two joules of energy per second to generate one coulomb of charge. Again, the power supply is using 2 joules/second, or 2 W.

## Calculating Power

◄ *OBJECTIVE 6*

You have seen that power is related to both current and voltage. The mathematical relationship among the three values is given as

$$P = VI \tag{4.4}$$

where $P$ = the power used, in watts

$V$ = the applied voltage, in volts

$I$ = the generated current, in amperes

Example 4.9 demonstrates the use of this equation.

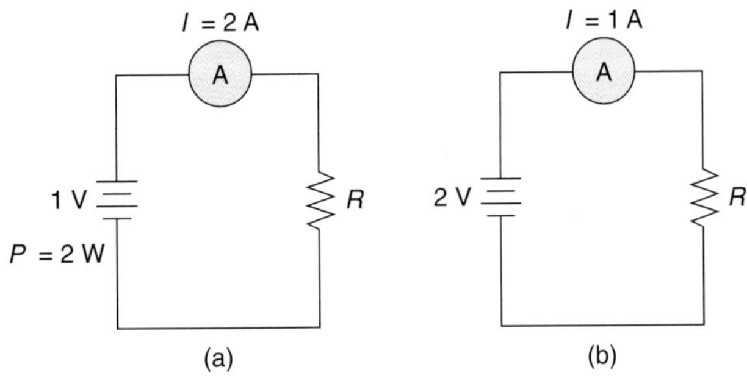


FIGURE 4.14

## Example 4.9

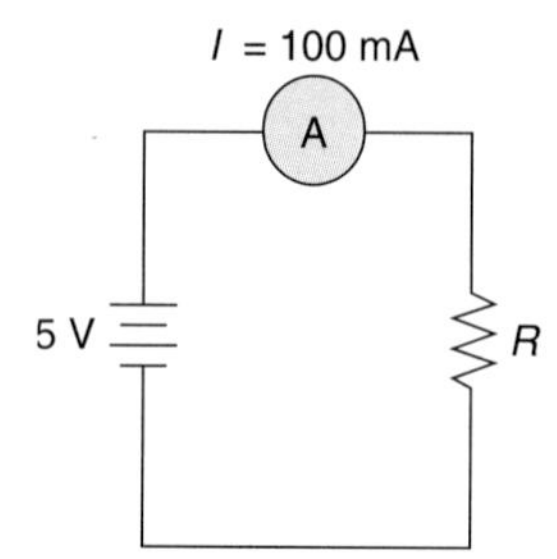


FIGURE 4.15

Calculate the power used by the voltage source in Figure 4.15.

***Solution:*** Using the values of voltage and current shown, the power used by the source is found as

$$P = VI = (5\text{V})(100\text{ mA}) = \mathbf{500\text{ mW}}$$

***PRACTICE PROBLEM 4.9***

A 12 V power supply generates 80 mA of current. Calculate the power used by the source.

## The Basis of Equation (4.4)

By definition, power is energy per unit time. Stated mathematically,

$$P = \frac{E}{t} \tag{4.5}$$

where $P$ = the power used, in watts

$E$ = energy, in joules

$t$ = time, in seconds

Equation (2.2) defined voltage as follows:

$$V = \frac{E}{Q}$$

This equation can be rewritten as

$$E = VQ$$

Substituting the above equation into equation (4.5), we get

$$P = \frac{VQ}{t}$$

or

$$P = V\frac{Q}{t}$$

Now, equation (2.1) defined current as

$$I = \frac{Q}{t}$$

Therefore, the last power equation can be rewritten as

$$P = VI$$

This is equation (4.4).

## Other Power Equations

We can use Ohm's law to derive two other useful power equations. By substituting $IR$ for the value of $V$ in equation (4.4), we obtain an equation that defines power in terms of current and resistance, as follows:

$$P = VI$$
$$= (IR)I$$

or

$$P = I^2R \qquad \textbf{(4.6)}$$

As the following example demonstrates, this equation can be used to calculate power when the values of current and resistance are known.

### EXAMPLE 4.10

Calculate the power used by the voltage source in Figure 4.16.

***Solution:*** Using the given values of current and resistance, the power used by the voltage source is found as

$$P = I^2R = (200\text{ mA})^2\,(1\text{ k}\Omega) = \mathbf{40\ W}$$

***PRACTICE PROBLEM 4.10***

A circuit like the one in Figure 4.16 has values of $I = 80$ mA and $R = 1.8$ kΩ. Calculate the amount of power used by the source.

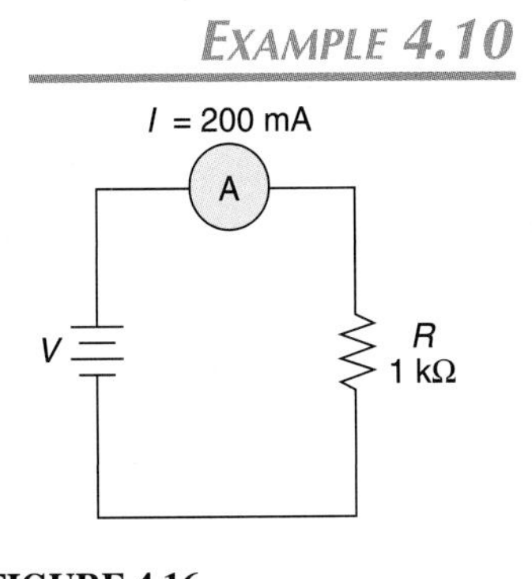


**FIGURE 4.16**

By substituting $V/R$ for the value of $I$ in equation (4.4), we obtain an equation that defines power in terms of voltage and resistance, as follows:

$$P = VI$$
$$= V\frac{V}{R}$$

or

$$P = \frac{V^2}{R} \qquad \textbf{(4.7)}$$

As the following example demonstrates, this equation can be used to calculate power when the values of voltage and resistance are known.

### EXAMPLE 4.11

Calculate the power used by the voltage source in Figure 4.17.

***Solution:*** Using the values shown, the power used by the voltage source is found as

$$P = \frac{V^2}{R} = \frac{(12\text{ V})^2}{10\text{ k}\Omega} = \mathbf{14.4\ mW}$$

***PRACTICE PROBLEM 4.11***

A circuit like the one in Figure 4.17 has values of $V = 10$ V and $R = 4.7$ kΩ. Calculate the power used by the source.

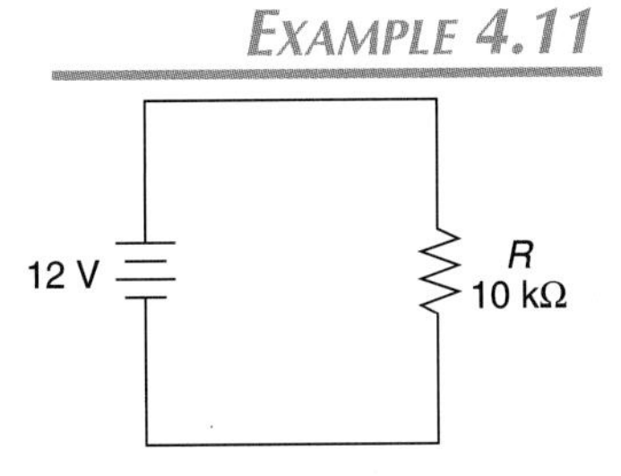


**FIGURE 4.17**

As you can see, power can be calculated using any two basic circuit properties and the appropriate power relationship. The basic power relationships are summarized as follows:

$$P = VI \qquad P = I^2R \qquad P = \frac{V^2}{R}$$

## Power and Heat

OBJECTIVE 7 ▶ You have been told that the power rating of a component is a measure of its ability to dissipate heat. At this point, we will look at the connection between power (as defined in this section) and heat.

In our basic science classes, we were taught that matter cannot be created or destroyed. The same principle applies to energy. While energy may change forms, it cannot be created or destroyed.

In any circuit, power is not created. It is merely transferred from one part of the circuit to another. This point is illustrated in Figure 4.18. As you can see, the dc power supply is drawing energy from the ac receptacle and transferring it to the resistor. Thus, the energy is merely being transferred from one place to another. It is not being created. At the same time, it cannot be destroyed, which leads us to the relationship between power and heat.

When energy is transferred to the resistor in Figure 4.18, the resistor absorbs that energy. Since this energy cannot be destroyed, the resistor converts it into another form: heat. Some other examples of energy conversion are as follows:

1. A toaster converts electrical energy into heat.
2. A light bulb converts electrical energy into light and heat.
3. A speaker converts electrical energy into mechanical energy (sound).

**Transducer**
Any device that converts energy from one form to another.

Note that any device designed to convert energy from one form to another is referred to as a **transducer.**

As you know, power is defined as energy used per unit time. If a device such as a resistor is absorbing a given amount of energy per unit time, it must be capable of converting that energy into heat at the same rate, which is why the power rating of a component is critical. If you exceed the power rating of a component, that component will not be able to convert the energy it is receiving quickly enough. As a result, the component will keep getting hotter, and will eventually be destroyed by the excessive heat.

OBJECTIVE 8 ▶ You were told in Chapter 3 that the power rating of a resistor is a parameter that indicates its maximum power dissipation capability. When choosing a resistor for a particular application, the power rating of the component should be much greater than the amount of power it is expected to absorb. This ensures that the resistor will be able to absorb the applied power without being destroyed. A practical application of this principle is demonstrated in Example 4.12.

**FIGURE 4.18**

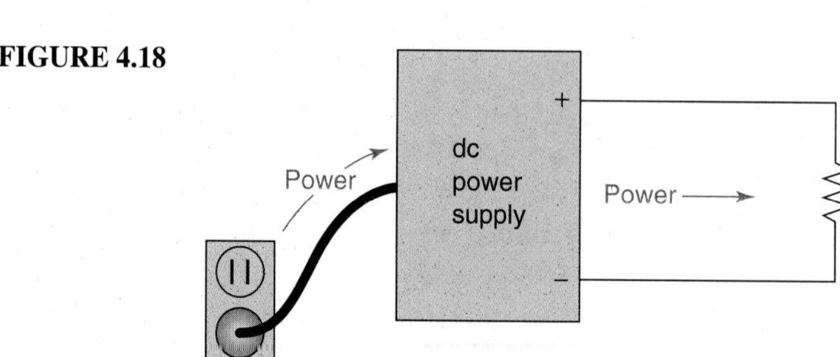

You want to build the circuit in Figure 4.19. Determine the minimum acceptable power rating for the resistor.

***Solution:*** The circuit has values of $V = 15$ V and $R = 1$ kΩ. Using these values, the power output from the voltage source is found as

$$P = \frac{V^2}{R} = \frac{(15\text{ V})^2}{1\text{ k}\Omega} = \mathbf{225\ mW}$$

Since all this power will be absorbed by the resistor, the power rating of the component must be greater than 225 mW. Thus, the resistor must have a power rating of at least ¼ W (250 mW).

***PRACTICE PROBLEM 4.12***

A circuit like the one in Figure 4.19 has values of $V = 18$ V and $R = 910\ \Omega$. Determine the minimum acceptable power rating for the resistor.

### EXAMPLE 4.12

*A Practical Consideration:* A common guideline for selecting a resistor is to use one with 2× the required power dissipation capability. For example, you would probably want to use a ½ watt resistor for the application in Example 4.12.

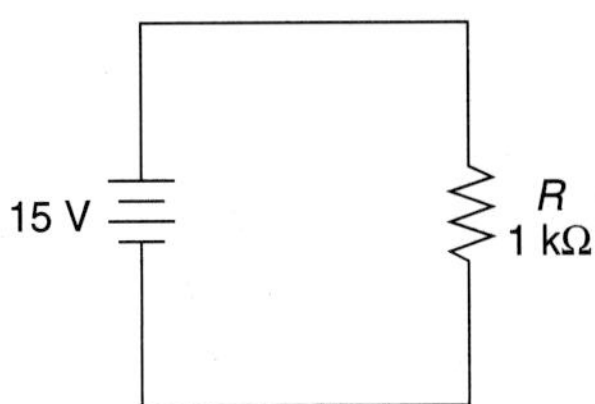


**FIGURE 4.19**

## Efficiency

◄ *OBJECTIVE 9*

The dc power supply in any circuit contains a variety of components that each use some amount of power. Therefore, the output power from the voltage source is always less than its input power.

The **efficiency** of a circuit or component is the ratio of its output power to its input power, given as a percentage. By formula,

**Efficiency**
The ratio of output power to input power, typically given as a percentage.

$$\eta = \frac{P_o}{P_i} \times 100 \qquad \mathbf{(4.8)}$$

where $\eta$ = the efficiency, written as a percentage

$P_o$ = the output power

$P_i$ = the input power

The following example shows how the efficiency of a dc power supply is calculated.

### EXAMPLE 4.13

A dc power supply has the following ratings: $P_o = 5$ W (maximum) and $P_i = 12$ W (maximum). Calculate the efficiency of the power supply.

***Solution:*** Using the values given, the efficiency of the power supply is found as

$$\eta = \frac{P_o}{P_i} \times 100 = \frac{5\text{ W}}{12\text{ W}} \times 100 = \mathbf{41.7\%}$$

This means that 41.7% of the power supply's input power is converted to output power. The other 58.3% of the input power is used by the dc power supply itself.

***PRACTICE PROBLEM 4.13***

A dc power supply has the following maximum ratings: $P_o = 20$ W and $P_i = 140$ W. Calculate the efficiency of the power supply.

An important point should be made about efficiency: It is not possible to have an efficiency greater than 100%. An efficiency greater than 100% would imply that the output power is greater than the input power, which is impossible. (*Remember:* energy cannot be created.) Also, since all electronic circuits absorb some amount of power, every practical efficiency rating is less than 100%.

## Energy Measurement

**Kilowatt-hour (kWh)**
A practical unit for measuring energy, equal to the amount of energy used by a 1000 W (1 kW) device run for one hour.

Every appliance in your home uses energy. While this energy cannot be measured directly, it can be determined by measuring the amount of power used and the time that it is used.

The practical measurement for the energy used by a utility customer is the **kilowatt-hour (kWh).** One kilowatt-hour is the amount of energy used by a 1000 W device run for one hour. The energy you use, in kilowatt-hours, is found as

$$E_{\text{kWh}} = \frac{P \cdot t}{1000} \qquad \textbf{(4.9)}$$

where $P$ = the power used, in watts

$t$ = the time, in hours

Note that the 1000 in the denominator is merely a conversion factor that converts watts into kilowatts. The following example demonstrates the use of equation (4.9).

### EXAMPLE 4.14

Calculate the power used to run 20 60-W lightbulbs for six hours.

***Solution:*** The combined power used by the bulbs is found as

$$20 \times 60 \text{ W} = 1200 \text{ W}$$

Using this combined power, the total energy used is found as

$$E_{\text{kWh}} = \frac{P \cdot t}{1000} = \frac{(1200 \text{ W})(6 \text{ h})}{1000} = \mathbf{7.2\ kWh}$$

***PRACTICE PROBLEM 4.14***

Calculate the total energy used by a 1500 W dishwasher, a 3600 W clothes dryer, and a 750 W microwave oven that are all used for 2 hours.

The service meter in your home measures energy in kilowatt-hours. Typically, dials on the meter are laid out as shown in Figure 4.20. Note that (from left to right), the dials indicate thousands of kWh, hundreds of kWh, and so on. For example, the dial shown in Figure 4.20 indicates that 1382.5 kWh of energy have been used. Also note that the dials alternate between clockwise and counterclockwise rotation.

**FIGURE 4.20**

### The Bottom Line

Power is the amount of energy supplied or absorbed by a circuit or component. Technically defined, it is a measure of energy used per unit time.

The basic unit of power is the watt (W). One watt of power is equal to one joule per second. When a voltage source uses one joule of energy per second to generate current, we say that it is using one watt of power.

Power can be calculated when any two of the following values are known: current, voltage, and resistance. The basic power equations are as follows:

$$P = VI \qquad P = I^2R \qquad P = \frac{V^2}{R}$$

These equations, along with Ohm's law, can be used to solve a variety of practical problems (as will be seen in the next section).

Energy cannot be created or destroyed. It is merely transferred from one device to another and converted from one form to another. When a voltage source provides a given amount of output power, that power is absorbed by the circuit resistance and is then dissipated in the form of heat.

If the power rating of a resistor is lower than the amount of power it is absorbing, it cannot dissipate the total heat generated. Thus, it continues to get hotter and is eventually destroyed.

The efficiency of a circuit or component is the ratio of its output power to its input power, given as a percentage. The higher the efficiency rating of a circuit or component, the less power it actually absorbs. In practice, no circuit or component is 100% efficient. All absorb some amount of power and therefore have efficiency ratings less than 100%.

Utility companies measure the energy you use in kilowatt-hours (kWh). One kilowatt-hour is the amount of energy that a 1000 W device uses when run for one hour.

### Section Review

1. What is power?
2. Explain the watt unit of power.
3. Describe the relationship between power and current.
4. Describe the relationship between power and voltage.
5. List the three basic power equations.
6. Discuss the relationship between power and heat.
7. What is a transducer?
8. What is efficiency? How is it calculated?
9. Why can't a circuit or device have a power rating greater than 100%?
10. What unit is used by power companies to measure the amount of energy you use? What does this unit equal?

## 4.3 MISCELLANEOUS TOPICS

In this section, we will discuss several topics that relate to basic circuits, including circuit problem solving, the effects of resistor tolerances, some frequently used terms, and two types of circuit faults.

## Solving Basic Circuit Problems

OBJECTIVE 10 ▶ In any circuit problem, you have two types of variables: those with known values (given in the problem) and those with unknown values. Any problem generally requires that you define one or more of the unknown variables in terms of the known variables. For example, consider the simple problem shown in Figure 4.21. To solve for the circuit current (the unknown value), we must define it in terms of voltage and resistance (the known values). In this case, the solution is simple. We just use Ohm's law in the form of

$$I = \frac{V}{R}$$

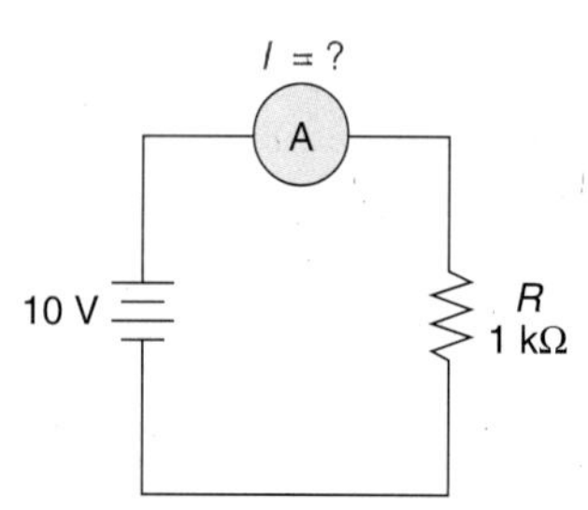


FIGURE 4.21

This equation defines our unknown value (current) in terms of our known values (voltage and resistance). Solving the equation solves the problem.

In many cases, the solution is not as obvious as it is in Figure 4.21. For example, consider the problem in Figure 4.22. In this case, we are asked to solve for the value of the resistor. We are given the values of applied voltage and power. Now, consider the equations we have used so far:

$$I = \frac{V}{R} \qquad V = IR \qquad R = \frac{V}{I}$$

$$P = \frac{V^2}{R} \qquad P = I^2R \qquad P = VI$$

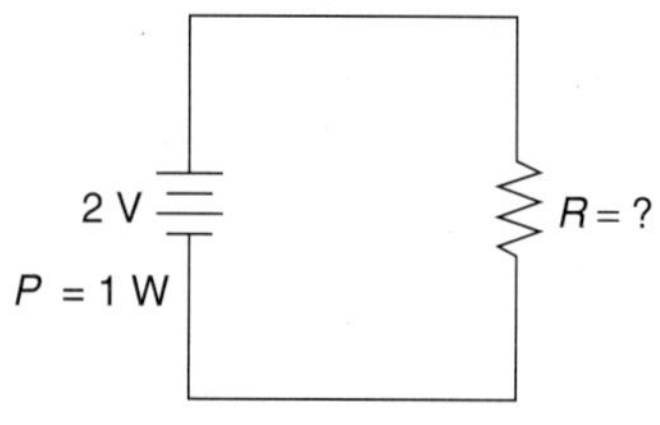


FIGURE 4.22

None of these equations defines resistance directly in terms of voltage and power. However, if you look closely at the equations, you will see that there is one equation that contains all three of our variables:

$$P = \frac{V^2}{R}$$

If we transpose this equation, we get

$$R = \frac{V^2}{P}$$

Now we have an equation that defines our unknown value (resistance) in terms of our known values (voltage and power), and we can solve the problem.

When you encounter any basic circuit problem, you should:

1. Find an equation that contains all the variables directly involved in the problem.
2. If necessary, transpose the equation to obtain one that defines the unknown value in terms of the known values.

The use of this procedure is demonstrated further in the following example.

### EXAMPLE 4.15

Calculate the value of current in the circuit shown in Figure 4.23.

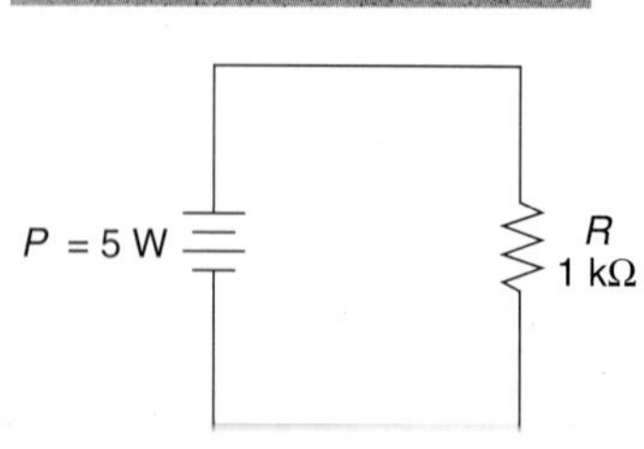


FIGURE 4.23

***Solution:*** We are trying to find a value of current using known values of power and resistance. The basic equation that contains all three of these properties is

$$P = I^2R$$

Transposing this equation to solve for *current,* we get

$$I = \sqrt{\frac{P}{R}}$$

In this case,

$$I = \sqrt{\frac{5\text{ W}}{1\text{ k}\Omega}} = \mathbf{70.7\text{ mA}}$$

***PRACTICE PROBLEM 4.15***

A circuit like the one shown in Figure 4.23 has the following values: $P = 2$ W and $R =$ 1.8 kΩ. Calculate the value of the applied voltage.

By following the procedure demonstrated in the example, you can solve for any one of the basic circuit properties (current, voltage, resistance, or power) when any two values are known.

## Resistor Tolerance

Up to this point, we have ignored the effect that resistor tolerance can have on the measured current in a basic circuit. As the following example illustrates, the actual current in a circuit will fall within a range of values determined by the tolerance of the resistor. ◀ *OBJECTIVE 11*

### EXAMPLE 4.16

Calculate the possible range of current values for the circuit in Figure 4.24.

***Solution:*** The color code of the resistor indicates that it is a 2 kΩ resistor with a 5% tolerance. The range of possible values for the component is found as

$$(2\text{ k}\Omega) \times 5\% = 100\ \Omega$$
$$2\text{ k}\Omega - 100\ \Omega = 1.9\text{ k}\Omega \quad \text{(minimum)}$$
$$2\text{ k}\Omega + 100\ \Omega = 2.1\text{ k}\Omega \quad \text{(maximum)}$$

The maximum circuit current is found using the minimum resistor value and Ohm's law, as follows:

$$I = \frac{V}{R} = \frac{10\text{ V}}{1.9\text{ k}\Omega} = \mathbf{5.26\text{ mA}}$$

The minimum circuit current is found using the maximum resistor value and Ohm's law, as follows:

$$I = \frac{V}{R} = \frac{10\text{ V}}{2.1\text{ k}\Omega} = \mathbf{4.76\text{ mA}}$$

Thus, the actual circuit current may fall anywhere between 4.76 mA and 5.26 mA.

**FIGURE 4.24**

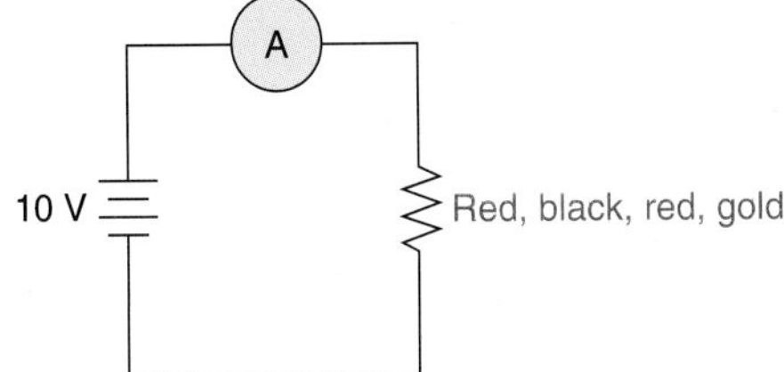


***PRACTICE PROBLEM 4.16***

A circuit like the one in Figure 4.24 has a 12 V source and a resistor that is color coded as follows: blue, gray, red, silver. Determine the range of current values for the circuit.

As you can see, the tolerance of a resistor can affect the actual value of current in a circuit. When working in the lab, keep in mind that the tolerance of a resistor may cause a measured current value to vary slightly from its predicted value.

## Circuit Loads

*OBJECTIVE 12* ▶

**Source**
The part of a circuit that supplies power.

**Load**
The part of a circuit that absorbs (uses) power.

The simplest circuit contains two components: one that supplies power and one that absorbs it. For example, the battery in Figure 4.25 is supplying power to the resistor. The resistor, in turn, is absorbing the power supplied by the battery.

In a circuit like the one in Figure 4.25, we generally refer to the components as the **source** and the **load.** The source supplies the power and the load absorbs (uses) it.

Every circuit or system, no matter how simple or complex, is designed to deliver power (in one form or another) to one or more loads. For example, consider the items shown in Figure 4.26. The stereo (a relatively complex system) is designed to deliver power to its load: the speakers. The table lamp (a relatively simple device) is designed to deliver power to its load: the lightbulb. In both cases, delivering power to the load is the purpose of the system or device.

In many schematics, the load on a circuit is represented as a single load resistor ($R_L$), as shown in Figure 4.27. The current through the load is called the load current ($I_L$), and the voltage across the load is referred to as the load voltage ($V_L$). The power absorbed by the load is referred to as the load power ($P_L$).

**Full load**
A load that draws maximum current from the source. (A full load occurs when load resistance is at its minimum value.)

When the resistance of a load is variable (as shown in Figure 4.27), the minimum load resistance is referred to as a **full load.** A full load draws maximum current from the source. As Ohm's law tells us, maximum current is produced when the load resistance is at its minimum value.

As you will see, circuits normally contain a single component that is referred to as the load. You will also find that our goal in analyzing these circuits is usually to determine the values of load resistance, current, voltage, and/or power.

**FIGURE 4.25**

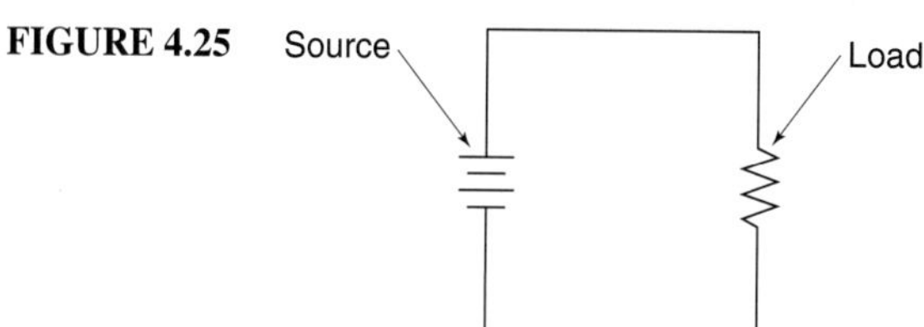

**FIGURE 4.26 Various electrical loads.**

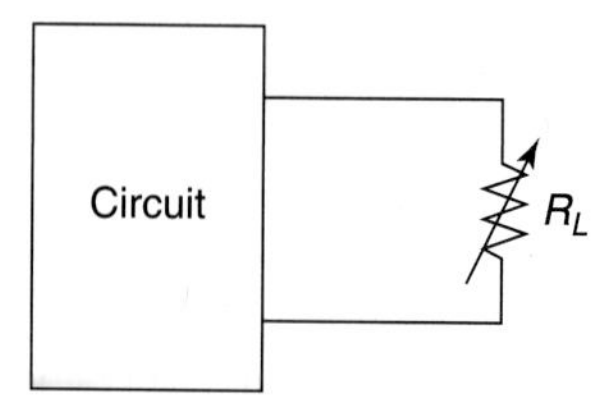


**FIGURE 4.27**

## *Circuit Faults: The Open Circuit*

◀ *OBJECTIVE 13*

Two common types of circuit faults are referred to as the *open circuit* and the *short circuit.* An **open circuit** is a physical break in a conduction path. Figure 4.28 shows the characteristics of an open circuit.

**Open circuit**
A physical break in a conduction path.

You already know that a fuse blocks current when it opens because it breaks the required path for the flow of charge. We get the same result if any component physically opens: the path for conduction is broken and the current drops to zero.

Since an open component is a physical break in a circuit, the resistance of the open component is infinite (for all practical purposes). Thus, if you attempt to measure the resistance of an open component, its resistance will be too high for the ohmmeter to measure. When an open exists in a basic circuit, the full applied voltage is felt across the open. This point is illustrated in Figure 4.29. If we were to measure the voltage across the open terminals of the battery, we would measure its rated value. When the voltmeter is connected across the open (as shown in Figure 4.29), it is effectively connected across the open terminals of the battery. Thus, we measure the full applied voltage across the open. As a summary, here are the characteristics of an open circuit:

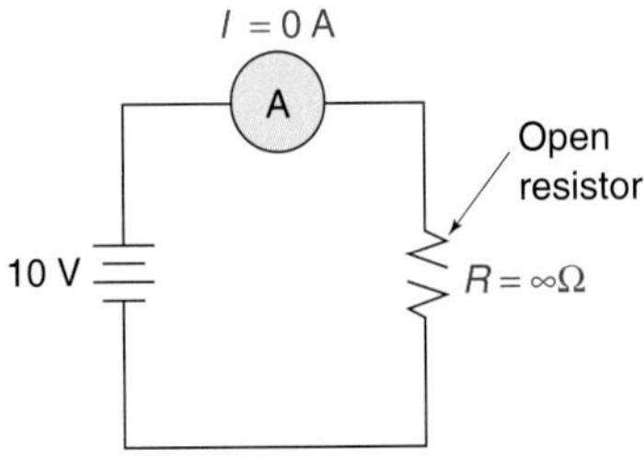


FIGURE 4.28

◀ *OBJECTIVE 14*

1. The circuit current drops to zero.
2. The resistance of the component is too high to measure.
3. The applied voltage is measured across the open.

Any component can open. Normally, an open is caused when the current through a component becomes too high, causing excessive power to be absorbed. This condition results in the component burning open. Note that an open in a circuit is repaired by replacing the open component.

## *Circuit Faults: The Short Circuit*

A short circuit is the exact opposite of an open. A **short circuit** is an extremely low resistance path between two points that does not normally exist. For example, consider the circuit shown in Figure 4.30. Here we have created a short by connecting a wire across the 1 kΩ resistor. Since current takes the path of least resistance (an old but true saying), we have effectively dropped the circuit resistance to 0.5 Ω, as shown in the figure. (This resistance value was chosen for the sake of discussion.)

**Short circuit**
An extremely low resistance path between two points that does not normally exist.

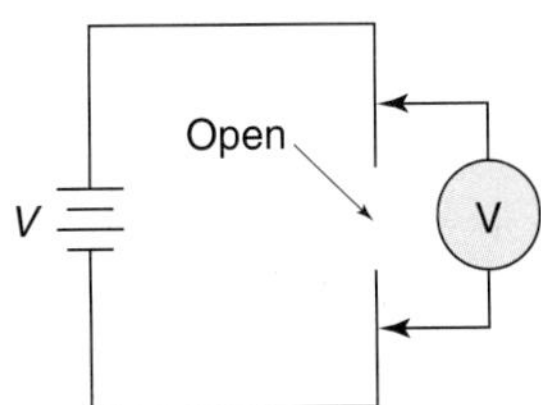


FIGURE 4.29

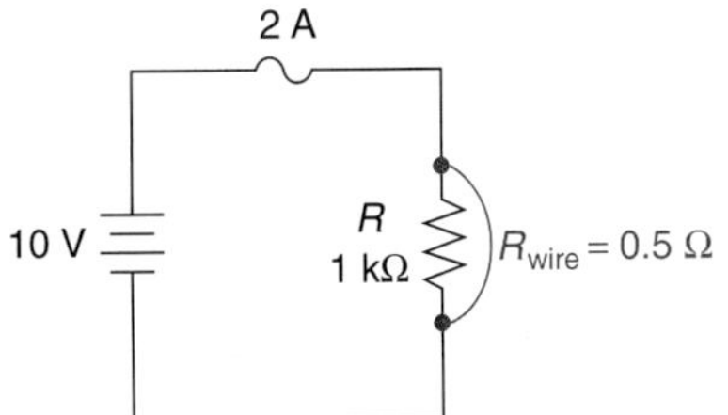


FIGURE 4.30

Ohm's law tells us that a decrease in resistance causes an increase in circuit current. For example, with the total short-circuit resistance of 0.5 Ω, the current in Figure 4.30 is found as:

$$I = \frac{V}{R} = \frac{10\text{ V}}{0.5\ \Omega} = \mathbf{20\ A}$$

In this case, the circuit current exceeds the rating of the fuse, and the fuse opens. At the same time, if there were no fuse in the circuit, the excessive current would eventually cause the short circuit current path (or the power source) to burn open. Thus, the current in a shorted circuit is extremely high until the fuse (or some other component) burns open as a result of the high current.

Ohm's law tells us that $V = IR$. Since the resistance of a shorted component is extremely low, the voltage across the shorted component is extremely low. As a summary, here are the characteristics of a short circuit:

1. The resistance of the shorted component is extremely low.
2. The current through the short circuit current path is extremely high. This high current may cause a circuit fuse (or some other component) to open.
3. The voltage across the shorted component is extremely low.

Common causes of apparently shorted resistors.

It should be noted that carbon-composition resistors do not internally short circuit. In other words, there is no internal fault that would cause the resistance of a carbon-composition resistor to drop to zero. When a carbon-composition resistor appears to be shorted, the short is usually caused by one of the conditions shown in Figure 4.31.

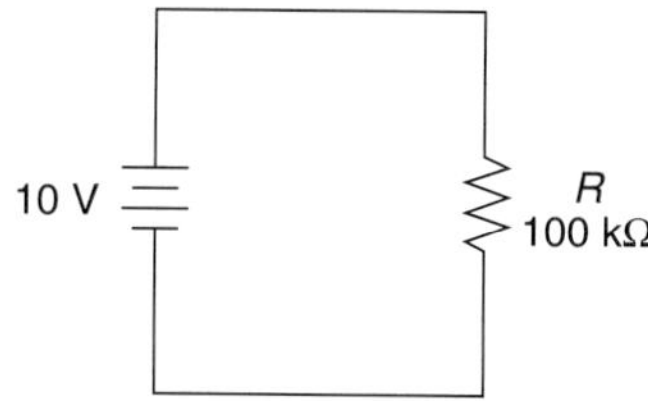


(a) R color code: brown, black, black, gold

*Note:* The solder bridge shown was built up for demonstration purposes. An actual solder bridge may be harder to see than the one shown here.

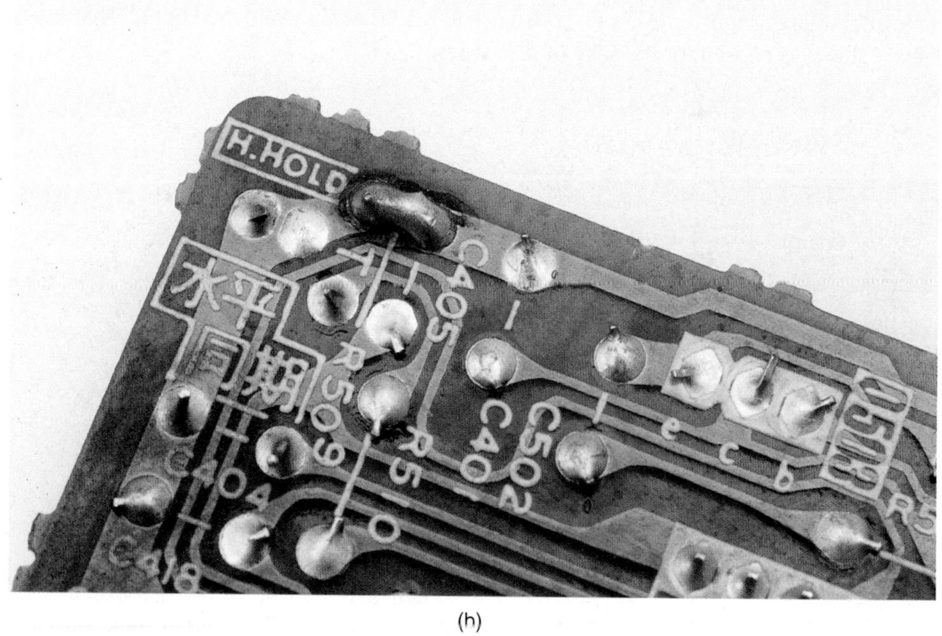


(b)

**FIGURE 4.31**

In Figure 4.31a, the schematic calls for a 100 kΩ resistor. However, the color code of the resistor indicates that its value is 10 Ω. In this case, the load resistor was accidentally replaced with one that had a much lower value. This condition would appear as a shorted component to the circuit.

Figure 4.31b shows another possible cause for a shorted resistor. Components are connected to a PC board using a compound called **solder.** Solder is a high-conductivity compound that has a low melting point. When a drop of melted solder is applied to the connection point between the resistor and the PC board, it cools and forms a solid electrical connection between the two points. (The procedure for soldering is covered in the lab manual that accompanies this text.) If the melted solder accidentally makes a connection between two points that shouldn't be connected, we have what is called a *solder bridge.* The solder bridge in Figure 4.31b forms a low-resistance path between the two sides of a resistor, effectively shorting the component. In this case, the short is repaired by removing the solder bridge.

**Solder**
A high-conductivity compound that has a relatively low melting point, used to affix components to each other and/or PC boards.

It should be noted that the other types of resistors (wire-wound, metal-film, integrated, etc.) can and do form internal short circuits. When one of these components shorts, replacing the component repairs the fault.

## One Final Note

We have now completed our coverage of circuits that contain only one resistive element. In Chapter 5, we will start to discuss circuits that contain more than one resistive element. As you will see, the circuits become more complex, but the principles and relationships that we have established so far still hold true.

### Section Review

1. Describe the procedure for selecting the proper equation to solve a basic circuit problem.
2. What effect does resistor tolerance have on a measured value of circuit current?
3. What is a source? What is a load?
4. What is a full load?
5. List the characteristics of an open circuit.
6. List the characteristics of a short circuit.

## Chapter Summary

Here is a summary of the major points made in this chapter:

1. Ohm's law states that current is directly proportional to voltage and inversely proportional to resistance.
2. The current through a fixed resistance:
   a. Increases when the applied voltage increases.
   b. Decreases when the applied voltage decreases.
3. The current generated by a fixed voltage:
   a. Decreases when the circuit resistance increases.
   b. Increases when the circuit resistance decreases.
4. Ohm's law can be used to predict the response of circuit current to a change in voltage or resistance.
5. A potentiometer wired as a variable resistor is called a rheostat.

6. When you adjust any circuit component, there is always some type of circuit response to the change.
7. Troubleshooting is the process of locating faults in an electronic circuit or system.
8. When troubleshooting, Ohm's law is used to determine the types of potential problems you are looking for.
9. Ohm's law is commonly used to:
   - a. Predict the response of a circuit to a change in voltage or resistance.
   - b. Calculate the value of current, voltage, or resistance in a circuit.
   - c. Determine the possible causes of a problem in a circuit.
10. Power is a measure of energy used per unit time.
11. Power is measured in watts (W). One watt of power is equal to 1 joule/second.
12. Power is directly proportional to both current and voltage.
13. In any circuit, power is not created. It is merely transferred from one part of the circuit to another.
14. Any device that converts energy from one form to another is called a transducer.
15. Resistors convert the power they absorb to heat.
16. When choosing a resistor for a given application, the power rating of the component must be greater than the anticipated power in the circuit.
17. Efficiency is the ratio of output power to input power, given as a percentage.
18. It is not possible to have an efficiency rating greater than 100%.
19. Utilities use the kilowatt-hour (kWh) as the unit of measure for energy. One kilowatt-hour is the amount of energy that a 1000 W appliance uses in one hour.
20. To solve basic circuit problems:
   - a. Find an equation that contains all the variables directly involved in the problem.
   - b. If necessary, rearrange the equation to obtain one that defines the unknown value in terms of known values.
21. Since resistors have tolerances, the measured current in a circuit actually falls within a range of values.
22. A source delivers power. A load absorbs (uses) power.
23. The purpose of a circuit or system is to deliver power (in one form or another) to a load.
24. A full load is one designed to draw maximum current from the source.
25. An open circuit is a break in the conduction path of a circuit. An open circuit has the following characteristics:
   - a. Circuit current is zero.
   - b. The resistance of the open component is too high to measure.
   - c. The applied voltage can be measured across the open.
26. A short circuit is a low-resistance condition that does not normally exist in a circuit. A short circuit has the following characteristics:
   - a. The resistance of the short circuit is extremely low.
   - b. The current through the short circuit is extremely high. This high current may cause a circuit fuse (or some other component) to burn open.
   - c. The voltage across the shorted component is very nearly zero.

## Equation Summary

| Equation Number | Equation | Section Number |
|---|---|---|
| (4.1) | $I = \frac{V}{R}$ | 4.1 |
| (4.2) | $V = IR$ | 4.1 |
| (4.3) | $R = \frac{V}{I}$ | 4.1 |
| (4.4) | $P = \frac{E}{t}$ | 4.2 |
| (4.5) | $P = \frac{V^2}{R}$ | 4.2 |
| (4.6) | $P = I^2R$ | 4.2 |
| (4.7) | $P = VI$ | 4.2 |
| (4.8) | $\eta = \frac{P_o}{P_i} \times 100$ | 4.2 |
| (4.9) | $E_{\mathrm{kWh}} = \frac{P \cdot t}{1000}$ | 4.2 |

## Key Terms

The following terms were introduced and defined in this chapter:

efficiency ($\eta$)
kilowatt-hour (kWh)
load
Ohm's law
open circuit
power ($P$)
rheostat
short circuit
solder
solder bridge
transducer
troubleshooting
watt (W)

## Practice Problems

1. For each combination of voltage and resistance, calculate the resulting current.
   a. $V = 12$ V, $R = 2.2$ kΩ
   b. $V = 80$ mV, $R = 1.8$ kΩ
   c. $V = 8$ V, $R = 470$ Ω
   d. $V = 16$ V, $R = 330$ kΩ
   e. $V = 120$ μV, $R = 6.8$ Ω
2. For each combination of voltage and resistance, calculate the resulting current.
   a. $V = 15$ V, $R = 4.7$ kΩ
   b. $V = 8.6$ V, $R = 510$ Ω
   c. $V = 22.8$ V, $R = 82$ kΩ
   d. $V = 10$ V, $R = 1.5$ MΩ
   e. $V = 2$ V, $R = 12$ Ω
3. For each combination of current and resistance, calculate the voltage required to generate the current.
   a. $I = 10$ mA, $R = 820$ Ω
   b. $I = 65$ mA, $R = 100$ Ω

c. $I = 130\ \mu A, R = 2.7\ k\Omega$

d. $I = 24\ mA, R = 1.1\ k\Omega$

e. $I = 800\ mA, R = 3.6\ \Omega$

**4.** For each combination of current and resistance, calculate the voltage required to generate the current.

a. $I = 20\ mA, R = 910\ \Omega$

b. $I = 33\ mA, R = 51\ \Omega$

c. $I = 14\ mA, R = 180\ \Omega$

d. $I = 60\ \mu A, R = 24\ k\Omega$

e. $I = 100\ \mu A, R = 16\ k\Omega$

**5.** For each combination of voltage and current, determine the resistance needed to limit the current to the given value.

a. $V = 6\ V, I = 4\ mA$

b. $V = 94\ V, I = 20\ mA$

c. $V = 11\ V, I = 500\ \mu A$

d. $V = 9\ V, I = 50\ mA$

e. $V = 33\ V, I = 330\ \mu A$

**6.** For each combination of voltage and current, calculate the resistance needed to limit the current to the given value.

a. $V = 8.8\ V, I = 24\ mA$

b. $V = 12\ V, I = 16\ mA$

c. $V = 28\ V, I = 8\ mA$

d. $V = 180\ mV, I = 260\ \mu A$

e. $V = 1\ V, I = 250\ nA$

**7.** For each combination of voltage and current, calculate the power supplied by the voltage source.

a. $V = 14\ V, I = 160\ mA$

b. $V = 120\ V, I = 40\ mA$

c. $V = 5\ V, I = 2.2\ A$

d. $V = 24\ V, I = 600\ \mu A$

e. $V = 15\ V, I = 850\ mA$

**8.** For each combination of voltage and current given in Problem 5, calculate the power supplied by the voltage source.

**9.** For each combination of current and resistance, calculate the power supplied by the voltage source.

a. $I = 25\ mA, R = 3.3\ k\Omega$

b. $I = 16\ mA, R = 12\ k\Omega$

c. $I = 4.6\ mA, R = 9.1\ k\Omega$

d. $I = 7.5\ mA, R = 11\ k\Omega$

e. $I = 420\ \mu A, R = 150\ k\Omega$

**10.** For each combination of current and resistance given in Problem 3, calculate the power supplied by the voltage source.

**11.** For each combination of voltage and resistance, calculate the power supplied by the voltage source.

a. $V = 15\ V, R = 4.7\ k\Omega$

b. $V = 8\ V, R = 47\ k\Omega$

c. $V = 120$ mV, $R = 14\ \Omega$

d. $V = 3.8$ V, $R = 510\ \Omega$

e. $V = 28$ V, $R = 120$ k$\Omega$

**12.** For each combination of voltage and resistance given in Problem 1, calculate the power supplied by the voltage source.

**13.** Calculate the minimum acceptable power rating for the resistor in Figure 4.32a.

**14.** Calculate the minimum acceptable power rating for the resistor in Figure 4.32b.

**15.** Calculate the minimum acceptable power rating for the resistor in Figure 4.32c.

**16.** Calculate the minimum acceptable power rating for the resistor in Figure 4.32d.

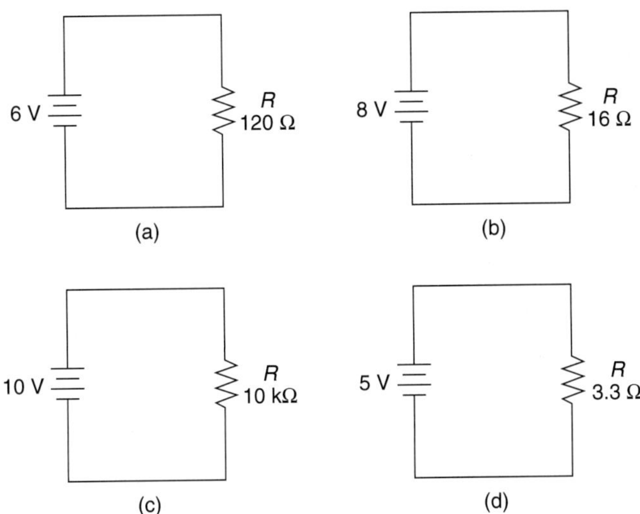


**FIGURE 4.32**

**17.** For each combination of input and output power, calculate the efficiency of the circuit.

a. $P_i = 2$ W, $P_o = 36$ mW

b. $P_i = 6$ W, $P_o = 445$ mW

c. $P_i = 8.5$ W, $P_o = 6.7$ W

d. $P_i = 100$ W, $P_o = 34$ W

e. $P_i = 15$ kW, $P_o = 600$ mW

**18.** Calculate the energy used (in kWh) to run twelve 150 W lightbulbs for eight hours.

**19.** Calculate the energy used (in kWh) to run a 500 W microwave oven and a 2400 W air conditioner for three hours.

**20.** Calculate the energy used to run the following combination of appliances for the times given: a 700 W microwave oven for 20 minutes, a 3800 W dishwasher for 30 minutes, a 2200 W air conditioner for 12 hours, and an 1800 W clothes dryer for 1 hour.

**21.** Complete the chart below.

| Current | Voltage | Resistance | Power |
|---|---|---|---|
| a. 10 mA | ____ | ____ | 4 W |
| b. ____ | 32 V | ____ | 16 mW |
| c. ____ | ____ | 3.3 k$\Omega$ | 231 mW |
| d. 15 mA | 45 V | ____ | ____ |
| e. 24 mA | ____ | 1.2 k$\Omega$ | ____ |

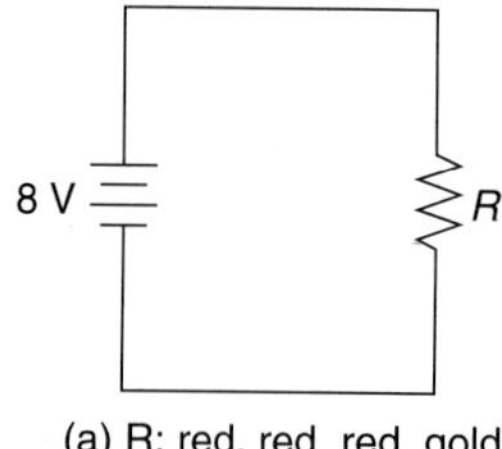

(a) R: red, red, red, gold

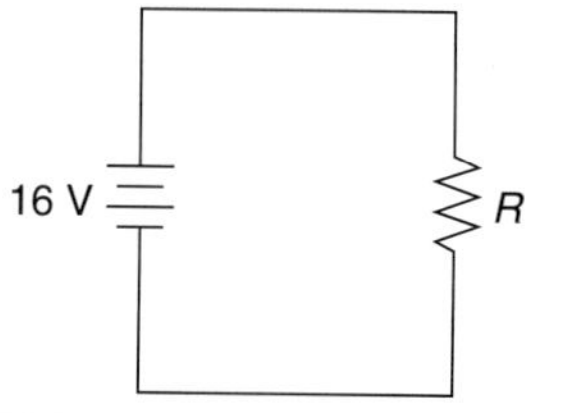

(b) R: yellow, violet, orange, silver

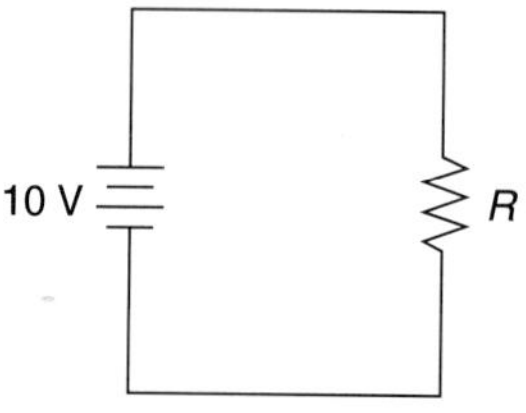

(c) R: orange, orange, red, gold

5 V
R

(d) R: brown, green, green, silver

FIGURE 4.33

22. Complete the chart below.

| | Current | Voltage | Resistance | Power |
|---|---|---|---|---|
| a. | ____ | ____ | 1 kΩ | 240 mW |
| b. | 50 mA | 2 V | ____ | ____ |
| c. | ____ | 8 V | 200 Ω | ____ |
| d. | ____ | 16 V | ____ | 800 mW |
| e. | 35 mA | ____ | ____ | 1.4 W |

23. Calculate the range of possible current values for the circuit in Figure 4.33a.
24. Calculate the range of possible current values for the circuit in Figure 4.33b.
25. Calculate the range of possible current values for the circuit in Figure 4.33c.
26. Calculate the range of possible current values for the circuit in Figure 4.33d.

## The Brain Drain

27. Calculate the minimum allowable setting for the potentiometer in Figure 4.34.
28. Calculate the maximum allowable setting for the voltage source in Figure 4.35.
29. Ohm's law states that:

    **a.** The current through a *fixed* resistance is directly proportional to voltage.

    **b.** The current generated by a *fixed* voltage is inversely proportional to resistance.

    Using the circuit in Figure 4.36 as a model, prove that Ohm's law doesn't apply when *both* voltage and resistance change.
30. Without substituting power ($P$) for either expression, prove the following:

$$\frac{V^2}{R} = I^2R$$

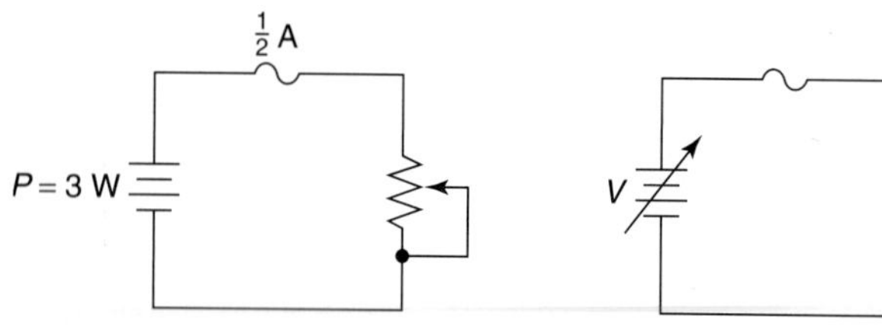

FIGURE 4.34

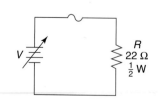

FIGURE 4.35

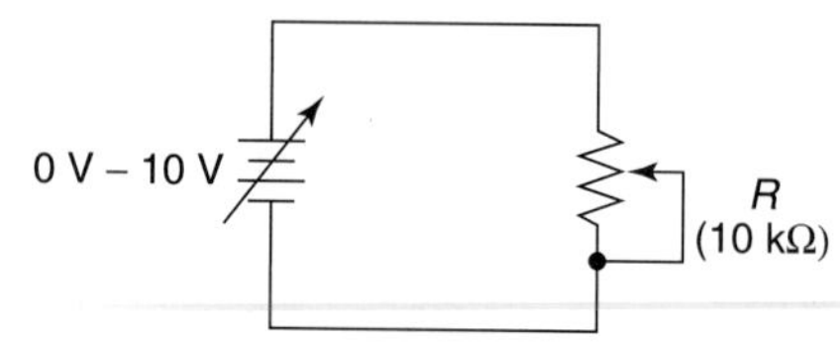

FIGURE 4.36

## Answers to the Example Practice Problems

**4.1** 180 mA
**4.2** The current in the circuit will be 1.11 A, so the fuse will blow.
**4.3** 1.65 V
**4.4** 13.2 V
**4.5** 300 Ω
**4.6** 60 Ω
**4.7** When $V = 3$ V, $I = 1.5$ mA (half the original value).
**4.8** When $R = 5$ kΩ, $I = 2$ mA (twice the original value).
**4.9** 960 mW
**4.10** 11.52 W
**4.11** 21.3 mW
**4.12** $P = 356$ mW, so the resistor must have a minimum rating of ½ W (500 mW).
**4.13** 14.3%
**4.14** 11.7 kWh
**4.15** 60 V
**4.16** 1.60 mA to 1.96 mA

## EWB Applications Problems

| Figure | EWB File Reference |
|---|---|
| 4.1 | EWBA4_1.ewb |
| 4.12 | EWBA4_12.ewb |
| 4.29 | EWBA4_29.ewb |
| 4.30 | EWBA4_30.ewb |

# 5

# SERIES CIRCUITS

## OBJECTIVES

*After studying the material in this chapter, you should be able to:*

1. State the current characteristic that distinguishes series circuits from parallel circuits.
2. Calculate any resistance value in a series circuit, given the other resistances.
3. Determine the total current in any series circuit.
4. Determine the potentiometer setting required to provide a specific amount of current in a series circuit.
5. Describe the voltage and power relationships in a series circuit.
6. State and explain Kirchhoff's voltage law.
7. Calculate the voltage from any point in a series circuit to ground.
8. Calculate the voltage across any resistor (or group of resistors) in a series circuit using the voltage-divider equation.
9. Perform a complete mathematical analysis of any series circuit.
10. List the fault symptoms associated with an open resistor in a series circuit.
11. Discuss the effect of measuring the voltage across an open component on the actual value of that voltage.
12. Calculate the voltage reading that will be obtained across an open component in a series circuit, given the internal resistance of the voltmeter used.
13. Discuss the symptoms associated with multiple opens in a series circuit.
14. List the fault symptoms associated with one or more shorted resistors in a series circuit.
15. Discuss the effects of component aging and stress on circuit operation.
16. Compare and contrast ideal and practical voltage sources.
17. Discuss the effects of source resistance on circuit voltage values.
18. Discuss the maximum power transfer theorem and calculate the maximum load power for a given series circuit.
19. Calculate the total voltage and current values for a series circuit containing two series-aiding or series-opposing voltage sources.
20. Compare and contrast earth ground with chassis ground.

Now that we have discussed the fundamental concepts of electricity and electronics, it is time to start applying these concepts to some common types of circuits. Two types of circuits found in almost every type of electronic system are the *series* circuit and the *parallel* circuit. These two types of circuits are shown in Figure 5.1. As you can see, the two circuits look very different from each other. In fact, they not only look different, they are almost opposites in terms of their operating characteristics. At the same time, most electronic circuits, no matter how complex, are designated as series circuits, parallel circuits, or some combination of the two.

In this chapter, we will discuss the operating characteristics of series circuits, along with some related topics. Parallel circuits are introduced in Chapter 6.

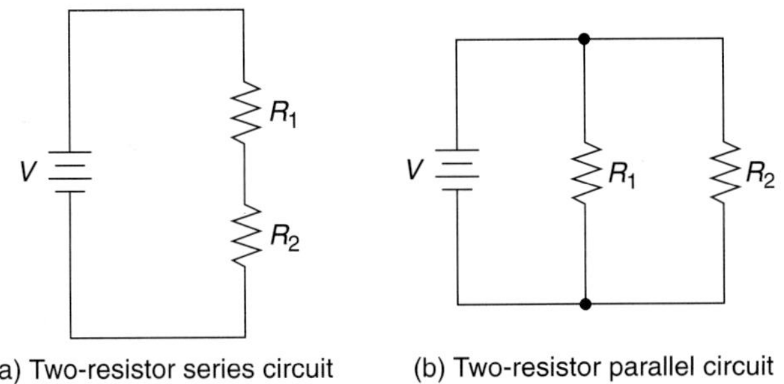


**FIGURE 5.1 The simplest series and parallel circuits.**

## 5.1 SERIES CIRCUIT CHARACTERISTICS

◀ *OBJECTIVE 1*

Simply defined, a **series circuit** is a circuit that contains only one current path. For example, consider the circuits shown in Figure 5.2. In each case, the current generated by the voltage source has only one path, and that path contains all the components in the circuit. In contrast, the parallel circuit in Figure 5.1b contains two current paths between the terminals of the voltage source; one through $R_1$ and one through $R_2$.

**Series circuit**
A circuit that contains only one current path.

### *Resistance Characteristics*

◀ *OBJECTIVE 2*

As shown in Figure 5.2, the current in a series circuit passes through all the resistors in the circuit. Since the circuit current passes through all the resistors, the total opposition to current is equal to the sum of the individual resistor values. By formula,

$$R_T = R_1 + R_2 + \ldots + R_n \quad \textbf{(5.1)}$$

where $R_T$ = the total circuit resistance

$R_n$ = the highest numbered resistor in the circuit

The total resistance in a series circuit is calculated as shown in Example 5.1.

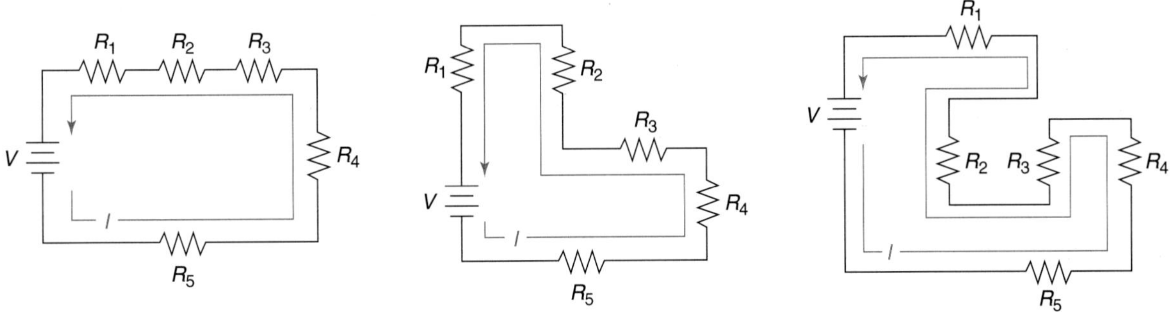


**FIGURE 5.2**

## EXAMPLE 5.1

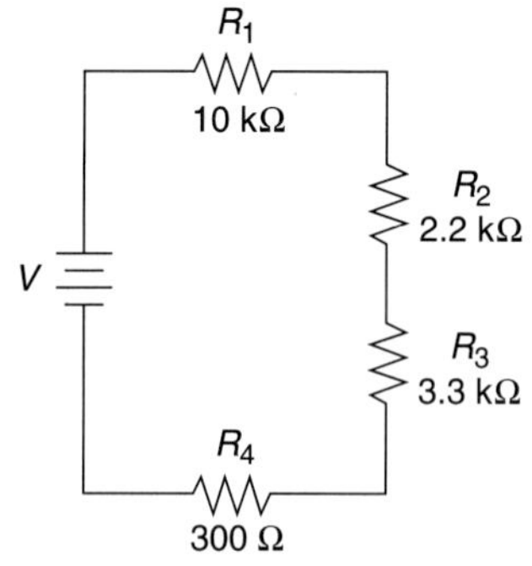

Calculate the total resistance in the circuit shown in Figure 5.3.

***Solution:*** Using equation (5.1) and the values given in the figure, the total resistance is found as

$$R_T = R_1 + R_2 + R_3 + R_4 = 10\ \text{k}\Omega + 2.2\ \text{k}\Omega + 3.3\ \text{k}\Omega + 300\ \Omega = \mathbf{15.8\ k\Omega}$$

***PRACTICE PROBLEM 5.1***

A circuit like the one shown in Figure 5.3 has the following values: $R_1 = 680\ \Omega$, $R_2 = 1.5\ \text{k}\Omega$, $R_3 = 470\ \Omega$, and $R_4 = 3.6\ \text{k}\Omega$. Calculate the total resistance in the circuit.

**FIGURE 5.3**

When you need to find the value of an unknown resistance in a series circuit, you can calculate it by subtracting the sum of the known resistances from the total circuit resistance. This technique is demonstrated in the following example.

## EXAMPLE 5.2

Determine the potentiometer setting required in Figure 5.4 to provide a total circuit resistance of 12 kΩ.

***Solution:*** The combined resistance of $R_1$ and $R_2$ is found as

$$\mathbf{R_1 + R_2} = 1.2\ \text{k}\Omega + 3.3\ \text{k}\Omega = \mathbf{4.5\ k\Omega}$$

$R_3$ must account for the difference between 4.5 kΩ and the desired total of 12 kΩ. Therefore, its adjusted value is found as

$$\mathbf{R_3} = R_T - (R_1 + R_2) = 12\ \text{k}\Omega - 4.5\ \text{k}\Omega = \mathbf{7.5\ k\Omega}$$

***PRACTICE PROBLEM 5.2***

A circuit like the one shown in Figure 5.4 has values of $R_1 = 47\ \text{k}\Omega$ and $R_2 = 110\ \text{k}\Omega$. Determine the potentiometer setting that will provide a total circuit resistance of 159 kΩ.

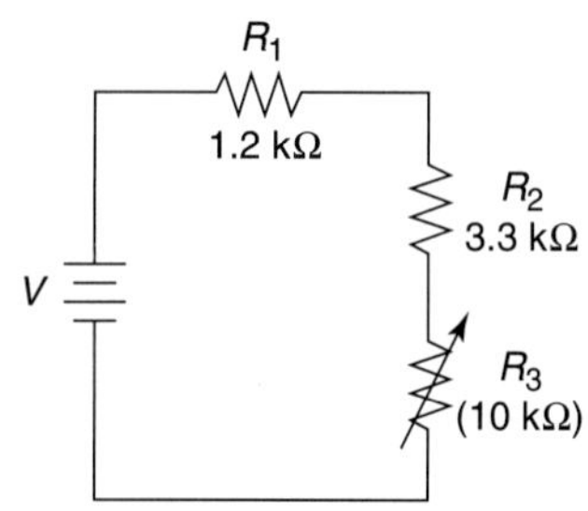


**FIGURE 5.4**

As a summary, the total resistance in a series circuit is equal to the sum of the individual resistor values. When the value of one resistor is unknown, it can be determined by subtracting the sum of the known resistor values from the total circuit resistance.

## Current Characteristics

Measuring current in a series circuit.

Since a series circuit contains only one current path, we can state that the current at any point in a series circuit must equal the current at every other point in the circuit. This principle is illustrated in Figure 5.5. As you can see, the ammeters are providing identical (1mA) readings. The fact that all the meters show the same reading makes sense when you consider the fact that current is the flow of charge. Since charge is leaving (and entering) the source at a rate of one milliampere, it must be flowing at the same rate at all points in the circuit.

Figure 5.5 also illustrates the fact that it makes no difference where you measure the current in a series circuit. Since the current is the same at all points, you will obtain the same reading regardless of where you place the meter in the circuit.

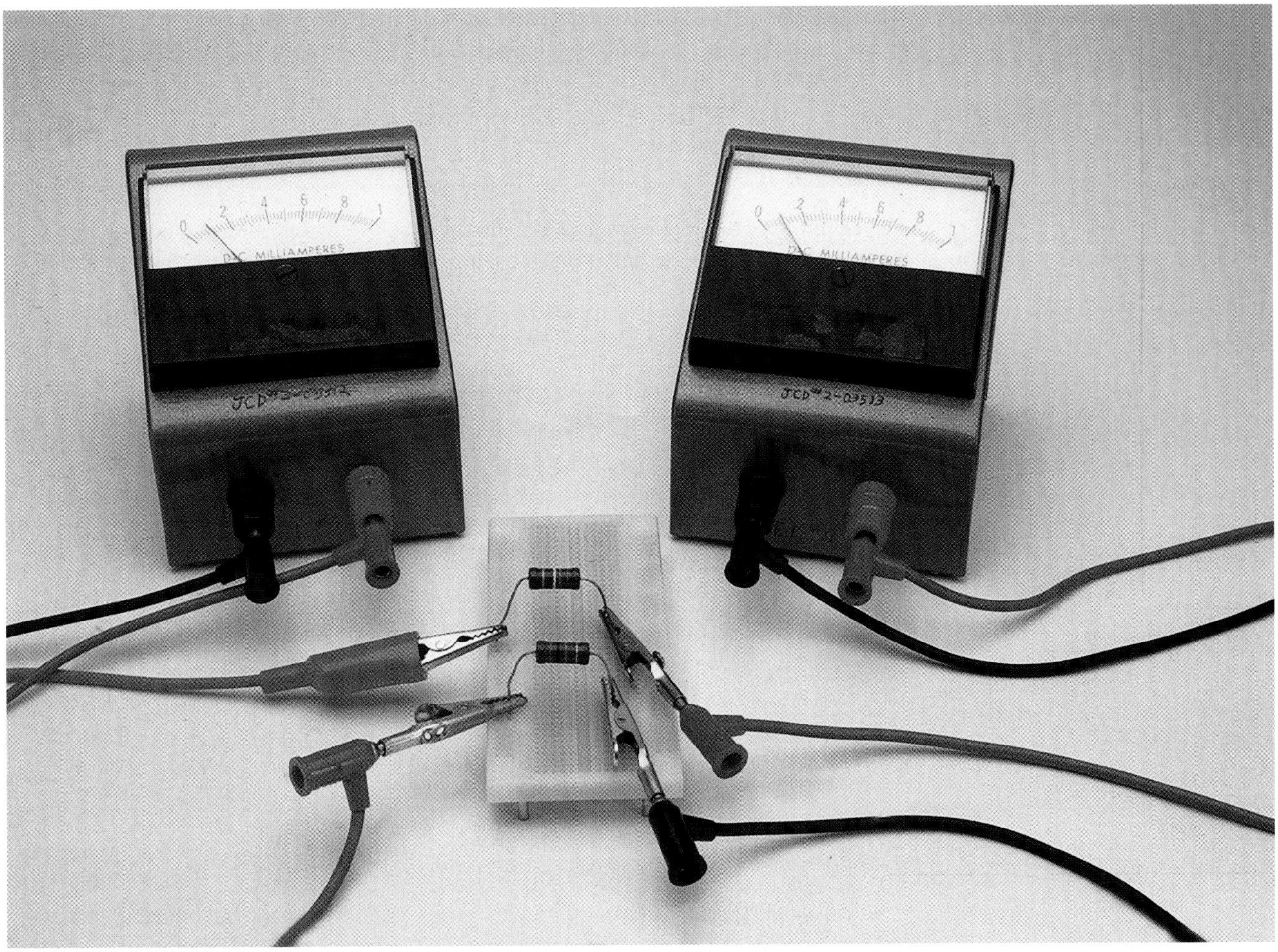

**FIGURE 5.5**

The actual value of current in a series circuit depends on the source voltage ($V_S$) and the total circuit resistance ($R_T$). When the source voltage and total circuit resistance are known, Ohm's law is used to calculate the total circuit current, as demonstrated in the following example.

◄ *OBJECTIVE 3*

## EXAMPLE 5.3

Calculate the total current ($I_T$) in the circuit shown in Figure 5.6.

***Solution:*** First, the total resistance in the circuit is found as

$$\boldsymbol{R_T} = R_1 + R_2 + R_3 = 1\text{ k}\Omega + 10\text{ k}\Omega + 3\text{ k}\Omega = \mathbf{14\text{ k}\Omega}$$

Now, using $R_T = 14\text{ k}\Omega$ and $V_S = 7\text{ V}$, the total circuit current is found as

$$\boldsymbol{I_T} = \frac{V_S}{R_T} = \frac{7\text{ V}}{14\text{ k}\Omega} = \mathbf{500\ \mu A}$$

Thus, the measured current at any point in the circuit would have a value of approximately 500 μA.

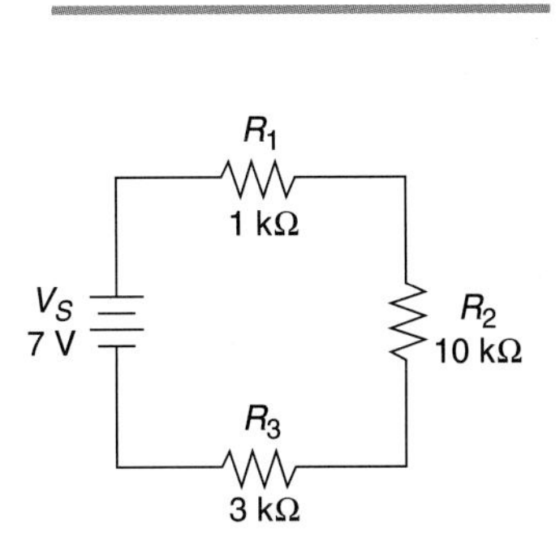


**FIGURE 5.6**

### *PRACTICE PROBLEM 5.3*

A series circuit like the one in Figure 5.6 has the following values: $R_1 = 12\text{ k}\Omega$, $R_2 = 1.5\text{ k}\Omega$, $R_3 = 7.5\text{ k}\Omega$, and $V_S = 15\text{ V}$. Calculate the total current in the circuit.

OBJECTIVE 4 ▶

In Example 5.2, you were shown how to determine the potentiometer setting needed to provide a specific amount of total resistance. The following example shows how a potentiometer is used to provide for a specific value of circuit current.

## EXAMPLE 5.4

The circuit in Figure 5.7 contains a potentiometer used to control the amount of circuit current. Determine the setting of $R_3$ needed to set the circuit current to 15 mA.

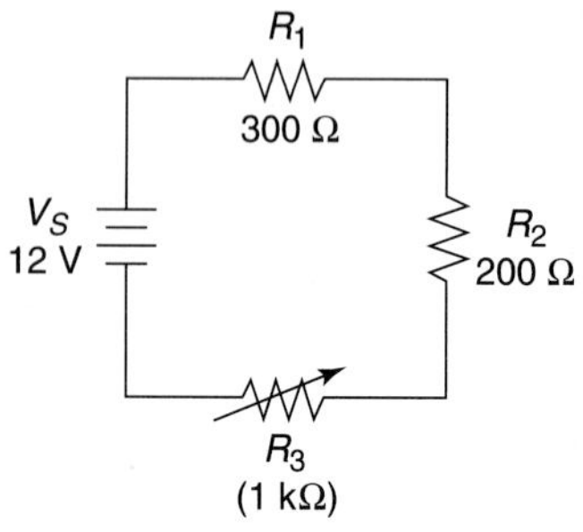


FIGURE 5.7

***Solution:*** First, Ohm's law is used to determine the total resistance required to limit the circuit current to 15 mA, as follows:

$$R_T = \frac{V_S}{I_T} = \frac{12\text{ V}}{15\text{ mA}} = \mathbf{800\ \Omega}$$

The sum of $R_1$ and $R_2$ is now subtracted from the value of $R_T$ to obtain the adjusted value of $R_3$, as follows:

$$R_3 = R_T - (R_1 + R_2) = 800\ \Omega - 500\ \Omega = \mathbf{300\ \Omega}$$

As you can see, adjusting the value of $R_3$ to 300 Ω will provide a total circuit resistance of 800 Ω. This total resistance will limit the circuit current to the desired value of 15 mA.

***PRACTICE PROBLEM 5.4***

A circuit like the one in Figure 5.7 has the following values: $V_S = 24$ V, $R_1 = 1.2$ kΩ, and $R_2 = 4.7$ kΩ. Determine the setting of $R_3$ needed to limit the circuit current to 2.5 mA.

As a summary, the current through a series circuit is equal at all points in the circuit and can be measured at any point in the circuit. The actual value of current in a series circuit is determined by the source voltage and the total circuit resistance.

## Voltage Characteristics

OBJECTIVE 5 ▶

Whenever current passes through a resistance, a difference of potential (voltage) is developed across that resistance, as given by the following relationship:

$$V = IR$$

Since the current in a series circuit passes through all the resistors, it would follow that a voltage would be developed across each resistor. For example, consider the circuit shown in Figure 5.8. The circuit current (2 mA) is shown to be passing through a 1 kΩ resistor and a 3 kΩ resistor. According to Ohm's law, the voltage across $R_1$ (which is designated as $V_1$) is found as

$$V_1 = I_T R_1 = (2\text{ mA})(1\text{ k}\Omega) = \mathbf{2\ V}$$

*An Alternate Approach:* The polarity of the voltage across each resistor can be determined by tracing from the component leads to the voltage source. For example, the bottom side of $R_2$ is connected to the (−) terminal of the battery. The top side of the component is connected (via $R_1$) to the (+) terminal of the voltage source.

The voltage across $R_2$ (which is designated as $V_2$) is found as

$$V_2 = I_T R_2 = (2\text{ mA})(3\text{ k}\Omega) = \mathbf{6\ V}$$

If we were to connect a series of voltmeters to this circuit (as shown in Figure 5.9), we would obtain the readings shown. Note that the polarities of the component voltages are determined by the direction of the circuit current. Since charge flows from negative to positive, the entry side of the component is more negative than the exit side. Thus, the component voltages are given the polarity signs shown in the figure.

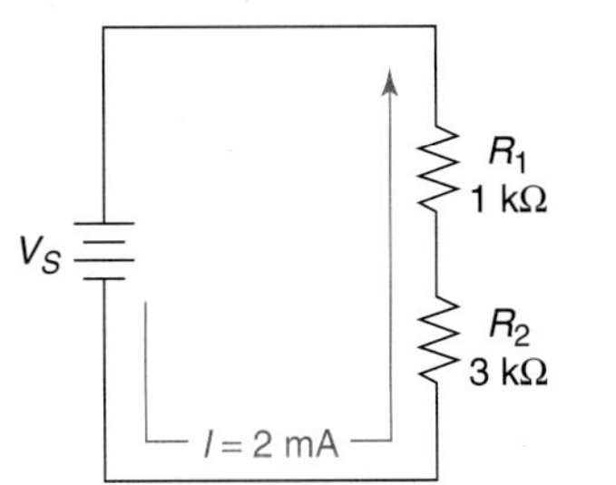


**FIGURE 5.8**

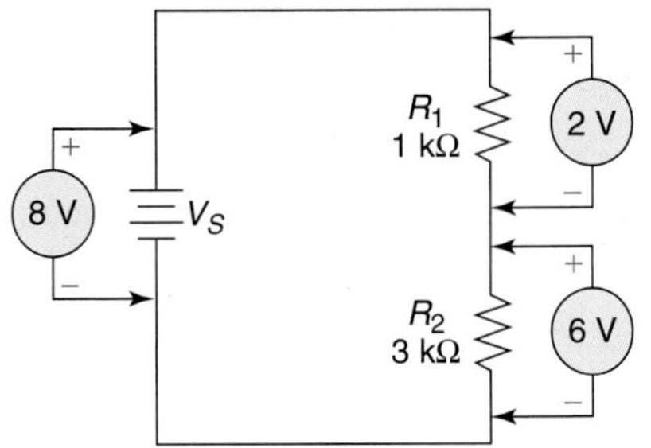


**FIGURE 5.9**

If you look closely at Figure 5.9, you'll see that the sum of the component voltages is equal to the source (or total) voltage. This relationship, which holds true for all series circuits, is given as

$$V_S = V_1 + V_2 + \ldots + V_n \quad \textbf{(5.2)}$$

where $V_S$ = the source (or total) voltage

$V_n$ = the voltage across the highest numbered resistor in the circuit

One application of this relationship is demonstrated in the following example.

*Reference:* The relationship given in equation (5.2) is referred to as Kirchhoff's voltage law. This law is discussed in detail in Section 5.2.

### *EXAMPLE 5.5*

Determine the value of the source voltage in Figure 5.10.

***Solution:*** Using the values given in the figure and equation (5.2), the value of the source voltage is found as

$$V_S = V_1 + V_2 + V_3 = 5\text{ V} + 12\text{ V} + 24\text{ V} = \mathbf{41\ V}$$

***PRACTICE PROBLEM 5.5***

A series circuit like the one shown in Figure 5.10 has the following values: $V_1 = 2$ V, $V_2 = 16$ V, and $V_3 = 8.8$ V. Determine the value of the source voltage.

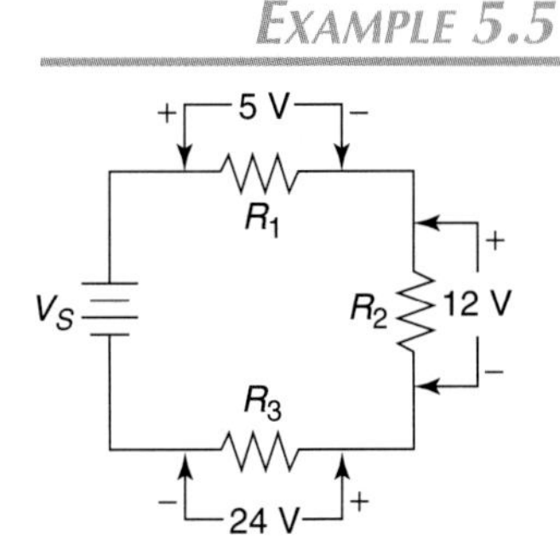


**FIGURE 5.10**

In the next section, we will take a closer look at the voltage relationships in series circuits. For now, remember that the sum of the component voltages in a series circuit is equal to the source (or total) voltage.

## *Power Characteristics*

Whenever current passes through a resistance, some amount of power is dissipated by the component, as given in the following relationship:

$$P = I^2R$$

Since the current in a series circuit passes through all the resistors, it would follow that they all dissipate some amount of power. Referring back to Figure 5.8, the power dissipated by $R_1$ (which is designated as $P_1$) is found as

$$\mathbf{P_1} = I^2R_1 = (2\text{ mA})^2(1\text{ k}\Omega) = \mathbf{4\ mW}$$

The power dissipated by $R_2$ (which is designated as $P_2$) is found as

$$\mathbf{P_2} = I^2R_2 = (2\text{ mA})^2(3\text{ k}\Omega) = \mathbf{12\ mW}$$

*Remember:* As you saw in Chapter 4, power can also be calculated using

$P = IV$ or $P = \dfrac{V^2}{R}$

The total power dissipated by the resistors in a series circuit is equal to the total power supplied by the source. By formula,

*Remember:* The law of conservation of energy states that energy cannot be created or destroyed. Therefore, the components in a circuit must use all the power provided by the source.

$$P_S = P_1 + P_2 + \ldots + P_n \tag{5.3}$$

where $P_S$ = the total power supplied to the circuit by the source

$P_n$ = the power dissipated by the highest numbered resistor in the circuit

For example, the total power supplied by the source in Figure 5.8 is found as

$$\boldsymbol{P_S} = P_1 + P_2 = 4 \text{ mW} + 12 \text{ mW} = \mathbf{16\ mW}$$

The following example verifies the relationship given in equation (5.3).

## EXAMPLE 5.6

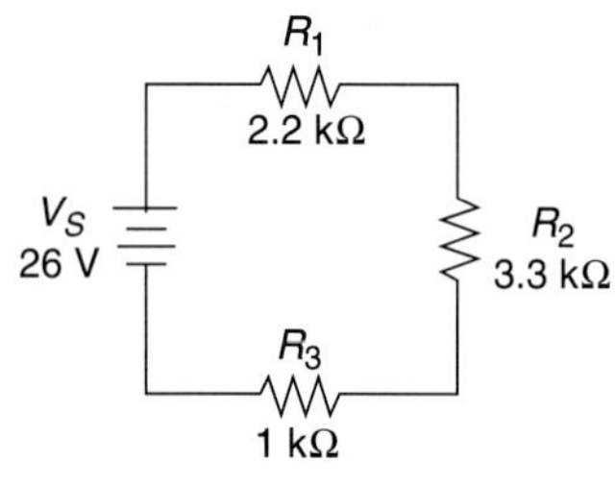


**FIGURE 5.11**

Verify that the total power dissipated by the resistors in Figure 5.11 is equal to the power supplied by the source.

***Solution:*** First, the total resistance in the circuit is found as

$$\boldsymbol{R_T} = R_1 + R_2 + R_3 = 2.2 \text{ k}\Omega + 3.3 \text{ k}\Omega + 1 \text{ k}\Omega = \mathbf{6.5\ k\Omega}$$

Next, the circuit current is found as

$$\boldsymbol{I_T} = \frac{V_S}{R_T} = \frac{26 \text{ V}}{6.5 \text{ k}\Omega} = \mathbf{4\ mA}$$

Now that we know the value of the circuit current, we can find the power dissipated by each of the resistors in the circuit as follows:

$$\boldsymbol{P_1} = I_T^2 R_1 = (4 \text{ mA})^2(2.2 \text{ k}\Omega) = \mathbf{35.2\ mW}$$
$$\boldsymbol{P_2} = I_T^2 R_2 = (4 \text{ mA})^2(3.3 \text{ k}\Omega) = \mathbf{52.8\ mW}$$
$$\boldsymbol{P_3} = I_T^2 R_3 = (4 \text{ mA})^2(1 \text{ k}\Omega) = \mathbf{16\ mW}$$

Adding the individual power dissipation values, the total power dissipated by the resistors is found as

$$\boldsymbol{P_S} = P_1 + P_2 + P_3 = 35.2 \text{ mW} + 52.8 \text{ mW} + 16 \text{ mW} = \mathbf{104\ mW}$$

The total power supplied by the source can also be found as

$$\boldsymbol{P_S} = I_T^2 R_T = (4 \text{ mA})^2(6.5 \text{ k}\Omega) = \mathbf{104\ mW}$$

or

$$\boldsymbol{P_S} = V_S I_T = (26 \text{ V})(4 \text{ mA}) = \mathbf{104\ mW}$$

or

$$\boldsymbol{P_S} = \frac{V_S^2}{R_T} = \frac{(26 \text{ V})^2}{6.5 \text{ k}\Omega} = \mathbf{104\ mW}$$

Regardless of the approach taken, the results are the same and verify the relationship given in equation (5.3).

### *PRACTICE PROBLEM 5.6*

A circuit like the one in Figure 5.11 has the following values: $V_S = 18$ V, $R_1 = 180\ \Omega$, $R_2 = 220\ \Omega$, $R_3 = 140\ \Omega$. Verify that the total power supplied by the source is equal to the sum of the resistor power dissipation values.

## Summary

A series circuit is one that contains a single path for current. The total resistance in a series circuit is equal to the sum of the individual resistor values. The total circuit current is the same at every point in the circuit and is generally found by dividing the source voltage by the total circuit resistance.

When current passes through the resistors in a series circuit, a voltage is developed across each resistor. The sum of these voltages is equal to the circuit's source voltage.

The current through the resistors in a series circuit also causes each component to dissipate some amount of power. The sum of the component power dissipation values is equal to the total power supplied by the source. The resistance, current, voltage, and power characteristics of series circuits are summarized in Figure 5.12.

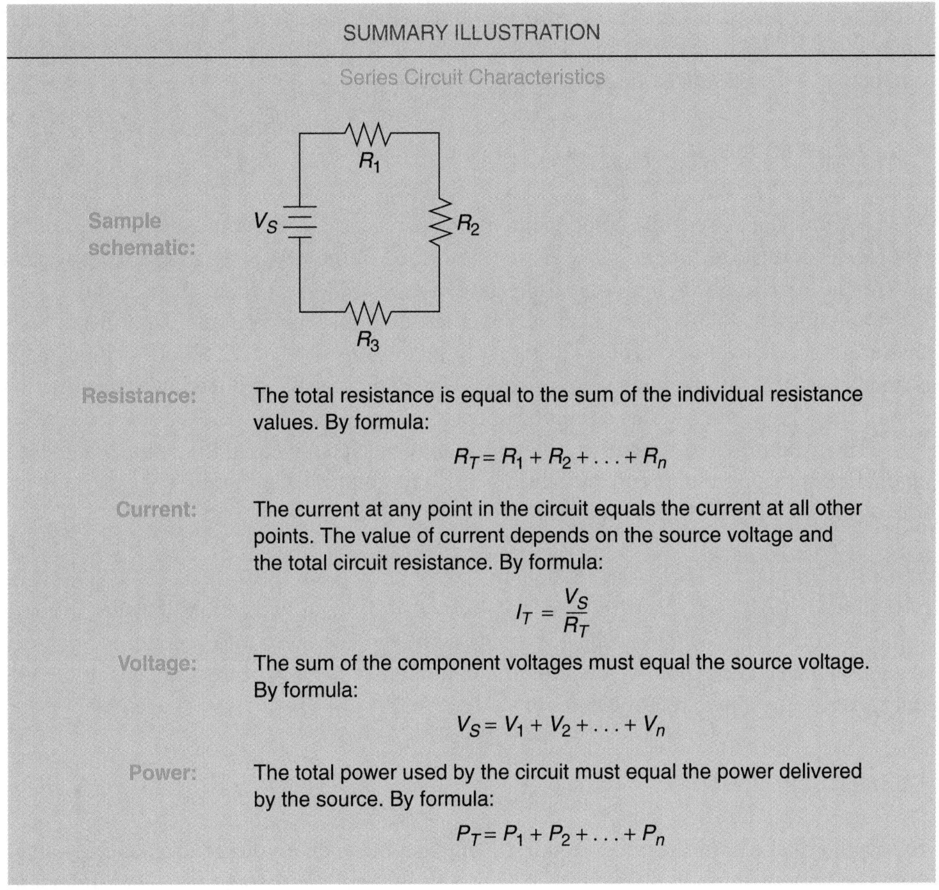


**FIGURE 5.12**

### Section Review

1. What is a series circuit?
2. Describe the relationship between the total resistance in a series circuit and the individual resistor values.
3. Describe the relationship among current values at various points in a series circuit.
4. What determines the value of current in a series circuit?
5. Describe the procedure for setting the current in a series circuit to a specific value using a potentiometer.
6. How do you determine the polarity of the voltage across a given resistor?

7. Describe the relationship between the total voltage applied to a series circuit and the individual component voltages.
8. Every resistor in a series circuit dissipates some amount of power. Why?
9. Describe the relationship between the power supplied by the source in a series circuit and the power dissipated by the individual components.
10. List four ways to calculate the source power in a series circuit.

## 5.2 VOLTAGE RELATIONSHIPS: KIRCHHOFF'S VOLTAGE LAW, VOLTAGE REFERENCES, AND THE VOLTAGE DIVIDER

In the last section, you were shown that the sum of the component voltages in a series circuit is equal to the source voltage. In this section, we will take a closer look at the relationship between the various voltages in a series circuit.

### Kirchhoff's Voltage Law

*OBJECTIVE 6* ▶

As you know, the sum of the component voltages in a series circuit must equal the source voltage. This relationship, which was first described in the 1840s by a German physicist named Gustav Kirchhoff, has come to be known as *Kirchhoff's voltage law.*

*A Practical Consideration:* Kirchhoff used the term *closed loop* to describe a series circuit. In practice, the term *loop* is also used to describe certain series-parallel connections (as you will learn in Chapter 7).

Kirchhoff's voltage law is actually worded like this: the algebraic sum of the voltages around a closed loop is zero. The term *closed loop,* as used in Kirchhoff's voltage law, is another way of describing a series circuit. Thus, Kirchhoff's voltage law states that the sum of the voltages in a series circuit must equal zero.

The meaning of Kirchhoff's voltage law may not seem clear at first, but it can be explained easily using the circuit in Figure 5.13. According to Kirchhoff's voltage law, the sum of the voltages shown must equal zero. By formula,

$$V_S + V_1 + V_2 = 0 \text{ V}$$

In Figure 5.13, $V_S = 10$ V, $V_1 = 2$ V, and $V_2 = 8$ V. It would seem that the sum of these voltages would be 20 V. However, Kirchhoff's voltage law takes into account the polarity of each voltage. Therefore, we need to assign a polarity to each of the voltages in the circuit to verify the above relationship.

### Assigning Voltage Polarities

*A Practical Consideration:* When working with Kirchhoff's voltage law, we need to distinguish between a voltage drop and a voltage rise. However, most technicians refer to any component voltage as a voltage drop, regardless of whether it is actually a drop or a rise.

To assign voltage polarities, we need to distinguish between a voltage drop and a voltage rise. The current through a given resistor produces a voltage rise across the component, meaning the exit side of the component is more positive than the entry side. For example, look at $R_2$ in Figure 5.13. As the current passes through the component counterclockwise, it goes from one potential to a more positive potential. Therefore, there is a voltage rise across the component. The same holds true for the current through $R_1$. At the same time,

**FIGURE 5.13**

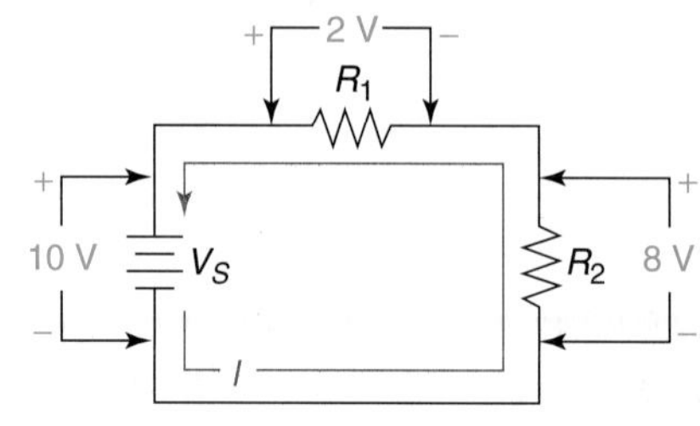

the current through the source experiences a voltage drop because it goes from one potential to a more negative potential.

In any Kirchhoff's loop equation, the polarity assigned to a given voltage is determined by whether it is a drop or a rise, as follows:

1. A voltage rise is represented as a positive value.
2. A voltage drop is represented as a negative value.

Using these guidelines, the voltages around the loop in Figure 5.13 are assigned as follows:

$$V_S = -10\text{ V} \qquad V_1 = +2\text{ V} \qquad V_2 = +8\text{ V}$$

If we put these values into our Kirchhoff's loop equation, we get

$$V_S + V_1 + V_2 = 0\text{ V}$$
$$-10\text{ V} + 2\text{ V} + 8\text{ V} = 0\text{ V}$$

which is true. Note that the result is the same no matter where we start in the circuit.

If we had taken the polarities of the voltages in Figure 5.13 into account when writing the Kirchhoff's loop equation for the circuit, the equation would have been written as

$$-V_S + V_1 + V_2 = 0\text{ V}$$

Adding $V_S$ to both sides of the equation gives us

$$V_1 + V_2 = V_S$$

Thus, Kirchhoff's voltage law is just another way of saying that the sum of the component voltages in a closed loop (series circuit) must equal the source voltage. The following example illustrates a practical application of this law.

### *EXAMPLE 5.7*

Find the value of $R_L$ in Figure 5.14.

***Solution:*** We know that the sum of the component voltages must equal the source voltage. Therefore, the voltage across $R_1$ must equal the difference between the source voltage (15 V) and the load voltage (6 V), as follows:

$$\mathbf{V_1} = V_S - V_L = 15\text{ V} - 6\text{ V} = \mathbf{9\text{ V}}$$

Since current is the same at all points in a series circuit, the value of $I_T$ can be found using the values of $V_1$ and $R_1$, as follows:

$$\mathbf{I_T} = \frac{V_1}{R_1} = \frac{9\text{ V}}{2.2\text{ k}\Omega} = \mathbf{4.09\text{ mA}}$$

Now that we know the values of $V_L$ and $I_T$, the resistance of the load can be found as

$$\mathbf{R_L} = \frac{V_L}{I_T} = \frac{6\text{ V}}{4.09\text{ mA}} = \mathbf{1.47\text{ k}\Omega}$$

***PRACTICE PROBLEM 5.7***

A circuit like the one in Figure 5.14 has the following values: $V_S = 22$ V, $V_L = 3$ V, and $R_1 = 8.2$ kΩ. Calculate the value of the circuit current.

**FIGURE 5.14**

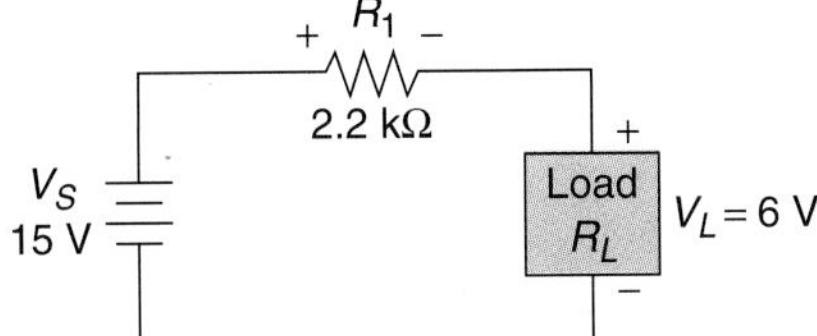

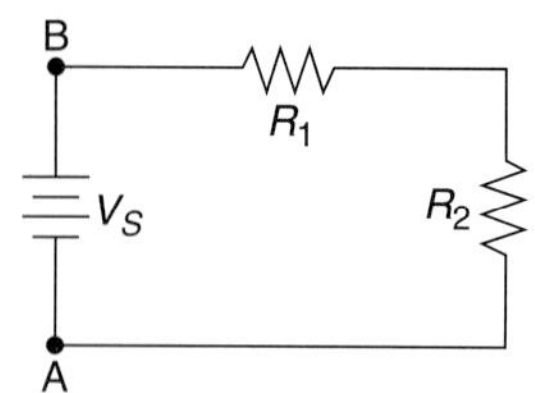


EWB

**FIGURE 5.15**

As you have seen, there are two forms of Kirchhoff's voltage law. The formal form states that the sum of the voltages around a closed loop must equal zero. The informal form states that the sum of the component voltages in a series circuit must equal the applied voltage. As you will learn, the informal form of Kirchhoff's law is used far more often than the formal form.

## Voltage References

Time and again, the topic of polarity has come up in our discussions on circuit voltages. In Chapter 3, it was stated that we must agree on a reference point if the polarity of a given voltage is to be agreed on. For example, consider the circuit in Figure 5.15. Is the source voltage positive or negative? If we reference the source voltage (point B) to point A, we say that the source voltage is positive with respect to the reference. If we reference the source voltage (point A) to point B, we say that the source voltage is negative with respect to the reference.

*OBJECTIVE 7* ▶

*Reference:* The topic of ground is discussed further in Section 5.4.

In any circuit, there is a point that serves as the reference for all voltages in the circuit. This point is referred to as *ground* (or *common*) and is designated by the symbol shown in Figure 5.16a. Note that ground is considered to be the 0 V point in a given circuit, and that all voltages in the circuit are referenced to that point. For example, the negative sides of the source and the resistor in Figure 5.16b are both connected to ground. Thus, $V_S$ and $V_1$ are both positive with respect to ground (the reference). The positive sides of the source and the resistor in Figure 5.16c are connected to ground, so $V_S$ and $V_1$ are both negative with respect to ground.

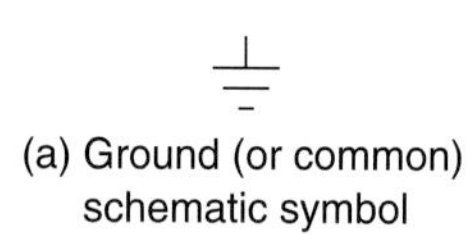
(a) Ground (or common) schematic symbol

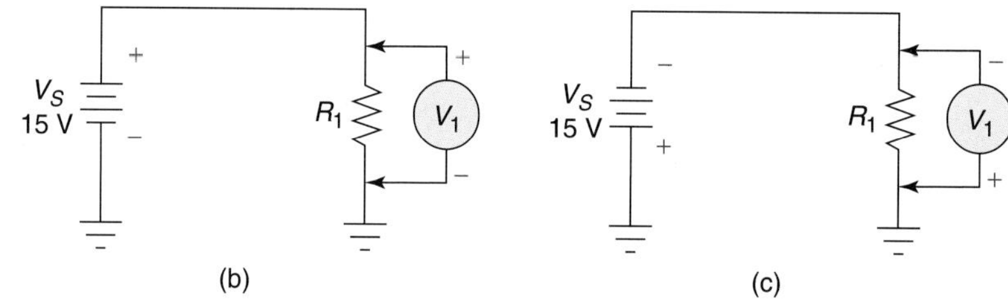

EWB

**FIGURE 5.16**

When the ground symbol appears more than once in a circuit, it indicates that the various points are physically connected, as shown in Figure 5.17. Thus, the ground symbols in a given schematic indicate that two or more points are:

1. Considered to be at a 0 V potential.
2. Physically connected to each other.

Note that the ground connection throughout a circuit is often referred to as a **ground-loop path.**

**Ground-loop path**
The conducting path between the ground connections throughout a circuit.

**FIGURE 5.17**

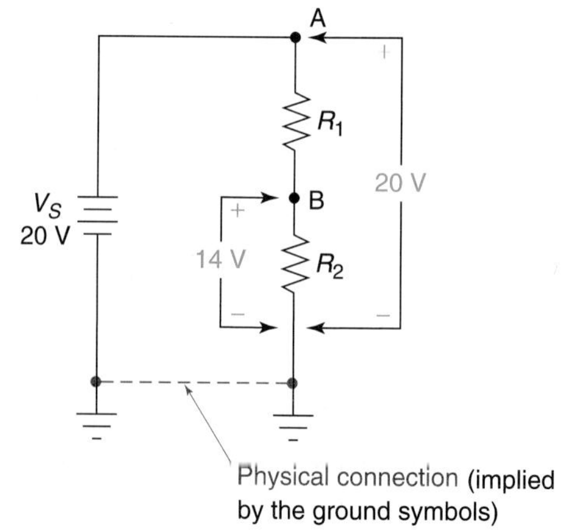

In the circuit shown in Figure 5.17, the voltage from point A to ground is equal to the source voltage, 20 V. The voltage from point B to ground is equal to 14 V. Generally, we say that the 6 V difference between the voltages at points A and B is dropped across $R_1$. The remaining 14 V is dropped across $R_2$. (*Remember:* The term *voltage drop* is commonly used to describe a change from one potential to a lower potential.)

## The Voltage Divider Relationship

The term *voltage divider* is often used to describe a series circuit because the source voltage is divided among the components in the circuit. For example, consider the circuits shown in Figure 5.18. In each case, the source voltage is divided among the resistors in the circuit. As is always the case, the sum of the resistor voltages in each series circuit is equal to the source voltage.

If you look closely at the circuits in Figure 5.18, you'll see that the ratio of any resistor voltage to the source voltage is equal to the ratio of that resistor's value to the total resistance. For example, in Figure 5.18a, $R_1$ is shown to be 1 kΩ. The total resistance in the circuit is equal to the sum of the two resistors, 2 kΩ. Thus, $R_1$ accounts for one-half of the total circuit resistance, and it is dropping one-half of the source voltage. By the same token:

1. In Figure 5.18b, $R_1$ (1 kΩ) accounts for one-third of the total circuit resistance (3 kΩ), and it is dropping one-third of the source voltage.
2. In Figure 5.18c, $R_1$ (1 kΩ) accounts for one-fourth of the total circuit resistance (4 kΩ), and it is dropping one-fourth of the source voltage.

As you can see, the ratio of $V_1$ to $V_S$ is equal to the ratio of $R_1$ to $R_T$ in each of the circuits shown in Figure 5.18. This relationship, which exists for all series circuits, is stated mathematically as follows:

$$\frac{V_n}{V_S} = \frac{R_n}{R_T} \tag{5.4}$$

where $R_n$ = the resistor of interest

$V_n$ = the voltage across $R_n$ (where $n$ is the component number)

The validity of equation (5.4) is demonstrated further in Figure 5.19. Here we have two circuits along with their calculated values of current, $V_1$ and $V_2$. As you can see, the values of $V_1$ and $V_2$ for the two circuits are equal, even though the actual resistor values are different. This is because the voltage values are determined by the resistance ratios rather than by the resistor values themselves. In both circuits, $R_1$ accounts for three-fourths of the total resistance and thus drops three-fourths of the source voltage.

*An Important Point:* The resistance and voltage ratios for the circuits are equal. However, the current in Figure 5.19b is significantly lower than the current in Figure 5.19a.

If we multiply both sides of equation (5.4) by $V_S$, we get the following useful equation:

◄ *OBJECTIVE 8*

$$V_n = V_S\frac{R_n}{R_T} \tag{5.5}$$

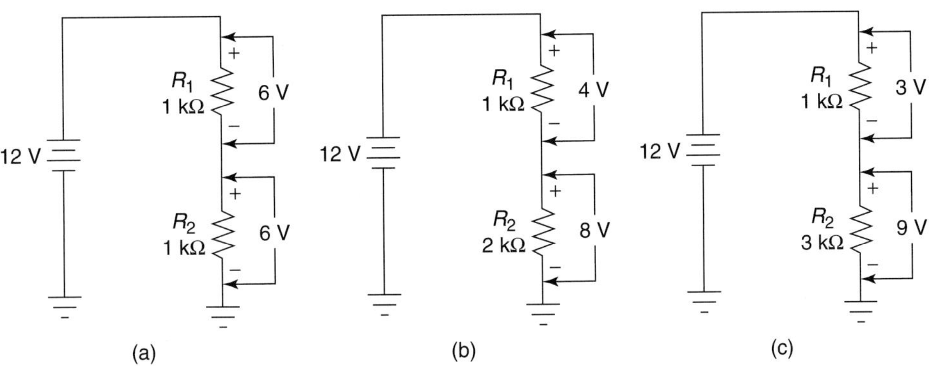


**FIGURE 5.18**

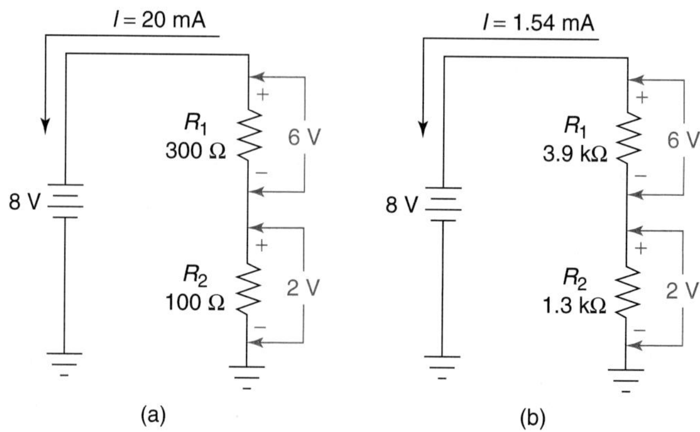


**FIGURE 5.19**

Equation (5.5), which is commonly referred to as the *voltage-divider equation,* allows us to calculate the voltage drop across a resistor without first calculating the value of the circuit current. The use of the voltage-divider equation is demonstrated in the following example.

## EXAMPLE 5.8

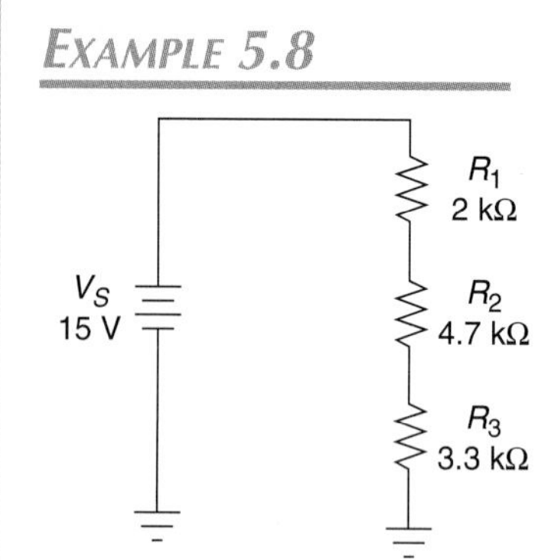


**FIGURE 5.20**

Determine the voltage across $R_3$ in Figure 5.20.

***Solution:*** Since we are trying to find the voltage across $R_3$, we will use $V_3$ and $R_3$ in place of $V_n$ and $R_n$ in the voltage-divider equation, as follows:

$$V_3 = V_S\frac{R_3}{R_T} = (15\text{ V})\frac{3.3\text{ k}\Omega}{10\text{ k}\Omega} = (15\text{ V})(0.33) = \mathbf{4.95\text{ V}}$$

***PRACTICE PROBLEM 5.8***

A circuit like the one in Figure 5.20 has values of $V_S = 12$ V, $R_1 = 1$ kΩ, $R_2 = 2.2$ kΩ, and $R_3 = 3.3$ kΩ. Using the voltage-divider equation, calculate the value of $V_2$.

Had we not used the voltage-divider equation, the problem in Example 5.8 would have been solved as follows:

$$I_T = \frac{V_S}{R_T} = \frac{15\text{ V}}{10\text{ k}\Omega} = \mathbf{1.5\text{ mA}}$$

and

$$V_3 = I_T R_3 = (1.5\text{ mA})(3.3\text{ k}\Omega) = \mathbf{4.95\text{ V}}$$

As you can see, we obtained the same result with fewer steps using the voltage-divider equation.

The following example demonstrates how the voltage-divider equation can be used, with Kirchhoff's voltage law, to calculate the values of the voltage drops in a circuit.

## EXAMPLE 5.9

Determine the values of $V_1$ and $V_2$ for the circuit in Figure 5.21.

***Solution:*** We'll start by calculating the value of $V_1$, as follows:

$$V_1 = V_S\frac{R_1}{R_T} = (14\text{ V})\frac{10\text{ k}\Omega}{25\text{ k}\Omega} = (14\text{ V})(0.4) = \mathbf{5.6\text{ V}}$$

Since the sum of the voltages must equal the source voltage, $V_2$ can be found as the difference between $V_S$ and $V_1$, as follows:

$$V_2 = V_S - V_1 = 14\text{ V} - 5.6\text{ V} = \mathbf{8.4\text{ V}}$$

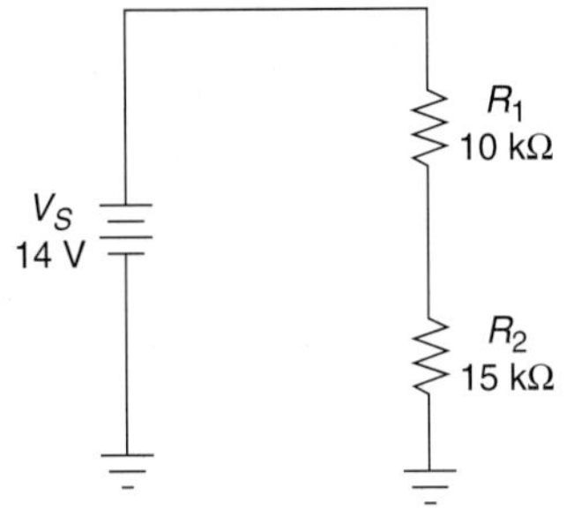


FIGURE 5.21

***PRACTICE PROBLEM 5.9***

A circuit like the one in Figure 5.21 has values of $V_S = 20$ V, $R_1 = 33$ kΩ, and $R_2 = 47$ kΩ. Determine the values of $V_1$ and $V_2$ for the circuit using the voltage-divider equation and Kirchhoff's voltage law.

Several points should be made regarding Example 5.9:

1. We could have used the voltage-divider equation to find $V_2$ and Kirchhoff's voltage law to find $V_1$. The results would have been the same. (Try it!)
2. Had we used the previously established method of finding $V_1$ and $V_2$, we would have started by calculating the values of $R_T$ and $I_T$. Multiplying each resistor value by $I_T$ would have given us the same values of $V_1$ and $V_2$ found in the example. The combination of the voltage-divider equation and Kirchhoff's voltage law has allowed us to determine the resistor voltages with less effort.

The voltage-divider equation can also be used to find the voltage across a group of resistors, as demonstrated in the following example.

## EXAMPLE 5.10

For the circuit shown in Figure 5.22, determine the value of the voltage from point A to ground ($V_A$).

***Solution:*** According to Kirchhoff's voltage law, the voltage from point A to ground must equal the sum of $V_3$ and $V_4$. Thus,

$$R_n = R_3 + R_4 = \mathbf{3\text{ k}\Omega}$$

Using this value in the voltage-divider equation, the value of $V_A$ is found as:

$$V_A = V_S\frac{R_n}{R_T} = (12\text{ V})\frac{3\text{ k}\Omega}{15\text{ k}\Omega} = \mathbf{2.4\text{ V}}$$

*Avoiding a Common Mistake: $V_A$ in Figure 5.22 equals the sum of the voltages between point A and ground ($V_3 + V_4$). It does not necessarily equal the sum of ($V_1 + V_2$). Make sure you identify the voltages of interest correctly when using the voltage divider equation.*

***PRACTICE PROBLEM 5.10***

A circuit like the one in Figure 5.22 has the following values: $V_S = 18$ V, $R_1 = 2.2$ kΩ, $R_2 = 4.7$ kΩ, $R_3 = 3.3$ kΩ, and $R_4 = 10$ kΩ. Determine the voltage from point A to ground.

FIGURE 5.22

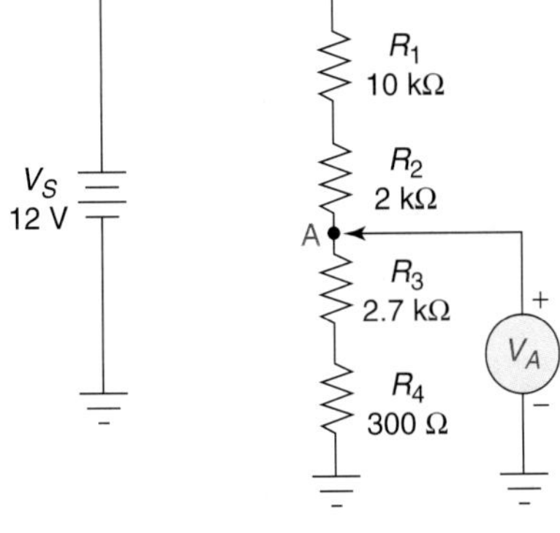

Again, had we used the previously established approach to the problem, we would have calculated the values of $R_T$, $I_T$, $V_3$, and $V_4$. Then we would have added the values of $V_3$ and $V_4$ to obtain the result found in Example 5.10. Once again, the voltage-divider equation has provided a more efficient method of solving the problem.

## One Final Note

In this section, we have established two extremely important relationships that apply to almost every series circuit: Kirchhoff's voltage law and the voltage-divider equation. As you will learn, these relationships are used almost as commonly as Ohm's law.

### Section Review

1. What does Kirchhoff's voltage law state?
2. When setting up a Kirchhoff's loop equation, how do you determine the polarity of each of the voltages?
3. What is the informal form of Kirchhoff's voltage law?
4. What is ground? What is its symbol?
5. What is indicated by the appearance of the ground symbol at two or more points in a circuit?
6. What is meant by the term *voltage drop?*
7. What is the voltage-divider relationship?
8. How does the voltage-divider equation simplify circuit voltage calculations?

# 5.3 SERIES CIRCUIT ANALYSIS, FAULT SYMPTOMS, AND TROUBLESHOOTING

At this point, you have been exposed to the basic operating characteristics of the series circuit. In this section, we will look at the complete analysis of a series circuit. We will also discuss the characteristic symptoms of several circuit faults and a basic approach to series circuit troubleshooting.

## Series Circuit Analysis

*OBJECTIVE 9* ▶ The complete analysis of a series circuit involves determining the values of $R_T$, $I_T$, and $P_T$, along with the resistor voltage and power values. The following example demonstrates the complete analysis of a series circuit.

### EXAMPLE 5.11

Perform the complete analysis of the circuit shown in Figure 5.23.

***Solution:*** First, the total resistance in the circuit is found as

$$R_T = R_1 + R_2 + R_3 = \mathbf{500\ \Omega}$$

Now, the total circuit current is found as

$$I_T = \frac{V_S}{R_T} = \frac{12\text{ V}}{500\ \Omega} = \mathbf{24\ mA}$$

The total power drawn from the source can now be found as

$$P_S = V_S I_T = (12\text{ V})(24\text{ mA}) = \mathbf{288\ mW}$$

We could use the voltage-divider equation to determine the voltage across each resistor. However, since we already know the value of the circuit current, we'll use Ohm's law, as follows:

$$\mathbf{V_1} = I_T R_1 = (24\text{ mA})(120\ \Omega) = \mathbf{2.88\ V}$$
$$\mathbf{V_2} = I_T R_2 = (24\text{ mA})(200\ \Omega) = \mathbf{4.8\ V}$$
$$\mathbf{V_3} = I_T R_3 = (24\text{ mA})(180\ \Omega) = \mathbf{4.32\ V}$$

The individual power dissipation values for the resistors can now be found as:

$$\mathbf{P_1} = V_1 I_T = (2.88\text{ V})(24\text{ mA}) = \mathbf{69.12\ mW}$$
$$\mathbf{P_2} = V_2 I_T = (4.8\text{ V})(24\text{ mA}) = \mathbf{115.2\ mW}$$
$$\mathbf{P_3} = V_3 I_T = (4.32\text{ V})(24\text{ mA}) = \mathbf{103.68\ mW}$$

This completes the analysis of the circuit.

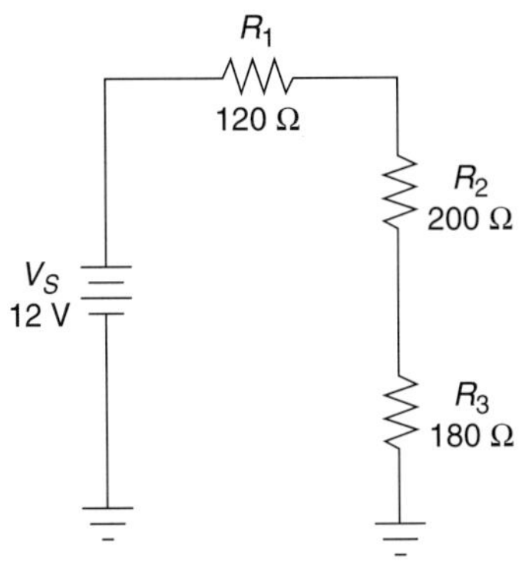


**FIGURE 5.23**

*An Alternate Approach*: The individual power values can be determined using a power divider equation that is similar to the voltage-divider equation. This power divider equation is

$$P_n = P_S \frac{R_n}{R_T}$$

***PRACTICE PROBLEM 5.11***

A circuit like the one in Figure 5.23 has the following values: $V_S = 13$ V, $R_1 = 330\ \Omega$, $R_2 = 220\ \Omega$, and $R_3 = 100\ \Omega$. Perform the complete analysis of the circuit.

The component voltage and power dissipation values found in the example can be verified by comparing their sums to the source voltage and total power, as follows:

$$\mathbf{V_1 + V_2 + V_3} = 2.88\text{ V} + 4.8\text{ V} + 4.32\text{ V} = \mathbf{12\ V}$$

and

$$\mathbf{P_1 + P_2 + P_3} = 69.12\text{ mW} + 115.2\text{ mW} + 103.68\text{ mW} = \mathbf{288\ mW}$$

Since $V_S = 12$ V and $P_T = 288$ mW (as found in the example), we have verified that the component voltage and power dissipation values are correct.

## *Open-Resistor Fault Symptoms*

◄ *OBJECTIVE 10*

When one resistor in a series circuit opens, the results of the open are as follows:

**1.** The total circuit current drops to zero.

**2.** The total source voltage is dropped across the open component.

These two symptoms can be explained with the help of Figure 5.24, where $R_2$ is shown as being open. As you know, there must be a complete path between the terminals of the voltage source to have any current. Since the path in the circuit has been broken, $I_T = 0$ A.

*Remember:* A series circuit has only one current path. Therefore, an open component prevents current throughout the entire circuit.

Ohm's law tells us that

$$V = IR$$

Since the current in Figure 5.24 is 0 A, the value of $V_1$ is found as

$$\mathbf{V_1} = I_T R_1 = (0\text{ A})(1\text{ k}\Omega) = \mathbf{0\ V}$$

Since $V_1 = 0$ V, Kirchhoff's voltage law tells us that all the source voltage must fall across the open component, as follows:

$$\mathbf{V_2} = V_S - V_1 = 12\text{ V} - 0\text{ V} = \mathbf{12\ V}$$

**FIGURE 5.24**

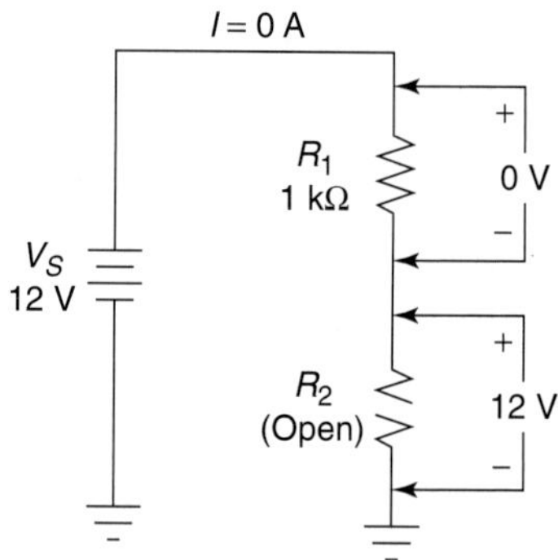


Thus, the voltage across an open component in a series circuit is equal to the source voltage. This holds true no matter how many resistors are in the circuit (assuming there is only *one* open component).

Here's another way of looking at it: picture the value of $R_2$ increasing in increments to a maximum value of $\infty\ \Omega$. As the value of $R_2$ is increased, the voltage drop across the component increases, as given by the relationship

$$V_2 = V_S\frac{R_2}{R_T}$$

The higher the value of $R_2$, the greater the value of $V_2$. As the value of $R_2$ approaches infinity, the value of $V_R$ approaches the value of $V_S$. Therefore, all the source voltage is measured across $R_2$ when the component's value reaches $\infty\ \Omega$.

You may be wondering why, with 0 A in the circuit, we cannot calculate a value of

$$\mathbf{V_2} = I_T R_2 = (0\text{ A})(R_2) = \mathbf{0\ V}$$

In mathematics, we are taught that

$$(a \times 0) = 0 \qquad \text{and} \qquad (a \times \infty) \text{ is undefined}$$

Since $I_T R_2 = (0\text{ A})(\infty\ \Omega)$, it is (by definition) undefined, meaning that it has no solution. This is why Ohm's law cannot be used to determine the value of $V_2$ for the circuit when $R_1$ is open.

## *Measuring the Voltage across an Open Component: A Practical Consideration*

*OBJECTIVE 11* ▶ When you measure the voltage across an open resistor in a series circuit with a voltmeter, the measured voltage may be slightly lower than the source voltage. The reason for this is illustrated in Figure 5.25.

When a voltmeter is connected across an open component (as shown in Figure 5.25), the meter completes the series circuit. As a result, some amount of current is present in the

**FIGURE 5.25**

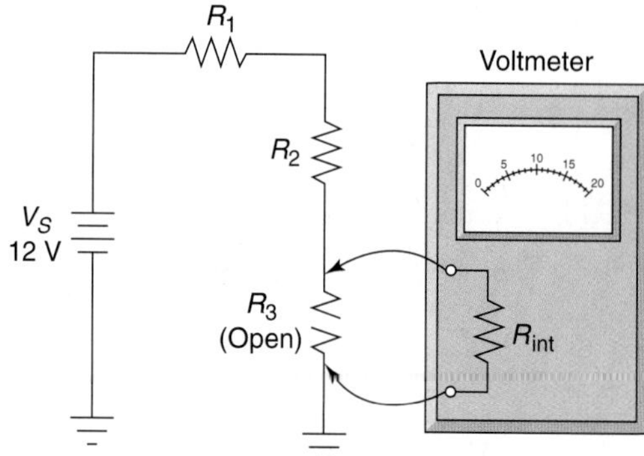

circuit, causing a voltage to be dropped across each of the other resistors. Since $V_1$ and $V_2$ (in this case) each have some measurable value, the voltage across the open is slightly less than the source voltage.

How much less? That depends on the values of the resistors in the circuit and the internal resistance of the meter. When the internal resistance of the meter ($R_{int}$) is much greater than the total of the other resistor values in the circuit, the presence of the meter has almost no effect on the measured voltage across the open. This point is illustrated in the following example.

## EXAMPLE 5.12

◀ *OBJECTIVE 12*

The voltage across the open resistor in Figure 5.26a is measured with a DMM that has an internal resistance ($R_{int}$) of 10 MΩ (a typical value). Determine the meter reading.

***Solution:*** When the meter is connected to the circuit, the circuit is completed. However, the total circuit is now equal to the sum of $R_1$ and the internal resistance of the meter, as follows:

$$R_T = R_1 + R_{int} = 10\ \text{k}\Omega + 10\ \text{M}\Omega = \mathbf{10.01\ M\Omega}$$

*Note:* Since $R_2$ is open, its rated value is not involved in the calculation of total resistance for the circuit when the meter is inserted.

With a total resistance of 10.01 MΩ, the voltage across $R_1$ can be found as

$$V_1 = V_S \frac{R_1}{R_T} = (10\ \text{V}) \frac{10\ \text{k}\Omega}{10.01\ \text{M}\Omega} = \mathbf{9.99\ mV}$$

With 9.99 mV dropped across $R_1$, the measured voltage across the open component is found as

$$V_2 = V_S - V_1 = 10\ \text{V} - 9.99\ \text{mV} = \mathbf{9.99\ V}$$

In this case, placing the voltmeter across the open resistor causes a very slight error in the reading.

***PRACTICE PROBLEM 5.12***

A circuit like the one in Figure 5.26a has values of $V_S = 20$ V and $R_1 = 12$ kΩ. Assuming that $R_2$ has opened, predict the value of $V_2$ that would be measured using a voltmeter with an internal resistance of 10 MΩ.

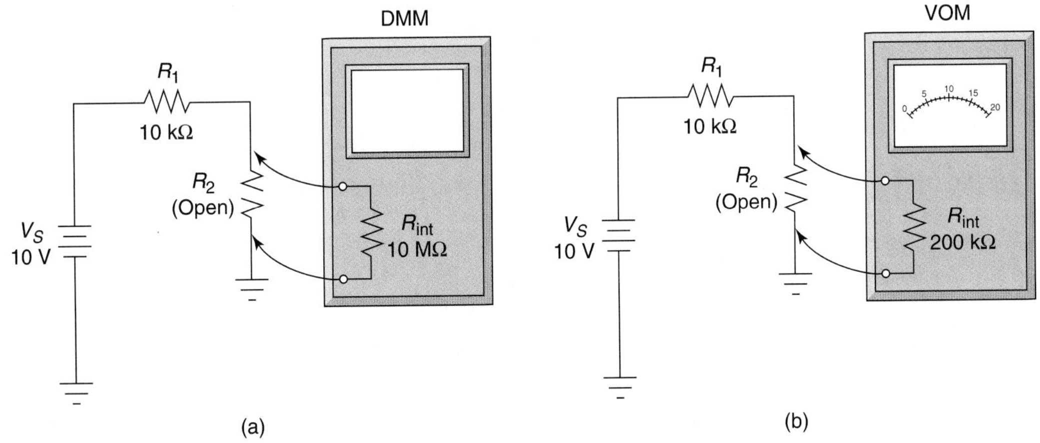

**FIGURE 5.26**

When a voltmeter with a relatively low internal resistance is used, the effect of connecting the meter across the open becomes more noticeable. This point is illustrated in the following example.

## EXAMPLE 5.13

The voltage across the open component in Figure 5.26b is measured with a VOM that has an internal resistance of 200 kΩ. Determine the value that will be read by the meter.

***Solution:*** With the VOM connected, the total circuit resistance is found as the sum of $R_1$ and the internal resistance of the meter, as follows:

$$R_T = R_1 + R_{\text{int}} = 10\text{ k}\Omega + 200\text{ k}\Omega = \mathbf{210\text{ k}\Omega}$$

*Note:* As was the case in Example 5.12, the rated value of $R_2$ is ignored in the calculation of total resistance because the component is open.

Using this value of $R_T$, the voltage across $R_1$ can now be found as

$$V_1 = V_S \frac{R_1}{R_T} = (10\text{ V}) \frac{10\text{ k}\Omega}{210\text{ k}\Omega} = \mathbf{476\text{ mV}}$$

Finally, the voltage measured across $R_2$ is found as the difference between $V_1$ and $V_S$, as follows:

$$V_2 = V_S - V_1 = 10\text{ V} - 476\text{ mV} = \mathbf{9.52\text{ V}} \quad \text{(approximately)}$$

While the error caused in this case is not critical, it is significantly greater than the one found in Example 5.12.

***PRACTICE PROBLEM 5.13***

Refer back to Practice Problem 5.12. Determine the meter reading if the voltmeter used has an internal resistance of 200 kΩ.

When taking voltage measurements, you should keep several points in mind:

1. The higher the internal resistance of a voltmeter, the more accurate the readings it will provide.
2. The internal resistance of a typical DMM is much higher than that of a typical VOM. For this reason, DMMs are usually preferred over VOMs for voltage readings.

You should also keep in mind the fact that connecting a voltmeter across an open component will complete the circuit and, in most cases, cause an error in the reading across the open component. (In fact, the act of taking any measurement has some effect on a circuit because the measuring device isn't part of the original circuit.)

## Multiple Opens

*OBJECTIVE 13* ▶ It is not uncommon for a fault in a series circuit to cause more than one component to open. When this happens:

1. The circuit current drops to zero, as always.
2. The source voltage is dropped across the entire circuit that lies between the most negative open terminal and the most positive open terminal.

For example, consider the circuit shown in Figure 5.27. In this case, $R_2$ and $R_4$ have both opened. As a result, the source voltage would be measured from the positive side of $R_2$ to the negative side of $R_4$. The reason for this is the fact that no voltage is dropped across $R_1$ or $R_5$ (since there is no current through either resistor). As a result, the sum of $V_2$, $V_3$, and $V_4$ must equal the supply voltage, in keeping with Kirchhoff's voltage law. The applied voltage is felt across the entire open circuit rather than across any one component.

Here's an interesting point: if we were to connect a voltmeter across the individual resistors in Figure 5.27, we would measure

$$V_2 \cong 0\text{ V} \qquad V_3 \cong 0\text{ V} \qquad V_4 \cong 0\text{ V}$$

FIGURE 5.27

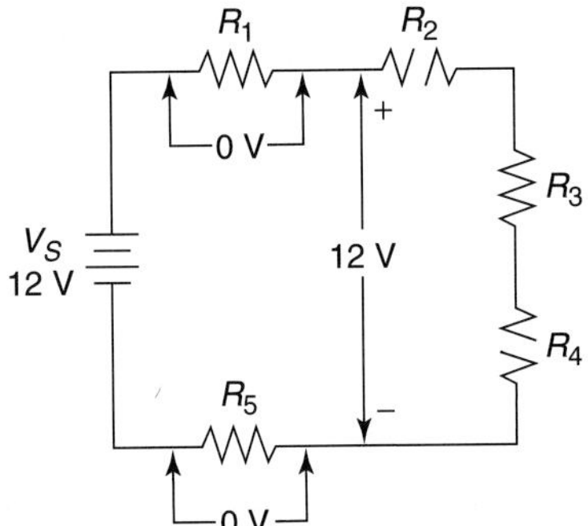


Adding these measured values together, we'd get $V_2 + V_3 + V_4 \cong 0$ V. This would indicate a conflict between the real and theoretical voltages in the circuit.

The problem in this case is the fact that a voltmeter must make a direct (or indirect) connection to the source with *both* leads to provide an accurate measurement. When we measure $V_2$, $V_3$, or $V_4$, one (or both) of the meter leads is isolated from the source by an open component. Therefore, each meter reading is meaningless. Only when measuring $(V_2 + V_3 + V_4)$ are both leads connected (indirectly) to the source, providing a meaningful reading.

## Shorted-Resistor Fault Symptoms

◄ *OBJECTIVE 14*

When a resistor in a series circuit is shorted, the results of the short are as follows:

1. The voltage across the shorted component drops to zero.
2. The total circuit resistance decreases, causing a proportional increase in the total circuit current.
3. The increase in circuit current results in a proportional increase in the voltage drops across the remaining resistors.

*Remember: I* is inversely proportional to *R*.

These characteristics are illustrated in Figure 5.28.

In Figure 5.28a, the normal circuit conditions are shown. In Figure 5.28b, $R_2$ is shown to be shorted. Assuming that $R_2 \cong 0\ \Omega$, the total circuit resistance drops from 4 kΩ to 2 kΩ. This drop in resistance causes the circuit current to double. The increase in circuit current causes both $V_1$ and $V_3$ to increase to 4 V. As you can see, the presence of a shorted

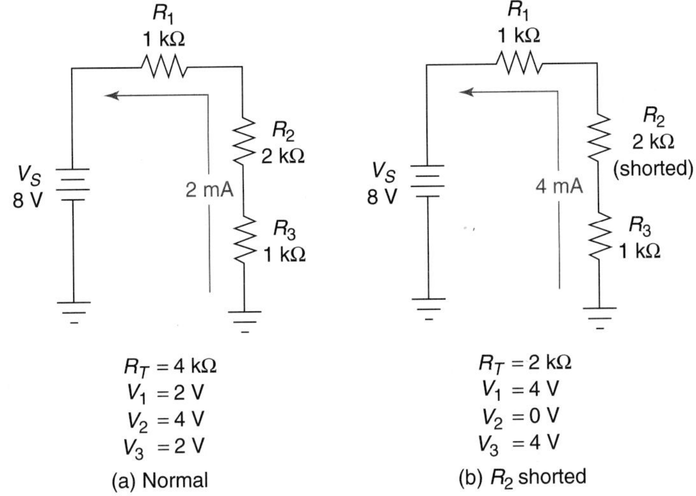


FIGURE 5.28

component has caused a decrease in the total resistance, an increase in the circuit current, and an increase in the voltage drops across the remaining components. At the same time, the voltage across the shorted component has dropped to 0 V.

It should be noted that a short in any circuit can result in other components being damaged from excessive current. For example, the increase in circuit current could cause:

1. Another component to burn open (if $I^2R$ exceeds the power rating of the component).
2. A circuit fuse (if present) to blow.

If a faulty circuit contains a burned component or an open fuse, there's a good chance that there is a shorted component somewhere in the circuit. When either of these conditions exists, the other components in the circuit should be tested before the circuit is assumed to be repaired.

## Component Aging and Stress

*OBJECTIVE 15* ▶ A shorted or open component is often referred to as a *catastrophic failure* because it is the complete (and sometimes violent) failure of the component. However, not all circuit failures are catastrophic in nature.

In many cases, the operation of a circuit can be affected to a lesser degree by the partial failure of an old or stressed component. For example, consider the circuit shown in Figure 5.29. In this case, $V_1$ should be 1 V, $V_2$ should be 10 V, and $V_3$ should be 15 V, as indicated by the source voltage and the rated values of the resistors. However, $V_1$ and $V_3$ are both higher than normal and $V_2$ is lower than normal. In this case, age and/or stress has caused the value of $R_2$ to decrease, even though the component hasn't failed completely. In a case like this, a series of resistance measurements may be required to determine which resistor is the cause of the problem.

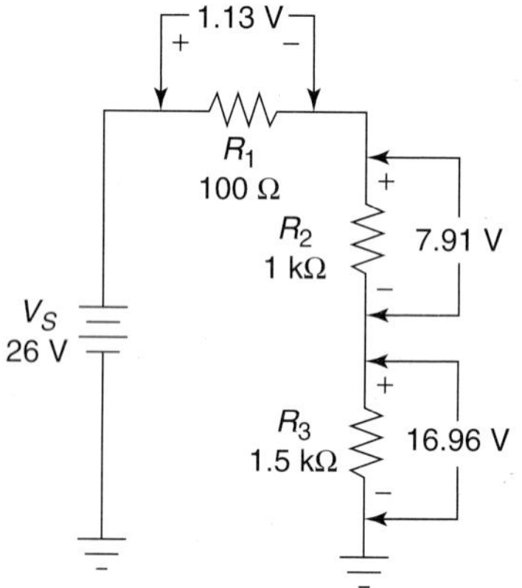


**FIGURE 5.29 Faulty voltage readings caused by a change in the value of $R_2$.**

## Series Circuit Troubleshooting

As you know, troubleshooting is the process of locating and repairing one or more faults in a circuit. In many cases, circuit faults can be detected using your senses (sight, smell, hearing, and touch). In many other cases, a given circuit must be tested using standard test equipment to locate any faults. Let's start by identifying some basic resistor faults that can usually be detected with your senses.

When a resistor opens, it is normally the result of excessive current and the heat it produces. Therefore, an open resistor will often show signs of excessive heat, such as:

1. Extremely dark color bands.
2. One or more black patches (the result of burning).
3. A crack in the body of the component.

When you see any of these signs, odds are that the component has burned open.

When the value of a resistor slowly decreases (due to component aging or stress), the current through the component slowly increases. This causes an increase in the amount of power that the component is dissipating, which increases the stress and eventually leads to complete failure of the component. However, even before the resistor fails completely, there may be signs that it is going bad:

1. The circuit containing the resistor may operate erratically (or outside of its design specifications): sometimes it may work and other times it may not.
2. The component may smell as if it is beginning to burn.
3. The component may be unusually hot.
4. The component may be discolored by excessive heat.

When a circuit begins to operate erratically or smell as if it is burning, the component is probably in the last stages of failing. The failing component will usually be extremely hot to the touch. If you find a component that is unusually hot, remove it from the circuit and check its resistance. *A word of caution: Before touching any circuit components, always remove power from the circuit to prevent any possibility of shock. After removing power, touch the component body (not the leads) lightly and quickly to protect yourself from getting burned.*

When there are no physical signs of damage to a circuit, any component failures must be diagnosed with the use of your test equipment. In most series dc circuits, any circuit faults can be diagnosed with the use of your DMM and your knowledge of how they operated under normal conditions. The following series of troubleshooting applications demonstrates this fact.

*A Word of Caution:* Avoid touching the leads of any component, even when you believe power has been removed from the circuit. Some components (other than resistors) can store a charge even after power has been removed from a circuit. If you touch the leads of any component, you run the risk of receiving an electrical shock from a charge-storing component.

## TROUBLESHOOTING APPLICATION 1

The circuit in Figure 5.30 was tested, giving the results shown in the diagram. As you can see, nearly all the supply voltage is dropped across $R_2$. In this case, the problem is easy to diagnose: $R_2$ is open and must be replaced. Note that the slight difference between $V_2$ and $V_S$ is caused by the internal resistance of the voltmeter (which completed the circuit).

**FIGURE 5.30**

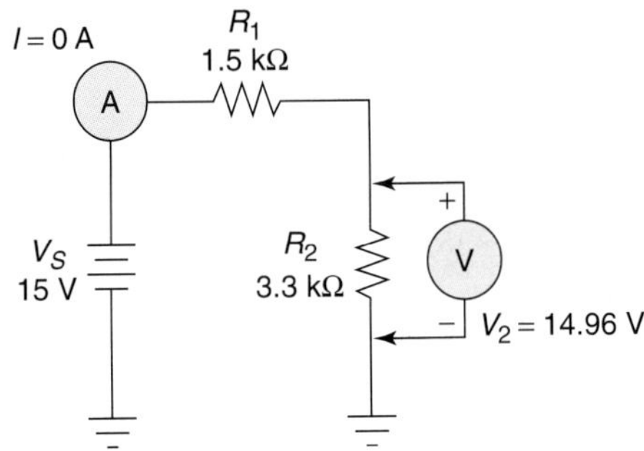


EWB

## TROUBLESHOOTING APPLICATION 2

The circuit in Figure 5.31a was tested, giving the results shown in the diagram. Again, the fact that the source voltage is dropped across a single component indicates that the component has opened. In this case, the faulty component is $R_3$. When $R_3$ is replaced, the circuit is retested, giving the results shown in Figure 5.31b. The 0 V reading across $R_1$ along with the abnormally high readings across $R_2$ and $R_3$ indicate that $R_1$ is shorted. Apparently, $R_1$ shorted and the increased current through $R_3$ caused that component to open.

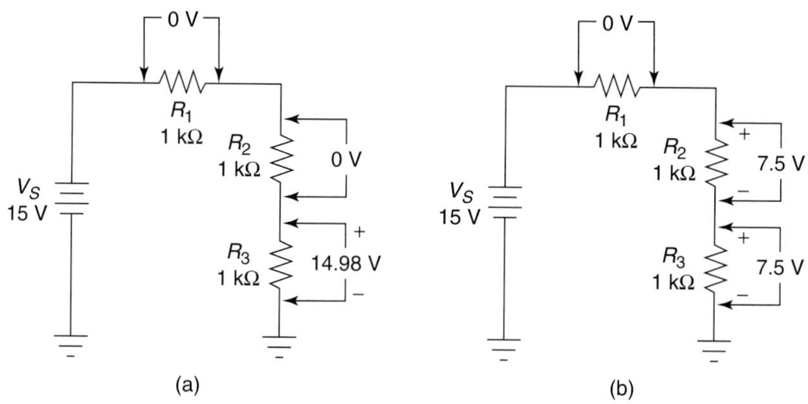


**FIGURE 5.31**

## Troubleshooting Application 3

The fuse in the circuit in Figure 5.32 has opened. When the fuse is replaced and power is applied, the new fuse blows immediately. In this case, there is no safe method of restoring operation to the circuit to take voltage measurements. Therefore, the only logical solution is to measure the circuit resistance. When the resistance is measured, $R_T$ is found to be 15 Ω. Since this reading is 10 Ω less than it should be, the indication is that $R_1$ is shorted. Replacing $R_1$ restores normal operation to the circuit.

*A Practical Consideration:* On many printed circuit boards, there is another tell-tale sign of a shorted component. In many cases, the increase in current (and therefore heat) caused by a short will cause one or more of the copper traces on the board to burn open and/or lift off the board itself. In most cases, a lifted or burned copper trace will lead you directly to the source of the short.

**FIGURE 5.32**

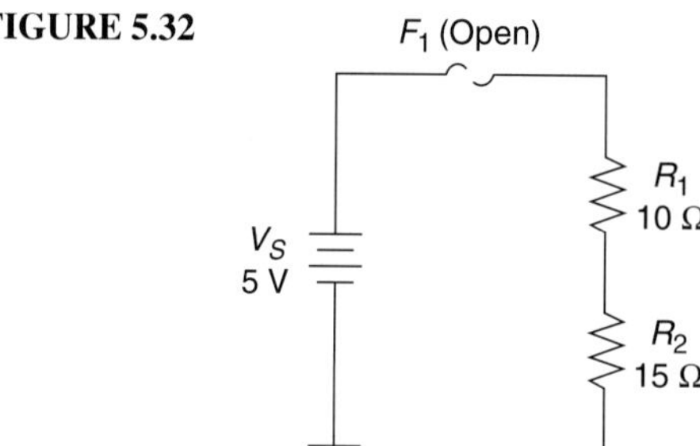


As you can see, troubleshooting involves:

1. Taking the appropriate circuit measurements.
2. Thinking about what those measurements tell you.

### One Final Note

Troubleshooting is a skill. Like any skill, it is not learned by reading but through experience. The troubleshooting cases in this section have been provided to show you some of the basics of troubleshooting. However, your experiences will show you that very few troubleshooting cases are as cut and dry as those presented here.

## Section Review

1. When performing the complete mathematical analysis of a circuit, how can you check your component voltage and power calculations quickly?
2. What are the symptoms of a single open resistor in a series circuit?
3. When a single resistor in a series circuit opens, the source voltage is dropped across the open component. Why?
4. Why is the measured voltage across an open component usually slightly less than the full source voltage?
5. Why are voltmeters with high internal resistance preferred over those with low internal resistance?
6. What are the symptoms associated with multiple opens in a series circuit?
7. What are the fault symptoms associated with a shorted resistor in a series circuit?
8. What can happen to the other components in a series circuit as the result of one component shorting?
9. What are the symptoms normally associated with component aging and stress?
10. What are the physical signs of a component that has been destroyed by excessive heat?
11. What precautions should be taken when touching circuit components?
12. What is involved in troubleshooting?

# 5.4 RELATED TOPICS

In this section, we will complete the chapter by discussing some miscellaneous topics that relate to series circuits.

## The Potentiometer As a Voltage Divider

A single potentiometer can be used as a variable voltage divider when connected as shown in Figure 5.33a. The voltage from the wiper arm to ground is determined by the setting of the potentiometer. For example, the potentiometer in Figure 5.33a is shown to be set so

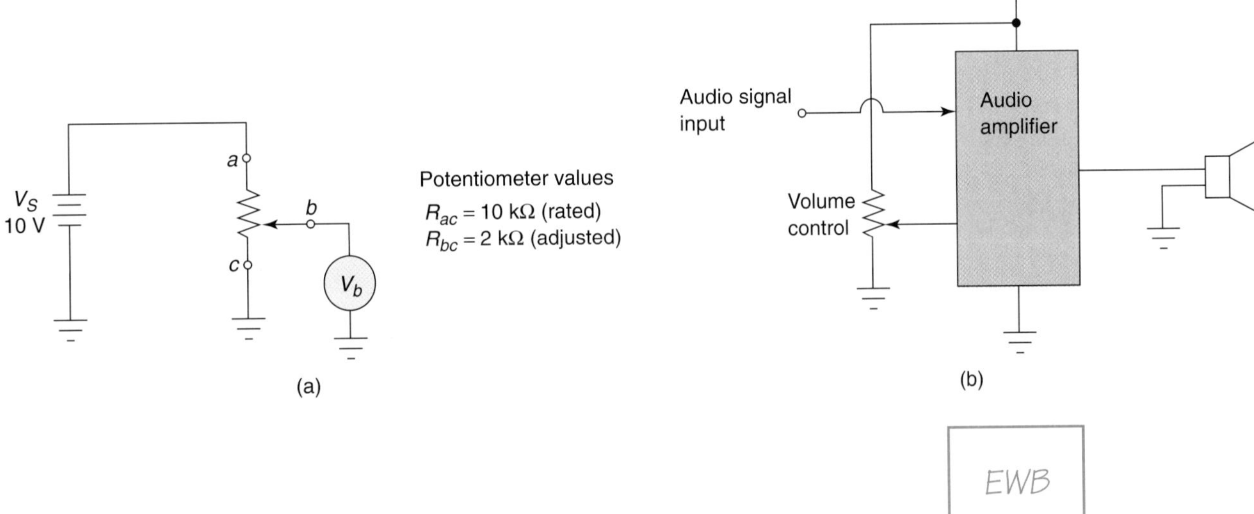


**FIGURE 5.33 The potentiometer as a voltage divider.**

that $R_{bc}$ is 2 kΩ. Since the outer terminal resistance of the potentiometer is 10 kΩ, the value of $V_b$ is found as

$$V_b = V_S\frac{R_{bc}}{R_{ac}} = (10\text{ V})\frac{2\text{ k}\Omega}{10\text{ k}\Omega} = \mathbf{2\text{ V}}$$

If the potentiometer is adjusted so that $R_{bc}$ equals 6 kΩ, $V_b$ changes as follows:

$$V_b = V_S\frac{R_{bc}}{R_{ac}} = (10\text{ V})\frac{6\text{ k}\Omega}{10\text{ k}\Omega} = \mathbf{6\text{ V}}$$

As you can see, a potentiometer wired as shown in Figure 5.33a can be used as a variable voltage divider. One application of such a circuit is shown in Figure 5.33b.

*Reference:* The purpose of this discussion is only to demonstrate an application of the potentiometer as a voltage divider. Amplifier operation is traditionally covered in a course on semiconductor devices.

The block in Figure 5.33b represents an audio amplifier. The purpose of an audio amplifier is to increase the power level (strength) of an audio signal input. The process of increasing the audio signal power is referred to as *amplification.* The volume control on the audio amplifier is a potentiometer. When the potentiometer setting is varied, the wiper arm voltage to the audio amplifier increases (or decreases). By design, the output power from the audio amplifier varies with the setting of the volume control.

## Source Resistance: A Practical Consideration

*OBJECTIVE 16* ▶

**Ideal voltage source**
A voltage source that maintains a constant output voltage regardless of the resistance of its load(s).

**No-load output voltage ($V_{NL}$)**
The output from a voltage source when its terminals are open. The maximum possible output from a voltage source.

The **ideal voltage source** maintains a constant output voltage regardless of the resistance of the load. For example, let's say that the voltage source in Figure 5.34a is an ideal voltage source. As shown in the figure, the voltage across the open terminals of the source is 10 V. Note that this open terminal voltage is referred to as the **no-load output voltage ($V_{NL}$).** When the various load resistances shown in Figure 5.34b are connected to the source, it maintains the same 10 V output. Thus, for the ideal voltage source,

$$V_{NL} = V_L$$

regardless of the value of $R_L$.

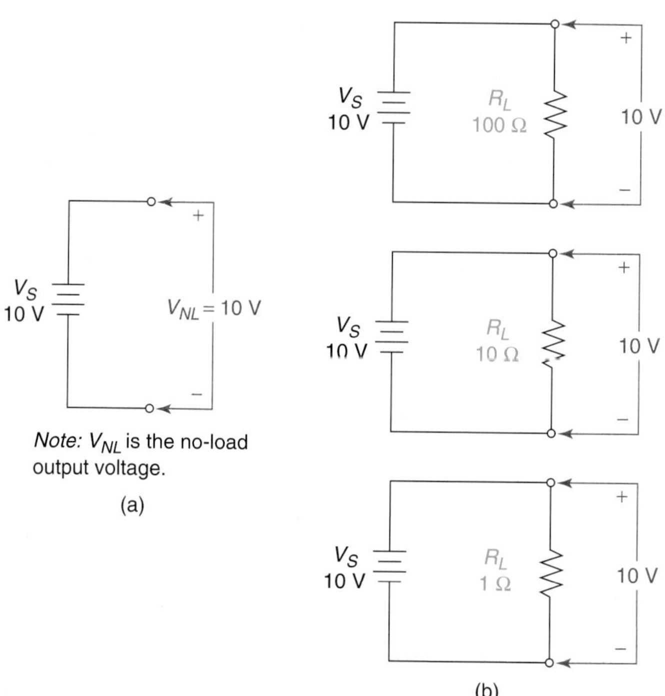


**FIGURE 5.34**

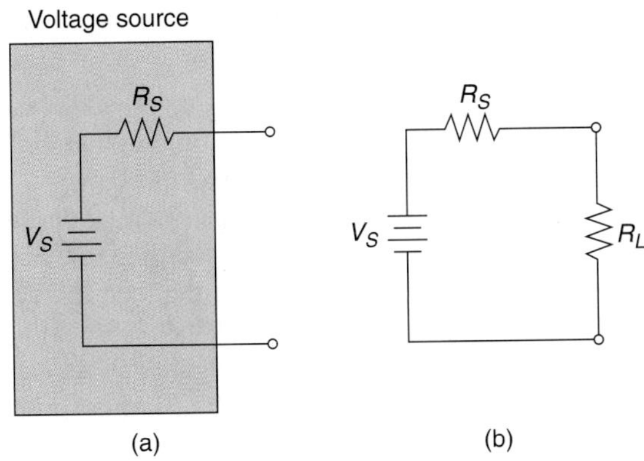


**FIGURE 5.35 Voltage source internal resistance.**

Unfortunately, the ideal voltage source does not exist at this time. For any practical voltage source, a decrease in load resistance results in a decrease in the source voltage. The reason for this lies in the fact that every voltage source has some amount of *internal resistance,* as represented by the resistor ($R_S$) in Figure 5.35a. When a load is connected to the source (as shown in Figure 5.35b), the load forms a voltage divider with the internal resistance of the source. This causes $V_L$ to be lower than the no-load output voltage ($V_{NL}$). This point is demonstrated in the following example.

◄ *OBJECTIVE 17*

## EXAMPLE 5.14

The no-load output voltage of the source in Figure 5.36 is 12 V. Calculate the values of $V_L$ for $R_L = 1\text{ k}\Omega$ and $R_L = 100\ \Omega$.

***Solution:*** The load resistance forms a voltage divider with the internal resistance of the source ($R_S$). When $R_L = 1\text{ k}\Omega$, $V_L$ is found as

$$V_L = V_S\frac{R_L}{R_T} = (12\text{ V})\frac{1\text{ k}\Omega}{1.05\text{ k}\Omega} = \mathbf{11.43\ V}$$

where

$$R_T = R_L + R_S$$

When $R_L = 100\ \Omega$, $V_L$ is found as

$$V_L = V_S\frac{R_L}{R_T} = (12\text{ V})\frac{100\ \Omega}{150\ \Omega} = \mathbf{8\ V}$$

As you can see, the output from the practical voltage source decreases as the load resistance decreases.

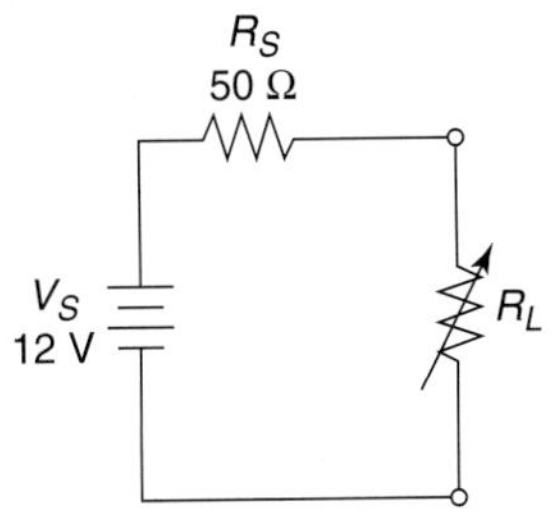

**FIGURE 5.36**

***PRACTICE PROBLEM 5.14***

A voltage source with 40 Ω of internal resistance has a no-load output of 14 V. Determine the values of $V_L$ when $R_L = 1\text{ k}\Omega$ and $R_L = 120\ \Omega$.

The internal resistance of most dc voltage sources is typically 50 Ω or less and thus does not present a major problem for loads in the low kΩ range or higher. However, you should be aware of the existence of source resistance for two reasons:

1. The internal resistance of the source forms a series circuit with the load and therefore can cause a significant drop in output voltage when a low-resistance load is present.
2. The internal resistance of the source determines the maximum amount of power that can be transferred from the source to its load.

At this point, we'll take a look at the relationship among source resistance, load resistance, and maximum load power.

## Maximum Power Transfer

OBJECTIVE 18 ▶

In Chapter 4, you were told that electronic systems are designed for the purpose of delivering power (in one form or another) to a given load. It follows that we would be interested in knowing the maximum possible output power for a given source.

**Maximum power transfer theorem**
Maximum power transfer from a source to its load occurs when the load and source resistances are equal.

The **maximum power transfer theorem** establishes a relationship among source resistance, load resistance, and maximum load power. According to this theorem, maximum power transfer from a source to its load occurs when the load resistance is equal to the source resistance. This means that load power reaches its maximum possible value when $R_L = R_S$. For example, consider the circuit shown in Figure 5.37a. According to the maximum power transfer theorem, the voltage source will transfer maximum power to its load when the potentiometer is set to 10 Ω (the value of $R_S$).

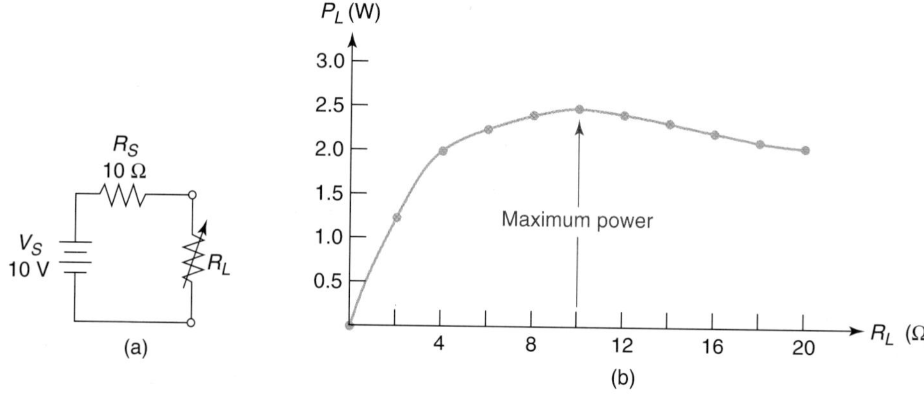


**FIGURE 5.37**

While the maximum power transfer theorem is not easy to derive mathematically, it can be demonstrated by performing a series of power calculations on the circuit in Figure 5.37a. In Table 5.1, we see the values of load voltage, current, and power that occur when the load in Figure 5.37a is varied from 0 Ω to 20 Ω (at 2 Ω increments). If you look at the load power ($P_L$) column of values, you'll see that it increases as we increase $R_L$ from 0 Ω to 10 Ω. Then, as we increase the values of $R_L$ above 10 Ω, load power starts to decrease again. Note that the maximum value of load power (2.50 W) occurs when $R_L$ is set to 10 Ω, the value of the source resistance.

If we plot a graph of $P_L$ versus $R_L$, we get the curve shown in Figure 5.37b. Again, you can see that maximum load power in this circuit occurs when $R_L = R_S$.

For each value of $R_L$ in Table 5.1, the other values were found using the following sequence of calculations:

1. $V_L = V_S \dfrac{R_L}{R_L + R_S}$
2. $I_L = \dfrac{V_L}{R_L}$
3. $P_L = V_L I_L$

**TABLE 5.1 Example Load Values for Figure 5.37**

| $R_L$ | $V_L$ | $I_L$ | $P_L$ |
|---|---|---|---|
| 0 Ω | 0 V | 1 A | 0 W |
| 2 Ω | 1.66 V | 833 mA | 1.38 W |
| 4 Ω | 2.86 V | 714 mA | 2.04 W |
| 6 Ω | 3.75 V | 625 mA | 2.34 W |
| 8 Ω | 4.44 V | 556 mA | 2.47 W |
| 10 Ω | 5.00 V | 500 mA | 2.50 W |
| 12 Ω | 5.45 V | 456 mA | 2.48 W |
| 14 Ω | 5.83 V | 417 mA | 2.43 W |
| 16 Ω | 6.15 V | 385 mA | 2.37 W |
| 18 Ω | 6.43 V | 357 mA | 2.30 W |
| 20 Ω | 6.67 V | 334 mA | 2.23 W |

The maximum power transfer theorem allows us to calculate easily the maximum possible load power for a circuit when the source resistance is known. This is demonstrated in the following example.

## EXAMPLE 5.15

Calculate the maximum possible load power for the circuit in Figure 5.38.

***Solution:*** The load power will be at maximum when $R_L$ is set to equal the value of $R_S$, 25 Ω. When the potentiometer is set to this value, the load voltage is one-half the no-load source voltage, and

$$P_L = \frac{V_L^2}{R_L} = \frac{(5\text{ V})^2}{25\ \Omega} = \mathbf{1\ W} \quad \text{(maximum)}$$

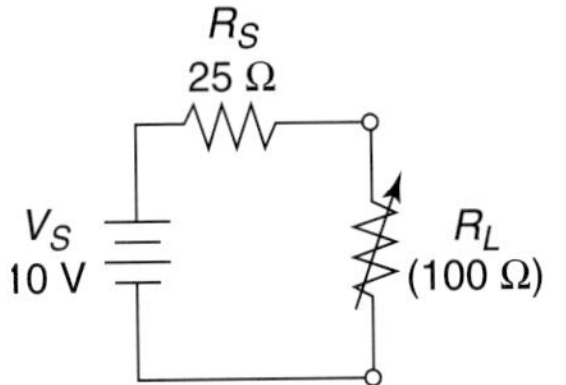


**FIGURE 5.38**

### *PRACTICE PROBLEM 5.15*

A circuit like the one in Figure 5.38 has values of $V_{NL} = 32$ V and $R_S = 40\ \Omega$. Calculate the maximum possible load power for the circuit.

An important point needs to be made: the maximum power transfer theorem holds true when the source resistance is fixed and the load resistance is variable. However, it fails to hold true when the situation is reversed. For example, consider the circuits shown in Figure 5.39. In Figure 5.39a, we have a voltage source with a fixed internal resistance and a variable load. The calculated values of $V_L$ and $P_L$ are provided for three settings of $R_L$. As you can see, maximum load power occurs when $R_L = 10\ \Omega$. Thus, the maximum power transfer theorem holds true when source resistance is fixed and load resistance is variable.

The maximum power transfer theorem fails to hold true when load resistance is fixed and source resistance is variable.

In Figure 5.39b, the load resistance is fixed at 10 Ω and the source resistance is shown to be variable. You would think that this load would receive maximum power when the source resistance is set to 10 Ω. However, this is not true. If $R_S$ is set to 10 Ω, the load voltage is 5 V (according to the voltage-divider equation), and the load power is found as

$$P_L = \frac{V_L^2}{R_L} = \frac{(5\text{ V})^2}{10\ \Omega} = \mathbf{2.5\ W}$$

If $R_S$ is set to 0.1 Ω, the voltage-divider equation gives us a load voltage of 9.9 V, and the load power is found as

$$P_L = \frac{V_L^2}{R_L} = \frac{(9.9\text{ V})^2}{10\ \Omega} = \mathbf{9.8\ W}$$

As you can see, maximum power transfer to a specific load resistance occurs when $R_S << R_L$. Thus, the maximum power transfer theorem fails to hold true when load resistance is fixed and source resistance is variable.

$R_S$
10 Ω
$V_S$
10 V
$R_L$
(100 Ω)

| For $R_L = 5\ \Omega$: | For $R_L = 10\ \Omega$: | For $R_L = 15\ \Omega$: |
|---|---|---|
| $V_L = 3.33$ V | $V_L = 5$ V | $V_L = 6$ V |
| $P_L = 2.22$ W | $P_L = 2.5$ W | $P_L = 2.4$ W |

(a)

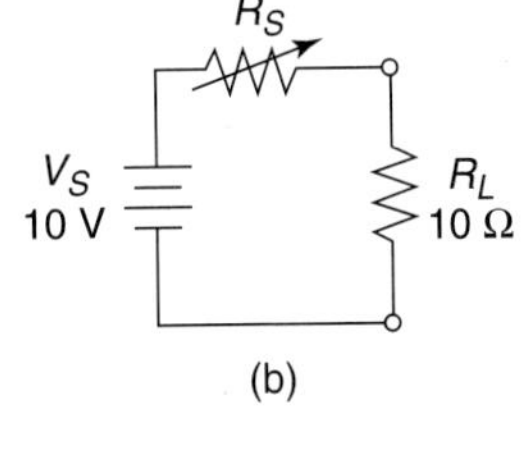


(b)

**FIGURE 5.39**

For a specific source resistance, maximum power transfer occurs when $R_L = R_S$. However, for a specific load resistance, maximum power transfer occurs when $R_S$ is set to its lowest possible value.

A few more points need to be made regarding maximum power transfer:

1. Maximum power transfer from a source to a load is often a major concern in circuit design. Since most loads have rated values (8 Ω stereo speakers, for example), circuit designers are generally concerned with minimizing source resistance.
2. Maximum load power does not necessarily correspond to maximum load voltage. This point is verified by the values listed in Table 5.1.
3. In Chapter 3, we discussed the concept of source efficiency. You may recall that the efficiency of a power supply is a ratio of its output power to its input power. Maximum power transfer does not mean that the power supply is 100% efficient. In fact, when maximum power transfer occurs in a circuit with a fixed source resistance, source efficiency is only 50% because $R_L$ and $R_S$ are equal in value. Therefore, each is using half the source input power.

## Series-Connected Voltage Sources

*OBJECTIVE 19* ▶

**Series-aiding voltage sources** Sources connected so that the current supplied by the individual sources are in the same direction.

**Series-opposing voltage sources** Sources connected so that the current supplied by the individual sources are in opposition to each other.

In later chapters, we will encounter situations where two (or more) voltage sources are connected in series. These voltage sources are connected either as series-aiding or series-opposing sources, as shown in Figure 5.40.

**Series-aiding voltage sources** (like those in Figure 5.40a) are connected so that the currents generated by the individual sources are in the same direction. As a result, each source is aiding the other(s) in generating current through the circuit. In contrast, **series-opposing voltage sources** (like those in Figure 5.40b) are connected so that the currents that would be generated by the individual sources are in opposition to each other. In effect, the voltage sources are opposing each other in the generation of circuit current.

***Series-Aiding Voltage Sources.*** When two voltage sources are connected as shown in Figure 5.40a, the total voltage and current in the circuit equals the sum of the individual source voltages and currents. By formula,

$$V_T = V_A + V_B \tag{5.6}$$

and

$$I_T = I_A + I_B \tag{5.7}$$

We are accustomed to adding series voltages, so equation (5.6) should appear familiar to you. However, equation (5.7) appears strange because current values are equal, not additive, throughout a series circuit. Here's the problem: equation (5.7) uses independent values of source current; that is, it assumes that the values of $I_A$ and $I_B$ are calculated as if they were the only current in the circuit. This point is illustrated in the following example.

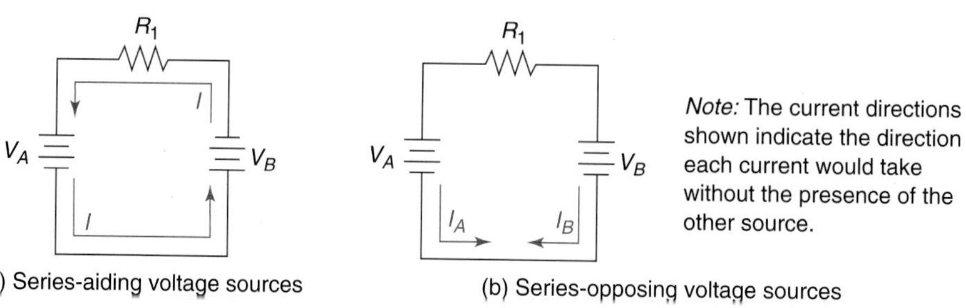

R1
I
VA
VB
I
R1
VA
VB
IA
IB
Note: The current directions shown indicate the direction each current would take without the presence of the other source.
(a) Series-aiding voltage sources
(b) Series-opposing voltage sources

**FIGURE 5.40 Series-connected voltage sources.**

Calculate the total current through the circuit in Figure 5.41a.

## EXAMPLE 5.16

***Solution:*** To solve the problem using equation (5.7), we have to treat the circuit as two single-source circuits. These circuits are shown in Figure 5.41b. Note that each source is replaced by a wire when removed from the circuit. (This represents the ideal value of $R_S = 0\ \Omega$ for the source.)

For the $V_A$ circuit, the total current is found as:

$$I_A = \frac{V_A}{R_T} = \frac{11\text{ V}}{550\ \Omega} = \mathbf{20\ mA}$$

For the $V_B$ circuit, the total current is found as

$$I_B = \frac{V_B}{R_T} = \frac{33\text{ V}}{550\ \Omega} = \mathbf{60\ mA}$$

Now, equation (5.7) can be used to find the total circuit current, as follows:

$$I_T = I_A + I_B = 20\text{ mA} + 60\text{ mA} = \mathbf{80\ mA}$$

***PRACTICE PROBLEM 5.16***

Using equation (5.7), calculate the total current for the circuit in Figure 5.41c.

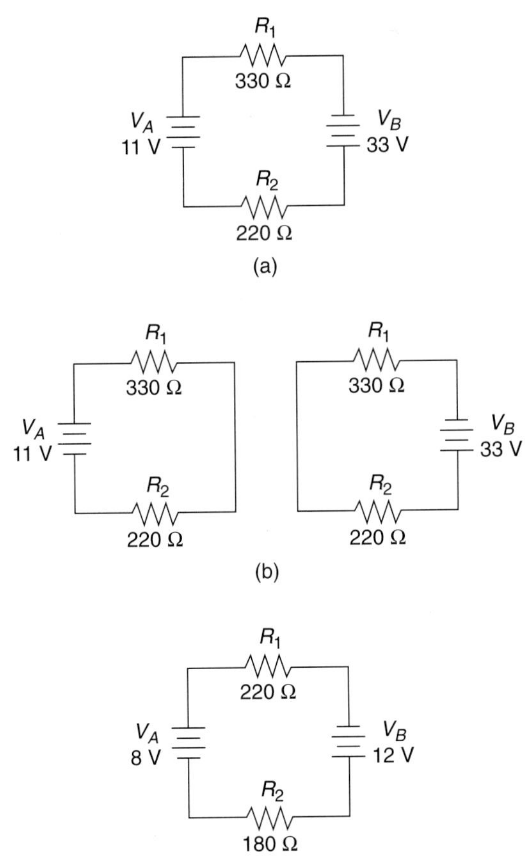


**FIGURE 5.41**

Equation (5.7) and Example 5.16 were included merely to illustrate the effects of the individual sources on the voltage and current values in a series circuit. In practice, a circuit with series-aiding sources is normally analyzed by adding the source voltages together and then solving for the total circuit current. This approach is demonstrated in the following example.

### EXAMPLE 5.17

Calculate the total current value for the circuit in Figure 5.41a.

***Solution:*** Since the sources are connected in series-aiding fashion, the total voltage can be found as

$$V_T = V_A + V_B = 11\text{ V} + 33\text{ V} = \mathbf{44\text{ V}}$$

Now, the total current can be found as

$$I_T = \frac{V_T}{R_T} = \frac{44\text{ V}}{550\ \Omega} = \mathbf{80\text{ mA}}$$

***PRACTICE PROBLEM 5.17***

Using the approach demonstrated in this example, recalculate the value of $I_T$ for the circuit in Figure 5.41c.

***Series-Opposing Voltage Sources.*** When two voltage sources are connected as shown in Figure 5.40b, the total voltage and current in the circuit equals the difference between the individual source voltages and currents. By formula,

$$V_T = |V_A - V_B| \tag{5.8}$$

and

$$I_T = |I_A - I_B| \tag{5.9}$$

Again, the current equation assumes that the circuit is analyzed as two single-source circuits. This point is illustrated in the following example.

### EXAMPLE 5.18

Using equation (5.9), calculate the total current for the circuit in Figure 5.42a.

***Solution:*** Again, the circuit is split into two single-source circuits. In each case, the source that was removed has been replaced by a wire (representing the ideal value of $R_S = 0\ \Omega$).

For the $V_A$ circuit, the total current is found as

$$I_A = \frac{V_A}{R_T} = \frac{8\text{ V}}{800\ \Omega} = \mathbf{10\text{ mA}}$$

For the $V_B$ circuit, the total current is found as

$$I_B = \frac{V_B}{R_T} = \frac{2\text{ V}}{800\ \Omega} = \mathbf{2.5\text{ mA}}$$

Finally, the total circuit current is found as

$$I_T = |I_A - I_B| = 10\text{ mA} - 2.5\text{ mA} = \mathbf{7.5\text{ mA}}$$

***PRACTICE PROBLEM 5.18***

Using equation (5.9), calculate the total current for the circuit in Figure 5.42c.

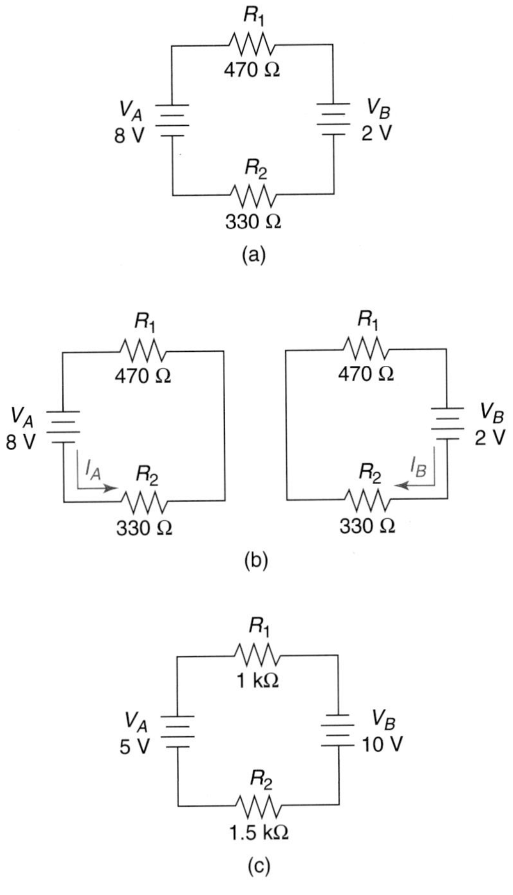


**FIGURE 5.42**

Again, we have taken the long approach to solving the circuit to demonstrate the relationship between the source currents. The analysis of a circuit with series-opposing sources is usually performed by finding the difference between the source voltages and using Ohm's law to determine the circuit current. This procedure is demonstrated in the following example.

### EXAMPLE 5.19

Using equation (5.8) and Ohm's law, recalculate the total current for the circuit in Figure 5.42a.

***Solution:*** The total voltage in the circuit is found as

$$\boldsymbol{V_T} = |V_A - V_B| = 8\text{ V} - 2\text{ V} = \mathbf{6\text{ V}}$$

Now, the total circuit current can be found as

$$\boldsymbol{I_T} = \frac{V_T}{R_T} = \frac{6\text{ V}}{800\ \Omega} = \mathbf{7.5\text{ mA}}$$

***PRACTICE PROBLEM 5.19***

Using equation (5.8) and Ohm's law, recalculate the value of $I_T$ for the circuit in Figure 5.42c.

***Determining the Direction of Circuit Current.*** The final step in solving for $I_T$ in a circuit with series-opposing sources is to determine the direction of the current. For example, in Figure 5.42a, two sources attempt to generate current in opposite directions. The question is, which direction does the total current take?

For any pair of series-opposing sources, the total current is in the direction of the source with the greater magnitude. In the circuit of Figure 5.42a, the magnitude of $V_A$ (8 V) is greater than the magnitude of $V_B$ (2 V). Therefore, the circuit current is in the direction indicated by source A. Note that the other source ($V_B$) is forced to conduct in the wrong direction. This is almost always the case for a given pair of series-opposing sources. The one exception is when the sources are equal in value. In this case,

$$V_T = |V_A - V_B| = \mathbf{0\ V}$$

and $I_T = 0$ A.

***One More Point.*** Batteries are not normally connected in series-opposing fashion. However, certain components (such as capacitors) are capable of storing a charge under specific circumstances. When this is the case, a charge-storing component can act as a series-opposing source to any dc voltage source in the circuit. When this occurs, the material presented here will apply. (You will see this later in the text.)

## Earth Ground Versus Chassis Ground

*OBJECTIVE 20* ▶

Earlier in the chapter, you were introduced to ground as a 0 V reference and common connection point in a circuit. There are actually two common types of ground connections: *earth ground* and *chassis ground*. The schematic symbols for each are shown in Figure 5.43.

As its name implies, earth ground provides a physical connection to the earth, via the wall outlet. This ensures that the ground in the circuit is at a 0 V potential with respect to the earth. Since the circuit is grounded to the earth, no difference of potential exists across you (or any piece of test equipment) that comes into contact with the ground connection.

*An Important Point:* Chassis ground should never be connected to any earth ground in the system. To do so could severely damage the system.

Chassis ground, on the other hand, is not returned to earth. In other words, chassis ground provides a 0 V reference for the components in the circuit, but it does not provide a connection between the circuit and the earth. As a result, there may be a difference of potential between the circuit ground and the earth. This difference of potential could be felt

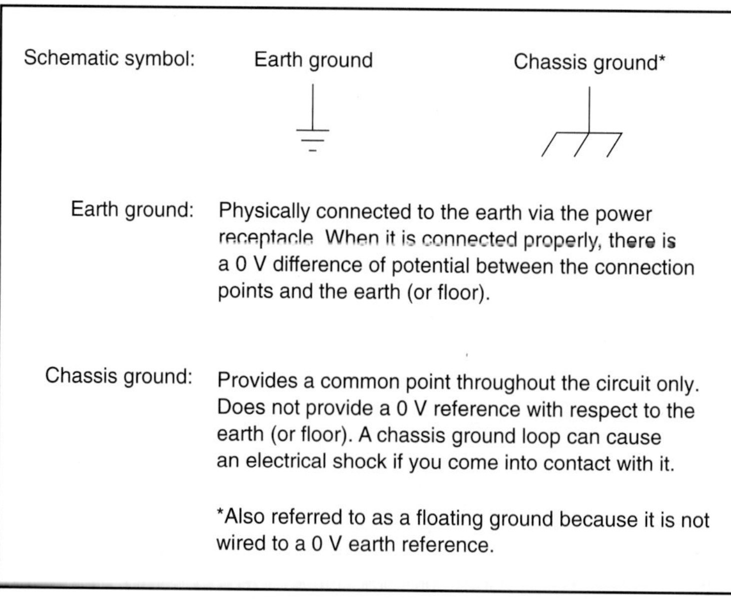


**FIGURE 5.43 Earth ground versus chassis ground.**

by you or any piece of test equipment that happened to come into contact with the circuit ground. You must always be careful when dealing with a chassis ground loop: it can hurt you if you happen to come into contact with it.

## Section Review

1. Explain how a potentiometer can be used as a variable voltage divider.
2. What happens to the output of an ideal voltage source when the load resistance changes?
3. What happens to the output of a practical voltage source when the load resistance changes?
4. Why do the output characteristics of the ideal and practical voltage sources differ?
5. When is maximum power transferred from a source to its load?
6. In terms of circuit current, how do series-aiding voltage sources differ from series-opposing voltage sources?
7. Why must you be careful when working on an electronics system that has a chassis ground?

## Chapter Summary

Here is a summary of the major points made in this chapter:

1. The two most basic types of circuits are the series circuit and the parallel circuit.
2. Most circuits can be classified as a series circuit, a parallel circuit, or some combination of the two.
3. A series circuit contains only one current path.
4. The total resistance in a series circuit is equal to the sum of the individual resistor values.
5. You can determine the value of an unknown resistance in a series circuit by subtracting the sum of the other resistors from the total circuit resistance.
6. The current at any point in a series circuit equals the current at every other point in the circuit.
7. Since current is constant throughout a series circuit, it makes no difference where in the circuit it is measured.
8. The actual value of current in a series circuit is determined by the source voltage and the total circuit resistance.
9. The current passing through the resistors in a series circuit causes a voltage to be developed across each resistor.
10. The polarity of the voltage across a given resistor in a series circuit is indicated by the direction of the current through the component.
11. The sum of the resistor voltages in a series circuit must equal the source voltage.
12. The current passing through the resistors in a series circuit causes each resistor to dissipate some amount of power.
13. The total power dissipated by the resistors in a series circuit is equal to the total power supplied by the source.
14. Kirchhoff's voltage law states that the sum of the voltages around a closed loop must equal zero. The term *closed loop* is used to describe a series circuit.
15. When setting up a Kirchhoff's loop equation, the polarities of the source and component voltages must be taken into account.

16. Kirchhoff's voltage law is just another way of saying that the sum of the component voltages in a series circuit must equal the total (or source) voltage.

17. Ground is the accepted reference for all voltages in a circuit.

18. The presence of the ground symbol in a schematic indicates that two or more points are:

    a. Considered to be at a 0 V potential.

    b. Physically connected to each other.

19. The term *voltage drop* is often used to describe the voltage across a given component in a circuit.

20. The term *voltage divider* is often used to describe a series circuit because the source voltage is divided among the resistors in the circuit.

21. The ratio of a given resistor voltage to the source voltage is equal to the ratio of the resistor's value to the total circuit resistance.

22. The voltage-divider equation allows you to calculate the voltage across a given resistor (or group of resistors) without calculating the value of the circuit current.

23. The complete mathematical analysis of a series circuit involves determining the values of $R_T$, $I_T$, and $P_T$, along with the individual resistor voltage and power values.

24. When one resistor in a series circuit opens:

    a. The circuit current drops to zero.

    b. The total source voltage is dropped across the open component.

25. The measured voltage across an open resistor is usually lower than the source voltage because the voltmeter completes the circuit. When the circuit is completed, the low circuit current causes some of the source voltage to be dropped across the other resistors in the circuit. This reduces the voltage across the open. The higher the internal resistance of the meter, the closer a voltage reading comes to its ideal value.

26. When multiple opens exist in a series circuit:

    a. The circuit current drops to zero.

    b. The source voltage is dropped across the entire circuit that lies between the most negative open terminal and the most positive open terminal.

27. When a resistor in a series circuit is shorted:

    a. The voltage across the shorted component drops to nearly 0 V.

    b. The total circuit current increases, causing the other resistor voltages to increase.

    c. The high circuit current can cause another component to burn open and/or the fuse in the circuit to blow.

28. Opens and shorts are referred to as catastrophic failures because they are complete (and sometimes violent) component failures.

29. When a resistor fails partially (due to component aging or stress), the voltage across the component falls gradually out of tolerance. Such a component will eventually fail completely.

30. When a resistor burns open, it will generally show signs of excessive heat, such as:

    a. Extremely dark color bands.

    b. One or more black patches (from burning).

    c. A crack in the body of the component.

31. Partial failure of a resistor usually has one or more of the following symptoms:

    a. The circuit containing the resistor may operate erratically.

    b. The component may smell as if it is beginning to burn.

    c. The component may be unusually hot.

    d. The component may be discolored.

32. Troubleshooting involves:
   a. Taking the appropriate circuit measurements.
   b. Thinking about what those measurements tell you.
33. A single potentiometer can be used as a variable voltage divider.
34. The ideal voltage source maintains a constant output voltage regardless of the resistance of the load.
35. The open-terminal output voltage from a source is called its no-load output voltage.
36. For a practical voltage source, a decrease in load resistance results in a decrease in load voltage.
   a. The cause of the decrease in load voltage is the internal resistance of the source.
   b. The load on a source forms a voltage divider with the source's internal resistance.
37. The maximum power transfer theorem states that maximum transfer of power from a source to its load occurs when the load resistance is equal to the source resistance.
   a. This theorem holds true when source resistance is fixed and load resistance is variable.
   b. This theorem does not hold true when source resistance is variable and load resistance is fixed. In this case, maximum power transfer occurs when $R_S$ is at its lowest possible value.
38. Series-aiding voltage sources are connected so that the currents generated by the individual sources are in the same direction. The total voltage equals the sum of the sources.
39. Series-opposing voltage sources are connected so that the currents that would be generated by the individual sources are in opposition to each other.
   a. The total voltage equals the difference between the sources.
   b. The circuit current is in the direction of the source with the greatest magnitude.
40. Earth ground provides a physical connection to earth. Chassis ground serves only as a 0 V reference within a circuit. There may be a difference of potential between a chassis ground and earth ground. Therefore, you must be careful not to assume that it is safe to touch a chassis ground connection.

## Equation Summary

| Equation Number | Equation | Section Number |
|---|---|---|
| **(5.1)** | $R_T = R_1 + R_2 + \ldots + R_n$ | 5.1 |
| **(5.2)** | $V_S = V_1 + V_2 + \ldots + V_n$ | 5.1 |
| **(5.3)** | $P_S = P_1 + P_2 + \ldots + P_n$ | 5.1 |
| **(5.4)** | $\frac{V_n}{V_S} = \frac{R_n}{R_T}$ | 5.2 |
| **(5.5)** | $V_n = V_S\frac{R_n}{R_T}$ | 5.2 |
| **(5.6)** | $V_T = V_A + V_B$ | 5.4 |
| **(5.7)** | $I_T = I_A + I_B$ | 5.4 |
| **(5.8)** | $V_T = \lvert V_A - V_B \rvert$ | 5.4 |
| **(5.9)** | $I_T = \lvert I_A - I_B \rvert$ | 5.4 |

## Key Terms

The following terms were introduced and defined in this chapter:

catastrophic failure
closed loop
ground
ideal voltage source
Kirchhoff's voltage law
maximum power transfer theorem
no-load output voltage
series-aiding voltage sources
series circuit
series-opposing voltage sources
source resistance
variable voltage divider
voltage divider
voltage drop

## Practice Problems

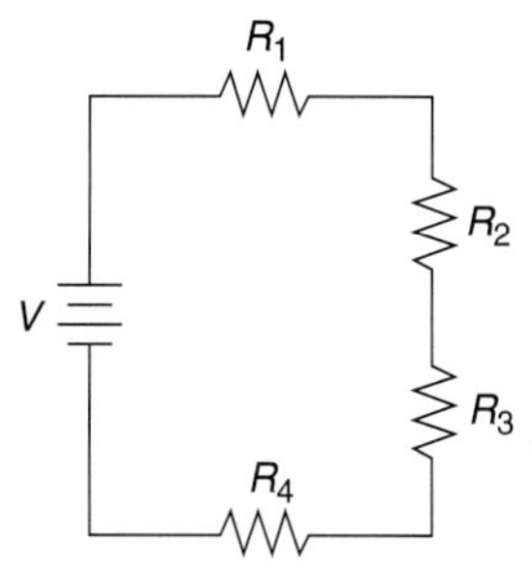


FIGURE 5.44

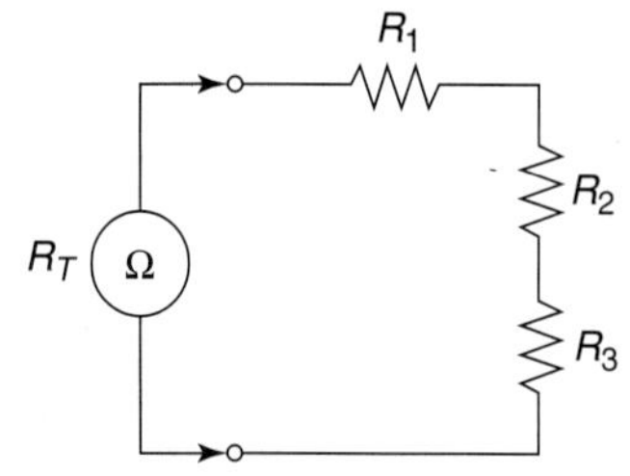


FIGURE 5.45

**1.** Each resistor combination below is connected as shown in Figure 5.44. For each combination, calculate the total circuit resistance.

| | $R_1$ | $R_2$ | $R_3$ | $R_4$ |
|---|---|---|---|---|
| a. | 1 kΩ | 220 Ω | 330 Ω | 1.1 kΩ |
| b. | 10 Ω | 18 Ω | 47 Ω | 200 Ω |
| c. | 150 Ω | 220 Ω | 820 Ω | 51 Ω |
| d. | 10 kΩ | 91 kΩ | 5.1 kΩ | 300 Ω |

**2.** Each resistor combination below is connected as shown in Figure 5.44. For each combination, calculate the total circuit resistance.

| | $R_1$ | $R_2$ | $R_3$ | $R_4$ |
|---|---|---|---|---|
| a. | 1 MΩ | 470 kΩ | 270 kΩ | 51 kΩ |
| b. | 22 kΩ | 39 kΩ | 12 kΩ | 75 kΩ |
| c. | 360 Ω | 1.1 kΩ | 68 Ω | 2.2 kΩ |
| d. | 8.2 kΩ | 3.3 kΩ | 9.1 kΩ | 5.1 kΩ |

**3.** The resistor combinations below are for a circuit like the one in Figure 5.45. In each case, determine the unknown resistor value.

| | $R_1$ | $R_2$ | $R_3$ | $R_T$ |
|---|---|---|---|---|
| a. | 1.1 kΩ | 330 Ω | ____ | 1.9 kΩ |
| b. | ____ | 47 kΩ | 91 kΩ | 165 kΩ |
| c. | 33 kΩ | ____ | 6.2 kΩ | 44.3 kΩ |
| d. | 27 Ω | 39 Ω | 82 Ω | ____ |

**4.** The resistor combinations below are for a circuit like the one in Figure 5.45. In each case, determine the unknown resistor value.

| | $R_1$ | $R_2$ | $R_3$ | $R_T$ |
|---|---|---|---|---|
| a. | 200 kΩ | 33 kΩ | ____ | 308 kΩ |
| b. | ____ | 82 kΩ | 75 kΩ | 204 kΩ |
| c. | 510 Ω | ____ | 68 Ω | 611 Ω |
| d. | 1.5 MΩ | 2.2 MΩ | 10 MΩ | ____ |

**5.** Determine the potentiometer setting required in Figure 5.46a to provide a total resistance of 52 kΩ.

**6.** Determine the potentiometer setting required in Figure 5.46b to provide a total resistance of 872 Ω.

**7.** Calculate the total current for the circuit in Figure 5.47a.

**8.** Calculate the total current for the circuit in Figure 5.47b.

**9.** Calculate the total current for the circuit in Figure 5.47c.

**10.** Calculate the total current for the circuit in Figure 5.47d.

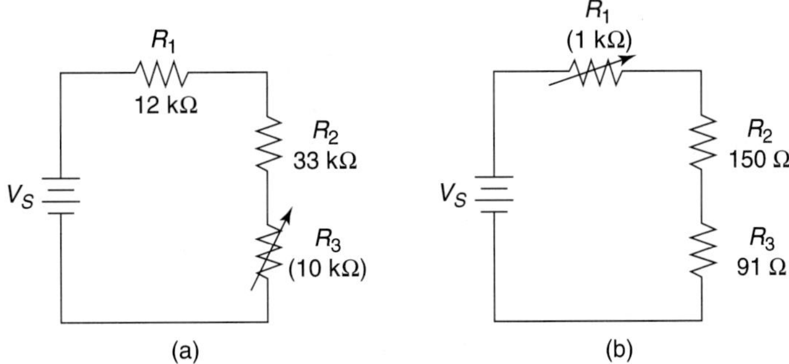


**FIGURE 5.46**

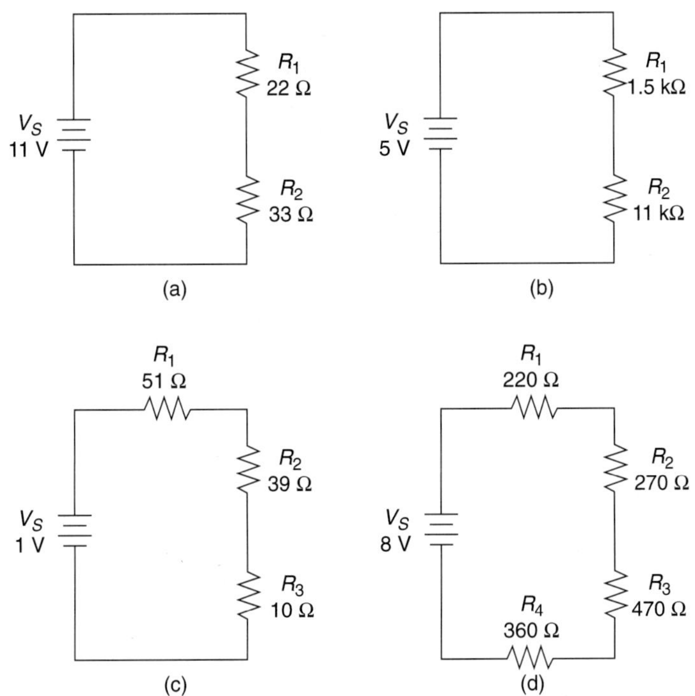


**FIGURE 5.47**

**11.** Determine the potentiometer setting needed to set the current in Figure 5.48a to 18 mA.

**12.** Determine the potentiometer setting needed to set the current in Figure 5.48b to 27 mA.

**13.** Determine the potentiometer setting needed to set the current in Figure 5.48c to 560 μA.

**14.** Determine the potentiometer setting needed to set the current in Figure 5.48d to 1.22 mA.

**15.** Calculate the value of the source voltage in Figure 5.49a.

**16.** Calculate the value of the source voltage in Figure 5.49b.

**17.** Calculate the value of the source voltage in Figure 5.49c.

**18.** Calculate the value of the source voltage in Figure 5.49d.

**19.** For the circuit in Figure 5.50a, verify that the total power dissipated by the resistors is equal to the power supplied by the source.

**20.** Repeat Problem 19 for the circuit in Figure 5.50b.

**21.** Repeat Problem 19 for the circuit in Figure 5.50c.

**22.** Repeat Problem 19 for the circuit in Figure 5.50d.

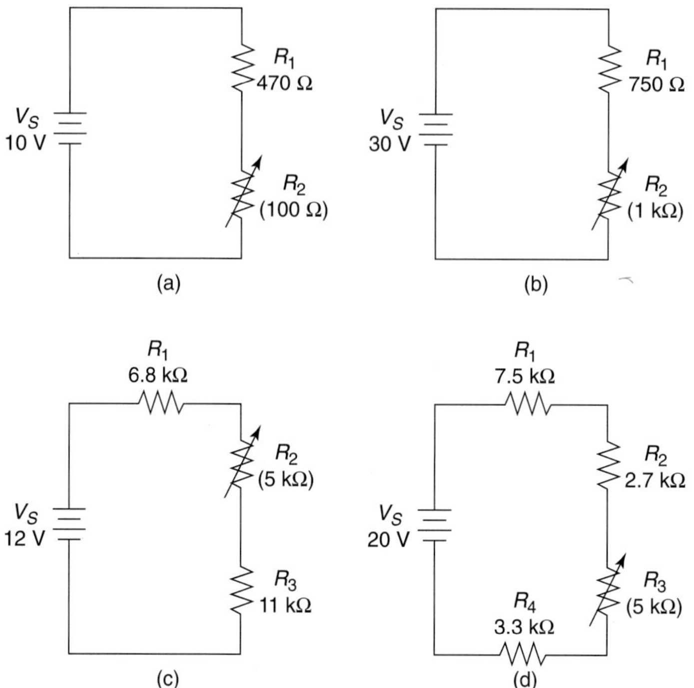


**FIGURE 5.48**

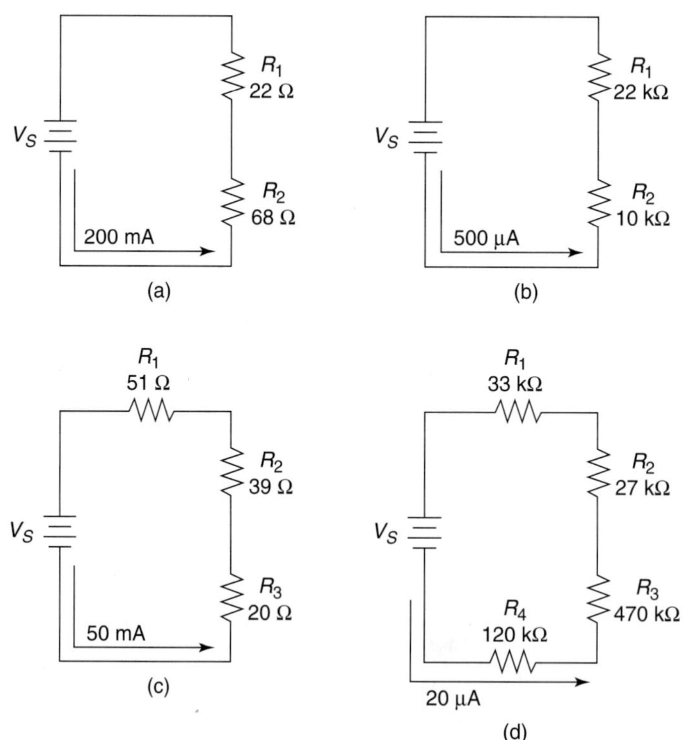


**FIGURE 5.49**

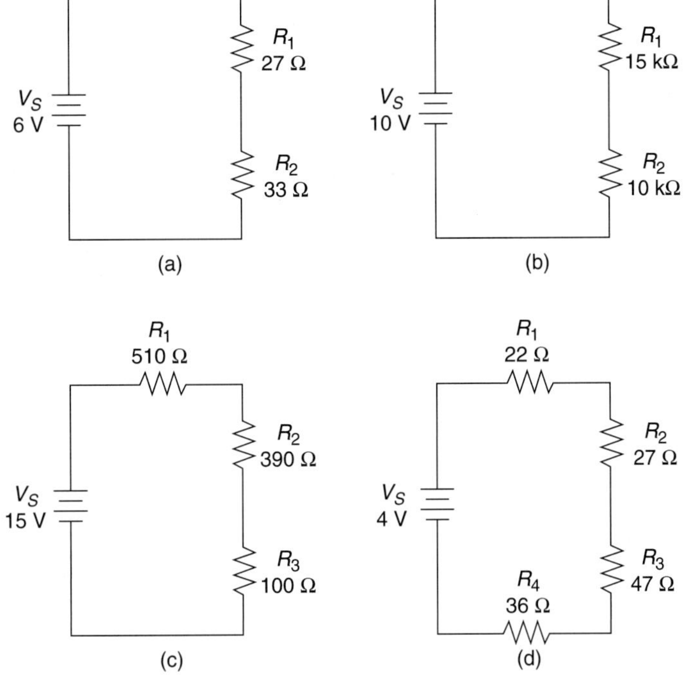


**FIGURE 5.50**

**23.** Write the Kirchhoff's loop equation for the circuit in Figure 5.51a and verify that the voltages add up to 0 V. (Begin at the negative side of the source and go clockwise around the loop.)

**24.** Repeat Problem 23 for the circuit in Figure 5.51b.

**25.** Calculate the value of the current in Figure 5.52a.

**26.** Calculate the value of the current in Figure 5.52b.

**27.** Calculate the value of the current in Figure 5.52c.

**28.** Calculate the value of the current in Figure 5.52d.

**29.** For the circuit in Figure 5.53a, calculate the voltage from point A to ground.

**30.** For the circuit in Figure 5.53b, calculate the voltage from point A to ground.

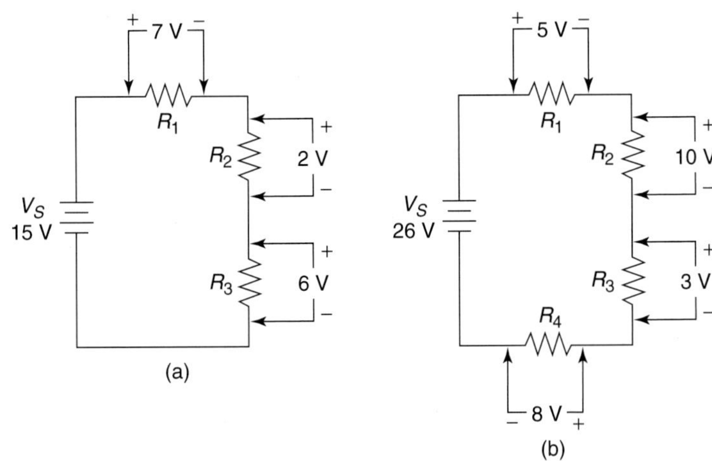


**FIGURE 5.51**

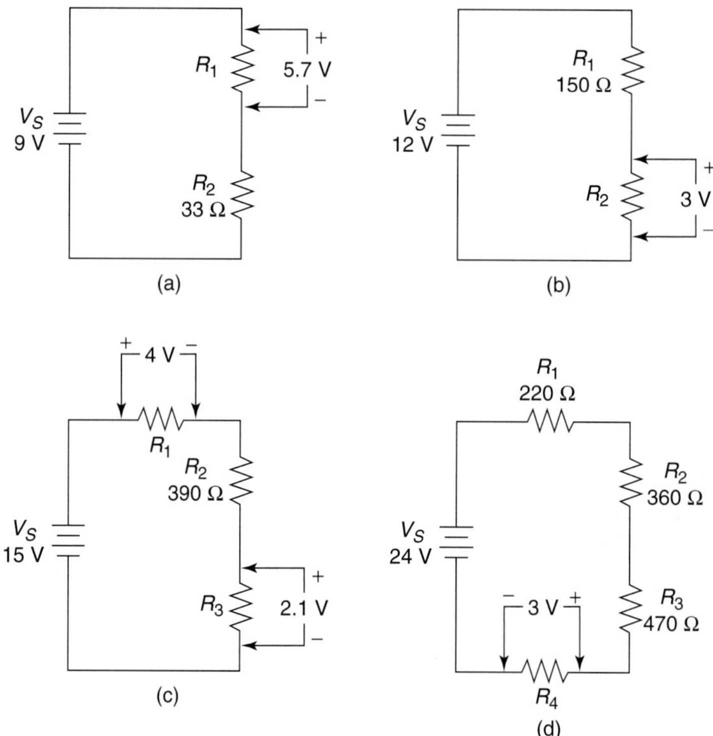


**FIGURE 5.52**

**31.** Using the voltage-divider equation, calculate the voltage across each of the resistors in Figure 5.54a.

**32.** Repeat Problem 31 for the circuit in Figure 5.54b.

**33.** Repeat Problem 31 for the circuit in Figure 5.54c.

**34.** Repeat Problem 31 for the circuit in Figure 5.54d.

**35.** Using the voltage-divider equation, determine the voltage from point A to ground for the circuit in Figure 5.55a.

**36.** Repeat Problem 35 for the circuit in Figure 5.55b.

**37.** Repeat Problem 35 for the circuit in Figure 5.55c.

**38.** Repeat Problem 35 for the circuit in Figure 5.55d.

**39.** Perform a complete mathematical analysis of the circuit in Figure 5.56a.

**40.** Perform a complete mathematical analysis of the circuit in Figure 5.56b.

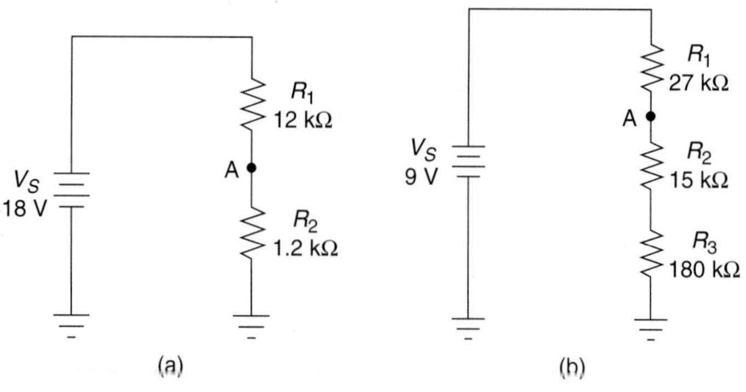


**FIGURE 5.53**

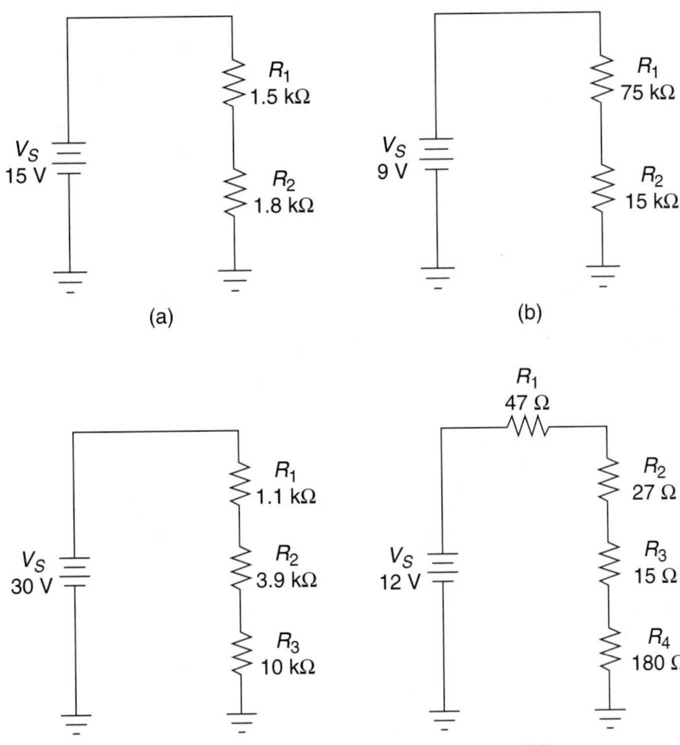


**FIGURE 5.54**

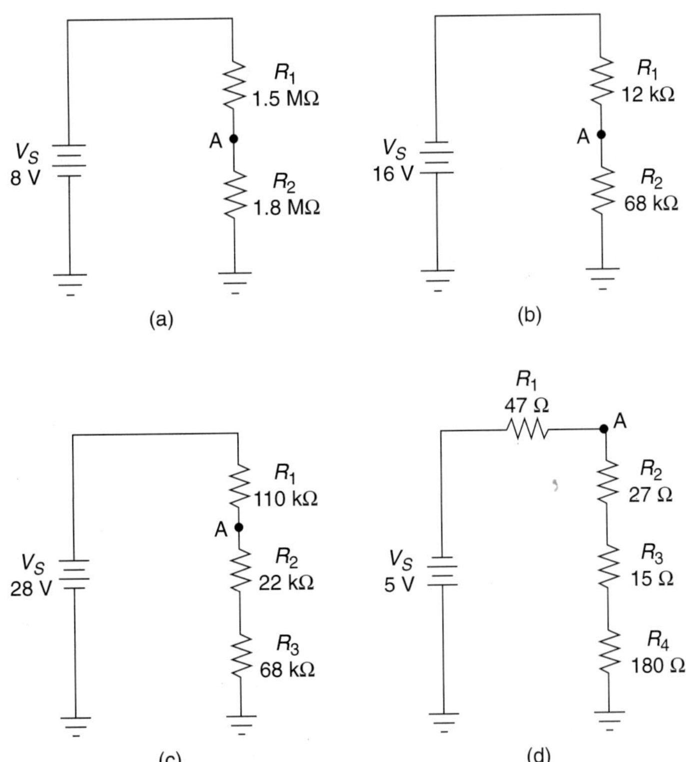


**FIGURE 5.55**

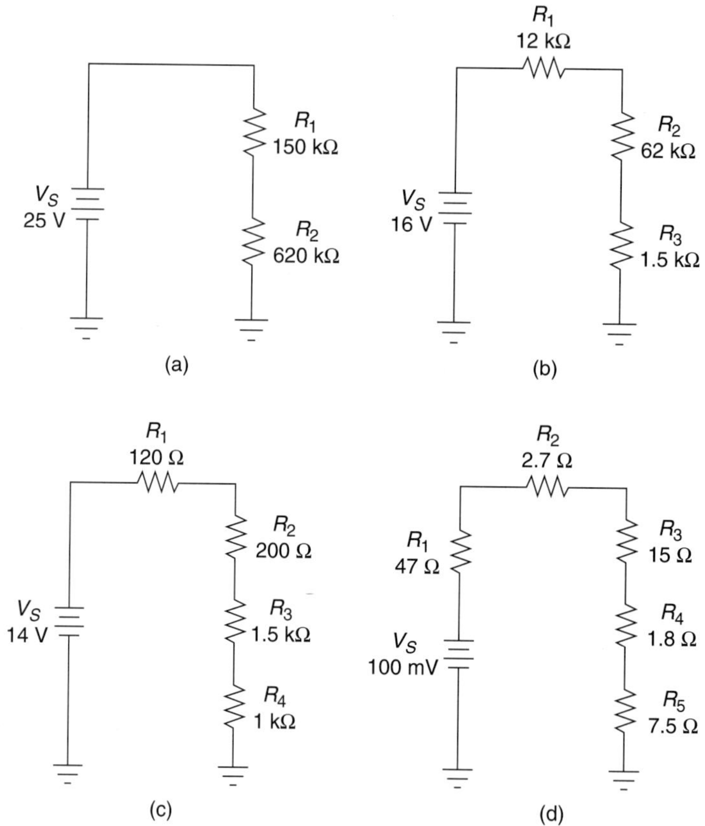


**FIGURE 5.56**

**41.** Perform a complete mathematical analysis of the circuit in Figure 5.56c.

**42.** Perform a complete mathematical analysis of the circuit in Figure 5.56d.

**43.** The voltage across the open resistor in Figure 5.57a is measured with a DMM that has an internal resistance of 10 MΩ. Determine the meter reading.

**44.** The voltage across the open resistor in Figure 5.57a is measured with a VOM that has an internal resistance of 200 kΩ. Determine the meter reading.

**45.** Repeat Problem 43 for the circuit in Figure 5.57b.

**46.** Repeat Problem 44 for the circuit in Figure 5.57b.

**47.** For each of the following values of $R_L$ (1 kΩ, 100 Ω, and 10 Ω), calculate the value of the load voltage for the circuit in Figure 5.58a.

**48.** Repeat Problem 47 for the circuit in Figure 5.58b.

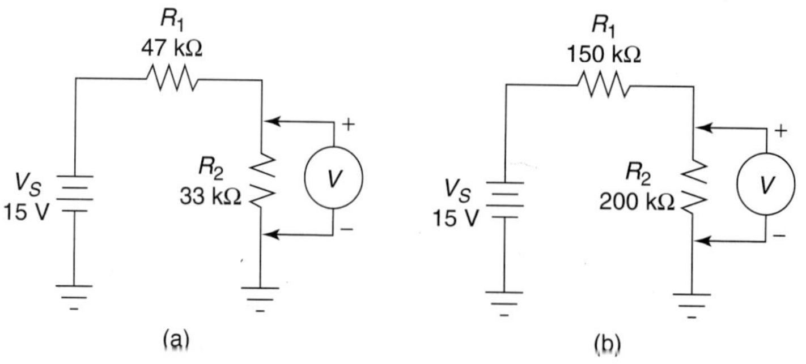


**FIGURE 5.57**

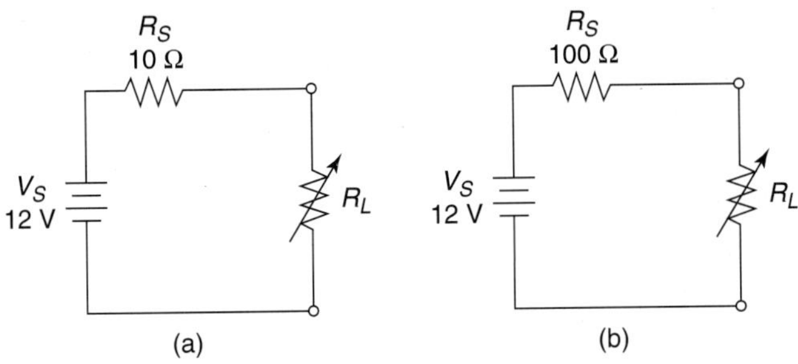


**FIGURE 5.58**

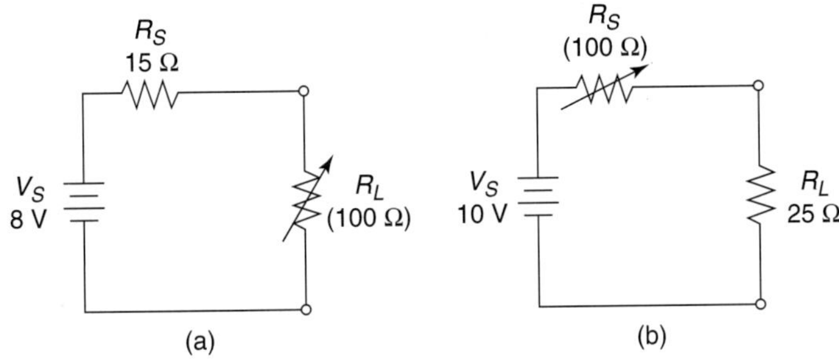


**FIGURE 5.59**

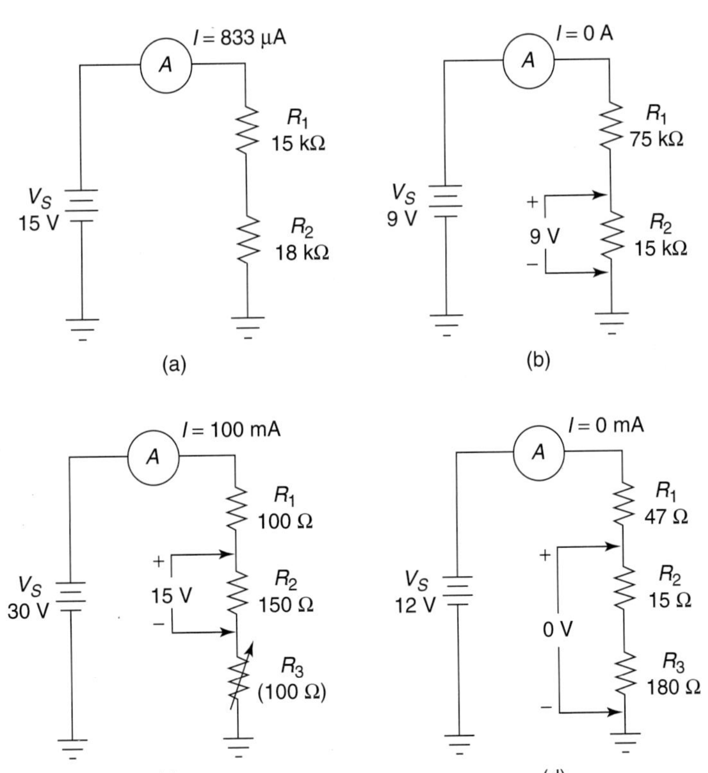


**FIGURE 5.60**

**49.** Calculate the maximum possible value of load power for the circuit in Figure 5.59a.

**50.** Calculate the maximum possible value of load power for the circuit in Figure 5.59b.

## Troubleshooting Practice Problems

**51.** Determine the fault (if any) indicated by the readings in Figure 5.60a.

**52.** Determine the fault (if any) indicated by the readings in Figure 5.60b.

**53.** Determine the fault (if any) indicated by the readings in Figure 5.60c.

**54.** Determine the fault (if any) indicated by the readings in Figure 5.60d.

## The Brain Drain

**55.** Each of the resistors in Figure 5.61 has a tolerance of 5%. Determine the minimum and maximum values of $I_T$ for the circuit (assuming that all the resistors are within tolerance).

**56.** For the circuit in Figure 5.62:

**a.** Calculate the values of $V_L$ for $R_L$ settings of 0 Ω to 1 kΩ, at 100 Ω increments.

**b.** Using your calculated values, plot a curve of $V_L$ versus $R_L$. (*Note:* The dependent variable $V_L$ is represented on the *y*-axis. The independent variable $R_L$ is represented on the *x*-axis.)

**c.** Determine (from your curve) the minimum allowable value of $R_L$, assuming that $V_L$ can drop as far as 10% below its maximum value.

## Answers to the Example Practice Problems

**5.1** 6.25 kΩ
**5.2** 2 kΩ
**5.3** 714 μA
**5.4** 3.7 kΩ
**5.5** 26.8 V
**5.6** $P_1 = 6$ W, $P_2 = 7.33$ W, $P_3 = 4.67$ W, $P_T = 18$ W
**5.7** 2.32 mA
**5.8** 4.06 V
**5.9** $V_1 = 8.25$ V, $V_2 = 11.75$ V
**5.10** 11.85 V
**5.11** $R_T = 650\ \Omega$, $I_T = 20$ mA, $P_T = 260$ mW, $V_1 = 6.6$ V, $V_2 = 4.4$ V, $V_3 = 2$ V, $P_1 =$ 132 mW, $P_2 = 88$ mW, $P_3 = 40$ mW
**5.12** 19.98 V
**5.13** 18.87 V
**5.14** When $R_L = 1$ kΩ, $V_L = 13.46$ V
When $R_L = 120\ \Omega$, $V_L = 10.5$ V
**5.15** 6.4 W (maximum)
**5.16** $I_T = 50$ mA

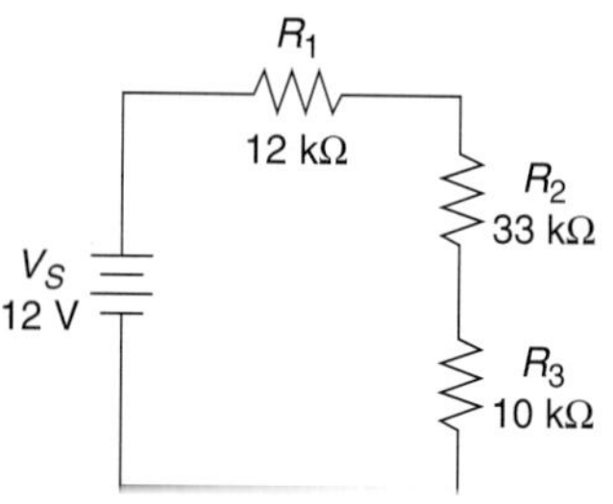


**FIGURE 5.61**

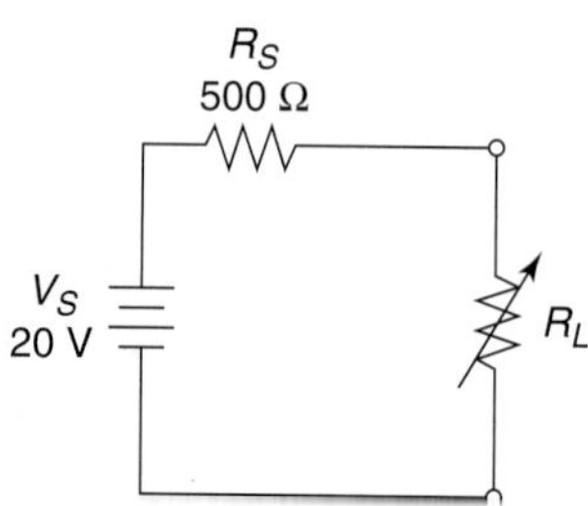


**FIGURE 5.62**

**5.17** $I_T = 50$ mA
**5.18** $I_T = 2$ mA
**5.19** $I_T = 2$ mA

---

## EWB Applications Problems

| Figure | EWB File Reference |
|---|---|
| 5.15 | EWBA5_15.ewb |
| 5.16 | EWBA5_16.ewb |
| 5.30 | EWBA5_30.ewb |
| 5.33 | EWBA5_33.ewb |

# 6
# PARALLEL CIRCUITS

## OBJECTIVES

*After studying the material in this chapter, you should be able to:*

1. Identify a parallel circuit.
2. Describe the current and voltage characteristics of parallel circuits.
3. Calculate any current value in a parallel circuit given the source voltage and the branch resistance values.
4. Contrast the current and voltage characteristics of series and parallel circuits.
5. Calculate the total resistance in any parallel circuit given the branch resistance values.
6. Describe Kirchhoff's current law and solve the Kirchhoff's current equation for a given parallel circuit.
7. Compare and contrast a voltage source and a current source.
8. Use the current-divider equations to solve for either branch current in a two-branch circuit.
9. Perform the complete mathematical analysis of any parallel circuit.
10. Describe the procedure used to locate an open branch in a parallel circuit.
11. Describe the symptoms associated with a shorted branch in a parallel circuit.
12. Describe how the fuse in a parallel circuit can indicate the type of fault you are looking for when troubleshooting.

In this chapter, we will continue our coverage of basic circuits by discussing the operating characteristics of parallel circuits. As you will see, these circuits are nearly the opposite of series circuits in many ways.

## 6.1 PARALLEL CIRCUIT CHARACTERISTICS

◀ *OBJECTIVE 1*

A **parallel circuit** is a circuit that provides more than one current path between any two points. Several parallel circuits are shown in Figure 6.1. As you can see, each circuit contains two or more paths for current. Note that each current path in a parallel circuit is referred to as a **branch.** For example, the circuit in Figure 6.1a contains two branches, the circuit in Figure 6.1b contains three branches, and so on.

**Parallel circuit**
A circuit that provides more than one current path between any two points.

**Branch**
A term used to identify each current path in a parallel circuit.

### *Current Characteristics*

◀ *OBJECTIVE 2*

Using the circuit in Figure 6.1a, let's take a look at the current action in a typical parallel circuit: the current leaves the negative terminal of the source. At the first connection point in the circuit (point A), the current splits. Part of the current passes through $R_1$ and the remainder passes through $R_2$. At the second connection point (point B), the currents recombine into a single current and return to the source.

The total current in a parallel circuit ($I_T$) is equal to the sum of the individual branch current values. By formula,

$$I_T = I_1 + I_2 + \ldots + I_n \tag{6.1}$$

*Reference:* The relationship given in equation (6.1) is referred to as Kirchhoff's current law. This law is discussed in detail in Section 6.2.

where $\quad I_n$ = the current through the highest numbered branch in the circuit

The total current in a parallel circuit is calculated as shown in Example 6.1.

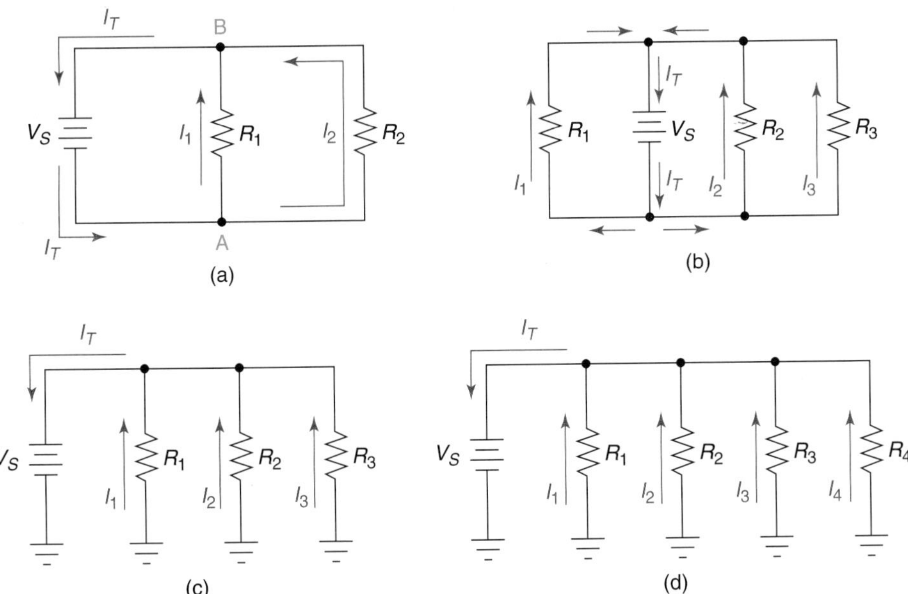


**FIGURE 6.1 Parallel circuits.**

*EXAMPLE 6.1*

The circuit in Figure 6.2 has the branch current values shown. Calculate the total current in the circuit.

***Solution:*** Using equation (6.1), the total circuit current is found as

$$I_T = I_1 + I_2 + I_3 = 5 \text{ mA} + 11 \text{ mA} + 19 \text{ mA} = \mathbf{35\ mA}$$

FIGURE 6.2

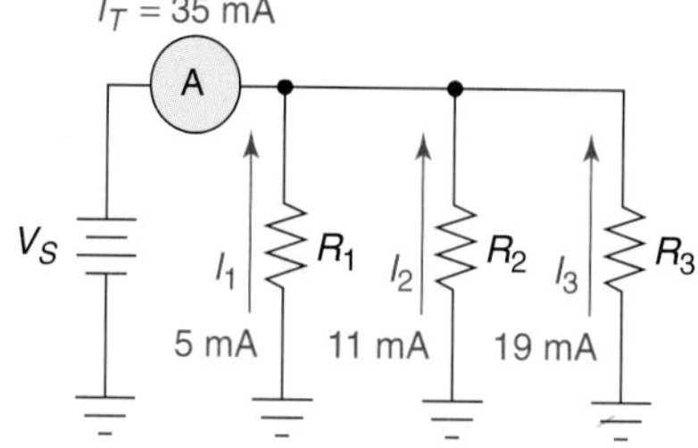


***PRACTICE PROBLEM 6.1***

A circuit like the one in Figure 6.2 has the following values: $I_1 = 6.4$ mA, $I_2 = 3.3$ mA, and $I_3 = 14.8$ mA. Calculate the total current in the circuit.

When you think about it, the result in Example 6.1 makes sense. The branch currents must come from the circuit's voltage source. Therefore, the total current supplied by the source *must* equal the sum of the branch currents.

## Voltage and Current Values

When two or more components are connected in parallel, the voltage across each component is equal to the voltage across all the others. This characteristic is illustrated in Figure 6.3.

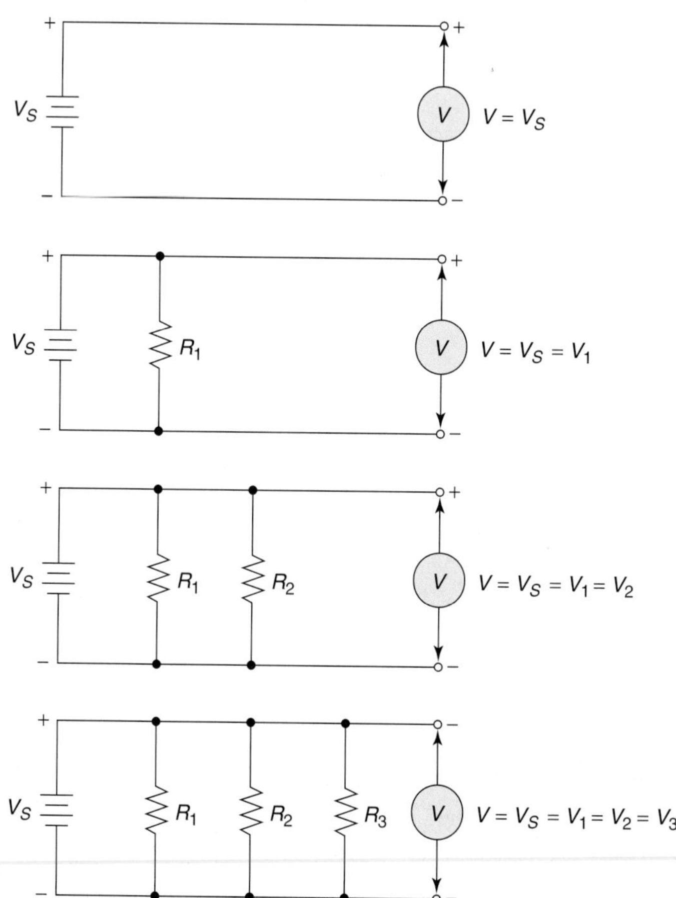


**FIGURE 6.3 The voltages across parallel components are equal.**

As the first circuit shows, the source voltage can be measured between the (+) and (−) conductors. Each time a resistor is added, one of its terminals is connected to the (+) conductor and the other is connected to the (−) conductor. Therefore, the voltage across each resistor is equal to the source voltage. By formula,

$$V_S = V_1 = V_2 = \ldots = V_n \tag{6.2}$$

where $V_n$ = the voltage across the highest numbered branch

The current through each branch of a parallel circuit is determined by the source voltage and the resistance of the branch. Therefore, we can use Ohm's law to calculate the value of any branch current as follows: ◄ *OBJECTIVE 3*

$$I_n = \frac{V_S}{R_n}$$

Once we have calculated the value of each branch current in a parallel circuit, we simply add those values to determine the total circuit current. This method of circuit current analysis is demonstrated Example 6.2.

## EXAMPLE 6.2

Calculate the total current in the circuit shown in Figure 6.4.

***Solution:*** First, we calculate the individual branch currents, as follows:

$$I_1 = \frac{V_S}{R_1} = \frac{5\text{ V}}{1\text{ k}\Omega} = \mathbf{5\text{ mA}}$$

$$I_2 = \frac{V_S}{R_2} = \frac{5\text{ V}}{2\text{ k}\Omega} = \mathbf{2.5\text{ mA}}$$

**FIGURE 6.4**

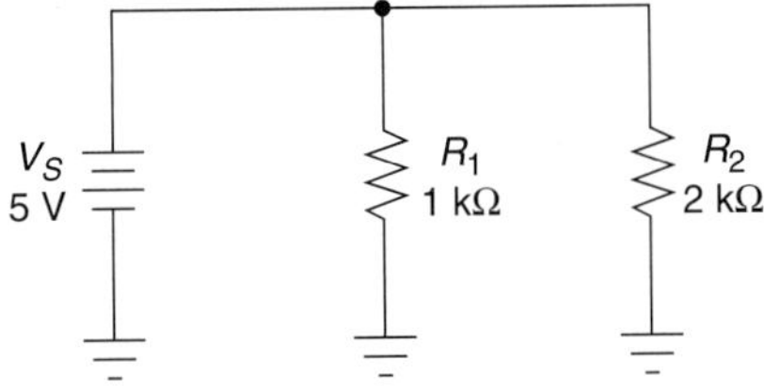


EWB

Now, we simply add these branch currents to obtain the value of the total circuit current, as follows:

$$I_T = I_1 + I_2 = 5\text{ mA} + 2.5\text{ mA} = \mathbf{7.5\text{ mA}}$$

***PRACTICE PROBLEM 6.2***

A circuit like the one in Figure 6.4 has the following values: $V_S = 12$ V, $R_1 = 240\ \Omega$, and $R_2 = 360\ \Omega$. Calculate the total current in the circuit.

Before we move on, let's take a moment to compare the voltage and current characteristics of parallel circuits with those of series circuits. These characteristics are summarized in Table 6.1. As you can see, the voltage and current characteristics of the parallel circuit are the opposite of those for the series circuit. In series circuits, currents are equal at all points and the sum of the component voltages equals the total (or source) voltage. In parallel circuits, voltages are equal across all branches and the sum of the branch currents equals the total current. ◄ *OBJECTIVE 4*

**TABLE 6.1 Comparison of Series and Parallel Circuit Characteristics**

| *Circuit Type* | *Voltage Characteristics* | *Current Characteristics* |
|---|---|---|
| Series | $V_T = V_1 + V_2 + \ldots + V_n$ | Equal at all points |
| Parallel | Equal across all branches | $I_T = I_1 + I_2 + \ldots + I_n$ |

You may recall that a series circuit is often referred to as a voltage divider because its source voltage is divided among the circuit resistors. By the same token, a parallel circuit is often referred to as a *current divider* because its total circuit current is divided among the circuit branches. The concept of a current divider is discussed in greater detail later in this chapter.

## Resistance Characteristics

In Example 6.2, the total current supplied by a 5 V source was found to be 7.5 mA. If we apply Ohm's law to these values, the total circuit resistance is found as

$$R_T = \frac{V_S}{I_T} = \frac{5\text{ V}}{7.5\text{ mA}} = 667\ \Omega$$

As you can see, this value is lower than the value of either resistor in the circuit.

In any parallel circuit, the total circuit resistance is always lower than any of the branch resistance values. This relationship is not as difficult to understand as it may seem. Consider the circuits shown in Figure 6.5. In Figure 6.5a, a 10 V source is generating 1 A of current through 10 Ω of resistance. If a second branch is added to the circuit (as shown in Figure 6.5b), the same source is now generating another 1 A of current through the second branch. As you can see, the total current generated has increased, even though the applied voltage has not changed. According to Ohm's law, this can happen only if the total resistance in the circuit decreases. Therefore, the addition of parallel branches must result in a decrease in circuit resistance.

*Another Approach:* Total current is greater than any branch current. Therefore, total resistance must be less than any branch resistance (since current and resistance are inversely proportional).

## Conductance Characteristics

To establish the relationship between the branch resistance values and the total resistance in a parallel circuit, we need to take a brief look at the *conductance* characteristics of parallel circuits. You may recall that conductance ($G$) is a measure of the ability of a component or circuit to conduct. In Figure 6.5, the addition of a parallel branch increased the circuit's ability to conduct. Therefore, the addition of the branch increased the conductance of the circuit.

*Don't Forget:* Total current increases when we add branches to a parallel circuit.

The total conductance ($G_T$) in a parallel circuit is equal to the sum of the branch conductance values. By formula,

$$G_T = G_1 + G_2 + \ldots + G_n \qquad \textbf{(6.3)}$$

where $G_n$ = the conductance of the highest numbered branch in the circuit

Example 6.3 demonstrates the use of this equation.

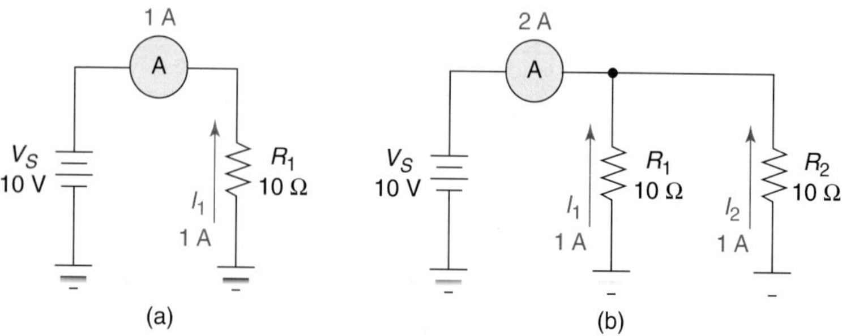


**FIGURE 6.5**

Calculate the total conductance for the circuits shown in Figure 6.5.

**EXAMPLE 6.3**

***Solution:*** In Chapter 2, conductance (in Siemens) was shown to equal the reciprocal of resistance. In Figure 6.5a, the total conductance is found as the reciprocal of the circuit resistance, as follows:

$$G_T = \frac{1}{R_T} = \frac{1}{10\ \Omega} = \mathbf{100\ mS}$$

In Figure 6.5b, each of the resistors has a conductance of 100 mS. Therefore, the total conductance is found as

$$G_T = G_1 + G_2 = 100\text{ mS} + 100\text{ mS} = \mathbf{200\ mS}$$

***PRACTICE PROBLEM 6.3***

A circuit like the one in Figure 6.5b has the following values: $R_1 = 100\ \Omega$ and $R_2 = 1\ \text{k}\Omega$. Calculate the conductance of each of the resistors and the total circuit conductance.

It may interest you to know that we have established another way in which series and parallel circuits can be viewed as opposites. In series circuits, the individual resistance values are additive; that is, they add up to equal the total resistance. In parallel circuits, conductance (which is the opposite of resistance) is additive.

## Calculating Total Resistance in a Parallel Circuit

◄ *OBJECTIVE 5*

We can use the conductance relationship given in equation (6.3) to develop an equation for the total resistance in a parallel circuit. Since resistance and conductance are reciprocal values, the total resistance of a parallel circuit can be found as

$$R_T = \frac{1}{G_T}$$

The total conductance ($G_T$) in a parallel circuit is equal to the sum of the branch conductance values. Therefore, the above equation can be rewritten as

$$R_T = \frac{1}{G_1 + G_2 + \ldots + G_n} \quad \textbf{(6.4)}$$

Finally, since $G = 1/R$, the relationship given in equation (6.4) can be rewritten as

$$R_T = \frac{1}{\frac{1}{R_1} + \frac{1}{R_2} + \ldots + \frac{1}{R_n}} \quad \textbf{(6.5)}$$

Equation (6.5) may look intimidating at first, but it's actually very easy to solve on any standard calculator. To solve the equation, you enter the values into your calculator in the same manner as you would any addition problem. However:

1. After entering each resistor value, you press the *reciprocal* $1/x$ key.
2. After all the values have been entered, press the = key followed by the reciprocal key.

The use of this key sequence is illustrated in Example 6.4.

## EXAMPLE 6.4

Determine the calculator key sequence you would use to solve for the value of $R_T$ in Figure 6.6.

**FIGURE 6.6**

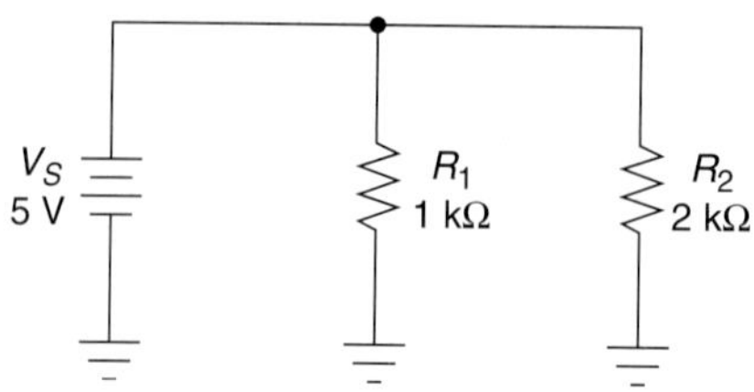


***Solution:*** The problem is solved using the following key sequence:

**1 [EE] 3** $\quad 1/x \quad + \quad$ **2 [EE] 3** $\quad 1/x \quad = \quad 1/x$

Using this key sequence, the calculator gives you a result of approximately 667 Ω.

***PRACTICE PROBLEM 6.4***

A circuit like the one in Figure 6.6 has the following values: $R_1 = 240\ \Omega$ and $R_2 = 360\ \Omega$. Solve for the value of $R_T$ for the circuit.

Another method commonly used to solve for the total resistance of a parallel circuit is the product-over-sum method. This method allows you to solve for the total resistance of a *two-branch* circuit as follows:

$$R_T = \frac{R_1 R_2}{R_1 + R_2} \tag{6.6}$$

The use of this equation (which is derived in Appendix E) is demonstrated in Example 6.5.

## EXAMPLE 6.5

Using equation (6.6), solve for the total resistance of the circuit in Figure 6.6.

***Solution:*** The value of $R_T$ for the circuit is found as

$$R_T = \frac{R_1 R_2}{R_1 + R_2} = \frac{(1\ \text{k}\Omega)(2\ \text{k}\Omega)}{1\ \text{k}\Omega + 2\ \text{k}\Omega} = \mathbf{667\ \Omega} \quad \text{(approximately)}$$

***PRACTICE PROBLEM 6.5***

Rework Practice Problem 6.4 using the product-over-sum method.

A couple of points should be made regarding equation (6.6):

1. The equation can be used only on parallel circuits *containing two branches.* As written, it will not work when there are more than two branches in the circuit.
2. Equation (6.6) actually requires more key strokes than equation (6.5). For this reason, you may find it easier to stick with equation (6.5) when solving for the total resistance of any two-branch parallel circuit.

There is another special-case method for finding the value of $R_T$. When a given number of branches *with equal resistances* are connected in parallel, the total resistance is equal to the value of one resistor divided by the number of branches. For example,

FIGURE 6.7

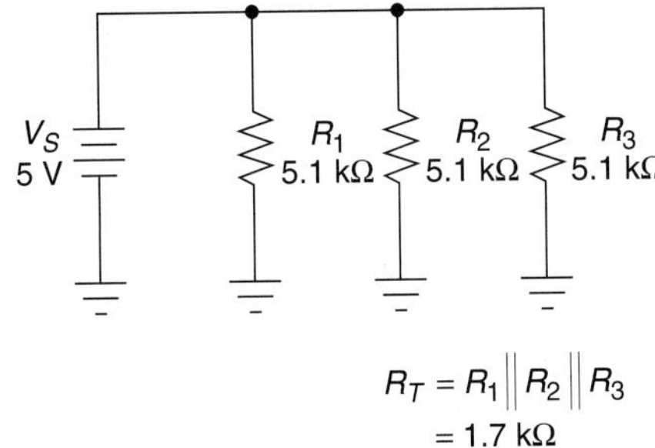


consider the circuit shown in Figure 6.7. Here we have three 5.1 kΩ resistors connected in parallel. The total resistance of this circuit can be found as

$$R_T = \frac{5.1\text{ k}\Omega}{3} = 1.7\text{ k}\Omega$$

Note that the value of one resistor (5.1 kΩ) is divided by the number of branches in the circuit (3). Again, this is a method that can be used only in a special case. All of the branch resistances must be equal in value for this method to be useful.

The equation $R_T = R_1 \| R_2 \| R_3$ shown in Figure 6.7 means that the total resistance is equal to the parallel combination of $R_1$, $R_2$, and $R_3$. Since there is more than one way to calculate the value of $R_T$ in many parallel circuits, we will use this short-hand notation in our future discussions. Just remember that the $\|$ symbol means that the resistor values are combined as parallel resistance values.

## Putting It All Together

There are three methods commonly used to calculate the total resistance of a parallel circuit. These methods and their applications are summarized in Table 6.2.

## Power Characteristics

The power characteristics of parallel circuits are nearly identical to those of series circuits. For example, the total power in a parallel circuit is found as the sum of the power dissipation values for the individual components in the circuit. Also, component power dissipation values in any parallel circuit are calculated using the same standard power relationships we used in series circuits.

As you know, power is a measure of the rate at which a component absorbs (or uses) energy. That rate does not take into account how the component is connected in a circuit. For example, a 100 Ω resistor with 100 mA of current through it uses 1 W of power, regardless of whether that component is in a series circuit or a parallel circuit. Therefore, the

**TABLE 6.2 Methods for Solving Total Parallel Resistance**

| *Equation* | *Applications* |
|---|---|
| $R_T = \dfrac{1}{\dfrac{1}{R_1} + \dfrac{1}{R_2} + \ldots + \dfrac{1}{R_n}}$ | All circuits |
| $R_T = \dfrac{R_1 R_2}{R_1 + R_2}$ | Circuits containing only two branches |
| $R_T = \dfrac{\text{value of one resistor}}{\text{number of branches}}$ | Circuits containing branches with equal resistance values |

*Don't Forget:* When the source voltage and total current are known, $R_T$ can also be found using Ohm's law.

total power used by a circuit is found as the sum of the individual power values, regardless of the circuit configuration.

In terms of power dissipation, there is one major difference between series and parallel circuits. In a series circuit, the higher the value of a resistor, the higher the percentage of the total power it dissipates. In a parallel circuit, the opposite is true; that is, the lower the value of branch resistance, the higher the percentage of the total power it dissipates. For example, the circuit in Figure 6.4 (Example 6.2) has the following power dissipation values:

$$P_1 = \frac{V_S^2}{R_1} = \frac{(5\text{ V})^2}{1\text{ k}\Omega} = 25\text{ mW}$$

and

$$P_2 = \frac{V_S^2}{R_2} = \frac{(5\text{ V})^2}{2\text{ k}\Omega} = 12.5\text{ mW}$$

As you can see, the smaller resistor is dissipating more power because power is inversely proportional to resistance when voltage is fixed.

## One Final Note

You have been shown the basic voltage, current, resistance, and power characteristics of parallel circuits. Figure 6.8 provides a comparison of a basic parallel circuit and a basic series circuit.

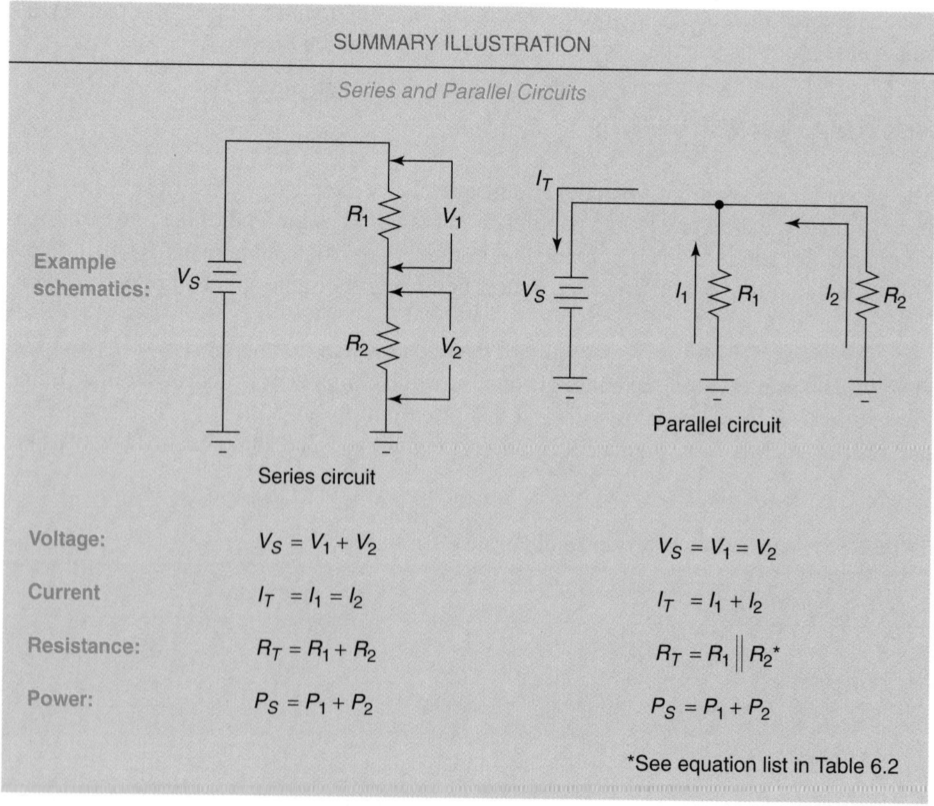


**FIGURE 6.8**

## Section Review

1. What is a branch?
2. Describe the current characteristics of a parallel circuit.
3. Describe the voltage characteristics of a parallel circuit.
4. Why are parallel circuits referred to as current dividers?
5. Why is the total resistance of a parallel circuit lower than any of its branch resistance values?
6. Compare and contrast series and parallel circuits in terms of their basic properties.

# 6.2 CURRENT RELATIONSHIPS: KIRCHHOFF'S CURRENT LAW, CURRENT SOURCES, AND THE CURRENT DIVIDER

In the last section, you were shown that the sum of the branch currents in a parallel circuit is equal to the total circuit current. In this section, we will take a closer look at the relationship between the various currents in a parallel circuit.

## Kirchhoff's Current Law

◀ *OBJECTIVE 6*

As you know, Kirchhoff's voltage law describes the relationship between the various voltages in a series circuit. In a similar fashion, Kirchhoff's current law describes the relationship between the various currents in a parallel circuit. According to **Kirchhoff's current law,** the algebraic sum of the currents entering and leaving a point must equal zero. In other words, the total current leaving any point must equal the total current entering the point. This relationship is illustrated in Figure 6.9. In Figure 6.9a, the total current shown entering the connection point ($I_T$) is 10 mA. The sum of the currents leaving the point ($I_1 + I_2$) is also 10 mA. In Figure 6.9b, the sum of the currents entering the connection point ($I_1 + I_2$) is 26 mA. The sum of the currents leaving the connection point ($I_3 + I_4$) is also 26 mA. In each circuit, the total current leaving the connection point is equal to the total current entering the point. Note that points connecting two or more current paths (like those in Figure 6.9) are referred to as **nodes.**

**Kirchhoff's current law**
A law which states that the algebraic sum of the currents entering and leaving a point must equal zero.

**Node**
A word used to describe any point that connects two or more current paths.

Kirchhoff's current equation for any node sets the algebraic sum of the currents equal to zero. For example, the Kirchhoff's current equation for Figure 6.9a is written as

$$I_T + I_1 + I_2 = 0\text{ A}$$

If we assign a positive value to the current entering the node ($I_T$) and a negative value to the currents leaving the node ($I_1$ and $I_2$), the current equation works, as follows:

$$I_T + I_1 + I_2 = 10\text{ mA} + (-2\text{ mA}) + (-8\text{ mA}) = 10\text{ mA} - 2\text{ mA} - 8\text{ mA} = 0\text{ A}$$

**FIGURE 6.9**

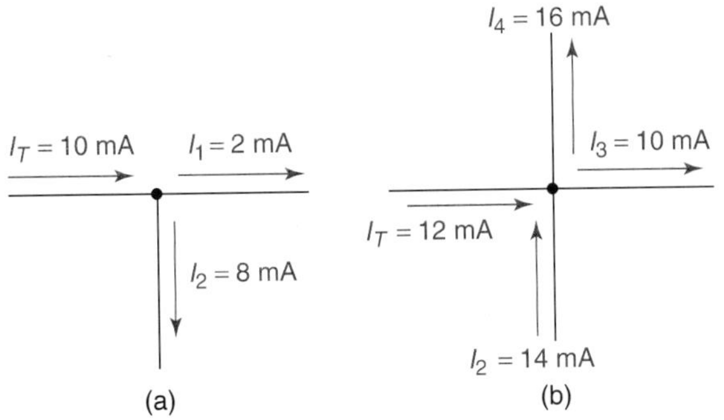

Note that the current equation for a given circuit is always solved by assigning positive and negative current values as we did in this case.

There are various practical applications for Kirchhoff's current law. One such application is demonstrated in Example 6.6.

## EXAMPLE 6.6

Determine the setting of the potentiometer required to set the total circuit current in Figure 6.10 to 28 mA.

***Solution:*** First, the value of $I_1$ is found using Ohm's law, as follows:

$$\boldsymbol{I_1} = \frac{V_S}{R_1} = \frac{12\text{ V}}{3\text{ k}\Omega} = \mathbf{4\text{ mA}}$$

To set $I_T$ to 28 mA, the sum of the branch currents must equal 28 mA. With $I_1 = 4$ mA, the value of $I_2$ is found as

$$\boldsymbol{I_2} = I_T - I_1 = 28\text{ mA} - 4\text{ mA} = \mathbf{24\text{ mA}}$$

Note that the above equation is simply a form of Kirchhoff's current law. Finally, Ohm's law is used to determine the potentiometer setting, as follows:

$$\boldsymbol{R_2} = \frac{V_S}{I_2} = \frac{12\text{ V}}{24\text{ mA}} = \mathbf{500\ \Omega}$$

Setting the potentiometer ($R_2$) to this value will set the total circuit current to 28 mA.

**FIGURE 6.10**

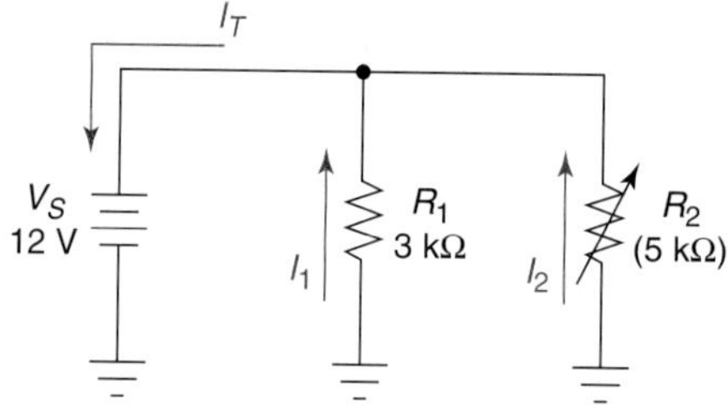


***PRACTICE PROBLEM 6.6***

A circuit like the one in Figure 6.10 has the following values: $V_S = 18$ V and $R_1 = 2$ kΩ. Determine the potentiometer setting required to set the total source current to 15 mA.

In Example 6.6, we used a potentiometer to determine the value of the total circuit current. There is a type of source, however, whose output current is fixed and relatively independent of the resistance values in the circuit. This type of source is referred to as a *current source.*

## Current Sources

*OBJECTIVE 7* ▶ Every circuit you have seen up to this point has contained a voltage source. The purpose of a dc voltage source is to provide an output voltage that remains relatively constant over a wide range of load resistance values. For example, consider the circuit shown in Figure 6.11. Ideally, the output from the dc voltage source is approximately 10 V, regardless of which load resistance value is used. When the load resistance changes from one value to another, the circuit responds with a change in current (in keeping with Ohm's law).

**Current source**
A source designed to provide an output current that remains relatively constant over a wide range of load resistance values.

A **current source** is a source designed to provide an output current value that remains relatively constant over a wide range of load resistance values. The schematic

FIGURE 6.11

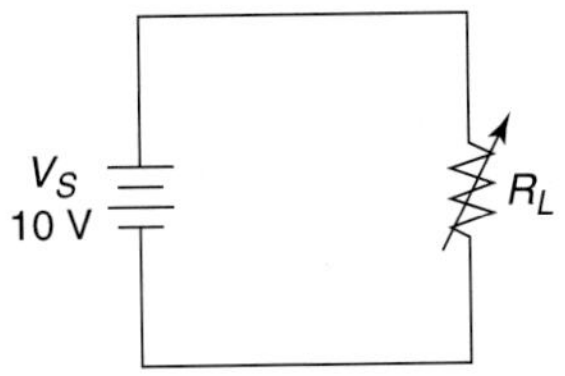


| $R_L$ | $V_L$ | $I_L$ |
|---|---|---|
| 100 Ω | 10 V | 100 mA |
| 1 kΩ | 10 V | 10 mA |
| 10 kΩ | 10 V | 1 mA |

EWB

symbol for a dc current source is shown in Figure 6.12a. Note that the output current value of the source is given and that the arrow in the symbol indicates the direction of the current.

Figure 6.12b contains a 10 mA current source and a variable load resistance. When the load resistance is changed, the circuit current remains at the rated value of the source. Thus, a change in the load resistance must result in a change in load voltage (in keeping with Ohm's law). This point is demonstrated further in Example 6.7.

*A Practical Consideration:* The relationship between source current and load resistance demonstrated in Example 6.7 is *ideal.* The ideal current source has *infinite* internal resistance, which keeps circuit current constant when the load resistance changes. (Since the source resistance is infinite, any change in load resistance has no effect on the total circuit resistance.) In practice, the internal resistance of a current source is less than infinite, so its output current *does* vary slightly when load resistance changes. This point is discussed further in Chapter 8.

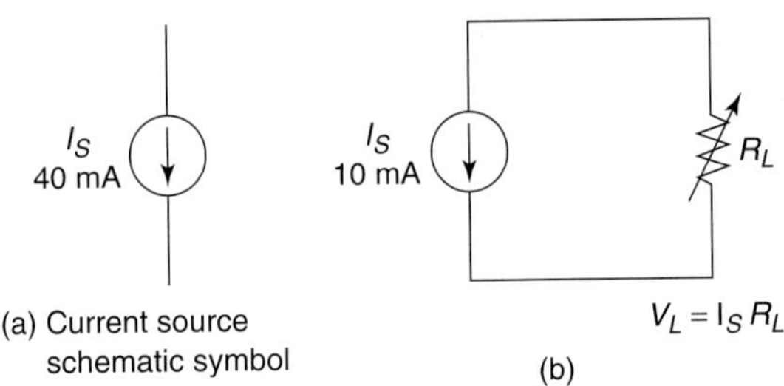


FIGURE 6.12

EWB

### EXAMPLE 6.7

Calculate the change in $V_L$ that occurs in Figure 6.13 if $R_L$ is adjusted from 1 kΩ to 4 kΩ.

***Solution:*** When $R_L$ = 1 kΩ, the value of $V_L$ is found as

$$V_L = I_S R_L = (1.5\text{ mA})(1\text{ k}\Omega) = \mathbf{1.5\text{ V}}$$

The value of the circuit current is fixed at 1.5 mA. Therefore, when $R_L$ changes to 4 kΩ, $V_L$ changes as follows:

$$V_L = I_S R_L = (1.5\text{ mA})(4\text{ k}\Omega) = \mathbf{6\text{ V}}$$

As you can see, the change in load resistance causes $V_L$ to change from 1.5 V to 6 V. The value of the circuit current (in this case) is independent of the load resistance.

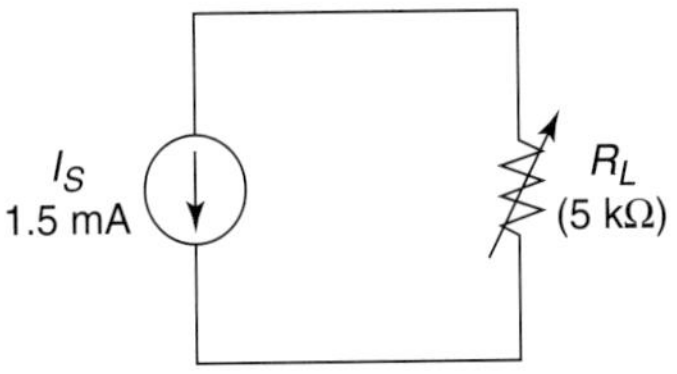

$I_S$ 1.5 mA — $R_L$ (5 kΩ)

FIGURE 6.13

***PRACTICE PROBLEM 6.7***

A circuit like the one in Figure 6.13 has a 100 μA current source. Calculate the change in $V_L$ that occurs if $R_L$ is changed from 12 kΩ to 32 kΩ.

When a current source is used in a parallel circuit, the process for determining the values of the branch currents can get a bit more complicated. This point is illustrated in Example 6.8.

## EXAMPLE 6.8

Calculate the values of $I_1$ and $I_2$ for the circuit in Figure 6.14.

**FIGURE 6.14**

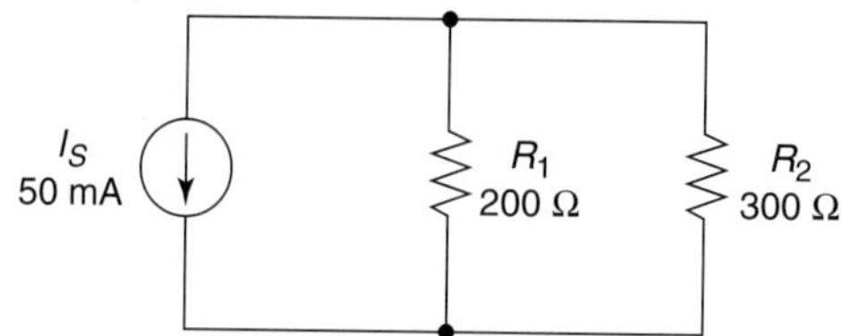


***Solution:*** Before we can find the values of $I_1$ and $I_2$, we need to know the value of the voltage across the two branches. To determine the value of the source voltage, we have to start by finding the value of $R_T$ for the circuit, as follows:

$$R_T = R_1 \| R_2 = (200\ \Omega) \| (300\ \Omega) = \mathbf{120\ \Omega}$$

Using Ohm's law, we can now find the value of the source voltage, as follows:

$$V_S = I_S R_T = (50\text{ mA})(120\ \Omega) = \mathbf{6\ V}$$

Now, using $V_S = 6$ V, the values of the branch currents can be calculated using Ohm's law, as follows:

$$I_1 = \frac{V_S}{R_1} = \frac{6\text{ V}}{200\ \Omega} = \mathbf{30\ mA}$$

and

$$I_2 = \frac{V_S}{R_2} = \frac{6\text{ V}}{300\ \Omega} = \mathbf{20\ mA}$$

Note that the sum of $I_1$ and $I_2$ is 50 mA: the given value of the output from the current source.

***PRACTICE PROBLEM 6.8***

A circuit like the one in Figure 6.14 has the following values: $I_S = 25$ mA, $R_1 = 100\ \Omega$, and $R_2 = 150\ \Omega$. Calculate the values of $I_1$ and $I_2$ for the circuit.

There is a simpler way to solve a problem like the one in Example 6.8. When a current source is used in a parallel circuit, we can treat the circuit as a *current divider.*

## Current Dividers

*OBJECTIVE 8* ▶ You may recall that a series circuit can be viewed as a voltage divider because the source voltage is divided among the resistors in the circuit. In a similar fashion, a parallel circuit can be viewed as a current divider because the source current is divided among the branches in the circuit.

A comparison of voltage dividers and current dividers is provided in Figure 6.15. Note the difference in the resistance ratios used in the divider equations. For the voltage divider, the source voltage is multiplied by the ratio of the component resistance to the total circuit resistance. For the current divider, the source current is multiplied by the ratio of the total circuit resistance to the component (or branch) resistance, as follows:

$$I_n = I_S \frac{R_T}{R_n} \tag{6.7}$$

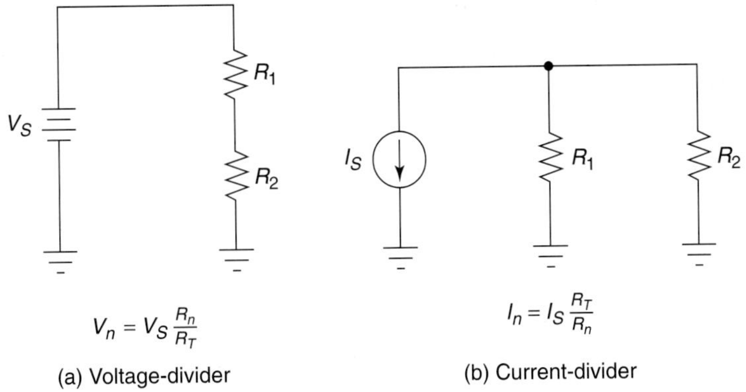


**FIGURE 6.15**

Again, we have a situation where series and parallel circuits have opposite characteristics. Example 6.9 demonstrates the use of equation (6.7).

## EXAMPLE 6.9

Calculate the branch currents for the circuit in Figure 6.16. (This is the same circuit we analyzed in Example 6.8.)

**FIGURE 6.16**

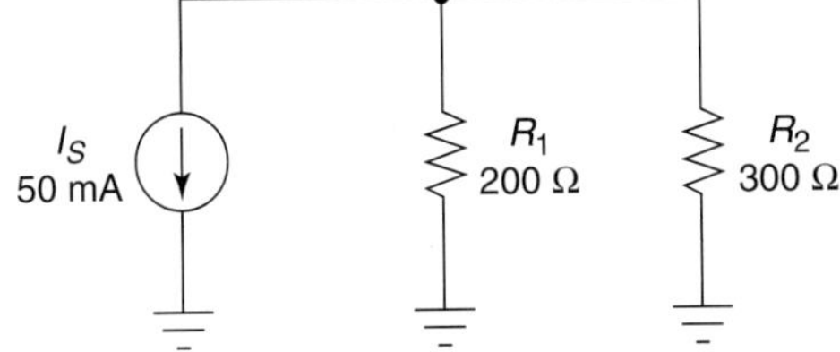


***Solution:*** First, the total resistance in the circuit is found as

$$\boldsymbol{R_T} = R_1 \| R_2 = 200\ \Omega \| 300\ \Omega = \mathbf{120\ \Omega}$$

Now, the value of $I_1$ can be found as

$$\boldsymbol{I_1} = I_S \frac{R_T}{R_1} = (50\text{ mA})\frac{120\ \Omega}{200\ \Omega} = \mathbf{30\ mA}$$

Finally, the value of $I_2$ can be found as

$$\boldsymbol{I_2} = I_S \frac{R_T}{R_2} = (50\text{ mA})\frac{120\ \Omega}{300\ \Omega} = \mathbf{20\ mA}$$

***PRACTICE PROBLEM 6.9***

Rework Practice Problem 6.8 using the technique demonstrated in this example.

In Example 6.9, we could have solved for either branch current value and then subtracted that value from the source current to obtain the value of the other branch current. For example, we could have solved for $I_1$, and then found $I_2$ as

$$I_2 = I_S - I_1$$

However, we solved the circuit as shown so that you can see how the current-divider equation works for both branch currents.

## The Basis for Equation (6.7)

We know that the voltages across parallel branches are equal. For either branch in Figure 6.15b,

$$V_n = V_S$$

or

$$I_n R_n = I_S R_T$$

If we divide both sides of the above equation by $R_n$, we get

$$I_n = I_S \frac{R_T}{R_n}$$

which is equation (6.7).

## Solving Current Dividers: Another Approach

Equation (6.7) can be modified to provide us with another easy method of determining the values of the branch currents in a two-branch circuit. For example, the branch current values for the circuit in Figure 6.15b can be found as

$$I_1 = I_S \frac{R_2}{R_1 + R_2} \qquad \textbf{(6.8)}$$

and

$$I_2 = I_S \frac{R_1}{R_1 + R_2} \qquad \textbf{(6.9)}$$

Note that in each of the above equations *the resistance of the other branch is used in the numerator of the resistance ratio.* If we want to determine the value of $I_1$, we use the resistance of the $I_2$ branch in the equation, and vice versa. We will discuss the reason for this in a moment. First, we'll look at how these equations are used.

### EXAMPLE 6.10

Using the current-divider equations, solve for the branch current values in Figure 6.16. (This figure is located in Example 6.9.)

***Solution:*** The value of $I_1$ is found as

$$I_1 = I_S \frac{R_2}{R_1 + R_2} = (50 \text{ mA}) \frac{300\ \Omega}{500\ \Omega} = \mathbf{30\ mA}$$

The value of $I_2$ is now found as

$$I_2 = I_S \frac{R_1}{R_1 + R_2} = (50 \text{ mA}) \frac{200\ \Omega}{500\ \Omega} = \mathbf{20\ mA}$$

If you look back at Examples 6.8 and 6.9, you'll see that these are the same results we obtained using the procedures established in those examples.

***PRACTICE PROBLEM 6.10***

Using the current-divider equations, rework Practice Problem 6.8.

Once again, we could have solved for either branch current value and then subtracted that value from the source current to obtain the value of the other branch current.

## The Basis for Equations (6.8) and (6.9)

In Figure 6.15, you were shown that the value of a given branch current can be found as

$$I_n = I_S \frac{R_T}{R_n}$$

For example, the value of $I_1$ in Figure 6.15b could be found as

$$I_1 = I_S \frac{R_T}{R_1} \qquad \text{or} \qquad I_1 = I_S\left(R_T \frac{1}{R_1}\right)$$

Using the product-over-sum method, the value of $R_T$ for the circuit would be found as

$$R_T = \frac{R_1 R_2}{R_1 + R_2}$$

If we use this fraction in place of $R_T$ in the equation for $I_1$, we get

$$I_1 = I_S\left(\frac{R_1 R_2}{R_1 + R_2} \times \frac{1}{R_1}\right) = I_S\left(\frac{R_1 R_2}{R_1(R_1 + R_2)}\right) = I_S \frac{R_2}{R_1 + R_2}$$

As you can see, this is equation (6.8). By performing the same manipulations on the current-divider equation for $I_2$, we would obtain equation (6.9).

## The Bottom Line

The relationship between the various currents in a parallel circuit is described by Kirchhoff's current law. In effect, this law states that the sum of the branch currents in a parallel circuit must equal the source current.

In most cases, the values of the branch currents in a parallel circuit can be found using Ohm's law. When the branch current values are known, they are simply added to find the value of the total circuit current.

A current source is one whose output current remains relatively constant for a wide range of load resistance values. When a current source is used in a parallel circuit, the process of determining the values of the branch currents is greatly simplified by using the current-divider equations.

### Section Review

1. What is Kirchhoff's current law?
2. What is a current source?
3. How does a circuit with a current source respond to a change in load resistance?
4. Describe how you would solve for the various current values in a parallel circuit containing a voltage source.
5. Describe how you would solve for the various current values in a parallel circuit containing a current source.

# 6.3 PARALLEL CIRCUIT ANALYSIS, FAULT SYMPTOMS, AND TROUBLESHOOTING

At this point, you have been exposed to the basic operating characteristics of parallel circuits. In this section, we will look at the complete analysis of a parallel circuit. We will also discuss the characteristic symptoms of several circuit faults and the basic approach to parallel circuit troubleshooting.

## Parallel Circuit Analysis

*OBJECTIVE 9* ▶ The complete analysis of a parallel circuit involves determining the values of $R_T$, $I_T$, and $P_T$, along with the individual branch current and power values. The following example demonstrates the complete procedure.

### EXAMPLE 6.11

EWB

Perform a complete mathematical analysis of the circuit in Figure 6.17.

**FIGURE 6.17**

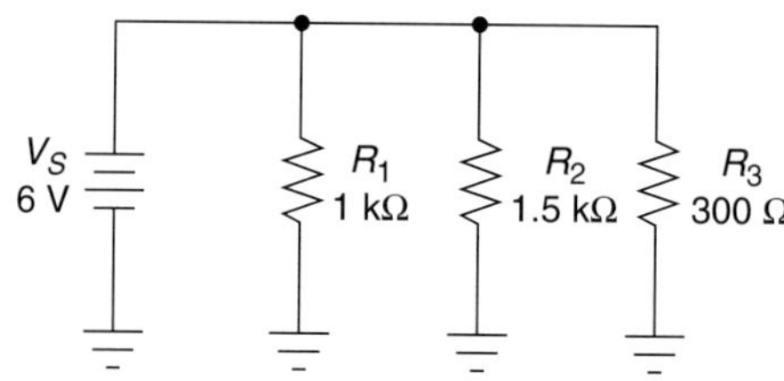


***Solution:*** First, we will solve for the total circuit resistance, as follows:

$$R_T = R_1 \| R_2 \| R_3 = 1\text{ k}\Omega \| 1.5\text{ k}\Omega \| 300\ \Omega = \mathbf{200\ \Omega}$$

Now we will solve for the individual branch currents, as follows:

$$I_1 = \frac{V_S}{R_1} = \frac{6\text{ V}}{1\text{ k}\Omega} = \mathbf{6\text{ mA}}$$

$$I_2 = \frac{V_S}{R_2} = \frac{6\text{ V}}{1.5\text{ k}\Omega} = \mathbf{4\text{ mA}}$$

and

$$I_3 = \frac{V_S}{R_3} = \frac{6\text{ V}}{300\ \Omega} = \mathbf{20\text{ mA}}$$

Using these branch current values and Kirchhoff's current law, the value of the source current is now found as

$$I_S = I_1 + I_2 + I_3 = 6\text{ mA} + 4\text{ mA} + 20\text{ mA} = \mathbf{30\text{ mA}}$$

Now we'll calculate the power dissipation value of each of the branch resistors, as follows:

$$P_1 = V_S I_1 = (6\text{ V})(6\text{ mA}) = \mathbf{36\text{ mW}}$$

$$P_2 = V_S I_2 = (6\text{ V})(4\text{ mA}) = \mathbf{24\text{ mW}}$$

and

$$P_3 = V_S I_3 = (6\text{ V})(20\text{ mA}) = \mathbf{120\text{ mW}}$$

Now we can use these power dissipation values to find the total power dissipation of the circuit, as follows:

$$P_T = P_1 + P_2 + P_3 = 36\text{ mW} + 24\text{ mW} + 120\text{ mW} = \mathbf{180\text{ mW}}$$

Incidentally, had we calculated $P_T$ as the product of the source voltage (6 V) and the total circuit current (30 mA), we would have obtained the same result: 180 mW.

***PRACTICE PROBLEM 6.11***

A circuit like the one in Figure 6.17 has the following values: $V_S = 10$ V, $R_1 = 2$ kΩ, $R_2 = 1$ kΩ, and $R_3 = 10$ kΩ. Perform a complete mathematical analysis of the circuit.

## Fault Symptoms: An Open Branch

◄ *OBJECTIVE 10*

You may recall that an open component in a series circuit can easily be found by measuring the voltage across each of the components in the circuit. The total voltage is dropped across the open component.

An open branch in a parallel circuit is more difficult to diagnose than an open resistor in a series circuit. The reason for this is illustrated in Figure 6.18. As you can see, the voltage readings in the two circuits are identical. The fact that $R_2$ has opened has had no effect on the voltage measured across that branch. Therefore, voltage readings will not help to diagnose an open component in a parallel circuit (as they will in a series circuit).

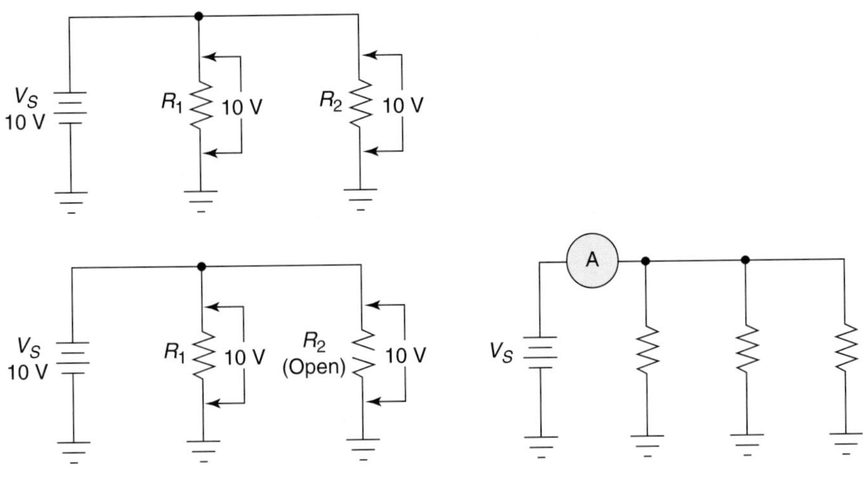


**FIGURE 6.18**

**FIGURE 6.19**

One way of locating an open branch is illustrated in Figure 6.19. An ammeter is connected as shown in the figure. Then one of the branches is disconnected. If the disconnected branch is already open, there will be no change in the ammeter reading. If the branch is not open, the ammeter reading will decrease when the branch is disconnected. This change is caused by the decrease in current that results when a branch is removed from a parallel circuit. (Remember: The more branches in a parallel circuit, the higher the total circuit current.)

If disconnecting a branch causes a decrease in the ammeter reading, reconnect the branch and repeat the process with the next branch. Proceed until you hit the branch that does not cause a decrease in the reading when disconnected. That branch is open.

*A Practical Consideration:* The procedure demonstrated in Figure 6.19 is not easy to perform when working with printed circuit (PC) boards. However, if it must be used to locate a fault, a branch can be effectively removed from the circuit by disconnecting one end of the component in the branch.

## A Practical Consideration

In most cases, a parallel circuit is made up of components other than resistors. For example, consider the circuit shown in Figure 6.20. This circuit consists of three dc motors connected in parallel to a voltage source. The source in this circuit provides the voltage required to drive each of the motors.

**FIGURE 6.20**

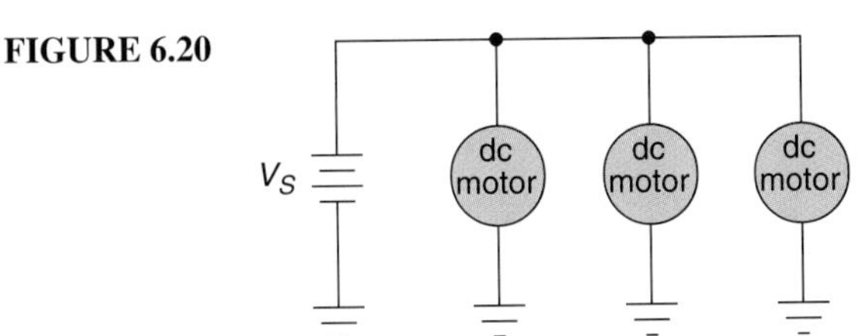

If a motor in Figure 6.20 opens, it simply stops working. At the same time, the other two motors continue to operate normally. Thus, there are no measurements, current or otherwise, required to find the open component. The motor that isn't running is the open component.

## Fault Symptoms: A Shorted Branch

*OBJECTIVE 11* ▶ When a branch in a parallel circuit shorts, the total circuit resistance drops to approximately 0 Ω, as illustrated in Figure 6.21a. You know that the total resistance in a parallel circuit must be lower than any of the individual branch resistance values. When $R_3$ in Figure 6.21a is shorted, the lowest branch resistance drops to nearly 0 Ω. Therefore, the total circuit resistance cannot be greater than the resistance of the shorted branch.

The low circuit resistance causes a drastic increase in the source current. In the case of the circuit in Figure 6.21a, this increase in current will cause one of two things to happen:

1. The resistor that is shorted will eventually burn open.
2. The source will burn out.

*A Practical Consideration:* Most voltage sources contain internal fuses (or circuits) designed to protect the source from a shorted output. This fuse (or circuit) in the voltage source may not be visible without opening the source.

If the shorted branch burns open, the circuit will take on the characteristics of an open circuit. However, since the resistor has burned, it will be easy to spot without the use of test equipment. If the source burns out, the circuit voltage drops to 0 V, the circuit current drops to 0 A, and the source must be replaced.

The circuit in Figure 6.21b contains a fuse (which is more realistic). If a branch shorts in this circuit, the fuse will open, preventing further damage. If the fuse is replaced before the short is corrected, the new fuse will also blow. Therefore, a circuit that continually blows fuses contains a shorted component or branch.

When a fuse is present in a parallel circuit, the only safe way to detect the source of the short is to disconnect each branch in the circuit and measure its resistance. The resistance of the shorted branch (as you know) will be nearly 0 Ω.

## A Word of Caution

Some technicians you meet may attempt to find a short like the one in Figure 6.21b by defeating the fuse; that is, they may connect a wire across the fuse holder to force the circuit to continue operating under conditions that would normally blow the fuse. The goal of this trick is to cause the shorted component to burn out, thus giving them an easy way to determine which branch is shorted. *This type of shortcut is extremely dangerous!* Not only can it cause a fire, it can also damage the circuit's power supply and may cause injury to the technician. For your own safety, always use the established troubleshooting techniques rather than potentially dangerous shortcuts.

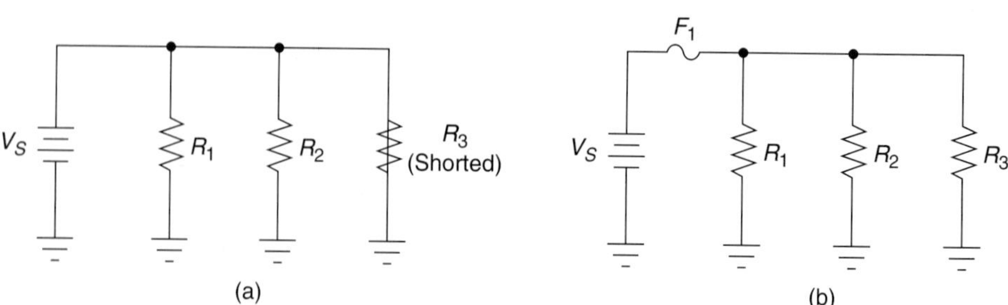


**FIGURE 6.21**

## Parallel Circuit Troubleshooting

The first step in troubleshooting a parallel circuit is to determine whether you are looking for a shorted branch or an open branch. In most cases, the type of problem you are looking for is identified by the circuit fuse or the power supply. For example, consider the circuits shown in Figure 6.22. ◄ *OBJECTIVE 12*

The circuit in Figure 6.22a contains an open branch. As you can see, the source voltage is normal and the fuse is intact. The circuit in Figure 6.22b contains a shorted branch, as indicated by the blown fuse.

The circuit in Figure 6.22c contains a shorted branch but does not have a fuse to indicate the problem. In this case, the output from the power supply is close to 0 V. The reason for this low output voltage is shown in Figure 6.22d. The shorted branch causes the total resistance of the parallel network to drop to zero. Therefore, the only resistance in the entire circuit is the internal resistance of the voltage source. According to Kirchhoff's voltage law, all the source voltage would have to be dropped across this internal resistance. At the same time, the increase in circuit current will cause the voltage source to get very hot.

As a summary, here are the primary indicators of the type of problem in a parallel circuit:

1. For a circuit that has a fuse:
   a. If the fuse is blown, you're probably looking for a shorted branch.
   b. If the fuse is good, you're probably looking for an open branch.
2. For a circuit that does not have a fuse:
   a. If the source voltage is normal, you're probably looking for an open branch.
   b. If the source voltage is low and the source is hot, you're looking for a shorted branch.

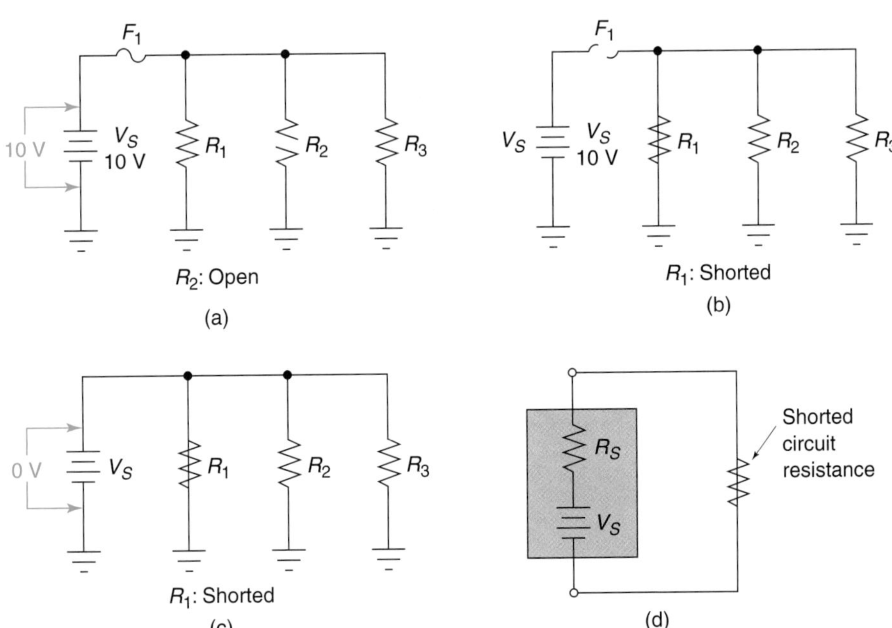


**FIGURE 6.22**

Let's tie all this information together by going through a few troubleshooting applications.

## TROUBLESHOOTING APPLICATION 1

The circuit in Figure 6.23 has a blown fuse. When the fuse is replaced, the new fuse quickly blows. In this case, the circuit current is extremely high, indicating that there is a shorted branch. The branches of the circuit are disconnected and the resistance of each branch is checked. The results of the resistance measurements are as follows: $R_1$ is measured at 998 Ω, $R_2$ is measured at 0 Ω, and $R_3$ is measured at 2.21 kΩ. From these readings, it is obvious that $R_2$ is shorted and must be replaced.

*AN OBSERVATION* You may be wondering why we bothered to check $R_3$ after $R_2$ was found to be shorted. The reason is that it is possible (although very unlikely) that more than one branch could be shorted. Therefore, the resistance of every branch should be checked.

## TROUBLESHOOTING APPLICATION 2

The circuit in Figure 6.24 has the readings shown. Motor 3 is not working and the fuse has not blown. In this case, an open is indicated by the good fuse and the normal power supply output voltage. Since two of the motors are working, the connection between the top of motor 2 and the top of motor 3 is checked and found to be open. Replacing the connection restores normal operation to the circuit.

*AN OBSERVATION* Why did we check the connection between the motors before checking the bad motor? If you look at the figure, you'll see that $V_1$ and $V_2$ are normal while $V_3 = 0$ V. The only condition that could cause this combination of voltage readings is an open between motor 3 and the other two motors. (A shorted motor would blow the fuse, isolating the supply voltage from all the motors.) The open in the motor 3 circuit isolates the supply voltage from motor 3 (causing $V_3$ to equal 0 V) while not affecting the values of $V_1$ and $V_2$.

As you can see, the approach to troubleshooting parallel circuits is very similar to the one used for series circuits. It involves taking the appropriate circuit measurements and thinking about what those measurements tell you.

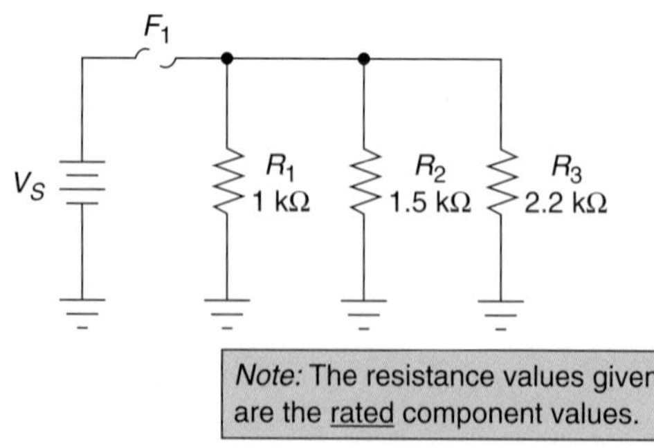


FIGURE 6.23

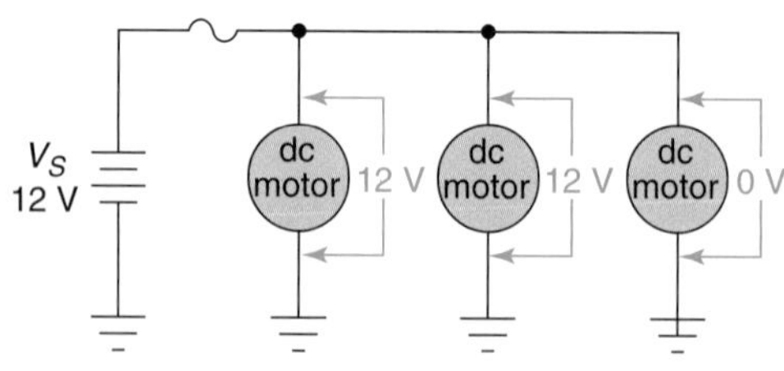


FIGURE 6.24

### Section Review

1. An open branch in a parallel circuit is more difficult to diagnose than an open resistor in a series circuit. Why?
2. List the steps you would take to find an open branch in a three-branch parallel circuit.

3. If a branch in a parallel circuit shorts, the total resistance drops to nearly 0 Ω. Why?
4. In a parallel circuit that does not contain a fuse, what are the possible outcomes of a shorted branch?
5. Why is it dangerous to bypass the fuse in any circuit?
6. List the steps you would take to find a shorted branch in a three-branch parallel circuit that does not contain a fuse.

## Chapter Summary

Here is a summary of the major points made in this chapter:

1. A parallel circuit is a circuit that provides more than one current path between the terminals of the source.
2. Each current path in a parallel circuit is referred to as a *branch.*
3. The total current in a parallel circuit is equal to the sum of the branch currents.
4. The voltage across each branch in a parallel circuit is equal to the voltage across all the other branches.
5. The current through each branch in a parallel circuit is determined by the source voltage and the resistance of the branch.
6. In terms of their voltage and current characteristics, parallel and series circuits are opposites:
    a. Currents are equal throughout series circuits and additive in parallel circuits.
    b. Voltages are additive in series circuits and equal across all branches in parallel circuits.
7. Parallel circuits are often referred to as current dividers because the source current is divided among the branches.
8. The total resistance in a parallel circuit is lower than any of the branch resistance values.
9. The total conductance in a parallel circuit is equal to the sum of the branch conductance values.
10. There are several means by which the total resistance of a parallel circuit can be calculated. The resistance equations for parallel circuits are summarized in Table 6.2.
11. The power characteristics of parallel circuits are nearly identical to those of series circuits.
    a. The total power in a parallel circuit is found as the sum of the branch power values.
    b. In parallel circuits, the lower-resistance branches dissipate the greatest amount of power. (In series circuits, the higher-value resistors dissipate the greatest amount of power.)
12. Kirchhoff's current law states that the algebraic sum of the currents entering and leaving a point must equal zero. In other words, it states that the total current leaving any point must equal the current entering the point.
13. A current source is one designed to provide an output current value that remains constant over a wide range of load resistance values. The schematic symbol for a dc current source indicates the value and direction of its output current.
14. In any circuit containing a current source, a change in load resistance results in a change in load voltage. Load current remains relatively fixed.
15. The current-divider equation is a valuable tool for calculating the values of the branch currents in a parallel circuit that contains a current source.

16. An open branch in a parallel circuit is more difficult to diagnose than an open resistor in a series circuit. An open branch is normally found using a technique that involves disconnecting the circuit branches while measuring the total circuit current.

17. When a parallel circuit contains a fuse, a shorted branch will cause the fuse to blow.

18. When a parallel circuit is constructed without a fuse, a shorted branch will generally cause one (or both) of the following to occur:

    a. The resistance in the shorted branch eventually burns open.

    b. The voltage source gets very warm and its output drops to an extremely low value.

19. Bypassing the fuse in any circuit is extremely dangerous and should never be attempted.

20. In a parallel circuit containing a fuse, the condition of the fuse is a good indicator of the type of problem you're looking for.

    a. If the fuse has blown, you're looking for a short.

    b. If the fuse is good, you're looking for an open.

## Equation Summary

| Equation Number | Equation | Section Number |
|---|---|---|
| **(6.1)** | $I_T = I_1 + I_2 + \ldots + I_n$ | 6.1 |
| **(6.2)** | $V_S = V_1 = V_2 = \ldots = V_n$ | 6.1 |
| **(6.3)** | $G_T = G_1 + G_2 + \ldots + G_n$ | 6.1 |
| **(6.4)** | $R_T = \dfrac{1}{G_1 + G_2 + \ldots + G_n}$ | 6.1 |
| **(6.5)** | $R_T = \dfrac{1}{\dfrac{1}{R_1} + \dfrac{1}{R_2} + \ldots + \dfrac{1}{R_n}}$ | 6.1 |
| **(6.6)** | $R_T = \dfrac{R_1 R_2}{R_1 + R_2}$ | 6.1 |
| **(6.7)** | $I_n = I_S \dfrac{R_T}{R_n}$ | 6.2 |
| **(6.8)** | $I_1 = I_S \dfrac{R_2}{R_1 + R_2}$ | 6.2 |
| **(6.9)** | $I_2 = I_S \dfrac{R_1}{R_1 + R_2}$ | 6.2 |

## Key Terms

The following terms were introduced and defined in this chapter:

| | | |
|---|---|---|
| branch | current source | parallel circuit |
| current divider | Kirchhoff's current law | |

## Practice Problems

1. The table below refers to Figure 6.25. For each combination of branch currents, find the total circuit current.

| | $I_1$ | $I_2$ | $I_3$ | $I_T$ |
|---|---|---|---|---|
| a. | 12 mA | 2.8 mA | 6.6 mA | ______ |
| b. | 5.5 mA | 360 μA | 1.04 mA | ______ |
| c. | 43 mA | 25 mA | 3.3 mA | ______ |
| d. | 500 μA | 226 μA | 1.08 mA | ______ |

2. The table below refers to Figure 6.25. For each combination of branch currents, find the total circuit current.

| | $I_1$ | $I_2$ | $I_3$ | $I_T$ |
|---|---|---|---|---|
| a. | 1.5 mA | 3.3 mA | 17 mA | ______ |
| b. | 480 μA | 20 mA | 5.1 mA | ______ |
| c. | 52 mA | 28 mA | 800 μA | ______ |
| d. | 10 mA | 112 mA | 34 mA | ______ |

3. Determine the values of the branch currents and the total current for the circuit in Figure 6.26a.
4. Determine the values of the branch currents and the total current for the circuit in Figure 6.26b.
5. Determine the values of the branch currents and the total current for the circuit in Figure 6.27a.
6. Determine the values of the branch currents and the total current for the circuit in Figure 6.27b.

FIGURE 6.25

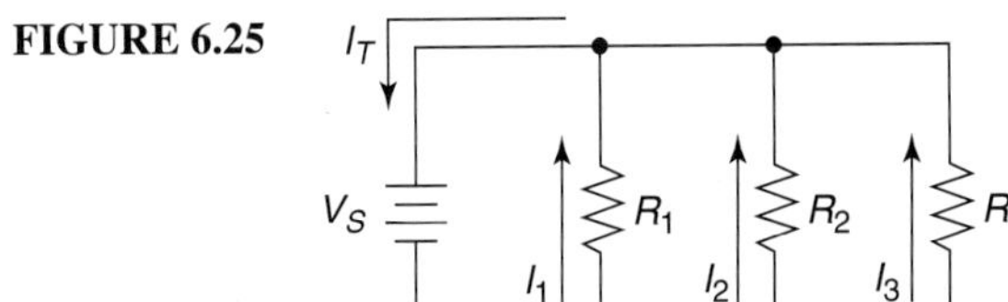


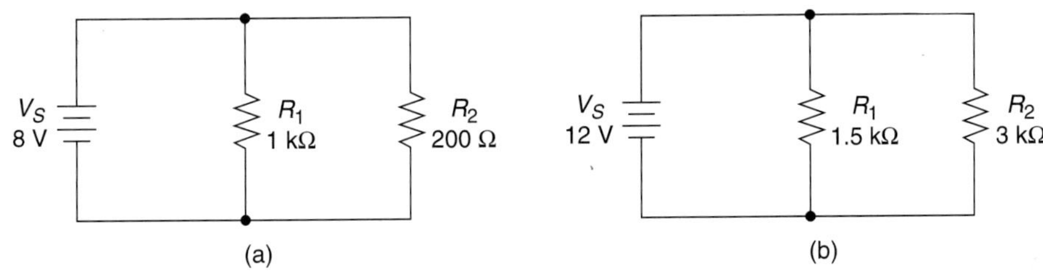


FIGURE 6.26

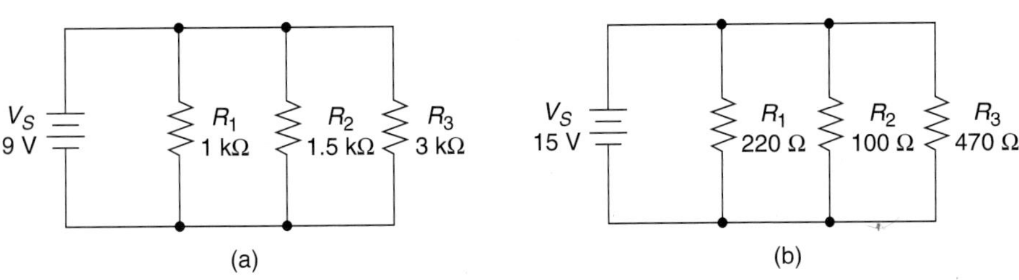


FIGURE 6.27

7. Calculate the value of $R_T$ for the circuit in Figure 6.26a.
8. Calculate the value of $R_T$ for the circuit in Figure 6.26b.
9. Calculate the value of $R_T$ for the circuit in Figure 6.27a.
10. Calculate the value of $R_T$ for the circuit in Figure 6.27b.
11. Calculate the value of $R_T$ for the circuit in Figure 6.28a.
12. Calculate the value of $R_T$ for the circuit in Figure 6.28b.
13. Calculate the value of $R_T$ for the circuit in Figure 6.29a.
14. Calculate the value of $R_T$ for the circuit in Figure 6.29b.
15. Determine the potentiometer setting required to set the total circuit current in Figure 6.30a to 50 mA.
16. Determine the potentiometer setting required to set the total circuit current in Figure 6.30b to 36 mA.
17. For the circuit in Figure 6.31a, determine the change in $V_L$ that occurs when $R_L$ is changed from 1.2 kΩ to 3.3 kΩ.
18. For the circuit in Figure 6.31b, determine the change in $V_L$ that occurs when $R_L$ is changed from 470 Ω to 220 Ω.

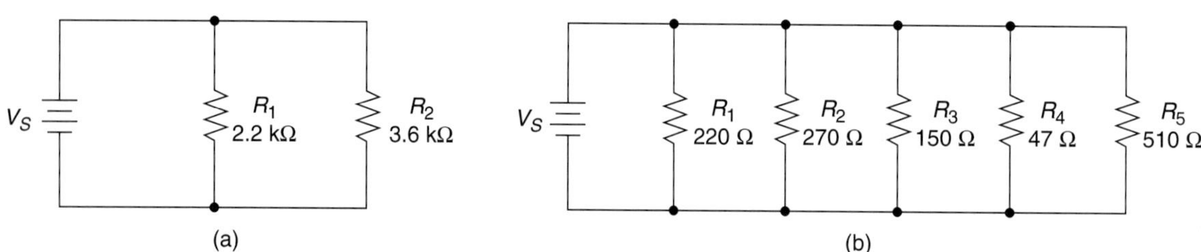


**FIGURE 6.28**

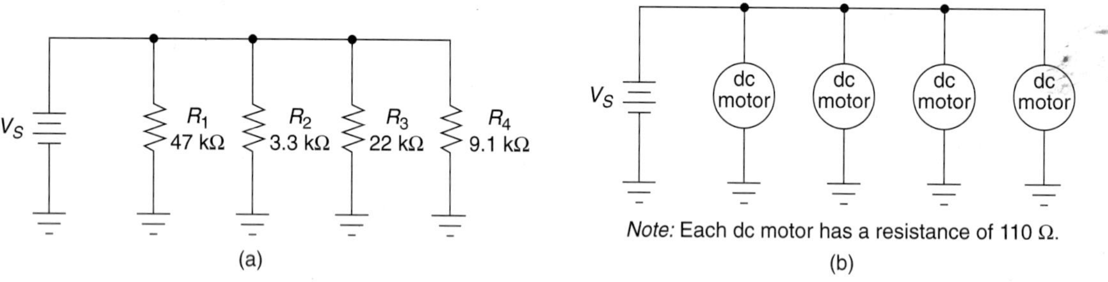


**FIGURE 6.29**

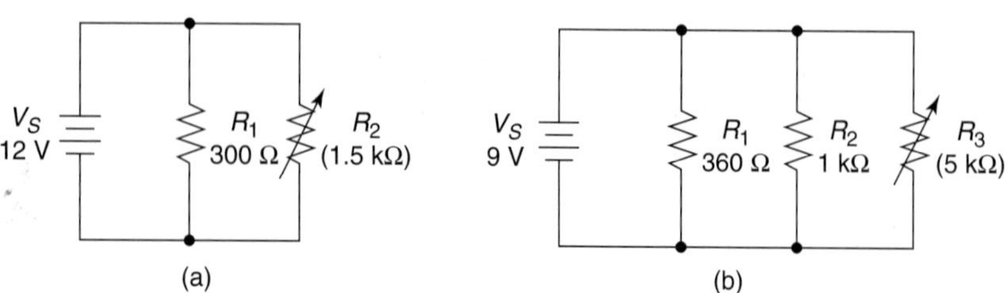


**FIGURE 6.30**

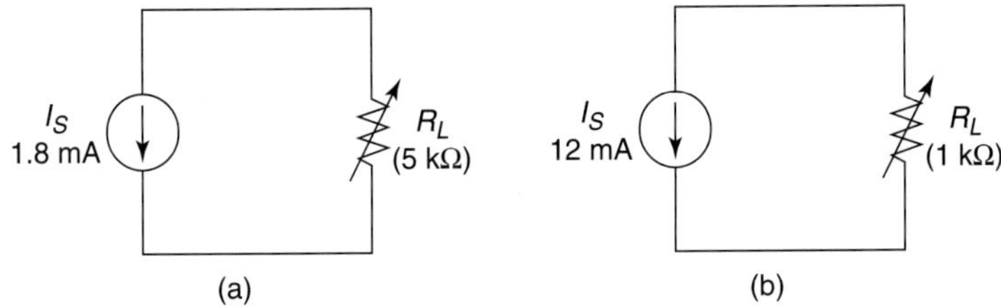


**FIGURE 6.31**

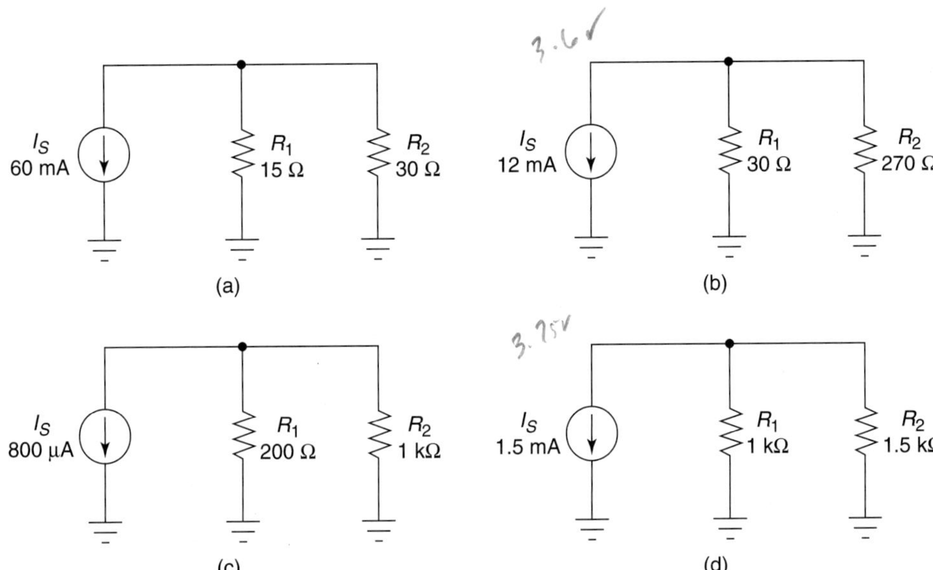


**FIGURE 6.32**

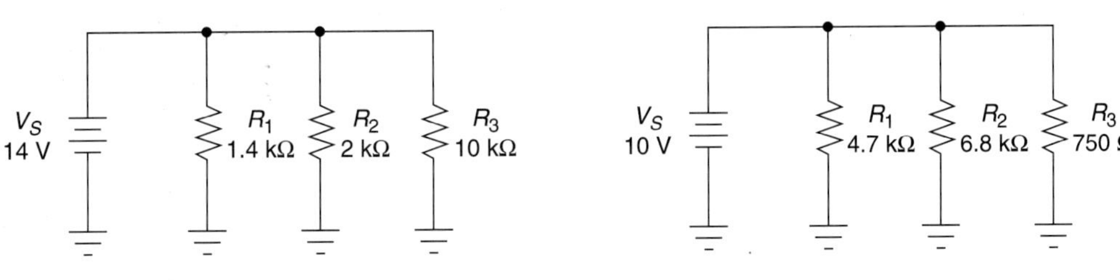


**FIGURE 6.33**

**FIGURE 6.34**

19. Determine the branch current values for the circuit in Figure 6.32a.
20. Determine the branch current values for the circuit in Figure 6.32b.
21. Determine the branch current values for the circuit in Figure 6.32c.
22. Determine the branch current values for the circuit in Figure 6.32d.
23. Perform a complete mathematical analysis of the circuit in Figure 6.33.
24. Perform a complete mathematical analysis of the circuit in Figure 6.34.
25. Perform a complete mathematical analysis of the circuit in Figure 6.35.
26. Perform a complete mathematical analysis of the circuit in Figure 6.36.

FIGURE 6.35

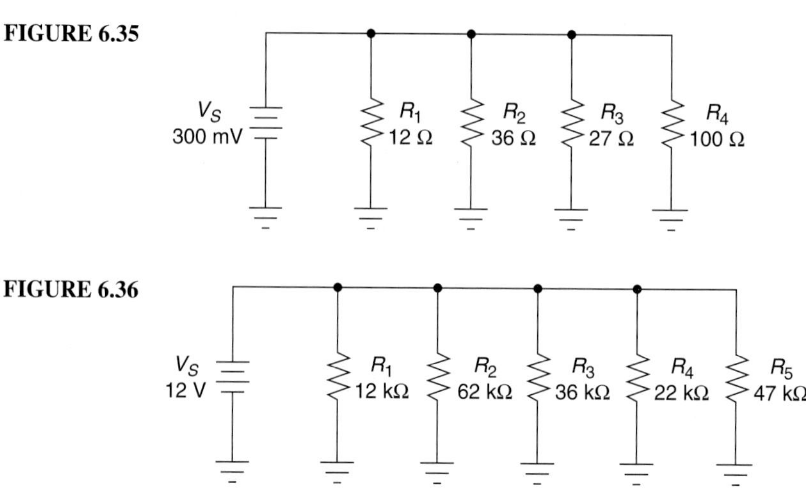


FIGURE 6.36

## Troubleshooting Practice Problems

**27.** Determine the fault (if any) indicated by the readings in Figure 6.37a.

**28.** Determine the fault (if any) indicated by the readings in Figure 6.37b.

**29.** Determine the fault (if any) indicated by the readings in Figure 6.38a.

**30.** Determine the fault (if any) indicated by the readings in Figure 6.38b.

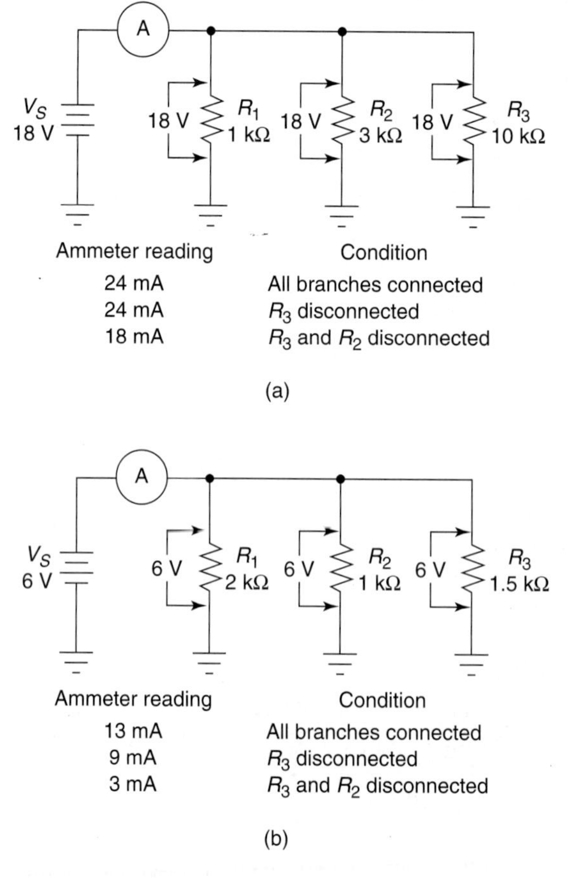


FIGURE 6.37

**FIGURE 6.38**

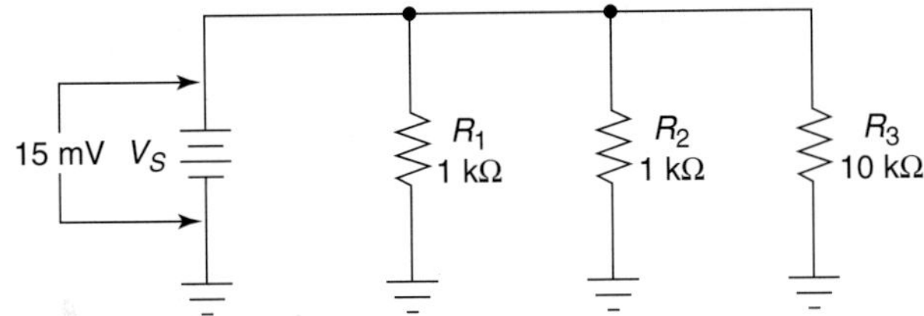


*Note:* The voltage source is rated at 10 V and is disconnected for all resistance measurements.

| Measured total resistance | Condition |
|---|---|
| 10 mΩ | All resistors connected |
| 10 mΩ | $R_3$ disconnected |
| 1 kΩ | $R_3$ and $R_2$ disconnected |

(a)

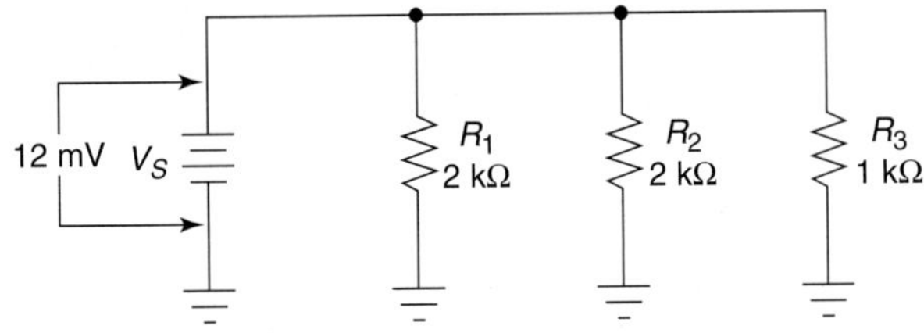


*Note:* The voltage source is rated at 15 V and is disconnected for all resistance measurements.

| Measured total resistance | Condition |
|---|---|
| 500 Ω | All resistors connected |
| 1 kΩ | $R_3$ disconnected |
| 2 kΩ | $R_3$ and $R_2$ disconnected |

(b)

---

*The Brain Drain*

**31.** For each of the circuits shown in Figure 6.39, find the missing value(s).

**32.** Using equation (6.6) as a base, derive the equations you would use to find the value of either branch resistance when the other branch resistance and the total resistance values are known.

**33.** In terms of circuit fault diagnosis, explain why motors (like those shown in Figure 6.29b) are always connected in parallel.

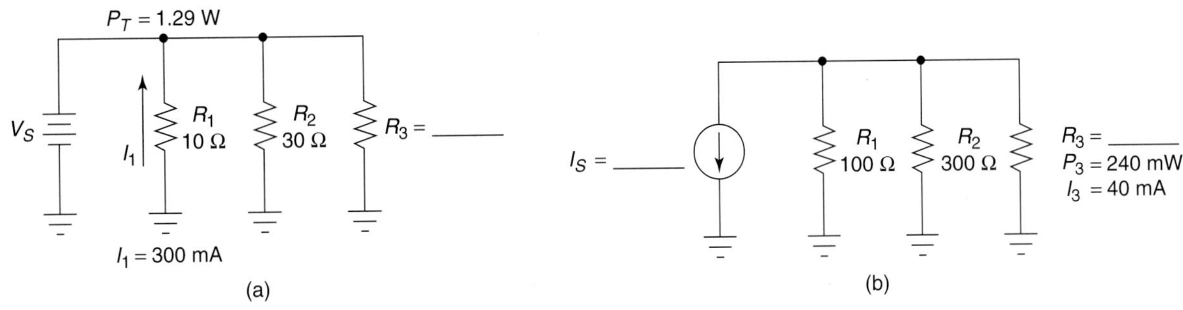


**FIGURE 6.39**

## Answers to the Example Practice Problems

**6.1.** 24.5 mA
**6.2.** 83.3 mA
**6.3.** 11 mS
**6.4.** 144 Ω
**6.5.** 144 Ω
**6.6.** 3 kΩ
**6.7.** $V_L$ changes from 1.2 V to 3.2 V.
**6.8.** $I_1 = 15$ mA, $I_2 = 10$ mA
**6.9.** See the answers to Problem 6.8.
**6.10.** See the answers to Problem 6.8.
**6.11.** $R_T = 625\ \Omega$, $I_1 = 5$ mA, $I_2 = 10$ mA, $I_3 = 1$ mA, $I_T = 16$ mA, $P_1 = 50$ mW, $P_2 = 100$ mW, $P_3 = 10$ mW, $P_T = 160$ mW

## EWB Applications Problems

| Figure | EWB File Reference |
|---|---|
| 6.4 | EWBA6_4.ewb |
| 6.11 | EWBA6_11.ewb |
| 6.12 | EWBA6_12.ewb |
| 6.17 | EWBA6_17.ewb |

# 7

# SERIES-PARALLEL CIRCUITS

## OBJECTIVES

*After studying the material in this chapter, you should be able to:*

1. Describe and analyze the operation of series circuits that are connected in parallel.
2. Describe and analyze the operation of parallel circuits that are connected in series.
3. Derive the series-equivalent or parallel-equivalent for any given series-parallel circuit.
4. Solve a given series-parallel circuit for any current, voltage, or power value.
5. List the steps used to solve for any value in a series-parallel circuit.
6. Discuss the effects of load resistance on the operation of a loaded voltage divider.
7. Discuss the concept of voltage divider stability.
8. Discuss voltmeter loading and its effects on voltage measurements.
9. Describe the construction and operation of the Wheatstone bridge.
10. Discuss the use of the Wheatstone bridge as a resistance measuring circuit.
11. Describe the construction and operation of a variable voltage divider.
12. List the basic series-parallel fault symptoms.
13. Describe the general approach to troubleshooting series-parallel circuits.

**Series-parallel circuit**
A circuit that contains both series and parallel elements.

**Combination circuit**
Another name for a series-parallel circuit.

All the circuits we've covered so far have contained components connected either in series or in parallel. In practice, most circuits contain a combination of both series and parallel connections. Circuits that contain both series and parallel elements are referred to as **series-parallel circuits** or **combination circuits.** Several examples of series-parallel circuits are shown in Figure 7.1. As you can see, some of the components in each circuit are clearly in series or parallel, and some are not. This holds true for any series-parallel circuit.

In this chapter, you will be shown how to analyze series-parallel circuits. You will also be introduced to some specific circuits that are commonly used in a variety of applications.

## 7.1 AN INTRODUCTION TO SERIES-PARALLEL CIRCUITS

Series-parallel circuits vary widely in their applications and complexity. For example, a stereo system can be viewed (loosely) as one extremely complex series-parallel circuit. At the same time, series-parallel circuits come in relatively simple configurations. For example, consider the circuits shown in Figure 7.2.

The circuit in Figure 7.2a contains two series circuits connected in parallel. One series circuit consists of $R_1$ and $R_2$, while the other consists of $R_3$ and $R_4$. The circuit in Figure 7.2b is simply two parallel circuits connected in series. $R_1$ and $R_2$ make up one of the parallel circuits, while $R_3$ and $R_4$ make up the other.

### Connecting Series Circuits in Parallel

*OBJECTIVE 1* ▶ Let's take a closer look at the circuit in Figure 7.2a. For the sake of discussion, the branches in the circuit are identified as shown in Figure 7.3. Since $R_1$ and $R_2$ form a series circuit, branch A retains all the characteristics of any series circuit. Therefore:

1. The currents through the two components are identical ($I_1 = I_2$).
2. The sum of the component voltages equals the source voltage ($V_1 + V_2 = V_S$).
3. The total resistance of the branch equals the sum of the resistor values ($R_A = R_1 + R_2$).

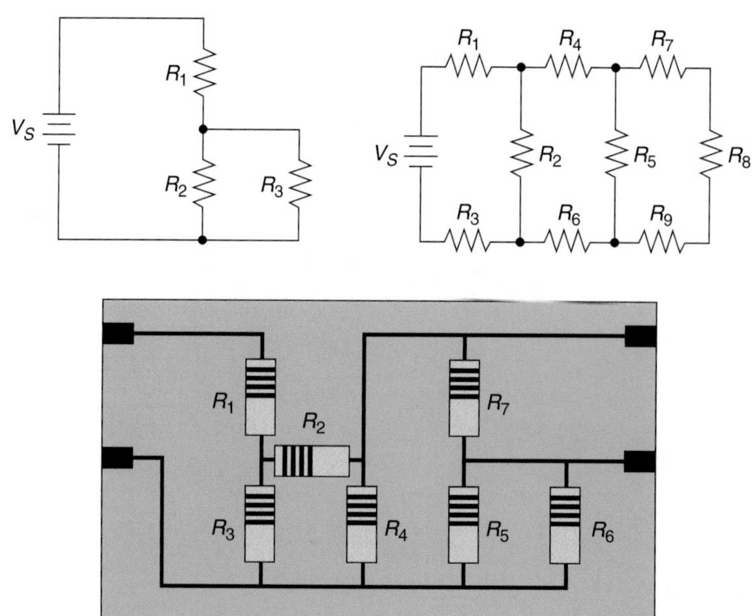


**FIGURE 7.1 Examples of series-parallel circuits.**

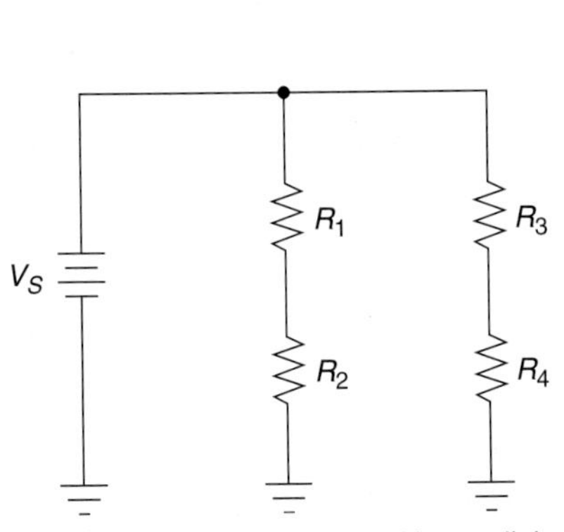


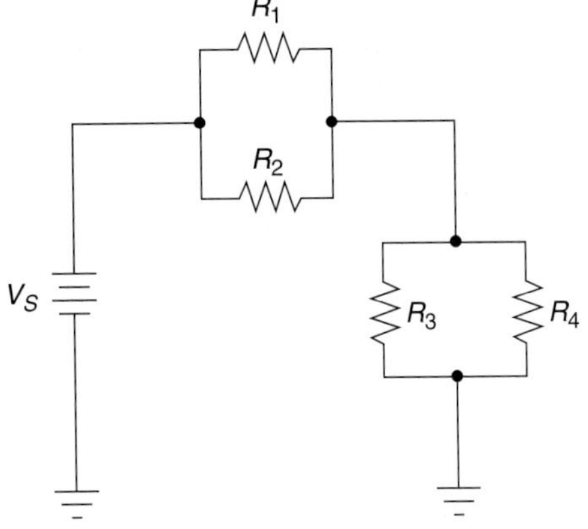


**FIGURE 7.2 Basic series-parallel configurations.**

Since $R_3$ and $R_4$ also form a series circuit:

1. $I_3 = I_4$
2. $V_3 + V_4 = V_S$
3. $R_B = R_3 + R_4$

Since these two branches form a parallel circuit, the combination of branch A and branch B retains the characteristics of any parallel circuit. Therefore:

1. The voltage across each branch is equal to the source voltage ($V_A = V_B = V_S$).
2. The sum of the branch currents is equal to the source current ($I_A + I_B = I_S$).
3. The total circuit resistance is equal to the parallel combination of $R_A$ and $R_B$ ($R_T = R_A \| R_B$).

**FIGURE 7.3**

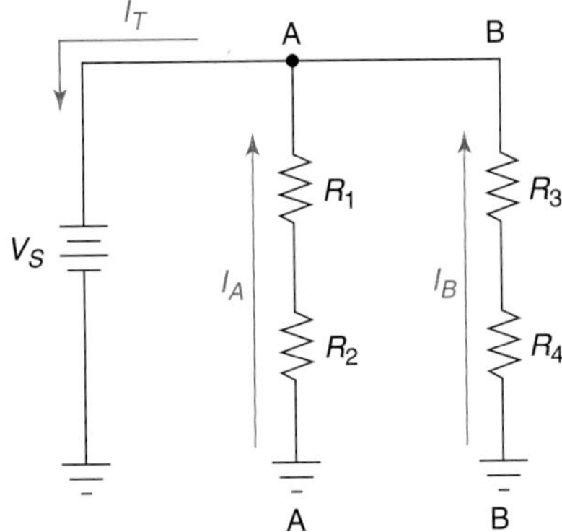


When two (or more) series circuits are connected in parallel, the analysis of the circuit begins with treating each branch as an individual series circuit. For each series circuit, the values of voltage current, resistance, and power are calculated. Then those values are used (as they would be in any parallel circuit) to determine the overall values for the circuit. This approach to analyzing the circuit is demonstrated in Example 7.1.

**EXAMPLE 7.1**

Perform a complete mathematical analysis of the circuit in Figure 7.4.

***Solution:*** We'll start with the analysis of branch A. The total resistance of this branch is found as

$$\boldsymbol{R_A} = R_1 + R_2 = 1\text{ k}\Omega + 1.5\text{ k}\Omega = \mathbf{2.5\text{ k}\Omega}$$

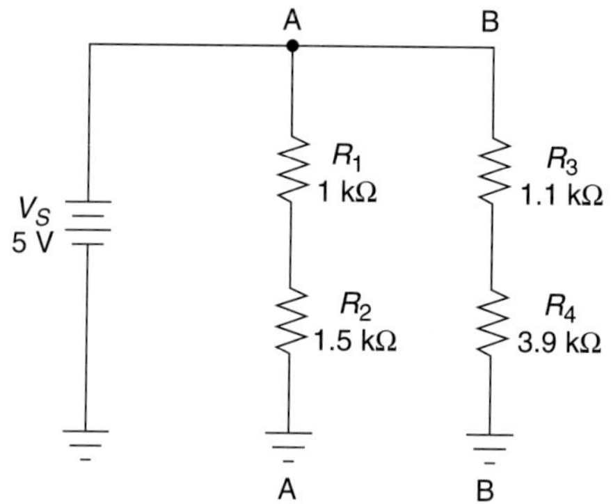


EWB

**FIGURE 7.4**

The current through branch A can now be found using Ohm's law, as follows:

$$\boldsymbol{I_A} = \frac{V_S}{R_A} = \frac{5\text{ V}}{2.5\text{ k}\Omega} = \mathbf{2\text{ mA}}$$

Since $I_A$ is the current through $R_1$ and $R_2$, the values of $V_1$ and $V_2$ can be found as

$$\boldsymbol{V_1} = I_A R_1 = (2\text{ mA})(1\text{ k}\Omega) = \mathbf{2\text{ V}}$$

and

$$\boldsymbol{V_2} = I_A R_2 = (2\text{ mA})(1.5\text{ k}\Omega) = \mathbf{3\text{ V}}$$

The power values for branch A can now be found as

$$\boldsymbol{P_1} = I_A V_1 = (2\text{ mA})(2\text{ V}) = \mathbf{4\text{ mW}}$$
$$\boldsymbol{P_2} = I_A V_2 = (2\text{ mA})(3\text{ V}) = \mathbf{6\text{ mW}}$$

and

$$\boldsymbol{P_A} = P_1 + P_2 = 4\text{ mW} + 6\text{ mW} = \mathbf{10\text{ mW}}$$

Now, the sequence of branch A calculations is repeated for branch B. First, the total branch resistance is found as

$$\boldsymbol{R_B} = R_3 + R_4 = 1.1\text{ k}\Omega + 3.9\text{ k}\Omega = \mathbf{5\text{ k}\Omega}$$

Next, the branch current is found as

$$\boldsymbol{I_B} = \frac{V_S}{R_B} = \frac{5\text{ V}}{5\text{ k}\Omega} = \mathbf{1\text{ mA}}$$

Using this value of $I_B$, we can calculate the resistor voltages in the branch as follows:

$$\boldsymbol{V_3} = I_B R_3 = (1\text{ mA})(1.1\text{ k}\Omega) = \mathbf{1.1\text{ V}}$$

and

$$\boldsymbol{V_4} = I_B R_4 = (1\text{ mA})(3.9\text{ k}\Omega) = \mathbf{3.9\text{ V}}$$

Finally, the branch power values are found as

$$\boldsymbol{P_3} = I_B V_3 = (1\text{ mA})(1.1\text{ V}) = \mathbf{1.1\text{ mW}}$$
$$\boldsymbol{P_4} = I_B V_4 = (1\text{ mA})(3.9\text{ V}) = \mathbf{3.9\text{ mW}}$$

and

$$\boldsymbol{P_B} = P_3 + P_4 = 1.1\text{ mW} + 3.9\text{ mW} = \mathbf{5\text{ mW}}$$

Since the branches are in parallel, the total circuit current is found as the sum of the branch currents, as follows:

$$I_T = I_A + I_B = 2 \text{ mA} + 1 \text{ mA} = \mathbf{3 \text{ mA}}$$

The total circuit resistance can be found as the parallel combination of $R_A$ and $R_B$, as follows:

$$R_T = R_A \| R_B = 2.5 \text{ k}\Omega \| 5 \text{ k}\Omega = \mathbf{1.67 \text{ k}\Omega}$$

Finally, the total circuit power can be found as the sum of the branch power values, as follows:

$$P_T = P_A + P_B = 10 \text{ mW} + 5 \text{ mW} = \mathbf{15 \text{ mW}}$$

***PRACTICE PROBLEM 7.1***

A circuit like the one in Figure 7.4 has the following values: $R_1 = 220\ \Omega$, $R_2 = 330\ \Omega$, $R_3 = 120\ \Omega$, $R_4 = 100\ \Omega$, and $V_S = 11$ V. Perform a complete mathematical analysis of the circuit.

## *Some Observations About the Circuit in Example 7.1*

Using the voltage and current values found in Example 7.1, we can make some important observations about series-parallel circuit characteristics. Figure 7.5a shows the voltage values we calculated for the circuit in Figure 7.4. Note that the voltage across branch A equals the voltage across branch B. These voltages *must* be equal in value because the branches are in parallel. At the same time, no two resistor voltages are equal because both branches are voltage dividers that split the source voltage independently. In other words, branch A divides $V_S$ differently than branch B does.

The current characteristics of the circuit are illustrated in Figure 7.5b. As is always the case with series circuits, the values of $I_1$ and $I_2$ are equal, as are the values of $I_3$ and $I_4$. The branch currents ($I_A$ and $I_B$) are not equal in this case. At the same time, the sum of $I_A$ and $I_B$ is equal to the total source current (as is the case in any parallel circuit).

The bottom line is this: each branch is a series circuit that retains the voltage and current characteristics of any series circuit. The combination of the branches, however, is a parallel circuit that retains the voltage and current characteristics of any parallel circuit.

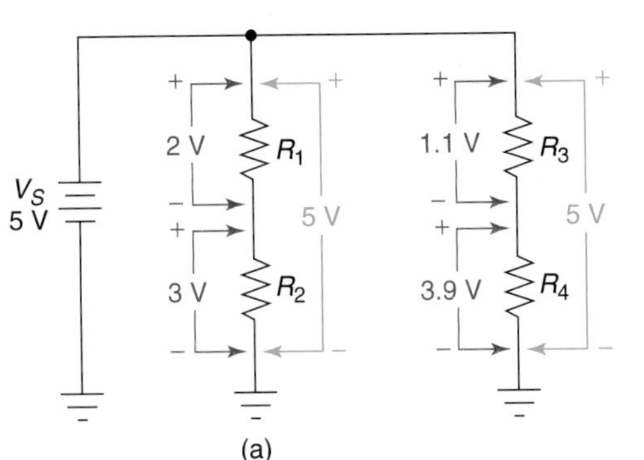


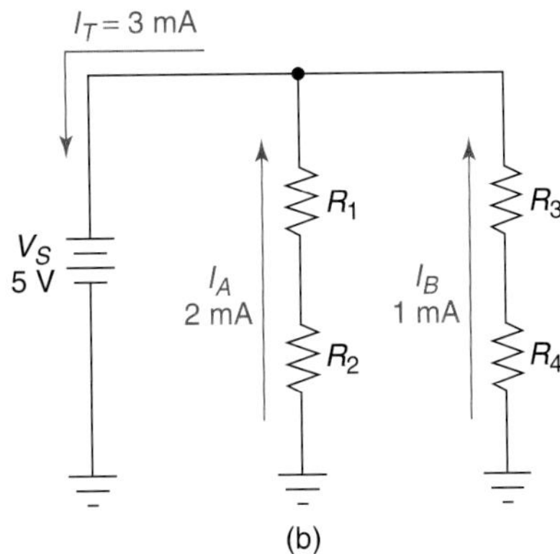


**FIGURE 7.5**

**FIGURE 7.6**

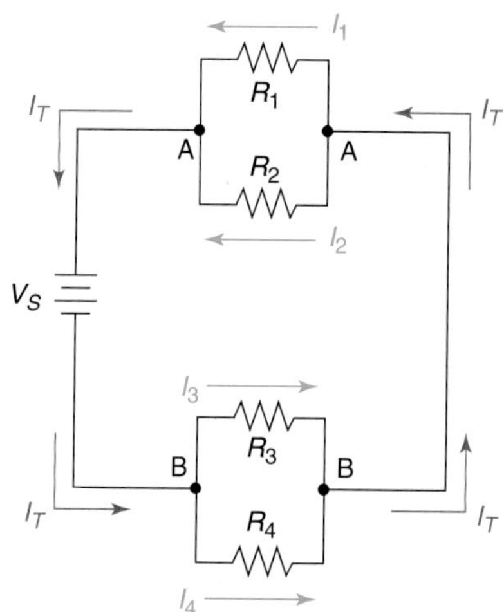


## Connecting Parallel Circuits in Series

*OBJECTIVE 2* ▶ Figure 7.6 shows two parallel circuits (identified as loop A and loop B) connected in series with a single voltage source. Loop A (which consists of $R_1$ and $R_2$) retains the characteristics of any parallel circuit. Therefore:

1. The voltages across the two components are equal ($V_1 = V_2$).
2. The sum of the component currents equals the total current through the loop ($I_1 + I_2 = I_T$).
3. The total resistance of the loop is equal to the parallel combination of the two resistor values ($R_A = R_1 \| R_2$).

Note that $R_A$ is used to denote the total resistance of loop A. By the same token, loop B (which consists of $R_3$ and $R_4$) retains the characteristics of any parallel circuit. Therefore:

1. $V_3 = V_4$
2. $I_3 + I_4 = I_T$
3. $R_B = R_3 \| R_4$

Since these two loops form a series circuit, the combination of loop A and loop B retains the characteristics of any series circuit. Therefore:

1. The sum of $V_A$ and $V_B$ equals the source voltage ($V_A + V_B = V_S$).
2. The currents through the two loops must equal the source current.
3. The total circuit resistance equals the sum of $R_A$ and $R_B$ ($R_T = R_A + R_B$).

When two or more parallel circuits are connected in series (as shown in Figure 7.6), the analysis of the circuit starts with representing each of the parallel circuits as a single resistance. Then the voltage across each equivalent resistance is determined. These voltage values are then used to determine the current values through the various resistors. This approach to analyzing the circuit is demonstrated in Example 7.2.

**EXAMPLE 7.2**

Perform a complete mathematical analysis of the circuit in Figure 7.7.

***Solution:*** First, the total resistance of loop A (the parallel combination of $R_1$ and $R_2$) is found as

$$R_A = R_1 \| R_2 = 200\ \Omega \| 300\ \Omega = \mathbf{120\ \Omega}$$

The total resistance of loop B is now found as

$$\boldsymbol{R_B} = R_3 \| R_4 = 240\ \Omega \| 360\ \Omega = \mathbf{144\ \Omega}$$

Using the calculated values of $R_A$ and $R_B$, we can redraw the original circuit as shown in Figure 7.8a. Using this equivalent circuit, we can calculate the voltage across each of the loops as follows:

$$\boldsymbol{V_A} = V_S \frac{R_A}{R_A + R_B} = (33\ \text{V}) \frac{120\ \Omega}{264\ \Omega} = \mathbf{15\ V}$$

and

$$\boldsymbol{V_B} = V_S \frac{R_B}{R_A + R_B} = (33\ \text{V}) \frac{144\ \Omega}{264\ \Omega} = \mathbf{18\ V}$$

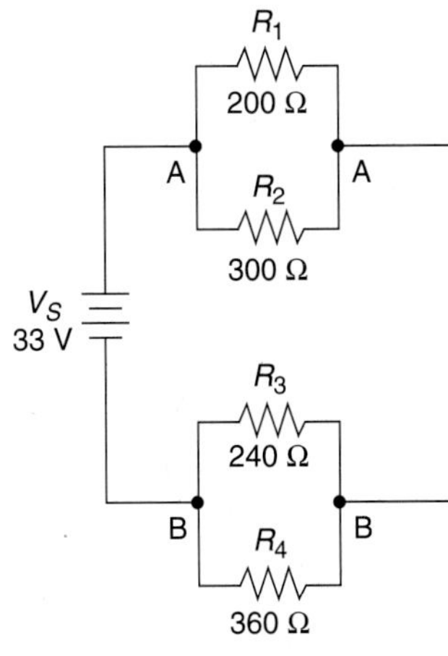


**FIGURE 7.7**

These calculations indicate that we have 15 V across loop A and 18 V across loop B, as shown in Figure 7.8b. We can now use $V_A$ to find the values of $I_1$ and $I_2$, as follows:

$$\boldsymbol{I_1} = \frac{V_A}{R_1} = \frac{15\ \text{V}}{200\ \Omega} = \mathbf{75\ mA}$$

and

$$\boldsymbol{I_2} = \frac{V_A}{R_2} = \frac{15\ \text{V}}{300\ \Omega} = \mathbf{50\ mA}$$

Using $V_B$, we can now find the values of $I_3$ and $I_4$, as follows:

$$\boldsymbol{I_3} = \frac{V_B}{R_3} = \frac{18\ \text{V}}{240\ \Omega} = \mathbf{75\ mA}$$

and

$$\boldsymbol{I_4} = \frac{V_B}{R_4} = \frac{18\ \text{V}}{360\ \Omega} = \mathbf{50\ mA}$$

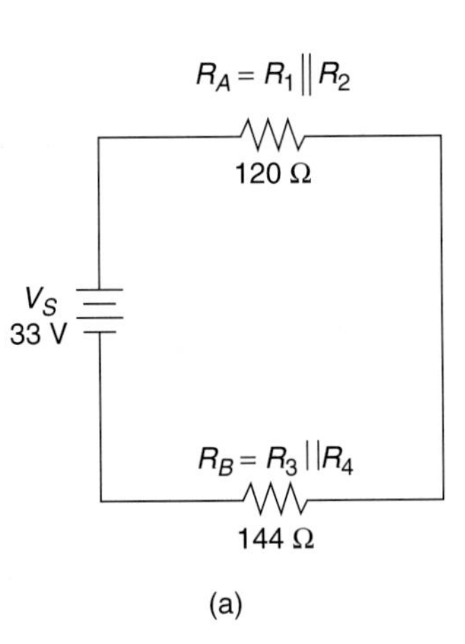


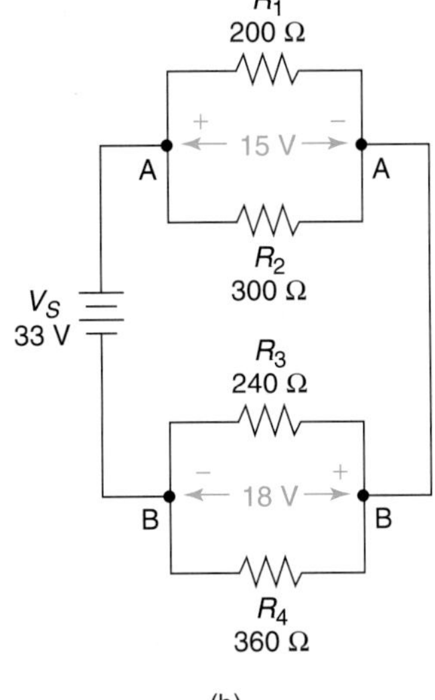


**FIGURE 7.8**

The power dissipated by each of the resistors in the circuit can now be found as

$$P_1 = V_A I_1 = (15\text{ V})(75\text{ mA}) = \mathbf{1.125\ W}$$
$$P_2 = V_A I_2 = (15\text{ V})(50\text{ mA}) = \mathbf{750\ mW}$$
$$P_3 = V_B I_3 = (18\text{ V})(75\text{ mA}) = \mathbf{1.350\ W}$$

and

$$P_4 = V_B I_4 = (18\text{ V})(50\text{ mA}) = \mathbf{900\ mW}$$

Finally, the total circuit current and power values can be found as

$$I_T = I_1 + I_2 = 75\text{ mA} + 50\text{ mA} = \mathbf{125\ mA}$$

and

$$P_T = P_1 + P_2 + P_3 + P_4 = 1.125\text{ W} + 750\text{ mW} + 1.350\text{ W} + 900\text{ mW} = \mathbf{4.125\ W}$$

***PRACTICE PROBLEM 7.2***

A circuit like the one in Figure 7.7 has the following values: $R_1 = 100\ \Omega$, $R_2 = 150\ \Omega$, $R_3 = 180\ \Omega$, $R_4 = 120\ \Omega$, and $V_S = 330$ mV. Perform a complete mathematical analysis of the circuit.

## *Some Observations About the Circuit in Example 7.2*

Using the voltage and current values found in Example 7.2, we can make some observations about parallel circuits that are connected in series. For example, Figure 7.9a shows the voltage values we calculated in Example 7.2. Note that the sum of the loop voltages equals the source voltage. Also note that each loop voltage is applied equally across two parallel resistors; that is, $V_A = V_1 = V_2$ and $V_B = V_3 = V_4$.

The current characteristics of the circuit in Example 7.2 are illustrated in Figure 7.9b. As is always the case with parallel circuits, $I_T = I_1 + I_2$ and $I_T = I_3 + I_4$. In fact, we could have calculated the total current in Example 7.2 using $I_T = I_3 + I_4$. The result would have been the same.

EWB

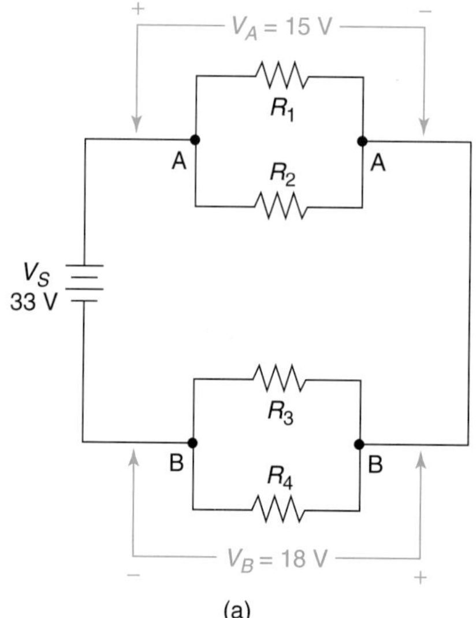


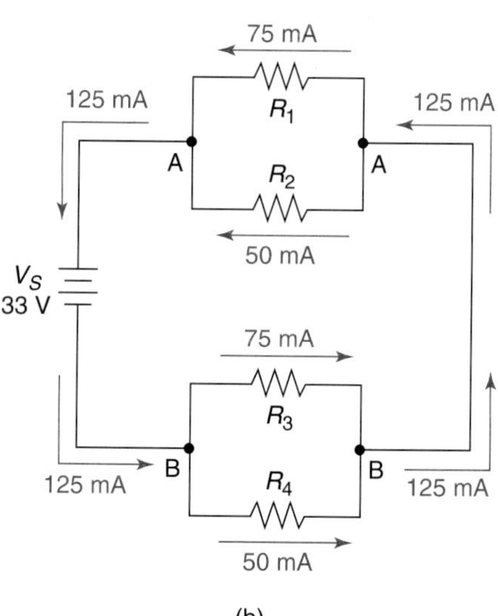


**FIGURE 7.9**

So here's the bottom line: each loop is a parallel circuit that retains all the voltage and current characteristics of any parallel circuit. At the same time, the overall circuit acts as a series circuit and retains all the voltage and current characteristics of any series circuit.

## Circuit Variations

Series-parallel circuits come in a near-infinite variety of configurations, from relatively simple to extremely complex. In this section, we have analyzed only two of the most basic configurations. Throughout the remainder of this chapter (and the next), we will look at a wide variety of series-parallel circuits. Our emphasis will be on establishing methods of analysis for those circuits. As you will see, any series-parallel circuit can be dealt with effectively when you:

1. Use the proper approach.
2. Remember the basic principles of series and parallel circuits.

### Section Review

1. List the current, voltage, and resistance characteristics of series circuits.
2. List the current, voltage, and resistance characteristics of parallel circuits.

*3. Refer to Figure 7.10a. Determine which (if any) of the following statements are true for the circuit shown.

   a. Under normal operating conditions, $V_3 = V_4$.
   b. $R_2$ is in series with $R_3$.
   c. The value of $V_S$ can be found by adding the values of $V_1$, $V_2$, and $V_4$.
   d. The total power ($P_T$) for the circuit can be found by adding the values of $P_1$, $P_2$, and $P_4$.

*4. Answer each of the following questions for the circuit shown in Figure 7.10b.

   a. Which two current values add up to equal the value of $I_4$?
   b. Which two current values add up to equal the value of $I_1$?
   c. Is $R_1$ in series with $R_2$? Explain your answer.
   d. What three voltage values could be added to determine the value of $V_S$?

*The answers to these questions are included with the practice problem answers at the end of the chapter.

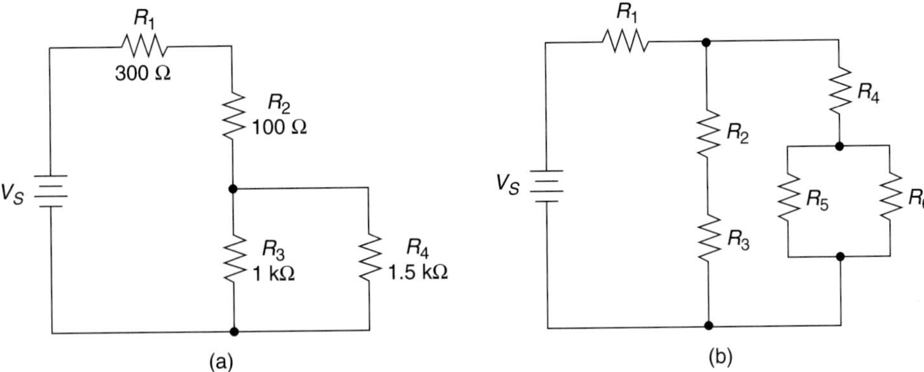


FIGURE 7.10

# 7.2 ANALYZING SERIES-PARALLEL CIRCUITS

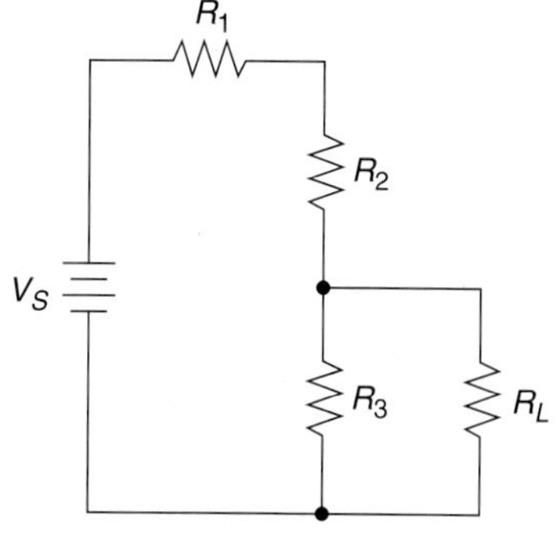


FIGURE 7.11

In most cases, the goal of any series-parallel circuit analysis is to determine the value of voltage, current, and/or power for one or more components. For example, consider the circuit shown in Figure 7.11. One practical circuit analysis goal would be to determine the value of load power for the circuit. Another might be to determine the load current response to a change in source voltage.

Finding a single component value in a series or parallel circuit is usually a simple, straightforward problem. However, this is not the case with series-parallel circuits. To calculate a single voltage, current, or power value in a series-parallel circuit, we usually find that we must first calculate several other values. For example, let's say we want to calculate the value of load power ($P_L$) for the circuit in Figure 7.11. To do so, we must solve for a series of values like those listed in Table 7.1. This series of calculations is demonstrated in Example 7.3.

**TABLE 7.1 The Series of Calculations Needed to Find the Value of $P_L$ for the Circuit in Figure 7.11**

| *To Find . . .* | *First Determine the Value of . . .* |
|---|---|
| Load power ($P_L$) | Load voltage ($V_L$) |
| Load voltage ($V_L$) | The equivalent resistance of the parallel circuit ($R_3 \| R_L$) |

## EXAMPLE 7.3

Calculate the value of load power for the circuit in Figure 7.12a.

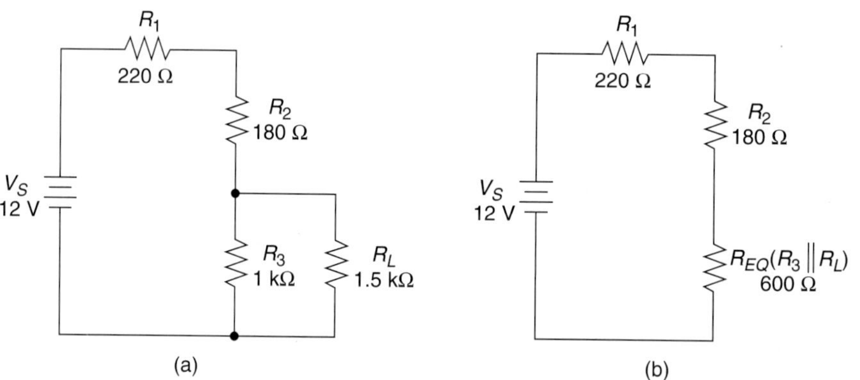


FIGURE 7.12

***Solution:*** As you can see, the circuit in Figure 7.12a is nearly identical to the one in Figure 7.11. The only difference is the added component values. The first step in calculating load power is to combine $R_3$ and $R_L$ into a single equivalent resistance ($R_{EQ}$), as follows:

$$R_{EQ} = R_3 \| R_L = 1\text{ k}\Omega \| 1.5\text{ k}\Omega = \mathbf{600\ \Omega}$$

If we replace the parallel circuit with its equivalent resistance, we get the circuit shown in Figure 7.12b. As you can see, we now have a simple series circuit. The voltage across the equivalent resistance ($V_{EQ}$) can now be found as

$$V_{EQ} = V_S \frac{R_{EQ}}{R_1 + R_2 + R_{EQ}} = (12\text{ V}) \frac{600\ \Omega}{1\text{ k}\Omega} = \mathbf{7.2\ V}$$

The voltage across the equivalent resistance ($V_{EQ}$) is equal to the voltage across the original parallel circuit; that is, $V_{EQ} = V_3 = V_L$. Therefore, we can find the value of load power as follows:

$$P_L = \frac{V_L^2}{R_L} = \frac{(7.2\text{ V})^2}{1.5\text{ k}\Omega} = \mathbf{34.56\text{ mW}}$$

*PRACTICE PROBLEM 7.3*

A circuit like the one in Figure 7.12a has the following values: $R_1 = 120\ \Omega$, $R_2 = 150\ \Omega$, $R_3 = 300\ \Omega$, $R_L = 100\ \Omega$, and $V_S = 6$ V. Calculate the value of load power for the circuit.

## Equivalent Circuits

◀ *OBJECTIVE 3*

The process of analyzing a series-parallel circuit usually begins with deriving a simplified equivalent of the original, like the one in Figure 7.12a. Equivalent circuits are derived by combining groups of parallel and/or series components to obtain an equivalent, but simpler, circuit. For example, consider the circuits shown in Figure 7.13. Here we have a series-parallel circuit and its series equivalent. The equivalent circuit shown was derived as follows:

1. $R_2$ and $R_3$ were combined into a single equivalent resistance ($R_{EQ1}$), just as we did in Example 7.3. The value of $R_{EQ1}$ was found as

$$R_{EQ1} = R_2 \| R_3 = 2\text{ k}\Omega \| 3\text{ k}\Omega = 1.2\text{ k}\Omega$$

2. $R_5$ and $R_L$ were combined into another single equivalent resistance ($R_{EQ2}$). The value of $R_{EQ2}$ was found as

$$R_{EQ2} = R_5 \| R_L = 2\text{ k}\Omega \| 18\text{ k}\Omega = 1.8\text{ k}\Omega$$

Using these values, the equivalent circuit was drawn as shown in Figure 7.13b.

The process of deriving an equivalent circuit ends when the original series-parallel circuit has been simplified to a **series-equivalent** or **parallel-equivalent** circuit; that is, a circuit that is made up entirely of components connected either in series or in parallel. This point is demonstrated in Examples 7.4 and 7.5.

**Series-equivalent circuit**
An equivalent circuit made up entirely of components connected in series.

**Parallel-equivalent circuit**
An equivalent circuit made up entirely of components connected in parallel.

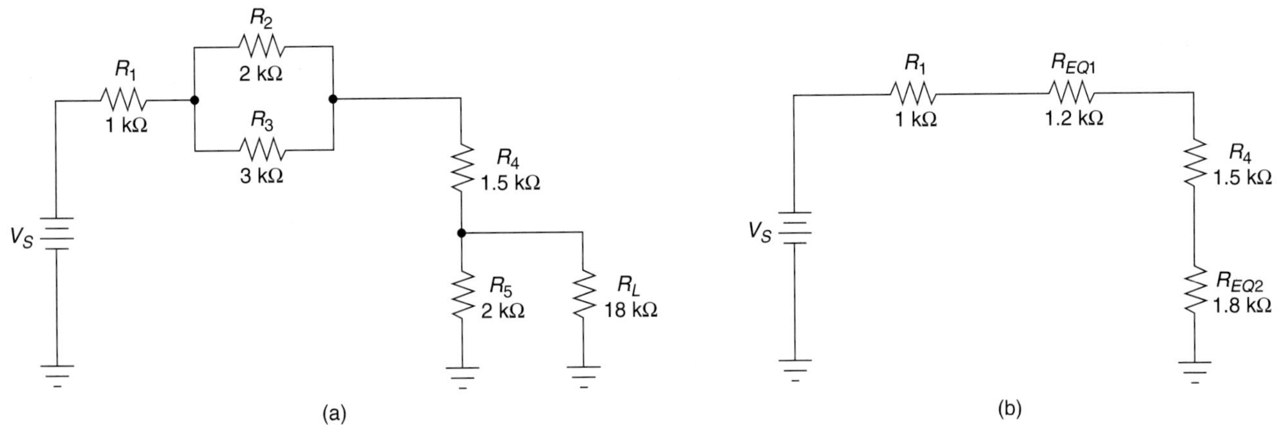


FIGURE 7.13

## EXAMPLE 7.4

Derive an equivalent for the series-parallel circuit shown in Figure 7.14a.

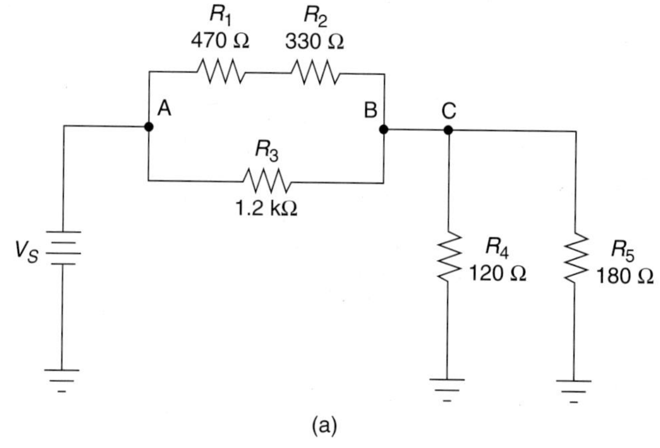


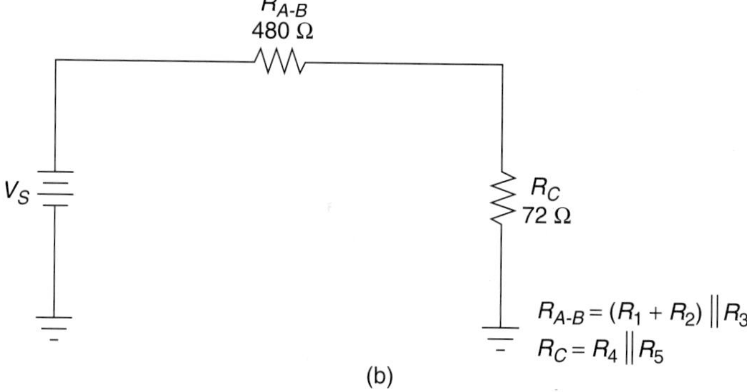


**FIGURE 7.14**

***Solution:*** The series-parallel circuit shown in Figure 7.14a can be reduced to a simple series-equivalent circuit by:

1. Combining the resistors between points A and B into a single equivalent resistance ($R_{A\text{-}B}$)
2. Combining the resistors between point C and ground into a single equivalent resistance ($R_C$)

Between points A and B, $R_1$ and $R_2$ form a series circuit that is in parallel with $R_3$. Therefore, the value of $R_{A\text{-}B}$ can be found as

$$\mathbf{R_{A\text{-}B}} = (R_1 + R_2) \| R_3 = (470\ \Omega + 330\ \Omega) \| 1.2\ \text{k}\Omega = 800\ \Omega \| 1.2\ \text{k}\Omega = \mathbf{480\ \Omega}$$

Between point C and ground, we have a parallel circuit made up of $R_4$ and the load. Therefore, the value of $R_C$ can be found as

$$\mathbf{R_C} = R_4 \| R_L = 120\ \Omega \| 180\ \Omega = \mathbf{72\ \Omega}$$

Using the calculated values of $R_{A\text{-}B}$ and $R_C$, the series-equivalent of the original circuit is drawn as shown in Figure 7.14b.

***PRACTICE PROBLEM 7.4***

Derive the series-equivalent for the circuit shown in Figure 7.15.

FIGURE 7.15

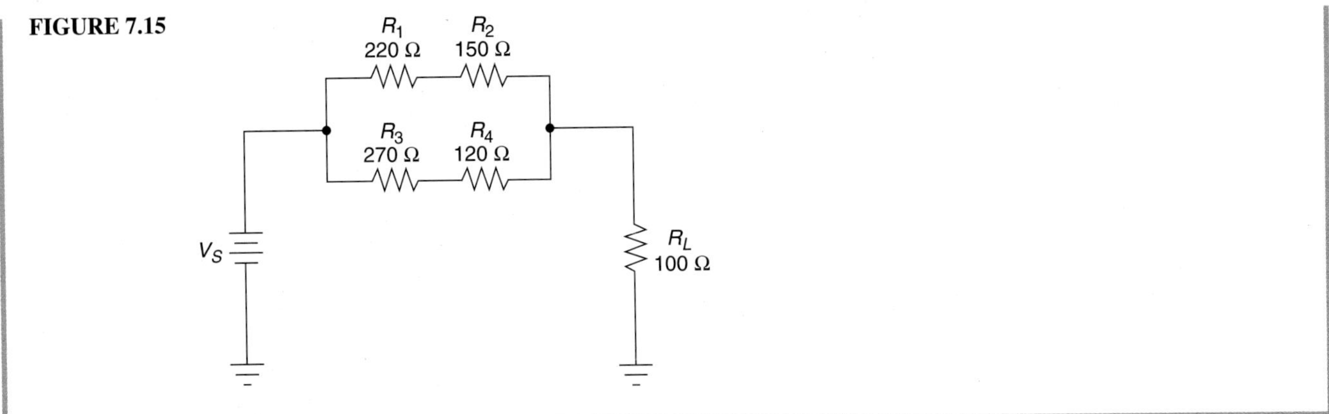


EXAMPLE 7.5

Derive the parallel-equivalent for the circuit shown in Figure 7.16a.

***Solution:*** The series-parallel circuit in Figure 7.16a can be reduced to a simple parallel-equivalent circuit by:

1. Combining the resistors between points A and B into a single equivalent resistance ($R_{A\text{-}B}$)
2. Combining the resistors between points C and D into a single equivalent resistance ($R_{C\text{-}D}$)

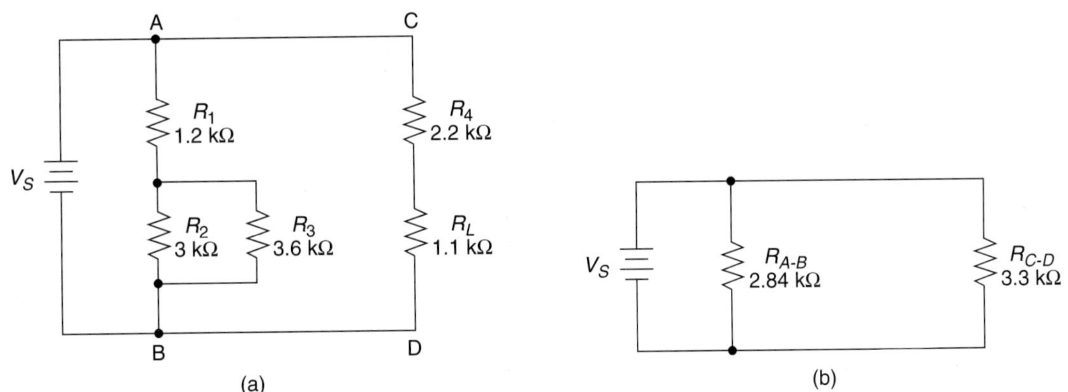


FIGURE 7.16

Between points A and B, $R_1$ is in series with the parallel combination of $R_2$ and $R_3$. Therefore, the value of $R_{A\text{-}B}$ can be found as

$$\boldsymbol{R_{A\text{-}B}} = R_1 + (R_2 \| R_3) = 1.2\ \text{k}\Omega + (3\ \text{k}\Omega \| 3.6\ \text{k}\Omega) = 1.2\ \text{k}\Omega + 1.64\ \text{k}\Omega = \mathbf{2.84\ k\Omega}$$

Between points C and D, $R_4$ is in series with the load resistance. Therefore, the value of $R_{C\text{-}D}$ can be found as

$$\boldsymbol{R_{C\text{-}D}} = R_4 + R_L = 2.2\ \text{k}\Omega + 1.1\ \text{k}\Omega = \mathbf{3.3\ k\Omega}$$

Using the calculated values of $R_{A\text{-}B}$ and $R_{C\text{-}D}$, the parallel-equivalent of the original circuit is drawn as shown in Figure 7.16b.

***PRACTICE PROBLEM 7.5***

Derive the parallel-equivalent for the circuit shown in Figure 7.17.

FIGURE 7.17

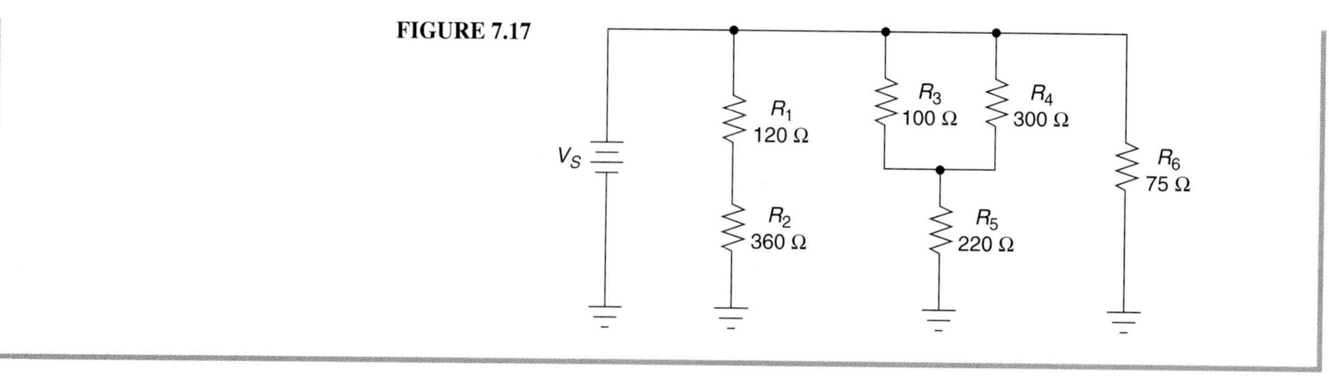


## Series-Parallel Circuit Analysis

*OBJECTIVE 4* ▶ In some circuit analysis problems, you must reduce an entire circuit to its series or parallel equivalent to determine a given current, voltage, and/or power value. In others, a given circuit value can be determined by reducing only a portion of the original circuit; that is, we can solve for a given value without reducing the entire circuit to a series or parallel equivalent.

The key to analyzing any series-parallel circuit is to know how far the circuit must be simplified to solve a given problem. For example, consider the circuit shown in Figure 7.18a. Let's say we want to determine the value of load power for this circuit. If we combine the load resistance and $R_3$ into a single equivalent resistance (as shown in Figure 7.18b), we have reduced the circuit enough to solve for the load power. This point is demonstrated in Example 7.6.

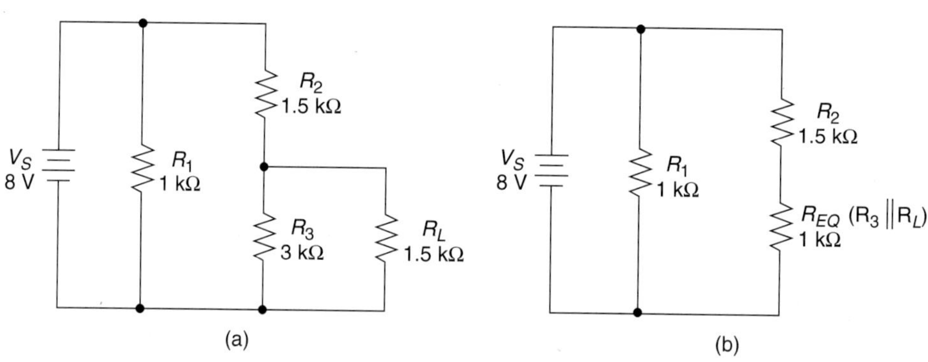


FIGURE 7.18

### EXAMPLE 7.6

Determine the value of load power ($P_L$) for the circuit in Figure 7.18a.

***Solution:*** As shown in Figure 7.18b,

$$R_{EQ} = R_3 \| R_L = 3\text{ k}\Omega \| 1.5\text{ k}\Omega = \mathbf{1\text{ k}\Omega}$$

Since the source voltage is applied to the series combination of $R_2$ and $R_{EQ}$, the voltage across $R_{EQ}$ can be found as

$$V_{EQ} = V_S \frac{R_{EQ}}{R_2 + R_{EQ}} = (8\text{ V})\frac{1\text{ k}\Omega}{2.5\text{ k}\Omega} = \mathbf{3.2\text{ V}}$$

As you know, $V_{EQ} = V_3 = V_L$. Therefore, the value of load power can be found as

$$P_L = \frac{V_L^2}{R_L} = \frac{(3.2\text{ V})^2}{1.5\text{ k}\Omega} = \mathbf{6.83\text{ mW}}$$

***PRACTICE PROBLEM 7.6***

A circuit like the one in Figure 7.18a has the following values: $V_S = 12$ V, $R_1 = 2$ kΩ, $R_2 = 3$ kΩ, $R_3 = 12$ kΩ, and $R_L = 18$ kΩ. Calculate the value of load power for the circuit.

In Example 7.6, we didn't need to reduce the entire circuit to solve the problem. We needed only to simplify the branch containing $R_2$, $R_3$, and the load. At the same time, calculating such values as total resistance ($R_T$), current ($I_T$), or power ($P_T$) would have required a complete reduction of the circuit to its simplest parallel equivalent.

## Example Analysis Problems

As you can see, the key to solving for any given value in a series-parallel circuit is knowing how far the circuit needs to be reduced to solve the problem. At this point, we will work through several analysis problems to further demonstrate this principle. The circuits we

**FIGURE 7.19**

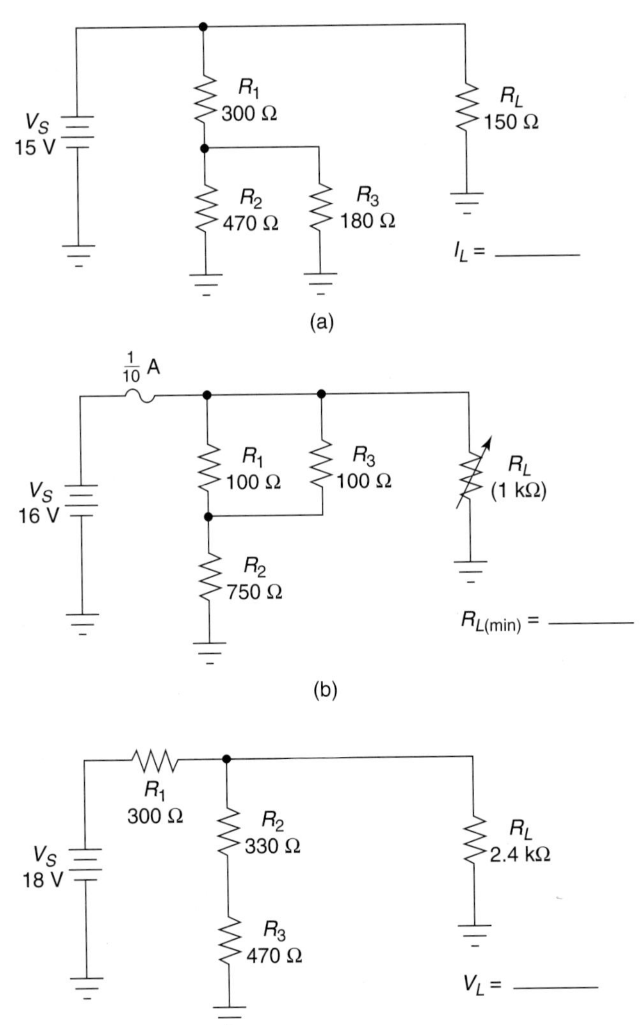


EWB

will analyze (and the values we will solve for) are shown in Figure 7.19. Before going through the examples, see if you can determine how far each of the circuits must be reduced to solve for the value.

## EXAMPLE 7.7

Determine the value of load current ($I_L$) for the circuit in Figure 7.19a.

***Solution:*** This problem is a lot easier to solve than it may seem at first. Since the load is in parallel with the rest of the circuit, the full value of $V_S$ is applied to the load; that is, $V_L = V_S$, and

$$I_L = \frac{V_L}{R_L} = \frac{15\ \text{V}}{150\ \Omega} = \mathbf{100\ mA}$$

As you can see, no circuit reduction was needed to solve for the value of $I_L$ in Figure 7.19a. Remember that, even though they are more complex, series-parallel circuits still operate by the same rules as other circuits. If you keep the basics in mind, many series-parallel circuit problems end up being simpler to solve than they first appear to be.

## EXAMPLE 7.8

Determine the minimum allowable resistance for the load in Figure 7.19b.

***Solution:*** We will solve this problem by determining the maximum allowable value of load current. The first step in doing so is to combine $R_1$, $R_2$, and $R_3$ into a single equivalent resistance ($R_{EQ}$), as shown in Figure 7.20. The value of $R_{EQ}$ shown in the figure was found as

$$R_{EQ} = (R_1 \| R_3) + R_2 = (100\ \Omega \| 100\ \Omega) + 750\ \Omega = \mathbf{800\ \Omega}$$

**FIGURE 7.20**

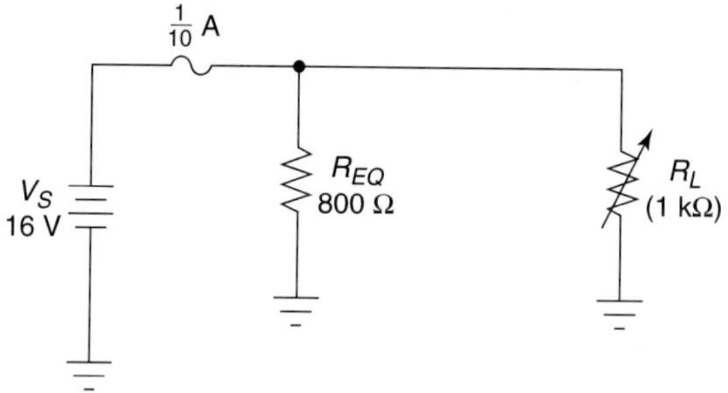


Now we have a simple parallel current problem like some of those we solved in Chapter 6. The maximum allowable current in the circuit (as determined by the fuse) is 100 mA. The value of the current through $R_{EQ}$ (which we will designate as $I_{EQ}$) is found using Ohm's law, as follows:

$$I_{EQ} = \frac{V_S}{R_{EQ}} = \frac{16\ \text{V}}{800\ \Omega} = \mathbf{20\ mA}$$

According to Kirchhoff's current law, the maximum allowable value of $I_L$ must equal the difference between $I_{EQ}$ and the maximum circuit current. Therefore, the maximum allowable value of load current is found as

$$I_{L(\text{max})} = I_T - I_{EQ} = 100\ \text{mA} - 20\ \text{mA} = \mathbf{80\ mA}$$

Finally, the minimum allowable setting for $R_L$ can be found using Ohm's law, as follows:

$$R_{L(\min)} = \frac{V_S}{I_{L(\max)}} = \frac{16\text{ V}}{80\text{ mA}} = \mathbf{200\ \Omega}$$

As long as $R_L$ is set to a value of 200 Ω or higher, the fuse in the circuit will not blow.

***PRACTICE PROBLEM 7.8***

A circuit like the one in Figure 7.19b has the following values: $V_S = 12$ V, $R_1 = 150\ \Omega$, $R_2 = 1.1\ \text{k}\Omega$, and $R_3 = 300\ \Omega$. If the fuse is rated at ⅛ A, what is the minimum allowable setting for $R_L$?

As you can see, the problem in Example 7.8 was solved using an equivalent circuit, Ohm's law, and Kirchhoff's current law. Again, a series-parallel circuit was reduced and solved using the basic circuit laws.

**EXAMPLE 7.9**

Determine the value of load voltage ($V_L$) for the circuit in Figure 7.19c.

***Solution:*** To solve this problem, we must reduce the circuit to its series-equivalent. This series equivalent is shown in Figure 7.21. The value of $R_{EQ}$ (which represents the total resistance of the parallel branches) was found as

$$R_{EQ} = (R_2 + R_3) \| R_L = (330\ \Omega + 470\ \Omega) \| \ 2.4\ \text{k}\Omega = \mathbf{600\ \Omega}$$

The voltage across this equivalent resistance (which we will designate as $V_{EQ}$) can be found using the voltage-divider equation, as follows:

$$V_{EQ} = V_S \frac{R_{EQ}}{R_1 + R_{EQ}} = (18\text{ V}) \frac{600\ \Omega}{900\ \Omega} = \mathbf{12\ V}$$

Since $V_{EQ}$ is the voltage across the parallel branches, the load voltage is equal to 12 V. By formula,

$$V_L = V_{EQ} = \mathbf{12\ V}$$

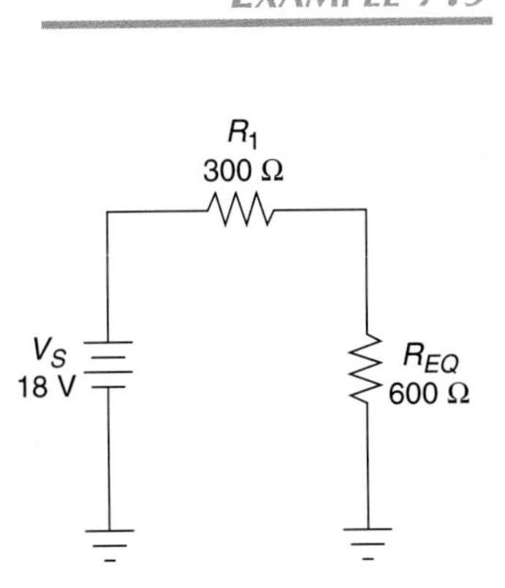


**FIGURE 7.21**

***PRACTICE PROBLEM 7.9***

A circuit like the one in Figure 7.19c has the following values: $V_S = 12$ V, $R_1 = 200\ \Omega$, $R_2 = 100\ \Omega$, $R_3 = 300\ \Omega$, and $R_L = 1.2\ \text{k}\Omega$. Determine the value of load voltage for the circuit.

Again, the problem was solved using an equivalent circuit and basic circuit relationships; in this case, the voltage-divider equation and the relationship between parallel voltages.

## Summary

◄ *OBJECTIVE 5*

As you can see, the analysis of most series-parallel circuits involves three steps:

1. Reduce the circuit to a series-equivalent or a parallel-equivalent, depending on the type of circuit.
2. Apply the basic series and parallel circuit principles and relationships to the equivalent circuit.
3. Use the values obtained from the equivalent circuit to solve the original for the desired value(s).

When you use this three-step approach, you will rarely have a problem solving for any single value in a series-parallel circuit.

**Section Review**

1. How are equivalent circuits derived?
2. What are the steps involved in solving most practical series-parallel circuit problems?

## 7.3 CIRCUIT LOADING

You will see many circuits and concepts time and again throughout your career. One important concept is the relationship between a load and the operation of its source circuit. In this section, we will discuss the effects that loads can have on circuit operation.

### Loaded Voltage Dividers

*OBJECTIVE 6* ▶ Most electronic circuits are designed for the purpose of supplying power (in one form or another) to one or more loads. For example, consider the circuit shown in Figure 7.22. Here we have a voltage divider used to supply the voltages required for the operation of two loads (designated as Load A and Load B). The voltage and current ratings for each

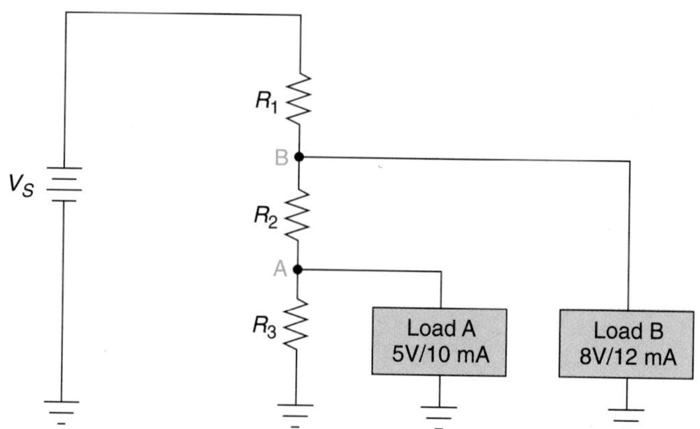


**FIGURE 7.22 A loaded voltage divider.**

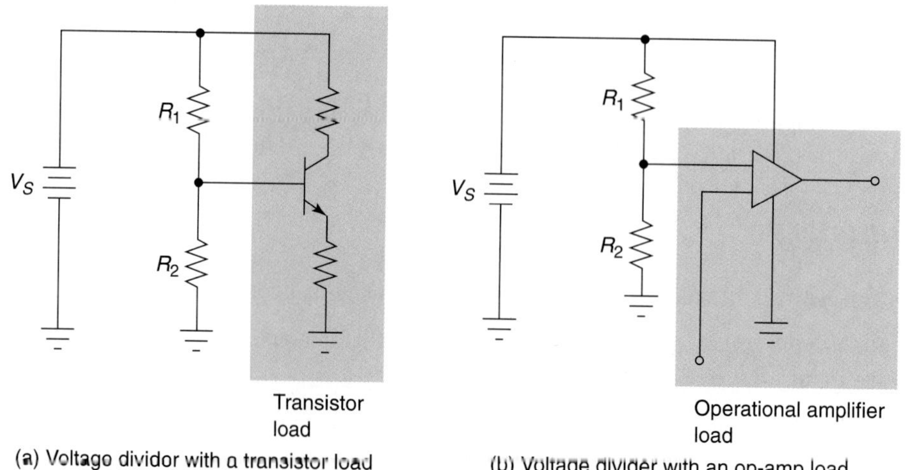


(a) Voltage divider with a transistor load (b) Voltage divider with an op-amp load

**FIGURE 7.23 Practical examples of loaded voltage dividers.**

load are given in the illustration. Assuming that the circuit is operating properly, the voltage divider provides the voltages and currents required by the loads.

In practice, voltage dividers are usually designed for the purpose of providing one or more specific load voltages (as shown in Figure 7.22). When used for this purpose, the voltage divider is referred to as a **loaded voltage divider.** Two practical examples of loaded voltage dividers are shown in Figure 7.23.

**Loaded voltage divider**
A voltage divider designed specifically to provide one (or more) specific load voltages.

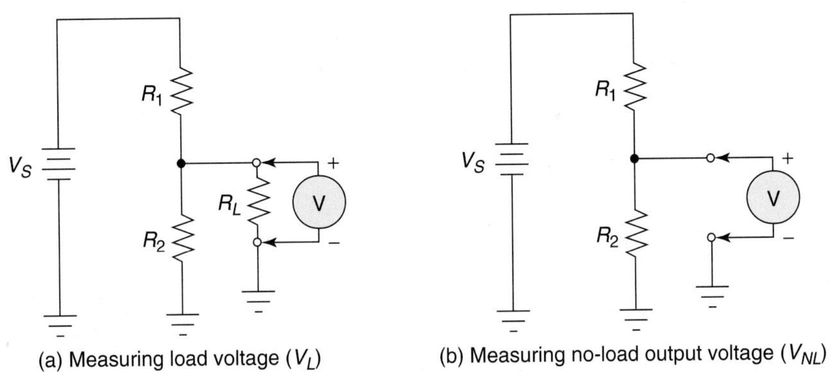


**FIGURE 7.24 Voltage divider output measurements.**

The operation and analysis of a loaded voltage divider is very similar to that of any other series-parallel circuit. However, one value of interest is somewhat unique to this circuit. This value is called the **no-load output voltage ($V_{NL}$).** The no-load output voltage is the voltage measured at the load terminals with the load removed. Note that the term *output voltage* is commonly used to describe the voltage at the load terminals of a circuit.

The measurement of $V_{NL}$ is illustrated in Figure 7.24b. As you can see, the load is removed from the circuit and the voltmeter is connected to the open load terminals.

**No-load output voltage ($V_{NL}$)**
The voltage measured at the load terminals with the load removed. For any voltage source, the maximum possible output voltage.

## The Relationship Between $V_{NL}$ and $V_L$

The no-load output voltage for a voltage divider is always greater than the value of $V_L$ for the circuit. The reason for this is illustrated in Figure 7.25. In Figure 7.25a, the total resistance from point A to ground ($R_{EQ}$) can be found as

The value of $V_L$ is always lower than the value of $V_{NL}$.

$$R_{EQ} = R_2 \| R_L$$

and the load voltage ($V_L$) can be found as

$$V_L = V_S \frac{R_{EQ}}{R_1 + R_{EQ}}$$

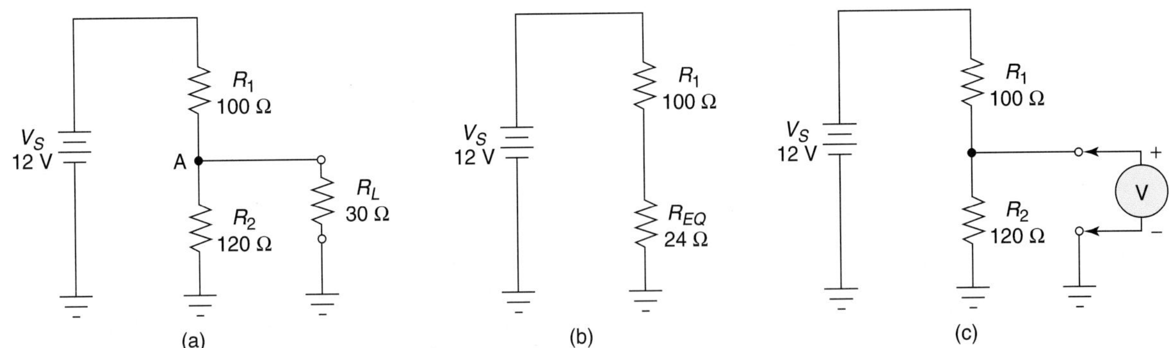


**FIGURE 7.25**

When the load is removed, the only resistance between point A and ground is $R_2$ (as shown in Figure 7.25b). Therefore, the value of $V_{NL}$ can be found as

$$V_{NL} = V_S \frac{R_2}{R_1 + R_2}$$

Since $R_2$ must be greater than the value of $R_{EQ}$, the value of $V_{NL}$ must be greater than the value of $V_L$. This point is demonstrated further in Example 7.10.

### EXAMPLE 7.10

Determine the values of $V_L$ and $V_{NL}$ for the circuit in Figure 7.25a.

***Solution:*** To find the value of $V_L$ for the circuit, we must start by combining $R_2$ and $R_L$ into a single equivalent resistance, as follows:

$$\boldsymbol{R_{EQ}} = R_2 \| R_L = 120\ \Omega \| 30\ \Omega = \mathbf{24\ \Omega}$$

Using this value, the circuit can now be redrawn as shown in Figure 7.25b. Using the voltage-divider equation, the voltage across $R_{EQ}$ can now be found as

$$\boldsymbol{V_{EQ}} = V_S \frac{R_{EQ}}{R_1 + R_{EQ}} = (12\ \text{V}) \frac{24\ \Omega}{124\ \Omega} = \mathbf{2.32\ V}$$

Since $V_{EQ}$ is across the parallel combination of $R_2$ and the load, $V_L = 2.32$ V. If we remove the load (as shown in Figure 7.25c), the voltmeter is measuring the voltage across $R_2$. Using the voltage-divider equation, this voltage can be found as

$$\boldsymbol{V_{NL}} = V_S \frac{R_2}{R_1 + R_2} = (12\ \text{V}) \frac{120\ \Omega}{220\ \Omega} = \mathbf{6.55\ V}$$

***PRACTICE PROBLEM 7.10***

A circuit like the one in Figure 7.25a has the following values: $V_S = 15$ V, $R_1 = 1$ k$\Omega$, $R_2 = 1.5$ k$\Omega$, and $R_L = 3$ k$\Omega$. Calculate the values of $V_L$ and $V_{NL}$ for the circuit.

As you can see, the value of $V_{NL}$ is significantly greater than the value of $V_L$. This relationship (which always holds true) can be used to explain several practical circuit relationships. The first of these is the relationship between load resistance and the operation of its source circuit.

## The Effects of Load Resistance on Circuit Operation

The results from Example 7.10 can be used to describe the effects of load resistance on the operation of a loaded voltage divider. For convenience, the results from the example are summarized as follows:

| Load Resistance | Output Voltage |
|---|---|
| $R_L = 30\ \Omega$ | $V_L = 2.32$ V |
| $R_L = \infty\ \Omega$ | $V_{NL} = 6.55$ V |

As you can see, load voltage varies directly with load resistance. As the resistance of the load increases, so does the circuit output voltage. Conversely, a decrease in load resistance causes a decrease in output voltage.

In many cases, the resistance of a given load is variable over a specified range. When a voltage divider has a variable load, the output voltage varies directly with the resistance of the load. This point is demonstrated in Example 7.11.

Determine the range of output voltages for the circuit in Figure 7.26a.

*EXAMPLE 7.11*

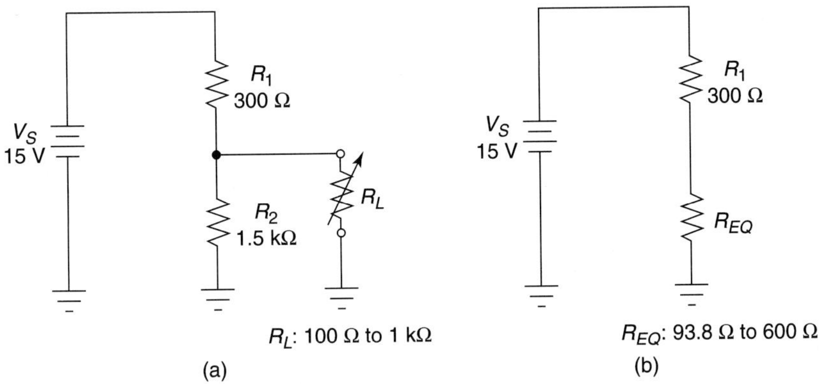


**FIGURE 7.26**

***Solution:*** The load is shown as being variable over a range of 1 kΩ to 100 Ω. When $R_L = 1\text{ k}\Omega$,

$$\boldsymbol{R_{EQ}} = R_2 \| R_L = 1.5\text{ k}\Omega \, \| 1\text{ k}\Omega = \mathbf{600\ \Omega}$$

and

$$\boldsymbol{V_L} = V_S \frac{R_{EQ}}{R_1 + R_{EQ}} = (15\text{ V})\frac{600\ \Omega}{900\ \Omega} = \mathbf{10\ V}$$

When $R_L = 100\ \Omega$,

$$\boldsymbol{R_{EQ}} = R_2 \| R_L = 1.5\text{ k}\Omega \, \| 100\ \Omega = \mathbf{93.8\ \Omega}$$

and

$$\boldsymbol{V_L} = V_S \frac{R_{EQ}}{R_1 + R_{EQ}} = (15\text{ V})\frac{93.8\ \Omega}{393.8\ \Omega} = \mathbf{3.57\ V}$$

***PRACTICE PROBLEM 7.11***

A circuit like the one in Figure 7.26a has the following values: $V_S = 12$ V, $R_1 = 220\ \Omega$, $R_2 = 330\ \Omega$, and $R_L = 50\ \Omega$ to $500\ \Omega$. Determine the range of output voltages for the circuit.

As you can see, a variable load can have a pretty drastic effect on the output from a voltage divider. However, voltage dividers are usually designed so that any variations in output voltage (due to a change in load resistance) are held to a minimum. Let's take a look at voltage divider stability and how it is accomplished.

## Voltage Divider Stability

◄ *OBJECTIVE 7*

**Voltage divider stability** The ability of a voltage divider to maintain a stable output voltage despite normal (or anticipated) variations in load demand.

When we talk about **voltage divider stability,** we are talking about the ability to maintain a stable load voltage despite normal (or anticipated) variations in load demand. For example, consider the circuit shown in Figure 7.27. The load in this circuit has ratings of $V_L =$

FIGURE 7.27

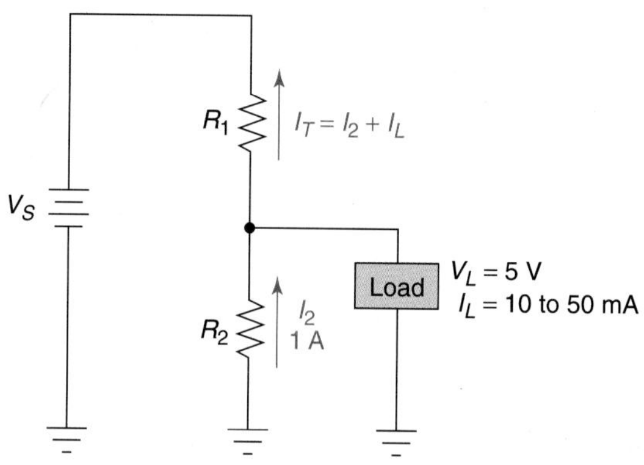


5 V and $I_L = 10$ mA to 50 mA. These ratings mean that the load resistance can be expected to vary between

$$R_L = \frac{5\text{ V}}{10\text{ mA}} = 500\ \Omega \qquad \text{and} \qquad R_L = \frac{5\text{ V}}{50\text{ mA}} = 100\ \Omega$$

The stability of the voltage divider is a measure of its ability to maintain a constant output voltage despite this possible change in the value of $R_L$.

How voltage divider stability is achieved.

Voltage divider stability is accomplished by designing the circuit so that the load current makes up a relatively small percentage of the total circuit current. For example, the circuit in Figure 7.27 is designed to have a value of $I_2 = 1$ A. Since the total current in the circuit is equal to the sum of $I_2$ and $I_L$, it has a range of

$$I_T = 1\text{ A} + 10\text{ mA} = 1.01\text{ A} \qquad \text{to} \qquad I_T = 1\text{ A} + 50\text{ mA} = 1.05\text{ A}$$

As you can see, the change in load current has no practical effect on the total circuit current. Since $V_1$ can be found as

$$V_1 = I_T R_1$$

the value of $V_1$ is also relatively stable against changes in $I_L$. This means that the load voltage must also be relatively stable, since the sum of $V_1$ and $V_L$ must equal the source voltage. The concept of voltage divider stability is demonstrated further in the following example.

## EXAMPLE 7.12

Calculate the normal range of $V_L$ values for the loaded voltage divider in Figure 7.28a.

***Solution:*** The load is shown to have the following ratings: $V_L = 12$ V and $I_L = 10$ mA to 30 mA. Using these values, the range of load resistance values can be found as follows:

$$R_{L(\min)} = \frac{V_L}{I_{L(\max)}} = \frac{12\text{ V}}{30\text{ mA}} = \mathbf{400\ \Omega} \qquad \text{and} \qquad R_{L(\max)} = \frac{V_L}{I_{L(\min)}} = \frac{12\text{ V}}{10\text{ mA}} = \mathbf{1.2\ k\Omega}$$

For reference, the series equivalent of the voltage divider is shown in Figure 7.28b. When $R_L = 400\ \Omega$,

$$R_{EQ} = R_2 \| R_L = 33\ \Omega \| 400\ \Omega = \mathbf{30.5\ \Omega}$$

and

$$V_L = V_S \frac{R_{EQ}}{R_1 + R_{EQ}} = (20\text{ V})\frac{30.5\ \Omega}{50.5\ \Omega} = \mathbf{12.08\ V}$$

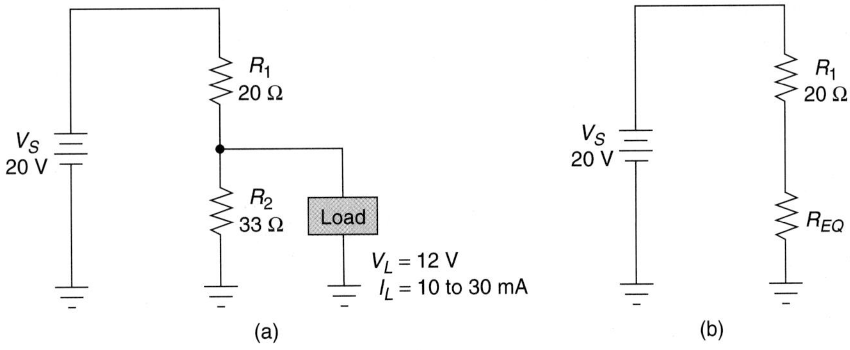


**FIGURE 7.28**

When $R_L = 1.2\text{ k}\Omega$,

$$R_{EQ} = R_2 \| R_L = 33\ \Omega \| 1.2\text{ k}\Omega = \mathbf{32.1\ \Omega}$$

and

$$V_L = V_S \frac{R_{EQ}}{R_1 + R_{EQ}} = (20\text{ V})\frac{32.1\ \Omega}{52.1\ \Omega} = \mathbf{12.32\ V}$$

***PRACTICE PROBLEM 7.12***

A circuit like the one in Figure 7.28a has the following values: $V_S = 18$ V, $R_1 = 68\ \Omega$, and $R_2 = 75\ \Omega$. The load on the circuit is rated at $V_L = 9$ V and $I_L = 2$ mA to 12 mA. Determine the normal range of $V_L$ values for the circuit.

As you can see, the circuit in Example 7.12 is very stable against changes in load resistance. When the load resistance increased by 200%, the load voltage increased by approximately 2%. In this case, the change in $V_L$ was negligible.

You may recall that voltage divider stability is achieved by designing the circuit so that the total circuit current is much greater than the load current. This relationship is accomplished so that the current through $R_2$ (which is referred to as the *bleeder current*) is much greater than the load current. By formula,

$$I_B >> I_L \tag{7.1}$$

where $I_B$ = the current through the component that is parallel to the load.

The validity of this relationship can be demonstrated with the results from Example 7.12. Using Ohm's law, we can calculate the following values for the circuit in the example:

$$I_L = \frac{V_L}{R_L} = \frac{12.32\text{ V}}{1.2\text{ k}\Omega} = 30.2\text{ mA} \quad \text{(maximum)}$$

and

$$I_2 = \frac{V_L}{R_2} = \frac{12.32\text{ V}}{33\ \Omega} = 366\text{ mA} \quad \text{(minimum)}$$

As you can see, $I_2 >> I_L$, and the circuit fulfills the requirement given in equation (7.1).

## Degree of Stability

Every loaded voltage divider provides some degree of load voltage stability (assuming the circuit has been designed properly). The degree of stability depends on the exact relationship between the bleeder current and the maximum load current. For example, consider the circuit shown in Figure 7.29. In this circuit, the load current is shown to have a range of

**FIGURE 7.29**

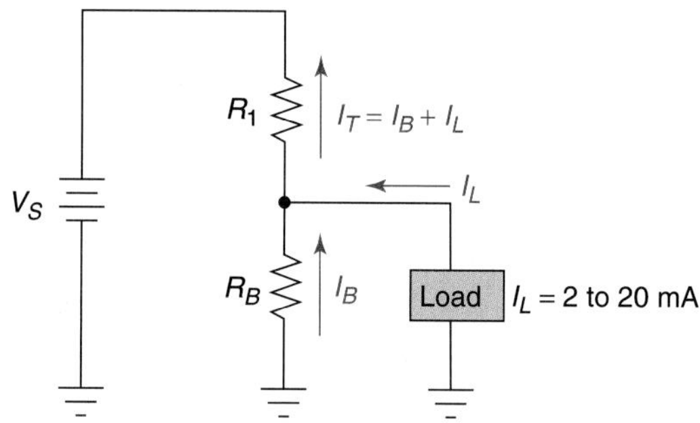


2 mA to 20 mA. In general, a loaded voltage divider is considered to be stable if the bleeder current is at least 10 times the maximum load current. By formula,

$$I_{B(\min)} \geq 10I_{L(\max)} \tag{7.2}$$

Therefore, for the circuit in Figure 7.29, the minimum acceptable value of bleeder current is 200 mA. As long as the circuit design provides a bleeder current equal to (or greater than) 200 mA, the output voltage will remain stable for the normal range of load resistance values.

*A Practical Consideration:* Generally, the bleeder resistor in a loaded voltage divider must have a higher power handling capability than the load because the bleeder and the load have the same voltage across them, but the bleeder is handling a much higher current. For example, if $I_B = 10I_L$, then the power dissipated by the bleeder resistor is 10 times the power dissipated by the load. Therefore, the bleeder must have a higher power rating than the load.

It should be noted that the greater the ratio of bleeder current to maximum load current, the greater the stability. At the same time, the higher the bleeder current, the higher the power that is used by the bleeder resistor. In other words, using a higher bleeder current provides better stability but wastes more power in the process. The engineer must consider this tradeoff when designing the circuit.

When load resistance is known, it is easy to determine the relative stability of a loaded voltage divider. All you need to do is compare the value of $R_{L(\min)}$ to the value of the bleeder resistor. Minimum stability, as defined in equation (7.2), is achieved when

$$R_{L(\min)} \geq 10R_B \tag{7.3}$$

where $R_B$ is the value of the bleeder resistor. If you look back to Example 7.12, you can see that the minimum load resistance was 400 Ω, which is greater than 10 times the value of the bleeder resistor ($R_2$ = 33 Ω).

## Open Loads

If the load on a voltage divider should open, it has the same impact as physically removing the load from the circuit. Therefore, the value of $V_L$ (for an open load) is approximately equal to the circuit's no-load output voltage ($V_{NL}$).

For some circuits, a voltmeter reading of $V_L = V_{NL}$ is a sure indicator of an open load. However, further testing may be needed in some circuits. For example, the circuit in Example 7.12 (Figure 7.28) was determined to have a normal output of $V_L$ = 12.08 V to 12.32 V. The no-load output voltage for this circuit can be found as

$$V_{NL} = V_S \frac{R_2}{R_1 + R_2} = (20\text{ V})\frac{33\ \Omega}{53\ \Omega} = 12.45\text{ V}$$

As you can see, there isn't much difference between the normal range of $V_L$ for this circuit and the value of $V_{NL}$. This is generally the case for any voltage divider designed for a stable output. In fact, the greater the stability of the circuit, the closer the normal value of $V_L$ will be to the value of $V_{NL}$. For this reason, you need to determine the relative stability of the voltage divider before assuming that a reading of $V_L \cong V_{NL}$ means the load is open.

## Voltmeter Loading

*OBJECTIVE 8* ▶

When a voltmeter is used to measure any voltage in a series circuit, the voltmeter may cause the reading to be significantly lower than the actual component voltage due to the effects of the meter's internal resistance, as shown in Figure 7.30. When the meter is

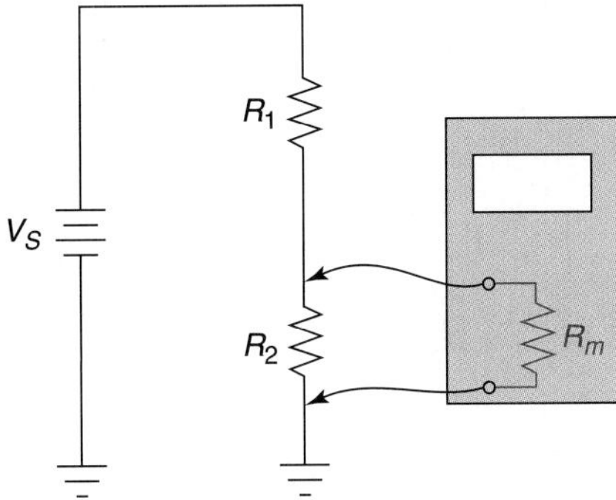


**FIGURE 7.30 Voltmeter loading.**

connected as shown in the figure, it essentially converts the series circuit into a loaded voltage divider, with the meter's internal resistance acting as the load. If the internal resistance of the meter is low enough, the impact on the voltage being measured can be fairly severe. This point is demonstrated in the following example.

## EXAMPLE 7.13

Determine the value of $V_2$ for the circuit in Figure 7.31a. Then determine the value of $V_2$ when the voltmeter is connected as shown in Figure 7.31b.

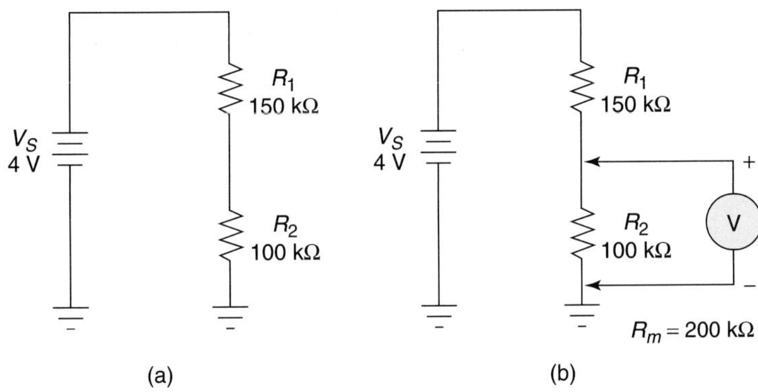


**FIGURE 7.31**

***Solution:*** The value of $V_2$ for the series circuit is found as

$$\mathbf{V_2} = V_S \frac{R_2}{R_1 + R_2} = (4 \text{ V}) \frac{100 \text{ k}\Omega}{250 \text{ k}\Omega} = \mathbf{1.6 \text{ V}}$$

When the meter is connected as shown in Figure 7.31b, $R_2$ and the internal resistance of the meter ($R_m$) form a parallel circuit. The resistance of this circuit is found as

$$\mathbf{R_{EQ}} = R_2 \| R_m = 100 \text{ k}\Omega \| 200 \text{ k}\Omega = \mathbf{66.7 \text{ k}\Omega}$$

Taking this equivalent resistance into account, the value of $V_2$ can be found as follows:

$$\mathbf{V_2} = V_S \frac{R_{EQ}}{R_1 + R_{EQ}} = (4 \text{ V}) \frac{66.7 \text{ k}\Omega}{216.7 \text{ k}\Omega} = \mathbf{1.23 \text{ V}}$$

***PRACTICE PROBLEM 7.13***

The value of $V_2$ for the circuit in Figure 7.31a is measured using a DMM with an internal resistance of 10 MΩ. Calculate the value of $V_2$ with the meter connected. Compare your result to the values obtained in the example.

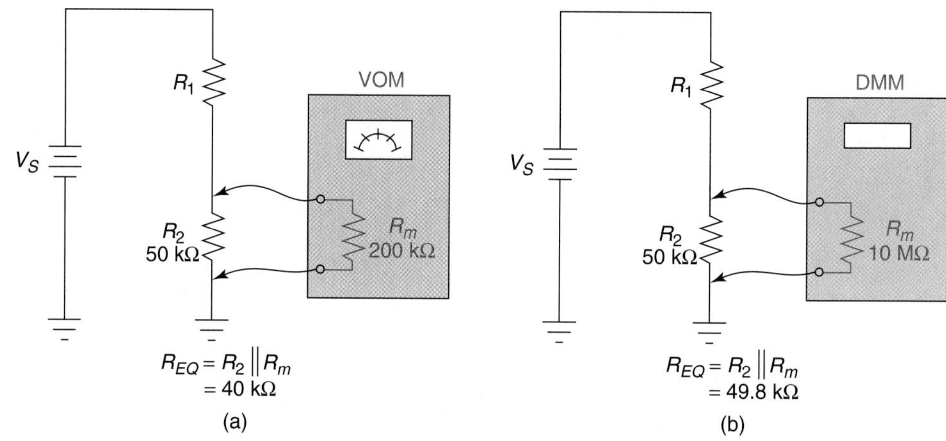


**FIGURE 7.32 VOM loading versus DMM loading.**

As you can see, connecting a voltmeter to a circuit can have a significant effect on the accuracy of the reading. This impact is reduced by using a meter with a high internal resistance. For example, consider the circuits shown in Figure 7.32. The internal resistance of the VOM in Figure 7.32a produces a parallel equivalent resistance of 40 kΩ. At the same time, the DMM in Figure 7.32b produces a parallel equivalent resistance of 49.5 kΩ. Since this value is approximately equal to the value of $R_2 = 50$ kΩ, the DMM presents almost no load on the circuit. Therefore, the reading in Figure 7.32b will be extremely accurate.

Why DMMs are preferred over VOMs for voltage measurements.

When you consider the impact of circuit loading, it is easy to understand why DMMs are preferred over VOMs for most voltage measurements. DMMs typically have much higher input resistance than VOMs. Therefore, voltage measurements taken with a DMM are much more accurate than comparable readings taken with a VOM.

## Summary

In this section, you have seen how the presence of a load can affect the operation of a voltage divider. When dealing with any loaded voltage divider, remember to look at the relationship between the value of the load resistance and that of the bleeder resistor. This relationship will tell you:

1. How stable the circuit is against changes in load resistance.
2. How close the normal value of $V_L$ will be to the no-load output voltage ($V_{NL}$).
3. How accurate any voltage reading will be (when the load is your voltmeter).

## Section Review

1. What is a loaded voltage divider?
2. What is the no-load output voltage of a voltage divider?
3. For a loaded voltage divider, what is the relationship between $V_L$ and $V_{NL}$?
4. For a loaded voltage divider, what is the relationship between output voltage and load resistance?
5. What is meant by the term *voltage divider stability*?
6. What design requirement must be met to produce a stable voltage divider?
7. What is the circuit recognition feature of a stable voltage divider?
8. Discuss the effects of voltmeter loading on series circuit voltage measurements.

# 7.4 THE WHEATSTONE BRIDGE

As you know, there are a near-infinite variety of series-parallel circuit configurations. In this section, we will concentrate on one specific circuit application: the **Wheatstone bridge.** As you will see, this circuit can be used to provide extremely accurate resistance measurements.

◄ *OBJECTIVE 9*

**Wheatstone bridge**
A circuit containing four resistors and a meter "bridge" that provides resistive measurements.

## Bridge Construction

The Wheatstone bridge consists of a dc voltage source, four resistors, and a meter, as shown in Figure 7.33. The resistors form two series circuits connected in parallel. The meter is connected as a "bridge" between the series circuits (which is how the circuit name was derived).

The meter shown in Figure 7.33 is a **galvanometer:** a current meter that indicates both the magnitude and the direction of a low-value current. As you will see, the galvanometer can be used to indicate the relationship between the voltages at points A and B in the circuit.

*Note:* The bridge between the resistive branches forms a unique connection; that is, it is connected in such a way that it is not in series, parallel, or series-parallel with any specific components. Even though it is approached as a series-parallel circuit, the Wheatstone bridge is actually an electrical anomaly. It is the only circuit that cannot be classified as series, parallel, or series-parallel.

**Galvanometer**
A current meter that indicates both the magnitude and direction of a low-value current.

## Bridge Operation

Each branch in the Wheatstone bridge is a voltage divider. As such, the values of $V_A$ and $V_B$ (which are identified in Figure 7.34) can be found as

$$V_A = V_S \frac{R_2}{R_1 + R_2} \tag{7.4}$$

and

$$V_B = V_S \frac{R_4}{R_3 + R_4} \tag{7.5}$$

*A Practical Consideration:* Equations (7.4) and (7.5) hold true only when the galvanometer is not connected. When the galvanometer is connected, another approach must be taken to analyze the circuit.

The operation of the Wheatstone bridge is based on the fact that it has three possible operating states, that is, three possible combinations of $V_A$ and $V_B$. These operating states are as follows:

$$V_A = V_B \qquad V_A > V_B \qquad V_A < V_B$$

Each of these operating states is illustrated in Figure 7.35.

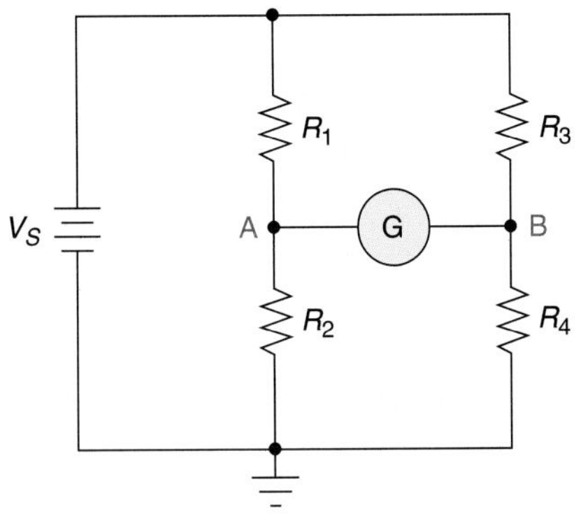

**FIGURE 7.33 The Wheatstone bridge.**

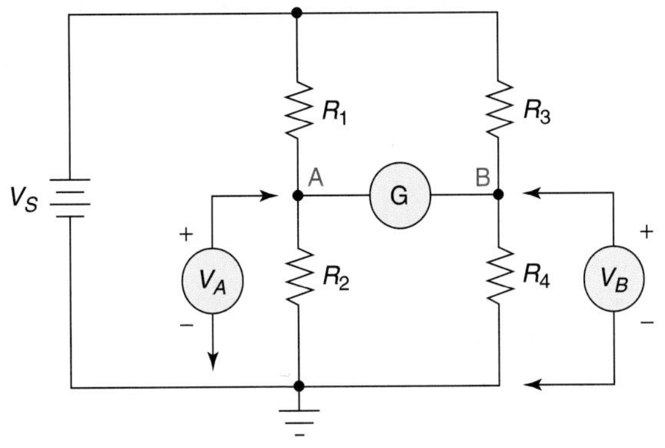

**FIGURE 7.34**

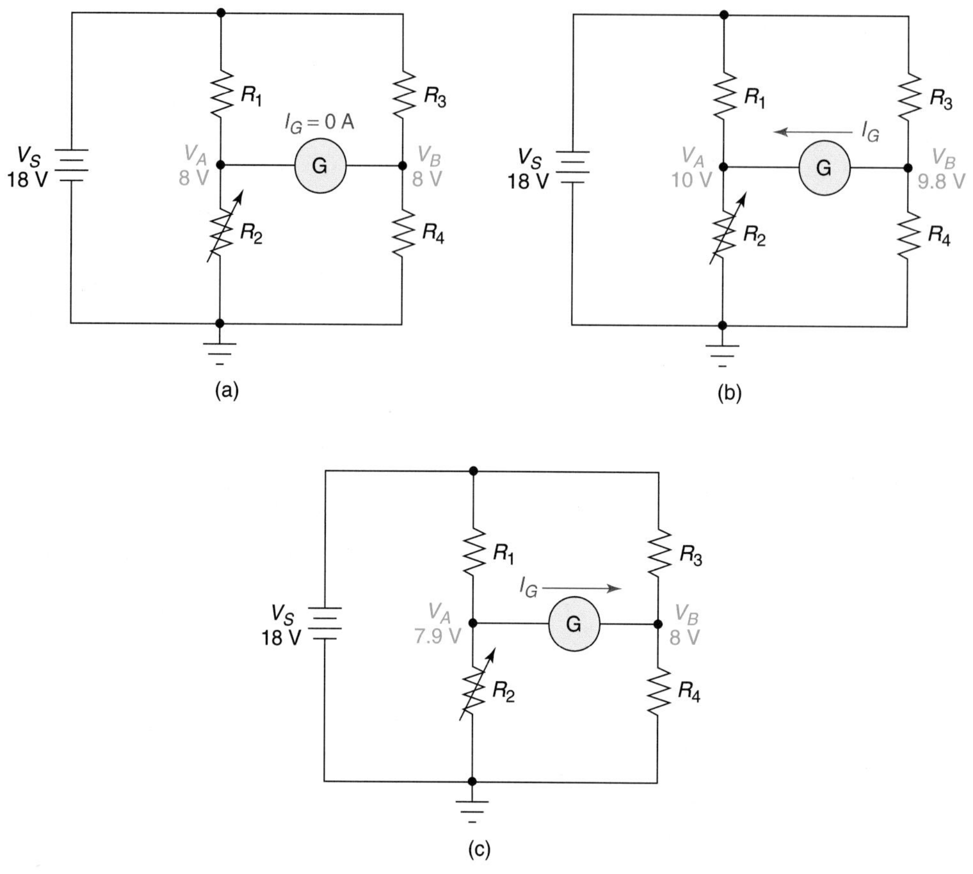


EWB

**FIGURE 7.35 Wheatstone bridge operation.**

In Figure 7.35, $R_2$ is shown to be a potentiometer. By adjusting the value of this component, we can set $V_A$ to any desired value (within limits). In Figure 7.35a, $R_2$ has been adjusted so that $V_A = V_B$. When these voltages are equal, there is no difference of potential across the galvanometer. Therefore, the galvanometer current ($I_G$) equals 0 A. When in this operating state, the bridge is said to be **balanced.**

**Balanced**
A term used to describe the bridge operating state when the galvanometer current is 0 A. A bridge is balanced when the voltages on the two sides of the meter are equal.

In Figure 7.35b, $R_2$ has been adjusted so that $V_A > V_B$. Since the voltages are no longer equal, there is a difference of potential (voltage) across the galvanometer. As a result, a measurable value of $I_G$ is generated in the direction shown in the figure. The galvanometer reading indicates the magnitude and direction of $I_G$.

In Figure 7.35c, $R_2$ has been adjusted so that $V_A < V_B$. Again, the voltages are not equal, and there is a difference of potential across the galvanometer. As a result, a measurable value of $I_G$ is generated in the direction shown in the figure. If you compare this operating state to the one shown in Figure 7.35b, you'll see that the voltage polarity across the galvanometer has reversed, which is why the currents are in opposite directions.

## Resistance Ratios

By definition, a Wheatstone bridge is balanced when $V_A = V_B$. For these voltages to be equal, the resistance ratios in equations (7.4) and (7.5) must be equal. In other words, a bridge is balanced only when

$$\frac{R_2}{R_1} = \frac{R_4}{R_3} \tag{7.6}$$

Note that this relationship can be derived by setting equations (7.4) and (7.5) equal to each other and simplifying the result. As Table 7.2 shows, each of the operating states is produced by a specific relationship between the resistance ratios. The relationships shown in Table 7.2 are important because they will help you to understand how the Wheatstone bridge can be used as a resistance measuring circuit.

**TABLE 7.2 Bridge Resistance Ratios and the Resulting Operating States**

| *Resistance Ratio* | *Resulting Operating State* |
|---|---|
| $\frac{R_2}{R_1} = \frac{R_4}{R_3}$ | $V_A = V_B$ |
| $\frac{R_2}{R_1} > \frac{R_4}{R_3}$ | $V_A > V_B$ |
| $\frac{R_2}{R_1} < \frac{R_4}{R_3}$ | $V_A < V_B$ |

## Measuring Resistance with a Wheatstone Bridge

OBJECTIVE 10

A Wheatstone bridge can be used as an effective resistance measuring circuit when constructed as shown in Figure 7.36. In the bridge circuit shown, the unknown resistance ($R_x$) has been placed in the $R_1$ position. Also, $R_2$ has been replaced by a calibrated decade box. You may recall that a decade box contains several series-connected potentiometers. By adjusting the individual potentiometers, the overall resistance of the decade box ($R_{db}$) can be set to any value within its limits, usually between 1 Ω and 9.999999 MΩ. (The value of $R_{db}$ for any setting is indicated by the calibrated readouts on the decade box.)

*Reference:* A decade box is shown in Figure 3.24.

When connected as shown in Figure 7.36, the value of the unknown resistance is found by balancing the bridge, that is, by adjusting the value of $R_{db}$ so that the galvanometer reads 0 A. When the bridge is balanced, the resistance ratios of the series circuits must be equal. Therefore, the value of $R_x$ can be found using a modified form of equation (7.6), as follows:

$$R_x = R_{db}\frac{R_3}{R_4} \quad \textbf{(7.7)}$$

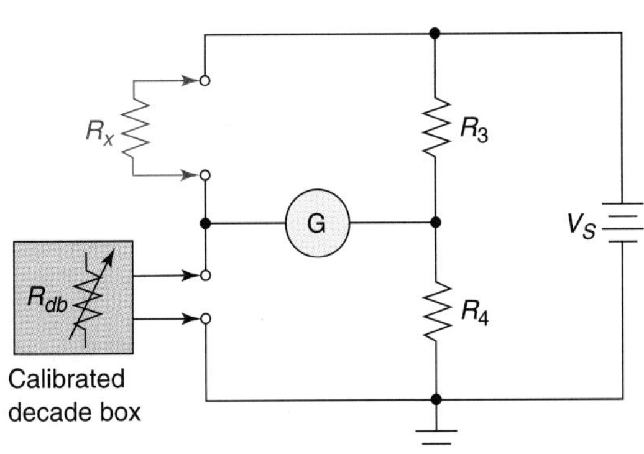


**FIGURE 7.36 Using a Wheatstone bridge to measure resistance.**

Note that this equation was derived by taking the reciprocals of the resistance ratios in equation (7.6), substituting $R_x$ and $R_{db}$ for $R_1$ and $R_2$, and solving for $R_x$. Example 7.14 demonstrates the procedure for measuring an unknown resistance with a Wheatstone bridge.

## EXAMPLE 7.14

Determine the value of $R_x$ for the circuit in Figure 7.37.

**FIGURE 7.37**

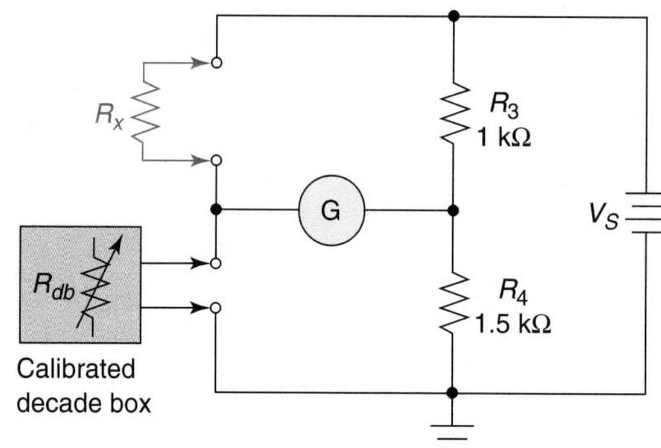


***Solution:*** The bridge is constructed and the calibrated decade box is adjusted until the bridge is balanced ($I_G = 0$ A). When the bridge is balanced, the readouts on the decade box indicate that the value of $R_{db}$ is 30 kΩ. Using the known resistance values, the value of $R_x$ is found as

$$\mathbf{R_x} = R_{db}\frac{R_3}{R_4} = (30\text{ k}\Omega)\frac{1\text{ k}\Omega}{1.5\text{ k}\Omega} = \mathbf{20\text{ k}\Omega}$$

***PRACTICE PROBLEM 7.14***

A circuit like the one in Figure 7.37 has values of $R_3 = 10$ kΩ and $R_4 = 20$ kΩ. The galvanometer indicates that the bridge is balanced when $R_{db}$ is set to 15 kΩ. What is the value of $R_x$?

When a Wheatstone bridge is used to measure resistance, the accuracy of the reading depends on two things:

1. How well the decade box is calibrated
2. The tolerance ratings of $R_3$ and $R_4$

If the decade box isn't calibrated, the value of $R_{db}$ indicated by the readout will not be accurate. This inaccuracy will lead to a significant error in the calculated value of $R_x$. At the same time, using resistors with high tolerance ratings may also cause a significant error in the calculated value of $R_x$. In other words, the more accurate the known values of $R_3$, $R_4$, and $R_{db}$, the more accurate the calculated value of $R_x$.

## A Circuit Modification

Calculating the value of $R_x$ is easier when a Wheatstone bridge is constructed as shown in Figure 7.38. In this circuit, $R_3 = R_4 = 10$ kΩ. When $R_3$ and $R_4$ are equal in value, the fraction in equation (7.4) equals 1, and the value of $R_x$ can be found as

$$R_x = R_{db} \quad (\text{when } R_3 = R_4) \tag{7.8}$$

This means that a bridge with equal values of $R_3$ and $R_4$ will balance when $R_{db}$ equals the value of $R_x$. This simplifies the overall procedure for measuring the value of an unknown resistance.

FIGURE 7.38

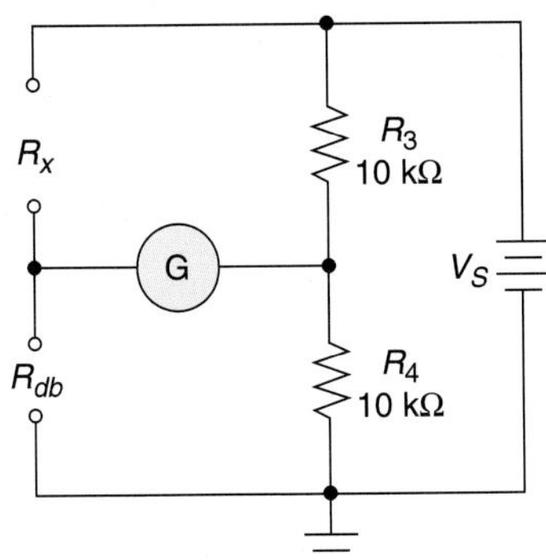


## One Final Note

In this section, the Wheatstone bridge has been presented as a circuit that you might see on a regular basis (like a loaded voltage divider). However, this is not usually the case. You will see circuits that resemble the Wheatstone bridge, but the circuit itself is used almost exclusively in meter circuits. While you may use a Wheatstone bridge on a regular basis, you generally will not see one when working on a given electronic system. As a reference, the characteristics of the Wheatstone bridge are summarized in Figure 7.39.

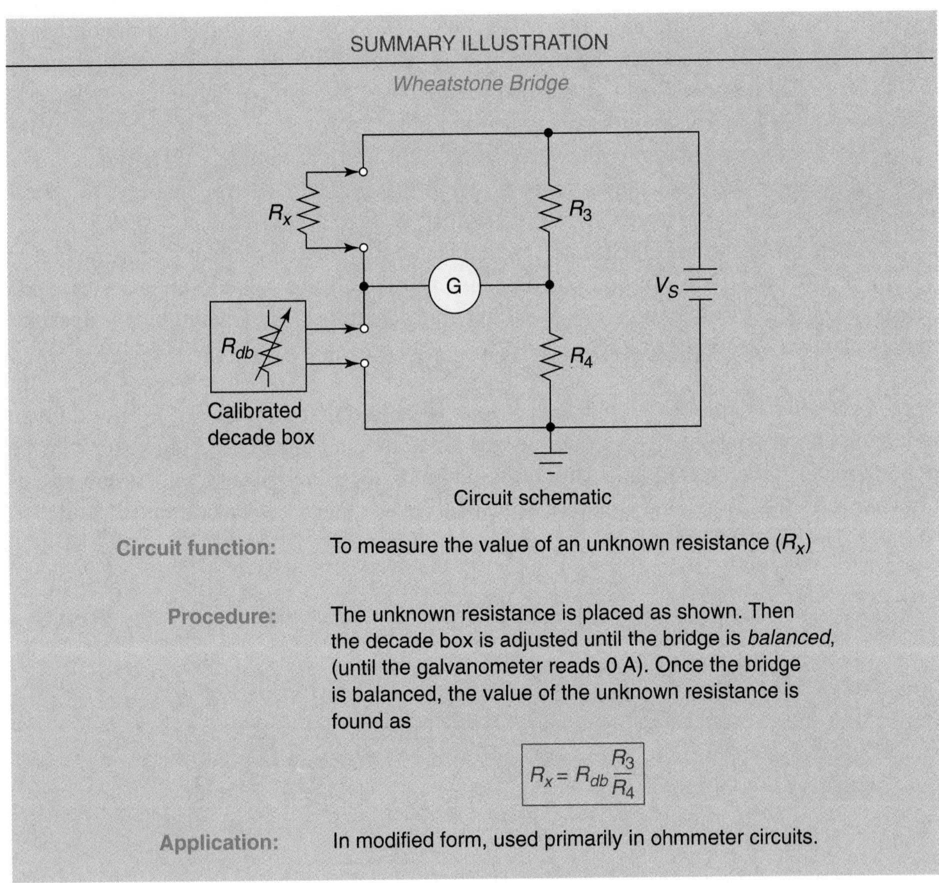

SUMMARY ILLUSTRATION
Wheatstone Bridge
$R_x$
$R_3$
G
$V_S$
$R_{db}$
$R_4$
Calibrated decade box
Circuit schematic
Circuit function: To measure the value of an unknown resistance ($R_x$)
Procedure: The unknown resistance is placed as shown. Then the decade box is adjusted until the bridge is balanced, (until the galvanometer reads 0 A). Once the bridge is balanced, the value of the unknown resistance is found as
$R_x = R_{db}\frac{R_3}{R_4}$
Application: In modified form, used primarily in ohmmeter circuits.


FIGURE 7.39

## Section Review

1. What is a galvanometer?
2. What are the three operating states of a Wheatstone bridge?
3. Describe a balanced bridge.
4. Describe the operating states of a Wheatstone bridge in terms of resistance ratios.
5. Describe the procedure for measuring resistance with a Wheatstone bridge.

# 7.5 VARIABLE VOLTAGE DIVIDERS, FAULT SYMPTOMS, AND TROUBLESHOOTING

In this section, we will briefly touch on one more common series-parallel circuit. We will also discuss some common series-parallel circuit fault symptoms and the general approach to troubleshooting these circuits.

## Variable Voltage Dividers

*OBJECTIVE 11* ► A potentiometer can be (and often is) used as a variable voltage divider. A variable voltage divider is one that can be varied to provide a specific load voltage. An example is shown in Figure 7.40. When the potentiometer setting is varied, it changes the voltage from the wiper arm to ground, that is, the voltage across $R_{bc}$. Since the load is in parallel with $R_{bc}$, the load voltage also changes.

When connected as shown in Figure 7.40, the load voltage can be varied to almost any value between 0 V and the value of $V_S$. For example, if we assume that the circuit in Figure 7.40 has a 10 V source, the range of load voltages for the circuit is approximately 0 V to 10 V. The range of load voltages is provided by varying the potentiometer from one extreme to the other. This point is illustrated in Figure 7.41.

When set to one extreme, the value of $R_{bc}$ will be approximately 0 Ω. When this is the case, we have the equivalent circuit shown in Figure 7.41a. As you can see, $R_{bc}$ is effectively shorting out the load. Therefore, the load voltage is approximately 0 V.

When set to the other extreme (as represented by the equivalent circuit in Figure 7.41b), $R_{ab}$ is effectively shorted out. Because of this, all the source voltage is dropped across the parallel combination of $R_{bc}$ and the load. Therefore, the load voltage is approximately equal to the source voltage.

***Potentiometer Resistance.*** In Chapter 3, you were told that the value of $R_{ac}$ for a given potentiometer is constant. (This principle was illustrated in Figure 3.18.) However, when a potentiometer is connected as a variable voltage divider, this is not true. As the potentiometer setting is varied, the effective value of $R_{ac}$ also changes. This point is illustrated in

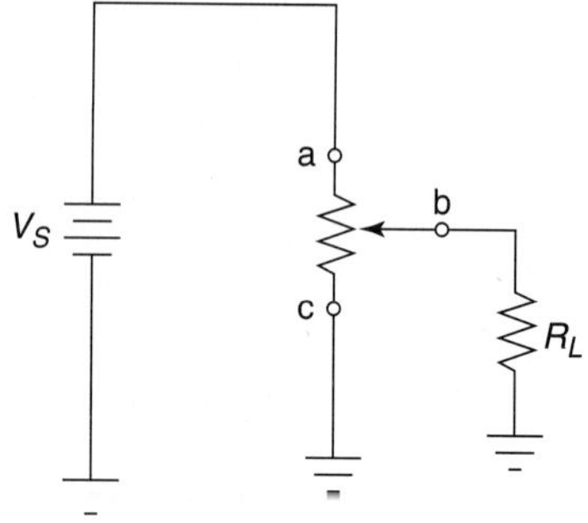


**FIGURE 7.40 A potentiometer as a variable voltage divider.**

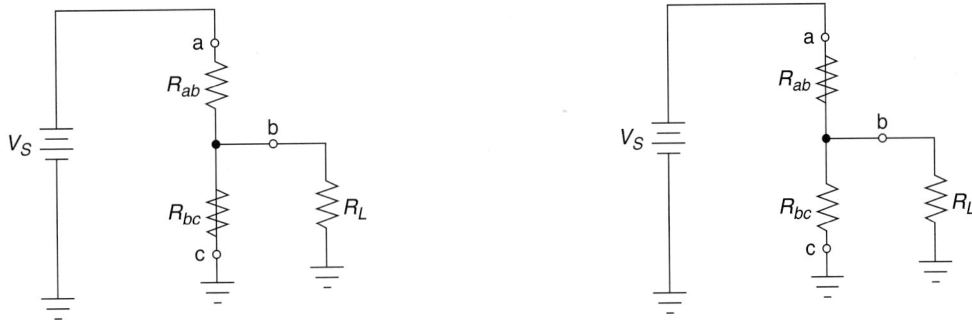


(a) The potentiometer is adjusted for minimum load voltage. (b) The potentiometer is adjusted for maximum load voltage.

**FIGURE 7.41 Equivalent circuits for a loaded voltage divider adjusted to its limits.**

Figure 7.42. Here we have a potentiometer rated at 10 kΩ with a 10 kΩ load. When the potentiometer is set to $R_{bc} = 0\ \Omega$, the value of $R_{ac}$ is equal to the value of $R_{ab}$, 10 kΩ. However, when set to the other extreme, the value of $R_{ac}$ is equal to the parallel combination of $R_{bc}$ and the load resistance. This parallel combination has a total resistance of 5 kΩ. Therefore, changing the setting on a loaded voltage divider causes the value of $R_{ac}$ to change.

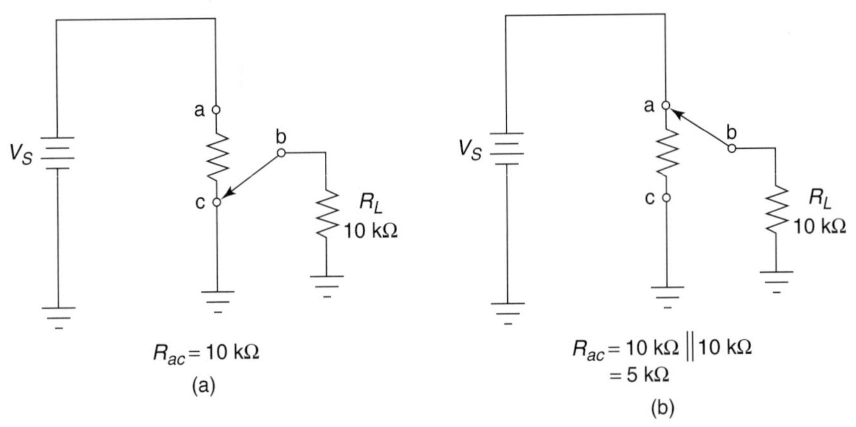


*Note:* The potentiometer symbols have been altered in this figure to illustrate a point. They are not normally drawn as shown here.

**FIGURE 7.42**

***Circuit Adjustments.*** The relationship between $R_{ac}$ and the potentiometer setting may limit the range over which the potentiometer can be varied. This restriction is illustrated in Figure 7.43. Here we have a loaded voltage divider with a 500 Ω potentiometer and a ⅛ A

**FIGURE 7.43**

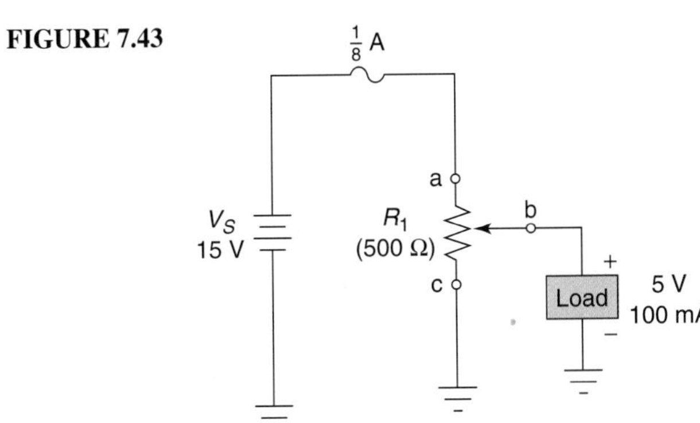

(125 mA) fuse. It would seem (from the potentiometer rating) that the circuit current would never be sufficient to blow the fuse. However, this is not the case. The load has a rating of $V_L = 5$ V at 100 mA. This means that the resistance of the load is approximately 50 Ω. If the potentiometer is varied so that $R_{bc} = 500\ \Omega$, the total circuit resistance drops to

$$R_{EQ} = R_{bc} \| R_L = 500\ \Omega \| 50\ \Omega = 45.5\ \Omega$$

At this setting, the total current through the circuit tries to increase to

$$I_T = \frac{V_S}{R_{EQ}} = \frac{15\text{ V}}{45.5\ \Omega} = 330\text{ mA}$$

which will cause the fuse to blow. As you can see, you must be careful when adjusting the potentiometer in a loaded voltage divider. The fact that $R_{ac}$ changes with the setting means that there may be a limit on how far the potentiometer can be varied in one direction. To be safe, never adjust a potentiometer any further than is required to adjust the load voltage to its proper value.

## Fault Symptoms in Series-Parallel Circuits

When troubleshooting a series-parallel circuit, it is important to remember the symptoms of open and shorted components in both series circuits and parallel circuits. As a review, these symptoms are given in Table 7.3. Even though series-parallel circuits are more complex, the symptoms associated with each of the faults listed in Table 7.3 remain the same. This is demonstrated in Examples 7.15 and 7.16.

**TABLE 7.3 Symptoms of Open and Shorted Components in Series and Parallel Circuits**

*OBJECTIVE 12* ▶

| *Circuit Type* | *Fault* | *Symptom(s)* |
|---|---|---|
| Series | Open | Circuit current drops to zero.<br>The applied voltage is dropped across the open component.<br>The voltage across each remaining component drops to 0 V. |
| | Short | Circuit current increases.<br>The voltage across the shorted component drops to 0 V.<br>The voltage across each remaining component increases. |
| Parallel | Open | Current through the open branch drops to zero.<br>Total circuit current decreases.<br>Other branches continue to operate normally. |
| | Short | The shorted branch causes the power supply fuse to blow.<br>The measured resistance across all branches is 0 Ω. (The faulty branch shorts out all parallel branches.)<br>The voltage measured across any branch is 0 V. |

### EXAMPLE 7.15

$R_1$ in Figure 7.44 is open. Determine the effect of this open on the voltage and current values in the circuit.

***Solution:*** Since $R_1$ is open, there is no current through that branch. Therefore, $I_A = 0$ A. With no current through the branch, no voltage is developed across $R_2$. Therefore, $V_2 = 0$ V and $V_1 = V_S$.

The problem in the circuit is isolated to the first branch; that is, it does not affect the operation of branch B. Therefore, the branch B values are all normal: $I_B = 4.8$ mA, $V_3 = 4.8$ V, and $V_4 = 7.2$ V.

Finally, the loss of $I_A$ causes the total current in the circuit to decrease. The total current is now equal to the value of $I_B$, 4.8 mA.

**FIGURE 7.44**

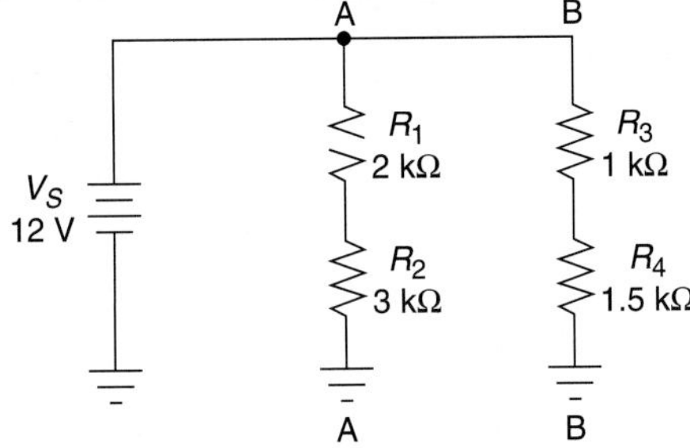


***PRACTICE PROBLEM 7.15***

Assume that $R_1$ in Figure 7.44 is normal and that $R_4$ is open. Determine the effect that this fault has on the voltage and current values in the circuit.

The symptoms of the problem in Example 7.15 are illustrated in Figure 7.45. Compare the readings in branch A to the series-open symptoms in Table 7.3. As you can see, the symptoms are exactly as described in the table:

1. The branch current has dropped to zero.
2. The applied voltage is dropped across the open component ($R_1$).
3. The voltage across the remaining component ($R_2$) is 0 V.

Compare the overall readings in the circuit to the parallel-open symptoms in Table 7.3. Again, the symptoms are exactly as described in the table:

1. Current through the open branch has dropped to zero.
2. The total circuit current has decreased.
3. The working branch has been unaffected by the open.

Even though the circuit is more complex than those covered in previous chapters, it still responds to the fault in a logical, predictable fashion. Let's go through another fault and its symptoms.

**FIGURE 7.45**

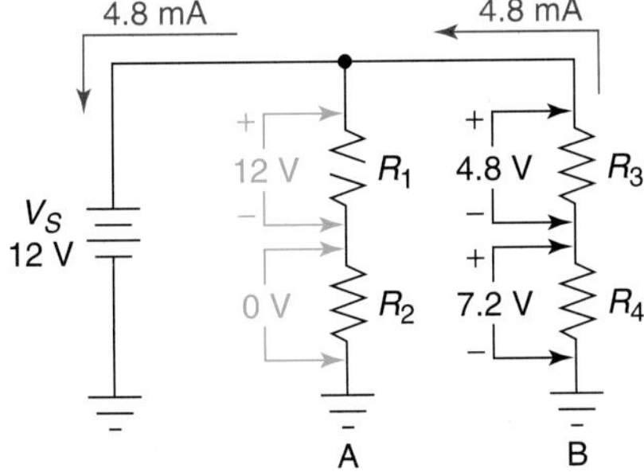


*EXAMPLE 7.16*

$R_2$ in Figure 7.46 is shorted. Determine the effect of this short on the voltage and current values in the circuit.

***Solution:*** Since $R_2$ is shorted, the combined resistance of $R_1$ and $R_2$ drops to approximately 0 Ω. Therefore, $V_{AB} = 0$ V and $V_{CD} = V_S$. The increase in $V_{CD}$ causes the current through loop B to increase from 125 mA (normal) to 230 mA. Since the loops are in series, the total circuit current also increases to 230 mA.

**FIGURE 7.46**

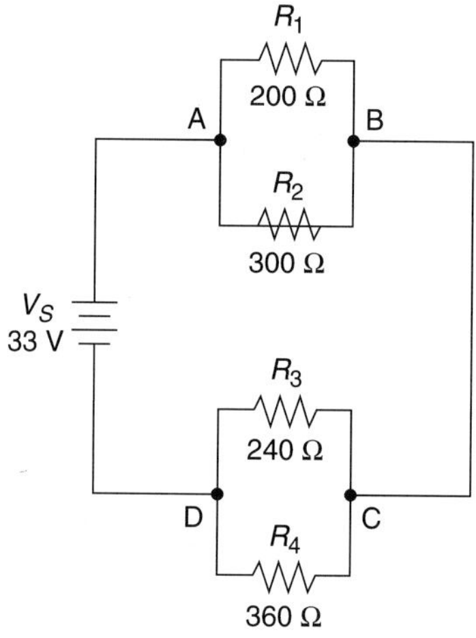


***PRACTICE PROBLEM 7.16***

Assume that $R_2$ in Figure 7.46 is normal and that $R_3$ is shorted. Determine the effect that this fault has on the voltage and current values in the circuit.

The symptoms of the problem in Example 7.16 are illustrated in Figure 7.47. Compare the loop A readings to the parallel-short symptoms in Table 7.3. As you can see, the symptoms are almost identical to those described in the table:

1. The combined resistance of the loop is 0 Ω.
2. The voltage across both branches is 0 V.

Compare the overall readings in the circuit to the series-short symptoms listed in Table 7.3. Again, the symptoms are identical to those listed in the table:

1. The total circuit current increased (from 125 mA to 230 mA).
2. The voltage across the shorted component (loop A) dropped to 0 V.
3. The voltage across the remaining component (loop B) increased.

Once again, a series-parallel circuit responds to a fault in a logical and predictable fashion.

**FIGURE 7.47**

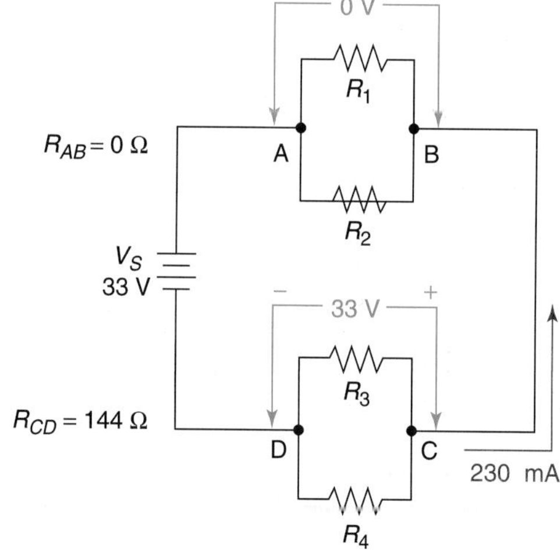

## *Series-Parallel Circuit Troubleshooting*

*OBJECTIVE 13*

In many practical situations, series-parallel circuit troubleshooting begins with determining whether or not a given fault is load-specific, that is, whether or not the fault is affecting more than one load. Determining whether or not a fault is load-specific can lead you directly to the most likely cause(s) of the fault. For example, consider the circuit shown in Figure 7.48. Let's assume that load 1 isn't working and the other two are. In this case, the problem must lie with the load itself and/or $R_4$. Why? Because these two components are the only ones unique to load 1. In other words, they are the only ones that could cause a problem in the load 1 circuit without affecting the other two loads. Using the same logic:

1. A malfunction unique to load 2 could be caused only by the load itself and/or $R_5$.
2. A malfunction unique to load 3 could be caused only by the load itself and/or $R_6$.

As you can see, load-specific faults are relatively easy to isolate.

Faults common to more than one load can also be easy to isolate when you take the time to identify the components common to those loads. Usually, one (or more) of these components is the source of the problem. For example, let's assume that loads 1 and 2 in Figure 7.48 are not working. The most likely cause of the problem would be $R_1$. Why? Because this component is part of the voltage divider that supplies power to both loads. Since the voltage source and the fuse also affect the operation of load 3 (which is working), they are not likely to be the cause of the problem. Therefore, we would start by testing $R_1$. If all three loads were inoperative, we would assume that the voltage source or the fuse was the cause of the problem.

As you can see, series-parallel circuit troubleshooting starts with observing symptoms and isolating the possible cause(s) of those symptoms. This process is demonstrated further in the following troubleshooting applications.

**FIGURE 7.48**

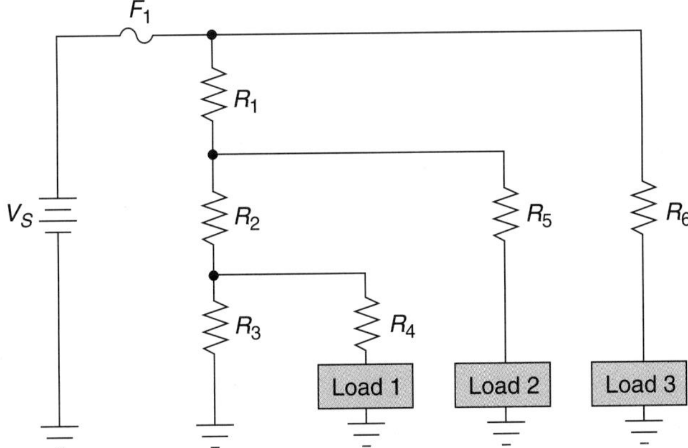


## TROUBLESHOOTING APPLICATION 1

The circuit in Figure 7.49 is designed to provide the operating voltages required by three loads. However, loads 2 and 3 are not working. Since B is the common point for these two loads, the voltage at this point is measured and found to be 0 V. Load 1 is working (indicating a voltage at point A), so $R_3$ is assumed to be open. Checking the resistance of this component verifies that it is open. Replacing the component restores normal operation to the circuit.

*AN OBSERVATION* For this circuit, loads 2 and 3 form the only load pair that could be inoperative. In other words, there is no single fault that could prevent loads 1 and 3 (or loads 1 and 2) from operating without affecting the third load.

FIGURE 7.49

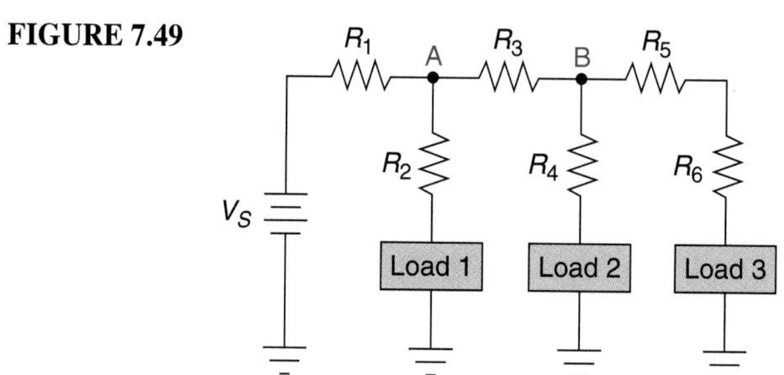


## TROUBLESHOOTING APPLICATION 2

Refer to Figure 7.50. Load 3 in this circuit is not operating. Since the other two loads are operating normally, the only suspect components are load 3 and $R_4$. The voltage across load 3 is measured and found to be equal to the voltage at point B, which is slightly higher than normal. This could occur only if the load were open. The load is replaced, restoring normal operation to the circuit.

*AN OBSERVATION* Had the problem been caused by an open $R_4$, the voltage across the load would have been 0 V because all of $V_B$ would have been dropped across the open resistor, as is the case in any series connection. Had the problem been a shorted $R_4$, the voltage at point B would have been slightly lower than normal because of the decreased parallel equivalent resistance between point B and ground.

FIGURE 7.50

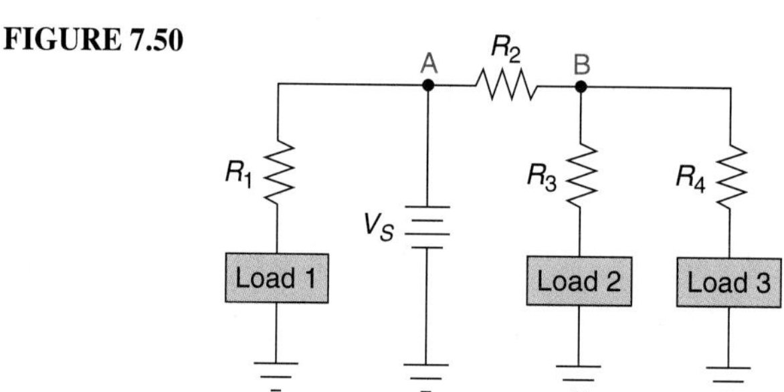


## TROUBLESHOOTING APPLICATION 3

None of the loads in Figure 7.50 are operating. In this case, the voltage source is the only component common to all the inoperative loads. It is assumed to be faulty and is replaced. Replacing the source restores normal operation to the circuit.

*AN OBSERVATION* In this case, it was not necessary to test the voltage source because no other component could have caused the symptoms. You will find that this happens fairly often when troubleshooting circuits and systems.

As you can see, practical troubleshooting starts before using any test equipment. You can generally save yourself a great deal of time and trouble by observing the operation of a circuit and thinking your way through the problem before actually testing the circuit.

## Section Review

1. Describe the use of a potentiometer as a variable voltage divider.
2. How does varying the setting of a variable voltage divider affect the effective value of $R_{ac}$ for the potentiometer?
3. Why should you be careful when adjusting the potentiometer setting in a variable voltage divider?
4. What fault symptoms need to be remembered when troubleshooting series-parallel circuits? Why should these symptoms be kept in mind?
5. What is the first step in troubleshooting a practical series-parallel circuit? Why is this step important?

## Chapter Summary

Here is a summary of the major points made in this chapter:

1. Circuits that contain both series and parallel elements are referred to as series-parallel circuits or combination circuits.
2. When series circuits are connected in parallel:
   - **a.** Each branch retains all the characteristics of any series circuit.
   - **b.** The combined branches retain all the characteristics of any parallel circuit.
3. When parallel circuits are connected in series:
   - **a.** Each loop is a parallel circuit that retains all the characteristics of any parallel circuit.
   - **b.** The combined loops retain all the characteristics of any series circuit.
4. Any series-parallel circuit can be dealt with effectively when you:
   - **a.** Use the proper approach.
   - **b.** Remember the basic principles of series circuits and parallel circuits.
5. In most cases, the goal of any series-parallel circuit analysis is to determine the value of voltage, current, and/or power for one or more components.
6. The process of analyzing a series-parallel circuit usually begins with deriving a simplified equivalent of the original circuit. The simplification process ends when the circuit has been reduced to either a series-equivalent or a parallel-equivalent circuit.
7. The complete analysis of most series-parallel circuits involves three steps:
   - **a.** Reduce the circuit to a series-equivalent or a parallel-equivalent, depending on the type of circuit.
   - **b.** Apply the basic series and parallel circuit principles and relationships to the equivalent circuit.
   - **c.** Use the values obtained from the equivalent circuit to solve the original for the desired values.
8. Voltage dividers are usually designed for the purpose of providing one or more specific load voltages. When used for this purpose, the voltage divider is referred to as a loaded voltage divider.
9. The no-load output voltage ($V_{NL}$) from a voltage divider is the voltage measured at the load terminals with the load removed.
   - **a.** The term *output voltage* is commonly used to describe the voltage at the load terminals of a circuit.
   - **b.** The value of $V_{NL}$ is always greater than the value of the load voltage ($V_L$) for the circuit.

10. For a loaded voltage divider, load voltage varies directly with load resistance.
11. Voltage dividers are usually designed so that any variations in output voltage (due to a change in load resistance) are held to a minimum.
12. Voltage divider stability refers to the circuit's ability to maintain a stable load voltage despite normal (or anticipated) variations in load demand.
    a. Stability is accomplished by designing a voltage divider so that the load current makes up a relatively small percentage of the total circuit current.
    b. The degree of stability depends on the exact relationship between the bleeder current and the maximum load current. The greater the ratio of bleeder current to load current, the greater the stability.
13. When designed for a high degree of stability, the values of $V_L$ and $V_{NL}$ for a loaded voltage divider are very nearly equal. In this case, a value of $V_L \cong V_{NL}$ may not indicate that a load is open. (In many circuits, $V_L \cong V_{NL}$ is a classic symptom of an open load.)
14. A voltmeter may provide a reading that is significantly lower than the actual component voltage.
    a. When a voltmeter is connected across a component in a series circuit, the circuit becomes a series-parallel circuit.
    b. The effect of meter loading depends on the meter's internal resistance. The lower the value of $R_m$, the greater the error in the voltage reading.
15. The Wheatstone bridge is a circuit that can be used to provide extremely accurate resistance measurements and is shown in Figure 7.33. The meter used is a galvanometer (a current meter that indicates both the magnitude and the direction of a low-value current).
16. When the resistance ratios in a Wheatstone bridge are equal, the bridge is said to be balanced. Balance in a Wheatstone bridge is indicated by a galvanometer reading of $I_G = 0$ A.
17. A Wheatstone bridge can be used as a resistance measuring circuit when constructed as shown in Figure 7.36. The value of an unknown resistance is measured as follows:
    a. The unknown resistance is placed in the position labeled $R_x$.
    b. The calibrated decade box is adjusted until the bridge is balanced.
    c. The reading on the decade box is used (with the other known resistance values in the circuit) to calculate the unknown value.
18. A potentiometer can be used as a variable voltage divider (see Figure 7.41).
    a. Varying the potentiometer setting allows the value of $V_L$ to vary between 0 V and the value of $V_S$.
    b. When its setting is varied, the effective value of $R_{ac}$ for the potentiometer may drop below its rated value.
    c. Because $R_{ac}$ can decrease, you must be careful when adjusting the potentiometer in a loaded voltage divider. Never adjust the component any further than needed to obtain the correct value of $V_L$.
19. When troubleshooting a series-parallel circuit, remember the symptoms of open and shorted components in both series and parallel circuits.
    a. These symptoms are provided in Table 7.3.
    b. Even though series-parallel circuits are more complex, the symptoms associated with each type of fault (listed in Table 7.3) remain the same.
20. In many practical situations, series-parallel circuit troubleshooting begins with determining whether or not a given fault is load-specific. Determining whether or not a fault is load-specific can lead you directly to the cause(s) of the problem.
21. Circuit troubleshooting starts with observing symptoms and isolating the possible cause(s) of those symptoms. (Determining whether or not the problem is load-specific is part of this process.)

## Equation Summary

| Equation Number | Equation | Section Number |
|---|---|---|
| **(7.1)** | $I_B >> I_L$ | 7.3 |
| **(7.2)** | $I_{B(\min)} \geq 10I_L$ | 7.3 |
| **(7.3)** | $R_{L(\min)} \geq 10R_B$ | 7.3 |
| **(7.4)** | $V_A = V_S \dfrac{R_2}{R_1 + R_2}$ | 7.4 |
| **(7.5)** | $V_B = V_S \dfrac{R_4}{R_3 + R_4}$ | 7.4 |
| **(7.6)** | $\dfrac{R_2}{R_1} = \dfrac{R_4}{R_3}$ | 7.4 |
| **(7.7)** | $R_x = R_{db} \dfrac{R_3}{R_4}$ | 7.4 |
| **(7.8)** | $R_x = R_{db}$ (when $R_3 = R_4$) | 7.4 |

## Key Terms

The following terms were introduced and defined in this chapter:

bleeder current ($I_B$)
circuit loading
combination circuit
galvanometer
loaded voltage divider
load-specific fault
no-load output voltage ($V_{NL}$)
parallel-equivalent circuit
series-equivalent circuit
series-parallel circuit
stability (voltage divider)
variable voltage divider
voltage divider stability
voltmeter loading
Wheatstone bridge

## Practice Problems

1. Perform a complete mathematical analysis of the circuit in Figure 7.51.
2. Perform a complete mathematical analysis of the circuit in Figure 7.52.
3. Perform a complete mathematical analysis of the circuit in Figure 7.53.
4. Perform a complete mathematical analysis of the circuit in Figure 7.54.
5. Perform a complete mathematical analysis of the circuit in Figure 7.55.
6. Perform a complete mathematical analysis of the circuit in Figure 7.56.
7. Perform a complete mathematical analysis of the circuit in Figure 7.57.
8. Perform a complete mathematical analysis of the circuit in Figure 7.58.
9. Derive the series-equivalent of the circuit in Figure 7.59a.

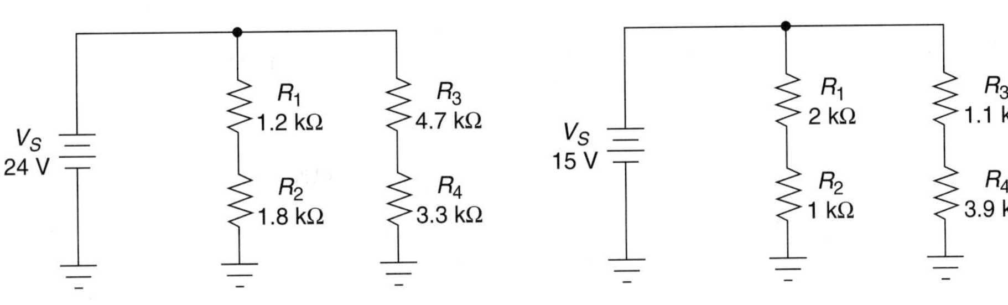


**FIGURE 7.51**

**FIGURE 7.52**

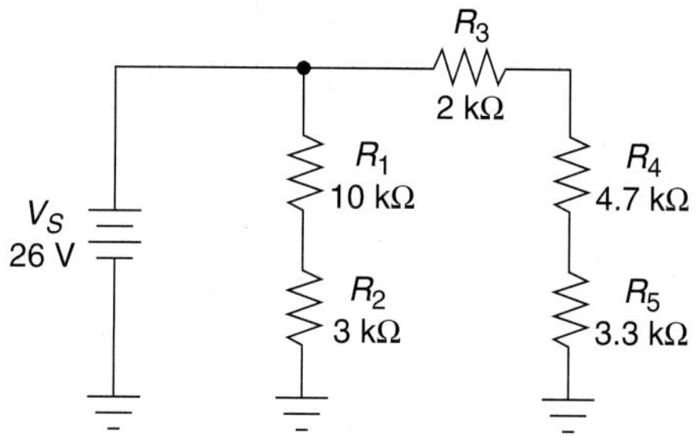

FIGURE 7.53

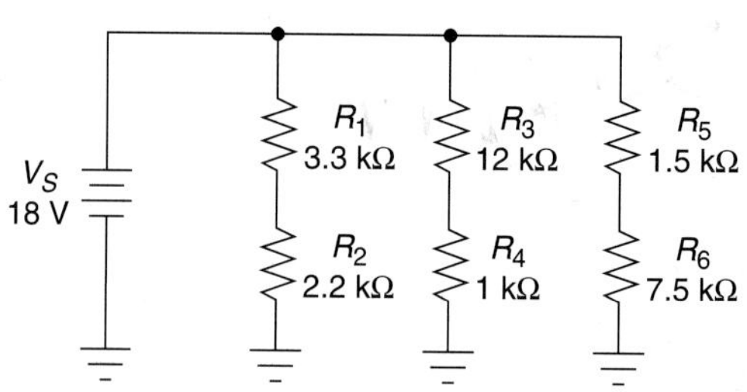


FIGURE 7.54

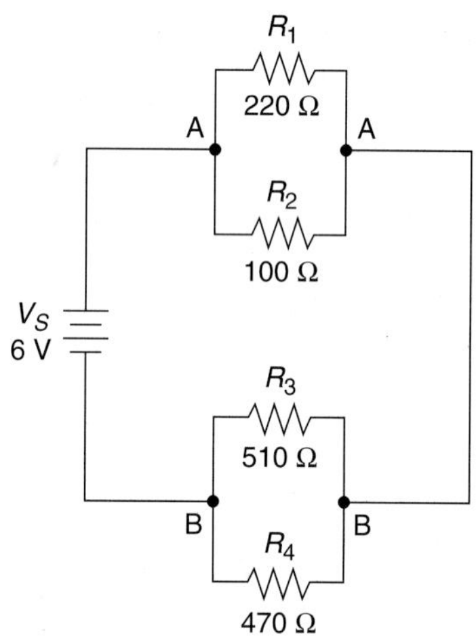

FIGURE 7.55

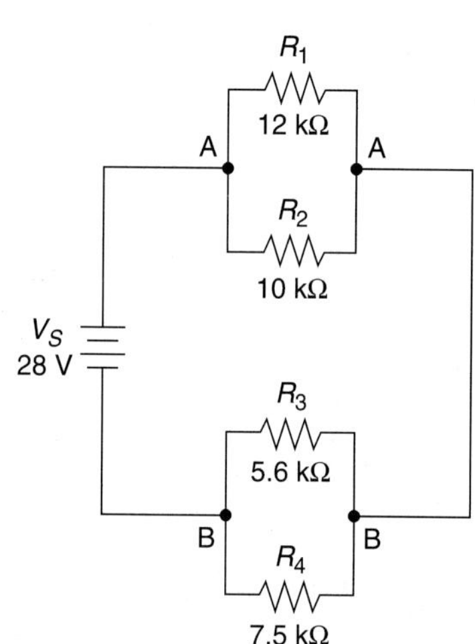


FIGURE 7.56

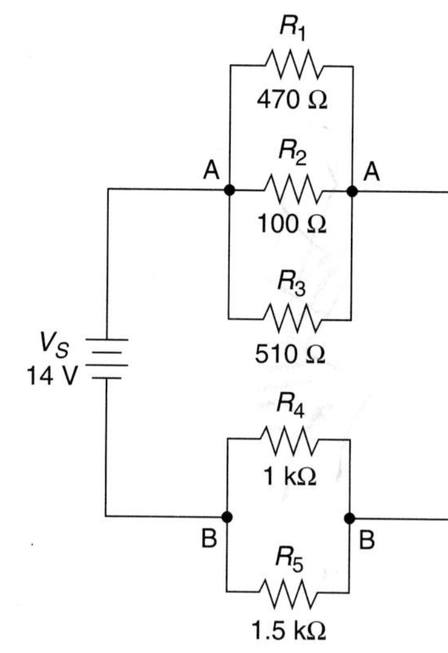


FIGURE 7.57

**10.** Derive the series-equivalent of the circuit in Figure 7.59b.

**11.** Derive the parallel-equivalent of the circuit in Figure 7.60a.

**12.** Derive the parallel-equivalent of the circuit in Figure 7.60b.

**13.** Determine the value of load voltage ($V_L$) for the circuit in Figure 7.59a.

**14.** Determine the value of load voltage ($V_L$) for the circuit in Figure 7.59b.

**15.** Determine the value of load current ($I_L$) for the circuit in Figure 7.60a.

FIGURE 7.58

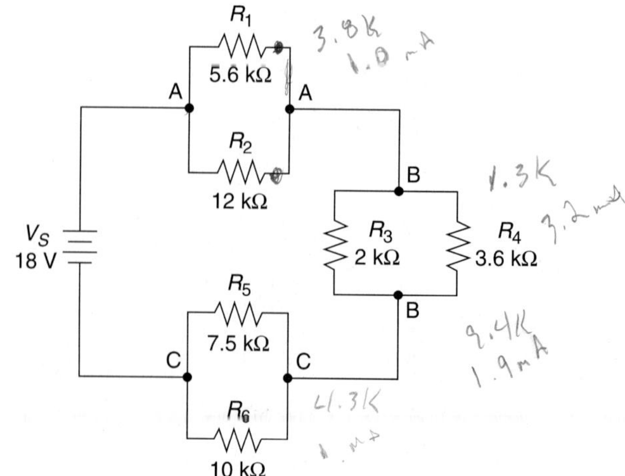

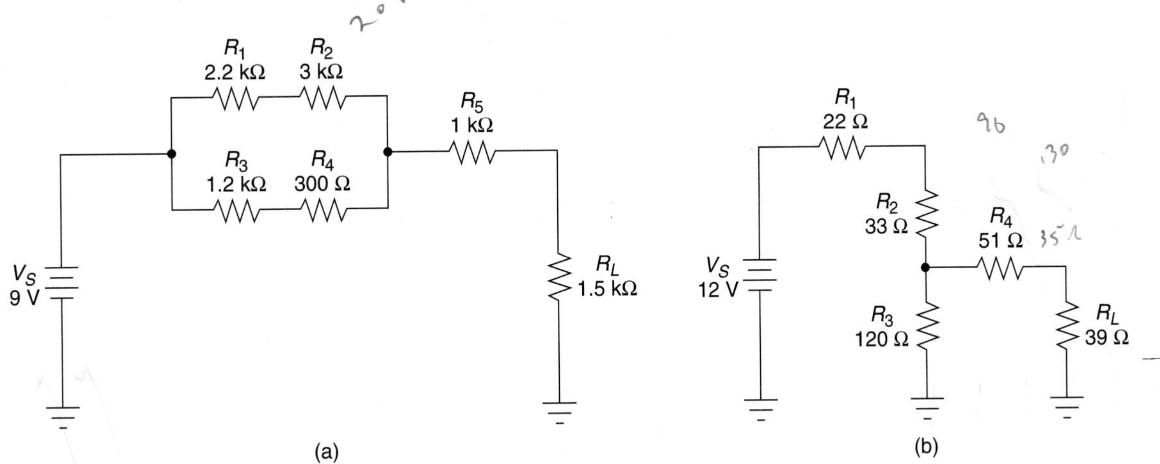


**FIGURE 7.59**

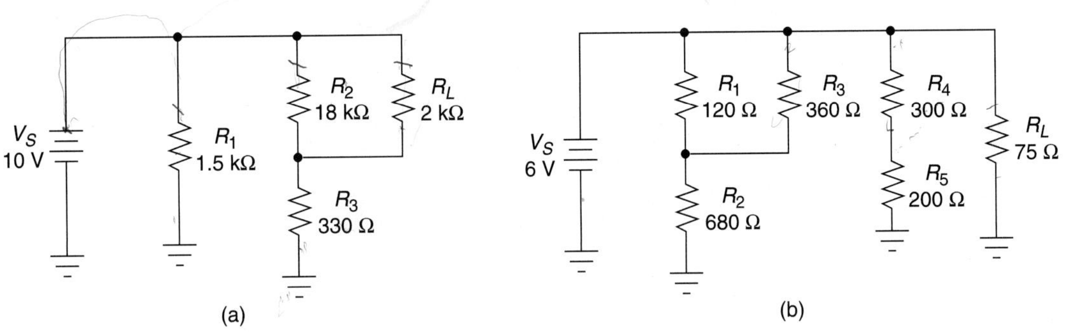


**FIGURE 7.60**

**16.** Determine the value of load current ($I_L$) for the circuit in Figure 7.60b.

**17.** Determine the value of load power ($P_L$) for the circuit in Figure 7.61a.

**18.** Determine the value of load power ($P_L$) for the circuit in Figure 7.61b.

**19.** Determine the values of $V_L$ and $V_{NL}$ for the circuit in Figure 7.62a.

**20.** Determine the values of $V_L$ and $V_{NL}$ for the circuit in Figure 7.62b.

**21.** Determine the normal range of output voltages for the circuit in Figure 7.63a.

**22.** Determine the normal range of output voltages for the circuit in Figure 7.63b.

**23.** Determine the normal range of output voltages for the circuit in Figure 7.64a.

**24.** Determine the normal range of output voltages for the circuit in Figure 7.64b.

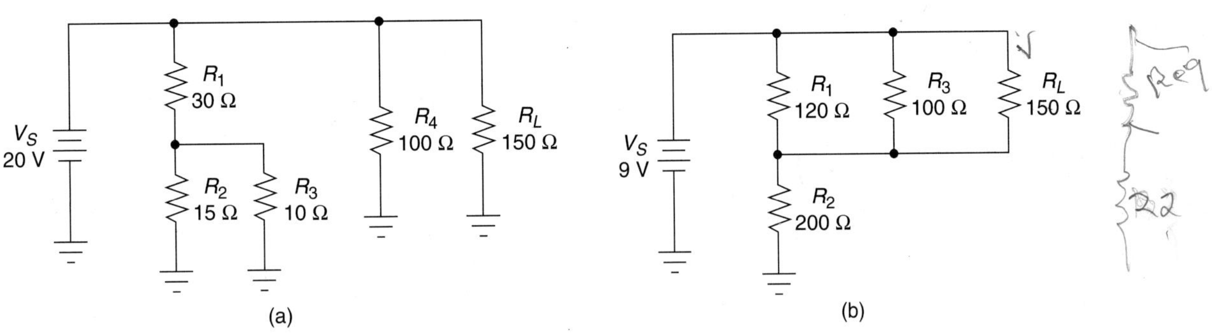


**FIGURE 7.61**

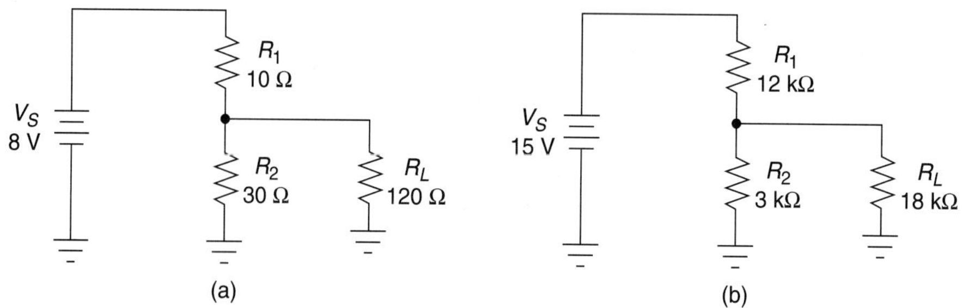


**FIGURE 7.62**

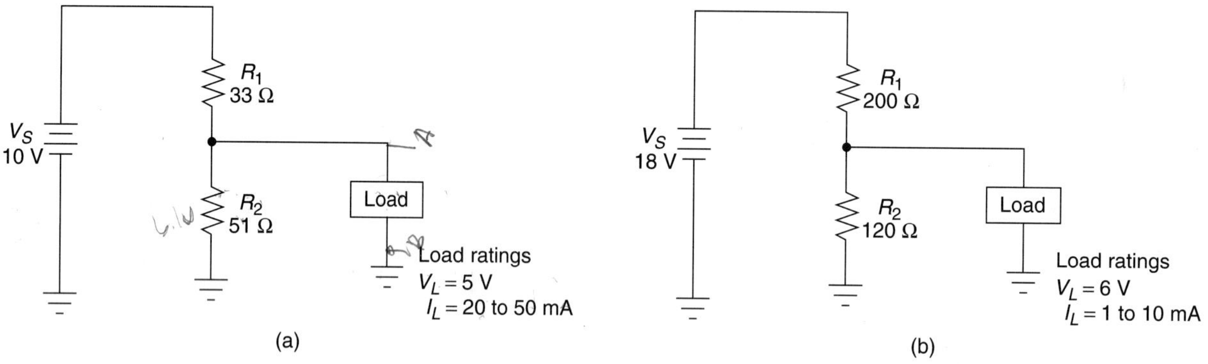


**FIGURE 7.63**

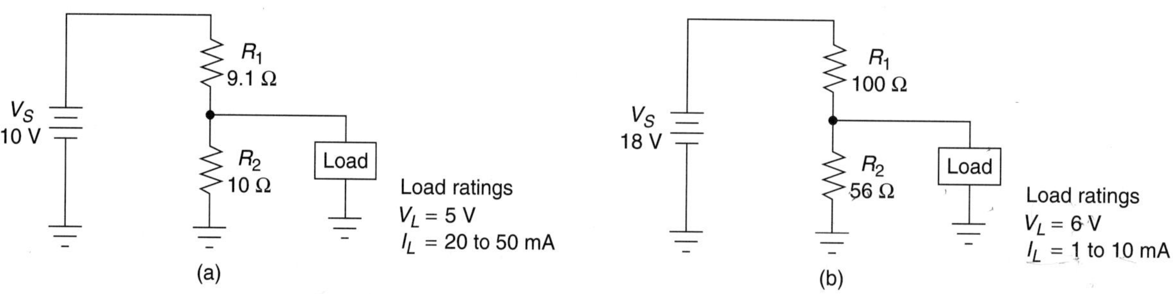


**FIGURE 7.64**

**25.** Determine the value of $V_2$ for the circuit in Figure 7.65a. Then determine the reading that will be produced when the voltmeter is connected as shown in Figure 7.65b.

**26.** Repeat Problem 25 for the circuit shown in Figure 7.66.

**27.** Determine the value of $R_x$ for the circuit in Figure 7.67a.

**28.** Determine the value of $R_x$ for the circuit in Figure 7.67b.

**FIGURE 7.65**

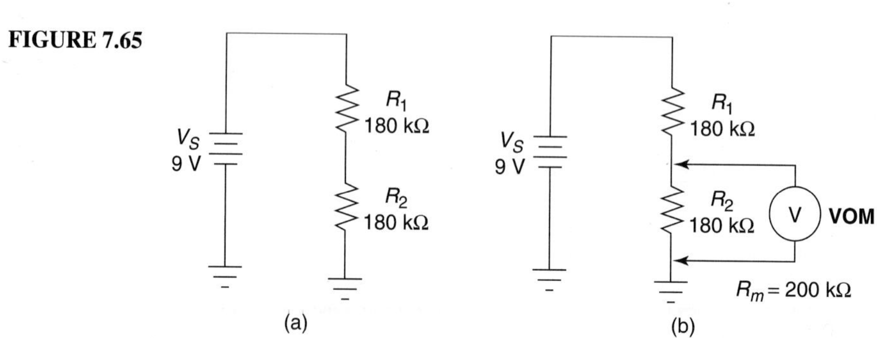

FIGURE 7.66

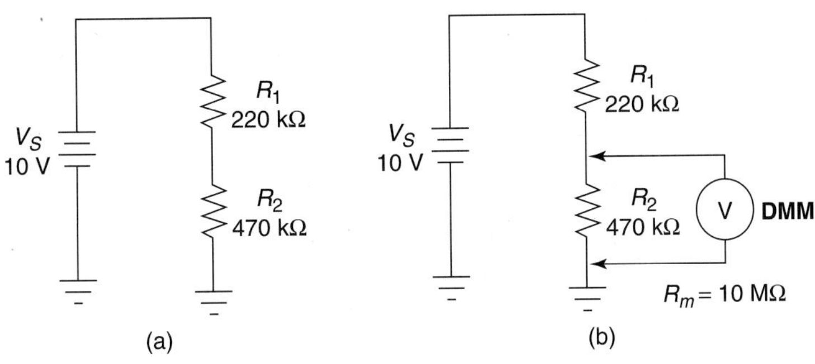


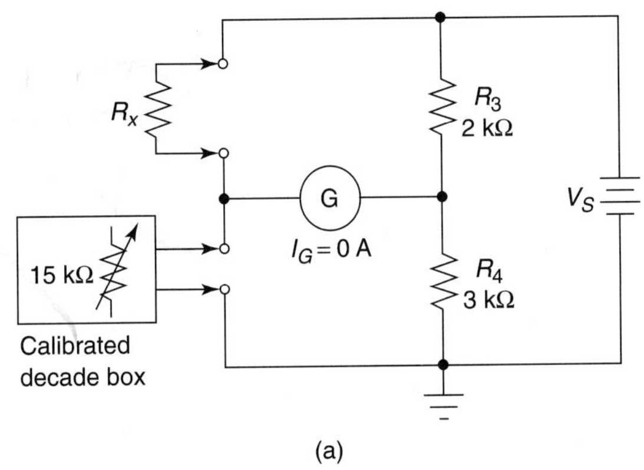


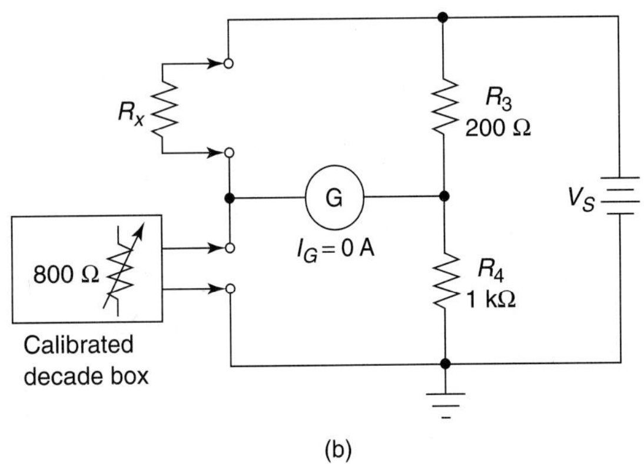


FIGURE 7.67

## Troubleshooting Practice Problems

**29.** Calculate the voltage and current values for the circuit in Figure 7.68. Then predict how these values would change if:

**a.** $R_1$ opened.

**b.** $R_3$ shorted.

**30.** Calculate the voltage and current values for the circuit in Figure 7.69. Then predict how these values would change if

**a.** $R_2$ opened.

**b.** $R_4$ shorted.

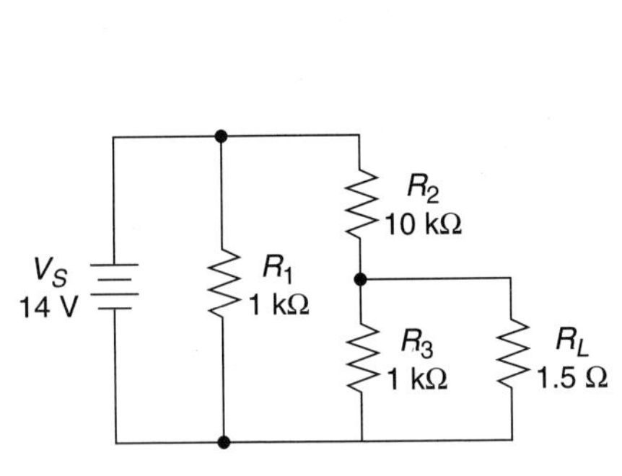


FIGURE 7.68

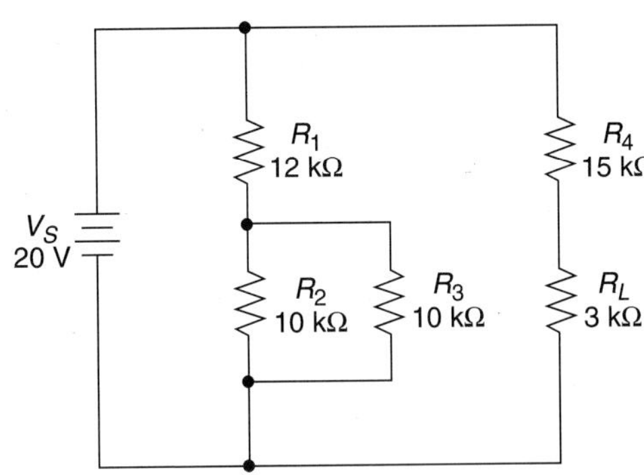


FIGURE 7.69

**FIGURE 7.70**

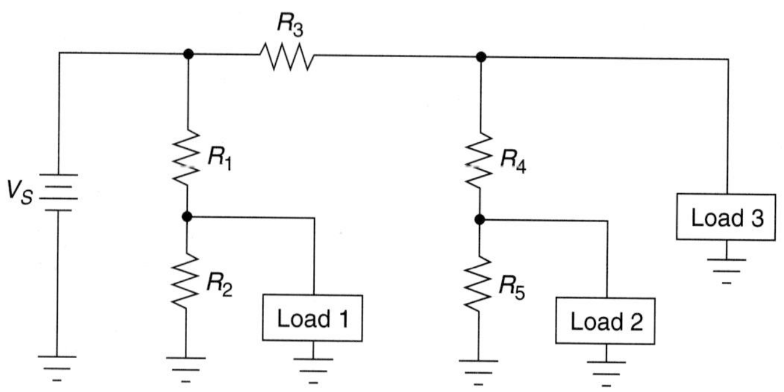


**FIGURE 7.71**

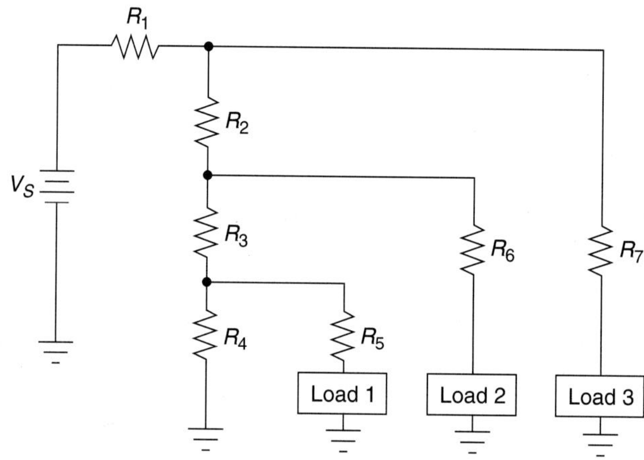


**31.** Refer to Figure 7.70. Determine the probable cause(s) of each of the following:

**a.** Load 1 is inoperative. The others are working normally.

**b.** Loads 2 and 3 are inoperative. Load 1 is working normally.

**32.** Refer to Figure 7.71. Determine the probable cause(s) of each of the following:

**a.** Load 2 is inoperative. The others are working normally.

**b.** None of the loads are working.

## The Brain Drain

**33.** For the circuit shown in Figure 7.72, determine the adjusted value of $R_2$ that will cause load 1 to run at its rated values.

**34.** Determine the value of $V_6$ for the circuit in Figure 7.73.

**35.** Figure 7.74 shows a simplified block diagram for a basic AM receiver. The power supply is a circuit that supplies the dc voltages required for each block to function. Each block performs a specific operation on the input signal received from the antenna. If you treat each block as a single component, is the receiver a series, parallel, or series-parallel circuit? Explain your answer.

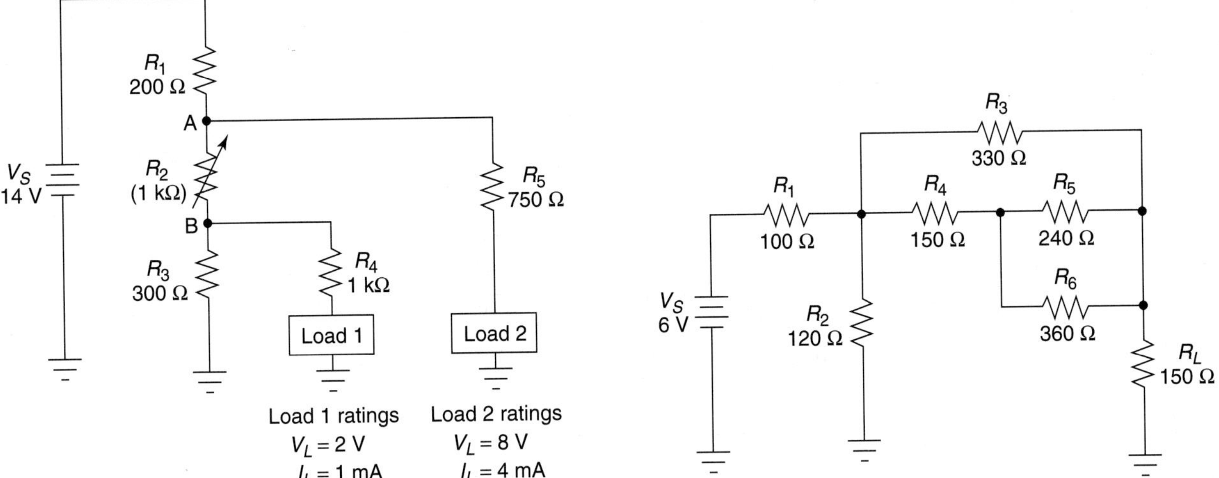


**FIGURE 7.72**

**FIGURE 7.73**

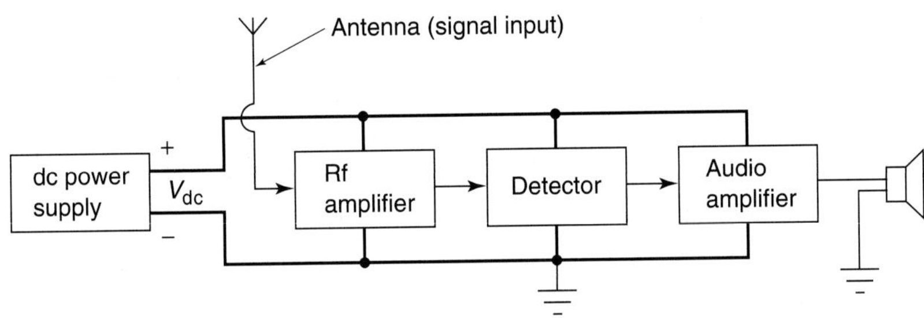


**FIGURE 7.74 A simplified AM receiver block diagram.**

## Answers to the Example Practice Problems

**7.1.** $I_A = 20$ mA, $V_1 = 4.4$ V, $V_2 = 6.6$ V, $P_1 = 88$ mW, $P_2 = 132$ mW, $I_B = 50$ mA, $V_3 = 6$ V, $V_4 = 5$ V, $P_3 = 300$ mW, $P_4 = 250$ mW, $I_T = 70$ mA, $R_T = 157\ \Omega$, $P_T = 770$ mW

**7.2.** $R_A = 60\ \Omega$, $R_B = 72\ \Omega$, $V_A = 150$ mV, $V_B = 180$ mV, $I_1 = 1.5$ mA, $I_2 = 1$ mA, $I_3 = 1$ mA, $I_4 = 1.5$ mA, $P_1 = 225\ \mu$W, $P_2 = 150\ \mu$W, $P_3 = 180\ \mu$W, $P_4 = 270\ \mu$W, $I_T = 2.5$ mA, $R_T = 132\ \Omega$, $P_T = 825\ \mu$W

**7.3.** 17 mW

**7.4.** See Figure 7.75.

**7.5.** See Figure 7.76.

**7.6.** 3.99 mW

**7.8.** 104.4 Ω

**7.9.** 7.2 V

**7.10.** $V_L = 7.5$ V, $V_{NL} = 9$ V

**7.11.** 1.98 V to 5.7 V

**7.12.** 9.01 V to 9.37 V

**7.13.** The DMM will provide a reading of 5.96 V. This is much closer to the calculated value of $V_2$.

**7.14.** 7.5 kΩ

**7.15.** $V_4 = 12$ V, $V_3 = 0$ V, $I_3 = I_4 = 0$ A, $I_T$ decreases to 2.4 mA

**7.16.** $V_{AB} = 33$ V, $V_{CD} = 0$ V, $I_S$ increases to 275 mA

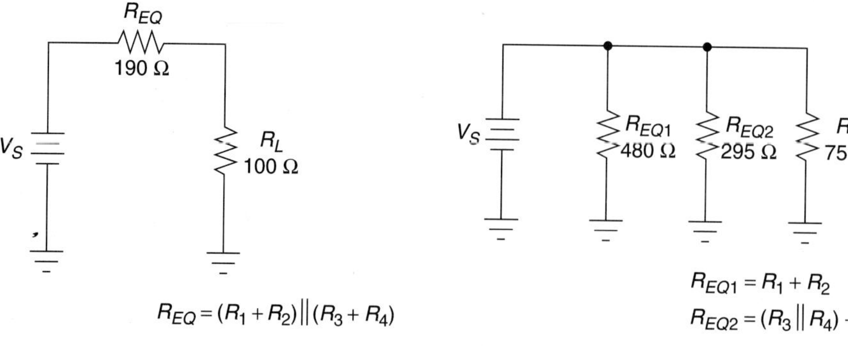


FIGURE 7.75

FIGURE 7.76

## Answers to the Section 7.1 Review Problems

**3.** a. True b. False c. True d. False

**4.**
a. $I_4 = I_5 + I_6$
b. $I_1 = I_2 + I_4$ or $I_1 = I_3 + I_4$
c. No, it is in series with the parallel combination of the two branches.
d. Any of the following combinations could be added to determine the value of $V_S$: $(V_1 + V_2 + V_3)$ or $(V_1 + V_4 + V_5)$ or $(V_1 + V_4 + V_6)$.

## EWB Applications Problems

| Figure | EWB File Reference |
|---|---|
| 7.4 | EWBA7_4.ewb |
| 7.9 | EWBA7_9.ewb |
| 7.19 | EWBA7_19.ewb |
| 7.35 | EWBA7_35.ewb |

# 8

# CIRCUIT ANALYSIS: THEOREMS AND CONVERSIONS

## OBJECTIVES

*After studying the material in this chapter, you should be able to:*

1. Explain how the superposition theorem is used to analyze multisource circuits.
2. Perform the analysis of a two-source circuit using the superposition theorem.
3. Discuss the characteristics of ideal and practical voltage and current sources.
4. Convert a voltage source to a current source and vice versa.
5. Describe load analysis.
6. Discuss Thevenin's theorem and the purpose it serves.
7. Describe the processes used to determine the values of Thevenin voltage ($V_{th}$) and Thevenin resistance ($R_{th}$) for a series-parallel circuit.
8. Derive the Thevenin equivalent of a given series-parallel circuit.
9. Use Thevenin's theorem to determine the range of output voltages for a series-parallel circuit with a variable load.
10. Use Thevenin's theorem to determine the maximum possible value of load power for a series-parallel circuit with a variable load.
11. Use Thevenin's theorem to analyze the operation of multisource, multiload, and bridge circuits.
12. Discuss Norton's theorem and the purpose it serves.
13. Describe the processes used to determine the values of Norton current ($I_n$) and Norton resistance ($R_n$) for a series-parallel circuit.
14. Analyze the operation of a series-parallel circuit using a Norton equivalent circuit.
15. Describe and perform Thevenin-to-Norton and Norton-to-Thevenin conversions.
16. Describe Millman's theorem and apply it to the analysis of a series-parallel circuit containing parallel voltage sources.
17. Describe delta ($\Delta$) and wye (Y) circuits.
18. Perform $\Delta$-to-Y and Y-to-$\Delta$ circuit conversions.

In Chapter 7, you were introduced to the fundamentals of series-parallel circuit operation and analysis. In this chapter, we will look at some analysis techniques used to solve some specific types of problems. These analysis techniques are generally referred to as network theorems. Network theorems are most commonly used to:

1. Analyze circuits that cannot be solved easily using the techniques established in Chapter 7.
2. Analyze series-parallel circuits that contain more than one voltage (or current) source.
3. Predict how a series-parallel circuit will respond to changes in load demand.

In this chapter, you will be shown how to use a variety of network theorems. As you will see, many of them simply provide alternate approaches to solving one of the problems listed above. You will also be shown how to perform a number of circuit conversions.

## 8.1 SUPERPOSITION

**Multisource circuit**
A circuit with more than one voltage and/or current sources.

All of the series-parallel circuits we've discussed up to this point have had only one voltage source. There are circuits, however, that contain more than one voltage (or current) source. A circuit of this type is referred to as a **multisource circuit.** Several examples of multisource circuits are shown in Figure 8.1.

When analyzing multisource circuits, the effects of each voltage (or current) source must be taken into account. In this section, we will cover one of several methods used to analyze multisource circuits. Some other commonly used (and more complex) methods are discussed in Chapter 9.

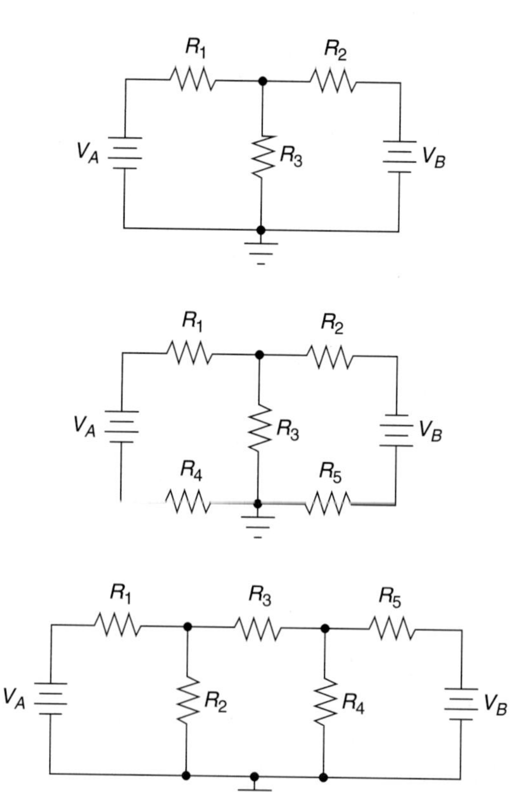


**FIGURE 8.1 Basic multisource circuit configurations.**

## The Superposition Theorem

◀ *OBJECTIVE 1*

The **superposition theorem** states that the response of a circuit to more than one source can be determined by analyzing the circuit's response to each source (alone) and combining the results. For example, consider the circuit shown in Figure 8.2. According to the superposition theorem, we can determine the component voltages in the circuit by:

**Superposition theorem**
The response of a circuit to more than one source can be determined by analyzing the circuit's response to each source (alone) and combining the results.

1. Analyzing the circuit as if $V_A$ were the only voltage source.
2. Analyzing the circuit as if $V_B$ were the only voltage source.
3. Combining the results from Steps 1 and 2.

**FIGURE 8.2**

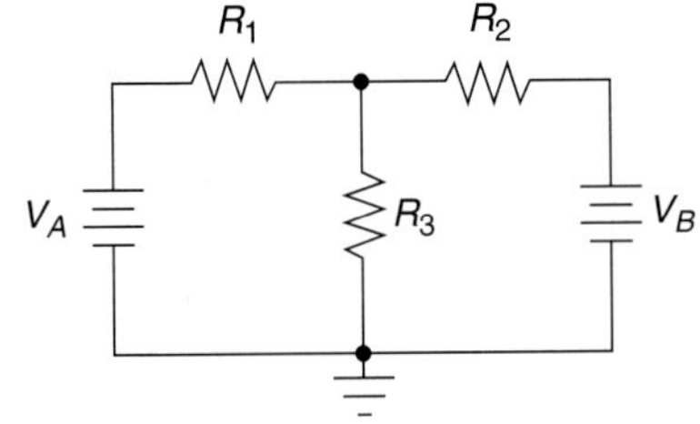


When using the superposition theorem to solve a multisource circuit, it is very helpful to draw the original circuit as several single-source circuits. For example, the circuit in Figure 8.2 would be redrawn as shown in Figure 8.3. Note that each circuit is identical to the original, except that one source has been removed. When drawn as shown, the analysis of the circuit for each voltage source becomes a problem like those we covered in Chapter 7. For example, the circuit in Figure 8.3a can be simplified to a series-equivalent by combining $R_2$ and $R_3$ into a parallel equivalent resistance. At that point, determining the component voltages is simple. The circuit in Figure 8.3b can be simplified in the same manner: $R_1$ and $R_3$ are combined into a parallel equivalent resistance, leaving us with a simple series circuit.

*Note:* When a voltage source is removed as shown in Figure 8.3, it is always replaced by a wire. This simulates the ideal value of $R_S = 0\ \Omega$ for the missing voltage source.

After calculating the voltage values for the circuits in Figure 8.3, we combine the values to obtain the actual circuit voltages. However, care must be taken to ensure that the

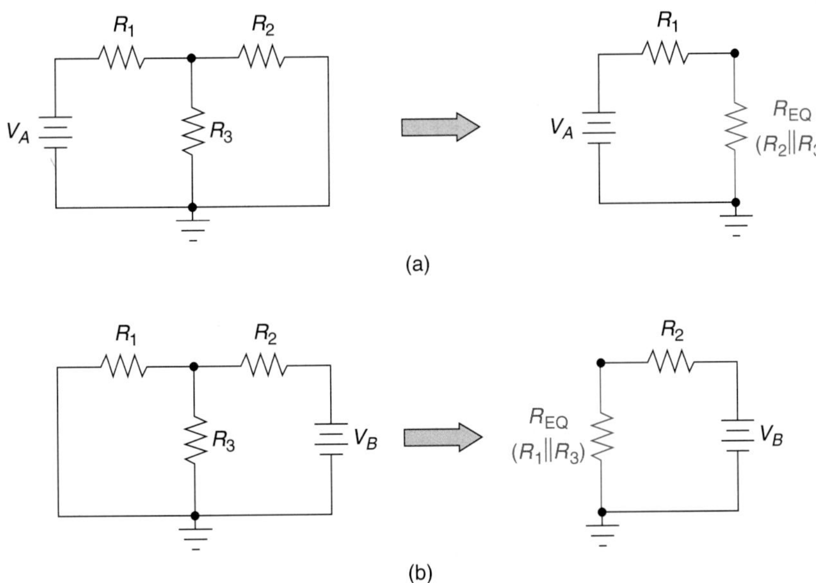


**FIGURE 8.3**

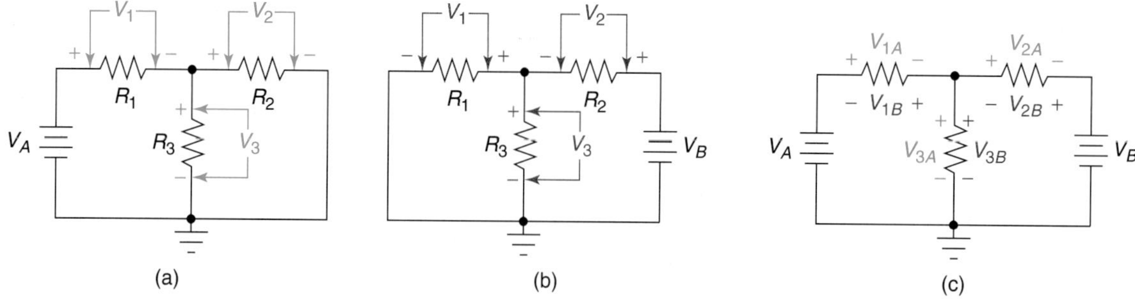


**FIGURE 8.4**

voltages are combined properly. This point is illustrated in Figure 8.4. Figure 8.4 (a and b) shows the polarities of the voltages produced by their respective sources. If we combine these voltages (as shown in Figure 8.4c), you can see that:

1. $V_{1A}$ and $V_{1B}$ are opposing voltages. Therefore, the actual component voltage equals the difference between the two.
2. $V_{2A}$ and $V_{2B}$ are also opposing voltages. Again, the actual component voltage equals the difference between the two.
3. $V_{3A}$ and $V_{3B}$ are aiding voltages. Therefore, the actual component voltage equals the sum of the two.

*OBJECTIVE 2* ▶ The superposition analysis of a simple multisource circuit is demonstrated in Example 8.1. As you will see, the polarities of the various voltages determine how they are combined in the final steps of the analysis.

## EXAMPLE 8.1

Determine the values of $V_1$, $V_2$, and $V_3$ for the circuit shown in Figure 8.5a.

***Solution:*** The first step is to split the circuit into two single-source circuits. These circuits (along with their series-equivalents) are shown in Figures 8.5b and 8.5c. For the circuit in Figure 8.5b, the value of $R_{EQ}$ is found as

$$\boldsymbol{R_{EQ}} = R_2 \| R_3 = 15\ \Omega \| 10\ \Omega = \mathbf{6\ \Omega}$$

Using this value in the voltage-divider equation, the value of $V_{EQ}$ can be found as

$$\boldsymbol{V_{EQ}} = V_A \frac{R_{EQ}}{R_{EQ} + R_1} = (12\text{ V})\frac{6\ \Omega}{36\ \Omega} = \mathbf{2\ V}$$

Since this voltage is across the parallel combination of $R_2$ and $R_3$, $V_{2A}$ and $V_{3A}$ both equal 2 V. The voltage across $R_1$ can now be found as

$$\boldsymbol{V_1} = V_A - V_{EQ} = 12\text{ V} - 2\text{ V} = \mathbf{10\ V}$$

At this point, we have calculated the following values:

$$V_{1A} = 10\text{ V} \qquad V_{2A} = 2\text{ V} \qquad V_{3A} = 2\text{ V}$$

Performing a similar analysis of the circuit in Figure 8.5c, we get the following values:

$$\boldsymbol{R_{EQ}} = R_1 \| R_3 = 30\ \Omega \| 10\ \Omega = \mathbf{7.5\ \Omega}$$

$$\boldsymbol{V_{EQ}} = V_B \frac{R_{EQ}}{R_{EQ} + R_1} = (9\text{ V})\frac{7.5\ \Omega}{22.5\ \Omega} = \mathbf{3\ V}$$

and

$$\boldsymbol{V_2} = V_B - V_{EQ} = 9\text{ V} - 3\text{ V} = \mathbf{6\ V}$$

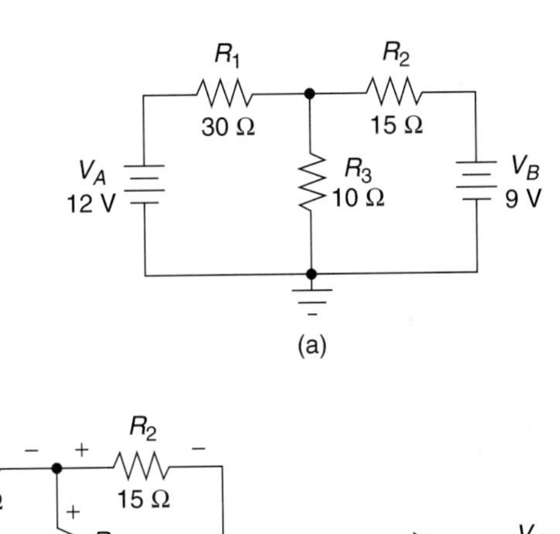

(a)

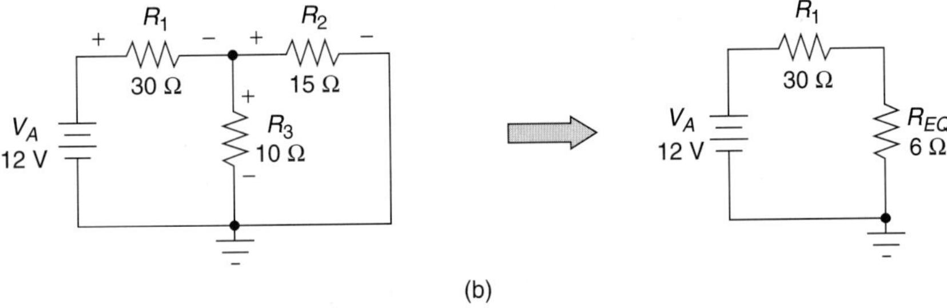

(b)

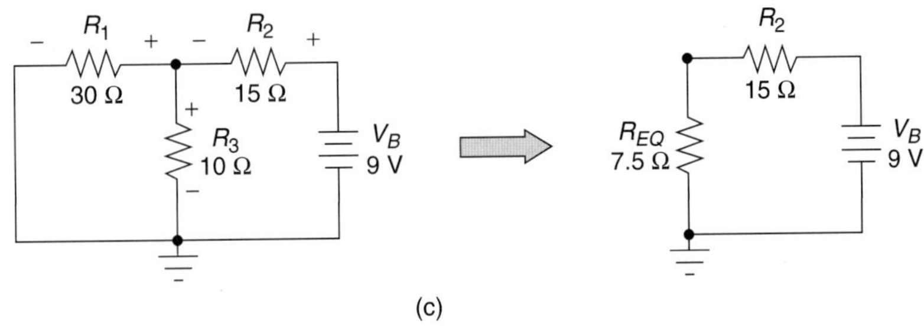

(c)

**FIGURE 8.5**

For the circuit in Figure 8.5c, we have calculated the following values:

$$V_{1B} = 3 \text{ V} \qquad V_{2B} = 6 \text{ V} \qquad V_{3B} = 3 \text{ V}$$

Finally, we can determine the actual component voltages by combining the A and B results. If you compare the polarities of $V_{1A}$ and $V_{1B}$ in Figure 8.5, you can see that they are opposing. Therefore,

$$\mathbf{V_1} = |V_{1A} - V_{1B}| = 10 \text{ V} - 3 \text{ V} = \mathbf{7\ V}$$

Figure 8.5 also shows $V_{2A}$ and $V_{2B}$ to be opposing. Therefore,

$$\mathbf{V_2} = |V_{2B} - V_{2A}| = 6 \text{ V} - 2 \text{ V} = \mathbf{4\ V}$$

Since $V_{3A}$ and $V_{3B}$ are aiding voltages, the actual value of $V_3$ is found as

$$\mathbf{V_3} = |V_{3A} + V_{3B}| = 2 \text{ V} + 3 \text{ V} = \mathbf{5\ V}$$

For the circuit in Figure 8.5a, the actual component voltages are as follows:

$$V_1 = 7 \text{ V} \qquad V_2 = 4 \text{ V} \qquad V_3 = 5 \text{ V}$$

***PRACTICE PROBLEM 8.1***

A circuit like the one in Figure 8.5a has the following values: $V_A = 14$ V, $V_B = 7$ V, $R_1 = 100\ \Omega$, $R_2 = 150\ \Omega$, and $R_3 = 150\ \Omega$. Calculate the values of $V_1$, $V_2$, and $V_3$ for the circuit.

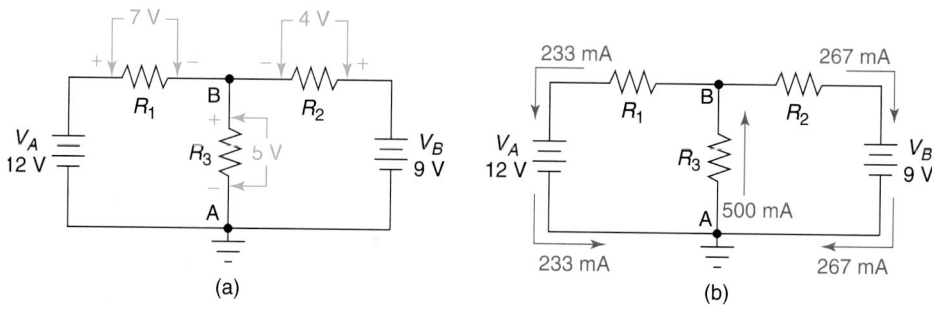


EWB

**FIGURE 8.6**

The results from Example 8.1 can be used to make several important points. Figure 8.6a shows the original circuit with the calculated component voltages included. First, note the polarities of $V_1$ and $V_2$. In both cases, the assumed polarity matches that of the greater voltage. Since $V_{1A}$ was greater than $V_{1B}$, the assumed polarity of $V_1$ matches that of $V_{1A}$. By comparing Figure 8.6 to Figure 8.5, you can see that the same holds true in the case of $V_2$. The polarity of $V_2$ matches the polarity of $V_{2B}$, which was greater than $V_{2A}$.

Figure 8.6a demonstrates another important point. As you know, Kirchhoff's voltage law states that series voltages must add up to equal the source voltage. If you look at the values in Figure 8.6a, you can see that

$$V_A = V_1 + V_3 \qquad \text{and} \qquad V_B = V_2 + V_3$$

Figure 8.6b shows the circuit currents and their values. Note that the value of each current was found using Ohm's law and the values shown in Figure 8.6a. Using the current values shown, the overall operation of the circuit can be described as follows:

1. $V_A$ generates a total current of 233 mA.
2. $V_B$ generates a total current of 267 mA.
3. At point A, the source currents combine. The sum of these currents (500 mA) passes through $R_3$.
4. At point B, the currents split. Each current returns to its source via the series resistor.

Figure 8.6b also demonstrates the fact that the circuit operates according to Kirchhoff's current law. As you can see, the total current entering a given point or component equals the total current leaving that point or component.

Even in a multisource circuit, the fundamental laws of circuit operation hold true. Using these laws provides you with a means of quickly checking the results of any circuit analysis problem.

Example 8.2 demonstrates how changing the polarity of one voltage source affects the superposition analysis of a multisource circuit. As you will see, the change in polarity is not considered until the final steps of the analysis.

## EXAMPLE 8.2

Determine the values of $V_1$, $V_2$, and $V_3$ for the circuit in Figure 8.7a.

***Solution:*** The first step (again) is to split the circuit into two single-source circuits. These circuits (along with their series-equivalents) are shown in Figure 8.7 (b and c). For Figure 8.7b, the value of $R_{EQ}$ is found as

$$\boldsymbol{R_{EQ}} = R_2 \| R_3 = 36\ \Omega \| 12\ \Omega = \mathbf{9\ \Omega}$$

The value of $V_{EQ}$ is now found as

$$\boldsymbol{V_{EQ}} = V_A \frac{R_{EQ}}{R_{EQ} + R_1} = (15\text{ V})\frac{9\ \Omega}{27\ \Omega} = \mathbf{5\ V}$$

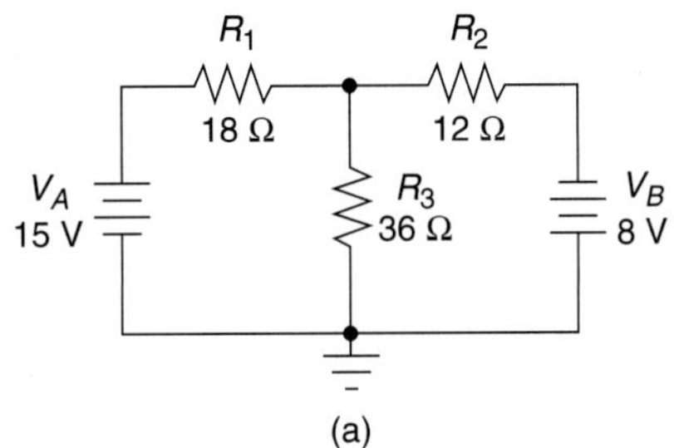


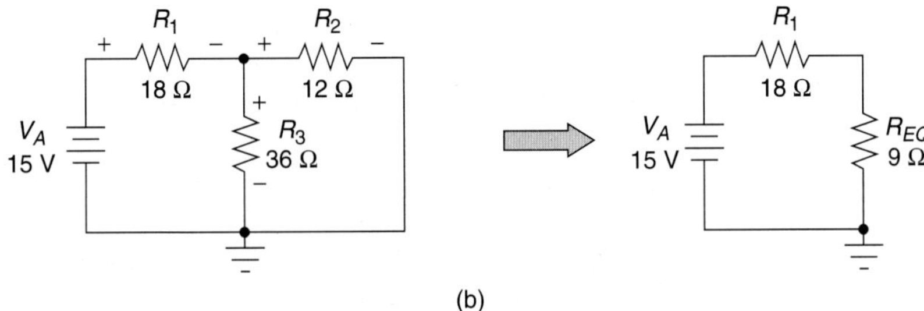


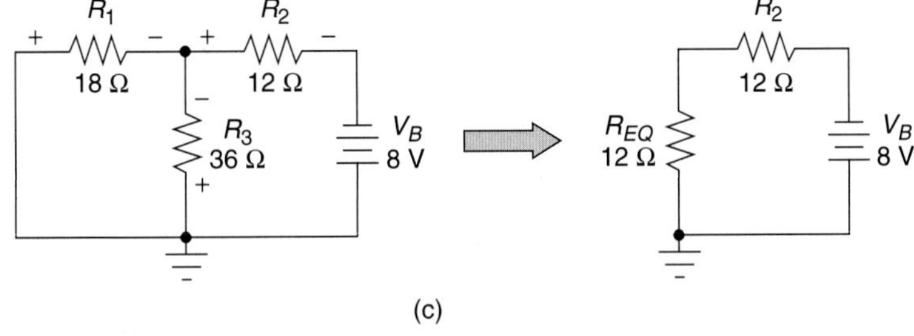


**FIGURE 8.7**

Since this voltage is across the parallel combination of $R_2$ and $R_3$, $V_{2A}$ and $V_{3A}$ both equal 5 V. The voltage across $R_1$ can be found as

$$\mathbf{V_{1A}} = V_A - V_{EQ} = 15\text{ V} - 5\text{ V} = \mathbf{10\text{ V}}$$

At this point, we know the following values for the circuit:

$$V_{1A} = 10\text{ V} \qquad V_{2A} = 5\text{ V} \qquad V_{3A} = 5\text{ V}$$

If you look at the circuit in Figure 8.7c, you can see that $V_B$ is a negative voltage. Even so, we will ignore the change in polarity for now. Performing an analysis like the one we just finished, we get the following values:

$$\mathbf{R_{EQ}} = R_1 \| R_3 = 18\ \Omega \| 36\ \Omega = \mathbf{12\ \Omega}$$

$$\mathbf{V_{EQ}} = V_B \frac{R_{EQ}}{R_{EQ} + R_2} = (8\text{ V})\frac{12\ \Omega}{24\ \Omega} = \mathbf{4\text{ V}}$$

and

$$\mathbf{V_{1B}} = V_B - V_{EQ} = 8\text{ V} - 4\text{ V} = \mathbf{4\text{ V}}$$

For the circuit in Figure 8.7c, we have calculated the following values:

$$V_{1B} = 4\text{ V} \qquad V_{2B} = 4\text{ V} \qquad V_{3B} = 4\text{ V}$$

Finally, we can determine the actual component voltages by combining the A and B results. This is the point where the polarity of $V_B$ affects the circuit analysis. If you compare the polarities of $V_{1A}$ and $V_{1B}$ in Figure 8.7, you can see that they are aiding. Therefore,

$$\mathbf{V_1} = V_{1A} + V_{1B} = 10\text{ V} + 4\text{ V} = \mathbf{14\text{ V}}$$

By the same token, $V_{2A}$ and $V_{2B}$ are aiding. Therefore,

$$\mathbf{V_2} = V_{2A} + V_{2B} = 5\text{ V} + 4\text{ V} = \mathbf{9\text{ V}}$$

Finally, $V_{3A}$ and $V_{3B}$ in Figure 8.7 are shown to be opposing. Therefore,

$$\mathbf{V_3} = V_{3A} - V_{3B} = 5\text{ V} - 4\text{ V} = \mathbf{1\text{ V}}$$

For the circuit in Figure 8.7a, the actual component voltages are as follows:

$$V_1 = 14\text{ V} \qquad V_2 = 9\text{ V} \qquad V_3 = 1\text{ V}$$

***PRACTICE PROBLEM 8.2***

A circuit like the one in Figure 8.7a has the following values: $V_A = 7$ V, $V_B = 21$ V, $R_1 = 150\ \Omega$, $R_2 = 100\ \Omega$, and $R_3 = 150\ \Omega$. Calculate the values of $V_1$, $V_2$, and $V_3$ for the circuit.

The results from Example 8.2 are illustrated in Figure 8.8. When you look at the circuit voltages in Figure 8.8a, it is fairly easy to see that

$$V_A = V_1 + V_3$$

It is not as easy to see how the second loop fulfills Kirchhoff's voltage law. It would seem on the surface that the voltages across $R_2$ and $R_3$ add up to 10 V. However, this is not the case. If we write the Kirchhoff's loop equation for loop B, we get

$$V_3 - V_2 + V_B = 0\text{ V}$$

This equation was derived by starting at the ground connection and going clockwise around loop B. The polarity of each voltage was determined (as usual) by referencing the voltage to the first polarity sign we encountered. The Kirchhoff's loop equation for loop B can be rewritten as

$$V_2 - V_3 = V_B$$

Substituting the values shown in Figure 8.8, we get

$$9\text{ V} - 1\text{ V} = 8\text{ V}$$

which is true. Again, the circuit has conformed to Kirchhoff's voltage law. It just takes a little more work to see the relationship in this case.

Figure 8.8b shows the circuit current values and directions. Using the currents shown (which were determined using the values in Figures 8.7a and 8.8a), we can describe the operation of the circuit as follows:

1. The current leaving $V_A$ is 778 mA.
2. At point A, the 778 mA current splits; 28 mA passes through $R_3$ and 750 mA passes through $V_B$.
3. At point B, the currents recombine, and the 778 mA returns to $V_A$ via $R_1$.

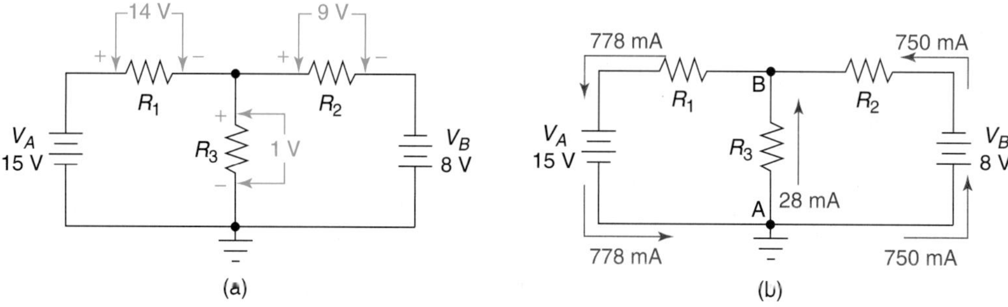


**FIGURE 8.8**

Again, a close look at the current values shows that the circuit operates as described by Kirchhoff's current law.

## One Final Note

In this section, you have been introduced to one of many ways to analyze multisource circuits. In Chapter 9, you will be introduced to several more analysis techniques commonly used on these types of circuits. As you will see, the techniques covered in Chapter 9 approach multisource circuits very differently.

**Section Review**

1. What is the superposition theorem? To what type of circuit does it apply?
2. Why is it helpful to redraw a multisource circuit as two (or more) single-source circuits when using the superposition theorem?
3. How can you perform a quick check of your analysis on a multisource circuit?

# 8.2 VOLTAGE AND CURRENT SOURCES

◄ *OBJECTIVE 3*

Some circuit analysis problems require the ability to convert a voltage source to an equivalent current source, or vice versa. In this section, we will briefly review the operating characteristics of voltage and current sources. Then we'll look at the methods used to convert each type of source to the other. As you will see, source conversions will play a role in some of our discussions later in this chapter.

## Voltage Sources

In Chapter 5, you were introduced to the operating characteristics of voltage sources. These characteristics are summarized in Figure 8.9. As you can see, the primary difference between ideal and practical voltage sources is the effect that source resistance has on load

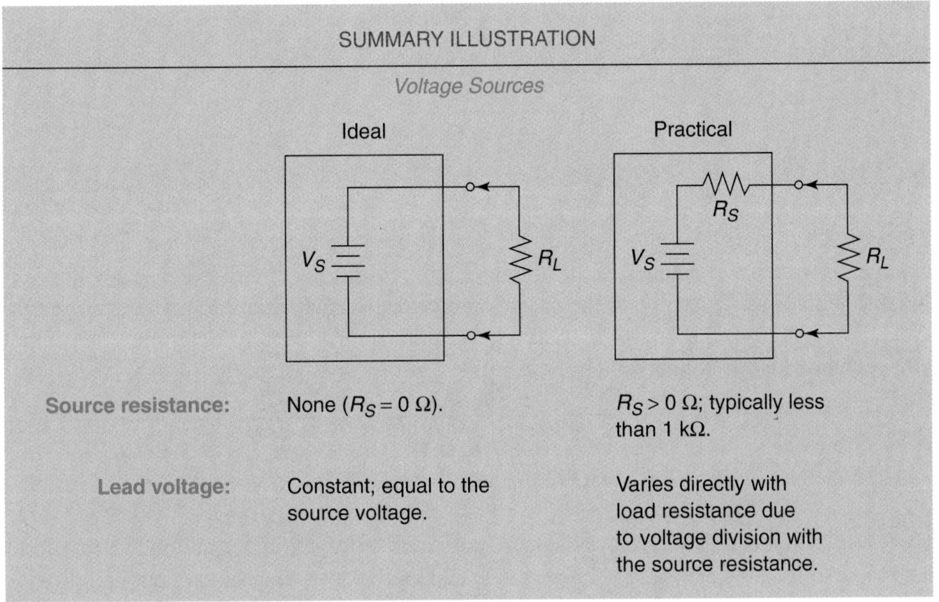


**FIGURE 8.9**

voltage. The ideal voltage source has a constant output, regardless of the value of $R_L$. The practical voltage source, on the other hand, has an output voltage that varies directly with the value of $R_L$, which means that a change in load resistance will cause a similar change in load voltage.

## Current Sources

In Chapter 6, you were introduced to current sources. Current source operating characteristics are summarized in Figure 8.10. Again, the primary difference between the ideal and practical sources is the effect of source resistance. In this case, source resistance affects the value of load current. The ideal current source produces a constant load current, regardless of the value of $R_L$. However, the output from a practical current source varies inversely with the load resistance; that is, as load resistance decreases, load current increases, and vice versa.

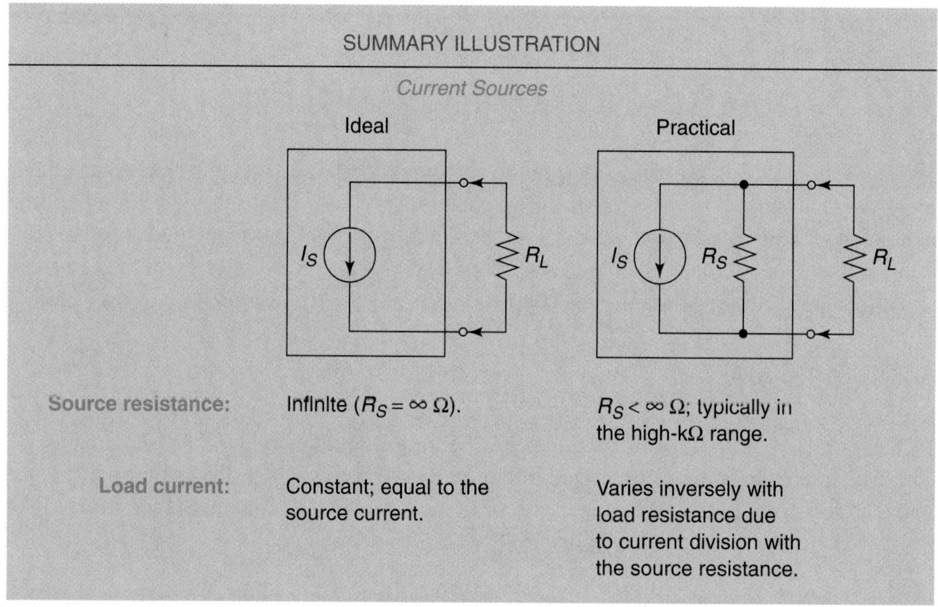


**FIGURE 8.10**

## Voltage and Current Sources: A Comparison

The characteristics of voltage and current sources are summarized in Figure 8.11. As you can see, the internal resistance of the voltage source causes the circuit to act as a voltage divider. The internal resistance of the current source causes that circuit to act as a current divider. In each case, the appropriate divider relationship is used to determine the output value when a load is connected.

## Equivalent Voltage and Current Sources

For every voltage source, there exists an equivalent current source (and vice versa). For any value of load resistance, a voltage source and its current source equivalent will provide the same output values. This point is demonstrated in Example 8.3.

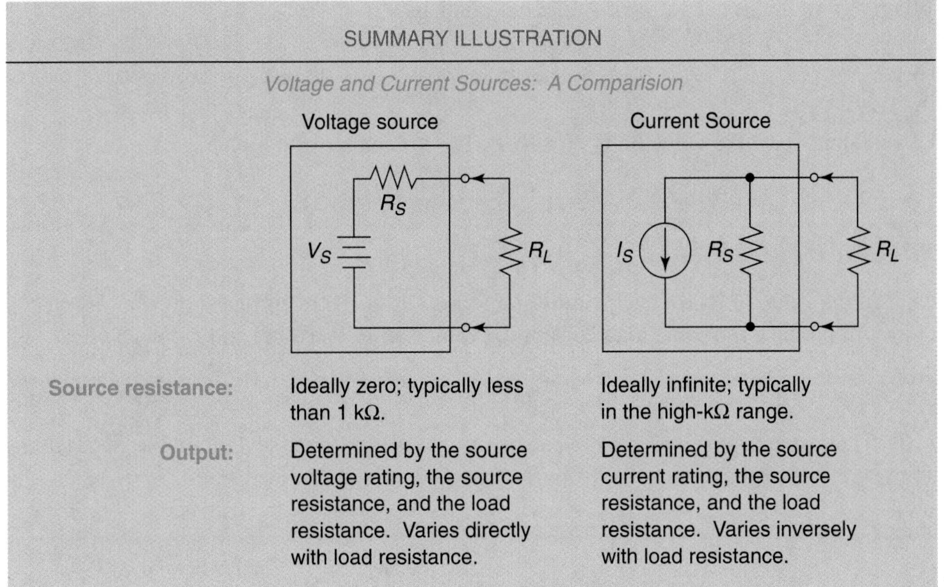


| | Voltage source | Current Source |
|---|---|---|
| Source resistance: | Ideally zero; typically less than 1 kΩ. | Ideally infinite; typically in the high-kΩ range. |
| Output: | Determined by the source voltage rating, the source resistance, and the load resistance. Varies directly with load resistance. | Determined by the source current rating, the source resistance, and the load resistance. Varies inversely with load resistance. |

**FIGURE 8.11**

*EXAMPLE 8.3*

Figure 8.12 shows a voltage source and its current source equivalent. Determine the value of $V_L$ provided by each of the circuits for values of $R_L = 400\ \Omega$ and $R_L = 100\ \Omega$.

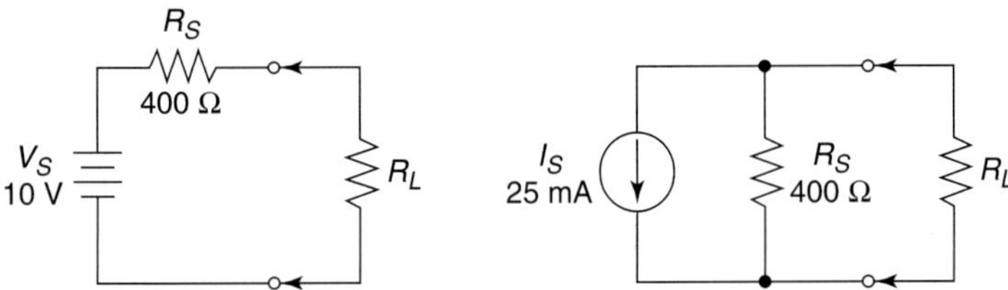


**FIGURE 8.12**

***Solution:*** First, we'll analyze the voltage source. When $R_L = 400\ \Omega$, the output from the voltage source can be found as

$$V_L = V_S \frac{R_L}{R_L + R_S} = (10\text{ V})\frac{400\ \Omega}{800\ \Omega} = \mathbf{5\ V}$$

When $R_L = 100\ \Omega$, the output from the voltage source can be found as

$$V_L = V_S \frac{R_L}{R_L + R_S} = (10\text{ V})\frac{100\ \Omega}{500\ \Omega} = \mathbf{2\ V}$$

Now, we'll take a look at the current source. Since it provides an output current, an extra step will be needed to determine the output voltage for each value of $R_L$. When $R_L = 400\ \Omega$, the output from the current source can be found as

$$I_L = I_S \frac{R_L \| R_S}{R_L} = (25\text{ mA})\frac{200\ \Omega}{400\ \Omega} = \mathbf{12.5\ mA}$$

Using Ohm's law, the value of $V_L$ (for $R_L = 400\ \Omega$) can now be found as

$$V_L = I_L R_L = (12.5\text{ mA})(400\ \Omega) = \mathbf{5\ V}$$

When $R_L = 100\ \Omega$, the value of $I_L$ can be found as

$$I_L = I_S \frac{R_L \| R_S}{R_L} = (25\text{ mA})\frac{80\ \Omega}{100\ \Omega} = \mathbf{20\ mA}$$

Using Ohm's law, the value of $V_L$ (for $R_L = 100\ \Omega$) can now be found as

$$V_L = I_L R_L = (20\text{ mA})(100\ \Omega) = \mathbf{2\ V}$$

***PRACTICE PROBLEM 8.3***

Change the value of $R_S$ for each circuit in Figure 8.12 to $100\ \Omega$. Then show that the circuits will produce the same load current for values of $R_L = 100\ \Omega$ and $R_L = 25\ \Omega$.

The results from Example 8.3 can be used to demonstrate several important relationships. As a reference, the results from the example are summarized as follows:

| *Voltage Source* | | *Current Source* | |
|---|---|---|---|
| Load Resistance ($R_L$) | Load Voltage ($V_L$) | Load Current ($I_L$) | Load Voltage ($V_L$) |
| 400 Ω | 5 V | 12.5 mA | 5 V |
| 100 Ω | 2 V | 20 mA | 2 V |

First, note the fact that the two circuits produced the same output voltage for each value of load resistance. This demonstrates the fact that they are equivalent circuits. Also, note that:

1. The value of $V_L$ for the voltage source varied directly with the value of $R_L$.
2. The value of $I_L$ for the current source varied inversely with the value of $R_L$.

These results verify the relationships given in Figures 8.9 and 8.10.

## Source Conversions

*OBJECTIVE 4* ▶ Source conversions are actually very simple. The relationships used to convert one type of source to the other are shown in Figure 8.13. As you can see, the value of $R_S$ does not change from one circuit to the other. When converting a voltage source to a current source (as shown in Figure 8.13a), the value of $I_S$ is found as

$$I_S = \frac{V_S}{R_S} \tag{8.1}$$

Once the value of $I_S$ has been calculated, the current source is drawn as shown in Figure 8.13a. Note that the direction of the arrow in the current source must match the direction of the current from the voltage source. The process for converting a voltage source to a current source is demonstrated further in Example 8.4.

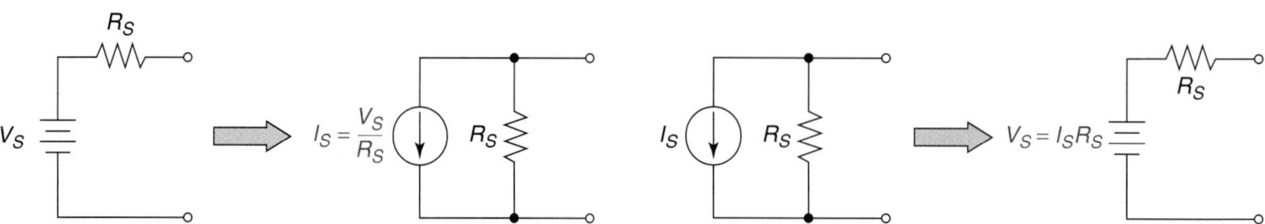


**FIGURE 8.13 Source conversions.**

## *Example 8.4*

Convert the voltage source in Figure 8.14a to an equivalent current source.

**FIGURE 8.14**

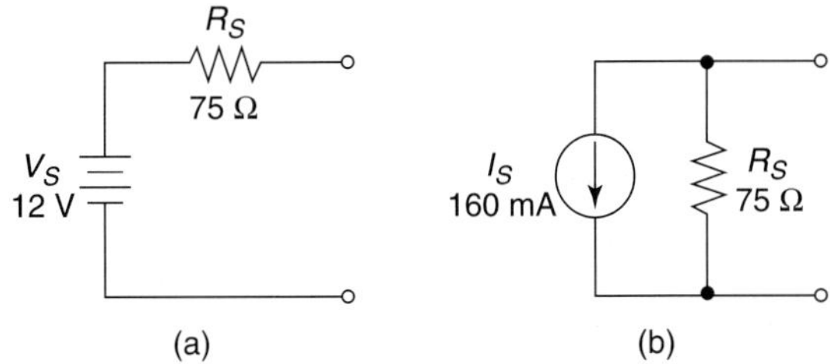


***Solution:*** The value of the source resistance, 75 Ω, remains the same as shown in the figure. The value of $I_S$ is found using equation (8.1), as follows:

$$I_S = \frac{V_S}{R_S} = \frac{12\text{ V}}{75\ \Omega} = \mathbf{160\ mA}$$

Using the values of $I_S = 160$ mA and $R_S = 75\ \Omega$, the current source is drawn as shown in Figure 8.14b.

### *PRACTICE PROBLEM 8.4*

A voltage source like the one in Figure 8.14a has values of $V_S = 8$ V and $R_S = 16\ \Omega$. Convert the voltage source to an equivalent current source. Then determine the output current provided by both circuits for a 32 Ω load.

When converting a current source to a voltage source (as shown in Figure 8.13), the value of source voltage is found as

$$V_S = I_S R_S \quad \textbf{(8.2)}$$

Once the value of $V_S$ is known, the voltage source is drawn as shown in Figure 8.13b. Again, care must be taken to ensure that the polarity of the voltage source matches the direction of current indicated in the current source. The process for converting a current source to a voltage source is demonstrated further in Example 8.5.

## *Example 8.5*

Convert the current source in Figure 8.15a to an equivalent voltage source.

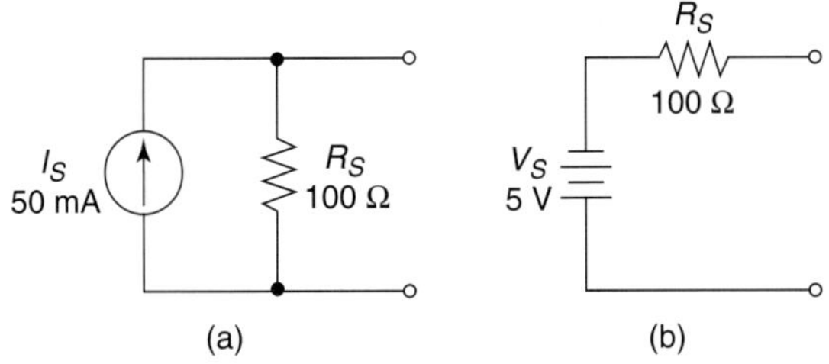


**FIGURE 8.15**

***Solution:*** The value of the source resistance, 100 Ω, remains the same as shown in the figure. The value of $V_S$ is found using equation (8.2), as follows:

$$V_S = I_S R_S = (50\text{ mA})(100\ \Omega) = \mathbf{5\ V}$$

Using the values of $V_S = 5$ V and $R_S = 100\ \Omega$, the voltage source is drawn as shown in Figure 8.15b. Note that the polarity of the voltage source matches that indicated by the current source.

***PRACTICE PROBLEM 8.5***

A current source like the one in Figure 8.15a has values of $I_S = 40$ mA and $R_S = 150\ \Omega$. Convert the current source to an equivalent voltage source. Then determine the output voltage provided by both circuits for a 100 Ω load.

Converting a voltage source to a current source (or vice versa) is not something that you will do regularly. However, later in this chapter you will see how source conversions can simplify certain circuit analysis problems.

## Section Review

1. In terms of source resistance, how does the ideal voltage source differ from the practical voltage source?
2. In terms of load voltage, how does the ideal voltage source differ from the practical voltage source?
3. In terms of source resistance, how does the ideal current source differ from the practical current source?
4. In terms of load current, how does the ideal current source differ from the practical current source?
5. Compare and contrast voltage sources and current sources.
6. Describe the process for converting a voltage source to a current source.
7. Describe the process for converting a current source to a voltage source.

## 8.3 THEVENIN'S THEOREM

*OBJECTIVE 5* ▶ In the last section, you saw how many electronic circuits experience a change in output voltage, current, and power when there is a change in load resistance. For example, Figure 8.11 showed that the output from a voltage (or current) source changes when the load resistance changes. Analyzing the effect that a change in load has on the output from a given circuit is referred to as **load analysis.**

**Load analysis**
A method of predicting the effect that a change in load has on the output from a given circuit.

The load analysis of a series or parallel circuit is relatively simple and straightforward. For example, if the load in Figure 8.16a changes, we can easily predict the change in load voltage using the voltage divider relationship. By the same token, the current divider relationship can be used to predict how the circuit in Figure 8.16b will respond to a change in load resistance.

**FIGURE 8.16**

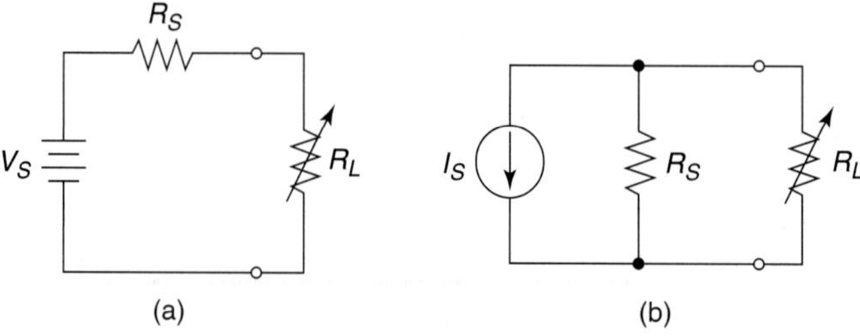

**FIGURE 8.17**

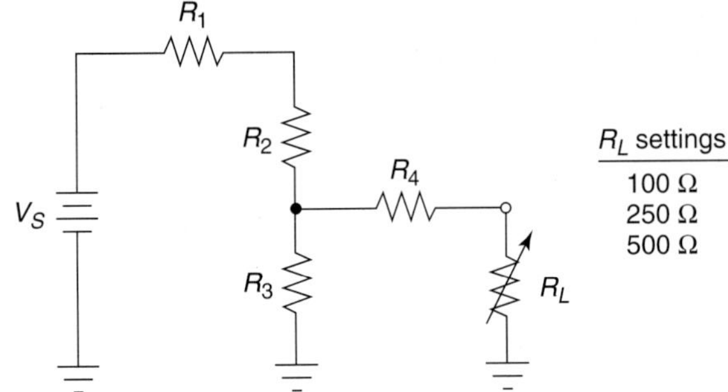


Using only the analysis techniques we have established so far, the load analysis of a series-parallel circuit could become a long and tedious process. For example, consider the circuit shown in Figure 8.17. Imagine trying to predict the value of load voltage produced by this circuit for the load resistance values shown. For each value of load resistance, we would have to derive a new series-equivalent, solve for the values in the equivalent circuit, and use those values to calculate the actual load voltage. Fortunately, the type of problem represented in Figure 8.17 can be simplified by using Thevenin's theorem.

◄ *OBJECTIVE 6*

**Thevenin's theorem**
A theorem stating that any resistive circuit or network can be represented as a voltage source in series with a source resistance.

**Thevenin's theorem** states that any resistive circuit or network, no matter how complex, can be represented as a voltage source in series with a source resistance. For example, the series-parallel circuit in Figure 8.18 can be represented by the Thevenin equivalent circuit shown in the figure. Note that the source voltage and resistance in Figure 8.18b are referred to as the Thevenin voltage ($V_{th}$) and the Thevenin resistance ($R_{th}$).

Later in this section, we'll discuss the process used to derive the Thevenin equivalent of a series-parallel circuit. First, let's take a look at how this theorem can be used to simplify the load analysis of series-parallel circuits.

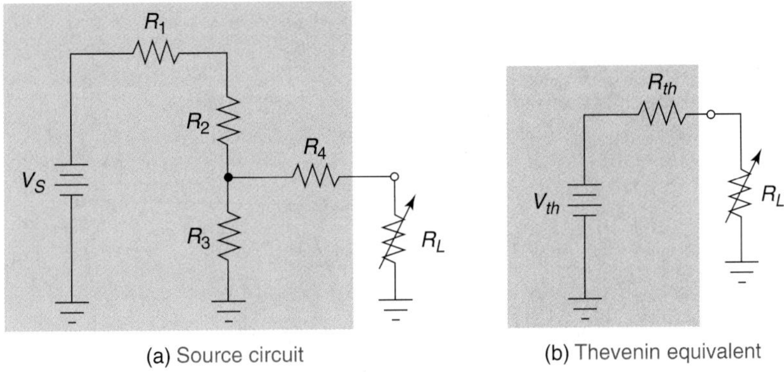


**FIGURE 8.18 A series-parallel circuit and its Thevenin equivalent.**

## The Purpose Served by Thevenin's Theorem

The strength of the Thevenin equivalent circuit lies in the fact that it produces the same output values as the original circuit for any given value of load resistance. For example, the circuits in Figure 8.18 (a and b) will produce the same values of load voltage, current, and power when their load resistance values are equal. This relationship is demonstrated further in Example 8.6.

## EXAMPLE 8.6

Figure 8.19 contains a loaded voltage divider and its Thevenin equivalent circuit. Determine the load voltage produced by each circuit for load resistance values of 100 Ω and 1 kΩ.

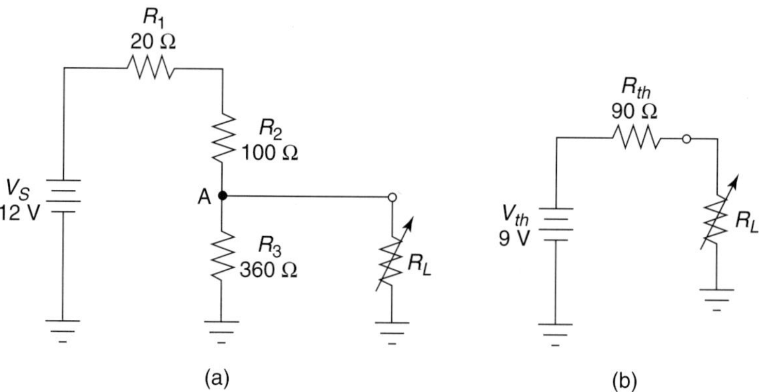


**FIGURE 8.19**

***Solution:*** We'll start with the loaded voltage divider shown in Figure 8.19a. When $R_L = 100\ \Omega$,

$$\boldsymbol{R_{EQ}} = R_3 \| R_L = 360\ \Omega \| 100\ \Omega = \mathbf{78.3\ \Omega}$$

and

$$\boldsymbol{V_{EQ}} = V_S \frac{R_{EQ}}{R_T} = (12\text{ V}) \frac{78.3\ \Omega}{198.3\ \Omega} = \mathbf{4.74\ V}$$

Therefore, $V_L = 4.74$ V when $R_L = 100\ \Omega$.

When $R_L = 1\text{ k}\Omega$,

$$\boldsymbol{R_{EQ}} = R_3 \| R_L = 360\ \Omega \| 1\text{ k}\Omega = \mathbf{264.7\ \Omega}$$

and

$$\boldsymbol{V_{EQ}} = V_S \frac{R_{EQ}}{R_T} = (12\text{ V}) \frac{264.7\ \Omega}{384.7\ \Omega} = \mathbf{8.26\ V}$$

Therefore, $V_L = 8.26$ V when $R_L = 1\text{ k}\Omega$.

Using the Thevenin equivalent circuit, we can solve the same problem as follows. When $R_L = 100\ \Omega$,

$$\boldsymbol{V_L} = V_{th} \frac{R_L}{R_L + R_{th}} = (9\text{ V}) \frac{100\ \Omega}{190\ \Omega} = \mathbf{4.74\ V}$$

When $R_L = 1\text{ k}\Omega$,

$$\boldsymbol{V_L} = V_{th} \frac{R_L}{R_L + R_{th}} = (9\text{ V}) \frac{1\text{ k}\Omega}{1.09\text{ k}\Omega} = \mathbf{8.26\ V}$$

***PRACTICE PROBLEM 8.6***

Figure 8.20 shows a series-parallel circuit and its Thevenin equivalent. Determine the output produced by each circuit for values of $R_L = 150\ \Omega$ and $R_L = 30\ \Omega$.

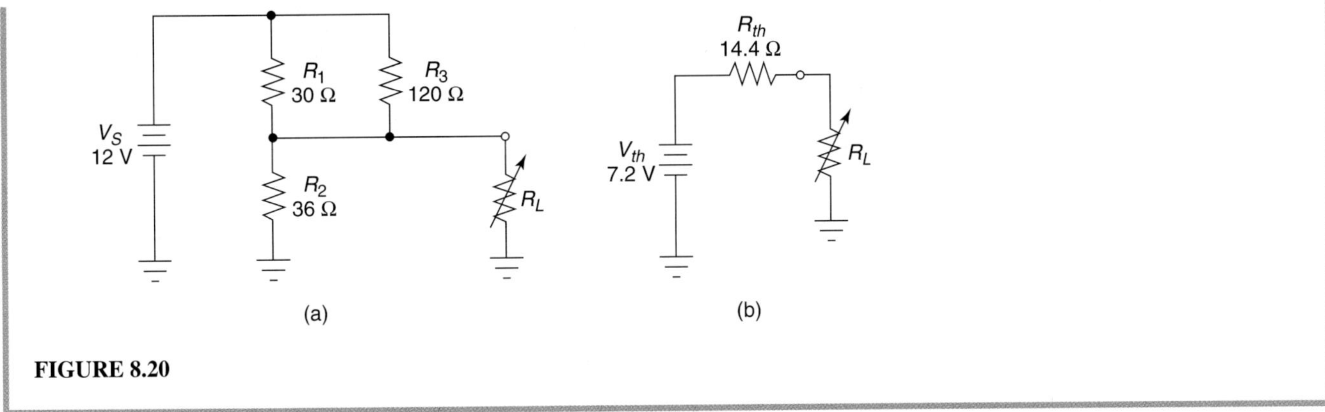


**FIGURE 8.20**

As you can see, the Thevenin equivalent for a series-parallel circuit produces the same output for any value of load resistance, which means that the load analysis of a series-parallel circuit (like the one in Figure 8.18) can be performed as follows:

1. Derive the Thevenin equivalent for the series-parallel circuit.
2. For each value of load resistance, determine the output from the Thevenin equivalent circuit using the voltage divider relationship.

The values obtained using this simple process are the same as those produced by the original circuit. As you can see, Thevenin's theorem can be an extremely powerful load analysis tool.

It should be stressed that the Thevenin equivalent of a series-parallel circuit is used to predict only the change in load values that result from a change in load resistance. If you want to know how any of the other circuit values change (such as $V_1$ or $V_2$, etc.), use the standard analysis techniques that you learned in Chapter 7.

## Determining the Value of $V_{th}$

◄ *OBJECTIVE 7*

**Thevenin voltage ($V_{th}$)** The voltage measured across the output terminals of a circuit with the load removed.

Deriving the Thevenin equivalent of a series-parallel circuit begins with determining the value of **Thevenin voltage ($V_{th}$)**. Although we've given it a new name, the Thevenin voltage for a series-parallel circuit is nothing more than its no-load output voltage ($V_{NL}$); that is, the voltage present at the output terminals of the circuit when the load is removed. (You may recall that we calculated the value of $V_{NL}$ for several circuits in Chapter 7.) The measurement of $V_{th}$ is illustrated in Figure 8.21. If you compare this measurement to the one shown in Figure 7.24, you'll see that $V_{th}$ and $V_{NL}$ are the same. The process for calculating the value of $V_{th}$ is demonstrated in Example 8.7.

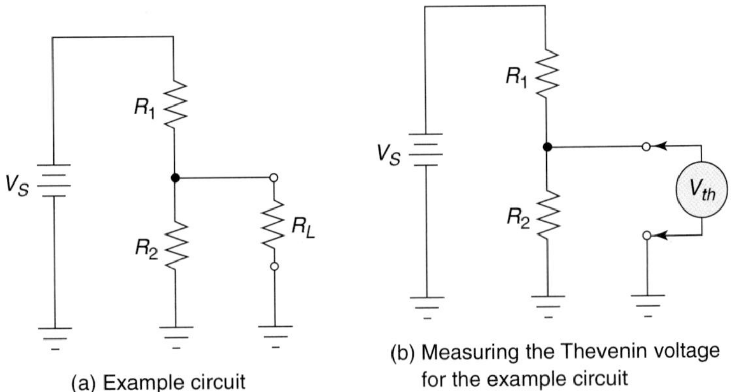


**FIGURE 8.21 Measuring Thevenin voltage.**

## EXAMPLE 8.7

Determine the value of $V_{th}$ for the series-parallel circuit shown in Figure 8.22a.

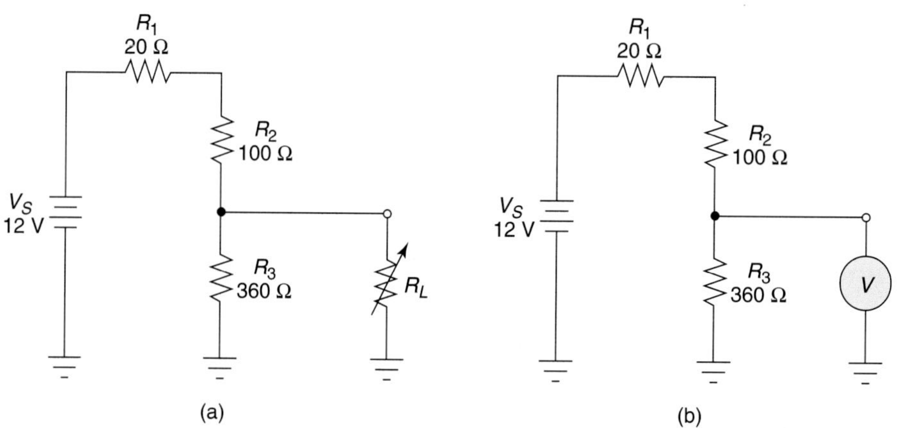


**FIGURE 8.22**

***Solution:*** Since $V_{th}$ is equal to the no-load output voltage, the first step is to remove the load. This gives us the circuit shown in Figure 8.22b. For this circuit, the voltage across the open load terminals is equal to the voltage across $R_3$. Therefore, $V_{th}$ can be found as

$$V_{th} = V_S \frac{R_3}{R_T} = (12 \text{ V}) \frac{360 \, \Omega}{480 \, \Omega} = \mathbf{9 \, V}$$

***PRACTICE PROBLEM 8.7***

Determine the value of $V_{th}$ for the circuit shown in Figure 8.23.

**FIGURE 8.23**

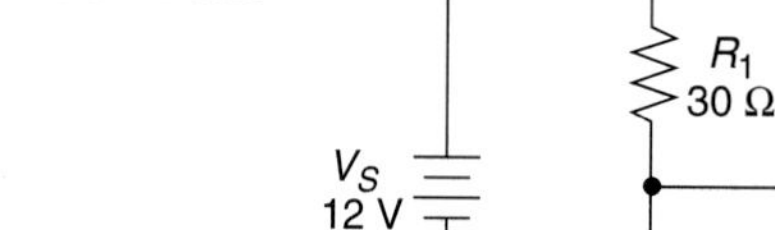

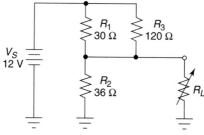

## Thevenin Resistance ($R_{th}$)

**Thevenin resistance ($R_{th}$)** The resistance measured across the output terminals of a circuit with the load removed. To measure $R_{th}$, the voltage source must be removed and replaced with a wire.

The second step in deriving a Thevenin equivalent circuit is determining its value of **Thevenin resistance ($R_{th}$).** In essence, $R_{th}$ can be thought of as the no-load output resistance ($R_{NL}$) of a circuit; that is, the resistance measured across the output terminals with the load removed. Figure 8.24 shows how this resistance can be measured. Note that the load has been removed and the source ($V_S$) has been replaced by a wire.

Why replace the voltage source with a wire? First, you know that the voltage source must be removed from the circuit before making any resistance measurement. If we want an accurate resistance measurement, we must replace the source with an equivalent resistance. In other words, if we assume that $V_S$ has an internal resistance of 0 Ω, then we must replace it with a resistance of approximately 0 Ω. Otherwise, the measured output resistance will not be accurate. Therefore, we replace the source with a wire.

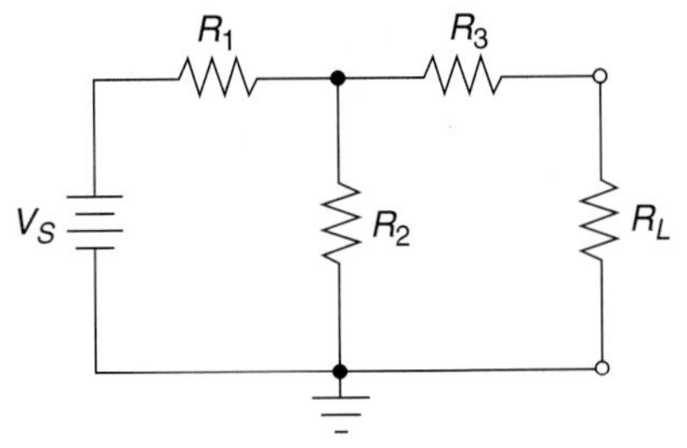


(a) Example circuit

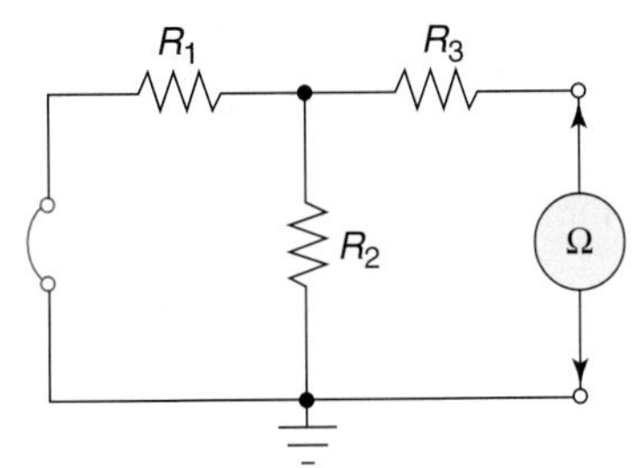


(b) Measuring the Thevenin resistance for the example circuit. (Note that the source has been replaced by a jumper wire.)

**FIGURE 8.24 Measuring Thevenin resistance.**

*A Practical Consideration:* The value of $R_S = 0\ \Omega$ represented by the wire is an ideal value. A practical voltage source always has some measurable amount of internal resistance. Therefore, the value of $R_{th}$ calculated (or measured) with the source shorted is always slightly lower than the actual value.

The currents shown in Figure 8.25 represent those generated by the meter. The currents indicate that (from the meter's perspective):

1. $R_1$ is in parallel with $R_2$.
2. $R_3$ is in series with the other two resistors.

Therefore, the Thevenin resistance of the circuit in Figure 8.25 is found as

$$R_{th} = (R_1 \| R_2) + R_3$$

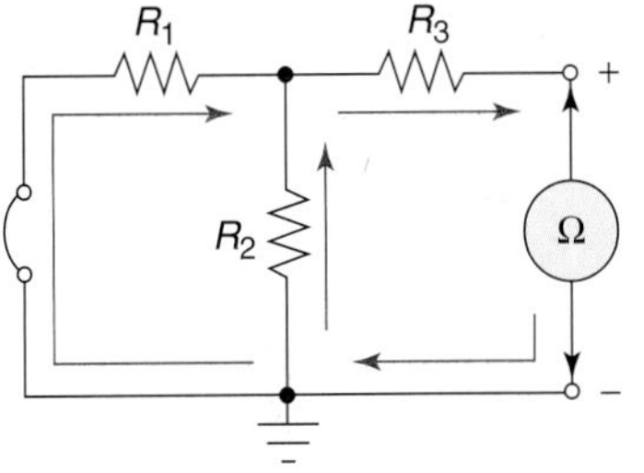


**FIGURE 8.25 Currents produced by the ohmmeter.**

Calculating circuit resistance from the viewpoint of the load is a process that usually takes a while to grow accustomed to. In all our previous experiences, we simplified circuits by combining resistor values from load to source. However, when we're analyzing a circuit from the viewpoint of the load (as shown in Figure 8.25), our perspective has changed. We are now combining resistance values from source to load.

With practice, you will be able to combine resistance values easily from any perspective. In the meantime, you will find it helpful to draw the circuit (as shown in Figure 8.25) and include the meter current arrows. These arrows will help you to determine which components are in series (or parallel). Example 8.8 demonstrates the complete procedure for calculating the Thevenin resistance of a series-parallel circuit.

*An Important Point:* The situation represented in Figure 8.25 is similar to the one we encountered using the superposition theorem. The ohmmeter acts as a voltage source when connected as shown. The value of $R_{th}$ is the total circuit resistance as seen by this source (the ohmmeter).

### EXAMPLE 8.8

Determine the value of Thevenin resistance for the circuit in Figure 8.26a.

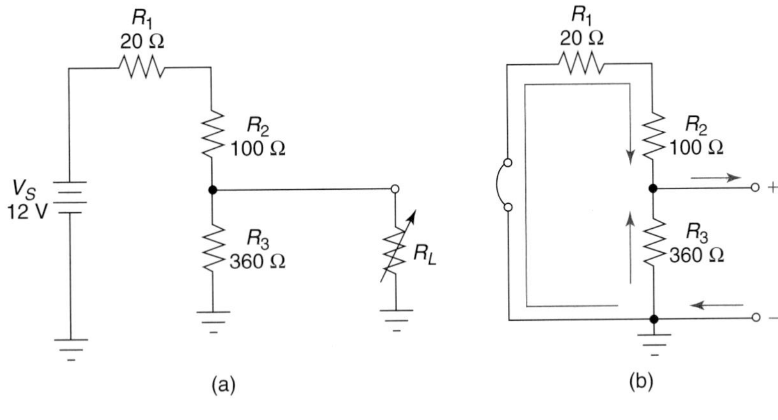


**FIGURE 8.26**

*Solution:* First, the circuit is drawn as shown in Figure 8.26b. Note that the load has been removed and the source has been replaced by a wire. The current arrows shown represent the current that would be produced by an ohmmeter connected to the open load terminals. These currents indicate that:

1. $R_1$ is in series with $R_2$.
2. $R_3$ is in parallel with the other two resistors.

Therefore, the Thevenin resistance for the circuit can be found as

$$\mathbf{R_{th}} = (R_1 + R_2)\|R_3 = 120\ \Omega\|360\ \Omega = \mathbf{90\ \Omega}$$

***PRACTICE PROBLEM 8.8***

Determine the Thevenin resistance for the circuit shown in Figure 8.23 (Example 8.7).

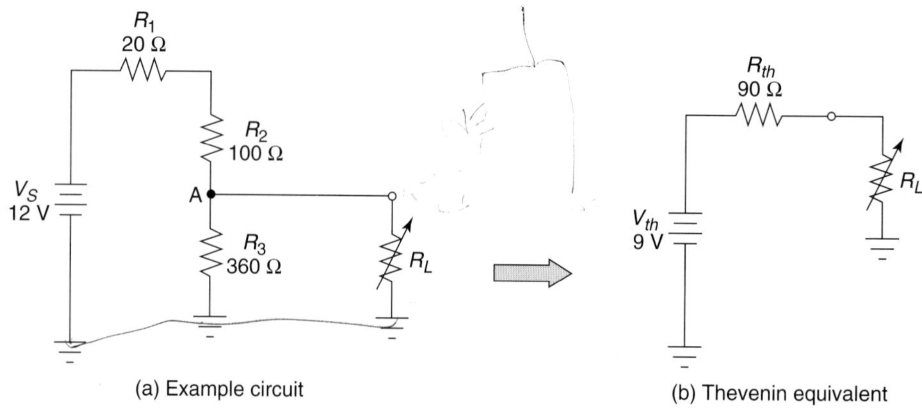


**FIGURE 8.27**

The results from the last two examples are illustrated in Figure 8.27. As you can see, the values of $V_{th}$ and $R_{th}$ were combined into a single Thevenin equivalent circuit. The relationship between the series-parallel circuit and its Thevenin equivalent was demonstrated in Example 8.6.

## Deriving Thevenin Equivalent Circuits

*OBJECTIVE 8* ► You have been shown how to determine the Thevenin equivalent voltage and resistance for a series-parallel circuit. At this point, we will put it all together by deriving the Thevenin equivalents for several example circuits.

### EXAMPLE 8.9

Derive the Thevenin equivalent of the circuit shown in Figure 8.28a.

*Solution:* We'll start by drawing the circuit with the load removed, as shown in Figure 8.28b. As you can see, the voltage across the open load terminals ($V_{th}$) is equal to the voltage across the parallel combination of $R_2$ and $R_3$. Combining these two resistors into a single equivalent resistance, we get the series-equivalent circuit shown. The value of $R_{EQ}$ for this circuit is found as

$$\mathbf{R_{EQ}} = R_2\|R_3 = 120\ \Omega\|360\ \Omega = \mathbf{90\ \Omega}$$

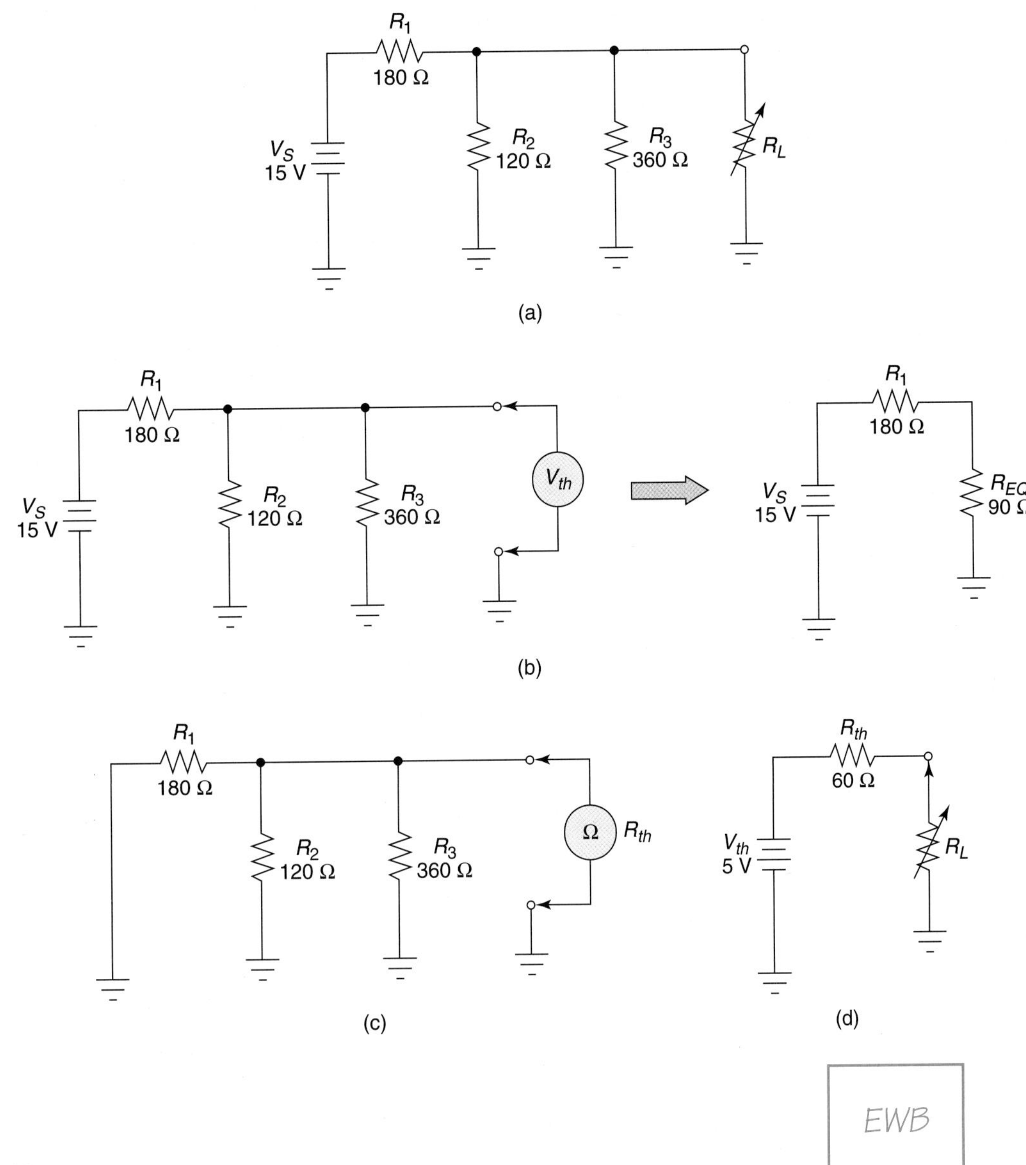


**FIGURE 8.28**

Since the Thevenin voltage is equal to the voltage across $R_{EQ}$, its value can be found as

$$\boldsymbol{V_{th}} = V_S \frac{R_{EQ}}{R_1 + R_{EQ}} = (15\text{ V})\frac{90\ \Omega}{270\ \Omega} = \mathbf{5\ V}$$

To determine the value of $R_{th}$, we need to replace the voltage source with a wire, as shown in Figure 8.28c. The resistance at the open output terminals ($R_{th}$) can now be found as

$$\boldsymbol{R_{th}} = R_1 \| R_2 \| R_3 = 180\ \Omega \| 120\ \Omega \| 360\ \Omega = \mathbf{60\ \Omega}$$

Using our calculated values of $V_{th}$ and $R_{th}$, the Thevenin equivalent of the original circuit is drawn as shown in Figure 8.28d.

***PRACTICE PROBLEM 8.9***

Derive the Thevenin equivalent of the circuit shown in Figure 8.29.

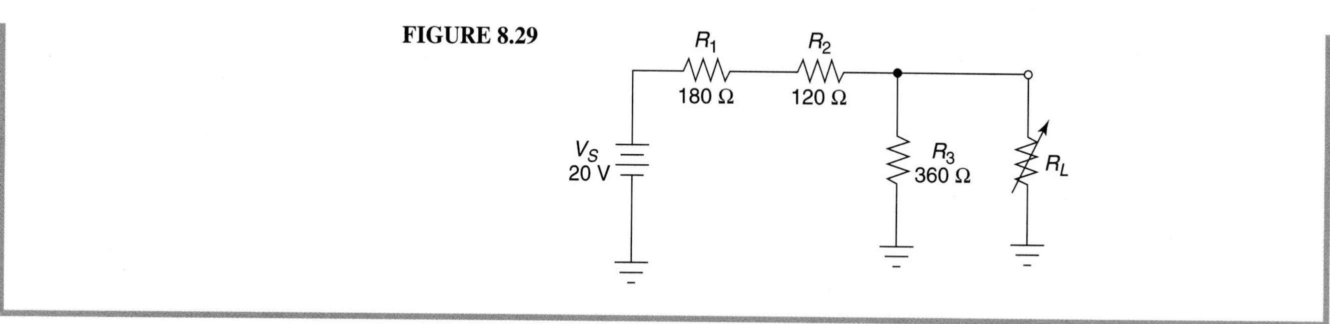


When one or more resistors are in series with the load, they have no effect on the value of $V_{th}$. This point is demonstrated in the following example.

## EXAMPLE 8.10

Derive the Thevenin equivalent of the circuit in Figure 8.30a.

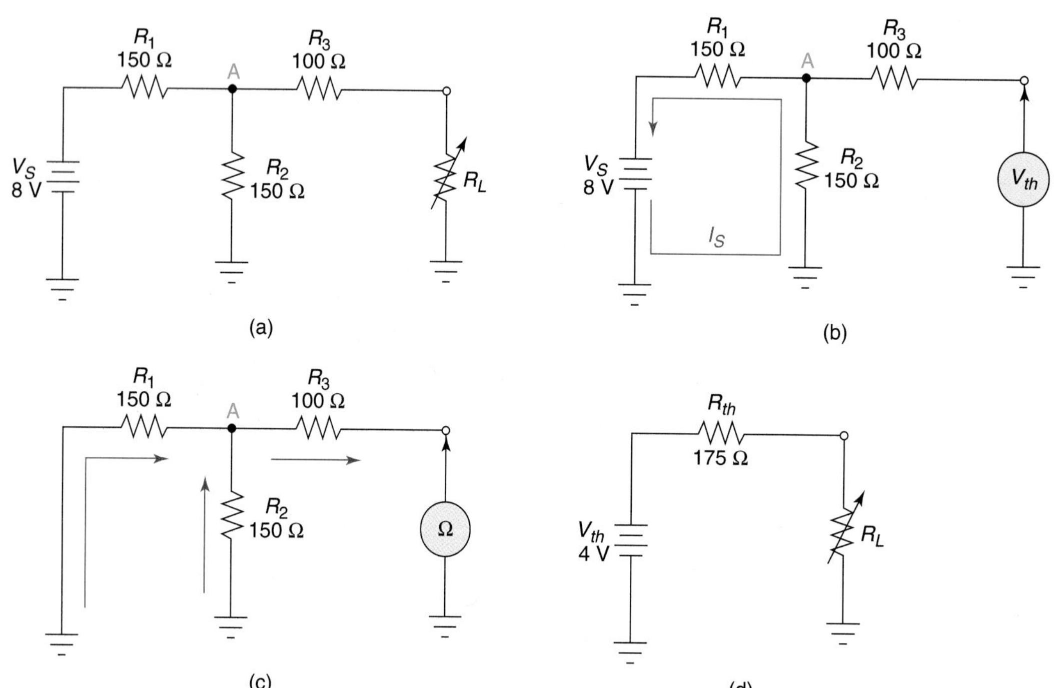


FIGURE 8.30

***Solution:*** First, the load is removed, giving us the circuit shown in Figure 8.30b. Note that the source current passes only through $R_1$ and $R_2$ because $R_3$ is not part of any complete current path. Since $I_3 = 0$ A, there is no voltage drop across $R_3$. Therefore, the voltage at the load terminal equals the voltage at point A; that is, $V_{th} = V_2$. The value of $V_{th}$ for Figure 8.30a can be found as

$$\boldsymbol{V_{th}} = V_S \frac{R_2}{R_1 + R_2} = (8\text{ V})\frac{150\ \Omega}{300\ \Omega} = \mathbf{4\ V}$$

The value of $R_{th}$ for the circuit can be found with the help of Figure 8.30c. The current arrows shown would be produced by an ohmmeter connected to the output terminals of the circuit. As these arrows indicate:

$$\boldsymbol{R_{th}} = (R_1 \| R_2) + R_3 = (150\ \Omega \| 150\ \Omega) + 100\ \Omega = \mathbf{175\ \Omega}$$

As you can see, $R_3$ (which is in series with the load) did not affect the value of $V_{th}$. At the same time, it did weigh into the value of $R_{th}$. The Thevenin equivalent of the original circuit is drawn as shown in Figure 8.30d.

***PRACTICE PROBLEM 8.10***

Derive the Thevenin equivalent of the circuit shown in Figure 8.31.

**FIGURE 8.31**

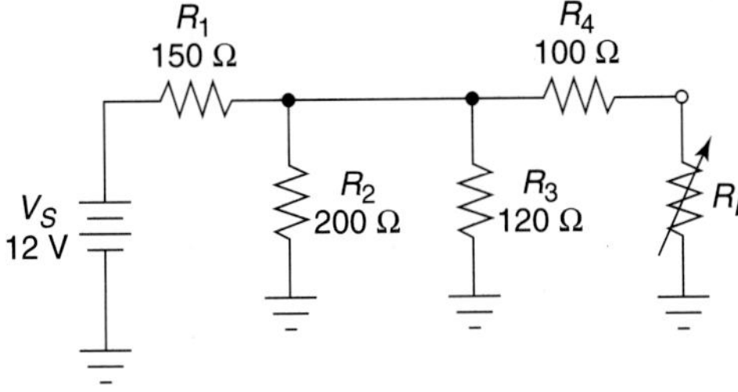


When a circuit contains branches in parallel with the source, one or more resistors in the circuit may have no effect on the values of $V_{th}$ and $R_{th}$. This point is demonstrated in the following example.

## EXAMPLE 8.11

Derive the Thevenin equivalent of the circuit shown in Figure 8.32a.

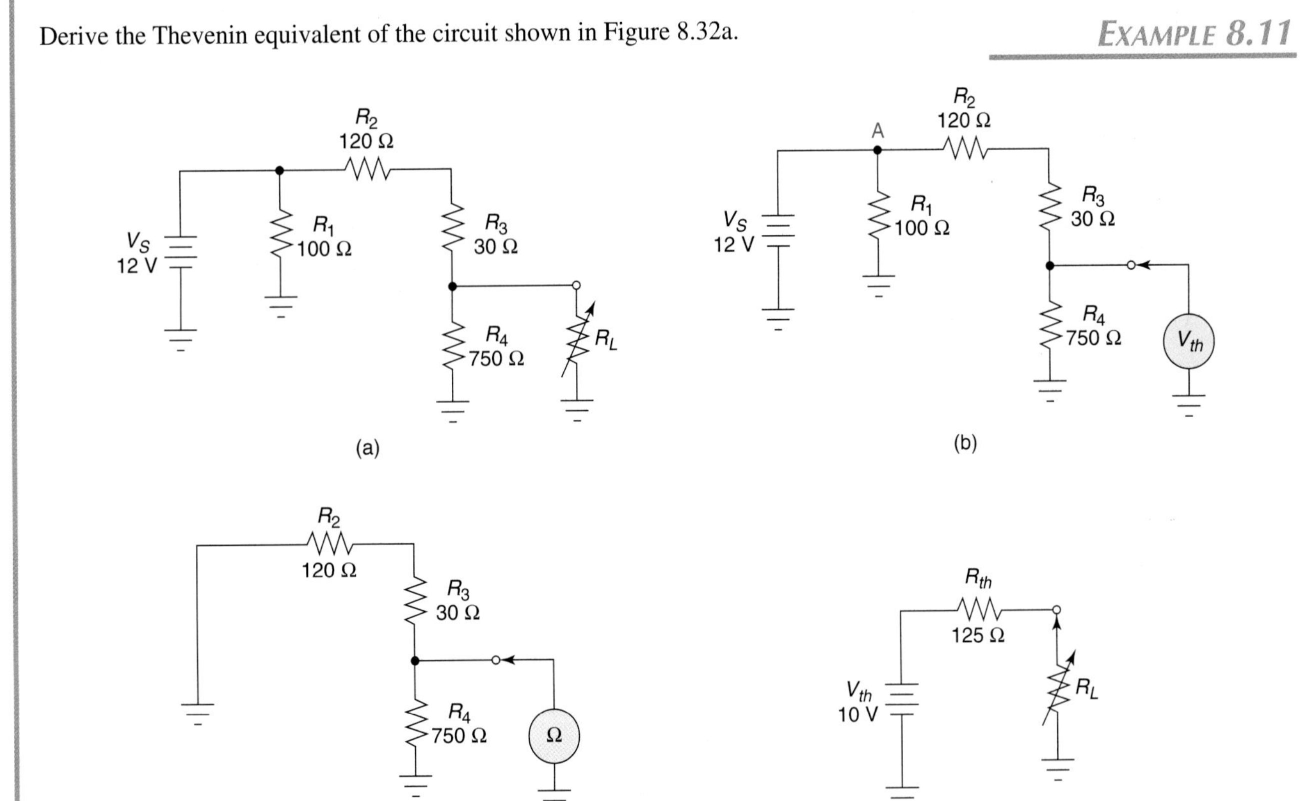


**FIGURE 8.32**

***Solution:*** As always, the first step is to remove the load (as shown in Figure 8.32b). In this circuit, $R_1$ is in parallel with the branch containing the other three resistors. Therefore, the voltage from point A to ground is equal to the source voltage, 12 V. This means that only $R_2$, $R_3$, and $R_4$ are involved in the calculation of $V_{th}$. Since $V_{th} = V_4$ in this circuit, its value can be found as

$$\boldsymbol{V_{th}} = V_S \frac{R_4}{R_2 + R_3 + R_4} = (12 \text{ V}) \frac{750 \ \Omega}{900 \ \Omega} = \mathbf{10 \ V}$$

When we short out the source in this circuit (to determine the value of $R_{th}$), the wire used to replace the source is in parallel with $R_1$. Therefore, $R_1$ is shorted and does not affect the value of Thevenin resistance for the circuit. (This is why $R_1$ is not shown in the resistance diagram in Figure 8.32c.) For the circuit shown, the series combination of $R_2$ and $R_3$ is in parallel with $R_4$. Therefore,

$$\boldsymbol{R_{th}} = (R_2 + R_3) \| R_4 = 150 \ \Omega \| 750 \ \Omega = \mathbf{125 \ \Omega}$$

Finally, using the calculated values of $V_{th}$ and $R_{th}$, the Thevenin equivalent of the original circuit is drawn as shown in Figure 8.32d.

***PRACTICE PROBLEM 8.11***

Derive the Thevenin equivalent for the circuit shown in Figure 8.33.

**FIGURE 8.33**

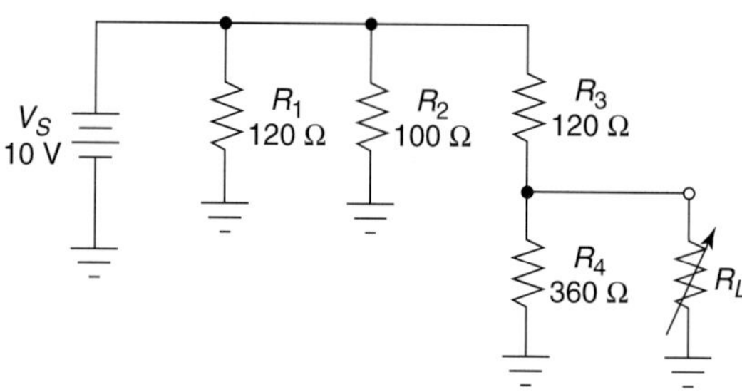


## One Final Note

Although it may seem complicated at first, deriving the Thevenin equivalents of series-parallel circuits becomes relatively simple with practice. Once you have become comfortable with using it, Thevenin's theorem can be used to solve circuit problems that would be extremely difficult to solve otherwise.

You were told at the beginning of this section that the strength of Thevenin's theorem can be seen in its circuit applications. In the next section, we will take a look at several of these applications.

### Section Review

1. What is load analysis?
2. What is Thevenin's theorem?
3. What purpose is served by Thevenin's theorem?
4. Using Thevenin's theorem, what is the procedure for performing the load analysis of a series-parallel circuit?
5. When used in circuit analysis, what is the limitation of Thevenin's theorem?
6. What is Thevenin voltage? How is it measured?
7. What is Thevenin resistance? How is it measured?
8. When measuring Thevenin resistance, why is the circuit voltage source replaced by a wire?

## 8.4 APPLICATIONS OF THEVENIN'S THEOREM

In the last section, you were shown how to derive the Thevenin equivalent of a series-parallel circuit. In this section, you will be shown how Thevenin equivalent circuits can be applied to a variety of circuit analysis problems.

### Load Voltage Ranges

Thevenin's theorem is most commonly used to predict the change in load voltage that will result from a change in load resistance. Since many loads are variable, it is important to be able to predict how a change in load will affect the output from the source circuit. ◄ *OBJECTIVE 9*

In many cases, it is desirable to know the normal range of output voltages for a circuit with a variable load. When the normal range of $V_L$ is known, it is easy to determine whether or not a particular load voltage reading indicates a problem in the circuit.

The range of load voltages for a circuit is found by determining the values of $V_L$ that correspond to the minimum and maximum load resistance values. An example of this type of load analysis problem is provided in Example 8.12.

**EXAMPLE 8.12**

The circuit in Figure 8.34a has a load with a range of $R_L = 10\ \Omega$ to $100\ \Omega$. Determine the normal range of $V_L$ values for the circuit.

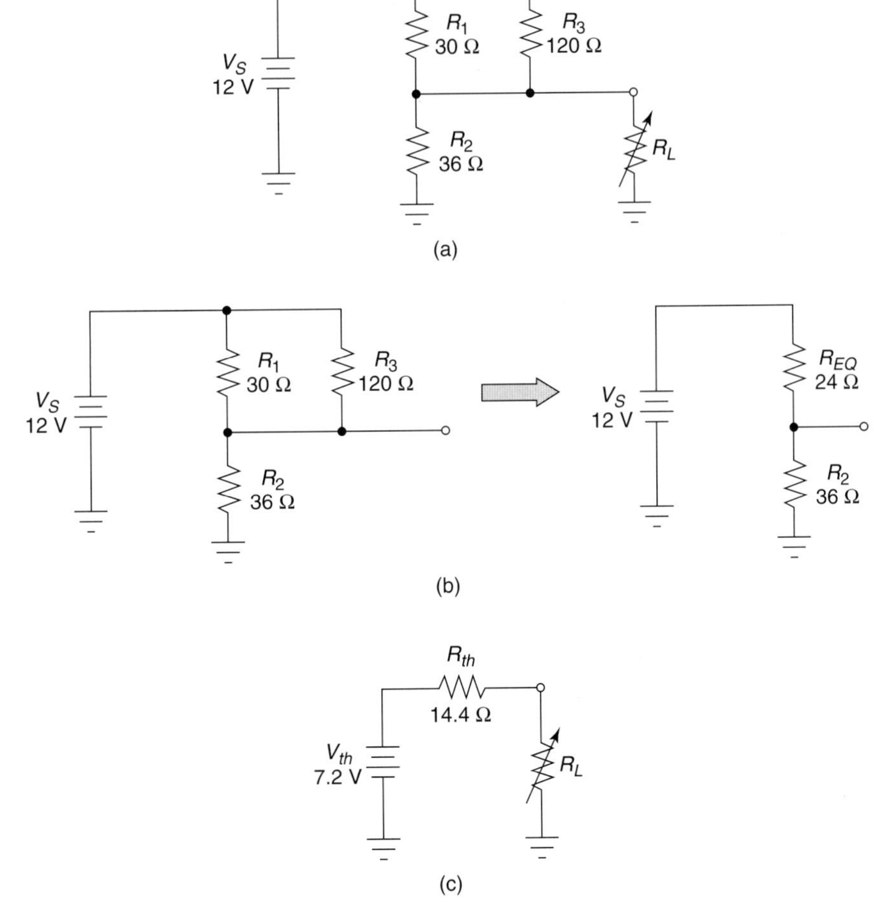


**FIGURE 8.34**

***Solution:*** First, the Thevenin equivalent for the circuit must be derived. When the load is removed, the circuit can be simplified to a series equivalent as shown in Figure 8.34b. The value of $R_{EQ}$ in the equivalent circuit was found as

$$\boldsymbol{R_{EQ}} = R_1 \| R_3 = 30\ \Omega \| 120\ \Omega = \mathbf{24\ \Omega}$$

Using the series equivalent, the value of $V_{th}$ (which equals the voltage across $R_2$) can be found as

$$\boldsymbol{V_{th}} = V_S \frac{R_2}{R_2 + R_{EQ}} = (12\ \text{V}) \frac{36\ \Omega}{60\ \Omega} = \mathbf{7.2\ V}$$

We can also use the series equivalent in Figure 8.34b to solve for $R_{th}$, as follows:

$$\boldsymbol{R_{th}} = R_2 \| R_{EQ} = 36\ \Omega \| 24\ \Omega = \mathbf{14.4\ \Omega}$$

These results are now combined in the Thevenin equivalent circuit shown in Figure 8.34c. When $R_L = 10\ \Omega$, the output from this circuit can be found as

$$\boldsymbol{V_L} = V_{th} \frac{R_L}{R_L + R_{th}} = (7.2\ \text{V}) \frac{10\ \Omega}{24.4\ \Omega} = \mathbf{2.95\ V}$$

When $R_L = 100\ \Omega$, the output from the circuit can be found as

$$\boldsymbol{V_L} = V_{th} \frac{R_L}{R_L + R_{th}} = (7.2\ \text{V}) \frac{100\ \Omega}{114.4\ \Omega} = \mathbf{6.29\ V}$$

These values indicate that the original circuit should always have an output voltage that falls somewhere between 2.95 V and 6.29 V. If $V_L$ is ever outside this range, there is a problem in the circuit.

***PRACTICE PROBLEM 8.12***

Refer to Practice Problem 8.11. Assume the circuit has a range of $R_L = 100\ \Omega$ to $500\ \Omega$. Determine the normal range of $V_L$ for the circuit.

## Maximum Power Transfer

*OBJECTIVE 10* ▶

In Chapter 5, you were introduced to the maximum power transfer theorem. As you may recall, this theorem states that maximum power transfer from a circuit to a variable load occurs when the load resistance equals the source resistance. The maximum power transfer theorem is summarized in Figure 8.35. (A more detailed discussion can be found in Section 5.4.)

When maximum power transfer occurs in a series-parallel circuit.

For a series-parallel circuit with a variable load, maximum power transfer occurs when the load resistance equals the Thevenin resistance of the circuit, which means that load power reaches its maximum possible value when $R_L = R_{th}$. Example 8.13 demonstrates the calculation of maximum load power for a series-parallel circuit.

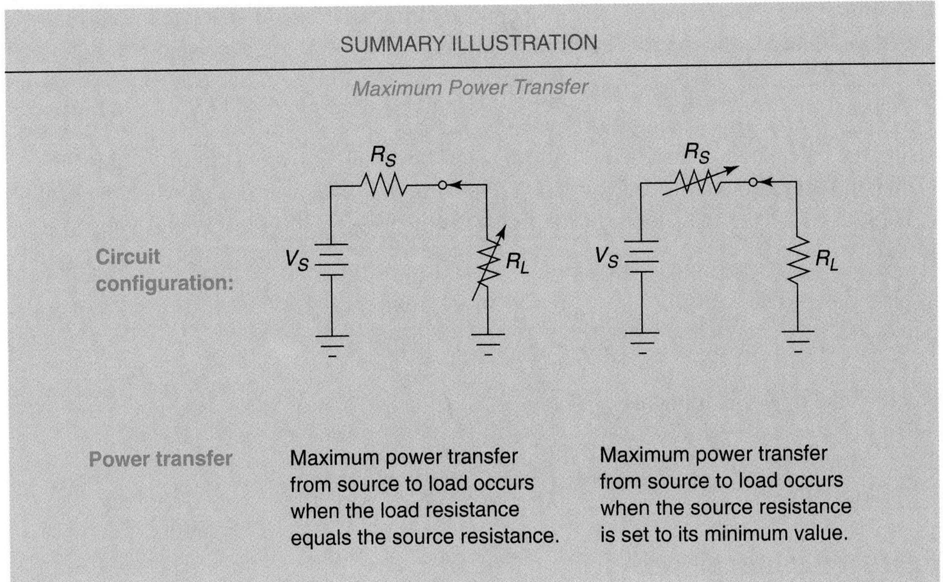

SUMMARY ILLUSTRATION
Maximum Power Transfer
Circuit configuration:
$R_S$
$V_S$
$R_L$
$R_S$
$V_S$
$R_L$
Power transfer
Maximum power transfer from source to load occurs when the load resistance equals the source resistance.
Maximum power transfer from source to load occurs when the source resistance is set to its minimum value.


**FIGURE 8.35**

## EXAMPLE 8.13

Calculate the maximum load power for the circuit in Figure 8.36a.

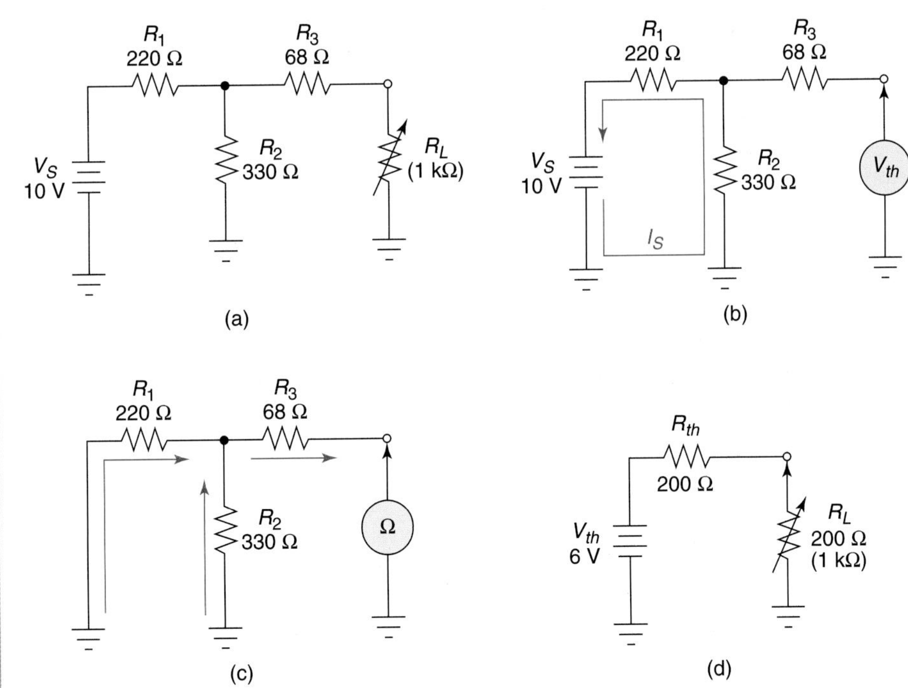


**FIGURE 8.36**

***Solution:*** The solution starts with deriving the Thevenin equivalent of the circuit. As shown in Figure 8.36b, the circuit current passes only through $R_1$ and $R_2$ when the load is removed. Therefore, $V_{th} = V_2$. For this circuit, the value of $V_{th}$ is found as:

$$\mathbf{V_{th}} = V_S \frac{R_2}{R_1 + R_2} = (10 \text{ V})\frac{330 \ \Omega}{550 \ \Omega} = \mathbf{6 \ V}$$

If an ohmmeter were connected to the open load terminal, it would generate the currents shown in Figure 8.36c. As these currents indicate:

$$R_{th} = (R_1 \| R_2) + R_3 = (220\ \Omega \| 330\ \Omega) + 68\ \Omega = \mathbf{200\ \Omega}$$

Using the calculated values of $V_{th}$ and $R_{th}$, the Thevenin equivalent circuit is drawn as shown in Figure 8.36d. Since $R_{th} = 200\ \Omega$, maximum load power occurs when the load is also set to 200 Ω (as shown in the figure). When $R_L = 200\ \Omega$, the load voltage is found as

$$V_L = V_{th}\frac{R_L}{R_L + R_{th}} = (6\text{ V})\frac{200\ \Omega}{400\ \Omega} = \mathbf{3\ V}$$

and the load power is found as

$$P_{L(\max)} = \frac{V_L^2}{R_L} = \frac{(3\text{ V})^2}{200\ \Omega} = \mathbf{45\ mW}$$

***PRACTICE PROBLEM 8.13***

Calculate the maximum load power for the circuit in Figure 8.37.

**FIGURE 8.37**

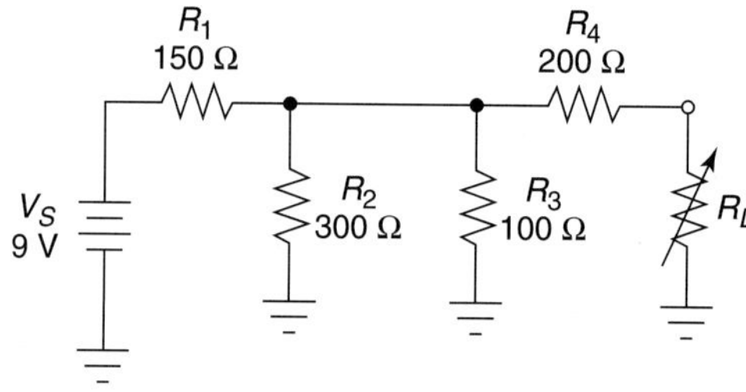


## Multisource Circuits

*OBJECTIVE 11* ▶ Earlier in the chapter, you were shown how to analyze a multisource circuit using the superposition theorem. You may recall that superposition analysis involved:

1. Determining the circuit response to each individual source.
2. Combining the results from Step 1.

How to derive the Thevenin equivalent of a multisource circuit.

Thevenin's theorem can be used (along with the superposition theorem) to predict the response a multisource circuit will have to a change in load resistance. For example, consider the circuit shown in Figure 8.38. The Thevenin equivalent of the circuit can be derived as follows:

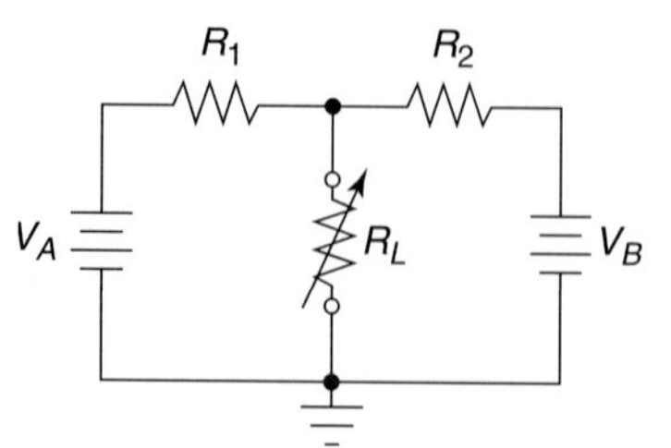


**FIGURE 8.38**

1. Remove the load (as always).
2. Using the superposition theorem, calculate the voltage across the open load terminals. This is the Thevenin voltage ($V_{th}$) for the circuit.
3. Short all voltage sources and calculate the resistance that would be measured across the load terminals. This is the Thevenin resistance ($R_{th}$) for the circuit.

This procedure is demonstrated in Example 8.14.

Predict the value of $V_L$ that will be produced by the circuit in Figure 8.39a for values of $R_L = 40\ \Omega$ and $R_L = 90\ \Omega$.

EXAMPLE 8.14

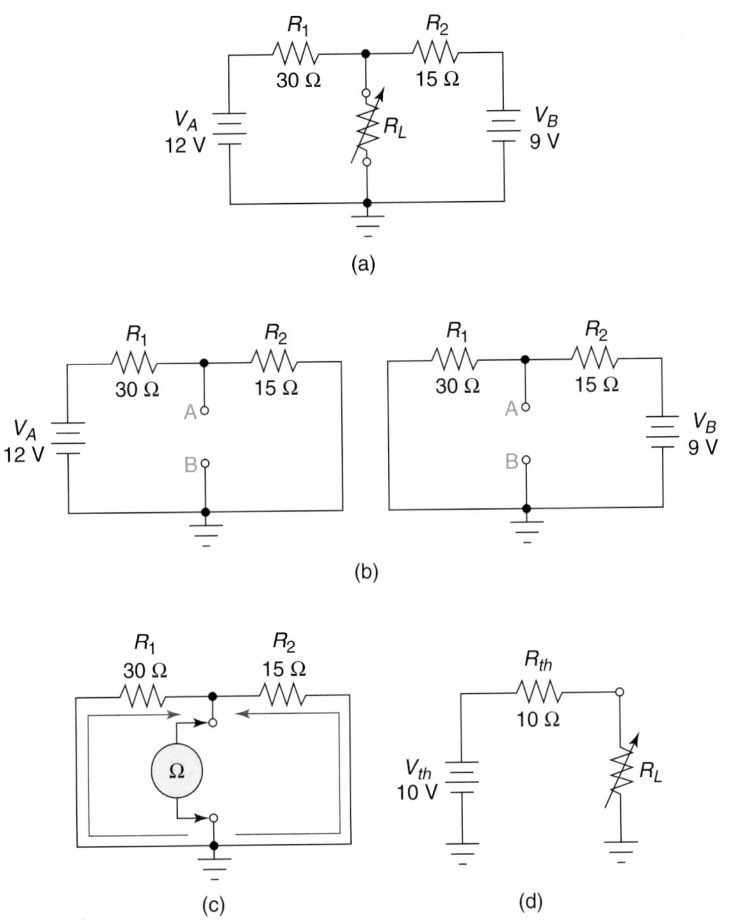


**FIGURE 8.39**

***Solution:*** First, we'll remove the load and $V_B$, as shown in Figure 8.39b. If you trace from load terminal B around to the right, you'll see that the open load terminals are connected directly across $R_2$. The value of $V_2$ produced by $V_A$ (with $V_B$ removed) is found as

$$\mathbf{V_{2A}} = V_A \frac{R_2}{R_1 + R_2} = (12\text{ V})\frac{15\ \Omega}{45\ \Omega} = \mathbf{4\ V}$$

Next, we'll replace $V_B$ and remove $V_A$ (as shown in Figure 8.39b). If you trace from load terminal B around to the left, you'll see that the load terminals are connected directly to both sides of $R_1$. The value of $V_1$ produced by $V_B$ (with $V_A$ removed) is found as

$$\mathbf{V_{1B}} = V_B \frac{R_1}{R_1 + R_2} = (9\text{ V})\frac{30\ \Omega}{45\ \Omega} = \mathbf{6\ V}$$

The value of $V_{th}$ is equal to the sum of these voltages, as follows:

$$\mathbf{V_{th}} = V_{1B} + V_{2A} = 6\text{ V} + 4\text{ V} = \mathbf{10\ V}$$

Figure 8.39c shows the circuit with both sources shorted out. The current arrows represent the current that would be generated by an ohmmeter connected to the load terminals. According to these arrows, $R_1$ and $R_2$ are in parallel. Therefore, the value of $R_{th}$ is found as

$$\mathbf{R_{th}} = R_1 \| R_2 = 30\ \Omega \| 15\ \Omega = \mathbf{10\ \Omega}$$

Using the calculated values of $V_{th}$ and $R_{th}$, the Thevenin equivalent of the multisource circuit is drawn as shown in Figure 8.39d. When $R_L = 40\ \Omega$, the value of $V_L$ can be found as

$$V_L = V_{th}\frac{R_L}{R_L + R_{th}} = (10\text{ V})\frac{40\ \Omega}{50\ \Omega} = \mathbf{8\ V}$$

When $R_L = 90\ \Omega$, the value of $V_L$ can be found as

$$V_L = V_{th}\frac{R_L}{R_L + R_{th}} = (10\text{ V})\frac{90\ \Omega}{100\ \Omega} = \mathbf{9\ V}$$

***PRACTICE PROBLEM 8.14***

Predict the values of $V_L$ that will be produced by the circuit in Figure 8.40 for $R_L = 20\ \Omega$ and $R_L = 50\ \Omega$.

**FIGURE 8.40**

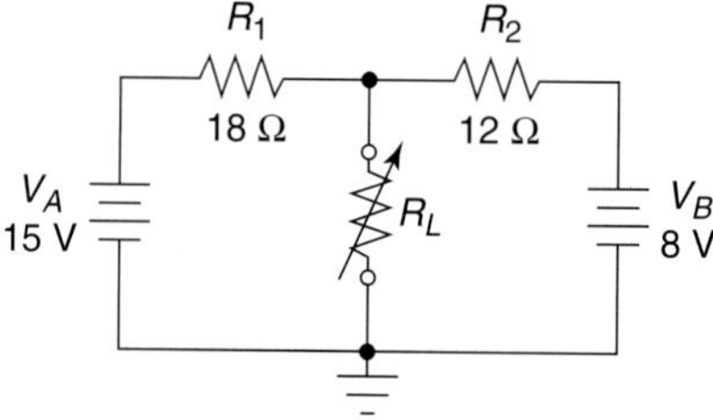


Once the Thevenin equivalent of a multisource circuit has been derived, it can also be used for maximum load power calculations. This point is illustrated in Example 8.15.

## EXAMPLE 8.15

Calculate the maximum value of load power for the circuit in Figure 8.39a.

***Solution:*** In Example 8.14, we calculated a Thevenin resistance of 10 Ω for the circuit. Therefore, maximum load power occurs when $R_L$ is set to 10 Ω. Using a value of $R_L = 10\ \Omega$ in our Thevenin equivalent circuit (Figure 8.39d), the load voltage is found as

$$V_L = V_{th}\frac{R_L}{R_L + R_{th}} = (10\text{ V})\frac{10\ \Omega}{20\ \Omega} = \mathbf{5\ V}$$

Now, the maximum load power can be found as

$$\mathbf{P_{L(max)}} = \frac{V_L^2}{R_L} = \frac{(5\text{ V})^2}{10\ \Omega} = \mathbf{2.5\ W}$$

***PRACTICE PROBLEM 8.15***

Calculate the maximum possible value of load power for the circuit in Figure 8.40.

As you can see, the combination of Thevenin's theorem and the superposition theorem can be used to reduce a multisource circuit to a Thevenin equivalent. The Thevenin equivalent circuit can then be used to:

1. Predict any output changes that result from a change in load resistance.
2. Determine the maximum possible value of load power for a variable load.

FIGURE 8.41

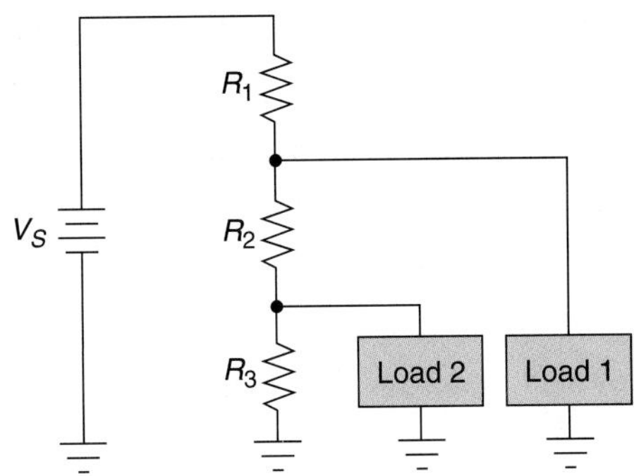


## Multiload Circuits

As you were told in Chapter 7, loaded voltage dividers often have more than one load. For example, the voltage divider in Figure 8.41 is being used to supply the operating voltages for two loads. When using Thevenin's theorem to analyze a circuit like the one in Figure 8.41, you need to keep several points in mind:

1. Each load has its own Thevenin equivalent circuit. Since they are connected to the circuit at different points, each load in Figure 8.41 has its unique Thevenin equivalent of the original circuit.
2. When deriving the Thevenin equivalent for each load, the resistance of the other load must be taken into account.

These points are illustrated in Figure 8.42. Here we have a loaded voltage divider and two equivalent circuits. Figure 8.42b shows the circuit with an open load 1 terminal. Figure

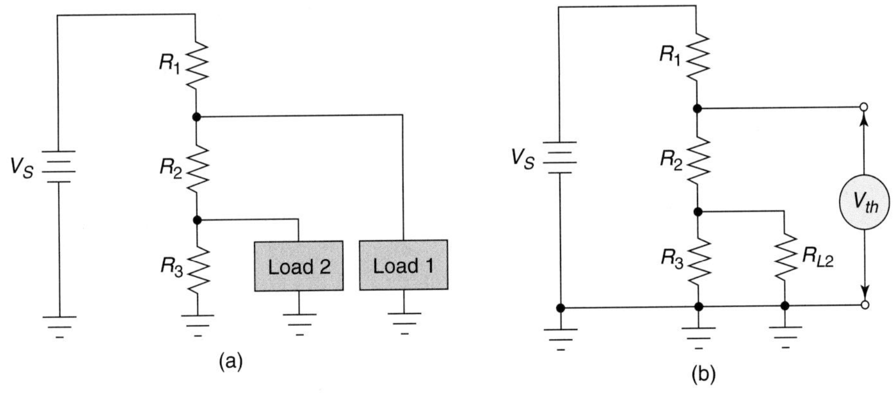


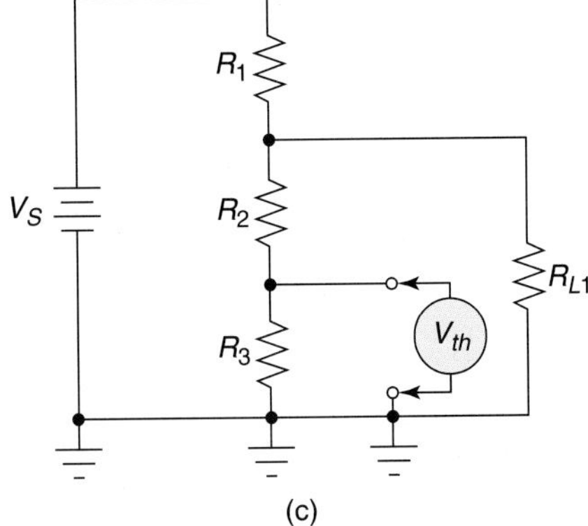


FIGURE 8.42

8.42c shows the circuit with an open load 2 terminal. As you can see, each load "sees" a unique source circuit; that is, the source circuit for load 1 is different than the one for load 2.

When analyzing the circuit in Figure 8.42a, a Thevenin equivalent must be derived for each load. Then these circuits can be used separately to perform the analyses of the loads. The analysis of one load in a multiload circuit is demonstrated in Example 8.16.

## EXAMPLE 8.16

Determine the maximum value of load power for load 1 in Figure 8.43a. Assume that load 2 is fixed at a value of $R_{L2} = 100\ \Omega$.

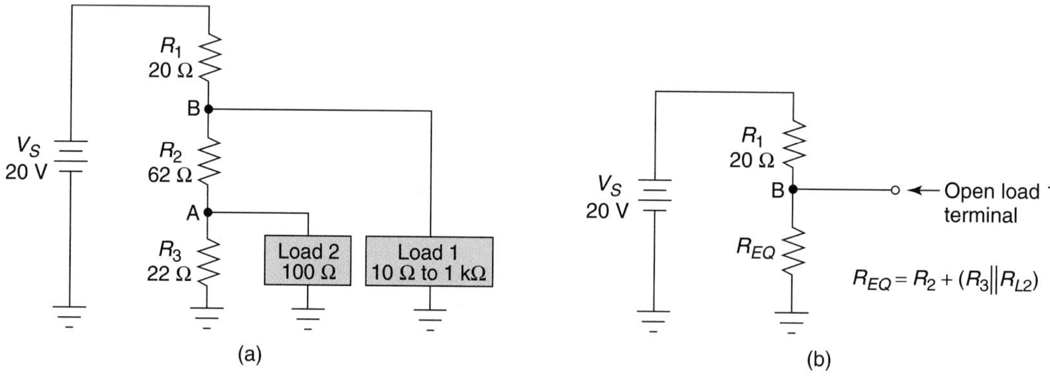


**FIGURE 8.43**

***Solution:*** The simplest approach to this problem is to combine $R_2$, $R_3$, and the resistance of load 2 ($R_{L2}$) into a single equivalent resistance (as shown in Figure 8.43b). The value of $R_{EQ}$ (which represents the total resistance from point B to ground) can be found as

$$\boldsymbol{R_{EQ}} = R_2 + (R_3 \| R_{L2}) = 62\ \Omega + (22\ \Omega \| 100\ \Omega) = 62\ \Omega + 18\ \Omega = \mathbf{80\ \Omega}$$

Now, using the value of $R_{EQ} = 80\ \Omega$, the voltage at the open load 1 terminal can be found as

$$\boldsymbol{V_{th1}} = V_S \frac{R_{EQ}}{R_1 + R_{EQ}} = (20\ \text{V}) \frac{80\ \Omega}{100\ \Omega} = \mathbf{16\ V}$$

Measured from the open load 1 terminal, the value of $R_{th}$ in Figure 8.43b is equal to the parallel combination of $R_1$ and $R_{EQ}$. Therefore:

$$\boldsymbol{R_{th1}} = R_1 \| R_{EQ} = 20\ \Omega \| 80\ \Omega = \mathbf{16\ \Omega}$$

This value indicates that maximum load power will be generated when $R_{L1}$ is set to 16 Ω. When $R_{L1} = 16\ \Omega$:

$$\boldsymbol{V_L} = V_{th} \frac{R_L}{R_L + R_{th}} = (16\ \text{V}) \frac{16\ \Omega}{32\ \Omega} = \mathbf{8\ V}$$

and

$$\boldsymbol{P_{L(\max)}} = \frac{V_L^2}{R_L} = \frac{(8\ \text{V})^2}{16\ \Omega} = \mathbf{4\ W}$$

***PRACTICE PROBLEM 8.16***

Assume that the positions of load 1 and load 2 in Figure 8.43a have been reversed. Calculate the maximum load power for load 1.

The analysis technique demonstrated in Example 8.16 can be used on any number of loads. However, there is one restriction: each load (other than the load of interest) must be assumed to have a fixed resistance. If you were to assume a range of values for more than one load, you wouldn't be able to determine set values of $V_{th}$ and $R_{th}$ for the load of interest. Without set values of $V_{th}$ and $R_{th}$, attempting to solve the problem would be pointless.

## Bridge Circuits

In Chapter 7, you were introduced to the Wheatstone bridge; a circuit that can be used to provide extremely accurate resistance measurements. In general, the term *bridge* is used to describe any circuit constructed as shown in Figure 8.44. Note that the load ($R_L$) forms the "bridge" between the two series circuits.

**FIGURE 8.44 Bridge configuration.**

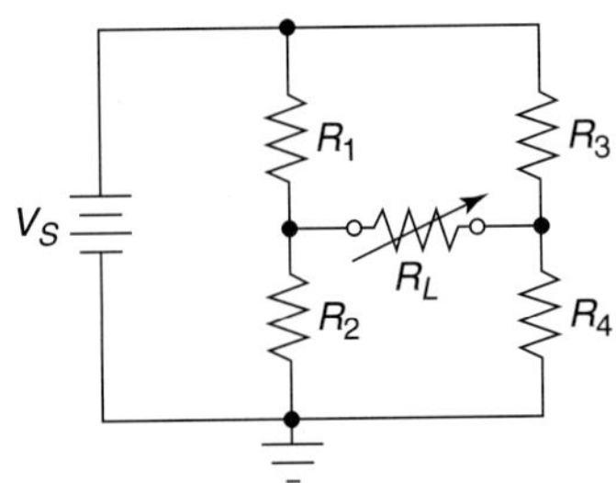


In a bridge circuit like the one shown in Figure 8.44, Thevenin's theorem can be used to predict the magnitude and polarity of the load voltage. Figure 8.45 helps to show how the Thevenin equivalent of a bridge circuit is derived. When the load is removed (as shown in Figure 8.45a), the voltage across the open load terminals is equal to the difference between $V_A$ and $V_B$. Therefore, the value of $V_{th}$ for the circuit is found as

$$V_{th} = |V_A - V_B| \quad \textbf{(8.3)}$$

The Thevenin resistance of a bridge circuit is measured as shown in Figure 8.45b. As usual, the ohmmeter is connected across the open load terminals. Note the highlighted line around the outside of the circuit. This line connects the bottom sides of $R_2$ and $R_4$ to the top sides of $R_1$ and $R_3$. Therefore, the circuit can be redrawn as shown in Figure 8.45c. As the currents in this circuit indicate:

$$R_{th} = (R_1 \| R_2) + (R_3 \| R_4) \quad \textbf{(8.4)}$$

Once the values of $V_{th}$ and $R_{th}$ are known, the Thevenin equivalent circuit can be used to make the same types of predictions as those we made for the other types of circuits. Deriving the Thevenin equivalent of a bridge circuit is demonstrated in Example 8.17.

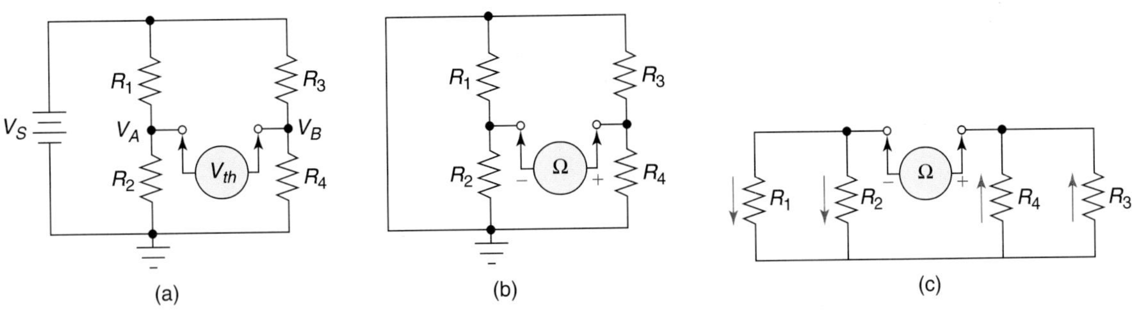


**FIGURE 8.45 Thevenin voltage and resistance.**

## EXAMPLE 8.17

Predict the values of $V_L$ that will be produced by the bridge in Figure 8.46a for values of $R_L = 100\ \Omega$ and $R_L = 300\ \Omega$.

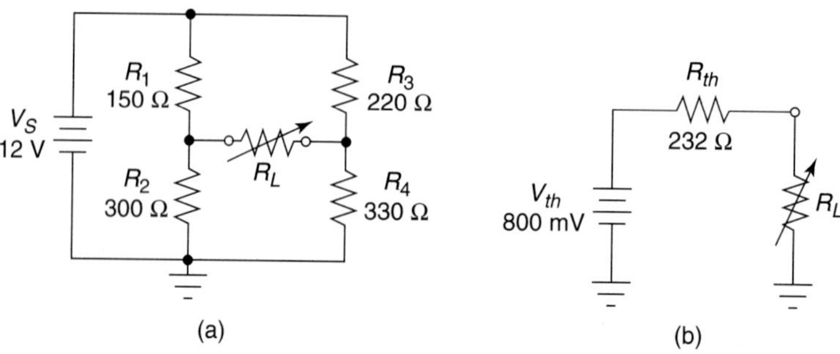


**FIGURE 8.46**

***Solution:*** To find the value of $V_{th}$, we need to determine the open-load values of $V_A$ and $V_B$. As you were shown in Chapter 7, these values are found as

$$\boldsymbol{V_A} = V_S \frac{R_2}{R_1 + R_2} = (12\ \text{V})\frac{300\ \Omega}{450\ \Omega} = \mathbf{8\ V}$$

and

$$\boldsymbol{V_B} = V_S \frac{R_4}{R_3 + R_4} = (12\ \text{V})\frac{330\ \Omega}{550\ \Omega} = \mathbf{7.2\ V}$$

Now, the value of $V_{th}$ can be found as

$$\boldsymbol{V_{th}} = V_A - V_B = 8\ \text{V} - 7.2\ \text{V} = \mathbf{800\ mV}$$

Using equation (8.4), the value of $R_{th}$ for the circuit can be found as

$$\boldsymbol{R_{th}} = (R_1 \| R_2) + (R_3 \| R_4) = (150\ \Omega \| 300\ \Omega) + (220\ \Omega \| 330\ \Omega) = 100\ \Omega + 132\ \Omega = \mathbf{232\ \Omega}$$

Using the calculated values of $V_{th}$ and $R_{th}$, the Thevenin equivalent circuit is constructed as shown in Figure 8.46b. When $R_L$ is adjusted to 100 Ω, the output from the circuit is found as

$$\boldsymbol{V_L} = V_{th} \frac{R_L}{R_L + R_{th}} = (800\ \text{mV})\frac{100\ \Omega}{332\ \Omega} = \mathbf{241\ mV}$$

When $R_L$ is adjusted to 300 Ω, the output from the circuit is found as

$$\boldsymbol{V_L} = V_{th} \frac{R_L}{R_L + R_{th}} = (800\ \text{mV})\frac{300\ \Omega}{532\ \Omega} = \mathbf{451\ mV}$$

***PRACTICE PROBLEM 8.17***

A circuit like the one in Figure 8.46a has the following values: $V_S = 10$ V, $R_1 = 200\ \Omega$, $R_2 = 200\ \Omega$, $R_3 = 100\ \Omega$, and $R_4 = 300\ \Omega$. Predict the value of $V_L$ that will be produced by the circuit for $R_L = 150\ \Omega$ and $R_L = 220\ \Omega$.

## One Final Note

In this section and in the last, you were shown that Thevenin's theorem can be a powerful load analysis tool. In many cases, it can be used to solve problems that would be extremely difficult to solve otherwise.

In the next section, you will be introduced to several other network theorems. While each has its own merits and applications, you will find that Thevenin's theorem is used more often than the rest of the network theorems combined.

**Section Review**

1. When does maximum power transfer occur in a series-parallel circuit with a variable load?
2. Describe the process used to predict the value of maximum load power for a series-parallel circuit with a variable load.
3. Describe how Thevenin's theorem can be used to perform the load analysis of a multi-source circuit.
4. Describe how Thevenin's theorem can be used to perform the load analysis of a multi-load circuit.
5. What restriction applies to the Thevenin analysis of a multiload circuit?
6. Describe the process you would use to predict the value of maximum load power for a bridge circuit with a variable load.

# 8.5 NORTON'S THEOREM

In this section, we will discuss another network theorem that can be used in the load analysis of a series-parallel circuit. As you will see, this theorem (called Norton's theorem) is very closely related to Thevenin's theorem.

## Norton's Theorem: An Overview

◄ *OBJECTIVE 12*

**Norton's theorem**
Any resistive circuit or network, no matter how complex, can be represented as a current source in parallel with a source resistance.

**Norton's theorem** states that any resistive circuit or network, no matter how complex, can be represented as a current source in parallel with a source resistance. A series-parallel circuit is shown with its Norton equivalent circuit in Figure 8.47. Note that the current source and resistance are referred to as the Norton current ($I_n$) and the Norton resistance ($R_n$), respectively.

Obviously, Norton's theorem is very similar to Thevenin's theorem. The primary difference is that Norton represents a complex circuit as a practical current source while Thevenin would represent the same circuit as a practical voltage source. Later in this section, we will look at the similarities and differences between these theorems in more detail.

When using Norton's theorem to perform the load analysis of a circuit, the first step is to derive the Norton equivalent circuit. Then the load is replaced and the equivalent circuit is analyzed as a current divider. Load analysis using Norton's theorem is demonstrated in several upcoming examples.

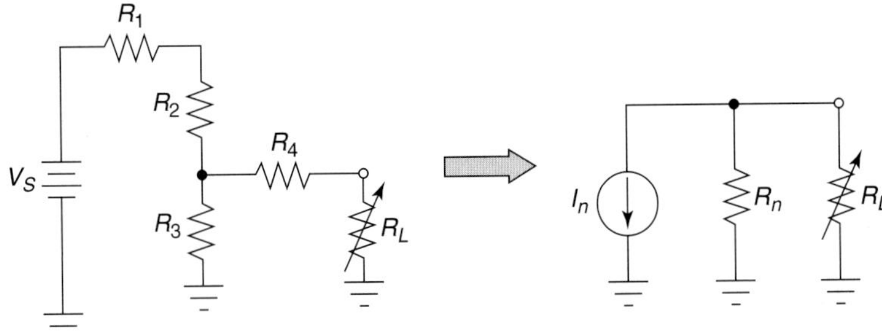


**FIGURE 8.47 A series-parallel circuit and its Norton equivalent circuit.**

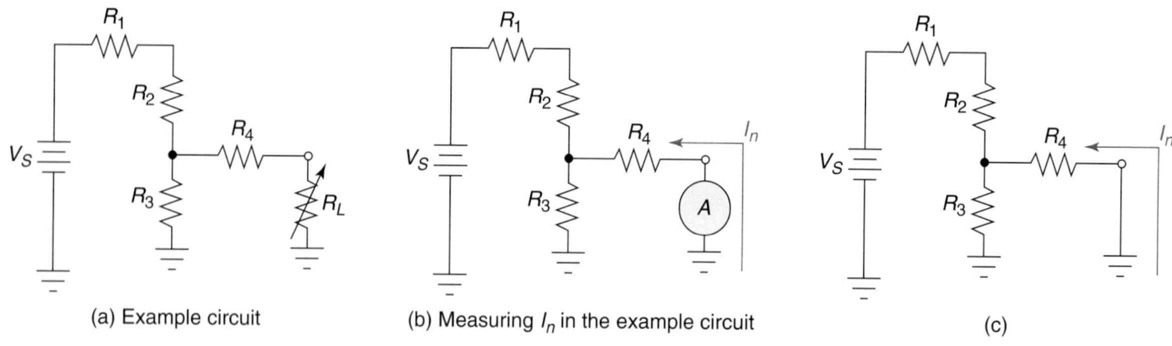


**FIGURE 8.48 Measuring Norton current in a series-parallel circuit.**

## Determining the Value of $I_n$

*OBJECTIVE 13* ▶ By definition, Norton current ($I_n$) is the current through the shorted load terminals. The measurement of $I_n$ is illustrated in Figure 8.48b. As you can see, the load has been removed and a current meter has been connected across the load terminals. Since the internal resistance of a current meter is 0 Ω (ideally), the current being measured in Figure 8.48b is equal to the Norton current for the circuit.

To calculate the value of $I_n$ for a series-parallel circuit, you short the load as shown in Figure 8.48c. Then you calculate the current through the shorted load terminals. This procedure is demonstrated in Example 8.18.

### EXAMPLE 8.18

Calculate the value of $I_n$ for the circuit in Figure 8.49a.

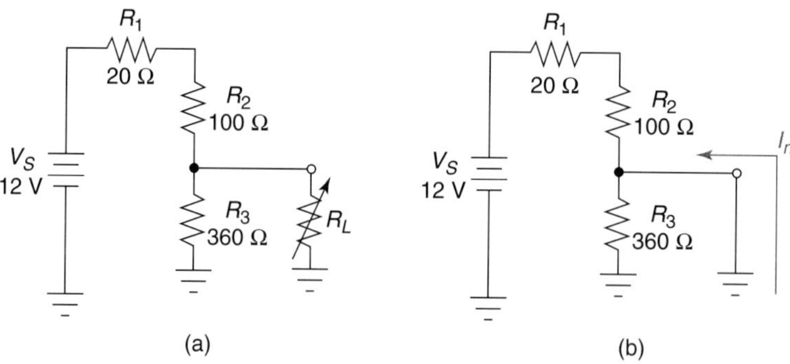


**FIGURE 8.49**

***Solution:*** The first step is to short the load as shown in Figure 8.49b. When the load is shorted, it also shorts out $R_3$. Therefore, the value of $I_n$ is determined by the source voltage and the series combination of $R_1$ and $R_2$, as follows:

$$I_n = \frac{V_S}{R_1 + R_2} = \frac{12\text{ V}}{120\ \Omega} = \mathbf{100\ mA}$$

***PRACTICE PROBLEM 8.18***

Calculate the value of $I_n$ for the circuit in Figure 8.50.

FIGURE 8.50

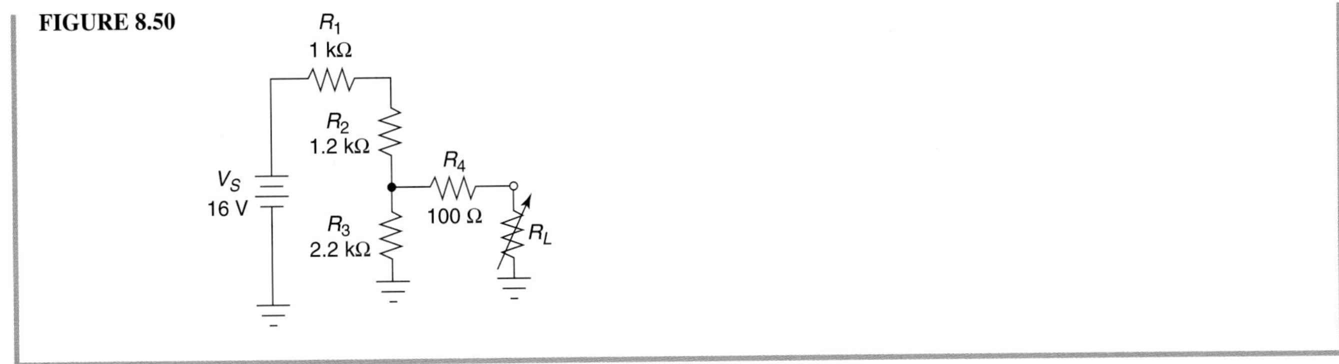


## Norton Resistance ($R_n$)

The next step in deriving a Norton equivalent circuit is determining the value of Norton resistance ($R_n$). The value of $R_n$ for a series-parallel circuit is measured and calculated exactly like Thevenin resistance:

1. The load is removed.
2. The source is replaced by its resistive equivalent.
3. The resistance (as measured at the open load terminals) is determined.

Example 8.19 demonstrates the fact that $R_n$ is calculated the same way as $R_{th}$.

EXAMPLE 8.19

Calculate the value of $R_n$ for the circuit shown in Figure 8.51a.

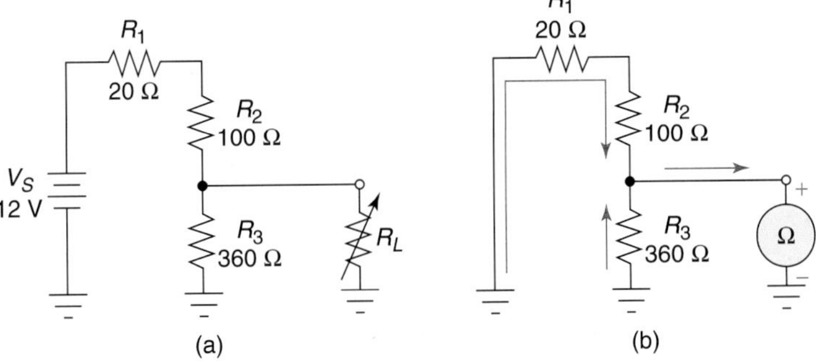


FIGURE 8.51

***Solution:*** The first step is to remove the load and short out the voltage source. This gives us the circuit shown in Figure 8.51b. The currents shown in the circuit are those that would be produced by an ohmmeter connected to the load terminals. As these arrows indicate, $R_n$ can be calculated as follows:

$$\boldsymbol{R_n} = (R_1 + R_2)\|R_3 = 120\ \Omega\|360\ \Omega = \mathbf{90\ \Omega}$$

***PRACTICE PROBLEM 8.19***

Calculate the value of $R_n$ for the circuit shown in Figure 8.50.

As you can see, there is no difference between $R_n$ and $R_{th}$. They are calculated (and measured) in exactly the same fashion.

## Load Analysis Using Norton Equivalent Circuits

*OBJECTIVE 14* ▶ Norton equivalent circuits are used in the same fashion as Thevenin equivalent circuits. Once the Norton equivalent of a circuit is derived, the load is placed in the equivalent circuit. Then, the current divider relationship is used to determine the value of load current ($I_L$). The following series of examples demonstrates how load analysis problems can be solved using Norton equivalent circuits.

### EXAMPLE 8.20

Predict the values of $I_L$ that will be produced by the circuit in Figure 8.52a for values of $R_L = 25\ \Omega$ and $R_L = 75\ \Omega$.

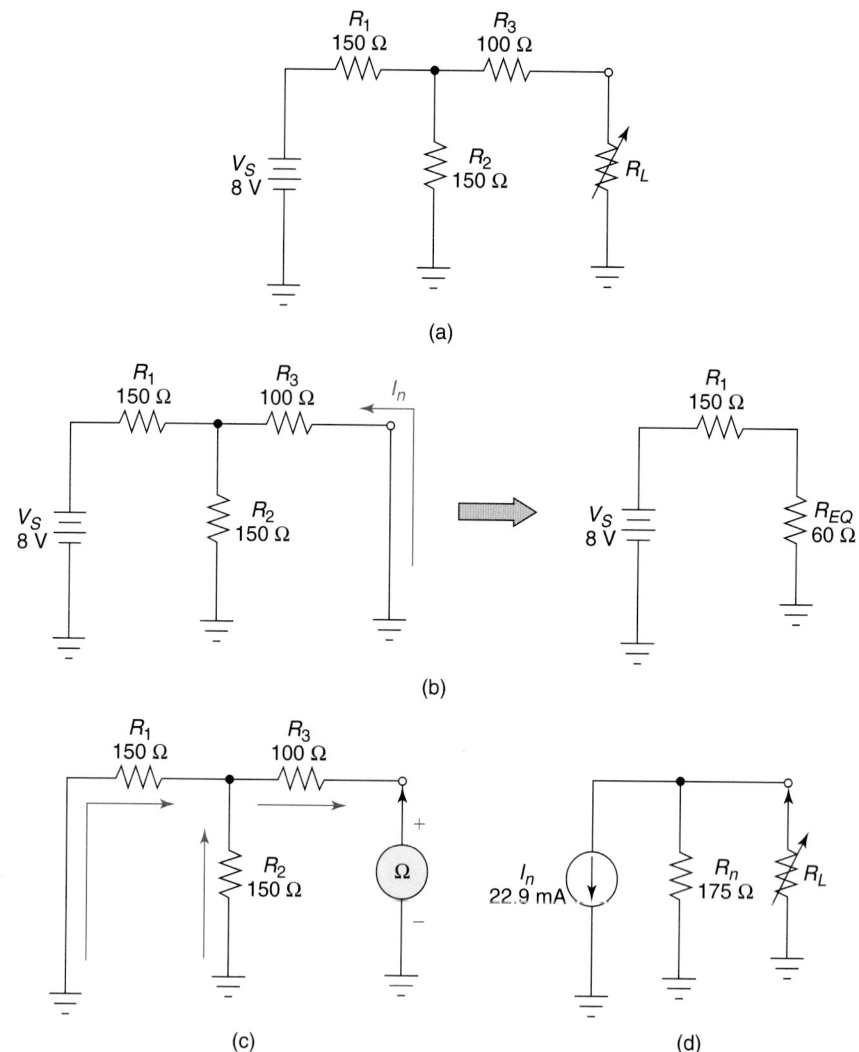


**FIGURE 8.52**

***Solution:*** To determine the Norton current for the circuit, we need to calculate the current through the shorted load terminals (as shown in Figure 8.52b). With the load shorted, the value of $I_n$ equals the current through $R_3$. Using the series equivalent circuit shown in Figure 8.52b, the value of $I_3$ can be found using the following sequence of calculations:

$$\boldsymbol{R_{EQ}} = R_2 \| R_3 = 150\ \Omega \| 100\ \Omega = \mathbf{60\ \Omega}$$

$$\boldsymbol{V_{EQ}} = V_S \frac{R_{EQ}}{R_1 + R_{EQ}} = (8\text{ V})\frac{60\ \Omega}{210\ \Omega} = \mathbf{2.29\ V}$$

and

$$\boldsymbol{I_3} = \frac{V_{EQ}}{R_3} = \frac{2.29\text{ V}}{100\ \Omega} = \mathbf{22.9\ mA}$$

Since this current passes through the shorted load terminals, the value of $I_n$ for the circuit is 22.9 mA. If we remove the load and replace the source in our original circuit, we get the circuit shown in Figure 8.52c. The currents shown in the figure are those that would be produced by an ohmmeter connected to the load terminals. As these arrows indicate, $R_n$ can be calculated as follows:

$$\boldsymbol{R_n} = (R_1 \| R_2) + R_3 = (150\ \Omega \| 150\ \Omega) + 100\ \Omega = \mathbf{175\ \Omega}$$

Now, using the calculated values of $I_n$ and $R_n$, the Norton equivalent circuit is constructed as shown in Figure 8.52d. When $R_L = 25\ \Omega$, the value of $I_L$ can be found using the current divider relationship, as follows:

$$\boldsymbol{I_L} = I_n \frac{R_L \| R_n}{R_L} = (22.9\text{ mA})\frac{25\ \Omega \| 175\ \Omega}{25\ \Omega} = \mathbf{20\ mA}$$

When $R_L = 75\ \Omega$, the value of $I_L$ can be found as

$$\boldsymbol{I_L} = I_n \frac{R_L \| R_n}{R_L} = (22.9\text{ mA})\frac{75\ \Omega \| 175\ \Omega}{75\ \Omega} = \mathbf{16\ mA}$$

***PRACTICE PROBLEM 8.20***

Predict the values of $I_L$ that will be produced by the circuit in Figure 8.53 for values of $R_L = 100\ \Omega$ and $R_L = 150\ \Omega$.

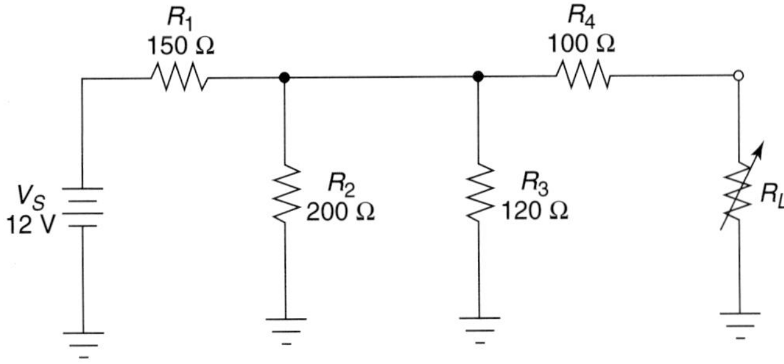


**FIGURE 8.53**

Now, let's see how Norton's theorem can be used to calculate the maximum possible load power for a series-parallel circuit with a variable load.

## EXAMPLE 8.21

Determine the maximum load power for the circuit in Figure 8.54a.

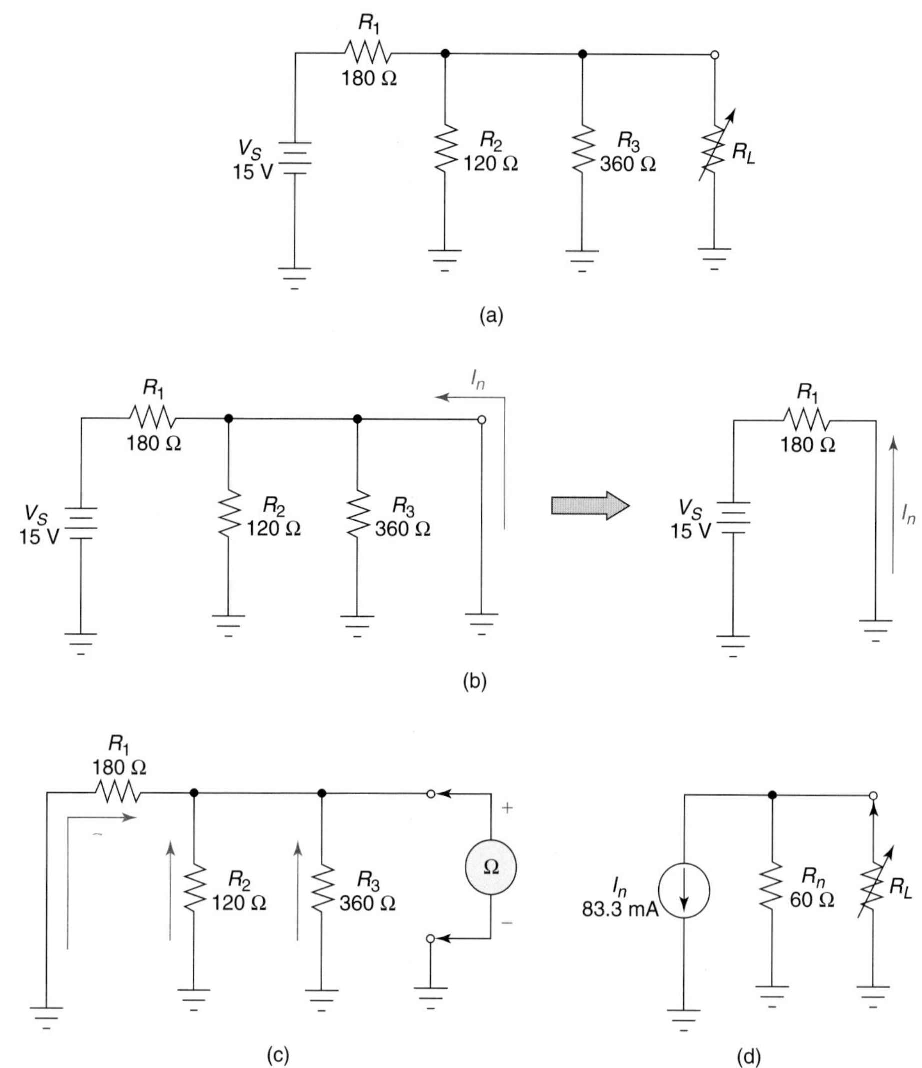


EWB

**FIGURE 8.54**

***Solution:*** The first step is to short out the load. In this case, shorting out the load also shorts out both $R_2$ and $R_3$ (as shown in Figure 8.54b). As a result, the current through the shorted load can be found simply as

$$\boldsymbol{I_n} = \frac{V_S}{R_1} = \frac{15\text{ V}}{180\ \Omega} = \mathbf{83.3\ mA}$$

If we remove the load and short the source, we get the circuit shown in Figure 8.54c. The arrows represent the current that would be generated by an ohmmeter connected to the load terminals. As these arrows indicate:

$$\boldsymbol{R_n} = R_1 \| R_2 \| R_3 = 180\ \Omega \| 120\ \Omega \| 360\ \Omega = \mathbf{60\ \Omega}$$

Using the values of $I_n$ and $R_n$, the Norton equivalent circuit is constructed as shown in Figure 8.54d. As you know, maximum power transfer to a variable load occurs when load resistance equals the source resistance. In this case, when $R_L = R_n$. When $R_L =$ 60 Ω, the load current can be found as

$$I_L = I_n \frac{R_L \| R_n}{R_L} = (83.3 \text{ mA}) \frac{60\ \Omega \| 60\ \Omega}{60\ \Omega} = \mathbf{41.65\ mA}$$

Finally, the maximum load power can be found as

$$P_{L(\max)} = I_L^2 R_L = (41.65 \text{ mA})^2 (60\ \Omega) = \mathbf{104.2\ mW}$$

***PRACTICE PROBLEM 8.21***

Calculate the maximum load power for the circuit shown in Figure 8.55.

**FIGURE 8.55**

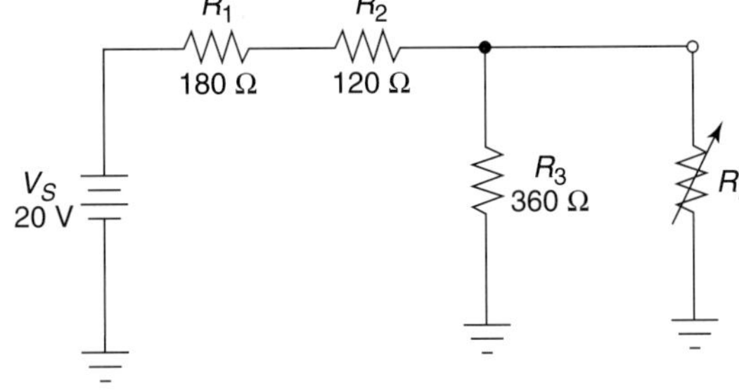


## *Norton-to-Thevenin and Thevenin-to-Norton Conversions*

*OBJECTIVE 15*

Many of the example circuits used in this section were also used to demonstrate Thevenin's theorem. The Norton and Thevenin equivalents derived in a pair of matched examples are shown (with the original circuit) in Figure 8.56.

The load calculations shown for each equivalent circuit were made using an assumed value of $R_L = 60\ \Omega$, the value of the source resistance. As you can see, the Norton and Thevenin equivalent circuits provide identical values of load voltage, current, and power. This proves that the Norton and Thevenin equivalents of a given circuit are also the equivalents of each other; that is, for every Norton equivalent circuit, there is a matching Thevenin equivalent circuit, and vice versa.

Using the techniques we discussed for source conversions, you can convert either type of equivalent circuit to the other. Both types of conversions are shown in Figure 8.57. As you can see, the conversion formulas used are the same as those you learned in Section 8.2. If we apply the Thevenin-to-Norton relationships to the Thevenin circuit in Figure 8.56, we get:

$$R_n = R_{th} = 60\ \Omega \qquad \text{and} \qquad I_n = \frac{V_{th}}{R_{th}} = \frac{5\text{ V}}{60\ \Omega} = 83.3 \text{ mA}$$

These are the same values as those shown in the Norton equivalent circuit in Figure 8.56.

## *One Final Note*

As your study of electronics continues, you'll find that there are analysis situations that lend themselves more toward one theorem than the other; that is, some circuits are better represented as voltage sources while others are better represented as current sources. This will become very evident when you study electronic devices: a group of components often represented using Thevenin and Norton equivalent circuits.

In most cases, the choice of theorem that you use for load analysis is up to you. Many technicians find it easier to become proficient at using one type of theorem and then

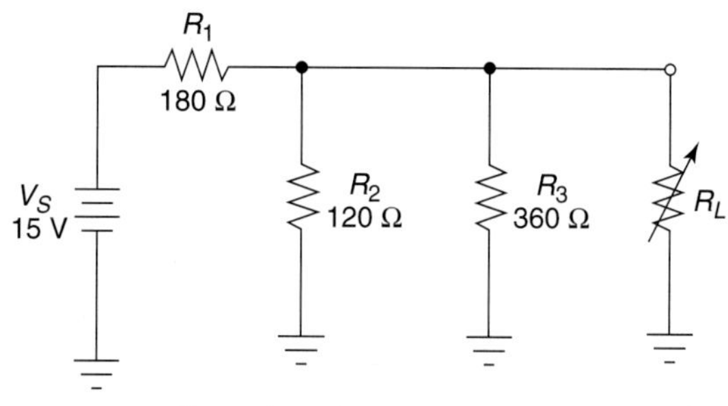


Example series-parallel circuit

Thevenin equivalent
(from example 8.9)

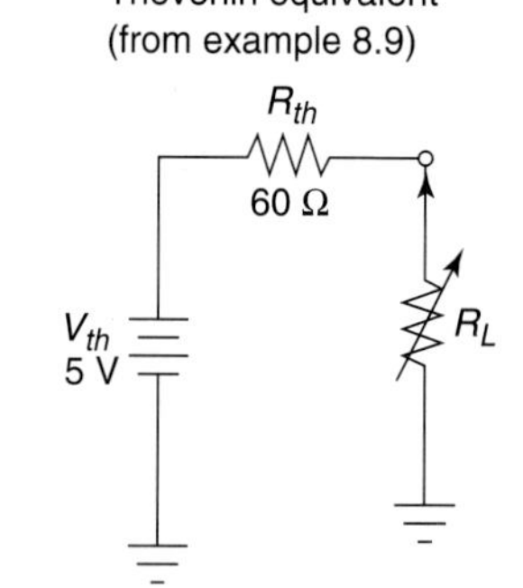


Load calculations

$$V_L = V_{th}\frac{R_L}{R_L + R_{th}} = (5\text{ V})\frac{60\ \Omega}{120\ \Omega} = 2.5\text{ V}$$

$$I_L = \frac{V_L}{R_L} = \frac{2.5\text{ V}}{60\ \Omega} = 41.67\text{ mA}$$

$$P_L = \frac{V_L^2}{R_L} = \frac{(2.5\text{ V})^2}{60\ \Omega} = 104.2\text{ mW}$$

Norton equivalent
(from example 8.21)

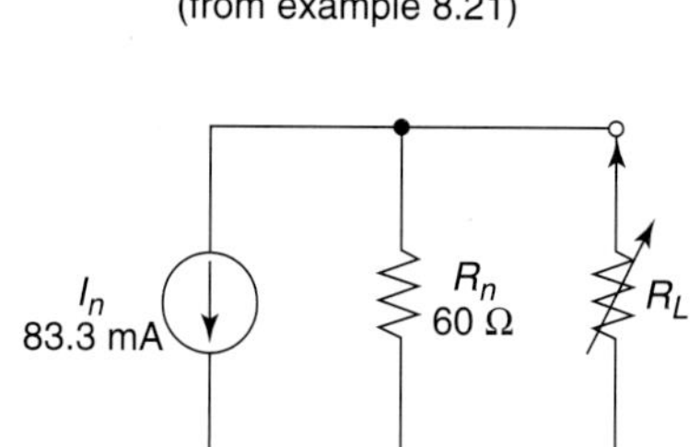


Load calculations

$$I_L = I_n\frac{R_L \| R_n}{R_L} = (83.3\text{ mA})\frac{30\ \Omega}{60\ \Omega} = 41.67\text{ mA}$$

$$V_L = I_L R_L = (41.67\text{ mA})(60\ \Omega) = 2.5\text{ V}$$

$$P_L = I_L^2 R_L = (41.67\text{ mA})^2(60\ \Omega) = 104.2\text{ mW}$$

**FIGURE 8.56 Comparing Norton and Thevenin equivalent circuits.**

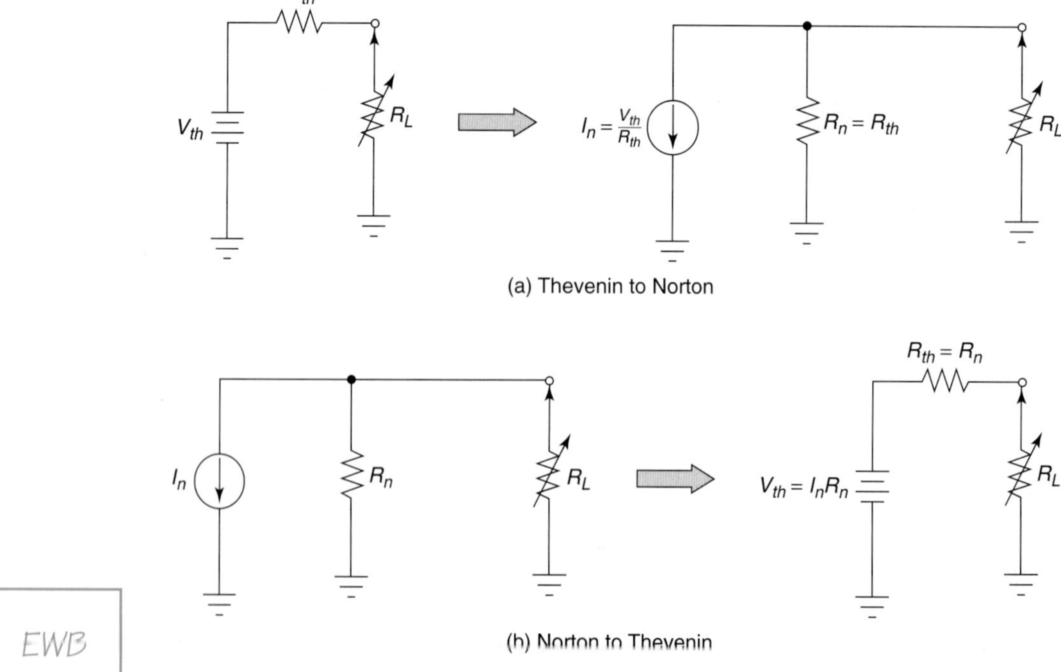


(a) Thevenin to Norton

(b) Norton to Thevenin

EWB

**FIGURE 8.57 Thevenin-to-Norton and Norton-to-Thevenin conversions.**

using source conversions to derive the other (when the need arises). Others find it beneficial to become proficient at both Norton's and Thevenin's theorems. Again, the choice is yours.

### Section Review

1. Describe the process for determining the value of Norton current for a series-parallel circuit.
2. Describe the process for determining the value of Norton resistance for a series-parallel circuit.
3. In terms of load analysis, how does Norton's theorem differ from Thevenin's theorem?
4. Describe the relationship between the Norton and Thevenin equivalents of a circuit.
5. How do you convert from one equivalent circuit (Norton or Thevenin) to the other?

## 8.6 OTHER NETWORK THEOREMS AND CONVERSIONS

There are some special-case network theorems and circuit conversions that you may need to use now and then. While these theorems and conversions are not used as often as the others we've discussed, there are occasions where they can greatly simplify the analysis of a circuit.

### Millman's Theorem

◄ *OBJECTIVE 16*

**Millman's theorem** states that any number of voltage sources connected in parallel can be represented as a single voltage source in series with a source resistance. The value of this theorem is demonstrated in Figure 8.58.

**Millman's theorem**
Any number of voltage sources connected in parallel can be represented as a single voltage source in series with a single resistance.

In Figure 8.58a, we have a group of voltage sources, each with its own values of $V_S$ and $R_S$. Connecting these sources in parallel with a load would give us the circuit shown in Figure 8.58b. As you know, parallel voltages must be equal. Therefore, there can be only one value of load voltage for the circuit in Figure 8.58b. Millman's theorem provides us with a means of determining the value of this voltage.

In essence, Millman's theorem states that a multisource circuit like the one in Figure 8.59 can be represented as shown in the figure. Note that the Millman values of voltage and resistance are labeled $V_m$ and $R_m$. Once the Millman equivalent of the original circuit is derived, it can be used to predict the value of $V_L$ in the same fashion as a Thevenin equivalent circuit.

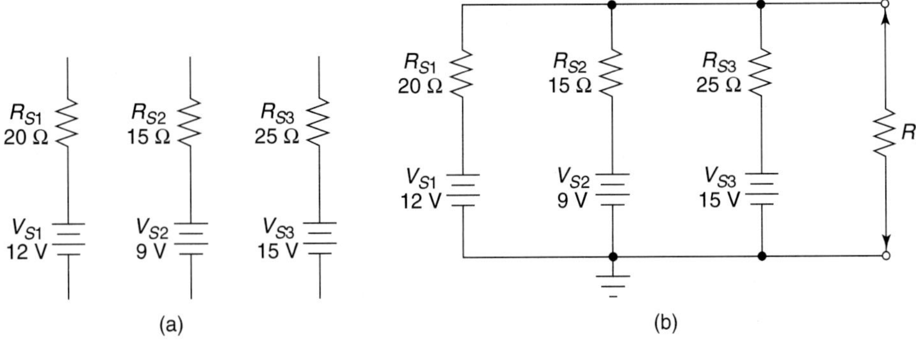


**FIGURE 8.58 Multiple sources connected in parallel.**

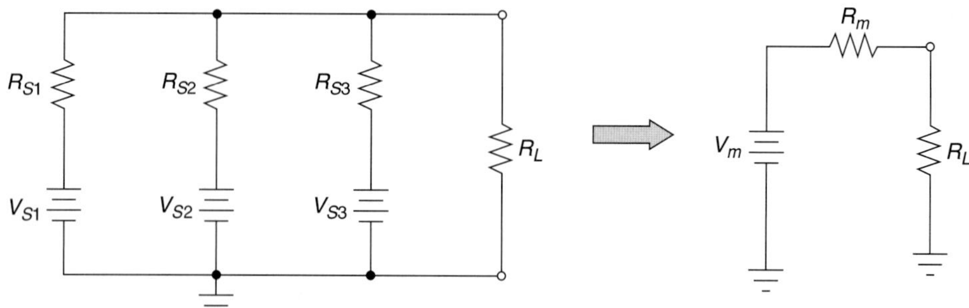


**FIGURE 8.59 A multisource circuit and its Millman equivalent.**

***Calculating the Values of $R_m$ and $V_m$.*** The Millman resistance for a circuit like the one in Figure 8.59 is actually very simple to calculate. If we remove the load and short out the voltage sources (as we always do for resistance calculations), we get the circuit shown in Figure 8.60. As you can see, this is nothing more than a group of parallel resistors. Therefore, $R_m$ is found using

$$R_m = R_{S1} \| R_{S2} \| \ldots \| R_{Sn} \tag{8.5}$$

where $R_{Sn}$ is the resistance of the highest numbered source.

The equation used to calculate the value of Millman voltage is as follows:

$$V_m = R_m\left(\frac{V_1}{R_1} + \frac{V_2}{R_2} + \ldots + \frac{V_n}{R_n}\right) \tag{8.6}$$

where $V_n$ and $R_n$ are the voltage and resistance ratings of the highest numbered branch in the circuit. The derivation of equation (8.6) is shown in Appendix E.

It is important to remember that equations (8.5) and (8.6) are both solved for the circuit with the load removed. Once the values of $R_m$ and $V_m$ are determined, the load is placed in the Millman equivalent circuit to determine the actual load voltage. Example 8.22 demonstrates the entire process.

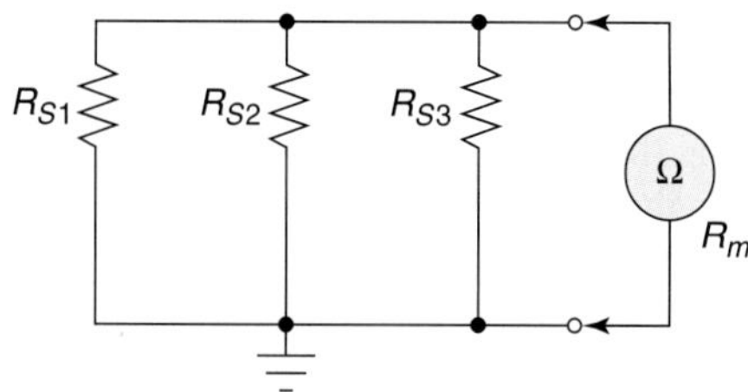


**FIGURE 8.60 Measuring the Millman equivalent resistance of the multisource circuit in Figure 8.59.**

## EXAMPLE 8.22

Determine the value of load voltage for the circuit in Figure 8.61a.

***Solution:*** First, we need to calculate the Millman resistance of the circuit. When we remove the load and short out the voltage sources, we get the circuit shown in Figure 8.61b. For this circuit,

$$\mathbf{R_m} = R_{S1} \| R_{S2} \| R_{S3} = 30\ \Omega \| 30\ \Omega \| 60\ \Omega = \mathbf{12\ \Omega}$$

Now, the Millman voltage for the circuit can be found as

$$\mathbf{V_m} = R_m\left(\frac{V_1}{R_1} + \frac{V_2}{R_2} + \frac{V_3}{R_3}\right) = (12\ \Omega)\left(\frac{3\ \text{V}}{30\ \Omega} + \frac{9\ \text{V}}{30\ \Omega} + \frac{12\ \text{V}}{60\ \Omega}\right) = \mathbf{7.2\ V}$$

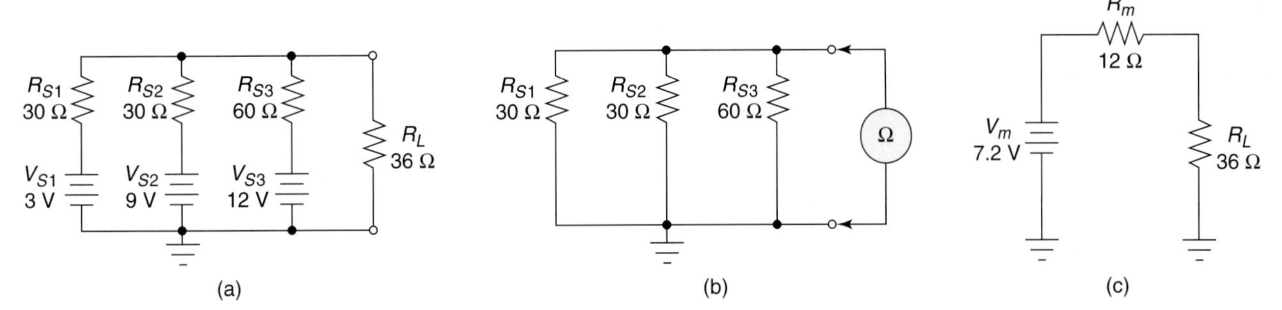


**FIGURE 8.61**

Using the calculated values of $V_m$ and $R_m$, the Millman equivalent circuit is constructed as shown in Figure 8.61c. For the load resistance shown,

$$V_L = V_m \frac{R_L}{R_L + R_m} = (7.2\text{ V})\frac{36\ \Omega}{48\ \Omega} = \mathbf{5.4\ V}$$

***PRACTICE PROBLEM 8.22***

Determine the value of load voltage for the circuit shown in Figure 8.62.

**FIGURE 8.62**

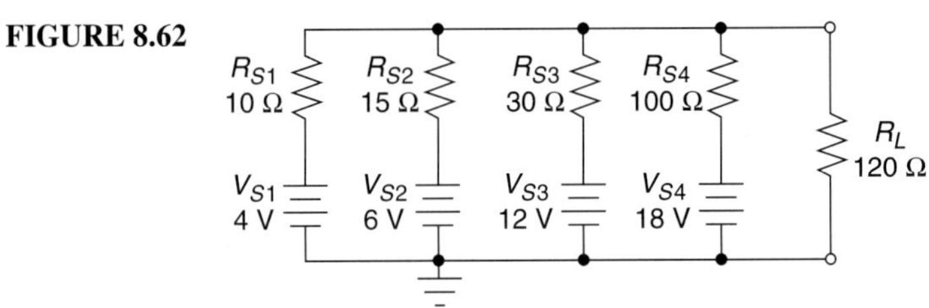


In the following example, one of the voltage sources is negative, and one of the branches contains only a resistance. Pay close attention to how these two branches affect the calculations of the Millman voltage and resistance.

EXAMPLE 8.23

Determine the value of load voltage for the circuit shown in Figure 8.63a.

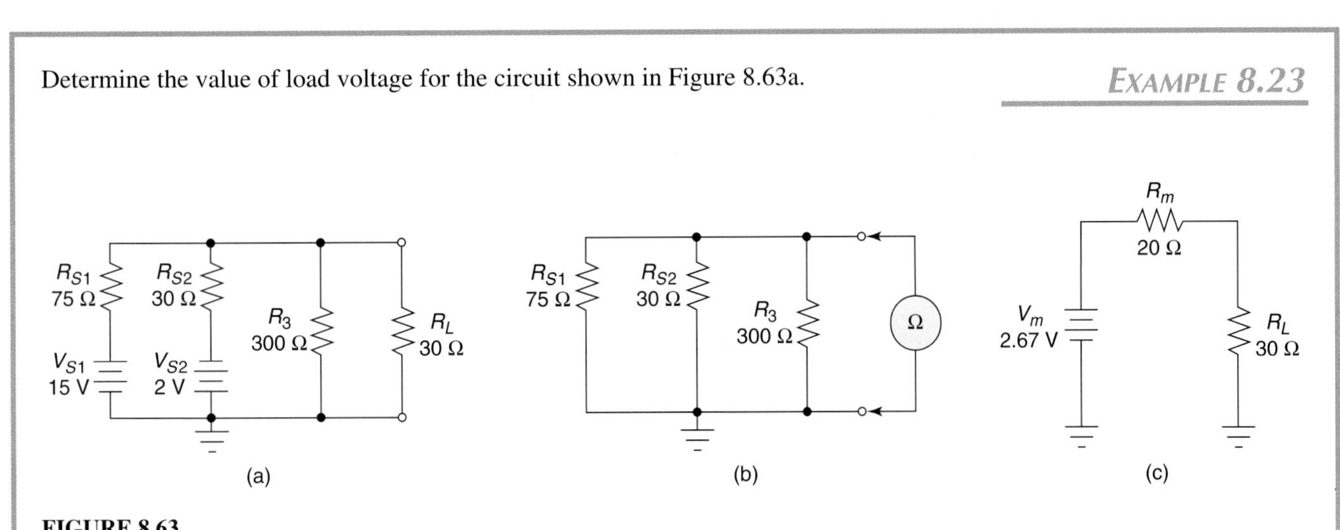


**FIGURE 8.63**

***Solution:*** Again, we'll start by calculating the Millman resistance of the circuit. When we remove the load and short out the voltage sources, we get the circuit shown in Figure 8.63b. For this circuit,

$$R_m = R_{S1} \| R_{S2} \| R_3 = 75\ \Omega \| 30\ \Omega \| 300\ \Omega = \mathbf{20\ \Omega}$$

Now, the value of $V_m$ can be found as

$$V_m = R_m\left(\frac{V_1}{R_1} + \frac{V_2}{R_2} + \frac{V_3}{R_3}\right) = (20\ \Omega)\left(\frac{15\text{ V}}{75\ \Omega} + \frac{-2\text{ V}}{30\ \Omega} + \frac{0\text{ V}}{300\ \Omega}\right) = \mathbf{2.67\ V}$$

Using the calculated values of $V_m$ and $R_m$, the Millman equivalent circuit is constructed as shown in Figure 8.63c. For the load resistance shown,

$$V_L = V_m \frac{R_L}{R_L + R_m} = (2.67\text{ V})\frac{30\ \Omega}{50\ \Omega} = \mathbf{1.6\ V}$$

***PRACTICE PROBLEM 8.23***

Calculate the value of load voltage for the circuit in Figure 8.64.

**FIGURE 8.64**

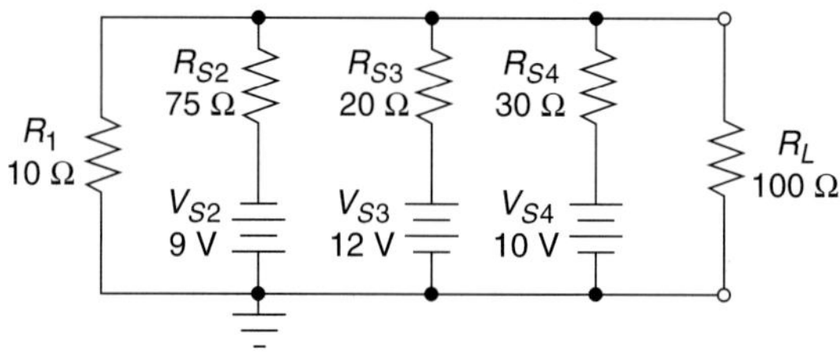


As the example demonstrated:

1. A negative voltage is entered as such in the $V_m$ equation.
2. A purely resistive branch is assumed to have a value of 0 V in the $V_m$ equation.

Once you have derived the Millman equivalent of a circuit, it can be used to solve any of the common load analysis problems, such as maximum power transfer, the response of the circuit to a change in load, and so on.

## Delta and Wye Circuits

*OBJECTIVE 17* ▶

*A Practical Consideration:* Delta and wye circuits are used extensively in ac power distribution applications.

Two series-parallel networks that we have ignored up until now are the *delta circuit* and the *wye circuit.* These circuits, which are shown in Figure 8.65a, are named for their resemblance to the Greek letter delta (Δ) and the letter Y. Sometimes, the circuits are drawn as shown in Figure 8.65b. When drawn as such, they are sometimes referred to as *Pi* (π) and *T* circuits. (Regardless of the names used, we're talking about the same two circuits.)

For every delta (Δ) circuit, there exists an equivalent wye (Y) circuit. This point is illustrated in Figure 8.66. Note that:

1. The circuits have been drawn in π and T form to help you follow the calculations shown in the figure.
2. The circuits have identical source voltages connected across their *x* and *z* terminals.
3. Each circuit has a load connected to the *y* and *z* terminals.

The calculations in the figure are those required to derive the Thevenin equivalent of each circuit. As you can see, the circuits have identical Thevenin equivalents. Therefore, each circuit must be the equivalent of the other.

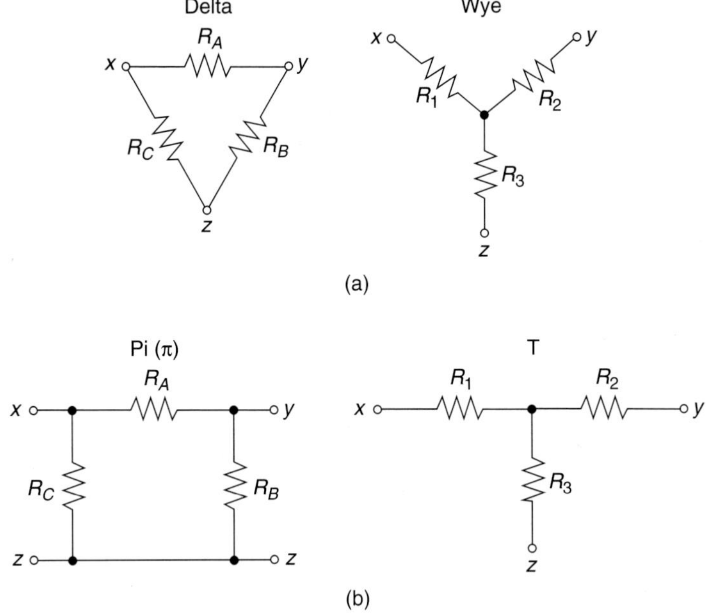


**FIGURE 8.65 Delta and wye circuits (with π and T representations).**

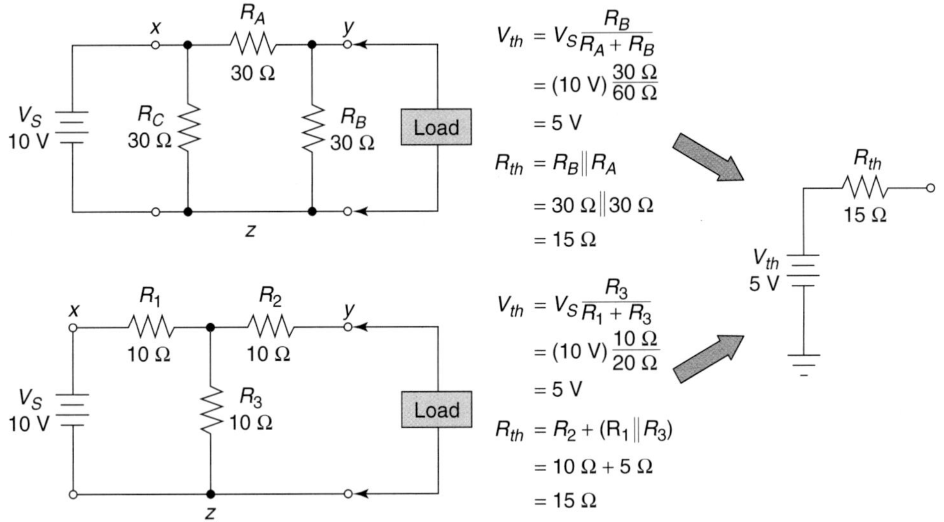


**FIGURE 8.66 Equivalent π and T circuits.**

***Circuit Conversions.*** Some circuit analysis problems can be made easier by converting a Δ circuit to a Y circuit, or vice versa. When you have a Δ circuit like the one in Figure 8.67, the values of the resistors in the equivalent Y circuit can be found using the following relationships: *OBJECTIVE 18*

$$R_1 = \frac{R_A R_C}{R_A + R_B + R_C} \tag{8.7}$$

$$R_2 = \frac{R_A R_B}{R_A + R_B + R_C} \tag{8.8}$$

and

$$R_3 = \frac{R_B R_C}{R_A + R_B + R_C} \tag{8.9}$$

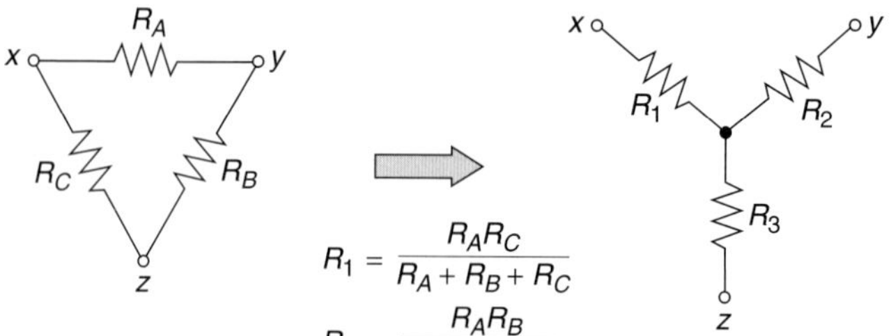

$$R_1 = \frac{R_A R_C}{R_A + R_B + R_C}$$

$$R_2 = \frac{R_A R_B}{R_A + R_B + R_C}$$

$$R_3 = \frac{R_B R_C}{R_A + R_B + R_C}$$

**FIGURE 8.67 Δ-to-Y conversion.**

These relationships are derived in Appendix E. Example 8.24 demonstrates the conversion of a Δ circuit to a Y circuit.

## EXAMPLE 8.24

Convert the circuit shown in Figure 8.68a to a Y circuit.

EWB

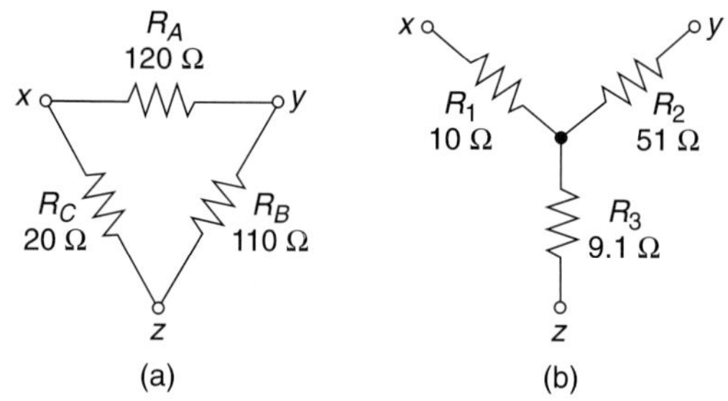

(a) $R_A$ 120 Ω, $R_B$ 110 Ω, $R_C$ 20 Ω

(b) $R_1$ 10 Ω, $R_2$ 51 Ω, $R_3$ 9.1 Ω

**FIGURE 8.68**

***Solution:*** The value of $R_1$ is found as

$$\boldsymbol{R_1} = \frac{R_A R_C}{R_A + R_B + R_C} = \frac{(120\ \Omega)(20\ \Omega)}{250\ \Omega} = \mathbf{9.6\ \Omega}$$

The value of $R_2$ is found as

$$\boldsymbol{R_2} = \frac{R_A R_B}{R_A + R_B + R_C} = \frac{(120\ \Omega)(110\ \Omega)}{250\ \Omega} = \mathbf{52.8\ \Omega}$$

The value of $R_3$ is found as

$$\boldsymbol{R_3} = \frac{R_B R_C}{R_A + R_B + R_C} = \frac{(110\ \Omega)(20\ \Omega)}{250\ \Omega} = \mathbf{8.8\ \Omega}$$

Using standard-value resistors, the Y circuit would be constructed as shown in Figure 8.68b.

### *PRACTICE PROBLEM 8.24*

A circuit like the one in Figure 8.68a has the following values: $R_A = 200\ \Omega$, $R_B = 120\ \Omega$, and $R_C = 30\ \Omega$. Calculate the values of the resistors for the equivalent Y circuit. Draw the equivalent circuit using the closest standard-value resistors.

The relationships given in equations (8.7) through (8.9) would be difficult to commit to memory, but there is a pattern that will make them relatively easy to remember. This pattern can be seen when we draw the Δ and Y circuits as shown in Figure 8.69. First, note that each relationship contains the sum of the three resistors in its denominator. The pattern used in the numerator of each relationship is as follows: each numerator contains the product of the two Δ resistors adjacent to the Y resistor. For example, the resistors in Figure 8.69 are positioned as follows:

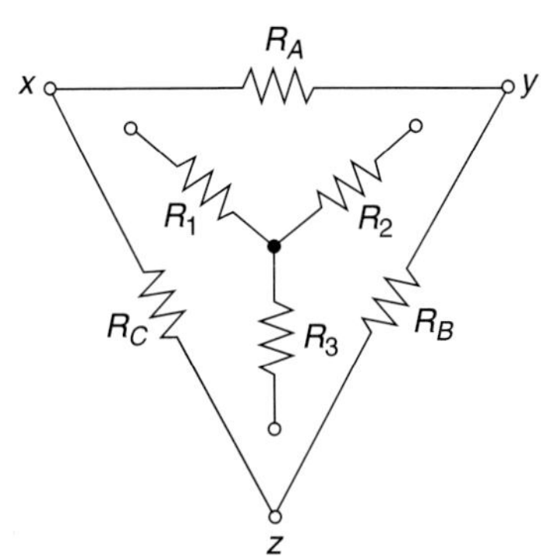


FIGURE 8.69

1. $R_1$ is adjacent to $R_A$ and $R_C$.
2. $R_2$ is adjacent to $R_A$ and $R_B$.
3. $R_3$ is adjacent to $R_B$ and $R_C$.

When you compare these statements to the relationships given, you can see the pattern in the numerator values.

When you have a Y circuit, the values of the resistors in the equivalent Δ circuit can be found using the following relationships:

$$R_A = \frac{R_1R_2 + R_2R_3 + R_1R_3}{R_3} \tag{8.10}$$

$$R_B = \frac{R_1R_2 + R_2R_3 + R_1R_3}{R_1} \tag{8.11}$$

and

$$R_C = \frac{R_1R_2 + R_2R_3 + R_1R_3}{R_2} \tag{8.12}$$

These relationships are also derived in Appendix E.

As before, there is a pattern in the conversion formulas that can be seen with the aid of Figure 8.69. First, the numerator of each equation contains the sum of the three possible two-resistor products. The denominator of each equation contains the resistance in the Y circuit that is opposite the resistor in the Δ circuit. For example, $R_3$ is opposite $R_A$ in Figure 8.69, and its value appears in the denominator for the $R_A$ equation. The same pattern can be seen by comparing the conversion formulas to the resistor positions in Figure 8.69. Example 8.25 demonstrates the conversion of a Y circuit to an equivalent Δ circuit.

## EXAMPLE 8.25

In the last example, we converted a Δ circuit to a Y circuit. We then showed the Y circuit as it would be built using standard-value resistors. Using the standard component values shown in Figure 8.68b, convert the circuit back to Δ form to see how close our actual circuit came to the original.

***Solution:*** Using the standard value components shown, the value of $R_A$ is found as

$$R_A = \frac{R_1R_2 + R_2R_3 + R_1R_3}{R_3}$$

$$= \frac{(10\ \Omega)(51\ \Omega) + (51\ \Omega)(9.1\ \Omega) + (10\ \Omega)(9.1\ \Omega)}{9.1\ \Omega}$$

$$\cong 117\ \Omega$$

The value of $R_B$ is found as

$$R_B = \frac{R_1R_2 + R_2R_3 + R_1R_3}{R_1}$$

$$= \frac{(10\ \Omega)(51\ \Omega) + (51\ \Omega)(9.1\ \Omega) + (10\ \Omega)(9.1\ \Omega)}{10\ \Omega}$$

$$\cong 107\ \Omega$$

Finally, the value of $R_C$ is found as

$$R_C = \frac{R_1R_2 + R_2R_3 + R_1R_3}{R_2}$$

$$= \frac{(10\ \Omega)(51\ \Omega) + (51\ \Omega)(9.1\ \Omega) + (10\ \Omega)(9.1\ \Omega)}{51\ \Omega}$$

$$\cong 21\ \Omega$$

When you compare these values to the circuit in Figure 8.68, you can see that our standard-value equivalent (Figure 8.68b) would work in place of the original circuit.

***PRACTICE PROBLEM 8.25***

Convert your standard-value Y circuit from Practice Problem 8.24 back to Δ form and compare your resistor values to the original values given in the problem.

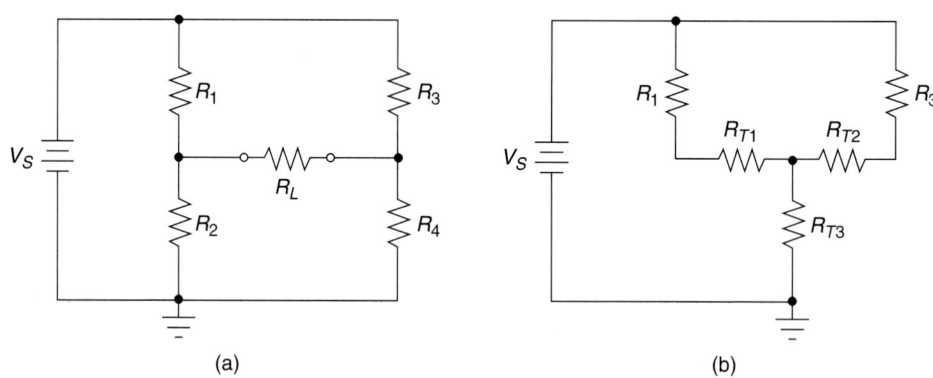


**FIGURE 8.70 Using Δ-to-Y conversion on a bridge circuit.**

***Δ-to-Y Conversion: An Application.*** You were shown earlier in this chapter how to analyze a bridge circuit using Thevenin's theorem. The analysis of a bridge circuit can be made even simpler using Δ-to-Y conversion, as shown in Figure 8.70. The combination of $R_2$, $R_L$, and $R_4$ in Figure 8.70a makes up the Δ circuit (drawn in the π form). When we convert this Δ circuit to a Y, we get the circuit shown in Figure 8.70b. As you can see, we now have two parallel branches in series with a single resistor. This circuit can be analyzed easily using the techniques we discussed in Chapter 7.

## One Final Note

As you were told earlier, Millman's theorem and the conversions shown in this section are not used that often. However, when you encounter situations like those shown in the examples, you will find the material in this section to be very useful.

### Section Review

1. What is Millman's theorem?
2. What does Millman's theorem have in common with Thevenin's theorem?
3. Describe the procedure used to derive the Millman equivalent of a multisource circuit.

4. Explain how each of the following is handled in a Millman conversion problem:
   a. A purely resistive branch, i.e., a branch containing only a resistor.
   b. A negative voltage source.
5. Describe the relationships used to:
   a. Derive the Y equivalent of a Δ circuit.
   b. Derive the Δ equivalent of a Y circuit.

## Chapter Summary

Here is a summary of the major points made in this chapter:

1. A multisource circuit is one that contains more than one voltage (or current) source. When analyzing a multisource circuit, the effects of all the sources must be taken into account.
2. The superposition theorem states that the response of a circuit to more than one source can be determined by analyzing the circuit's response to each source (alone) and combining the results.
3. The primary difference between the ideal and practical voltage source is the effect that source resistance has on load voltage.
   a. The output from the ideal voltage source remains constant, despite variations in load resistance.
   b. The output from a practical voltage source varies directly with changes in load resistance.
4. The primary difference between the ideal and practical current source is the effect that source resistance has on load current.
   a. The output from the ideal current source remains constant, despite variations in load resistance.
   b. The output from a practical current source varies inversely with changes in load resistance.
5. For every voltage source, there exists an equivalent current source (and vice versa).
   a. Two sources are considered to be equivalents when they provide the same output values for any given value of load resistance.
   b. Example 8.4 demonstrates the process for converting a voltage source to an equivalent current source.
   c. Example 8.5 demonstrates the process for converting a current source to an equivalent voltage source.
6. Analyzing the effect that a load has on the output from a given circuit is referred to as load analysis.
7. Thevenin's theorem states that any resistive circuit or network, no matter how complex, can be represented as a voltage source in series with a source resistance.
   a. Any series-parallel circuit can be represented as a Thevenin equivalent circuit. The components of this circuit are called the Thevenin voltage ($V_{th}$) and the Thevenin resistance ($R_{th}$).
   b. The strength of the Thevenin equivalent circuit lies in the fact that it produces the same output values as the original circuit for any given value of load resistance.
8. When using Thevenin's theorem, the load analysis of a series-parallel circuit is performed as follows:
   a. Derive the Thevenin equivalent for the series-parallel circuit.
   b. For each value of load resistance, determine the output from the Thevenin equivalent circuit using the voltage divider relationship. The values obtained equal the actual outputs from the original circuit.

9. To determine the value of $V_{th}$:
   a. Remove the load.
   b. Calculate the voltage across the open load terminals. This voltage is $V_{th}$.
10. Thevenin voltage ($V_{th}$) is the no-load output voltage ($V_{NL}$) of the original circuit.
11. To determine the value of $R_{th}$:
   a. Remove the load and replace the voltage source with a wire (to represent the ideal source resistance of 0 Ω).
   b. Calculate the resistance that would be measured across the open load terminals. This resistance is $R_{th}$.
12. Thevenin resistance ($R_{th}$) is the no-load output resistance ($R_{NL}$) of the original circuit.
13. Thevenin's theorem is most commonly used to predict the change in load voltage that will result from a change in load resistance.
   a. In many cases, it is desirable to know the normal range of output voltages for a circuit. When the range is known, it is easy to determine whether or not a voltage reading indicates a problem with the circuit.
   b. The range of load voltages for a circuit is found by determining the values of $V_L$ that correspond to the minimum and maximum values of $R_L$.
14. For a series-parallel circuit with a variable load, maximum power transfer occurs when $R_L$ is set to the value of $R_{th}$.
15. The load analysis of a multisource circuit is performed using Thevenin's theorem along with the superposition theorem (as demonstrated in Example 8.14).
16. In a multiload circuit, no two loads have the same Thevenin equivalent circuit.
   a. To derive the Thevenin equivalent circuit for each load, you must assume fixed values of resistance for the other loads.
   b. The Thevenin analysis of a multiload circuit is demonstrated in Example 8.16.
17. Thevenin's theorem can be used to analyze the operation of a bridge circuit, as demonstrated in Example 8.17.
18. Norton's theorem states that any resistive circuit or network, no matter how complex, can be represented as a current source in parallel with a source resistance.
   a. Any series-parallel circuit can be represented as a Norton equivalent circuit. The components of this circuit are called the Norton current ($I_n$) and the Norton resistance ($R_n$).
   b. In essence, Norton's theorem is a current-source version of Thevenin's theorem.
19. To determine the value of $I_n$:
   a. Short the load terminals.
   b. Calculate the value of current through the shorted load terminals. This current is $I_n$.
20. The value of $R_n$ is found using the same procedure as the one given for finding $R_{th}$. (See number 11 in this summary.)
21. Load analysis using Norton's theorem is demonstrated in Example 8.20.
22. For every Norton equivalent circuit, there is a matching Thevenin equivalent circuit (and vice versa).
23. Using the appropriate source conversion technique, a Thevenin equivalent circuit can be converted to a Norton equivalent circuit (and vice versa).
24. Millman's theorem states that any number of voltage sources connected in parallel can be represented as a single voltage source in series with a source resistance. Once the Millman equivalent of a multisource circuit is derived, it can be used to determine the value of load voltage in the same manner as a Thevenin equivalent circuit.

25. Delta and wye circuits are named for their resemblance to the Greek letter delta (Δ) and the letter Y. When drawn as shown in Figure 8.65b, the circuits are commonly referred to as Pi (π) and T circuits.

26. For every Δ circuit, there exists an equivalent Y circuit (and vice versa).
    a. Δ-to-Y conversion is demonstrated in Example 8.24.
    b. Y-to-Δ conversion is demonstrated in Example 8.25.

## Equation Summary

| Equation Number | Equation | Section Number |
|---|---|---|
| (8.1) | $I_S = \frac{V_S}{R_S}$ | 8.2 |
| (8.2) | $V_S = I_S R_S$ | 8.2 |
| (8.3) | $V_{th} = \lvert V_A - V_B \rvert$ | 8.4 |
| (8.4) | $R_{th} = (R_1 \| R_2) + (R_3 \| R_4)$ | 8.4 |
| (8.5) | $R_m = R_{S1} \| R_{S2} \| \ldots \| R_{Sn}$ | 8.6 |
| (8.6) | $V_m = R_m\left(\frac{V_1}{R_1} + \frac{V_2}{R_2} + \ldots + \frac{V_n}{R_n}\right)$ | 8.6 |
| (8.7)* | $R_1 = \frac{R_A R_C}{R_A + R_B + R_C}$ | 8.6 |
| (8.8)* | $R_2 = \frac{R_A R_B}{R_A + R_B + R_C}$ | 8.6 |
| (8.9)* | $R_3 = \frac{R_B R_C}{R_A + R_B + R_C}$ | 8.6 |
| (8.10)* | $R_A = \frac{R_1 R_2 + R_2 R_3 + R_1 R_3}{R_3}$ | 8.6 |
| (8.11)* | $R_B = \frac{R_1 R_2 + R_2 R_3 + R_1 R_3}{R_1}$ | 8.6 |
| (8.12)* | $R_C = \frac{R_1 R_2 + R_2 R_3 + R_1 R_3}{R_2}$ | 8.6 |

*See the circuit model in Figure 8.69 for component positions.

## Key Terms

The following terms were introduced and defined in this chapter:

delta (Δ) circuit
load analysis
Millman equivalent circuit
Millman's theorem
multisource circuit
network theorem
Norton current ($I_n$)
Norton equivalent circuit
Norton resistance ($R_n$)
Norton's theorem
pi (π) circuit
superposition theorem
T circuit
Thevenin equivalent circuit
Thevenin resistance ($R_{th}$)
Thevenin's theorem
Thevenin voltage ($V_{th}$)
wye (Y) circuit

## Practice Problems

1. Determine the values of $V_1$, $V_2$, and $V_3$ for the circuit in Figure 8.71a.
2. Determine the values of $V_1$, $V_2$, and $V_3$ for the circuit in Figure 8.71b.
3. Determine the values of $V_1$, $V_2$, and $V_3$ for the circuit in Figure 8.72a.
4. Determine the values of $V_1$, $V_2$, and $V_3$ for the circuit in Figure 8.72b.
5. Determine the values of $V_1$ through $V_4$ for the circuit in Figure 8.73a.
6. Determine the values of $V_1$ through $V_4$ for the circuit in Figure 8.73b.
7. Using a value of $R_L = 100\ \Omega$, determine whether or not the circuits in Figure 8.74a are equivalent circuits.
8. Using a value of $R_L = 2.2\ \text{k}\Omega$, determine whether or not the circuits in Figure 8.74b are equivalent circuits.
9. Convert the voltage source in Figure 8.75a to an equivalent current source.

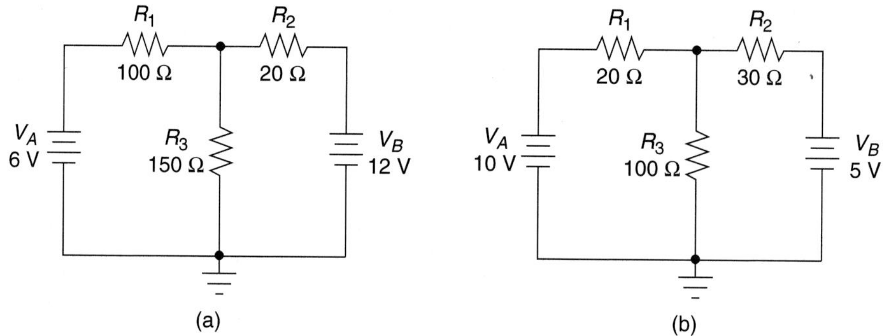


**FIGURE 8.71**

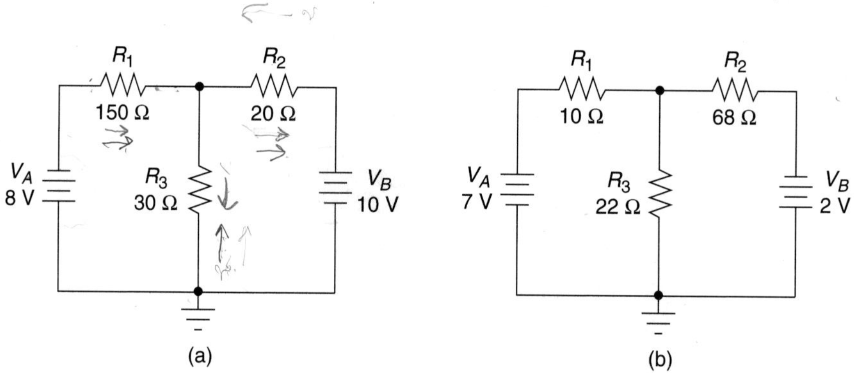


**FIGURE 8.72**

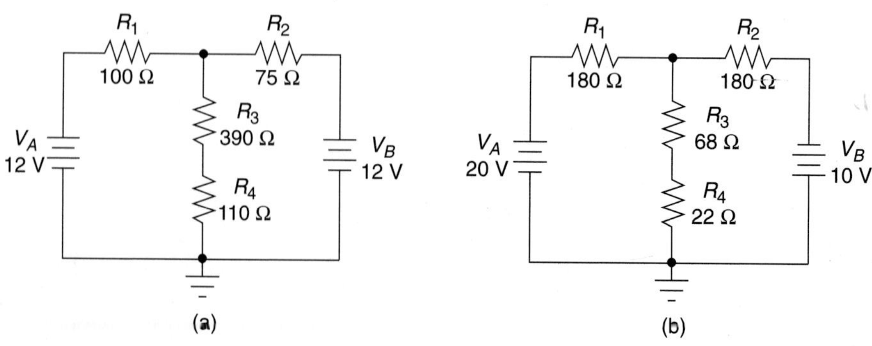


**FIGURE 8.73**

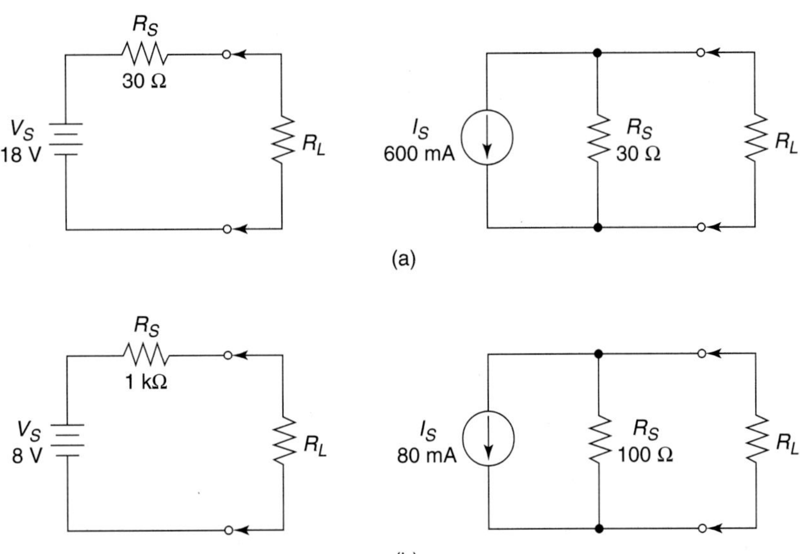


**FIGURE 8.74**

10. Convert the voltage source in Figure 8.75b to an equivalent current source.
11. Derive the current source equivalent of the voltage source in Figure 8.75c. Then calculate the values of $V_L$ produced by each circuit for values of $R_L = 100\ \Omega$ and $R_L = 330\ \Omega$.
12. Derive the current source equivalent of the voltage source in Figure 8.75d. Then calculate the values of $I_L$ produced by each circuit for values of $R_L = 180\ \Omega$ and $R_L = 1\ \text{k}\Omega$.
13. Convert the current source in Figure 8.76a to an equivalent voltage source.
14. Derive the voltage source equivalent of the current source in Figure 8.76b. Then calculate the values of $V_L$ produced by each circuit for values of $R_L = 75\ \Omega$ and $R_L = 150\ \Omega$.
15. Derive the Thevenin equivalent of the circuit in Figure 8.77a.

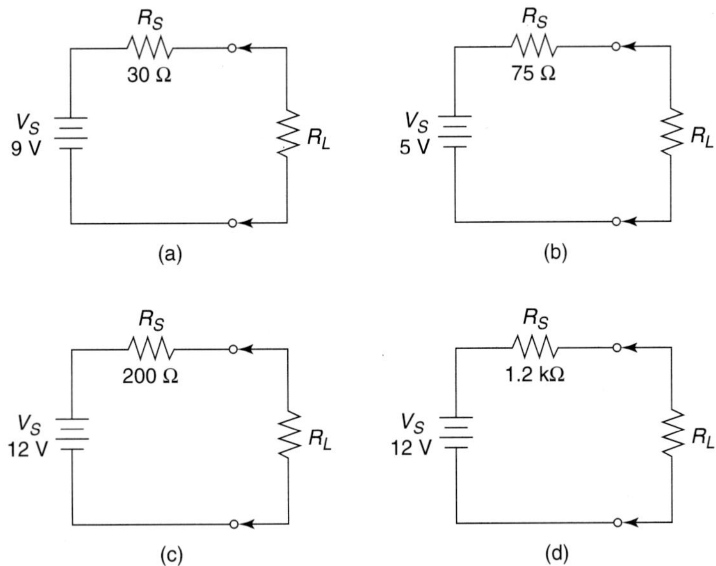


**FIGURE 8.75**

FIGURE 8.76

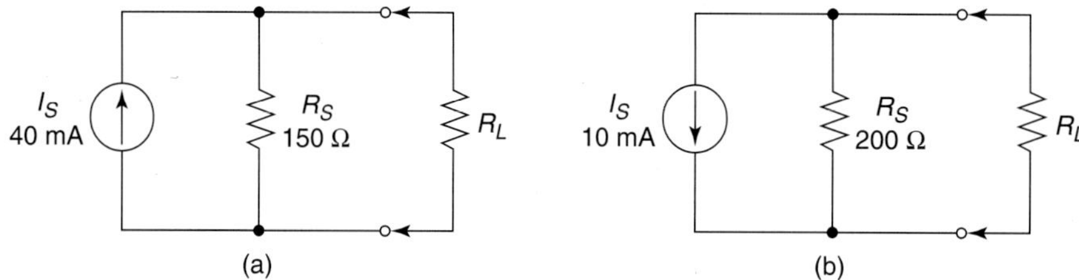


FIGURE 8.77

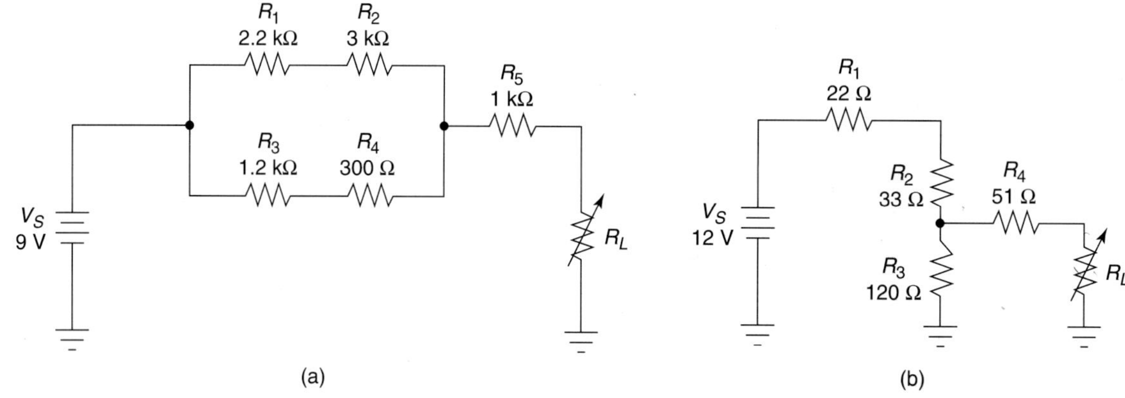


16. Derive the Thevenin equivalent of the circuit in Figure 8.77b.
17. Derive the Thevenin equivalent of the circuit in Figure 8.78a.
18. Derive the Thevenin equivalent of the circuit in Figure 8.78b.
19. The circuit in Figure 8.79a has a load with a range of $R_L$ = 100 Ω to 1 kΩ. Using the Thevenin equivalent of the circuit, determine its normal range of load voltage values.
20. The circuit in Figure 8.79b has a load with a range of $R_L$ = 10 Ω to 100 Ω. Using the Thevenin equivalent of the circuit, determine its normal range of load voltage values.

FIGURE 8.78

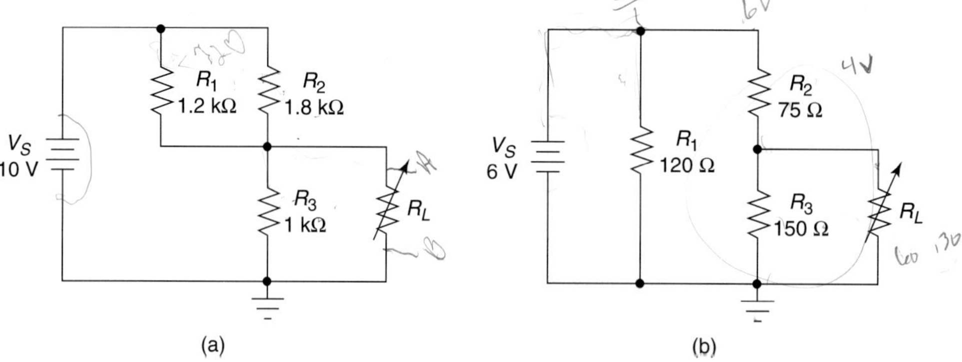


FIGURE 8.79

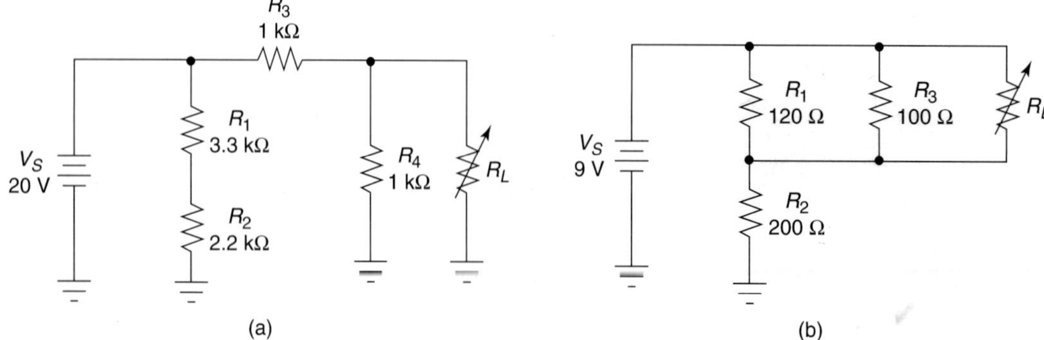

FIGURE 8.80

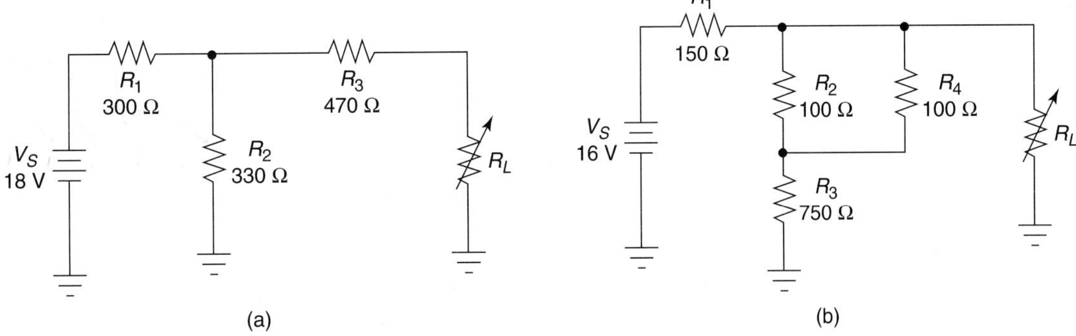


21. Calculate the maximum possible load power for the circuit in Figure 8.80a.
22. Calculate the maximum possible load power for the circuit in Figure 8.80b.
23. Calculate the normal range of output voltage values for the circuit in Figure 8.81a.
24. Calculate the normal range of output voltage values for the circuit in Figure 8.81b.
25. Calculate the maximum possible load power for the circuit in Figure 8.82a.
26. Calculate the maximum possible load power for the circuit in Figure 8.82b.
27. Derive the Thevenin equivalent circuit for load 1 in Figure 8.83.
28. Derive the Thevenin equivalent circuit for load 2 in Figure 8.83.
29. Determine the maximum possible load power for the bridge circuit in Figure 8.84a.
30. Determine the maximum possible load power for the bridge circuit in Figure 8.84b.
31. Derive the Norton equivalent for the circuit in Figure 8.77a.

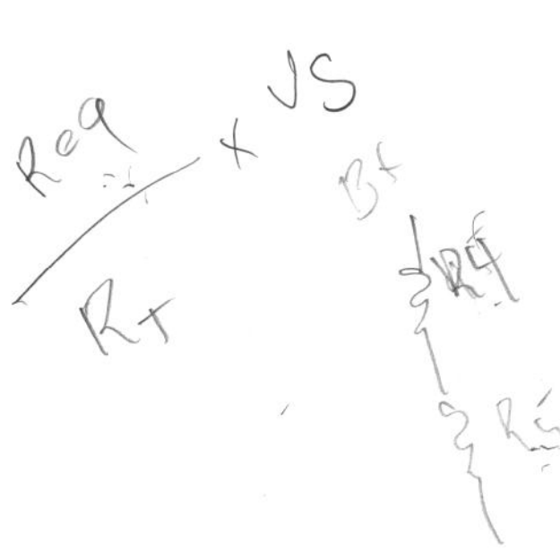

FIGURE 8.81

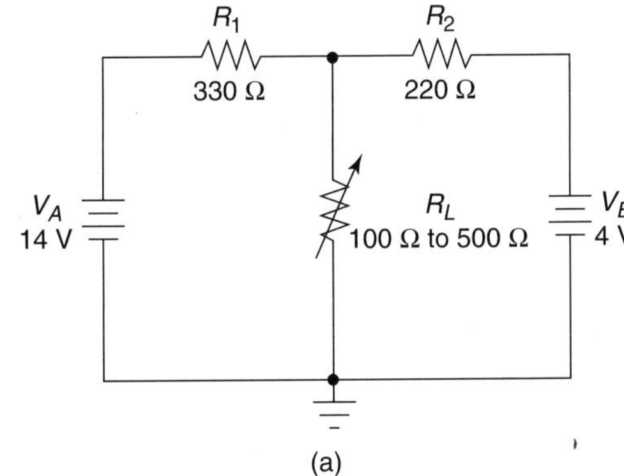


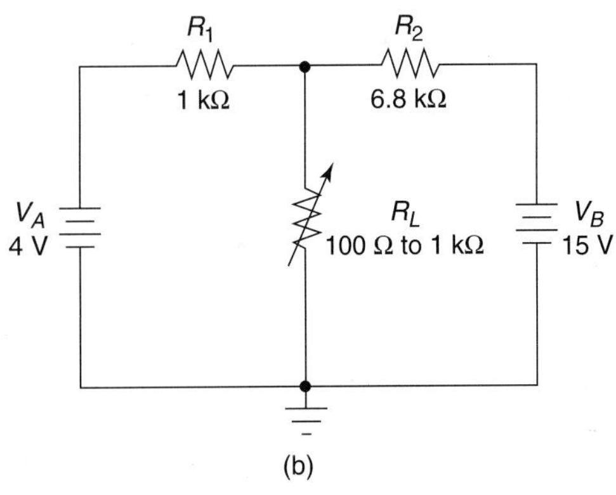


FIGURE 8.82

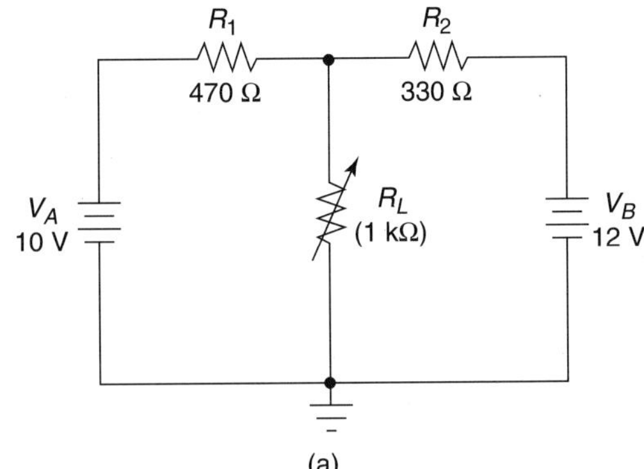


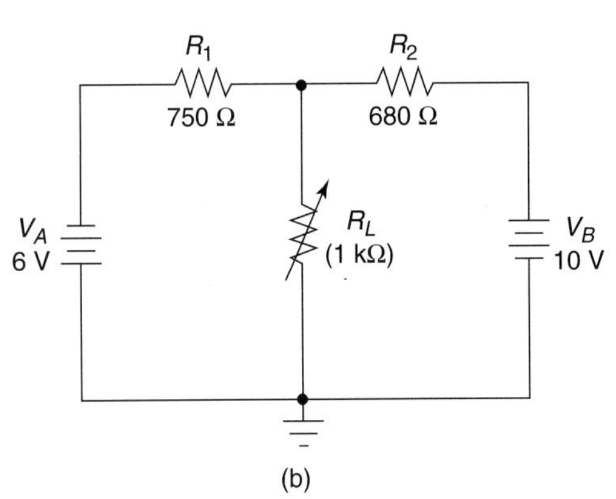

FIGURE 8.83

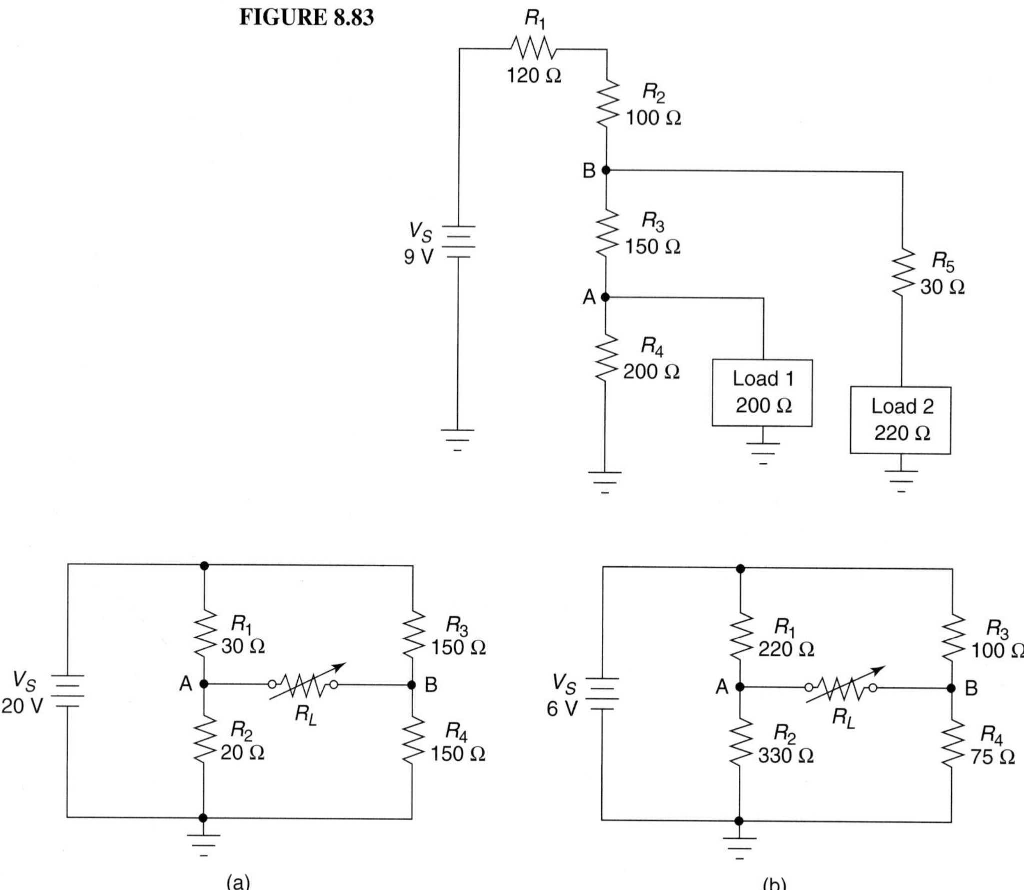


FIGURE 8.84

**32.** Derive the Norton equivalent for the circuit in Figure 8.77b.

**33.** Derive the Norton equivalent for the circuit in Figure 8.78a.

**34.** Derive the Norton equivalent for the circuit in Figure 8.78b.

**35.** Using the Norton equivalent for the circuit in Figure 8.85a, determine its normal range of load current values.

**36.** Using the Norton equivalent for the circuit in Figure 8.85b, determine its normal range of load current values.

**37.** Determine the value of $V_L$ for the circuit in Figure 8.86a.

**38.** Determine the value of $V_L$ for the circuit in Figure 8.86b.

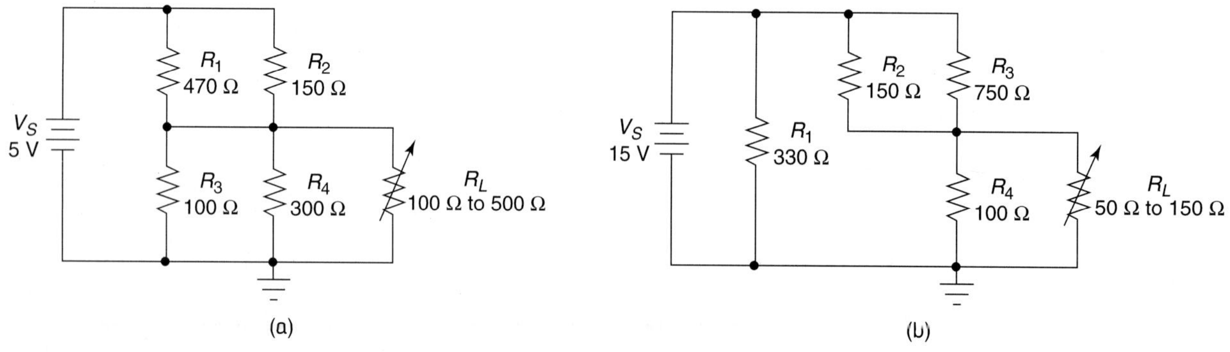


FIGURE 8.85

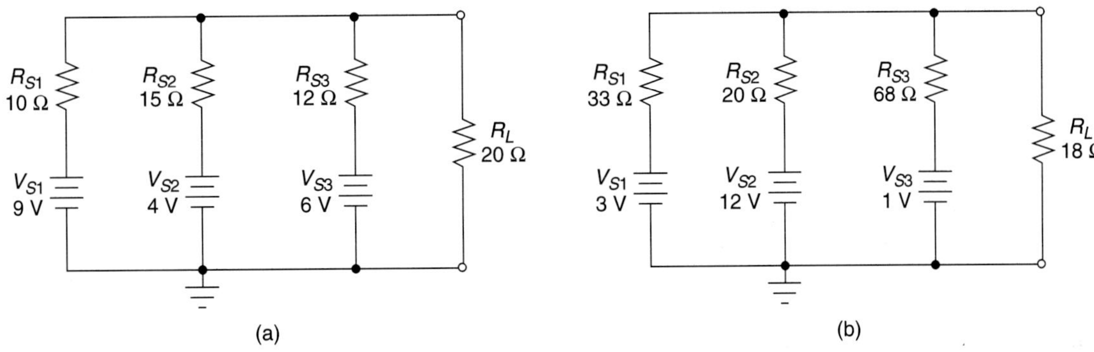


**FIGURE 8.86**

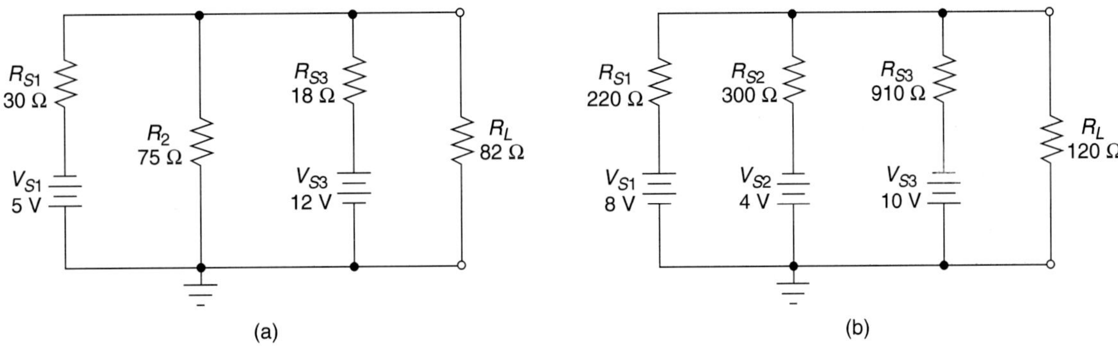


**FIGURE 8.87**

**39.** Determine the value of $V_L$ for the circuit in Figure 8.87a.

**40.** Determine the value of $V_L$ for the circuit in Figure 8.87b.

**41.** Determine the maximum possible load power for the circuit in Figure 8.88a.

**42.** Determine the normal range of output voltages for the circuit in Figure 8.88b.

**FIGURE 8.88**

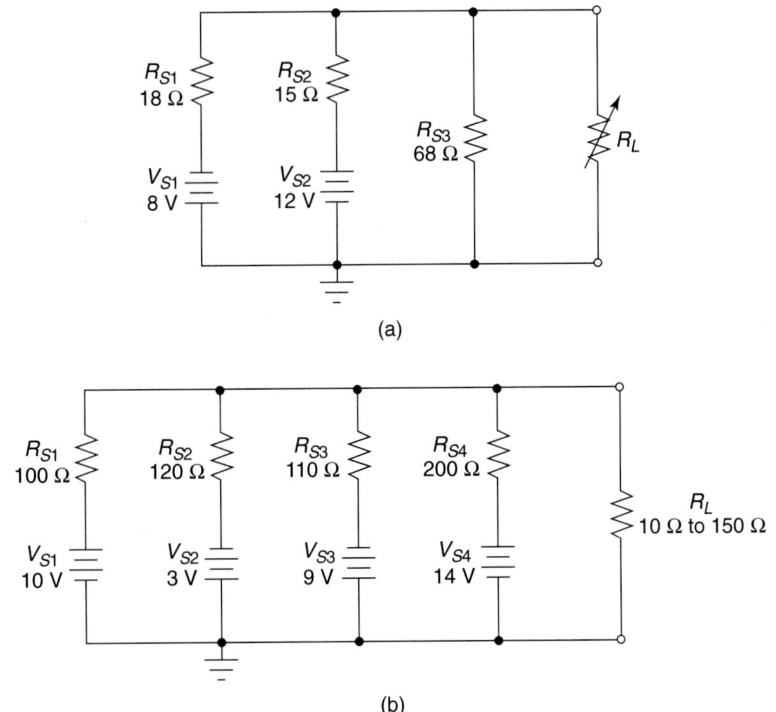

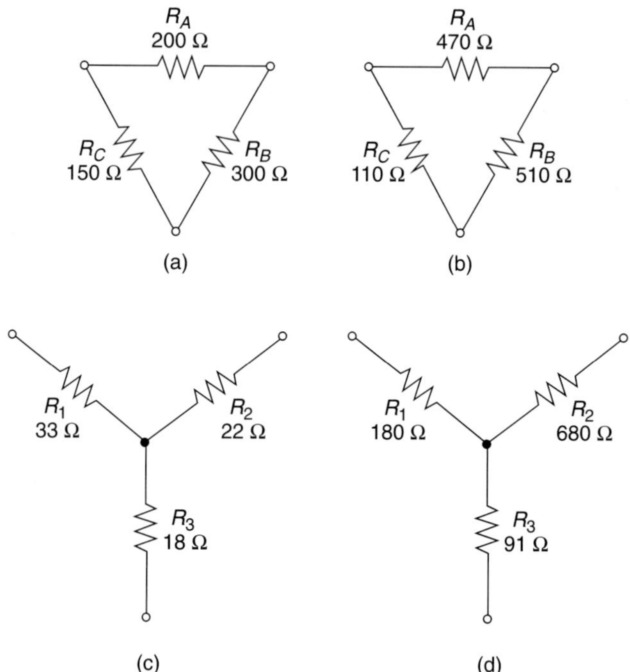


**FIGURE 8.89**

**43.** Perform a Δ-to-Y conversion on the circuit in Figure 8.89a.

**44.** Perform a Δ-to-Y conversion on the circuit in Figure 8.89b.

**45.** Perform a Y-to-Δ conversion on the circuit in Figure 8.89c.

**46.** Perform a Y-to-Δ conversion on the circuit in Figure 8.89d.

## The Brain Drain

**47.** Calculate the range of $V_L$ values for each load in Figure 8.90.

**48.** Calculate the range of load voltage values for the circuit in Figure 8.91.

**49.** Determine the maximum possible load power for the circuit shown in Figure 8.92.

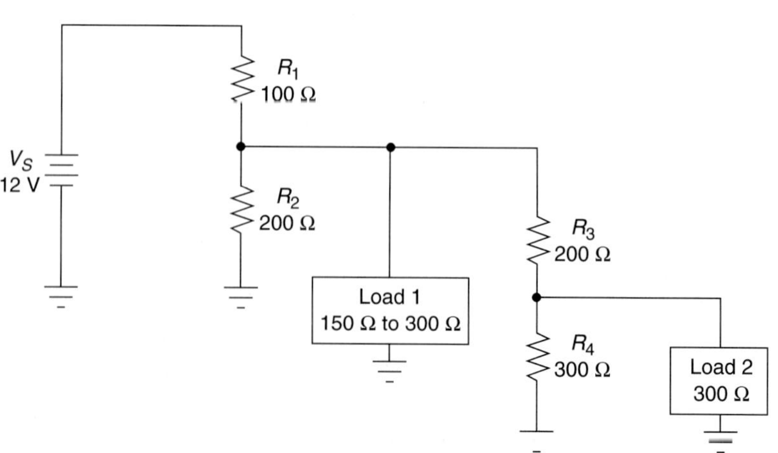


**FIGURE 8.90**

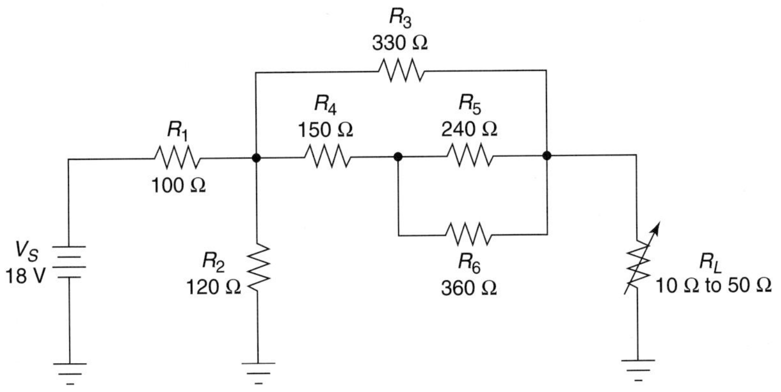


**FIGURE 8.91**

**FIGURE 8.92**

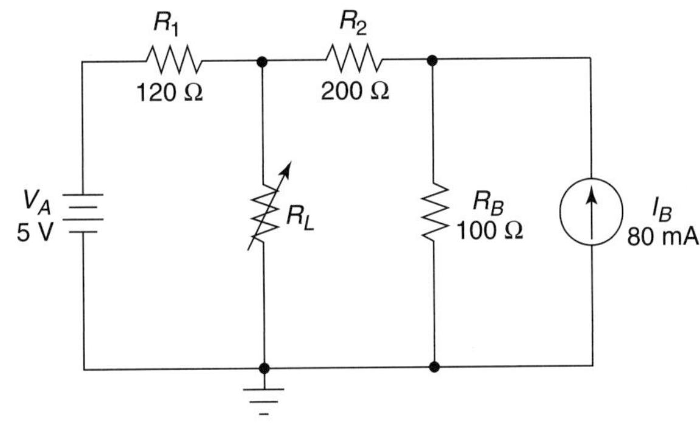


## Answers to the Example Practice Problems

**8.1.** $V_1 = 6$ V, $V_2 = 1$ V, $V_3 = 8$ V
**8.2.** $V_1 = 14$ V, $V_2 = 14$ V, $V_3 = 7$ V
**8.3.** For $R_L = 100\ \Omega$: $I_L = 50$ mA. For $R_L = 25\ \Omega$: $I_L = 80$ mA.
**8.4.** $I_S = 500$ mA, $R_S = 16\ \Omega$. For $R_L = 32\ \Omega$: $I_L = 167$ mA.
**8.5.** $V_S = 6$ V, $R_S = 150\ \Omega$. For $R_L = 100\ \Omega$: $V_L = 2.4$ V.
**8.6.** For $R_L = 150\ \Omega$: $V_L = 6.57$ V. For $R_L = 30\ \Omega$: $V_L = 4.86$ V.
**8.7.** 7.2 V
**8.8.** 14.4 Ω
**8.9.** $V_{th} = 10.9$ V, $R_{th} = 164\ \Omega$
**8.10.** $V_{th} = 4$ V, $R_{th} = 150\ \Omega$
**8.11.** $V_{th} = 7.5$ V, $R_{th} = 90\ \Omega$
**8.12.** 3.95 V to 6.36 V
**8.13.** 9 mW
**8.14.** When $R_L = 20\ \Omega$: $V_L = 7.94$ V. When $R_L = 50\ \Omega$: $V_L = 9.44$ V.
**8.15.** 4.05 W
**8.16.** 193 mW
**8.17.** When $R_L = 150\ \Omega$: $V_L = 1.15$ V. When $R_L = 220\ \Omega$: $V_L = 1.39$ V.
**8.18.** 6.7 mA
**8.19.** 1.2 kΩ
**8.20.** When $R_L = 100\ \Omega$: $I_L = 16$ mA. When $R_L = 150\ \Omega$: $I_L = 13.35$ mA.
**8.21.** 181.8 mW
**8.22.** 6.32 V
**8.23.** 3.94 V
**8.24.** $R_1 = 17.14\ \Omega$, $R_2 = 68.6\ \Omega$, and $R_3 = 10.28\ \Omega$. Use $R_1 = 18\ \Omega$, $R_2 = 68\ \Omega$, and $R_3 = 10\ \Omega$.
**8.25.** $R_A = 208\ \Omega$, $R_B = 115.8\ \Omega$, $R_C = 30.6\ \Omega$

## EWB Applications Problems

| Figure | EWB File References |
|---|---|
| 8.6 | EWBA8_6.ewb |
| 8.28 | EWBA8_28.ewb |
| 8.54 | EWBA8_54.ewb |
| 8.57 | EWBA8_57.ewb |
| 8.68 | EWBA8_68.ewb |

# 9 COMPLEX ANALYSIS TECHNIQUES

## OBJECTIVES

*After studying the material in this chapter, you should be able to:*

1. Solve simultaneous equations using the substitution method.
2. Solve simultaneous equations using the alter-and-add method.
3. Write the determinant matrix for a pair of two-variable simultaneous equations.
4. Write and solve the matrix equation for a 2-by-2 matrix.
5. Solve a pair of two-variable simultaneous equations using determinants.
6. Derive and solve 3-by-3 matrices.
7. Derive the branch current equations for a two-loop multisource circuit.
8. Solve for the currents in a two-loop multisource circuit using branch current analysis.
9. Compare and contrast branch current analysis and mesh current analysis.
10. Derive the mesh current equations for a multisource circuit.
11. Solve for the currents in a two-loop multisource circuit using mesh current analysis.
12. Solve a three-loop multisource circuit using determinants and mesh current analysis.
13. Compare and contrast node voltage analysis with the current methods.
14. Solve for the voltages in a two-loop multisource circuit using node voltage analysis.

In Chapter 8, you were introduced to the network theorems and conversions that are most commonly used to analyze the operation of multisource circuits. In this chapter, you will be introduced to some other analysis techniques that are used on multisource circuits. As you will see, some of these analysis techniques can be applied to circuits and networks that would be difficult to analyze using the methods covered in Chapter 8.

Two of the analysis techniques covered in this chapter involve the use of *simultaneous equations.* For this reason, we'll begin with a review of these equations and the methods most commonly used to solve them. For those who are already comfortable with simultaneous equations and their solutions, Section 9.1 can be skipped without loss of continuity.

## 9.1 SOLVING SIMULTANEOUS EQUATIONS

**Simultaneous equations**
Two or more equations used to describe relationships that occur at the same time. Any group of linear simultaneous equations has one common solution.

Two (or more) equations used to describe relationships that occur at the same time are referred to as **simultaneous equations.** For example, refer to the circuit shown in Figure 9.1. Kirchhoff's voltage law tells us that

$$V_A = V_1 + V_3 \qquad \text{and} \qquad V_B = V_2 + V_3$$

These two equations are considered to be simultaneous equations because:

1. They describe the operation of the same circuit.
2. They have a common solution.

The first statement indicates that you would never have a situation where the $V_A$ equation would be true and the $V_B$ equation would be false (or vice versa). For any set of values, both equations hold true or both are false.

Simultaneous equations like those for Figure 9.1 have only one common solution; that is, only one set of numbers will make both equations true. For example, in Chapter 8, we analyzed the circuit in Figure 9.2. At that time, we used the superposition theorem to calculate the voltage values shown in the figure. With the supply voltage and resistance values given, there is only one possible combination of $V_1$, $V_2$, and $V_3$ that will make both of the equations for the figure true. That combination of values is shown in the circuit.

When solving simultaneous equations, we are determining the single combination of values that will make all the equations true. There are several commonly used approaches

FIGURE 9.1

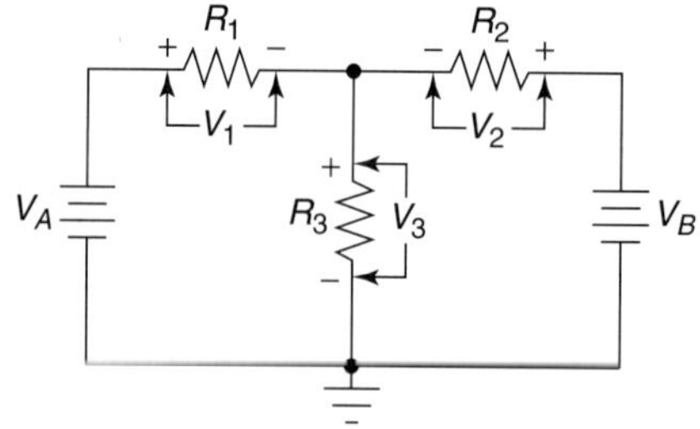


EWB

FIGURE 9.2

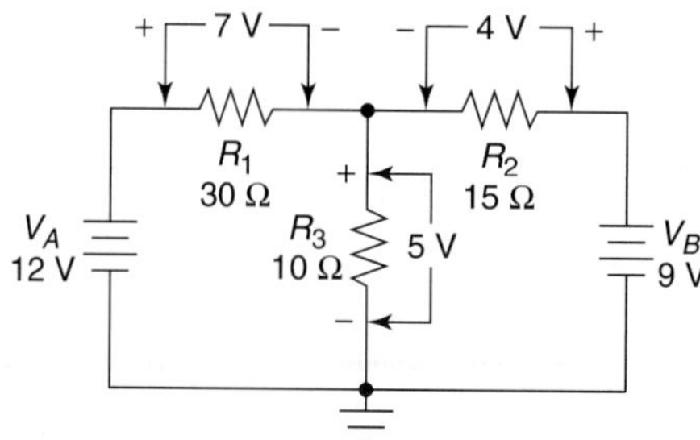

to solving simultaneous equations. We will discuss these approaches in this section. The rest of the chapter will be spent discussing the different methods used to derive simultaneous equations for multisource circuits.

## The Substitution Method

The **substitution method** involves combining two simultaneous equations so that they become one single-variable equation. To demonstrate this procedure, we'll assume that we are solving the following simultaneous equations:

◂ *OBJECTIVE 1*

**Substitution method**
A method of solving two simultaneous equations where the equations are combined so that they become one single-variable equation.

$$\text{Equation 1:} \quad x + y = 14$$
$$\text{Equation 2:} \quad 2x - y = 10$$

To find the common solution to these equations, the substitution method proceeds as follows.

**Step 1:** Rewrite equation 1 so that it "defines" one of the variables.

This means that we want to transpose the equation to isolate one of the variables. If we transpose the equation to isolate $x$, we get:

$$x = 14 - y$$

**Step 2:** Substitute the equation from Step 1 for the defined variable in equation 2.

In this step, we take (14 − y) and substitute it for $x$ in equation 2, as follows:

$$2x - y = 10$$
$$2(14 - y) - y = 10$$

**Step 3:** Solve the new single-variable equation.

$$2(14 - y) - y = 10$$
$$28 - 2y - y = 10$$
$$28 - 3y = 10$$
$$3y = 18$$
$$y = 6$$

**Step 4:** Substitute the value found in Step 3 into equation 1 and solve for the other variable.

Now that we know $y = 6$, we can solve for the value of $x$ as follows:

$$x + y = 14$$
$$x + 6 = 14$$
$$x = 8$$

We now know that the common solution to the two equations is $x = 8$ and $y = 6$. This is the only combination of values that will make *both* of the original equations true.

The substitution method of solving simultaneous equations is demonstrated further in Example 9.1.

**EXAMPLE 9.1**

The equations listed below describe the voltage relationships in a multisource circuit. Using the substitution method, determine the values of $V_1$ and $V_2$.

$$\text{Equation 1:} \quad 10V_1 + 5V_2 = 40\text{ V}$$
$$\text{Equation 2:} \quad V_1 + V_2 = 6\text{ V}$$

***Solution:*** First, we will transpose equation 2 to define $V_1$ as follows:

$$V_1 = 6\text{ V} - V_2$$

Now, we will substitute $(6\text{ V} - V_2)$ for $V_1$ (in equation 1) as follows:

$$10V_1 + 5V_2 = 40\text{ V}$$
$$10(6\text{ V} - V_2) + 5V_2 = 40\text{ V}$$

We can now solve for the value of $V_2$, as follows:

$$10(6\text{ V} - V_2) + 5V_2 = 40\text{ V}$$
$$60\text{ V} - 10V_2 + 5V_2 = 40\text{ V}$$
$$60\text{ V} - 5V_2 = 40\text{ V}$$
$$5V_2 = 20\text{ V}$$
$$V_2 = 4\text{ V}$$

Now that we know the value of $V_2$, we can use equation 2 to find the value of $V_1$, as follows:

$$V_1 + V_2 = 6\text{ V}$$
$$V_1 + 4\text{ V} = 6\text{ V}$$
$$V_1 = 2\text{ V}$$

Thus, the circuit voltage values are $V_1 = 2\text{ V}$ and $V_2 = 4\text{ V}$. Again, this is the only combination of $V_1$ and $V_2$ that makes both of the equations true.

***PRACTICE PROBLEM 9.1***

A multisource circuit is described by the following simultaneous equations: $V_1 - V_2 = 8\text{ V}$ and $V_1 + 3V_2 = 36\text{ V}$. Using the substitution method, determine the values of $V_1$ and $V_2$.

The substitution method is the simplest approach to solving simultaneous equations where one variable is easily defined by the other. However, it can get a bit more difficult when the variables are not so easily defined. For example, consider the following equations:

| | |
|---|---|
| Equation 1: | $6x - 7y = 23$ |
| Equation 2: | $7x + 11y = 44$ |

If we were to transpose equation 1 to define $x$, we would get:

$$x = \frac{23 + 7y}{6}$$

Substituting $\frac{23 + 7y}{6}$ into equation 2 would produce a mathematical nightmare. Fortunately, there is a simpler method of solving more complex simultaneous equations like these.

## The Alter-and-Add Method

*OBJECTIVE 2* ▶ The alter-and-add method of solving simultaneous equations is based on an important principle: *when two equations are true, then the sum of those equations must also be true.* For example, assume that we have the following simultaneous equations:

$$4x + 3y = 10 \qquad \text{and} \qquad 7x + 6y = 19$$

We can add these two equations together to produce a third equation, as follows:

| | |
|---|---|
| Equation 1: | $4x + 3y = 10$ |
| Equation 2: | $7x + 6y = 19$ |
| Sum: | $11x + 9y = 29$ |

Assuming that the initial equations are true, then the sum of those equations ($11x + 9y = 29$) must also be true.

The principle described above is the basis of the **alter-and-add method** of solving simultaneous equations. This method involves adding two equations so that one of the variables is eliminated. This is normally accomplished by altering one (or both) of the equations. For example, let's say that we want to find the common solution to the following simultaneous equations:

**Alter-and-add method** A method of solving two simultaneous equations where the equations are added so that one of the variables is eliminated. This normally involves altering one (or both) of the equations.

Equation 1: $5x - 6y = 15$
Equation 2: $3x + 3y = 42$

The alter-and-add method of solving these equations proceeds as follows.

**Step 1:** Alter equation 2 to change the coefficient of $y$ from 3 to 6.

This is accomplished by multiplying both sides of the equation by (2), as follows:

$$(3x + 3y = 42) \times (2) \rightarrow 6x + 6y = 84$$

**Step 2:** Add the altered form of equation 2 to equation 1.

| | |
|---|---|
| Equation 1: | $5x - 6y = 15$ |
| Altered Equation 2: | $6x + 6y = 84$ |
| Sum: | $11x = 99$ |

**Step 3:** Solve for the variable in the sum equation.

$$11x = 99$$
$$x = 9$$

**Step 4:** Using the known value, solve for the remaining unknown value.

$$5x - 6y = 15$$
$$5(9) - 6y = 15$$
$$45 - 6y = 15$$
$$6y = 30$$
$$y = 5$$

We now know that the common solution to the two equations is $x = 9$ and $y = 5$. This is the only combination of values that will make both of the original equations true.

*A Practical Consideration:* When you have correctly solved for one variable, you could substitute that value into any of the step equations and it would work. However, if you made a mistake in altering (or adding) the equations, the result would be wrong. By substituting your results into both of the original equations, you'll ensure that your results are correct.

As you can see, equation 2 was altered so that the $y$ terms would cancel when the equations were added together.

The key to successfully using the alter-and-add method is learning to recognize how the equations can be altered to eliminate a variable. In the following examples, you will see a number of different situations and some alterations that can be used in solving the equations.

### *Example 9.2*

Determine the common solution to the following simultaneous equations:

Equation 1: $6x + 8y = 60$
Equation 2: $4x + 2y = 30$

***Solution:*** For these equations, we can get the $y$ terms to cancel if we multiply equation 2 by $(-4)$ and add the result to equation 1. First, equation 2 is altered as follows:

$$(4x + 2y = 30) \times (-4) \rightarrow -16x - 8y = -120$$

This altered equation is now added to equation 1:

$$\begin{array}{rcr} 6x + 8y & = & 60 \\ -16x - 8y & = & -120 \\ \hline -10x & = & -60 \end{array}$$

and

$$x = 6$$

The value of $x = 6$ can now be used to solve for $y$, as follows:

$$\begin{aligned} 6x + 8y &= 60 \\ 6(6) + 8y &= 60 \\ 36 + 8y &= 60 \\ 8y &= 24 \\ y &= 3 \end{aligned}$$

Thus, the common solution for the equations is $x = 6$ and $y = 3$.

***PRACTICE PROBLEM 9.2***

Determine the common solution to the following simultaneous equations: $3x + 12y = 42$ and $9x + 6y = 6$.

In some cases, it is necessary to alter *both* of the original equations to find their common solution. A problem of this type is solved in Example 9.3.

### EXAMPLE 9.3

Determine the common solution to the following simultaneous equations:

$$\begin{aligned} &\text{Equation 1:} \quad 3x - 2y = 5 \\ &\text{Equation 2:} \quad 4x + 5y = 22 \end{aligned}$$

***Solution:*** For these equations, we can get the $x$ terms to cancel if we multiply equation 1 by $(-4)$ and equation 2 by $(3)$, as follows:

$$\begin{aligned} (3x - 2y = 5) \times (-4) &\rightarrow -12x + 8y = -20 \\ (4x + 5y = 22) \times (3) &\rightarrow 12x + 15y = 66 \end{aligned}$$

Now, we can add the altered equations to eliminate the $x$ terms, as follows:

$$\begin{array}{r} -12x + 8y = -20 \\ 12x + 15y = 66 \\ \hline 23y = 46 \end{array}$$

and

$$y = 2$$

Substituting $y = 2$ into equation 1 allows us to solve for $x$, as follows:

$$\begin{aligned} 3x - 2y &= 5 \\ 3x - 2(2) &= 5 \\ 3x - 4 &= 5 \\ 3x &= 9 \\ x &= 3 \end{aligned}$$

Thus, the common solution to the equations is $x = 3$ and $y = 2$.

***PRACTICE PROBLEM 9.3***

Determine the common solution to the following simultaneous equations: $3x + 2y = 28$ and $2x + 5y = 48$.

**TABLE 9.1 Terms and the Multipliers That Can Be Used to Eliminate Them**

| | *Term* | *Multiplier* | *Result* | *Multiplier* | *Result* |
|---|---|---|---|---|---|
| Equation 1: | $5x$ | $(-8)$ | $-40x$ | $(8)$ | $40x$ |
| Equation 2: | $8x$ | $(5)$ | $40x$ | $(-5)$ | $-40x$ |
| Equation 1: | $3y$ | $(-2)$ | $-6y$ | $(2)$ | $6y$ |
| Equation 2: | $2y$ | $(3)$ | $6y$ | $(-3)$ | $-6y$ |

*Note:* When the coefficients in the original equations are mixed (i.e., one positive and one negative), then the multipliers used are both positive (or negative).

As you can see, we had to come up with a common multiple of the $x$ term coefficients (3 and 4) to eliminate the $x$ variables. The easiest way to determine the multipliers that will eliminate a given variable is to multiply each equation by the coefficient of the term in the other. For example, consider the following equations:

Equation 1: $5x + 3y = 37$
Equation 2: $8x + 2y = 34$

Table 9.1 breaks the equations down by terms and the two sets of multipliers that can be used to eliminate each. Note that one of the multipliers in each pair is negative. The negative multiplier is used so that the altered equations will contain equal-value positive and negative coefficients. That way, the coefficients will result in zero when the altered equations are added.

## Determinants

The third commonly used method of solving simultaneous equations involves the use of **determinants.** A determinant is a value found using the coefficients of the variables in the equations. For example, in the following equations:

**Determinant**
For an equation containing two or more variables, a value found using the coefficients of the variables.

$$5x + 3y = 37$$
$$8x + 2y = 34$$

the coefficients of the variables are:

1. 5 and 8 (for $x$)
2. 3 and 2 (for $y$)

To solve for the determinant of the equations, the coefficients are first written in the form of a **matrix.** A matrix is a row-column pattern like the one shown in Figure 9.3. As you can see, the coefficients are positioned in the matrix exactly as they are in the equations.

◀ *OBJECTIVE 3*

**Matrix**
A row-column pattern containing the coefficients of an equation.

The matrix in Figure 9.3 is referred to as a *2-by-2 matrix* or an *order-2 matrix.* Both of these names indicate that the matrix contains two rows and two columns. By the same token, a 3-by-3 (or order-3) matrix contains three rows and three columns. (Matrices always contain an equal number of rows and columns.)

**FIGURE 9.3 Matrix construction.**

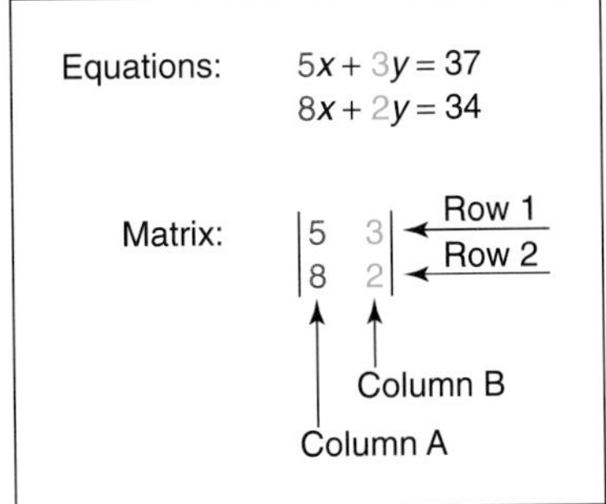

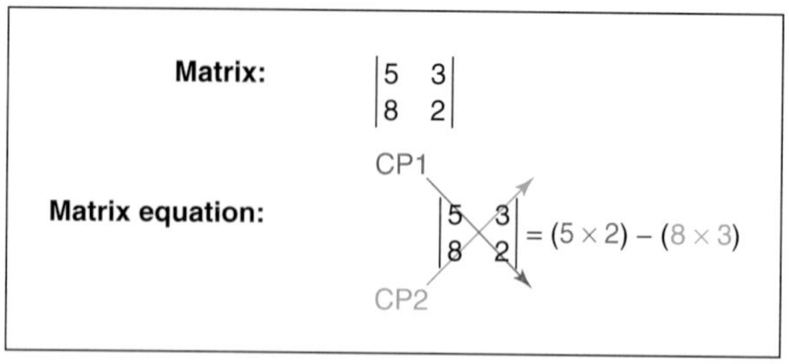

Matrix: $\begin{vmatrix} 5 & 3 \\ 8 & 2 \end{vmatrix}$

Matrix equation: $\begin{vmatrix} 5 & 3 \\ 8 & 2 \end{vmatrix} = (5 \times 2) - (8 \times 3)$ (CP1, CP2)

FIGURE 9.4 **Writing the matrix equation.**

| Equations | Determinant solution |
|---|---|
| $4x + 3y = 15$<br>$2x + 5y = 25$ | $\begin{vmatrix} 4 & 3 \\ 2 & 5 \end{vmatrix} = (4 \times 5) - (2 \times 3) = 14$ |
| $5V_1 + 4V_2 = 20$ V<br>$2V_1 + 3V_2 = 16$ V | $\begin{vmatrix} 5 & 4 \\ 2 & 3 \end{vmatrix} = (5 \times 3) - (2 \times 4) = 7$ |
| $7x + 8y = 54$<br>$5x + 10y = 60$ | $\begin{vmatrix} 7 & 8 \\ 5 & 10 \end{vmatrix} = (7 \times 10) - (5 \times 8) = 30$ |

FIGURE 9.5 **Example equations and their determinants.**

*Note:* In any group of simultaneous equations, there must be one equation for each variable. Since each column represents a variable and each row represents an equation, a matrix always has an equal number of rows and columns.

*OBJECTIVE 4* ▶

**Matrix equation**
An equation used to solve for the value of a matrix.

After drawing a matrix, a **matrix equation** is derived as shown in Figure 9.4. As you can see, the values in the matrix are split into two cross-products (labeled CP1 and CP2). The equation is then written as the difference between these cross-products. Note that CP1 comes first in the matrix equation. For consistency, the down cross-product always comes before the up product in the matrix equation. In other words, the up cross-product is always subtracted from the down cross-product.

The determinant for the matrix in Figure 9.4 is the solution to the matrix equation. Thus, for the matrix shown, the determinant ($D$) is found as

$$D = \begin{vmatrix} 5 & 3 \\ 8 & 2 \end{vmatrix} = (5 \times 2) - (8 \times 3) = -14$$

Figure 9.5 shows several equation pairs and the calculation of the determinant for each. Note the order of the cross-products in the matrix equations: the up cross-product is always subtracted from the down cross-product.

*OBJECTIVE 5* ▶ ***Simultaneous Equations Have More Than One Matrix.*** A matrix like the one in Figure 9.3 is referred to as a determinant matrix. For any pair of equations like the one shown in the figure, there are actually three matrices. This point is illustrated in Figure 9.6. Note that the determinant matrix is formed as described earlier in this section.

The matrix for each variable is found by replacing the coefficients of the variable with the constants (sums) given in the equations. For the $x$-matrix in Figure 9.6:

1. The coefficients of $x$ (5 and 8) have been replaced with the constants, 37 and 34, respectively.
2. The $y$ coefficients have been left unchanged.

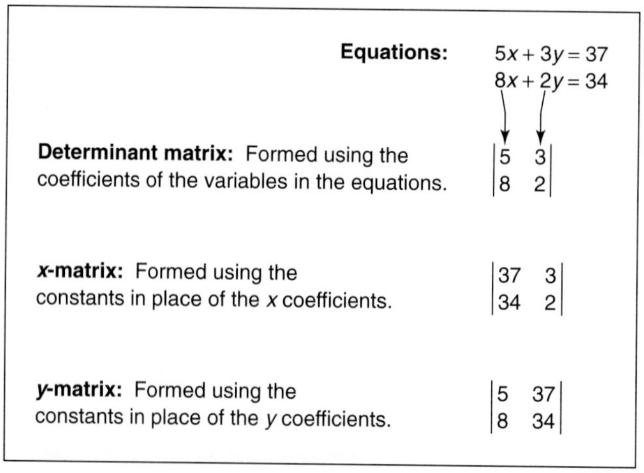

**Equations:** $5x + 3y = 37$, $8x + 2y = 34$

**Determinant matrix:** Formed using the coefficients of the variables in the equations. $\begin{vmatrix} 5 & 3 \\ 8 & 2 \end{vmatrix}$

***x*-matrix:** Formed using the constants in place of the $x$ coefficients. $\begin{vmatrix} 37 & 3 \\ 34 & 2 \end{vmatrix}$

***y*-matrix:** Formed using the constants in place of the $y$ coefficients. $\begin{vmatrix} 5 & 37 \\ 8 & 34 \end{vmatrix}$

FIGURE 9.6 **An equation pair and its matrices.**

For the $y$-matrix:

1. The coefficients of $y$ (3 and 2) have been replaced by the constants, 37 and 34, respectively.
2. The $x$ coefficients have been left unchanged.

***Solving Simultaneous Equations Using Determinants.*** Once you learn how to derive the various matrices for a pair of simultaneous equations, it becomes fairly quick and easy to solve the equations for their common solution. The procedure is as follows:

1. Derive and solve the matrix for the determinant and each of the variables.
2. Solve for each variable by dividing its matrix solution by the determinant.

For example, the determinant for the equations in Figure 9.6 was found earlier as:

$$D = \begin{vmatrix} 5 & 3 \\ 8 & 2 \end{vmatrix} = (5 \times 2) - (8 \times 3) = -14$$

The solution to the $x$-matrix in Figure 9.6 (which we will designate as $x'$) can be found as:

$$x' = \begin{vmatrix} 37 & 3 \\ 34 & 2 \end{vmatrix} = (37 \times 2) - (34 \times 3) = -28$$

The solution to the $y$-matrix in Figure 9.6 (which we will designate as $y'$) can be found as

$$y' = \begin{vmatrix} 5 & 37 \\ 8 & 34 \end{vmatrix} = (5 \times 34) - (8 \times 37) = -126$$

Using these matrix solutions, the values of $x$ and $y$ can be found as

$$x = \frac{x'}{D} = \frac{-28}{-14} = 2 \qquad \text{and} \qquad y = \frac{y'}{D} = \frac{-126}{-14} = 9$$

The combination of $x = 2$ and $y = 9$ is the common solution to the equations given in Figure 9.6. Example 9.4 further demonstrates the entire process for solving simultaneous equations using determinants.

**EXAMPLE 9.4**

Find the common solution to the following simultaneous equations:

| | |
|---|---|
| Equation 1: | $2x + 3y = 26$ |
| Equation 2: | $3x + 4y = 37$ |

***Solution:*** Using the coefficients of the variables, the determinant matrix ($D$) is written and solved as follows:

$$D = \begin{vmatrix} 2 & 3 \\ 3 & 4 \end{vmatrix} = (2 \times 4) - (3 \times 3) = -1$$

Substituting the constants in place of the $x$-coefficients gives us the $x$-matrix. This matrix is derived and solved as follows:

$$x' = \begin{vmatrix} 26 & 3 \\ 37 & 4 \end{vmatrix} = (26 \times 4) - (37 \times 3) = -7$$

Substituting the constants in place of the $y$-coefficients gives us the $y$-matrix. This matrix is derived and solved as follows:

$$y' = \begin{vmatrix} 2 & 26 \\ 3 & 37 \end{vmatrix} = (2 \times 37) - (3 \times 26) = -4$$

The value of $x$ can now be found as

$$\begin{aligned} x &= \frac{x'}{D} \\ &= \frac{-7}{-1} \\ &= 7 \end{aligned}$$

The value of $y$ can now be found as

$$\begin{aligned} y &= \frac{y'}{D} \\ &= \frac{-4}{-1} \\ &= 4 \end{aligned}$$

Thus, the combination of $x = 7$ and $y = 4$ is the common solution to the equations.

***PRACTICE PROBLEM 9.4***

Using determinants, find the common solution to the following simultaneous equations: $7x + 3y = 36$ and $5x + 5y = 20$.

In all of our examples up to this point, the coefficients of the variables (as well as the constants) have all been positive values. Whenever a coefficient (or constant) is negative, it must be written as such in the matrices. This point is demonstrated in Example 9.5.

## EXAMPLE 9.5

Find the common solution to the following simultaneous equations:

$$\begin{aligned} &\text{Equation 1:} \quad 4x - 7y = -53 \\ &\text{Equation 2:} \quad 3x + 3y = 51 \end{aligned}$$

***Solution:*** First, the matrices are derived and solved as follows:

$$D = \begin{vmatrix} 4 & -7 \\ 3 & 3 \end{vmatrix} = (4 \times 3) - [3 \times (-7)] = 33$$

$$x' = \begin{vmatrix} -53 & -7 \\ 51 & 3 \end{vmatrix} = (-53 \times 3) - [51 \times (-7)] = 198$$

$$y' = \begin{vmatrix} 4 & -53 \\ 3 & 51 \end{vmatrix} = (4 \times 51) - [3 \times (-53)] = 363$$

Now, the values of $x$ and $y$ can be found as

$$\begin{aligned} x &= \frac{x'}{D} & \text{and} \qquad y &= \frac{y'}{D} \\ &= \frac{198}{33} & &= \frac{363}{33} \\ &= 6 & &= 11 \end{aligned}$$

Thus, the combination of $x = 6$ and $y = 11$ is the common solution to the equations.

**PRACTICE PROBLEM 9.5**

Solve for the common solution to the following simultaneous equations: $4x - 2y = 12$ and $-6x + 7y = -6$.

## Deriving and Solving 3-by-3 Matrices

Some circuit analysis problems involve more than two equations. For example, the circuit in Figure 9.7 contains three loops, each having its own equation. Solving such a problem would be extremely difficult using any of the techniques we covered earlier. ◄ *OBJECTIVE 6*

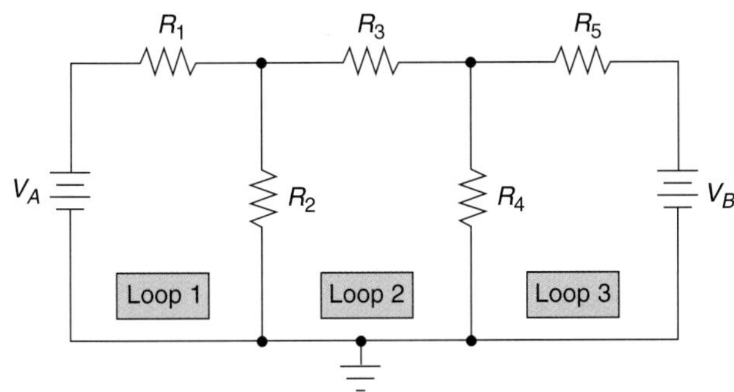


**FIGURE 9.7 A three-loop multisource circuit.**

The determinant method of solving simultaneous equations can be used to solve more complex groups of equations. For example, here are three three-variable simultaneous equations:

$$4x + 2y + 3z = 41$$
$$3x + 2y + 4z = 46$$
$$x + 3y + 2z = 34$$

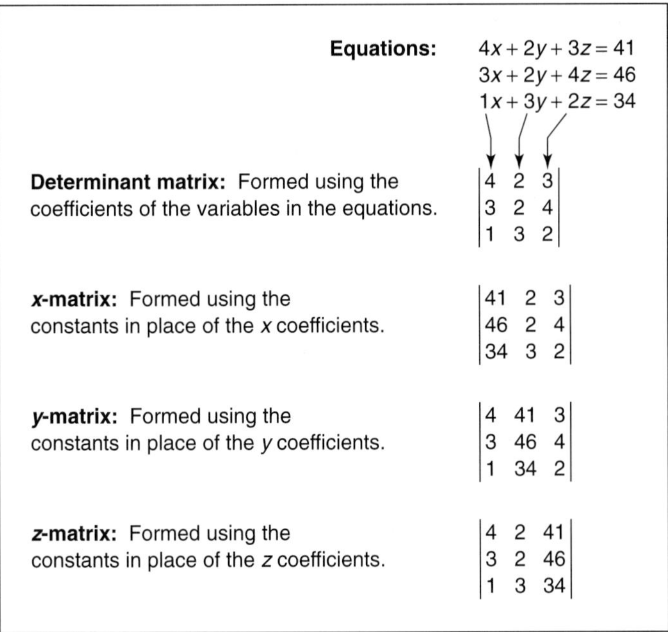

Equations:
4x + 2y + 3z = 41
3x + 2y + 4z = 46
1x + 3y + 2z = 34
Determinant matrix: Formed using the coefficients of the variables in the equations.
|4 2 3|
|3 2 4|
|1 3 2|
x-matrix: Formed using the constants in place of the x coefficients.
|41 2 3|
|46 2 4|
|34 3 2|
y-matrix: Formed using the constants in place of the y coefficients.
|4 41 3|
|3 46 4|
|1 34 2|
z-matrix: Formed using the constants in place of the z coefficients.
|4 2 41|
|3 2 46|
|1 3 34|


**FIGURE 9.8 3-by-3 matrices.**

Step 1: Expand the matrix.

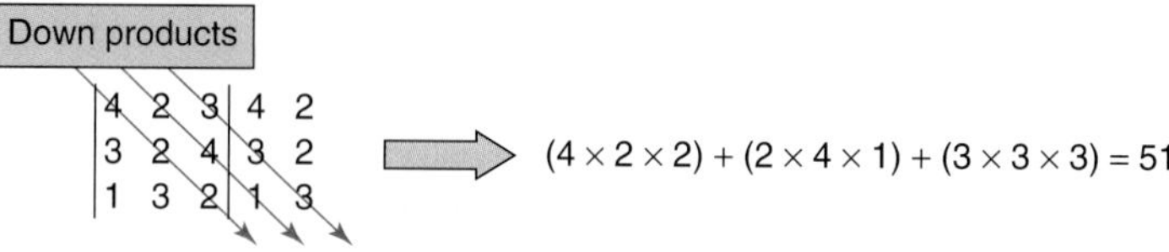


Step 2: Find the sum of the down products.

Down products

$$\begin{vmatrix} 4 & 2 & 3 \\ 3 & 2 & 4 \\ 1 & 3 & 2 \end{vmatrix} \begin{matrix} 4 & 2 \\ 3 & 2 \\ 1 & 3 \end{matrix} \Rightarrow (4 \times 2 \times 2) + (2 \times 4 \times 1) + (3 \times 3 \times 3) = 51$$

Step 3: Find the sum of the up products.

$$\begin{vmatrix} 4 & 2 & 3 \\ 3 & 2 & 4 \\ 1 & 3 & 2 \end{vmatrix} \begin{matrix} 4 & 2 \\ 3 & 2 \\ 1 & 3 \end{matrix} \Rightarrow (1 \times 2 \times 3) + (3 \times 4 \times 4) + (2 \times 3 \times 2) = 66$$

Up products

Step 4: Subtract the up sum from the down sum.

$D = 51 - 66 = -15$

**FIGURE 9.9 Solving a 3-by-3 matrix.**

These equations are similar (in form) to the equations that would be used to analyze the circuit in Figure 9.7. The determinant method discussed earlier can be modified to solve this type of problem.

The equations above are solved using 3-by-3 (or order-3) matrices. These matrices are written for the determinant and the variables as shown in Figure 9.8. Note that the procedure for writing each matrix is exactly as the procedure described for a 2-by-2 matrix. The only difference is the size of the matrix.

Solving a 3-by-3 matrix takes several steps (as shown in Figure 9.9). The first step is to expand the matrix by repeating the first two columns outside the matrix. After expanding the matrix, three down products are identified and added as shown in the figure. Then, three up products are identified and added. Finally, the determinant is found by subtracting the sum of the up products from the sum of the down products.

When solving three equations like those in Figure 9.8, all the matrices are derived and solved. Then the results are combined (as before) to determine the values of the variables. The complete process is demonstrated in Example 9.6.

### *Example 9.6*

Find the common solution to the simultaneous equations shown in Figure 9.8.

***Solution:*** The determinant matrix was solved ($D = -15$) in Figure 9.9. The $x$-matrix in Figure 9.8 is expanded and solved as follows:

$$x' = \begin{vmatrix} 41 & 2 & 3 \\ 46 & 2 & 4 \\ 34 & 3 & 2 \end{vmatrix} \quad \begin{matrix} 41 & 2 \\ 46 & 2 \\ 34 & 3 \end{matrix}$$

$$\begin{aligned} &= [(41 \times 2 \times 2) + (2 \times 4 \times 34) + (3 \times 46 \times 3)] \\ &\quad -[(34 \times 2 \times 3) + (3 \times 4 \times 41) + (2 \times 46 \times 2)] \\ &= 850 - 880 \\ &= -30 \end{aligned}$$

The $y$-matrix in Figure 9.8 is expanded and solved as follows:

$$y' = \begin{vmatrix} 4 & 41 & 3 \\ 3 & 46 & 4 \\ 1 & 34 & 2 \end{vmatrix} \begin{matrix} 4 & 41 \\ 3 & 46 \\ 1 & 34 \end{matrix}$$

$$\begin{aligned} &= [(4 \times 46 \times 2) + (41 \times 4 \times 1) + (3 \times 3 \times 34)] \\ &\quad - [(1 \times 46 \times 3) + (34 \times 4 \times 4) + (2 \times 3 \times 41)] \\ &= 838 - 928 \\ &= -90 \end{aligned}$$

The $z$-matrix in Figure 9.8 is expanded and solved as follows:

$$z' = \begin{vmatrix} 4 & 2 & 41 \\ 3 & 2 & 46 \\ 1 & 3 & 34 \end{vmatrix} \begin{matrix} 4 & 2 \\ 3 & 2 \\ 1 & 3 \end{matrix}$$

$$\begin{aligned} &= [(4 \times 2 \times 34) + (2 \times 46 \times 1) + (41 \times 3 \times 3)] \\ &\quad - [(1 \times 2 \times 41) + (3 \times 46 \times 4) + (34 \times 3 \times 2)] \\ &= 733 - 838 \\ &= -105 \end{aligned}$$

We can now combine the matrix solutions to solve for the variables, as follows:

$$x = \frac{x'}{D} = \frac{-30}{-15} = 2 \qquad y = \frac{y'}{D} = \frac{-90}{-15} = 6 \qquad z = \frac{z'}{D} = \frac{-105}{-15} = 7$$

Thus, the combination of $x = 2$, $y = 6$, and $z = 7$ is the common solution to the three equations shown in Figure 9.8.

***PRACTICE PROBLEM 9.6***

Find the common solution to the following simultaneous equations:

$$\begin{aligned} 6x + 2y + z &= 3 \\ 3x + 5y + 2z &= 17 \\ 4x + 4y + 3z &= 19 \end{aligned}$$

As you can see, the key to solving three three-variable equations is learning how to write, expand, and solve the matrices. The rest of the problem is solved just like the problems we solved earlier in this section.

Now that we have reviewed the commonly used methods of solving simultaneous equations, we will see how to apply these equations to the analysis of multisource circuits.

### Section Review

1. What are simultaneous equations?
2. What is the substitution method of solving simultaneous equations?
3. List the steps involved in the substitution method.
4. What is the alter-and-add method of solving simultaneous equations?
5. List the steps involved in the alter-and-add method.
6. What is a determinant?
7. What is a matrix?

8. What values are included in the determinant matrix for a given set of equations?
9. What values are included in the matrix for a given variable?
10. Which method of solving two simultaneous equations do you prefer? Why?

# 9.2 BRANCH CURRENT ANALYSIS

In Chapter 8, you were shown how to analyze a multisource circuit using the superposition theorem. In this section, you will be shown the branch current method of multisource circuit analysis.

## Branch Currents

The circuit in Figure 9.10 contains two branches, or loops. Loop 1 is made up of $V_A$, $R_1$ and $R_3$. Loop 2 is made up of $V_B$, $R_2$, and $R_3$. Note that $R_3$ is the only component common to both loops.

**Branch current**
A term used to describe the current generated through each loop of a multisource circuit.

The current through each loop is referred to as a **branch current.** In Figure 9.10, the branch currents are identified as $I_1$ and $I_2$. As you can see, the current through $R_3$ is equal to the sum of the branch currents. By formula,

$$I_3 = I_1 + I_2$$

**Branch current analysis**
A method of analyzing multisource circuits by redefining the circuit voltages as products of resistor values and branch currents.

Since $R_3$ is common to both loops, $I_1$ and $I_2$ are also common to both loops. As you will see, this is the basis of **branch current analysis,** which is a method of analyzing multisource circuits by redefining the circuit voltages as products of resistor values and branch currents. Basically, branch current analysis involves:

1. Writing the branch current equation for each loop in the circuit. Since the branch current equations describe the operation of a single circuit, they are (by definition) simultaneous equations.
2. Solving the simultaneous branch current equations using one of the techniques you were shown in Section 9.1.

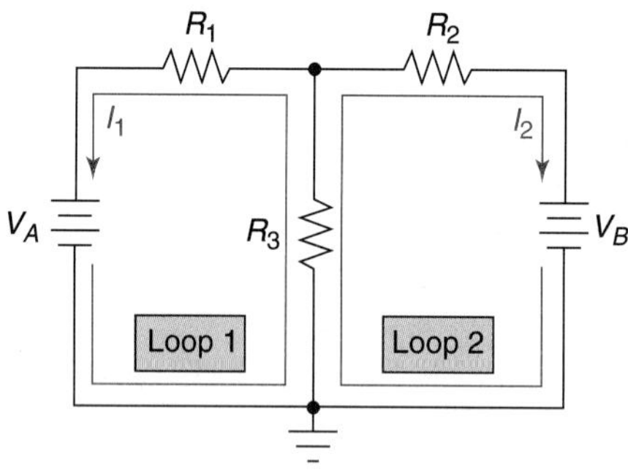


**FIGURE 9.10 Branch currents.**

## Branch Current Equations

*OBJECTIVE 7* ▶ Branch current equations are derived using Kirchhoff's voltage law and Ohm's law. For example, take a look at the circuit in Figure 9.11. Deriving the branch current equation for loop 1 begins with writing the Kirchhoff's voltage equation for that loop, as follows:

$$V_1 + V_3 = V_A \qquad \text{or} \qquad V_1 + V_3 = 8\text{ V}$$

FIGURE 9.11

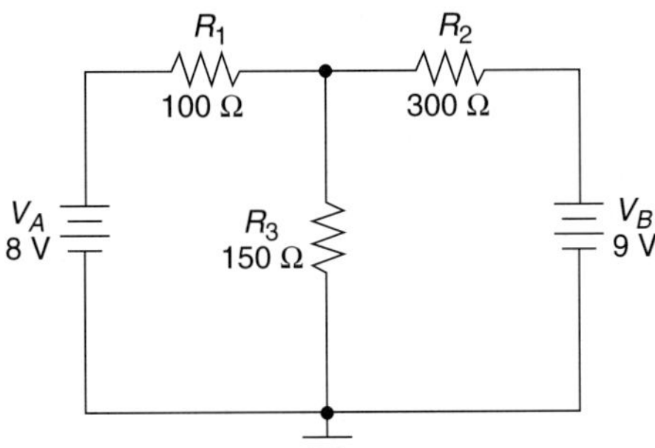


EWB

The current paths for the circuit in Figure 9.11 are identical to those shown in Figure 9.10; that is:

1. The current through $R_1$ equals $I_1$.
2. The current through $R_3$ equals $(I_1 + I_2)$.

Therefore, the Kirchhoff's voltage equation for loop 1 can be rewritten as

$$I_1R_1 + (I_1 + I_2)R_3 = 8 \text{ V}$$

If we substitute the resistor values into the above equation, we get:

$$100I_1 + 150(I_1 + I_2) = 8 \text{ V}$$

Before we continue, there are a couple of points that should be made about this last equation. First, note that the resistor values have been placed in front of the current variables. This simply allows us to view them in a more familiar form ($100I_1$ is more familiar to us than $I_1100$). Second, the unit of measure for the resistors (Ω) is not carried in the equation. It is understood, however, that the numeric values are resistor values.

The Kirchhoff's equation for loop 1 can now be simplified as follows:

$$100I_1 + 150I_1 + 150I_2 = 8 \text{ V} \qquad \text{or} \qquad 250I_1 + 150I_2 = 8 \text{ V}$$

This is the branch current equation for loop 1. Note that it defines the operation of loop 1 in terms of the two branch currents. This is our goal in writing the equation. Using Figure 9.10 as a reference, the branch current equation for loop 2 can be derived as follows:

$$\begin{aligned} V_2 + V_3 &= V_B \\ V_2 + V_3 &= 9 \text{ V} \\ I_2R_2 + (I_1 + I_2)R_3 &= 9 \text{ V} \\ 300I_2 + 150(I_1 + I_2) &= 9 \text{ V} \\ 300I_2 + 150I_1 + 150I_2 &= 9 \text{ V} \\ 150I_1 + 450I_2 &= 9 \text{ V} \end{aligned}$$

We now have two simultaneous equations that describe the operation of the circuit. These equations are

$$\begin{aligned} &\text{Loop 1:} \qquad 250I_1 + 150I_2 = 8 \text{ V} \\ &\text{Loop 2:} \qquad 150I_1 + 450I_2 = 9 \text{ V} \end{aligned}$$

## Solving the Circuit

Once we have the branch current equations, we can easily solve for the circuit currents and voltages. The equations for the circuit in Figure 9.11 can easily be solved using the alter-and-add method, as follows:

$$\begin{aligned} (250I_1 + 150I_2 = 8 \text{ V}) \times (-3) \rightarrow -750I_1 - 450I_2 &= -24 \text{ V} \\ 150I_1 + 450I_2 &= 9 \text{ V} \\ \hline -600I_1 \qquad\qquad &= -15 \text{ V} \\ I_1 \qquad\qquad &= 25 \text{ mA} \end{aligned}$$

Substituting 25 mA for $I_1$ in the loop 1 equation gives us:

$$250I_1 + 150I_2 = 8\text{ V}$$
$$250(25\text{ mA}) + 150I_2 = 8\text{ V}$$
$$6.25\text{ V} + 150I_2 = 8\text{ V}$$
$$150I_2 = 1.75\text{ V}$$
$$I_2 = 11.67\text{ mA}$$

Now we can solve for the component voltages as follows:

$$V_1 = I_1R_1 = (25\text{ mA})(100\ \Omega) = 2.5\text{ V}$$
$$V_2 = I_2R_2 = (11.67\text{ mA})(300\ \Omega) = 3.5\text{ V}$$
$$V_3 = (I_1 + I_2)R_3 = (36.67\text{ mA})(150\ \Omega) = 5.5\text{ V}$$

The values calculated for the circuit in Figure 9.11 are shown in Figure 9.12. As you can see, the component voltages add up to the source voltage in each loop, and the currents throughout the circuit obey Kirchhoff's current law. Therefore, our results must be correct.

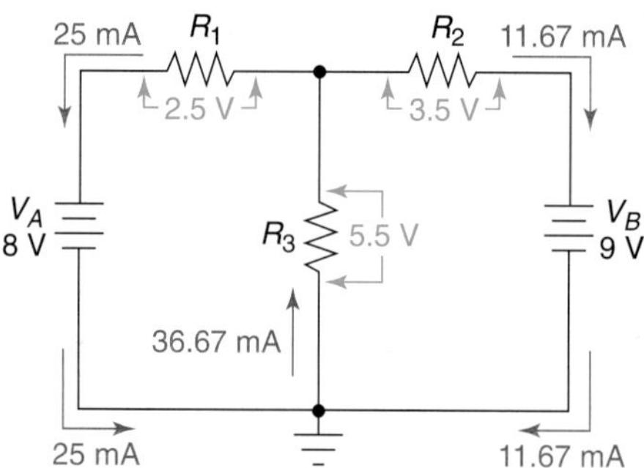


**FIGURE 9.12 The current and voltage values for the circuit in Figure 9.11.**

## *Writing Branch Current Equations: The Observation Method*

Now that you have seen how branch current equations are derived, you can be shown a simpler and faster method of writing these equations. This observation method is illustrated in Figure 9.13. Figure 9.13a shows the original circuit (from Figure 9.11). Figure 9.13b shows only loop 1 (with $I_2$ included). If you look closely at the loop 1 circuit, you'll see that

1. $I_1$ passes through a total of 250 Ω, so we'll use $250I_1$ in our loop equation.
2. $I_2$ passes through a total of 150 Ω, so we'll use $150I_2$ in our loop equation.

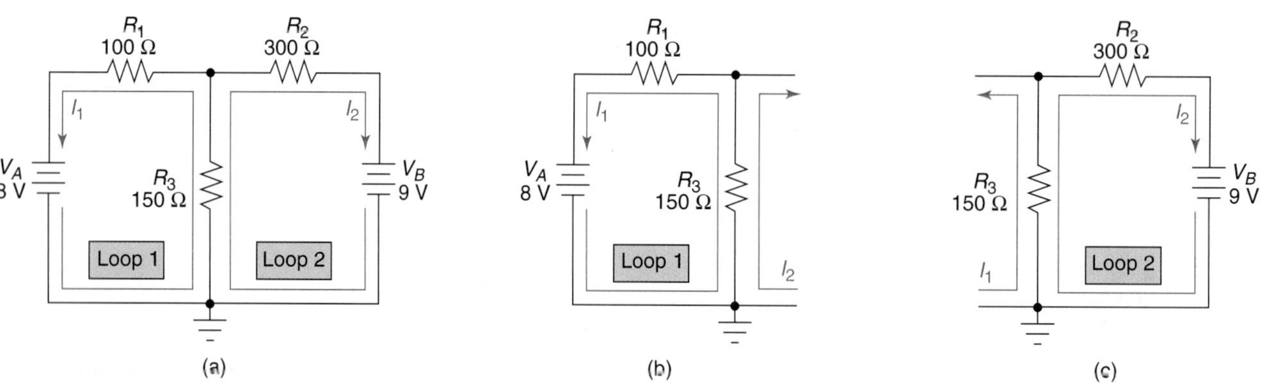


**FIGURE 9.13**

Using these values and the supply voltage, the loop 1 equation is written as $250I_1 + 150I_2 = 8\text{ V}$. This is the same equation we derived earlier for this circuit. By the same token, the loop 2 circuit in Figure 9.13c shows that:

1. $I_1$ passes through a total of 150 Ω, giving us $150I_1$.
2. $I_2$ passes through a total of 450 Ω, giving us $450I_2$.

Using these values and the supply voltage, the loop 2 equation is written as $150I_1 + 450I_2 = 9\text{ V}$, the same equation we derived earlier. As you can see, this method of writing the branch current equations is much faster than going through the mathematical derivations.

## Branch Current Analysis: Example Problems

Now that you have seen the overall process of branch current analysis, we'll go through a series of example problems to help you get more comfortable with this analysis technique. Our first example problem is very similar to the one we solved earlier.

◄ *OBJECTIVE 8*

**EXAMPLE 9.7**

Determine the current and component voltage values for the circuit in Figure 9.14a.

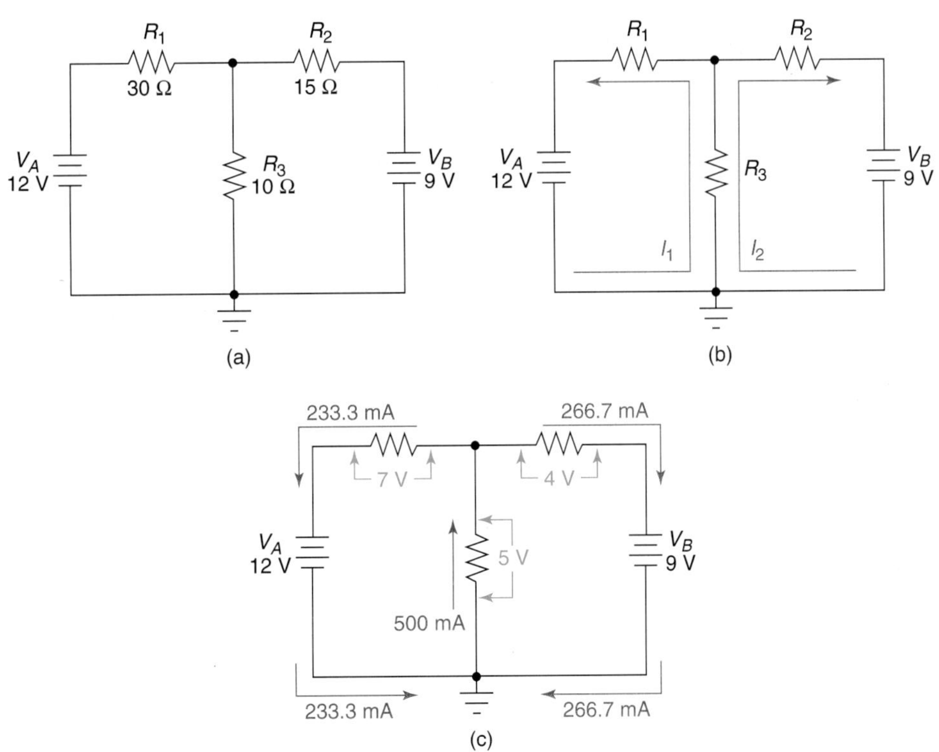


**FIGURE 9.14**

***Solution:*** The currents are identified as shown in Figure 9.14b. We'll use the observation method of writing the loop equations. In loop 1:

1. $I_1$ passes through a total of 40 Ω, giving us $40I_1$.
2. $I_2$ passes through a total of 10 Ω, giving us $10I_2$.

Therefore, the loop 1 equation is $40I_1 + 10I_2 = 12$ V.
In loop 2:

1. $I_1$ passes through a total of 10 Ω, giving us $10I_1$.
2. $I_2$ passes through a total of 25 Ω, giving us $25I_2$.

Therefore, the loop 2 equation is $10I_1 + 25I_2 = 9$ V.

Using the determinant method of solving the equations, the value of $I_1$ can be found as

$$I_1 = \frac{I_1\text{-matrix}}{D\text{-matrix}} = \frac{\begin{vmatrix} 12 & 10 \\ 9 & 25 \end{vmatrix}}{\begin{vmatrix} 40 & 10 \\ 10 & 25 \end{vmatrix}} = \frac{(12 \times 25) - (9 \times 10)}{(40 \times 25) - (10 \times 10)} = \frac{210}{900} = 233.3 \text{ mA}$$

The value of $I_2$ can be found as

$$I_2 = \frac{I_2\text{-matrix}}{D\text{-matrix}} = \frac{\begin{vmatrix} 40 & 12 \\ 10 & 9 \end{vmatrix}}{\begin{vmatrix} 40 & 10 \\ 10 & 25 \end{vmatrix}} = \frac{(40 \times 9) - (10 \times 12)}{(40 \times 25) - (10 \times 10)} = \frac{240}{900} = 266.7 \text{ mA}$$

Now, using the resistor values shown in the figure, the component voltages can be found as

$$V_1 = I_1R_1 = (233.3 \text{ mA})(30\ \Omega) = 7 \text{ V}$$
$$V_2 = I_2R_2 = (266.7 \text{ mA})(15\ \Omega) = 4 \text{ V}$$
$$V_3 = (I_1 + I_2)R_3 = (500 \text{ mA})(10\ \Omega) = 5 \text{ V}$$

The calculated current and voltage values are shown in Figure 9.14c. As you can see, the circuit follows Kirchhoff's current and voltage laws.

***PRACTICE PROBLEM 9.7***

A circuit like the one in Figure 9.14 has the following values: $V_A = 14$ V, $V_B = 21$ V, $R_1 = 100\ \Omega$, $R_2 = 150\ \Omega$, and $R_3 = 150\ \Omega$. Calculate the current and component voltage values for the circuit.

In some cases, the branch current analysis of a circuit provides a negative current value. When this occurs, the negative value simply indicates that the current is in the opposite direction of the one we assumed. This point is demonstrated in Example 9.8.

## EXAMPLE 9.8

Calculate the current and component voltage values for the circuit in Figure 9.15a.

***Solution:*** The currents are assumed to have the directions shown in Figure 9.15b. Again, we'll use the observation method of deriving the loop equations. In loop 1:

1. $I_1$ passes through a total of 250 Ω, giving us $250I_1$.
2. $I_2$ passes through a total of 150 Ω, giving us $150I_2$.

Therefore, the loop 1 equation is $250I_1 + 150I_2 = 18$ V.

In loop 2:

1. $I_1$ passes through a total of 150 Ω, giving us $150I_1$.
2. $I_2$ passes through a total of 300 Ω, giving us $300I_2$.

Therefore, the loop 2 equation is $150I_1 + 300I_2 = 7$ V.

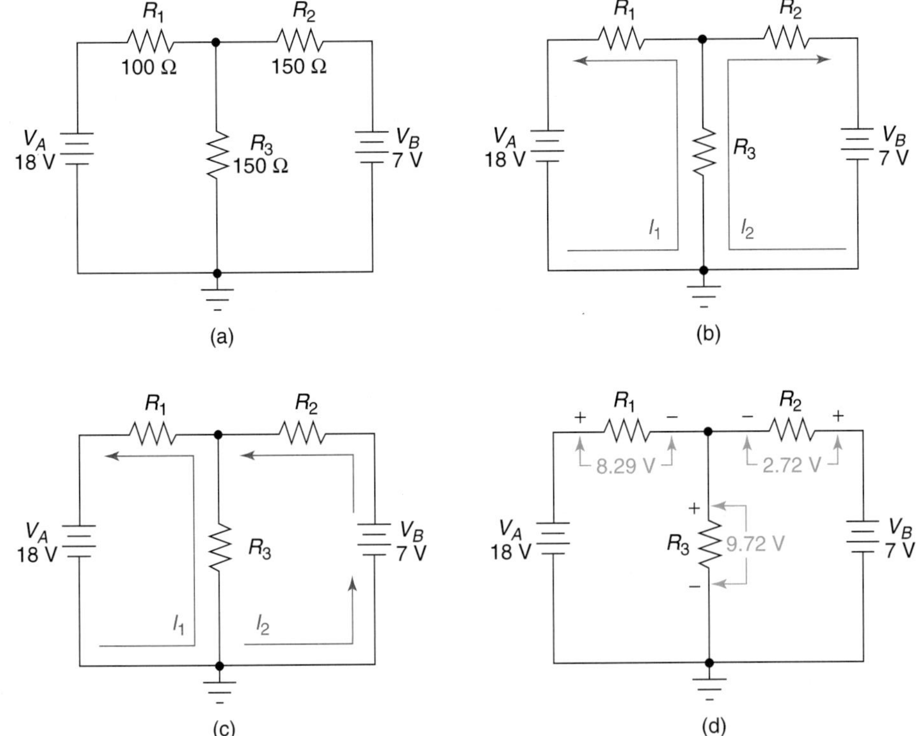


**FIGURE 9.15**

Using the determinant method of solving the equations, the value of $I_1$ can be found as

$$I_1 = \frac{I_1\text{-matrix}}{D\text{-matrix}} = \frac{\begin{vmatrix} 18 & 150 \\ 7 & 300 \end{vmatrix}}{\begin{vmatrix} 250 & 150 \\ 150 & 300 \end{vmatrix}} = \frac{(18 \times 300) - (7 \times 150)}{(250 \times 300) - (150 \times 150)} = \frac{4350}{52{,}500} = 82.9 \text{ mA}$$

The value of $I_2$ can be found as

$$I_2 = \frac{I_2\text{-matrix}}{D\text{-matrix}} = \frac{\begin{vmatrix} 250 & 18 \\ 150 & 7 \end{vmatrix}}{\begin{vmatrix} 250 & 150 \\ 150 & 300 \end{vmatrix}} = \frac{(250 \times 7) - (150 \times 18)}{(250 \times 300) - (150 \times 150)} = \frac{-950}{52{,}500}$$

$$= -18.1 \text{ mA}$$

The negative value of $I_2$ indicates that our assumed direction for that current was incorrect. The actual current directions are shown in Figure 9.15c. As shown,

$$I_1 = 82.9 \text{ mA}$$
$$I_2 = 18.1 \text{ mA}$$
$$I_3 = I_1 - I_2 = 64.8 \text{ mA}$$

Note that $I_3$ equals the difference between the branch currents because they are in opposite directions. Using the calculated current values, the component voltages can be found as

$$V_1 = I_1R_1 = (82.9 \text{ mA})(100\ \Omega) = 8.29 \text{ V}$$
$$V_2 = I_2R_2 = (18.1 \text{ mA})(150\ \Omega) = 2.72 \text{ V}$$
$$V_3 = I_3R_3 = (64.8 \text{ mA})(150\ \Omega) = 9.72 \text{ V}$$

These voltages are shown in Figure 9.15d. As you can see, our results are consistent with Kirchhoff's current and voltage laws.

*PRACTICE PROBLEM 9.8*

A circuit like the one in Figure 9.15a has the following values: $V_A = 7$ V, $V_B = 21$ V, $R_1 = 150\ \Omega$, $R_2 = 100\ \Omega$, and $R_3 = 300\ \Omega$. Calculate the current and voltage values for the circuit.

You may be wondering why $I_2$ is in the direction shown in Figure 9.15c. The reason is that the value of $V_A$ is sufficient to force $V_B$ to conduct in the "wrong" direction, which is why our assumed current direction was incorrect.

In the next section, we will discuss another method of current analysis called mesh current analysis. As you will see, negative current values frequently occur when this method of analyzing circuits is used. However, it can be applied to some situations where branch current analysis cannot be used.

### Section Review

1. What are branch currents?
2. Describe the branch current analysis procedure.
3. Describe the observation method of writing branch current equations.
4. What does a negative current result indicate?

## 9.3 MESH CURRENT ANALYSIS

*OBJECTIVE 9* ▶ Branch current analysis (like most circuit analysis techniques) has its limitations. While it works very well for two-loop multisource circuits, it doesn't work well for circuits containing more than two loops. For example, consider the circuit shown in Figure 9.16. It is easy to predict the directions of the loop 1 and loop 3 currents, but what about the loop 2 current? Branch current analysis does not address this problem. A circuit like the one in Figure 9.16 can be solved using mesh current analysis. As you will see, mesh current analysis is *almost* identical to branch current analysis.

### *Mesh Currents*

**Mesh**
A term often used to describe a multisource circuit that contains more than one loop.

A multisource circuit containing more than one loop is often referred to as a **mesh.** In practice, the term is generally used to describe multisource circuits containing one or more loops that are not connected directly to a voltage (or current) source. (For example, loop 2 in Figure 9.16 is not connected directly to either of the voltage sources.)

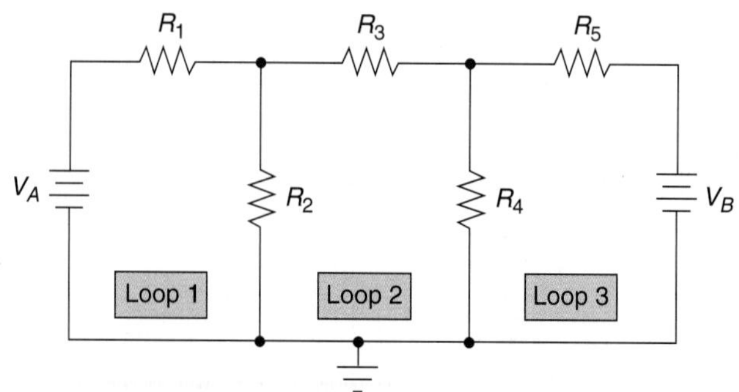


**FIGURE 9.16**

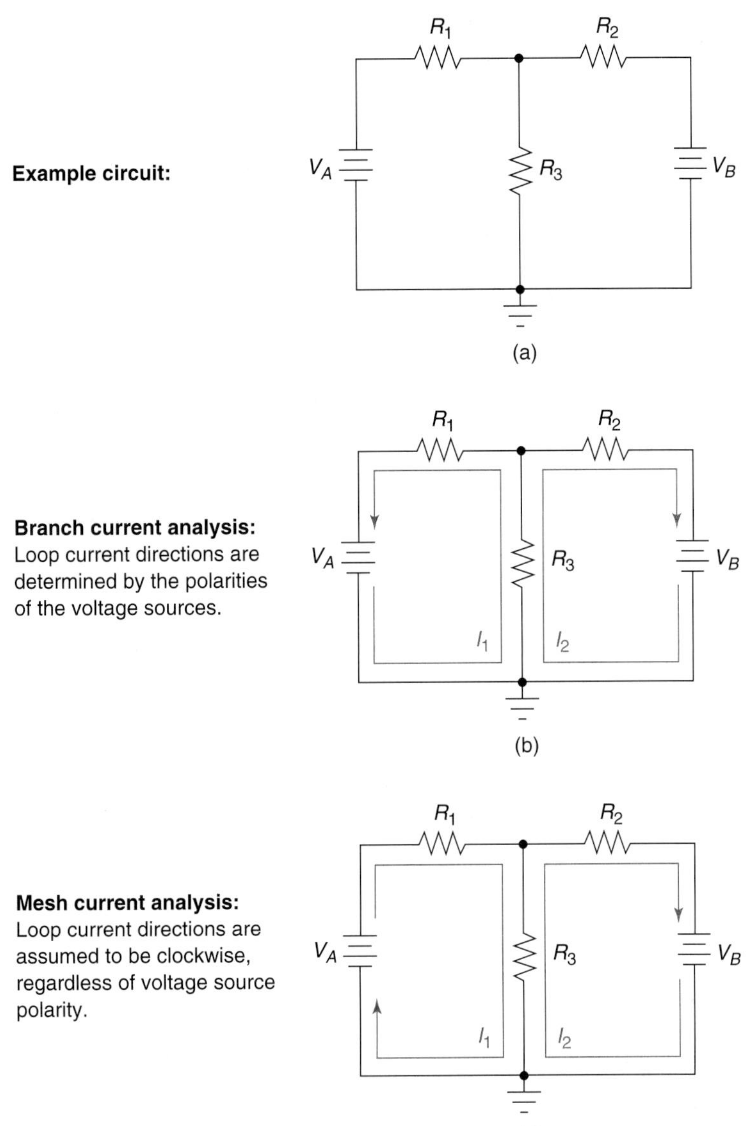


**FIGURE 9.17 Assigning current directions.**

For all practical purposes, **mesh current analysis** can be viewed as a modified form of branch current analysis. The primary difference is the assumed directions of the loop currents. Let's say that we're analyzing the operation of the circuit in Figure 9.17a. When using branch current analysis, the direction of each loop current is assigned according to the polarity of the voltage source. Thus, branch current analysis assumes the current directions shown in Figure 9.17b. Mesh current analysis, on the other hand, assumes that each loop current is in a clockwise direction, regardless of the polarity of the voltage source. Thus, mesh current analysis assumes the current directions shown in Figure 9.17c.

**Mesh current analysis** A form of branch current analysis where all loop currents are assumed to be in a clockwise direction, regardless of voltage source polarities.

The strength of mesh current analysis can be seen if we refer back to the circuit in Figure 9.16. For branch current analysis, the polarity of each source must be considered when assigning the current directions. However, since loop 2 is not connected directly to any voltage source, we have no basis for assigning a direction to the current in this loop. In mesh current analysis, however, all loop currents are assumed to be clockwise. Thus, the loop 2 current for the circuit in Figure 9.16 would be assigned the direction shown in Figure 9.18. As you can see, mesh current analysis provides us with the foundation needed to solve the circuit.

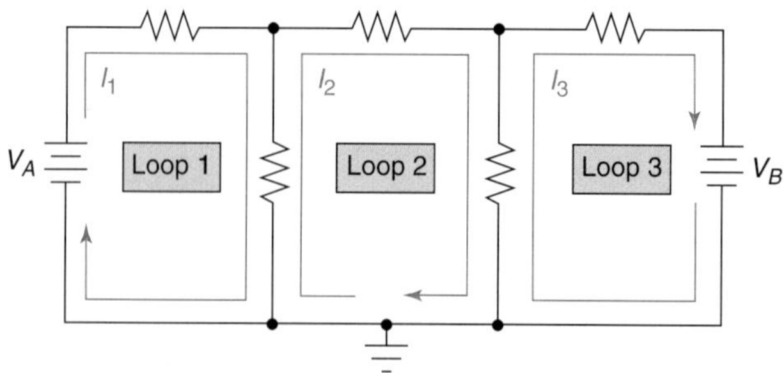


**FIGURE 9.18 Mesh currents in a three-loop circuit.**

## *Mesh Current Equations*

*OBJECTIVE 10* ▶

Since mesh current analysis assumes that all loop currents are in a clockwise direction, it presents us with a unique problem: the current direction in a given loop may not match the polarity of the voltage source. For example, the direction of $I_1$ in Figure 9.18 does not match the polarity of $V_A$; that is, given the polarity of $V_A$, we would expect $I_1$ to be in the opposite direction. Even so, the current is assumed to be in a clockwise direction.

When writing mesh equations, the relationship between the polarity of the source and the assumed direction of current must be taken into account, as follows:

1. If the polarity of the source matches the assumed direction of current, the source voltage is written as a positive value in the mesh equation.
2. If the polarity of the sources does not match the assumed direction of current, the source voltage is written as a negative value in the mesh equation.

For example, the polarity of $V_A$ in Figure 9.19 does not match the assumed direction of $I_1$. Therefore, the source voltage is entered as (−12 V) in the mesh equation for loop 1. With this information, the loop 1 equation is written as

$$500I_1 - 200I_2 = -12\text{ V}$$

Note that the equation is written in terms of the difference between the loop currents because the loop currents are always assumed to be in a clockwise direction. Therefore, $I_1$ and $I_2$ are in opposite directions through $R_3$, and $I_3$ is equal to the difference between them $(I_1 - I_2)$.

Since the assumed direction of $I_2$ is consistent with the polarity of $V_B$, the source voltage is entered as a positive value (10 V) in the mesh equation for loop 2. With this in mind, the loop 2 equation is written as

$$320I_2 - 200I_1 = 10\text{ V}$$

Again, note the positive value of the source voltage, and the fact that the equation is written as the difference between the loop currents.

**FIGURE 9.19**

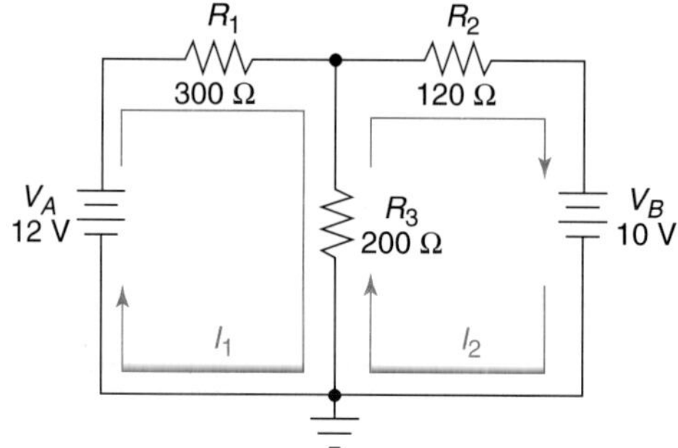

## *Solving the Circuit*

Once the mesh equations are written for a given circuit, they are solved in the same manner as branch currents. This is demonstrated in Example 9.9. ◀ *OBJECTIVE 11*

**EXAMPLE 9.9**

Determine the values of $I_1$ and $I_2$ for the circuit in Figure 9.19.

***Solution:*** We already determined that the circuits have the following mesh equations:

$$\text{Loop 1:} \quad 500I_1 - 200I_2 = -12\text{ V}$$
$$\text{Loop 2:} \quad -200I_1 + 320I_2 = 10\text{ V}$$

Note that the order of the terms in the loop 2 equation has been reversed to align the like terms in the two equations. Using the alter-and-add approach, the value of $I_2$ can be found as follows:

$$\begin{array}{lrr} \text{Loop 2:} & (-200I_1 + 320I_2 = 10\text{ V}) \times (2.5) \rightarrow -500I_1 + 800I_2 = & 25\text{ V} \\ \text{Loop 1:} & 500I_1 - 200I_2 = & -12\text{ V} \\ \hline \text{Sum:} & 600I_2 = & 13\text{ V} \end{array}$$

Dividing both sides of the sum equation by 600 gives us:

$$I_2 = 21.67\text{ mA}$$

The value of $I_2$ can now be used to solve for the value of $I_1$, as follows:

$$-200I_1 + 320I_2 = 10\text{ V}$$
$$-200I_1 + (320)(21.67\text{ mA}) = 10\text{ V}$$
$$-200I_1 + 6.93\text{ V} = 10\text{ V}$$
$$-200I_1 = 3.07\text{ V}$$

or

$$I_1 = -15.33\text{ mA}$$

Thus, the calculated values for the circuit are $I_1 = -15.33$ mA and $I_2 = 21.67$ mA.

***PRACTICE PROBLEM 9.9***

Using mesh current analysis, determine the values of $I_1$ and $I_2$ for the circuit in Figure 9.20.

**FIGURE 9.20**

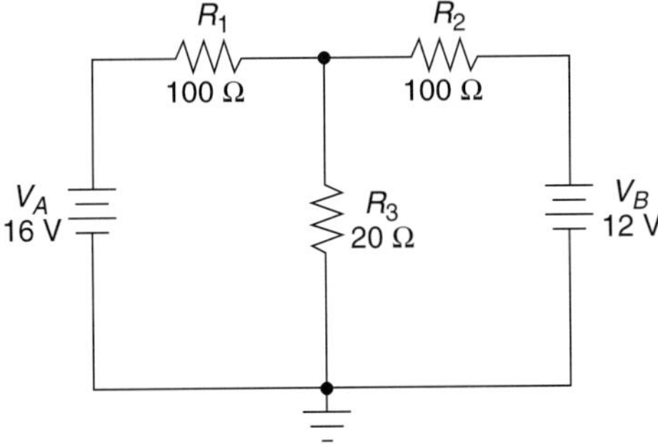


The negative result for $I_1$ in Example 9.9 indicates that the current in loop 1 is actually in the opposite direction of the one assumed; that is, $I_1$ actually goes counterclockwise around the loop, not clockwise as we assumed. With this information, the currents in the

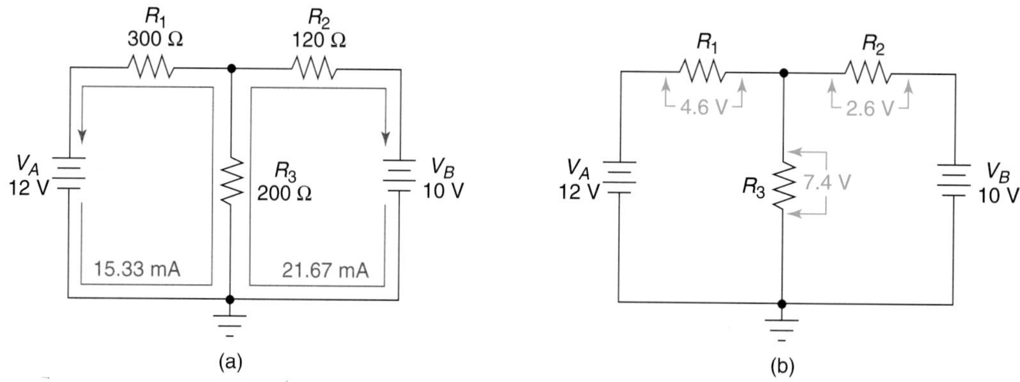


**FIGURE 9.21**

circuit are as shown in Figure 9.21a. Using these current values, the component voltages can be found as:

$$V_1 = I_1R_1 = (15.35 \text{ mA})(300\ \Omega) = 4.6 \text{ V}$$
$$V_2 = I_2R_2 = (21.67 \text{ mA})(120\ \Omega) = 2.6 \text{ V}$$
$$V_3 = (I_1 + I_2)R_3 = (37 \text{ mA})(200\ \Omega) = 7.4 \text{ V}$$

The component voltages are shown in Figure 9.21b. As you can see, each loop in the circuit conforms to Kirchhoff's voltage law.

## Solving a Three-Loop Circuit

*OBJECTIVE 12* ▶ As you know, mesh current analysis can be used to solve for the current values in a three-loop circuit. Let's say that we want to solve for the values of $I_1$, $I_2$, and $I_3$ in Figure 9.22. The first step is to draw the loop currents in a clockwise direction as shown in the figure. The loop 1 equation can now be written as:

$$\text{Loop 1:} \qquad 25I_1 - 15I_2 + 0I_3 = -20 \text{ V}$$

Note that $I_3$ is included in the equation as $(0I_3)$, which indicates that $I_3$ does not pass through any of the components in loop 1. Also note the negative value of $V_A$, a result of the fact that the polarity of $V_A$ does not match the assumed direction of $I_1$.

Loop 2 is not connected directly to any voltage source. However, Kirchhoff's voltage law tells us that the sum of the voltages around the loop must equal zero. Therefore, the loop 2 equation can be written as follows:

$$\text{Loop 2:} \qquad 50I_2 - 15I_1 - 15I_3 = 0 \text{ V}$$

Note that both $I_1$ and $I_3$ weigh into this equation because both currents pass through one of the resistors in loop 2.

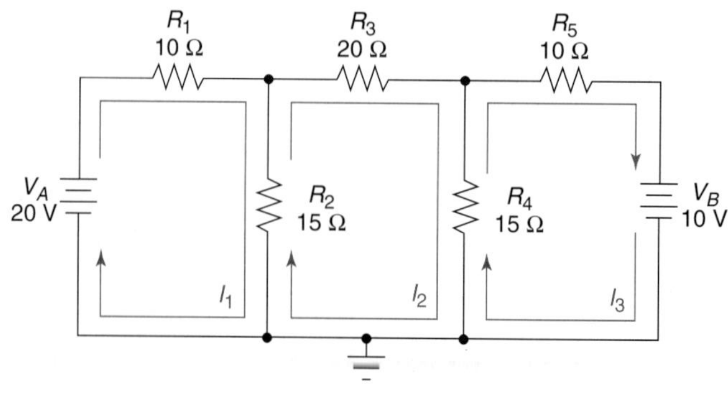


EWB

**FIGURE 9.22**

Finally, the current through loop 3 is consistent with the polarity of $V_B$. Therefore, the loop 3 equation can be written as follows:

$$\text{Loop 3:} \qquad 25I_3 - 15I_2 + 0I_1 = 10\text{ V}$$

In this case, the loop 1 current is entered as $0I_1$ because this current does not pass through any of the components in loop 3.

Once the three loop equations have been written, they can be solved as shown in Example 9.10. Note that solving this problem requires the use of 3-by-3 matrices.

**EXAMPLE 9.10**

Determine the values of $I_1$, $I_2$, and $I_3$ for the circuit in Figure 9.22.

***Solution:*** We'll start by rearranging the current terms in the equations, as follows:

$$\begin{array}{ll} \text{Loop 1:} & 25I_1 - 15I_2 + 0I_3 = -20\text{ V} \\ \text{Loop 2:} & -15I_1 + 50I_2 - 15I_3 = 0\text{ V} \\ \text{Loop 3:} & 0I_1 - 15I_2 + 25I_3 = 10\text{ V} \end{array}$$

For the above equations, the $D$-matrix is written and solved as follows:

$$D = \left|\begin{array}{rrr} 25 & -15 & 0 \\ -15 & 50 & -15 \\ 0 & -15 & 25 \end{array}\right| \begin{array}{rr} 25 & -15 \\ -15 & 50 \\ 0 & -15 \end{array}$$

$$\begin{aligned} &= (31{,}250 + 0 + 0) - (0 + 5625 + 5625) \\ &= 31{,}250 - 11{,}250 \\ &= 20{,}000 \end{aligned}$$

The $I_1$-matrix is now written and solved as follows:

$$I_1' = \left|\begin{array}{rrr} -20 & -15 & 0 \\ 0 & 50 & -15 \\ 10 & -15 & 25 \end{array}\right| \begin{array}{rr} -20 & -15 \\ 0 & 50 \\ 10 & -15 \end{array}$$

$$\begin{aligned} &= (-25{,}000 + 2250 + 0) - (0 - 4500 + 0) \\ &= -22{,}750 + 4500 \\ &= -18{,}250 \end{aligned}$$

The $I_2$-matrix is now written and solved as follows:

$$I_2' = \left|\begin{array}{rrr} 25 & -20 & 0 \\ -15 & 0 & -15 \\ 0 & 10 & 25 \end{array}\right| \begin{array}{rr} 25 & -20 \\ -15 & 0 \\ 0 & 10 \end{array}$$

$$\begin{aligned} &= (0 + 0 + 0) - (0 - 3750 + 7500) \\ &= 0 - 3750 \\ &= -3750 \end{aligned}$$

The $I_3$-matrix is now written and solved as follows:

$$I_3' = \left|\begin{array}{rrr} 25 & -15 & -20 \\ -15 & 50 & 0 \\ 0 & -15 & 10 \end{array}\right| \begin{array}{rr} 25 & -15 \\ -15 & 50 \\ 0 & -15 \end{array}$$

$$\begin{aligned} &= (12{,}500 + 0 - 4500) - (0 + 0 + 2250) \\ &= 8000 - 2250 \\ &= 5750 \end{aligned}$$

Now that we have solved each matrix, the current values for the circuit can be solved as follows:

$$I_1 = \frac{I_1'}{D} = \frac{-18{,}250}{20{,}000} = -912.5 \text{ mA} \qquad I_2 = \frac{I_2'}{D} = \frac{-3750}{20{,}000} = -187.5 \text{ mA} \qquad I_3 = \frac{I_3'}{D} = \frac{5750}{20{,}000} = 287.5 \text{ mA}$$

***PRACTICE PROBLEM 9.10***

Calculate the values of $I_1$, $I_2$, and $I_3$ for the circuit in Figure 9.23.

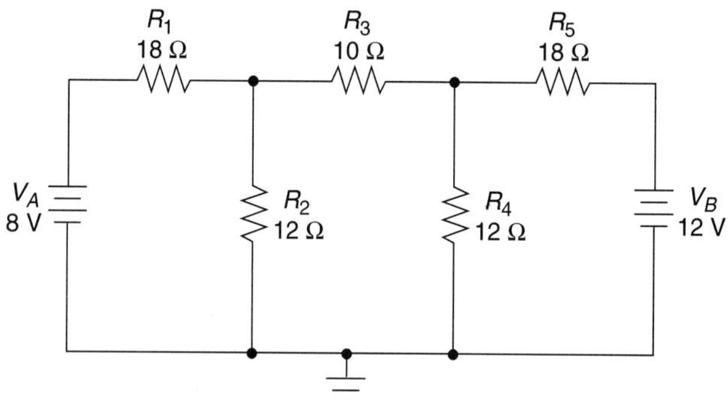


**FIGURE 9.23**

The negative values of $I_1$ and $I_2$ indicate that these currents are actually in the opposite direction of the one we assumed. Thus, the three currents are as shown in Figure 9.24a. Using the current values shown, we can calculate the component voltages as follows:

$$\begin{aligned}
V_1 &= I_1R_1 = (912.5 \text{ mA})(10\ \Omega) = 9.125 \text{ V} \\
V_2 &= (I_1 - I_2)R_2 = (725 \text{ mA})(15\ \Omega) = 10.875 \text{ V} \\
V_3 &= I_2R_3 = (187.5 \text{ mA})(20\ \Omega) = 3.75 \text{ V} \\
V_4 &= (I_2 + I_3)R_4 = (475 \text{ mA})(15\ \Omega) = 7.125 \text{ V} \\
V_5 &= I_3R_5 = (287.5 \text{ mA})(10\ \Omega) = 2.875 \text{ V}
\end{aligned}$$

The calculated component voltages are shown in Figure 9.24b. As you can see, the sum of the component voltages in loop 1 equals $V_A$. The sum of the component voltages in loop 3 equals $V_B$. The sum of the component voltages in loop 2 equals 0 V.

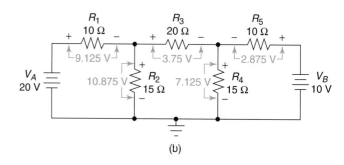

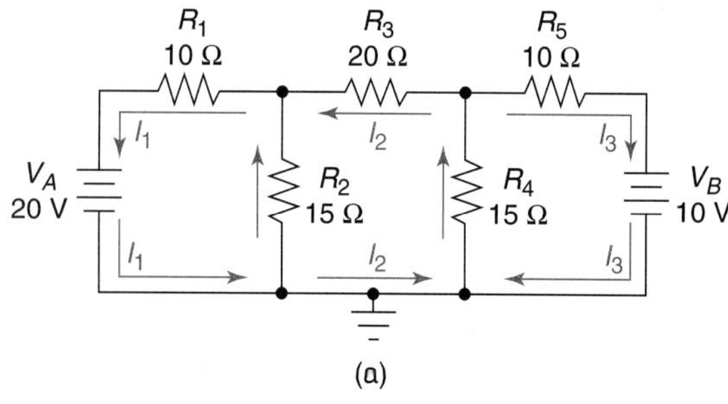


(a)

(b)

**FIGURE 9.24**

## One Final Note

We have now completed our coverage of the two current analysis techniques. In the next section, we will discuss a method for analyzing multisource circuits using voltage relationships.

**Section Review**

1. What is the primary limitation of branch current analysis?
2. What is the term *mesh* generally used to describe?
3. What is the primary difference between mesh current analysis and branch current analysis?
4. How does the relationship between voltage polarity and current direction affect the loop equations in mesh current analysis?

# 9.4 NODE VOLTAGE ANALYSIS

◀ *OBJECTIVE 13*

In this section, we will discuss the third (and simplest) of the circuit analysis techniques, node voltage analysis. As you will see, this method of circuit analysis does not require the use of simultaneous equations. At the same time, its application is limited to two-loop multisource circuits.

## Nodes

Any connection point common to two or more branches in a circuit is referred to as a **node.** For example, points x and y in each of the circuits in Figure 9.25 are nodes.

The term **node voltage** is used to identify the voltage from one node to the other. For each circuit in Figure 9.25, the voltage from x to y is the node voltage. In Figure 9.25a, the node voltage is equal to $V_3$. In Figure 9.25b, the node voltage is equal to $(V_3 + V_4)$.

**Node voltage analysis** is a circuit analysis technique that defines the node voltage in terms of the other voltages in the circuit. As you will see, this approach results in a single-variable circuit equation that can easily be solved using simple algebra.

**Node**
Any connection point common to two or more branches.

**Node voltage**
The voltage from one node to another.

**Node voltage analysis**
A circuit analysis technique where the node voltage is defined in terms of the other voltages in the circuit.

## Node Voltage Analysis

◀ *OBJECTIVE 14*

Node voltage analysis begins with writing a current equation for the component(s) between the nodes. For example, the current through $R_3$ in Figure 9.26 can be found as

$$I_3 = I_1 + I_2$$

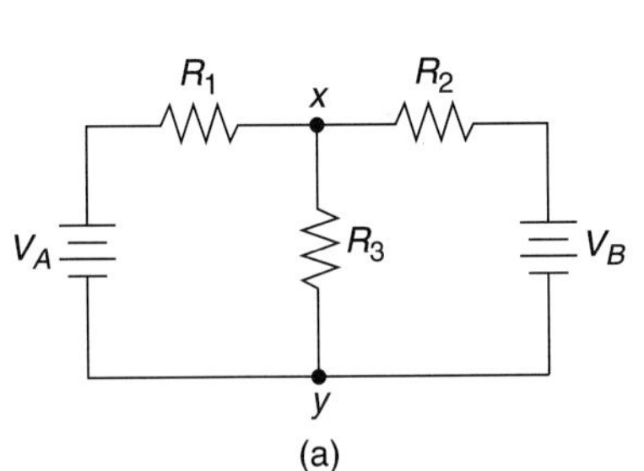


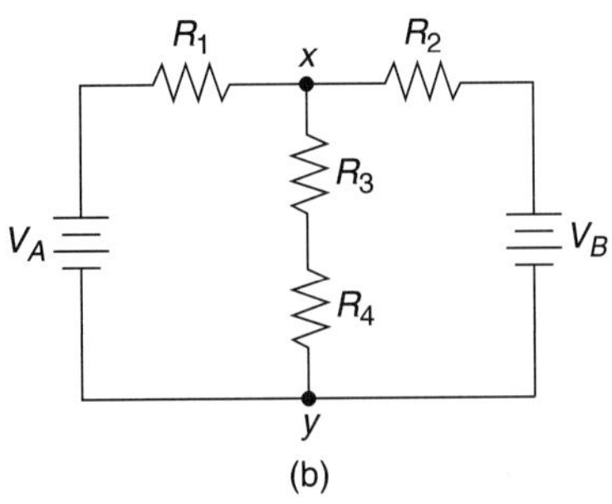


**FIGURE 9.25**

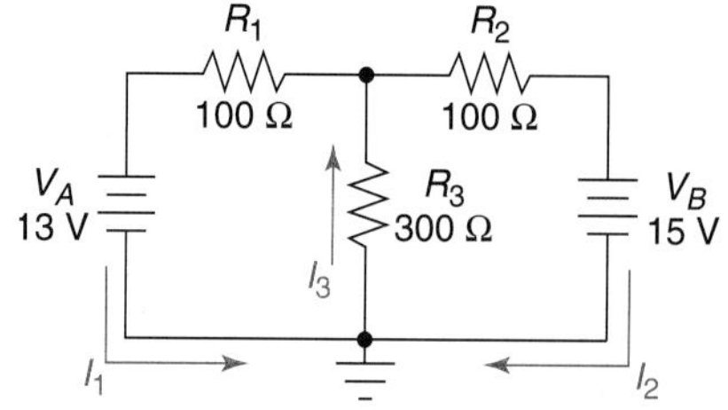


FIGURE 9.26

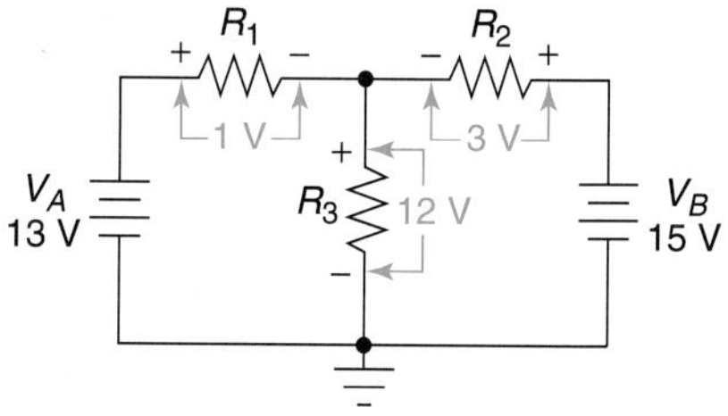


FIGURE 9.27

Now, Ohm's law is used to rewrite the current equation, as follows:

$$\frac{V_3}{R_3} = \frac{V_1}{R_1} + \frac{V_2}{R_2}$$

or (substituting the resistor values):

$$\frac{V_3}{300} = \frac{V_1}{100} + \frac{V_2}{100}$$

Note that the units of measure for the resistors (Ω) have been left out to simplify the equation.

Now we want to eliminate the resistor values from the equation by rewriting it so that the three terms have a common denominator, as follows:

$$\frac{V_3}{300} = \frac{3V_1}{300} + \frac{3V_2}{300}$$

As you can see, the common denominator (300) was achieved by multiplying the $V_1$ and $V_2$ terms by 1 in the form of $\frac{3}{3}$. Once we have a common denominator, it can be dropped, giving us:

$$V_3 = 3V_1 + 3V_2$$

According to Kirchhoff's voltage law, $V_1$ and $V_2$ can be found as

$$\begin{aligned} V_1 &= V_A - V_3 \\ &= 13\text{ V} - V_3 \end{aligned} \qquad \text{and} \qquad \begin{aligned} V_2 &= V_B - V_3 \\ &= 15\text{ V} - V_3 \end{aligned}$$

If we substitute these equations for $V_1$ and $V_2$ in the node equation, we get:

$$V_3 = 3(13\text{ V} - V_3) + 3(15\text{ V} - V_3)$$

We now have a single-variable equation that can be solved easily as follows:

$$\begin{aligned} V_3 &= 3(13\text{ V} - V_3) + 3(15\text{ V} - V_3) \\ V_3 &= 39\text{ V} - 3V_3 + 45\text{ V} - 3V_3 \\ V_3 &= 84\text{ V} - 6V_3 \\ 7V_3 &= 84\text{ V} \\ V_3 &= 12\text{ V} \end{aligned}$$

Once we know the value of $V_3$, the values of $V_1$ and $V_2$ can be found as

$$\begin{aligned} V_1 &= V_A - V_3 \\ &= 13\text{ V} - 12\text{ V} \\ &= 1\text{ V} \end{aligned} \qquad \text{and} \qquad \begin{aligned} V_2 &= V_B - V_3 \\ &= 15\text{ V} - 12\text{ V} \\ &= 3\text{ V} \end{aligned}$$

The voltage values calculated for the circuit are shown in Figure 9.27.

As you have been shown, node voltage analysis allows us to calculate the voltage values in a two-loop circuit without the use of simultaneous equations. The complete procedure is demonstrated further in Example 9.11.

Calculate the values of $V_1$, $V_2$, and $V_3$ for the circuit in Figure 9.28.

EXAMPLE 9.11

EWB

**FIGURE 9.28**

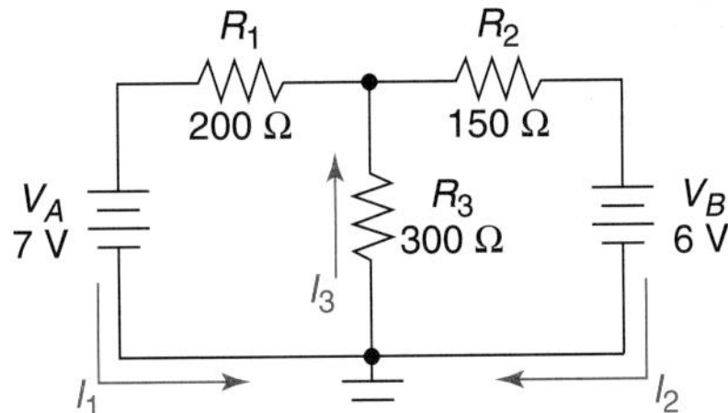


***Solution:*** The current between the nodes can be found as

$$I_3 = I_1 + I_2 \qquad \text{or} \qquad \frac{V_3}{R_3} = \frac{V_1}{R_1} + \frac{V_2}{R_2}$$

Substituting the resistor values gives us

$$\frac{V_3}{300} = \frac{V_1}{200} + \frac{V_2}{150}$$

The lowest common denominator for these three terms is 600. Using this value, the equation is rewritten as

$$\frac{2V_3}{600} = \frac{3V_1}{600} + \frac{4V_2}{600}$$

The denominators can be dropped, giving us:

$$2V_3 = 3V_1 + 4V_2$$

According to Kirchhoff's law:

$$\begin{aligned} V_1 &= V_A - V_3 \\ &= 7\text{ V} - V_3 \end{aligned} \qquad \text{and} \qquad \begin{aligned} V_2 &= V_B - V_3 \\ &= 6\text{ V} - V_3 \end{aligned}$$

Substituting these equations into the node equation, we get

$$\begin{aligned} 2V_3 &= 3V_1 + 4V_2 \\ 2V_3 &= 3(7\text{ V} - V_3) + 4(6\text{ V} - V_3) \end{aligned}$$

Solving for $V_3$, we get:

$$\begin{aligned} 2V_3 &= 3(7\text{ V} - V_3) + 4(6\text{ V} - V_3) \\ 2V_3 &= 21\text{ V} - 3V_3 + 24\text{ V} - 4V_3 \\ 2V_3 &= 45\text{ V} - 7V_3 \\ 9V_3 &= 45\text{ V} \\ V_3 &= 5\text{ V} \end{aligned}$$

Once the value of $V_3$ is known, the other component voltages are found as

$$\begin{aligned} V_1 &= V_A - V_3 \\ &= 7\text{ V} - 5\text{ V} \\ &= 2\text{ V} \end{aligned} \qquad \text{and} \qquad \begin{aligned} V_2 &= V_B - V_3 \\ &= 6\text{ V} - 5\text{ V} \\ &= 1\text{ V} \end{aligned}$$

The calculated component voltages are shown in Figure 9.29a. As you can see, our calculated values are consistent with Kirchhoff's voltage law.

***PRACTICE PROBLEM 9.11***

Using node voltage analysis, calculate the values of $V_1$, $V_2$, and $V_3$ for the circuit in Figure 9.29b.

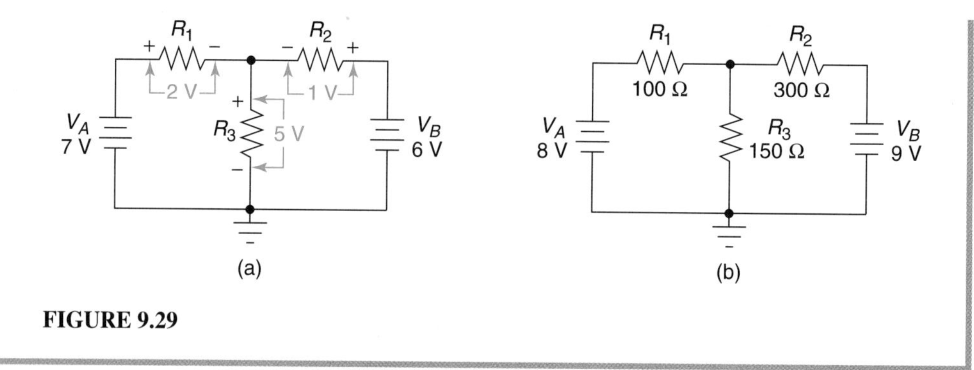


**FIGURE 9.29**

## One Final Note

In this chapter, you have been introduced to three complex analysis techniques. Since multisource circuits like those presented in this chapter are not very common, you probably won't be using any of these techniques all that often. However, learning them does serve several purposes. First, it helps to demonstrate the interdependence of the values in any circuit; that is, it helps to demonstrate how few of the variables in a given circuit are independent. Second, in the event that you do need to perform an analysis of the type shown in this chapter, you will know how to go about it. Of course, the method used in such a case is generally up to you.

## Chapter Summary

Here is a summary of the major points made in this chapter:

1. Two (or more) equations used to describe relationships that occur at the same time are referred to as simultaneous equations.
2. In electronics applications, simultaneous equations:
   a. Describe the operation of the same circuit.
   b. Have only one common solution.
3. The substitution method of solving simultaneous equations involves combining the equations so that they become one single-variable equation. The step-by-step procedure is outlined starting on page 319.
4. The alter-and-add method of solving simultaneous equations is based on the following principle: when two equations are true, the sum of those equations must also be true.
5. The alter-and-add method involves adding two simultaneous equations together so that one of the variables is eliminated. The step-by-step procedure is outlined starting on page 320.
6. The key to successfully using the alter-and-add method is learning to recognize how the equations can be altered to eliminate a variable.
7. A determinant is a value found using the coefficients of the variables in a group of simultaneous equations.
8. To solve for the determinant of a group of equations, the coefficients are first written in the form of a matrix.
   a. A matrix is a column-row structure (see Figure 9.3).
   b. The most common matrices are the 2-by-2 and 3-by-3 matrices. These are also referred to as order-2 and order-3 matrices.
9. After drawing the matrix for a group of equations, a matrix equation is derived and solved (see Figure 9.4).

10. The determinant for a matrix is the solution to the matrix equation.
11. A group of simultaneous equations has more than one matrix.
    a. The determinant matrix is written using the variable coefficients.
    b. The matrix for each variable is written using the constants in place of the coefficients for the variable (see Figure 9.6).
12. The value of each variable in a group of equations is found as follows:
    a. Derive and solve the determinant matrix.
    b. Derive and solve the variable matrix.
    c. Divide the result from Step b by the result from Step a.
13. A 3-by-3 matrix is derived and solved as shown in Figures 9.8 and 9.9.
14. Each loop in a multisource circuit is referred to as a branch. The current through each loop is referred to as a branch current.
15. Branch current analysis involves:
    a. Writing the branch current equation for each loop in the circuit.
    b. Solving the simultaneous branch current equations using one of the techniques outlined in this chapter.
16. Branch current analysis (like any other technique) has its limitations: it doesn't work well on three-loop circuits.
17. A multisource circuit containing more than one loop is often referred to as a mesh. The term *mesh* is often used to indicate that one or more loops are not tied directly to either voltage (or current) source.
18. Mesh current analysis is nearly identical to branch current analysis. In mesh current analysis, loop currents are always assumed to be in a clockwise direction, regardless of the polarity of the source.
19. When writing mesh equations, the relationship between the assumed direction of current and the polarity of the voltage source must be taken into account.
    a. When they match, the source voltage is written as a positive value in the mesh equation.
    b. When they do not match, the source voltage is written as a negative value in the mesh equation.
20. Mesh current analysis can be used to solve three-loop circuits (see Example 9.10).
21. Any connection point common to two or more branches in a circuit is referred to as a node.
22. The term *node voltage* is used to identify the voltage between two nodes.
23. *Node voltage analysis* does not require the use of simultaneous equations.
24. The procedure for *node voltage analysis* is described in detail beginning on page 343.

## Key Terms

The following terms were introduced and defined in this chapter:

alter-and-add method
branch
branch current
branch current analysis
determinant
matrix
matrix equation
mesh
mesh current analysis
node
node voltage
node voltage analysis
observation method
order-3 matrix
order-2 matrix
simultaneous equations
substitution method
3-by-3 matrix
2-by-2 matrix

## Practice Problems

1. Using the substitution method, find the common solution to each pair of simultaneous equations.

   a. $5x + 3y = 29$
      $2x - y = 5$

   b. $3x - 7y = 21$
      $x + 8y = 69$

   c. $12x - 5y = 19$
      $x + 20y = 22$

   d. $3x + 12y = 48$
      $2x - y = -21$

2. Using the substitution method, find the common solution to each pair of simultaneous equations.

   a. $3I_1 - 5I_2 = 3$ mA
      $I_1 - 2I_2 = 2$ mA

   b. $15I_1 - I_2 = 56$ mA
      $I_1 - 3I_2 = 36$ mA

   c. $3V_1 + V_2 = 22$ V
      $5V_1 + 18V_2 = 53$ V

   d. $10V_1 + 2V_2 = 66$ V
      $-V_1 - V_2 = -15$ V

3. Using the alter-and-add method, find the common solution to each pair of simultaneous equations.

   a. $4x + 3y = 3$
      $3x + 6y = 36$

   b. $-5x - 7y = 29$
      $3x - 4y = -1$

   c. $3x - 4y = 17.5$
      $2x + 5y = -7.5$

   d. $15x + 12y = 21$
      $-5x + 4y = 17$

4. Using the alter-and-add method, find the common solution to each pair of simultaneous equations.

   a. $3V_1 + 2V_2 = 40$ V
      $10V_1 - 4V_2 = 0$ V

   b. $3I_1 + 2I_2 = 39$ mA
      $10I_1 - 4I_2 = 2$ mA

   c. $10V_1 - 2V_2 = 22$ V
      $-4V_1 + 4V_2 = -20$ V

   d. $6I_1 + 8I_2 = 69$ mA
      $8I_1 + 16I_2 = 116$ mA

5. Calculate the determinant for each pair of equations in Problem 1.

6. Calculate the determinant for each pair of equations in Problem 2.

7. Using determinants, find the common solution to each pair of simultaneous equations listed in Problem 3.

8. Using determinants, find the common solution to each pair of simultaneous equations listed in Problem 4.

*Note:* In the chapter discussions, problems were developed specifically to provide whole-number answers. Problems 9 and 10 do not work out as neatly as the in-chapter problems.

9. Using determinants, find the common solution to each group of simultaneous equations.

   a. $2x + 4y - 3z = 27$
      $3x - 2y + 5z = -20$
      $5x + y + 3z = 0$

   b. $4x + 2y + 12z = 82$
      $7x - 5y + 7z = 57$
      $2x + 3y - 2z = -29$

10. Using determinants, find the common solution to each group of simultaneous equations.

    a. $3V_1 + 2V_2 + 2V_3 = 42$ V
       $2V_1 + 3V_2 - 5V_3 = -82$ V
       $5V_1 - 4V_2 + V_3 = 34$ V

    b. $3I_1 + 2I_2 = 34$ mA
       $-2I_1 + 8I_2 - 3I_3 = 31$ mA
       $-5I_2 + 20I_3 = 100$ mA

11. Using branch current analysis, calculate the loop current and component voltage values for the circuit in Figure 9.30a.

12. Using branch current analysis, calculate the loop current and component voltage values for the circuit in Figure 9.30b.

13. Using branch current analysis, calculate the loop current and component voltage values for the circuit in Figure 9.30c.

14. Using branch current analysis, calculate the loop current and component voltage values for the circuit in Figure 9.30d.

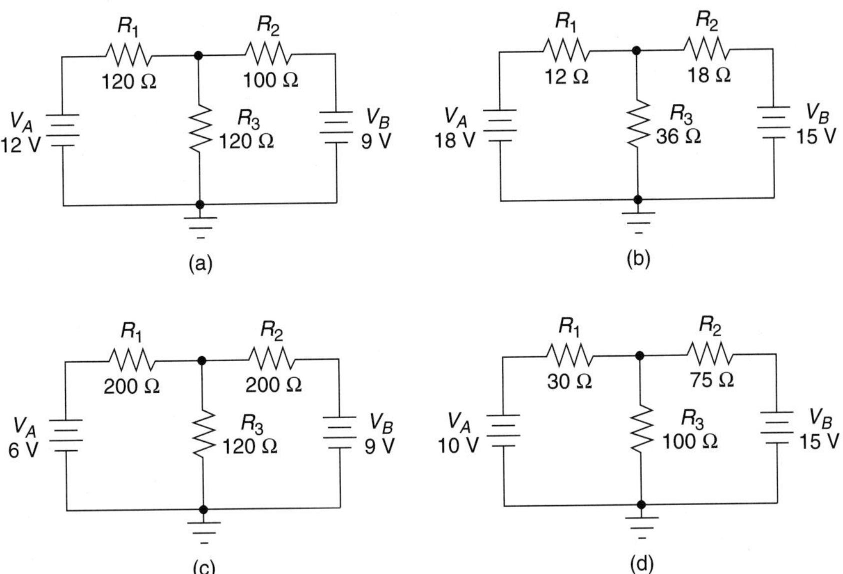


**FIGURE 9.30**

15. Using mesh current analysis, calculate the loop current and component voltage values for the circuit in Figure 9.31a.
16. Using mesh current analysis, calculate the loop current and component voltage values for the circuit in Figure 9.31b.
17. Using mesh current analysis, calculate the loop current and component voltage values for the circuit in Figure 9.31c.
18. Using mesh current analysis, calculate the loop current and component voltage values for the circuit in Figure 9.31d.
19. Using mesh current analysis, calculate the loop current and component voltage values for the circuit in Figure 9.32a.
20. Using mesh current analysis, calculate the loop current and component voltage values for the circuit in Figure 9.32b.

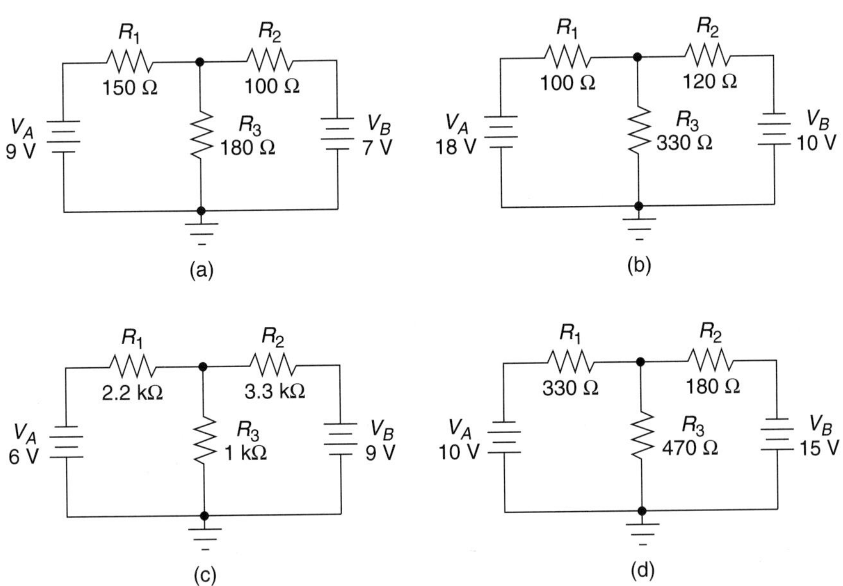


**FIGURE 9.31**

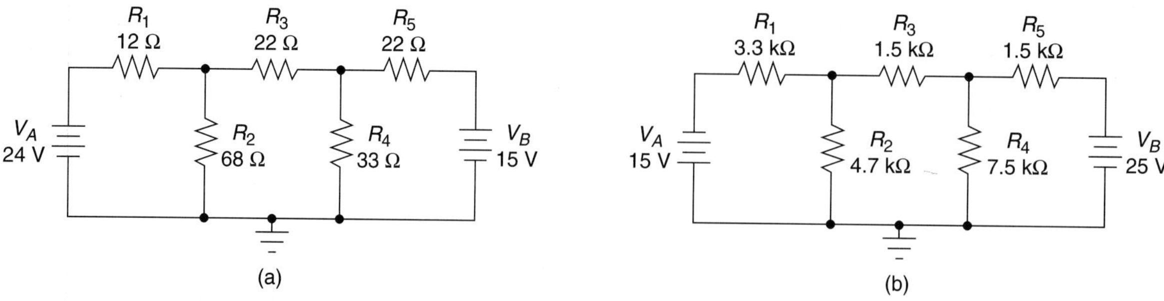


**FIGURE 9.32**

**21.** Using node voltage analysis, calculate the component voltage values for the circuit in Figure 9.30a.

**22.** Using node voltage analysis, calculate the component voltage values for the circuit in Figure 9.30b.

**23.** Using node voltage analysis, calculate the component voltage values for the circuit in Figure 9.30c.

**24.** Using node voltage analysis, calculate the component voltage values for the circuit in Figure 9.30d.

## The Brain Drain

**25.** Using calculated circuit values, prove that $V_2 + V_3 + V_4 = 0$ V for the circuit in Figure 9.33.

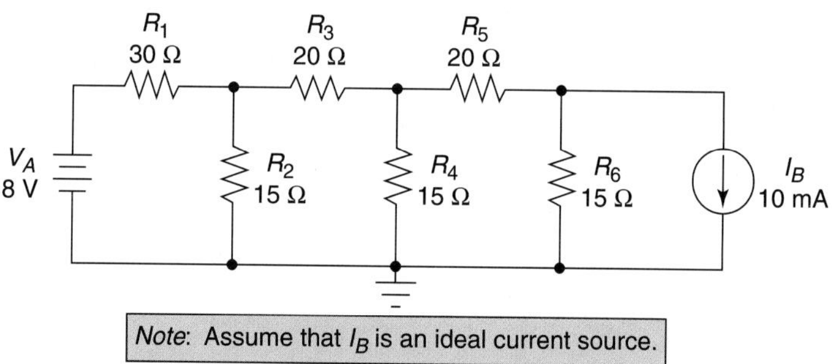


EWB

**FIGURE 9.33**

## Answers to the Example Practice Problems

**9.1.** $V_1 = 15$ V, $V_2 = 7$ V
**9.2.** $x = -2, y = 4$
**9.3.** $x = 4, y = 8$
**9.4.** $x = 6, y = -2$
**9.5.** $x = 4.5, y = 3$
**9.6.** $x = -1, y = 2, z = 5$
**9.7.** $I_1 = 20$ mA, $I_2 = 60$ mA, $V_1 = 2$ V, $V_2 = 9$ V, $V_3 = 12$ V
**9.8.** $I_1 = 38.9$ mA, $I_2 = 81.7$ mA, $I_3 = 42.8$ mA
$V_1 = 5.84$ V, $V_2 = 8.17$ V, $V_3 = 12.84$ V
**9.9.** $I_1 = -120$ mA, $I_2 = 80$ mA
**9.10.** $I_1 = -240.4$ mA, $I_2 = 65.7$ mA, $I_3 = 426.2$ mA
**9.11.** $V_1 = 2.5$ V, $V_2 = 3.5$ V, $V_3 = 5.5$ V

## EWB Applications Problems

| Figure | EWB File Reference |
|---|---|
| 9.2 | EWBA9_2.ewb |
| 9.11 | EWBA9_11.ewb |
| 9.22 | EWBA9_22.ewb |
| 9.28 | EWBA9_28.ewb |
| 9.33 | EWBA9_33.ewb |

# 10

# AN INTRODUCTION TO MAGNETISM

## OBJECTIVES

*After studying the material in this chapter, you should be able to:*

1. Describe magnetic force.
2. Identify the poles of a magnet and state the relationships between them.
3. List and define the common units of magnetic flux.
4. Contrast the following: magnetic fields, magnetic flux, and flux density.
5. List and describe the common units of flux density.
6. Calculate the flux density at any point in a magnetic field, given the cross-sectional area and the total flux.
7. Compare and contrast permeability with relative permeability.
8. Given the permeability of a material, calculate its relative permeability.
9. Discuss the domain theory of the source of magnetism.
10. Describe the magnetic induction methods of producing a magnet.
11. Define retentivity and discuss its relationship to permeability.
12. Define reluctance and discuss its relationship to permeability.
13. List and describe the magnetic classifications of materials.
14. Describe the relationship between current and a resulting magnetic field.
15. Compare the basic magnetic and electric quantities.
16. Discuss the relationships that exist among the following: magnetomotive force, ampere-turns, coil current, core permeability, and coil flux density.
17. Define and discuss hysteresis.
18. Describe magnetic shielding.
19. List and describe the various shapes of magnets.
20. Describe the proper care and handling of magnets.

Magnetism plays an important role in the operation of many electronic components and systems. For example, audio and video systems, electric motors, and many analog meters use the properties of magnetism to some extent.

In this chapter, we will discuss many of the basic properties of magnetism. A complete study of the subject would require several volumes, so our discussion will be limited to those principles that relate directly to the study of electricity and electronics.

## 10.1 MAGNETISM: AN OVERVIEW

As you know, a magnet can attract (or repel) another magnet or any piece of iron. The force that a magnet exerts on the objects around it is referred to as **magnetic force.**

◄ *OBJECTIVE 1*

**Magnetic force**
The force that a magnet exerts on the objects around it.

When exposed to magnetic force, iron filings line up as shown in Figure 10.1a. For this reason, magnetic force is commonly represented as a series of lines. These lines form closed loops around (and through) the magnet, as illustrated in Figure 10.1b.

*Note:* The pattern shown in Figure 10.1a was produced using a circular magnet.

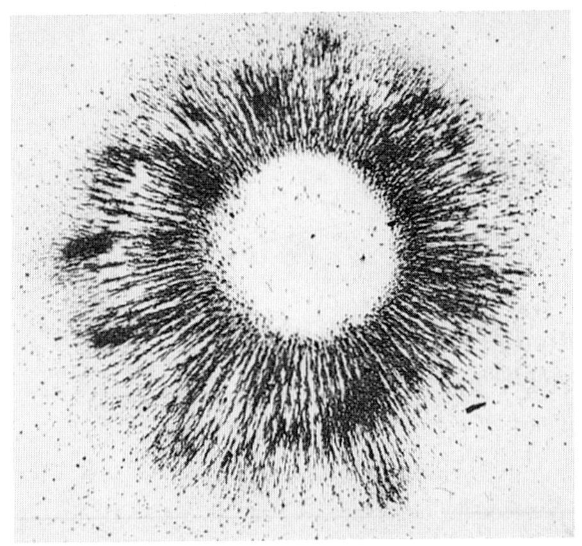

(a) Iron filings lining up along lines of magnetic force

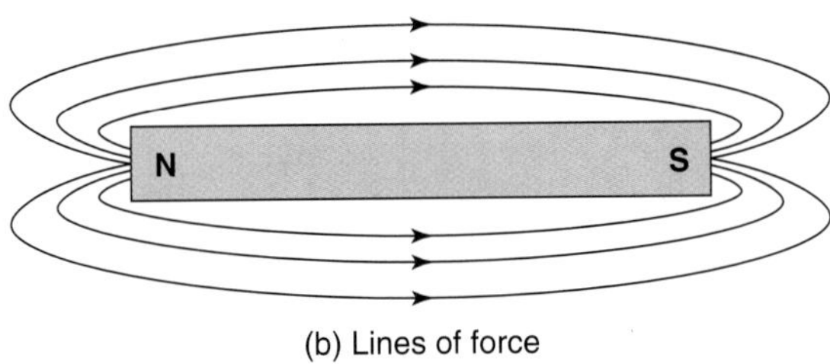


(b) Lines of force

**FIGURE 10.1**

### Magnetic Poles

The points where magnetic lines of force leave (and return to) a magnet are referred to as the **poles.** By convention, the lines of force are assumed to emanate from the **north-seeking pole (N)** and return to the magnet via its **south-seeking pole (S).** Within the magnet, the lines of force continue from (S) to (N).

◄ *OBJECTIVE 2*

**Poles**
The points where magnetic lines of force leave (and return to) a magnet. The poles are referred to as the north-seeking pole (N) and the south-seeking pole (S).

The terms used to describe the poles of a magnet are derived as shown in Figure 10.2. The earth itself is one huge magnet, with *magnetic north* and *magnetic south* poles. If we suspend a magnet, the magnet turns so that one pole points to magnetic north and the other points to magnetic south. The poles are then identified as the north-seeking and south-seeking poles.

FIGURE 10.2 Pole identification.

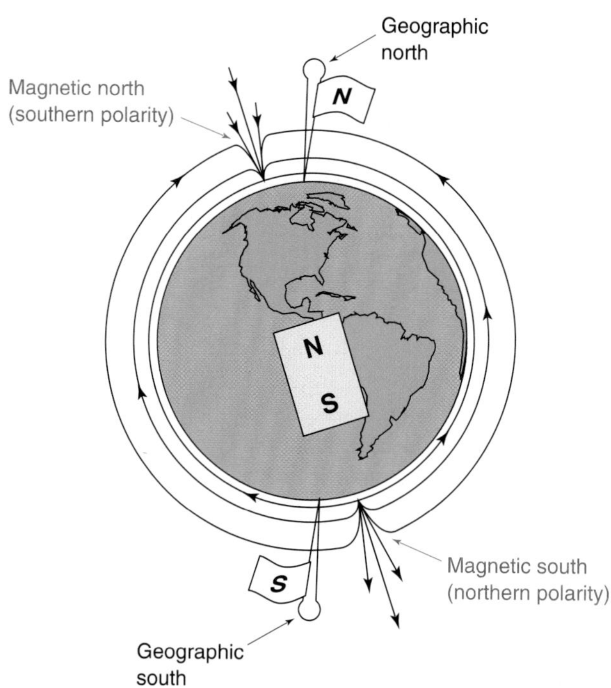


## Like Poles and Unlike Poles

You may recall that like charges repel (and unlike charges attract) each other. Magnetic poles have similar effects on each other; that is:

1. Like poles repel each other.
2. Unlike poles attract each other.

These interactions between the poles are illustrated in Figure 10.3.

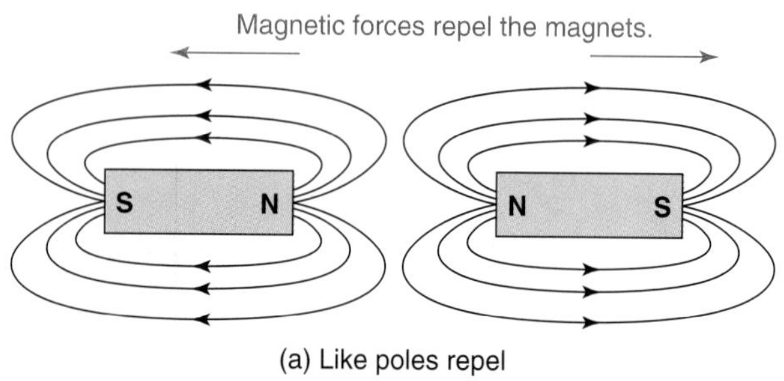


(a) Like poles repel

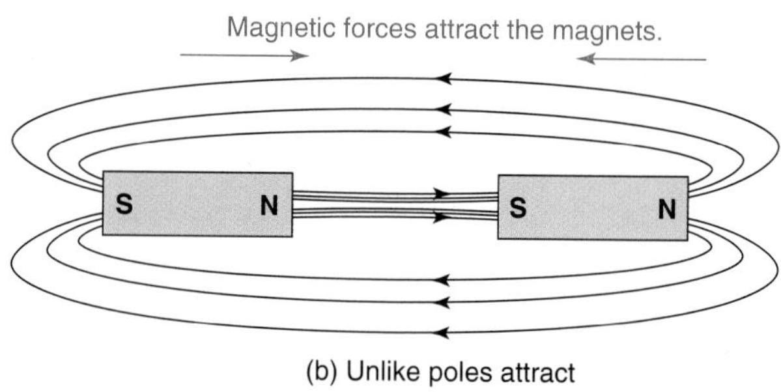


(b) Unlike poles attract

**FIGURE 10.3 Pole interactions.**

## Magnetic Flux

*OBJECTIVE 3* ▶

**Magnetic flux**
A term used to describe the total lines of force produced by a magnet.

The lines of force produced by a magnet are collectively referred to as **magnetic flux.** There are two fundamental units of measure for magnetic flux. These units are identified and defined as follows:

| Unit of Measure | Unit System | Value (Defined) |
|---|---|---|
| maxwell (Mx) | cgs | 1 Mx = 1 line of force |
| weber (Wb) | SI | 1 Wb = $1 \times 10^8$ Mx |

The maxwell (Mx) is easy to visualize. Each magnet represented in Figure 10.3 has six lines of force. Therefore, the magnetic flux (in maxwells) for each magnet is given as

$$\Phi = 6 \text{ Mx}$$

where $\Phi$ (the Greek letter *phi*) is used to represent magnetic flux.

The maxwell is not a practical unit of measure. Even small magnets produce flux in the thousands of maxwells (and higher). On the other hand, the weber (Wb) is often too large a unit of measure to be useful. In such cases, the **micro-weber (μWb)** is the preferred unit of measure. The amount of flux contained in one micro-weber can be determined as follows

**Micro-weber (μWb)** A practical unit of measure of magnetic flux, equal to 100 maxwells (lines of force).

$$\begin{aligned} 1 \ \mu\text{Wb} &= (1 \times 10^{-6}) \text{ Wb} \\ &= (1 \times 10^{-6})(1 \times 10^{8} \text{ Mx}) \\ &= 1 \times 10^{2} \text{ Mx} \\ &= 100 \text{ Mx} \end{aligned}$$

Therefore, one micro-weber contains 100 lines of flux.

Magnetic characteristics are typically measured in both cgs (centimeter-gram-second) units and SI (Système International) units. Throughout the remainder of this chapter, we will focus on the SI units of measure for magnetic quantities because they are used more often than cgs units. The cgs units, along with any applicable conversions, will be identified in the margins.

## Flux Density

◄ *OBJECTIVE 4*

The area of space surrounding a magnet contains magnetic flux. This area of space is referred to as a **magnetic field.** A two-dimensional representation of a magnetic field is shown in Figure 10.4. It should be noted that the field is actually a three-dimensional space (similar to a cylinder) that surrounds the magnet.

**Magnetic field** The area of space surrounding a magnet that contains magnetic flux.

**FIGURE 10.4 Magnetic field.**

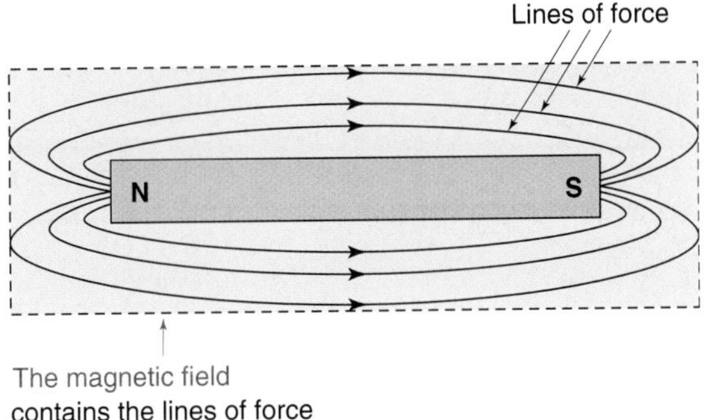


The strength of the force produced by a magnet is a function of the field size and the amount of flux it contains: the greater the amount of flux per unit area, the stronger the magnetic force. **Flux density** is a measure of flux per unit area, and therefore, it is an indicator of magnetic strength. (In fact, flux density is often referred to as field strength.) By formula, flux density is found as

**Flux density** The amount of flux per unit area. Flux density is an indicator of field strength.

$$B = \frac{\Phi}{A} \tag{10.1}$$

where $B$ = the flux density
$\Phi$ = the amount of flux
$A$ = the cross-sectional area containing the flux

The concept of flux density is illustrated in Figure 10.5. As you can see, the area in Figure 10.5a has six lines of flux passing through it. An area of equal size shown in Figure 10.5b has only three lines of flux. Since flux density is a measure of flux per unit area, the flux density in Figure 10.5a is twice that in Figure 10.5b.

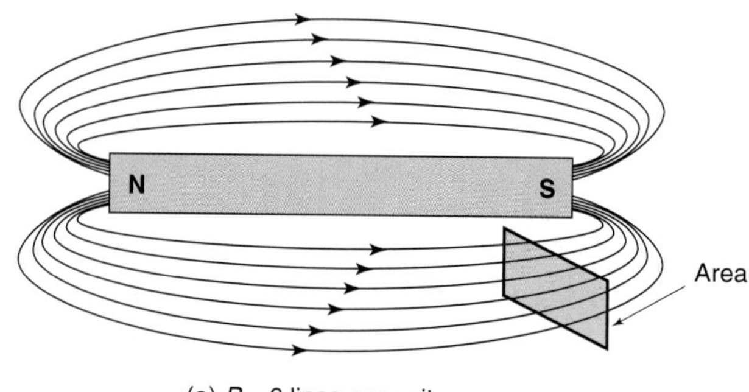


(a) $B = 6$ lines per unit area

N
S
Area

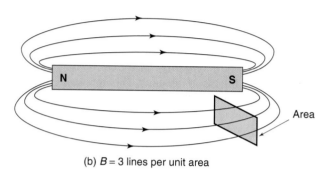

(b) $B = 3$ lines per unit area

**FIGURE 10.5 Flux density.**

*OBJECTIVE 5* ▶

As you know, the SI unit of magnetic flux is the weber. The SI defines flux density in terms of webers per square meter. The SI unit of measure for flux density, the tesla (T) is defined as follows:

*Note:* The cgs unit of measure for flux density is the *gauss.* One gauss equals 1 maxwell per square centimeter ($1\ \text{G} = 1\ \text{Mx/cm}^2$).

| Unit of Measure | Value (Defined) |
|---|---|
| tesla (T) | $1\ \text{T} = 1\ \dfrac{\text{Wb}}{\text{m}^2}$ |

The calculation of flux density (in teslas) is demonstrated in the following example.

## EXAMPLE 10.1

Calculate the flux density for the shaded area in Figure 10.6a. Assume that each line of force represents 40 μWb of flux.

*OBJECTIVE 6* ▶

***Solution:*** The shaded area is shown to be one square centimeter ($1\ \text{cm}^2$), which is equal to $1 \times 10^{-4}$ square meters. Therefore, the flux density represented in Figure 10.6a is found as

$$\begin{aligned} \boldsymbol{B} &= \frac{\Phi}{A} \\ &= \frac{(3)(40 \times 10^{-6}\ \text{Wb})}{1 \times 10^{-4}\ \text{m}} \\ &= 1.2 \times 10^{-1}\ \text{Wb/m}^2 \\ &= \mathbf{1.2 \times 10^{-1}\ T} \end{aligned}$$

***PRACTICE PROBLEM 10.1***

Calculate the flux density for the shaded area in Figure 10.6b. Assume that each line of force represents 75 μWb of flux.

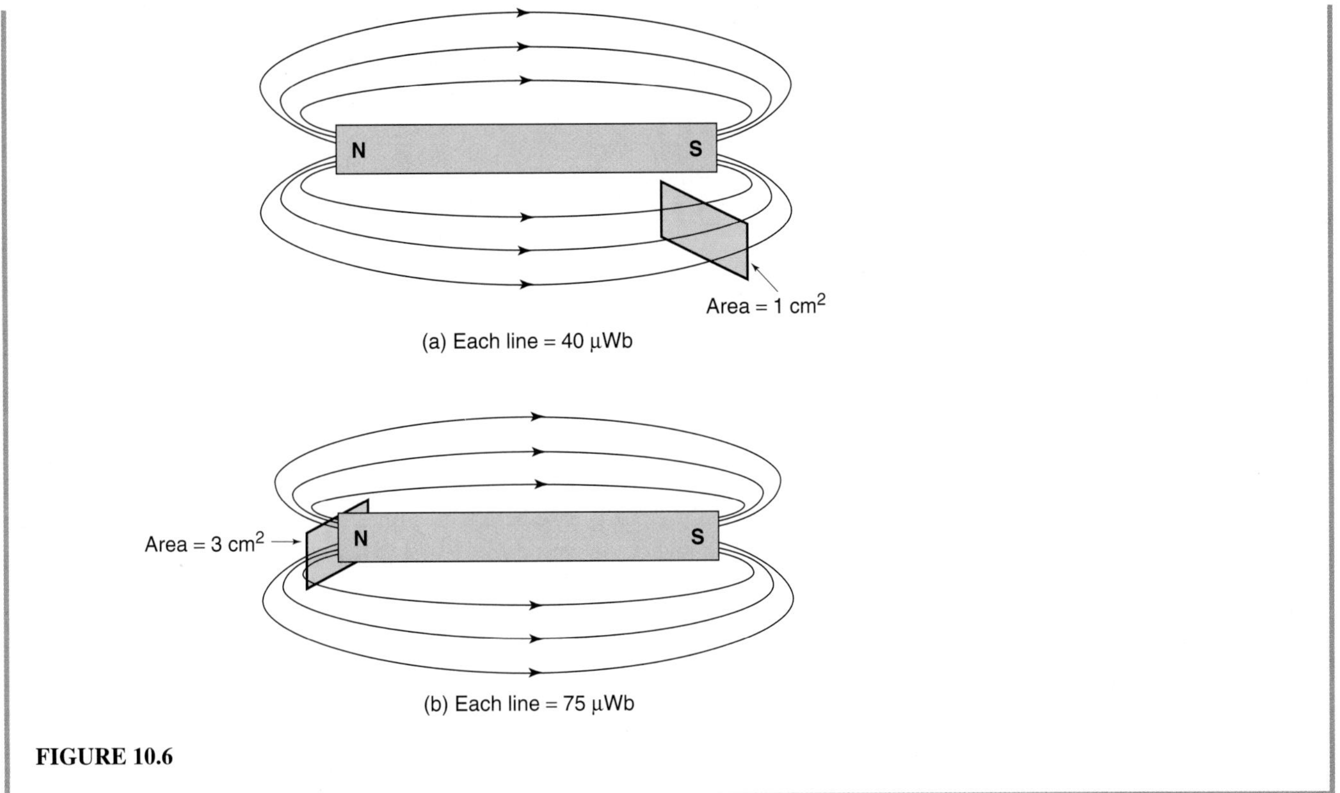


**FIGURE 10.6**

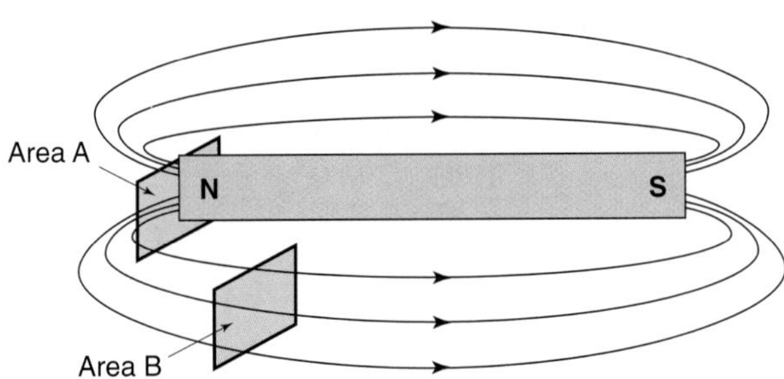


**FIGURE 10.7 Flux density ($B$) is greatest at the poles.**

For any magnet, the maximum value of flux density can be measured at the poles. This point is illustrated in Figure 10.7. As you can see, there is twice as much flux in area A as in area B. Since all lines of flux emanate from (and return to) the magnetic poles, the flux density must be greater at the poles than anywhere else in the magnetic field.

## Section Review

1. What is magnetic force?
2. How is magnetic force commonly represented?
3. What are the poles of a magnet? What are they called?
4. How do like and unlike poles affect each other?
5. What is magnetic flux?
6. List and define the common units of measure for magnetic flux.
7. What is a magnetic field?

8. What is flux density? What does it indicate?
9. List and define the units of measure for flux density.

## 10.2 MAGNETIC CHARACTERISTICS OF MATERIALS

Some materials respond to the presence of a magnetic force while others do not. In this section, we will discuss the magnetic characteristics and classifications of materials.

### Permeability

*OBJECTIVE 7* ▶

**Permeability** is a measure of the ease with which lines of magnetic force are established within a material. The higher the permeability of a given material, the easier it is to produce magnetic lines of force within the material.

**Permeability**
A measure of how easily magnetic lines of force are established within a material.

Materials with high permeability have the ability to concentrate magnetic lines of force. This point is illustrated in Figure 10.8. The bar shown is made of soft iron, a material with a permeability much higher than that of the surrounding air. Since the iron bar accepts the lines of force more readily, they align themselves in the iron bar as shown.

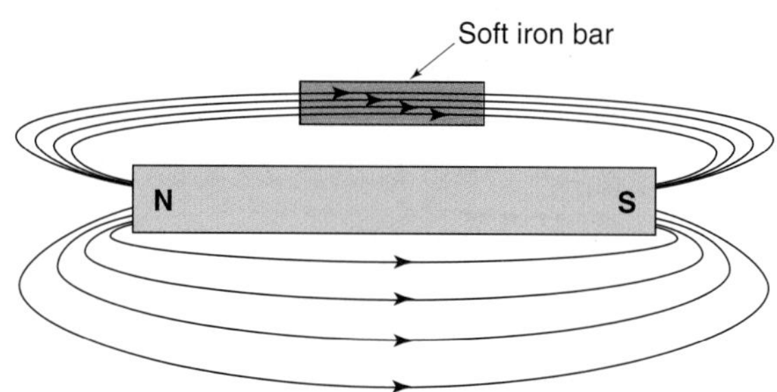


**FIGURE 10.8 The effect of a high-permeability material on magnetic flux.**

Since the lines of force in Figure 10.8 are drawn to the iron bar, the flux density at any point in the bar is greater than in the surrounding air. For example, assume that the cross-sectional areas shown in Figure 10.9 represent one square centimeter (1 $cm^2$). As you can see, the flux density in the iron bar (area A) is twice that of area B. Thus, high permeability materials can be used to increase flux density.

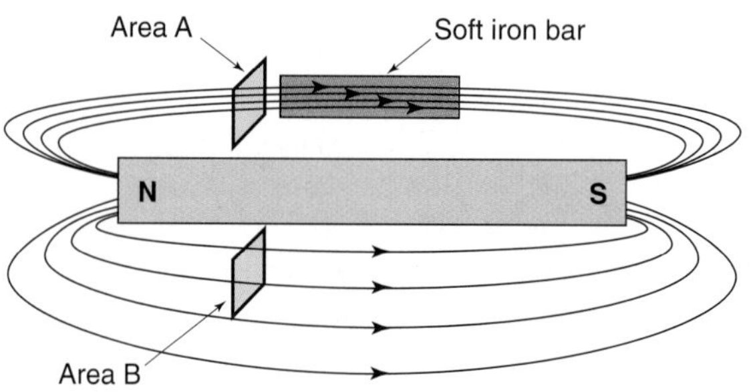


**FIGURE 10.9**

## Relative Permeability

In general, materials are rated according to their **relative permeability ($\mu_r$):** the ratio of the material's permeability to that of free space. In SI units, the permeability of free space ($\mu_0$) is known to be

**Relative permeability ($\mu_r$)** The ratio of a material's permeability to that of free space.

$$\mu_0 = 4\pi \times 10^{-7} \frac{\text{Wb}}{\text{A} \cdot \text{m}}$$

where Wb/A · m (webers per ampere-meter) is the unit of measure for permeability

*Note:* The value of $4\pi \times 10^{-7}$ given in the permeability relationship is often written as $1.26 \times 10^{-6}$. The two values are the same.

The *ampere-meter* is discussed in detail later in this chapter. (For now, we simply want to establish the fact that permeability has a unit of measure.)

When the permeability of a material is known, its relative permeability is found as

◀ *OBJECTIVE 8*

$$\mu_r = \frac{\mu_m}{\mu_0} \qquad \textbf{(10.2)}$$

where $\mu_r$ = the relative permeability of the material
$\mu_m$ = the permeability of the material, in Wb/A · m
$\mu_0$ = the permeability of free space, in Wb/A · m

Since relative permeability is a ratio of one permeability value to another, it has no unit of measure. The concept of relative permeability is illustrated further in the following example.

### EXAMPLE 10.2

The permeability of a piece of iron is approximately $2.5 \times 10^{-4}$ Wb/A · m. Calculate the relative permeability of this material.

***Solution:*** Using the values given for iron and free space, the relative permeability of the iron is found as

$$\mu_r = \frac{\mu_m}{\mu_0} = \frac{2.5 \times 10^{-4} \frac{\text{Wb}}{\text{A} \cdot \text{m}}}{4\,\pi \times 10^{-7} \frac{\text{Wb}}{\text{A} \cdot \text{m}}} = \mathbf{199}$$

This result indicates that the permeability of the iron is approximately 200 times that of *an equal volume* of free space.

Relative permeability is important because it equals the ratio of flux density within a material to flux density in free space; that is, it tells us how much flux density is increased by a given material relative to the space surrounding it. For example, if we assume that the flux density in the space surrounding an iron bar equals $2 \times 10^{-3}$ teslas, the flux density in the iron bar is 200 times as great, or $4 \times 10^{-1}$ teslas.

Why $\mu_r$ is important.

## The Source of Magnetism: Domain Theory

The domain theory is one of many possible explanations for the source of magnetism. To understand this theory, we need to take another look at the makeup of the atom.

◀ *OBJECTIVE 9*

As you know, all atoms contain electrons that orbit a nucleus. It is believed that every electron in a given atom has both an electric charge and a magnetic field. The electric charge is always negative (and is offset by a matching positive charge in the nucleus). The polarity of the magnetic field, however, depends on the direction the electron spins on its own axis.

Each electron in an atom spins in a clockwise or counterclockwise direction. For example, let's assume that each of the electrons in Figure 10.10 is spinning in the direction

**FIGURE 10.10 Electron spin.**

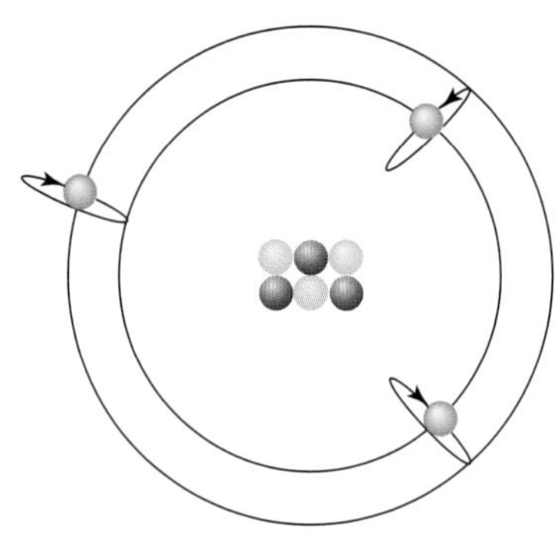

indicated. When an atom contains an equal number of electrons spinning in each direction, their magnetic fields cancel each other. As a result, the atom does not generate magnetic force. When the electrons spinning in one direction outnumber the ones spinning in the other direction, the atom acts as a source of magnetic force. The polarity of the magnetic force depends on the net spin direction of the electrons.

**Domain theory**
One possible explanation for the source of magnetism. This theory states that atoms with like magnetic fields join together to form magnetic domains, each acting like a magnet.

**Domain theory** states that atoms with like magnetic fields join together to form magnetic domains. Each domain acts as a magnet, with two poles and a magnetic field. In a nonmagnetized material, the domains are positioned randomly as shown in Figure 10.11a. As a result, the magnetic fields produced by the domains cancel each other, and the material has no net magnetic force. However, if the domains can be aligned (as shown in Figure 10.11b), the material generates magnetic force.

**FIGURE 10.11**

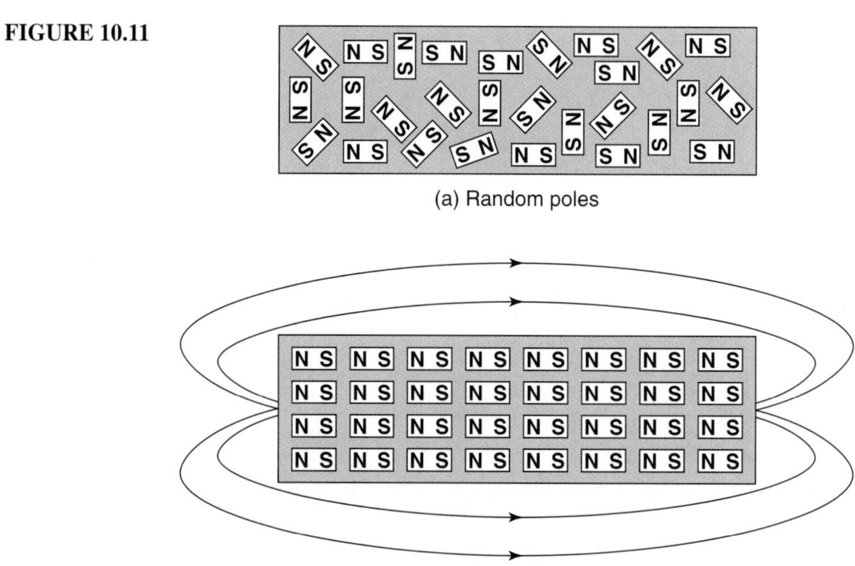

(a) Random poles

(b) Aligned poles and resulting flux

## *Magnetic Induction*

*OBJECTIVE 10* ▶

**Induction**
The process of producing an artificial magnet.

Most magnets are artificially produced; that is, the magnetic domains are aligned using an external force. The process of producing a magnet in this fashion is generally referred to as **induction.** Three methods of producing a magnet by induction are illustrated in Figure 10.12.

In Figure 10.12a, a bar of steel is stroked with a magnet. The flux generated by the magnet aligns the magnetic domains in the steel. As a result, the steel becomes magnetized. It is also possible to magnetize an object simply by placing it within the field

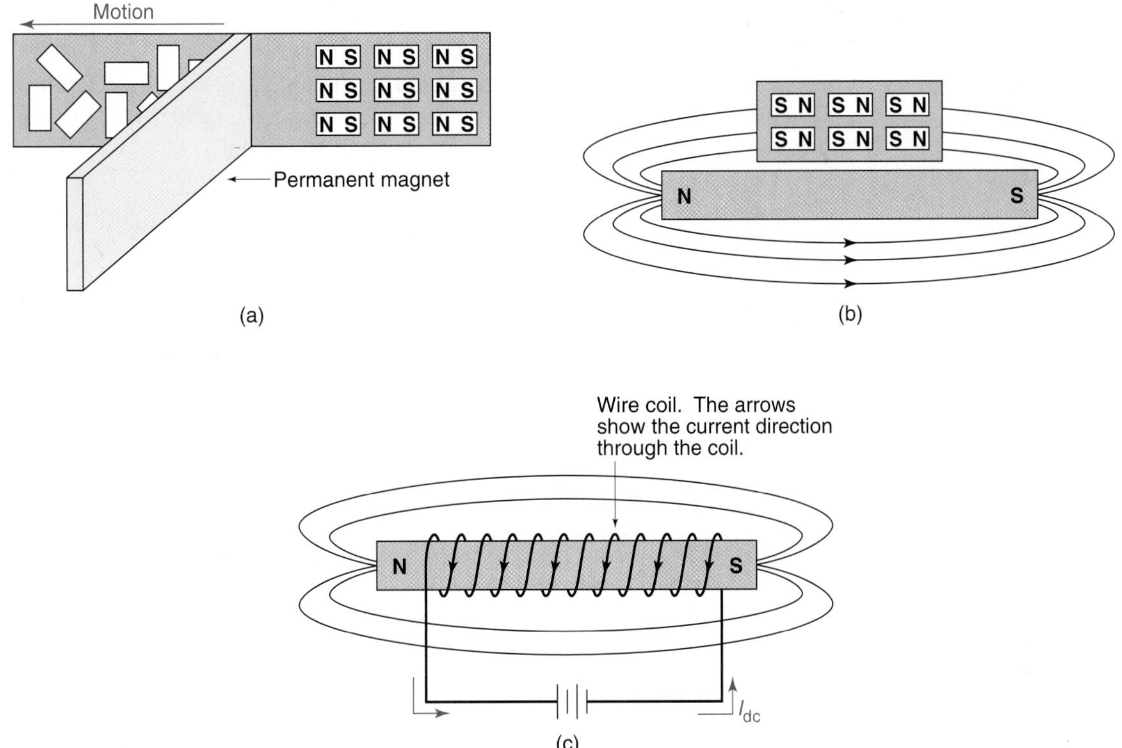


**FIGURE 10.12 Aligning magnetic domains.**

of a magnet. This method of producing a magnet is illustrated in Figure 10.12b. In this case, the flux produced by the magnet aligns the domains in the iron bar. As a result, the iron bar becomes magnetized. Note that the poles in the iron bar are opposite those in the magnet.

The methods described above are referred to as **magnetic induction** because an external magnetic force is used to align the domains. An iron bar can also be magnetized using *a strong* dc current, as shown in Figure 10.12c. In the next section, we will discuss the relationship between electric current and magnetic fields. For now, we will simply state that a strong dc current can be used to align magnetic domains. Thus, the iron bar in Figure 10.12c is magnetized by the dc current through the coil.

**Magnetic induction**
Using an external magnetic force to align the magnetic domains within a material, thereby magnetizing the material.

The higher the permeability of a material, the easier it is to align its magnetic domains, and thus, become magnetized. Since high-permeability materials have the ability to concentrate magnetic flux, it follows that the increased flux density aligns the domains in the material better. On the other hand, low-permeability materials do not concentrate magnetic flux. As a result, their domains do not align themselves as well.

## Retentivity

◄ *OBJECTIVE 11*

High-permeability materials are magnetized easily, but they lose most of their magnetic strength quickly. On the other hand, low-permeability materials are difficult to magnetize, but they tend to retain most of their magnetic strength for a long time. The ability of an artificial magnet to retain its magnetic characteristics is referred to as **retentivity.** High-retentivity materials, such as hardened steel, are used to produce permanent magnets. Once magnetized, they remain magnetic. Low-retentivity materials, such as soft iron, act as temporary magnets; that is, they lose most of their magnetic strength soon after being isolated from an external source of magnetic force.

**Retentivity**
The ability of a material to retain its magnetic characteristics after a magnetizing force is removed.

## Reluctance

OBJECTIVE 12 ▶

**Reluctance**
The opposition that a material presents to magnetic lines of force (the magnetic equivalent of resistance).

As you know, the opposition to current in a dc circuit is called resistance. The opposition that a material presents to magnetic lines of force is called **reluctance.** In essence, reluctance can be viewed as the magnetic resistance of a material.

The reluctance of a material can be found as

$$\Re = \frac{\ell}{\mu A} \tag{10.3}$$

where
$\Re$ = the reluctance of the material
$\ell$ = the length of the material
$\mu$ = the permeability of the material
$A$ = the cross-sectional area of the material

*Note:* In the next section, you will see that current (in amperes) can be used to generate flux (in webers) through a material. The unit of A/Wb indicates that the amount of current required to generate a given amount of flux is directly proportional to the reluctance of the material.

The equation for magnetic reluctance is very similar to equation (2.5), which is used to find the resistance of a material.

Equation (10.3) demonstrates the fact that reluctance varies inversely with permeability. This makes sense when you consider that:

1. Permeability is a measure of how easily a material passes magnetic flux.
2. Reluctance is a measure of opposition to magnetic flux.

Reluctance is typically given as a ratio. For example, the SI unit of measure for reluctance is *amperes per weber* (A/Wb). In the cgs system, reluctance is measured in gilberts per maxwell (Gb/Mx).

## Material Classifications

OBJECTIVE 13 ▶

All materials can be classified according to their magnetic characteristics. The magnetic classifications of materials and their characteristics are identified in Table 10.1. As Table 10.1 illustrates:

1. High-permeability (ferromagnetic) materials are considered to be magnetic, while all others are generally considered to be nonmagnetic.
2. Retentivity varies inversely with permeability and directly with reluctance; that is, the more difficult it is to induce magnetism in a material, the longer the magnetic characteristics are retained. By the same token, the easier it is to induce magnetism, the faster the magnetic characteristics fade.

As you will learn, high retentivity is a desirable characteristic in certain applications and a liability in others.

**TABLE 10.1 Magnetic Classifications of Materials**

| *Classification* | *Ferromagnetic* | *Paramagnetic* | *Diamagnetic* |
|---|---|---|---|
| Relative permeability ($\mu_r$): | Much greater than 1 | Slightly greater than 1 | Less than 1 |
| Retentivity: | Low | Moderate | High |
| Reluctance: | Low | Moderate | High |
| Considered to be . . . | Magnetic | Slightly magnetic | Nonmagnetic |
| Examples: | Iron | Aluminum | Copper |
| | Steel | Air | Silver |
| | Nickel | Platinum | Gold |
| | Cobalt | | Zinc |
| | Alnico[a] | | |
| | Permalloy[a] | | |

[a]These are alloys. Alnico is made primarily of aluminum, nickel, and cobalt. Permalloy is made primarily of iron and nickel.

## Section Review

1. What is permeability?
2. What effect does a high-permeability material have on magnetic flux?
3. What is relative permeability?
4. According to domain theory, what determines the magnetic characteristics of an atom?
5. What is a magnetic domain?
6. List and describe the methods of magnetic induction.
7. What is retentivity?
8. What types of materials are used to produce permanent magnets? Temporary magnets?
9. What is reluctance?
10. What is the relationship between reluctance and permeability? Explain your answer.
11. Using the information contained in Table 10.1, compare and contrast the magnetic characteristics of aluminum and iron.

# 10.3 ELECTROMAGNETISM

*OBJECTIVE 14*

In 1820, the Danish physicist Hans Christian Oersted discovered that an electric current produces a magnetic field. Using a structure similar to the one shown in Figure 10.13a, he found that the compass always aligned itself at 90° angles to a current-carrying wire. When the current was in the direction shown in Figure 10.13b, the compass aligned itself as shown. When the current was reversed (as shown in Figure 10.13c), the compass reversed direction.

**FIGURE 10.13**

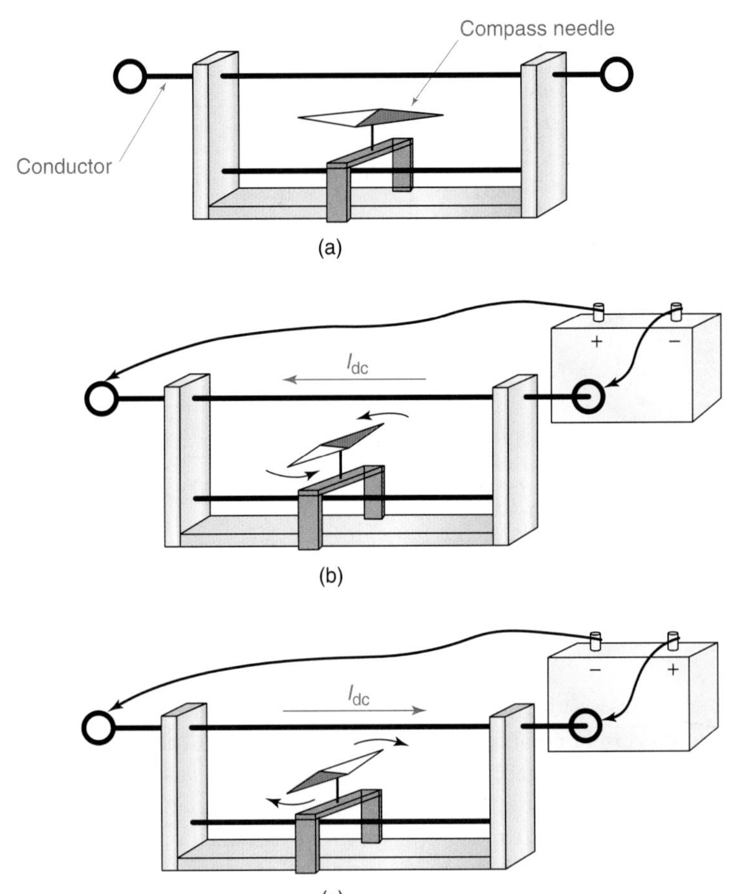

As a result of Oersted's work, it is known that a current carrying wire generates lines of magnetic force that circle the wire as shown in Figure 10.14. As the figure shows, reversing the direction of the current causes the polarity of the flux to reverse.

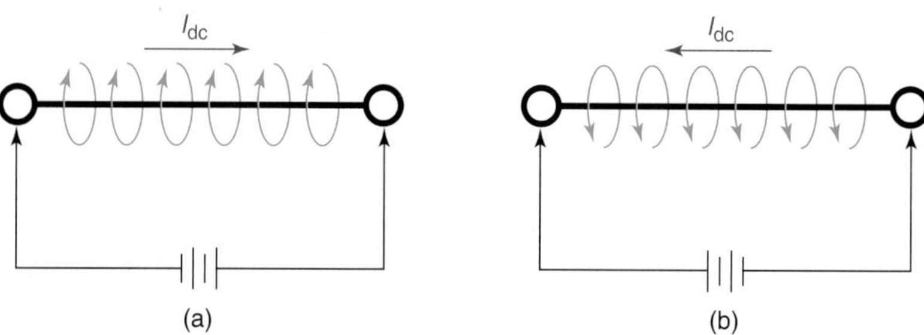


**FIGURE 10.14**

## The Left-Hand Rule

**Left-hand rule**
A memory aid that helps you remember the relationship between current direction and the direction of the resulting magnetic field.

The **left-hand rule** is a memory aid designed to help you remember the relationship between current direction and the direction of the resulting magnetic field. The left-hand rule is illustrated in Figure 10.15. As you can see, the fingers indicate the direction taken by the flux when the thumb points in the direction of the current. If the current is reversed, the left hand is also reversed. When repositioned, the fingers indicate that the flux direction has also changed.

**FIGURE 10.15 Left-hand rule.**

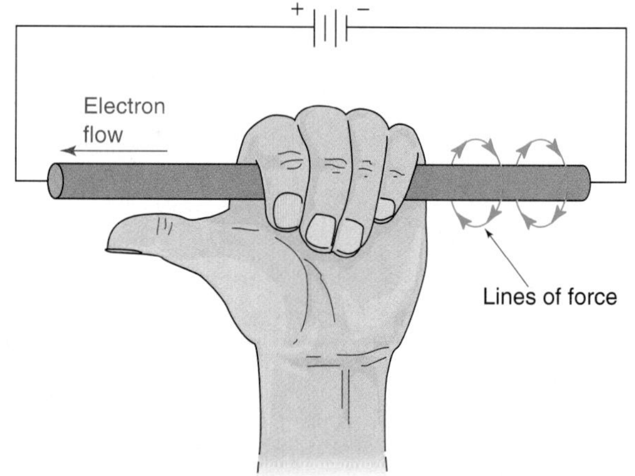


## The Coil

**Coil**
A wire wrapped into a series of loops for the purpose of concentrating magnetic lines of force.

**Core**
The space within the turns of a coil.

If we take a wire and wrap it into a series of loops, called a **coil,** the lines of force generated by the current combine to form a magnetic field. The magnetic field produced by a coil is illustrated in Figure 10.16. Note that the space inside the coil is called the **core.** If the core is made of an iron bar, the lines of flux are concentrated in the bar. As described in the last section, the iron bar becomes magnetic by induction. (However, the magnetic characteristics fade quickly after power is removed from the coil because of iron's low retentivity.)

**FIGURE 10.16 Magnetic field generated by the current through a coil.**

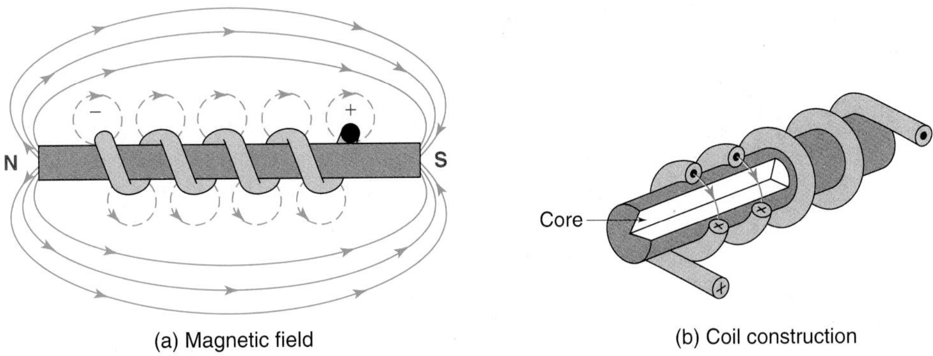


The left-hand rule can be applied to a coil as shown in Figure 10.17. In this case, the thumb indicates the direction of the magnetic flux when the fingers point in the direction of the coil current. As shown, the direction of the flux reverses when the current direction through the coil reverses.

**FIGURE 10.17 Left-hand rule for generators.**

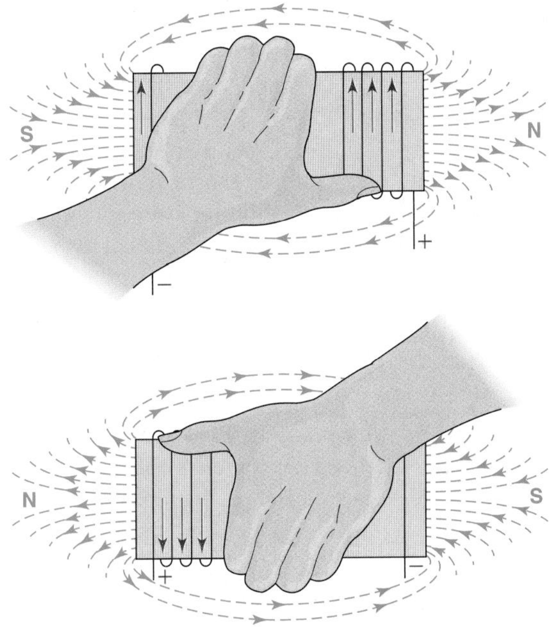


Figure 10.18 further demonstrates the fact that a coil can be used to produce a magnet. When current is drawn through the coil, the iron core becomes magnetic and lifts the iron filings off the table. The strength of the magnet depends (in part) on the amount of magnetomotive force (mmf) produced by the current through the coil.

**FIGURE 10.18 Iron filings lifted off the table by a magnet.**

## Magnetomotive Force (mmf)

OBJECTIVE 15 ▶

**Magnetomotive force (mmf)**
The force that produces magnetic flux. The magnetic equivalent of electromotive force.

In Figure 10.18, the current through the coil generates magnetic flux through the reluctance of the iron core. Generating magnetic flux through any value of reluctance requires a certain amount of **magnetomotive force (mmf).** As Table 10.2 indicates, mmf is to magnetic circuits what electromotive force (emf) is to electric circuits.

*Note:* In cgs units:
Magnetomotive force is measured in *gilberts* (Gb).
Flux is measured in *maxwells* (Mx).
Reluctance is measured in *gilberts per maxwell* (Gb/Mx).

**TABLE 10.2 A Comparison Between Basic Electric and Magnetic Quantities**

| | *Electric Circuit* | *Magnetic Circuit* |
|---|---|---|
| Force | Electromotive force (emf) (measured in volts) | Magnetomotive force (mmf) (measured in ampere-turns) |
| Flow | Current ($I$) (measured in amperes) | Flux ($\phi$) (measured in webers) |
| Opposition | Resistance ($R$) (measured in ohms) | Reluctance ($\Re$) (measured in ampere-turns/weber) |

OBJECTIVE 16 ▶

**Rowland's law**
Magnetic flux is directly proportional to mmf and inversely proportional to reluctance.

Just as Ohm's law defines the relationship between emf, current, and resistance, **Rowland's law** defines the relationship between mmf, flux, and reluctance. Rowland's law states that magnetic flux is directly proportional to magnetomotive force and inversely proportional to reluctance. By formula,

$$\Phi = \frac{F}{\Re} \tag{10.4}$$

where $\Phi$ = the magnetic flux, in webers (Wb)
$F$ = the magnetomotive force, in ampere-turns (A · t)
$\Re$ = the reluctance, in ampere-turns per weber (A · t/Wb)

Equation (10.4) is useful because it shows that there are two ways to increase the amount of flux produced by a magnet like the one in Figure 10.16. The amount of flux produced by the magnet can be increased by:

1. Increasing the amount of magnetomotive force.
2. Using a lower reluctance material as the core.

If you refer back to Figure 10.16, you'll see that the iron-core coil contains more flux than the air-core coil. Since iron has less reluctance than air, more flux is generated by the coil with the iron coil. As you will see, two methods can be used to increase the amount of magnetomotive force produced by the coil.

## Ampere-Turns

**Ampere-turns (A · t)**
The product of coil current and the number of turns in the component. Usually represented as $NI$ in equations ($N$ = number of turns, $I$ = coil current).

The magnetomotive force produced by a coil is proportional to both the current through the coil and the number of turns. These two factors are combined into a single value called the **ampere-turn (A · t).** The ampere-turn value for a given coil equals the product of the current and the number of turns, as illustrated in Figure 10.19. As you can see, the value of A · t for a coil can be varied by changing either the amount of current or the number of

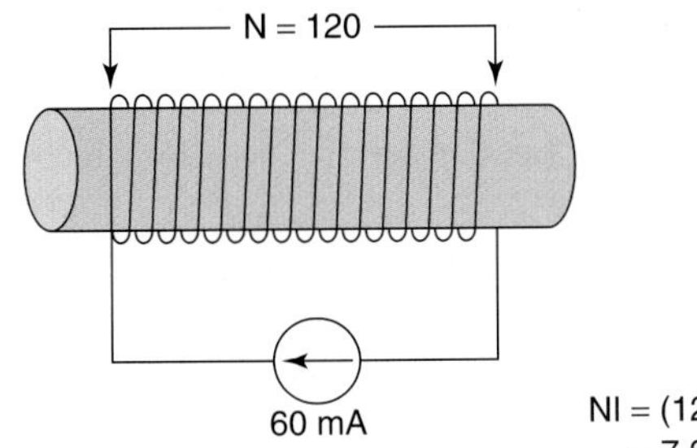


**FIGURE 10.19 Ampere-turns value for a coil.**

turns in the coil. Note that ampere-turns are commonly represented using $NI$ ($N$ for "number of turns" and $I$ for "current").

The amount of magnetomotive force produced by a coil equals the ampere-turns value of the coil. By formula,

$$\text{mmf} = NI \tag{10.5}$$

*Note:* In the cgs system, magnetomotive force (in gilberts) is found as

$$\text{mmf} = 1.2566\,NI$$

The flux density produced by a coil depends on three factors: the permeability of the core, the ampere-turns of the coil, and the length of the coil. When these values are known, the flux density produced by a coil (in teslas) is found as

$$B = \frac{\mu_m NI}{\ell} \tag{10.6}$$

where
$B$ = the flux density, in teslas
$\mu_m$ = the permeability of the core material, in Wb/A · m
$NI$ = the ampere-turns value of the coil
$\ell$ = the length of the coil (in meters)

The use of this equation is demonstrated in Example 10.3.

## EXAMPLE 10.3

The coil in Figure 10.20 has an air core. Calculate the flux density (in teslas) for the coil.

***Solution:*** As stated earlier, the permeability of air is approximately $4\pi \times 10^{-7}$ Wb/A · m. The length of the coil is shown to be 5 cm, which is equal to $5 \times 10^{-2}$ m. Using these values in equation (10.6), the flux density produced by the coil is found as

$$\begin{aligned} B &= \frac{\mu_m NI}{\ell} \\ &= \frac{\left(4\pi \times 10^{-7}\,\frac{\text{Wb}}{\text{A}\cdot\text{m}}\right)(100)(200 \times 10^{-3}\,\text{A})}{5 \times 10^{-2}\,\text{m}} \\ &= 502.7 \times 10^{-6}\,\frac{\text{Wb}}{\text{m}^2} \\ &= \mathbf{502.7\ \mu T} \end{aligned}$$

**FIGURE 10.20**

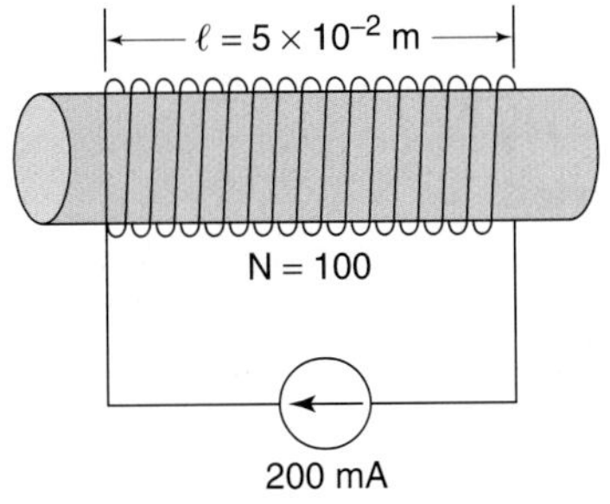


*A Practical Consideration:* When dealing with coils, the unit of measure for permeability is sometimes written as webers per ampere-turn meters (Wb/At · m). The reason for doing so is mathematical. If "turns" is considered a unit of measure, then permeability must be given in (Wb/At · m) for the units to resolve themselves in the calculation of $B$. However, if turns are treated as a multiplier without units, then the conventional unit of measure (Wb/A · m) is used. In either case, the value of the measurement is the same.

*PRACTICE PROBLEM 10.3*

The air-core coil in Figure 10.20 is replaced with an iron-core coil having the same dimensions. Calculate the flux density produced by the coil.

If you take another look at equation (10.6), you'll see that:

1. The values of $N$ and $\ell$ are determined by the physical characteristics of the coil.
2. The value of $\mu_m$ is determined by the type of material found in the core.

As such, the values of $N$, $\ell$, and $\mu_m$ cannot be changed easily for a given coil. Therefore, the coil current ($I$) is the value that is normally adjusted to vary flux density. The effect of varying current on the value of flux density is demonstrated in Example 10.4.

### EXAMPLE 10.4

The current through the air-core coil in Figure 10.20 is increased to 250 mA. Calculate the resulting change in flux density for the coil.

***Solution:*** Using a value of $I$ = 250 mA, the flux density for the coil is found as

$$B = \frac{\mu_m NI}{\ell}$$

$$= \frac{\left(4\pi \times 10^{-7} \frac{\text{Wb}}{\text{A} \cdot \text{m}}\right)(100)(250 \times 10^{-3}\text{ A})}{5 \times 10^{-2}\text{ m}}$$

$$= 628.3 \times 10^{-6} \frac{\text{Wb}}{\text{m}^2}$$

$$= \mathbf{628.3\ \mu T}$$

In Example 10.3, we calculated a value of $B$ = 502.6 μT for $I$ = 200 mA. Using this value, the change in flux density is found as

$$\Delta B = 628.3\ \mu\text{T} - 502.6\ \mu\text{T} = \mathbf{125.7\ \mu T}$$

*PRACTICE PROBLEM 10.4*

The current through the coil in Practice Problem 10.3 is decreased to 150 mA. Calculate the change in flux density that results from the change in current. (Compare the percent of change in current to the percent of change in flux density.)

## Hysteresis

*OBJECTIVE 17* ▶

**Hysteresis**
The time lag between the removal of a magnetizing force and the drop in flux density.

**Hysteresis curve**
A curve representing flux density ($B$) as a function of the field intensity ($H$) of a magnetizing force.

As you know, most materials lose much of their magnetic strength (flux density) when the magnetizing force is removed. For example, the iron core in Figure 10.21 experiences a significant drop in flux density when current is removed from the coil. The time required for the flux density to drop after the magnetizing force is removed depends on the retentivity of the material. The time lag between the removal of a magnetizing force and the drop in flux density is referred to as **hysteresis.**

The principle of hysteresis is commonly represented using the **hysteresis curve** shown in Figure 10.22a. This curve represents flux density ($B$) as a function of the field intensity ($H$) of the magnetizing force. Figure 10.22b shows what happens when a current is applied to a coil like the one in Figure 10.21. The field intensity ($H$) of the magnetizing force reaches its maximum value ($+H_M$), as does the flux density through the core. When

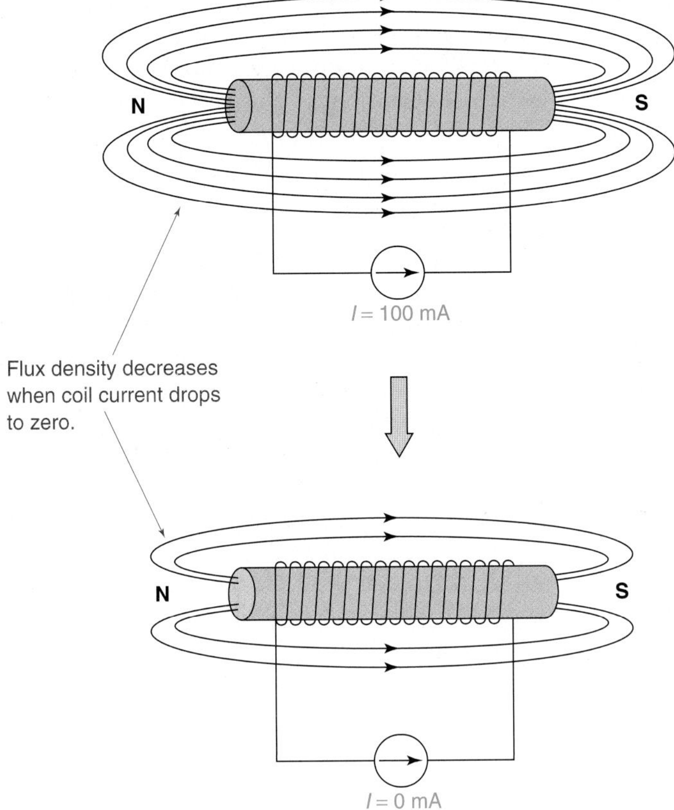


**FIGURE 10.21 Hysteresis.**

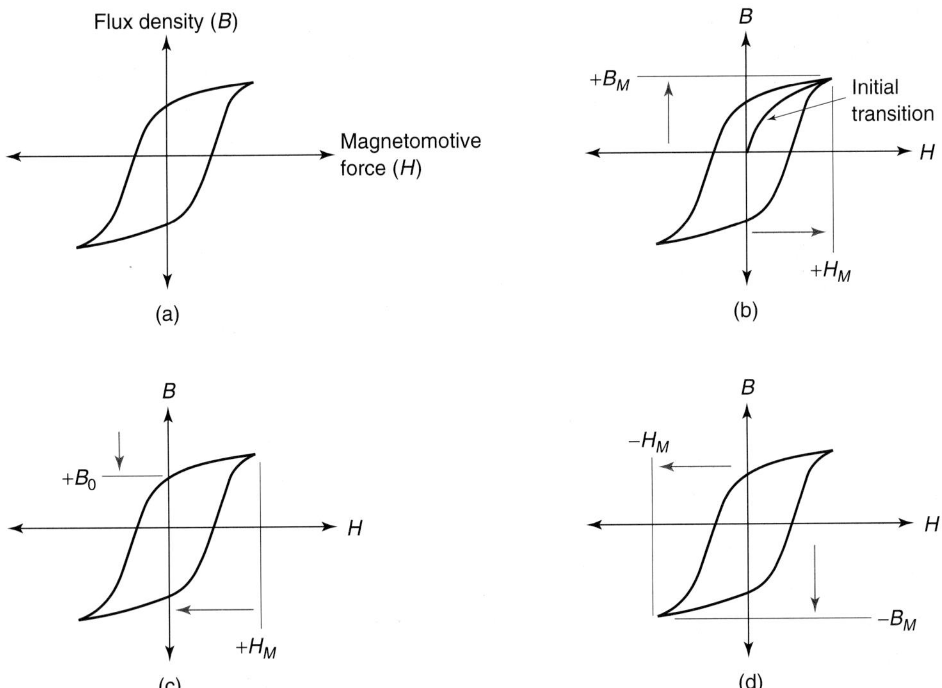


**FIGURE 10.22 Hysteresis curves.**

power is removed from the coil, the value of $H$ drops to zero. However, there is some **residual flux density** in the core, represented by the positive value of $(+B_0)$ in the curve (Figure 10.22c). The time required for the flux density to drop as well as the amount of flux density depend on the retentivity of the material. The higher the retentivity of the material:

**Residual flux density**
The flux density that remains in a material after any mmf is removed.

*A Practical Consideration:* When switching from one magnetic polarity to another, energy must be used to overcome any residual flux in the material. This loss of energy (called hysteresis loss) is discussed in Chapter 15.

1. The longer it takes for $B$ to drop to $+B_0$.
2. The greater the value of $+B_0$.

If the current through the coil is reversed, the polarity of the magnetomotive force and the resulting flux density reverses. This condition is represented in Figure 10.22d. Again, if power is removed, magnetomotive force drops to zero and the residual flux density drops to the value represented on the curve. Note that the magnetomotive force required to return the value of flux density ($B$) to zero is referred to as the **coercive force.**

**Coercive force**
The magnetomotive force required to return the value of flux density within a material to zero.

Note that the two halves of the hysteresis curve are simply mirror images. One half represents the response of the core to a current generated in one direction. The other half shows the response of the core when the coil current reverses direction.

## Section Review

1. What relationship did Oersted discover between magnetism and current?
2. What happens to the polarity of a magnetic field when current is reversed?
3. What is a coil? What is the core of a coil?
4. What is magnetomotive force (mmf)?
5. What is Rowland's law? Which law of electrical circuits does it resemble?
6. What is an ampere-turn?
7. What is the relationship between ampere-turns and flux density?
8. In a practical sense, how is the amount of magnetomotive force produced by a given coil usually varied?
9. What is hysteresis?
10. What is residual flux density?
11. Using the hysteresis curve, describe what happens when power is applied to and removed from a coil.

# 10.4 RELATED TOPICS

In this section, we will briefly discuss several topics that relate to magnetism. Among the topics we will cover are magnetic shielding, common shapes of magnets, and the proper care and storage of magnets.

## Magnetic Shielding

*OBJECTIVE 18* ▶

Some instruments, like analog meter movements, are sensitive to magnetic flux; that is, exposure to magnetic flux causes them to operate improperly. When such an instrument must be placed in a magnetic field, **magnetic shielding** is used to insulate the instrument from the effects of the magnetic flux.

**Magnetic shielding**
Insulating an instrument (or material) from magnetic flux by diverting the lines of force around the instrument (or material).

It should be noted that there is no known magnetic insulating material. Flux passes through any material, regardless of its reluctance. Because of this, magnetic shielding is accomplished by diverting lines of flux around the object to be shielded. The means by which this is accomplished is illustrated in Figure 10.23.

Figure 10.23 shows the lines of force between two magnetic poles. As you can see, a soft iron ring has been placed between the poles. Rather than follow a straight-line path between the poles, the lines of force bend to pass through the high-permeability material. If a

FIGURE 10.23 **Magnetic shielding.**

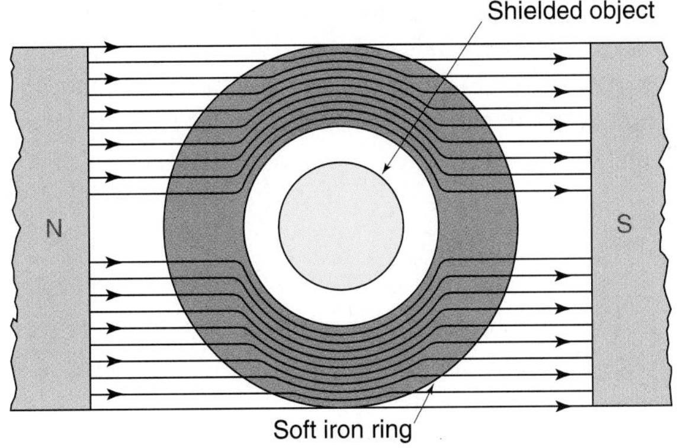


magnetically sensitive device is placed in the middle of the iron ring, the lines of force are diverted around the device. As a result, it is shielded from the flux.

## *Air Gaps*

The space directly between two magnetic poles is called an **air gap.** Figure 10.24 shows the air gap for two types of magnets. The narrower the air gap between two poles, the stronger the force of attraction between them. With a narrow air gap, there is little room for the lines of force between two poles to spread out. As a result, a narrow air gap produces a narrow magnetic field with relatively high flux density (which is why the force of attraction between two magnets increases when you move them closer together).

**Air gap**
The area of free space (if any) between two magnetic poles.

**FIGURE 10.24 Air gaps of horseshoe and round magnets.**

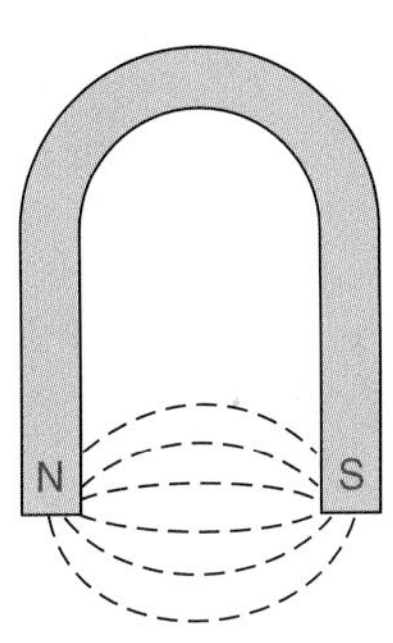


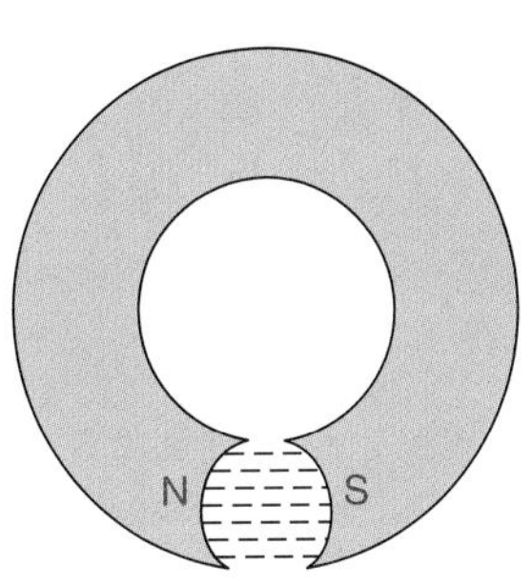


## *Ring Magnets*

**Ring magnets** are unique because they do not have any identifiable poles or air gaps. They are most commonly used in analog meter movements. A ring magnet can be produced in one of several fashions, as shown in Figure 10.25.

◄ *OBJECTIVE 19*

**Ring magnet**
A magnet that forms a closed loop and therefore has no identifiable poles. There is no magnetic flux in the center of a ring magnet.

**FIGURE 10.25 Ring magnets.**

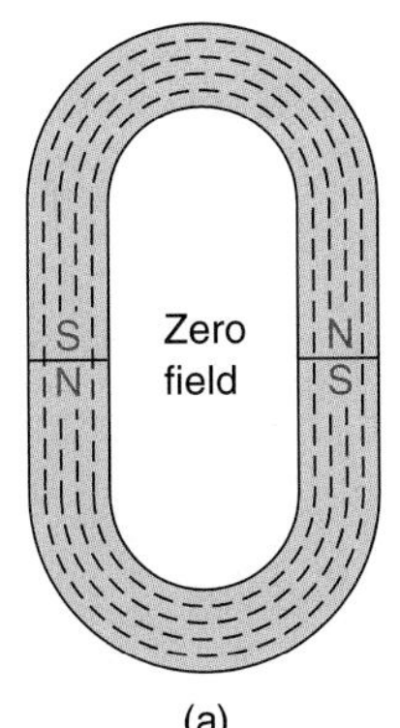


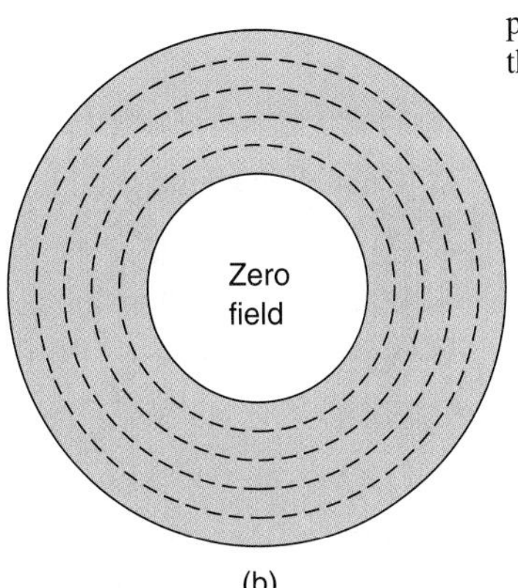

Figure 10.25a shows a ring magnet that is actually made up of two horseshoe magnets. When connected as shown, the lines of flux remain in the ring magnet. Therefore, there is no magnetic field in the center area of the magnet.

When constructed as shown in Figure 10.25b, a ring magnet has no identifiable poles. Rather, a continuous series of magnetic flux circles are produced within the magnet.

## Care and Storage

*OBJECTIVE 20* ▶ Over time, even a "permanent" magnet can lose most of its magnetic strength if not cared for properly. The most common causes of loss of magnetic strength for a permanent magnet are:

1. Allowing the magnet to be jarred or dropped.
2. Exposing the magnet to sufficiently high temperatures.
3. Storing a magnet improperly.

When a magnet is jarred or dropped, the impact can knock the magnetic domains out of alignment. As a result, magnetic strength is lost. Excessive heat can also cause misalignment of the domains, resulting in a loss of magnetic strength.

When storing magnets, care must be taken to ensure that the flux produced by the magnet is not lost externally. This is accomplished by storing one or more magnets in one of the ways shown in Figure 10.26. Each of the storing methods shown has the same effect as constructing a ring magnet; that is, the flux generated by the magnet(s) stays within a loop, rather than being lost externally.

**FIGURE 10.26 Storing magnets.**

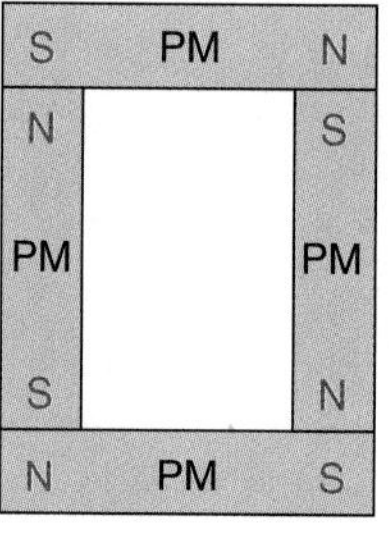


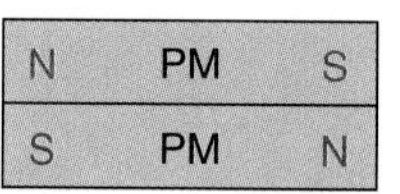


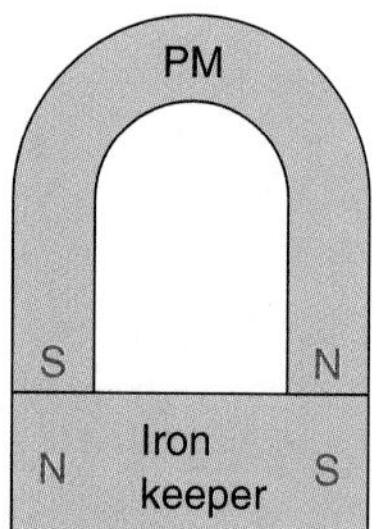


PM: permanent magnet

### Section Review

1. What is magnetic shielding? How is it accomplished?
2. What is an air gap?
3. What is the relationship between air gap width and magnetic strength?
4. What characteristics distinguish a ring magnet from other types of magnets?
5. What is the most common application for ring magnets?
6. List the most common causes of loss of magnetic strength for a permanent magnet.
7. Describe the proper storage of a magnet.

Here is a summary of the major points made in this chapter:

1. The force that a magnet exerts on the objects around it is referred to as magnetic force.
2. The points where magnetic lines of force leave (and return to) a magnet are referred to as the poles.
   - **a.** The two poles are the north-seeking (N) and south-seeking (S) poles.
   - **b.** Like poles repel each other. Unlike poles attract.
3. The lines of force produced by a magnet are referred to collectively as magnetic flux.
4. There are two fundamental units of measure for flux.
   - **a.** The maxwell (Mx): 1 Mx = 1 line of flux.
   - **b.** The weber (Wb): 1 Wb = $1 \times 10^8$ Mx.

   The maxwell is the cgs unit of measure. The weber is the SI unit of measure.
5. The micro-weber ($\mu$Wb), which equals 100 Mx, is a commonly used unit of measure for flux.
6. The area of space that surrounds a magnet and contains magnetic flux is called the magnetic field.
7. Flux density is a measure of flux per unit area. Flux density is an indicator of magnetic strength.
8. The SI unit of flux density is the tesla (T). One tesla equals one weber per square meter.
9. In cgs units, flux density is measured in gauss (G), where 1 G = 1 Mx/cm$^2$.
10. Flux density is greatest at the poles of a magnet.
11. *Permeability* ($\mu$) is a measure of how easily magnetic lines of force are established within a material. Permeability is (essentially) the magnetic equivalent of conductance.
12. Materials with high permeability have the ability to concentrate lines of magnetic force.
    - **a.** High-permeability materials can be used to increase flux density.
    - **b.** Permeability is measured in Wb/A · m.
13. Relative permeability ($\mu_r$) is the ratio of a material's permeability to that of free space ($\mu_0$). Since $\mu_r$ is a ratio, it has no units.
14. Relative permeability is important because it equals the ratio of flux density within a material to flux density in free space.
15. Domain theory states that atoms with like magnetic fields join together to form magnetic domains.
    - **a.** Every atom has both an electric field and a magnetic field. The polarity of the field depends on the direction the electron spins on its own axis.
    - **b.** Each domain acts as a magnet, with two poles and a magnetic field.
    - **c.** In nonmagnetized materials, the domains are positioned randomly, so their magnetic fields cancel.
    - **d.** In magnetic materials, the domains can be aligned to produce a magnet.
16. Induction is the process of producing an artificial magnet. An external source of magnetism is used to align the magnetic domains in a material.
17. There are three methods of induction, as shown in Figure 10.12.
    - **a.** Induction can be accomplished using the force of an external magnet or an electric current.
    - **b.** The domains in high-permeability materials are relatively easy to align because these materials concentrate magnetic flux.

18. The ability of an artificial magnet to retain its magnetic characteristics is called retentivity.

    a. High-permeability materials have low retentivity. Low-permeability materials have high retentivity.

    b. High-retentivity (low-permeability) materials are used to produce permanent magnets.

    c. Low-retentivity (high-permeability) materials are used to produce temporary magnets.

19. The opposition that a material presents to magnetic lines of force is called reluctance.

    a. Reluctance is (essentially) the magnetic equivalent of resistance.

    b. Reluctance varies inversely with permeability.

    c. In the SI, reluctance is measured in ampere-turns per weber (A · t/Wb).

20. All materials fall within one of three magnetic classifications.

    a. Ferromagnetic materials have high permeability and are considered to be magnetic.

    b. Paramagnetic materials have relatively low permeability and are not considered to be magnetic.

    c. Diamagnetic materials have very low permeability and are not considered to be magnetic.

    The characteristics of these magnetic classifications are summarized in Table 10.1.

21. In 1820, Oersted discovered that an electric current produces a magnetic field. The magnetic field resulting from a current exists at 90° angles to the current.

22. The left-hand rule is a memory aid designed to help you remember the relationship between current direction and the direction of the resulting magnetic field. The left-hand rule is illustrated in Figure 10.16.

23. A wire wrapped into a series of loops is called a coil. The area in the center of the coil is called the core.

24. The lines of force produced by the current through a coil join to form a magnetic field with relatively high flux density. If the core contains an iron bar, the bar becomes magnetic by induction.

25. The left-hand rule can be applied to a coil, as illustrated in Figure 10.18.

26. Generating magnetic flux through any value of reluctance requires a certain amount of magnetomotive force (mmf). Magnetomotive force is to magnetic circuits what electromotive force (emf) is to electric circuits.

27. Rowland's law defines the relationship between mmf, reluctance, and magnetic flux.

    a. Magnetic flux is directly proportional to mmf and inversely proportional to reluctance.

    b. Rowland's law is similar in structure to Ohm's law.

28. The magnetomotive force produced by a coil is proportional to both the current through the coil and the number of turns (loops) contained in the coil.

    a. These two factors are combined in the ampere-turns value, which equals the product of the two.

    b. Ampere-turns are commonly represented as $NI$ ($N$ for the number of turns and $I$ for the current).

29. The flux density produced by a coil depends on:

    a. The permeability of the core.

    b. The number of coil turns.

    c. The current through the coil.

    d. The length of the coil.

30. Of the values listed above, coil current is the value that is normally varied to control the flux density produced by a coil. Core permeability, coil turns, and coil length are physical characteristics that are not easily changed for a given coil.

31. Materials lose much of their magnetic strength (flux density) when a magnetizing force is removed.
    a. The time lag between removing a mmf and the resulting loss in flux density is referred to as hysteresis.
    b. The principle of hysteresis is commonly represented using the hysteresis curve (Figure 10.22).
    c. The flux density that remains when a magnetizing force is removed is called residual flux density.
    d. The amount of residual flux density depends on the resistivity of the material.

32. Magnetic shielding is used to protect sensitive instruments from magnetic flux.
    a. There is no known magnetic insulating material.
    b. Shielding is accomplished by diverting magnetic flux around the sensitive instrument, as shown in Figure 10.23.

33. The space directly between two magnetic poles is called an air gap.
    a. The narrower the air gap, the stronger the force of attraction (or repulsion) between the poles.
    b. Ring magnets are unique because they do not have any identifiable poles or air gaps.

34. The most common causes of loss of magnetic strength for a permanent magnet are:
    a. Allowing the magnet to be jarred or dropped.
    b. Exposing the magnet to sufficiently high temperatures.
    c. Storing a magnet improperly.

35. When storing a magnet, care must be taken to ensure that the flux produced by the magnet is not lost externally. Magnets should be stored as shown in Figure 10.26.

## Equation Summary

| Equation Number | Equation | Section Number |
|---|---|---|
| (10.1) | $B = \frac{\Phi}{A}$ | 10.1 |
| (10.2) | $\mu_r = \frac{\mu_m}{\mu_0}$ | 10.2 |
| (10.3) | $\Re = \frac{\ell}{\mu A}$ | 10.2 |
| (10.4) | $\Phi = \frac{F}{\Re}$ | 10.3 |
| (10.5) | $\text{mmf} = NI$ | 10.3 |
| (10.6) | $B = \frac{\mu_m NI}{\ell}$ | 10.3 |

## Key Terms

The following terms were introduced and defined in this chapter:

air gap
ampere-turns
coercive force
coil
core
diamagnetic
domain theory
ferromagnetic
flux density
gauss
hysteresis
hysteresis curve
induction
magnetic domain
magnetic field
magnetic flux
magnetic force
magnetic induction
magnetic north
magnetic shielding
magnetic south
magnetomotive force (mmf)
maxwell (Mx)
micro-weber ($\mu$Wb)
north-seeking pole
paramagnetic
permanent magnet
permeability ($\mu$)
poles
relative permeability ($\mu_r$)
reluctance
residual flux density
retentivity
ring magnet
Rowland's law
south-seeking pole
temporary magnet
tesla (T)
weber (Wb)

## Practice Problems

1. Calculate the flux density for the designated area in Figure 10.27a. Assume that each line of force represents 120 $\mu$Wb of flux.
2. Calculate the flux density for the shaded area in Figure 10.27b. Assume that each line of force represents 200 $\mu$Wb of flux.
3. A magnet with a cross-sectional area of 1.5 $cm^2$ generates 300 $\mu$Wb of flux at its poles. Calculate the flux density at the poles of the magnet.
4. A magnet with a cross-sectional area of 3 $cm^2$ generates 1200 $\mu$Wb of flux at its poles. Calculate the flux density at the poles of the magnet.
5. The permeability of a piece of low-carbon steel is determined to be $7.54 \times 10^{-4}$ Wb/A · m. Calculate the relative permeability of this material.

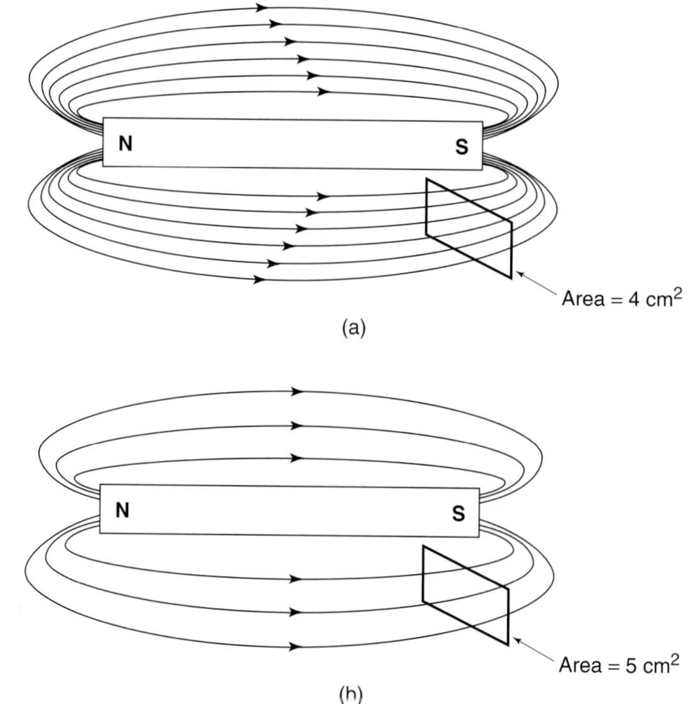


**FIGURE 10.27**

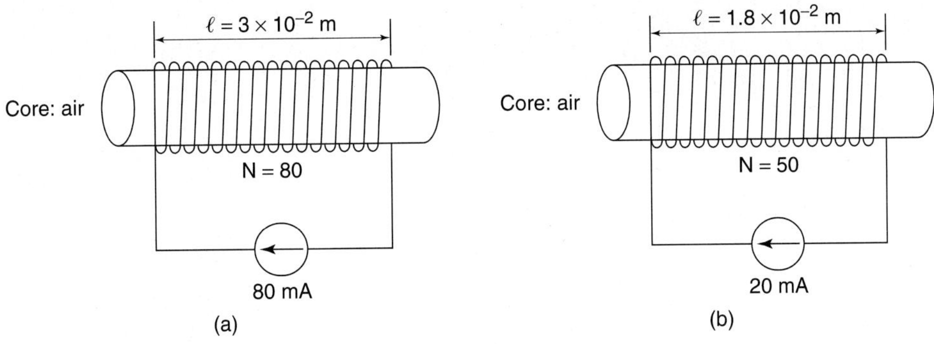


**FIGURE 10.28**

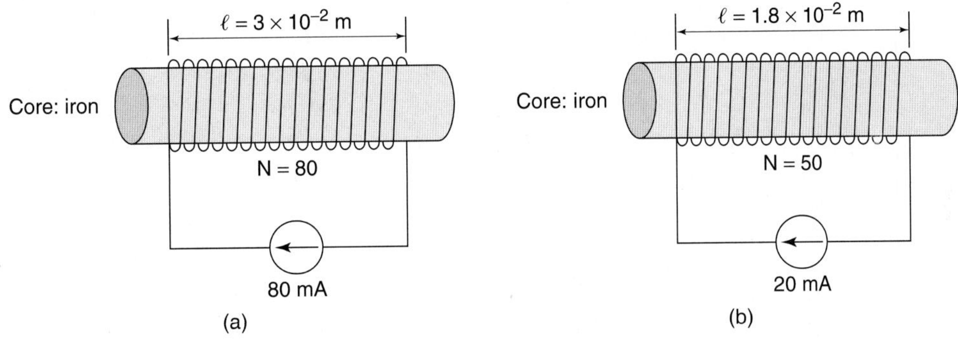


**FIGURE 10.29**

6. Calculate the relative permeability of free space.
7. The permeability of a piece of permalloy is determined to be 0.1 Wb/A · m. Calculate the relative permeability of this material.
8. The permeability of a piece of ingot iron is determined to be $3.77 \times 10^{-3}$ Wb/A · m. Calculate the relative permeability of this material.
9. Calculate the flux density produced by the coil in Figure 10.28a.
10. Calculate the flux density produced by the coil in Figure 10.28b.
11. Calculate the flux density produced by the coil in Figure 10.29a.
12. Calculate the flux density produced by the coil in Figure 10.29b.

---

## Answers to the Example Practice Problems

**10.1** 1.5 T
**10.3** $1 \times 10^{-1}$ T
**10.4** $2.5 \times 10^{-2}$ T

# 11

# ac CHARACTERISTICS

## OBJECTIVES

*After studying the material in this chapter, you should be able to:*

1. Describe alternating current (ac).
2. Describe the makeup of a sinusoidal waveform.
3. Describe the relationship between cycle time and frequency.
4. Determine the cycle time and frequency of a waveform displayed on an oscilloscope.
5. Define and identify each of the following magnitude-related values: peak, peak-to-peak, instantaneous, full-cycle average, and half-cycle average.
6. Describe the relationship between average ac power and rms values.
7. Describe the relationship between peak and rms values.
8. Describe how each of the magnitude-related values is measured.
9. Describe the magnetic induction of current.
10. Compare and contrast phase and time measurements.
11. Discuss phase angles and how they are measured.
12. Calculate the instantaneous value at any point on a sine wave using the degree approach.
13. Describe a sine wave in terms of radians and angular velocity.
14. Calculate the instantaneous value at any point on a sine wave using the radian approach.
15. Compare and contrast the degree and radian approach to measuring phase.
16. Compare and contrast static and dynamic values.
17. Discuss the effects of a dc offset on sine wave measurements.
18. Describe wavelength and calculate its value for any waveform.
19. Discuss the relationship between a waveform and its harmonics.
20. Describe and analyze rectangular waveforms.
21. Describe sawtooth waveforms.

In Chapter 2, you were told that there are two types of current: direct current (dc) and alternating current (ac). In dc circuits, charge flows in only one direction. Up to this point, we have limited our discussions to dc circuits and characteristics.

The receptacles (wall outlets) in your home provide power in the form of alternating current and voltage. It follows then, that everything you connect to the receptacles must rely on this alternating current and voltage for their operation. Refrigerators, electric stoves, stereo systems, computers, televisions, and so on, all derive their power from an alternating current (ac) source. Therefore, to understand the operation of these appliances and systems, you must be thoroughly familiar with the characteristics of alternating current and voltage.

In this chapter, we will focus primarily on ac measurements, sources, and characteristics. Then, in future chapters, we will analyze the operation of various ac circuits. As your studies of electricity and electronics continue, you will see that most systems contain both direct and alternating currents and voltages.

## 11.1 ALTERNATING CURRENT (ac): OVERVIEW AND TIME MEASUREMENTS

◄ *OBJECTIVE 1*

Generally, the term **alternating current** (or **ac**) is used to describe any current that periodically changes direction. For example, consider the current versus time relationship shown in Figure 11.1. The horizontal axis of the graph is used to represent time ($t$). The vertical axis of the graph is used to represent both the **magnitude** (value) and the direction of the current. As the figure shows, the current builds to a peak value in one direction and returns to zero ($t_0$ to $t_1$). It then builds to a peak value in the other direction and returns to zero ($t_1$ to $t_2$). Note that the current not only changes direction, it is constantly changing in magnitude.

**Alternating current (ac)** A term used to describe any current that periodically changes direction.

**Magnitude** Another word for quantity or value.

A graph of the relationship between magnitude and time is referred to as a **waveform.** The waveform in Figure 11.1 is called a sinusoidal waveform, or sine wave. (The basis for this name is discussed in Section 11.3.)

**Waveform** A graph of the relationship between a current (or voltage) and time.

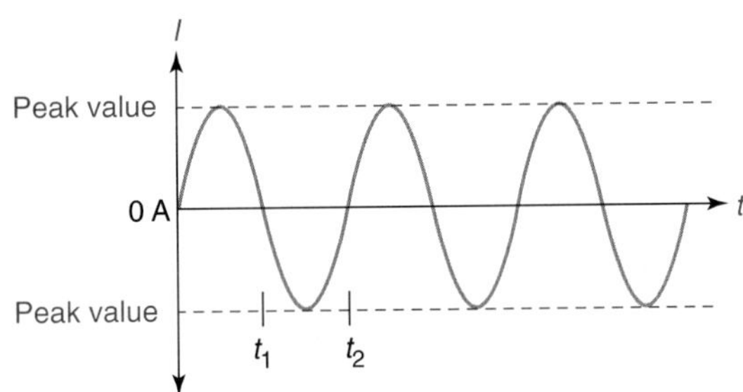


**FIGURE 11.1 Alternating current (ac).**

There are many common ac waveforms. Each of these waveforms is generated by some type of ac generator or ac source. Some common ac waveforms are identified in Figure 11.2, along with the schematic symbols for their sources.

In this chapter, we will focus on the most common ac waveform: the sine wave. Unless indicated otherwise, any ac waveform discussed in this chapter is assumed to be sinusoidal. When other waveforms are discussed, they will be identified by name.

### Basic ac Operation

The operation of a simple ac circuit is illustrated in Figure 11.3. Figure 11.3a highlights the operation of the circuit during the positive transition of the ac source. Figure 11.3b highlights the operation of the circuit during the negative transition of the source. Using

EWB

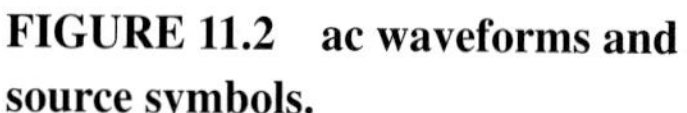

**FIGURE 11.2 ac waveforms and source symbols.**

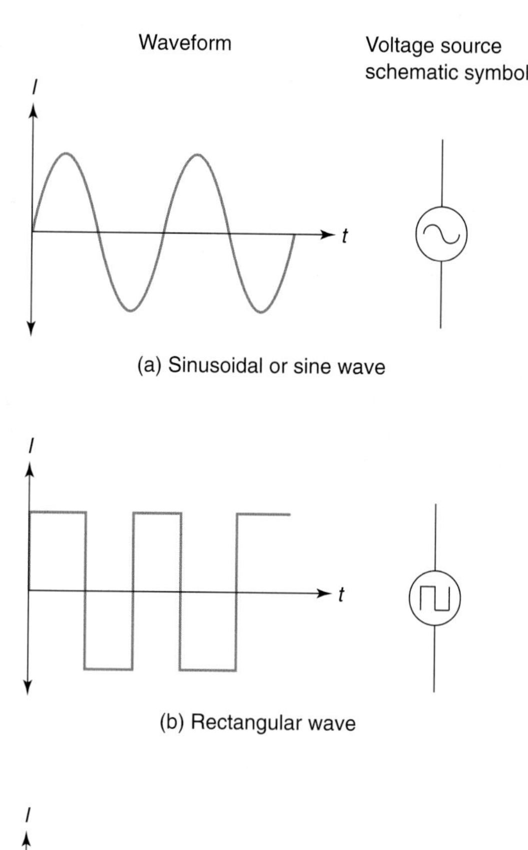


(a) Sinusoidal or sine wave

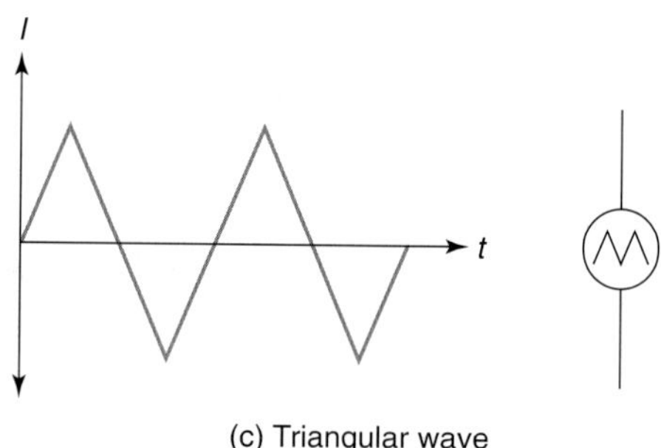


(b) Rectangular wave

(c) Triangular wave

the time periods shown in the figure, the operation of the circuit can be described as follows:

| Time Period | Circuit Operation |
|---|---|
| $t_0$ to $t_1$ | The output from the voltage source increases from 0 V to its positive peak value (+V peak). As the voltage increases, the current also increases (from 0 A) to a peak value, as shown in the figure. Note the voltage polarity and current direction shown. |
| $t_1$ to $t_2$ | The output from the voltage source returns to 0 V. As the voltage decreases, so does the value of the circuit current. However, neither the voltage polarity nor the current direction have changed (yet). |
| $t_2$ to $t_3$ | The output from the voltage source is again increasing in magnitude. However, the polarity of the source voltage has now changed. As the source voltage increases to its negative peak value (−V peak), the circuit current also increases to its peak value. Note that both the voltage polarity and the current direction have reversed (as shown in the circuit). |
| $t_3$ to $t_4$ | The output from the voltage source returns again to 0 V. As the voltage decreases, so does the value of the circuit current. |

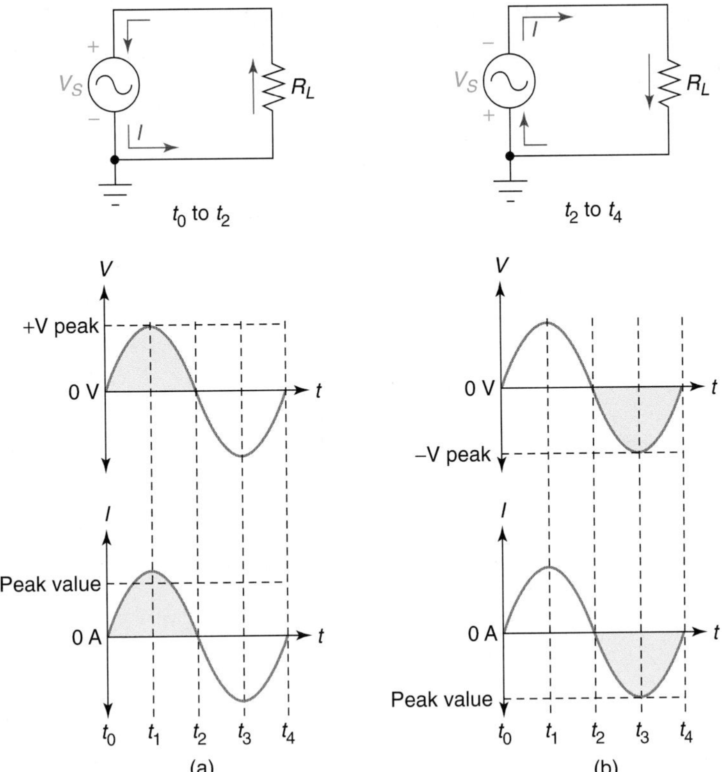


**FIGURE 11.3 Basic ac operation.**

In this description of ac circuit operation, we referred to both source voltage and current. Throughout this chapter, you will see the terms *alternating current* and *alternating voltage.* Unless indicated otherwise, you should assume that these terms describe a sinusoidal waveform like the one in Figure 11.3.

## Alternations and Cycles

◄ *OBJECTIVE 2*

The positive and negative transitions described in Figure 11.3 are referred to as **alternations.** The complete transition through one positive alternation and one negative alternation is referred to as a **cycle.** These terms are used to identify the parts of the waveform in Figure 11.4. Since two alternations make up a cycle, each alternation is commonly referred to as a half-cycle.

Most waveforms are made up of continuous cycles. The alternations and cycles of several common wave shapes are identified in Figure 11.5. As you can see, one cycle consists of a negative alternation and a positive alternation, regardless of the shape of the waveform.

**Alternation**
A term used to describe the positive and negative transitions of a waveform. For sine waves, each alternation is also referred to as a *half-cycle.*

**Cycle**
The complete transition through one positive alternation and one negative alternation of a waveform.

**FIGURE 11.4 Waveform makeup.**

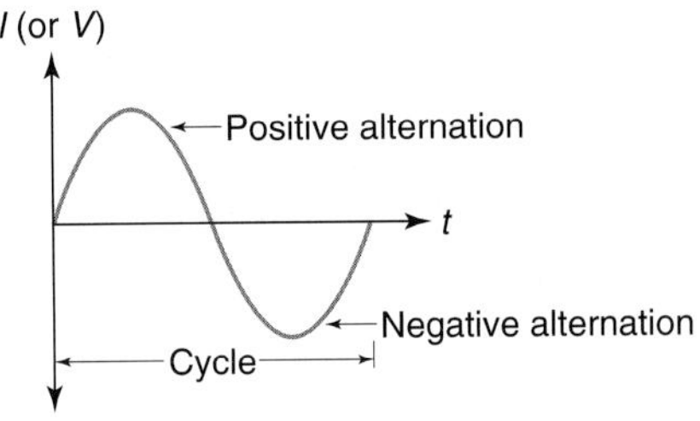

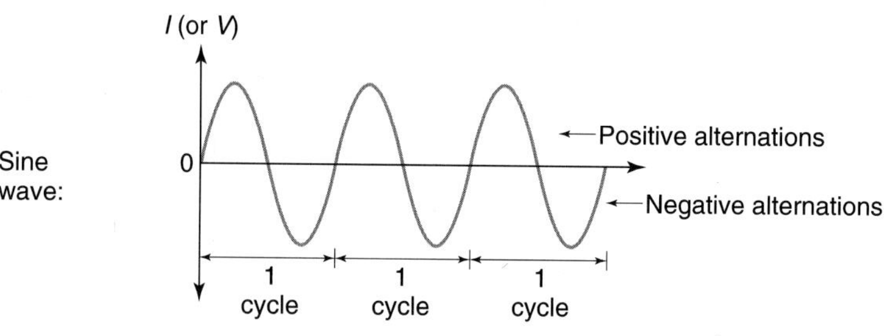


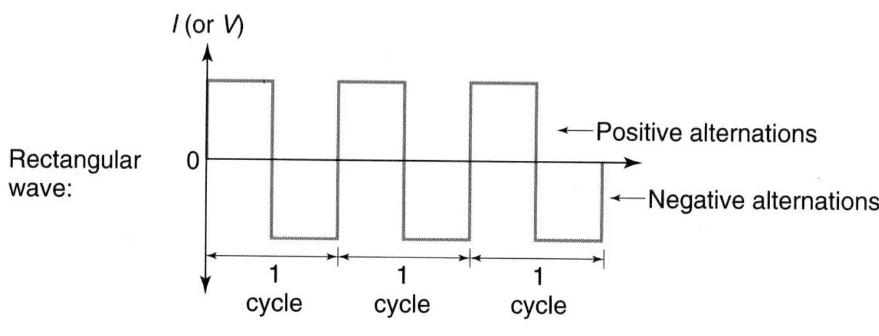


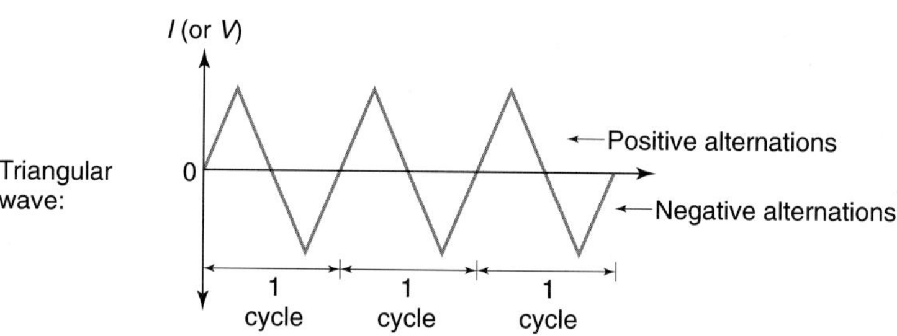


**FIGURE 11.5 Alternations and cycles of common wave shapes.**

*OBJECTIVE 3* ▶

## *Cycle Time (Period) and Frequency*

**Signal**
A term commonly used in reference to a waveform.

**Cycle time**
The time required to complete one cycle of an ac signal (also referred to as *period*).

It takes some measurable amount of time for any ac waveform (or **signal**) to complete a cycle. The time required to complete one cycle of an ac signal is referred to as its **cycle time,** or *period.* Example 11.1 demonstrates the concept of cycle time.

### EXAMPLE 11.1

Determine the cycle time of the waveform in Figure 11.6.

**FIGURE 11.6**

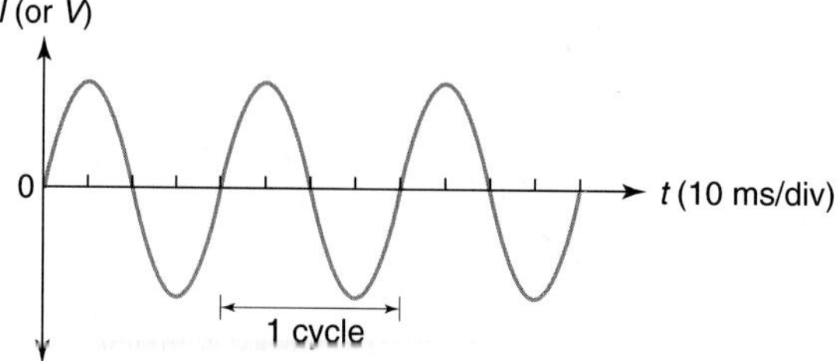

***Solution:*** As shown, each division on the horizontal axis of the graph represents a time period of 10 ms. One cycle of the waveform is four divisions in length. Therefore, the cycle time ($T_C$) of the waveform is found as

$$T_C = (4 \text{ divisions}) \times \frac{10 \text{ ms}}{\text{div}} = \mathbf{40 \text{ ms}}$$

***PRACTICE PROBLEM 11.1***

Calculate the cycle time of the waveform shown in Figure 11.7.

**FIGURE 11.7**

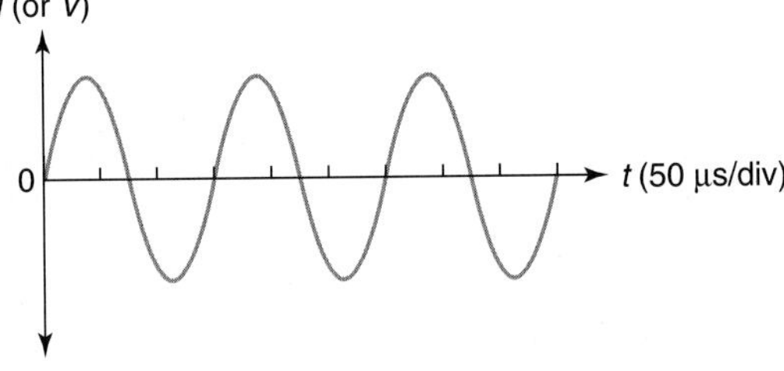


The period of a waveform can be measured from any given point on the waveform to the identical point in the next cycle. This concept is illustrated in Figure 11.8. As you can see, cycle time can be measured:

1. From one positive peak to the next positive peak.
2. From one negative peak to the next negative peak.
3. From the start of one cycle to the start of the next cycle.

In Figure 11.8, each of these sets of points span four divisions. With a horizontal scale of 5 ms/div, the total cycle time (regardless of the measurement points) works out to be

$$T_C = (4 \text{ div}) \times \frac{5 \text{ ms}}{\text{div}} = 20 \text{ ms}$$

The **frequency** of a waveform is a measure of the rate at which the cycles repeat themselves, in cycles per second. The concept of frequency is illustrated in Figure 11.9. The waveform shown has a cycle time of 200 ms. Therefore, the cycle repeats itself five times every second, and the frequency ($f$) of the waveform is given as

**Frequency**
A measure of the rate at which the cycles of a waveform repeat themselves, in cycles per second.

$$f = 5 \text{ cycles per second}$$

The unit of measure for frequency is **hertz (Hz).** One hertz is equal to one cycle per second. Therefore, the frequency of the waveform in Figure 11.9 would be written as

**Hertz (Hz)**
The unit of measure of frequency, equal to one cycle per second.

$$f = 5 \text{ Hz}$$

**FIGURE 11.8**

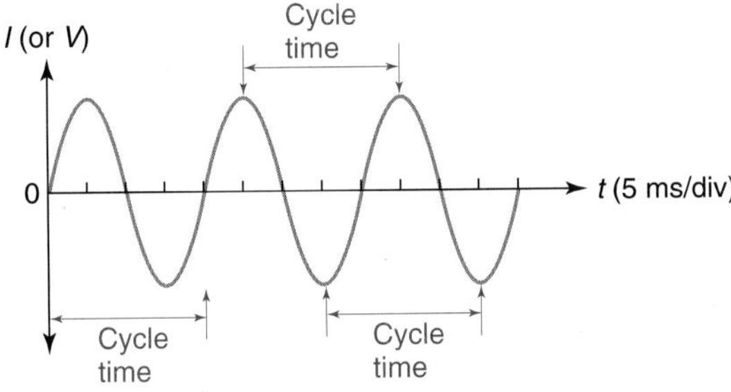

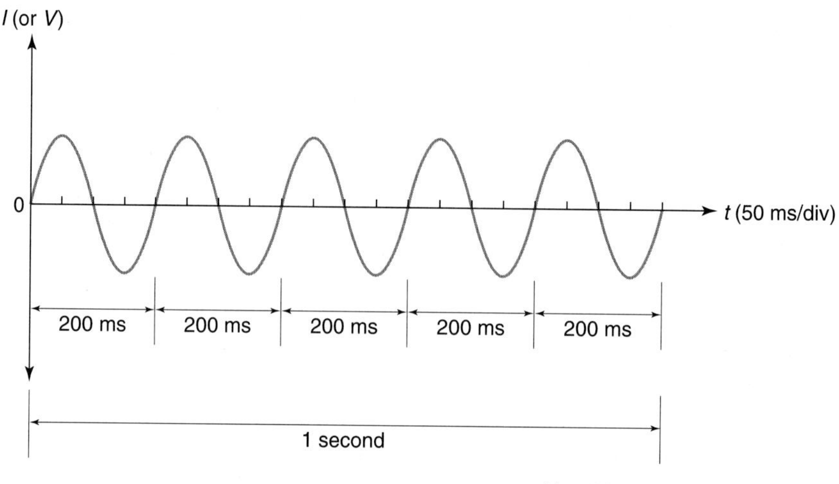


**FIGURE 11.9 Cycle time and frequency.**

When the cycle time of a waveform is known, the frequency of the waveform can be found as

$$f = \frac{1}{T_C} \tag{11.1}$$

where $T_C$ = the cycle time of the waveform in seconds.

The use of this equation is demonstrated in Example 11.2.

## EXAMPLE 11.2

Calculate the frequency of a waveform with a 100 ms cycle time.

***Solution:*** With a cycle time of 100 ms, the frequency can be found as

$$f = \frac{1}{T_C} = \frac{1}{100 \times 10^{-3}\text{s}} = \mathbf{10\ Hz}$$

This means that a waveform with a 100 ms cycle time repeats itself ten times each second.

***PRACTICE PROBLEM 11.2***

A sine wave has a cycle time of 40 ms. Calculate the frequency of the waveform.

When the frequency of a waveform is known, the cycle time can be found as

$$T_C = \frac{1}{f} \tag{11.2}$$

The use of this relationship is demonstrated in Example 11.3.

**EXAMPLE 11.3**

Calculate the cycle time of a sine wave with a frequency of 400 Hz.

***Solution:*** With a frequency of 400 Hz, the cycle time of the waveform is found as

$$T_C = \frac{1}{f} = \frac{1}{400 \text{ Hz}} = \mathbf{2.5\ ms}$$

This means that a 400 Hz waveform completes one cycle every 2.5 ms.

***PRACTICE PROBLEM 11.3***

Calculate the cycle time of a 1 kHz sine wave.

Equations (11.1) and (11.2) indicate that frequency and cycle time are inversely proportional. This relationship is verified further in Example 11.4.

**EXAMPLE 11.4**

The ac source in Figure 11.10 is shown to have an output frequency range of 1 kHz to 5 kHz. Calculate the range of cycle time values for the source.

**FIGURE 11.10**

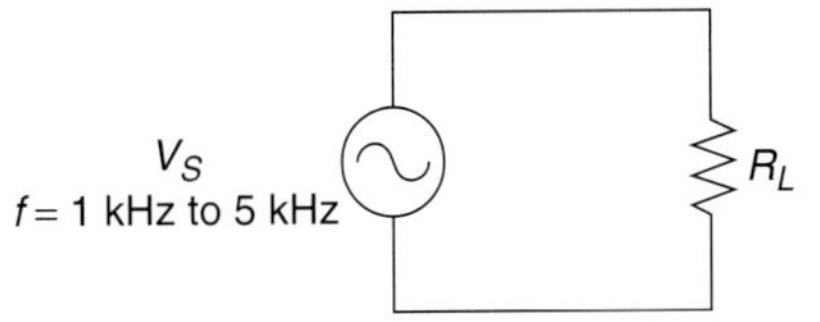

***Solution:*** When the output frequency from the source is 1 kHz, the cycle time is found as

$$T_C = \frac{1}{f} = \frac{1}{1 \times 10^3 \text{ Hz}} = \mathbf{1\ ms}$$

When the output frequency is adjusted to 5 kHz, the cycle time becomes

$$T_C = \frac{1}{f} = \frac{1}{5 \times 10^3 \text{ Hz}} = \mathbf{0.2\ ms\ (200\ \mu s)}$$

As you can see, when frequency increases by a factor of 5, cycle time decreases by the same factor.

***PRACTICE PROBLEM 11.4***

An ac source like the one in Figure 11.10 has an output frequency range of 100 Hz to 1.5 kHz. Determine the range of cycle time values for the circuit.

## Oscilloscope Time and Frequency Measurements

◄ *OBJECTIVE 4*

An **oscilloscope** is an instrument that provides a visual representation of a voltage waveform. This visual display can be used to make various magnitude and time-related measurements. An oscilloscope is shown in Figure 11.11.

**Oscilloscope**
An instrument that provides a visual representation of a voltage waveform.

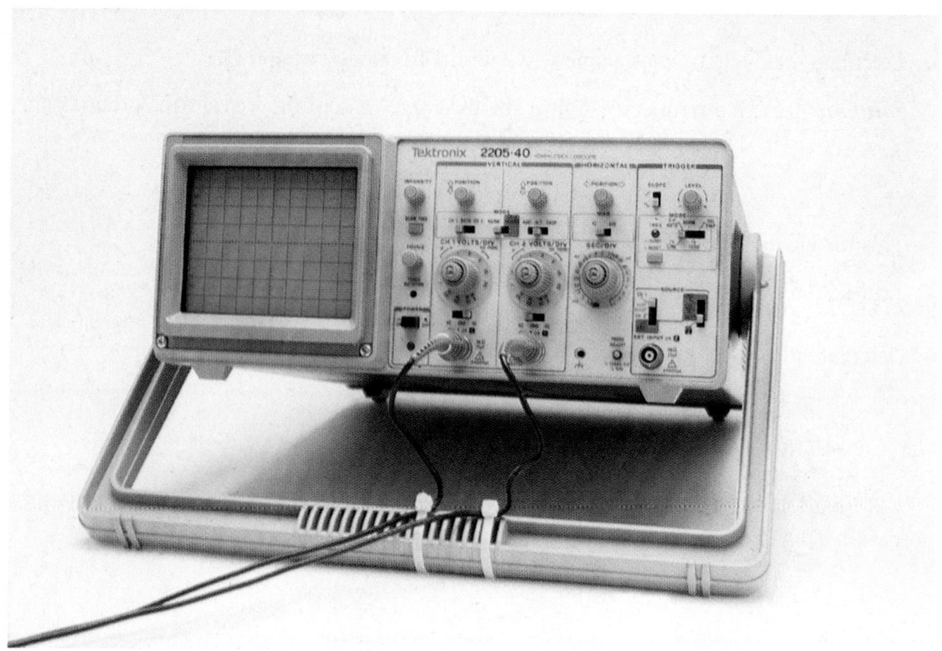


**FIGURE 11.11 Oscilloscope.**

The *x*- and *y*-axes on the screen of an oscilloscope are divided into a series of major divisions and minor divisions, as shown in Figure 11.12. The divisions along the *x*-axis are used to measure time, while the divisions along the *y*-axis are used to measure voltage. For now, we are interested only in time-related measurements. (Voltage-related measurements are addressed in the next section.)

**Time base**
The oscilloscope control used to set the period of time represented by the space between adjacent major divisions along the horizontal axis.

The setting of the **time base** control on the oscilloscope determines the amount of time represented by the space between each pair of major divisions along the *x*-axis. This point is illustrated in Figure 11.13a. If the time base control is set to 5 ms/div, the time intervals have the values shown. In each case, the value of the time interval equals the product of the time base setting and the number of major divisions.

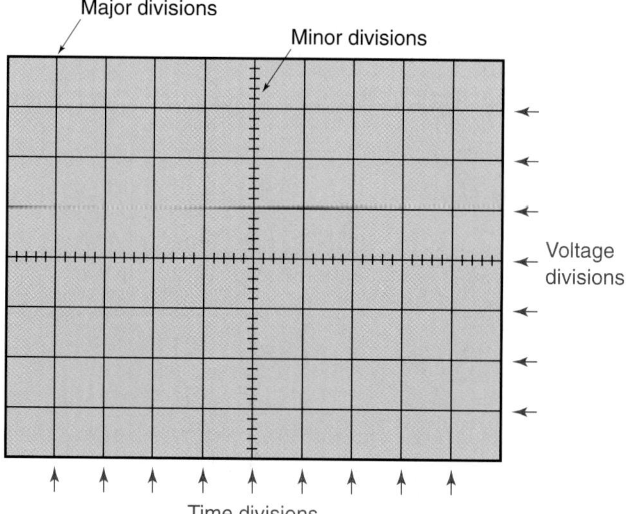


**FIGURE 11.12 Oscilloscope display grid.**

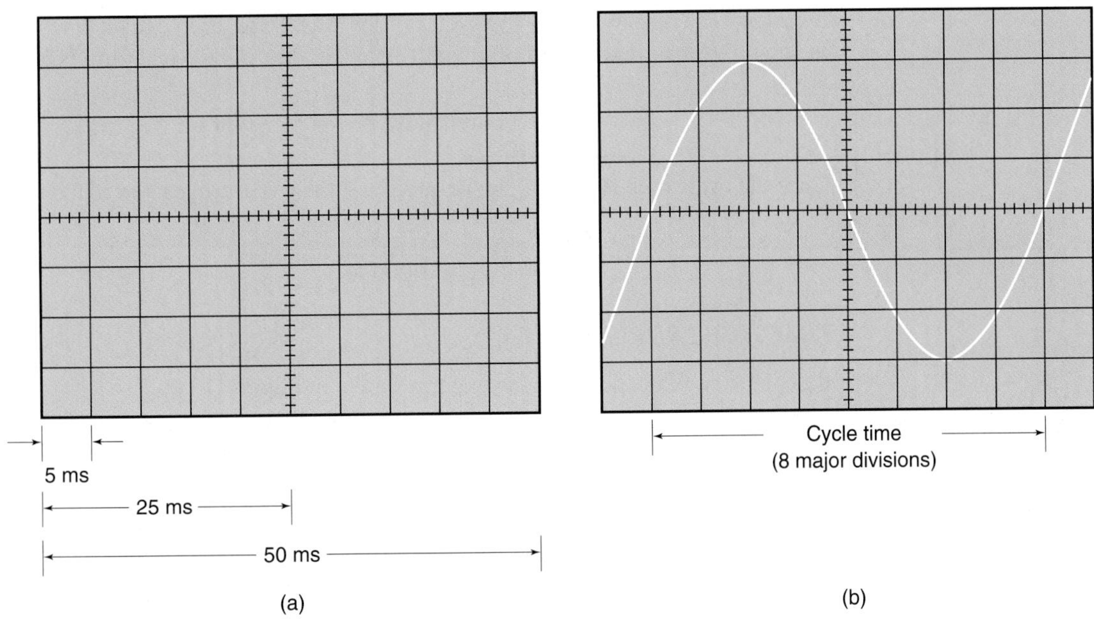


**FIGURE 11.13 Measuring cycle time.**

Calculating cycle time.

Figure 11.13b shows how the cycle time of a sine wave is measured with an oscilloscope. As shown in the figure, one cycle of the waveform is 8 major divisions in length. If the time base control is set to 5 ms/div, the cycle time of the waveform is found as

$$T_C = (8 \text{ divisions}) \times \frac{5 \text{ ms}}{\text{div}} = 40 \text{ ms}$$

The frequency of a waveform can be determined by measuring its cycle time and then calculating its frequency. This technique is demonstrated in Example 11.5.

## EXAMPLE 11.5

Determine the frequency of the waveform shown in Figure 11.14a.

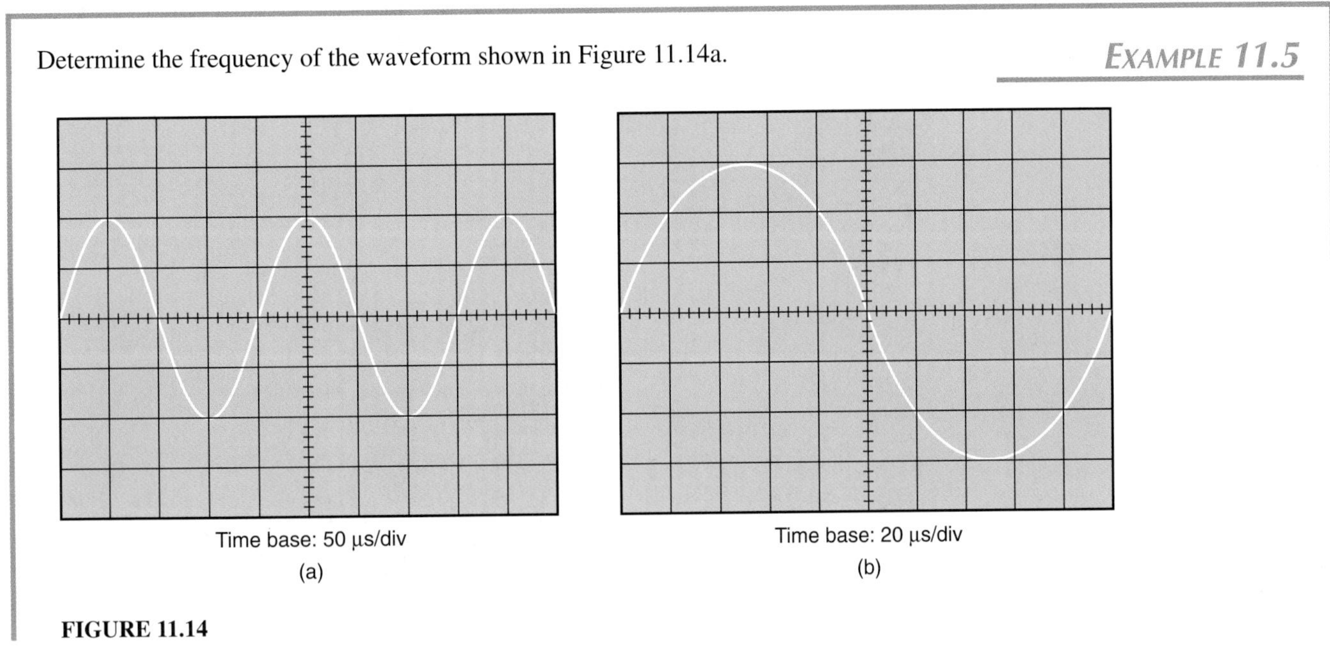


**FIGURE 11.14**

***Solution:*** The display shows 2½ cycles of the waveform, each cycle being 4 divisions in length. With a time base of 50 μs/div, the cycle time of the waveform is found as

$$T_C = (4 \text{ divisions}) \times \frac{50\ \mu\text{s}}{\text{div}} = \mathbf{200\ \mu s}$$

Once the cycle time is known, the frequency of the waveform can be found as

$$f = \frac{1}{T_C} = \frac{1}{200\ \mu\text{s}} = \mathbf{5\ kHz}$$

***PRACTICE PROBLEM 11.5***

Calculate the frequency of the waveform shown in Figure 11.14b.

*An Analogy:* Changing the time base setting on the oscilloscope is like changing the power setting on a telescope. Changing the telescope power setting makes the object appear closer (or further away); it doesn't change the actual distance to the object. By the same token, changing the time base setting on an oscilloscope changes only the display, not the waveform values.

When you work Practice Problem 11.5, you get a result of 5 kHz: the same value calculated for the waveform in Figure 11.14a. The fact that these two waveforms have the same frequency demonstrates an important characteristic of the time base control on the oscilloscope: when you vary the time base setting, you do not change the frequency of the waveform, only the time represented by each division along the horizontal axis of the display. Since the oscilloscope displays voltage as a function of time, changing the time base changes only the waveform display, not the waveform itself.

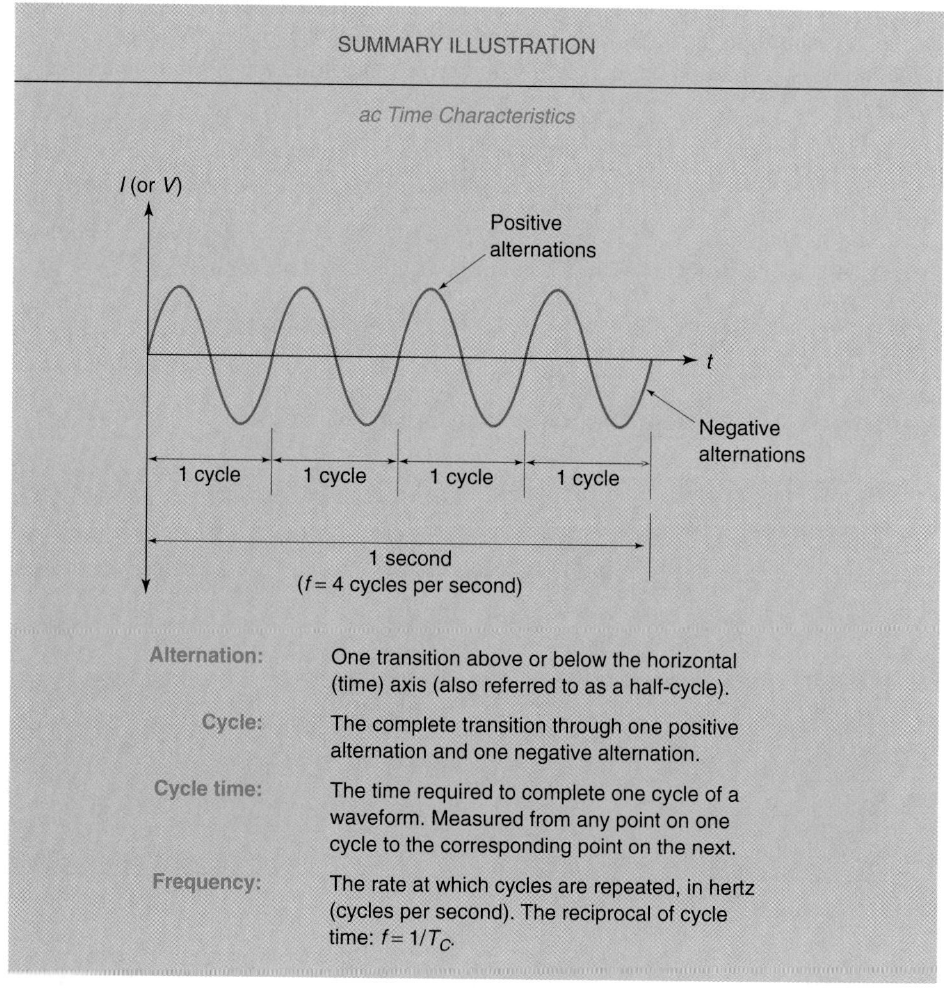


**FIGURE 11.15**

## Summary of Time-Related Characteristics

Up to this point, we have focused primarily on the time-related characteristics of ac waveforms. These characteristics (and terms) are summarized in Figure 11.15. In the next section, we'll look at several magnitude-related characteristics.

### Section Review

1. What is the term alternating current used to describe?
2. What is a waveform?
3. List the three most common waveforms.
4. What is the relationship between alternations and cycles?
5. What is cycle time? By what other name is it known?
6. Where on a sine wave can cycle time be measured?
7. What is frequency? What is its unit of measure?
8. Describe the relationship between cycle time and frequency.
9. What is an oscilloscope?
10. What is adjusted using the time base control on an oscilloscope?
11. How do you measure cycle time with an oscilloscope?

# 11.2 MAGNITUDE VALUES AND MEASUREMENTS

There are many different ways to describe and measure the magnitude of an ac waveform. In this section, we will discuss the common methods for describing and measuring the magnitude of a given waveform.

◄ *OBJECTIVE 5*

## Peak Values

Every ac waveform has two **peak values,** as shown in Figure 11.16. For the voltage waveform, the peak values are ±10 V. For the current waveform, the peak values are ±1 mA.

A **true ac waveform** is a waveform with the peak values equal in magnitude (like those shown in Figure 11.16). Later in this chapter, you will be shown some waveforms that are not true ac.

**Peak value**
The maximum value reached by either alternation of a waveform.

**True ac waveform**
An ac waveform with peak values that are equal in magnitude.

## Peak-to-Peak Values

The **peak-to-peak value** of a waveform equals the difference between its positive and negative peak values. For the voltage waveform in Figure 11.16, the difference between the peak values is shown to be 20 V. For the current waveform, the difference between the peak values is shown to be 2 mA.

The peak-to-peak value of a pure ac waveform is twice its peak value. By formula,

$$V_{PP} = 2\,V_{pk} \quad \text{(for pure ac)} \tag{11.3}$$

and

$$I_{PP} = 2\,I_{pk} \quad \text{(for pure ac)} \tag{11.4}$$

These relationships can be verified by comparing the peak and peak-to-peak values shown in Figure 11.16. In each case, the peak-to-peak value is twice the peak value.

**Peak-to-peak value**
The difference between the peak values.

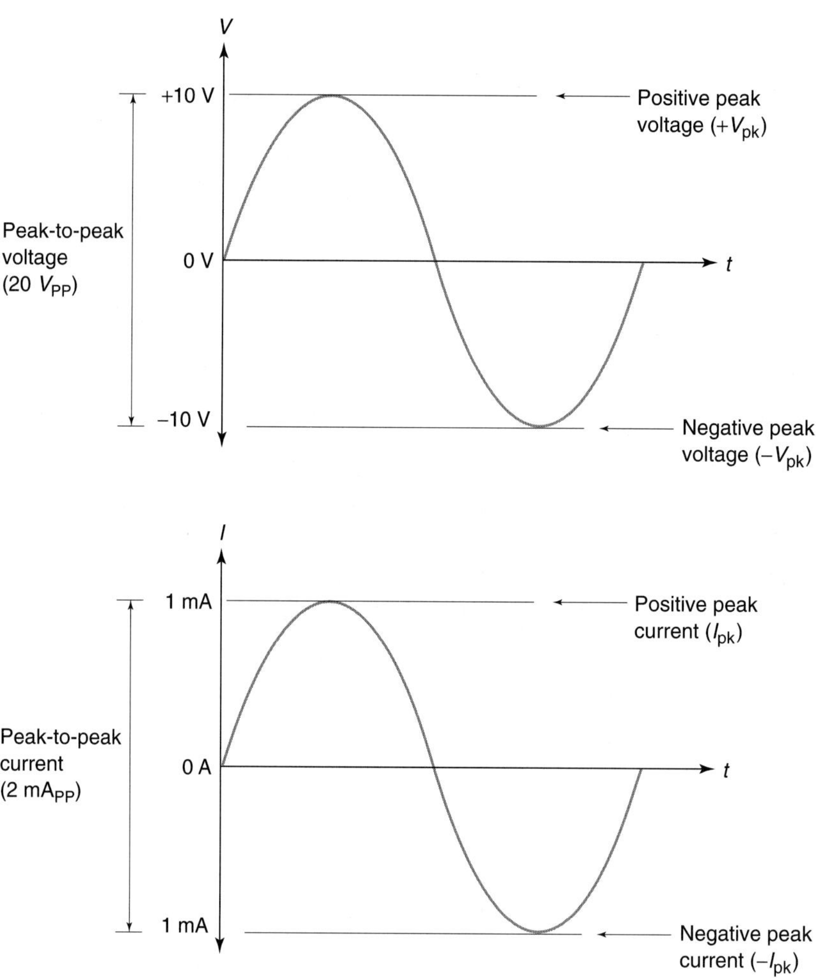


**FIGURE 11.16 Peak and peak-to-peak values.**

## Instantaneous Values

**Instantaneous value**
The magnitude of a voltage or current at a specified point in time.

An **instantaneous value** is the magnitude of a voltage or current at a specified point in time. For example, consider the voltage sine wave shown in Figure 11.17. Two time intervals have been specified ($t_1$ and $t_2$). At each time interval, a corresponding instantaneous voltage has been identified. These voltages are labeled $v_1$ and $v_2$, respectively.

In the next section, you will learn how to calculate the actual value of any instantaneous voltage (or current). For now, we simply want to establish these basic principles and practiccs:

1. For every time interval in a waveform, there is one instantaneous value of voltage (or current).
2. Since we can divide one cycle of a waveform into an infinite number of time intervals, each cycle contains an infinite number of instantaneous values.
3. Instantaneous values are identified using lowercase (rather than uppercase) letters. For example, the positive peak voltage in Figure 11.17 is labeled using uppercase letters ($+V_{pk}$) because its value remains constant from one cycle to the next. However, the instantaneous voltages are labeled using lowercase letters, indicating that their values change from one instant to the next.

As you will learn, instantaneous values are used to define other types of voltage and current measurements.

**FIGURE 11.17 Instantaneous voltages.**

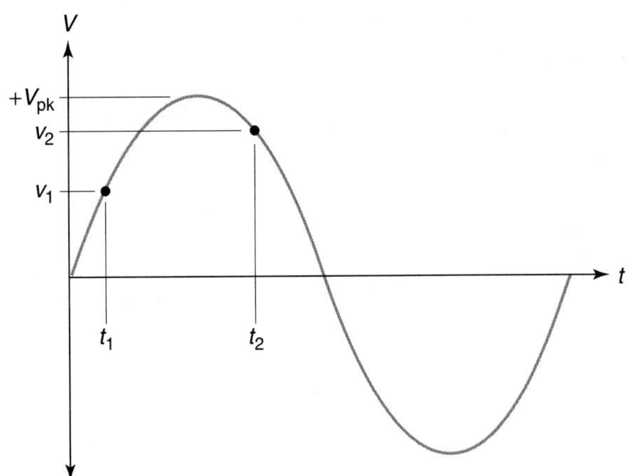


## Full-Cycle Average

The **full-cycle average** of a waveform is the average of all its instantaneous values of voltage (or current) throughout one complete cycle. The full-cycle average of a voltage waveform is determined as shown in Figure 11.18.

**Full-cycle average** The average of all the instantaneous values of a waveform voltage (or current) throughout one complete cycle. Sometimes referred to as *full-wave average.*

The full-cycle average of a pure ac waveform is always zero. The reason for this can be seen by comparing the positive and negative $v_1$ and $v_2$ values highlighted in Figure 11.18. As these values indicate, every positive instantaneous value has a negative equivalent. When these two voltages are added, the result is 0 V. Therefore, every positive instantaneous value is negated by a negative equivalent, and the overall average of the waveform is zero.

So, why bother? Later in this chapter when we discuss waveforms that are not pure ac, you will see that the full-cycle average can have a value other than zero.

**FIGURE 11.18 Determining the full-wave average of a waveform.**

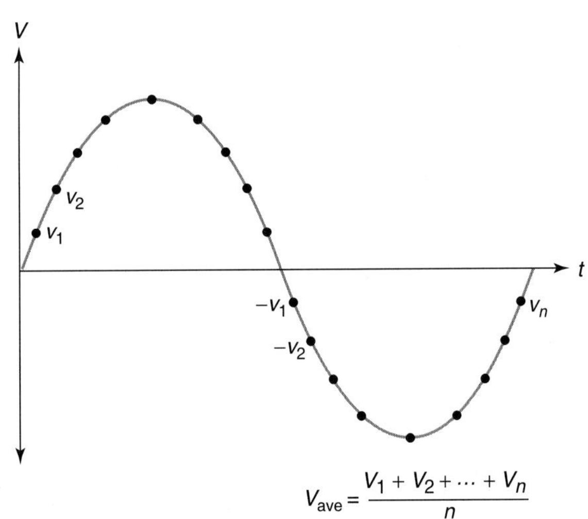


$$V_{ave} = \frac{V_1 + V_2 + \cdots + V_n}{n}$$

## Half-Cycle Average

In some applications, it is helpful to know the half-cycle average of a sine wave (or some other waveform). The **half-cycle average** of a waveform is the average of all its instantaneous values of voltage (or current) through either of its alternations (half-cycles). The half-cycle average of a waveform can be determined as shown in Figure 11.19a. As you can see, all of the instantaneous values are taken from the positive alternation. Therefore, the half-cycle average always has some measurable value.

**Half-cycle average** The average of all the instantaneous values of a waveform voltage (or current) throughout one alternation of a cycle. Sometimes referred to as *half-wave average.*

FIGURE 11.19 **Determining the half-wave average of a sine wave.**

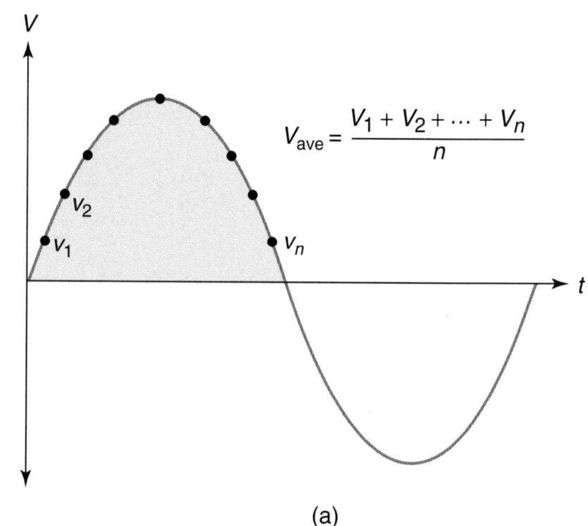


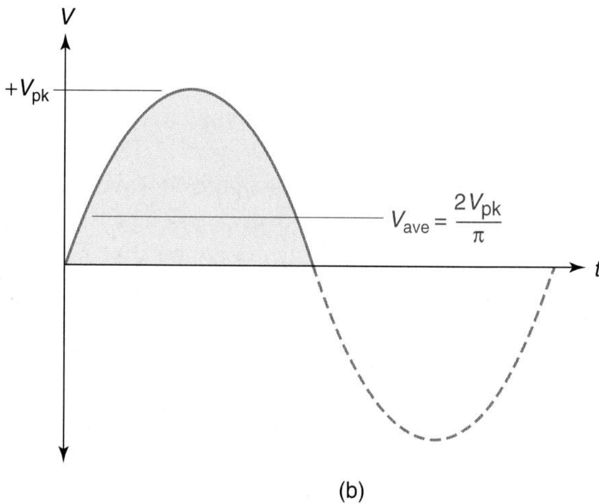


It is not necessary to measure and average a series of instantaneous values to determine the value of $V_{ave}$ for a sine wave. In Appendix E, the following equation is derived for finding the half-cycle average voltage of a pure ac sine wave:

$$V_{ave} = \frac{2\ V_{pk}}{\pi} \quad \text{(for a pure ac sine wave)} \tag{11.5}$$

Since $\frac{2}{\pi} \cong 0.637$, the above equation is often written as

$$V_{ave} \cong 0.637\ V_{pk} \quad \text{(for a pure ac sine wave)} \tag{11.6}$$

The process for determining the value of $V_{ave}$ for a sine wave is demonstrated in Example 11.6.

## EXAMPLE 11.6

Calculate the half-cycle average voltage ($V_{ave}$) for the sine wave in Figure 11.20a.

***Solution:*** The waveform is shown to have a peak value of +15 V. Using this value in equation (11.5), the value of $V_{ave}$ is found as

$$\mathbf{V_{ave}} = \frac{2\ V_{pk}}{\pi} = \frac{30\text{ V}}{\pi} = \mathbf{9.55\ V}$$

**FIGURE 11.20**

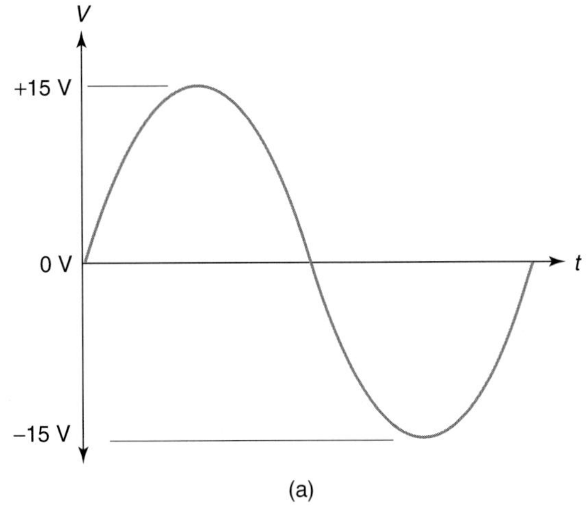


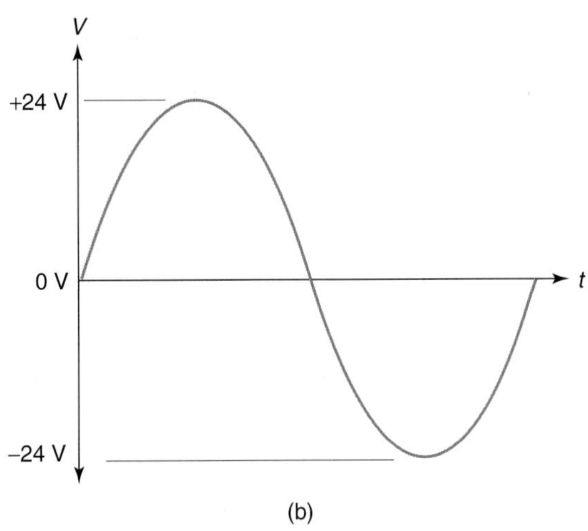


***PRACTICE PROBLEM 11.6***

Calculate the value of $V_{\text{ave}}$ for the waveform in Figure 11.20b.

## Average ac Power and rms Values

*OBJECTIVE 6*

In Chapter 4, you were told that electrical and electronic systems are designed for the purpose of delivering power (in one form or another) to a given load. You were also told that load power can be found as

$$P_L = \frac{V^2}{R_L}$$

Figure 11.21a shows a dc circuit and an ac circuit. Calculating the load power for the dc circuit is relatively simple: we just use the values of $V_S$ and $R_L$ in the proper power equation. For example, if the dc circuit shown has values of $V_S = 12$ V and $R_L = 100\ \Omega$, the load power for the circuit is found as

$$P_L = \frac{V_S^2}{R_L} = \frac{(12\text{ V})^2}{100\ \Omega} = 1.44\text{ W}$$

Calculating the load power for the ac circuit in Figure 11.21a gets a bit more complicated because the source voltage is changing constantly. For example, let's say that the voltage waveform in Figure 11.21b represents the output from the ac source. Since the

**FIGURE 11.21**

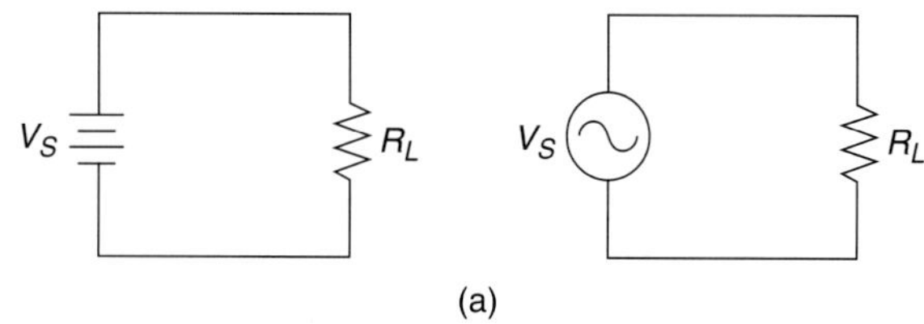


(a)

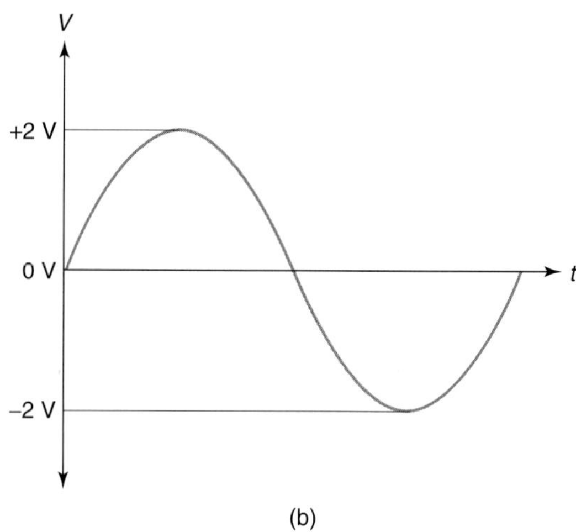


(b)

value of $V_S$ is changing constantly, so is the value of load power (according to the power equation above).

When dealing with ac circuits, we are interested in the average ac power generated throughout each cycle. The concept of average ac power is illustrated in Figure 11.22a. If we assume that the curve shown represents the variations in load power over one cycle of load voltage, **average ac power** ($P_{ac}$) for the load is the value that falls midway between 0 W and the peak value. (The basis for positioning the curve as shown will become clear as this discussion continues.)

**Average ac power**
The average power generated over the course of each complete cycle of an ac waveform. On a graph of power, the value that falls halfway between 0 W and peak power.

As you know, power is proportional to the square of voltage. If we assume that $R_L$ in Figure 11.22a has a value of 1 Ω, then the load power for the circuit can be found as

$$P_L = \frac{V_L^2}{1\ \Omega}$$

or

$$P_L = V_L^2 \quad (\text{for } R_L = 1\ \Omega)$$

This relationship indicates that the $P_L$ and $V_L^2$ curves for this circuit are identical. Keeping this relationship in mind, look at the waveforms shown in Figure 11.22b. The $V_L^2$ curve in the figure was derived by squaring the values in the source voltage ($V_S$) curve. As you can see, all the values in the $V_L^2$ curve are positive because squaring any value (positive or negative) gives us a positive result.

According to the curve, load power equals the value of $P_{ac}$ when

$$V_L^2 = \frac{V_{pk}^2}{2}$$

which is the average value of the $V_L^2$ waveform. Another way of expressing this relationship is

*Note:* Equation (11.7) is application-specific, meaning that it is written for a specific circuit. Equation (11.8) provides a general-purpose form of the equation.

$$P_L = P_{ac} \qquad \text{when} \qquad V_L = \sqrt{\frac{V_{pk}^2}{2}} \tag{11.7}$$

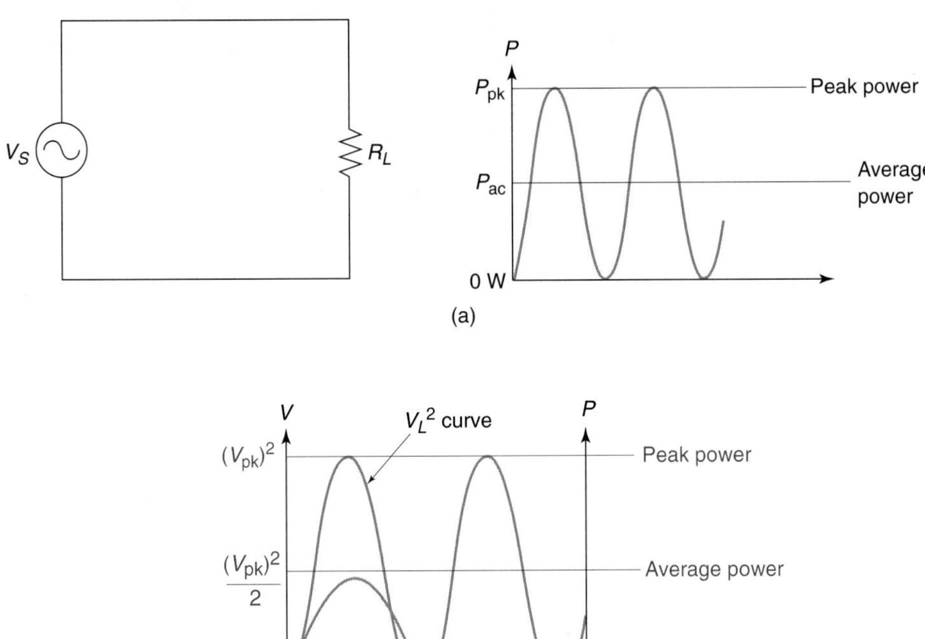


**FIGURE 11.22 Sine wave voltage and power curves.**

Equation (11.7) defines what is called the root-mean-square (rms) value of voltage for the circuit in Figure 11.22. The **root-mean-square (rms) value** of a voltage (or current) waveform is the value that, when used in the appropriate power equation, will give you the average ac power of the waveform.

**Root-mean-square (rms) value** The value of voltage (or current) that, when used in the appropriate power equation, gives the average ac power of the waveform.

The waveforms in Figure 11.23 illustrate the basis for the term *root-mean-square*. First, the instantaneous voltages in the sine wave (labeled *V* curve) are squared. This gives us the $V^2$ curve shown in the figure.

The average (or **mean**) value of the $V^2$ curve is the value that lies halfway between its peaks. Since the curve has peak values of $V_{pk}^2$ and 0 V, the mean is found as

**Mean** Another word for average.

$$\text{mean} = \frac{V_{pk}^2}{2}$$

**FIGURE 11.23 The rms value of a voltage sine wave.**

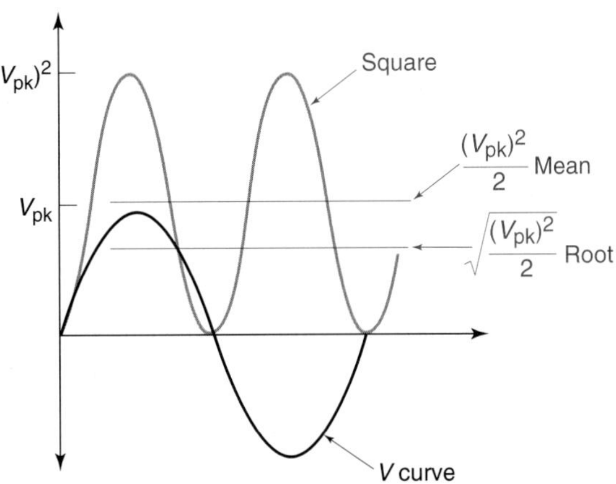

Finally, we take the square root of the mean, which gives us

$$V_{rms} = \sqrt{\frac{V_{pk}^2}{2}} \tag{11.8}$$

As you can see, this is the same equation we used to determine the value of $V_L$ that provides a value of $P_L = P_{ac}$. Therefore, *the rms value of a voltage (or current) is the value we use to calculate average ac power.*

*OBJECTIVE 7* ▶ Equation (11.8) establishes the relationship between peak and rms voltage. However, it can be greatly simplified, as follows:

$$V_{rms} = \sqrt{\frac{V_{pk}^2}{2}} = \frac{\sqrt{V_{pk}^2}}{\sqrt{2}} = \frac{V_{pk}}{\sqrt{2}}$$

And finally, since $1/\sqrt{2} \cong 0.707$, the relationship between peak and rms voltage can be written as

$$V_{rms} = 0.707\ V_{pk} \tag{11.9}$$

By the same token, the rms value of a current sine wave can be found as

$$I_{rms} = 0.707\ I_{pk} \tag{11.10}$$

Example 11.7 demonstrates the use of these equations for determining rms values.

## EXAMPLE 11.7

Calculate the rms values of voltage and current for the circuit in Figure 11.24a.

EWB

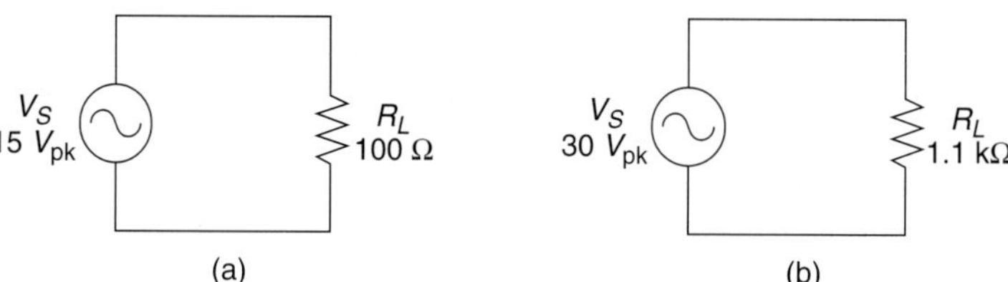


**FIGURE 11.24**

***Solution:*** Using the peak value shown, the rms voltage is found as

$$\mathbf{V_{rms}} = 0.707\ V_{pk} = (0.707)(15\text{ V}) = \mathbf{10.6\ V}$$

The peak current through the circuit can be found using Ohm's law, as follows:

$$\mathbf{I_{pk}} = \frac{V_{pk}}{R_L} = \frac{15\text{ V}}{100\ \Omega} = \mathbf{150\ mA}$$

The rms current can now be found as

$$\mathbf{I_{rms}} = 0.707\ I_{pk} = (0.707)(150\text{ mA}) = \mathbf{106\ mA}$$

(Of course, we could have simply divided the value of $V_{rms}$ by the value of $R_L$ to find the value of rms current. The result would have been the same.)

***PRACTICE PROBLEM 11.7***

Calculate the rms values of voltage and current for the circuit in Figure 11.24b.

As stated earlier, ac power calculations are always made using rms values. Example 11.8 demonstrates the process for determining average load power when the value of peak voltage is known.

**EXAMPLE 11.8**

A circuit like those in Figure 11.24 has values of $V_{pk} = 12$ V and $R_L = 330\ \Omega$. Calculate the value of load power ($P_L$) for the circuit.

***Solution:*** The load power is found using the rms value of source voltage. This value can be found as

$$V_{\mathbf{rms}} = 0.707\ V_{pk} = (0.707)(12\text{ V}) = \mathbf{8.48\ V}$$

Once the peak voltage has been converted to an rms value, the load power can be found as

$$P_L = \frac{V_{rms}^2}{R_L} = \frac{(8.48\text{ V})^2}{330\ \Omega} = \mathbf{218\ mW}$$

***PRACTICE PROBLEM 11.8***

A circuit like those in Figure 11.24 has values of $V_{pk} = 10$ V and $R_L = 1.5\text{ k}\Omega$. Calculate the average load power for the circuit.

There are several more points that need to be made regarding rms values and their applications:

Some important points regarding rms values and their applications.

1. Power has always been used as the basis of comparison for ac and dc; that is, two sources (one dc and one ac) are considered equivalents when they deliver the same amount of power to a given load.

   Average ac power is the equivalent of dc power. Since rms values are used to calculate average ac power, they are also an indicator of equivalent dc power. In other words, a sine wave with a given rms value delivers the same amount of power to a load as a dc source with the same value. (For example, a 20 $V_{rms}$ sine wave provides the same amount of power to a given load as a 20 $V_{dc}$ source.)
2. Since power is a primary concern with any electrical or electronic system, rms values are used more often than any of the other magnitude-related measurements (such as peak, average, and so on).
3. When a magnitude-related measurement is not specifically identified (as peak, peak-to-peak, or average) it is automatically assumed to be an rms value. For example, the receptacles in your home are identified as being 120 $V_{ac}$ or 240 $V_{ac}$. Since these values are not identified as peak, peak-to-peak, or average, we know they are rms values.

## Measuring Magnitude-Related Values

◄ *OBJECTIVE 8*

You have been introduced to peak, peak-to-peak, average, and rms values. At this point, we need to look at how the various waveform values are measured in practice.

***Measuring Peak and Peak-to-Peak Voltages.*** In the last section, you were shown how the oscilloscope is used to measure cycle time and frequency. You were also told that the oscilloscope can be used to make magnitude-related measurements.

Both peak and peak-to-peak voltages can be measured using an oscilloscope. These measurements are illustrated in Figure 11.25. (The measurement of cycle time is shown to provide a basis of comparison between time and magnitude measurements.)

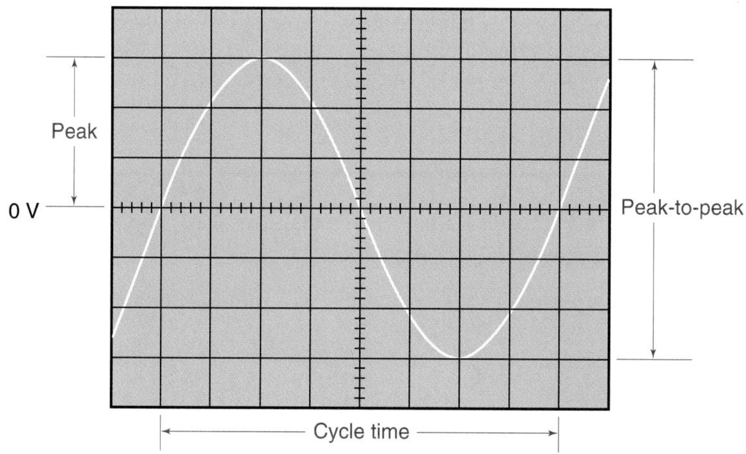


**FIGURE 11.25 Sine wave voltage (and time) measurements.**

**Vertical sensitivity**
The oscilloscope control used to set the amount of voltage represented by the space between adjacent major divisions along the vertical axis. Usually referred to as the volts per division (V/div) control.

Voltage measurements are made along the vertical axis ($y$-axis) of the display. The value represented by each pair of major divisions is controlled by the **vertical sensitivity** control. This point is illustrated in Figure 11.26a. Note that the vertical sensitivity control is simply labeled *volts/division* on most oscilloscopes. With a vertical sensitivity of 5 V per

To establish the 0 V reference at the center of the grid:

1. Set the ac/gnd/dc switch to the gnd position.
2. Turn the vertical position control until the trace rides on the center line of the grid.
3. Return the ac/gnd/dc switch to its original setting.

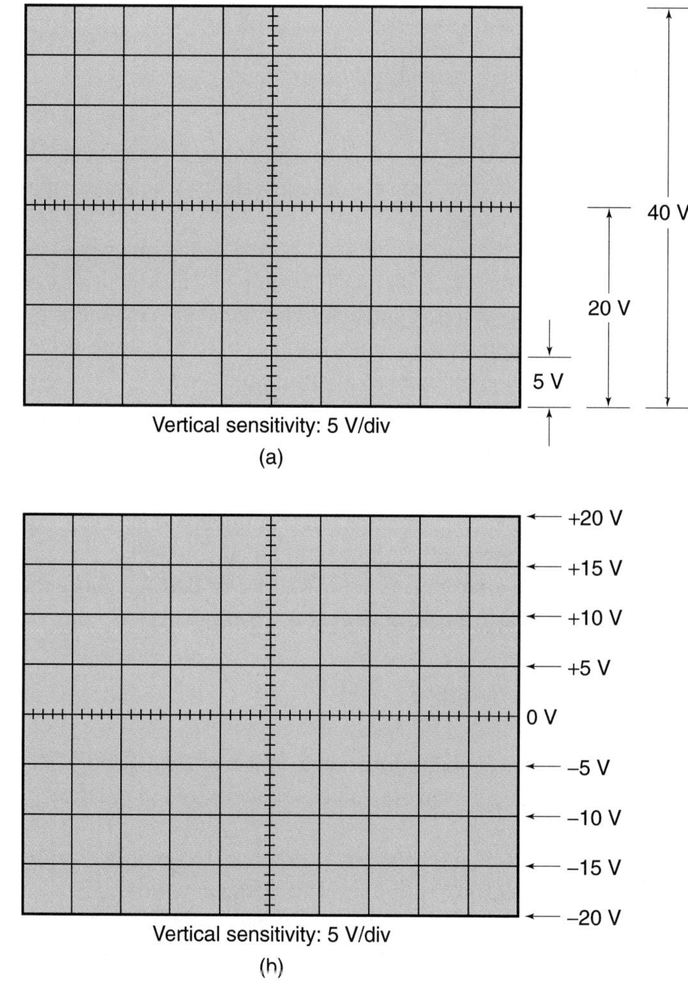


**FIGURE 11.26 Vertical sensitivity (volts per division).**

division, the voltage ranges have the values shown. If the volts/div setting is changed, the voltage represented by each range changes.

When the center of the grid is established as 0 V, the divisions above the center line represent positive voltage values and the divisions below the center line represent negative voltage values. This point is illustrated in Figure 11.26b. As Example 11.9 demonstrates, it is easy to measure the peak value of a sine wave when the 0 V reference on the grid is established.

### EXAMPLE 11.9

The center of the grid in Figure 11.27a has been established as the 0 V reference. Determine the peak value of the sine wave.

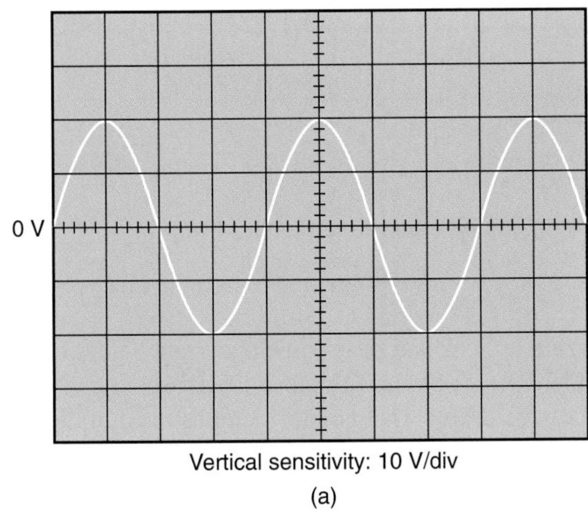


Vertical sensitivity: 10 V/div

(a)

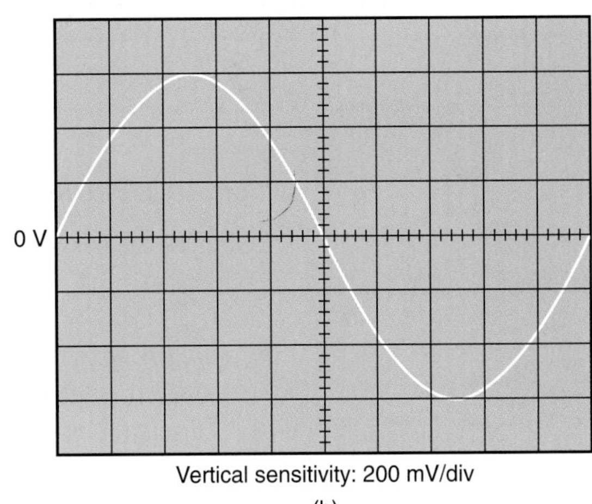


Vertical sensitivity: 200 mV/div

(b)

**FIGURE 11.27**

***Solution:*** The positive peak is two major divisions above the 0 V line. With a vertical sensitivity of 10 V per division, the peak voltage is found as

$$V_{\mathbf{pk}} = (2 \text{ divisions}) \times \frac{10 \text{ V}}{\text{div}} = \mathbf{20\ V}$$

***PRACTICE PROBLEM 11.9***

Determine the peak voltage for the sine wave displayed in Figure 11.27b.

As Example 11.10 demonstrates, peak-to-peak voltage is measured using the same basic technique used to measure peak voltage. The only difference is that we do not need to establish the 0 V point on the grid to measure peak-to-peak voltage because we are *not* referencing our measurement to the 0 V point.

### EXAMPLE 11.10

Determine the peak-to-peak voltage displayed in Figure 11.28a.

***Solution:*** The waveform is shown to be approximately five major divisions in height (from peak to peak). With a vertical sensitivity of 50 mV/div, the peak-to-peak voltage is found as

$$V_{\mathbf{PP}} = (5 \text{ divisions}) \times \frac{50 \text{ mV}}{\text{div}} = \mathbf{250\ mV}$$

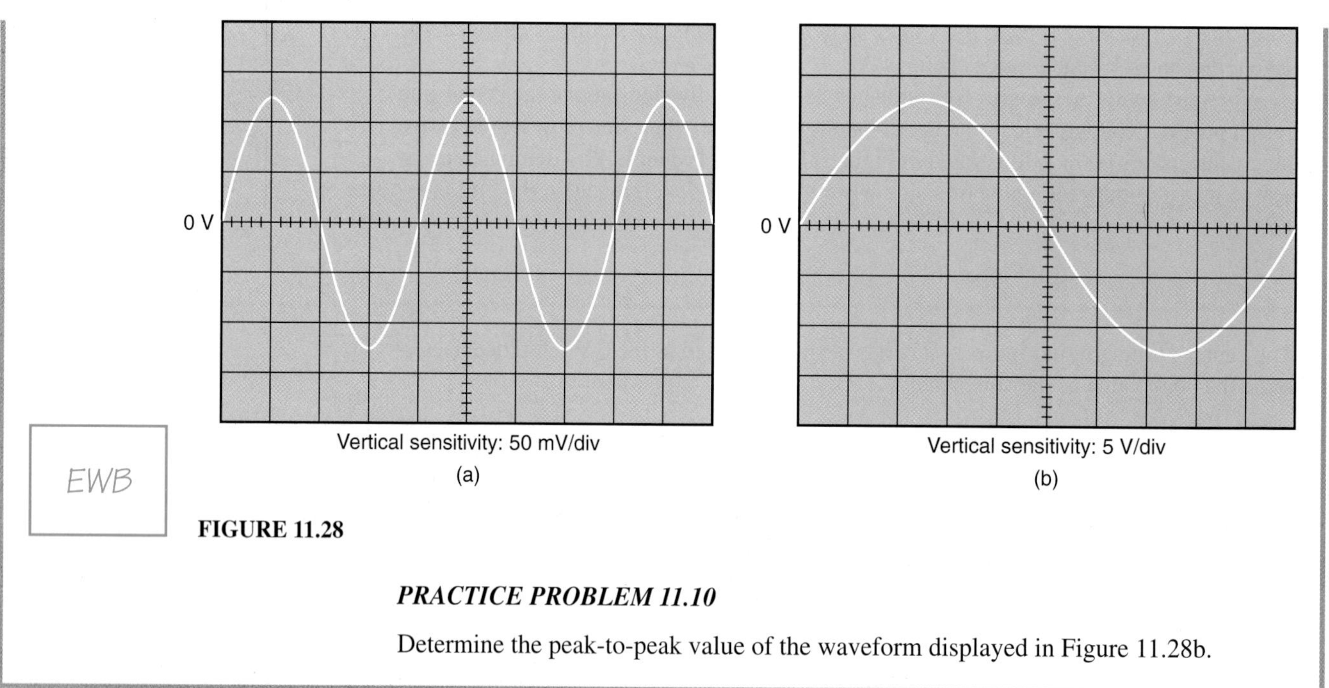


**FIGURE 11.28**

***PRACTICE PROBLEM 11.10***

Determine the peak-to-peak value of the waveform displayed in Figure 11.28b.

***Peak and Peak-to-Peak Current.*** Peak and peak-to-peak current values cannot be measured directly using any standard pieces of test equipment. Even so, knowing the value of peak current for an ac signal can be useful. This point is demonstrated in Example 11.11.

## EXAMPLE 11.11

The load in Figure 11.29 represents a current-sensitive device that has an absolute current limit of 500 mA. The internal resistance of the device is 10 Ω. Determine whether the value of the series resistor ($R_S$) is sufficient to protect the load from excessive current.

**FIGURE 11.29**

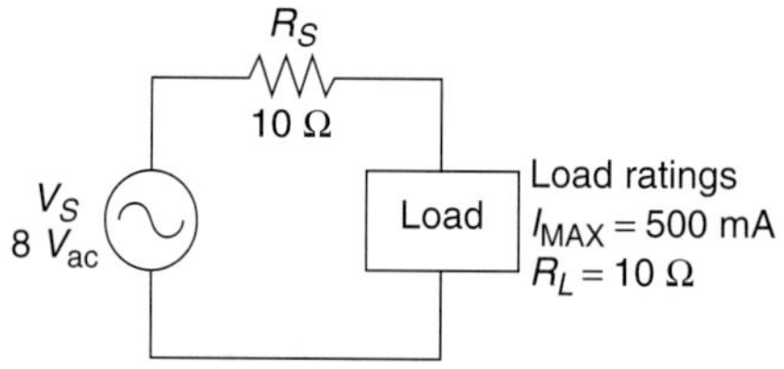

***Solution:*** The source has an rms value of 8 $V_{ac}$. The total resistance in the circuit is found as $R_S + R_L = 20\ \Omega$. Using this value, the circuit current can be found as

$$I = \frac{V_S}{R_S + R_L} = \frac{8\ V_{ac}}{20\ \Omega} = \mathbf{400\ mA}$$

It would appear that the value of $R_S$ is sufficient to protect the load. However, the current value we calculated is an rms value. Using a transposed form of equation (11.9), the peak current can be found as

$$I_{pk} = \frac{I_{rms}}{0.707} = \frac{400\ \text{mA}}{0.707} = \mathbf{566\ mA}$$

This current value occurs only for a brief instant at the sine wave peaks, so it may not seriously affect the load. At the same time, we cannot be certain that the load will *not* be affected by the peak current values. Therefore, a higher value series resistor must be used.

As you can see, peak current can be a practical value, even though it cannot be measured using any standard pieces of test equipment. Peak-to-peak current, on the other hand, is a conceptual value that does not play a role in practical situations.

***Measuring Average Voltage and Current.*** Earlier, you were told that the full-cycle average of a pure ac voltage is 0 V. You were also told that the half-cycle average of a pure ac voltage is the average of all the instantaneous values in one alternation of the waveform.

The half-cycle average of any ac voltage can be measured using a dc voltmeter. There is, however, one requirement: the waveform must be altered so its alternations are all positive or all negative. The reason for this is illustrated in Figure 11.30a. If we attempt to read the value of $V_{ave}$ as shown, both alternations of the source voltage are applied to the meter. As a result, the meter reads the average of the entire cycle, which is 0 V. So, if we want to read the average of either alternation, we have to ensure that the dc meter "sees" only positive alternations or negative alternations, not both. This can be accomplished using a circuit called a rectifier.

A **rectifier** is a circuit that converts ac to pulsating dc; that is, it converts the ac to a series of alternations that are all positive or all negative. The **full-wave rectifier** in Figure 11.30b essentially converts the negative input alternations to positive alternations. The **half-wave rectifier** in Figure 11.30c simply eliminates the negative alternations at its input.

**Rectifier**
A circuit that converts ac to pulsating dc.

**Full-wave rectifier**
Converts negative alternations to positive alternations (or vice versa) to provide a single-polarity output.

**Half-wave rectifier**
Eliminates the negative (or positive) alternations to provide a single-polarity output.

Once an ac voltage has been rectified, a dc voltmeter can be used to measure its half-cycle average. For example, if a dc voltmeter is connected to the output of the full-wave rectifier in Figure 11.30b, the meter reading will equal the half-cycle average of the positive alternation of the input waveform. By formula,

$$V_{dc} = \frac{2\,V_{pk}}{\pi} \cong 0.637\,V_{pk} \quad \text{(full-wave rectified)} \tag{11.11}$$

If you compare this equation to equations (11.5) and (11.6), you'll see that they all define the dc average of one alternation of a voltage sine wave.

Now, compare the output from the half-wave rectifier (Figure 11.30c) to that of the full-wave rectifier. As you can see, the half-wave rectifier provides one output transition for every two provided by the full-wave rectifier. For this reason, a dc voltmeter con-

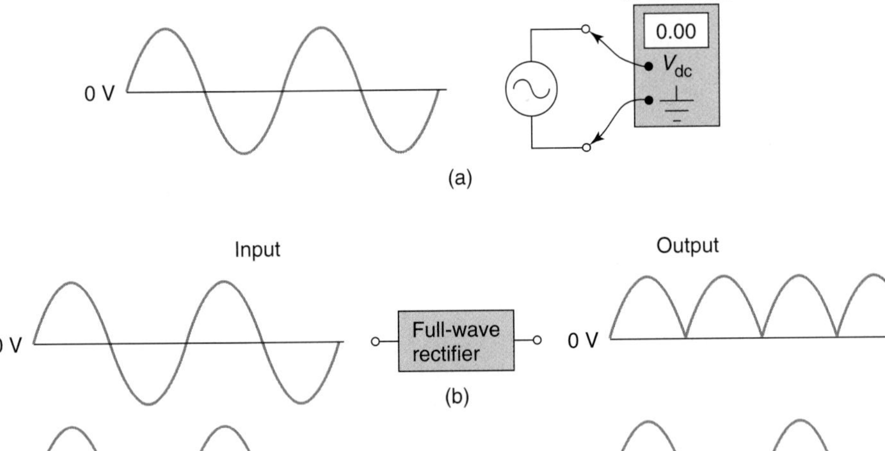


**FIGURE 11.30 Rectification.**

*A Practical Consideration:* The rectifiers in Figure 11.30 are positive rectifiers, meaning that their outputs consist of positive alternations. Negative rectifiers (not shown) provide outputs that consist of negative alternations.

EWB

nected to a half-wave rectifier reads half the value provided by the full-wave rectifier. By formula,

$$V_{dc} = \frac{V_{pk}}{\pi} \cong 0.318\ V_{pk} \quad \text{(half-wave rectified)} \qquad \textbf{(11.12)}$$

If you compare equation (11.12) with equation (11.11), you'll see that the output from a half-wave rectifier is half that of a full-wave rectifier.

***Rectifier Construction and Operation.*** Both half-wave and full-wave rectifiers contain semiconductor components called *diodes.* Rectifier construction and operation are beyond the scope of this discussion. (You will learn about diodes and their role in rectifier operation when you study semiconductor devices and circuits.) For now, we are interested only in establishing the need to rectify a sine wave in order to measure its half-cycle average.

***Measuring Average Current.*** Once rectified, the half-cycle average current of a sine wave can be measured using a dc ammeter. Assuming the sine wave is full-wave rectified, a dc ammeter will provide a reading equal to the half-cycle average current.

The value of half-cycle current for a sine wave is found using the same basic relationship that we used to find half-cycle average voltage. Therefore, the value read by a dc ammeter that is connected to the output of a full-wave rectifier can be found as

$$I_{dc} = \frac{2\ I_{pk}}{\pi} \cong 0.637\ I_{pk} \quad \text{(full-wave rectified)} \qquad \textbf{(11.13)}$$

If the sine wave is half-wave rectified, the value measured by a dc ammeter will be one-half that of a full-wave rectified waveform. By formula,

$$I_{dc} = \frac{I_{pk}}{\pi} \cong 0.318\ I_{pk} \quad \text{(half-wave rectified)} \qquad \textbf{(11.14)}$$

Again, the output from the half-wave rectifier is shown to be half that of the output from a full-wave rectifier.

**Effective value**
Another name for an rms value. It is derived from the fact that an rms value provides the same heating effect as its dc equivalent.

***Measuring Effective Values.*** The rms values of an ac waveform are commonly referred to as **effective values.** The name stems from the fact that rms values produce the same heating effect as their equivalent dc values.

Effective (rms) voltage is measured using an ac voltmeter, while effective current is measured using an ac ammeter. On most DMMs, switching from a dc scale to an ac scale is simply a matter of using the selector switch to select the type of measurement desired. On others, you may also need to switch your probe to a specially designated input.

## Summary

In this section, you have been introduced to peak, peak-to-peak, average, and effective (rms) values, along with the means by which each is measured. The various sine wave voltage values are summarized in Figure 11.31. Even though voltages are emphasized in the figure, sine wave currents are labeled and defined in a similar fashion. Sine wave voltage and current values are measured using various types of test equipment. Voltage and current measurements are summarized in Table 11.1. Note the absence of peak current and peak-to-peak current. These have been left out of the table because they cannot be measured directly using any standard piece of test equipment.

You have now been introduced to the time, voltage, and current characteristics of sinusoidal waveforms. In the next section, we will take a closer look at sine waves and the methods used to generate them.

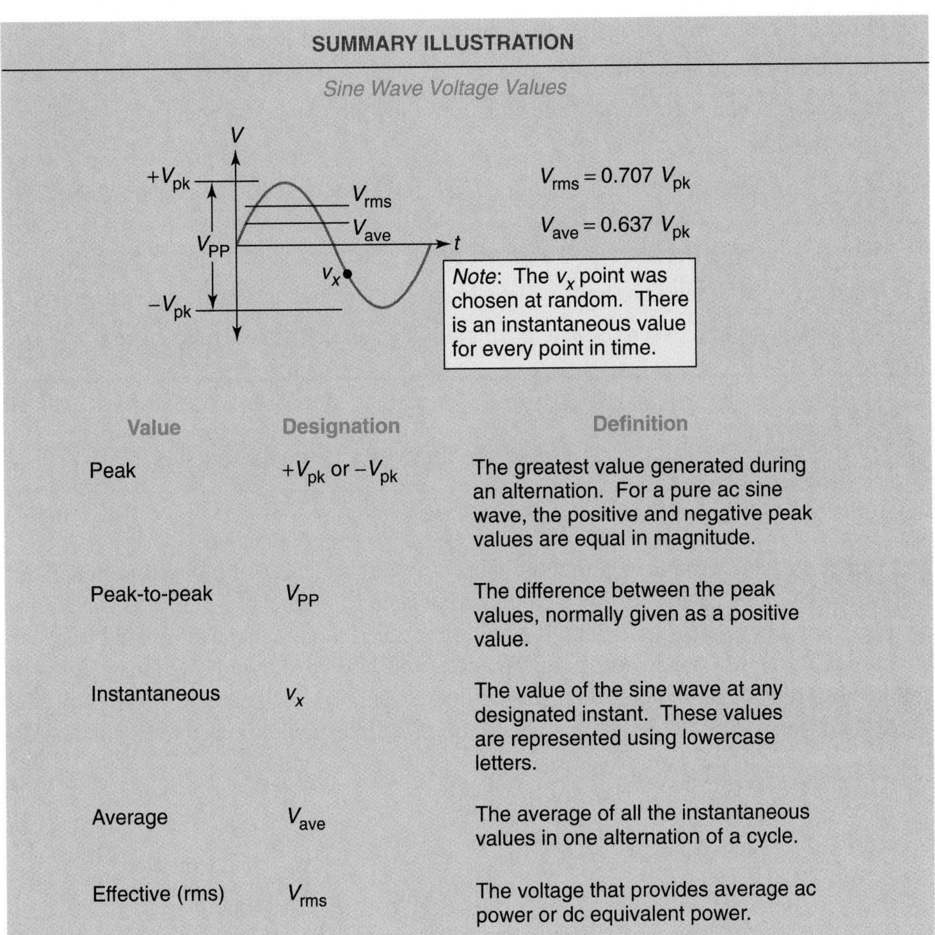

SUMMARY ILLUSTRATION
Sine Wave Voltage Values
V
$+V_{pk}$
$V_{rms}$
$V_{ave}$
t
$V_{PP}$
$v_x$
$-V_{pk}$
$V_{rms} = 0.707\ V_{pk}$
$V_{ave} = 0.637\ V_{pk}$
Note: The $v_x$ point was chosen at random. There is an instantaneous value for every point in time.
Value
Designation
Definition
Peak
$+V_{pk}$ or $-V_{pk}$
The greatest value generated during an alternation. For a pure ac sine wave, the positive and negative peak values are equal in magnitude.
Peak-to-peak
$V_{PP}$
The difference between the peak values, normally given as a positive value.
Instantaneous
$v_x$
The value of the sine wave at any designated instant. These values are represented using lowercase letters.
Average
$V_{ave}$
The average of all the instantaneous values in one alternation of a cycle.
Effective (rms)
$V_{rms}$
The voltage that provides average ac power or dc equivalent power.


**FIGURE 11.31**

**TABLE 11.1 Sine Wave Values and Measurements**

| *Value* | *. . . Is Measured Using a(n) . . .* |
|---|---|
| Peak voltage | oscilloscope[a] |
| Peak-to-peak voltage | oscilloscope[a] |
| Instantaneous voltage | oscilloscope |
| Half-cycle average voltage | dc voltmeter[b] |
| Half-cycle average current | dc ammeter[b] |
| Effective (rms) voltage | ac voltmeter |
| Effective (rms) current | ac ammeter |

[a]See Figure 11.25.

[b]Must be full-wave rectified; see Figure 11.30.

## Section Review

1. What is the relationship between the two peak values of a pure ac waveform?
2. How do you calculate the peak-to-peak value of a waveform?
3. What are instantaneous values? How are they designated?
4. What is the full-wave average of a waveform? What is its value for any pure ac waveform?
5. What is the half-wave average of a sine wave?
6. Why is average ac power an important consideration?

7. What is the root-mean-square (rms) value of a sine wave?
8. What is the relationship between rms voltage (and current) and dc equivalent power?
9. In terms of voltage (or current) designations, how do you know when a given value is an rms value?
10. Describe the use of the oscilloscope to measure peak voltages and peak-to-peak voltages.
11. Why is a rectifier needed to measure the half-cycle average of a pure ac sine wave?
12. What are used to measure half-cycle average voltage and half-cycle average current?
13. What are used to measure effective (rms) voltage and effective (rms) current?

## 11.3 SINE WAVES: PHASE MEASUREMENTS AND INSTANTANEOUS VALUES

In Chapter 10, you were shown that the current through a coil can be used to generate a magnetic field. As a review, this principle is illustrated in Figure 11.32. Just as current can be used to produce a magnetic field, the field produced by a magnet can be used to generate current through a conductor. In this section, you will be shown how a magnetic field can be used to generate a sinusoidal alternating current. You will also be shown how sine wave characteristics affect its instantaneous values and the methods used to calculate those values.

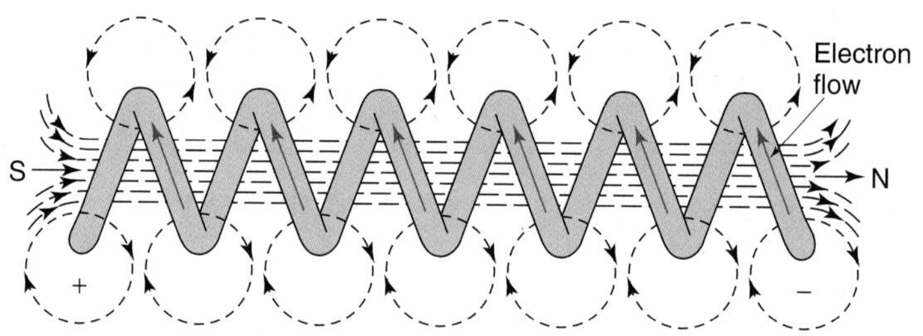


**FIGURE 11.32 Magnetic induction.**

### Magnetic Induction of Current

OBJECTIVE 9 ▶ If a conductor is cut by perpendicular lines of magnetic force, maximum current is induced in the conductor. This principle is illustrated in Figure 11.33. In Figure 11.33a, the conductor is at a 90° angle to the lines of force. When the conductor is moved downward through the lines of force, a current is generated through the conductor in the direction shown.

In Figure 11.33b, the direction of conductor motion has reversed. As the conductor moves upward through the lines of force, a current is generated in the direction shown. Note that reversing the direction of motion reverses the direction of the conductor current.

In Figures 11.33a and b, the conductor, the lines of force, and the direction of conductor motion are all at 90° angles to each other. This relationship is illustrated further in Figure 11.33c. As shown, the conductor and the lines of force are at a 90° angle on the horizontal plane. The directions of conductor motion are at 90° angles to the plane and, therefore, at 90° angles to the conductor and lines of force. As long as the angles are maintained at 90°, maximum current is generated in the wire. However, if any one of the angles changes, the amount of current through the conductor decreases. For example, consider the conductor motion illustrated in Figure 11.33d. When the conductor motion is in the direction of the flux, the current generated in the conductor drops to 0 A. This drop in current happens because the conductor is no longer cutting the lines of force.

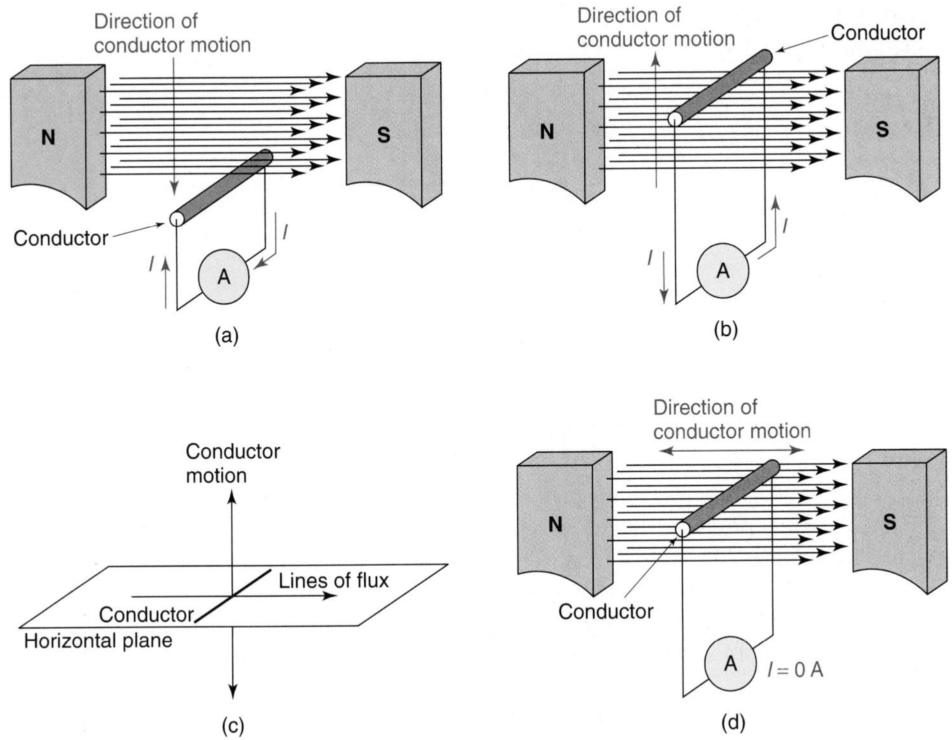


**FIGURE 11.33 Current generated by magnetic induction.**

## The Left-Hand Rule for Generators

In Chapter 10, you were introduced to the left-hand rule. You may recall that the left-hand rule is a memory aid designed to help you remember the relationship between a current and its associated magnetic field. We can use a variation on the left-hand rule to help determine the direction of the current generated in a conductor passing through a magnetic field. The application of the left-hand rule in this instance is illustrated in Figure 11.34.

## Generating a Sine Wave

A sine wave can be generated by rotating a loop conductor through a stationary magnetic field. A simplified sine wave generator is illustrated in Figure 11.35. As you can see, there is a loop conductor (or **rotor**) positioned in the magnetic field. Each end of the rotor is

**Rotor**
The rotating part in a motor or generator. In Figure 11.35, the conducting loop is the rotor.

**FIGURE 11.34 Left-hand rule.**

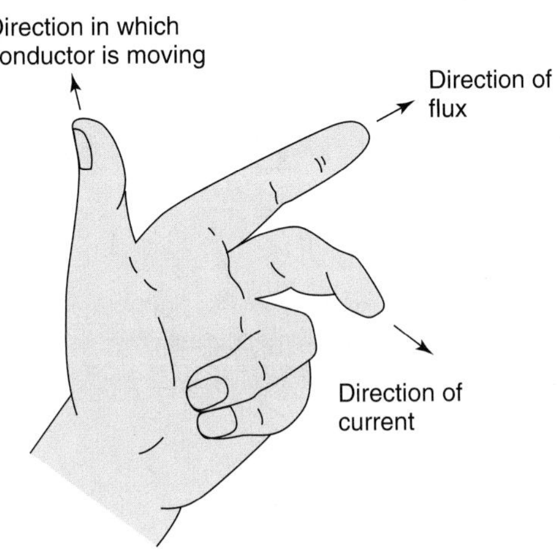

FIGURE 11.35 **Simplified sine wave generator.**

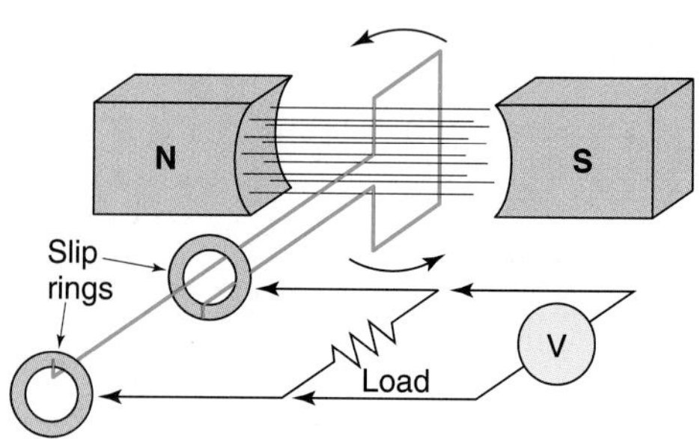


connected to a slip ring. When an ammeter (or some other load) is connected across the slip rings, the circuit is completed, providing a path for any current produced by the generator.

As it rotates through the magnetic field, a current is induced through the rotor. The amount of current produced and its direction are determined by the angle of the conductor, relative to the magnetic lines of force. This point is illustrated in Figure 11.36.

Figure 11.36 illustrates the operation of the simplified generator shown in Figure 11.35. The voltage curves shown represent the load voltage that would be produced by any generated current. To simplify the illustration, the load circuitry (from Figure 11.35) has been omitted.

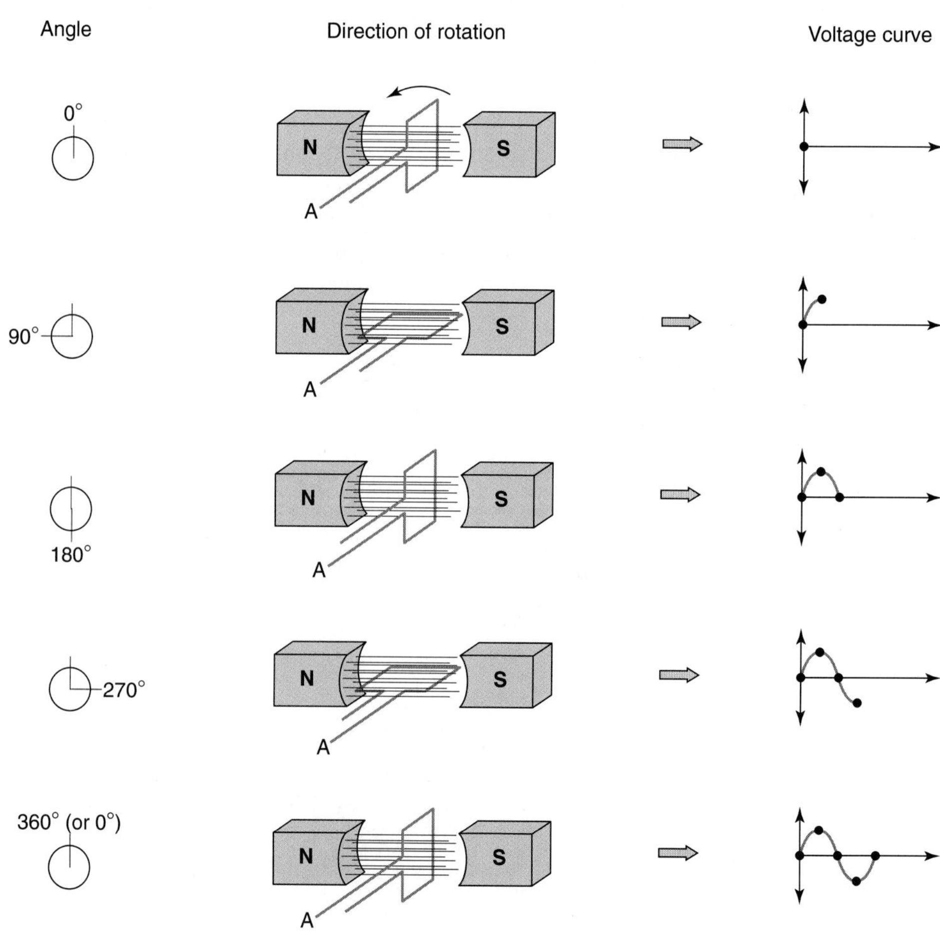


**FIGURE 11.36 Generating a sine wave.**

When side A of the conductor is at 0°, its motion is parallel to the lines of force. Since it is not cutting the lines of force, the current through the conductor is 0 A, and there is no voltage across the load. When A has rotated to the 90° point, it is:

1. Perpendicular to the lines of force.
2. Traveling in a downward direction.

Since the angle of the conductor motion (relative to the lines of force) is 90°, maximum current is induced through the conductor. For the sake of discussion, this current is assumed to generate a positive load voltage, as shown in the voltage curve.

When side A of the conductor reaches 180°, it is once again parallel to the lines of force. As a result, the conductor current returns to 0 A and the load voltage returns to 0 V. When side A of the conductor has rotated to the 270° point, it is:

1. Perpendicular (again) to the lines of force.
2. Traveling in an upward direction.

Once again, current through the conductor reaches a maximum value. However, since the direction has changed (from downward to upward), the direction of the conductor current has reversed. As a result, the polarity of the load voltage has reversed. When side A of the conductor reaches 360° (its starting position), the load current and voltage both return to zero, and the cycle repeats itself.

## Phase

◄ *OBJECTIVE 10*

If you compare each curve in Figure 11.36 with its corresponding angle of rotation, you'll see that:

1. The positive peak occurs when the angle of rotation is 90°.
2. The waveform returns to 0 V when the angle of rotation reaches 180°.
3. The negative peak occurs when the angle of rotation reaches 270°.
4. The waveform returns to 0 V (again) when the angle of rotation reaches 360° (or 0°).

As the angle diagrams in Figure 11.36 indicate, these angles are all relative to the starting point of the waveform. This is illustrated further in Figure 11.37. Note that the horizontal axis of the graph represents the angles of rotation, designated by the Greek letter θ (theta).

Any point on a waveform can be identified by its phase. The **phase** of a given point is its position relative to the start of the waveform and is usually given in degrees. (Later,

**Phase**
The position of a given point on a waveform relative to the start of the waveform, usually expressed in degrees.

**FIGURE 11.37 Degree measurements.**

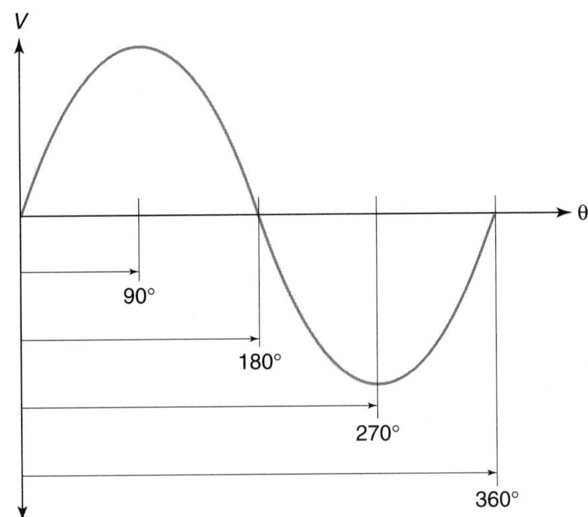

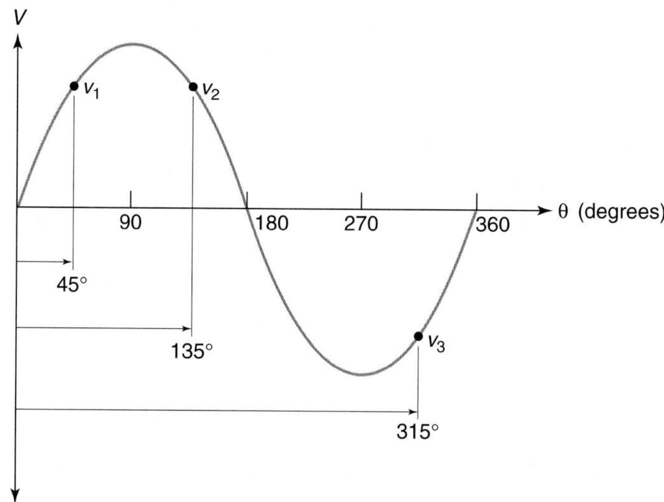


**FIGURE 11.38 Example phase measurements.**

you'll be shown another method of expressing phase.) Several points are identified by phase in Figure 11.38.

Sine waves are often described in terms of phase for several reasons:

Why phase is used to describe the points on a waveform.

1. Phase measurements allow us to identify points on a waveform independent of cycle time and frequency.
2. Phase plays an important role in the instantaneous values of sine wave voltage and current.

The first of these points is illustrated in Figure 11.39. The two waveforms shown have frequencies of 250 Hz and 10 kHz. The 250 Hz waveform reaches its positive peak after 1 ms. The 10 kHz waveform reaches the same point after only 25 μs. Even so, both waveforms reach a positive peak when $\theta = 90°$. Therefore, the phase measurement provides us with a means of identifying corresponding points on the two waveforms independent of time and frequency.

We will discuss the relationship between phase and instantaneous values of voltage and current later in this section. At this point, we will take a closer look at the relationship between phase and time measurements.

## Phase and Time Measurements

Phase and time measurements are related as shown in Figure 11.40. As the figure implies, the ratio of phase ($\theta$) to 360° equals the ratio of instantaneous time ($t$) to cycle time ($T_C$). By formula,

$$\frac{\theta}{360°} = \frac{t}{T_C} \tag{11.15}$$

where
$\theta$ = the phase of the point ($v_x$)
360 = the number of degrees in one complete cycle
$t$ = the time from the start of the cycle to ($v_x$)
$T_C$ = the time required for one complete cycle

If we apply this relationship to the 90° point on the waveform in Figure 11.39b, we get

$$\frac{90°}{360°} = \frac{25\ \mu\text{s}}{100\ \mu\text{s}}$$

which is true. The same can be demonstrated for any identified point on either of the waveforms shown.

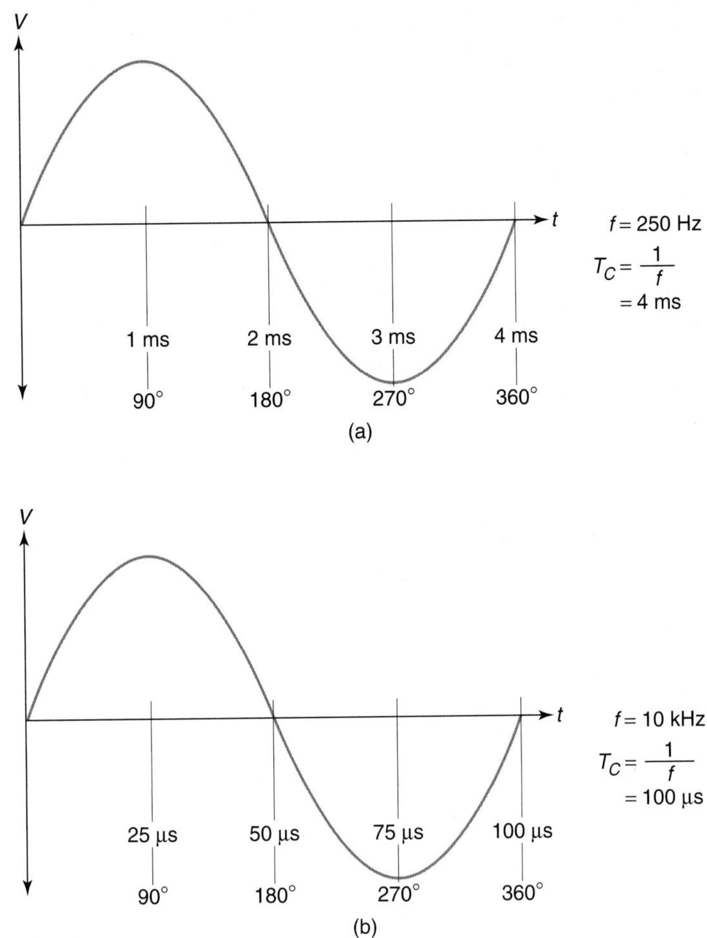


**FIGURE 11.39 Phase and time measurements.**

There is an important point that needs to be made at this time. Earlier, you were told that phase is independent of time and frequency. This is true when using phase to describe identical points on two or more waveforms. However, within each individual waveform, the phase/time ratio holds true.

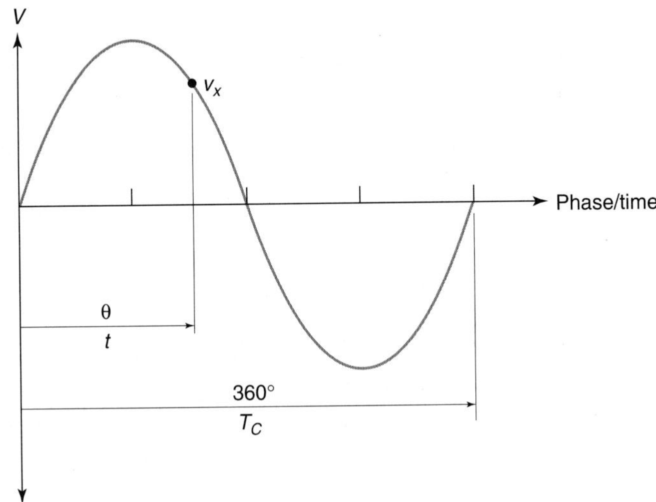


**FIGURE 11.40 Phase/time conversions.**

When the phase and cycle time of a waveform are known, the value of $t$ can be found as

$$t = T_C \frac{\theta}{360°} \tag{11.16}$$

Example 11.12 demonstrates the use of this relationship.

## EXAMPLE 11.12

Assume that the waveform shown in Figure 11.40 has the following values: $\theta = 140°$ and $f = 15$ kHz. Determine the time from the start of the cycle until $\theta$ is reached.

***Solution:*** First, we need to determine the cycle time of the waveform, as follows:

$$T_C = \frac{1}{f} = \frac{1}{15\text{ kHz}} = \mathbf{66.67\ \mu s}$$

The time required to reach an angle of 140° can now be found as

$$t = T_C \frac{\theta}{360°} = (66.67\ \mu\text{s})\frac{140°}{360°} \cong \mathbf{25.9\ \mu s}$$

***PRACTICE PROBLEM 11.12***

Assume that the waveform in Figure 11.40 has the following values: $\theta = 135°$ and $f = 250$ Hz. Determine the time from the start of the cycle until $\theta$ is reached.

Equation (11.15) can also be modified as follows to provide an equation for $\theta$:

$$\theta = (360°)\frac{t}{T_C} \tag{11.17}$$

This relationship allows us to determine the phase (in degrees) for any point on an oscilloscope display. The application of the equation is demonstrated in Example 11.13.

## EXAMPLE 11.13

Determine the value of $\theta$ for the point highlighted in Figure 11.41a.

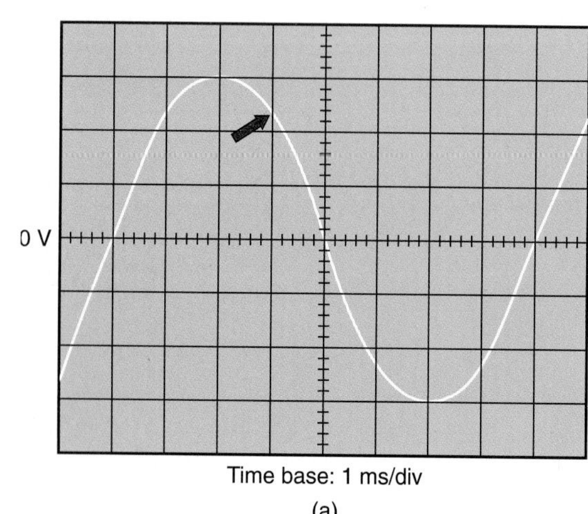


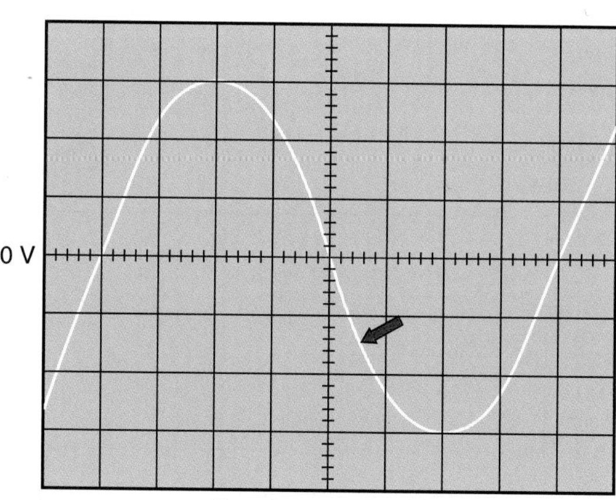


**FIGURE 11.41**

***Solution:*** With a time base of 1 ms/div, the cycle time of the waveform is found as

$$T_C = (8\text{ div})\frac{1\text{ ms}}{\text{div}} = \mathbf{8\text{ ms}}$$

The highlighted point occurs 3 divisions into the waveform. Therefore, the value of $t$ for the point is found as

$$t = (3\text{ div})\frac{1\text{ ms}}{\text{div}} = \mathbf{3\text{ ms}}$$

Now, the phase measured at the point is found as

$$\theta = (360°)\frac{t}{T_C} = (360°)\left(\frac{3\text{ ms}}{8\text{ ms}}\right) = \mathbf{135°}$$

***PRACTICE PROBLEM 11.13***

Determine the value of θ for the point highlighted in Figure 11.41b.

## Phase Angles

◄ *OBJECTIVE 11*

There are many cases where a circuit contains two (or more) sine waves of equal frequency that do not reach their peaks at the same time. Waveforms of this nature are said to be *out of phase.* Figure 11.42a shows an example of two waveforms that are out of phase. As the figure implies, the two waveforms have the same cycle time, but they do not reach their peak values and zero-crossings at the same time. The phase difference between two

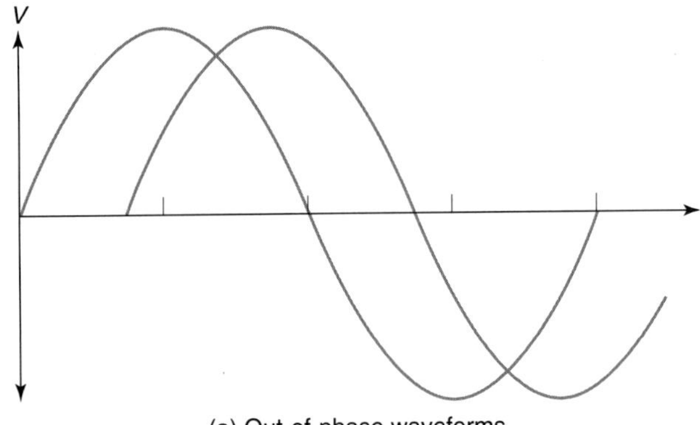


(a) Out-of-phase waveforms

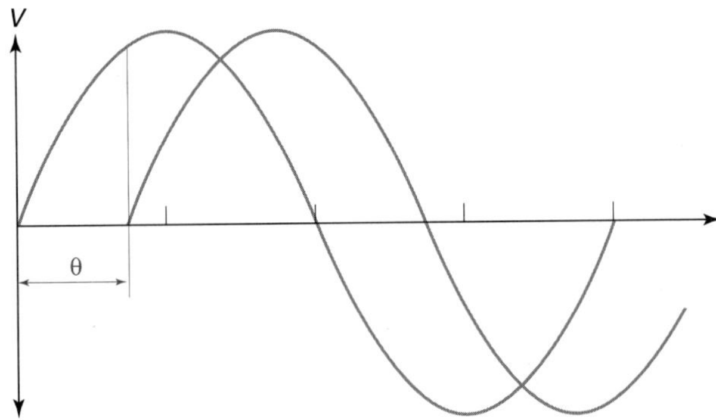


(b) The phase angle (θ) between two sine waves

**FIGURE 11.42 Phase relationships.**

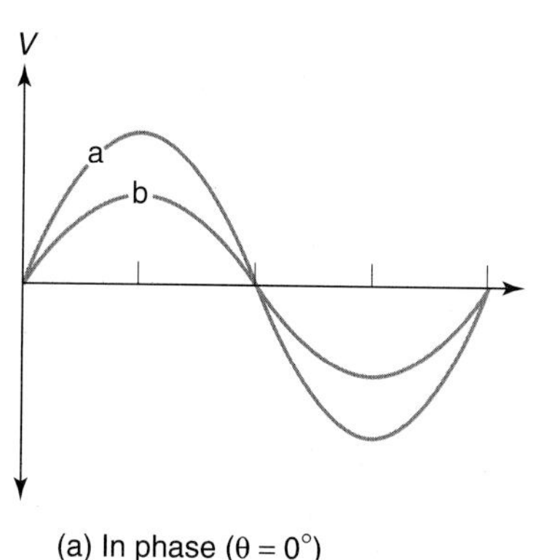

(a) In phase (θ = 0°)

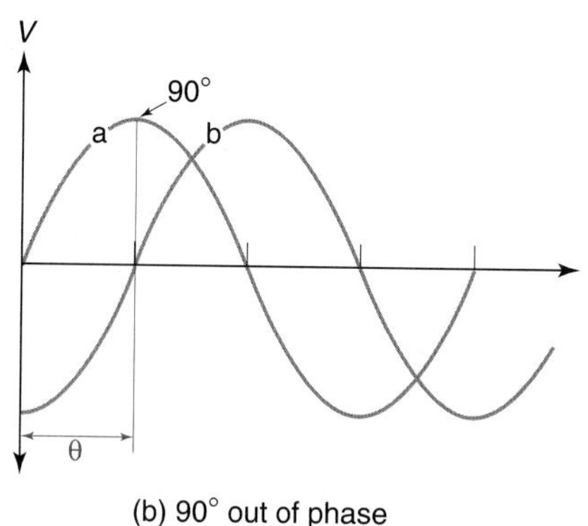

(b) 90° out of phase

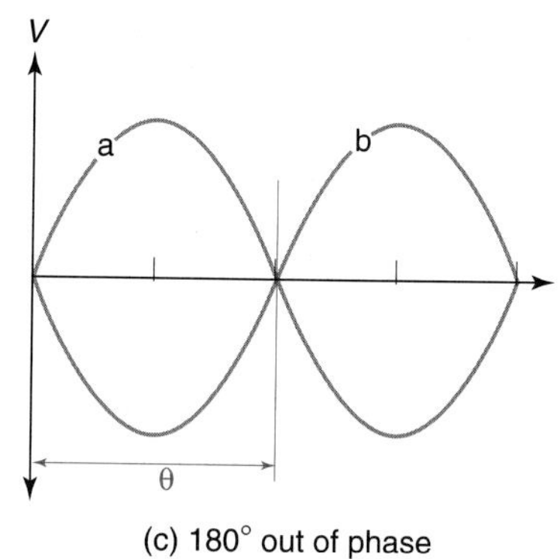

(c) 180° out of phase

**FIGURE 11.43**

**Phase angle**
The phase difference between two or more waveforms.

waveforms is referred to as their **phase angle.** The phase angle between two waveforms is identified in Figure 11.42b.

Sometimes it is easy to determine the phase relationship between a pair of waveforms. For example, consider the waveform pairs shown in Figure 11.43. The waveforms in Figure 11.43a are in phase, which is indicated by the fact that their peaks and zero crossings occur at the same time (even though they have different peak values). The waveforms in Figure 11.43b are 90° out of phase. Two waveforms are 90° out of phase when the 0° point of one corresponds (in time) to the 90° point of the other. The waveforms in Figure 11.43c are 180° out of phase, which is indicated by the fact that one waveform reaches its positive peak at the same time that the other reaches its negative peak.

The waveforms in Figure 11.43a demonstrate another important point. The phase angle between two (or more) waveforms is strictly a function of time. The peak voltage (or current) values do not determine the phase relationship between the waveforms.

**Dual-trace oscilloscope**
An oscilloscope with two independent traces that can be used to compare two waveforms. Each trace has its own set of vertical controls.

## Measuring Phase Angles

When there is any question about the phase angle between two waveforms, it can be determined using a **dual-trace oscilloscope** and equation (11.17). This technique for measuring the phase angle between two waveforms is demonstrated in Example 11.14.

### EXAMPLE 11.14

A dual-trace oscilloscope is used to measure the phase angle between two sine waves. The oscilloscope display is shown in Figure 11.44a. Determine the phase angle (θ) displayed.

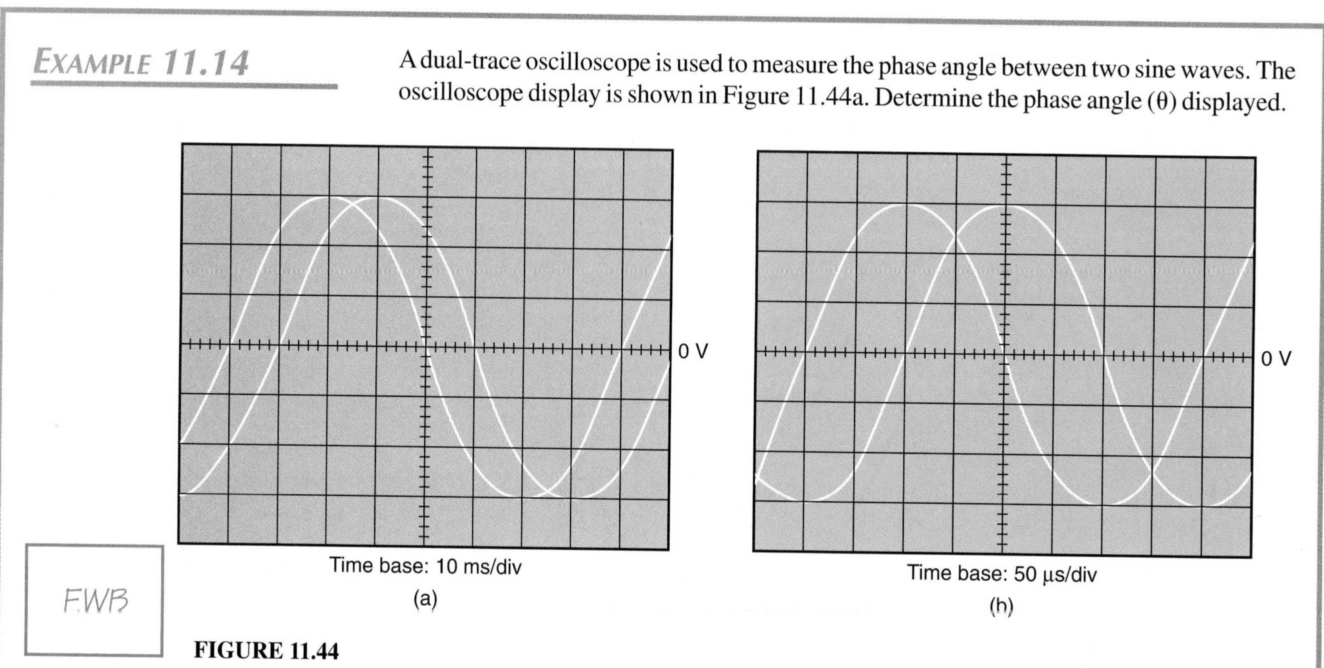

**FIGURE 11.44**

***Solution:*** The cycle time for each of the waveforms is found as

$$T_C = (8 \text{ divisions}) \times \frac{10 \text{ ms}}{\text{div}} = \mathbf{80\ ms}$$

If you look at the zero-crossing points for the two waveforms, you'll see that the waveforms are separated by 1 major division. Therefore, the value of time between the waveforms $t$ is found as

$$t = (1 \text{ division}) \times \frac{10 \text{ ms}}{\text{div}} = \mathbf{10\ ms}$$

Using the values of $T_C = 80$ ms and $t = 10$ ms in equation (11.17), the phase angle between the waveforms can be found as

$$\theta = (360°)\frac{t}{T_C} = (360°)\frac{10 \text{ ms}}{80 \text{ ms}} = \mathbf{45°}$$

***PRACTICE PROBLEM 11.14***

Determine the phase angle between the waveforms shown in Figure 11.44b.

In later chapters, you will be shown some practical situations involving phase angles and their measurement. For now, we are simply interested in establishing the following:

1. Two (or more) waveforms of equal frequency that do not peak at the same time are said to be *out of phase*.
2. Any phase difference between two waveforms of equal frequency is called a phase angle.
3. The phase angle between two waveforms can be determined using a dual-trace oscilloscope and equation (11.17).

## Instantaneous Values

◀ *OBJECTIVE 12*

Earlier in this section, you were told that an instantaneous value is the magnitude of a voltage (or current) at a specified point in time. For example, $v_1$ in Figure 11.45a is an instantaneous value. Now that we have discussed phase relationships, we are ready to look at the method used to determine any instantaneous value.

The instantaneous value of a sine wave voltage (or current) is proportional to the peak value of the waveform and the sine of the phase angle. This relationship is given as:

$$v = V_{pk} \sin \theta \quad \mathbf{(11.18)}$$

Example 11.15 demonstrates the use of this equation.

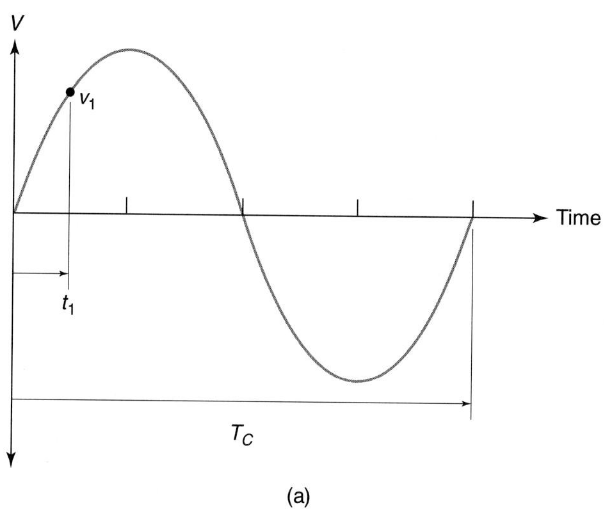

(a)

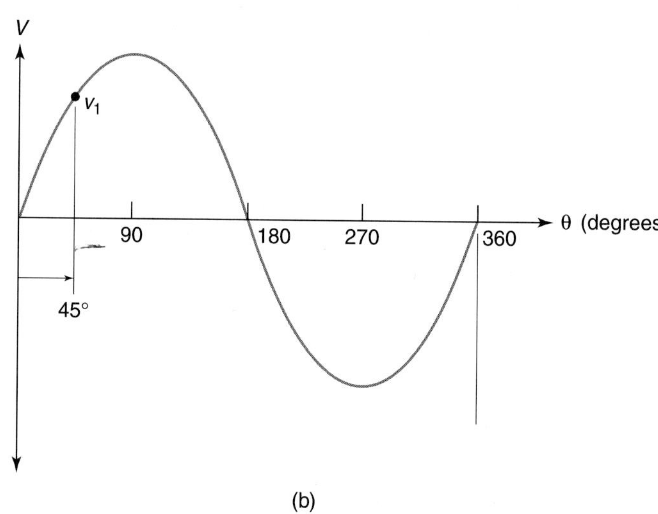

(b)

**FIGURE 11.45**

## EXAMPLE 11.15

Determine the value of $v_1$ in Figure 11.45b. Assume that the waveform has a peak value of 10 V.

***Solution:*** As shown in the figure, $v_1$ occurs when $\theta = 45°$. With a peak value of 10 V, the value of $v_1$ is found as

$$\begin{aligned} \mathbf{v_1} &= V_{pk} \sin \theta \\ &= (10\ V_{pk})(\sin 45°) \\ &= (10\ V_{pk})(0.707) \\ &= \mathbf{7.07\ V} \end{aligned}$$

Thus, $v_1 = 7.07$ V when $\theta = 45°$.

### *PRACTICE PROBLEM 11.15*

A sine wave has a peak value of 12 V. Determine the instantaneous value of the waveform when $\theta = 60°$.

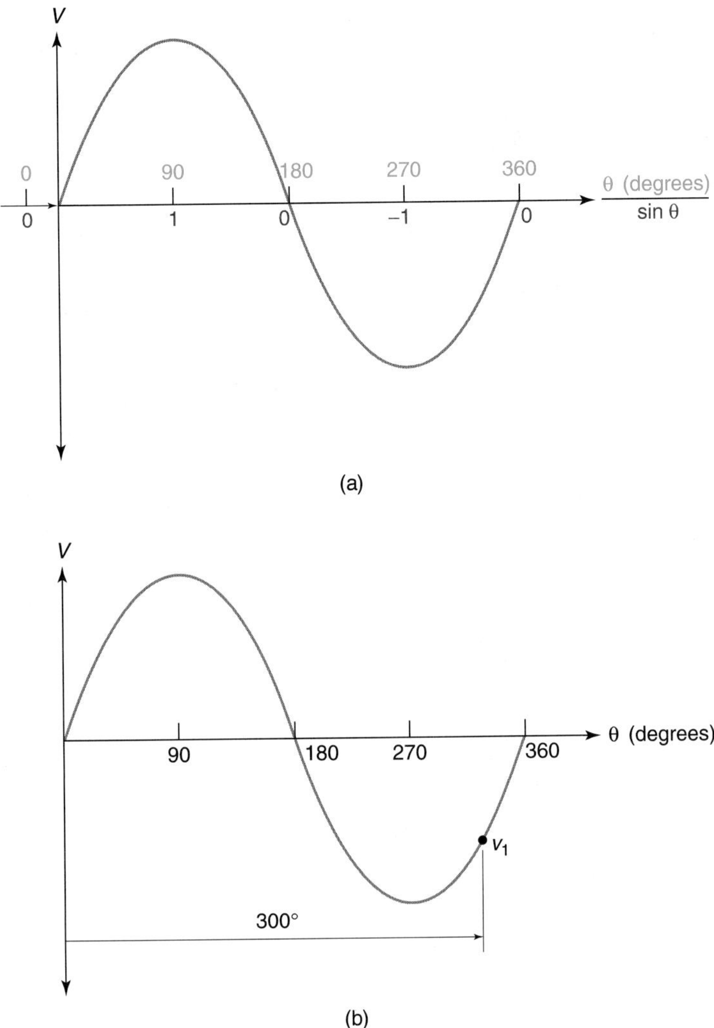


**FIGURE 11.46**

The peak value of a sine wave (like those in Figure 11.45) is a constant; that is, the peak value does not change over the course of a cycle. However, the phase angle does change from one instant to the next. Since the instantaneous values are proportional to the sine of the phase angle (sin θ), the waveform is said to vary at a sinusoidal rate. (This is where the name *sine wave* comes from.)

The basis for the term *sine wave.*

Figure 11.46a shows a sine wave divided into 90° increments. At each increment, the value of sin θ is shown. From the sin θ values shown, we can draw the following conclusions:

*Note:* The sine values shown at each increment in Figure 11.46a can be verified by entering the angle into your calculator and pressing the "sin" key.

1. During the positive alternation of the waveform, sin θ is positive. Therefore, $V_{pk} \sin \theta$ is also positive.
2. During the negative alternation of the waveform, sin θ is negative. Therefore, $V_{pk} \sin \theta$ is also negative.

This means that equation (11.18) will yield the correct polarity for any instantaneous value in either alternation. This is demonstrated further in Example 11.16.

## EXAMPLE 11.16

Determine the value of $v_1$ in Figure 11.46b. Assume that the waveform has a peak value of 15 V.

***Solution:*** Using equation (11.18), the value of $v_1$ is found as

$$\begin{aligned} v_1 &= V_{pk} \sin \theta \\ &= (15\ V_{pk})(\sin 300°) \\ &= (15\ V_{pk})(-0.866) \\ &\cong \mathbf{-13\ V} \end{aligned}$$

As you can see, sin θ automatically provides the correct voltage polarity for any angle.

***PRACTICE PROBLEM 11.16***

A sine wave has a peak value of 30 V. Determine the instantaneous value of the waveform when $\theta = 200°$.

To determine the instantaneous value of a waveform at a specified time interval, we must determine the phase angle at that time interval. This point is demonstrated in Example 11.17.

## EXAMPLE 11.17

Determine the value of $v_1$ in Figure 11.47a.

**FIGURE 11.47**

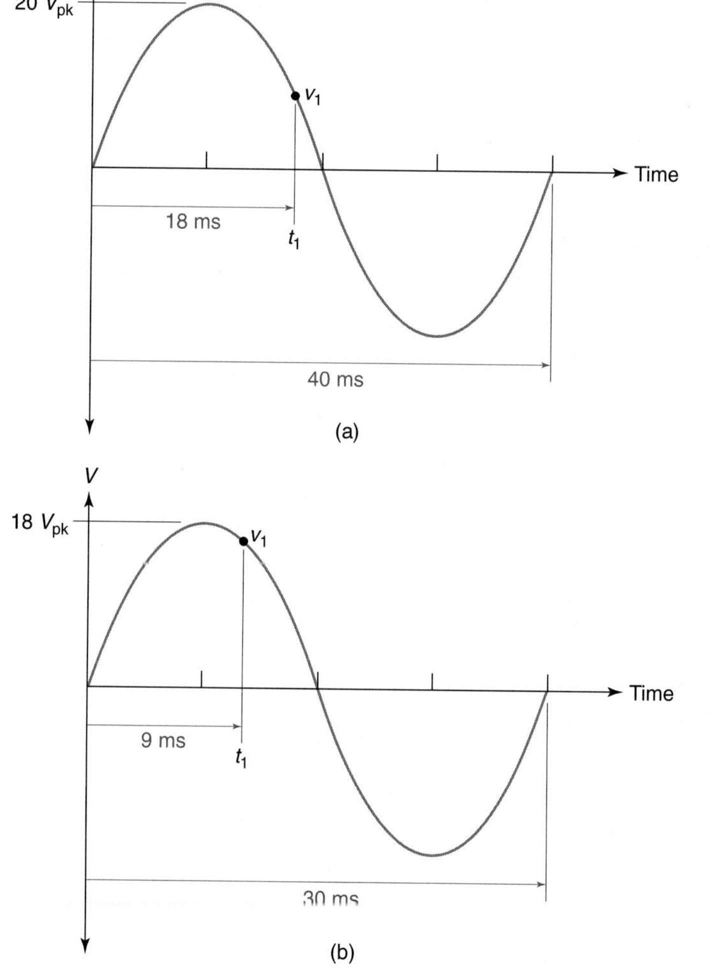

***Solution:*** First, we need to determine the phase angle of the waveform at $t_1$. Using equation (11.17), the value of θ is found as

$$\boldsymbol{\theta} = (360°)\frac{t_1}{T_C} = (360°)\frac{18 \text{ ms}}{40 \text{ ms}} = \mathbf{162°}$$

Now, the value of $v_1$ can be found as

$$\begin{aligned} v_1 &= V_{pk} \sin \theta \\ &= (20 \text{ V})(\sin 162°) \\ &= (20 \text{ V})(0.309) \\ &= \mathbf{6.18 \text{ V}} \end{aligned}$$

***PRACTICE PROBLEM 11.17***

Determine the value of $v_1$ in Figure 11.47b.

## Phase Measurements: The Radian Method

◄ *OBJECTIVE 13*

So far, we have used degrees for all our phase measurements. Another phase measurement technique involves the use of radians. At this point, we will discuss radian measurements and calculations.

**Radian**
The angle formed at the center of a circle by two radii and an arc of equal length. (*Note: radii* is the plural form of *radius.*)

***Radians.*** Technically defined, a **radian** is the angle formed at the center of a circle by two radii and an arc of equal length. A radian is formed as shown in Figure 11.48. The two radii ($r_1$ and $r_2$) are separated by an arc ($r_a$). The lengths of $r_1$, $r_2$, and $r_a$ are all equal. The angle formed by $r_1$ and $r_2$ (which is labeled θ in the figure) is one radian.

The radian is independent of the size of the circle. For example, if we were to double the size of the circle in Figure 11.48, the angle designated by θ would not change. This is due to the fact that $r_1$, $r_2$, and $r_a$ would all increase by the same factor. Thus, the number of degrees contained in a radian is the same for every circle.

So, how many degrees are there in one radian? In geometry, we learned that the circumference of a circle can be found as

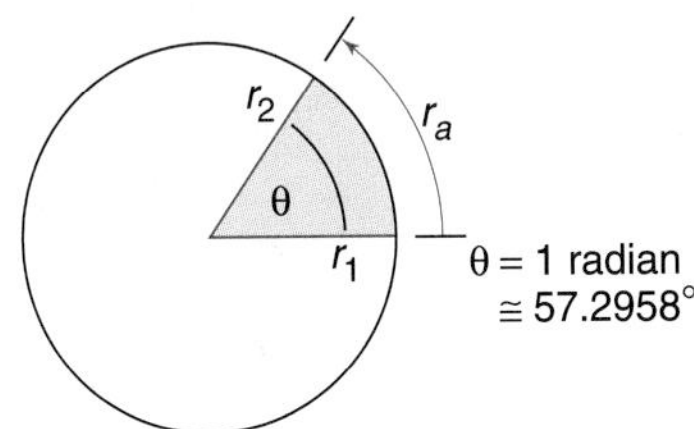


**FIGURE 11.48 Vector rotation equal to one radian.**

$$\ell = 2\pi r$$

where

$\ell$ = the length of the circumference
$r$ = the radius of the circle

The length of the circumference ($\ell$) varies from one circle to another. However, the number of degrees (360) does not. If we rewrite the equation in terms of degrees, we get

$$360° = 2\pi r$$

If we solve this equation for *r*, we get:

$$r = \frac{360°}{2\pi} \cong 57.2958°$$

*An Important Point:* There are 360° in a sine wave. Since there are 2π radians in 360°, there are:

1. π radians in each alternation of a sine wave.
2. 2π radians in each complete cycle of a waveform.

***Angular Velocity.*** Earlier in this section, you were shown how a sine wave can be generated by rotating a conducting loop (or rotor) through a permanent magnetic field. The speed of rotation for the rotor is referred to as **angular velocity.** The angular velocity (in radians per second) can be found as

**Angular velocity**
The speed of rotation for the rotor in a sine wave generator. The rate at which the phase of a sine wave changes and, therefore, the rate at which its instantaneous values change.

$$\omega = 2\pi f \qquad \textbf{(11.19)}$$

where

ω = the angular velocity, in radians per second (ω is the lowercase form of Ω)
2π = the number of radians in one cycle
$f$ = the number of cycles per second (frequency)

You may be wondering how $2\pi f$ works out to be a measure of velocity (distance per unit time). It is easy to see that $2\pi f$ is a measure of velocity when we consider that

$$2\pi = \text{radians per cycle} \qquad f = \text{cycles per second}$$

Therefore, the product of these values can be written as

$$2\pi f = \frac{\text{rads}}{\text{cycle}} \times \frac{\text{cycles}}{\text{second}} = \frac{\text{rads}}{\text{second}}$$

Since 1 radian corresponds to a length of vector rotation (shown as $r_a$ in Figure 11.48), radians per second is a measure of distance per unit time (velocity).

Angular velocity is of interest because it indicates the rate at which the phase of a sine wave changes and, therefore, the rate at which its instantaneous values change. This can be seen by referring back to Figure 11.36. If the rotor is spinning at a set rate, then the phase of the sine wave must be changing at the same rate. Since the instantaneous values of the sine wave are proportional to the phase, they must also be changing at the same rate. Therefore, the angular velocity tells us the rate at which the phase and instantaneous values of a sine wave change. As you will learn in upcoming chapters, this rate of change is used in many ac circuit calculations.

*OBJECTIVE 14* ▶ ***Instantaneous Values.*** Earlier, you were shown that an instantaneous value of sine wave voltage can be found as

$$v = V_{pk} \sin \theta$$

When you know the frequency of the waveform, the instantaneous value at a designated time interval ($t$) can also be found using

$$v = V_{pk} \sin \omega t \qquad \textbf{(11.20)}$$

where $\omega = 2\pi f$ (the angular velocity, in radians)
$t =$ the designated time interval from the start of the cycle

Figure 11.49 will help to clarify the relationship between the two methods for calculating an instantaneous value. Phase measurements designate specific points on a sine wave. For example, the 90° point is identified in Figure 11.49a. When the point is identified, calculating the instantaneous value is a relatively simple problem. The question is: How can the product of $\omega t$ also designate a specific point on a sine wave?

Earlier, you were told that angular velocity is measured in radians per second. This is easier to see if we rewrite $\omega$ as follows:

$$\omega = 2\pi f = 2\pi \frac{1}{T_C}$$

Using the rewritten form of $\omega$, the product $\omega t$ can be written as

$$\omega t = \left(2\pi \frac{1}{T_C}\right) t = 2\pi \frac{t}{T_C}$$

This relationship is shown in Figure 11.49b. Now, consider the following: $2\pi$ is the number of radians in one complete cycle. (The quarter-cycle increments of $2\pi$ are shown in the figure.) The ratio $t/T_C$ is the same one we used to designate instantaneous points in equation (11.17). If we were to assume (as an example) that $t/T_C = 1/4$, then the product $\omega t$ would have a value of

$$\omega t = 2\pi \frac{1}{4} = \frac{2\pi}{4} = \frac{\pi}{2}$$

As shown in Figure 11.49b, this designates the same point as the one selected in Figure 11.49a. Thus, the product $\omega t$ always specifies a point on a sine wave. Knowing where the point is and the peak value allows us to calculate the instantaneous value of the waveform.

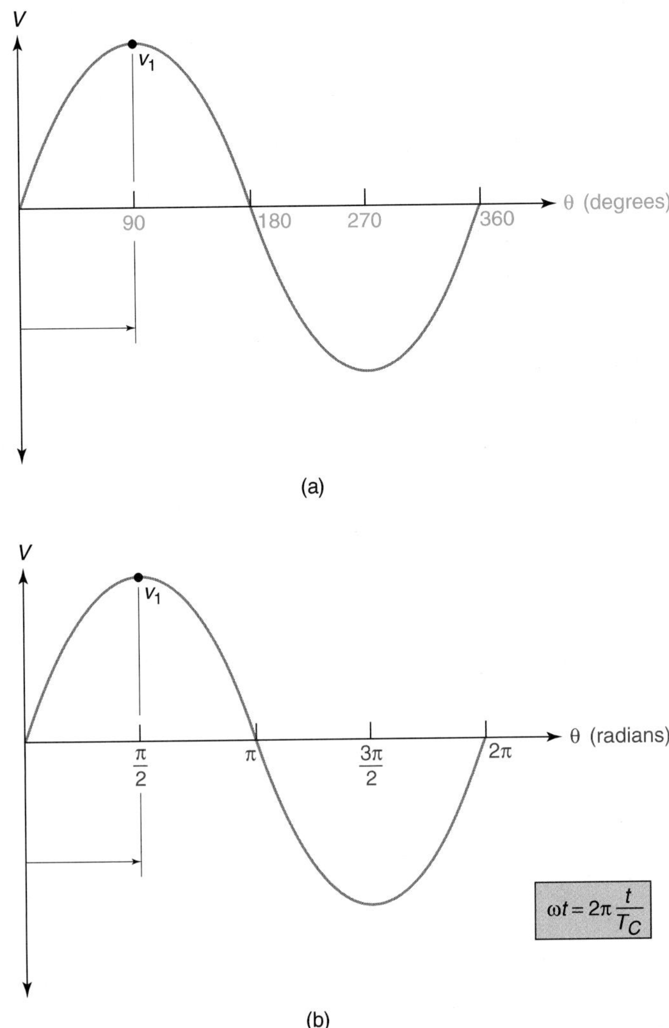


**FIGURE 11.49 Comparing degree and radian measurements.**

Equation (11.20) will always provide the same result as $V_{pk} \sin \theta$ *provided your calculator is set for the radian mode of operation.* This calculator mode converts your radian input to degrees automatically. As a result, the sine function provides the correct result. If the calculator is left in the degree mode of operation, your result will be incorrect. This point is illustrated further in Example 11.18.

*A Practical Consideration:* Your calculator must be set for the radian mode of operation when using equation (11.20). For most calculators, the mode of operation is set using a key labeled DRG. If your calculator does not have a key with this label, check the operating manual to determine which key to use.

## EXAMPLE 11.18

The waveform in Figure 11.50a has the values shown. Calculate the value of $v_1$ using both the degree and radian methods.

***Solution:*** The waveform shown is the same one we analyzed in Example 11.17, where we calculated the following values:

$$\theta = 162° \qquad \text{and} \qquad v_1 = 6.18 \text{ V}$$

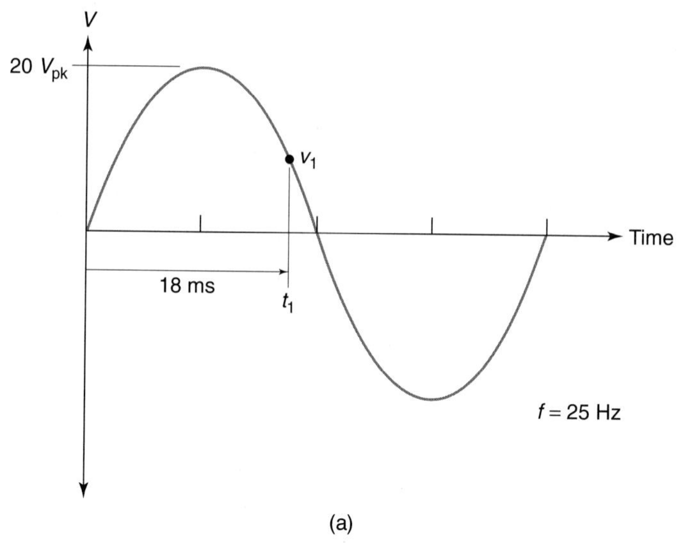


(a)

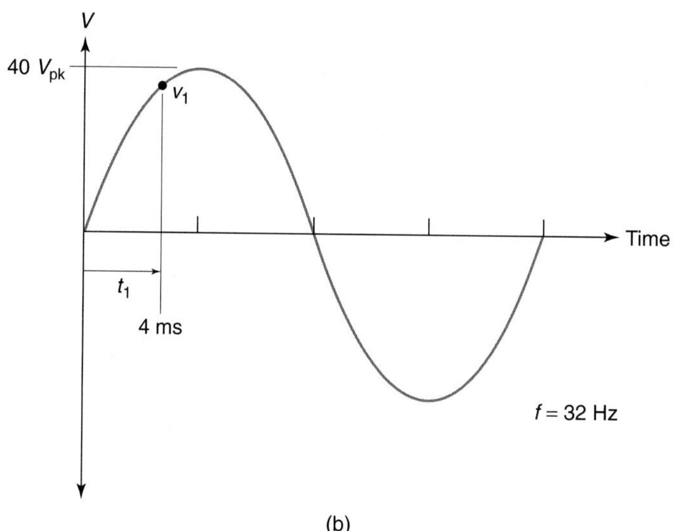


(b)

**FIGURE 11.50**

After switching your calculator to the radian mode of operation, the value of $v_1$ can be found as

$$\begin{aligned} \mathbf{v_1} &= V_{pk} \sin \omega t \\ &= (20\text{ V}) \sin (2\pi \times 25\text{ Hz} \times 18\text{ ms}) \\ &= (20\text{ V}) \sin (28.274) \\ &= (20\text{ V})(0.3090) \\ &= \mathbf{6.18\text{ V}} \end{aligned}$$

As you can see, we obtained the same value of $v_1$ as we did in Example 11.17. Remember, though, that the problem was solved correctly using the calculator's radian mode of operation. Had we attempted to use the degree mode of operation, we would have obtained

$$\mathbf{v_1} = (20\text{ V}) \sin (28.274) = (20\text{ V})(0.4737) = \mathbf{9.474\text{ V}}$$

which is incorrect.

***PRACTICE PROBLEM 11.18***

Calculate the value of $v_1$ for the waveform in Figure 11.50b using both the degree and radian approaches. (A model of the degree approach can be found in Example 11.17.)

## *Putting It All Together: Phase Measurements and Calculations*

The information in this section has dealt with phase measurements and calculations. Before ending this section, let's take a moment to put this information into perspective.

A sine wave is made up of an infinite number of instantaneous values. Certain circuit analysis problems require that you be able to calculate an instantaneous value of sine wave voltage (or current). To calculate one of these values, you must be able to determine: ◄ *OBJECTIVE 15*

1. The peak value of the waveform.
2. The position of the point relative to the start of the waveform cycle, that is, its phase. Phase is measured in either degrees or radians.

The degree and radian approaches to solving for an instantaneous voltage are compared in Figure 11.51. As you can see, the degree approach is preferable when a phase value (in degrees) is known. However, if the waveform frequency and the instant of time ($t$) are known, you may find the radian approach easier. Examples 11.19 and 11.20 present applications for each approach.

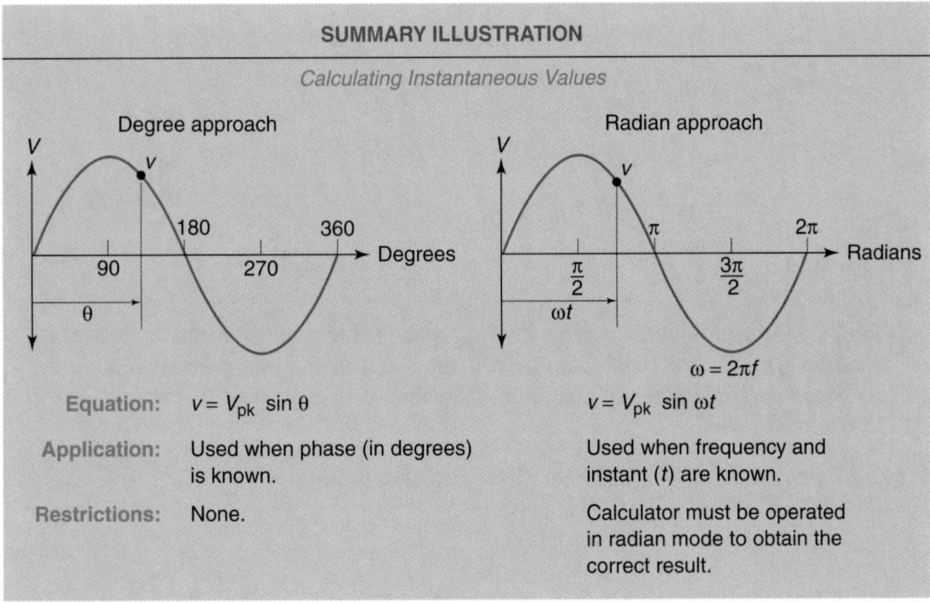

SUMMARY ILLUSTRATION
Calculating Instantaneous Values
Degree approach
V
v
180
360
Degrees
90
270
θ
Radian approach
V
v
π
2π
Radians
π/2
3π/2
ωt
ω = 2πf
Equation: v = Vpk sin θ
v = Vpk sin ωt
Application: Used when phase (in degrees) is known.
Used when frequency and instant (t) are known.
Restrictions: None.
Calculator must be operated in radian mode to obtain the correct result.


**FIGURE 11.51**

***EXAMPLE 11.19***

The waveforms in Figure 11.52a are 30° out of phase. Determine the instantaneous value of waveform A when waveform B is at 0°.

***Solution:*** Since we know the phase angle between the waveforms, the degree approach is the better of the two. Using the values shown in the figure, we get

$$v = V_{pk} \sin \theta = (12\text{ V}) \sin (30°) = \mathbf{6\text{ V}}$$

***PRACTICE PROBLEM 11.19***

The waveforms in Figure 11.52b are 75° out of phase. Determine the instantaneous value of waveform A when waveform B is at 10°. (Be careful on this one.)

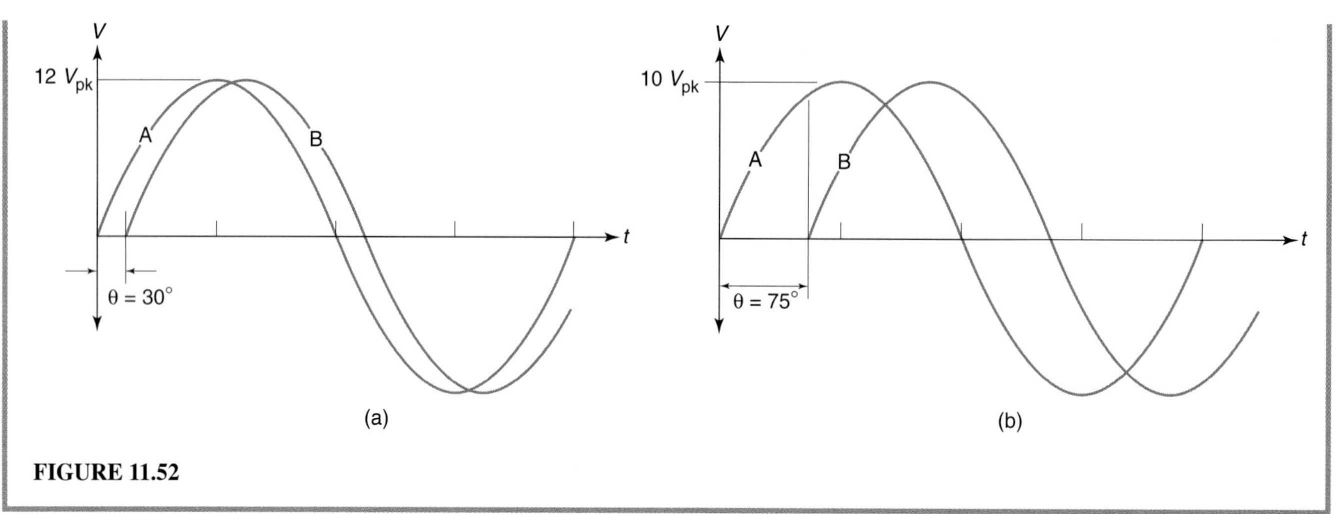


**FIGURE 11.52**

## *EXAMPLE 11.20*

Refer to Figure 11.53a. Determine the instantaneous value that the source voltage reaches 120 μs after the start of each cycle.

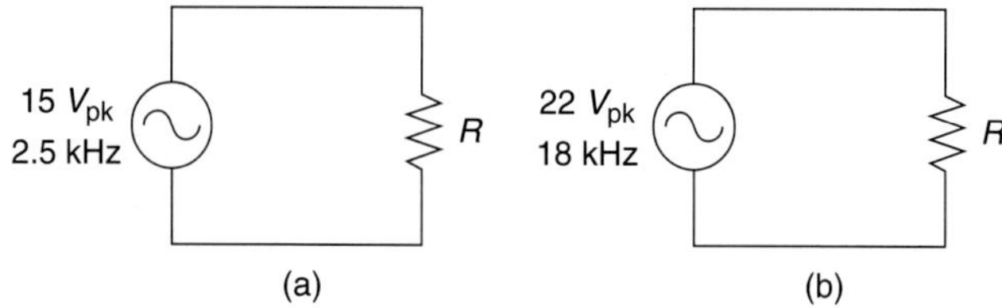


**FIGURE 11.53**

***Solution:*** In this problem, we know the frequency of the sine wave and the instant ($t$) of interest. Therefore, the radian approach is the better of the two. After switching the calculator to the radian mode of operation, the value of $v$ is found as

$$\begin{aligned} v &= V_{pk} \sin \omega t \\ &= (15 \text{ V}) \sin (2\pi \times 2.5 \text{ kHz} \times 120\ \mu\text{s}) \\ &= (15 \text{ V}) \sin (1.885) \\ &= (15 \text{ V})(0.9511) \\ &= \mathbf{14.27\ V} \end{aligned}$$

***PRACTICE PROBLEM 11.20***

Refer to Figure 11.53b. Determine the instantaneous value that the source voltage reaches 30 μs after the start of each cycle.

As you can see, the approach used to solve an instantaneous value problem depends on the known values. In future chapters, we will select our approach to each problem based on the information provided.

## *Section Review*

1. Briefly describe how current can be generated by magnetic induction.
2. Briefly describe the operation of the simple ac generator in Figure 11.35.
3. What is the phase of a point on a waveform?

4. Why are sine waves commonly described in terms of phase?
5. How can phase and time measurements be related?
6. What is meant by the term *phase angle*?
7. Describe the relationship between phase angle and each of the following: frequency, cycle time, peak voltage (or current).
8. Describe the process for measuring phase angles with a dual-trace oscilloscope.
9. What factors determine the instantaneous value of a sine wave voltage (or current)? Which of these values is a constant for a given waveform? Which varies from one point on the waveform to the next?
10. What is a radian?
11. How many radians are contained in one alternation of a sine wave? In one complete cycle?
12. What is angular velocity? Why is it significant?
13. Compare the degree and radian approaches to measuring phase.
14. When using the radian approach to calculating instantaneous values, what adjustment must be made to your calculator?
15. When would you use the degree approach to calculating an instantaneous value? When would you use the radian approach?

## 11.4 RELATED TOPICS

The purpose of this chapter has been to introduce you to some of the basic principles of ac. In this section, we will complete the chapter by discussing various topics that relate to ac waveforms and their measurements. The topics in this section do not necessarily relate directly to each other, so they can be covered in any order desired.

### *Static and Dynamic Values*

◄ *OBJECTIVE 16*

Most ac circuit values are classified as being either static or dynamic. **Static values** are constant; that is, they do not change during the normal operation of the circuit. For example, the resistor in Figure 11.54a has a value that remains constant at $R_L = 1\ \text{k}\Omega$. Since this value does not change during the normal operation of the circuit, it is considered to be static. Note that static values are usually independent of other circuit values. On the other

**Static values**
Values that are constant. Static values do not change as a result of normal circuit operation.

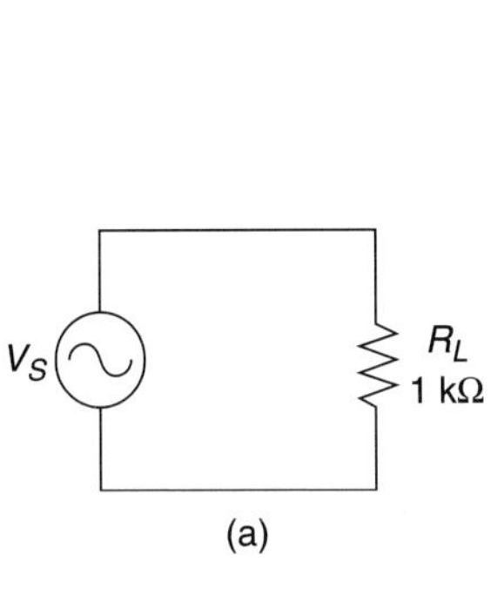

(a)

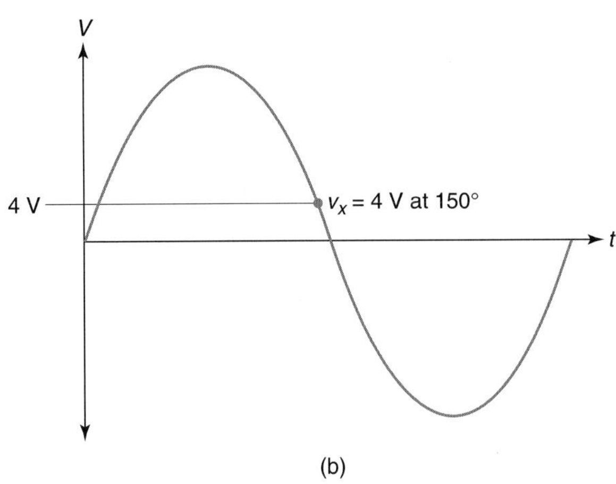

(b)

**FIGURE 11.54**

**Dynamic values**
Values that change as a direct result of normal circuit operation. Dynamic values always have specified conditions.

hand, **dynamic values** change during the normal operation of the circuit. For example, the instantaneous value of a sine wave is dynamic because it changes from one instant to the next.

A dynamic value always has a specified condition, that is, a condition under which the measurement is valid. For example, look at the waveform in Figure 11.54b. If we want to indicate the value of $v_x$, we must include both the value and the phase angle because the waveform has two 4 V points in the positive half-cycle. Therefore, we must indicate to which 4 V point we are referring. For the waveform shown, the value of $v_x$ could be written as

$$v_x = 4\text{ V}\angle 150^\circ$$

where 4 V is the value and $\angle 150^\circ$ indicates the phase angle where the measurement is valid. (Incidentally, the equation above is written in a form called *polar notation.* We will discuss this notation, and others, in Chapter 12.)

*OBJECTIVE 17* ▶

## dc Offsets

Earlier in the chapter, you were told that a pure ac waveform is one that has:

1. Equal positive and negative peak values.
2. An average value of 0 V (or 0 A).

**dc offset**
A term used to describe a dc value (other than zero) that acts as the reference for a sine wave (or some other waveform).

In some cases, an ac waveform contains a **dc offset;** that is, the waveform is centered around some value other than zero. A voltage sine wave with a dc offset is shown in Figure 11.55a. As the figure implies, a sine wave with a dc offset has:

1. Unequal positive and negative peak values.
2. An average value equal to the value of the dc offset (rather than zero).

**Amplifier**
A circuit that can be used to increase the power level of a sine wave (or some other waveform).

Various sources can generate dc offsets. For example, an **amplifier** can introduce a dc offset as shown in Figure 11.55b. The purpose of the amplifier is to increase the power level of the sine wave. It does so by transferring power from its dc source ($+V_{dc}$) to the

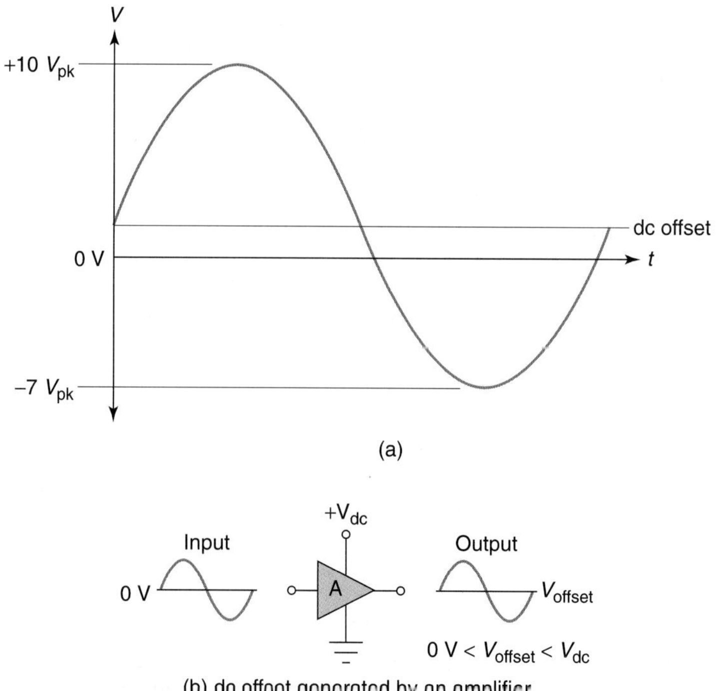


**FIGURE 11.55 dc offsets.**

waveform. In the process, it also introduces a dc offset into the sine wave. Depending on the design of the amplifier, the dc offset voltage falls somewhere within the range shown in the figure.

## Determining the Value of Offset Voltage

The dc offset in any sine wave can be measured directly with a dc voltmeter. For example, to measure the dc offset introduced by an amplifier, we would connect a dc voltmeter to the amplifier output as shown in Figure 11.56. Note that the dc offset reading on the meter (+5 V) falls halfway between the peak values of the output waveform. This relationship, which always holds true, can be expressed as

$$V_{\text{offset}} = \frac{+V_{\text{pk}} + (-V_{\text{pk}})}{2} \tag{11.21}$$

Using the values in the figure, this relationship can be verified as follows:

$$V_{\text{offset}} = \frac{+20\text{ V} + (-10\text{ V})}{2} = \frac{+10\text{ V}}{2} = +5\text{ V}$$

In upcoming chapters, you won't see many problems involving dc offsets. However, when you study solid-state devices and digital electronics, you will find that dc offsets are very much a part of the picture.

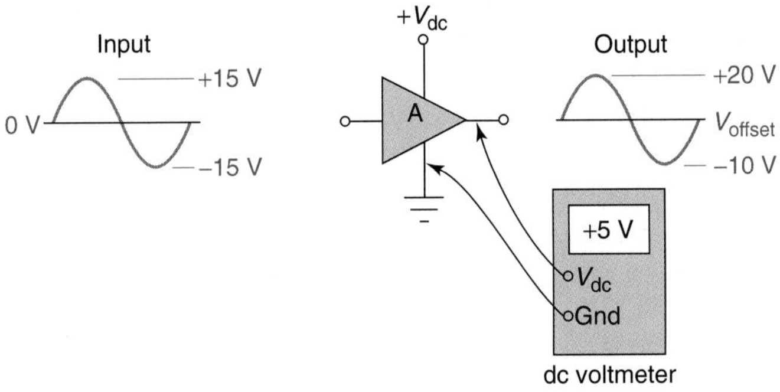


**FIGURE 11.56 Measuring a dc offset.**

EWB

## Wavelength

◀ *OBJECTIVE 18*

**Electromagnetic wave**
A waveform that consists of perpendicular electric and magnetic fields.

When you study communications electronics, you will learn how a sine wave can be transmitted through space as a series of electromagnetic waves. An **electromagnetic wave** is a waveform that consists of perpendicular electric and magnetic fields. A sinusoidal electromagnetic wave is illustrated in Figure 11.57. As the figure implies, the electric field ($y$-axis) and magnetic field ($x$-axis) are always at 90° angles. Note that both fields are perpendicular to the direction of travel for the waveform (along the $z$-axis of the graph).

**FIGURE 11.57 Electromagnetic wave.**

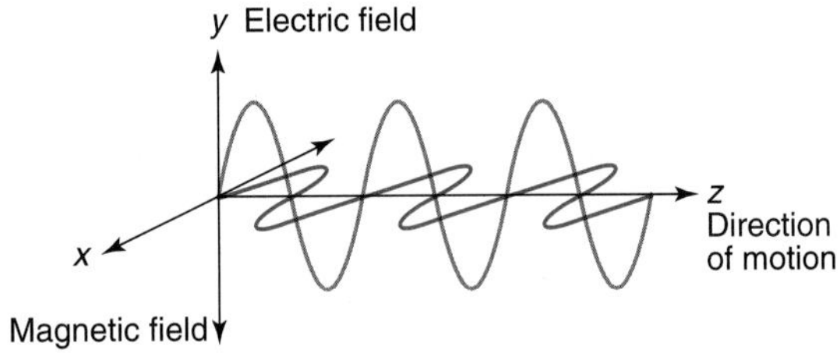

**Wavelength** ($\lambda$)
The physical length of a transmitted waveform.

Every cycle of a transmitted waveform has a measurable length, called its **wavelength.** The wavelength for one cycle of a waveform is determined by:

1. The speed at which it travels through space.
2. The cycle time of the waveform.

In a vacuum, all electromagnetic waves travel at the velocity of electromagnetic radiation, which is commonly referred to as the *speed of light.* This velocity is approximately 186,000 miles per second. This and other useful speed of light measurements are listed in Table 11.2.

*A Memory Aid:* Many technicians find it easier to remember the speed of light in either of the following forms:

300 m/μs    or    984 ft/μs

**TABLE 11.2 Commonly Used Speed of Light Measurements**

| *Value* | *Unit Identification* |
|---|---|
| $3 \times 10^8$ m/s | Meters per second |
| $3 \times 10^{10}$ cm/s | Centimeters per second |
| $1.86 \times 10^5$ mi/s | Miles per second |
| $9.84 \times 10^8$ ft/s | Feet per second |
| $1.18 \times 10^{10}$ in/s | Inches per second |

The concept of wavelength is illustrated in Figure 11.58. The waveform shown is transmitted into space from an antenna. In free space, the waveform travels at a speed of $3 \times 10^8$ meters per second. With a cycle time of 1 second, point A on the waveform will have traveled $3 \times 10^8$ meters by the time point B leaves the source. Thus, the wave shown in Figure 11.58 would have a wavelength of $3 \times 10^8$ meters.

The wavelength of a transmitted waveform is represented using the Greek letter *lambda* ($\lambda$). The value of $\lambda$ for a given waveform can be found as the product of velocity and cycle time. By formula,

$$\lambda = cT_C \tag{11.22}$$

where $c$ = the speed of light
$T_C$ = the cycle time of the waveform

For the waveform in Figure 11.58, the wavelength can be calculated as

$$\lambda = cT_C = (3 \times 10^8 \text{ m/s})(1 \text{ s}) = 3 \times 10^8 \text{ m}$$

This calculation matches the description of the waveform given earlier.

There is another (and more commonly used) method of calculating wavelength. Since cycle time and frequency are reciprocal functions, wavelength can be found as

$$\lambda = c\frac{1}{f}$$

or

$$\lambda = \frac{c}{f} \tag{11.23}$$

where $f$ = the frequency of the transmitted waveform.

**FIGURE 11.58 Wavelength.**

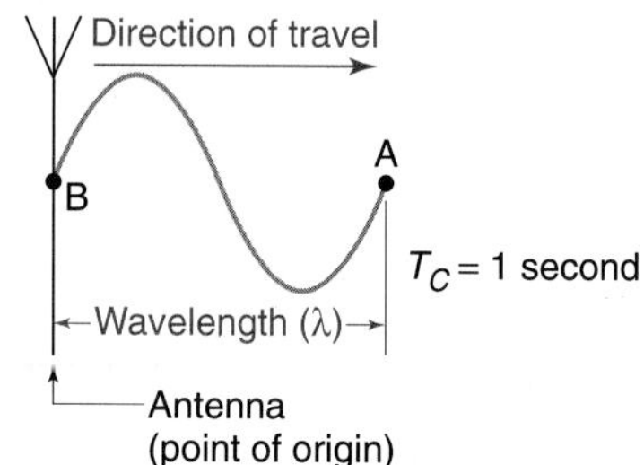

When calculating wavelength, the unit of measure for $\lambda$ is determined by the unit of measure for its velocity. For example, if you use $c = 3 \times 10^{10}$ cm/s in your calculation, the value of $\lambda$ obtained will be in centimeters. If you use $c = 9.84 \times 10^{8}$ ft/s, the value of $\lambda$ will be in feet, and so on. Therefore, the value used for $c$ is determined by the desired unit of measure for $\lambda$. This point is demonstrated in Example 11.21.

### EXAMPLE 11.21

Calculate the wavelength (in meters) for a 150 MHz sine wave.

***Solution:*** Since the desired unit of measure for $\lambda$ is meters, we will use $c = 3 \times 10^{8}$ m/s in equation (11.23), as follows:

$$\lambda = \frac{c}{f} = \frac{3 \times 10^{8} \text{ m/s}}{150 \times 10^{6} \text{ Hz}} = \mathbf{2\ m}$$

***PRACTICE PROBLEM 11.21***

Calculate the wavelength of a 225 MHz sine wave in centimeters, feet, and inches.

As equation (11.23) indicates, wavelength and frequency are inversely proportional. This relationship is illustrated further in Example 11.22.

### EXAMPLE 11.22

A sine wave has a frequency that is variable from 150 MHz to 400 MHz. Calculate the range of wavelengths for the waveform.

***Solution:*** In Example 11.21, we calculated a 2 meter wavelength for a 150 MHz signal. For a 400 MHz signal, the wavelength (in meters) is found as

$$\lambda = \frac{c}{f} = \frac{3 \times 10^{8} \text{ m/s}}{400 \times 10^{6} \text{ Hz}} = \mathbf{0.75\ m}$$

As you can see, an increase in frequency (150 MHz to 400 MHz) results in a decrease in wavelength (2 m to 0.75 m). Therefore, wavelength and frequency are inversely proportional.

***PRACTICE PROBLEM 11.22***

The frequency of a sine wave is variable from 200 MHz to 250 MHz. Calculate the range of wavelengths (in feet) for the waveform.

As you continue your studies, you'll see that wavelength is an important consideration in various communications applications. Some examples of wavelength-sensitive devices are antennas, coaxial cables, fiber optic cables, and opto-electronic devices.

## Harmonics

◄ *OBJECTIVE 19*

A **harmonic** is a whole-number multiple of a given frequency. For example, a 2 kHz sine wave has harmonics of

$$2 \text{ kHz} \times 2 = 4 \text{ kHz} \qquad 2 \text{ kHz} \times 3 = 6 \text{ kHz}$$
$$2 \text{ kHz} \times 4 = 8 \text{ kHz} \qquad 2 \text{ kHz} \times 5 = 10 \text{ kHz}$$

and so on. Note that a group of related frequencies like the one above is referred to as a **harmonic series.** The reference frequency in a harmonic series (2 kHz in this case) is referred to as the **fundamental frequency.**

**Harmonic**
A whole-number multiple of a given frequency.

**Harmonic series**
A group of related harmonic frequencies.

**Fundamental frequency**
The lowest frequency in a harmonic series.

**TABLE 11.3 Frequency Identifiers**

| *Fundamental Frequency* | *Multiplier* | *Harmonic Frequency* | *Identifier* |
|---|---|---|---|
| 100 Hz | 2 | 200 Hz | 2nd-order harmonic |
| 100 Hz | 3 | 300 Hz | 3rd-order harmonic |
| 100 Hz | 4 | 400 Hz | 4th-order harmonic |
| . | | | |
| . | | | |
| . | | | |
| 100 Hz | 10 | 1 kHz | 10th-order harmonic |

***n*-order harmonics** The frequencies (other than the fundamental) in a harmonic series.

**Octave** A frequency multiplier equal to 2.

The fundamental frequency is the lowest frequency in any harmonic series. The other frequencies are generally identified as ***n*-order harmonics.** This is illustrated in Table 11.3.

Two of the frequency relationships in Table 11.3 are often identified using other names. When frequency changes by a factor of 2, the change is referred to as one **octave.** Thus, the 2nd-order harmonic is said to be one octave above the fundamental frequency. By the same token, the 4th-order harmonic is:

1. One octave above the 2nd-order harmonic ($200 \text{ Hz} \times 2 = 400 \text{ Hz}$).
2. Two octaves above the fundamental frequency ($100 \text{ Hz} \times 2 \times 2 = 400 \text{ Hz}$).

**Decade** A frequency multiplier equal to 10.

When frequency changes by a factor of 10, the change is referred to as one **decade.** Thus, the 10th-order harmonic is said to be one decade above the fundamental frequency. It would appear on the surface that one decade is the same as five octaves. However, this is not the case. The frequency that is five octaves above 100 Hz would be found as

$$\begin{aligned} f &= 100 \text{ Hz} \times 2 \times 2 \times 2 \times 2 \times 2 \quad \text{(doubling frequency 5 times)} \\ &= 3.2 \text{ kHz} \end{aligned}$$

As you can see, we found the frequency that is five octaves above the fundamental by doubling the frequency five times. The frequency that is $n$ octaves above a given frequency ($f_0$) can be found as

$$f = 2^n f_0 \tag{11.24}$$

Had we used this relationship for the 100 Hz frequency in Table 11.3, the result would have been

$$\begin{aligned} f &= 2^n f_0 \\ &= 2^5 (100 \text{ Hz}) \\ &= (32)(100 \text{ Hz}) \\ &= 3.2 \text{ kHz} \end{aligned}$$

The frequency that is $n$ decades above a given frequency can be found as

$$f = 10^n f_0 \tag{11.25}$$

As you can see, the only thing that has changed is the number that is raised to the $n$ power.

In Chapter 19, we will focus on circuits whose outputs are frequency dependent; that is, their outputs vary directly (or inversely) with the frequency of operation. At that time, you will be shown how octave and decade frequency intervals are often used to describe circuit frequency response.

*OBJECTIVE 20* ▶

## Nonsinusoidal Waveforms: Rectangular Waves

**Rectangular waves** Waveforms that alternate between two dc levels.

**Rectangular waves** are waveforms that alternate between two dc levels. An example of a rectangular waveform is shown in Figure 11.59. The waveform shown alternates between the $-V_{dc}$ and $+V_{dc}$ levels.

**FIGURE 11.59 Rectangular waveform.**

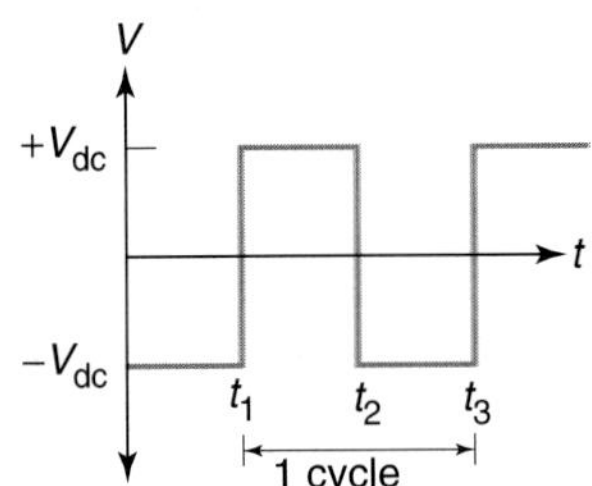


Figure 11.60 shows the response of a circuit to a rectangular-wave input. From $t_1$ to $t_2$, the source has the polarity and current direction shown in Figure 11.60b. From $t_2$ to $t_3$, the source polarity and current direction are reversed, as shown in Figure 11.60c. Even though the rectangular wave is made up of alternating dc levels, it produces current that alternates directions. Therefore, it is classified correctly as an ac waveform.

*A Practical Consideration:* The exact values of $-V_{dc}$ and $+V_{dc}$ aren't important at this point. As long as one value is positive and the other is negative, the waveform qualifies as an ac waveform.

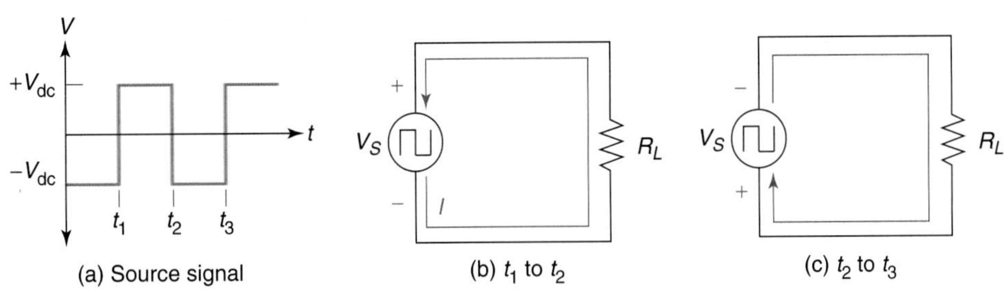


**FIGURE 11.60 Circuit response to a rectangular input.**

***Terminology and Time Measurements.*** The parts of a rectangular waveform are identified in Figure 11.61. The positive half-cycle is referred to as the **pulse width.** The negative half-cycle is referred to as the **space width.** The sum of the pulse width and space width equals the cycle time of the waveform.

**Pulse width**
A term generally used to describe the positive alternation of a rectangular waveform.

**Space width**
A term generally used to describe the negative alternation of a rectangular waveform.

***Duty Cycle.*** Circuits designed to respond to rectangular waveforms are generally referred to as pulse and switching circuits, or digital circuits. Because of certain design factors, many digital circuits have a limit on the ratio of pulse width to cycle time. This ratio, which is commonly given as a percentage, is referred to as **duty cycle.** By formula,

**Duty cycle**
The ratio of pulse width to cycle time, usually given as a percentage.

$$\text{duty cycle } (\%) = \frac{\text{PW}}{T_C} \times 100 \qquad \textbf{(11.26)}$$

where PW = the pulse width of the circuit input
$T_C$ = the cycle time of the circuit input

Example 11.23 demonstrates the calculation of duty cycle for a rectangular waveform.

**FIGURE 11.61 Pulse width, space width, and cycle time.**

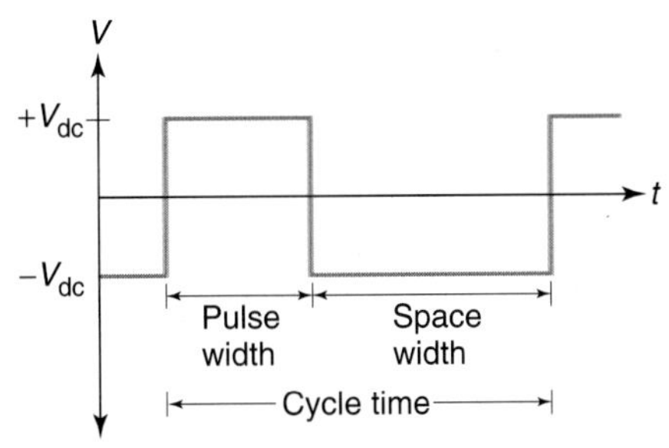


*A Practical Consideration:* When you study digital electronics, you'll learn how the pulse width and space width polarities can be the opposite of those shown here; that is, the pulse width can be negative and the space width can be positive.

## EXAMPLE 11.23

Determine the duty cycle of the waveform shown in Figure 11.62a.

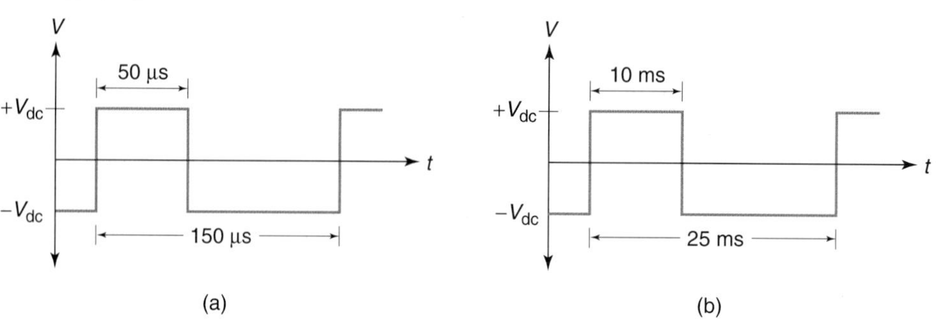


**FIGURE 11.62**

***Solution:*** Using the values shown in the figure, the duty cycle is found as

$$\textbf{duty cycle} = \frac{\text{PW}}{T_C} \times 100 = \frac{50\ \mu\text{s}}{150\ \mu\text{s}} \times 100 = \mathbf{33.3\%}$$

This value indicates that the pulse width makes up 33.3% of the total cycle.

***PRACTICE PROBLEM 11.23***

Calculate the duty cycle of the waveform shown in Figure 11.62b.

When you study digital electronics, you will see that the duty cycle rating of a component is a limit. If the duty cycle limit of a component is exceeded, the component may respond erratically.

**Square wave**
A special-case rectangular wave with equal pulse width and space width values. The duty cycle of any square wave is 50%.

**Harmonic synthesis**
Generating a square wave by adding a fundamental frequency to an infinite number of its odd-order harmonics.

***Square Waves.*** The **square wave** is a special-case rectangular waveform that has equal pulse width and space width values. A square wave is shown in Figure 11.63. Since the values of PW and SW are equal, the duty cycle of a square wave is always 50%.

There is another characteristic that separates the square wave from other rectangular waveforms: it is possible to generate a square wave by adding a sine wave to an infinite number of its odd-order harmonics. This method of generating a square wave, which is referred to as **harmonic synthesis,** is illustrated in Figure 11.64. In Figure 11.64a, a fundamental ($f$) is added to its 3rd-order harmonic ($3f$). The result is similar to the sum wave shown. Adding the fundamental to $3f$ and $5f$ produces the sum wave in Figure 11.64b. If we were to continue adding higher odd-order harmonics, the variations in the sum wave would cancel out and we would have a square wave. Note that the resulting square wave has the same frequency as the fundamental.

**FIGURE 11.63 Square wave measurements.**

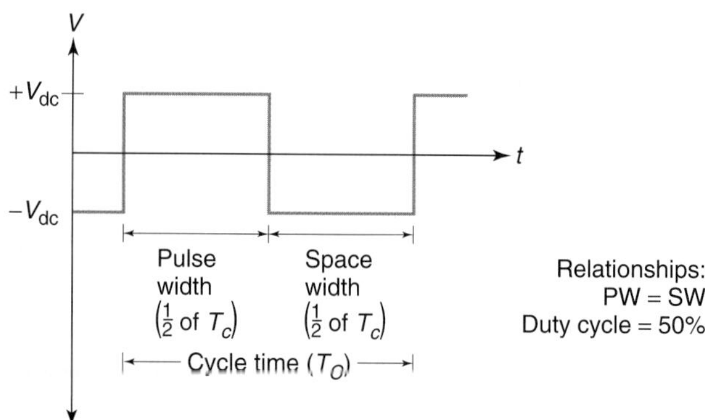

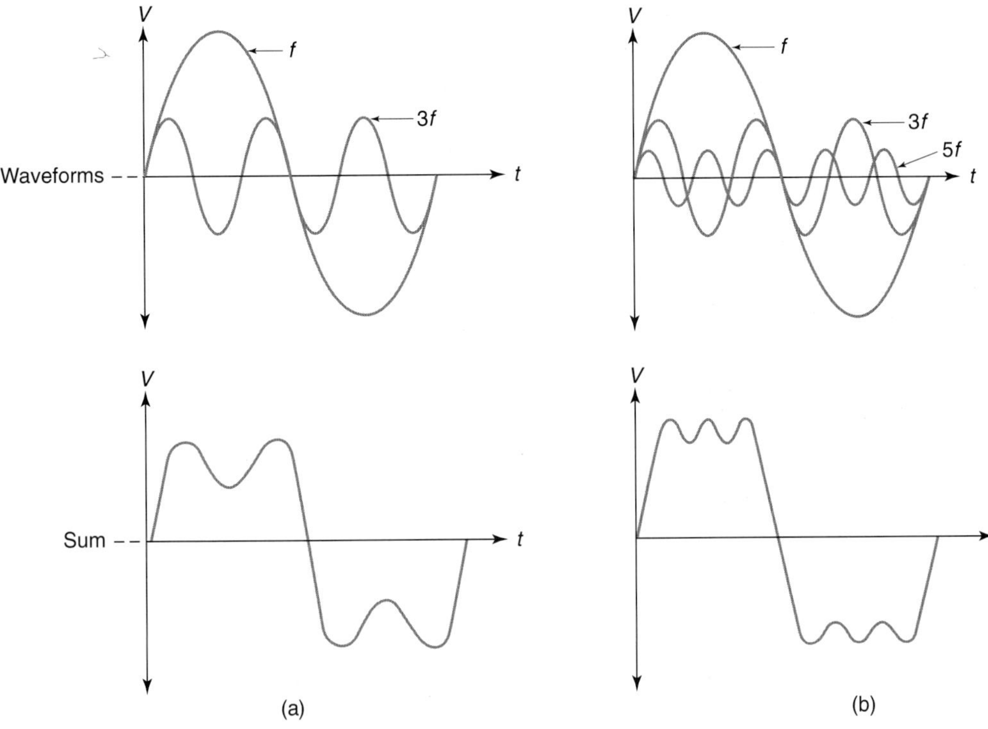


**FIGURE 11.64 Harmonic synthesis.**

EWB

By definition, a fundamental sine wave frequency must be added to an infinite number of odd-order harmonics to produce a square wave. In practice, only the first five odd-order harmonics are required; that is, $f + 3f + 5f + 7f + 9f + 11f$ produces a square wave that is difficult to distinguish from the ideal square wave.

It should be noted that square waves are not commonly generated by adding sinusoidal waveforms. However, you should know that every square wave contains frequencies that are much higher than the frequency of the waveform itself. For example, a 1 kHz square wave contains frequencies of 1 kHz, 3 kHz, 5 kHz, 7 kHz, 9 kHz, 11 kHz, and so on. This point will become important when you begin to work with filters.

***Symmetrical and Nonsymmetrical Waveforms.*** A **symmetrical waveform** is one with cycles that are made up of identical halves. For example, look at the square wave in Figure 11.65a. In this case:

**Symmetrical waveform**
A waveform with cycles made up of identical halves. Waveforms can be symmetrical in time and/or amplitude.

1. PW = SW, so the waveform is symmetrical in time.
2. The waveform varies equally above and below 0 V, so it is symmetrical in amplitude.

By definition, a square wave is symmetrical in time. However, a square wave may be nonsymmetrical in amplitude; that is, it may vary unequally above and below 0 V. For example, consider the waveform shown in Figure 11.65b. The waveform has peak values of +5 V and −15 V. When the peak values of a rectangular waveform are unequal, then the amplitude of the waveform is considered to be nonsymmetrical.

A square wave with equal positive and negative peak values has an average value of 0 V. For a square wave with nonsymmetrical peak values, the average voltage can be found as

*Note:* Equation (11.27) is waveform-specific: it applies only to square waves. Equation (11.28) provides a general-purpose form of the equation.

$$V_{ave} = \frac{+V_{pk} + (-V_{pk})}{2} \tag{11.27}$$

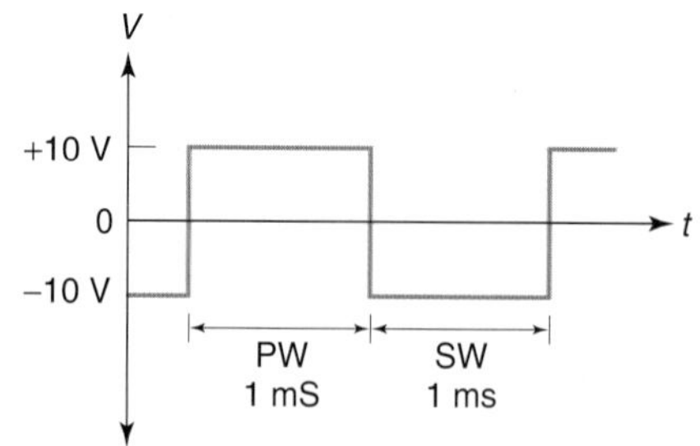

(a) Symmetrical time and amplitude

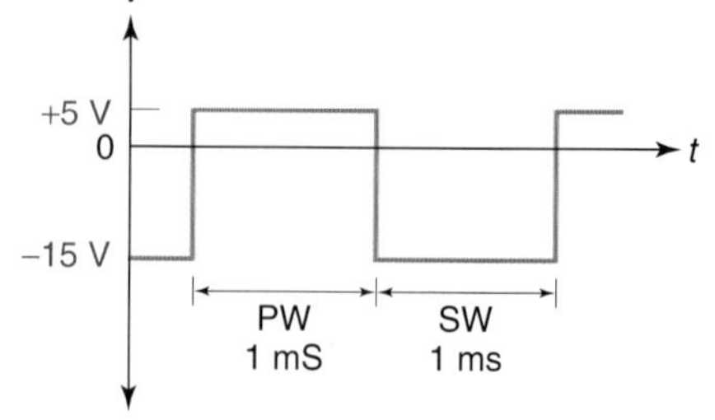

(b) Symmetrical time, nonsymmetrical amplitude

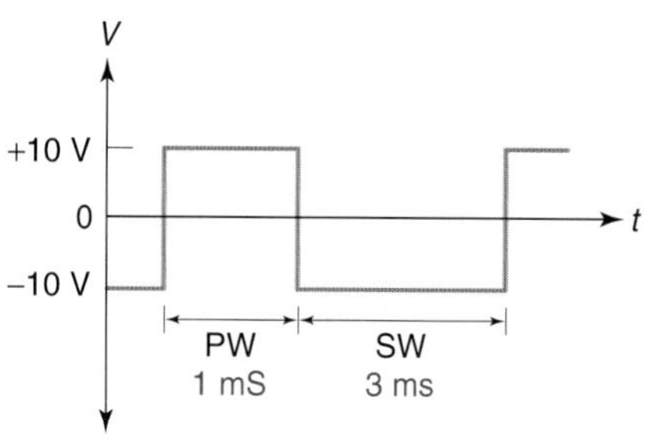

(c) Nonsymmetrical time, symmetrical amplitude

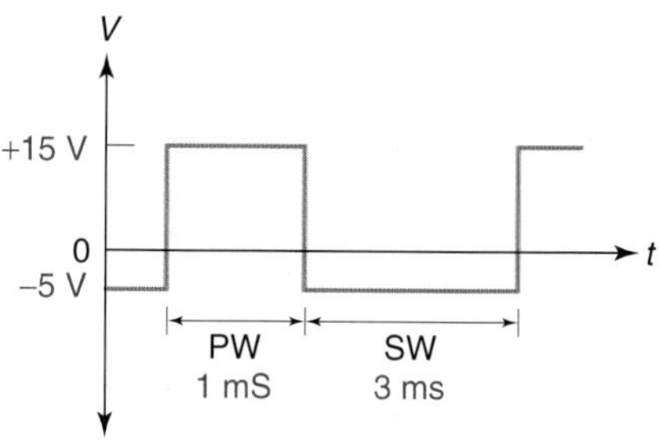

(d) Nonsymmetrical time and amplitude

**FIGURE 11.65 Symmetrical and nonsymmetrical waveforms.**

As you can see, this is the same relationship we use to determine the average value of a sine wave with a dc offset. For the waveform in Figure 11.65b, the average value can be found as

$$V_{ave} = \frac{+V_{pk} + (-V_{pk})}{2} = \frac{+5\text{ V} + (-15\text{ V})}{2} = \frac{-10\text{ V}}{2} = -5\text{ V}$$

When you first glance at equation (11.27), it may seem that $[(+V_{pk}) + (-V_{pk})]$ is the same thing as the peak-to-peak value of the waveform. However, this is not the case. The peak-to-peak value of a waveform equals the difference between the peak value, not the sum of the peak values. Had we used the peak-to-peak value of Figure 11.65b in equation (11.27), the result would have been +10 V. This result is impossible, given the fact that the positive peak value is only +5 V.

Figure 11.65c shows a waveform that has symmetrical peak values, but is nonsymmetrical in time. For this type of waveform, the average value is found as

$$V_{ave} = \frac{(\text{PW})(+V_{pk}) + (\text{SW})(-V_{pk})}{T_C} \tag{11.28}$$

This equation considers the difference in time spent at each peak value. For the waveform in Figure 11.65c, the average value can be found as

$$\begin{aligned} V_{ave} &= \frac{(\text{PW})(+V_{pk}) + (\text{SW})(-V_{pk})}{T_C} \\ &= \frac{(1\text{ ms})(+10\text{ V}) + (3\text{ ms})(-10\text{ V})}{4\text{ ms}} \\ &= \frac{(+10\text{ V}\cdot\text{ms}) - (30\text{ V}\cdot\text{ms})}{4\text{ ms}} \\ &= \frac{-20\text{ V}\cdot\text{ms}}{4\text{ ms}} \\ &= -5\text{ V} \end{aligned}$$

Equation (11.28) can also be used to find the average value of a rectangular waveform that is nonsymmetrical in both time and amplitude. For example, look at the waveform in Figure 11.65d. For this waveform, the average value can be found as

$$V_{\text{ave}} = \frac{(\text{PW})(+V_{\text{pk}}) + (\text{SW})(-V_{\text{pk}})}{T_C}$$

$$= \frac{(1\text{ ms})(+15\text{ V}) + (3\text{ ms})(-5\text{ V})}{4\text{ ms}}$$

$$= \frac{(15\text{ V}\cdot\text{ms}) - (15\text{ V}\cdot\text{ms})}{4\text{ ms}}$$

$$= 0\text{ V}$$

As this result indicates, a rectangular waveform can have an average of 0 V and still be considered nonsymmetrical in amplitude. Even though the average works out to 0 V, the peaks still vary unequally above and below 0 V.

***Measuring the Value of $V_{ave}$ for a Rectangular Waveform.*** The value of $V_{\text{ave}}$ for any rectangular waveform can be measured easily using an oscilloscope. The procedure is as follows:

1. Set the ac/gnd/dc switch for the oscilloscope input to the ac position.
2. Adjust the vertical position control so that the positive peak falls on any major horizontal division line.
3. Switch the ac/gnd/dc setting to dc. Note:
   a. The direction that the waveform shifts.
   b. The number of divisions that the waveform shifts.

   (These can be seen by comparing the new position of the positive peak to its original position.)
4. Determine the value of $V_{\text{ave}}$ as follows:
   a. The direction of the shift indicates the polarity of $V_{\text{ave}}$. If the waveform shifts up, $V_{\text{ave}}$ is positive. If the waveform shifts down, it is negative.
   b. The number of divisions shifted times the V/div setting indicates the magnitude of $V_{\text{ave}}$. (*Note:* If $V_{\text{ave}} = 0$ V, the waveform will not shift when the ac/gnd/dc setting is changed.)

Figure 11.66 demonstrates this technique. Figure 11.66a shows the waveform with the ac/gnd/dc switch in the ac position. The vertical sensitivity is set to 5 V/div. When the ac/gnd/dc setting is changed to dc, the waveform shifts to the position shown in Figure 11.66b. Note that it has shifted in the positive direction, indicating that the value of $V_{\text{ave}}$ is positive. Since the waveform shifted 1½ major divisions, the value of $V_{\text{ave}}$ can be found as

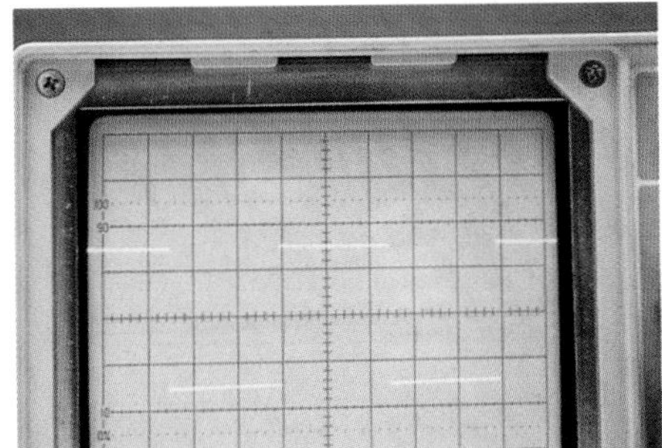

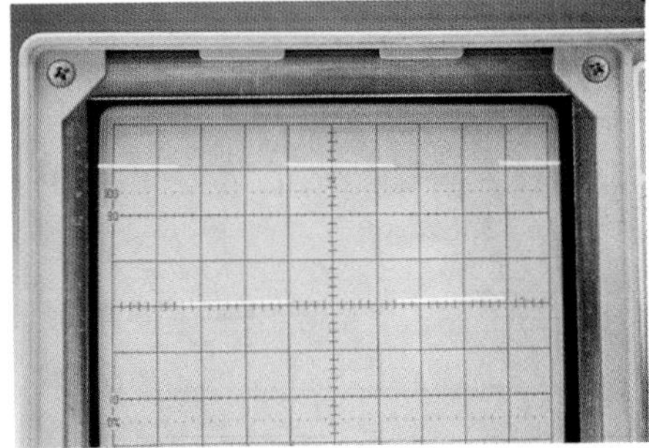

**FIGURE 11.66 Measuring average voltage with an oscilloscope.**

$$V_{ave} = (1.5 \text{ divisions})\frac{5 \text{ V}}{\text{div}} = 7.5 \text{ V}$$

As your oscilloscope skills develop, you'll see that the value of $V_{ave}$ for any type of waveform can be measured using this technique. For example, it can be used to measure the average value of a sine wave with a dc offset.

OBJECTIVE 21 ▸

## Nonsinusoidal Waveforms: Sawtooth Waves

**Sawtooth**
A term used to describe a waveform that changes constantly at a linear rate.

**Ramp**
Another name for a sawtooth. The term is also used to describe the actual transition from one peak to the other.

**Slope**
The rate at which a value changes. For a voltage sawtooth, slope is measured in volts per unit time.

The term **sawtooth** is used to describe any waveform that changes constantly at a linear rate; that is, it is made up of straight-line changes back and forth between its peak values. A sawtooth waveform is shown in Figure 11.67a. Note the straight-line changes from $-V_{pk}$ to $+V_{pk}$ and back. Also note that the waveform does not level out at its peak values (like a rectangular waveform). Because of its shape, the sawtooth waveform is sometimes referred to as a **ramp.**

The **slope** of a sawtooth waveform is the rate at which its value changes, measured in volts per unit time. Slope is measured as shown in Figure 11.67b. Note that the voltage represented changes by 6 V over a period of 4 ms. As the calculations show, this translates into a slope of 1.5 V/ms.

The slopes of the positive-going and negative-going transitions of a sawtooth are not necessarily equal. This point is demonstrated by the waveform in Figure 11.67a. For this waveform, the slope of the negative-going transition is much greater than that of the positive-going transition (the higher the slope, the greater the change per unit time).

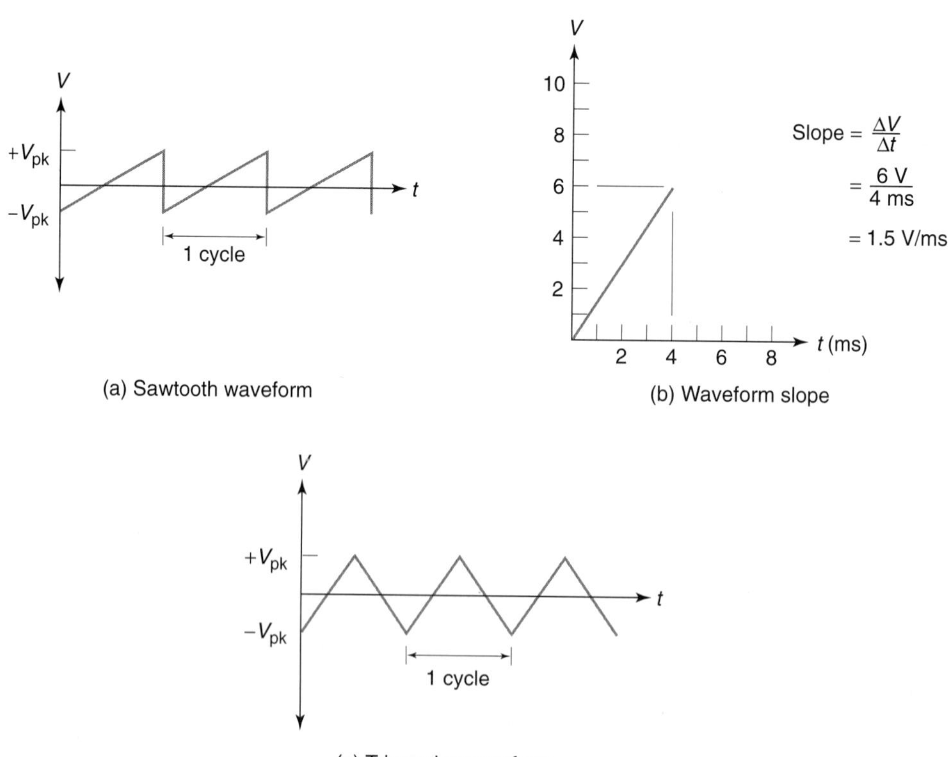


**FIGURE 11.67 Sawtooth waveforms.**

## Triangular Waveforms

The **triangular waveform** is a symmetrical sawtooth waveform; that is, the positive-going and negative-going transitions are mirror images of each other. A triangular waveform is shown in Figure 11.67c. Note that the slopes of the transitions are always equal in magnitude. For example, if the positive-going transition in Figure 11.66c has a value of 2 V/ms, then the negative-going transition has a slope of −2 V/ms.

**Triangular waveform** A symmetrical sawtooth waveform. In a triangular waveform, the slopes of the transitions are equal.

### Section Review

1. What is a static value? What is a dynamic value?
2. How can you distinguish a dynamic value from a static value?
3. What is a dc offset? What effect does it have on the average value of a sine wave?
4. How do you measure the average value of a sine wave with a dc offset?
5. What is an electromagnetic wave?
6. What is wavelength?
7. What factors determine the wavelength of a given waveform?
8. List the commonly used speed-of-light values.
9. When calculating wavelength, what determines the unit of measure for the result?
10. What is a harmonic?
11. In a string of harmonic frequencies, which is the fundamental?
12. What is an octave? What is a decade?
13. What is a rectangular waveform?
14. Draw a rectangular waveform and identify each of the following: pulse width, space width, and cycle time.
15. What is duty cycle?
16. What is a square wave?
17. Describe harmonic synthesis.
18. In a practical sense, how many sinusoidal frequencies are needed to produce a square wave?
19. What is a symmetrical waveform?
20. Describe the various possible combinations of symmetrical and nonsymmetrical measurements. (See Figure 11.64.)
21. Describe the process for measuring the average value of a rectangular waveform using an oscilloscope.
22. What is a sawtooth waveform? By what other name is it known?
23. What is the slope of a ramp?
24. What is a triangular waveform?

## Chapter Summary

Here is a summary of the major points made in this chapter:

1. The term *alternating current* is used to describe any current that changes direction periodically.
2. A graph of the relationship between any magnitude and time is referred to as a waveform. On a waveform display, magnitude is measured along the vertical axis. Time is measured along the horizontal axis.

3. The positive and negative transitions of a waveform are referred to as alternations.
4. The complete transition through one positive alternation and one negative alternation is referred to as a cycle. Since two alternations make up one cycle, each alternation is commonly referred to as a half-cycle.
5. The time required to complete one cycle of an ac signal is referred to as its cycle time or period.
6. Cycle time can be measured from any given point on the waveform to the identical point in the next cycle.
7. The frequency of a waveform is a measure of the rate at which the cycles repeat themselves, measured in cycles per second.
8. The unit of measure for frequency is hertz (Hz). One hertz is equal to one cycle per second.
9. Frequency and cycle time are inversely proportional. Each can be calculated as the reciprocal of the other:

$$T_C = \frac{1}{f} \qquad \text{and} \qquad f = \frac{1}{T_C}$$

10. An oscilloscope is a piece of test equipment that provides a visual representation of a voltage waveform.
    a. The visual display can be used to make various magnitude-related and time-related measurements.
    b. The setting of the time base control determines the amount of time represented by the space between each pair of major divisions along the $x$-axis. (See Figure 11.13.)
    c. When you vary the time base setting on the oscilloscope, you change only the waveform display, not the waveform itself.
11. The maximum value that a waveform reaches in each transition is called the peak value of that transition.
12. A pure ac waveform is one with peak values equal in magnitude.
13. The peak-to-peak value of a waveform is equal to the difference between its peak values. For a pure ac waveform, the peak-to-peak value is twice its peak value.
14. An instantaneous value is the magnitude of a waveform at a specified point in time.
    a. For every time interval in a waveform, there is only one instantaneous value.
    b. A sine wave can be divided into an infinite number of time intervals, so there are an infinite number of instantaneous values in each cycle.
    c. Instantaneous values are represented using lowercase letters (as opposed to constant values, which are represented using uppercase letters).
15. The full-cycle average of a waveform is the average of all its instantaneous values through one complete cycle.
    a. The full-cycle average of a pure ac waveform is always zero.
    b. For a waveform that is not pure ac, the full-cycle average can have a value other than zero.
16. The half-cycle average of a waveform is the average of all its instantaneous values of voltage (or current) through one alternation of the input cycle.
17. When dealing with ac circuits, we are interested in the average ac power generated throughout each cycle.
18. The rms (root-mean-square) value of a voltage (or current) waveform is the value that, when used in the appropriate power equation, will provide the average ac power of the waveform. (The origin of the term *rms* is illustrated in Figure 11.23.)
19. The rms value of any sine wave is equal to 70.7% of its peak value.

20. Power has always been used as the basis of comparison for ac and dc. Two sources, one dc and one ac, are considered equivalent when they deliver the same amount of power to a given load.
    a. Average ac power is the equivalent of dc power.
    b. Since rms values are used to calculate average ac power, they also indicate equivalent dc power values.
21. When a magnitude-related value is not specifically identified (as peak, peak-to-peak, or average), it is automatically assumed to be an rms value.
22. Peak and peak-to-peak voltages are measured with an oscilloscope.
    a. The value represented by each major vertical division on the oscilloscope is determined by the setting of the vertical sensitivity control.
    b. To measure peak voltage, the 0 V position must be established on the oscilloscope grid. Voltages above the 0 V line are positive. Voltages below the 0 V line are negative.
    c. Peak and peak-to-peak measurements are illustrated in Examples 11.9 and 11.10.
23. Peak and peak-to-peak currents cannot be measured directly by any standard piece of test equipment.
    a. These values are usually determined using voltage measurements and Ohm's law. (See Example 11.11.)
    b. Peak-to-peak current is a conceptual value that does not play a major role in any practical situation.
24. The half-cycle average of a sine wave can be measured using a dc voltmeter and a circuit called a rectifier.
    a. A rectifier is a circuit that converts ac to pulsating dc.
    b. The output from a full-wave rectifier equals the half-cycle average of its sine wave input. (See Figure 11.30.)
    c. The output from a half-wave rectifier equals ½ of the output from a full-wave rectifier; that is, it equals one-half of the half-cycle average of its sine wave input.
    d. Rectifier construction and operation are covered in a course on semiconductor devices and circuits.
25. Half-cycle average current can be measured using a dc ammeter and a rectifier. Half-wave and full-wave rectifiers have the same effect on current measurements as they do on the half-cycle voltage measurements.
26. The rms values of a sine wave are also referred to as *effective* values.
27. Effective values are measured using ac voltmeters and ac ammeters.
28. The tools used to measure the various ac values are summarized in Table 11.1.
29. Just as current can be used to produce a magnetic field, the field produced by a magnet can be used to generate current through a conductor.
    a. If a conductor is cut by perpendicular lines of magnetic force, current is induced in the wire.
    b. The direction of current depends on the direction of motion that the conductor takes through the magnetic field. (See Figure 11.33.)
30. A sine wave can be generated by rotating a loop conductor through a stationary magnetic field. (See Figure 11.36.)
31. The phase of a given point is its position relative to the start of the waveform, usually given in degrees.
    a. Phase measurements allow us to identify points on a waveform independent of cycle time and frequency.
    b. Phase plays an important role in the instantaneous values of sine wave voltage and current.

32. The ratio of phase ($\theta$) to 360° equals the ratio of instantaneous time $t$ to cycle time $T_C$.

33. There are many cases where a circuit contains two (or more) sine waves of equal frequency that do not reach their peaks at the same time.
    a. Waveforms of this nature are said to be out of phase.
    b. The phase difference between waveforms that are out of phase is referred to as their phase angle.

34. Phase angles can be measured using a dual-trace oscilloscope. (See Example 11.14.)

35. The instantaneous value of a sine wave voltage (or current) is proportional to the peak value of the waveform and the sine of the phase angle.

36. Since instantaneous values are proportional to the sine of the phase angle ($\sin \theta$), the waveform is said to vary at a sinusoidal rate. (This is where the term *sine wave* comes from.)

37. A radian is the angle formed at the center of a circle by two radii and an arc of equal length. (See Figure 11.48.)

38. Regardless of the size of a circle, one radian equals 57.2958°.

39. There are $\pi$ radians in 180° and $2\pi$ radians in 360°.

40. The term *angular velocity* is used to describe:
    a. The speed of rotation for a rotor generating a sine wave.
    b. The rate at which the phase of a sine wave changes and, therefore, the rate at which its instantaneous values change.

41. When calculating instantaneous values, make sure that your calculator is set for the proper mode of operation, as follows:

| **Equation** | **Calculator Mode** |
|---|---|
| $v = V_{pk} \sin \theta$ | Degrees |
| $v = V_{pk} \sin \omega t$ | Radian |

Otherwise, your result will be incorrect.

42. The degree approach to calculating an instantaneous value is used when the phase ($\theta$) is known.

43. The radian approach to calculating an instantaneous value is used when the frequency (or cycle time) and specific time interval are known.

44. Static values are constant. Dynamic values change during the normal operation of a circuit. Dynamic values are always measured under one or more specified conditions.

45. Sine waves sometimes contain a dc offset.

46. A sine wave with a dc offset has:
    a. Unequal positive and negative peak values.
    b. An average value equal to the value of the dc offset (rather than zero).

47. The dc offset in any sine wave can be measured directly with a dc voltmeter.

48. An electromagnetic wave is a waveform that consists of perpendicular electric and magnetic fields. (See Figure 11.57.)

49. Every cycle of a transmitted waveform has a measurable length, called its wavelength. The wavelength ($\lambda$) of one cycle is determined by:
    a. The speed at which the waveform travels through space.
    b. The cycle time of the waveform.

50. All waveforms travel at the velocity of electromagnetic radiation, which is commonly referred to as the speed of light.
    a. The common units of measure are listed in Table 11.2.
    b. The unit of measure for wavelength ($\lambda$) is determined by the unit used to measure its velocity.

51. A harmonic is a whole-number multiple of a given frequency.
    a. The lowest frequency in a harmonic series is referred to as the fundamental frequency.
    b. The remaining frequencies in a harmonic series are referred to as *n*-order harmonics. (See Table 11.3.)
52. When frequency changes by a factor of 2, the change is referred to as one octave.
53. When frequency changes by a factor of 10, the change is referred to as one decade.
54. A rectangular wave is one that alternates between two dc levels.
55. The alternations of a rectangular wave are referred to as the pulse width (PW) and space width (SW). The sum of PW and SW equals cycle time.
56. The duty cycle of a rectangular wave is the ratio of input pulse width to cycle time, given as a percentage. Many digital circuits have restrictions on the duty cycle of any input waveform.
57. A square wave is a special-case rectangular wave that has equal PW and SW values.
58. A square wave can be generated by harmonic synthesis. This process involves adding a fundamental frequency to an infinite number of its odd-order harmonics.
    a. In practice, only the fundamental frequency and the first five odd-order harmonics are required to produce a square wave.
    b. Square waves are not commonly generated using this technique. However, the frequency makeup of square waves is important in filtering.
59. A symmetrical waveform is one with cycles made up of identical half-cycles. There are two types of symmetry: time and amplitude.
60. The value of $V_{\text{ave}}$ for a rectangular waveform can be measured using an oscilloscope. The technique is outlined on page 433.
61. A sawtooth is a waveform that changes constantly at a linear rate.
62. The slope of a sawtooth waveform is the rate at which its value changes.
63. A triangular wave is a symmetrical sawtooth waveform.

| Equation Number | Equation | Section Number | Equation Summary |
|---|---|---|---|
| **(11.1)** | $f = \dfrac{1}{T_C}$ | 11.1 | |
| **(11.2)** | $T_C = \dfrac{1}{f}$ | 11.1 | |
| **(11.3)** | $V_{PP} = 2\,V_{\text{pk}}$ | 11.2 | Equations (11.3) through (11.6) apply to pure ac waveforms only. |
| **(11.4)** | $I_{PP} = 2\,I_{\text{pk}}$ | 11.2 | |
| **(11.5)** | $V_{\text{ave}} = \dfrac{2\,V_{\text{pk}}}{\pi}$ | 11.2 | |
| **(11.6)** | $V_{\text{ave}} \cong 0.637\,V_{\text{pk}}$ | 11.2 | |
| **(11.7)** | $P_L = P_{\text{ac}} \quad \text{when} \quad V_L = \sqrt{\dfrac{V_{\text{pk}}^2}{2}}$ | 11.2 | |
| **(11.8)** | $V_{\text{rms}} = \sqrt{\dfrac{V_{\text{pk}}^2}{2}}$ | 11.2 | |
| **(11.9)** | $V_{\text{rms}} = 0.707\,V_{\text{pk}}$ | 11.2 | |

| | | |
|---|---|---|
| **(11.10)** | $I_{rms} = 0.707\ I_{pk}$ | 11.2 |
| **(11.11)** | $V_{dc} = \dfrac{2\ V_{pk}}{\pi} \cong 0.637\ V_{pk}$ | 11.2 |
| **(11.12)** | $V_{dc} = \dfrac{V_{pk}}{\pi} \cong 0.318\ V_{pk}$ | 11.2 |
| **(11.13)** | $I_{dc} = \dfrac{2\ I_{pk}}{\pi} \cong 0.637\ I_{pk}$ | 11.2 |
| **(11.14)** | $I_{dc} = \dfrac{I_{pk}}{\pi} \cong 0.318\ I_{pk}$ | 11.2 |
| **(11.15)** | $\dfrac{\theta}{360°} = \dfrac{t}{T_C}$ | 11.3 |
| **(11.16)** | $t = T_C \dfrac{\theta}{360°}$ | 11.3 |
| **(11.17)** | $\theta = (360°)\dfrac{t}{T_C}$ | 11.3 |
| **(11.18)** | $v = V_{pk} \sin \theta$ | 11.3 |
| **(11.19)** | $\omega = 2\pi f$ | 11.3 |
| **(11.20)** | $v = V_{pk} \sin \omega t$ | 11.3 |
| **(11.21)** | $V_{offset} = \dfrac{+V_{pk} + (-V_{pk})}{2}$ | 11.4 |
| **(11.22)** | $\lambda = cT_C$ | 11.4 |
| **(11.23)** | $\lambda = \dfrac{c}{f}$ | 11.4 |
| **(11.24)** | $f = 2^n f_0$ | 11.4 |
| **(11.25)** | $f = 10^n f_0$ | 11.4 |
| **(11.26)** | $\text{duty cycle (\%)} = \dfrac{\text{PW}}{T_C} \times 100$ | 11.4 |
| **(11.27)** | $V_{ave} = \dfrac{+V_{pk} + (-V_{pk})}{2}$ | 11.4 |
| **(11.28)** | $V_{ave} = \dfrac{(\text{PW})(+V_{pk}) + (\text{SW})(-V_{pk})}{T_C}$ | 11.4 |

## Key Terms

The following terms were introduced and defined in this chapter:

alternating current (ac)
amplifier
angular velocity
average ac power
cycle time
dc offset
decade
digital circuit
dual-trace oscilloscope
duty cycle
dynamic value
electromagnetic wave
frequency
full-cycle average
full-wave rectifier
fundamental frequency
half-cycle average
half-wave rectifier

harmonic
harmonic series
harmonic synthesis
magnitude
mean
$n$-order harmonic
octave
oscilloscope
peak-to-peak value
peak value
phase
phase angle
pulse width
pure ac waveform
radian
ramp
rectangular wave
rectifier
root-mean-square (rms)
sawtooth wave
signal
sine wave
sinusoidal waveform
space width
square wave
static value
symmetrical waveform
time base
triangular waveform
vertical sensitivity
wavelength

## Practice Problems

1. Calculate the cycle time of the waveform in Figure 11.68a.
2. Calculate the cycle time of the waveform in Figure 11.68b.
3. Calculate the cycle time of the waveform in Figure 11.69a.
4. Calculate the cycle time of the waveform in Figure 11.69b.
5. Calculate the frequency of each waveform in Figure 11.68.
6. Calculate the frequency of each waveform in Figure 11.69.
7. Calculate the frequency of the waveform in Figure 11.70a.
8. Calculate the frequency of the waveform in Figure 11.70b.

**FIGURE 11.68**

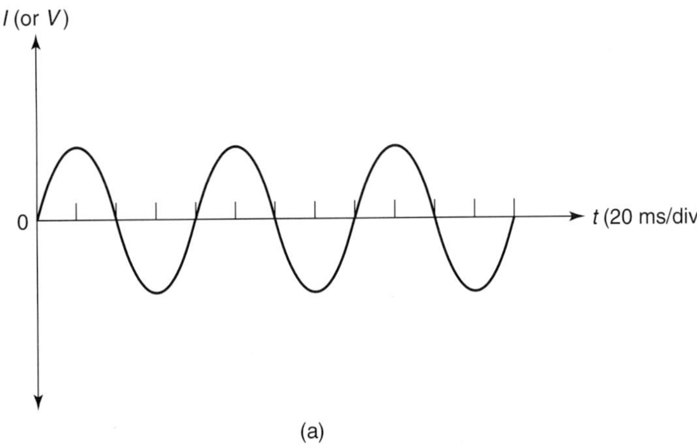


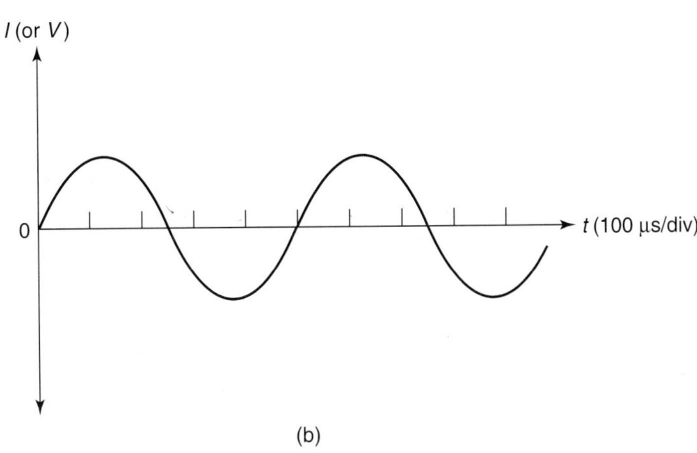

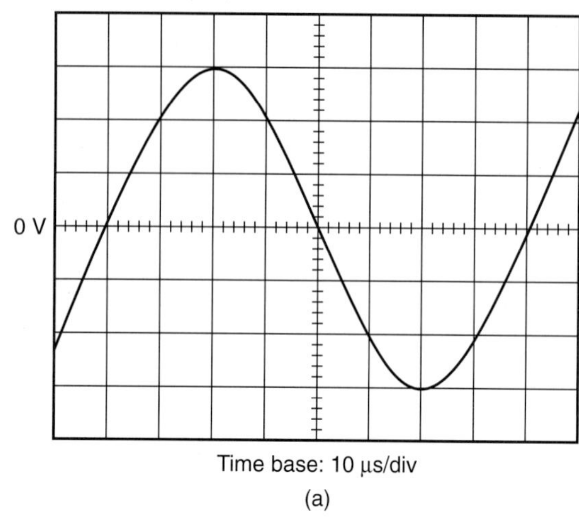


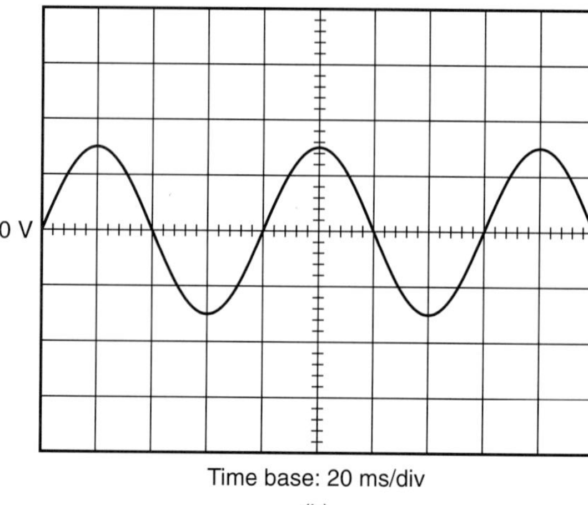


**FIGURE 11.69**

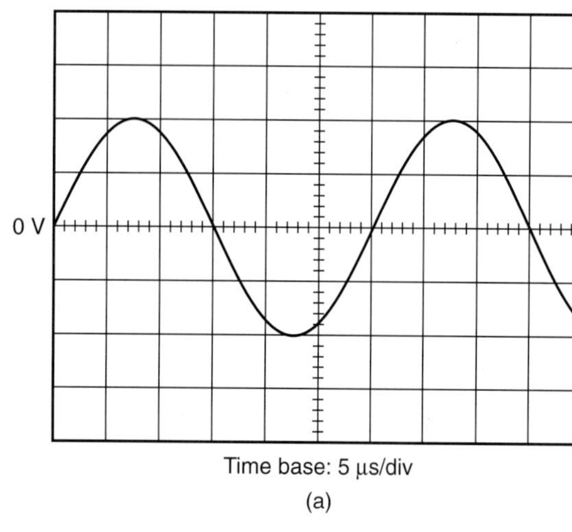


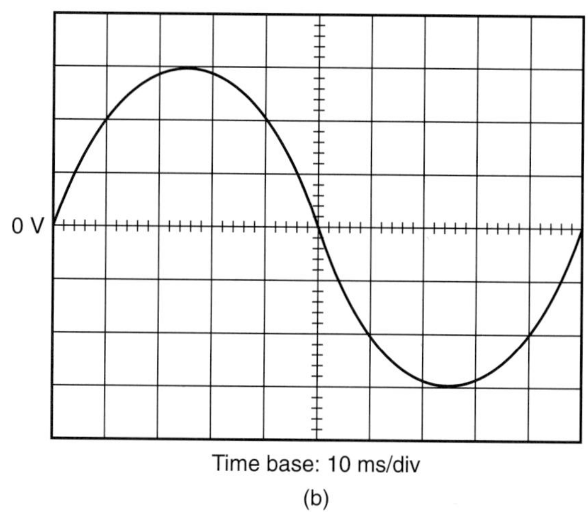


**FIGURE 11.70**

**9.** Calculate the cycle time that corresponds to each of the following frequencies:

**a.** 500 Hz

**b.** 2.5 kHz

**c.** 1 MHz

**d.** 75 kHz

**10.** Calculate the range of cycle time values for each of the following frequency ranges:

**a.** 100 Hz to 300 Hz

**b.** 1 kHz to 5 kHz

**c.** 20 Hz to 20 kHz

**d.** 1.5 MHz to 10 MHz

**11.** A sine wave has peak values of $\pm 12\ V_{pk}$. Calculate the waveform's half-cycle average voltage.

**12.** A sine wave has peak values of $\pm 20\ V_{pk}$. Calculate the waveform's half-cycle average voltage.

**13.** Determine the peak-to-peak and half-cycle average voltages for the waveform in Figure 11.71a.

FIGURE 11.71

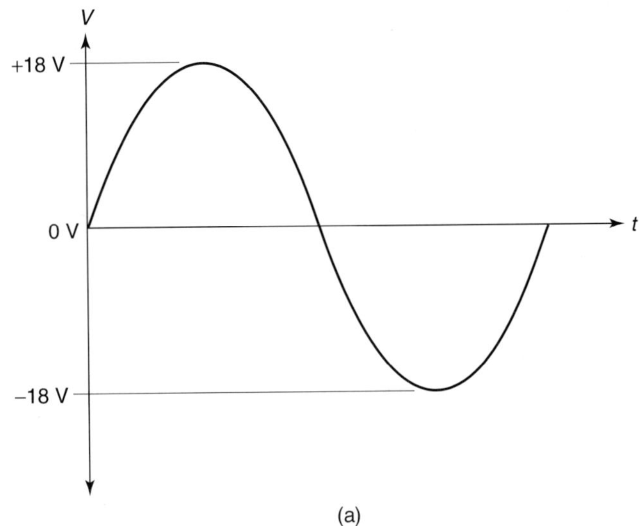

(a)

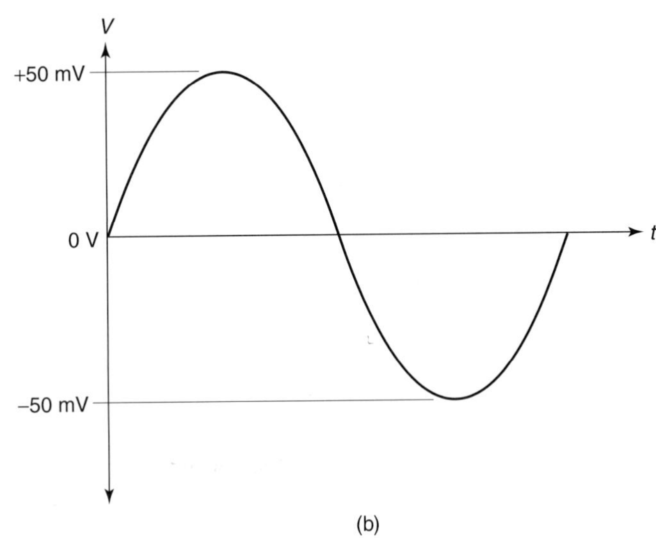

(b)

14. Determine the peak-to-peak and half-cycle average voltages for the waveform in Figure 11.71b.
15. Calculate the rms values of load voltage and current for the circuit in Figure 11.72a.
16. Calculate the rms values of load voltage and current for the circuit in Figure 11.72b.
17. Calculate the rms values of load voltage and current for the circuit in Figure 11.73a.
18. Calculate the rms values of load voltage and current for the circuit in Figure 11.73b.
19. Calculate the load power for the circuit in Figure 11.74a.
20. Calculate the load power for the circuit in Figure 11.74b.

FIGURE 11.72

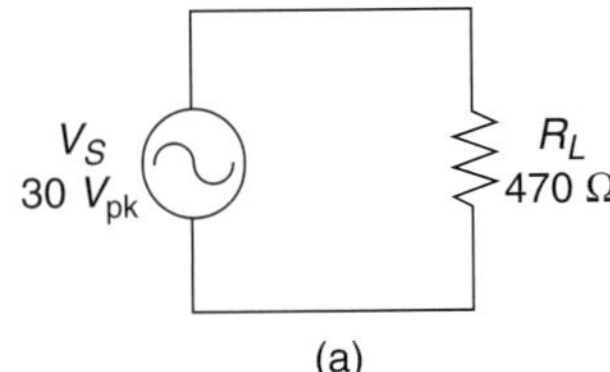

(a)

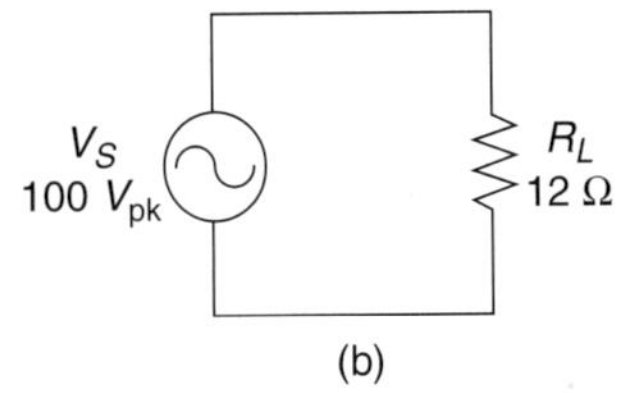

(b)

FIGURE 11.73

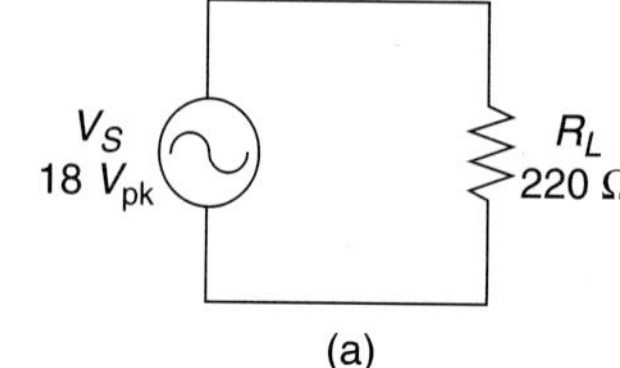


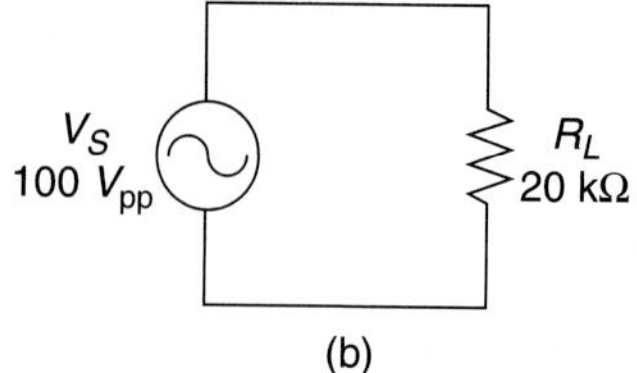


FIGURE 11.74

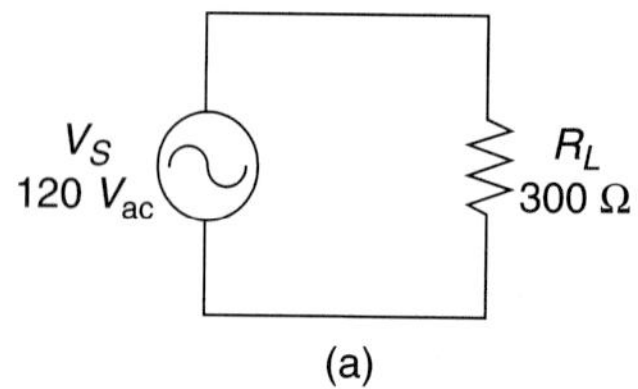


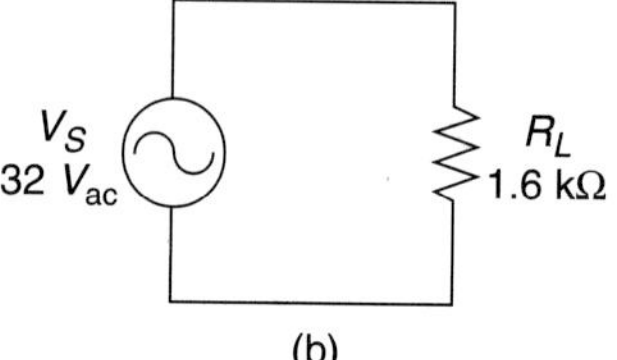


**21.** Determine the peak and peak-to-peak values of the waveform in Figure 11.69a. Assume that the vertical sensitivity setting of the oscilloscope is 50 mV/div.

**22.** Determine the peak and peak-to-peak values of the waveform in Figure 11.69b. Assume that the vertical sensitivity setting of the oscilloscope is 100 μV/div.

**23.** Determine the peak and peak-to-peak values of the waveform in Figure 11.70a. Assume that the vertical sensitivity setting of the oscilloscope is 10 V/div.

**24.** Determine the peak and peak-to-peak values of the waveform in Figure 11.70b. Assume that the vertical sensitivity setting of the oscilloscope is 200 mV/div.

**25.** For a 20 kHz sine wave:

**a.** $\theta = 30°$ when $t =$ ______.

**b.** $\theta = 150°$ when $t =$ ______.

**c.** $\theta =$ ______ when $t = 25$ μs.

**d.** $\theta =$ ______ when $t = 10$ μs.

**26.** For a 1.8 kHz sine wave:

**a.** $\theta = 200°$ when $t =$ ______.

**b.** $\theta = 15°$ when $t =$ ______.

**c.** $\theta =$ ______ when $t = 300$ μs.

**d.** $\theta =$ ______ when $t = 55$ μs.

**27.** Complete the table below.

| *Frequency* | *Cycle Time* | *Instantaneous Time (t)* | *Phase (θ)* |
|---|---|---|---|
| ______ | 10 ms | 2 ms | ______ |
| 120 kHz | ______ | ______ | 60° |
| ______ | 45 ms | ______ | 100° |
| ______ | ______ | 10 ms | 90° |

**28.** Complete the table below.

| *Frequency* | *Cycle Time* | *Instantaneous Time (t)* | *Phase (θ)* |
|---|---|---|---|
| ______ | 100 μs | 40 μs | ______ |
| 2 MHz | ______ | ______ | 200° |
| ______ | 50 ms | ______ | 300° |
| ______ | ______ | 1.25 ms | 60° |

FIGURE 11.75

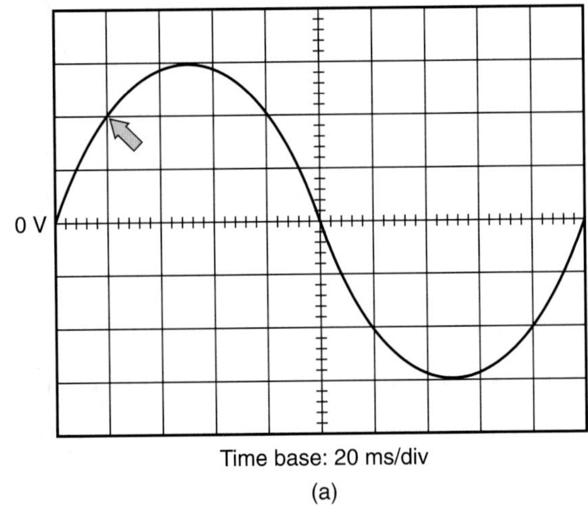


Time base: 20 ms/div

(a)

0 V

Time base: 5 μs/div

(b)

FIGURE 11.76

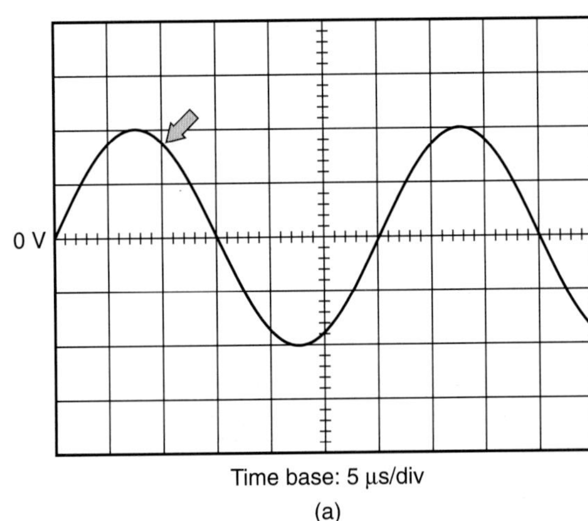


Time base: 5 μs/div

(a)

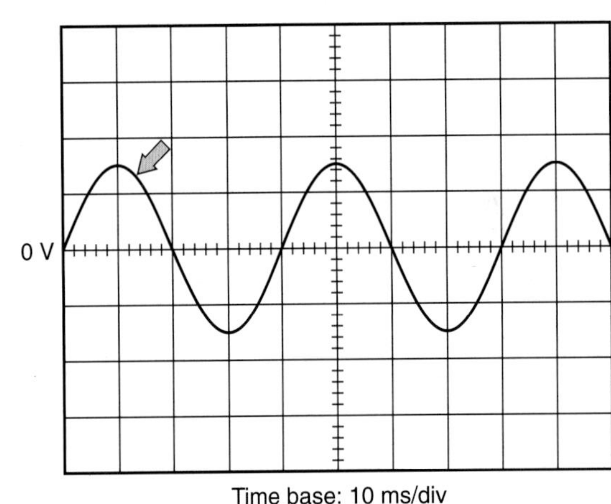


Time base: 10 ms/div

(b)

29. Determine the value of θ for the highlighted point in Figure 11.75a.
30. Determine the value of θ for the highlighted point in Figure 11.75b.
31. Determine the value of θ for the highlighted point in Figure 11.76a.
32. Determine the value of θ for the highlighted point in Figure 11.76b.
33. Determine the phase angle between the sine waves in Figure 11.77a.
34. Determine the phase angle between the sine waves in Figure 11.77b.

FIGURE 11.77

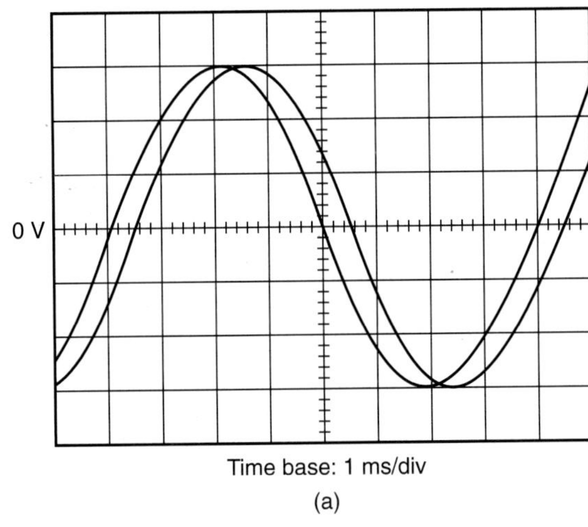


Time base: 1 ms/div

(a)

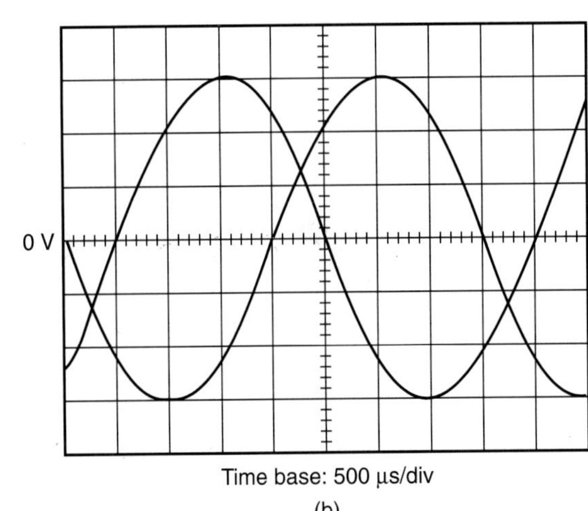


Time base: 500 μs/div

(b)

**FIGURE 11.78**

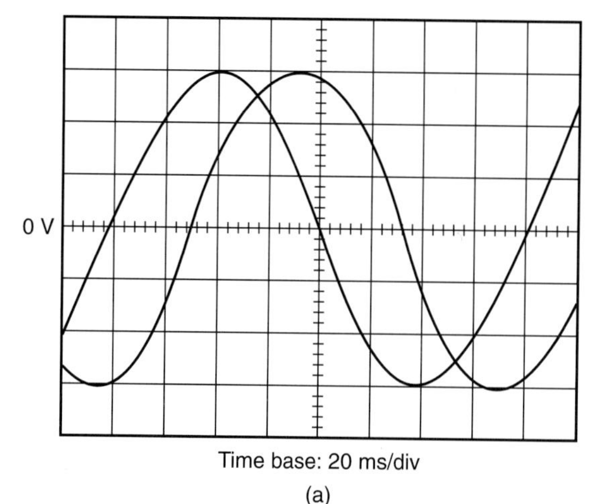

(a)

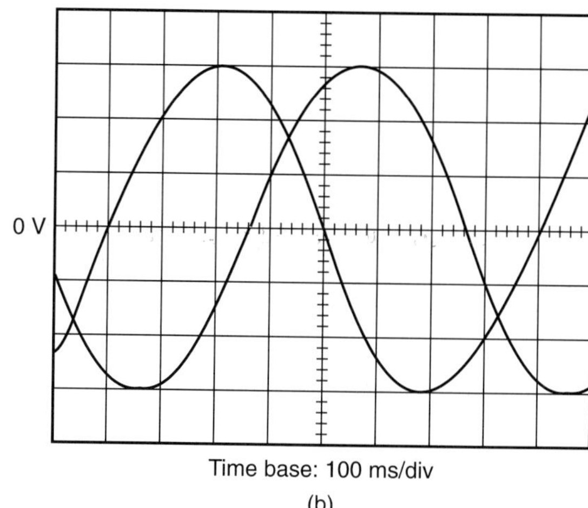

(b)

**35.** Determine the phase angle between the sine waves in Figure 11.78a.

**36.** Determine the phase angle between the sine waves in Figure 11.78b.

**37.** For a $\pm 15\ V_{pk}$ sine wave:

**a.** $v =$ ______ when $\theta = 60°$.

**b.** $v =$ ______ when $\theta = 150°$.

**c.** $v =$ ______ when $\theta = 240°$.

**d.** $v =$ ______ when $\theta = 350°$.

**38.** For a $\pm 170\ V_{pk}$ sine wave:

**a.** $v =$ ______ when $\theta = 45°$.

**b.** $v =$ ______ when $\theta = 190°$.

**c.** $v =$ ______ when $\theta = 10°$.

**d.** $v =$ ______ when $\theta = 180°$.

**39.** A sine wave has the following values: $T_C = 200\ \mu s$ and $V_{pk} = 15$ V. Determine the instantaneous voltage when $t = 120\ \mu s$.

**40.** A sine wave has the following values: $T_C = 16$ ms and $V_{pk} = 170$ V. Determine the instantaneous voltage when $t = 4$ ms.

**41.** A sine wave has the following values: $f = 1.5$ MHz and $V_{pk} = 50$ mV. Determine the instantaneous voltage when $t = 67$ ns. (Use the degree method.)

**42.** A sine wave has the following values: $f = 200$ kHz and $V_{pk} = 5$ V. Determine the instantaneous voltage when $t = 2.5\ \mu s$. (Use the degree method.)

**43.** Rework Problem 41 using the radian method.

**44.** Rework Problem 42 using the radian method.

**45.** The waveform in Figure 11.79a has the values shown. Calculate the value of $v_1$ using both the degree and radian methods.

**46.** The waveform in Figure 11.79b has the values shown. Calculate the value of $v_1$ using both the degree and radian methods.

**47.** The waveforms in Figure 11.80a are 45° out of phase. Determine the instantaneous value of waveform A when waveform B is at 0°.

**48.** The waveforms in Figure 11.80b are 60° out of phase. Determine the instantaneous value of waveform A when waveform B is at 20°.

**49.** A sine wave has peak values of +12 V and −18 V. Determine the dc offset in the waveform.

**50.** A sine wave has peak values of +32 V and −2 V. Determine the dc offset in the waveform.

**FIGURE 11.79**

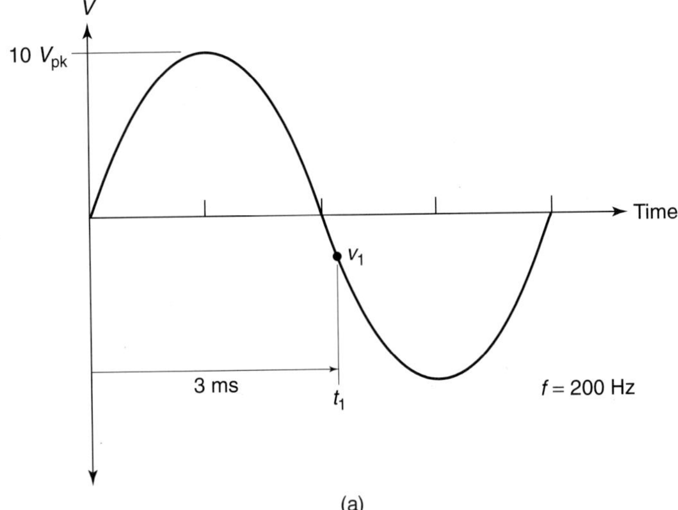


(a)

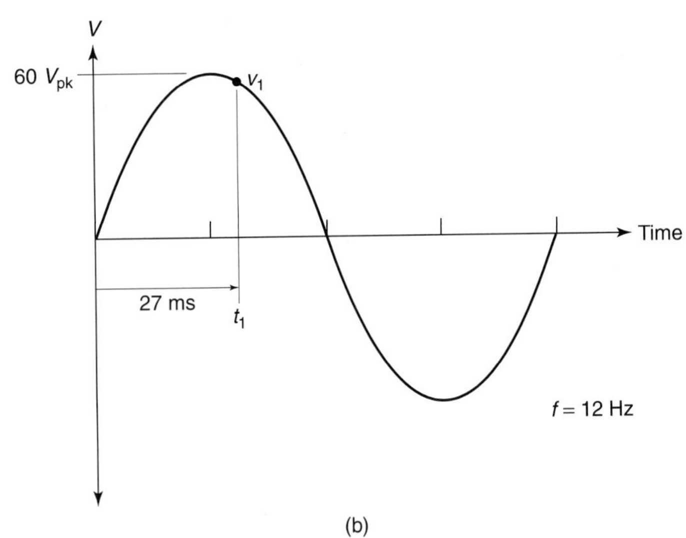


(b)

51. A sine wave has peak values of −8 V and −22 V. Determine the dc offset in the waveform.
52. A sine wave has peak values of +48 V and +12 V. Determine the dc offset in the waveform.
53. Calculate the wavelength (in centimeters) for a 300 MHz sine wave.
54. Calculate the wavelength (in meters) for a 220 kHz sine wave.
55. Calculate the wavelength (in miles) for a 60 Hz sine wave.
56. Calculate the wavelength (in meters) for a 500 kHz sine wave.
57. A sine wave has a frequency range of 15 kHz to 55 kHz. Calculate the range of wavelength values (in miles) for the waveform.
58. A sine wave has a frequency range of 30 MHz to 100 MHz. Calculate the range of wavelength values (in meters) for the waveform.
59. Determine the frequency that is 4 octaves above 1 kHz.
60. Determine the frequency that is 12 octaves above 300 Hz.
61. Determine the frequency that is 1.5 decades above 12 kHz.
62. Determine the frequency that is 3 decades above 10 Hz.

**FIGURE 11.80**

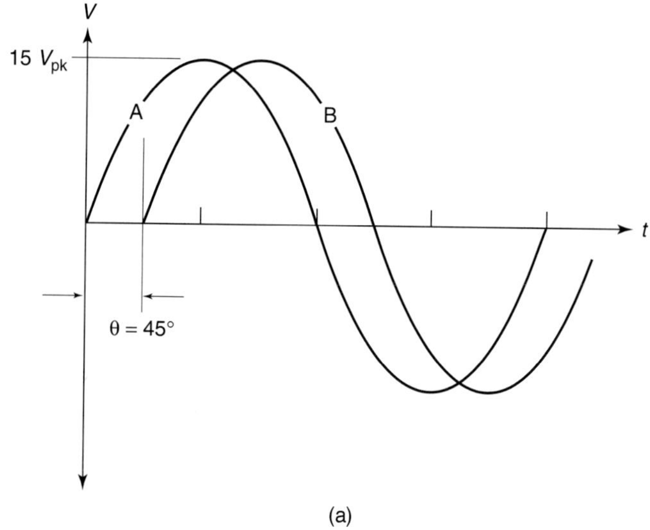


(a)

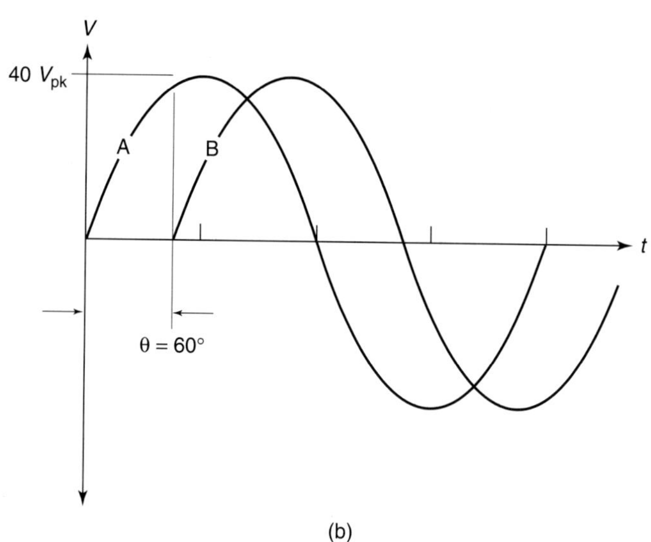


(b)

**63.** Determine the duty cycle of the waveform in Figure 11.81a.

**64.** Determine the duty cycle of the waveform in Figure 11.81b.

**65.** Determine the duty cycle of the waveform in Figure 11.82a.

**66.** Determine the duty cycle of the waveform in Figure 11.82b.

**67.** Determine the average value of the waveform in Figure 11.83a.

**68.** Determine the average value of the waveform in Figure 11.83b.

**69.** Determine the average value of the waveform in Figure 11.84a.

**70.** Determine the average value of the waveform in Figure 11.84b.

FIGURE 11.81

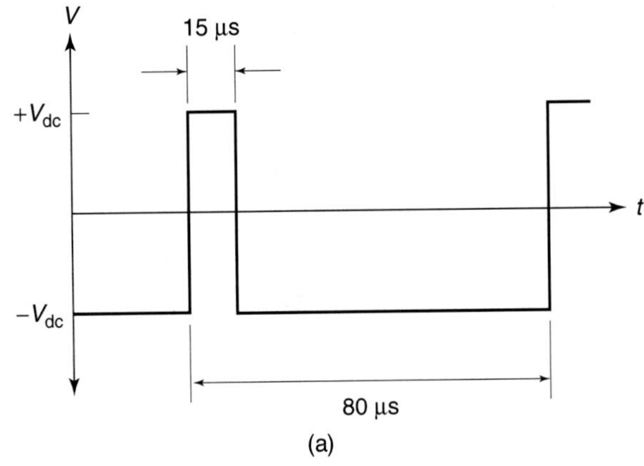

(a)

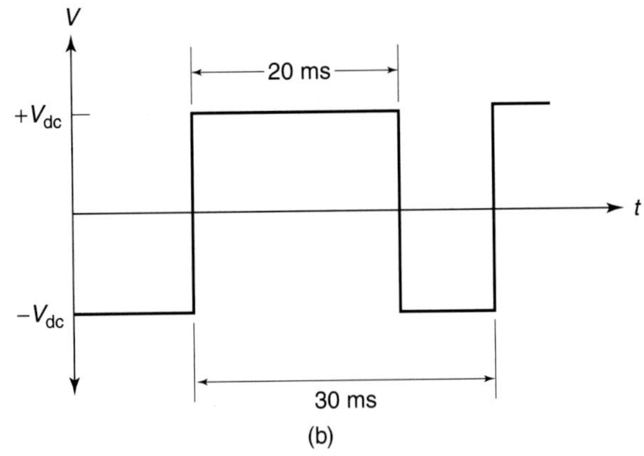

(b)

FIGURE 11.82

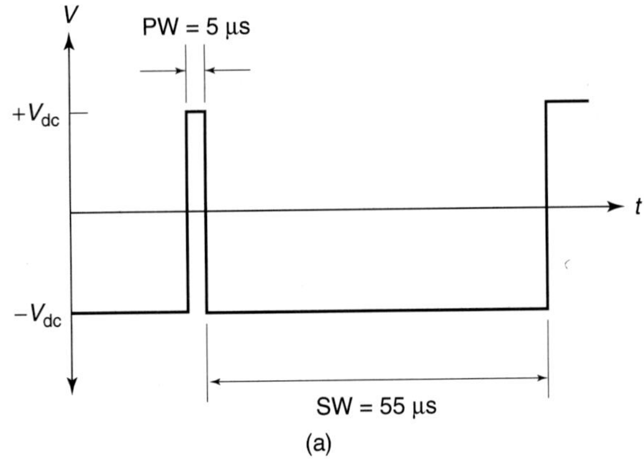

(a)

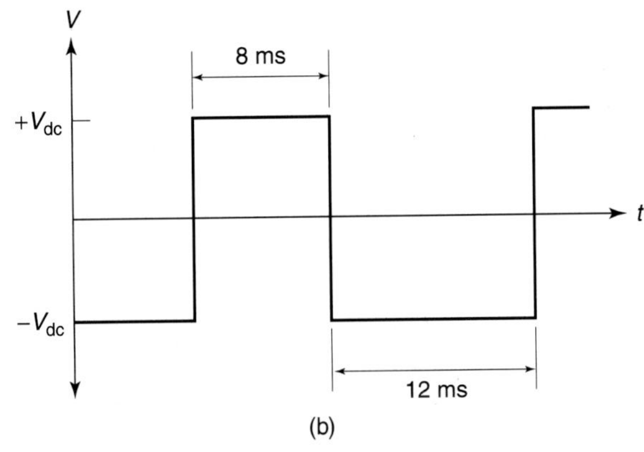

(b)

**FIGURE 11.83**

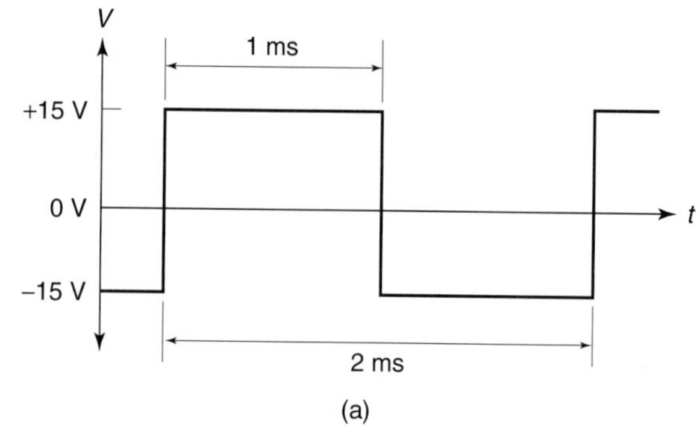

(a)

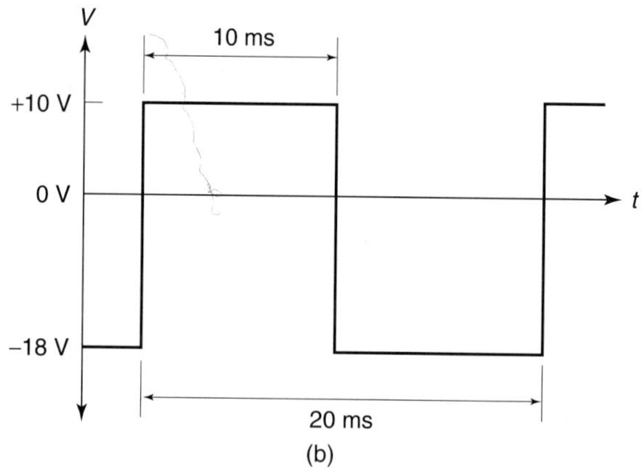

(b)

**FIGURE 11.84**

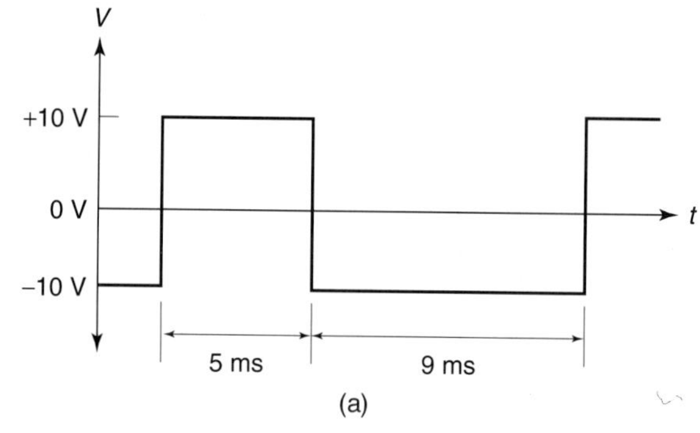

(a)

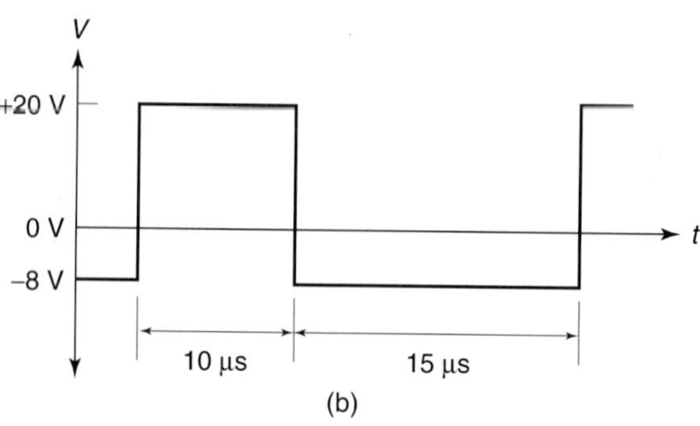

(b)

## The Brain Drain

71. A sine wave has the following values: $V_{pk} = +12$ V and $v = +4.8$ V when $t = 50$ μs. Determine the two lowest possible frequencies of the waveform. [*Hint:* $\sin \theta = \sin(180° - \theta)$.]

72. A sine wave takes 50 μs to reach a phase of 60°. Calculate the wavelength (in meters) of the waveform.

73. A sine wave has a value of $T_C = 25$ μs. Determine the wavelength (in centimeters) of its 5th-order harmonic.

74. A rectangular waveform has a 40% duty cycle and a space width of 60 μs. Determine its wavelength (in centimeters).

75. Determine the negative peak value of the waveform in Figure 11.85.

**FIGURE 11.85**

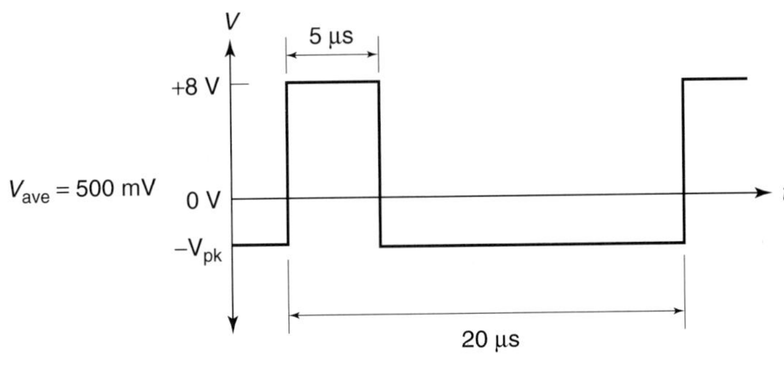


## Answers to the Example Practice Problems

**11.1.** 150 μs
**11.2.** 25 Hz
**11.3.** 1 ms
**11.4.** 10 ms to 667 μs
**11.5.** 5 kHz
**11.6.** 15.28 V
**11.7.** $V_{rms} = 21.21$ V, $I_{rms} = 19.28$ mA
**11.8.** 33.3 mW
**11.9.** 600 m$V_{pk}$
**11.10.** 25 $V_{PP}$
**11.12.** 1.5 ms
**11.13.** Approximately 203°
**11.14.** 90°
**11.15.** 10.4 V
**11.16.** −10.26 V
**11.17.** 17.1 V
**11.18.** For both methods, $v_1 = 28.8$ V
**11.19.** $v = 9.96$ V ($\theta = 85°$)
**11.20.** −5.47 V
**11.21.** 133 cm, 4.37 ft, 52.4 in.
**11.22.** 4.92 to 3.94 ft
**11.23.** 40%

## EWB Applications Problems

| Figure | EWB File Reference |
|---|---|
| 11.2 | EWBA11_2.ewb |
| 11.24 | EWBA11_24.ewb |
| 11.28 | EWBA11_28.ewb |
| 11.30 | EWBA11_30.ewb |
| 11.44 | EWBA11_44.ewb |
| 11.56 | EWBA11_56.ewb |
| 11.64 | EWBA11_64.ewb |

12

# POLAR AND RECTANGULAR NOTATION

## OBJECTIVES

*After studying the material in this chapter, you should be able to:*

1. Compare and contrast vector and scalar quantities.
2. Describe how magnitude and direction are represented using a vector.
3. Describe the polar notation method of expressing the value of a vector.
4. Express any angle as either a positive or a negative value.
5. Multiply and divide polar values.
6. Describe the processes used to add and subtract polar values.
7. Describe the coordinate notation method of expressing the value of a point.
8. Describe the rectangular notation method of expressing the value of a point.
9. Discuss the meaning of the *j*-operator.
10. State the result of any multiple *j*-operation.
11. Add and subtract complex numbers.
12. Multiply and divide complex numbers.
13. Identify the real and imaginary components of complex numbers.
14. Explain the concept of a trig function as a ratio of line lengths.
15. Convert any polar value to rectangular form.
16. Convert any rectangular value to polar form.

You were told in Chapter 11 that many ac circuit calculations involve phase angles measured in either degrees or radians. To perform calculations involving angles, we need to use more complex methods of representing values. In this chapter, we will discuss two methods for representing values containing angles: polar notation and rectangular notation. We will also take a look at basic operations involving polar and rectangular notation.

# 12.1 VECTORS

**Vector**
Any value that has both magnitude and an angle (or direction).

**Scalar**
A value that has only magnitude.

In Chapter 11, you were shown that some ac values have both a magnitude and an angle. For example, the instantaneous values shown on the sine wave in Figure 12.1a all have

*OBJECTIVE 1* ▶ magnitude and phase (an angle). Any value that has both magnitude and an angle (or direction) is referred to as a **vector**. Each of the instantaneous voltages in Figure 12.1a is a vector. In contrast, a value that has magnitude only is referred to as a **scalar**. The value of $R_L = 1\ \text{k}\Omega$ in Figure 12.1b is a scalar because resistance has no angle (or direction) associated with it.

It should be noted that both the magnitude and direction of a vector must always be specified. If the angle is omitted, the value is meaningless. (It would be like telling someone how far he or she needs to go to get somewhere without specifying which direction to take.)

## Plotting a Vector

*OBJECTIVE 2* ▶ Any vector can be plotted as a line on a graph. For example, let's say that we want to plot a line representing $v_1$ in Figure 12.2a. This line would be plotted as shown in Figure 12.2b. As the figure demonstrates:

1. The length ($\ell$) of the line represents the magnitude of $v_1$.
2. The angle between the line and the positive horizontal ($\theta$) represents the phase of $v_1$.

This is the standard approach to graphing the values associated with any vector.

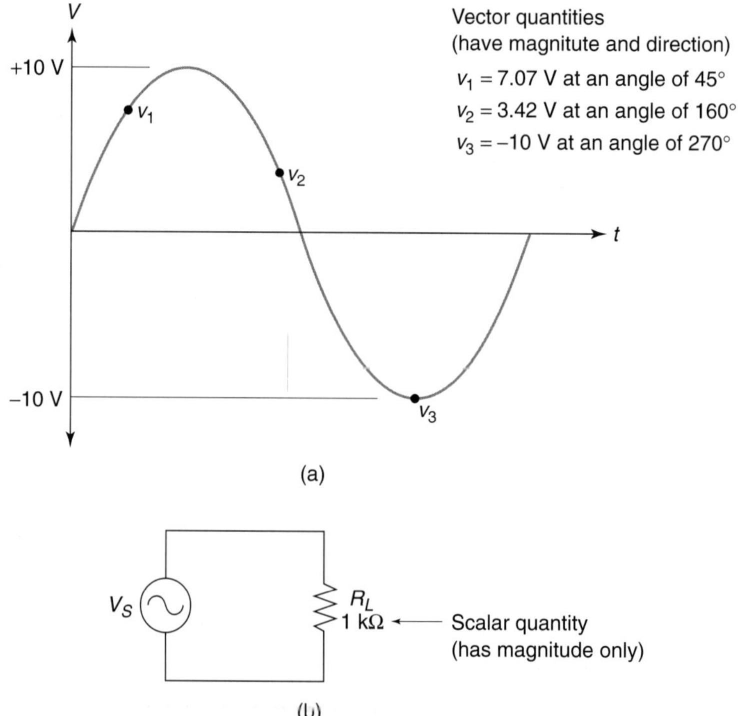


**FIGURE 12.1 Examples of vector and scalar quantities.**

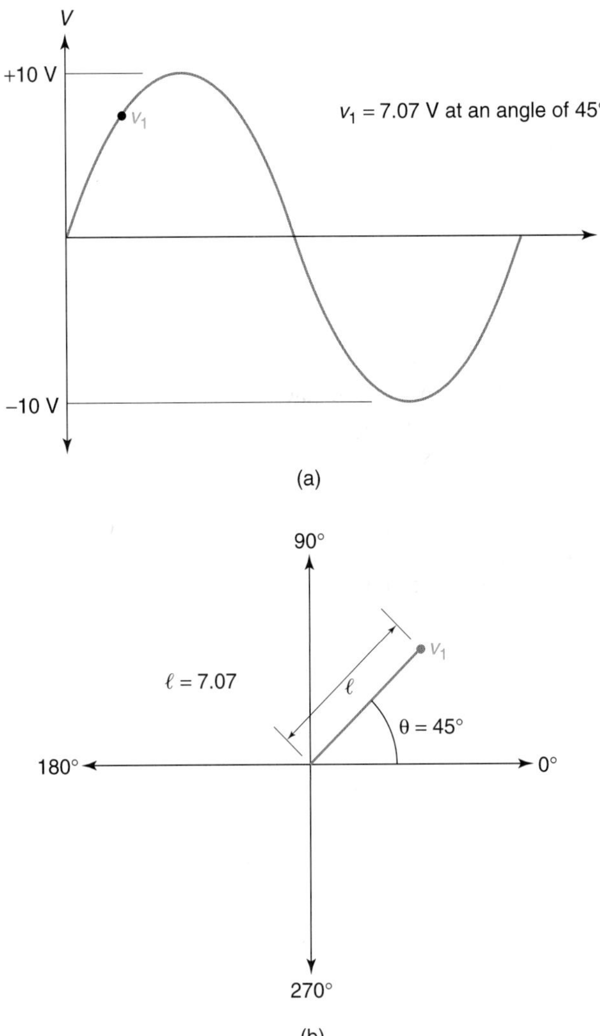


**FIGURE 12.2 Vector representation of a point.**

There is an important point that needs to be made at this time. According to *Webster's New World Dictionary,* a vector is:

1. A physical quantity with both magnitude and direction.
2. A directed line segment representing such a physical quantity.

The first definition describes $v_1$ in Figure 12.2a, while the second definition describes the line drawn in Figure 12.2b. This may seem confusing at first, but the point and the line represent the same value, which is why the term can be applied to both. To avoid any confusion, we will use the terms *vector quantity* to refer to a point (like $v_1$) and *vector* to refer to the line representing the point.

## Representing Vector Quantities

Many of the values in ac circuit calculations are vector quantities. As such, we need to be able to indicate clearly both their magnitude and direction. One method would be to draw their vectors as shown in Figure 12.2b. However, you'd find this extremely tiresome after a while.

In the next two sections, you will be introduced to several vector notations. Each of these provides a method of representing vector quantities without plotting the actual vectors. The first vector notation we will discuss is referred to as polar notation.

## Section Review

1. What is a vector quantity?
2. What is a scalar quantity?
3. When writing a vector quantity, what must be specified?
4. Describe the process for plotting a vector.

# 12.2 POLAR NOTATION

*OBJECTIVE 3* ▶

**Polar notation**
A method of expressing the value of a vector as a magnitude followed by an angle.

In **polar notation,** the value of a vector is expressed as a magnitude followed by an angle. Figure 12.3 shows several vectors. The polar notation for each vector appears with the graph.

Figure 12.3a shows a vector that is 8 units in length at a 45° angle from the positive horizontal (0° line). Therefore, its value is expressed as

$$v = 8\angle 45^\circ$$

(8 = Magnitude; 45° = Angle)

This is the form of any value written in polar notation.

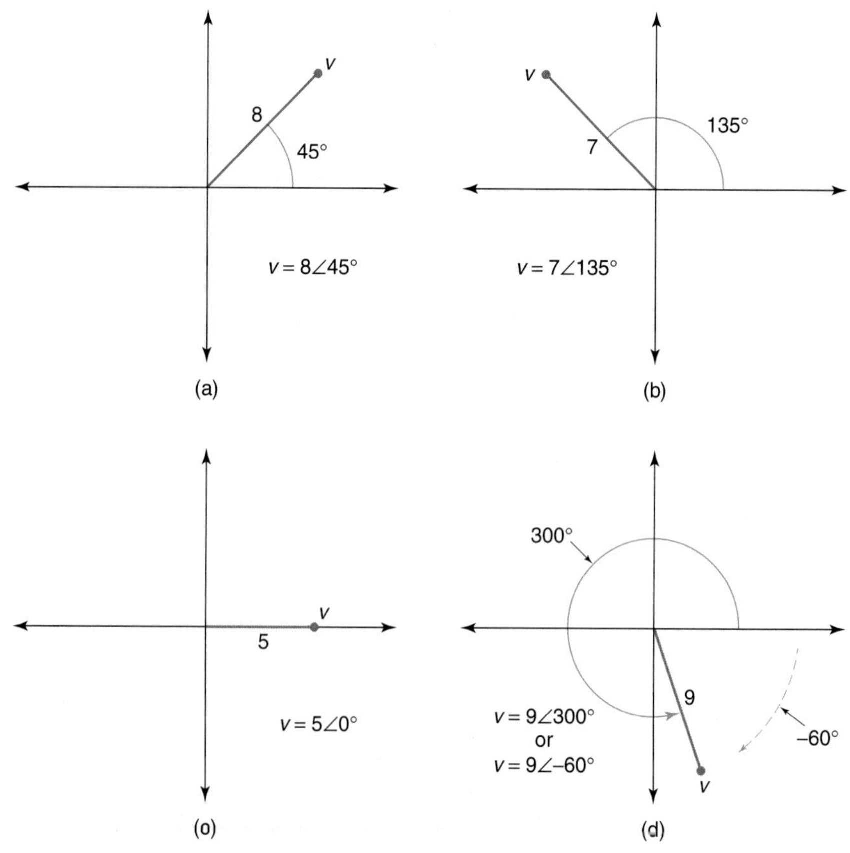


**FIGURE 12.3 Vectors expressed in polar notation.**

EWB

There are a few points that need to be made regarding the vectors in Figure 12.3. In Figure 12.3c, the vector is positioned on the positive horizontal of the graph. When this is the case, the vector is normally expressed as:

$$v = 5\angle 0° \qquad \text{or} \qquad v = 5$$

This demonstrates the only case where an angle is *not* specified in polar notation. When this occurs, the value of the angle is automatically assumed to be 0°.

Figure 12.3d demonstrates the fact that there are two possible representations for any angle. A positive angle indicates that the vector angle is measured in a counterclockwise direction from the positive horizontal. For the vector shown, this angle is 300°. A negative angle indicates that the vector angle is measured in a clockwise direction from the positive horizontal. For the vector shown, this angle is −60°. Note that the difference between the angles is 360°, the value of one complete rotation. ◄ *OBJECTIVE 4*

At this point, we will look at some basic mathematical operations involving polar values. Our purpose is to establish some procedures. In upcoming chapters, you will encounter various applications for these procedures.

## Multiplying and Dividing Polar Values

When multiplying one polar value by another, the result equals the product of the magnitudes followed by the sum of the angles. By formula, ◄ *OBJECTIVE 5*

$$(x\angle\theta_1)(y\angle\theta_2) = (xy)\angle(\theta_1 + \theta_2) \tag{12.1}$$

where $x, y$ = the magnitudes

$\theta_1, \theta_2$ = the angles

Example 12.1 demonstrates the relationship given in equation (12.1).

**EXAMPLE 12.1**

Determine the product of $8\angle 35°$ and $12\angle 45°$.

***Solution:*** Using equation (12.1), the product of the values is found as

$$(8\angle 35°)(12\angle 45°) = (8 \times 12)\angle(35° + 45°) = \mathbf{96\angle 80°}$$

***PRACTICE PROBLEM 12.1***

Determine the product of the following values: $6\angle 12°$, $5\angle 118°$, and $2\angle -20°$.

When dividing one polar value by another, the result equals the quotient of the magnitudes followed by the difference between the angles. By formula,

$$\frac{x\angle\theta_1}{y\angle\theta_2} = \frac{x}{y}\angle(\theta_1 - \theta_2) \tag{12.2}$$

Equation (12.2) may appear complicated, but all it says is that you divide the magnitudes and subtract the angles. This is demonstrated in Example 12.2.

**EXAMPLE 12.2**

Determine the quotient of the following fraction:

$$\frac{63\angle 60°}{12\angle 25°}$$

***Solution:*** Using the relationship given in equation (12.2), the fraction is solved as follows:

$$\frac{\mathbf{63\angle 60°}}{\mathbf{12\angle 25°}} = \frac{63}{12}\angle(60° - 25°) = \mathbf{5.25\angle 35°}$$

***PRACTICE PROBLEM 12.2***

Determine the quotient of the following fraction:

$$\frac{18\angle 120°}{100\angle -30°}$$

Example 12.3 shows how equations (12.1) and (12.2) can be used together to solve more complex problems.

## EXAMPLE 12.3

Simplify the following fraction:

$$\frac{(5\angle 30°)(12\angle 60°)}{(15\angle 40°)(5\angle 5°)}$$

***Solution:*** Using equations (12.1) and (12.2), the fraction is simplified as follows:

$$\frac{\mathbf{(5\angle 30°)(12\angle 60°)}}{\mathbf{(15\angle 40°)(5\angle 5°)}} = \frac{(5 \times 12)\angle(30° + 60°)}{(5 \times 15)\angle(40° + 5°)}$$

$$= \frac{60\angle 90°}{75\angle 45°}$$

$$= \frac{60}{75}\angle(90° - 45°)$$

$$= \mathbf{0.8\angle 45°}$$

***PRACTICE PROBLEM 12.3***

Simplify the following fraction:

$$\frac{(18\angle 35°)(5\angle -15°)}{(3\angle 12°)(10\angle -32°)}$$

## Adding and Subtracting Polar Values

*OBJECTIVE 6* ▶

*An Alternate Approach:* Many people find another approach to adding graphed vectors simpler than the method shown in Figure 12.4. A line is drawn from the end of each vector to form a parallelogram. The sum vector is then drawn from the origin of the graph (0, 0) to the opposite angle of the parallelogram.

Polar addition is more complicated than polar multiplication or division. The procedure for adding two polar values is illustrated in Figure 12.4. Figure 12.4a shows two polar values and the vectors they represent. To add the values, vector B is positioned so that it begins at the end of vector A. The sum vector is then drawn as shown in Figure 12.4b. The length of the sum vector equals the magnitude of the result. The angle of the sum vector (which is measured as shown) equals the angle of the result.

Polar subtraction is nearly identical to addition. The only difference is that the vector for the subtrahend is inverted. Polar subtraction is illustrated in Figure 12.5. The vectors used are the same as those in Figure 12.4. Figure 12.5b shows how the vectors are combined to subtract vector B from vector A. Since vector B is the subtrahend, it is inverted (meaning that it is rotated by 180°). The inverted vector is then added to vector A, and the

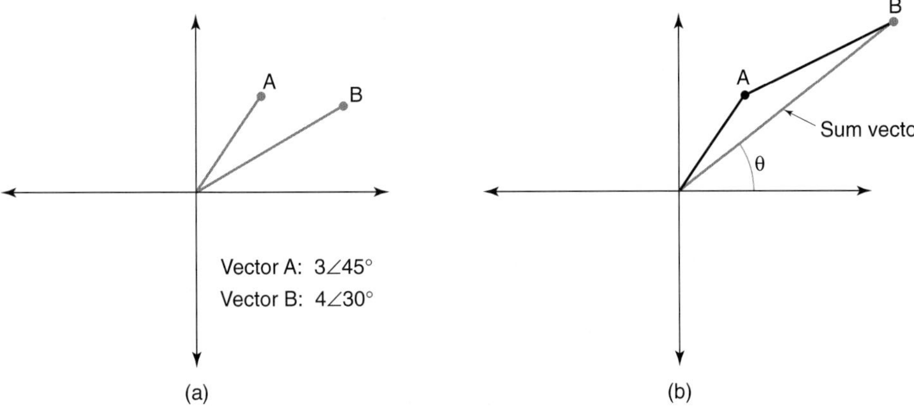


**FIGURE 12.4 Vector addition.**

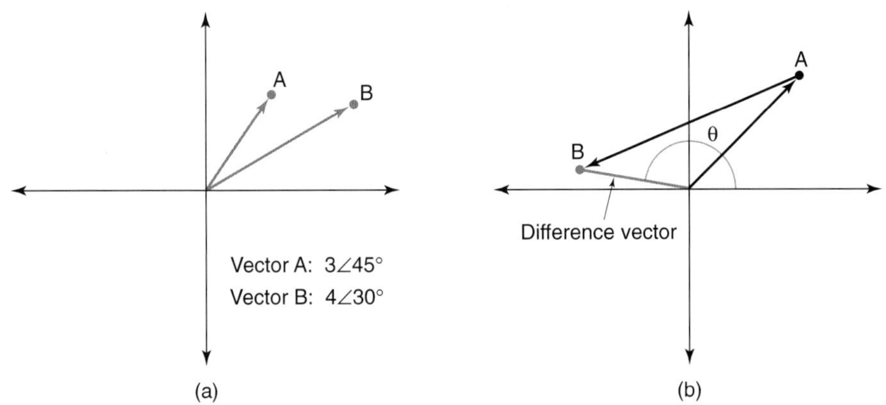


**FIGURE 12.5 Vector subtraction.**

difference vector is plotted. The length of the difference vector equals the magnitude of the result, and the angle relative to the positive horizontal is the angle of the result.

Obviously, polar addition and subtraction are far more involved than multiplication or division. There is another approach to adding and subtracting vector quantities that is far simpler. This approach involves the use of rectangular notation. As you will see, vector quantities expressed in rectangular notation can be added directly and easily.

*An Alternate Approach:* The alternate approach described for addition can be used for subtraction as well. After the subtrahend vector is rotated by 180°, the parallelogram is drawn. The difference vector is then drawn just as the sum vector is drawn for addition.

### Section Review

1. Describe the process for multiplying two polar values.
2. Describe the process for dividing one polar value by another.
3. Describe the process for adding two polar values.
4. Describe the process for subtracting one polar value from another.

## 12.3 RECTANGULAR NOTATION

There are several ways to express the value of any point on a graph. For example, let's say we want to express the value of the point shown in Figure 12.6a. We can express the value using polar notation, as shown in Figure 12.6b. If we assume that $z$ represents the length of the vector, then the vector quantity is expressed as $z\angle\theta$.

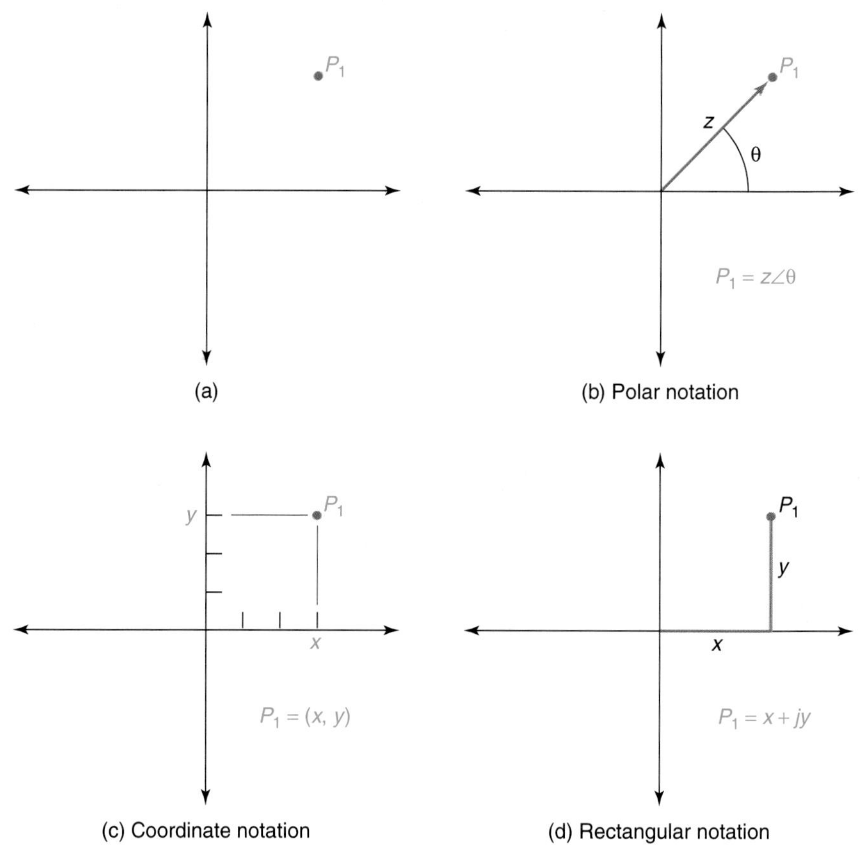


**FIGURE 12.6 Polar, coordinate, and rectangular notation.**

*OBJECTIVE 7* ▶

**Coordinate notation**
A method of expressing the value of a point in terms of its $x$-axis and $y$-axis values (or coordinates).

Another method that we can use to express the value of $P_1$, called **coordinate notation,** is illustrated in Figure 12.6c. This method simply identifies the point in terms of its $x$-axis and $y$-axis values, or *coordinates*. (In essence, this is the approach used to identify points on a map.) If Figure 12.6c has values of $x = 6$ and $y = 3$, then coordinate notation expresses the value of the point as

$$P_1 = (6, 3)$$

x-coordinate (↓ 6); y-coordinate (↑ 3)

Note that every point on the graph has a unique combination of coordinates; that is, every point has its own $(x, y)$ combination.

*OBJECTIVE 8* ▶

**Rectangular notation**
A variation on coordinate notation where the $x$- and $y$-coordinates are written in the form of $x + jy$.

**Rectangular notation** can be viewed as a variation on coordinate notation. In this case, the $x$- and $y$-coordinates are written in the form of an expression. For example, look at the graph in Figure 12.6d. The $x$ and $y$ values for the points are represented as vectors (rather than coordinates). These vectors are combined in the form of

$$x + jy$$

If we assume (again) that $P_1$ has values of $x = 6$ and $y = 3$, then rectangular notation expresses the point as

$$P_1 = 6 + j3$$

For now, we will assume that the $j$ in the expression merely identifies the $y$ value. (We will more accurately define $j$ later in this section.)

The graph in Figure 12.7 shows a series of points. Using the $x$ and $y$ values given, we can express each of the points in coordinate and rectangular notation. The values of each point are listed in the figure.

**FIGURE 12.7 Example points and their representations.**

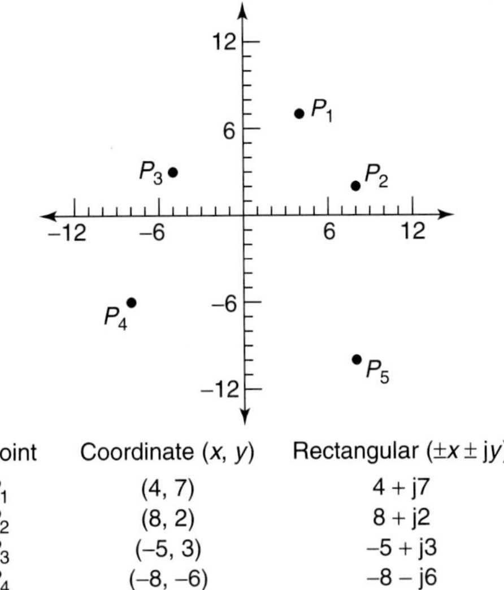


| Point | Coordinate (x, y) | Rectangular (±x ± jy) |
|---|---|---|
| $P_1$ | (4, 7) | 4 + j7 |
| $P_2$ | (8, 2) | 8 + j2 |
| $P_3$ | (−5, 3) | −5 + j3 |
| $P_4$ | (−8, −6) | −8 − j6 |
| $P_5$ | (8, −10) | 8 − j10 |

If you compare the $P_4$ and $P_5$ values with the first three, you'll see that the $+y$ and $-y$ values are distinguished by the sign preceding the *j;* that is:

1. A positive $y$ value is expressed in the form of $+jy$.
2. A negative $y$ value is expressed in the form of $-jy$.

This is always the case with rectangular notation.

## The j-Operator

The third column heading in Figure 12.7 indicates that every rectangular notation value can be expressed in the form of

◄ *OBJECTIVE 9*

$$\pm x \pm jy$$

The *j* that appears in this expression is actually referred to as the ***j*-operator**. As this name implies, *j* represents an operation (just like the signs $+$, $-$, $\times$, and $\div$). The *j*-operator represents a *vector rotation of 90°*. A positive *j*-operator indicates that the vector rotates in a positive direction. A negative *j*-operator indicates that the vector rotates in a negative direction. Both of these operations are illustrated in Figure 12.8.

***j*-operator** A symbol that indicates a vector rotation of 90°.

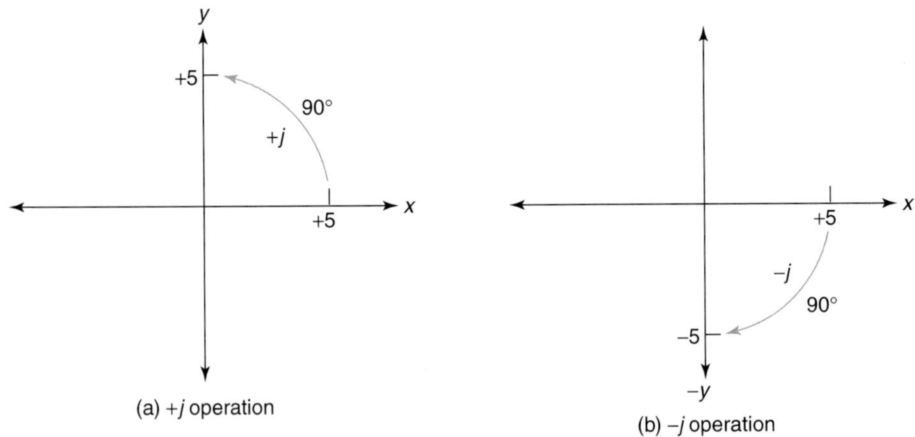


**FIGURE 12.8 Positive and negative *j*-operations.**

**FIGURE 12.9**

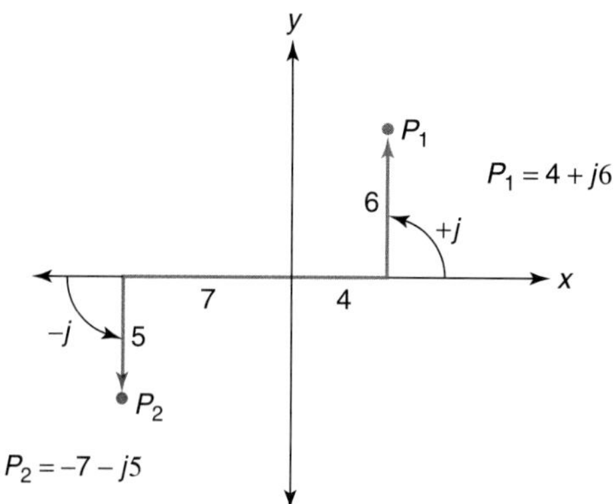


To understand the $+j$ operation, picture a vector lying on the $x$-axis of Figure 12.8a. Imagine this vector rotating in the direction indicated. A 90° rotation in this direction would place the vector on the positive $y$-axis. By the same token, imagine rotating the same vector as shown in Figure 12.8b. This $-j$ rotation would position the vector on the negative $y$-axis. Now, let's relate these operations to values expressed in rectangular notation.

When a value is expressed in rectangular notation, it is expressed as the sum of two vectors. For example, consider the points shown in Figure 12.9. $P_1$ is represented as the sum of the following:

1. A vector 4 units in length along the positive $x$-axis of the graph.
2. A vector 6 units in length at a 90° angle to the first. The direction of this vector is indicated by the $+j$-operator in the expression.

$P_2$ in the figure is represented as the sum of the following:

1. A vector 7 units in length along the negative $x$-axis of the graph.
2. A vector 5 units in length at a 90° angle to the first. The direction of this vector is indicated by the $-j$-operator in the expression.

As you will see, representing the points on a graph in this fashion allows us to add and subtract their values easily and directly: something we couldn't do using polar notation.

One more point: in rectangular notation, every value is written in the form of $(x \pm jy)$. Any number written in this form is referred to as a **complex number**.

**Complex number**
Any number written in the form of $(x \pm jy)$.

## *Multiple j-Operations*

*OBJECTIVE 10* ▶ Before beginning to solve problems involving complex numbers, we need to establish some relationships involving multiple $j$-operations. As you will see, these operations affect the outcomes of some of our calculations. Multiple $j$-operations are illustrated in Figure 12.10.

Figure 12.10a illustrates the $j^2$-operation. If we start with a vector on the $x$-axis and rotate that vector 90°, we get a value of

$$y = jx$$

If that vector is then rotated another 90°, the result equals $-x$ (as shown in the figure). By formula,

$$-x = jy$$

Since $y = jx$, the above equation can be rewritten as

$$-x = j(jx)$$

or

$$j^2x = -x \quad \textbf{(12.3)}$$

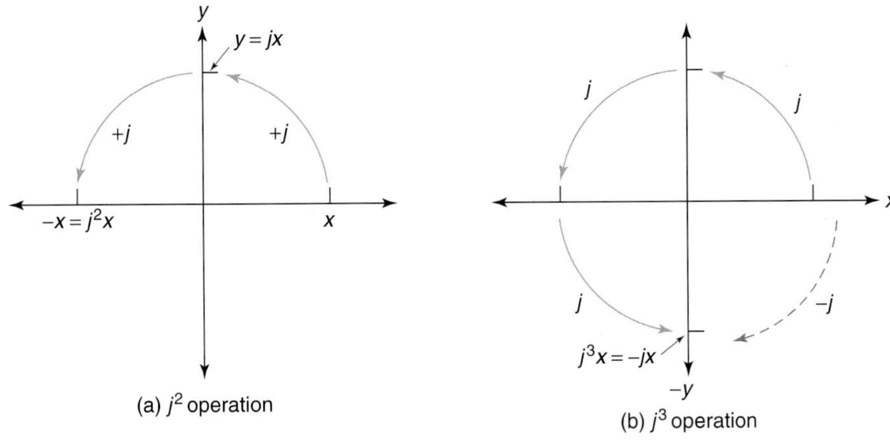


**FIGURE 12.10 Multiple *j*-operations.**

This is a transposed version of the equation shown in Figure 12.10a. Here are some examples of this relationship:

$$j^2 5 = -5 \qquad j^2 10 = -10 \qquad j^2(-2) = 2$$

This relationship will be used when we deal with multiplication and division of complex numbers.

Equation (12.3) can be developed further to provide another interesting relationship. If we divide both sides of the equation by $x$, we get

$$\frac{j^2 x}{x} = \frac{-x}{x}$$

or

$$j^2 = -1 \tag{12.4}$$

At the end of this section, we'll take equation (12.4) one step further to establish another interesting characteristic of the $j$-operator.

Figure 12.10b shows the result of rotating our vector an additional 90°. As you can see, three $j$-rotations have the same effect as a single rotation in the opposite direction. Therefore,

$$j^3 x = -jx \tag{12.5}$$

This relationship also plays a role in multiplying and dividing complex numbers.

## Adding and Subtracting Complex Numbers

◄ *OBJECTIVE 11*

Why complex numbers can be added directly to each other.

The ease of adding and subtracting complex numbers is a natural result of the way in which the values are represented. This point is illustrated in Figure 12.11.

When two (or more) vectors are positioned so that their sum forms a straight line, the vectors can be added directly regardless of the notation used to express the vectors. For example, let's assume we have the two vectors described in Figure 12.11a. Since these vectors have the same angle (θ) they can be added directly. The sum of the vectors in this case would be $7\angle\theta$, as shown in the figure. (The reason that adding polar values is normally so complicated is the fact that their angles usually are *not* the same.)

Now, let's apply this principle to the complex numbers represented in Figure 12.11b. Since the $x$ values are in the same direction, they can be added directly. The same applies to the $jy$ values. Therefore, two (or more) numbers expressed in complex form can be added directly; that is, the $x$ values are added to each other and the $y$ values are added to each other. The addition of two complex numbers is demonstrated in Example 12.4.

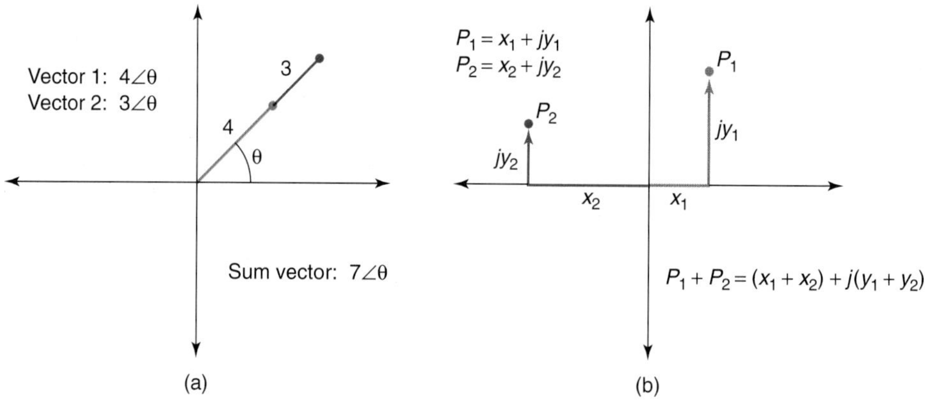


**FIGURE 12.11 Adding polar and rectangular vectors.**

## EXAMPLE 12.4

The vectors in Figure 12.11b have values of $P_1 = 4 + j6$ and $P_2 = -7 + j4$. Calculate the sum of the vectors.

***Solution:*** Since we are dealing with complex numbers, the values can be added directly as follows:

$$\begin{array}{ll} P_1: & 4 + j6 \\ P_2: & \underline{-7 + j4} \\ (P_1 + P_2): & -3 + j10 \end{array}$$

***PRACTICE PROBLEM 12.4***

Calculate the sum of the following vectors: $V_1 = 3 + j4$, $V_2 = 5 - j7$, and $V_3 = -2 + j4$.

The principle that allows us to add one complex number directly to another also allows us to subtract one complex number directly from another. Subtraction involving complex numbers is demonstrated in Example 12.5.

## EXAMPLE 12.5

Two vectors have the following values: $V_1 = 2 - j4$ and $V_2 = -2 + j5$. Calculate the value of $(V_1 - V_2)$.

***Solution:*** Since we are dealing with complex numbers, the value of $V_2$ can be subtracted directly from the value of $V_1$ as follows:

$$\begin{array}{ll} V_1: & 2 - j4 \\ -V_2: & \underline{-[-2 + j5]} \\ (V_1 - V_2): & 4 - j9 \end{array}$$

***PRACTICE PROBLEM 12.5***

Refer back to Practice Problem 12.4. For the values given, calculate the following: $(V_1 - V_2)$ and $(V_1 + V_2 - V_3)$.

As you can see, addition and subtraction are much easier to perform when vectors are expressed in complex form. However, the same cannot be said for the multiplication and division of complex numbers.

## Multiplying Complex Numbers

◀ *OBJECTIVE 12*

In algebra, we learn that a **binomial** is an expression written as the sum of (or difference between) two terms. By definition, complex numbers are binomials. The product of two binomials in the form of $(a + b)$ and $(c + d)$ is found by multiplying each term in the first expression by both terms in the second, as follows:

**Binomial**
An expression written as the sum of (or difference between) two terms.

$$\begin{aligned}(a + b)(c + d) &= a(c + d) + b(c + d)\\ &= ac + ad + bc + bd\end{aligned}$$

Example 12.6 shows how this approach is used to find the product of two complex numbers.

### EXAMPLE 12.6

Find the product of the following vectors:

$$V_1 = 3 + j6 \qquad \text{and} \qquad V_2 = 5 + j2$$

***Solution:*** Using the technique shown above, the product of the vectors can be found as:

$$\begin{aligned}\text{Step 1:}\quad V_1V_2 &= (3 + j6)(5 + j2)\\ \text{Step 2:}\quad &= 3(5 + j2) + j6(5 + j2)\\ \text{Step 3:}\quad &= 15 + j6 + j30 + j^2 12\\ \text{Step 4:}\quad &= 15 + j36 - 12\\ \text{Step 5:}\quad &= 3 + j36\end{aligned}$$

***PRACTICE PROBLEM 12.6***

Find the product of the following vectors:

$$V_1 = 4 + j3 \qquad \text{and} \qquad V_2 = 3 + j9$$

One of the steps in Example 12.6 warrants further discussion. If you compare Step 3 to Step 4, you'll see that

$$+j^2 12 = -12$$

This change is based on equation (12.4), which states that $j^2 = -1$. Using this relationship, the value of $-12$ (above) was derived as follows:

$$+j^2 12 = (+j^2)(12) = (-1)(12) = -12$$

The value ($j^2$) shows up in almost every multiplication and division problem involving complex numbers. Whenever this occurs, simply change the sign and drop the $j^2$ from the result, for example:

$$+j^2 5 = -5 \qquad -j^2 8 = +8$$

Note that the second example is based on the fact that $-j^2 = -(j^2) = -(-1) = +1$. In each case, the sign was changed and the $j^2$ was dropped.

As you can see, multiplying vectors in complex form is more involved than multiplying vectors in polar form. As we will demonstrate, the same holds true for vector division.

## Dividing Complex Numbers

To divide one complex number by another, we must first determine what is called the **conjugate** of the denominator. For any complex number, the conjugate is found by changing the sign of the $j$-operator. Table 12.1 shows several complex numbers and their conjugates.

**Conjugate**
For any binomial $(a \pm b)$, the value found by changing the sign between the variables from $+$ to $-$ (or vice versa).

**TABLE 12.1 Conjugates of Complex Numbers**

| *Complex Number* | *Conjugate* |
|---|---|
| $5 + j3$ | $5 - j3$ |
| $6 - j12$ | $6 + j12$ |
| $x + jy$ | $x - jy$ |

When a complex number is multiplied by its conjugate, the result is always a single number. This fact is based on the following rule of algebra:

$$(a + b)(a - b) = a^2 - b^2 \tag{12.6}$$

Example 12.7 demonstrates how this rule affects the outcome of multiplying a complex number by its conjugate.

## EXAMPLE 12.7

Find the product of $(6 + j5)$ and its conjugate.

***Solution:*** Using equation (12.6), the product is found as

$$\mathbf{(6 + j5)(6 - j5)} = 6^2 - (j5)^2 = 36 - (j^2 25) = 36 + 25 = \mathbf{61}$$

Again, we applied equation (12.4) to change $(-j^2 25)$ to $+25$ in the third step of the calculation.

***PRACTICE PROBLEM 12.7***

Find the product of $(8 - j6)$ and its conjugate.

Multiplying a complex number by its conjugate eliminates the *j*-operator.

As you can see, the result of the operation was a single number. This is always the case when a complex number is multiplied by its conjugate. Now, let's apply this principle to complex number division. Let's say that we have two vectors: $V_1 = 5 + j3$ and $V_2 = 2 + j4$. If we want to divide the first vector by the second, we start with

$$\frac{V_1}{V_2} = \frac{5 + j3}{2 + j4}$$

As written, the fraction cannot be simplified any further. However, the fraction can be simplified if we multiply both the numerator and the denominator by the conjugate of the denominator. By doing this, the denominator will be reduced to a single value that can be divided into each of the values in the numerator. Here is the step-by-step process:

**Step 1** Multiply the numerator and the denominator by the conjugate of the denominator.

$$\frac{5 + j3}{2 + j4} = \frac{(5 + j3)(2 - j4)}{(2 + j4)(2 - j4)}$$

Since we have multiplied both the numerator and the denominator by the same value $(2 - j4)$, we have not actually changed the value of the original fraction.

**Step 2** Solve the products in the numerator and the denominator.

$$\frac{(5 + j3)(2 - j4)}{(2 + j4)(2 - j4)} = \frac{10 - j20 + j6 - j^2 12}{2^2 - (j4)^2}$$

$$= \frac{10 - j14 + 12}{4 + 16}$$

$$= \frac{22 - j14}{20}$$

**Step 3** Divide each value in the numerator by the single value in the denominator.

$$\frac{22 - j14}{20} = \frac{22}{20} - \frac{j14}{20} = 1.1 - j0.7$$

Thus,

$$\frac{V_1}{V_2} = 1.1 - j0.7$$

As you can see, dividing one complex number by another is relatively complicated. Example 12.8 demonstrates the procedure from start to finish.

**EXAMPLE 12.8**

$V_1 = 6 - j3$ and $V_2 = 5 + j12$. Solve for the value of $V_1/V_2$.

***Solution:*** For the vectors given, the value of the fraction can be found as

$$\begin{aligned}\frac{\mathbf{V_1}}{\mathbf{V_2}} &= \frac{6 - j3}{5 + j12} \\ &= \frac{(6 - j3)(5 - j12)}{(5 + j12)(5 - j12)} \\ &= \frac{30 - j72 - j15 + j^2 36}{5^2 - (j12)^2} \\ &= \frac{30 - j87 - 36}{25 + 144} \\ &= \frac{-6 - j87}{169} \\ &= \mathbf{-0.036 - j0.515}\end{aligned}$$

***PRACTICE PROBLEM 12.8***

$V_1 = 4 + j8$ and $V_2 = 3 + j2$. Solve for the value of $V_1/V_2$.

## Polar Notation versus Rectangular Notation

You have now been shown how addition, subtraction, multiplication, and division are performed using both polar notation and rectangular notation. If you glance through the examples that have been presented up to this point, you'll see that:

1. Addition and subtraction are easier to perform on values written in rectangular form.
2. Multiplication and division are easier to perform on values written in polar notation.

In the next section, you'll be shown how to convert polar values to rectangular form and vice versa. This will enable you to solve easily any problem involving vectors, regardless of the notation used to express their values.

## Imaginary Numbers

◄ *OBJECTIVE 13*

To close this section, we will look at one more interesting characteristic of the $j$-operator. It should be noted that this characteristic is purely theoretical and therefore should be treated as such. According to equation (12.4):

$$j^2 = -1$$

If we were to take the square root of each side of the equation, we would obtain:

$$j = \sqrt{-1} \quad \textbf{(12.7)}$$

**Imaginary number**
A term used to describe any number containing the $j$-operator. The name is based on the relationship $j = \sqrt{-1}$.

Since a negative number cannot have a square root, the value expressed for $j$ in equation (12.7) is impossible. For this reason, the value of any number containing the $j$-operator is commonly referred to as an **imaginary number.**

In most references, the terms in a complex number are identified as follows:

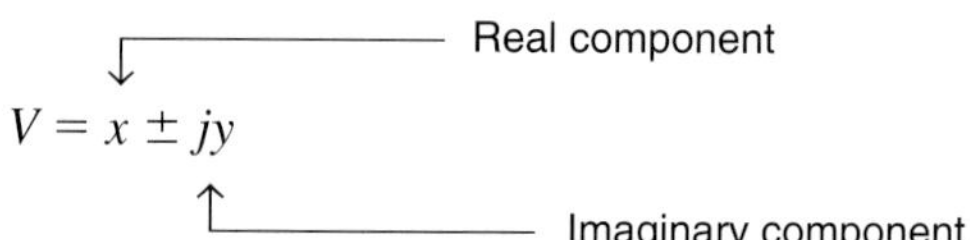


While the $jy$ term may be referred to as imaginary, you'll learn in upcoming chapters how these values have very real applications in electronics.

## Section Review

1. How is the value of a point represented in coordinate notation?
2. What is rectangular notation?
3. How would a $y$ value of $+8$ be expressed in rectangular notation? How would a $y$ value of $-8$ be expressed?
4. What does the $j$-operator represent?
5. In rectangular notation, how are values expressed?
6. What is a complex number?
7. Describe the process for adding (or subtracting) two complex numbers.
8. What is a binomial?
9. Describe the process for multiplying two complex numbers.
10. When multiplying two complex numbers, how do you treat any $j^2$ term in the result?
11. What is the conjugate of a complex number?
12. Describe the process for dividing one complex number by another.
13. List the notation of choice for each of the following: addition and subtraction, multiplication and division.
14. What is an imaginary number? Where does the term come from?

# 12.4 VECTOR NOTATION CONVERSIONS

You have been shown how certain operations are easier to perform using polar values while others are easier to perform using complex numbers. In this section, you will learn how vectors expressed in either notation (polar or rectangular) can be converted to the opposite form. This will enable you to perform any mathematical operation regardless of the form of the vector notation.

## Equivalent Polar and Rectangular Representations

Every point on a graph has both a polar representation and a rectangular representation. For example, Figure 12.12 shows a point ($P_1$) with both of its representations. Since the polar and rectangular representations shown describe the same point, we can say that

$$z\angle\theta = x + jy$$

for the point.

The combination of the polar and rectangular vectors for a given point always forms a right triangle.

If we combine the polar and rectangular representations in Figure 12.12, we get the right triangle shown in Figure 12.12c. Since the representations form a right triangle, the relationships between them must adhere to the principles of every right triangle. In other words, the relationships between ($z\angle\theta$) and ($x + jy$) are the same as those for any right triangle.

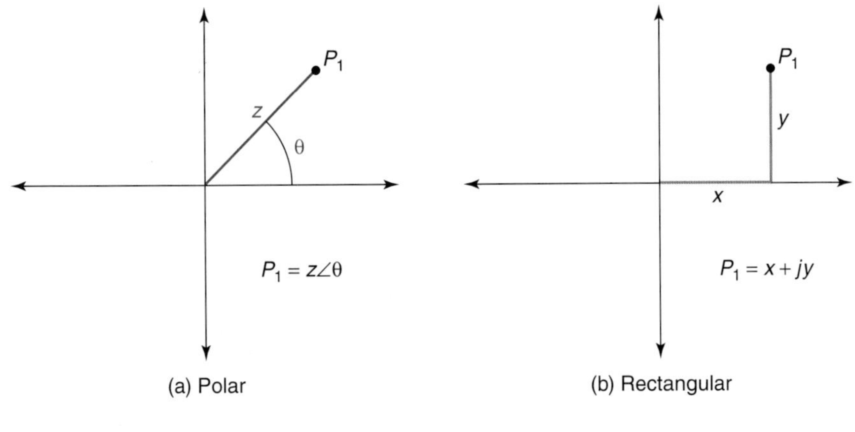


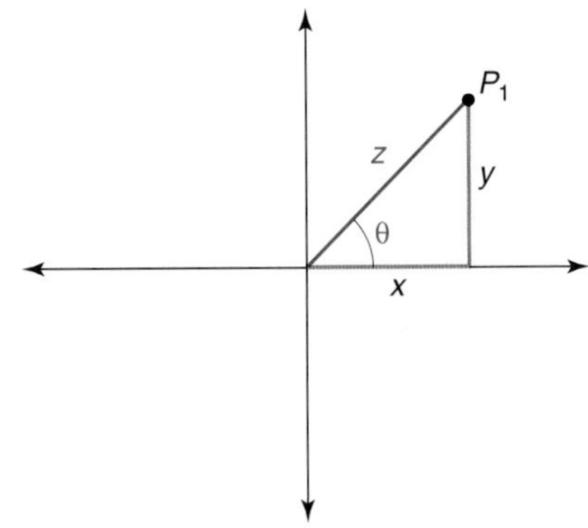


**FIGURE 12.12 Equivalent polar and rectangular representations.**

## *Right Triangle Relationships: The Pythagorean Theorem and Trigonometric Functions*

There are four fundamental relationships used to describe every right triangle. These relationships are listed in Figure 12.13.

**Theorem**
A relationship among a group of variables that always holds true.

**Pythagorean theorem**
A theorem stating that, for any right triangle, the square of the hypotenuse is equal to the sum of the squares of the other two sides.

A **theorem** can be viewed (loosely) as a relationship among a group of variables that always holds true. The **Pythagorean theorem** describes the relationship between the line lengths in any right triangle. It states that, for any right triangle, the square of the hypotenuse is equal to the sum of the squares of the other two sides. By formula,

$$z^2 = x^2 + y^2 \qquad \textbf{(12.8)}$$

◄ *OBJECTIVE 14*

Each trigonometric function shown in Figure 12.13 is used to relate angle θ to two sides of the triangle. If you look closely at the trig functions listed in the figure, you'll see that they contain every possible combination of two sides in the triangle. For future reference, we will identify the trig functions as follows:

$$\sin\theta = \frac{y}{z} \qquad \textbf{(12.9)}$$

$$\cos\theta = \frac{x}{z} \qquad \textbf{(12.10)}$$

and

$$\tan\theta = \frac{y}{x} \qquad \textbf{(12.11)}$$

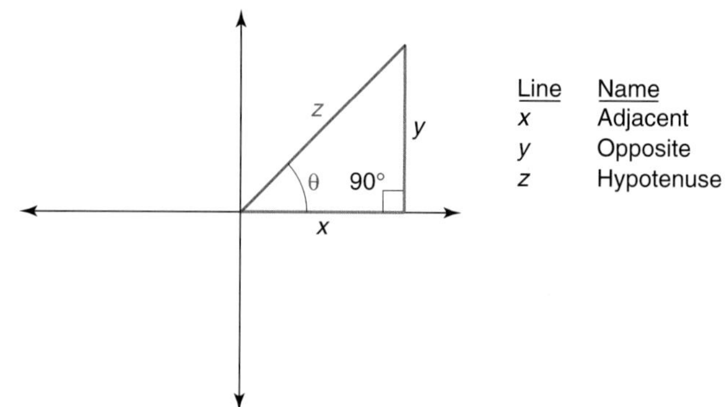


Pythagorean theorem: $z^2 = x^2 + y^2$

Trigonometric functions

| | | |
|---|---|---|
| Sine: | $\sin\theta =$ | $\dfrac{\text{opposite}}{\text{hypotenuse}} = \dfrac{y}{z}$ |
| Cosine: | $\cos\theta =$ | $\dfrac{\text{adjacent}}{\text{hypotenuse}} = \dfrac{x}{z}$ |
| Tangent: | $\tan\theta =$ | $\dfrac{\text{opposite}}{\text{adjacent}} = \dfrac{y}{x}$ |

**FIGURE 12.13 Right-triangle relationships.**

where $x =$ the side adjacent to angle $\theta$
$y =$ the side opposite angle $\theta$
$z =$ the hypotenuse (the side opposite the 90° angle)

Each trig function is nothing more than a ratio. For example, $\sin\theta$ for a given right triangle equals the ratio of the length of $y$ to the length of $z$. The reason we are interested in the length ratio is demonstrated in Figure 12.14. Here we have two triangles with the same angle $\theta$. For a given angle $\theta$, the ratio of $y$ to $z$ is constant, even though the actual line lengths may change. For the two triangles in Figure 12.14, $y_1 \neq y_2$ and $z_1 \neq z_2$. However,

$$\frac{y_1}{z_1} = \frac{y_2}{z_2}$$

Since the ratios are constant, the sine function for any angle is constant, regardless of the actual line lengths. For example,

$$\sin 30° = 0.5$$

This means that the ratio of $y$ to $z$ equals 0.5 when $\theta = 30°$, regardless of the actual lengths of $y$ and $z$.

**FIGURE 12.14 Side length ratios are constant for a given angle θ.**

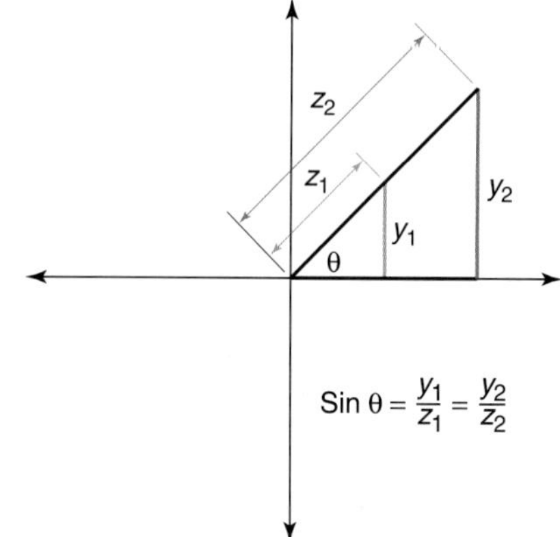

The constant-ratio characteristic of the sine function also holds true for the cosine and tangent functions. As you will see, this characteristic of trig functions makes them very useful when converting polar values to rectangular form.

## Polar-to-Rectangular Conversion

By transposing equations (12.9) and (12.10), we obtain the relationships needed to convert a polar value to rectangular form. These relationships are as follows: ◄ *OBJECTIVE 15*

$$x = z \cos \theta \qquad \textbf{(12.12)}$$

and

$$y = z \sin \theta \qquad \textbf{(12.13)}$$

Example 12.9 demonstrates the process for converting a vector value from polar form to rectangular form.

**EXAMPLE 12.9**

A vector has a value of $V_1 = 15\angle 60°$. Convert this value to rectangular form.

***Solution:*** Using equation (12.12), the value of $x$ can be found as

$$x = z \cos \theta = (15) \cos 60° = (15)(0.5) = \mathbf{7.5}$$

Using equation (12.13), the value of $y$ can be found as

$$y = z \sin \theta = (15) \sin 60° = (15)(0.866) \cong \mathbf{13}$$

Therefore, $15\angle 60° = 7.5 + j13$.

***PRACTICE PROBLEM 12.9***

A vector has a value of $V_1 = 20\angle 25°$. Convert this value to rectangular form.

Figure 12.4 showed how complicated it can be to determine the sum of two vectors written in polar form. Example 12.10 shows how we can simplify the problem by converting the polar values to rectangular form.

**EXAMPLE 12.10**

Two vectors are described as follows: $V_1 = 10\angle 45°$ and $V_2 = 20\angle -30°$. Calculate the value of $(V_1 + V_2)$.

***Solution:*** First, $V_1$ is converted to rectangular form as follows:

$$\begin{aligned} x &= 10 \cos 45° \\ &= (10)(0.707) \\ &= \mathbf{7.07} \end{aligned} \qquad \text{and} \qquad \begin{aligned} y &= 10 \sin 45° \\ &= (10)(0.707) \\ &= \mathbf{7.07} \end{aligned}$$

Thus, the first vector can be expressed as $V_1 = 7.07 + j7.07$. The $x$ and $y$ values for the second vector can be found as

$$\begin{aligned} x &= 20 \cos -30° \\ &= (20)(0.866) \\ &= \mathbf{17.32} \end{aligned} \qquad \text{and} \qquad \begin{aligned} y &= 20 \sin -30° \\ &= (20)(-0.5) \\ &= \mathbf{-10} \end{aligned}$$

Thus, the second vector can be expressed as $V_2 = 17.32 - j10$. Now, the sum of the vectors can be found as follows:

$$\begin{array}{ll} V_1: & 7.07 + j7.07 \\ V_2: & \underline{17.32 - j10.00} \\ (V_1 + V_2): & 24.39 - j2.93 \end{array}$$

***PRACTICE PROBLEM 12.10***

Two vectors are described as follows: $V_1 = 25\angle 60°$ and $V_2 = 8\angle -10°$. Calculate the value of $(V_1 + V_2)$.

Figure 12.15 shows how vectors $V_1$ and $V_2$ appear when plotted on a graph. If we were to place $V_2$ at the end of $V_1$ and plot the sum vector, it would be positioned as shown by the $(V_1 + V_2)$ vector in the diagram. As the arrows in the figure indicate, the sum vector has values of

$$x \cong 24 \qquad \text{and} \qquad y \cong 3$$

These values correspond closely to the vector sum of $(V_1 + V_2) = 24.39 - j2.93$ found in the example. However, we need to be able to relate the result to our original vector values. This means that we must be able to convert a rectangular value to polar form.

**FIGURE 12.15 Vectors for Example 12.10.**

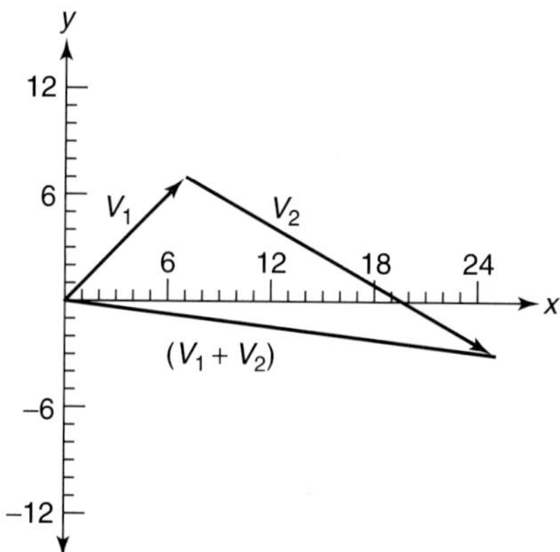


## Rectangular-to-Polar Conversion

*OBJECTIVE 16* ▶ Rectangular-to-polar conversion is performed using modified versions of the Pythagorean theorem and the tangent function. Given a complex number in the form of $(x + jy)$, the length of the vector can be found as

$$z = \sqrt{x^2 + y^2} \qquad \textbf{(12.14)}$$

Note that equation (12.14) is derived from the Pythagorean theorem as follows:

$$z^2 = x^2 + y^2 \qquad \sqrt{z^2} = \sqrt{x^2 + y^2} \qquad z = \sqrt{x^2 + y^2}$$

The value found using this equation is the length of $z\angle\theta$.

To determine the value of $\angle\theta$, we can use the inverse-tangent function. This function can be expressed as

$$\theta = \tan^{-1}\frac{y}{x} \qquad \textbf{(12.15)}$$

where $x, y$ = the components of the complex number
$\tan^{-1}$ = the inverse-tangent function

The inverse-tangent reverses the tangent operation. (The tangent of the angle gives us the ratio, so the inverse-tangent of the ratio gives us the angle.) Example 12.11 shows how equations (12.14) and (12.15) are used to convert a complex number to polar form.

### EXAMPLE 12.11

Convert the result from Example 12.10 back to polar form.

***Solution:*** In Example 12.10, we calculated a value of $(V_1 + V_2) = 24.39 - j2.93$. The length of the result vector is found as

$$z = \sqrt{x^2 + y^2} = \sqrt{(24.39)^2 + (-2.93)^2} = \sqrt{603.5} \cong \mathbf{24.6}$$

Note that the $y$ value is negative. Now, $\angle\theta$ can be found as

$$\boldsymbol{\theta} = \tan^{-1}\frac{y}{x} = \tan^{-1}\frac{-2.93}{24.39} = \tan^{-1}(-0.12) \cong \mathbf{-7°}$$

Thus, the result vector (expressed in polar form) has a value of $24.6\angle-7°$ (or $24.6\angle353°$).

***PRACTICE PROBLEM 12.11***

Convert the result from Practice Problem 12.10 to polar form.

*A Practical Consideration:* Equation (12.14) gives us what is called the geometric sum of the vectors. The geometric sum of any two vectors is always greater than the value of either vector and less than the algebraic sum of the two; that is,

$z > x \qquad z > y \qquad z < (x + y)$

Knowing this provides you with a method of quickly determining whether or not a calculated value of $z$ is in the ballpark.

If you compare the result of Example 12.11 with the sum vector in Figure 12.15, you'll see that:

1. The vector is close to 24 units in length.
2. The vector lies just below the positive horizontal (which corresponds to an angle of $-7°$).

## Putting It All Together

In upcoming chapters, we will be confronted with calculations involving both polar and complex numbers. When working these types of problems, you will find that notation conversions are often used to simplify the process. The relationships used to convert each vector notation to the other are summarized in Figure 12.16.

## One Final Note

Most scientific calculators can convert polar numbers to complex form (and vice versa). The keys used to initiate this function are usually labeled as P → R and R → P. Once you have learned how to perform the conversions manually (and learned *why* they work), you may find it faster to use the conversion functions of your calculator. Check the operator's manual for specific directions on using the notation conversion functions.

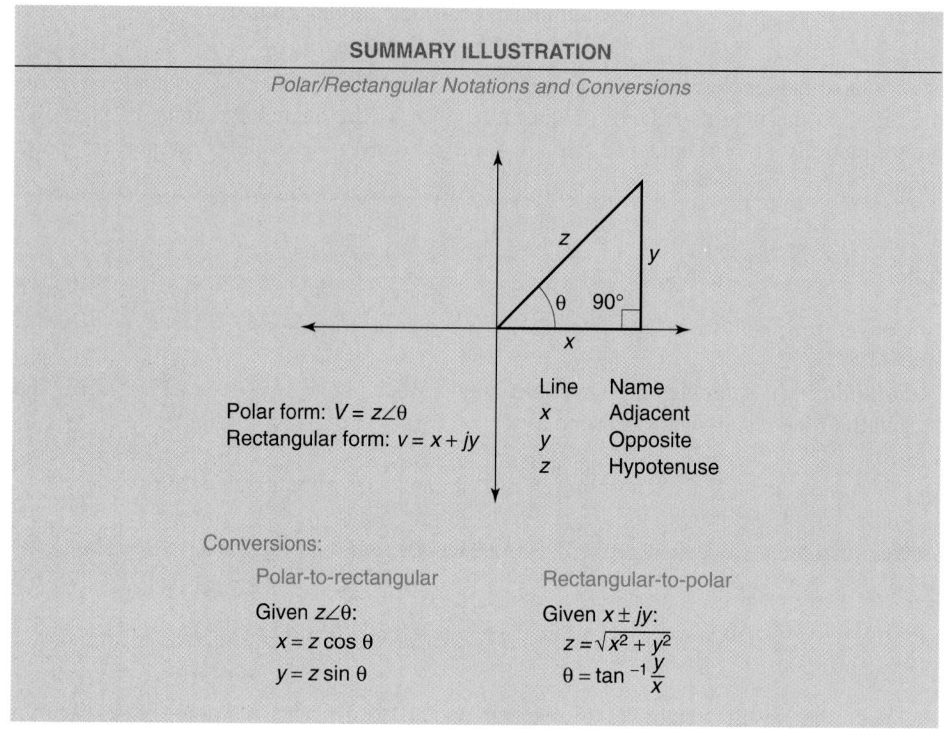

SUMMARY ILLUSTRATION
Polar/Rectangular Notations and Conversions
z
y
θ 90°
x
Line Name
x Adjacent
y Opposite
z Hypotenuse
Polar form: V = z∠θ
Rectangular form: v = x + jy
Conversions:
Polar-to-rectangular
Given z∠θ:
x = z cos θ
y = z sin θ
Rectangular-to-polar
Given x ± jy:
z = √(x² + y²)
θ = tan⁻¹ y/x


**FIGURE 12.16**

## Section Review

1. Demonstrate how the polar and rectangular values for a given point form a right triangle.
2. What is a theorem?
3. What is the relationship given in the Pythagorean theorem?
4. Explain the concept of a trig function as a ratio of line lengths.
5. List the relationships used to convert from polar form to rectangular form.
6. List the relationships used to convert from rectangular form to polar form.

## Chapter Summary

Here is a summary of the major points made in this chapter:

1. Any value that has both magnitude and an angle (or direction) is referred to as a vector. In contrast, a value with magnitude only is referred to as a scalar.
2. Both the magnitude and direction of a vector must be specified for the value to have any meaning.
3. In polar notation, the value of a vector is expressed as a magnitude followed by an angle.
   a. The polar representation of a vector is written in the following form:

$$V = z\angle\theta$$

Magnitude ($z$), Angle ($\theta$)

where $V$ is the variable used to represent the line.

   b. When an angle is not specified in a polar value, the angle is assumed to be 0°.
   c. A positive angle is measured in a counterclockwise direction from the positive horizontal of the graph.
   d. A negative angle is measured in a clockwise direction from the positive horizontal of the graph.
4. The product of two polar values equals the product of the magnitudes followed by the sum of the angles (see Example 12.1).
5. The quotient of two polar values equals the quotient of the magnitudes followed by the difference between the angles (see Example 12.2).
6. Polar addition is performed by positioning the vectors so that the second vector begins at the end of the first and drawing a sum vector (see Figure 12.4).
7. Polar subtraction is performed by rotating the subtrahend vector 180° and adding the result to the other vector (see Figure 12.5).
8. The procedures used for polar addition and subtraction are far more involved than those used for polar multiplication or division. Vector values can be added and subtracted more easily when the vector quantities are expressed in rectangular notation.
9. Coordinate notation expresses a vector quantity using its $x$-axis and $y$-axis values, or coordinates.
   a. The coordinate representation of a vector quantity is written in the form of

$$P = (x, y)$$

x-coordinate (arrow to $x$); y-coordinate (arrow to $y$)

   where $P$ is the variable used to represent the point.
   b. Every point on a graph has a unique combination of coordinates.
10. Rectangular notation can be viewed as a variation on coordinate notation. In this case, the $x$- and $y$-coordinates are written in the form of an expression.
   a. The rectangular representation of a vector is written in the form of

$$V = x \pm jy$$

   where $V$ is the variable used to represent the vector
   b. A positive $y$ value is expressed in the form of $+jy$.
   c. A negative $y$ value is expressed in the form of $-jy$.
11. The $j$ that appears in a rectangular expression is actually referred to as the $j$-operator.
12. The $j$-operator represents a vector rotation of 90°.
   a. A positive $j$-operator indicates that the vector rotates in a positive direction.
   b. A negative $j$-operator indicates that the vector rotates in a negative direction.
13. In rectangular notation, a vector quantity is expressed as the sum of two vectors. Any number written in rectangular form is referred to as a complex number.
14. The $j^2$-operation indicates a vector rotation of 180°. This operation reverses the sign of the value, as follows:

$$j^2 (y) = -y \qquad j^2(-y) = y$$

15. The $j^3$-operation indicates a vector rotation of 270°. This operation has the same effect as a $-j$ vector rotation, as follows:

$$j^3(y) = -jy$$

16. Two (or more) complex numbers can be added directly. The $x$ values are added to each other and the $y$ values are added to each other (see Example 12.4).

17. One complex number can be subtracted directly from another.
18. A binomial is an expression written as the sum of (or difference between) two variables. Complex numbers are binomials by definition.
19. The product of two binomials is found by multiplying each term in the first expression by both terms in the second (see Example 12.6).
20. Dividing complex numbers involves the conjugate of the divisor.
    a. The conjugate of a complex number is found by changing the sign of the $j$-operator.
    b. When a complex number is multiplied by its conjugate, the result is always a single number (see Example 12.7).
21. To simplify a fraction containing complex numbers:
    a. Multiply the numerator and the denominator by the conjugate of the denominator.
    b. Solve the products in the numerator and the denominator.
    c. Divide each value in the numerator by the result in the denominator.

    This process is demonstrated in Example 12.8.
22. Comparing polar and rectangular operations demonstrates that:
    a. Addition and subtraction are easier to perform using rectangular notation.
    b. Multiplication and division are easier to perform using polar notation.
23. Since $j^2 = -1$, as shown in equation (12.4), $j = \sqrt{-1}$.
    a. A negative number cannot have a square root. For this reason, any number containing the $j$-operator is called an imaginary number.
    b. In many references, the values in a complex number are identified as follows:

Real component

$$V = x \pm jy$$

Imaginary component

24. Every point on a graph has both a polar representation and a rectangular representation.
    a. For any given point, the combination of the polar and rectangular vectors forms a right triangle.
    b. The relationship between the polar and rectangular representations of a point is determined by the characteristics of every right triangle.
25. A theorem is a relationship among a group of variables that always holds true. The Pythagorean theorem states that, for any right triangle, the square of the hypotenuse is equal to the sum of the squares of the other two sides.
26. Each of the three basic trigonometric (trig) functions relates angle $\theta$ to two sides of the triangle. Each trig function is nothing more than a ratio.
27. Polar-to-rectangular conversion involves the sine and cosine functions (see Example 12.9).
28. Rectangular-to-polar conversion is performed using modified versions of the Pythagorean theorem and the tangent function (see Example 12.11).

## Equation Summary

| Equation Number | Equation | Section Number |
|---|---|---|
| (12.1) | $(x\angle\theta_1)(y\angle\theta_2) = (xy)\angle(\theta_1 + \theta_2)$ | 12.2 |
| (12.2) | $\dfrac{x\angle\theta_1}{y\angle\theta_2} = \dfrac{x}{y}\angle(\theta_1 - \theta_2)$ | 12.2 |

| | | |
|---|---|---|
| **(12.3)** | $j^2x = -x$ | 12.3 |
| **(12.4)** | $j^2 = -1$ | 12.3 |
| **(12.5)** | $j^3x = -jx$ | 12.3 |
| **(12.6)** | $(a + b)(a - b) = a^2 - b^2$ | 12.3 |
| **(12.7)** | $j = \sqrt{-1}$ | 12.3 |
| **(12.8)** | $z^2 = x^2 + y^2$ | 12.4 |
| **(12.9)** | $\sin \theta = \dfrac{y}{z}$ | 12.4 |
| **(12.10)** | $\cos \theta = \dfrac{x}{z}$ | 12.4 |
| **(12.11)** | $\tan \theta = \dfrac{y}{x}$ | 12.4 |
| **(12.12)** | $x = z \cos \theta$ | 12.4 |
| **(12.13)** | $y = z \sin \theta$ | 12.4 |
| **(12.14)** | $z = \sqrt{x^2 + y^2}$ | 12.4 |
| **(12.15)** | $\theta = \tan^{-1} \dfrac{y}{x}$ | 12.4 |

## Key Terms

The following terms were introduced and defined in this chapter:

| | | |
|---|---|---|
| binomial | imaginary number | rectangular notation |
| complex number | *j*-operator | scalar |
| conjugate | polar notation | theorem |
| coordinate notation | Pythagorean theorem | vector |

## Practice Problems

1. Solve for each product:
   **a.** $(5\angle-20°)(-2\angle14°)$
   **b.** $(3.3\angle45°)(12\angle144°)$
   **c.** $(5\angle77°)(3\angle28°)(-2\angle-80°)$
   **d.** $(12)(-3\angle280°)(-15\angle88°)$
2. Solve for each product:
   **a.** $(9\angle-30°)(2\angle-14°)$
   **b.** $(20\angle-55°)(0.5\angle55°)$
   **c.** $(-8\angle75°)(3\angle-32°)(-2\angle90°)$
   **d.** $(12\angle-12°)(-8\angle80°)(-1.5\angle68°)$
3. Simplify the following fractions:
   **a.** $\dfrac{15\angle40°}{3\angle25°}$ **b.** $\dfrac{0.28\angle240°}{-3\angle88°}$
   **c.** $\dfrac{36\angle-10°}{50\angle20°}$ **d.** $\dfrac{6}{24\angle-90°}$

4. Simplify the following fractions:

a. $\frac{288\angle 78^\circ}{-3\angle -12^\circ}$ b. $\frac{188\angle 90^\circ}{44\angle -300^\circ}$

c. $\frac{36\angle -10^\circ}{72}$ d. $\frac{15\angle -90^\circ}{75\angle -90^\circ}$

5. Simplify the following fractions:

a. $\frac{(12\angle 18^\circ)(-14\angle 12^\circ)}{(80\angle 45^\circ)(2\angle 15^\circ)}$

b. $\frac{(7\angle -48^\circ)(-9\angle 72^\circ)}{(20\angle 65^\circ)(-5\angle 18^\circ)}$

6. Simplify the following fractions:

a. $\frac{(0.8\angle 300^\circ)(180\angle 72^\circ)}{(-12\angle 45^\circ)(5\angle 20^\circ)(0.5\angle 90^\circ)}$

b. $\frac{1}{(10\angle 25^\circ)(-2\angle 95^\circ)}$

7. Solve for each sum:

a. $(5 + j3) + (-3 + j6) + (2 - j8)$

b. $(12 + j22) + (6 - j9) + (8 + j5)$

8. Solve for each sum:

a. $(-24 + j33) + (3.6 + j9) + (-2 - j40)$

b. $(10 + j10) + 5 + (6 - j9) + j5$

9. Solve each of the following:

a. $(5 + j3) - (-3 + j6) - (2 - j8)$

b. $(9 + j2) - (6 - j14) + (-8 - j5)$

10. Solve each of the following:

a. $(-24 + j33) - (3.6 + j9) - (-2 - j40)$

b. $(-10 + j10) - 15 - (-5 - j20) + j44$

11. Solve for each product:

a. $(3 + j9)(2 - j2)$

b. $(12 - j5)(3 - j8)$

c. $(-12 + j2)(5 - j5)$

d. $(3 + j5)(3 - j5)$

12. Solve for each product:

a. $(0.83 + j3.3)(4 - j0.9)$

b. $(40 - j18)(2 - j5)$

c. $(-12 + j2)(-12 - j2)$

d. $(3 + j5)(3 - j5)(4 + j1)$

13. Find the product of $(8 + j2)$ and its conjugate.

14. Find the product of $(-4 + j2)$ and its conjugate.

15. Find the product of $(15 - j8)$ and its conjugate.

16. Find the product of $(-6 - j9)$ and its conjugate.

17. Simplify the following fractions:

a. $\frac{5 + j15}{6 + j2}$

b. $\frac{-18 + j30}{-2 + j2}$

c. $\frac{12 - j12}{3 + j3}$

d. $\frac{10 - j12}{25 + j5}$

18. Simplify the following fractions:

a. $\frac{1}{5 + j5}$

b. $\frac{-10 + j0.5}{-1 + j10}$

c. $\frac{12 - j12}{12 + j12}$

d. $\frac{-15 + j45}{0 + j5}$

19. Convert each of the following to a complex number.

a. $25\angle 45°$

b. $50\angle -60°$

c. $18\angle 60°$

d. $100\angle 90°$

20. Solve each problem and express the answer in rectangular form.

a. $(9\angle -30°)(2\angle -14°)$

b. $\frac{188\angle 90°}{44\angle -300°}$

c. $(20\angle -55°) + (0.5\angle 55°)$

d. $\frac{(12\angle 18°)(-14\angle 12°)}{(80\angle 45°)(2\angle 15°)}$

21. Rewrite each fraction in Problem 17 in polar form.

22. Rewrite each fraction in Problem 18 in polar form.

---

## The Brain Drain

23. Solve the problem below. Give your answer in rectangular form.

$$\left(\frac{44\angle 30° + 10\angle -30°}{5 + j33}\right) \times \left(\frac{1}{10 - j5}\right) =$$

24. Graph the result vector for the problem below. Provide the values of $z$ and $\angle\theta$ on your graph.

$$\frac{(3 + j3)(5 - j10) + 20\angle 45°}{12 + j10 - 5\angle -30°} =$$

---

## Answers to the Example Practice Problems

**12.1** $60\angle 110°$
**12.2** $0.18\angle 150°$
**12.3** $3\angle 40°$
**12.4** $6 + j1$
**12.5** $V_1 - V_2 = -2 + j11$, $V_1 + V_2 - V_3 = 10 - j7$
**12.6** $-15 + j45$
**12.7** 100
**12.8** $2.15 + j1.23$
**12.9** $18.1 + j8.45$
**12.10** $20.3 + j20.3$
**12.11** $28.7\angle 44.8°$

# 13

# INDUCTORS

## OBJECTIVES

*After studying the material in this chapter, you should be able to:*

1. Describe the effect of varying current on a magnetic field.
2. List and discuss the first three of Faraday's laws of induction.
3. Describe the concept of self-induction.
4. Explain the concept of counter emf.
5. Define the henry (H) unit of inductance.
6. Calculate the value of an inductor (in henries) given its physical dimensions and core permeability.
7. Explain the phase relationship between inductor current and voltage.
8. Discuss mutual inductance and list the factors that affect its value.
9. Discuss coefficient of coupling, its normal range of values, and the factors that affect its values.
10. Calculate the total inductance for any group of inductors connected in series.
11. Describe the possible effects of mutual inductance on total series inductance.
12. Describe the procedure for measuring the coefficient of coupling between any two inductors.
13. Calculate the total inductance for any group of inductors connected in parallel.
14. Discuss inductive reactance ($X_L$) and its source.
15. Calculate the value of $X_L$ for any inductor operated at a specified frequency.
16. Perform basic circuit calculations involving inductive reactance.
17. Describe, compare, and contrast the following: apparent power, true power, and reactive power.
18. Discuss the concept of inductor quality ($Q$) and calculate its value for a given inductor operating at a specified frequency.
19. Compare and contrast iron-core and air-core inductors.
20. List and describe the various types of inductors.

You have seen how magnetic flux can be used to induce a current through a conductor and vice versa. In this chapter, we will take the relationship between magnetism and current one step further by discussing an electrical characteristic called inductance. Technically defined, **inductance** is the ability of a component with a changing current to induce a voltage across itself or a nearby circuit by generating a changing magnetic field. Based on its overall effect, inductance is sometimes described as:

**Inductance**
The ability of a component with a changing current to induce a voltage across itself or a nearby circuit by generating a changing magnetic field.

1. The ability of a component to oppose a change in current.
2. The ability of a component to store energy in an electromagnetic field.

As you will learn, both of these descriptions are valid.

An **inductor** is a component designed to provide a specific measure of inductance. Inductors can store energy in an electromagnetic field as well as oppose any change in current. These characteristics make the inductor useful in various applications, some of which will be introduced later in this chapter.

**Inductor**
A component designed to provide a specific amount of inductance.

## 13.1 INTRODUCTION TO INDUCTANCE

The descriptions of inductance given in the opening indicate that we are dealing (once again) with the relationship between current and magnetism. For this reason, we will start our discussion on inductance by reviewing some principles of magnetism.

### The Relationship between Current and Magnetism: A Review

In Chapter 10, you were shown that a current-carrying coil generates magnetic flux, as shown in Figure 13.1. As a review, here are some of the principles we established regarding the relationship between current and the resulting flux:

1. The current through a wire produces magnetic lines of force.
2. When a wire is wrapped in a series of loops, called a coil, the lines of force generated by the current combine to form a magnetic field.
3. The strength of the magnetic field is determined by the amount of magnetomotive force (mmf) produced by the current through the coil.
4. The mmf produced by the current through a coil is proportional to the level of current and the number of turns (loops) in the coil. These two factors are combined into a single value called the ampere-turn (At).

**FIGURE 13.1 Current-carrying coil and the resulting magnetic flux.**

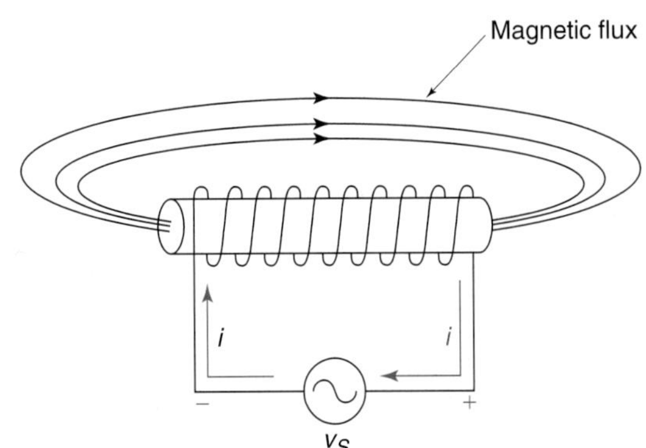

5. Flux density ($B$) is a measure of the concentration of a magnetic field. The flux density generated by a current-carrying coil is:
   a. Proportional to the permeability of the core material.
   b. Proportional to the ampere-turns value of the coil.
   c. Inversely proportional to the length of the coil.
6. The polarity of the flux generated by a current is determined by the current direction; that is, if the current through a coil reverses direction, so do the poles of the resulting magnetic field.

As you will see, all these factors play a role in the principle of inductance.

## The Effect of Varying Current on a Magnetic Field

*OBJECTIVE 1* ▶ A coil like the one in Figure 13.1 is referred to as an inductor. When current passes through an inductor, magnetic flux is generated as shown in the figure. In Chapter 10, you learned that flux density varies directly with the ampere-turns value of the coil. This relationship was given as

*Reference:* This equation was introduced in Chapter 10 as equation (10.6).

$$B = \frac{\mu_m NI}{\ell} \tag{13.1}$$

where
$B$ = the flux density, in webers per square meter ($\text{Wb/m}^2$)
$\mu_m$ = the permeability of the core material
$NI$ = the ampere-turns value of the component
$\ell$ = the length of the coil, in meters

According to this relationship:

1. An increase in inductor current causes an increase in flux density.
2. A decrease in inductor current causes a decrease in flux density.

In other words, flux density varies directly with inductor current.

The effect of varying inductor current on the resulting magnetic field is illustrated in Figure 13.2. As the inductor current increases, flux density increases and the magnetic field expands outward from the component. As the inductor current decreases, flux density decreases and the magnetic field collapses back into the component. To understand the impact of this changing magnetic field, we have to take a brief look at several relationships, called Faraday's laws of induction.

## Faraday's Laws of Induction

*OBJECTIVE 2* ▶ In 1831, an English scientist named Michael Faraday discovered that a magnetic field could be used to generate (or *induce*) a voltage across a coil, through a process called *electromagnetic induction.* Faraday's observations on the results of his experiments have come to be known as Faraday's laws of induction. The first three of these laws can be paraphrased as follows:

Law 1: To induce a voltage across a wire, there must be relative motion between the wire and the magnetic field.

Law 2: The voltage induced is proportional to the rate of change in magnetic flux encountered by the wire.

Law 3: When a wire is cut by $10^8$ perpendicular lines of force per second, one volt is induced across that wire.

*Note:* Based on Faraday's third law, we can now redefine the *weber* as the amount of flux that, when passing through a perpendicular coil in one second, induces 1 V across that coil.

Faraday's first law states that induction takes place when there is relative motion between the conductor and the magnetic field. In Chapter 11, you were shown how current is induced when a conductor moves through a stationary magnetic field. In the case of the

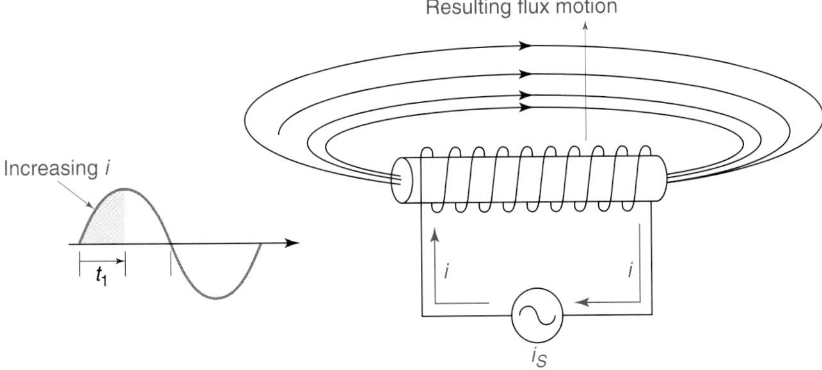

(a) An increase in current causes the magnetic field to expand outward.

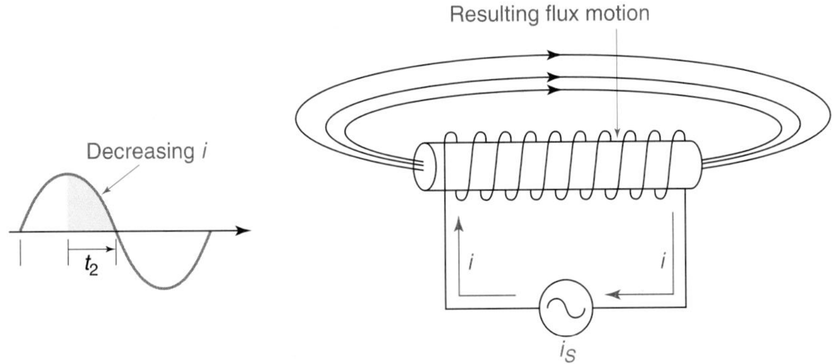

(b) A decrease in current causes the magnetic field to collapse inward.

**FIGURE 13.2 Effects of changing current on magnetic flux.**

inductor, the expanding and collapsing magnetic field passes through a stationary component. Even so, there is relative motion between the inductor and the magnetic field. Therefore, Faraday's first law of induction applies: the changing magnetic field induces a voltage across the inductor. The more rapid the change in the magnetic field, the greater the voltage induced.

Faraday's second law states that induced voltage is proportional to the rate of change in magnetic flux. This law can be expressed mathematically as

$$V = N\frac{d\phi}{dt} \tag{13.2}$$

where $N$ = the number of turns in the wire

$\frac{d\phi}{dt}$ = the instantaneous rate of change in flux

This relationship is illustrated in Figure 13.3, where a magnetic field is expanding outward from the core of an inductor. As it cuts through the inductor, it induces a voltage across the inductor. The magnitude of the induced voltage is proportional to the number of turns in the coil and the rate of change in flux. For example, if 1 weber ($10^8$ lines) of magnetic flux passes through the inductor per second, a 1 V difference of potential is induced across each turn in the component. If 1 Wb of flux cuts through the inductor every 100 ms, the rate of change in flux (per turn) is found as

$$\frac{d\phi}{dt} = \frac{1 \text{ Wb}}{100 \text{ ms}} = 10 \text{ Wb/s}$$

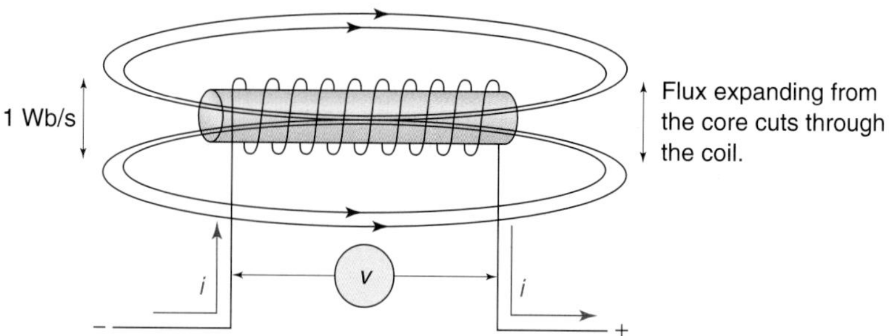
1 Wb/s
Flux expanding from
the core cuts through
the coil.
V
i
i
−
+

**FIGURE 13.3 One volt is induced across each turn when flux cuts through the coil at a rate of 1 weber per second.**

Therefore, 10 V is induced across each turn in the component (1 V for each Wb/s). Remember: it is the rate of change in magnetic flux that determines the voltage induced across each of the turns.

*OBJECTIVE 3* ▶

We have now established the following principles of operation for a current-carrying inductor:

1. The current through an inductor generates magnetic flux.
2. The amount of flux varies directly with inductor current.
3. As inductor current increases, the magnetic field expands.
4. As inductor current decreases, the magnetic field collapses.
5. As the magnetic field expands and collapses, it cuts through the stationary inductor, which produces relative motion between the two.
6. The relative motion caused by the changing magnetic field induces a voltage across the inductor.

**Self-inductance**
The ability of an inductor with a changing current to generate a changing magnetic field that, in turn, induces a voltage across the component.

This sequence indicates that a change in inductor current generates a moving magnetic field, which in turn induces a voltage across the inductor. This principle is referred to as **self-inductance.** To understand the impact of self-inductance, we have to take a look at another law of electromagnetic induction called Lenz's law.

## Lenz's Law

*OBJECTIVE 4* ▶

**Lenz's law**
An induced voltage always opposes its source (in keeping with the law of conservation of energy).

In 1834, a Russian physicist named Heinrich Lenz derived the relationship between a magnetic field and the voltage it induces. This relationship, which has come to be known as **Lenz's law,** can be paraphrased as follows: *an induced voltage always opposes its source;* that is, the polarity of the voltage induced by a magnetic field opposes the change in current that produced the magnetic field.

Lenz's law is illustrated in Figure 13.4. In Figure 13.4a:

1. An increase in inductor current causes the magnetic field to expand.
2. As the magnetic field expands, it cuts through the coil, inducing a voltage.
3. The polarity of the voltage is such that it opposes the increase in current.

In Figure 13.4b:

1. A decrease in inductor current causes the magnetic field to collapse.
2. As the magnetic field collapses, it cuts through the coil, inducing a voltage across the component. (Note that the polarity of the voltage has reversed.)
3. The polarity of the induced voltage is such that it opposes the decrease in current.

**Counter emf**
A term used to describe the voltage induced across an inductor by its own changing magnetic field. (The term serves as a reminder that the induced voltage opposes the change in current that generated the changing magnetic field to begin with.)

As you can see, the polarity of the induced voltage always opposes the change in coil current. Because it always opposes the change in coil current, the induced voltage is referred to as **counter emf.**

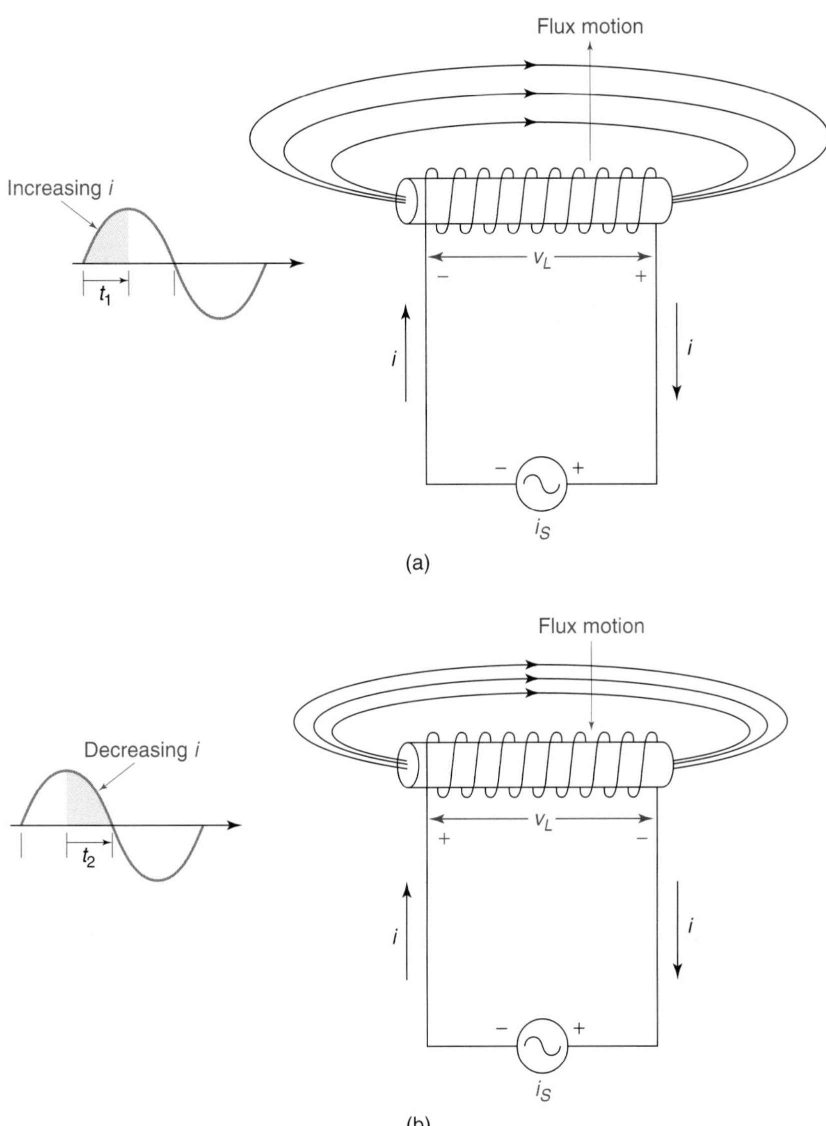


**FIGURE 13.4 An inductor opposes any change in current.**

## *Lenz's Law and Conservation of Energy*

The relationship between induced voltage and inductor current makes sense when you think of the component action in terms of the law of conservation of energy. If the voltage induced across an inductor aided the change in current, here's what would happen. The increasing current through the inductor (Figure 13.4a) would cause the magnetic field to expand. The expanding magnetic field would induce a voltage that would generate an additional increase in coil current. As the inductor current increased, the magnetic field would continue to expand, inducing more voltage, generating more current, and so on. Thus, any increase in current would end up generating an ever-increasing current, which is impossible.

## Induced Voltage

OBJECTIVE 5 ▶

As you know, a voltage is induced across an inductor when the current through the component changes. This relationship can be expressed as

*Note: di/dt* is simply a way of expressing an instantaneous rate of change in current. For example, if current is changing at a rate of 1 amp per second at an instant in time, then

$$\frac{di}{dt} = 1\text{ A/s}$$

$$-v_L = L\frac{di}{dt} \quad \textbf{(13.3)}$$

where $v_L$ = the instantaneous value of induced voltage
$L$ = the inductance of the coil, measured in henries (H)
$\frac{di}{dt}$ = the instantaneous rate of change in inductor current

Note that the minus sign in equation (13.3) is in keeping with Lenz's law, which defines induced voltage as a counter emf.

The **henry (H),** which is the unit of measure of inductance, is easy to define when we transpose equation (13.3) as follows:

*Note:* Since a unit of measure is neither positive nor negative, the minus sign in equation (13.3) is dropped from equation (13.4).

$$L = \frac{v_L}{\frac{di}{dt}} \quad \textbf{(13.4)}$$

**Henry (H)**
The unit of measure of inductance. One henry is the amount of inductance that produces a 1 V difference of potential when current changes at a rate of 1 A/s.

This equation indicates that inductance is measured in volts per rate of change in current. With this in mind, take a look at the inductor shown in Figure 13.5. The current through the inductor is shown to be changing at a rate of one ampere per second. This rate of change is inducing 1 V across the inductor. When a change of 1 A/s induces 1 V, the value of the inductance is said to be 1 henry (1 H). The relationship between inductance and inductor current and voltage is demonstrated further in Example 13.1.

**FIGURE 13.5 Henry (H) unit of inductance.**

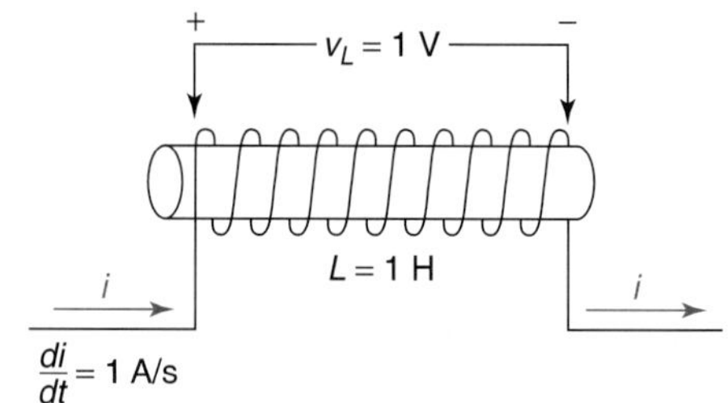


### EXAMPLE 13.1

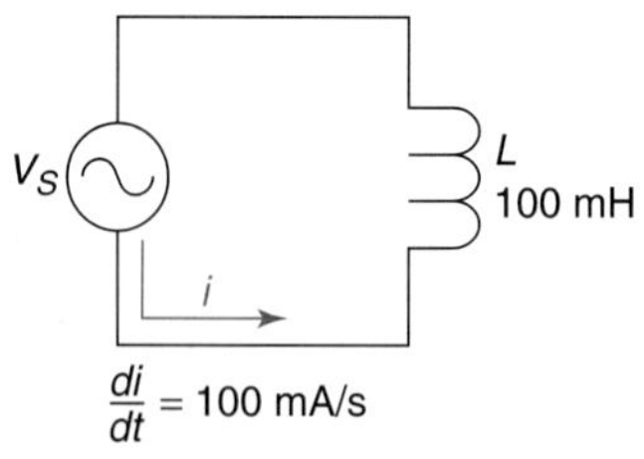


**FIGURE 13.6**

EWB

The schematic symbol for an air-core inductor is shown in Figure 13.6. For the component shown, calculate the value of $v_L$ when the current changes at a rate of 100 mA per second.

***Solution:*** The inductor is shown to have a value of 100 mH. Using equation (13.3), the value of the induced voltage can be found as

$$-v_L = L\frac{di}{dt} = (100\text{ mH})(100\text{ mA/s}) = \mathbf{10\text{ mV}}$$

***PRACTICE PROBLEM 13.1***

The current through a 33 mH inductor changes at a rate of 200 mA/s. Calculate the value of the induced voltage.

**FIGURE 13.7 Inductor physical dimensions.**

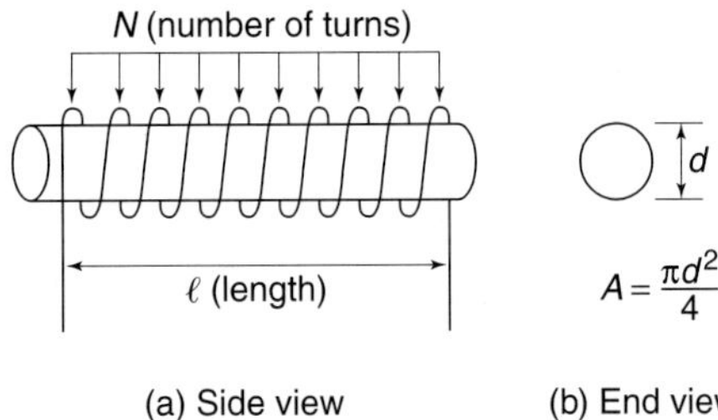


Equation (13.4) defines inductance in terms of induced voltage and rate of change in current. It is also possible to approximate the value of an inductor based on its physical characteristics, as follows:

$$L \cong \frac{\mu_m N^2 A}{\ell} \tag{13.5}$$

*Note:* Equation (13.5) holds true only if the length of the coil is at least 10 times the value of the core diameter (which is generally the case).

where $\mu_m$ = the permeability of the inductor core
$N^2$ = the square of the number of turns
$A$ = the cross-sectional area of the inductor core, in square meters ($m^2$)
$\ell$ = the length of the coil, in meters (m)

The physical characteristics used in equation (13.4) are illustrated in Figure 13.7. Note that the core of the inductor (when viewed from either end) is in the shape of a circle. Therefore, the area ($A$) of the core can be found in the same manner as for any circle. By formula,

$$A = \frac{\pi d^2}{4} \tag{13.6}$$

Example 13.2 shows how the value of an inductor can be approximated when its physical dimensions are known.

◄ *OBJECTIVE 6*

## EXAMPLE 13.2

The air-core inductor in Figure 13.8 has the physical dimensions shown. Calculate the value of the inductor. (The permeability of air was given as $4\pi \times 10^{-7}$ in Chapter 10.)

***Solution:*** With a diameter 0.5 cm, the cross-sectional area of the core is found as

$$A = \frac{\pi d^2}{4} = \frac{\pi(0.5 \times 10^{-2}\text{m})^2}{4} = \mathbf{1.96 \times 10^{-5}\ m^2}$$

Now, the value of the inductor can be found as

$$L \cong \frac{\mu_m N^2 A}{\ell} \cong \frac{(4\pi \times 10^{-7})(100)^2(1.96 \times 10^{-5}\text{m}^2)}{8 \times 10^{-2}\text{m}} \cong \mathbf{3.1\ \mu H}$$

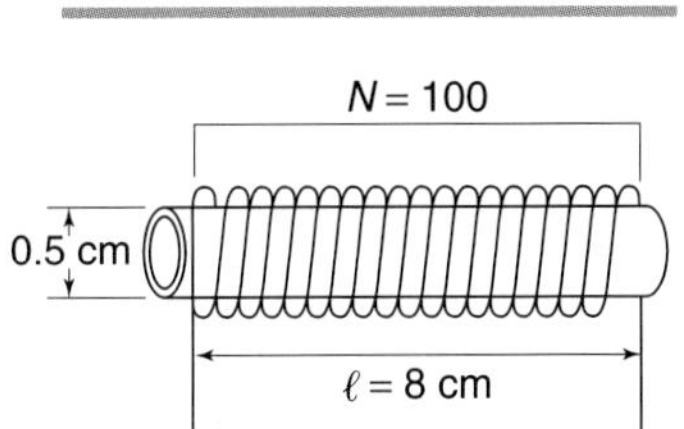


**FIGURE 13.8**

***PRACTICE PROBLEM 13.2***

An air-core inductor has the following dimensions: $d$ = 0.25 cm, $N$ = 150, and $\ell$ = 5 cm. Calculate the value of the component.

## Putting It All Together

An inductor is a component designed to provide a specific amount of inductance. When the current through an inductor changes, it produces a changing magnetic field. This field expands as inductor current increases and collapses as inductor current decreases. In the process of expanding and collapsing, the magnetic field cuts through the coil, inducing a

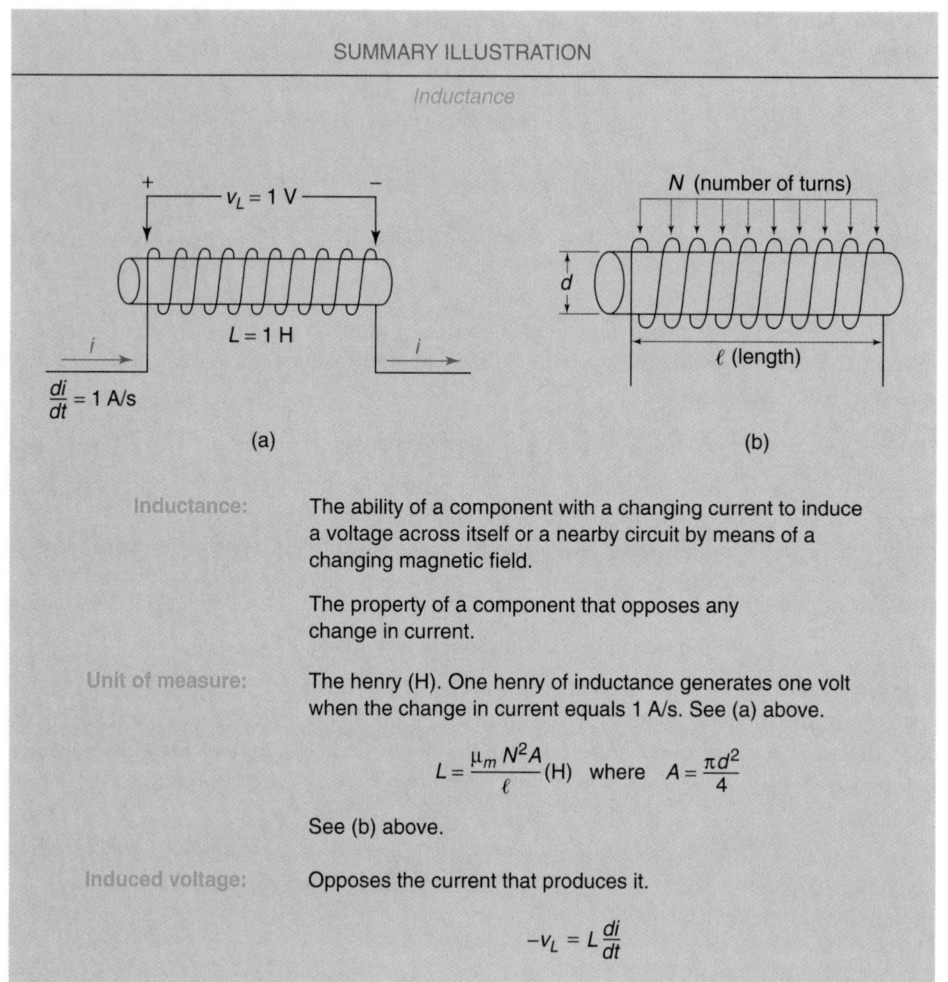
SUMMARY ILLUSTRATION
Inductance
+
$v_L = 1$ V
−
$L = 1$ H
i
i
$\frac{di}{dt} = 1$ A/s
(a)
N (number of turns)
d
ℓ (length)
(b)
Inductance:
The ability of a component with a changing current to induce a voltage across itself or a nearby circuit by means of a changing magnetic field.
The property of a component that opposes any change in current.
Unit of measure:
The henry (H). One henry of inductance generates one volt when the change in current equals 1 A/s. See (a) above.
$L = \frac{\mu_m N^2 A}{\ell}$ (H) where $A = \frac{\pi d^2}{4}$
See (b) above.
Induced voltage:
Opposes the current that produces it.
$-v_L = L\frac{di}{dt}$

**FIGURE 13.9**

voltage across the component. In keeping with the law of conservation of energy, the polarity of the induced voltage opposes the change in current that caused it.

Inductance is measured in volts per rate of change in current. The unit of measure of inductance is the henry (H). When the current through 1 H of inductance changes at a rate of 1 A/s, the difference of potential induced across the coil is 1 V.

The value of an inductor is determined by its physical dimensions. Equation (13.4) shows that inductance is:

1. Directly proportional to core permeability, the square of the number of turns, and the cross-sectional area of the core.
2. Inversely proportional to the length of the component.

Many of the properties of inductance are summarized in Figure 13.9.

## Section Review

1. List the three commonly accepted descriptions of inductance.
2. What is an inductor?
3. What is the relationship between flux density and core permeability? Between flux density and the ampere-turns of a coil? Between flux density and coil length?
4. How does a change in coil current affect the magnetic field generated by the component?

5. How is Faraday's first law of induction fulfilled by a coil and its magnetic field?
6. What is Lenz's law?
7. Explain Lenz's law in terms of the law of conservation of energy.
8. What is the unit of measure for inductance? In terms of volts per change in current, what does this unit equal?
9. Describe the relationship between the physical dimensions of an inductor and its inductance.

## 13.2 THE PHASE RELATIONSHIP BETWEEN INDUCTOR CURRENT AND VOLTAGE

In any purely resistive circuit, current and voltage are directly proportional (as described by Ohm's law). In a purely inductive circuit, the relationship between current and voltage gets a bit more complicated because the induced voltage is directly proportional to the rate of change in current. The relationship between inductor voltage and current was given earlier in the chapter as ◀ *OBJECTIVE 7*

$$-v_L = L\frac{di}{dt}$$

Assuming that the value of $L$ is constant, the $v_L$ equation indicates that induced voltage:

1. Reaches its maximum value when the rate of change in current ($di/dt$) reaches its maximum value.
2. Reaches its minimum value when the rate of change in current ($di/dt$) reaches its minimum value.

As you will see, the relationship between inductor voltage and $di/dt$ creates a phase angle between inductor current and voltage.

### Sine Wave Values of di/dt

Figure 13.10 shows a current sine wave. At first glance, you'd think that $di/dt$ reaches its maximum value at the peaks. However, the opposite is true. For any current sine wave:

$$\frac{di}{dt} \cong 0 \qquad \text{when} \qquad i = I_{pk}$$

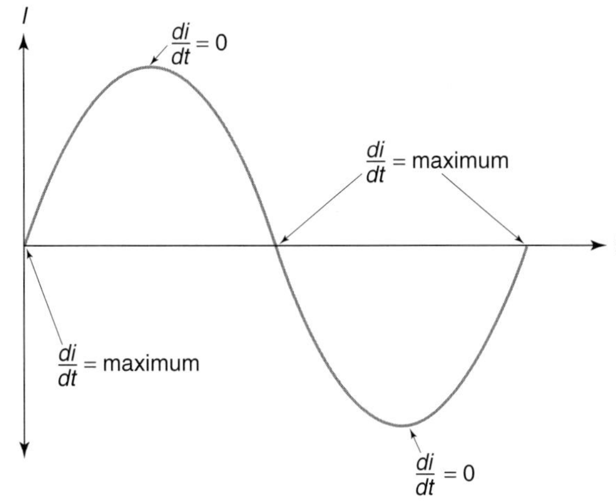


**FIGURE 13.10 Sine wave values of *di/dt*.**

This characteristic is based on the fact that $di/dt$ is an instantaneous rate of change in current. At the instant when current reaches either of its peak values, there is no change in current. At that instant, $di/dt$ drops to approximately zero. As an analogy, picture a ball thrown straight up into the air. When it reaches its maximum height, it must stop for just an instant before beginning to fall. At that instant, the rate of change in the ball's position is zero.

Assuming that $di/dt$ drops to approximately 0 A/s at each of the peaks, it must reach its maximum value somewhere between the peaks. For any current sine wave:

$$\frac{di}{dt} = \text{(its maximum value)} \qquad \text{when} \qquad i = 0$$

The two $di/dt$ relationships given for the waveform in Figure 13.10 indicate that $di/dt$ varies inversely with the instantaneous magnitude of current, which is why the current peaks and zero crossings in Figure 13.10 are labeled as shown.

## The Phase Relationship Between Inductor Voltage and Current

As you know, the voltage induced across a coil at any instant is directly proportional to the value of $di/dt$, as given in the relationship

$$-v_L = L\frac{di}{dt}$$

You also know that $di/dt$ varies inversely with the instantaneous inductor current (as shown in Figure 13.10). These relationships are summarized in Table 13.1.

**TABLE 13.1 The Relationships Among Inductor Instantaneous Values**

| *Current (i)* | *di/dt* | *Induced Voltage ($v_L$)* |
|---|---|---|
| Minimum ($i = 0$ A) | Maximum | Maximum ($v_L = V_{pk}$) |
| Maximum ($i = I_{pk}$) | Minimum | Minimum ($v_L = 0$ V) |

The voltage and current values listed in Table 13.1 indicate that inductor voltage and current are 90° out of phase. This relationship is illustrated in Figure 13.11. Table 13.2 describes the relationship between the inductor current and voltage waveforms.

As shown in Figure 13.11, there is a 90° phase shift between the voltage and current waveforms. If we use the voltage waveform as a reference, the inductor current begins its positive half-cycle 90° after the inductor voltage. Note that this relationship is normally described in one of two ways:

1. Voltage *leads* current by 90°.
2. Current *lags* voltage by 90°.

You will see that the terms *lead* and *lag* are commonly used when describing the phase relationships in ac circuits.

**TABLE 13.2 The Current and Voltage Relationship at Each Time Interval in Figure 13.11**

| *Time* | *Description* |
|---|---|
| $t_1$ | Current is at the zero crossing, $di/dt$ is at its maximum value, and inductor voltage is at its maximum value ($v_L = V_{pk}$). |
| $t_2$ | Current reaches its positive peak, $di/dt = 0$ A/s, and $v_L = 0$ V. |
| $t_3$ | Current crosses zero again, $di/dt$ and $v_L$ reach their peak (maximum) values. |
| $t_4$ | Current reaches its negative peak, $di/dt = 0$ A/s, and $v_L = 0$ V. |

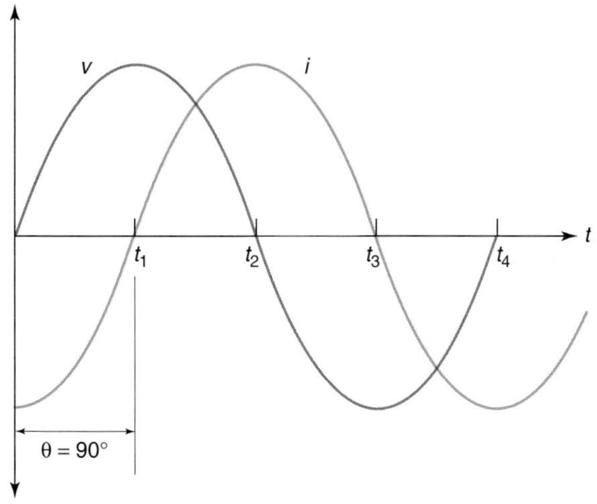

**FIGURE 13.11 Phase relationship between inductor voltage and current.**

## An Important Point

The current versus voltage relationship described in this section applies only to a purely inductive circuit, that is, a circuit that does not contain any significant amount of resistance. If we add one or more resistors to an inductive circuit, we have what is called a resistive-inductive (*RL*) circuit. As you will learn in Chapter 14, the phase relationship between circuit current and source voltage changes when resistance is added to an inductive circuit.

### Section Review

1. What is the relationship between the voltage induced across a coil and the rate of change in current ($di/dt$)?
2. What is the relationship between $di/dt$ and the instantaneous value of inductor current?
3. What is the relationship between the voltage induced across a coil and the instantaneous value of inductor current?
4. Using the above relationships, describe the phase relationship shown in Figure 13.11.
5. Define the terms *lead* and *lag*.

# 13.3 MUTUAL INDUCTANCE

◀ *OBJECTIVE 8*

When one inductor is placed in close proximity to another, the flux produced by each coil can induce a voltage across the other. When the magnetic field produced by one coil induces a voltage across another coil, we have what is called **mutual inductance.** The concept of mutual inductance is illustrated in Figure 13.12. In Figure 13.12a, a coil is shown producing a magnetic field as the result of a change in current. As you know, this magnetic field expands and collapses as the current changes. If a second coil is placed in the magnetic field (as shown in Figure 13.12b), the flux from the first coil cuts through the turns of the second coil. As a result, a voltage is induced across the second coil.

**Mutual inductance**
The inductance produced when the magnetic field generated by one inductor induces a voltage across another inductor in close proximity.

The amount of mutual inductance between two (or more) coils depends on several factors. First, and perhaps most important, is the distance between the coils. To have

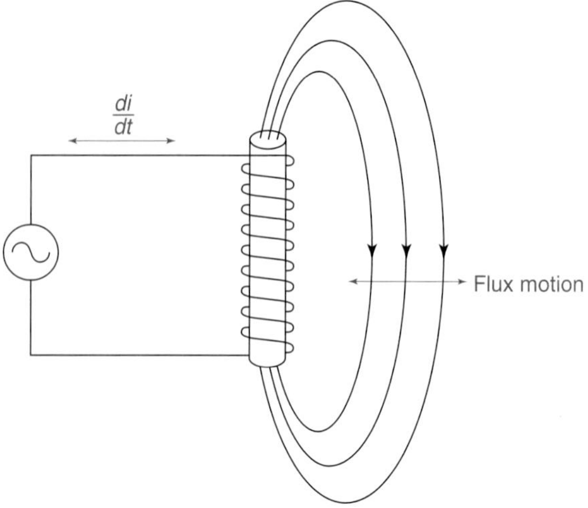


(a) A changing current produces a changing magnetic field.

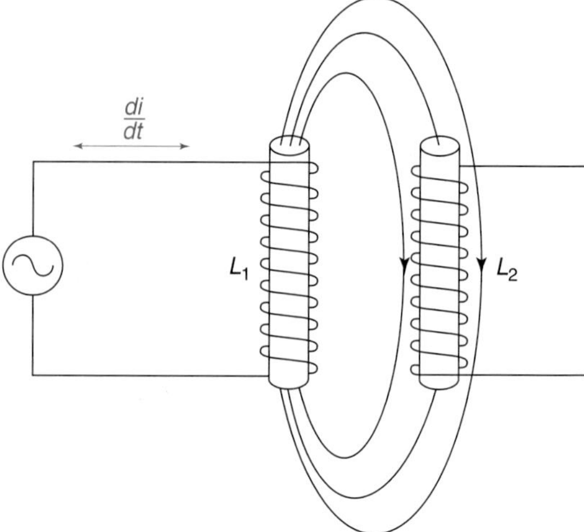


(b) A changing magnetic field induces a voltage across a second inductor.

**FIGURE 13.12 Mutual inductance.**

mutual inductance, the coils must be near enough to each other for the flux from one coil to pass through the turns of the other. If the coils are far enough apart, the flux from one won't pass through the turns of the other, and there is no mutual inductance.

Since induced voltage is affected by the physical dimension of a coil and the permeability of the core, these two factors also affect mutual inductance. For example, consider the two coils in Figure 13.12b. A high-inductance coil typically has:

1. A relatively high number of turns per unit length.
2. A high-permeability core.

A low-inductance coil typically has:

1. A relatively low number of turns per unit length.
2. A low-permeability core.

If $L_2$ in Figure 13.12b is a high-inductance coil, the flux will induce more voltage across the component than it will across a low-inductance coil.

The relationship between mutual inductance and the dimensions of the coils involved is expressed in the following equation:

$$L_M = \frac{\mu_m N_1 N_2 A}{\ell} \quad \textbf{(13.7)}$$

where $L_M$ = the mutual inductance between the coils, in henries (H)
$\mu_m$ = the permeability of the cores
$N_1, N_2$ = the number of turns in each of the coils
$A$ = the cross-sectional area of the cores, in square meters
$\ell$ = the length of the coils, in meters

Note that this equation assumes the following:

**Unity coupling**
A term used to describe a situation where 100% of the flux generated by one coil cuts through a parallel coil.

1. The two coils have identical cores and equal lengths.
2. There is **unity coupling** between the coils, which means that:
   a. The coils are physically parallel to each other.
   b. 100% of the flux from the first coil cuts through the second coil; that is, the coils are close enough to each other for all the flux from the first coil to pass through the second.

Note that two or more components (or circuits) are said to be **coupled** when energy is transferred from one to the other. In this case, energy is transferred between inductors via the lines of force that pass from one to the other.

**Coupling** When energy is transferred between two or more components (or circuits).

If any of the conditions listed are not met, equation (13.7) cannot be used to determine the value of mutual inductance between two coils. In this case, $L_M$ is best determined through actual circuit measurement. This point will be discussed in more detail later in this section.

## *Coefficient of Coupling (k)*

◄ *OBJECTIVE 9*

The **coefficient of coupling (*k*)** between two (or more) coils is a measure of the degree of coupling that takes place between the coils. The coefficient of coupling for a pair of coils can be found as

**Coefficient of coupling (*k*)** A measure of the degree of coupling that takes place between two (or more) coils.

$$k = \frac{\Phi_2}{\Phi_1} \quad \textbf{(13.8)}$$

where $\Phi_1$ = the amount of flux generated by coil $L_1$

$\Phi_2$ = the amount of $\Phi_1$ that passes through coil $L_2$ at a 90° angle to the turns of the coil

For example, let's say that two coils are positioned as shown in Figure 13.13a. Assuming that all the flux from coil $L_1$ passes through coil $L_2$:

*Note:* Since it is a ratio, *k* has no units of measure.

$$\Phi_2 = \Phi_1 \quad \text{and} \quad k = 1$$

This case, by definition, is unity coupling. Note that a value of $k = 1$ cannot be achieved in practice because the flux generated by the first coil expands in all directions. Therefore, it is impossible for all the flux to pass through coil 2, and *k* always has a value less than one.

*A Practical Consideration:* Unity coupling cannot be achieved in practice. However, by wrapping two coils around the same form, values of $k > 0.9$ can be achieved.

The greater the distance between two coils, the lower the value of *k*. For example, in Figure 13.13b, very little of the flux from coil 1 is reaching coil 2. Therefore:

$$\Phi_2 << \Phi_1 \quad \text{and} \quad k << 1$$

If the coils in Figure 13.13b are spaced far enough from each other, none of $\Phi_1$ will pass through coil 2. In this case, $k = 0$, which means that there is no mutual inductance between the coils.

In the case of 13.13c, the flux from coil 1 passes through coil 2. However, since the coils are perpendicular to each other, the lines of flux are parallel to the turns of coil 2. Therefore, no voltage is induced in the second coil, and $k = 0$. Once again, there is no mutual inductance between the coils.

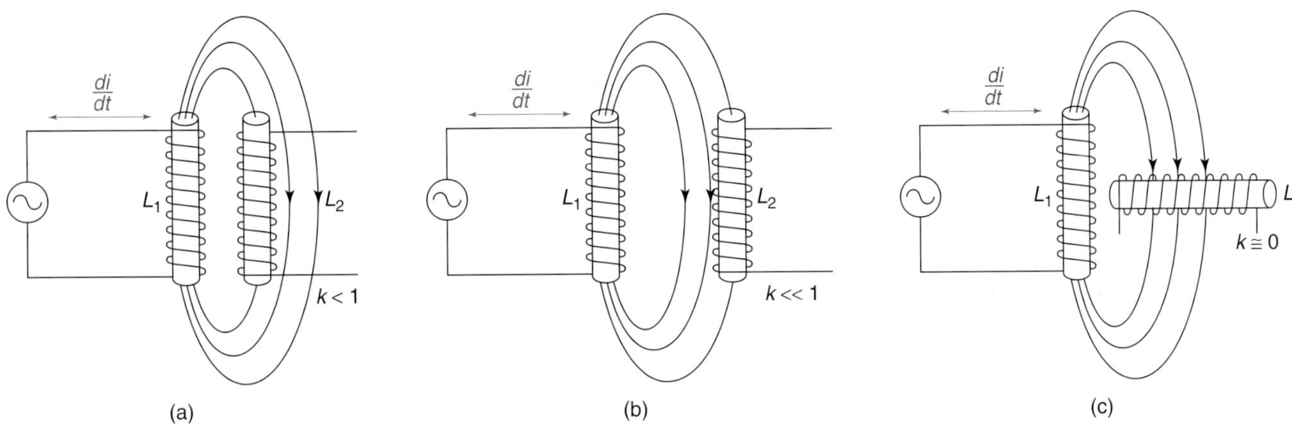


**FIGURE 13.13 Factors affecting coefficient of coupling.**

It is important to note that the coefficient of coupling ($k$) between two coils is not measured easily in practice. However, the effect of $k$ on mutual inductance can be expressed as:

$$L_M = k\sqrt{L_1 L_2} \qquad \textbf{(13.9)}$$

where $k$ = the coefficient of coupling between the coils

$\sqrt{L_1 L_2}$ = the geometric average of the coil inductance values, in henries

Later in this section, you will be shown how equation (13.9) can be used to help measure the coefficient of coupling between two coils.

## Series-Connected Coils

*OBJECTIVE 10* ▶ When $k = 0$ for a group of series-connected inductors, the total inductance of the circuit can be found as

$$L_T = L_1 + L_2 + \ldots + L_n \qquad \textbf{(13.10)}$$

where $L_n$ = the highest numbered inductor in the circuit

As you can see, total inductance (in this case) is found in the same fashion as total resistance in a series circuit. The use of this equation is demonstrated in Example 13.3.

### EXAMPLE 13.3

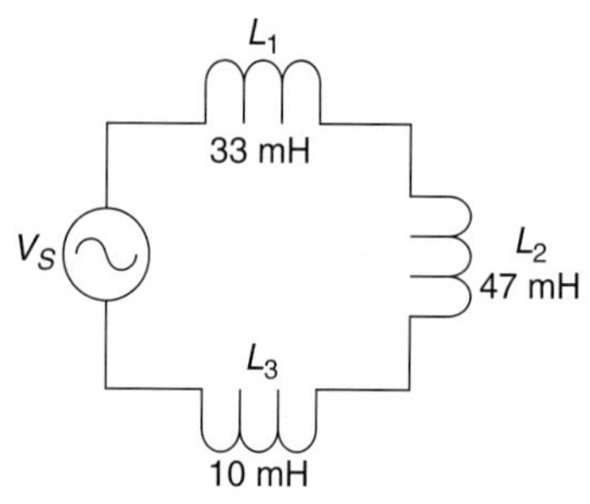


**FIGURE 13.14**

The coils in Figure 13.14 are spaced sufficiently apart so that there is no mutual inductance between them ($k = 0$). Calculate the total inductance in the circuit.

***Solution:*** Since $k = 0$, we can simply add the individual component values, as follows:

$$L_T = L_1 + L_2 + L_3 = 33 \text{ mH} + 47 \text{ mH} + 10 \text{ mH} = \mathbf{90\ mH}$$

***PRACTICE PROBLEM 13.3***

A circuit like the one in Figure 13.14 has the following values: $L_1 = 100\ \mu\text{H}$, $L_2 = 3.3$ mH, and $L_2 = 330\ \mu\text{H}$. Assuming that $k = 0$ throughout the circuit, calculate the value of $L_T$.

*OBJECTIVE 11* ▶ When $k \neq 0$ for series-connected inductors, their mutual inductance becomes a factor in the total circuit inductance. The total inductance of two coils that have some mutual inductance between them is found as

$$L_T = L_1 + L_2 \pm 2L_M \qquad \textbf{(13.11)}$$

As you can see, the value of $2L_M$ is added to (or subtracted from) the sum of $L_1$ and $L_2$. Whether the value of $2L_M$ is added or subtracted depends on how the components are connected, as illustrated in Figure 13.15.

When the inductors are connected as shown in Figure 13.15a, the magnetic poles are such that the flux produced by the coils combine. As a result, the overall flux is increased, which has the same effect as increasing the inductance. In this case, the inductors are in a **series-aiding** configuration, and

**Series-aiding**
A two-inductor series connection where the mutual inductance results in a total inductance greater than the sum of the individual component values.

$$L_T = L_1 + L_2 + 2L_M$$

When connected as shown in Figure 13.15b, the flux generated by each coil opposes the flux generated by the other. This results in a reduction in the overall flux, which has the

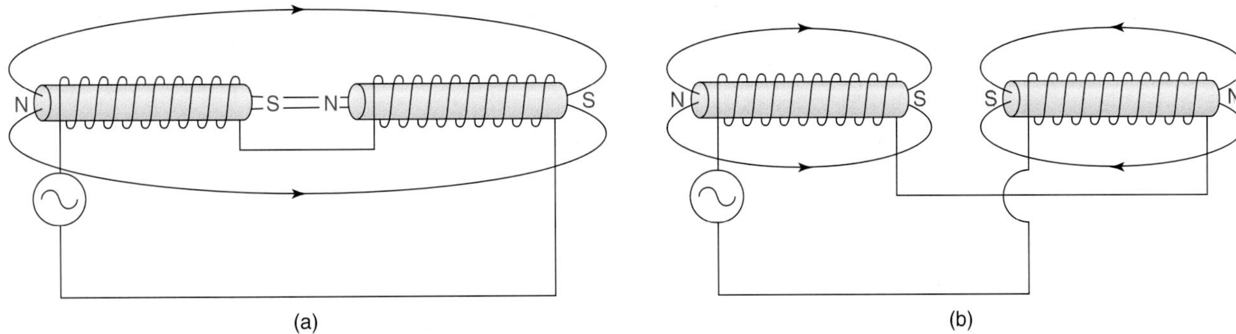


**FIGURE 13.15 Series-aiding and series-opposing connections.**

same effect as decreasing the total inductance. In this case, the inductors are in a **series-opposing** configuration, and

$$L_T = L_1 + L_2 - 2L_M$$

The effects of mutual inductance on series inductance calculations are demonstrated in the following examples. Example 13.4 demonstrates the effect that mutual inductance has on the total inductance of a series-aiding configuration.

**Series-opposing**
A two-inductor series connection where the mutual inductance results in a total inductance less than the sum of the individual component values.

EXAMPLE 13.4

The inductors in Figure 13.16 are connected in series-aiding fashion. Assume that the coefficient of coupling between the two is 0.6, and calculate the total series inductance in the circuit.

**FIGURE 13.16**

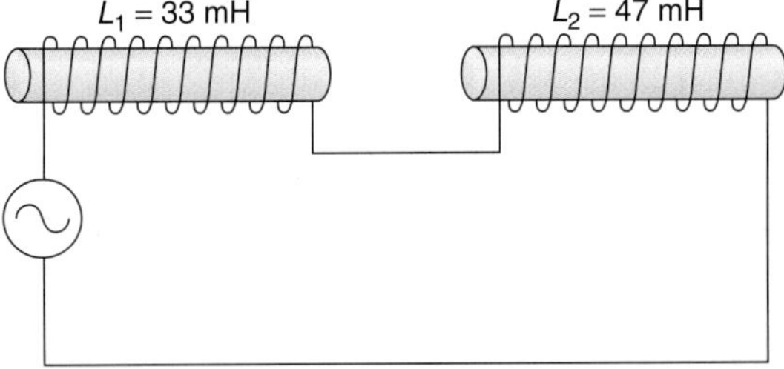


***Solution:*** First, the value of $L_M$ is found as

$$L_M = k\sqrt{L_1 L_2} = (0.6)\sqrt{(33 \text{ mH})(47 \text{ mH})} = (0.6)(39.4 \text{ mH}) = \mathbf{23.6\ mH}$$

Since the inductors are connected in series-aiding fashion, the total inductance is found as

$$\begin{aligned} L_T &= L_1 + L_2 + 2L_M \\ &= 33 \text{ mH} + 47 \text{ mH} + 2(23.6 \text{ mH}) \\ &= 33 \text{ mH} + 47 \text{ mH} + 47.2 \text{ mH} \\ &= \mathbf{127.2\ mH} \end{aligned}$$

As you can see, this value is significantly higher than the sum of the individual inductors. This drastic an increase is entirely possible when inductors are connected in series-aiding fashion.

***PRACTICE PROBLEM 13.4***

A circuit like the one in Figure 13.16 has the following values: $L_1 = 10$ mH, $L_2 = 22$ mH, and $k = 0.4$. Calculate the total inductance in the circuit.

The mutual inductance generated by series-opposing inductors can significantly reduce the total inductance in the circuit. This point is demonstrated in Example 13.5.

## EXAMPLE 13.5

The inductors in Figure 13.17 are connected in series-opposing fashion. Assuming that the coefficient of coupling between the two is 0.55, calculate the total inductance in the circuit.

**FIGURE 13.17**

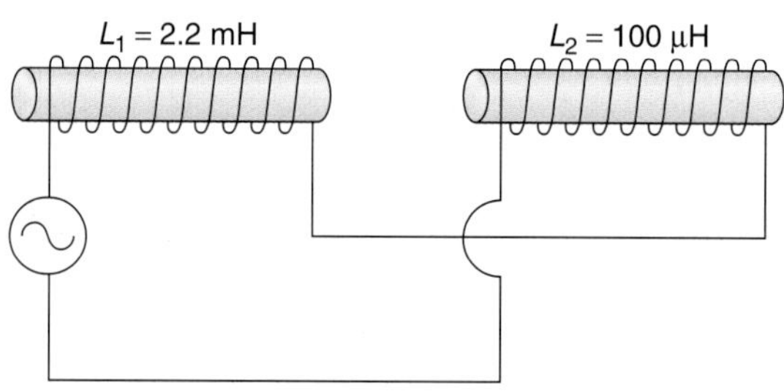


***Solution:*** Again, we begin by calculating the value of $L_M$ as follows:

$$\begin{aligned} L_M &= k\sqrt{L_1 L_2} \\ &= (0.55)\sqrt{(2.2\ \text{mH})(100\ \mu\text{H})} \\ &= (0.55)(469\ \mu\text{H}) \\ &= \mathbf{258\ \mu H} \end{aligned}$$

Since the inductors are connected in series-opposing fashion, the total inductance is found as

$$\begin{aligned} L_T &= L_1 + L_2 - 2L_M \\ &= 2.2\ \text{mH} + 100\ \mu\text{H} - 2(258\ \mu\text{H}) \\ &= 2.2\ \text{mH} + 100\ \mu\text{H} - 516\ \mu\text{H} \\ &= \mathbf{1.784\ mH} \end{aligned}$$

As you can see, the value of $L_T$ in this case is actually lower than the value of the largest inductor.

***PRACTICE PROBLEM 13.5***

A circuit like the one in Figure 13.17 has the following values: $L_1 = 4.7$ mH, $L_2 =$ 1 mH, and $k = 0.3$. Calculate the total inductance in the circuit.

*OBJECTIVE 12* ▸

## Measuring the Coefficient of Coupling ($k$)

**LCR bridge**
A piece of test equipment used to measure inductance ($L$), capacitance ($C$), and resistance ($R$).

It is possible to measure the coefficient of coupling between two inductors using an **LCR bridge.** An LCR bridge is a piece of equipment used to measure inductance ($L$), capacitance ($C$), and resistance ($R$).

If you have access to an LCR bridge, the mutual inductance between two series-connected inductors can be determined as follows:

1. Measure the value of each inductor.
2. Measure the total inductance of the pair.
3. Subtract the sum of the measured inductor values from the measured total inductance. If the result is zero, there is no mutual inductance in the circuit.

4. If the result from Step 3 does not equal zero, the mutual inductance between the inductors can be determined using the following equation:

$$L_M = \frac{L_T - (L_1 + L_2)}{2} \tag{13.12}$$

Note that equation (13.12) is merely a transposed form of equation (13.11). If the result of this equation is negative, the inductors are connected in series-opposing fashion. If the result is positive, they are connected in series-aiding fashion.

Once you have determined the actual values of $L_1$, $L_2$, and $L_M$, the coefficient of coupling can be found using the following equation:

$$k = \frac{|L_M|}{\sqrt{L_1 L_2}} \tag{13.13}$$

*A Practical Consideration:* Since $k$ is the ratio of one positive value to another, it is always positive, regardless of the circuit configuration. This is why we use the absolute value of $L_M$ in equation (13.12).

In this case, we are using a transposed form of equation (13.9). The following example demonstrates the entire procedure for measuring mutual inductance and the coefficient of coupling.

## EXAMPLE 13.6

Determine the coefficient of coupling for the circuit in Figure 13.18.

***Solution:*** The values of $L_1$, $L_2$, and $L_T$ for the circuit were measured using an LCR bridge. These measured values were as follows:

$$L_1 = 3.45 \text{ mH} \qquad L_2 = 1.15 \text{ mH} \qquad L_T = 5.8 \text{ mH}$$

Since $L_T \neq L_1 + L_2$, we know that there is mutual inductance between the components. The amount of mutual inductance is now found as:

$$\begin{aligned} L_M &= \frac{L_T - (L_1 + L_2)}{2} \\ &= \frac{5.8 \text{ mH} - (3.45 \text{ mH} + 1.15 \text{ mH})}{2} \\ &= \frac{1.2 \text{ mH}}{2} \\ &= \mathbf{0.6\ mH} \end{aligned}$$

Since the value of $L_M$ is positive, we know that the inductors are connected in series-aiding fashion. Finally, the coefficient of coupling between the inductors can be found as

$$\begin{aligned} k &= \frac{|L_M|}{\sqrt{L_1 L_2}} \\ &= \frac{0.6 \text{ mH}}{\sqrt{(3.45 \text{ mH})(1.15 \text{ mH})}} \\ &= \frac{0.6 \text{ mH}}{1.99 \text{ mH}} \\ &= \mathbf{0.3} \end{aligned}$$

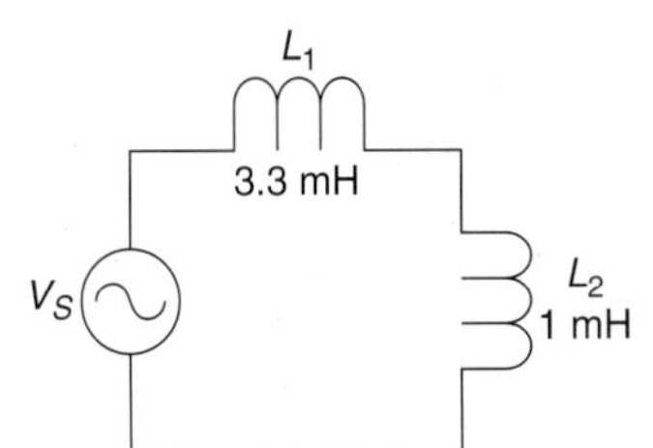


**FIGURE 13.18**

***PRACTICE PROBLEM 13.6***

An LCR bridge was used to measure the inductance values in a circuit like the one in Figure 13.18. The readings obtained were as follows: $L_1 = 2.29$ mH, $L_2 = 358$ μH, and $L_T = 3.1$ mH. Determine the coefficient of coupling (if any) between the inductors.

## Parallel-Connected Coils

OBJECTIVE 13 ▶ When $k = 0$ for a group of parallel-connected coils, the total inductance of the circuit can be found as

$$L_T = \frac{1}{\dfrac{1}{L_1} + \dfrac{1}{L_2} + \ldots + \dfrac{1}{L_n}} \tag{13.14}$$

where $L_n$ = the value of the highest numbered inductor in the circuit

As you can see, total inductance is calculated in the same fashion as total resistance in a parallel circuit. This means that we can also calculate total parallel inductance using

$$L_T = \frac{L_1 L_2}{L_1 + L_2} \tag{13.15}$$

for an inductive circuit containing only two branches. For a circuit containing any number of branches with equal inductances, the total inductance can be found using

$$L_T = \frac{L}{n} \tag{13.16}$$

If you compare equations (13.15) and (13.16) with the resistance equations given in Chapter 6, you'll see that they are identical formats used under the same conditions. The following series of examples demonstrates the applications of equations (13.14) through (13.16).

### EXAMPLE 13.7

The circuit in Figure 13.19 has a value of $k = 0$. Calculate the total inductance in the circuit.

**FIGURE 13.19**

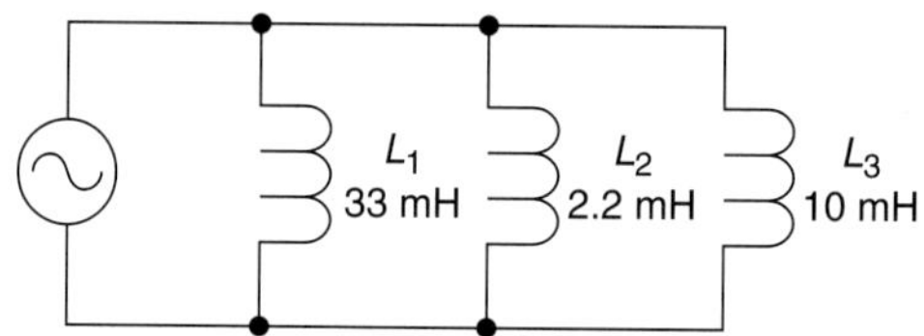


***Solution:*** Using equation (13.14), the total inductance in the circuit can be found as

$$\begin{aligned} L_T &= \frac{1}{\dfrac{1}{L_1} + \dfrac{1}{L_2} + \dfrac{1}{L_3}} \\ &= \frac{1}{\dfrac{1}{33 \text{ mH}} + \dfrac{1}{2.2 \text{ mH}} + \dfrac{1}{10 \text{ mH}}} \\ &= \frac{1}{0.5848 \text{ mH}} \\ &= \mathbf{1.71 \text{ mH}} \end{aligned}$$

***PRACTICE PROBLEM 13.7***

A circuit like the one in Figure 13.19 has the following values: $L_1 = 330\ \mu\text{H}$, $L_2 = 1$ mH, $L_3 = 470\ \mu\text{H}$, and $k = 0$. Calculate the total inductance in the circuit.

If you compare the result in Example 13.7 with the given branch values, you'll see another similarity between parallel inductive and resistive circuits. In any parallel circuit, the total inductance is lower than any of the branch values.

## EXAMPLE 13.8

Calculate the total inductance in the circuit shown in Figure 13.20. Assume that $k = 0$ for the circuit.

**FIGURE 13.20**

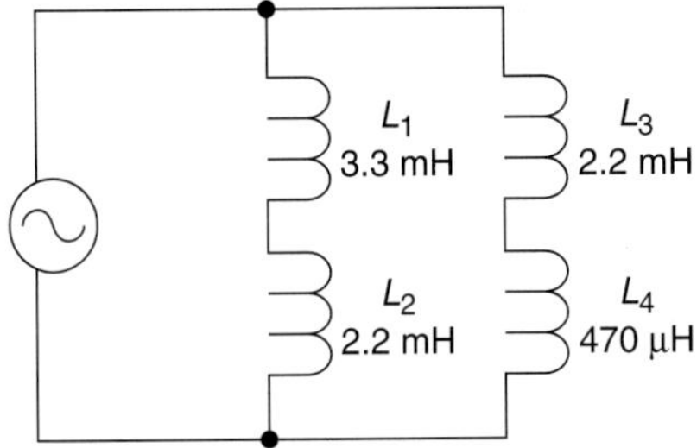


***Solution:*** The total inductance in the first branch ($L_{T(1)}$) is found as

$$L_{T(1)} = L_1 + L_2 = 3.3 \text{ mH} + 2.2 \text{ mH} = \mathbf{5.5 \text{ mH}}$$

The total inductance in the second branch ($L_{T(2)}$) is found as

$$L_{T(2)} = L_3 + L_4 = 2.2 \text{ mH} + 470 \ \mu\text{H} = \mathbf{2.67 \text{ mH}}$$

Now, using equation (13.15), the total inductance in the circuit can be found as

$$L_T = \frac{L_{T(1)}L_{T(2)}}{L_{T(1)} + L_{T(2)}} = \frac{(5.5 \text{ mH})(2.67 \text{ mH})}{5.5 \text{ mH} + 2.67 \text{ mH}} \cong \mathbf{1.8 \text{ mH}}$$

In this case, $L_T$ is lower than the total inductance of either branch.

### *PRACTICE PROBLEM 13.8*

A circuit like the one in Figure 13.20 has the following values: $L_1 = 10$ mH, $L_2 = 1$ mH, $L_3 = 4.7$ mH, $L_4 = 3.3$ mH, and $k = 0$. Calculate the total inductance in the circuit using equation (13.15).

## EXAMPLE 13.9

An inductive circuit contains six 220 mH inductors wired in parallel. Assume that $k = 0$ and determine the total inductance in the circuit.

***Solution:*** The total inductance (in this case) can be found as

$$L_T = \frac{L}{n} = \frac{220 \text{ mH}}{6} = \mathbf{36.67 \text{ mH}}$$

### *PRACTICE PROBLEM 13.9*

An inductive circuit contains four 330 mH inductors wired in parallel. Assume that $k = 0$ and determine the total inductance in the circuit.

If two parallel inductors are wired in close proximity to each other, there may be some measurable amount of mutual inductance between them. In this case, the total inductance in the circuit can be found as

$$L_T = \frac{(L_1 \pm L_M)(L_2 \pm L_M)}{(L_1 \pm L_M) + (L_2 \pm L_M)} \tag{13.17}$$

The following example demonstrates the effect of mutual inductance on total parallel inductance.

## EXAMPLE 13.10

The inductors in Figure 13.21 are wired in parallel-aiding fashion with a value of $k = 0.5$. Determine the total inductance in the circuit.

**FIGURE 13.21**

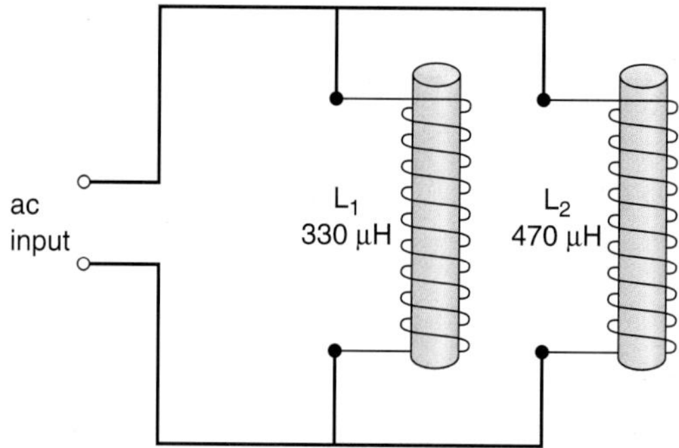


***Solution:*** If the circuit had a value of $k = 0$, the total inductance would be found as

$$\boldsymbol{L_T} = \frac{L_1 L_2}{L_1 + L_2} = \frac{(330\ \mu\text{H})(470\ \mu\text{H})}{330\ \mu\text{H} + 470\ \mu\text{H}} = \mathbf{194\ \mu H}$$

However, since $k = 0.5$, there is mutual inductance between the coils. The value of this mutual inductance is found as

$$\boldsymbol{L_M} = k\sqrt{L_1 L_2} = (0.5)\sqrt{(330\ \mu\text{H})(470\ \mu\text{H})} = (0.5)(394\ \mu\text{H}) = \mathbf{197\ \mu H}$$

With 197 μH of mutual inductance, the total inductance in the circuit is found as

$$\begin{aligned}\boldsymbol{L_T} &= \frac{(L_1 + L_M)(L_2 + L_M)}{(L_1 + L_M) + (L_2 + L_M)} \\ &= \frac{(330\ \mu\text{H} + 197\ \mu\text{H})(470\ \mu\text{H} + 197\ \mu\text{H})}{(330\ \mu\text{H} + 197\ \mu\text{H}) + (470\ \mu\text{H} + 197\ \mu\text{H})} \\ &= \frac{(527\ \mu\text{H})(667\ \mu\text{H})}{527\ \mu\text{H} + 667\ \mu\text{H}} \\ &= \mathbf{294\ \mu H}\end{aligned}$$

***PRACTICE PROBLEM 13.10***

A circuit like the one in Figure 13.21 has the following values: $L_1 = 1$ mH and $L_2 = 22$ mH. Determine the total inductance in the circuit when $k = 0$ and when $k = 0.25$.

## A Practical Consideration

Calculating the total inductance in any series or parallel network is rarely as complicated as presented here. In the course of circuit design and layout, the potential effects of mutual inductance are usually taken into account and avoided; that is, coils are normally positioned to reduce (or eliminate) any mutual inductance. Therefore, you will rarely be required to account for mutual inductance in the analysis of any circuit.

So why spend so much energy calculating the effects of mutual inductance on total inductance? You may find at times that there is a discrepancy between measured component values and total inductance. When this is the case, you shouldn't be surprised or confused. The probable cause is some amount of mutual inductance that is increasing (or reducing) the total inductance in the circuit.

### Section Review

1. What is mutual inductance?
2. List the factors that determine the amount of mutual inductance between two (or more) coils.
3. What is unity coupling? What do we mean when we say that two (or more) components are electrically coupled?
4. What is indicated by the coefficient of coupling ($k$) between two coils?
5. How do you determine the total inductance in a series circuit when $k = 0$? When $k \neq 0$?
6. What type of series connection is indicated by each of the following conditions?
   a. $k = 0$
   b. $k$ has a positive value.
   c. $k$ has a negative value.
7. Describe how an LCR bridge can be used to measure the coefficient of coupling between two inductors.
8. How do you determine the total inductance in a parallel circuit when $k = 0$? When $k \neq 0$?

## 13.4 INDUCTIVE REACTANCE ($X_L$)

◄ *OBJECTIVE 14*

Any component that has a voltage across it and a current through it is providing some measurable opposition to that current. For example, let's say the resistor in Figure 13.22a is limiting the circuit current to 1 mA, as indicated by the ac ammeter. The opposition to current (resistance) provided by the resistor is found as $R = V/I$, in this case, 4 kΩ.

Like a resistor, an inductor with a sine wave input provides a measurable amount of opposition to current. For example, consider the inductive waveforms and circuit shown in Figure 13.22b. As the waveforms indicate, there is a 90° phase difference between inductor current and voltage. At the same time, each waveform has a measurable rms (or effective) value. If we were to measure these rms values, we could determine the total opposition to current provided by the inductor. Using the rms values shown in the figure, this opposition could be found as

$$\text{Opposition} = \frac{V_{\text{rms}}}{I_{\text{rms}}} = \frac{10\text{ V}}{1\text{ mA}} = 10\text{ k}\Omega$$

### Inductor Resistance

As you know, an inductor is a wire wrapped into a series of loops. If we were to measure the resistance of a coil, we would obtain an extremely low reading (since wire has little resistance). For example, if the coil in Figure 13.22 was made up of 2 ft of 30 gage wire, its resistance could be approximated as

$$R = (\text{length}) \times (\text{ohms/unit length}) = (2\text{ ft}) \times \frac{103.2\ \Omega}{1000\text{ ft}} \cong 0.21\ \Omega$$

*Reference:* The value of 103.2 Ω/1000 ft for 30 gage wire was obtained from Table 3.1.

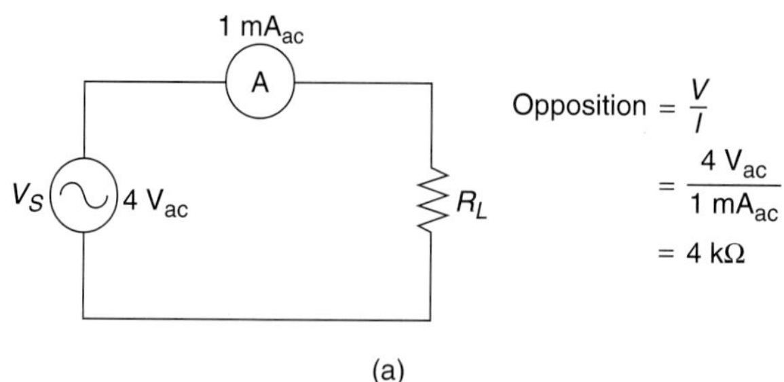

(a)

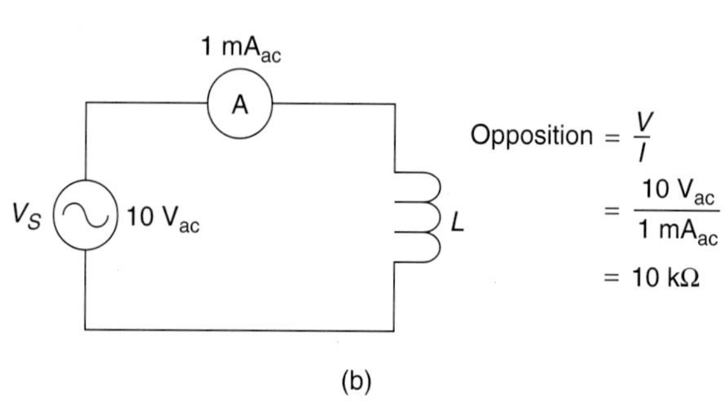

(b)

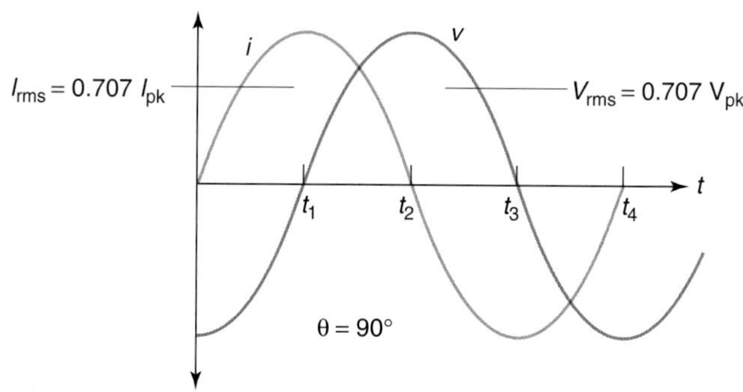


**FIGURE 13.22**

This is significantly lower than the 10 kΩ of opposition that we calculated using Ohm's law. Assuming that all the calculations are correct, there is only one way to account for the difference between our calculated values of 0.21 Ω and 10 kΩ: the inductor is providing some type of opposition to current other than resistance. This "other" opposition to current is known as **inductive reactance ($X_L$).**

## Reactance (X)

Earlier in this chapter, you were told that inductance can be defined as the ability of a component to oppose a change in current. In an inductive circuit with a sinusoidal input (like the one in Figure 13.22b), the current provided by the source is constantly changing. The opposition (in ohms) that an inductor presents to this changing current is referred to as *inductive reactance* ($X_L$). Note that the letter $X$ is used to designate reactance, and the subscript $L$ is used to identify the reactance as inductive. (In Chapter 16, you will be introduced to another type of reactance.)

**Inductive reactance ($X_L$)** The opposition (in ohms) that an inductor presents to a changing current.

## Calculating the Value of $X_L$

You have already been shown one way to calculate the value of $X_L$ for an inductor. When the rms values of inductor current and voltage are known, the reactance of the component can be found as

$$X_L = \frac{V_{rms}}{I_{rms}} \tag{13.18}$$

The use of this equation was demonstrated in Figure 13.22b.

*OBJECTIVE 15* ▶

There is another (and more commonly used) method for calculating the reactance of an inductor. When the value of an inductor and the frequency of its sine wave source are known, the reactance of the inductor can be found as

$$X_L = 2\pi f L \tag{13.19}$$

We will discuss the basis for this equation in a moment. Its use is demonstrated in Example 13.11.

**EXAMPLE 13.11**

Calculate the value of $X_L$ for the inductor in Figure 13.23.

EWB

**FIGURE 13.23**

***Solution:*** The circuit is shown to have values of $L = 1$ mH and $f = 50$ kHz. Using these values in equation (13.18), we get

$$X_L = 2\pi fL = 2\pi(50\text{ kHz})(1\text{ mH}) \cong \mathbf{314\ \Omega}$$

***PRACTICE PROBLEM 13.11***

A 470 mH inductor is operated at a frequency of 20 kHz. Calculate the reactance provided by the component.

Equation (13.19) may appear strange at first, but you'll see that it is based on several principles introduced earlier in the chapter. In terms of applications, equation (13.19) is usually easier to use than equation (13.18); all you need to know is the value of the inductor and the frequency of operation. Equation (13.18), on the other hand, requires that you calculate (or measure) the circuit values of voltage and current. In many cases, this is easier said than done.

*A Practical Consideration:* Nearly all electronics certification exams (like the FCC, CET, and NARTE exams) require that you know the relationship in equation (13.18). You may want to commit the equation (and its meaning) to memory.

## The Basis for Equation (13.19)

Equation (13.19) is based on the following relationships, which were established earlier in the chapter:

1. Induced voltage is proportional to the rate of change in current.
2. Induced voltage is proportional to inductance.

The instantaneous value of a sinusoidal current is proportional to the sine of the phase ($\theta$). Therefore, the rate of change in current is proportional to the rate of change in phase. You may recall that the rate of change in phase for a sine wave equals its angular velocity, found as $\omega = 2\pi f$. Based on these relationships, we can say that *induced voltage varies directly with angular velocity.* By formula,

$$V_{\text{rms}} \propto \omega \qquad \text{or} \qquad V_{\text{rms}} \propto 2\pi f$$

As the above equation indicates, *induced voltage is directly proportional to the operating frequency of the circuit.* It is also directly proportional to the value of coil inductance. You may recall that

$$-v_L = L\frac{di}{dt}$$

This equation indicates that the greater the inductance of a coil, the greater the voltage produced at a given rate of change in current. Since $V_{\text{rms}}$ is a function of all the values of $v_L$ in a cycle, we can say that

$$V_{\text{rms}} \propto L$$

Earlier, we said that

$$X_L = \frac{V_{rms}}{I_{rms}}$$

This relationship indicates that, for any given value of $I_{rms}$, inductive reactance is directly proportional to inductor voltage. Since $V_{rms}$ is directly proportional to operating frequency and inductance, $X_L$ must also be directly proportional to these values.

Having established the direct proportionality among $X_L$, operating frequency, and inductance, we can combine these factors into a single relationship:

$$X_L = \omega L \qquad \text{or} \qquad X_L = 2\pi f L$$

which is equation (13.19).

## $X_L$ and Ohm's Law

*OBJECTIVE 16* ► In an inductive circuit, inductive reactance ($X_L$) can be used with Ohm's law to determine the total circuit current, just like resistance. The analysis of a simple inductive circuit is demonstrated in Example 13.12.

### EXAMPLE 13.12

Calculate the total current for the circuit shown in Figure 13.24.

***Solution:*** Using the values shown, the circuit current is found as

$$\boldsymbol{I} = \frac{V_S}{X_L} = \frac{12\ \text{V}_{ac}}{1\ \text{k}\Omega} = \mathbf{12\ mA}$$

FIGURE 13.24

***PRACTICE PROBLEM 13.12***

A circuit like the one in Figure 13.24 has the following values: $V_S = 18\ \text{V}_{ac}$ and $X_L = 5\ \text{k}\Omega$. Calculate the total circuit current.

When the values of inductance, source voltage, and operating frequency for a circuit are known, the circuit current can be calculated as demonstrated in Example 13.13.

### EXAMPLE 13.13

Calculate the total current for the circuit in Figure 13.25a.

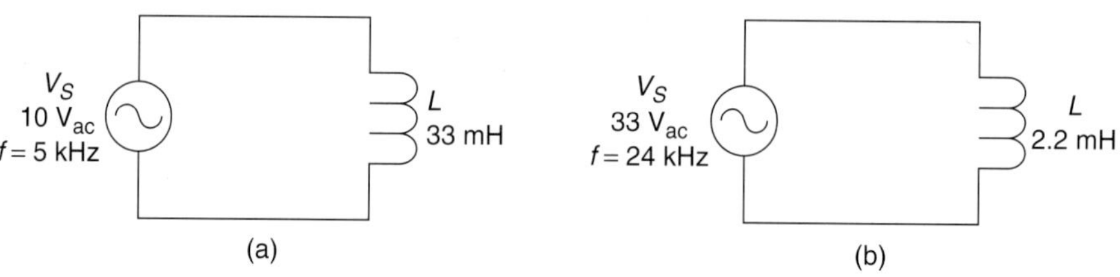


FIGURE 13.25

***Solution:*** First, the reactance of the coil is found as

$$X_L = 2\pi f L = 2\pi(5\ \text{kHz})(33\ \text{mH}) = \mathbf{1.04\ k\Omega}$$

Now, using $X_L$ and the value of the applied voltage, the circuit current is found as

$$I = \frac{V_S}{X_L} = \frac{10\ \text{V}_{\text{ac}}}{1.04\ \text{k}\Omega} = \mathbf{9.62\ mA}$$

***PRACTICE PROBLEM 13.13***

Calculate the value of current for the circuit in Figure 13.25b.

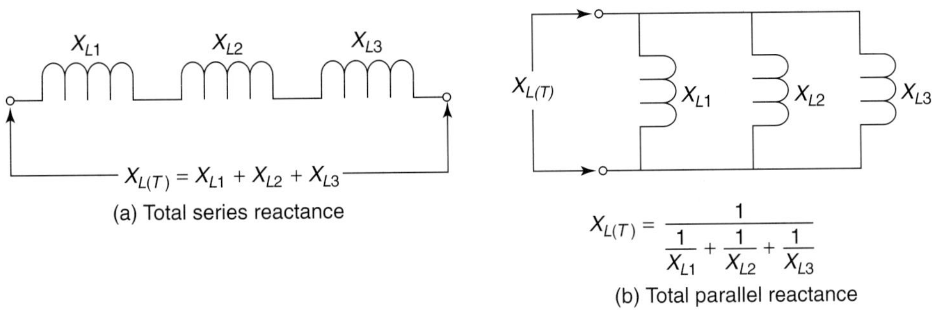


EWB

**FIGURE 13.26 Total series and parallel reactance.**

## Series and Parallel Reactances

When any number of reactive components are connected in series or parallel, the total reactance is found in the same fashion as total resistance. This point is illustrated in Figure 13.26. (Since we have used similar equations for resistance calculations, we do not need to rework them here.)

## Resistance, Reactance, and Impedance (Z)

Resistance is usually a static value; that is, the value of resistance in a circuit usually does not vary when circuit conditions change. (When you study electronic devices, you will learn of several exceptions to this statement.) Reactance, on the other hand, is a dynamic value. As you know, the reactance of an inductor does change when circuit operating frequency changes. Therefore, $X_L$ is a dynamic value. Although they are both oppositions to current (measured in ohms), resistance and reactance cannot be added algebraically to determine the total opposition to current in a circuit.

In Chapter 14, you will learn how to combine resistance and reactance into a single value called **impedance (Z).** Technically defined, impedance is the total opposition to current in an ac circuit, consisting of resistance and/or reactance. As you will learn, impedance is found as the geometric (vector) sum of resistance and reactance, expressed in ohms.

**Impedance**
The total opposition to current in an ac circuit, consisting of resistance and/or reactance. The geometric sum of resistance and reactance, expressed in ohms.

### Section Review

1. How is the opposition to current provided by an inductor measured?
2. Why do inductors have extremely low resistance values?
3. What is inductive reactance ($X_L$)?
4. What two relationships are commonly used to calculate the value of $X_L$?
5. Which of the relationships in Question 4 is used most often? Why?

6. In your own words, explain the basis of the following equation: $X_L = 2\pi fL$.
7. How is total current determined in a basic inductive circuit?
8. How is total reactance calculated in a series circuit? In a parallel circuit?
9. What is impedance (Z)?

# 13.5 RELATED TOPICS

In this section, we will complete the chapter by discussing some miscellaneous topics related to inductors, their applications, and ratings.

## Apparent Power ($P_{APP}$)

*OBJECTIVE 17* ►

When used in an ac circuit, any inductor has measurable values of voltage and current. If these values are plugged into the $P = IV$ equation, they provide a numeric result (as does any combination of voltage and current). For the circuit in Figure 13.27, this power equation yields a result of

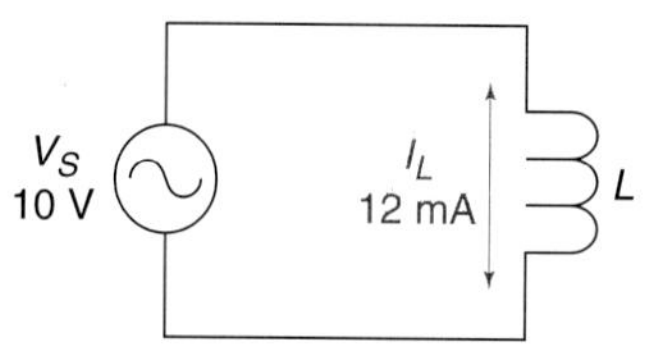


**FIGURE 13.27**

$$P = V_S I_L = (10\text{ V})(12\text{ mA}) = 120\text{ mW}$$

The only problem is that the result of this calculation is not a true indicator of energy used. As such, it is referred to as **apparent power ($P_{APP}$)**. The term *apparent power* is used because most of the energy is actually stored in the electromagnetic field generated by the component. Only a small portion of the value obtained is actually dissipated by the component. This point can be explained with the help of Figure 13.28.

**Apparent power ($P_{APP}$)**
The product of alternating voltage and current in a reactive circuit. The term is used because only a portion of the value obtained is dissipated by the reactive components.

**Winding resistance ($R_w$)**
The dc resistance of the wire used to make a coil.

As you know, power is dissipated whenever current passes through any measurable amount of resistance. With this in mind, consider the inductor shown in Figure 13.28. If we assume that the **winding resistance ($R_w$)** of the coil has a value of $R_w = 2\ \Omega$, then the power dissipated by the component can be found as

$$P = I_L^2 R_w = (100\text{ mA})^2(2\ \Omega) = 20\text{ mW}$$

At the same time, the inductor voltage and current values shown in the figure would indicate that the component is dissipating

$$P = I_L V_L = (100\text{ mA})(20\text{ V}) = 2\text{ W}$$

**FIGURE 13.28**

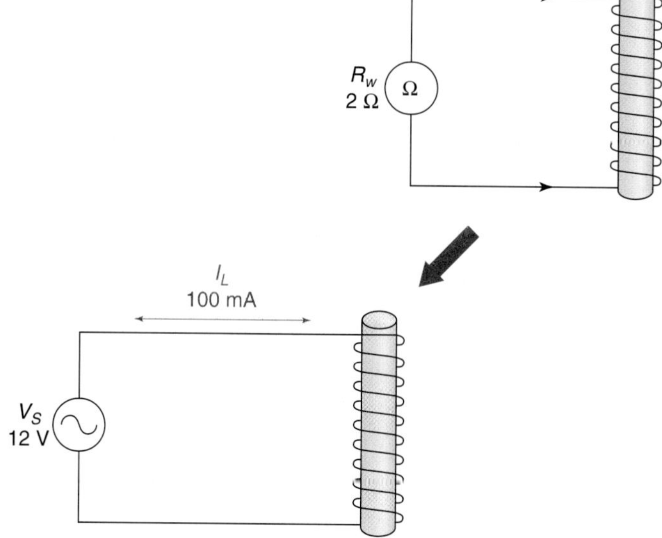

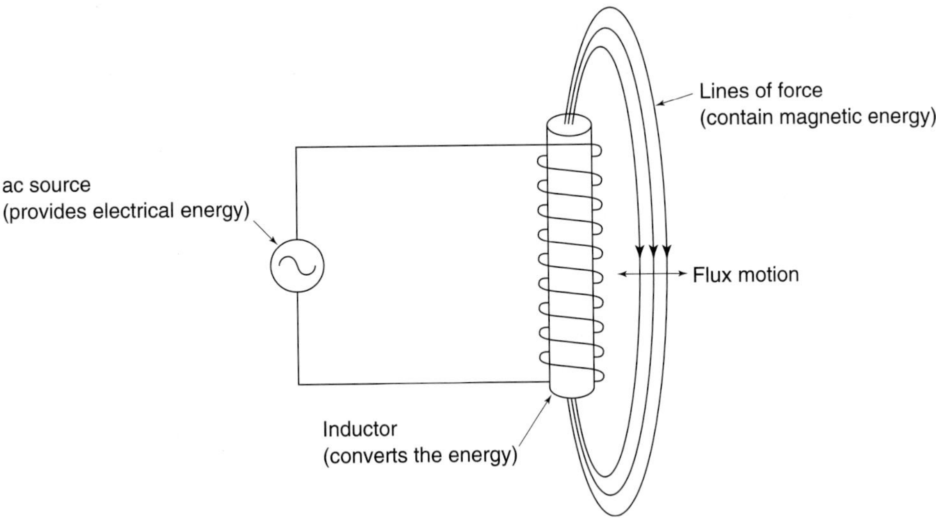


**FIGURE 13.29 An inductor converts electrical energy to magnetic energy.**

This value implies that the inductor is using energy at a rate of 2 joules per second. Since we know that the inductor is dissipating only 20 mW (which equals a rate of 0.02 joules per second), there is an obvious discrepancy between the two calculations.

*Remember:* One watt equals one joule per second.

The discrepancy lies not in our calculations, but rather in our interpretation of the results. We assume that the value of 2 W represents an amount of energy used each second. However, only a small portion (20 mW) represents the amount of energy used. The rest represents the rate at which the inductor transfers energy to the magnetic field. This transfer of energy can be explained with the help of Figure 13.29. The field generated by the inductor contains magnetic energy. Since energy cannot be created (or destroyed), this magnetic energy had to come from somewhere. The energy in the magnetic field comes from the ac source, via the inductor. In a sense, the inductor can be viewed as an energy converter because it converts electrical energy into magnetic energy.

*A Practical Consideration:* In most cases, the true power dissipation for an inductor makes up a very small portion of the apparent power. This is why inductors normally remain relatively cool to the touch, even when used in high-current applications.

We have now established three power values that can be calculated for any inductive circuit. The terms used to identify them are listed and defined in Table 13.3.

There are two important points that need to be made at this time. First, since reactive (or imaginary) power doesn't represent a true power value, it has its own unit of measure. To distinguish reactive power from true power, it is measured in a unit called **volt-amperes-reactive,** or **VAR.** By the same token, only a portion of apparent power is dissipated, so it also has its own unit of measure. As shown in the table, apparent power is measured in **volt-amperes (VA).** You will see calculations involving both VAR and VA in Chapter 14.

**Volt-amperes-reactive (VAR)** The unit of measure for reactive (imaginary) power. The unit is used to distinguish reactive power from resistive (true) power.

**Volt-amperes (VA)** The unit of measure for apparent power. This unit is used to distinguish apparent power from resistive and reactive power.

The second point deals with what happens to the energy transferred to the magnetic field. When the field collapses (due to a polarity change in the ac source), the energy that it

**TABLE 13.3 Inductor Power Values**

| *Term* | *Definition* |
|---|---|
| Resistive power ($P_{Rw}$) | The power actually dissipated by the winding resistance of an inductor. Also referred to as true power. Resistive power is measured in watts (W). |
| Reactive power ($P_X$) | A value that indicates the rate at which energy is transferred to an inductor's magnetic field. Also referred to as imaginary power. Reactive power is measured in *volt-amperes-reactive* (VAR). |
| Apparent power ($P_{APP}$) | The combination of resistive and reactive power. Only a small portion of $P_{APP}$ is dissipated by the inductor. The rest is transferred to the magnetic field by the inductor. Apparent power is measured in *volt-amperes* (VA). |

*Note:* Reactive and resistive power values cannot be added together algebraically. The method used to combine reactive and resistive power values is discussed in Chapter 14.

contains is returned to the circuit and, thus, to the ac source. As a result, the net reactive power drawn from the source is 0 VAR. The only power loss experienced by the source is the power used by any resistance in the circuit, which is why this power loss is referred to as true power.

In practice, every inductor has some measurable amount of winding resistance and therefore dissipates some amount of power. As you will see, the amount of winding resistance determines (in part) the quality of an inductor.

## *Inductor Quality* (*Q*)

*OBJECTIVE 18* ▶

**Quality** (*Q*)
A numeric value that indicates how close an inductor comes to having the power characteristics of the ideal inductor. For an inductor, the ratio of reactive power to true power. Sometimes referred to as the *figure of merit* of an inductor.

The ideal inductor (if it could be produced) would have no winding resistance. As a result, all the energy drawn from the ac source would be transferred to the inductor's magnetic field. The **quality** (*Q*) rating of an inductor is a numeric value that indicates how close the inductor comes to the power characteristics of the ideal component. The $Q$ of a given inductor can be found as the ratio of reactive power to true power for the component. By formula,

$$Q = \frac{P_X}{P_{Rw}} \tag{13.20}$$

where $P_X$ = the reactive power of the component, measured in VAR
$P_{Rw}$ = the true power dissipation of the component, measured in W

For the ideal inductor, $R_w = 0\ \Omega$ and, therefore, $P_{Rw} = 0$ W. Thus, the ideal inductor has a $Q$ of

The ideal value of $Q$.

$$Q = \frac{P_X}{P_{Rw}} = \frac{P_X}{0\text{ W}} = \infty$$

Since the ideal value of $Q$ is infinite, it follows that the higher its value of $Q$, the closer an inductor comes to having the power characteristics of the ideal component.

It takes very little winding resistance to have a significant impact on the quality of an inductor. This point is demonstrated in Example 13.14.

### EXAMPLE 13.14

The inductor in Figure 13.30a has a value of $R_w = 4\ \Omega$. Calculate the value of $Q$ for the component.

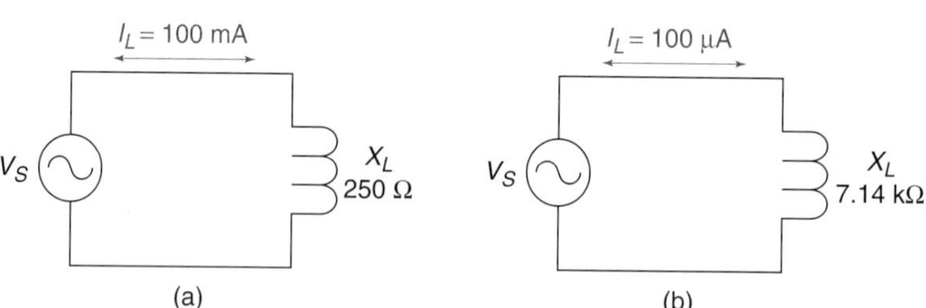


**FIGURE 13.30**

***Solution:*** The value of the reactive power ($P_X$) can be found using the circuit current and inductive reactance, as follows:

$$P_X = I_L^2 X_L = (100\text{ mA})^2(250\ \Omega) = \mathbf{2.5\ W}$$

The true power ($P_{Rw}$) can be found in a similar fashion, as follows:

$$P_{Rw} = I_L^2 R_w = (100\text{ mA})^2(4\ \Omega) = \mathbf{40\ mW}$$

Now the $Q$ of the inductor can be found as:

$$Q = \frac{P_X}{P_{Rw}} = \frac{2.5\text{ W}}{40\text{ mW}} = \mathbf{62.5}$$

As you can see, it took only 4 Ω of winding resistance to significantly reduce the $Q$ of the inductor from its ideal value of $Q = \infty$.

***PRACTICE PROBLEM 13.14***

The inductor in Figure 13.30b has a value of $R_w = 6.8\ \Omega$. Calculate the $Q$ of the component.

## Calculating the Value of Q

We can combine the equations used in Example 13.14 to develop a simpler means of calculating the value of $Q$ for an inductor. In the example, we calculated the values of reactive and true power as follows:

$$P_X = I_L^2 X_L \qquad \text{and} \qquad P_{Rw} = I_L^2 R_w$$

If we substitute these relationships into equation (13.20), we get

$$Q = \frac{P_X}{P_{Rw}} = \frac{I_L^2 X_L}{I_L^2 R_w}$$

or

$$Q = \frac{X_L}{R_w} \qquad \textbf{(13.21)}$$

This equation provides a much faster means of calculating the value of $Q$, as demonstrated in Example 13.15.

***EXAMPLE 13.15***

Using equation (13.21), recalculate the $Q$ of the inductor in Example 13.14.

***Solution:*** In Example 13.14, we were given values of $X_L = 250\ \Omega$ and $R_w = 4\ \Omega$. Using these values in equation (13.20), the $Q$ of the inductor can be found as

$$Q = \frac{X_L}{R_w} = \frac{250\ \Omega}{4\ \Omega} = \mathbf{62.5}$$

As you can see, this approach is considerably faster than using the circuit power values to determine the value of $Q$.

***PRACTICE PROBLEM 13.15***

Verify that equation (13.21) provides the same result you obtained in Practice Problem 13.14.

Equation (13.21) not only provides a faster means of calculating the value of $Q$, it also demonstrates a very important relationship. Since inductive reactance varies directly with operating frequency (as given in the relationship $X_L = 2\pi fL$), the $Q$ of an inductor also varies directly with the operating frequency of the component. This relationship is demonstrated in the following example.

## EXAMPLE 13.16

The ac source in Figure 13.31a is variable over a frequency range of 20 Hz to 20 kHz. Calculate the range of $Q$ values for the inductor.

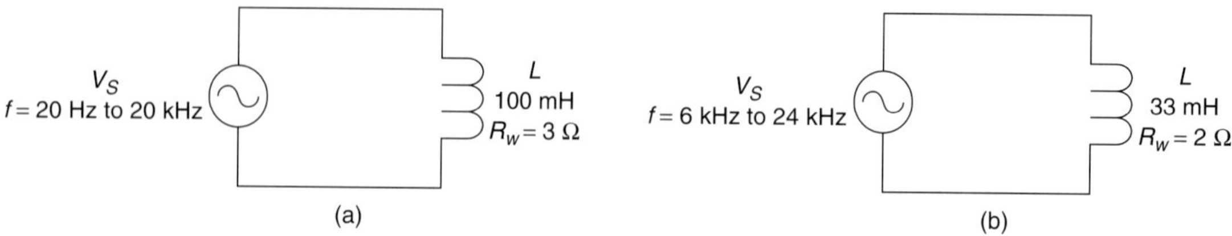


**FIGURE 13.31**

***Solution:*** When $f = 20$ Hz, the reactance of the inductor can be found as

$$X_L = 2\pi fL = 2\pi(20\text{ Hz})(100\text{ mH}) = \mathbf{12.6\ \Omega}$$

Now, the $Q$ of the inductor is found as

$$Q = \frac{X_L}{R_w} = \frac{12.6\ \Omega}{3\ \Omega} = \mathbf{4.2}$$

When $f = 20$ kHz, the reactance of the inductor can be found as

$$X_L = 2\pi fL = 2\pi(20\text{ kHz})(100\text{ mH}) = \mathbf{12.6\ k\Omega}$$

and the $Q$ of the inductor can be found as

$$Q = \frac{X_L}{R_w} = \frac{12.6\ \text{k}\Omega}{3\ \Omega} = \mathbf{4200} \qquad (4.2 \times 10^3)$$

Thus, when the ac source is varied across its frequency range, the $Q$ of the inductor varies between 4.2 and 4200.

***PRACTICE PROBLEM 13.16***

The ac source in Figure 13.31b is variable over a frequency range of 6 kHz to 24 kHz. Calculate the range of $Q$ values for the inductor.

**Frequency response**
The response that any reactive circuit has to a change in operating frequency.

In Chapter 19, you will be shown the role that the $Q$ of an inductor plays in the **frequency response** of any circuit containing the component. As you will see, the $Q$ of the inductor determines (in part) the response that any reactive (inductive) circuit has to a change in operating frequency. The relationship between operating frequency and inductor $Q$ is also important because it helps to explain why certain types of inductors are preferred over others for use in circuits that operate within specified frequency limits.

## Types of Inductors

Many types of inductors are designed for specific applications. Although these inductors each have unique characteristics, they all work according to the principles covered in this chapter. The purpose of this section is merely to point out the characteristics that make each type of inductor unique and, therefore, better suited than others for some specific application(s).

***Iron-Core versus Air-Core Inductors.*** Most inductors have cores made of either iron or air. Iron-core and air-core inductors are represented using the schematic symbols) shown in Figure 13.32. Iron-core inductors are better suited for use in low-frequency applications, such as audio and dc power supply circuits, because:

◄ *OBJECTIVE 19*

1. The dc winding resistance of an iron-core inductor is much lower than that of an equal-value air-core inductor.
2. The $Q$ of an iron-core inductor is much higher than that of an equal-value air-core inductor.

These characteristics are based on the relative permeability of iron.

In Chapter 10, you were told that relative permeability ($\mu_r$) is a ratio of the permeability of a material to that of air. For iron, $\mu_r \cong 200$ (all other factors being equal). With this in mind, look at the following relationship:

$$L = \frac{\mu_m N^2 A}{\ell}$$

Since iron has approximately 200 times the permeability of air, it takes far fewer turns (and, therefore, a shorter length of wire) to produce a value of inductance when an iron core is used. Since the resistance of any wire ($R_w$, in this case) varies directly with its length, *the winding resistance of an iron-core inductor is always significantly less than that of an equal-value air-core inductor.* As a result, the iron-core inductor:

1. Has a much higher value of $Q$ at low frequencies than does an equal-value air-core inductor.
2. Tends to be much smaller in size than an equal-value air-core inductor.

The relatively low winding resistance of an iron-core inductor makes it suitable for use in high-current applications, such as dc power supply filter circuits. With its low winding resistance, an iron-core inductor dissipates much less power than an equal-value air-core inductor. (Power dissipation is *always* an important consideration in high-current circuitry.) At the same time, iron cores can experience significant power losses when operated at higher frequencies, making air-core inductors more suitable for such applications. (The power losses that can occur in iron cores are discussed in more detail in Chapter 15.)

***Air Cores.*** The turns of an air-core inductor are self-supporting or they are physically supported by a nonmagnetic form, such as a ceramic tube.

***Toroids.*** A **toroid** is a coil wrapped around a doughnut-shaped magnetic core, as shown in Figure 13.33. Because of its shape, nearly all the flux produced by the coil remains in the core; that is, very little of the flux is lost to the air surrounding the coil. As a result, toroids have:

◄ *OBJECTIVE 20*

**Toroid**
An inductor with a doughnut-shaped core.

1. Inductance values that are much greater than their physical size would indicate.
2. Extremely high quality ($Q$) ratings.
3. Extremely accurate rated values.

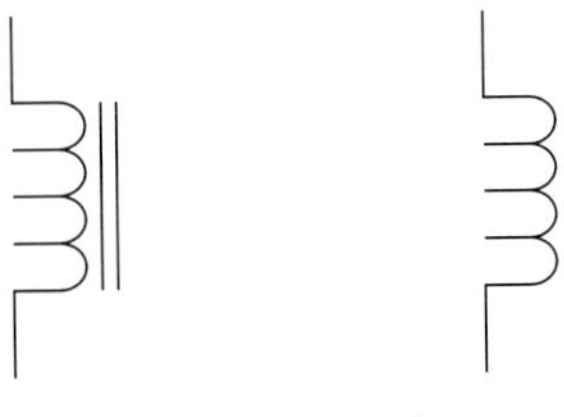


**FIGURE 13.32 Iron-core and air-core inductor symbols.**

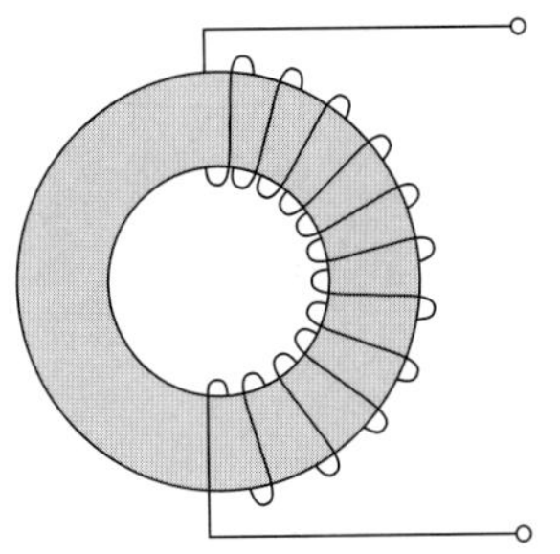

**FIGURE 13.33 Toroid.**

**Choke**
A low-resistance inductor designed to provide a high reactance within a designated frequency range.

***Chokes.*** A low-resistance inductor designed to provide a high opposition to current (reactance) within a designated frequency range is referred to as a **choke.** For example, look at the simplified dc power supply filter shown in Figure 13.34. The overall operation of a dc power supply is beyond the scope of this textbook and is covered in a course on solid-state circuits. However, we can state the following with regard to the filter:

1. The input from the rectifier is a varying dc current with a relatively high average value, typically greater than 1 A.
2. The input is varying above and below its average (dc) value at a frequency of approximately 120 Hz (as shown in Figure 13.34).

Because of the low dc resistance of the choke, little power is dissipated by the component. This characteristic is especially desirable in a dc power supply (which is designed to provide the greatest possible dc output power). At the same time, the choke is designed to have a high value of reactance at the frequency of the current variations (120 Hz). This high reactance effectively blocks the variations from the filter output. As a result, the filter output is a straight-line dc current that equals the dc average of the input current (as shown in Figure 13.34). As described here, the filter choke has:

1. Low winding (dc) resistance.
2. High reactance at the frequency of operation.

These are the characteristics of any inductor designed for use as a choke.

The two most common types of chokes are filter chokes and radio-frequency chokes (more commonly known as RF chokes). Typical filter and RF chokes are shown in Figure 13.35.

As you have seen, the filter choke is designed for operation in low-frequency circuits, typically operating at a frequency lower than 1 kHz. The RF choke is designed for operation within the radio-frequency range: approximately 30 kHz to 30 GHz. In Figure 13.35, the filter choke has an iron core and the RF choke has an air core. This is consistent with our previous discussion on core materials and frequency.

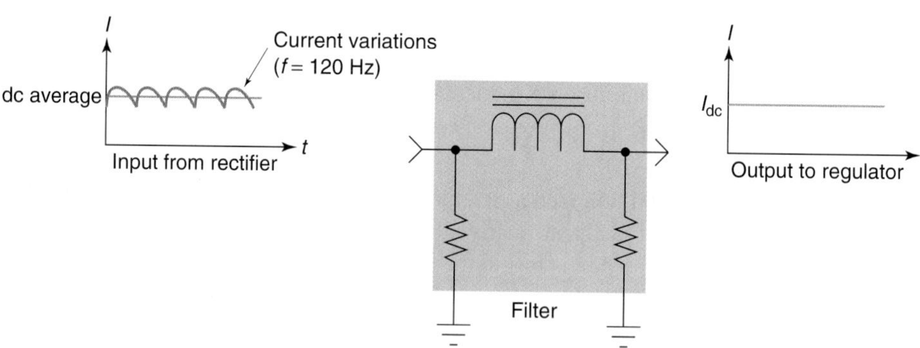


**FIGURE 13.34 Filter choke input and output currents.**

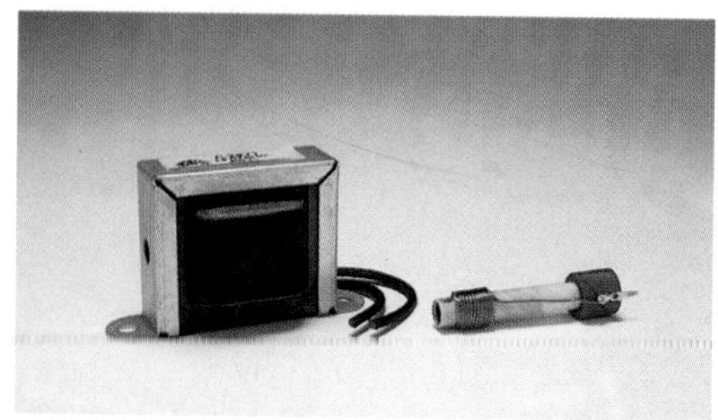

**FIGURE 13.35 A filter choke and an RF choke.**

## Section Review

1. Why is the term *apparent power* used to describe the product of inductor current and voltage?
2. Describe the inductor as an energy converter.
3. What terms are commonly used to describe resistive power and reactive power?
4. What are the units of measure for reactive power and apparent power?
5. List two characteristics of the ideal inductor.
6. What is the quality ($Q$) rating of an inductor?
7. What is the $Q$ of an ideal inductor? Explain the basis of this value.
8. What is the relationship between inductor $Q$ and winding resistance? Between inductor $Q$ and operating frequency?
9. Why are iron-core inductors better suited than air-core inductors for low-frequency applications?
10. Why are air-core inductors better suited than iron-core inductors for high-frequency applications?
11. What is a toroid? What are the significant characteristics of this component?
12. What is a choke?

## Chapter Summary

Here is a summary of the major points made in this chapter:

1. Inductance is commonly defined in one of three ways:
   a. The ability of a component with a changing current to induce a voltage across itself (or a nearby circuit) by generating a changing magnetic field.
   b. The ability of a component to oppose a change in current.
   c. The ability of a component to store energy in an electromagnetic field.
2. An inductor is a component designed to provide a specific amount of inductance.
3. When current passes through an inductor, magnetic flux is generated as shown in Figure 13.1.
   a. An increase in inductor current causes an increase in flux density. As flux density increases, the magnetic field expands outward from the component.
   b. A decrease in inductor current causes a decrease in flux density. As flux density decreases, the magnetic field collapses back into the component.
4. According to Faraday's first law of induction, there must be relative motion between a conductor and a magnetic field to induce a current through (or voltage across) the conductor. In the case of a coil, the relative motion is provided by an expanding and collapsing magnetic field.
5. One weber is the amount of flux that, when passing through a perpendicular coil in one second, induces 1 V across each turn of the component.
6. A change in inductor current generates a moving magnetic field that, in turn, induces a voltage across the inductor. This principle is known as self-inductance.

7. Lenz's law states that any induced voltage always opposes its source.
   a. Since it always opposes the change in coil current that produced it, induced voltage is referred to as counter emf.
   b. An induced voltage must oppose its source for an inductor to work within the laws of conservation of energy.
8. The unit of measure of inductance is the henry (H).
   a. Inductance is measured in volts induced per rate of change in current.
   b. When a change of 1 A/s induces 1 V, the value of inductance equals 1 H.
9. For any inductor, the value of inductance is:
   a. Directly proportional to core permeability, the square of the number of turns, and the cross-sectional area of the core.
   b. Inversely proportional to the length of the component.
10. For any sinusoidal current, the instantaneous rate of change in current ($di/dt$) is:
    a. Maximum at the zero crossings.
    b. Approximately zero at the positive and negative peaks.
11. Since $v_L$ is directly proportional to $di/dt$, the value of $v_L$ is:
    a. Maximum when current is at the zero crossings.
    b. Approximately zero when current is at its positive and negative peaks. Therefore, current and voltage in a purely inductive circuit are 90° out of phase.
12. The relationship between current and voltage in a purely inductive circuit is usually described in either of two ways:
    a. Voltage leads current by 90°.
    b. Current lags voltage by 90°.
13. When the magnetic field produced by one coil induces a voltage across another coil, we have what is called mutual inductance.
14. The amount of mutual inductance between two (or more) coils depends on:
    a. The distance between the coils.
    b. The physical characteristics of each coil.
    c. The angle between the coils.
15. The coefficient of coupling ($k$) between two (or more) coils is a measure of the degree of coupling that takes place between the coils.
    a. Unity coupling ($k = 1$) is the ideal value.
    b. Unity coupling (which cannot be achieved in practice) could occur only if identical coils were placed in parallel, with 100% of the flux from the first passing through the second.
16. When coils are connected in series, the total inductance is equal to the sum of the individual component values (assuming that there is no mutual inductance between the coils).
17. The effect of mutual inductance on series-connected inductors depends on whether the inductors are connected in series-aiding or series-opposing fashion.
    a. For a series-aiding connection: $L_T = L_1 + L_2 + 2L_M$
    b. For a series-opposing connection: $L_T = L_1 + L_2 - 2L_M$
    c. The effects of these connections are illustrated in Examples 13.4 and 13.5.
18. The coefficient of coupling between two inductors can be measured using an LCR bridge. (The procedure demonstrated in Example 13.6.)

19. Assuming that $k = 0$, the total inductance of parallel-connected coils can be found in the same fashion as total parallel resistance (see equations (13.13) through (13.15)).
    a. Like resistance, total parallel inductance is lower than any of the branch values.
    b. When $k \neq 0$, any total inductance calculation must account for the mutual inductance (see equation (13.16) and Example 13.10).
20. The opposition (in ohms) that an inductor presents to a changing current is referred to as inductive reactance ($X_L$).
    a. $X_L$ is directly proportional to the value of an inductor.
    b. $X_L$ is directly proportional to the inductor's operating frequency.
21. In any inductive ac circuit, Ohm's law can be used to calculate any one of the following values (given the other two): $V_L$, $I_L$, or $X_L$. (Ohm's law calculations for inductive ac circuits are demonstrated in Examples 13.12 and 13.13.)
22. Total series reactance is found in the same fashion as total series resistance.
23. Total parallel reactance is found in the same fashion as total parallel resistance.
24. Impedance is the total opposition to current in an ac circuit, consisting of resistance and/or reactance.
    a. Resistance is a static value.
    b. Reactance is a dynamic value.
    c. Resistance and reactance cannot be added algebraically to find impedance. (You will be shown how to combine resistance and reactance into a single impedance value in Chapter 14.)
25. The power value calculated using the rms values of voltage and current in an inductive circuit is referred to as apparent power.
    a. The term *apparent power* is used because most of the energy is transferred to the electromagnetic field of the inductor. Apparent power is measured in volt-amperes (VA).
    b. Apparent power consists of reactive power and true (resistive) power.
    c. Reactive power actually indicates the rate at which energy is transferred to the inductor's electromagnetic field. It is not a power dissipation value and, therefore, is measured in volt-amperes-reactive (VAR) rather than watts.
    d. True power is the power dissipated by the winding resistance of a coil. In most cases, true power makes up a small percentage of the apparent power.
26. The ideal inductor (if it could be produced) would have no winding resistance. As a result, all the energy drawn from the source would be transferred to the component's electromagnetic field.
27. The quality ($Q$) of an inductor is a numeric value that indicates how close the component comes to having the power characteristics of the ideal inductor.
    a. The higher the $Q$ of an inductor, the closer it comes to having the power characteristics of the ideal component.
    b. The $Q$ of an inductor equals the ratio of inductive reactance to winding resistance.
    c. Since $X_L$ varies with frequency, so does the $Q$ of an inductor (see Example 13.16).
28. Iron-core inductors are better suited than air-core inductors for low-frequency applications.
29. Compared to equal-value air-core inductors, iron-core inductors tend to have:
    a. Lower winding resistance.
    b. A higher value of $Q$.

30. A toroid is a coil wrapped around a doughnut-shaped magnetic core. Toroids have:
    a. High inductance values for their physical size.
    b. Extremely high $Q$ ratings.
    c. Extremely accurate rated values.
31. A low-resistance inductor designed to provide a high opposition to current (reactance) within a designated frequency range is called a choke.
    a. The two most common types of chokes are filter chokes and RF chokes.
    b. Filter chokes are used in low-frequency applications.
    c. RF (radio-frequency) chokes are designed for use in the radio-frequency range: approximately 30 kHz to 30 GHz.

## Equation Summary

| Equation Number | Equation | Section Number |
|---|---|---|
| **(13.1)** | $B = \dfrac{\mu_m NI}{\ell}$ | 13.1 |
| **(13.2)** | $V = N\dfrac{d\phi}{dt}$ | 13.1 |
| **(13.3)** | $-v_L = L\dfrac{di}{dt}$ | 13.1 |
| **(13.4)** | $L = \dfrac{v_L}{\dfrac{di}{dt}}$ | 13.1 |
| **(13.5)** | $L \cong \dfrac{\mu_m N^2 A}{\ell}$ | 13.1 |
| **(13.6)** | $A = \dfrac{\pi d^2}{4}$ | 13.1 |
| **(13.7)** | $L_M = \dfrac{\mu_m N_1 N_2 A}{\ell}$ | 13.3 |
| **(13.8)** | $k = \dfrac{\Phi_2}{\Phi_1}$ | 13.3 |
| **(13.9)** | $L_M = k\sqrt{L_1 L_2}$ | 13.3 |
| **(13.10)** | $L_T = L_1 + L_2 + \ldots + L_n$ | 13.3 |
| **(13.11)** | $L_T = L_1 + L_2 \pm 2L_M$ | 13.3 |
| **(13.12)** | $L_M = \dfrac{L_T - (L_1 + L_2)}{2}$ | 13.3 |
| **(13.13)** | $k = \dfrac{\lvert L_M \rvert}{\sqrt{L_1 L_2}}$ | 13.3 |
| **(13.14)** | $L_T = \dfrac{1}{\dfrac{1}{L_1} + \dfrac{1}{L_2} + \ldots + \dfrac{1}{L_n}}$ | 13.3 |

| | | |
|---|---|---|
| **(13.15)** | $L_T = \dfrac{L_1 L_2}{L_1 + L_2}$ | 13.3 |
| **(13.16)** | $L_T = \dfrac{L}{n}$ | 13.3 |
| **(13.17)** | $L_T = \dfrac{(L_1 \pm L_M)(L_2 \pm L_M)}{(L_1 \pm L_M) + (L_2 \pm L_M)}$ | 13.3 |
| **(13.18)** | $X_L = \dfrac{V_{\text{rms}}}{I_{\text{rms}}}$ | 13.4 |
| **(13.19)** | $X_L = 2\pi f L$ | 13.4 |
| **(13.20)** | $Q = \dfrac{P_X}{P_{Rw}}$ | 13.5 |
| **(13.21)** | $Q = \dfrac{X_L}{R_w}$ | 13.5 |

## Key Terms

The following terms were introduced and defined in this chapter:

apparent power
choke
coefficient of coupling ($k$)
counter emf
coupling
filter choke
frequency response
henry (H)
imaginary power
impedance ($Z$)
inductance
inductive reactance ($X_L$)
inductor
lag
LCR bridge
lead
Lenz's law
mutual inductance
quality ($Q$)
radio-frequency (RF) choke
reactance ($X$)
reactive power
resistive power
self-inductance
toroid
true power
unity coupling
volt-amperes (VA)
volt-amperes-reactive (VAR)
winding resistance

## Practice Problems

1. Calculate the value of $v_L$ developed across a 47 mH coil when the instantaneous rate of change in component current equals 200 mA/s.
2. Calculate the value of $v_L$ developed across a 4.7 H coil when the instantaneous rate of change in component current equals 1.5 A/s.
3. Calculate the value of $v_L$ developed across a 100 mH coil when the instantaneous rate of change in component current equals 25 mA/ms. (*Hint:* Convert the current change into A/s.)
4. Calculate the value of $v_L$ developed across a 470 μH coil when the instantaneous rate of change in component current equals 2 mA/μs.
5. Calculate the value of the air-core inductor shown in Figure 13.36a.
6. Calculate the value of the air-core inductor shown in Figure 13.36b.

**FIGURE 13.36**

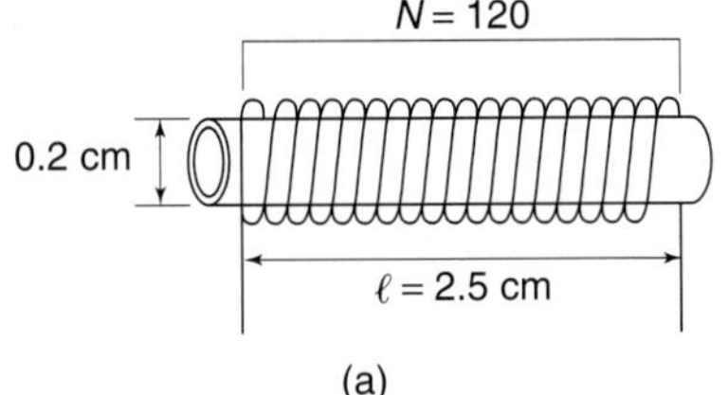


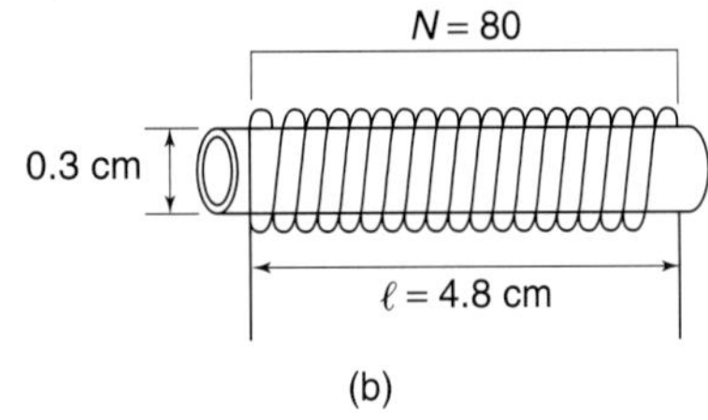

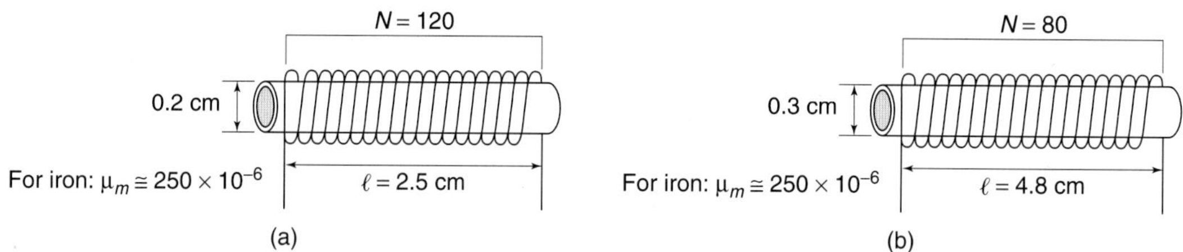


**FIGURE 13.37**

7. Calculate the value of the iron-core inductor shown in Figure 13.37a.
8. Calculate the value of the iron-core inductor shown in Figure 13.37b.
9. The coils in Figure 13.38a have a value of $k = 0$. Calculate the total inductance in the circuit.
10. The coils in Figure 13.38b have a value of $k = 0$. Calculate the total inductance in the circuit.
11. The inductors in Figure 13.39a are connected in series-aiding fashion. Assume that the coefficient of coupling between the two is 0.25 and calculate the total inductance in the circuit.
12. The inductors in Figure 13.39b are connected in series-opposing fashion. Assume that the coefficient of coupling between the two is 0.33 and calculate the total inductance in the circuit.

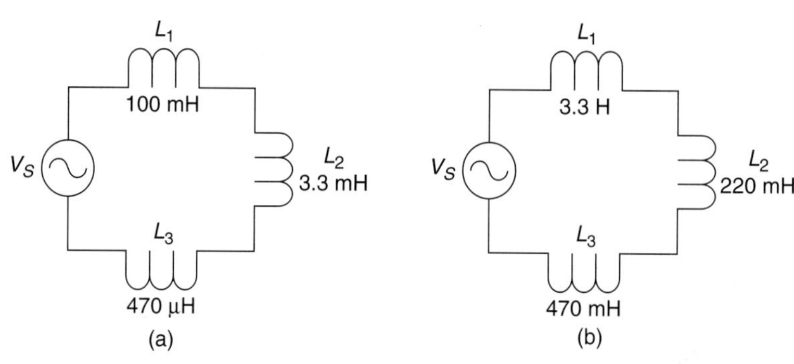


**FIGURE 13.38**

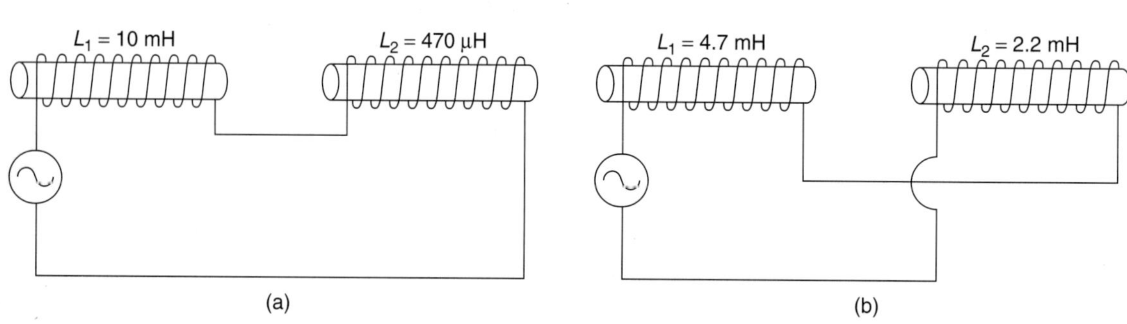


**FIGURE 13.39**

13. Determine the coefficient of coupling for the circuit in Figure 13.40a.
14. Determine the coefficient of coupling for the circuit in Figure 13.40b.
15. The circuit in Figure 13.41a has a value of $k = 0$. Calculate the total inductance in the circuit.

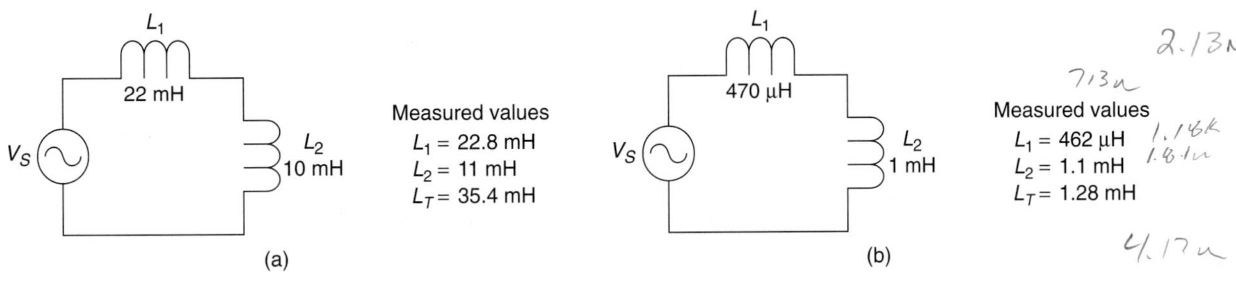


**FIGURE 13.40**

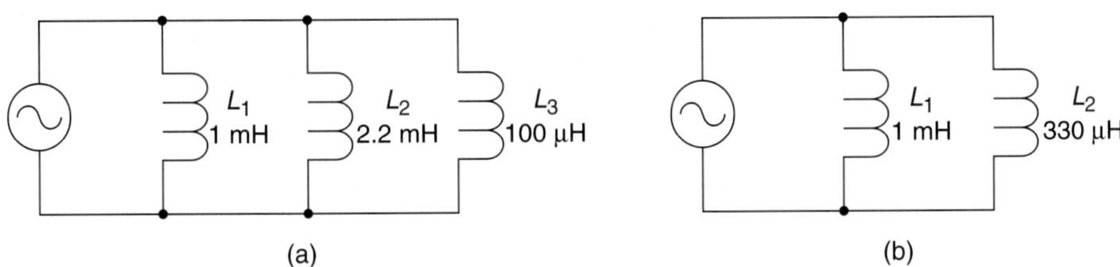


**FIGURE 13.41**

16. The inductors in Figure 13.41b are wired in parallel-aiding fashion with a value of $k$ = 0.7. Calculate the total inductance in the circuit.
17. Calculate the value of $X_L$ for the circuit in Figure 13.42a.
18. Calculate the value of $X_L$ for the circuit in Figure 13.42b.
19. The following table lists values for an inductive circuit. Complete the table.

| $V_{rms}$ | $I_{rms}$ | $X_L$ |
|---|---|---|
| 12 $V_{ac}$ | ______ | 2 kΩ |
| ______ | 30 mA | 180 Ω |
| 20 $V_{ac}$ | 480 μA | ______ |

20. The following table lists values for an inductive circuit. Complete the table.

| $V_{rms}$ | $I_{rms}$ | $X_L$ |
|---|---|---|
| 180 $mV_{ac}$ | ______ | 200 Ω |
| ______ | 40 μA | 18 MΩ |
| 2 $V_{ac}$ | 100 μA | ______ |

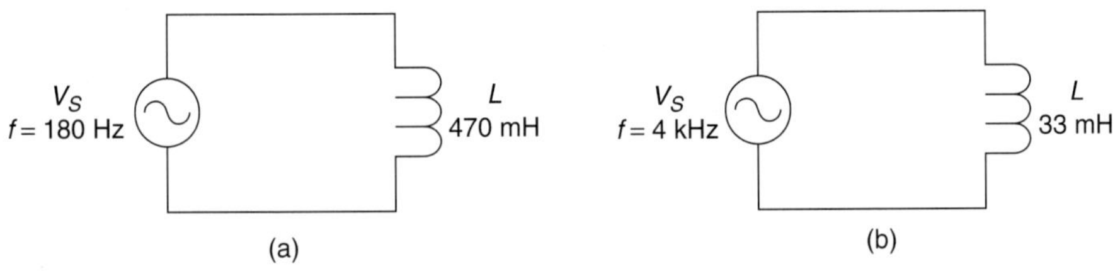


**FIGURE 13.42**

21. Determine the operating frequency of the circuit in Figure 13.43a.
22. Determine the operating frequency of the circuit in Figure 13.43b.
23. Calculate the $Q$ of the inductor in Figure 13.44a.
24. Calculate the $Q$ of the inductor in Figure 13.44b.
25. Calculate the $Q$ of the inductor in Figure 13.45a.
26. Calculate the $Q$ of the inductor in Figure 13.45b.
27. The source in Figure 13.46a is variable over a frequency range of 10 kHz to 50 kHz. Calculate the range of $Q$ values for the inductor.
28. The source in Figure 13.46b is variable over a frequency range of 100 Hz to 8 kHz. Calculate the range of $Q$ values for the inductor.

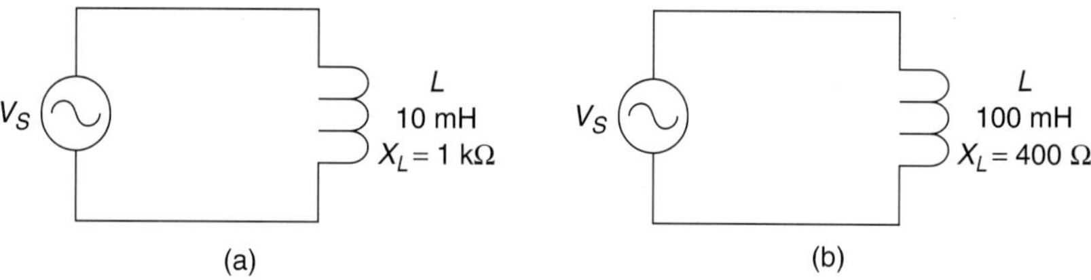


**FIGURE 13.43**

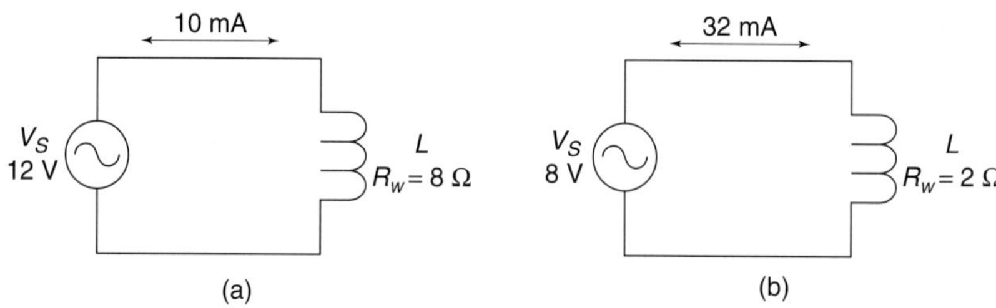


**FIGURE 13.44**

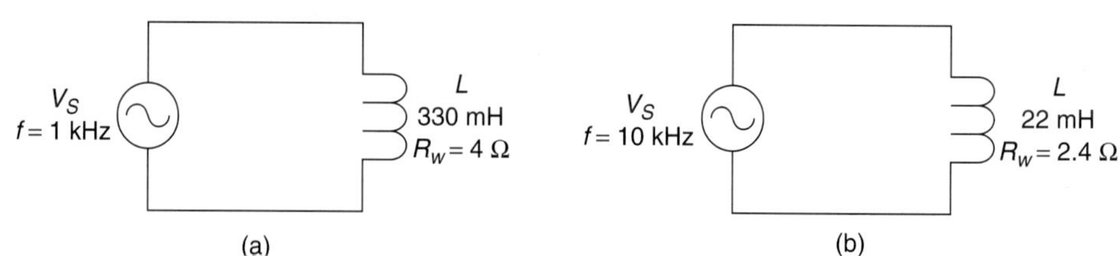


**FIGURE 13.45**

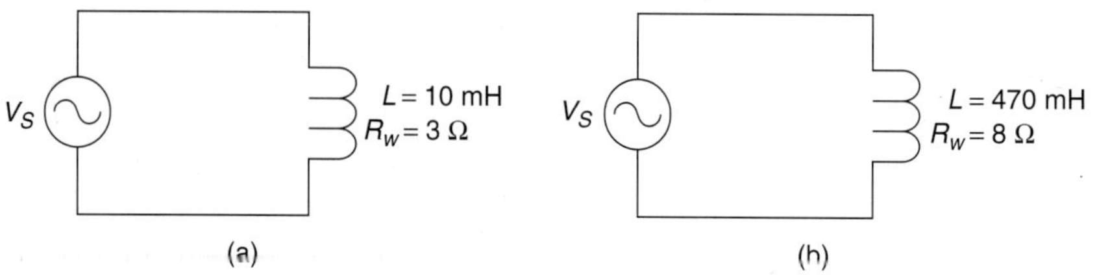


**FIGURE 13.46**

## The Brain Drain

29. Determine the frequency of operation that will provide a current of 100 mA through the circuit in Figure 13.47. (*Note:* $k = 0$)
30. The circuit current in Figure 13.48 has a measured value of 100 mA. Assume that the circuit has a value of $k = 0$ and determine the value of $L_2$.
31. The operating frequency of the circuit in Figure 13.49 is variable. Determine the values of $I_{rms}$ and $f$ when the inductor has a value of $Q = 25$.

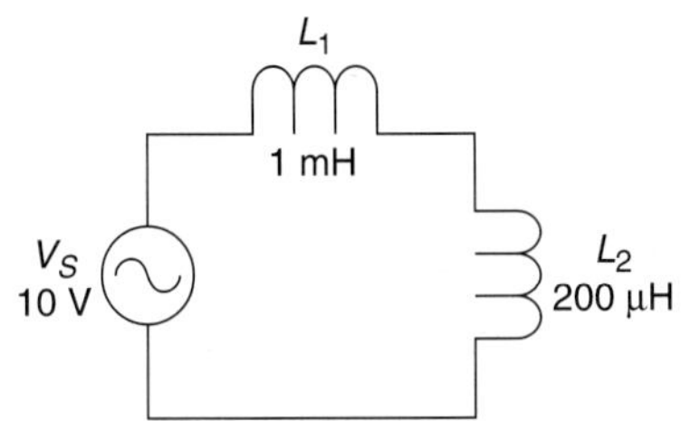


FIGURE 13.47

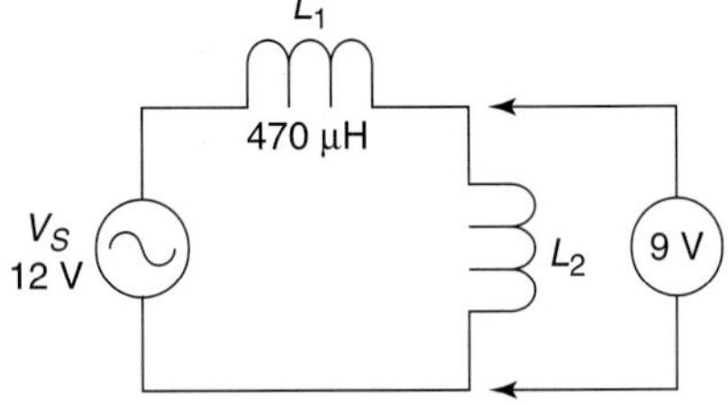


FIGURE 13.48

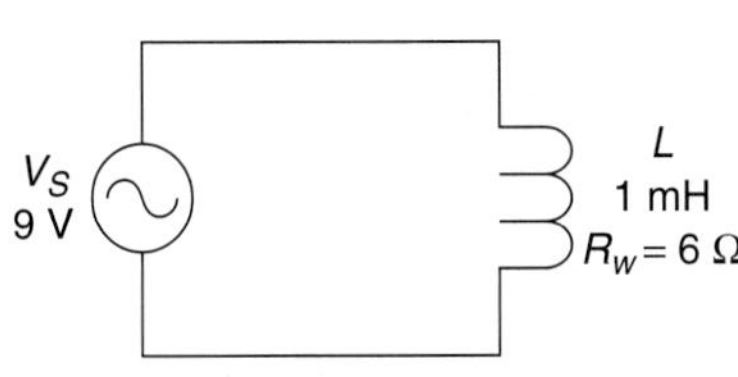


FIGURE 13.49

## Open to Discussion

32. You have an unlimited length of 20 gage wire and you need a 5 H inductor. If you had to construct the inductor yourself (using air as a core):
    a. What physical dimensions would it have?
    b. What value of $Q$ would it have at an operating frequency of 1 kHz?
    (*Note:* The resistance of 20 gage wire is provided in Chapter 3.)

## Answers to the Example Practice Problems

**13.1.** 6.6 mV
**13.2.** 2.78 μH
**13.3.** 3.73 mH
**13.4.** 43.9 mH
**13.5.** 4.4 mH
**13.6.** $k \cong 0.25$
**13.7.** 162.4 μH
**13.8.** 4.63 mH
**13.9.** 82.5 mH
**13.10.** 957 μH (when $k = 0$), 1.99 mH (when $k = 0.25$)
**13.11.** 59 kΩ
**13.12.** 3.6 mA
**13.13.** 99.5 mA
**13.14.** 1050
**13.16.** 622 to 2488

## EWB Applications Problems

| Figure | EWB File Reference |
|---|---|
| 13.6 | EWBA13_6.ewb |
| 13.11 | EWBA13_11.ewb |
| 13.23 | EWBA13_23.ewb |
| 13.26 | EWBA13_26.ewb |

# 14

# RESISTIVE-INDUCTIVE (*RL*) CIRCUITS

## OBJECTIVES

*After studying the material in this chapter, you should be able to:*

1. Compare and contrast series resistive, inductive, and resistive-inductive (*RL*) circuits.
2. Describe the voltage and phase angle characteristics and measurements for a series *RL* circuit.
3. Describe the impedance and current characteristics of a series *RL* circuit.
4. Perform a complete mathematical analysis of any series *RL* circuit.
5. Describe and analyze the frequency response of a series *RL* circuit.
6. Perform all relevant power calculations for a series *RL* circuit.
7. Compare and contrast parallel resistive, inductive, and resistive-inductive (*RL*) circuits.
8. Calculate any current or impedance value for a parallel *RL* circuit.
9. Measure the current phase angle of a parallel *RL* circuit.
10. Perform a complete mathematical analysis of any parallel *RL* circuit.
11. Describe and analyze the frequency response of a parallel *RL* circuit.
12. Perform a complete mathematical analysis of a series-parallel *RL* circuit.
13. Describe and explain the current and voltage waveforms found in an *RL* switching circuit.
14. Describe the universal current curve for an *RL* switching circuit and use it to approximate the value of the circuit current at a given time interval.
15. Calculate the value of circuit current at any point on a rise or decay curve using the universal curve equations.
16. Calculate the time constant and transition time for a series *RL* circuit.
17. State and explain the definition of the term *time constant.*
18. Calculate the time required for the current in an *RL* switching circuit to reach a designated value.

Up to this point, we have limited our discussions to circuits that are purely resistive or purely inductive. The purpose of limiting ourselves to so-called pure circuits has been to establish thoroughly the operating characteristics of each type of component. In practice, many circuits contain both resistive and reactive components. In this chapter, we will begin our analysis of such circuits by studying the combined effects of resistance and inductive reactance.

## 14.1 SERIES *RL* CIRCUITS

A **resistive-inductive (*RL*) circuit** is one that contains any combination of resistors and inductors. Several examples of basic *RL* circuits are shown in Figure 14.1.

**Resistive-inductive (*RL*) circuit** A circuit that contains any combination of resistors and inductors.

As we progress through this chapter, we will develop the current, voltage, power, and impedance characteristics of *RL* circuits. First, we will compare *RL* series circuits to purely resistive and purely inductive series circuits. The goal is to show the similarities and differences among the three.

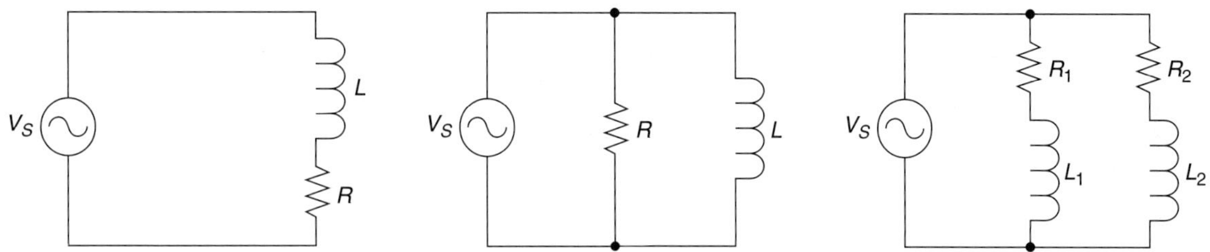


**FIGURE 14.1 Examples of basic *RL* circuits.**

### Series Circuits: A Comparison

◄ *OBJECTIVE 1*

In Chapters 5 and 13, we established the fundamental characteristics of resistive and inductive series circuits. As you know, these characteristics are as follows:

1. Current is the same at all points throughout the circuit.
2. The applied voltage equals the sum of the individual component voltages.
3. The total impedance (resistance or reactance) equals the sum of the individual component impedances.

These relationships are provided as equations in Figure 14.2.

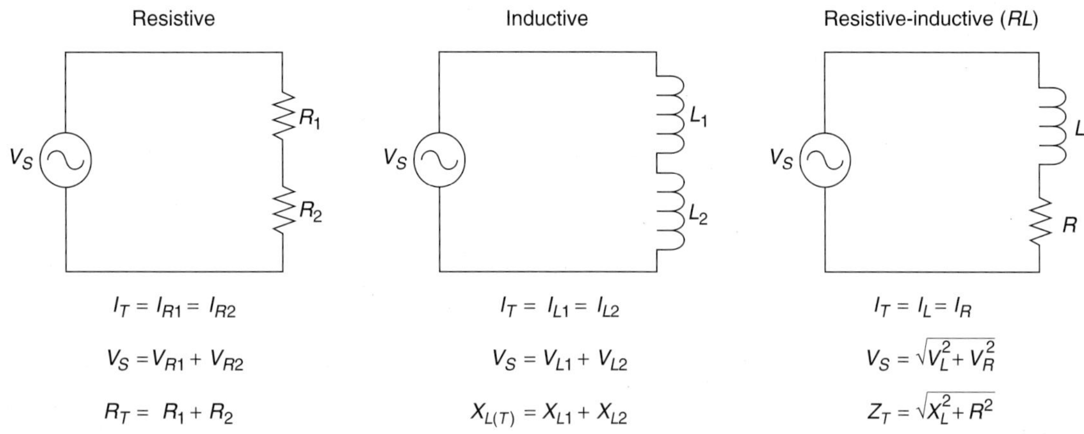


**FIGURE 14.2 Series circuit characteristics.**

The relationships listed for the resistive and inductive circuits should all be familiar. In each case, the current is constant throughout the circuit, and the component voltages add up to equal the source voltage ($V_S$). For the resistive circuit, the total impedance (opposition to current) is equal to the sum of the resistor values. For the inductive circuit, the total impedance is equal to the sum of the reactance ($X_L$) values.

As we progress through this section, we will discuss each relationship listed for the *RL* circuit in Figure 14.2. First, let's compare them to those given for the other two circuits. As the $I_T$ equation shows, current is constant throughout the series *RL* circuit, just as it is in the other series circuits. The voltage and impedance relationships appear to be radically different, however, than those for the resistive and inductive circuits. To understand the change, we need to establish the difference between the *algebraic sum* and the *geometric sum* of two values.

**Phasor**
A vector used to represent a value that constantly changes phase (such as, an instantaneous sine wave value). As such, a phasor is described as a vector having constant angular velocity. (*Note:* The distinction between phasors and vectors is made here for technical accuracy. Many technicians refer to both types of lines as vectors.)

***Algebraic and Geometric Sums.*** As you begin working with circuits that contain both resistive and reactive components, you'll find that many values must be represented using **phasors** (rotating vectors). As such, these values must be added together geometrically rather than algebraically. Given two variables, *a* and *b,* the algebraic and geometric sums of the variables are found as follows:

$$\text{Algebraic sum:} \quad c = a + b$$
$$\text{Geometric sum:} \quad c = \sqrt{a^2 + b^2}$$

Algebraic addition should not require any clarification, as you have always added values in this fashion. Geometric addition may seem new, but you were actually introduced to this type of addition in Chapter 12, where we used geometric addition to add vectors forming a 90° angle. The geometric sum gave us the length of the hypotenuse that formed a right triangle with the vectors.

The reason for using geometric addition in resistive-reactive circuits will be made clear as we continue through this chapter. For now, we need to establish only that any equation in the form of $c = \sqrt{a^2 + b^2}$ provides a geometric sum of the variables. With this in mind, let's go back to the *RL* circuit in Figure 14.2.

***RL Circuit Voltage and Impedance.*** In a series *RL* circuit, the geometric sum of the component voltages equals the source voltage. The geometric sum of resistance and reactance equals the total impedance (opposition to current) in the circuit. Even though we are using a different type of addition, the *RL* circuit still operates according to the rules we have established for all series circuits:

1. Component voltages add (geometrically) to equal the source voltage.
2. Resistance and reactance add (geometrically) to equal the total impedance.

Another important characteristic of *RL* circuits is that the source voltage and total impedance each have a phase angle that is normally expressed as part of the value. For example, the circuit in Figure 14.3 was analyzed using the principles and techniques that you will learn later in this section. As you can see, the source voltage and total impedance each have a phase angle. (For reasons that will be explained later, the phase angle for the current is assumed to be 0°.)

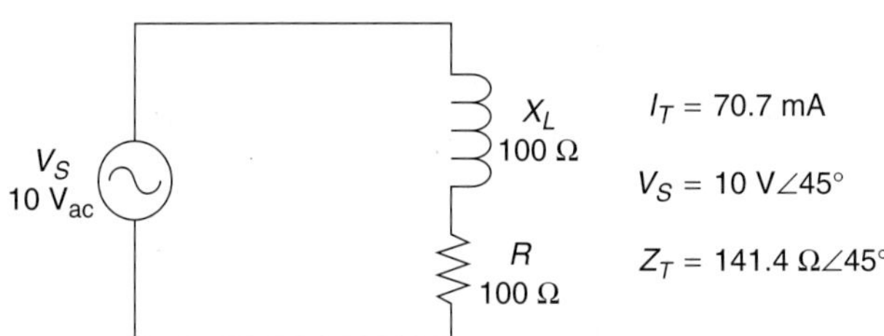


**FIGURE 14.3** ***RL* circuit values (with phase angles).**

## *Series Voltages*

In Chapter 13, you were told that voltage leads current by 90° in an inductive circuit. In a resistive circuit, voltage and current are always in phase. These two phase relationships are illustrated in Figure 14.4. If we combine the circuits in Figure 14.4, we get the series circuit and waveforms shown in Figure 14.5. Since current is constant throughout a series circuit, the inductor and resistor voltage waveforms are referenced to a single current waveform. As you can see: ◀ *OBJECTIVE 2*

1. The inductor voltage leads the circuit current by 90°.
2. The resistor voltage is in phase with the circuit current.

Therefore, inductor voltage leads resistor voltage by 90°. This phase relationship holds true for every series *RL* circuit with an ac source.

Since $V_L$ leads $V_R$ by 90°, the values of the component voltages are represented using phasors, as shown in Figure 14.6a. By plotting the value of $V_R$ on the $x$-axis (0°) and the value of $V_L$ on the $y$-axis (90°), the phase relationship between the two is represented in the graph.

Like any other series circuit, the source voltage in an *RL* circuit must equal the sum of the component voltages. Since the component voltages in an *RL* circuit are phasor values, they must be added geometrically. As you learned in Chapter 12, the sum of two phasors at a 90° angle can be found as

$$z = \sqrt{x^2 + y^2}$$

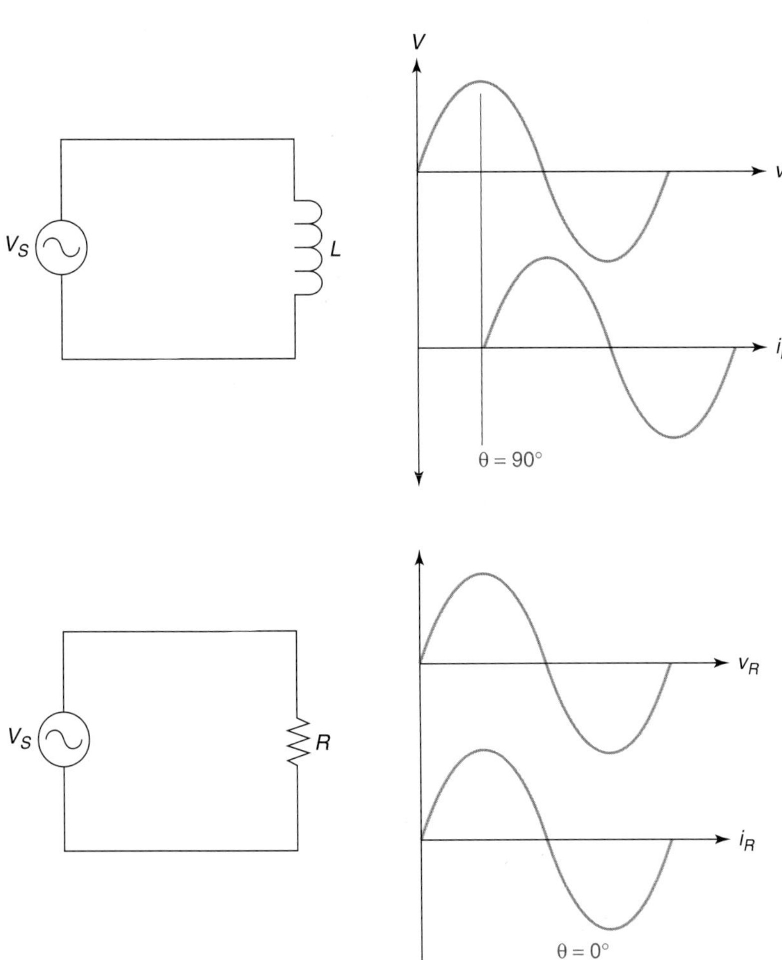


**FIGURE 14.4 Inductive and resistive circuit phase relationships.**

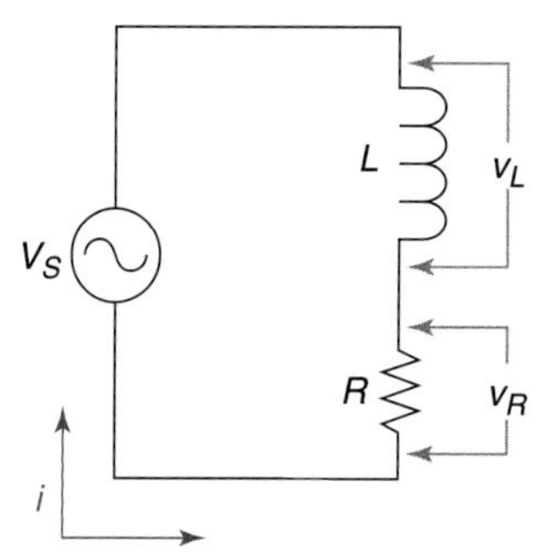


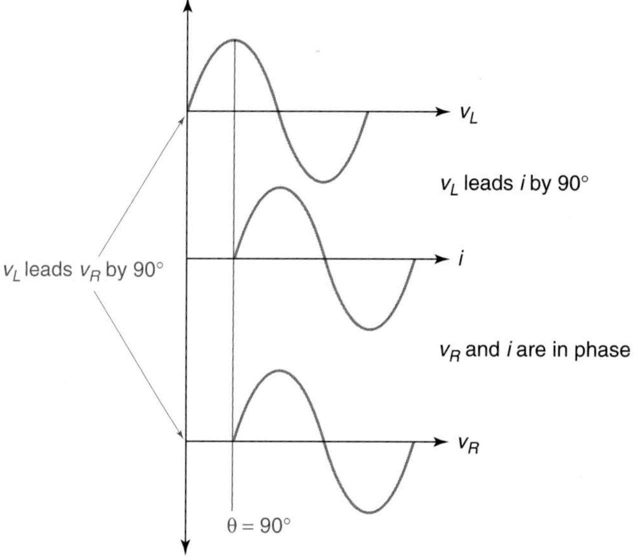


**FIGURE 14.5 Voltage phase relationships in a series *RL* circuit.**

where $z$ = the geometric sum of the two phasor magnitudes
$x$ and $y$ = the magnitudes of the phasors to be added

Relating this equation to Figure 14.6b, the sum of $V_L$ and $V_R$ can be found as

$$V_S = \sqrt{V_L^2 + V_R^2} \tag{14.1}$$

The source voltage phase angle.

The phase angle of $V_S$ is *the phase difference between the source voltage and the circuit current.* In Chapter 12, you were shown that this phase angle can be found as

$$\theta = \tan^{-1}\frac{y}{x}$$

where $\tan^{-1}$ = the inverse tangent of the fraction
$y$ = the magnitude of the $y$-axis value
$x$ = the magnitude of the $x$-axis value

**FIGURE 14.6 Vector (geometric) addition of $V_L$ and $V_R$.**

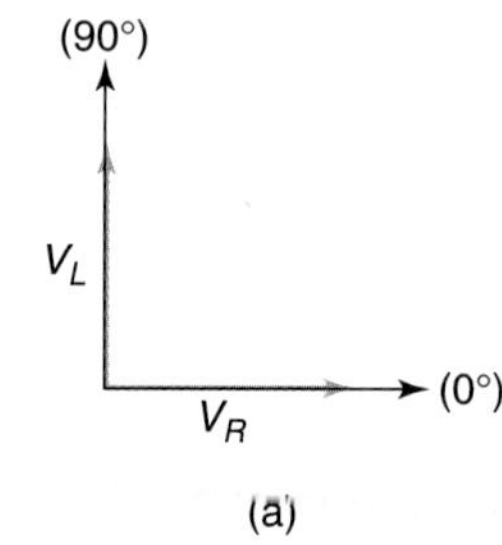


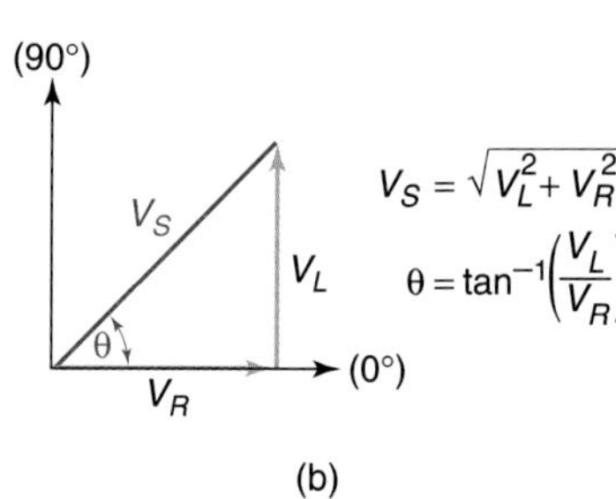

Relating this equation to Figure 14.6b, the phase angle of the source voltage can be found as

$$\theta = \tan^{-1}\frac{V_L}{V_R} \quad \textbf{(14.2)}$$

Example 14.1 demonstrates the calculation of source voltage for a series *RL* circuit.

**EXAMPLE 14.1**

The component voltages shown in Figure 14.7a were measured using an ac voltmeter. Calculate the source voltage for the circuit.

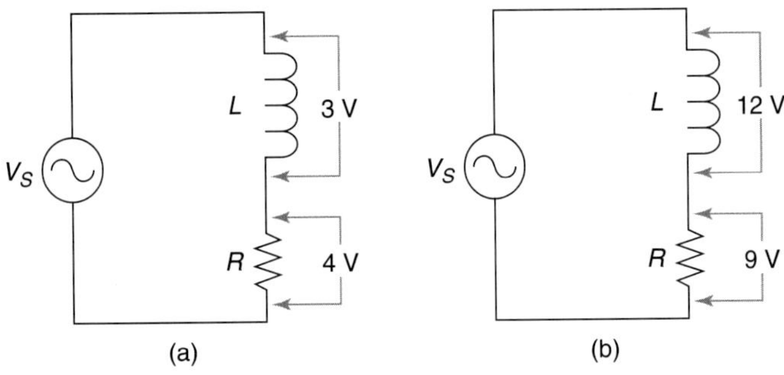


**FIGURE 14.7**

***Solution:*** Using equation (14.1), the magnitude of the source voltage is found as

$$V_S = \sqrt{V_L^2 + V_R^2} = \sqrt{(3\text{ V})^2 + (4\text{ V})^2} = \sqrt{25\text{ V}^2} = \mathbf{5\text{ V}}$$

The phase angle of the source voltage (relative to the circuit current) can be found as

$$\theta = \tan^{-1}\left(\frac{V_L}{V_R}\right) = \tan^{-1}\left(\frac{3\text{ V}}{4\text{ V}}\right) = \tan^{-1}(0.75) \cong \mathbf{36.9^\circ}$$

Therefore, the source voltage for the circuit is 5 V∠36.9°.

***PRACTICE PROBLEM 14.1***

Calculate the value of $V_S$ for the circuit in Figure 14.7b.

For any *RL* circuit, the limits on the phase angle between the source voltage and circuit current are given as follows:

$$0^\circ < \theta < 90^\circ \quad \textbf{(14.3)}$$

This range of values for $\theta$ makes sense when you consider that:

1. The phase angle in a purely resistive circuit is 0°.
2. The phase angle in a purely inductive circuit is 90°.

A circuit containing both resistance and inductance is neither purely resistive nor purely inductive. Therefore, it must have a phase angle that is greater than 0° and less than 90°. As you will see, the value of $\theta$ for a given series *RL* circuit depends on the ratio of inductive reactance to resistance.

## Measuring the Source Phase Angle ($\theta$)

Since circuit current is in phase with resistor voltage, $\theta$ can be measured by measuring the phase angle between the source voltage ($V_S$) and the resistor voltage ($V_R$). This can be accomplished by connecting a dual-trace oscilloscope to the circuit, as shown in Figure 14.8a.

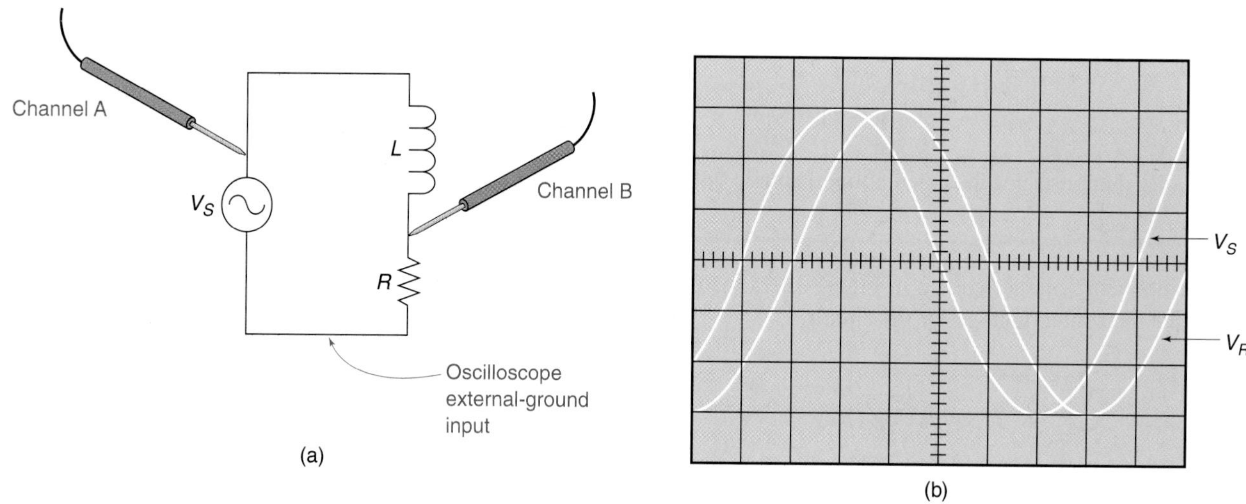


EWB

**FIGURE 14.8**

When the oscilloscope settings are adjusted properly, the display will resemble the one shown in Figure 14.8b. In Chapter 11, you were shown how this type of display can be used to calculate the phase angle, as follows:

$$\theta = (360°)\frac{t}{T_C} \tag{14.4}$$

where $t$ = the time between the zero crossings of the two waveforms
$T_C$ = the cycle time

As a review, the process for measuring θ with an oscilloscope is demonstrated in Example 14.2.

## EXAMPLE 14.2

Calculate the phase angle between the waveforms shown in Figure 14.9a.

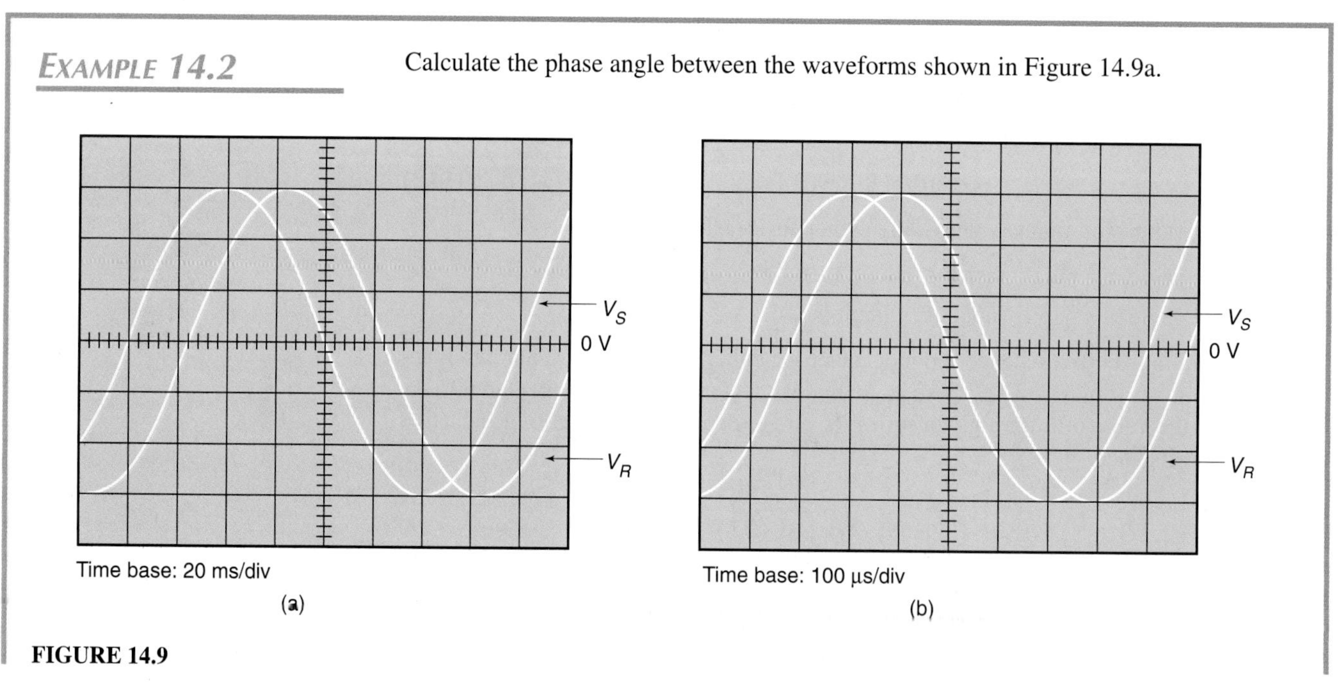


**FIGURE 14.9**

***Solution:*** There are 1.2 major divisions between the time that $V_S$ and $V_R$ cross the 0 V line. With a time base setting of 20 ms/div, the value of $t$ is found as

$$t = (1.2 \text{ div})\frac{20 \text{ ms}}{\text{div}} = \mathbf{24 \text{ ms}}$$

Each waveform is 8 major divisions in length. Therefore, the cycle time of each waveform can be found as

$$T_C = (8 \text{ div})\frac{20 \text{ ms}}{\text{div}} = \mathbf{160 \text{ ms}}$$

Finally, the phase angle between the waveforms can be found as

$$\theta = (360°)\frac{t}{T_C} = (360°)\frac{24 \text{ ms}}{160 \text{ ms}} = \mathbf{54°}$$

This result indicates that the source voltage is leading the circuit current by 54°.

***PRACTICE PROBLEM 14.2***

Calculate the phase angle between the waveforms shown in Figure 14.9b.

A simpler way to calculate θ from an oscilloscope display.

It's time to show you a shortcut for calculating the value of θ from an oscilloscope display. Since the time base setting is constant for the $t$ and $T_C$ measurements, it is not really necessary to calculate the actual time values. Instead, we can solve equation (14.4) using the number of divisions that make up the $t$ and $T_C$ intervals. For example, the time intervals in Figure 14.9a were indicated as follows:

$$t \rightarrow 1.2 \text{ divisions} \qquad T_C \rightarrow 8 \text{ divisions}$$

Had we used these values in place of the real-time values in equation (14.4), we would have calculated a phase angle of

$$\theta = (360°)\frac{t \text{ divisions}}{T_C \text{ divisions}} = (360°)\frac{1.2 \text{ div}}{8 \text{ div}} = 54°$$

which is the same result we obtained in the example. (If used correctly, this technique provides the same result for Practice Problem 14.2 that you obtained earlier. Try it!)

## Series Impedance

◄ *OBJECTIVE 3*

$X_L$ has a phase angle relative to series circuit current.

In an *RL* circuit, the total impedance (opposition to alternating current) is made up of resistance and inductive reactance. Like series voltages, the values of $R$ and $X_L$ for a series circuit are phasor quantities that must be added geometrically. The reason for this can be explained with the help of Figure 14.10.

In Chapter 13, you were told that inductive reactance can be found using Ohm's law, as follows:

$$X_L = \frac{V_L}{I_L}$$

Since Chapter 13 dealt with purely inductive circuits, it was not necessary to consider any phase angle associated with $X_L$. Now that we are dealing with *RL* circuits, we need to establish the fact that there *is* a phase angle associated with inductive reactance. This phase angle can be determined by including the phase angle of $V_L$ in the Ohm's law calculation of $X_L$. For the circuit in Figure 14.10:

$$X_L = \frac{V_L\angle 90°}{I\angle 0°} = \frac{V_L}{I}\angle 90°$$

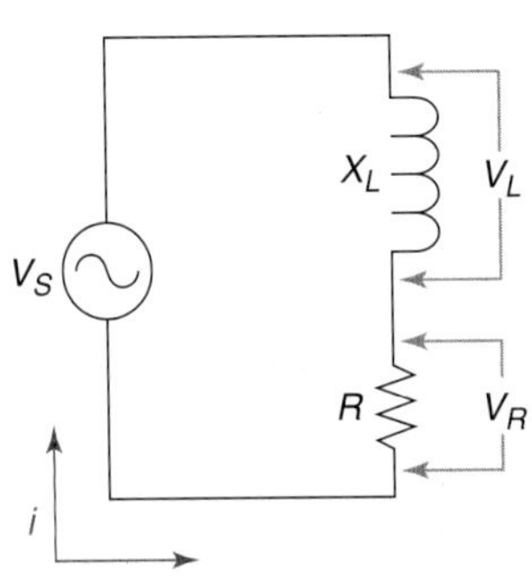


**FIGURE 14.10 Series *RL* voltages and circuit current.**

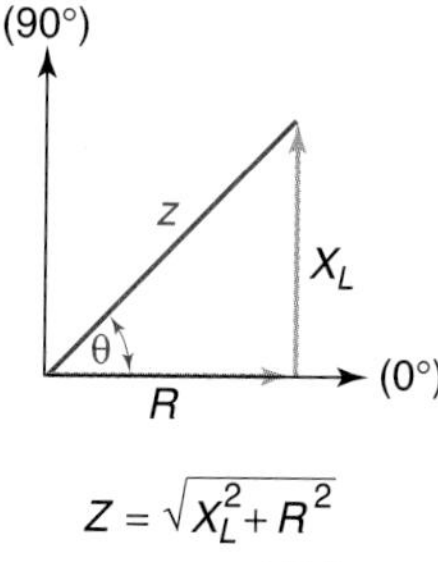


**FIGURE 14.11 Combining series resistance and inductive reactance.**

While this result does not include a magnitude, it *does* indicate that inductive reactance leads circuit current by 90°. By the same token, we can demonstrate that $R$ (Figure 14.10) has a 0° phase angle. Using the value of $\theta = 0°$ for $V_R$, the phase angle of $R$ is found as:

$$R = \frac{V_R \angle 0°}{I \angle 0°} = \frac{V_R}{I} \angle 0°$$

As these calculations indicate, $X_L$ leads the circuit current by 90°, while resistance and circuit current are in phase. Therefore, inductive reactance leads resistance by 90°. This means that $X_L$ and $R$ must be treated as phasor values and plotted as shown in Figure 14.11. As the figure indicates, the magnitude of $Z$ equals the geometric sum of $X_L$ and $R$. By formula,

$$Z = \sqrt{X_L^2 + R^2} \quad \textbf{(14.5)}$$

The phase angle of the circuit impedance (relative to the circuit current) is found in a similar fashion as the phase angle of the source voltage. By formula,

$$\theta = \tan^{-1} \frac{X_L}{R} \quad \textbf{(14.6)}$$

Example 14.3 demonstrates the procedure for determining the total impedance in a series *RL* circuit.

## EXAMPLE 14.3

Calculate the total impedance (including the phase angle) for the circuit in Figure 14.12a.

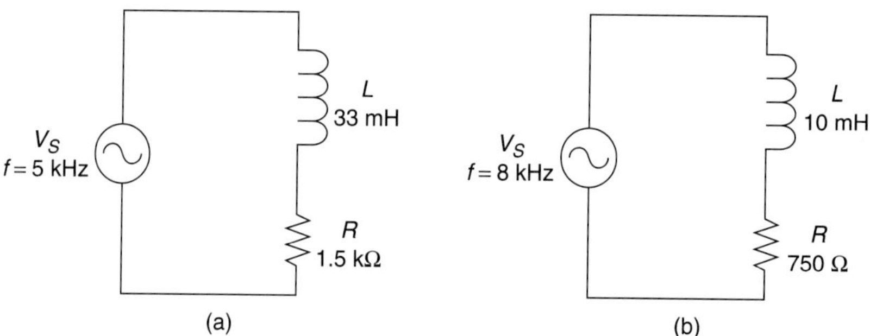


**FIGURE 14.12**

***Solution:*** Using the values shown in the figure, the value of $X_L$ is found as

$$X_L = 2\pi f L = 2\pi(5 \text{ kHz})(33 \text{ mH}) \cong \mathbf{1.04\ k\Omega}$$

Now, the magnitude of the circuit impedance can be found as

$$Z = \sqrt{X_L^2 + R^2} = \sqrt{(1.04\text{ k}\Omega)^2 + (1.5\text{ k}\Omega)^2} = \mathbf{1.82\ k\Omega}$$

and the phase angle can be found as

$$\theta = \tan^{-1}\left(\frac{X_L}{R}\right) = \tan^{-1}\left(\frac{1.04\text{ k}\Omega}{1.5\text{ k}\Omega}\right) = \tan^{-1}(0.693) = \mathbf{34.7°}$$

These results indicate that the circuit has a total impedance of 1.82 kΩ at an angle of 34.7° (relative to the circuit current).

***PRACTICE PROBLEM 14.3***

Calculate the total impedance (including the phase angle) for the circuit in Figure 14.12b.

## Resolving Series Circuit Phase Angles

Like numeric values, any group of phase angles associated with a given circuit must resolve themselves when the circuit analysis is complete. In other words, the phase angles must combine in such a way so that *every* circuit value conforms to the basic laws of circuit operation, such as Ohm's law and Kirchhoff's law. With this in mind, look at the circuit in Figure 14.13.

The circuit shown in the figure is basically the same circuit that we analyzed in Example 14.3. For the sake of discussion, a circuit current value of 1 mA has been added, and Ohm's law has been used to calculate the component voltages. According to equations (14.1) and (14.2),

$$V_S = \sqrt{V_L^2 + V_R^2} = \sqrt{(1.04\text{ V})^2 + (1.5\text{ V})^2} = 1.82\text{ V}$$

and

$$\theta = \tan^{-1}\left(\frac{V_L}{V_R}\right) = \tan^{-1}\left(\frac{1.04\text{ V}}{1.5\text{ V}}\right) = \tan^{-1}(0.693) = 34.7°$$

Therefore, the source voltage has a value of 1.82 V∠34.7°. In Example 14.3, we calculated a total circuit impedance of 1.82 kΩ∠34.7°. If we use these two values, Ohm's law provides a circuit current value of

$$I = \frac{V_S}{Z_T} = \frac{1.82\text{ V}\angle 34.7°}{1.82\text{ k}\Omega\angle 34.7°} = 1\text{ mA}\angle 0°$$

which agrees with our assumed value of circuit current. As you can see, the phase angles for the voltage, current, and impedance values all conform to basic circuit principles. This will always be the case when a circuit is analyzed properly.

## Calculating Series Circuit Current

When you use Ohm's law to calculate the total current in a series *RL* circuit, you may see cases where it appears that the circuit current has a negative phase angle. For example, consider the circuit in Figure 14.14. The analysis of this circuit would begin by finding the value of $X_L$, as follows:

$$X_L = 2\pi fL = 2\pi(10\text{ kHz})(200\text{ mH}) = 12.6\text{ k}\Omega$$

The next step would be to determine the magnitude and phase angle of the circuit impedance, as follows:

$$Z_T = \sqrt{X_L^2 + R^2} = \sqrt{(12.6\text{ k}\Omega)^2 + (15\text{ k}\Omega)^2} = 19.6\text{ k}\Omega$$

and

$$\theta = \tan^{-1}\left(\frac{X_L}{R}\right) = \tan^{-1}\left(\frac{12.6\text{ k}\Omega}{15\text{ k}\Omega}\right) = \tan^{-1}(0.84) = 40°$$

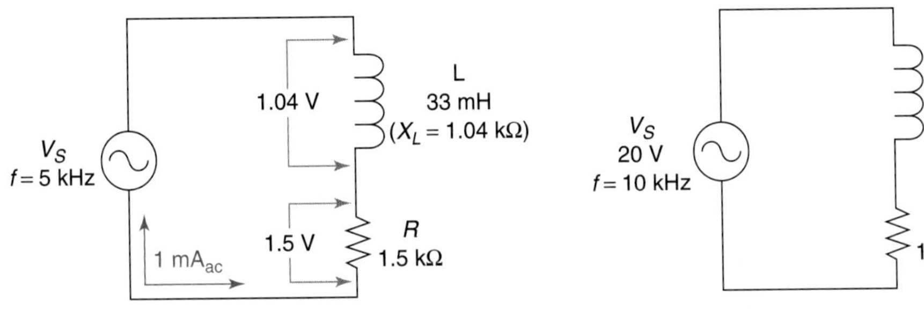


**FIGURE 14.13**

**FIGURE 14.14**

With a circuit impedance of 19.6 kΩ∠40°, Ohm's law provides us with a circuit current of

$$I = \frac{V_S}{Z_T} = \frac{20\text{ V}\angle 0°}{19.6\text{ k}\Omega\angle 40°} = 1.02\text{ mA}\angle -40°$$

As you can see, the calculated circuit current includes a negative phase angle.

The meaning of a negative current phase angle.

Although the negative phase angle of the circuit current may seem out of place, it really isn't. The negative phase angle in our result indicates that the circuit current lags the source voltage by 40°. As you learned in Chapter 13, this is the same thing as saying that the source voltage leads the circuit current by 40°. Therefore, we could express the phase relationship between the source voltage and circuit current as either:

$$V_S = 20\text{ V}\angle 0° \qquad\qquad V_S = 20\text{ V}\angle 40°$$
$$\text{or}$$
$$I = 1.02\text{ mA}\angle -40° \qquad\qquad I = 1.02\text{ mA}\angle 0°$$

These two sets of numbers say the same thing. The only difference is that the first set of values references the phase angle to the source voltage, and the second references the phase angle to the circuit current. Since current is constant throughout a series circuit, using current as the 0° reference is the preferred method. Therefore, we would rewrite our results as

$$V_S = 20\text{ V}\angle 40° \qquad I = 1.02\text{ mA}\angle 0°$$

You will see this adjustment made in several of the upcoming examples.

## Series Circuit Analysis

OBJECTIVE 4 ▶ You have been shown how the voltage, current, and impedance values are calculated for a basic series *RL* circuit. At this point, we will work through several circuit analysis problems. The goal is to demonstrate the approaches taken to various series *RL* circuit analysis problems. We'll start by working through a "straightforward" circuit analysis.

### EXAMPLE 14.4

Determine the voltage, current, and impedance values for the circuit in Figure 14.15a.

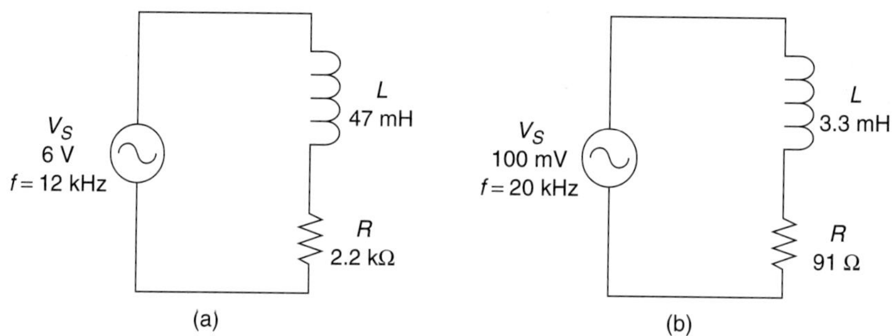


**FIGURE 14.15**

***Solution:*** First, the value of $X_L$ is found as

$$X_L = 2\pi fL = 2\pi(12\text{ kHz})(47\text{ mH}) = \mathbf{3.54\text{ k}\Omega}$$

The magnitude of the circuit impedance can now be found as

$$Z_T = \sqrt{X_L^2 + R^2} = \sqrt{(3.54\text{ k}\Omega)^2 + (2.2\text{ k}\Omega)^2} = \mathbf{4.17\text{ k}\Omega}$$

The impedance has a phase angle of

$$\boldsymbol{\theta} = \tan^{-1}\left(\frac{X_L}{R}\right) = \tan^{-1}\left(\frac{3.54\text{ k}\Omega}{2.2\text{ k}\Omega}\right) = \mathbf{58.1°}$$

Using the values of $Z_T$ and $V_S$, the circuit current is found as

$$\boldsymbol{I} = \frac{V_S}{Z_T} = \frac{6\text{ V}\angle 0°}{4.17\text{ k}\Omega\angle 58.1°} = \mathbf{1.44\text{ mA}\angle -58.1°}$$

Remember that the negative phase angle is relative to the source voltage. Since current is used as the 0° reference in a series circuit, the values of $V_S$ and $I$ are written as

$$V_S = 6\text{ V}\angle 58.1° \qquad I = 1.44\text{ mA}\angle 0°$$

Note that the phase angles were determined by adding 58.1° to the phase angles of the source voltage and circuit current.

Now, we can calculate the magnitudes of $V_L$ and $V_R$ as follows:

$$\boldsymbol{V_L} = I \cdot X_L = (1.44\text{ mA})(3.54\text{ k}\Omega) = \mathbf{5.10\text{ V}}$$

and

$$\boldsymbol{V_R} = I \cdot R = (1.44\text{ mA})(2.2\text{ k}\Omega) = \mathbf{3.17\text{ V}}$$

At this point, we have calculated all the circuit values. To verify our results, we will use the component voltages to calculate the magnitude and phase angle of the source. If everything has been calculated correctly, we will obtain the value of $V_S = 6\text{ V}\angle 58.1°$. First, the magnitude of the source is found as

$$\boldsymbol{V_S} = \sqrt{V_L^2 + V_R^2} = \sqrt{(5.10\text{ V})^2 + (3.17\text{ V})^2} = \mathbf{6\text{ V}}$$

Now, the source phase angle (relative to the circuit current) is found as

$$\boldsymbol{\theta} = \tan^{-1}\left(\frac{V_L}{V_R}\right) = \tan^{-1}\left(\frac{5.10\text{ V}}{3.17\text{ V}}\right) = \mathbf{58.1°}$$

These results match our known source values, so we know we have solved the circuit correctly.

***PRACTICE PROBLEM 14.4***

Determine the voltage, current, and impedance values for the circuit in Figure 14.15b. Verify your results using the component voltages (as we did in this example).

Using the voltage-divider equation.

There are always instances when we are interested in calculating only the component voltages in a series *RL* circuit. We can calculate the component voltages using the voltage-divider equation, just as we did in dc circuits. However, when dealing with *RL* circuits, the voltage-divider equation must be modified as follows:

$$V_n = V_S \frac{Z_n}{Z_T} \tag{14.7}$$

where $Z_n$ = the magnitude of $R$ or $X_L$

$Z_T$ = the geometric sum of $R$ and $X_L$

Example 14.5 demonstrates the use of the voltage-divider relationship in series *RL* circuit analysis.

## EXAMPLE 14.5

Calculate the component voltages for the circuit in Figure 14.16a.

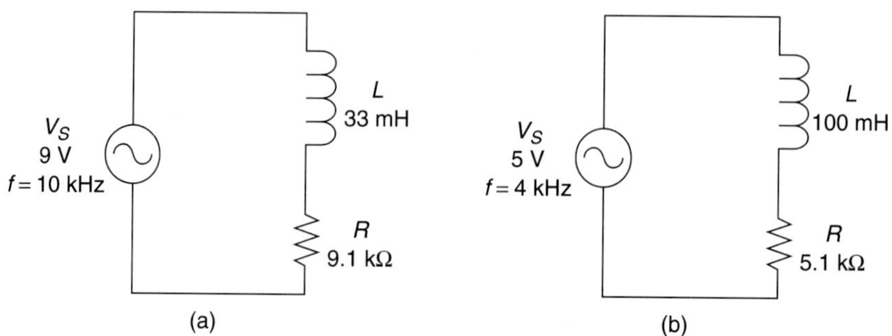


**FIGURE 14.16**

***Solution:*** First, the value of $X_L$ is found as

$$X_L = 2\pi fL = 2\pi(10 \text{ kHz})(33 \text{ mH}) = \mathbf{2.07\ k\Omega}$$

and the magnitude of the circuit impedance is found as

$$Z_T = \sqrt{X_L^2 + R^2} = \sqrt{(2.07 \text{ k}\Omega)^2 + (9.1 \text{ k}\Omega)^2} = \mathbf{9.33\ k\Omega}$$

Now, the magnitude of the inductor voltage can be found as

$$V_L = V_S \frac{X_L}{Z_T} = (9 \text{ V})\frac{2.07 \text{ k}\Omega}{9.33 \text{ k}\Omega} = \mathbf{2.00\ V}$$

and the magnitude of the resistor voltage can be found as

$$V_R = V_S \frac{R}{Z_T} = (9 \text{ V})\frac{9.1 \text{ k}\Omega}{9.33 \text{ k}\Omega} = \mathbf{8.78\ V}$$

***PRACTICE PROBLEM 14.5***

Calculate the component voltages for the circuit in Figure 14.16b.

If the component voltages calculated in Example 14.5 appear to be incorrect, you may be forgetting that the source voltage equals their geometric sum. The component voltages calculated in the example *do* add up to equal the source voltage, as follows:

$$V_S = \sqrt{V_L^2 + V_R^2} = \sqrt{(2.00 \text{ V})^2 + (8.78 \text{ V})^2} = 9 \text{ V}$$

When a series *RL* circuit contains more than one inductor and/or resistor, the analysis of the circuit takes a few more steps. This point is demonstrated in Example 14.6.

## EXAMPLE 14.6

Calculate the values of $Z_T$, $V_{L1}$, $V_{L2}$, and $V_R$ for the circuit in Figure 14.17a.

**FIGURE 14.17**

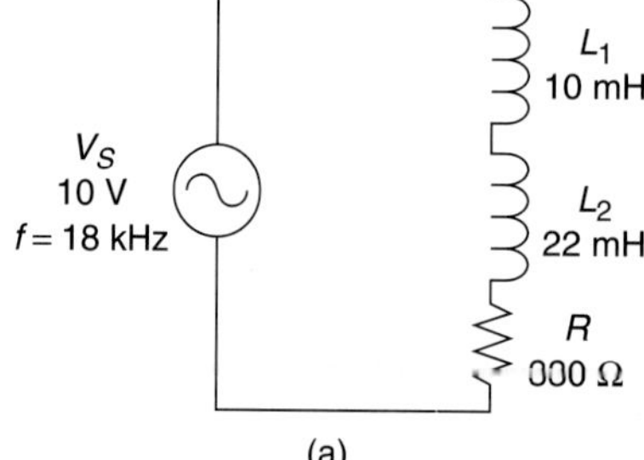


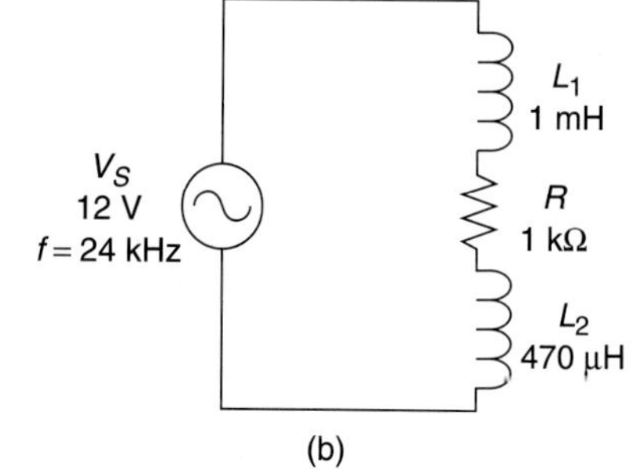

**Solution:** First, we must determine the reactance of each inductor, as follows:

$$X_{L1} = 2\pi f L_1 = 2\pi(18\text{ kHz})(10\text{ mH}) = \mathbf{1.13\ k\Omega}$$

and

$$X_{L2} = 2\pi f L_2 = 2\pi(18\text{ kHz})(22\text{ mH}) = \mathbf{2.49\ k\Omega}$$

To calculate the total circuit impedance, we need to determine the total circuit reactance. Since $X_{L1}$ and $X_{L2}$ are in phase with each other, the total reactance equals the algebraic sum of their values, as follows:

$$X_{L(T)} = X_{L1} + X_{L2} = 1.13\text{ k}\Omega + 2.49\text{ k}\Omega = \mathbf{3.62\ k\Omega}$$

*An Alternate Approach:* To calculate the total reactance, we could have found $L_T = L_1 + L_2$, and then used $L_T$ in the inductive reactance equation. The result would have been the same.

Now, using the values of $X_{L(T)}$ and $R$, the magnitude of the circuit impedance can be found as

$$Z_T = \sqrt{X_{L(T)}^2 + R^2} = \sqrt{(3.62\text{ k}\Omega)^2 + (330\ \Omega)^2} = \mathbf{3.64\ k\Omega}$$

Once we know the magnitude of the circuit impedance, we can use the voltage-divider equation to calculate the magnitudes of the component voltages, as follows:

$$V_{L1} = V_S\frac{X_{L1}}{Z_T} = (10\text{ V})\frac{1.13\text{ k}\Omega}{3.64\text{ k}\Omega} = \mathbf{3.10\ V}$$

$$V_{L2} = V_S\frac{X_{L2}}{Z_T} = (10\text{ V})\frac{2.49\text{ k}\Omega}{3.64\text{ k}\Omega} = \mathbf{6.84\ V}$$

and

$$V_R = V_S\frac{R}{Z_T} = (10\text{ V})\frac{330\ \Omega}{3.64\text{ k}\Omega} = \mathbf{0.91\ V}$$

***PRACTICE PROBLEM 14.6***

Calculate the values of $Z_T$, $V_{L1}$, $V_{L2}$, and $V_R$ for the circuit in Figure 14.17b. (*Note:* The placement of the components does not affect the analysis of the circuit.)

Once again, we can verify our results by using the component voltages to calculate the magnitude of the source voltage. Since $V_{L1}$ and $V_{L2}$ are in phase, they can be added together algebraically, as follows:

$$V_{L(T)} = V_{L1} + V_{L2} = 3.10\text{ V} + 6.84\text{ V} = 9.94\text{ V}$$

Then, using the values of $V_{L(T)}$ and $V_R$, the magnitude of the source voltage can be found as

$$V_S = \sqrt{V_{L(T)}^2 + V_R^2} = \sqrt{(9.94\text{ V})^2 + (0.91\text{ V})^2} \cong 10\text{ V}$$

Since our result was approximately 10 V (the given value of the source), we can assume that all our calculations are correct.

In Example 14.6, we used the total circuit reactance ($X_{L(T)}$) to calculate the circuit impedance. By the same token, the value of $X_{L(T)}$ would be used to calculate the phase angle of the circuit impedance (relative to the circuit current). For example, the phase angle of $Z_T$ in Figure 14.17a is found as

$$\theta = \tan^{-1}\left(\frac{X_{L(T)}}{R}\right) = \tan^{-1}\left(\frac{3.62\text{ k}\Omega}{330\ \Omega}\right) = 84.8°$$

When working with circuits containing multiple inductors and/or resistors, just remember the following:

1. Any two (or more) values, such as two or more reactances, that are in phase can be added algebraically.

2. Any two (or more) values, such as reactance and resistance, that are not in phase must be added geometrically.
3. Any overall circuit value, such as the magnitude of $Z_T$ or $\theta$, is calculated using the appropriate total values. (For example, $V_S$ would be calculated using $V_{L(T)}$ and $V_{R(T)}$.)

OBJECTIVE 5 ▶

## Series RL Circuit Frequency Response

**Frequency response** Any changes that occur in a circuit as a result of a change in operating frequency.

The term **frequency response** is used to describe any changes that occur in a circuit as a result of a change in operating frequency. For example, consider the circuit shown in Figure 14.18. Table 14.1 shows the circuit response to an increase in operating frequency.

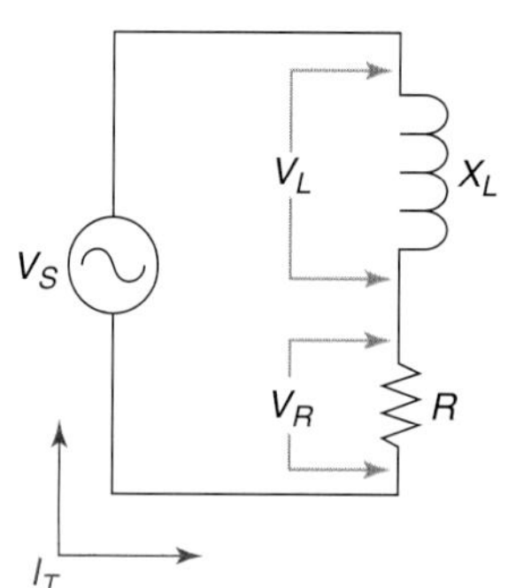


FIGURE 14.18

**TABLE 14.1 Series *RL* Circuit Response to an Increase in Operating Frequency**

| *Cause and Effect* | *Relevant Equation* |
|---|---|
| 1. The increase in frequency causes $X_L$ to increase. | $X_L = 2\pi fL$ |
| 2. The increase in $X_L$ causes: | |
| a. $\theta$ to increase. | $\theta = \tan^{-1}\frac{X_L}{R}$ |
| b. $Z_T$ to increase. | $Z_T = \sqrt{X_L^2 + R^2}$ |
| 3. The increase in $Z_T$ causes $I_T$ to decrease. | $I_T = \frac{V_S}{Z_T}$ |
| 4. The increase in $X_L$ causes $V_L$ to increase. | $V_L = I_T X_L$ |
| 5. The increase in $V_L$ causes $V_R$ to decrease. | $V_R = \sqrt{V_S^2 - V_L^2}$ |

Here's what happens if the operating frequency decreases:

1. The decrease in operating frequency causes $X_L$ to decrease.
2. The decrease in $X_L$ causes $\theta$ to decrease.
3. The decrease in $X_L$ also causes $Z_T$ to decrease.
4. The decrease in $Z_T$ causes $I_T$ to increase.
5. The increase in $X_L$ causes $V_L$ to decrease.
6. The increase in $V_L$ causes $V_R$ to increase.

Example 14.7 demonstrates the effects of a change in operating frequency on series *RL* circuit values.

### EXAMPLE 14.7

The voltage source in Figure 14.19a has the frequency limits shown. Calculate the circuit voltage, current, and impedance values at the frequency limits of the source.

FIGURE 14.19

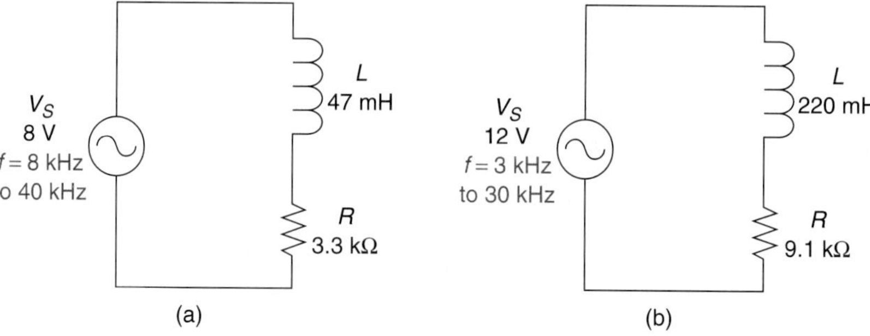

***Solution:*** When the operating frequency of the circuit is 8 kHz, the reactance of the inductor is found as

$$X_L = 2\pi fL = 2\pi(8\text{ kHz})(47\text{ mH}) = \mathbf{2.36\ k\Omega}$$

The magnitude of the circuit impedance is found as

$$Z_T = \sqrt{X_L^2 + R^2} = \sqrt{(2.36\text{ k}\Omega)^2 + (3.3\text{ k}\Omega)^2} = \mathbf{4.06\ k\Omega}$$

and the value of $\theta$ is found as

$$\theta = \tan^{-1}\left(\frac{X_L}{R}\right) = \tan^{-1}\left(\frac{2.36\text{ k}\Omega}{3.3\text{ k}\Omega}\right) = \mathbf{35.6°}$$

The magnitude of the circuit current is found as

$$I_T = \frac{V_S}{Z_T} = \frac{8\text{ V}}{4.06\text{ k}\Omega} = \mathbf{1.97\ mA}$$

Now, the magnitudes of the component voltages can be found as

$$V_L = I_T X_L = (1.97\text{ mA})(2.36\text{ k}\Omega) = \mathbf{4.65\ V}$$

and

$$V_R = I_T R = (1.97\text{ mA})(3.3\text{ k}\Omega) = \mathbf{6.50\ V}$$

When the operating frequency increases to 40 kHz, the reactance of the inductor increases to

$$X_L = 2\pi fL = 2\pi(40\text{ kHz})(47\text{ mH}) = \mathbf{11.8\ k\Omega}$$

This increase in $X_L$ gives us a circuit impedance of

$$Z_T = \sqrt{X_L^2 + R^2} = \sqrt{(11.8\text{ k}\Omega)^2 + (3.3\text{ k}\Omega)^2} = \mathbf{12.3\ k\Omega}$$

at a phase angle of

$$\theta = \tan^{-1}\left(\frac{X_L}{R}\right) = \tan^{-1}\left(\frac{11.8\text{ k}\Omega}{3.3\text{ k}\Omega}\right) = \mathbf{74.4°}$$

The increased circuit impedance reduces the magnitude of the circuit current to

$$I_T = \frac{V_S}{Z_T} = \frac{8\text{ V}}{12.3\text{ k}\Omega} = \mathbf{650\ \mu A}$$

Finally, the magnitudes of the component voltages at this operating frequency can be found as

$$V_L = I_T X_L = (650\ \mu\text{A})(11.8\text{ k}\Omega) = \mathbf{7.67\ V}$$

and

$$V_R = I_T R = (650\ \mu\text{A})(3.3\text{ k}\Omega) = \mathbf{2.15\ V}$$

***PRACTICE PROBLEM 14.7***

The voltage source in Figure 14.19b has the frequency limits shown. Calculate the circuit voltage, current, and impedance values at the frequency limits of the source.

The values obtained in Example 14.7 are listed in Table 14.2. When you compare the results with the cause-and-effect statements in Table 14.1, you'll see that the circuit responded to the change in frequency exactly as described in that table.

Throughout the remainder of this text, we will touch on the concept of circuit frequency response. As you will learn, frequency response is an integral part of any discussion on ac circuits.

**TABLE 14.2 A Comparison of the Two Sets of Results in Example 14.7**

| *Variable* | *Value at f = 8 kHz* | *Value at f = 40 kHz* | *Effect* |
|---|---|---|---|
| $X_L$ | 2.36 kΩ | 11.8 kΩ | Increased |
| $Z_T$ | 4.06 kΩ | 12.3 kΩ | Increased |
| θ | 35.6° | 74.4° | Increased[a] |
| $I_T$ | 1.97 mA | 650 μA | Decreased |
| $V_L$ | 4.65 V | 7.67 V | Increased |
| $V_R$ | 6.50 V | 2.15 V | Decreased |

[a]Effect refers to magnitude.

## One Final Note

You have been introduced to a variety of relationships in this section. Series *RL* circuits are relatively easy to analyze when you keep in mind the fact that many values must be added geometrically rather than algebraically. For future reference, the characteristics of series *RL* circuits are summarized in Figure 14.20.

You may have noticed that we did not address the subject of component power in this section. Since power calculations apply to both series and parallel *RL* circuits, they are introduced in a separate section.

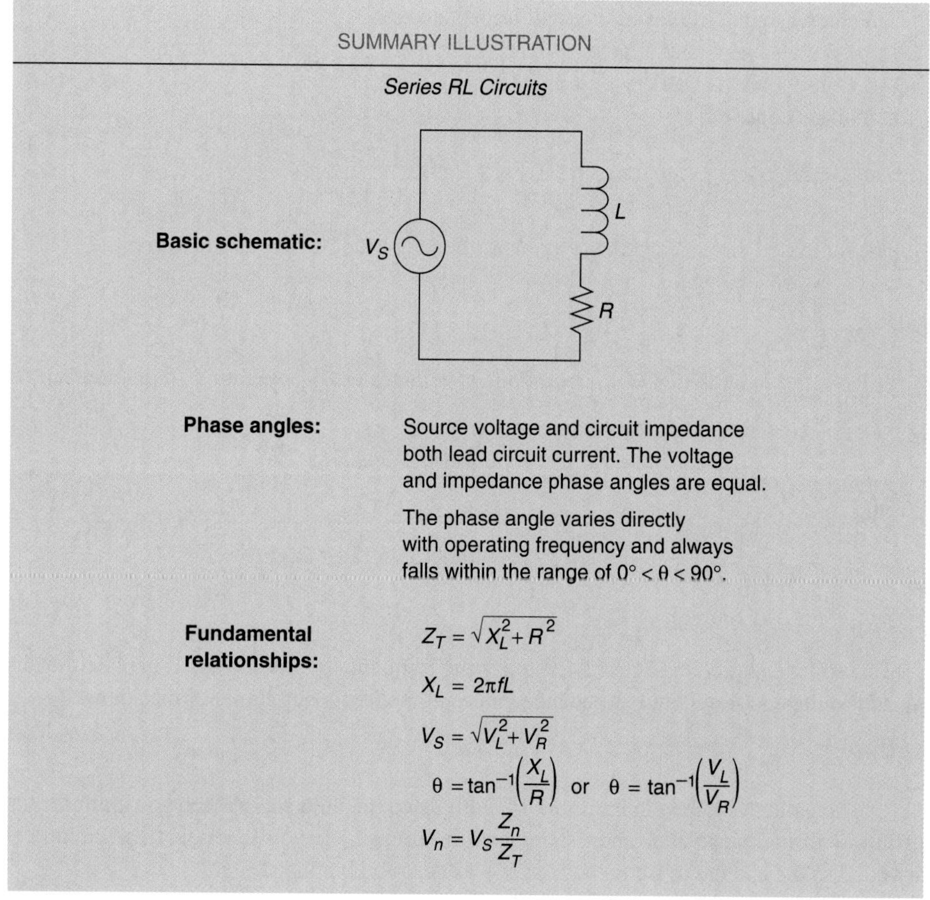
SUMMARY ILLUSTRATION
Series RL Circuits
Basic schematic:
$V_S$
L
R
Phase angles:
Source voltage and circuit impedance both lead circuit current. The voltage and impedance phase angles are equal.
The phase angle varies directly with operating frequency and always falls within the range of 0° < θ < 90°.
Fundamental relationships:
$Z_T = \sqrt{X_L^2 + R^2}$
$X_L = 2\pi fL$
$V_S = \sqrt{V_L^2 + V_R^2}$
$\theta = \tan^{-1}\left(\frac{X_L}{R}\right)$ or $\theta = \tan^{-1}\left(\frac{V_L}{V_R}\right)$
$V_n = V_S \frac{Z_n}{Z_T}$

**FIGURE 14.20**

## Section Review

1. What are the three fundamental characteristics of every series circuit?
2. What is the geometric sum of two variables? What does it represent?
3. In a series *RL* circuit, what is the phase relationship between $V_L$ and $V_R$?
4. In a series *RL* circuit, what does the phase angle of the source voltage represent?
5. Describe the process for measuring the phase angle between the source voltage and current for a series *RL* circuit.
6. In a series *RL* circuit, what is the phase relationship between inductive reactance and resistance?
7. In a series *RL* circuit, what does the phase angle of the circuit impedance represent?
8. What relationship must exist between the various phase angles in a series *RL* circuit?
9. When using Ohm's law to calculate the circuit current in a series *RL* circuit, what is indicated by a negative phase angle in the result? How is this phase angle normally resolved?
10. How is the voltage-divider equation modified for use in series *RL* circuits?
11. Describe the effect that a change in operating frequency has on the voltage, current, and impedance values in a series *RL* circuit.

# 14.2 POWER CHARACTERISTICS AND CALCULATIONS

◄ *OBJECTIVE 6*

Resistive-reactive circuit analysis involves all types of power calculations. Significant values of apparent power, true power, and reactive power are present in most resistive-reactive circuits. In this section, we will discuss the fundamental power characteristics and calculations for series *RL* circuits. As you will learn, the principles covered in this section apply to parallel *RL* circuits as well.

## ac Power Values: A Brief Review

In Chapter 13, we discussed apparent power ($P_{APP}$), resistive power ($P_R$), and reactive power ($P_X$) in terms of inductor operation. Each of these values is described (in terms of *RL* circuit values) in Table 14.3. The values of resistive and reactive power for a series *RL* circuit are calculated as demonstrated in Example 14.8.

**TABLE 14.3 Power Values in Resistive-Reactive Circuits**

| *Value* | *Definition* |
|---|---|
| Resistive power ($P_R$) | The power dissipated by the resistance in an *RL* circuit. Also known as true power. |
| Reactive power ($P_X$) | The value found using $P = I^2X_L$. Also known as imaginary power. The energy (per unit time) found using ($I^2X_L$) is actually used by the inductor to build its electromagnetic field and is returned to the circuit when the field collapses. $P_X$ is measured in volt-amperes-reactive (VAR) to distinguish it from actual power dissipation values. |
| Apparent power ($P_{APP}$) | The combination of resistive (true) power and reactive (imaginary) power. Measured in volt-amperes (VA) because its value does not represent a true power-dissipation value. |

## EXAMPLE 14.8

The circuit in Figure 14.21a has the values shown. Assuming that the inductor is an ideal component, calculate the values of resistive and reactive power for the circuit.

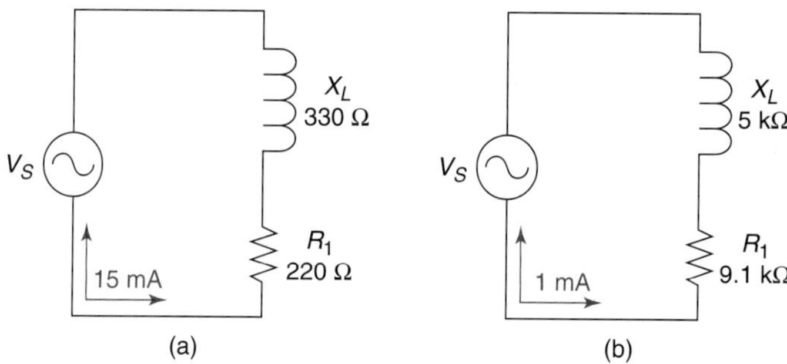


**FIGURE 14.21**

***Solution:*** Using the values shown, the reactive power is found as

$$P_X = I^2X_L = (15\text{ mA})^2(330\ \Omega) = \mathbf{74.3\ mVAR}$$

Since the inductor is assumed to be ideal, we do not need to account for any winding resistance ($R_w$) the component may have. Therefore, we can simply use the value of $R$ in our resistive power calculation, as follows:

$$P_R = I^2R_1 = (15\text{ mA})^2(220\ \Omega) = \mathbf{49.5\ mW}$$

***PRACTICE PROBLEM 14.8***

Calculate the values of $P_X$ and $P_R$ for the circuit in Figure 14.21b. Assume that the inductor is an ideal component.

The winding resistance of the coil ($R_w$) affects the calculated value of resistive power. However, the effect of $R_w$ on the value of resistive power is usually negligible. This point is demonstrated in Example 14.9.

## EXAMPLE 14.9

The coil in Figure 14.21 (Example 14.8) has a winding resistance of 0.8 Ω. Taking this value of $R_w$ into account, recalculate the value of resistive power for the circuit.

***Solution:*** The total resistance in the circuit is equal to the sum of $R_1$ and $R_w$. Thus, the resistive power in the circuit is found as

$$P_R = I^2(R_w + R_1) = (15\text{ mA})^2(220.8\ \Omega) = \mathbf{49.7\ mW}$$

As you can see, there is little difference between this value and the value of 49.5 mW found in Example 14.8. For this reason, $R_w$ is usually ignored in resistive power calculations for $RL$ circuits.

***PRACTICE PROBLEM 14.9***

Refer to Practice Problem 14.8. Assume that the coil has a winding resistance of 1 Ω and recalculate the resistive power in the circuit.

Remember that resistive power is dissipated by the resistance in an $RL$ circuit. Reactive power, on the other hand, is merely stored in the electromagnetic field of the inductor and returned to the circuit when the field collapses.

## Calculating Apparent Power ($P_{APP}$)

Apparent power in a series $RL$ circuit equals the geometric sum of resistive power and reactive power. This relationship can be explained with the aid of Figure 14.22. The graph in Figure 14.22a shows the phase relationship between $X_L$ and $R$. As indicated, $X_L$ has a 90° phase angle relative to $R$. If we consider this phase angle in our $P_X$ equation, we get

$$P_X = I^2X_L = (I^2\angle 0°)(X_L\angle 90°) = (I^2X_L)\angle 90°$$

While the result does not commit to a magnitude, it does indicate that reactive power has a phase angle of 90° (relative to the circuit current). If we perform the same type of calculation for $P_R$, we get

$$P_R = I^2R = (I^2\angle 0°)(R\angle 0°) = (I^2R)\angle 0°$$

Again, the result does not commit to a magnitude. However, it does indicate that resistive power is in phase with the circuit current. Since reactive power leads circuit current by 90°, it also leads resistive power by the same angle. Therefore, apparent power equals the geometric sum of the magnitudes of $P_X$ and $P_R$ as shown in Figure 14.22b. By formula,

$$P_{APP} = \sqrt{P_X^2 + P_R^2} \qquad \textbf{(14.8)}$$

where $P_{APP}$ = the apparent power, measured in volt-amperes (VA)

The phase angle of apparent power (relative to the circuit current) is found in the same fashion as the phase angle for impedance and source voltage. By formula,

$$\theta = \tan^{-1}\frac{P_X}{P_R} \qquad \textbf{(14.9)}$$

Example 14.10 demonstrates the use of these equations.

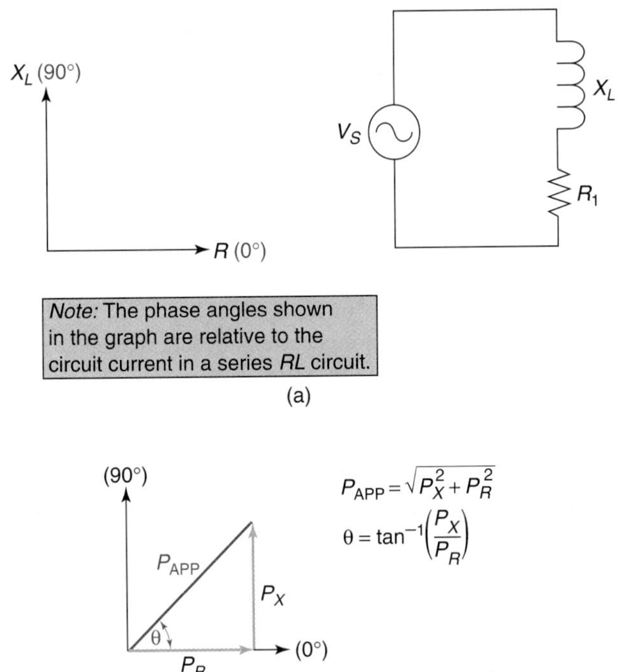


**FIGURE 14.22 Power phase relationships in a series $RL$ circuit.**

## EXAMPLE 14.10

Calculate the value of apparent power (including the phase angle) for the circuit in Example 14.8.

***Solution:*** In Example 14.8, we calculated the following magnitudes: $P_X = 74.3$ mVAR and $P_R = 49.5$ mW. Using these values in equation (14.8), the magnitude of apparent power is found as

$$\boldsymbol{P_{\mathrm{APP}}} = \sqrt{P_X^2 + P_R^2} = \sqrt{(74.3\text{ mVAR})^2 + (49.5\text{ mW})^2} = \mathbf{89.3\ VA}$$

and the phase angle is found as

$$\boldsymbol{\theta} = \tan^{-1}\left(\frac{P_X}{P_R}\right) = \tan^{-1}\left(\frac{74.3\text{ mVAR}}{49.5\text{ mW}}\right) = \tan^{-1}(1.501) = \mathbf{56.3^\circ}$$

***PRACTICE PROBLEM 14.10***

Calculate the value of $P_{\mathrm{APP}}$ (and the phase angle) for the circuit in Practice Problem 14.8.

When you calculate the value of apparent power for a circuit, you need to remember that only a portion of its value is dissipated by the circuit. (This fact is the reason why it has its own unit of measure.) At this point, we will discuss a multiplier that can be used to determine the portion of apparent power dissipated by the resistance in a circuit.

## Power Factor

**Power factor (PF)** The ratio of resistive power to apparent power, equal to cos θ.

The **power factor (PF)** for an *RL* circuit is the ratio of resistive power to apparent power. In other words, it is the ratio of actual power dissipation to apparent power. For any *RL* circuit, the power factor can be found as

$$\text{PF} = \cos\theta \tag{14.10}$$

The relationship given in equation (14.10) can be derived from the power triangle shown in Figure 14.23. As you learned in Chapter 12, the cosine of θ equals the ratio of the length of the adjacent (adj) side of the triangle to the length of the hypotenuse. By formula,

$$\cos\theta = \frac{\text{adj}}{\text{hyp}}$$

If we apply this relationship to the power triangle, we get

$$\cos\theta = \frac{P_R}{P_{\mathrm{APP}}} \tag{14.11}$$

which, by definition, is the power factor for the circuit.

If we transpose equation (14.11), we get a relationship that can be used to determine the actual power dissipation in an *RL* circuit, as follows:

$$P_R = P_{\mathrm{APP}}\cos\theta \tag{14.12}$$

Example 14.11 demonstrates a practical application of this relationship.

**FIGURE 14.23 Power factor.**

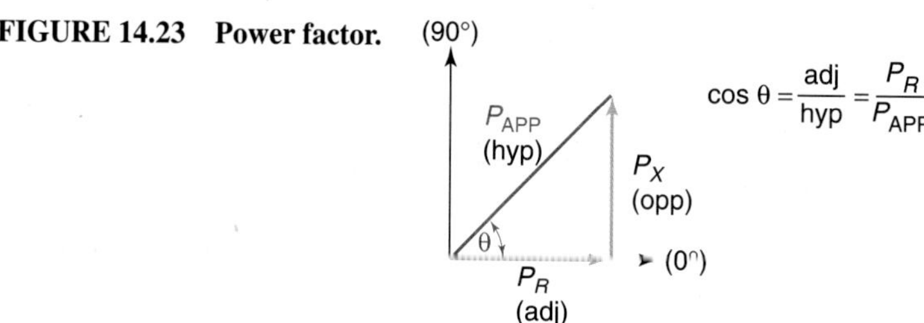

The waveforms shown in Figure 14.24a were obtained from a series *RL* circuit. The total circuit current was measured at 100 mA. Determine the values of apparent power and resistive power (dissipation) for the circuit.

## EXAMPLE 14.11

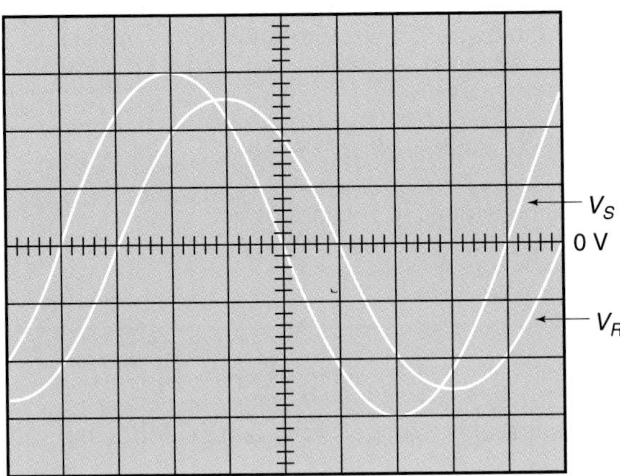


Time base: 20 ms/div
Vertical sensitivity: 5 V/div

(a)

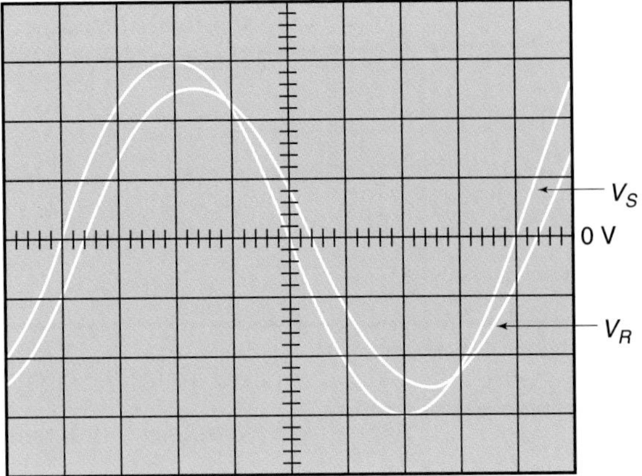


Time base: 100 μs/div
Vertical sensitivity: 2.5 V/div

(b)

**FIGURE 14.24**

***Solution:*** First, we need to obtain several values from the oscilloscope display. With a vertical sensitivity of 5 V/div, the source is shown to have a value of 15 $V_{pk}$. Since we are dealing with power calculations, we must convert this value to rms, as follows:

$$V_S = 0.707\ V_{pk} = (0.707)(15\ V_{pk}) = \mathbf{10.6\ V_{rms}}$$

Combining this value with the measured circuit current, we get an apparent power of

$$P_{APP} = V_S I_T = (10.6\ \text{V})(100\ \text{mA}) = \mathbf{1.06\ VA}$$

Now we need to determine the phase angle in the circuit. As the display shows, the relevant time periods are represented as follows:

$$t \rightarrow 1\ \text{div} \qquad T_C \rightarrow 8\ \text{div}$$

and

$$\theta = 360°\frac{t}{T_C} = 360°\frac{1\ \text{div}}{8\ \text{div}} = \mathbf{45°}$$

Now we can determine the resistive power (dissipation) in the circuit as follows:

$$P_R = P_{APP}\cos\theta = (1.06\ \text{VA})(\cos 45°) = (1.06\ \text{VA})(0.707) \cong \mathbf{750\ mW}$$

This means that approximately 750 mW of the power drawn from the source is dissipated by the circuit.

***PRACTICE PROBLEM 14.11***

The waveforms in Figure 14.24b were measured for a series *RL* circuit. The measured circuit current is 200 mA. Calculate the values of apparent power and resistive power (dissipation) for the circuit.

## *Impedance Applications of Cos θ*

Using (cos θ) to calculate circuit resistance.

The power factor (cos θ) for a circuit can also be used to calculate the value of *resistance* for a series *RL* circuit when the corresponding value of $Z_T$ is known. This principle is based on the fact that the impedance triangle for a series *RL* circuit is congruent to its power triangle. (This can be seen by comparing Figure 14.11 with Figure 14.22b.) When the total impedance of a series *RL* circuit is known, the total circuit resistance can be found as

$$R_T = Z_T \cos\theta \qquad \textbf{(14.13)}$$

An application for this equation is demonstrated in Example 14.12.

### EXAMPLE 14.12

Determine the circuit resistance for the circuit analyzed in Example 14.11.

***Solution:*** In Example 14.11, we determined that the circuit had the following values:

$$V_S = 10.6\text{ V} \qquad I_T = 100\text{ mA} \qquad \theta = 45°$$

The magnitude of the circuit impedance can be determined using Ohm's law, as follows:

$$Z_T = \frac{V_S}{I_T} = \frac{10.6\text{ V}}{100\text{ mA}} = \mathbf{106\ \Omega}$$

Now, the total resistance in the circuit can be found as

$$R_T = Z_T \cos\theta = (106\ \Omega)(\cos 45°) = (106\ \Omega)(0.707) \cong \mathbf{75\ \Omega}$$

***PRACTICE PROBLEM 14.12***

Calculate the total resistance for the circuit described in Practice Problem 14.11.

## *Putting It All Together: Series RL Circuit Analysis*

Now that we have covered the power characteristics and calculations associated with series *RL* circuits, we will close out the section with a complete circuit analysis problem.

### EXAMPLE 14.13

Perform a complete analysis of the series *RL* circuit shown in Figure 14.25a.

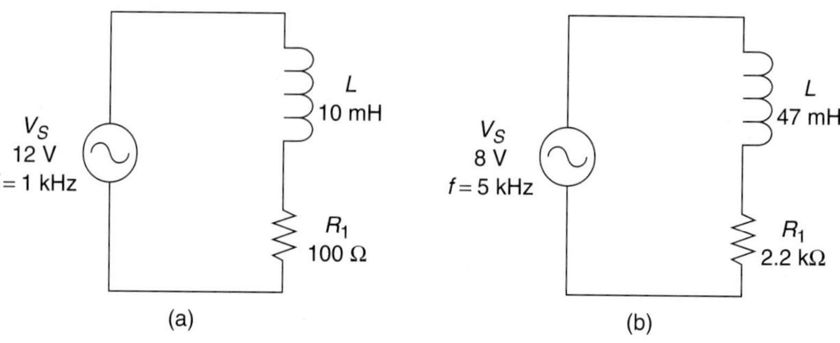


**FIGURE 14.25**

***Solution:*** The reactance of the inductor is found as

$$X_L = 2\,\pi f L = 2\,\pi(1\text{ kHz})(10\text{ mH}) = \mathbf{62.8\ \Omega}$$

This, combined with the value of $R_1$, gives us a circuit impedance of

$$Z_T = \sqrt{X_L^2 + R^2} = \sqrt{(62.8\ \Omega)^2 + (100\ \Omega)^2} = \mathbf{118\ \Omega}$$

at a phase angle of:

$$\theta = \tan^{-1}\frac{X_L}{R} = \tan^{-1}\frac{62.8\ \Omega}{100\ \Omega} = \tan^{-1}(0.628) = \mathbf{32.1^\circ}$$

The magnitude of the circuit current can now be found as

$$I_T = \frac{V_S}{Z_T} = \frac{12\text{ V}}{118\ \Omega} = \mathbf{102\text{ mA}}$$

So far, we know that the circuit has the following values:

$$Z_T = 118\ \Omega \qquad V_S = 12\text{ V} \qquad I_T = 102\text{ mA} \qquad \theta = 32.1^\circ$$

As you have been told, $\theta$ is the phase angle for both impedance and the source voltage, relative to the circuit current. Therefore, the values of $Z_T$ and $V_S$ are written more precisely as

$$Z_T = 118\ \Omega\angle 32.1^\circ \qquad \text{and} \qquad V_S = 12\text{ V}\angle 32.1^\circ$$

Using the voltage-divider equation, the voltage across the resistor can be found as

$$V_R = V_S\frac{R_1}{Z_T} = (12\text{ V}\angle 32.1^\circ)\frac{100\ \Omega\angle 0^\circ}{118\ \Omega\angle 32.1^\circ} = \mathbf{10.2\text{ V}\angle 0^\circ}$$

and the voltage across the inductor can be found as

$$V_L = V_S\frac{X_L}{Z_T} = (12\text{ V}\angle 32.1^\circ)\frac{62.8\ \Omega\angle 90^\circ}{118\ \Omega\angle 32.1^\circ} = \mathbf{6.39\text{ V}\angle 90^\circ}$$

Finally, the circuit power values can be found as

$$P_R = I_T V_R = (102\text{ mA})(10.2\text{ V}) = \mathbf{1.04\text{ W}}$$

$$P_X = I_T V_L = (102\text{ mA})(6.39\text{ V}) = \mathbf{652\text{ mVAR}}$$

and

$$P_{APP} = \sqrt{P_X^2 + P_R^2} = \sqrt{(652\text{ mVAR})^2 + (1.04\text{ W})^2} = \mathbf{1.23\text{ VA}}$$

This completes the analysis of the circuit.

***PRACTICE PROBLEM 14.13***

Perform a complete analysis of the circuit shown in Figure 14.25b.

You may have noticed that the power factor of the circuit was not included in our analysis. While the power factor can be used to perform several calculations (as demonstrated in this section), there are other ways to determine those same values. In Example 14.13, you did not see the power factor because it was not needed.

## One Final Note

We have completed the basic analysis of series *RL* circuits. Later in this chapter, we will discuss some of the basic faults that can occur in series *RL* circuits, along with their symptoms. At this point, we will move on to parallel *RL* circuits.

## Section Review

1. Compare and contrast the following: apparent power, resistive power, and reactive power.
2. Why is the winding resistance ($R_w$) of a coil normally ignored in the resistive power calculations for an *RL* circuit?
3. State and explain the phase relationship between reactive power ($P_X$) and resistive power ($P_R$).
4. What is the power factor for an *RL* circuit? What does it tell you?
5. How is the power factor of an *RL* circuit determined?

# 14.3 PARALLEL *RL* CIRCUITS

**Parallel *RL* circuit**
A circuit that contains one or more resistors in parallel with one or more inductors. Each branch in a parallel *RL* circuit contains only one component (either resistive or inductive).

A **parallel *RL* circuit** is one that contains one or more resistors in parallel with one or more inductors. Several parallel *RL* circuits are shown in Figure 14.26. As you can see, none of the circuit branches contains more than one component. (If they did, we'd be dealing with a series-parallel *RL* circuit.)

## *Parallel Circuits: A Comparison*

*OBJECTIVE 7* ►

As we did with series circuits, we will start by comparing a parallel *RL* circuit to purely resistive and purely inductive parallel circuits. A comparison among these circuits is provided in Figure 14.27.

In Chapters 6 and 13, we established the fundamental characteristics of resistive and inductive parallel circuits. As you know, these characteristics are:

1. All branch voltages equal the source voltage.
2. The circuit current equals the sum of the branch currents.
3. The total impedance (resistance or reactance) is lower than the lowest branch value and can be found as shown in Figure 14.27.

*A Practical Consideration:* The product-over-sum method of calculating total impedance is emphasized here because the reciprocal method is extremely tedious when applied to parallel *RL* circuits. (Impedance calculations are demonstrated later in this section.)

The relationships listed for the resistive and inductive circuits in Figure 14.27 should all be familiar. In each case, all the branch voltages equal the source voltage, and the total circuit current equals the sum of the branch currents. The total impedance for the resistive and inductive circuits can be found using the *product-over-sum* method shown.

The current and impedance relationships shown for the *RL* circuit in Figure 14.27 are explained in detail later in this section. Even so, you can see that the total current in a parallel *RL* circuit equals the geometric sum of the branch currents. By the same token, the calculation of total impedance involves the geometric sum of inductive reactance and resistance (in the denominator of the equation). As you may have guessed, these geometric relationships exist because of the presence of phase angles within the circuit.

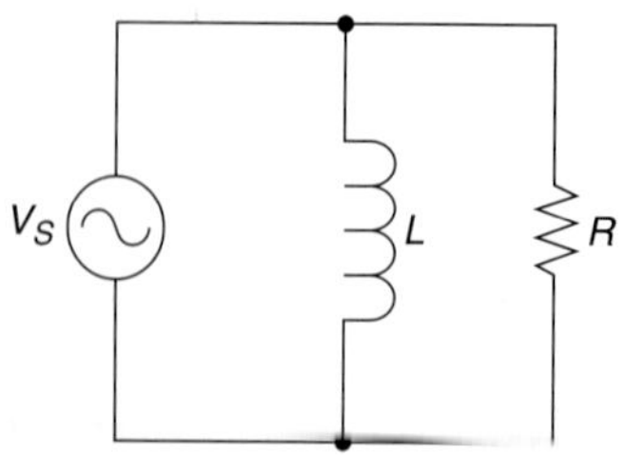

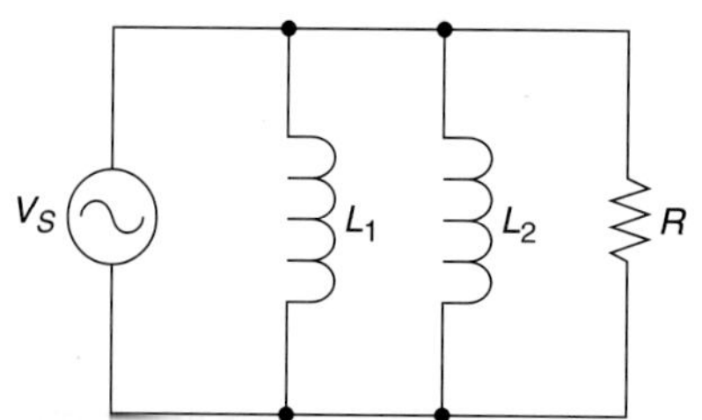

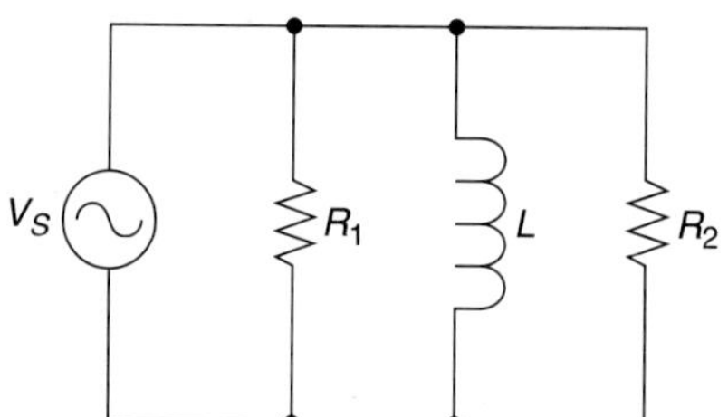


**FIGURE 14.26 Some basic parallel *RL* circuits.**

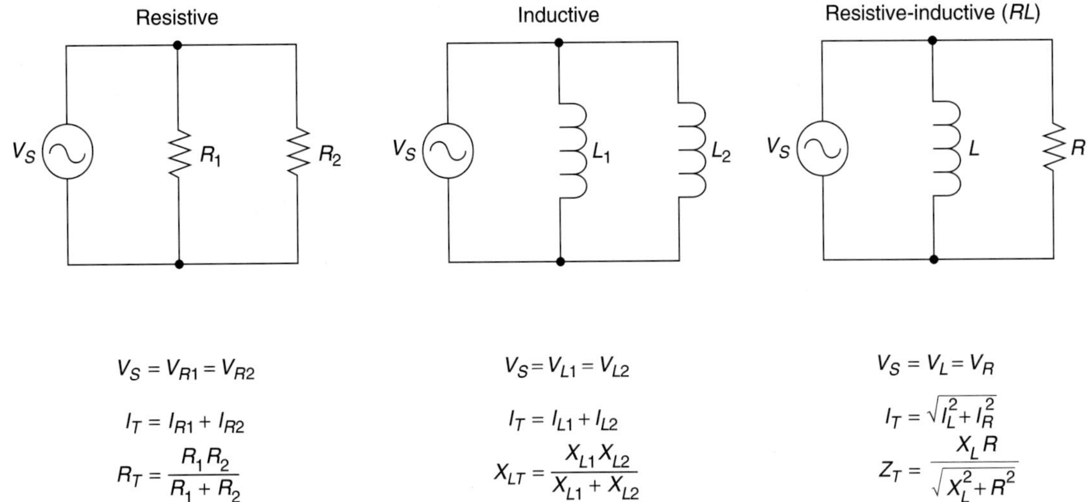


**FIGURE 14.27 Parallel circuit characteristics.**

## Branch Currents

Current lags voltage by 90° in an inductive circuit. At the same time, current and voltage in a resistive circuit are always in phase. You saw these two relationships in Figure 14.4. When combined in parallel, a resistor and an inductor produce the waveforms shown in Figure 14.28.

◀ *OBJECTIVE 8*

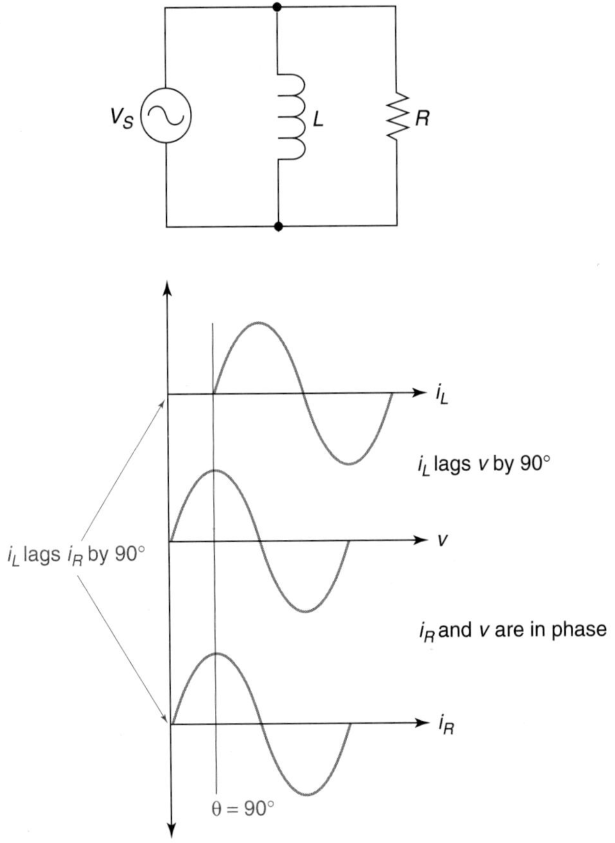


**FIGURE 14.28 Current phase relationships in a parallel *RL* circuit.**

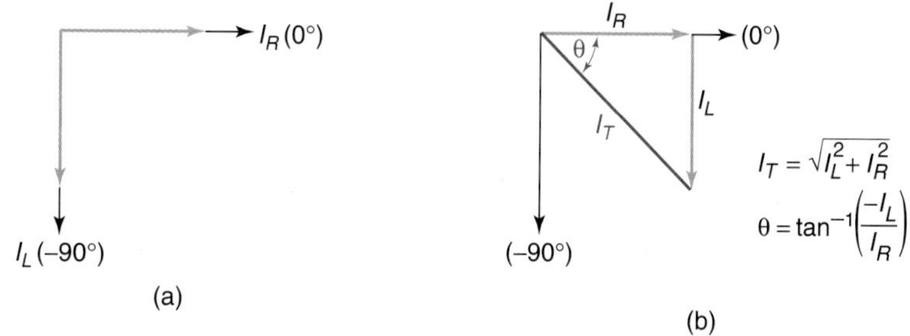


**FIGURE 14.29 Vector (geometric) addition of $I_L$ and $I_R$.**

Since voltage is constant across all branches in a parallel circuit, the inductor and resistor currents are referenced to a single voltage waveform. As you can see:

1. The inductor current lags the circuit voltage by 90°.
2. The resistor current is in phase with the circuit voltage.

Therefore, inductor current lags resistor current by 90°. This phase relationship holds true for every parallel *RL* circuit with an ac voltage source.

Since $I_L$ lags $I_R$ by 90°, the values of the component currents are represented using phasors, as shown in Figure 14.29a. By plotting the value of $I_R$ on the *x*-axis (0°) and the value of $I_L$ on the negative *y*-axis (−90°), the phase relationship between the two is represented in the graph.

As in any parallel circuit, the sum of $I_L$ and $I_R$ must equal the total circuit current. In this case, the phasor relationship between the two values requires that they be added together geometrically. By formula,

$$I_T = \sqrt{I_L^2 + I_R^2} \qquad \textbf{(14.14)}$$

The circuit current ($I_T$) phase angle.

The phase angle of $I_T$ in a parallel *RL* circuit is the phase difference between the circuit current and the source voltage. As shown in Figure 14.29b, the value of θ can be found as

$$\theta = \tan^{-1}\frac{-I_L}{I_R} \qquad \textbf{(14.15)}$$

Since voltage is equal across all branches in a parallel circuit, $V_S$ is used as the reference for all phase angle measurements. Therefore, the value of θ for the circuit current indicates its phase relative to the source voltage. Example 14.14 demonstrates the calculations of total current and θ for a parallel *RL* circuit.

## EXAMPLE 14.14

The current values shown in Figure 14.30a were measured as shown. Calculate the total current for the circuit.

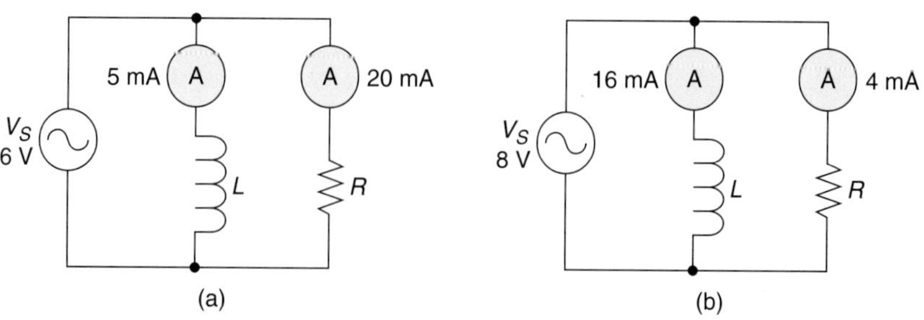


**FIGURE 14.30**

***Solution:*** Using equation (14.14), the magnitude of the circuit current is found as

$$I_T = \sqrt{I_L^2 + I_R^2} = \sqrt{(5\text{ mA})^2 + (20\text{ mA})^2} = \mathbf{20.6\ mA}$$

Since $X_L$ and $R$ do not have the same phase angle, the equal-value branches approach to calculating total impedance cannot be used on a parallel *RL* circuit. At the same time, the phase angles of $X_L$ and $R$ make the reciprocal method extremely long and tedious. For this reason, the product-over-sum approach to calculating circuit impedance is preferred for parallel *RL* circuits. Using the product-over-sum format, the magnitude of parallel *RL* circuit impedance can be found as

$$Z_T = \frac{X_L R}{\sqrt{X_L^2 + R^2}} \tag{14.18}$$

Note that the denominator of the fraction contains the geometric sum of the variables. Example 14.16 demonstrates the use of this equation.

*EXAMPLE 14.16*

Calculate the magnitude of $Z_T$ for the circuit in Figure 14.31a.

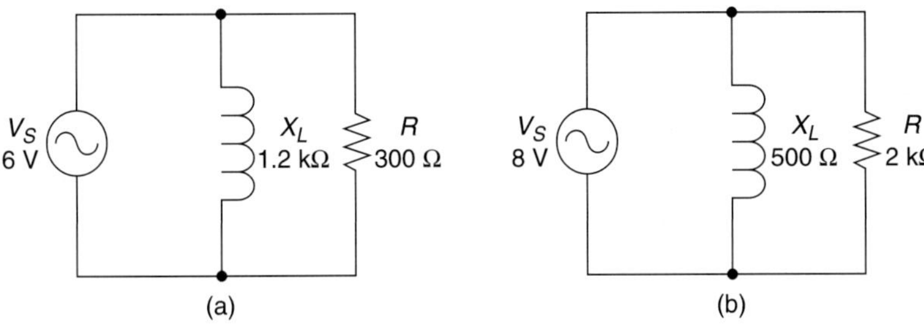


**FIGURE 14.31**

***Solution:*** Using the values shown in equation (14.18), the magnitude of $Z_T$ is found as

$$Z_T = \frac{X_L R}{\sqrt{X_L^2 + R^2}} = \frac{(1.2\text{ k}\Omega)(300\ \Omega)}{\sqrt{(1.2\text{ k}\Omega)^2 + (300\ \Omega)^2}} = \mathbf{291\ \Omega}$$

To provide a basis for comparison, the values used in this example match those found for the circuit in Example 14.15. If you compare the results of these two examples, you'll see that they are the same.

***PRACTICE PROBLEM 14.16***

Calculate the value of $Z_T$ for the circuit in Figure 14.31b. Compare your answer to the one obtained in Practice Problem 14.15.

***Calculating the Impedance Phase Angle.*** The phase angle of the circuit impedance for a parallel *RL* circuit can be found as

$$\theta = \tan^{-1}\frac{R}{X_L} \tag{14.19}$$

As you can see, the fraction in equation (14.19) is the reciprocal of the one used to calculate the impedance phase angle in a series *RL* circuit. We will discuss the reason for the change in a moment. First, we'll verify the validity of the equation.

## EXAMPLE 14.17

Calculate the phase angle of the circuit impedance for the circuit in Figure 14.31a (Example 14.16).

***Solution:*** The circuit has values of $X_L = 1.2\ \text{k}\Omega$ and $R = 300\ \Omega$. Using these values in equation (14.19), the value of θ for the circuit impedance is found as

$$\boldsymbol{\theta} = \tan^{-1}\left(\frac{R}{X_L}\right) = \tan^{-1}\left(\frac{300\ \Omega}{1.2\ \text{k}\Omega}\right) = \tan^{-1}(0.25) = \mathbf{14^\circ}$$

***PRACTICE PROBLEM 14.17***

Calculate the phase angle of the circuit impedance for the circuit in Figure 14.31b.

In Example 14.15, we calculated the phase angle of the circuit impedance using Ohm's law. When you compare the results of the two examples, you'll see that they are the same.

***The Basis of Equation (14.19).*** The current triangle in Figure 14.29b was used to illustrate the current relationships in a parallel *RL* circuit. As shown in the figure,

$$\theta = \tan^{-1}\left(\frac{-I_L}{I_R}\right)$$

This relationship was given earlier as equation (14.15).

The phase angle of circuit impedance is the positive-equivalent of the current phase angle. For this reason, we will rewrite equation (14.15) as follows:

$$\theta = \tan^{-1}\left(\frac{I_L}{I_R}\right)$$

(When written in this form, the equation provides a positive result, which is consistent with the impedance phase angle.) If we rewrite each current value in the equation in the form of *V/R,* we get

$$\theta = \tan^{-1}\left(\frac{\frac{V_S}{X_L}}{\frac{V_S}{R}}\right)$$

$$= \tan^{-1}\left(\frac{V_S}{X_L} \times \frac{R}{V_S}\right)$$

Finally, the values of $V_S$ cancel out, leaving

$$\theta = \tan^{-1}\left(\frac{R}{X_L}\right)$$

which is equation (14.19).

## Measuring the Current Phase Angle (θ)

*OBJECTIVE 9* ▶

**Current-sensing resistor** A low-value resistor used to produce a low-value voltage that is in phase with the circuit current.

Measuring the current phase angle of a parallel *RL* circuit is a bit more involved than measuring the phase angle in a series *RL* circuit because oscilloscopes are designed to display voltage waveforms (rather than current waveforms). Therefore, we have to modify any parallel *RL* circuit if we want to measure the current phase angle.

To display a current waveform on an oscilloscope effectively, we use a **current-sensing resistor.** A current-sensing resistor is a low-value resistor used to produce a

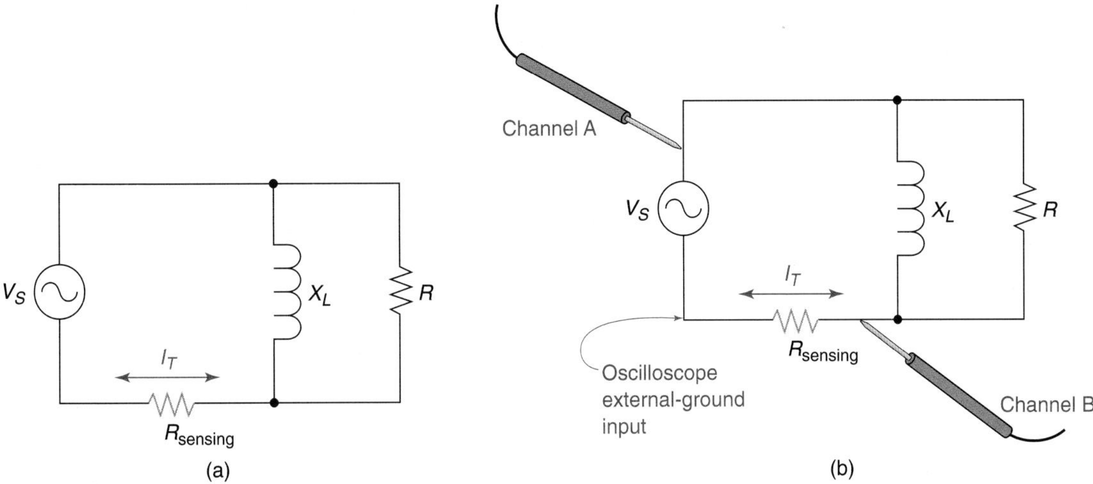


**FIGURE 14.32 Measuring θ with a current-sensing resistor.**

low-value voltage that is in phase with the circuit current. This low-value voltage can be displayed on an oscilloscope to provide a relatively accurate representation of the current's *phase*. A parallel *RL* circuit with an added current-sensing resistor is shown in Figure 14.32.

As you can see, the current-sensing resistor is placed in the path of the current we want to display. When the total circuit current ($I_T$) passes through the current-sensing resistor, a voltage is developed across the resistor. This voltage is in phase with the circuit current. When an oscilloscope is connected to the circuit (as shown in Figure 14.32b), channel A displays the source voltage and channel B displays the voltage across the sensing resistor. Since this voltage is in phase with the circuit current, the oscilloscope display shows the phase angle between the source voltage and the circuit current. When properly connected and calibrated, the oscilloscope provides a display similar to the one shown in Figure 14.33.

## Choosing the Value of the Current-Sensing Resistor

Why $R_{sensing}$ should have as low a value as possible.

Whenever you add a component to a circuit, you alter that circuit and the results of any tests performed on the circuit are affected. For example, when a current-sensing resistor is added as shown in Figure 14.32:

1. The circuit is changed from a parallel circuit to a series-parallel circuit.
2. The total impedance in the circuit is increased.
3. The total resistance in the circuit is increased, which causes the phase angle to change.

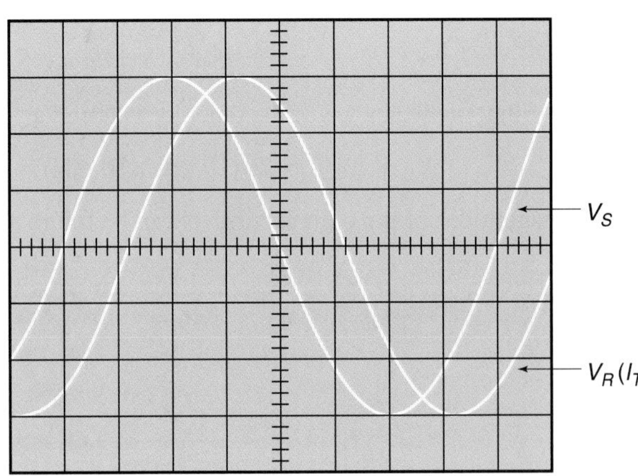


**FIGURE 14.33 A typical display of the current phase angle in a parallel *RL* circuit with a current-sensing resistor.**

*A Practical Consideration:* The magnitude of the voltage developed across the sensing resistor (Figure 14.33) is much lower than that of the source voltage. A display like the one shown is obtained by decreasing the *volts/div* setting for the channel displaying $V_R$.

EWB

**FIGURE 14.34 Selecting a current-sensing resistor.**

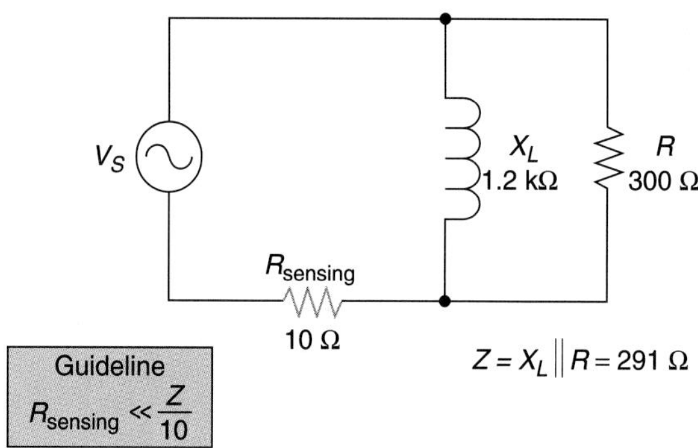


*A Practical Consideration:* Even though we want to use a low-value sensing resistor, the value of the component must be sufficient to generate a measurable voltage at the value of circuit current. If you cannot obtain a voltage reading across a sensing resistor, try using a higher value component (while not exceeding the maximum limit).

The key to using a current-sensing resistor successfully is to ensure that the value of the resistor is much lower than the value of total impedance for the circuit. For example, consider the circuit shown in Figure 14.34. The impedance of the parallel circuit is 291 Ω. A 10 Ω resistor has been selected as the current-sensing resistor.

Since $R_{sensing} << Z/10$, the component will not significantly affect the total impedance in the circuit or the current phase angle. In fact, the resistor in Figure 14.34 will cause a decrease in the phase angle of less than 1°. (You will learn how to calculate this change for yourself when we discuss series-parallel *RL* circuits in Section 14.4.)

## Parallel Circuit Analysis

*OBJECTIVE 10* ▶ In this section, you have been shown how current and impedance values are calculated for a parallel *RL* circuit. Parallel circuit power calculations are identical to those for series circuits, so they have already been covered as well. At this point, we will tie all these calculations together by working through several circuit analysis problems.

### EXAMPLE 14.18

Perform a complete analysis of the circuit in Figure 14.35a.

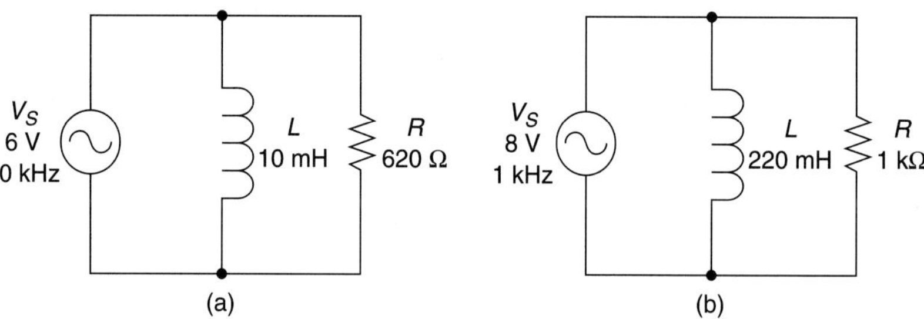


**FIGURE 14.35**

***Solution:*** First, the reactance of the inductor is found as

$$X_L = 2\pi fL = 2\pi(10\text{ kHz})(10\text{ mH}) = \mathbf{628\ \Omega}$$

Now, the magnitudes of the branch currents can be found as

$$I_L = \frac{V_S}{X_L} = \frac{6\text{ V}}{628\ \Omega} = \mathbf{9.55\ mA}$$

and

$$I_R = \frac{V_S}{R} = \frac{6\text{ V}}{620\ \Omega} = \mathbf{9.68\ mA}$$

The magnitude of the circuit current can now be found as

$$I_T = \sqrt{I_L^2 + I_R^2} = \sqrt{(9.55 \text{ mA})^2 + (9.68 \text{ mA})^2} = \mathbf{13.6 \text{ mA}}$$

and the current phase angle (relative to the source voltage) can be found as

$$\theta = \tan^{-1}\left(\frac{-I_L}{I_R}\right) = \tan^{-1}\left(\frac{-9.55 \text{ mA}}{9.68 \text{ mA}}\right) = \mathbf{-44.6°}$$

Once the total current is known, the circuit impedance can be found as

$$Z = \frac{V_S}{I_T} = \frac{6 \text{ V} \angle 0°}{13.6 \text{ mA} \angle -44.6°} = \mathbf{441 \ \Omega \angle 44.6°}$$

The value of reactive power ($P_X$) for the circuit can be found as

$$P_X = I_L^2 X_L = (9.55 \text{ mA})^2 (628 \ \Omega) = \mathbf{57.3 \text{ mVAR}}$$

and the value of resistive power dissipation ($P_R$) can be found as

$$P_R = I_R^2 R = (9.68 \text{ mA})^2 (620 \ \Omega) = \mathbf{58.1 \text{ mW}}$$

Once the values of reactive and resistive power are known, the value of apparent power ($P_{\text{APP}}$) can be found as

$$P_{\text{APP}} = \sqrt{P_X^2 + P_R^2} = \sqrt{(57.3 \text{ mVAR})^2 + (58.1 \text{ mW})^2} = \mathbf{81.6 \text{ mVA}}$$

This completes the analysis of the circuit.

***PRACTICE PROBLEM 14.18***

Perform a complete analysis of the circuit shown in Figure 14.35b.

There are several approaches we could take for checking the accuracy of the results in Example 14.18. Perhaps the simplest method is to use the power factor (cos θ) to calculate the value of resistive power, as follows:

$$P_R = P_{\text{APP}} \cos \theta = (81.6 \text{ mVA})(\cos 44.6°) = (81.6 \text{ mVA})(0.712) = 58.1 \text{ mW}$$

Since this result matches the value of $P_R$ found in the example, we can assume that the values found are accurate.

When a parallel *RL* circuit contains more than one inductor and/or resistor, the analysis of the circuit takes a few more steps. This point is demonstrated in Example 14.19.

## EXAMPLE 14.19

Perform a complete analysis on the circuit in Figure 14.36a.

**FIGURE 14.36**

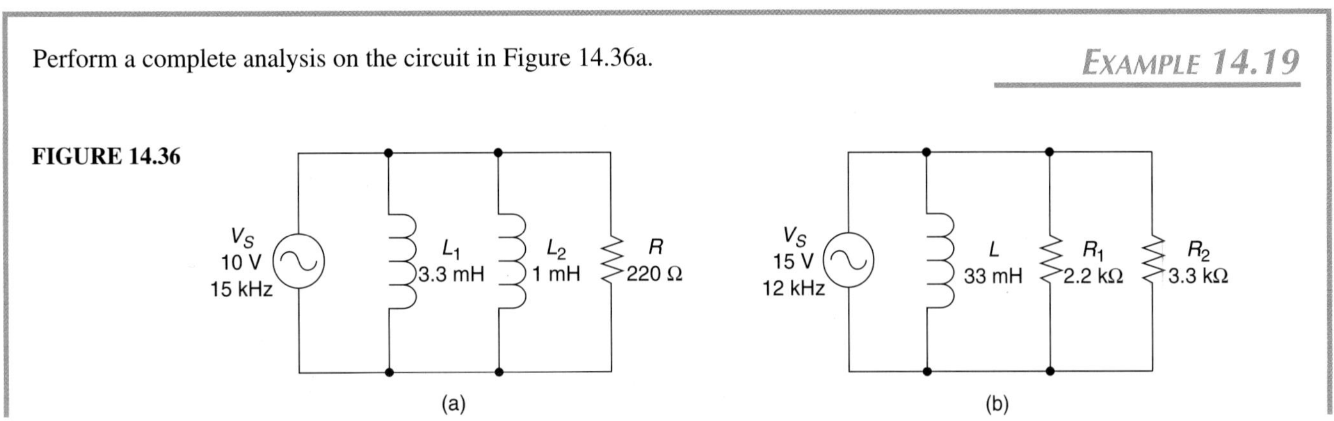

***Solution:*** First, the values of $X_L$ for the circuit are found as

$$\boldsymbol{X_{L1}} = 2\pi f L_1 = 2\pi(15\text{ kHz})(3.3\text{ mH}) = \mathbf{311\ \Omega}$$

and

$$\boldsymbol{X_{L2}} = 2\pi f L_2 = 2\pi(15\text{ kHz})(1\text{ mH}) = \mathbf{94.25\ \Omega}$$

Now, the magnitudes of the branch currents can be found as

$$\boldsymbol{I_{L1}} = \frac{V_S}{X_{L1}} = \frac{10\text{ V}}{311\ \Omega} = \mathbf{32.2\ mA}$$

$$\boldsymbol{I_{L2}} = \frac{V_S}{X_{L2}} = \frac{10\text{ V}}{94.25\ \Omega} = \mathbf{106.1\ mA}$$

and

$$\boldsymbol{I_R} = \frac{V_S}{R} = \frac{10\text{ V}}{220\ \Omega} = \mathbf{45.5\ mA}$$

Since $I_{L1}$ and $I_{L2}$ are in phase, their values can be added algebraically, as follows:

$$\boldsymbol{I_{L(T)}} = I_{L1} + I_{L2} = 32.2\text{ mA} + 106.1\text{ mA} = \mathbf{138.3\ mA}$$

Now, the total inductive current is used (with $I_R$) to find the magnitude of the circuit current, as follows:

$$\boldsymbol{I_T} = \sqrt{I_{L(T)}^2 + I_R^2} = \sqrt{(138.3\text{ mA})^2 + (45.5\text{ mA})^2} \cong \mathbf{146\ mA}$$

and the current phase angle can be found as

$$\boldsymbol{\theta} = \tan^{-1}\left(\frac{-I_{L(T)}}{I_R}\right) = \tan^{-1}\left(\frac{-138.3\text{ mA}}{45.5\text{ mA}}\right) = \mathbf{-71.8^\circ}$$

Once the total circuit current is known, the circuit impedance can be found as:

$$\boldsymbol{Z_T} = \frac{V_S}{I_T} = \frac{10\text{ V}\angle 0^\circ}{146\text{ mA}\angle -71.8^\circ} = \mathbf{68.5\ \Omega\angle 71.8^\circ}$$

The values of reactive power for the circuit can be found as:

$$\boldsymbol{P_{X(L1)}} = I_{L1}^2 X_{L1} = (32.2\text{ mA})^2(311\ \Omega) = \mathbf{322\ mVAR}$$

and

$$\boldsymbol{P_{X(L2)}} = I_{L2}^2 X_{L2} = (106.1\text{ mA})^2(94.25\ \Omega) = \mathbf{1.06\ VAR}$$

Since the inductor values are in phase with each other, the total reactive power can be found as

$$\boldsymbol{P_{X(T)}} = P_{X(L1)} + P_{X(L2)} = 322\text{ mVAR} + 1.06\text{ VAR} = \mathbf{1.38\ VAR}$$

The resistive power in the circuit can be found as

$$\boldsymbol{P_R} = I_R^2 R = (45.5\text{ mW})^2(220\ \Omega) = \mathbf{455\ mW}$$

Now, using resistive power and total reactive power, the value of apparent power for the circuit can be found as

$$\boldsymbol{P_{APP}} = \sqrt{P_{X(T)}^2 + P_R^2} = \sqrt{(1.38\text{ VAR})^2 + (455\text{ mW})^2} = \mathbf{1.45\ VA}$$

This completes the analysis of the circuit.

***PRACTICE PROBLEM 14.19***

Perform a complete analysis on the circuit in Figure 14.36b.

## Parallel Circuit Frequency Response

The frequency response of a parallel $RL$ circuit is quite different than that of a series $RL$ circuit. For example, consider the circuit shown in Figure 14.37. Table 14.5 shows the circuit response to an increase in operating frequency. ◄ *OBJECTIVE 11*

**TABLE 14.5 Parallel *RL* Circuit Response to an Increase in Operating Frequency**

| *Cause and Effect* | *Relevant Equation* |
|---|---|
| 1. The increase in frequency causes $X_L$ to increase. | $X_L = 2\pi fL$ |
| 2. The increase in $X_L$ causes the magnitude of: | |
| a. $I_L$ to decrease. | $I_L = \frac{V_S}{X_L}$ |
| b. $\theta$ to decrease. | $\theta = \tan^{-1}\frac{R}{X_L}$ |
| c. $Z_T$ to increase. | $Z_T = \frac{X_L R}{\sqrt{X_L^2 + R^2}}$ |
| 3. The decrease in $I_L$ causes $I_T$ to decrease. | $I_T = \sqrt{I_L^2 + I_R^2}$ |

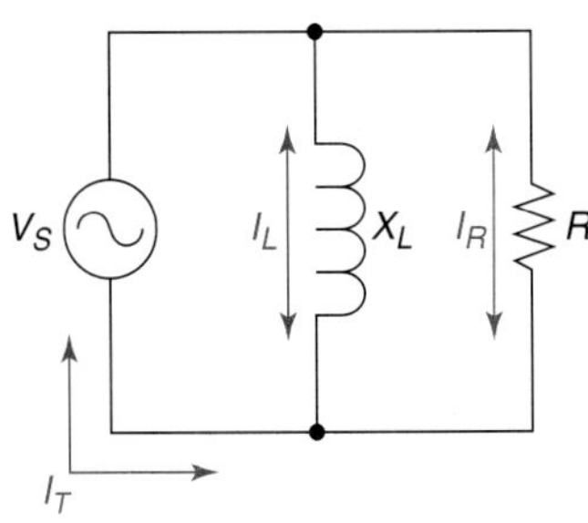


**FIGURE 14.37**

By the same token, here's what happens if the operating frequency decreases:

1. The decrease in operating frequency causes $X_L$ to decrease.
2. The decrease in $X_L$ causes:
   a. $I_L$ to increase.
   b. $\theta$ to increase.
   c. $Z_T$ to decrease.
3. The increase in $I_L$ causes $I_T$ to increase.

Example 14.20 demonstrates the effects of a change in operating frequency on parallel $RL$ circuits.

### EXAMPLE 14.20

The voltage source in Figure 14.38a has the frequency limits shown. Calculate the circuit current and impedance values at the frequency limits of the source.

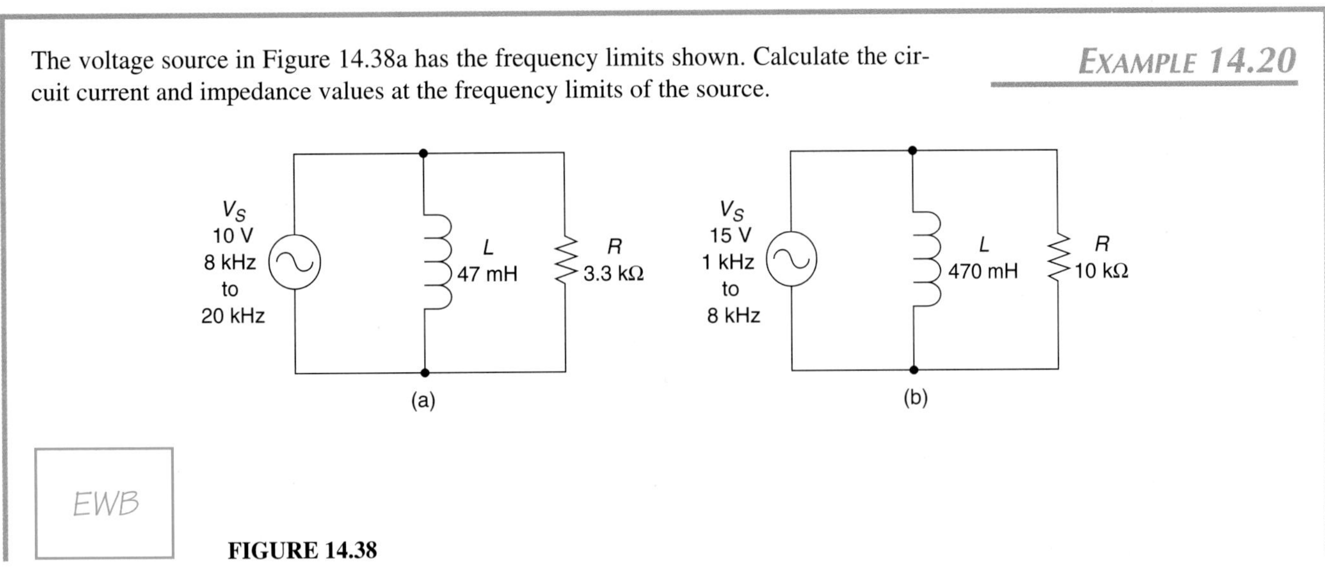


**FIGURE 14.38**

***Solution:*** When the operating frequency is set to 8 kHz, the reactance of the inductor is found as

$$X_L = 2\pi fL = 2\pi(8 \text{ kHz})(47 \text{ mH}) = \mathbf{2.36 \text{ k}\Omega}$$

The magnitudes of the branch currents are found as

$$I_L = \frac{V_S}{X_L} = \frac{10 \text{ V}}{2.36 \text{ k}\Omega} = \mathbf{4.24 \text{ mA}}$$

and

$$I_R = \frac{V_S}{R} = \frac{10 \text{ V}}{3.3 \text{ k}\Omega} = \mathbf{3.03 \text{ mA}}$$

Combining these values, the magnitude of the circuit current is found as

$$I_T = \sqrt{I_L^2 + I_R^2} = \sqrt{(4.24 \text{ mA})^2 + (3.03 \text{ mA})^2} = \mathbf{5.21 \text{ mA}}$$

and the current phase angle is found as

$$\theta = \tan^{-1}\left(\frac{-I_L}{I_R}\right) = \tan^{-1}\left(\frac{-4.24 \text{ mA}}{3.03 \text{ mA}}\right) = \mathbf{-54.4^\circ}$$

Finally, the circuit impedance is found as

$$Z_T = \frac{V_S}{I_T} = \frac{10 \text{ V} \angle 0^\circ}{5.21 \text{ mA} \angle -54.4^\circ} = \mathbf{1.92 \text{ k}\Omega \angle 54.4^\circ}$$

When the operating frequency is increased to 20 kHz, the reactance of the inductor is found as

$$X_L = 2\pi fL = 2\pi(20 \text{ kHz})(47 \text{ mH}) = \mathbf{5.91 \text{ k}\Omega}$$

The magnitudes of the branch currents are found as

$$I_L = \frac{V_S}{X_L} = \frac{10 \text{ V}}{5.91 \text{ k}\Omega} = \mathbf{1.69 \text{ mA}}$$

and

$$I_R = \frac{V_S}{R} = \frac{10 \text{ V}}{3.3 \text{ k}\Omega} = \mathbf{3.03 \text{ mA}}$$

Combining these values, the magnitude of the circuit current is found as

$$I_T = \sqrt{I_L^2 + I_R^2} = \sqrt{(1.69 \text{ mA})^2 + (3.03 \text{ mA})^2} = \mathbf{3.47 \text{ mA}}$$

and the current phase angle is found as

$$\theta = \tan^{-1}\left(\frac{-I_L}{I_R}\right) = \tan^{-1}\left(\frac{-1.69 \text{ mA}}{3.03 \text{ mA}}\right) = \mathbf{-29.2^\circ}$$

Finally, the circuit impedance is found as

$$Z_T = \frac{V_S}{I_T} = \frac{10 \text{ V} \angle 0^\circ}{3.47 \text{ mA} \angle -29.2^\circ} = \mathbf{2.88 \text{ k}\Omega \angle 29.2^\circ}$$

***PRACTICE PROBLEM 14.20***

The voltage source in Figure 14.38b has the frequency limits shown. Calculate the circuit current and impedance values at the frequency limits of the source.

**TABLE 14.6 A Comparison of the Two Sets of Results in Example 14.20**

| *Variable* | *Value at f = 8 kHz* | *Value at f = 20 kHz* | *Effect* |
|---|---|---|---|
| $X_L$ | 2.36 kΩ | 5.91 kΩ | Increased |
| $I_L$ | 4.24 mA | 1.69 mA | Decreased |
| $I_T$ | 5.21 mA | 3.47 mA | Decreased |
| θ | −54.4° | −29.2° | Decreased[a] |
| $Z_T$ | 1.92 kΩ | 2.88 kΩ | Increased |

[a]Effects refers to magnitude.

The values obtained in Example 14.20 are listed in Table 14.6. When you compare the results with the cause-and-effect statements in Table 14.5, you'll see that the circuit responded to the change in frequency exactly as described in that table.

## One Final Note

In this section, you have been introduced to parallel *RL* circuit relationships and analysis. For future reference, the basic characteristics of parallel *RL* circuits are summarized in Figure 14.39.

We have taken one approach to analyzing parallel *RL* circuits in this section. Another approach emphasizes the use of **susceptance** and **admittance;** the reciprocals of reactance and impedance, respectively. This approach to analyzing the operation of parallel *RL* circuits is introduced and demonstrated in Appendix F.

**Susceptance**
The reciprocal of reactance.

**Admittance**
The reciprocal of impedance.

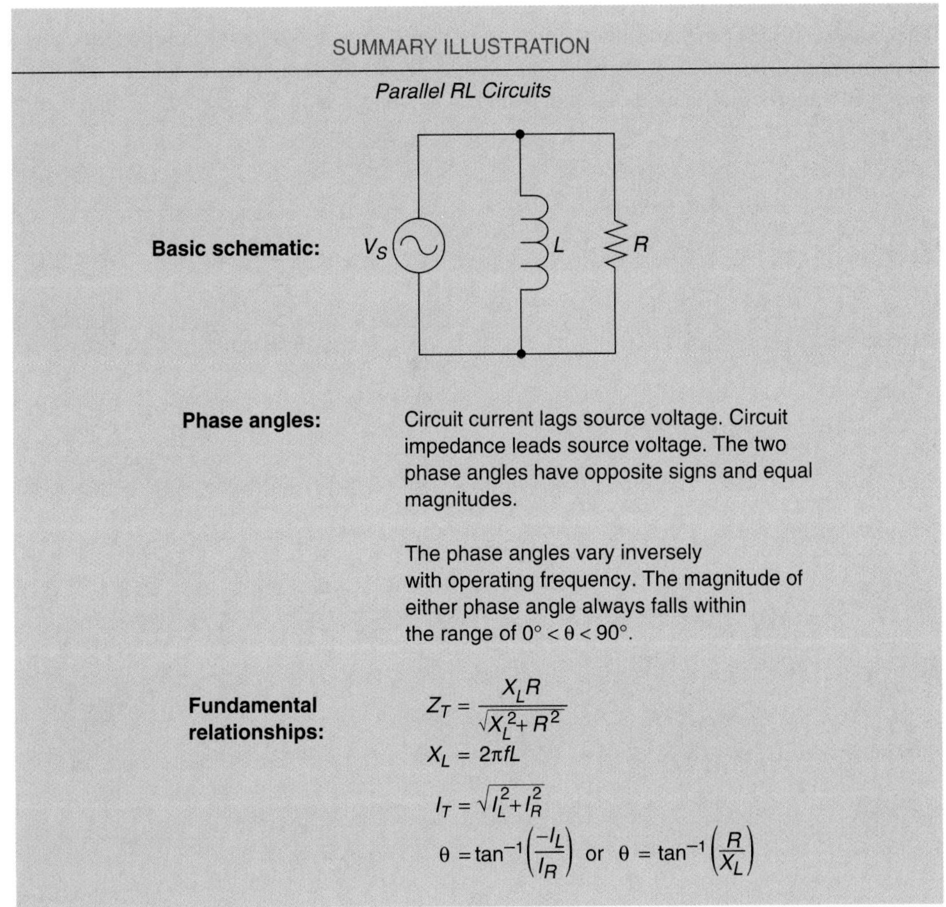

SUMMARY ILLUSTRATION
Parallel RL Circuits
Basic schematic:
$V_S$
L
R
Phase angles:
Circuit current lags source voltage. Circuit impedance leads source voltage. The two phase angles have opposite signs and equal magnitudes.
The phase angles vary inversely with operating frequency. The magnitude of either phase angle always falls within the range of 0° < θ < 90°.
Fundamental relationships:
$Z_T = \frac{X_L R}{\sqrt{X_L^2 + R^2}}$
$X_L = 2\pi f L$
$I_T = \sqrt{I_L^2 + I_R^2}$
$\theta = \tan^{-1}\left(\frac{-I_L}{I_R}\right)$ or $\theta = \tan^{-1}\left(\frac{R}{X_L}\right)$


**FIGURE 14.39**

## Section Review

1. What are the three fundamental characteristics of any parallel circuit?
2. In a parallel *RL* circuit, what is the phase relationship between $I_L$ and $I_R$?
3. In a parallel *RL* circuit, what does the current phase angle represent?
4. Why does the current in a parallel *RL* circuit have a negative phase angle?
5. What is the simplest approach to calculating the value of $Z_T$ for a parallel *RL* circuit?
6. When $X_L = R$ in a parallel *RL* circuit, can the equal value branches approach to calculating $Z_T$ be used? Explain your answer.
7. Describe the process for measuring the current phase angle in a parallel *RL* circuit.
8. When adding a current-sensing resistor to a parallel *RL* circuit, why must its value be kept below $Z_T/10$?
9. Describe the response of a parallel *RL* circuit to an increase in operating frequency.
10. Describe the response of a parallel *RL* circuit to a decrease in operating frequency.

## 14.4 SERIES-PARALLEL CIRCUIT ANALYSIS

*OBJECTIVE 12* ▶ You were first introduced to series-parallel circuit analysis in Chapter 7. At that time, you learned that analyzing a series-parallel circuit is simply a matter of:

1. Combining series components and values according to the rules of series circuits.
2. Combining parallel components and values according to the rules of parallel circuits.

The analysis of a series-parallel circuit with both resistive and reactive components gets a bit complicated because of the phase angles involved. For example, consider the circuit shown in Figure 14.40a. This circuit, which was analyzed in Example 14.1, has a total

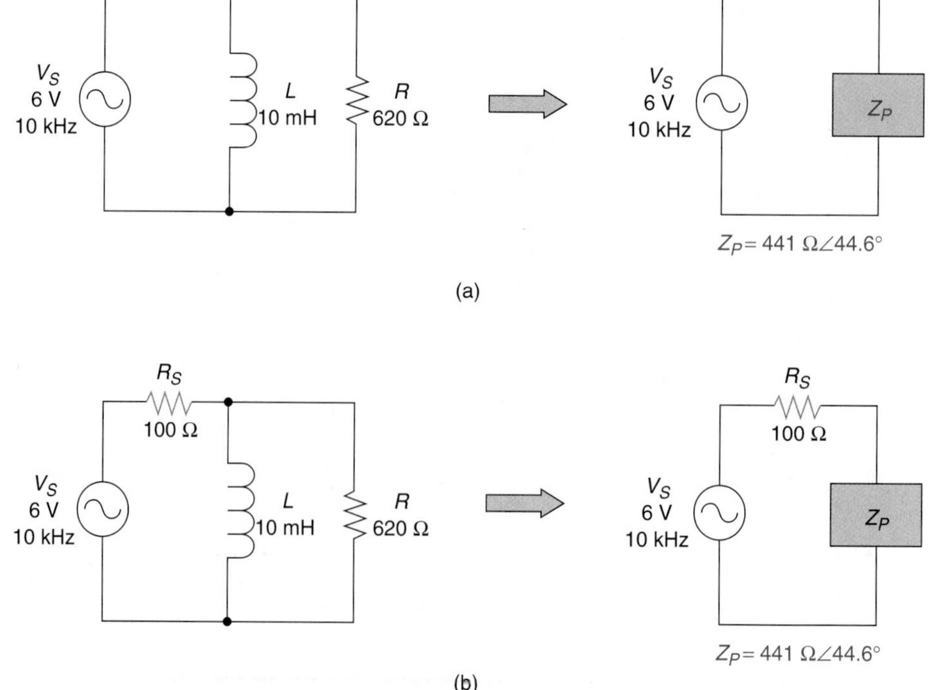


**FIGURE 14.40**

impedance of $Z_T = 432\ \Omega\angle 44.6°$. The circuit impedance is represented as a block ($Z_P$) in the equivalent circuit shown.

If we add a series resistor ($R_S$) to the circuit, we have the series-parallel circuit shown in Figure 14.40b. As the equivalent of this circuit shows, $R_S$ is in series with $Z_P$. The phase difference between $R_S$ and $Z_P$ makes it difficult to add their polar notation values. However, they can be added easily when these values are converted to rectangular notation.

## RL Circuit Values: Polar and Rectangular Forms

In Chapter 12, you were introduced to polar and rectangular representations of phasors. At that time, you were shown that any phasor can be represented in either polar form or rectangular form. These two forms and their conversion equations are summarized in Figure 14.41.

Earlier in this chapter, you were shown how $X_L$, $R$, and $Z$ are plotted as an impedance triangle. A comparison of an $RL$ impedance triangle with the one in Figure 14.41 is shown in Figure 14.42. As you can see, the triangles are identical. When the impedance of an $RL$ circuit is given in polar form, it can be converted to rectangular form using

$$R = Z \cos \theta \quad \textbf{(14.20)}$$

and

$$X_L = Z \sin \theta \quad \textbf{(14.21)}$$

The result is then written in the form $R + jX_L$. This conversion is demonstrated in Example 14.21.

*Note:* By now, you probably know that the conversions discussed here can be performed using any standard scientific calculator. However, manual conversions are discussed in this section to reinforce the relationships between the values.

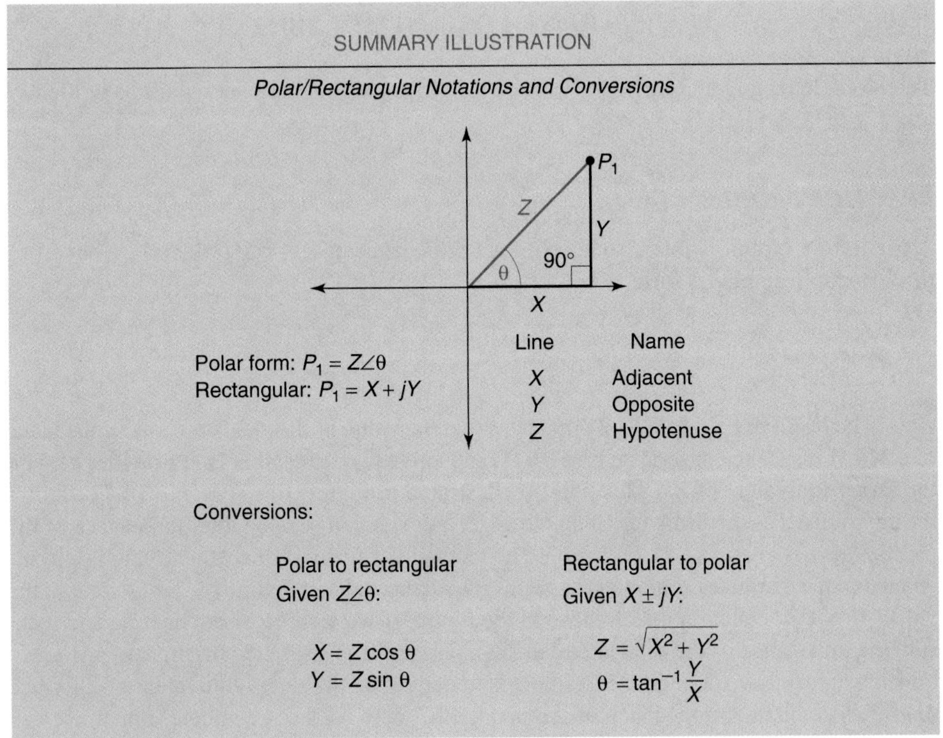


**FIGURE 14.41**

*Note:* If any of the relationships shown in Figure 14.41 are unfamiliar, you may want to review the appropriate discussion in Chapter 12 before continuing.

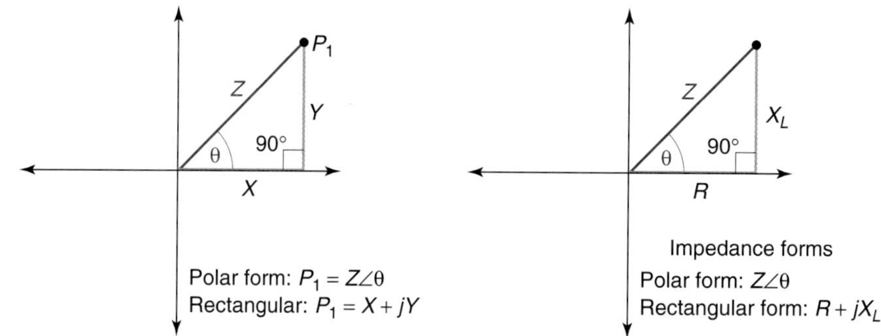


Conversion equations for the impedance triangle:

| Given: $Z\angle\theta$ | Given: $R + jX_L$ |
|---|---|
| $R = Z\cos\theta$ | $Z = \sqrt{X_L^2 + R^2}$ |
| $X_L = Z\sin\theta$ | $\theta = \tan^{-1}\dfrac{X_L}{R}$ |

**FIGURE 14.42**

## EXAMPLE 14.21

The parallel combination of $R$ and $L$ in Figure 14.40b is shown to have a value of $Z_P = 441\ \Omega\angle 44.6°$. Convert this value to rectangular form.

***Solution:*** The value of $R$ is found as

$$\boldsymbol{R} = Z\cos\theta = (441\ \Omega)(\cos 44.6°) = \mathbf{314\ \Omega}$$

The value of $X_L$ is found as

$$\boldsymbol{X_L} = Z\sin\theta = (441\ \Omega)(\sin 44.6°) = \mathbf{310\ \Omega}$$

Therefore, in rectangular form, the impedance of the parallel network would be written as $Z_P = (314 + j310)\ \Omega$.

***PRACTICE PROBLEM 14.21***

A parallel $RL$ circuit is found to have a total impedance of $Z_P = 560\ \Omega\angle 60°$. Convert this value to rectangular form.

Using rectangular notation provides us with the *series equivalent* of any impedance value written in polar notation.

The result in Example 14.21 indicates that the value of $Z_P$ equals the total impedance of a 314 Ω resistance in series with a 310 Ω inductive reactance; that is, it provides us with the series equivalent of $Z_P$. This point is illustrated in Figure 14.43. Figure 14.43a shows the circuit used as the basis for Example 14.21. As you can see, the total impedance of the parallel network ($Z_P$) is 441 Ω∠44.6°. The rectangular form of this value (which was calculated in the example) is drawn as a series impedance ($Z_S$) in Figure 14.43b. Note that the values in the $Z_S$ equation match those in the figure. If we were to disconnect $R_S$ and measure the impedance of the shaded circuit ($Z_S$), we would measure approximately 441 Ω. In effect, what we have done in the example is to determine the series equivalent of $Z_P$. Once the series equivalent impedance of the parallel $RL$ network has been determined, we can solve easily for the total circuit impedance, as demonstrated in Example 14.22.

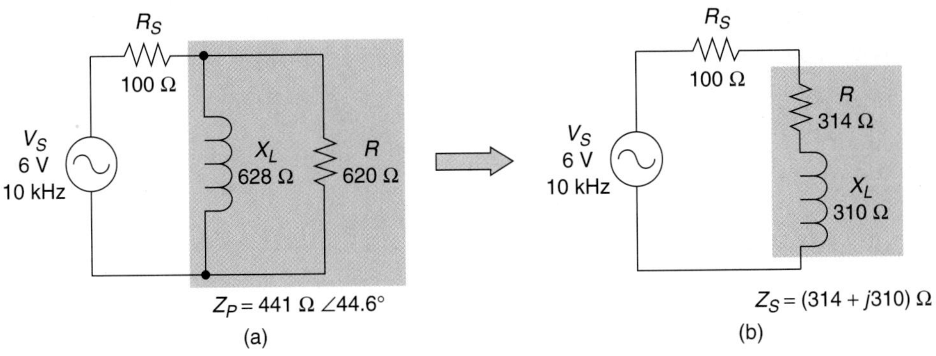


$Z_P = Z_S$

EWB

**FIGURE 14.43 Equivalent parallel and series impedances.**

**EXAMPLE 14.22**

Determine the total impedance (in polar form) for the circuit in Figure 14.43b.

***Solution:*** First, the total resistance in the circuit is found as

$$R_T = R_S + R = 100\ \Omega + 314\ \Omega = \mathbf{414\ \Omega}$$

Now, $R_T$ and $X_L$ can be combined to determine the magnitude of the circuit impedance, as follows:

$$Z_T = \sqrt{X_L^2 + R_T^2} = \sqrt{(310\ \Omega)^2 + (414\ \Omega)^2} = \mathbf{517\ \Omega}$$

and the phase angle of the circuit impedance can be found as

$$\theta = \tan^{-1}\left(\frac{X_L}{R_T}\right) = \tan^{-1}\left(\frac{310\ \Omega}{414\ \Omega}\right) = \mathbf{36.8°}$$

Thus, the total impedance of the original series-parallel circuit is 517 $\Omega\angle 36.8°$.

***PRACTICE PROBLEM 14.22***

A circuit like the one in Figure 14.43b has the following values: $R_S = 220\ \Omega$, $R = 872\ \Omega$, and $X_L = 444\ \Omega$. Calculate the total circuit impedance.

In Examples 14.21 and 14.22, we derived the series equivalent for the circuit in Figure 14.43a. The remainder of the circuit analysis would involve solving for the component current, voltage, and power values. Example 14.23 demonstrates the process from start to finish.

## EXAMPLE 14.23

Perform a complete analysis on the circuit shown in Figure 14.44a.

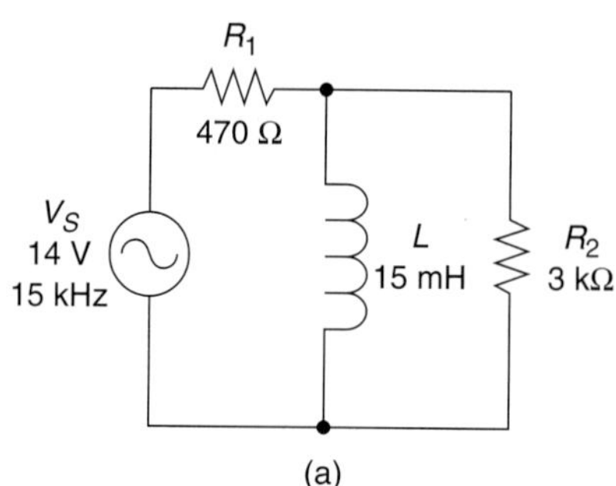


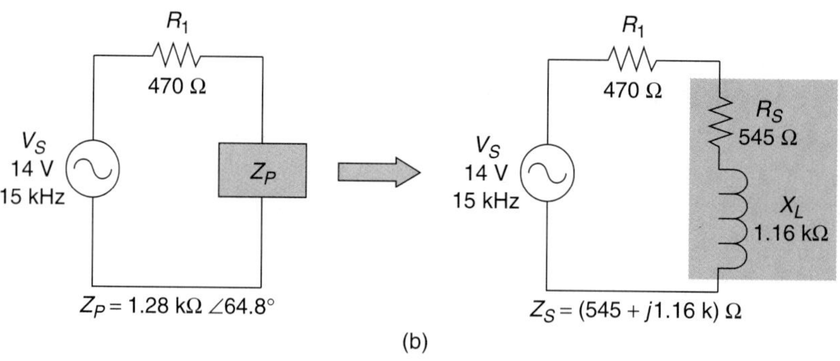


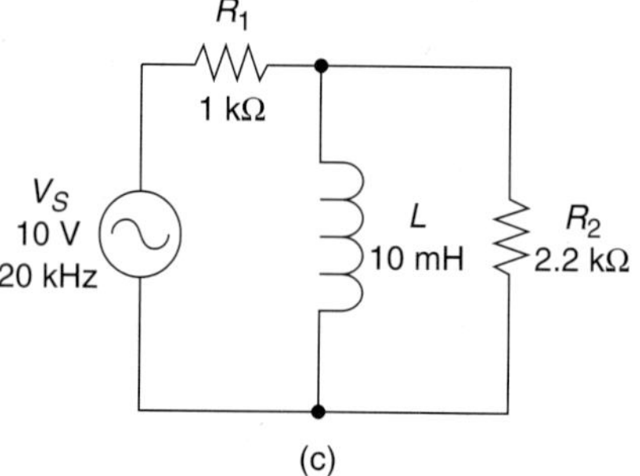


**FIGURE 14.44**

***Solution:*** First, the reactance of the inductor is found as

$$X_L = 2\,\pi fL = 2\,\pi(15\text{ kHz})(15\text{ mH}) = \mathbf{1.41\ k\Omega}$$

The magnitude of the parallel circuit impedance can be found as

$$Z_P = \frac{X_L R_2}{\sqrt{X_L^2 + R_2^2}} = \frac{(1.41\text{ k}\Omega)(3\text{ k}\Omega)}{\sqrt{(1.41\text{ k}\Omega)^2 + (3\text{ k}\Omega)^2}} = \mathbf{1.28\ k\Omega}$$

Since $L$ and $R_2$ are connected in parallel, the phase angle of $Z_P$ is found as

$$\theta = \tan^{-1}\left(\frac{R_2}{X_L}\right) = \tan^{-1}\left(\frac{3\text{ k}\Omega}{1.41\text{ k}\Omega}\right) = \mathbf{64.8^\circ}$$

We have determined that the parallel combination of $L$ and $R_2$ has a total impedance of 1.28 kΩ∠64.8°. This impedance is represented as a block ($Z_P$) in Figure 14.44b. The series equivalent of $Z_P$ can now be found using

$$R = Z\cos\theta = (1.28\text{ k}\Omega)(\cos 64.8^\circ) = \mathbf{545\ \Omega}$$

and

$$X_L = Z \sin \theta = (1.28\ \text{k}\Omega)(\sin 64.8°) = \mathbf{1.16\ k\Omega}$$

These values are included as $Z_S$ in the series equivalent circuit (Figure 14.44b).

Once the series equivalent circuit has been derived, the resistances in the circuit can be combined as follows:

$$\boldsymbol{R_T} = R_1 + R_S = 470\ \Omega + 545\ \Omega \cong \mathbf{1.02\ k\Omega}$$

The magnitude of the circuit impedance can be found as

$$\boldsymbol{Z_T} = \sqrt{X_L^2 + R_T^2} = \sqrt{(1.16\ \text{k}\Omega)^2 + (1.02\ \text{k}\Omega)^2} = \mathbf{1.54\ k\Omega}$$

and its phase angle can be found as

$$\boldsymbol{\theta} = \tan^{-1}\left(\frac{X_L}{R_T}\right) = \tan^{-1}\left(\frac{1.16\ \text{k}\Omega}{1.02\ \text{k}\Omega}\right) = \mathbf{48.7°}$$

Once the circuit impedance is determined, the circuit current can be found as

$$\boldsymbol{I_T} = \frac{V_S}{Z_T} = \frac{14\ \text{V}}{1.54\ \text{k}\Omega\angle 48.7°} = \mathbf{9.09\ mA\angle -48.7°}$$

This result indicates that current lags the source voltage by 48.7°. Earlier in the chapter, you were told that current is considered to be the zero degree reference in a series circuit. Since we are dealing with a series equivalent circuit, the phase angles are adjusted as follows:

$$V_S = 14\ \text{V}\angle 48.7° \qquad \text{and} \qquad I_T = 9.09\ \text{mA}\angle 0°$$

Now, the voltage across $R_1$ can be found as

$$\boldsymbol{V_1} = I_T R_1 = (9.09\ \text{mA})(470\ \Omega) = \mathbf{4.27\ V}$$

Note that the phase angles were left out of the $V_1$ calculation because both are known to be 0°. However, phase angles must be included when calculating the voltage across the parallel circuit impedance ($Z_P$). This voltage ($V_P$) can be found as

$$\boldsymbol{V_P} = I_T Z_P = (9.09\ \text{mA}\angle 0°)(1.28\ \text{k}\Omega\angle 64.8°) = \mathbf{11.6\ V\angle 64.8°}$$

Now that we know the voltage across the parallel circuit, we can calculate the values of $I_L$ and $I_{R2}$. Again, we must include the phase angles in our calculations, as follows:

$$\boldsymbol{I_L} = \frac{V_P}{X_L} = \frac{11.6\ \text{V}\angle 64.8°}{1.41\ \text{k}\Omega\angle 90°} = 8.23\ \text{mA}\angle(64.8° - 90°) = \mathbf{8.23\ mA\angle -25.2°}$$

and

$$\boldsymbol{I_{R2}} = \frac{V_P}{R_2} = \frac{11.6\ \text{V}\angle 64.8°}{3\ \text{k}\Omega\angle 0°} = \mathbf{3.87\ mA\angle 64.8°}$$

That completes the circuit currents and voltages. The apparent power for the circuit can be found using the values of $I_T$ and $Z_T$, as follows:

$$\boldsymbol{P_{APP}} = I_T^2 Z_T = (9.09\ \text{mA}\angle 0°)^2(1.54\ \text{k}\Omega\angle 48.7°) = \mathbf{127\ mVA\angle 48.7°}$$

Given the complexity of the circuit, the easiest way to calculate the actual power dissipation (that is, resistive power) is to use the power factor, as follows:

$$\boldsymbol{P_R} = P_{APP} \cos \theta = (127\ \text{mVA})(\cos 48.7°) = (127\ \text{mVA})(0.660) = \mathbf{83.8\ mW}$$

This completes the analysis of the circuit.

***PRACTICE PROBLEM 14.23***

Perform a complete analysis on the circuit shown in Figure 14.44c.

## Putting It All Together: Some Observations of Example 14.23

To make sense of the results in Example 14.23, we need to combine them and see how they fit together. For ease of discussion, Figure 14.45 includes the currents and voltages calculated in the example. (Note that the source has been given a phase angle relative to the circuit current, as found in the example.)

First, let's take a look at the parallel circuit ($L$ and $R_2$). Earlier in the chapter, you were given the following relationships for a parallel $RL$ circuit:

1. The voltages across the branches are equal.
2. $I_R$ leads $I_L$ by 90°.
3. The sum of the branch currents equals the total circuit current.

The first of these relationships is shown in the figure. The voltage across each branch equals the value of $V_P$ calculated in the example. The phase relationship between $I_L$ and $I_{R2}$ can be verified by finding the difference between their phases, as follows:

$$\theta = 64.8° - (-25.2°) = 64.8° + 25.2° = 90°$$

As you can see, the second relationship given for parallel $RL$ circuits also holds true. The third relationship can be verified by converting each of the branch currents to rectangular form. Using the conversion techniques covered in this section, the branch currents in the figure can be written as follows:

$$I_L = (7.45 - j3.50)\text{ mA} \qquad I_{R2} = (1.64 + j3.50)\text{ mA}$$

Adding these currents together, we get

$$\begin{aligned} I_L &= 7.45 - j3.50\text{ mA} \\ I_{R2} &= \underline{1.64 + j3.50\text{ mA}} \\ & \phantom{=}\ 9.09 + j0.00\text{ mA} = I_T \end{aligned}$$

The value of ($+j0.00$ mA) indicates that the total circuit current contains no reactive component. Therefore, the phase angle of the circuit current is 0°, and the branch currents add up to a value of $I_T = 9.09\text{ mA}\angle 0°$. This value matches the one shown in the figure.

We have shown that the characteristics of the parallel $RL$ circuit match those given earlier in the chapter. Now, let's look at the series combination of $R_1$ and ($L \| R_2$). Earlier in the chapter, you were given the following relationships for a series $RL$ circuit:

1. The currents through the series components are equal.
2. The source voltage equals the sum of the series component voltages.

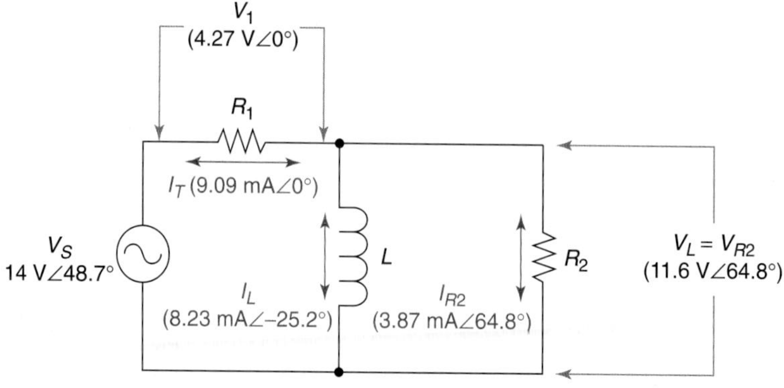


**FIGURE 14.45 The currents and voltages calculated in Example 14.23.**

We have already shown the first of these relationships to be true. The second can be demonstrated by converting $V_{R1}$ and the voltage across the parallel circuit ($V_P$) to rectangular form, as follows:

$$V_{R1} = (4.27 + j0)\text{ V} \qquad \text{and} \qquad V_P = (4.94\text{ V} + j10.5)\text{ V}$$

Adding these voltages together, we get

$$\begin{aligned} V_{R1} &= 4.27 + j0.00\text{ V} \\ V_P &= \underline{4.94 + j10.5\text{ V}} \\ &\phantom{=} 9.21 + j10.5\text{ V} = V_S \end{aligned}$$

We can verify this result by converting it to polar form, as follows:

$$V_S = \sqrt{(9.21\text{ V})^2 + (10.5\text{ V})^2} = 13.97\text{ V} \cong 14\text{ V}$$

and calculating the phase angle, as follows:

$$\theta = \tan^{-1}\left(\frac{10.5\text{ V}}{9.21\text{ V}}\right) = \tan^{-1}(1.140) = 48.7°$$

As you can see, all the relationships given earlier for series and parallel *RL* circuits held true for the circuit in Example 14.23. Although the numbers seem complicated, the results are always as expected (provided no mistakes are made).

## A Look Ahead

Up to this point, we have concentrated on the operation of *RL* circuits with sinusoidal input signals. In the next section, we will look at the response that series *RL* circuits have to rectangular inputs. More specifically, we will concentrate on the effect that a series *RL* circuit has on a square-wave input. As you will see, an *RL* circuit responds quite differently to a square wave than it does to a sine wave.

**Section Review**

1. How is the impedance of an *RL* circuit converted from polar form to rectangular form?
2. When a parallel *RL* impedance is converted from polar to rectangular form, what type of circuit conversion takes place?

# 14.5 SQUARE WAVE RESPONSE: *RL* TIME CONSTANTS

At the end of Chapter 11, you were introduced to rectangular waveforms. You may recall that a rectangular waveform is made up of alternating dc levels. As a review, rectangular and square-wave measurements are provided in Figure 14.46. The waveforms in the figure alternate between two dc levels identified as 0 V and +V. The time that each waveform spends at the +V level is referred to as the pulse width (PW). The time that each waveform spends at the 0 V level is referred to as the space width (SW). The combination of the PW and SW is the cycle time of the waveform. The square wave (Figure 14.46b) is a special-case rectangular waveform that has equal PW and SW values.

Any circuit with a square (or rectangular) wave input is generally referred to as a **switching circuit,** since the input switches back and forth between dc levels. In this section, we will concentrate on the characteristics of series *RL* switching circuits. Since these circuits operate using alternating dc levels, we will need to establish several new principles of *RL* circuit operation, most notably, the *RL time constant.*

**Switching circuit**
A term generally used to describe any circuit with a square (or rectangular) wave input.

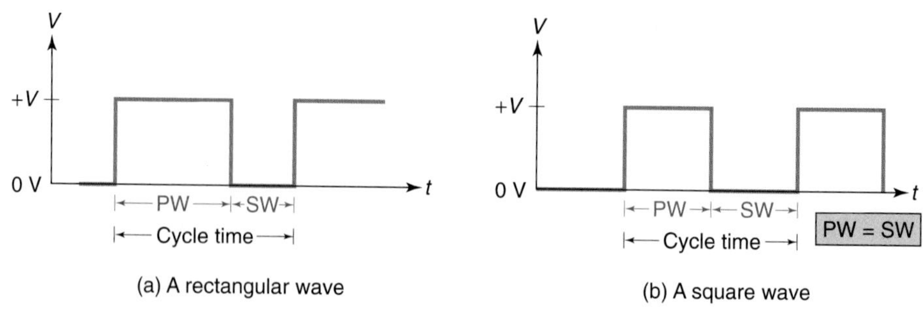


**FIGURE 14.46 Rectangular and square wave measurements.**

OBJECTIVE 13 ▶

## Voltage and Current Transitions

**Transition**
In a switching circuit, the change from one dc level to another.

In a rectangular waveform, the change from one dc level to the other is referred to as a **transition.** For example, the voltage and current transitions in a purely resistive circuit are shown in Figure 14.47. The waveform transitions occur at the time increments shown at the bottom of the waveforms ($t_1$ through $t_4$). The transitions that occur at $t_1$ and $t_3$ are called positive-going transitions. The transitions at $t_2$ and $t_4$ are called negative-going transitions. Note that the $V_R$ and $I$ waveforms are identical to the $V_S$ waveform. This is *not* the case when an inductor is added to the circuit.

## RL Circuit Waveforms

When a square wave is applied to a series *RL* circuit, the $V_R$ and current waveforms are altered as shown in Figure 14.48. (We'll look at the $V_L$ waveform in a moment.) At $t_1$, the source voltage makes the transition from 0 V to +V. Since an inductor opposes any change in current, the circuit current cannot rise to its maximum value at the same rate as the source voltage (like it did in the purely resistive circuit). Rather, current rises to its maximum value over the time period shown ($t_1$ to $t_2$). Since $V_R$ is always in phase with the circuit current, it also rises over the time period shown. (The transitions described here are repeated between $t_3$ and $t_4$.)

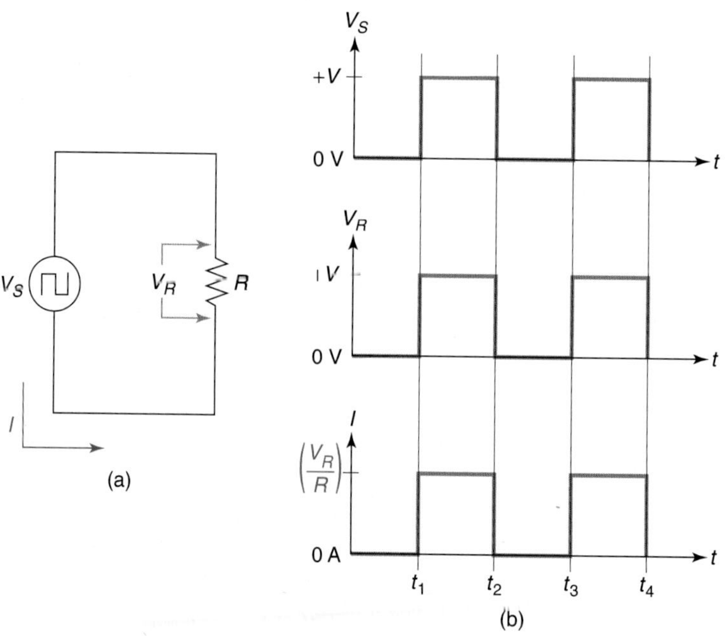


**FIGURE 14.47 Voltage and current waveforms in a purely resistive dc circuit.**

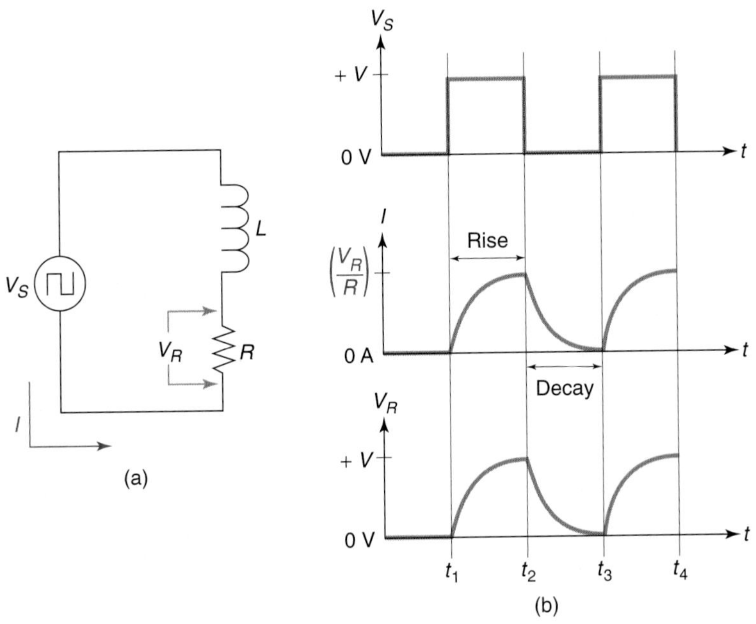


EWB

**FIGURE 14.48 Transition delay in an *RL* circuit.**

At $t_2$, the source voltage makes the transition back to 0 V. Again, the inductor prevents the circuit current from changing at the same rate as the source voltage. In this case, the current decay occurs over the time period shown ($t_2$ to $t_3$). As a result, $V_R$ decays slowly over the same time period.

## *Inductor Voltage ($V_L$)*

The $V_R$ and current waveforms in Figure 14.48 are explained easily in terms of the inductor's opposition to a change in current. This opposition slows the change in current, producing the waveforms shown. However, the waveform for the inductor voltage ($V_L$) requires a more thorough explanation. As Figure 14.49 shows, the $V_L$ waveform has a shape that you haven't seen before. (For the sake of discussion, Figure 14.49 assumes that the source has a peak value of +5 V.)

The shape of the $V_L$ waveform can be explained easily in terms of Kirchhoff's voltage law. As you know, this law states that the sum of the component voltages must equal the source voltage. Since the source in Figure 14.49 generates alternating dc voltages, the circuit operates according to dc principles. This means that the source voltage equals the algebraic sum of $V_L$ and $V_R$. By formula,

$$V_S = V_L + V_R$$

or

$$V_L = V_S - V_R \qquad \textbf{(14.22)}$$

where $V_S$, $V_L$, and $V_R$ are all dc voltage levels. With this relationship in mind, let's look at the voltage waveforms shown in Figure 14.49.

At $t_1$, the source makes the transition from 0 V to +5 V. At the instant of this transition, $V_R = 0$ V (since the rise in current is delayed by the inductor). Therefore, the value of $V_L$ at $t_1$ is found as

$$V_L = V_S - V_R = +5\text{ V} - 0\text{ V} = +5\text{ V}$$

This value corresponds to the peak value of $V_L$ shown at $t_1$ on the waveform.

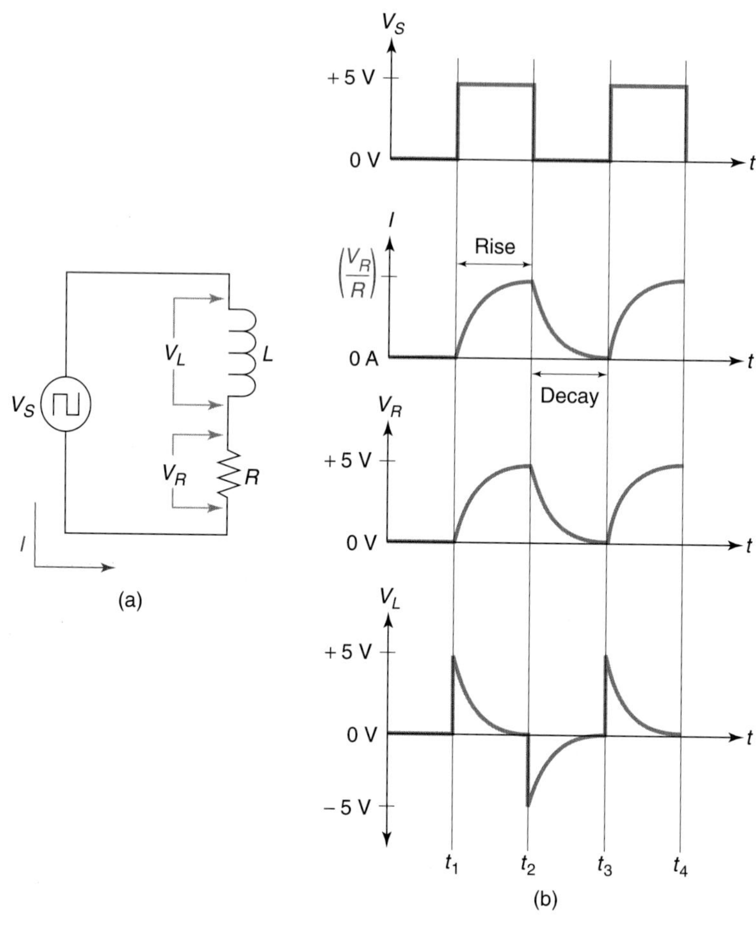


EWB

**FIGURE 14.49** ***RL*** **circuit waveforms.**

Immediately prior to $t_2$, the source is shown to be at +5 V. During the $t_1 \rightarrow t_2$ transition, $V_R$ has increased from 0 V to +5 V. At the same time, $V_L$ has decreased, as follows:

$$V_L = V_S - V_R = +5\text{ V} - 5\text{ V} = 0\text{ V}$$

Again, the calculated value corresponds to the appropriate point on the $V_L$ waveform. At $t_2$, the source makes the transition from +5 V to 0 V. At the instant this transition is completed, $V_R = +5$ V (since the current decay is delayed by the inductor). Therefore, the value of $V_L$ at $t_2$ is found as

$$V_L = V_S - V_R = 0\text{ V} - 5\text{ V} = -5\text{ V}$$

This value corresponds to the abrupt transition from 0 V to −5 V that $V_L$ makes at $t_2$. During the $t_2 \rightarrow t_3$ transition, $V_R$ decays from +5 V to 0 V. At the same time, $V_L$ decays, as follows:

$$V_L = V_S - V_R = 0\text{ V} - 0\text{ V} = 0\text{ V}$$

From then on, the cycle repeats itself.

## More on the RL Circuit Waveforms

Several more points need to be made regarding the waveforms in Figure 14.49. First, it should be noted that equation (14.22) holds true at any instant on the $V_R$ and $V_L$ waveforms. Even though we limited our discussion to the times immediately before and after the transitions, $V_R$ and $V_L$ change at the same rate. Therefore, their sum always equals the

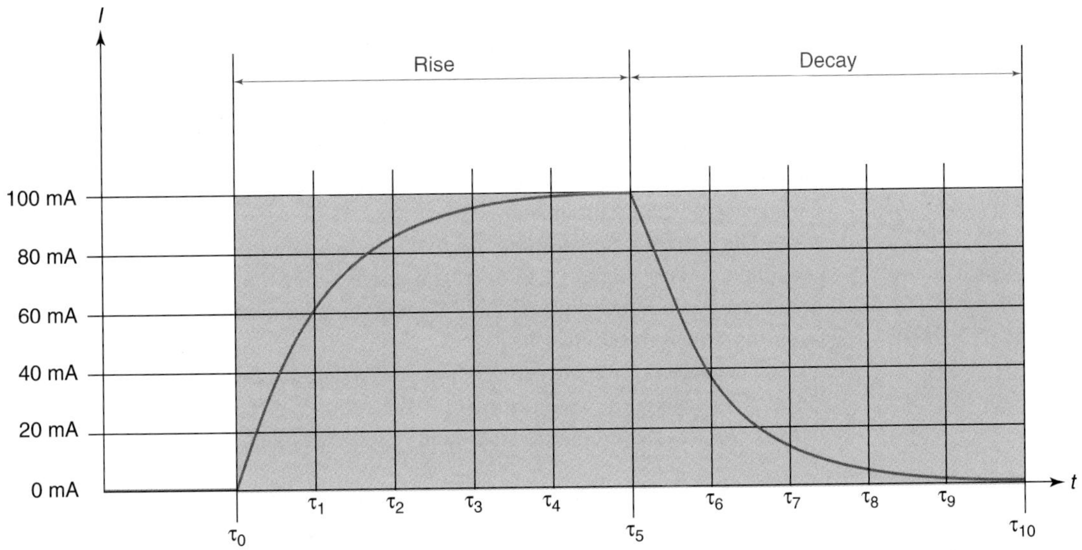


**FIGURE 14.50 A close-up of the current waveform for Figure 14.49.**

supply voltage. For example, at a point in time between $t_1$ and $t_2$, $V_R$ reaches a value of 3 V. At that time, $V_L$ has a value of

$$V_L = V_S - V_R = 5\text{ V} - 3\text{ V} = 2\text{ V}$$

This relationship between the circuit voltages holds true at every point on the voltage waveforms.

The second point we need to make deals with the rate at which current makes its transitions. As Figure 14.49 shows, the circuit current does not change at a **linear** (constant) rate. Rather, the rate of change in current decreases as the transition progresses. A close-up of the current waveform (Figure 14.50) can be used to illustrate this concept. (For the sake of discussion, we assume that the circuit current has a peak value of 100 mA.)

**Linear**
A term used to describe any value that changes at a constant rate.

Each half of the waveform in Figure 14.50 is divided into five equal time intervals. For reasons you will be shown later, each time interval is identified using the Greek letter tau, $\tau$. The estimated change in current ($\Delta I$) during each of the first five time intervals is provided in Table 14.7. As you can see, the change in current decreases from one time interval to the next during the rising portion of the curve. The same principle holds true for the decay portion of the curve. The estimated values of $\Delta I$ for the decay portion of the curve are listed in Table 14.8.

*Note:* The reason for using five equal time periods will be explained later in this section.

When you compare the values of $\Delta I$ in these two tables, you can see that the two halves of the curve are symmetrical; that is, the rate of rise in any time interval matches the corresponding rate of decay. This relationship holds true for any *RL* switching circuit.

**TABLE 14.7 Estimated Values of $\Delta I$ During the Rise Portion of the Curve in Figure 14.50**

| *Time Interval* | *Estimated $\Delta I$* |
|---|---|
| $\tau_0 \to \tau_1$ | 62 mA |
| $\tau_1 \to \tau_2$ | 22 mA |
| $\tau_2 \to \tau_3$ | 10 mA |
| $\tau_3 \to \tau_4$ | 2 mA |
| $\tau_4 \to \tau_5$ | <2 mA |

**TABLE 14.8 Estimated Values of $\Delta I$ During the Decay Portion of the Curve in Figure 14.50**

| *Time Interval* | *Estimated $\Delta I$* |
|---|---|
| $\tau_5 \to \tau_6$ | −62 mA |
| $\tau_6 \to \tau_7$ | −22 mA |
| $\tau_7 \to \tau_8$ | −10 mA |
| $\tau_8 \to \tau_9$ | −2 mA |
| $\tau_9 \to \tau_{10}$ | <−2 mA |

## The Universal Curve

OBJECTIVE 14 ▶ The shape of the curve in Figure 14.50 is a function of the inductor's opposition to a change in circuit current. This means that *the wave shape is independent of the peak value of the circuit current.* For example, if we doubled the peak source voltage in Figure 14.49, the peak current generated through the circuit would also double. However, the *shape* of the curve would remain unchanged.

**Universal curve**
One that can be used to predict the operation of any specified type of circuit.

The current characteristics of an *RL* switching circuit can be represented using what is called a **universal curve.** A universal curve is one that can be used to predict the operation of any specified type of circuit. The universal curve for the current in an *RL* switching circuit is shown in Figure 14.51.

The $x$-axis of the universal curve is divided into the same time intervals that we saw earlier. The $y$-axis, however, measures the ratio of circuit current ($I_t$) to the peak source current ($I_{pk}$). For example, at $\tau_1$, the curve indicates that

$$\frac{I_t}{I_{pk}} \cong 0.62$$

At $\tau_2$, the curve indicates that

$$\frac{I_t}{I_{pk}} \cong 0.85$$

and so on.

When the peak source current is known, the curve can be used to approximate the total circuit current at any time interval:

$$I_t = \left(\frac{I_t}{I_{pk}}\right) I_{pk} \tag{14.23}$$

where $\left(\frac{I_t}{I_{pk}}\right)$ = the current ratio obtained from the curve

$I_{pk}$ = the known peak source current

For example, the current ratio at $\tau_4$ in Figure 14.51 is approximately 0.95. If we know that an *RL* circuit has a value of $I_{pk}$ = 80 mA, then the circuit current at $\tau_4$ can be found as

$$I_t = \left(\frac{I_t}{I_{pk}}\right) I_{pk} = (0.95)(80 \text{ mA}) = 76 \text{ mA}$$

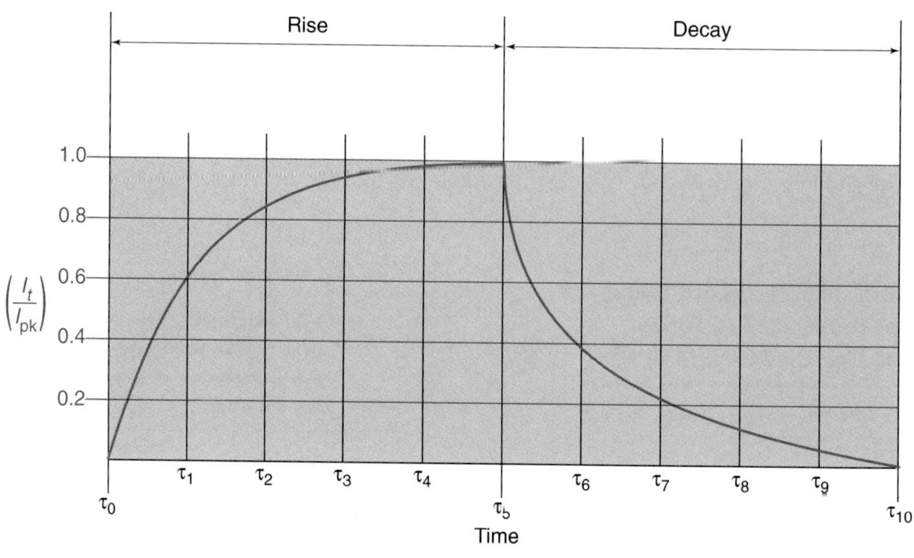


**FIGURE 14.51 The universal current curve.**

It should be noted that the curve provides only approximate values of circuit current at a given time interval. If a more exact value of circuit current is desired, it must be calculated using the *curve equation.*

## The Rise Curve Equation

The universal curve is an **exponential curve:** its rate of change is a function of a variable exponent. For example, the rise portion of the universal curve in Figure 14.52 is actually a plot of the equation

**Exponential curve** A curve whose rate of change is a function of a variable exponent.

$$\frac{I_t}{I_{pk}} = 1 - e^{-x} \tag{14.24}$$

where $e$ = the base of the natural log system (approximately 2.71828)
$x$ = the variable exponent

The only value on the right-hand side of the equation that is variable is $x$. Therefore, the shape of the curve must be a function of this variable exponent. By definition, that makes the curve an exponential curve.

Before demonstrating the relationship between equation (14.24) and the rise portion of the curve, we need to establish the calculator sequence used to solve the equation.

◄ *OBJECTIVE 15*

***Solving Equation (14.24).*** The $e^x$ function is the inverse of the natural log (ln) function. On most scientific calculators, the natural log key is labeled *ln*. This key has an INV (or 2nd) function of $e^x$ (as labeled above the key). To perform this function for a given value of $x$, the following key sequence is used:

[$x$] [INV] [ln]

When solving for $e^{-x}$, as we are in equation (14.24), the sign of $x$ must be assigned before performing the inverse ln function. Therefore, the following key sequence is used to solve for $e^{-x}$:

[$x$] [+/−] [INV] [ln]

For $x = 1$, equation (14.24) is solved using the following key sequence:

[1] [−] [1] [+/−] [INV] [ln] [=]

(the keys [1] [+/−] [INV] [ln] give $e^{-1}$)

which provides a result of 0.632.

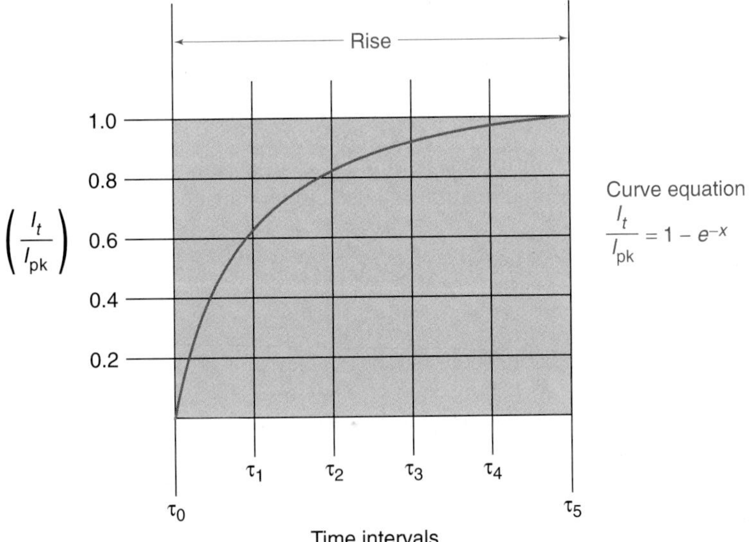


**FIGURE 14.52 The rise portion of the universal curve.**

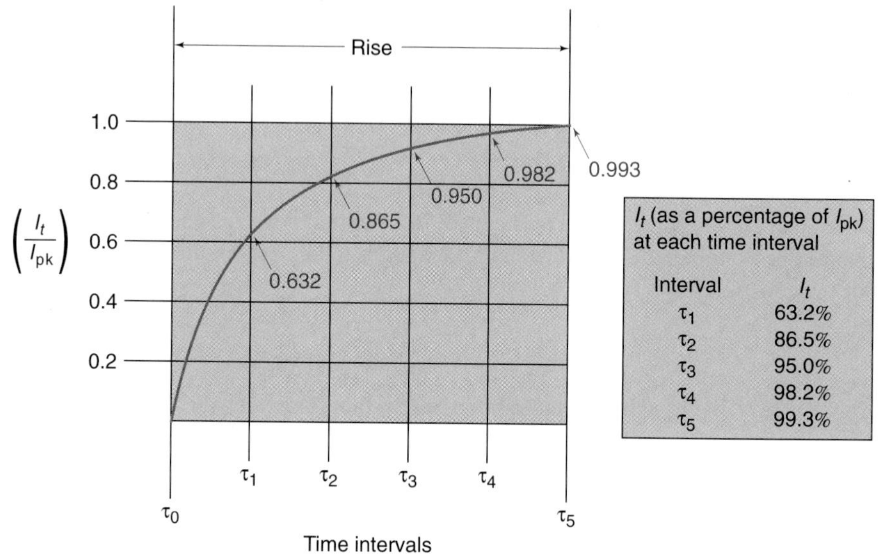


**FIGURE 14.53 The calculated current ratio at each time interval.**

**TABLE 14.9 Results of Equation (14.24) for Increasing Values of $x$**

| *Value of $x$* | *Equation (14.24) Result* |
|---|---|
| 1 | 0.632 |
| 2 | 0.865 |
| 3 | 0.950 |
| 4 | 0.982 |
| 5 | 0.993 |

***Plotting the Universal Curve.*** If we solve equation (14.24) for whole number values of $x = 1 \rightarrow 5$, we get the results listed in Table 14.9. If we assume that the values of $x$ in Table 14.9 correlate to the time intervals along the $x$-axis in Figure 14.53, then the curve forms a plot of the results. Using the values shown on the curve, we can express the value of $I_t$ at each time interval as a percentage of the peak circuit current ($I_{pk}$). For example, at $t_1$:

$$\frac{I_t}{I_{pk}} = 0.632$$

Therefore,

$$I_t = 0.632\, I_{pk}$$

This equation indicates that $I_t$ reaches 63.2% of its peak value at $t_1$. The value of $I_t$ (as a percentage of $I_{pk}$) for each time interval is listed in the figure. Based on the percentages shown, we can say that current increases from 0% to 99.3% of its peak value over the course of the five time intervals.

## The Decay Curve

The decay curve is simply a mirror image of the rise curve. Like the rise curve, it is an exponential curve. The exponential relationship used to plot the decay curve is

$$\frac{I_t}{I_{pk}} = e^{-x} \tag{14.25}$$

**TABLE 14.10 Results of Equation (14.25) for Increasing Values of $x$**

| *Value of $x$* | *Equation (14.25) Result* |
|---|---|
| 1 | 0.368 |
| 2 | 0.135 |
| 3 | 0.050 |
| 4 | 0.018 |
| 5 | 0.007 |

If we solve this relationship for whole number values of $x = 1 \rightarrow 5$, we get the results listed in Table 14.10.

The results listed in Table 14.10 are identified on the decay curve in Figure 14.54. Again, each time interval corresponds to a value of $x$ from the table. As was the case with the rise curve, we can express the value of $I_t$ at each time interval in Figure 14.54 as a percentage of the peak circuit current ($I_{pk}$). The value of $I_t$ (as a percentage of $I_{pk}$) for each time interval is listed in the figure. As you can see, the current decreases from approximately 100% to 0.7% of its peak value over the course of the five time intervals.

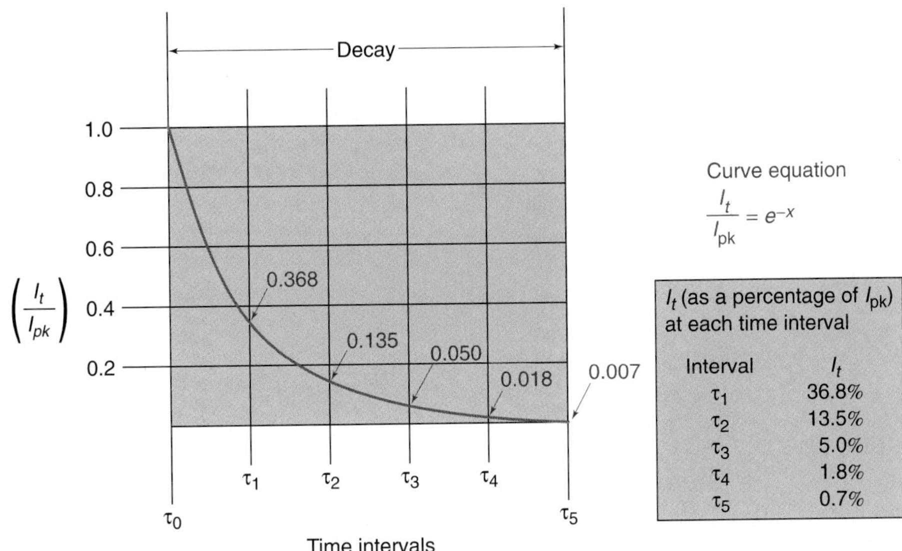


**FIGURE 14.54 The decay curve.**

## The Time Intervals

So far, we have established two important principles regarding the time intervals on the rise and decay curves:

1. Each time interval corresponds to a whole number value of $x$ in each curve equation.
2. It takes five of these time intervals to complete a rise or decay transition.

At this point, we will turn our attention to the time intervals on the curves. As you will see, each time interval in the current curves actually has a *real-time* value determined by the circuit values of inductance and resistance.

## The RL Time Constant ($\tau$)

◄ *OBJECTIVE 16*

Each time interval in the universal curve represents a real-time value referred to as a **time constant ($\tau$).** The term *constant* is used because $\tau$ is independent of the circuit current magnitude and operating frequency. For a given series *RL* switching circuit, the value of the *RL* time constant is found as

**Time constant ($\tau$)**
A time interval on any universal rise (or decay) curve that is constant and independent of magnitude and operating frequency. (*Note:* A more precise definition is provided later in this section.)

$$\tau = \frac{L}{R} \tag{14.26}$$

where $\tau$ = the duration of each time interval, in seconds
$L$ = the total series inductance
$R$ = the total series resistance

Example 14.24 demonstrates the use of this equation in determining the time constant for a series *RL* circuit with a square-wave input.

## EXAMPLE 14.24

Determine the time constant for the circuit in Figure 14.55a.

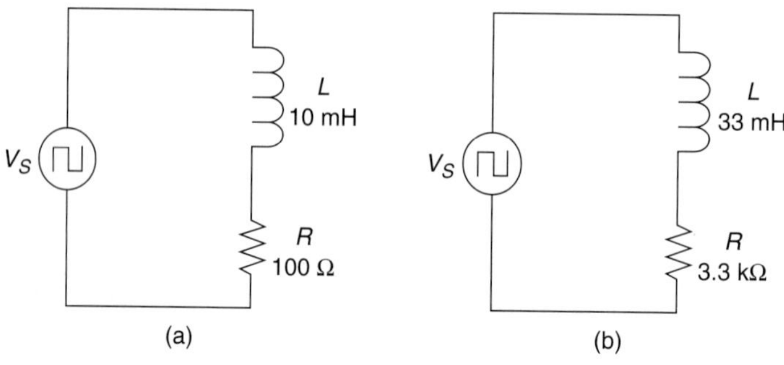


EWB

**FIGURE 14.55**

***Solution:*** Using equation (14.26), the time constant for the circuit is found as

$$\tau = \frac{L}{R} = \frac{10\text{ mH}}{100\ \Omega} = \mathbf{100\ \mu s}$$

***PRACTICE PROBLEM 14.24***

Calculate the value of the time constant for the circuit in Figure 14.55b.

Figure 14.56 shows the rise curve for the circuit in the example. As you can see, each time interval represents the value of the circuit time constant (100 μs). The total time required for the circuit current to complete its rise is approximately equal to five time constants. By formula,

$$T = 5\tau \tag{14.27}$$

where $T$ = the time required for the circuit current to rise to its peak value, in seconds
$\tau$ = the time constant of the circuit, in seconds

Example 14.25 demonstrates the process for determining the total time required for the current in an *RL* circuit to rise from zero to its peak value.

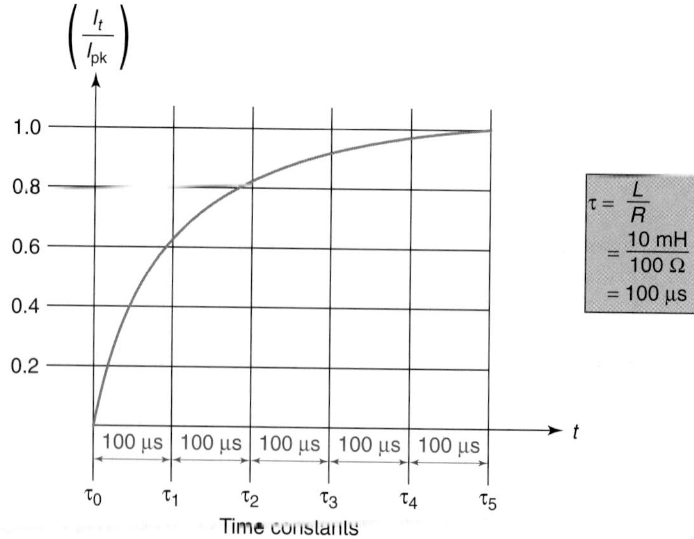


**FIGURE 14.56 The rise curve for the circuit in Example 14.24.**

Determine the time required for $I_t$ to reach its peak value in the circuit shown in Figure 14.57a.

**EXAMPLE 14.25**

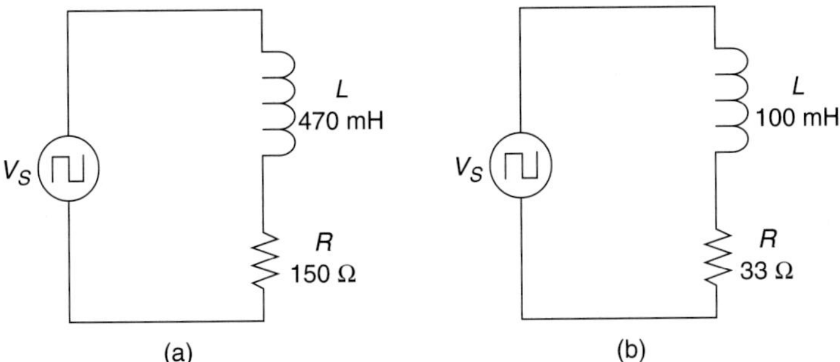


**FIGURE 14.57**

***Solution:*** First, the value of the circuit time constant is found as

$$\tau = \frac{L}{R} = \frac{470 \text{ mH}}{150\ \Omega} = \mathbf{3.13\ ms}$$

It takes five time constants for the rise portion of the curve to complete its transition. Therefore, the total time required for the transition is found as

$$T = 5\tau = (5)(3.13 \text{ ms}) = \mathbf{15.67\ ms}$$

***PRACTICE PROBLEM 14.25***

Determine the time required for $I_t$ to reach its peak value in the circuit shown in Figure 14.57b.

## The Basis for Equation (14.26)

You may be wondering how dividing inductance by resistance can yield a value measured in seconds. In Chapter 13, you were shown that the voltage across an inductor can be found as

$$v_L = L\frac{di}{dt}$$

where $dt$ is measured in seconds. During the rise (or decay) portion of the current curve, the source provides the circuit with a dc input. For a dc input, the $V_L$ equation can be rewritten as

$$\Delta V = L\frac{\Delta I}{\Delta t}$$

*Note: dx* is used to represent an *instantaneous rate of change* in a variable ($x$), and is normally used in *ac* calculations. In *dc* calculations, a rate of change in a variable ($x$) over a period of time is represented as $\Delta X$.

Rewriting the above equation for $L$, we get

$$\frac{(\Delta V)(\Delta t)}{\Delta I} = L \qquad \text{or} \qquad \left(\frac{\Delta V}{\Delta I}\right)(\Delta t) = L$$

Since $\dfrac{\Delta V}{\Delta I} = R$, the above equation can be rewritten as

$$(R)(\Delta t) = L$$

and, dividing both sides of the equation by $R$, we get

$$\Delta t = \frac{L}{R}$$

Using $\tau$ to represent the change in time ($\Delta t$), we can rewrite the above equation as

$$\tau = \frac{L}{R}$$

which is equation (14.26).

## Another Look at the Universal Curve Equations

Earlier in this section, you learned that the universal curve is actually a plot of the following equations:

| Rise Curve | Decay Curve |
|---|---|
| $\dfrac{I_t}{I_{pk}} = 1 - e^{-x}$ | $\dfrac{I_t}{I_{pk}} = e^{-x}$ |

At this point, we are ready to take a closer look at the variable exponent ($x$) in each of these equations. The variable ($x$) in the curve equations is actually a ratio of two time values. This ratio is given as

$$x = \frac{t}{\tau} \tag{14.28}$$

where
$x$ = the exponent of $e$ in the curve equations
$t$ = the time from the start of the curve
$\tau$ = the duration of one time constant

For example, refer back to Figure 14.56. As shown, the duration of one time constant in the curve is 100 μs. The time interval labeled $t_3$ occurs 300 μs after the start of the curve (which is labeled $\tau_0$). Therefore, the power of $e$ at this time interval is found as

$$x = \frac{t}{\tau} = \frac{300\ \mu\text{s}}{100\ \mu\text{s}} = 3$$

This result is used to calculate the current ratio at $\tau_3$, as follows:

$$\frac{I_t}{I_{pk}} = 1 - e^{-x} = 1 - e^{-3} = 0.95$$

Using equation (14.28) allows us to calculate the current ratio at any point on the rise (or decay) curve. This point is demonstrated in Example 14.26.

## EXAMPLE 14.26

The circuit in Figure 14.58a is shown to have a peak current of 100 mA. Determine the value of the circuit current 1.5 ms after the start of the decay cycle.

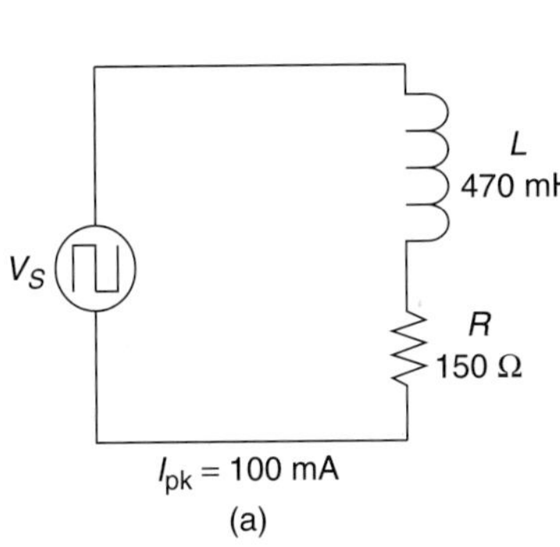


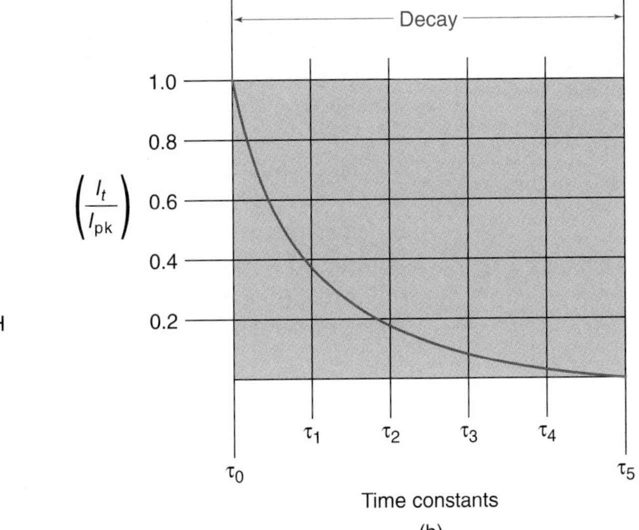


**FIGURE 14.58**

***Solution:*** In Example 14.25, we calculated a value of $\tau = 3.13$ ms for this circuit. At $t = 1.5$ ms, the power of $e$ is found as

$$x = \frac{t}{\tau} = \frac{1.5\text{ ms}}{3.13\text{ ms}} = \mathbf{0.48}$$

Now, the current ratio at the designated time is found as

$$\frac{I_t}{I_{pk}} = e^{-x} = e^{-0.48} = \mathbf{0.619}$$

Using this current ratio, the value of the circuit current at $t = 1.5$ ms is found as

$$I_t = 0.619 I_{pk} = (0.619)(100\text{ mA}) = \mathbf{61.9\ mA}$$

When you compare the current ratio (0.619) to the curve in Figure 14.58, you can see that it is very close to the value provided by the curve at the halfway point between $\tau_0$ and $\tau_1$.

***PRACTICE PROBLEM 14.26***

Refer back to Practice Problem 14.25. Assume that the peak circuit current for the circuit is 80 mA and determine the value of $I_t$ when $t = 10.6$ ms.

As you can see, the time ratio in the universal curve equation allows us to determine the circuit current at any point during the decay (or rise) cycle. (We could not make this type of calculation when we assumed that $x$ simply represented a whole number time interval.)

## Defining the Time Constant ($\tau$)

OBJECTIVE 17 ▶

**Time constant ($\tau$)**
The time required for the current in a switching circuit to increase (or decrease) by 63.2% of its maximum possible $\Delta I$.

You have been shown how to calculate the value of the time constant for a series *RL* circuit. You have also been shown how this value is used to calculate the current at any point on the rise (or decay) curve. However, we haven't really discussed what a time constant *is*. Technically defined, a **time constant** is the time required for the current in a switching circuit to increase (or decrease) by 63.2% of its maximum possible $\Delta I$. This definition can be explained with the aid of Figure 14.59.

The rise curve shown in the figure was derived using the approach demonstrated in Example 14.26; that is, the current ratio for each time interval was determined and then used to calculate the value of $I_t$. At any time interval, the maximum possible $\Delta I$ equals the difference between its value at the start of the time interval and its value at the end of the transition. For example, the circuit current at $\tau_0$ in the figure is 0 mA. The current at the end of the transition is shown to be approximately 100 mA. Therefore, the maximum possible $\Delta I$ (at $\tau_0$) is found as

$$\Delta I_{max} = 100 \text{ mA} - 0 \text{ mA} = 100 \text{ mA} \quad (\text{at } \tau_0)$$

During the first time constant, the circuit current increases by 63.2% of $\Delta I_{max}$. Therefore, the current at $\tau_1$ is found as

$$I_t = I_{\tau 0} + 0.632(\Delta I_{max}) = 0 \text{ mA} + 0.632(100 \text{ mA}) = 0 \text{ mA} + 63.2 \text{ mA} = 63.2 \text{ mA}$$

From $\tau_1$ to $\tau_2$, the current again changes by 63.2% of $\Delta I_{max}$. However, the current at the start of the time interval has a value of $I_t = 63.2$ mA. Therefore, the value of $\Delta I_{max}$ is found as

$$\Delta I_{max} = 100 \text{ mA} - 63.2 \text{ mA} = 36.8 \text{ mA} \quad (\text{at } \tau_1)$$

Using this value of $\Delta I_{max}$, the current at $\tau_2$ is found as

$$\begin{aligned} I_t = I_{\tau 1} + 0.632(\Delta I_{max}) &= 63.2 \text{ mA} + 0.632(36.8 \text{ mA}) \\ &= 63.2 \text{ mA} + 23.3 \text{ mA} \\ &= 86.5 \text{ mA} \end{aligned}$$

As you can see, this value corresponds to the value shown on the curve when $t = \tau_2$. Continuing this series of calculations provides the results given in Table 14.11. In each case, the circuit current at the end of the time interval matches the value shown in the curve (Figure 14.59). As the current calculations demonstrate, the current change from one time

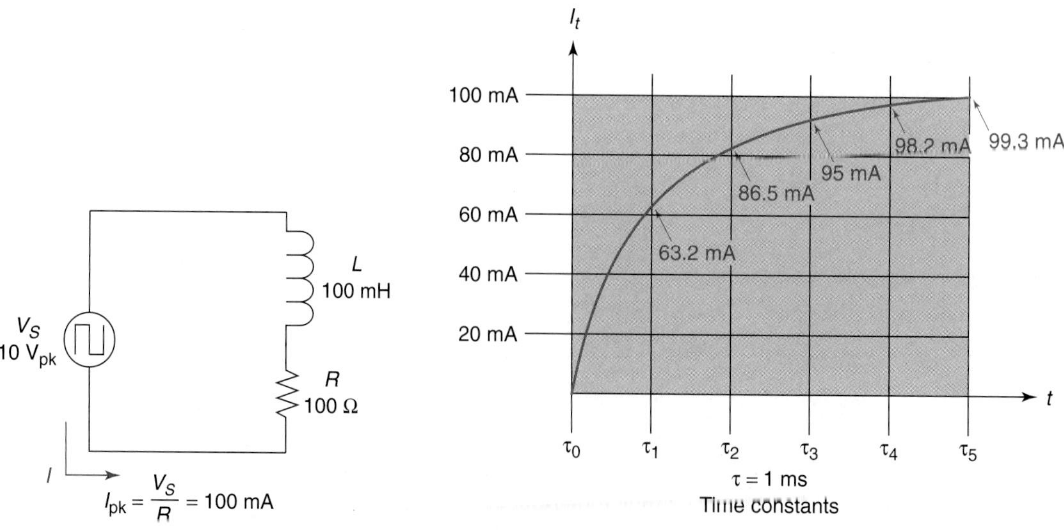


**FIGURE 14.59 Current values at each time constant of an *RL* circuit.**

**TABLE 14.11 Current Calculations for the Circuit in Figure 14.59**

| *Time Interval* | *Current Values* |
|---|---|
| $\tau_2 \to \tau_3$ | $I_{\tau 2} = 86.5$ mA<br>$\Delta I_{max} = 100$ mA $-$ 86.5 mA = 13.5 mA<br>$I_{\tau 3} = 86.5$ mA + 0.632(13.5 mA) = 95 mA |
| $\tau_3 \to \tau_4$ | $I_{\tau 3} = 95$ mA<br>$\Delta I_{max} = 100$ mA $-$ 95 mA = 5 mA<br>$I_{\tau 4} = 86.5$ mA + 0.632(5 mA) = 98.2 mA |
| $\tau_4 \to \tau_5$ | $I_{\tau 4} = 98.2$ mA<br>$\Delta I_{max} = 100$ mA $-$ 98.2 mA = 1.8 mA<br>$I_{\tau 5} = 86.5$ mA + 0.632(1.8 mA) = 99.3 mA |

constant to the next equals 63.2% of the maximum possible current transition. This is the basis for the definition of the time constant.

The current relationship described here also applies to the decay curve of an *RL* switching circuit. However, the current during each time interval decreases by 63.2% of the maximum possible current transition. Demonstrating this principle through calculations (like those in Table 14.11) is left as a problem at the end of the chapter in the section called "The Brain Drain."

## Time Calculations

Sometimes it is desirable to know how long it will take for the current in an *RL* switching circuit to reach a designated value. For example, take a look at the delay circuit shown in Figure 14.60. This circuit uses the current rise of the *RL* circuit to provide a delay between the positive-going transitions of $V_S$ and $V_{out}$. ◄ *OBJECTIVE 18*

The purpose served by a delay circuit like the one in Figure 14.60 is covered in a course on *digital* (switching) circuits. However, we can use the circuit to demonstrate the concept of a *signal delay.* The switching circuit is designed to switch between two output levels: 0 V and +5 V. The output from the circuit depends on the voltages at its signal inputs (*A* and *B*). The only time that the output from the switching circuit is at +5 V is when both of its input voltages ($V_A$ and $V_B$) are greater than or equal to +3 V. With this relationship in mind, let's look at the waveforms in Figure 14.60.

At the instant that $V_S$ makes its positive-going transition from 0 V to +5 V, there is no voltage across the resistor. As the circuit current increases, $V_R$ also increases (since $V_R = I_t R$). When $V_R$ reaches +3 V, the switching circuit has the following input voltages:

$$V_A = V_S = +5\text{ V} \qquad V_B = V_R = +3\text{ V}$$

These two voltages meet the requirements for a +5 V output from the switching circuit, so $V_{out}$ makes its positive-going transition. The time difference between the positive-going transitions of $V_S$ and $V_{out}$ is shown as the signal delay in the waveforms.

The analysis of a circuit like the one in Figure 14.60 requires that you know how long it takes for $V_R$ to reach its desired level. The time required for $V_R$ to reach a specified level can be found as

$$t = -\tau \ln\left(1 - \frac{V_R}{V_S}\right) \qquad \textbf{(14.29)}$$

where ln is the natural log function. Equation (14.29) is derived in Appendix E. Its use is demonstrated in Example 14.27.

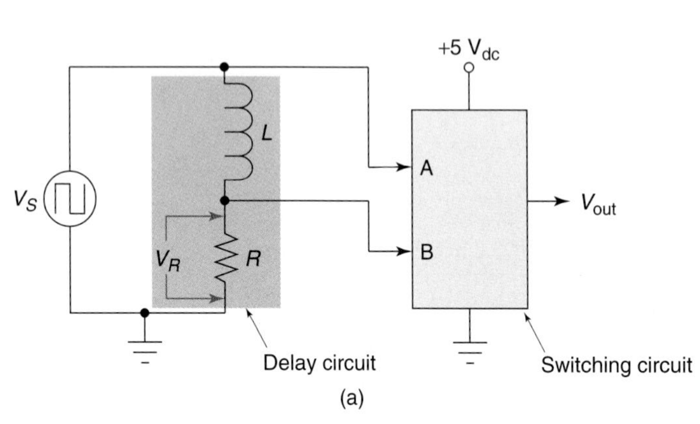


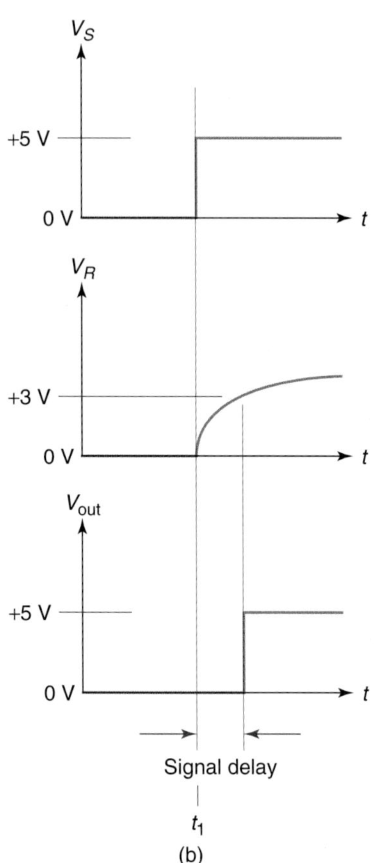


**FIGURE 14.60 An *RL* delay circuit and waveforms.**

## EXAMPLE 14.27

The circuit in Figure 14.60a has values of $L = 330$ mH and $R = 1.5$ kΩ. Determine the duration of the signal delay shown in the waveforms.

***Solution:*** First, the time constant of the *RL* circuit is found as

$$\tau = \frac{L}{R} = \frac{330 \text{ mH}}{1.5 \text{ k}\Omega} = \mathbf{220\ \mu s}$$

As the waveforms show, $V_{out}$ makes its transition when $V_R = +3$ V and $V_S = +5$ V. The time required for the $V_R$ transition can be found as

$$t = -\tau \ln\left(1 - \frac{V_R}{V_S}\right) = -(220\ \mu\text{s}) \ln\left(1 - \frac{3\text{ V}}{5\text{ V}}\right) = -(220\ \mu\text{s}) \ln(0.4) \cong \mathbf{202\ \mu s}$$

This result indicates that $V_{out}$ will make its positive-going transition approximately 202 μs after $V_S$ makes its positive-going transition.

### *PRACTICE PROBLEM 14.27*

A circuit like the one in Figure 14.60a has the following values: $L = 470$ mH, $R = 4.7$ kΩ, and $V_S = +10\ V_{pk}$. Determine the time required for $V_R$ to reach a value of $+8$ V.

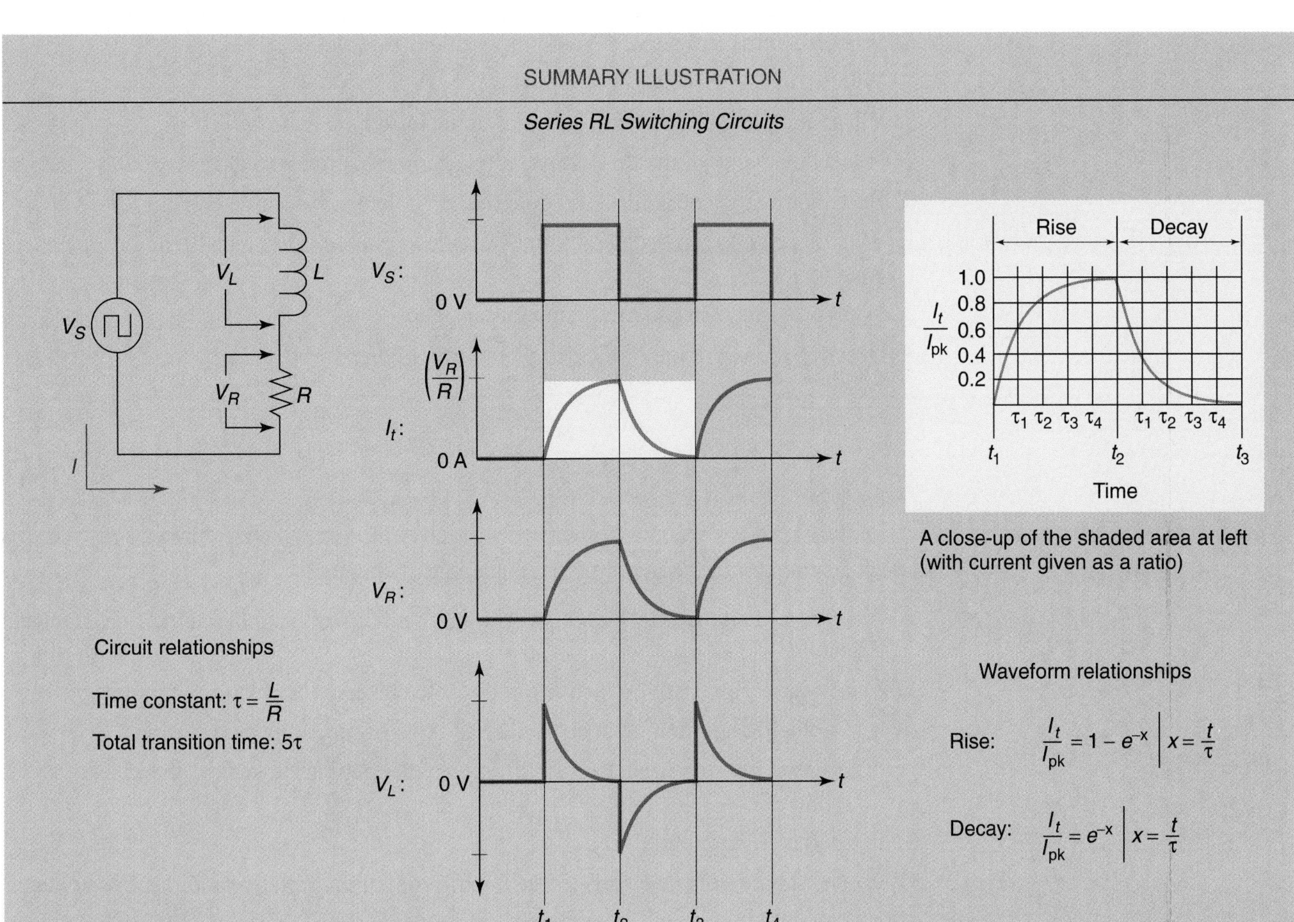


EWB

**FIGURE 14.61**

## One Final Note

This section introduced you to some of the basic principles of *RL* switching circuits and principles. Many of the principles and relationships introduced in this section are summarized in Figure 14.61. As your electronics education continues, you'll learn that the principles covered in this section are involved in a wide variety of applications.

### Section Review

1. Define each of the following terms: rectangular waveform, pulse width, space width, square wave, switching circuit.
2. What is a transition? What are the two types of transitions for a rectangular waveform?
3. Refer to Figure 14.48. What causes the shape of the $I$ and $V_R$ transitions?
4. Explain the shape of the $V_L$ waveform in Figure 14.49.
5. What is the relationship between the shape of a current rise (or decay) curve and the peak circuit current?

6. What is a universal curve? What purpose does it serve?
7. What is an exponential curve?
8. What is implied by the word *constant* in the term *time constant*?
9. How is the value of the time constant in a series *RL* circuit found?
10. How many time constants are required to complete a rise (or decay) curve?
11. What is actually represented by the variable exponent ($x$) in the rise and decay curve equations?
12. What is a time constant? Explain your answer.
13. Describe the operation of the delay circuit in Figure 14.60.

## Chapter Summary

Here is a summary of the major points made in this chapter:

1. A resistive-inductive (*RL*) circuit is one that contains any combination of resistors and inductors.
2. Many values in resistive-reactive circuits must be added as phasor quantities.
   a. Geometric addition must be used to add two or more phasor quantities.
   b. For two phasors ($a$ and $b$) at 90° angles, the geometric sum of the phasors ($c$) is found as $c = \sqrt{a^2 + b^2}$.
3. In a series *RL* circuit:
   a. The source voltage equals the geometric sum of the resistive and reactive voltages.
   b. The total impedance equals the geometric sum of resistance and reactance.
4. The values of source voltage and total impedance in a series *RL* circuit each have a phase angle that is normally expressed as part of the value.
5. Inductor voltage leads resistor voltage by 90° in a series *RL* circuit.
6. The source phase angle ($\angle\theta$) is the phase difference between the source voltage and the circuit current.
   a. Source voltage leads current in a series *RL* circuit.
   b. The source phase angle falls within the limits of $0° < \theta < 90°$.
7. Since circuit current is in phase with resistor voltage, $\theta$ can be measured by measuring the phase difference between $V_S$ and $V_R$. (The process for measuring the source phase angle is demonstrated in Example 14.2.)
8. The total impedance in a series circuit equals the geometric sum of $X_L$ and $R$.
9. The impedance phase angle in a series *RL* circuit equals the voltage phase angle.
10. Inductive reactance leads circuit current by 90° in a series *RL* circuit.
11. The term *frequency response* is used to describe any changes that occur in a circuit as a result of a change in operating frequency. The responses of a series *RL* circuit to an increase in operating frequency are listed in Table 14.1. (The responses of a series *RL* circuit to a decrease in operating frequency are described immediately following the table.)
12. In any resistive-reactive circuit:
   a. Resistive power (or true power) is measured in watts. $P_R$ is the power dissipated in the circuit.
   b. Reactive power (or imaginary power) is measured in volt-amperes-reactive (VAR).
   c. Apparent power (the geometric sum of resistive and reactive power) is measured in volt-amperes (VA).

13. The effect of inductor winding resistance ($R_w$) on the value of apparent power is usually negligible.
14. Apparent power leads the circuit current in a series *RL* circuit.
15. The power factor for a series *RL* circuit is the ratio of resistive power to apparent power.
    a. The power factor for an *RL* circuit is found as PF $= \cos\theta$.
    b. When PF and $Z_T$ are known, total circuit resistance can be found as $R_T = Z_T \cos\theta$.
16. The complete mathematical analysis of a series *RL* circuit is demonstrated in Example 14.13.
17. A parallel *RL* circuit contains one or more resistors in parallel with one or more inductors, each branch containing only one component.
18. In a parallel *RL* circuit:
    a. The total circuit current equals the geometric sum of the currents through the resistive and the reactive branches.
    b. The inductor current lags the source voltage by 90°.
    c. The resistor current is in phase with the circuit voltage.
19. The current phase angle ($\angle\theta$) in a parallel *RL* circuit is the phase difference between $I_T$ and $V_S$.
    a. Since voltage is assumed to have an angle of 0° in a parallel circuit, the current phase angle is always negative.
    b. For a parallel *RL* circuit, the current phase angle (relative to the source voltage) has limits of $-90° < \theta < 0°$.
20. The product-over-sum method for calculating total impedance is preferred for parallel *RL* circuits. (The geometric sum of $X_L$ and *R* is used in the denominator.)
21. The impedance phase angle in a parallel *RL* circuit is always the negative equivalent of the current phase angle.
22. The impedance phase angles for series and parallel *RL* circuits are calculated using reciprocal fractions, as follows:

| Series | Parallel |
|---|---|
| $\theta = \tan^{-1}\dfrac{X_L}{R}$ | $\theta = \tan^{-1}\dfrac{R}{X_L}$ |

23. Measuring the current phase angle in a parallel *RL* circuit requires the use of a sensing resistor.
    a. A sensing resistor is a low-value resistor used to produce a low-value voltage that is in phase with the circuit current.
    b. To prevent the sensing resistor from seriously affecting the current phase angle, the value of the resistor should be selected according to the following guideline: $R_{sensing} << Z/10$. (Remember that the value of the sensing resistor must be sufficient to generate a measurable voltage at the value of circuit current.)
24. A complete mathematical analysis of a parallel *RL* circuit is provided in Example 14.18.
25. The responses of a parallel *RL* circuit to an increase in operating frequency are listed in Table 14.5. The responses to a decrease in operating frequency are provided immediately following the table.
26. Analyzing a series-parallel circuit is simply a matter of:
    a. Combining series components and values according to the rules of series circuits.
    b. Combining parallel components and values according to the rules of parallel circuits.

27. The analysis of a series-parallel circuit with resistive and reactive components is more complicated because of the phase angles involved.
    a. Geometric addition is used to determine the sum of two phasors that are at 90° angles.
    b. Rectangular notation is used when adding two phasors that are at any angle other than 90° (or 0°).
28. To add two phasors that are at angles other than 90° (or 0°):
    a. Convert both phasors to rectangular form.
    b. Add the two rectangular values.
    c. Convert the result back to polar form (if needed).
29. When the total impedance of a parallel *RL* circuit is converted to rectangular form, the new value is the series equivalent of the parallel impedance network.
30. Rectangular waveforms consist of alternating dc levels.
31. The alternations of a rectangular waveform are referred to as the pulse width (PW) and the space width (SW).
    a. The combination of PW and SW is the cycle time of the waveform.
    b. A square wave is a rectangular waveform with equal PW and SW values.
32. Any circuit with a square (or rectangular) wave input is generally referred to as a switching circuit.
33. The change from one dc level to another in a switching circuit is referred to as a transition.
34. In an *RL* switching circuit, the inductor opposes any change in current. As a result, there is a measurable delay between the transition of the source voltage and the complete transition of the circuit current (see Figure 14.48).
35. The current transitions in an *RL* switching circuit are referred to as the rise and decay curves.
36. The $V_L$ curve in an *RL* switching circuit is shown in Figure 14.49.
37. A linear waveform is one that changes at a constant rate. The rise and decay curves for an *RL* switching circuit are not linear because the rate of change decreases as the transition progresses.
38. A universal curve is one that can be used to predict the operation of a specified type of circuit.
    a. The universal rise curve for an *RL* switching circuit is an exponential curve because its value is determined by a variable exponent.
    b. The equation for the rise portion of the universal curve (Figure 14.52) is $I_t/I_{pk} = 1 - e^{-x}$. In this case, $x$ is the variable exponent that determines the value of the current ratio.
39. The decay curve is simply a mirror image of the rise curve.
    a. The equation for the decay portion of the universal curve is $I_t/I_{pk} = e^{-x}$.
    b. Again, $x$ is the variable exponent that determines the value of the current ratio.
40. Each time interval in the universal curve represents a real-time value referred to as a time constant ($\tau$). The term *constant* is used because its value is independent of the circuit current magnitude and operating frequency.
41. Current rises (or decays) to its final value in five time constants.
42. The exponent ($x$) in the curve equations is actually a time ratio. This ratio is given as $x = t/\tau$, where $t =$ the time from the start of the transition.
43. Technically defined, a time constant is the time required for the current in a switching circuit to increase (or decrease) by 63.2% of its maximum possible transition.

| Equation Number | Equation | Section Number | Equation Summary |
|---|---|---|---|
| **(14.1)** | $V_S = \sqrt{V_L^2 + V_R^2}$ | 14.1 | |
| **(14.2)** | $\theta = \tan^{-1} \dfrac{V_L}{V_R}$ | 14.1 | |
| **(14.3)** | $0° < \theta < 90°$ | 14.1 | |
| **(14.4)** | $\theta = (360°)\dfrac{t}{T_C}$ | 14.1 | |
| **(14.5)** | $Z = \sqrt{X_L^2 + R^2}$ | 14.1 | |
| **(14.6)** | $\theta = \tan^{-1} \dfrac{X_L}{R}$ | 14.1 | |
| **(14.7)** | $V_n = V_S \dfrac{Z_n}{Z_T}$ | 14.1 | |
| **(14.8)** | $P_{\text{APP}} = \sqrt{P_X^2 + P_R^2}$ | 14.2 | |
| **(14.9)** | $\theta = \tan^{-1} \dfrac{P_X}{P_R}$ | 14.2 | |
| **(14.10)** | $\text{PF} = \cos\theta$ | 14.2 | |
| **(14.11)** | $\cos\theta = \dfrac{P_R}{P_{\text{APP}}}$ | 14.2 | |
| **(14.12)** | $P_R = P_{\text{APP}} \cos\theta$ | 14.2 | |
| **(14.13)** | $R_T = Z_T \cos\theta$ | 14.2 | |
| **(14.14)** | $I_T = \sqrt{I_L^2 + I_R^2}$ | 14.3 | |
| **(14.15)** | $\theta = \tan^{-1} \dfrac{-I_L}{I_R}$ | 14.3 | |
| **(14.16)** | $-90° < \theta < 0°$ | 14.3 | |
| **(14.17)** | $Z_T = \dfrac{V_S}{I_T}$ | 14.3 | |
| **(14.18)** | $Z_T = \dfrac{X_L R}{\sqrt{X_L^2 + R^2}}$ | 14.3 | |
| **(14.19)** | $\theta = \tan^{-1} \dfrac{R}{X_L}$ | 14.3 | |
| **(14.20)** | $R = Z\cos\theta$ | 14.4 | |
| **(14.21)** | $X_L = Z\sin\theta$ | 14.4 | |
| **(14.22)** | $V_L = V_S - V_R$ | 14.5 | |
| **(14.23)** | $I_t = \left(\dfrac{I_t}{I_{\text{pk}}}\right) I_{\text{pk}}$ | 14.5 | |

| | | |
|---|---|---|
| **(14.24)** | $\frac{I_t}{I_{pk}} = 1 - e^{-x}$ (rise curve) | 14.5 |
| **(14.25)** | $\frac{I_t}{I_{pk}} = e^{-x}$ (decay curve) | 14.5 |
| **(14.26)** | $\tau = \frac{L}{R}$ | 14.5 |
| **(14.27)** | $T = 5\tau$ | 14.5 |
| **(14.28)** | $x = \frac{t}{\tau}$ | 14.5 |
| **(14.29)** | $t = -\tau \ln\left(1 - \frac{V_R}{V_S}\right)$ | 14.5 |

## Key Terms

The following terms were introduced and defined in this chapter:

| | | |
|---|---|---|
| current-sensing resistor | linear | rise |
| decay | power factor (PF) | time constant ($\tau$) |
| exponential curve | resistive-inductive (*RL*) circuit | transition |
| frequency response | | universal curve |

## Practice Problems

1. The component voltages shown in Figure 14.62a were taken using an ac voltmeter. Determine the magnitude and phase angle of $V_S$ for the circuit.
2. The component voltages shown in Figure 14.62b were taken using an ac voltmeter. Determine the magnitude and phase angle of $V_S$ for the circuit.
3. Calculate the phase angle between the waveforms in Figure 14.63a.
4. Calculate the phase angle between the waveforms in Figure 14.63b.
5. Calculate the magnitude and phase angle of the circuit impedance for Figure 14.64a.
6. Calculate the magnitude and phase angle of the circuit impedance for Figure 14.64b.
7. Calculate the values of $Z_T$, $I_T$, $V_L$, $V_R$, and $\theta$ for the circuit in Figure 14.65a.
8. Calculate the values of $Z_T$, $I_T$, $V_L$, $V_R$, and $\theta$ for the circuit in Figure 14.65b.
9. Calculate the values of $Z_T$, $I_T$, $V_{L1}$, $V_{L2}$, $V_R$, and $\theta$ for the circuit in Figure 14.66a.
10. Calculate the values of $Z_T$, $I_T$, $V_{L1}$, $V_{L2}$, $V_{R1}$, $V_{R2}$, and $\theta$ for the circuit in Figure 14.66b.

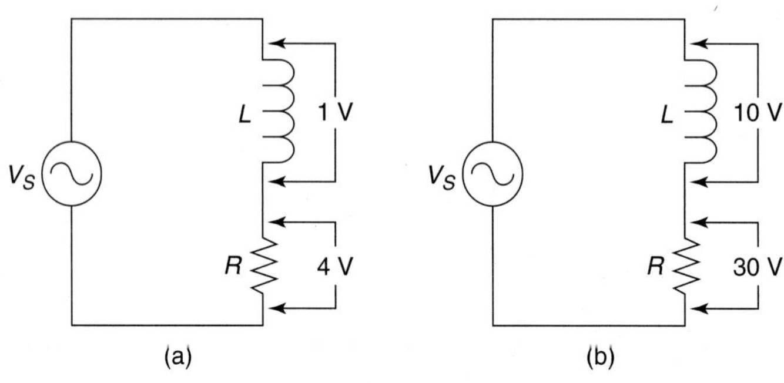


**FIGURE 14.62**

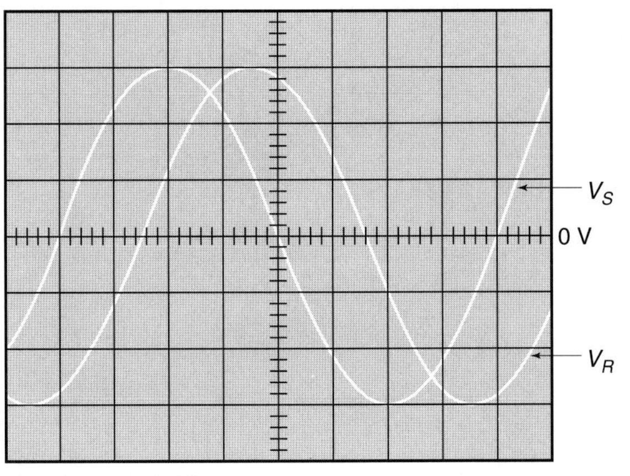


Time base: 5 ms/div

(a)

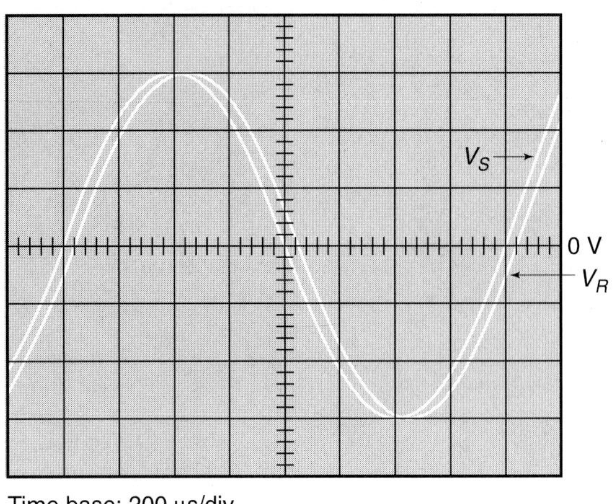


Time base: 200 μs/div

(b)

**FIGURE 14.63**

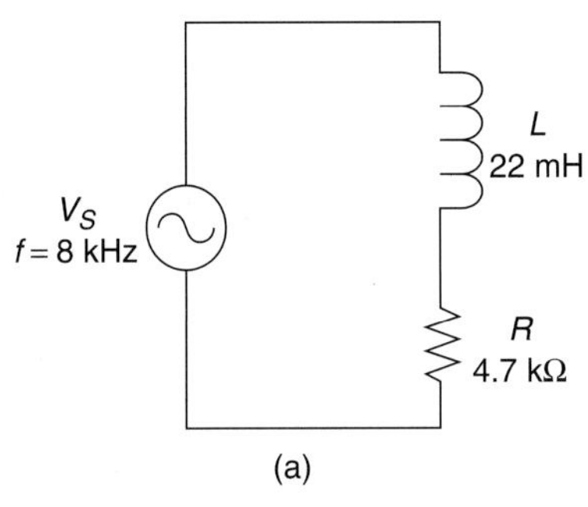


(a)

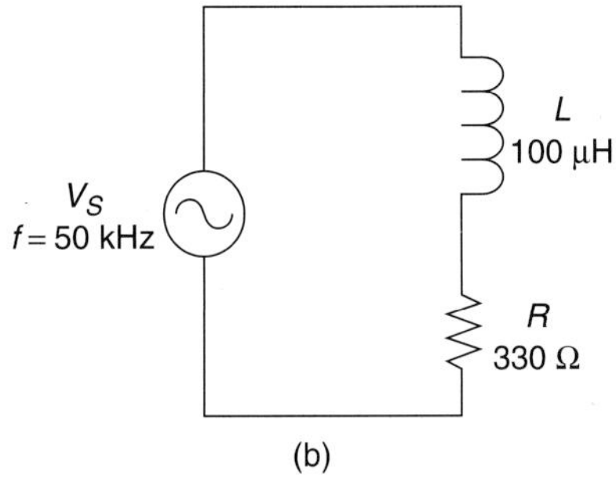


(b)

**FIGURE 14.64**

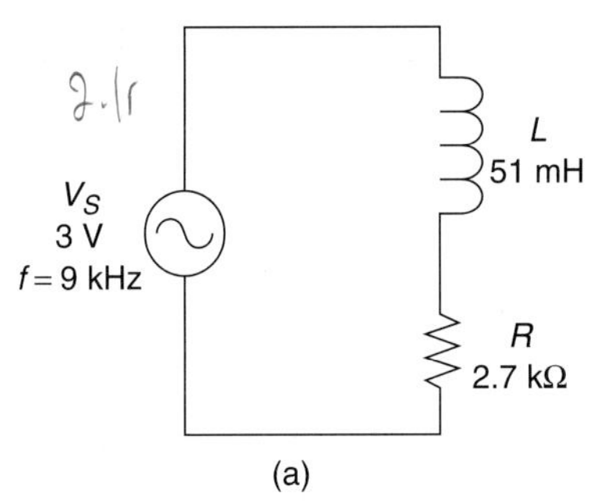


(a)

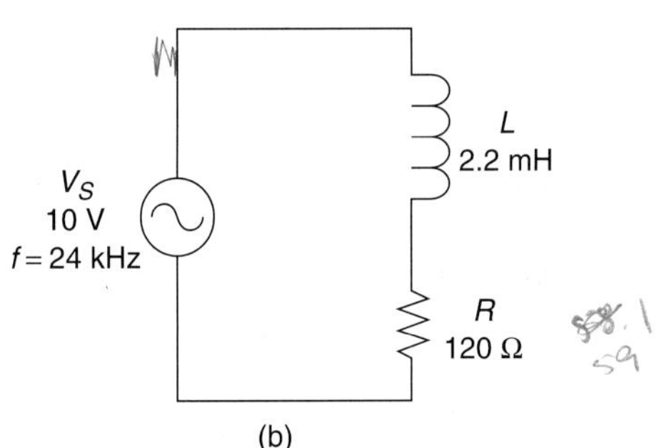


(b)

**FIGURE 14.65**

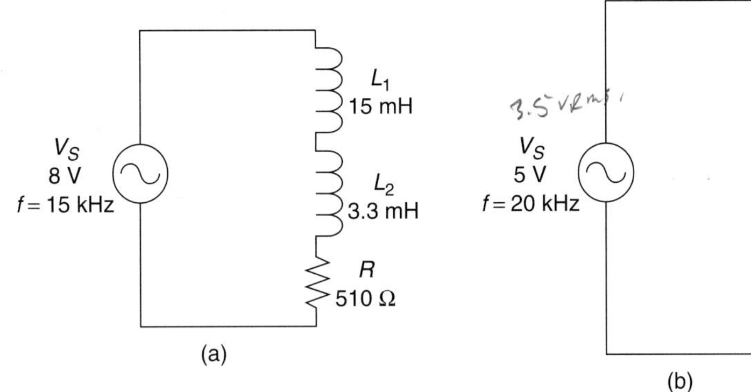


(b)

**FIGURE 14.66**

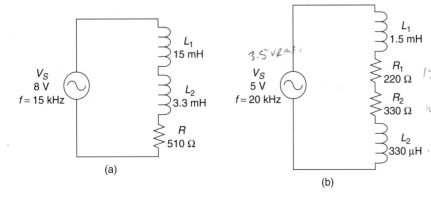

FIGURE 14.67

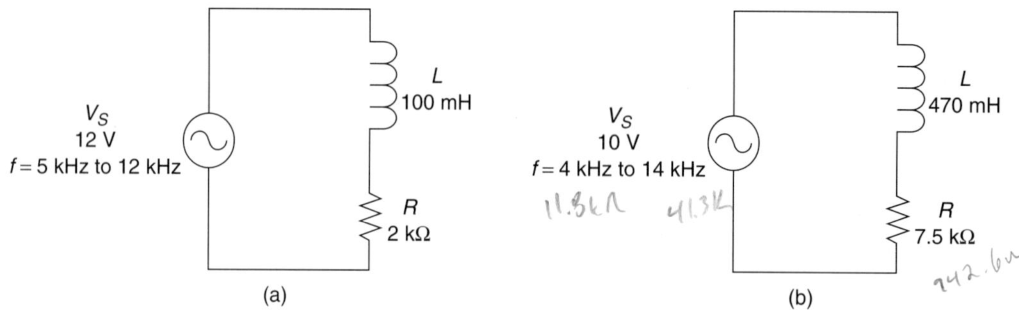


**11.** The voltage source in Figure 14.67a has the frequency limits shown. Calculate the range of $Z_T$, $V_L$, $V_R$, and $\theta$ values for the circuit.

**12.** The voltage source in Figure 14.67b has the frequency limits shown. Calculate the range of $Z_T$, $V_L$, $V_R$, and $\theta$ values for the circuit.

**13.** Refer to Figure 14.65a. Assume that the inductor is an ideal component and calculate the values of resistive and reactive power for the circuit.

**14.** Refer to Figure 14.65b. Assume that the inductor is an ideal component and calculate the values of resistive and reactive power for the circuit.

**15.** Refer to Figure 14.67a. Assume that the inductor has a winding resistance of 10 Ω and calculate the value of resistive power for the circuit when it is operating at its maximum source frequency.

**16.** Refer to Figure 14.67b. Assume that the inductor has a winding resistance of 800 mΩ and calculate the value of resistive power for the circuit when it is operating at its minimum source frequency.

**17.** Calculate the magnitude and phase angle of apparent power for the circuit described in Problem 13.

**18.** Calculate the magnitude and phase angle of apparent power for the circuit described in Problem 14.

**19.** Refer to Figure 14.63a. The current for this circuit was measured at 100 μA. Determine the values of $P_{APP}$, $\theta$, and $P_R$ for the circuit. Assume $V_S$ (only) is being measured at a 100 mV/div vertical sensitivity.

**20.** Refer to Figure 14.63b. The current for this circuit was measured at 50 mA. Determine the values of $P_{APP}$, $\theta$, and $P_R$ for the circuit. Assume $V_S$ (only) is being measured at a 2 V/div vertical sensitivity.

**21.** Determine the value of $R$ for the circuit in Problem 19.

**22.** Determine the value of $R$ for the circuit in Problem 20.

**23.** Perform a complete ac analysis of the circuit shown in Figure 14.68a.

**24.** Perform a complete ac analysis of the circuit shown in Figure 14.68b.

FIGURE 14.68

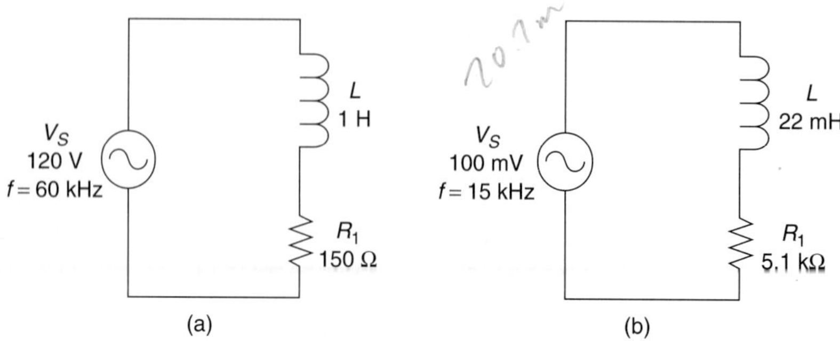

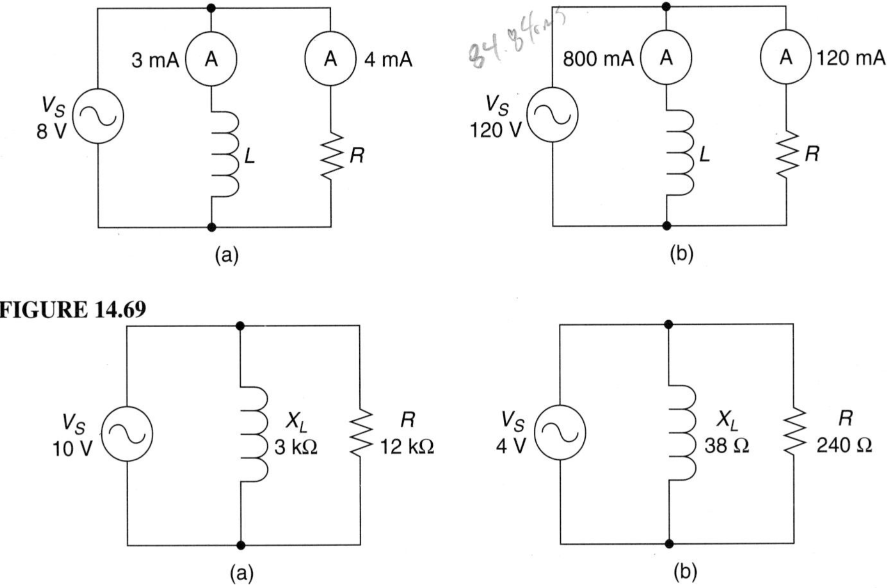


**FIGURE 14.69**

**FIGURE 14.70**

**25.** A parallel *RL* circuit has values of $I_L = 30$ mA and $I_R = 50$ mA. Calculate the magnitude and phase angle of the circuit current.

**26.** A parallel *RL* circuit has values of $I_L = 240$ mA and $I_R = 180$ mA. Calculate the magnitude and phase angle of the circuit current.

**27.** Calculate the total impedance for the circuit shown in Figure 14.69a.

**28.** Calculate the total impedance for the circuit shown in Figure 14.69b.

**29.** Calculate the magnitude and phase angle of the circuit impedance in Figure 14.70a.

**30.** Calculate the magnitude and phase angle of the circuit impedance in Figure 14.70b.

**31.** Perform a complete ac analysis of the circuit in Figure 14.71a.

**32.** Perform a complete ac analysis of the circuit in Figure 14.71b.

**33.** Perform a complete ac analysis of the circuit in Figure 14.72a.

**34.** Perform a complete ac analysis of the circuit in Figure 14.72b.

**35.** The voltage source in Figure 14.73a has the frequency limits shown. Calculate the circuit current and impedance values at the frequency limits of the source.

**36.** The voltage source in Figure 14.73b has the frequency limits shown. Calculate the circuit current and impedance values at the frequency limits of the source.

**37.** Refer to Problem 29. Express the circuit impedance in rectangular form.

**38.** Refer to Problem 30. Express the circuit impedance in rectangular form.

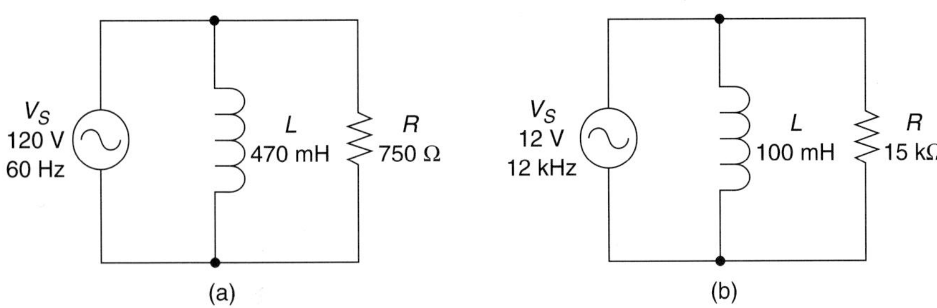


**FIGURE 14.71**

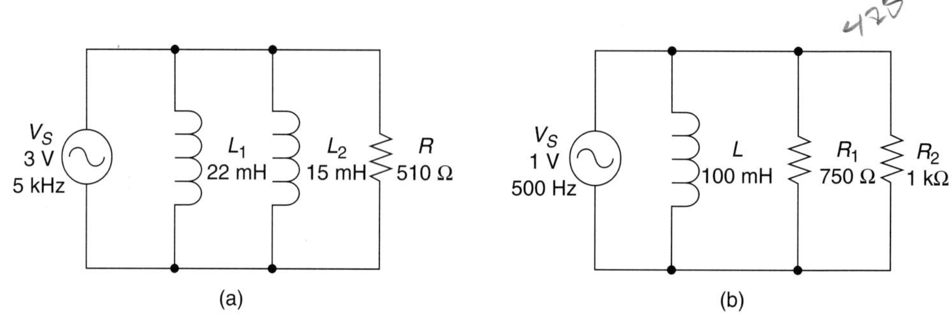


FIGURE 14.72

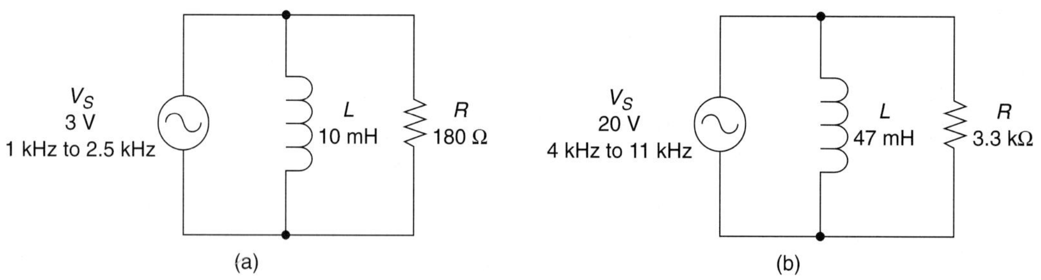


FIGURE 14.73

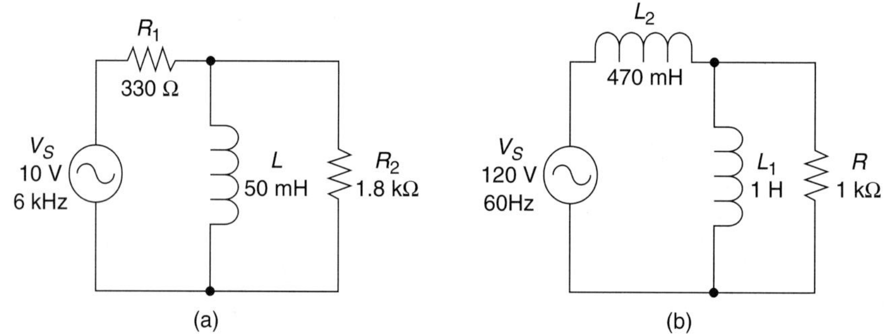


FIGURE 14.74

**39.** Perform a complete ac analysis on the circuit shown in Figure 14.74a.

**40.** Perform a complete ac analysis on the circuit shown in Figure 14.74b.

**41.** Determine the time constant for the circuit in Figure 14.75a.

**42.** Determine the time constant for the circuit in Figure 14.75b.

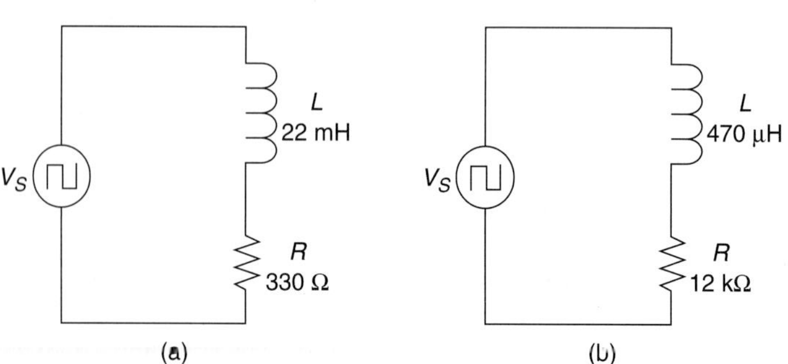


FIGURE 14.75

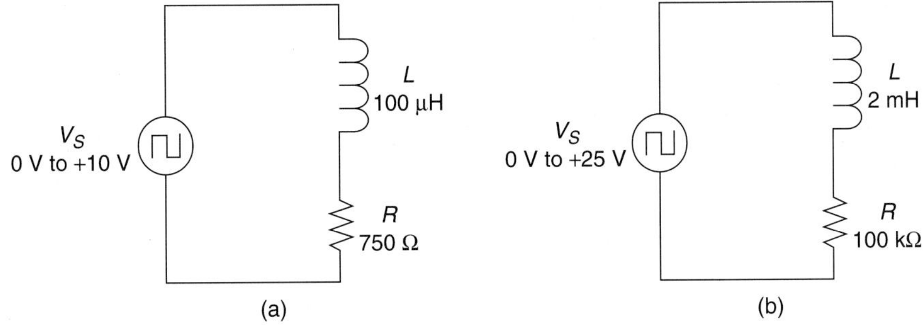


**FIGURE 14.76**

**43.** Determine the time required for the current in Figure 14.76a to reach its peak value.

**44.** Determine the time required for the current in Figure 14.76b to reach its peak value.

**45.** An *RL* switching circuit has values of $R = 150\ \Omega$, $L = 330$ mH, and $I_{pk} = 20$ mA. Determine the value of the circuit current ($I_t$) at the following times:

**a.** 5 ms after the start of the current rise.

**b.** 150 μs after the start of the current decay.

**46.** An *RL* switching circuit has values of $R = 33\ \Omega$, $L = 100\ \mu$H, and $I_{pk} = 75$ mA. Determine the value of the circuit current ($I_t$) at the following times:

**a.** 1.4 μs after the start of the current rise.

**b.** 2 μs after the start of the current decay.

**47.** Refer to Figure 14.76a. Determine the time required for $V_R$ to reach +4 V on the rise curve.

**48.** Refer to Figure 14.76b. Determine the time required for $V_R$ to reach +18 V on the rise curve.

---

*The Brain Drain*

**49.** Determine the potentiometer setting that will provide a circuit current of 50 mA in Figure 14.77. [*Hint:* Transpose equation (14.5) to provide an equation for *R*.]

**50.** Determine the *Q* of the inductor in Figure 14.78. [*Hint:* Determine the winding resistance of the coil.)

**51.** Using the decay curve and curve equation in Figure 14.54 and a series of calculations like those in Table 14.11, show that current decreases by 63.2% of its maximum $\Delta I$ during each time interval.

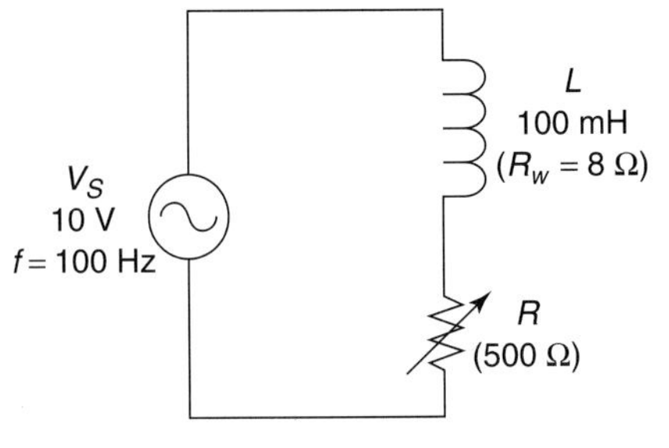

L
100 mH
($R_w$ = 8 Ω)
$V_S$
10 V
$f$ = 100 Hz
R
(500 Ω)

**FIGURE 14.77**

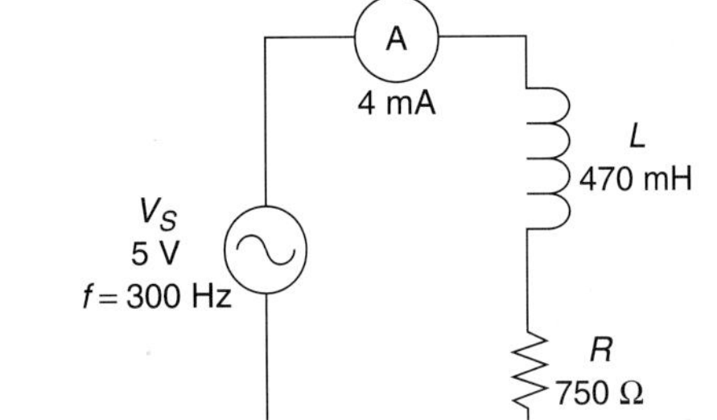


**FIGURE 14.78**

FIGURE 14.79

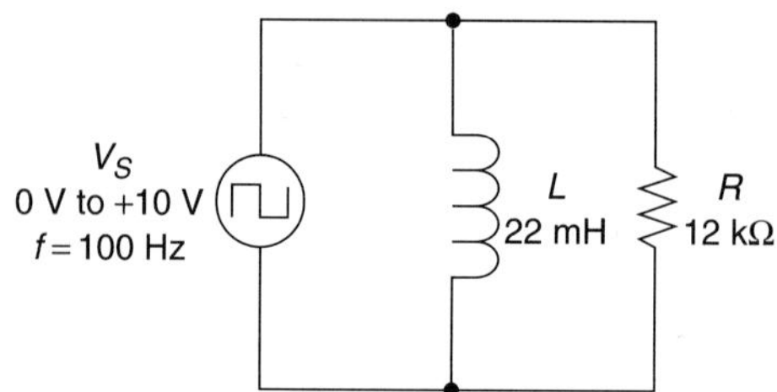


## Open for Discussion

52. Figure 14.79 shows a parallel *RL* switching circuit. This type of circuit was not discussed in this chapter simply because it would never be used in practice. Can you explain why?

## Answers to the Example Practice Problems

**14.1** 15 V∠53°
**14.2** 36°
**14.3** 903 Ω∠33.8°
**14.4** $X_L$ = 415 Ω, $Z_T$ = 425 Ω∠77.6°, $I_T$ = 235 μA, $V_L$ = 98 mV, $V_R$ = 21 mV
**14.5** $V_L$ = 2.21 V, $V_R$ = 4.49 V
**14.6** $Z_T$ = 1.02 kΩ∠12.5°, $V_{L1}$ = 1.77 V, $V_{L2}$ = 830 mV, $V_R$ = 11.7 V
**14.7** For $f$ = 3 kHz: $Z_T$ = 10 kΩ∠24.5°, $I_T$ = 1.2 mA, $V_L$ = 4.98 V, $V_R$ = 10.9 V. For $f$ = 30 kHz: $Z_T$ = 42.5 kΩ∠77.6°, $I_T$ = 283 μA, $V_L$ = 11.7 V, $V_R$ = 2.6 V.
**14.8** $P_X$ = 5 mVAR, $P_R$ = 9.1 mW
**14.9** 9.101 mW
**14.10** 10.38 VA∠28.8°
**14.11** $P_{APP}$ = 1.06 VA, $P_R$ = 1.01 W
**14.12** 25.2 Ω
**14.13** $X_L$ = 1.48 kΩ, $Z_T$ = 2.65 kΩ∠33.9°, $I_T$ = 3.02 mA, $V_S$ = 8 V∠33.9°, $V_R$ = 6.64 V, $V_L$ = 4.47 V, $P_R$ = 20.1 mW, $P_X$ = 13.5 mVAR, $P_{APP}$ = 24.2 mVA
**14.14** 16.49 mA∠−75.96°
**14.15** 485 Ω∠75.96°
**14.16** 485 Ω
**14.17** 75.96°
**14.18** $X_L$ = 1.38 kΩ, $I_L$ = 5.79 mA, $I_R$ = 1.5 mA, $I_T$ = 9.88 mA∠−35.9°, $Z_T$ = 810 Ω∠35.9°, $P_X$ = 46.3 mVAR, $P_R$ = 64 mW, $P_{APP}$ = 79 mVA
**14.19** $X_L$ = 2.49 kΩ, $I_L$ = 6.03 mA, $I_{R1}$ = 6.82 mA, $I_{R2}$ = 4.55 mA, $I_{RT}$ = 11.37 mA, $I_T$ = 12.87 mA∠−25.1°, $Z_T$ = 1.17 kΩ∠25.1°, $P_X$ = 90.5 mVAR, $P_{R1}$ = 102.3 mW, $P_{R2}$ = 68.3 mW, $P_{RT}$ = 170.6 mW, $P_{APP}$ = 193 mVA
**14.20** For $f$ = 1 kHz: $X_L$ = 2.95 kΩ, $I_L$ = 5.08 mA, $I_R$ = 1.5 mA, $I_T$ = 5.3 mA∠−73.5°, $Z_T$ = 2.83 kΩ∠73.5°. For $f$ = 8 kHz: $X_L$ = 23.6 kΩ, $I_L$ = 635 μA, $I_R$ = 1.5 mA, $I_T$ = 1.63 mA∠−22.9°, $Z_T$ = 9.2 kΩ∠22.9°
**14.21** (280 + $j$485) Ω
**14.22** 1.18 kΩ∠22.13°
**14.23** $X_L$ = 1.26 kΩ, $Z_P$ = 2.53 kΩ∠60.2°, $Z_T$ = 3.15 kΩ∠44.16°, $I_T$ = 3.17 mA∠−44.16°, $V_{R1}$ = 3.18 V, $V_P$ = 6.83 V, $I_L$ = 5.42 mA, $P_L$ = 37 mVAR, $I_{R2}$ = 3.1 mA, $P_{R2}$ = 21.2 mW, $P_{APP}$ = 31.7 mVA∠44.16°, $P_R$ = 22.7 mW
**14.24** 3.3 ms
**14.25** 16.5 seconds
**14.26** 37.9 mA
**14.27** 161 μs

## EWB Applications Problems

| Figure | EWB File Reference |
|---|---|
| 14.8 | EWBA14_8.ewb |
| 14.34 | EWBA14_34.ewb |
| 14.38 | EWBA14_38.ewb |
| 14.43 | EWBA14_43.ewb |
| *14.48 | EWBA14_48.ewb |
| *14.49 | EWBA14_49.ewb |
| 14.55 | EWBA14_55.ewb |
| *14.61 | EWBA14_61.ewb |

*These figures share a common file.

# 15 TRANSFORMERS

## OBJECTIVES

*After studying the material in this chapter, you should be able to:*

1. State the purpose served by a transformer.
2. Describe the construction of a typical transformer.
3. List the three basic types of transformers.
4. Briefly describe transformer operation in terms of electromagnetic induction.
5. Describe the turns ratio of a transformer and the purpose it serves.
6. Perform the voltage calculations for a transformer using its turns ratio.
7. Identify a transformer using its turns ratio.
8. Discuss the relationship between transformer input power and output power.
9. List and describe the power losses that occur within a typical transformer.
10. Perform the current calculations for a transformer using its turns ratio.
11. For each type of transformer (step-up, step-down, and isolation), describe the relationship between the input and output values of voltage, current, and power.
12. Describe the effect that a load has on transformer current values.
13. Describe the relationship between the input and output impedances of a transformer.
14. Calculate the impedance values for a transformer.
15. Perform a complete mathematical analysis of a loaded transformer.
16. Discuss the use of a transformer as an impedance matching component.
17. Discuss the voltage and current ratings of a transformer.
18. Describe the construction and operation of center-tapped and multiple-output transformers.
19. Discuss the operation of the autotransformer, along with its applications and limitations.
20. Discuss the need for transformer shielding.
21. List the common transformer faults and their symptoms.

In Chapter 13, you were introduced to the concept of mutual inductance: the process where the magnetic field produced by one coil induces a voltage across another coil. In this chapter, you will be introduced to the transformer, a component that operates on this principle.

# 15.1 TRANSFORMERS: AN OVERVIEW

◄ *OBJECTIVE 1*

A **transformer** is a two-coil component that uses electromagnetic induction to pass an ac signal from its input to its output, while providing dc isolation between the two. The ac and dc input/output characteristics of a transformer are illustrated in Figure 15.1. As you can see:

**Transformer**
A two-coil component that uses electromagnetic energy to pass an ac signal from its input to its output, while providing dc isolation between the two. (One exception to this definition is introduced in Section 15.4.)

1. An ac input to the transformer passes through to the output.
2. A dc input to the transformer is prevented from reaching the output.

When a component (or circuit) allows any signal or potential to pass from one point to another, it is said to provide *coupling* between the points. When a component (or circuit) prevents a signal or potential from passing between points, it is said to provide *isolation* between the two. Therefore, we can say that a transformer is a component that provides ac coupling and dc isolation between its input and output terminals. As you will see, this combination of characteristics makes the transformer extremely useful in many applications, such as power transmission and communications systems.

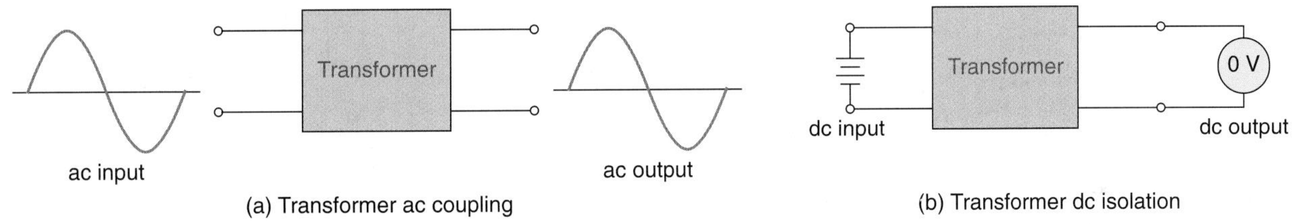


**FIGURE 15.1 Transformer input/output functions.**

## Transformer Construction

◄ *OBJECTIVE 2*

A transformer is made up of two coils, called the **primary** and the **secondary**. Several types of transformers are shown in Figure 15.2. Each of the transformers shown contains the primary and secondary coils, and a core that is typically made of air or iron. Note that the primary serves as the input to the transformer, and the secondary serves as its output.

**Primary**
The transformer coil that serves as the component input.

**Secondary**
The transformer coil that serves as the component output.

## Schematic Symbols

Figure 15.3 shows several commonly used transformer schematic symbols. In each case, the coils are identified by the labels on the input and output **terminals** (leads). $P_1$ and $P_2$ are the primary terminals. $S_1$ and $S_2$ are the secondary terminals.

**Terminals**
Another name for the leads on a component.

## Voltage Classifications

◄ *OBJECTIVE 3*

A transformer is commonly described in terms of the relationship between its input and output voltages. For example, Figure 15.4 shows three types of transformers, each with a unique relationship between the input and output voltages.

**FIGURE 15.2 Basic transformers.**

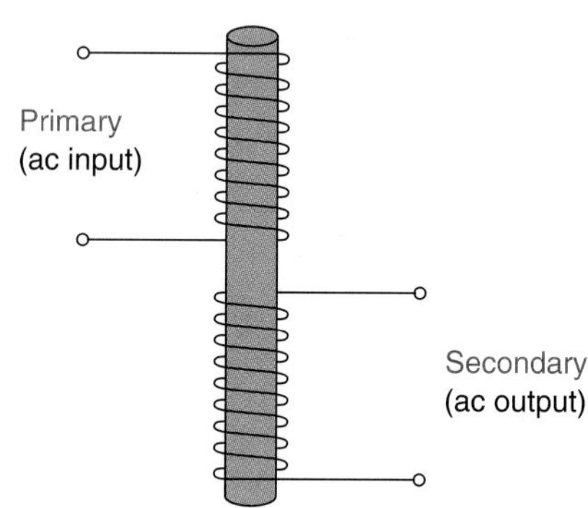


(a) Rf (radio frequency) transformer

*Note:* The primary and secondary designations for the autotransformer can be reversed. (Autotransformers are discussed in Section 15.4.)

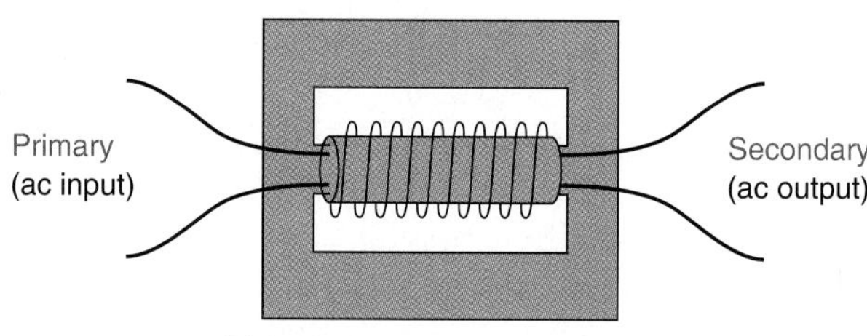


(b) Audio-frequency transformer

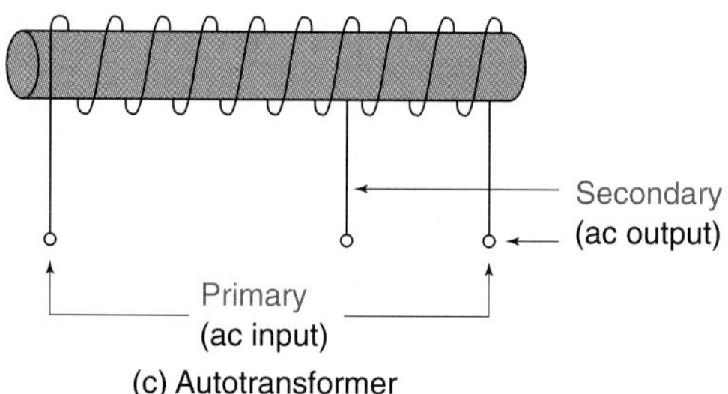


(c) Autotransformer

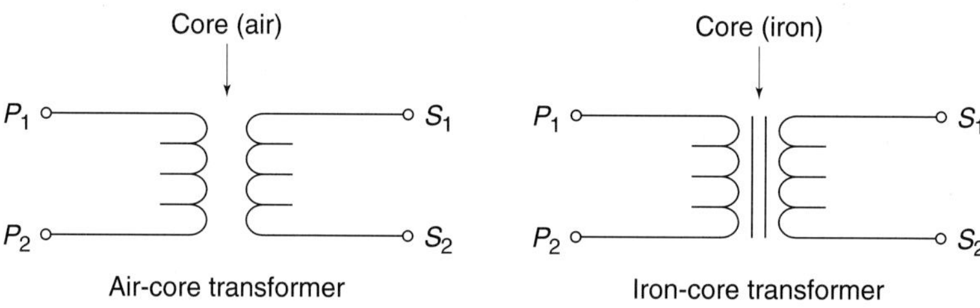


*Note:* The symbols for some other transformers are introduced (as needed) later in the chapter.

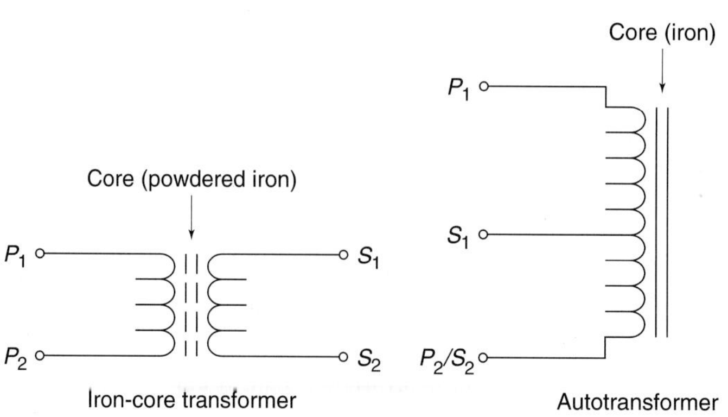


**FIGURE 15.3 Some common transformer schematic symbols.**

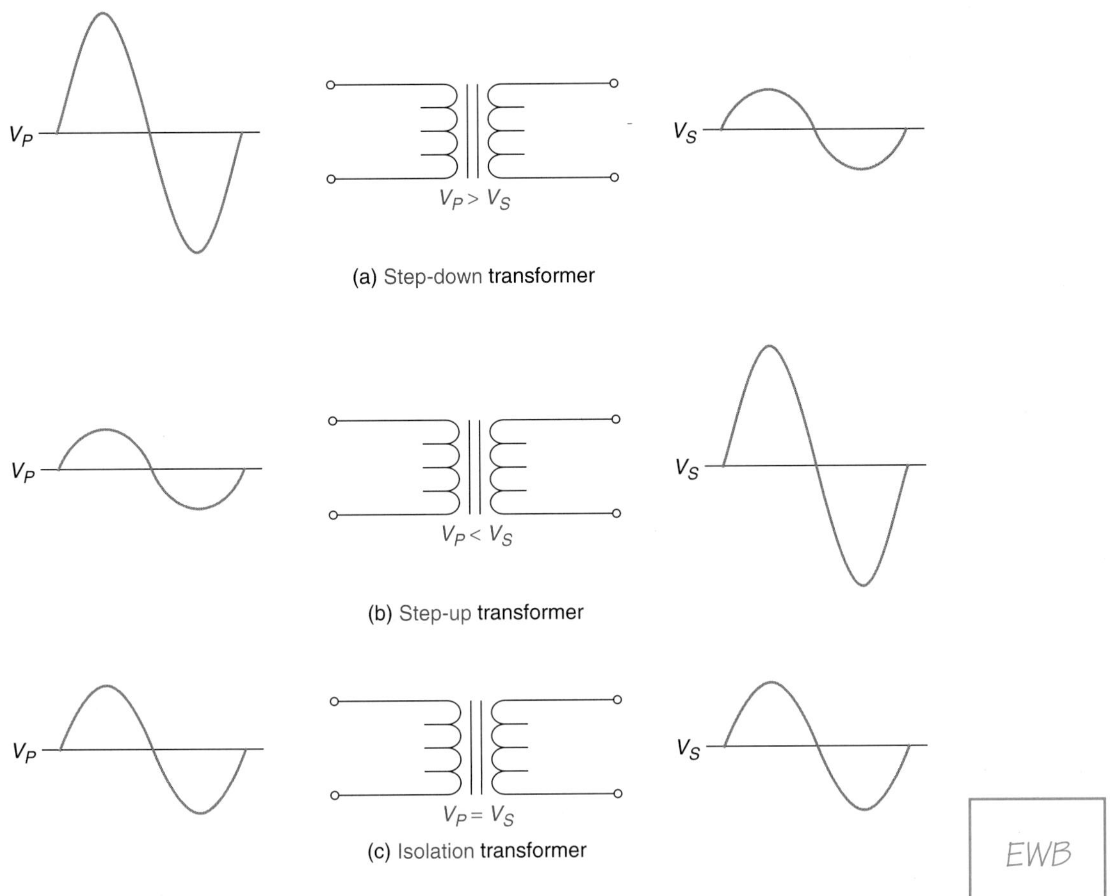


EWB

**FIGURE 15.4**

In Figure 15.4a, the secondary (output) voltage is less than the primary (input) voltage. This type of transformer is called a **step-down transformer.** In contrast, the **step-up transformer** shown in Figure 15.4b provides an output voltage that is greater than its input voltage. The **isolation transformer** shown in Figure 15.4c has equal input and output voltages. (Later in this section, we will discuss the input/output power and current relationships for each type of transformer.)

**Step-down transformer** A transformer with a secondary voltage that is less than the primary voltage.

**Step-up transformer** A transformer with a secondary voltage that is greater than the primary voltage.

**Isolation transformer** A transformer with equal secondary and primary voltages.

## Transformer Operation

◄ *OBJECTIVE 4*

Transformer operation is based on the principle of electromagnetic induction. This principle is illustrated (as a review) in Figure 15.5a. When there is a changing current in the $L_1$ circuit, a changing magnetic field is generated. This changing magnetic field cuts through $L_2$, inducing a voltage across that coil. It is important to note that:

1. Any voltage waveform induced across $L_2$ will have the same shape as the $L_1$ waveform. For example, if a sine wave is used to generate the changing current through $L_1$, the voltage induced across $L_2$ is also sinusoidal.
2. The inductors are not physically connected. Therefore, any dc voltage applied to $L_1$ is not coupled to $L_2$.

As these statements indicate, the two-inductor circuit couples an ac signal from one inductor to the other, while providing dc isolation between the two.

Figure 15.5b shows how this principle of electromagnetic induction applies to a transformer. When an ac input is applied to the primary, a changing current is generated

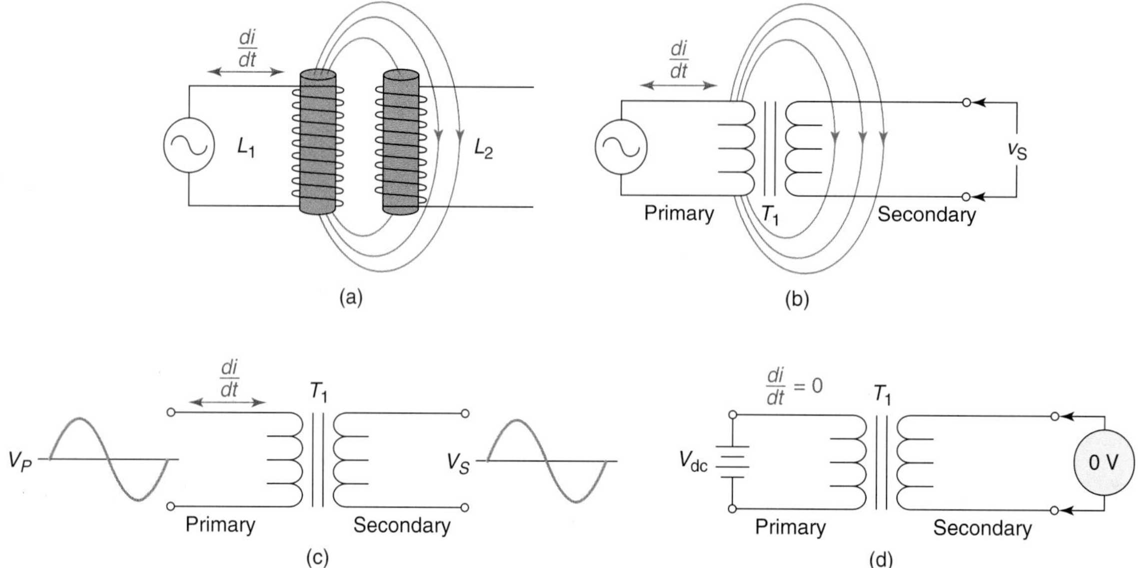


**FIGURE 15.5**

throughout the primary circuit. The resulting magnetic field cuts through the secondary coil, inducing a voltage across the secondary. This voltage is felt at the output terminals of the component ($S_1$ and $S_2$). Assuming that the ac input to the transformer is a sine wave, the output is also a sine wave, as shown in Figure 15.5c. As you can see, the ac input has been effectively coupled from the input of the transformer to its output.

If a dc voltage is applied to the primary of the transformer (as shown in Figure 15.5d), a changing current is not generated in the primary coil. As a result, the primary coil does not generate a changing magnetic field, no voltage is induced across the secondary coil, and the output from the component is 0 V. This is how the transformer provides dc isolation.

*OBJECTIVE 5* ▶

## Turns Ratio

**Turn**
Another name for a single loop of wire in a coil.

**Turns ratio**
The ratio of primary turns to secondary turns for a given transformer.

Each loop of wire in a coil is referred to as a **turn**. The **turns ratio** of a transformer is the ratio of primary turns to secondary turns. For example, the transformer in Figure 15.6 has 320 turns in the primary and 80 turns in the secondary. Using these values, the turns ratio of the component is found as

$$\frac{N_P}{N_S} = \frac{320 \text{ turns}}{80 \text{ turns}} = \frac{4}{1}$$

*A Practical Consideration:* Transformers are often described in terms of their turns ratios. For example, the transformer in Figure 15.6 may be referred to as a four-to-one transformer.

**FIGURE 15.6 Transformer turns ratio.**

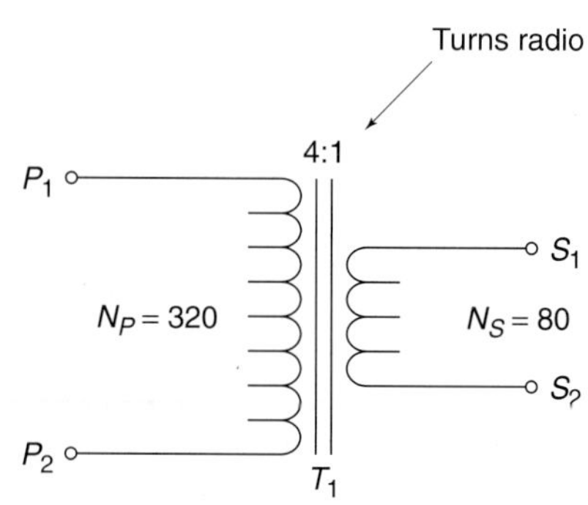

Note that the turns ratio of a transformer is normally written in the form shown in Figure 15.6. Also note that the turns ratio of a transformer is a function of its construction: in most cases, it cannot be changed.

The turns ratio of a transformer is important because it determines the ratio of input voltage to output voltage. By formula,

$$\frac{N_P}{N_S} = \frac{V_P}{V_S} \tag{15.1}$$

For the transformer in Figure 15.6, this equation indicates that the voltage across the primary is four times the voltage across the secondary when the component has an ac input. As such, the transformer shown in the figure is a step-down transformer.

In the upcoming sections, you will learn how the turns ratio of a transformer is used to determine more than just its output voltage. Current, impedance, and power calculations can also be made using this transformer rating.

**Section Review**

1. What is a transformer?
2. Describe the ac and dc input/output relationships for a transformer.
3. Describe the construction of a typical transformer.
4. Which coil serves as a transformer's input? Which serves as its output?
5. List and describe the three basic types of transformers.
6. Describe the process by which an ac input to the primary of a transformer causes a voltage to be induced across the secondary circuit.
7. What is the turns ratio of a transformer?
8. What is the relationship between the turns ratio of a transformer and its input/output voltage ratio?

# 15.2 TRANSFORMER VOLTAGE, CURRENT, AND POWER

You have seen that the relationship between primary and secondary voltage for a transformer is a function of the turns ratio of the component. In this section, we will discuss the voltage, current, and power characteristics of transformers. As you will see, the relationship between primary and secondary current is also a function of the component turns ratio. Power, on the other hand, is simply a function of voltage and current (as it has always been).

## Secondary Voltage ($V_S$)

◄ OBJECTIVE 6

If we solve equation (15.1) for secondary voltage ($V_S$), we get the following useful relationship:

$$V_S = V_P \frac{N_S}{N_P} \tag{15.2}$$

Equation (15.2) allows us to determine the secondary voltage when the primary voltage and turns ratio of a transformer are known. This point is demonstrated in Example 15.1.

## EXAMPLE 15.1

Determine the secondary voltage for the step-down transformer shown in Figure 15.7a.

**FIGURE 15.7**

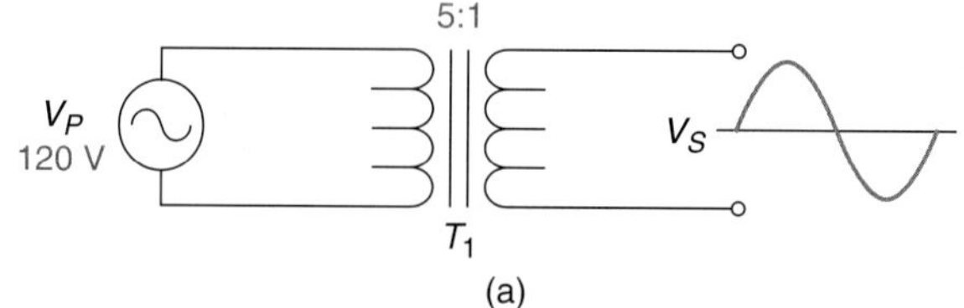

(a)

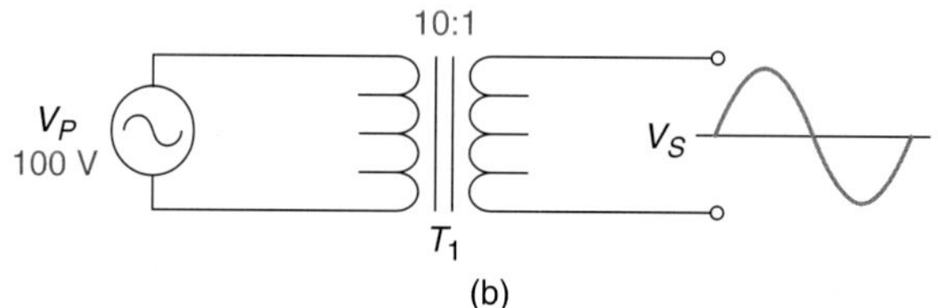

(b)

***Solution:*** The transformer shown has a 120 $V_{ac}$ input and a 5:1 turns ratio. Using these values, the ac secondary voltage is found as

$$V_S = V_P\frac{N_S}{N_P} = (120\ \text{V}_{\text{ac}})\frac{1}{5} = \mathbf{24\ V_{ac}}$$

***PRACTICE PROBLEM 15.1***

Determine the value of $V_S$ for the transformer shown in Figure 15.7b.

The step-down transformer in Example 15.1 causes a decrease in voltage from primary to secondary. The increase in voltage (from primary to secondary) produced by a step-up transformer is demonstrated in Example 15.2.

## EXAMPLE 15.2

Determine the secondary voltage for the step-up transformer shown in Figure 15.8a.

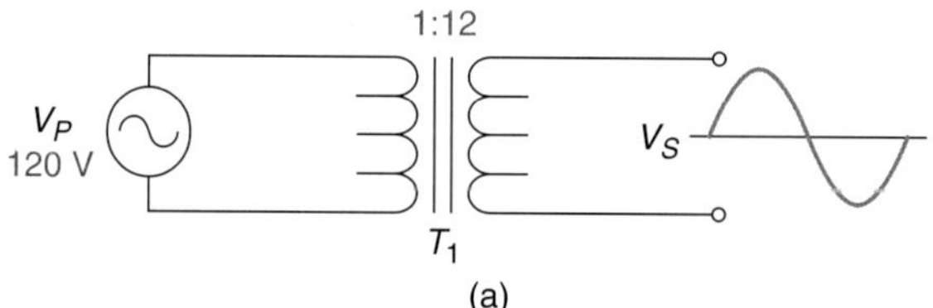

(a)

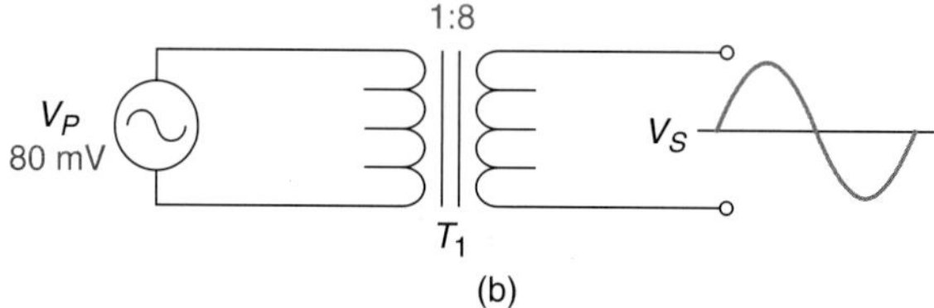

(b)

**FIGURE 15.8**

***Solution:*** The transformer shown has a 120 $V_{ac}$ input and a 1:12 turns ratio. Using these values, the secondary voltage is found as

$$V_S = V_P\frac{N_S}{N_P} = (120\ V_{ac})\frac{12}{1} = \mathbf{1.44\ kV}$$

***PRACTICE PROBLEM 15.2***

Determine the value of $V_S$ for the transformer shown in Figure 15.8b.

As you can see, the output voltage for the step-up transformer in Example 15.2 is significantly higher than its input voltage.

## Identifying a Transformer by Its Turns Ratio

◀ *OBJECTIVE 7*

If we compare the components analyzed in Examples 15.1 and 15.2, we can establish a means of identifying a transformer by its turns ratio. For convenience, the values of interest from the examples are summarized as follows:

| *Example* | *Type of Transformer* | $N_P$ | $N_S$ | $N_P$ *Versus* $N_S$ |
|---|---|---|---|---|
| 15.1 | Step-down | 5 | 1 | $N_P > N_S$ |
| 15.2 | Step-up | 1 | 12 | $N_P < N_S$ |

As these results indicate, the type of transformer is determined by the relationship between the number of primary and secondary turns. When $N_P > N_S$, the component is a step-down transformer. When $N_P < N_S$, the component is a step-up transformer.

As you were told earlier, an isolation transformer has equal input and output voltages. This input/output voltage relationship can exist only when $N_P = N_S$. Any transformer with a turns ratio of 1:1 is an isolation transformer.

## The Phase Relationship Between $V_P$ and $V_S$

The phase relationship between $V_P$ and $V_S$ depends on the wiring of the transformer. Some transformers are wired so that the primary and secondary voltages are in phase, while others are wired so that they are 180° out of phase. The input/output phase relationship for a transformer is commonly identified using **polarity dots** like those shown in Figure 15.9. Note that the input and output voltages are assumed to be in phase when no polarity dots are shown on a transformer symbol.

**Polarity dots**
Dots placed in the symbol of a transformer to indicate the phase relationship between the transformer input and output.

## Power Transfer

◀ *OBJECTIVE 8*

Under ideal conditions, all the power applied to the primary (input) of a transformer is transferred to the secondary (output). By formula,

$$P_P = P_S \quad \text{(ideal)} \qquad \textbf{(15.3)}$$

or

$$I_P V_P = I_S V_S \quad \text{(ideal)} \qquad \textbf{(15.4)}$$

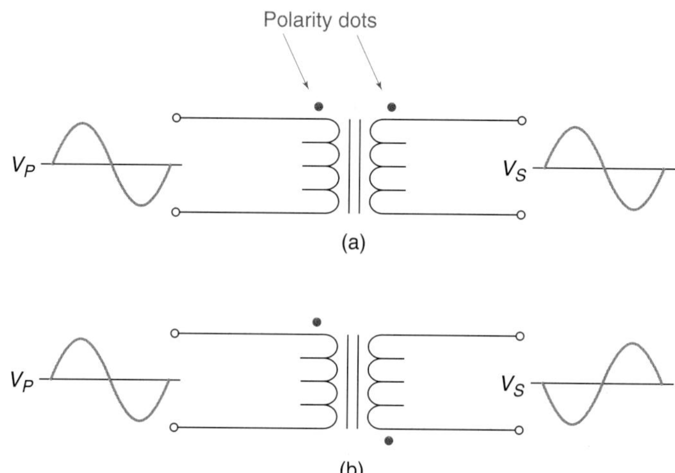


**FIGURE 15.9 Polarity dots.**

As you know, power is a measure of energy used (or transferred) per second. With this in mind, look at the transformer represented in Figure 15.10. Under ideal conditions, the resistance of the primary is 0 Ω, so it does not dissipate any power. Therefore, all the energy supplied to the transformer (by the source) is transferred to the magnetic flux. Assuming that $k = 1$ (which is also an ideal condition), all the energy in the magnetic flux is transferred to the secondary coil. The coil in the secondary then converts this energy back into electrical energy, which is delivered to the load. Thus, all the transformer input power is transferred (via the magnetic flux) to the load.

In Chapter 3, we discussed the concept of efficiency. You may recall that efficiency is the ratio of output power to input power, given as a percentage. The efficiency of a transformer is found as

$$\eta = \frac{P_S}{P_P} \times 100 \qquad (\%) \tag{15.5}$$

where
$\eta$ = the efficiency of the transformer (written as a percentage)
$P_S$ = the secondary (output) power
$P_P$ = the primary (input) power

Since the ideal transformer has equal input and output power values, its efficiency is 100%.

*OBJECTIVE 9* ▶ ***The Practical Transformer.*** In practice, no transformer is 100% efficient. The value of secondary power is always slightly lower than the value of primary power because of power losses that occur in any practical transformer. At this point, we will take a brief look at these losses.

**Copper loss**
Power dissipated by the winding resistance of the transformer coils. (Often called $I^2R$ *loss*.)

***Copper Loss.*** All wire has some measurable amount of resistance. Even though this resistance is very low (less than 100 Ω), it still causes some amount of power to be dissipated in both the primary and secondary coils. Note that **copper loss** is often referred to as $I^2R$ *loss*.

**FIGURE 15.10 Ideal power transfer.**

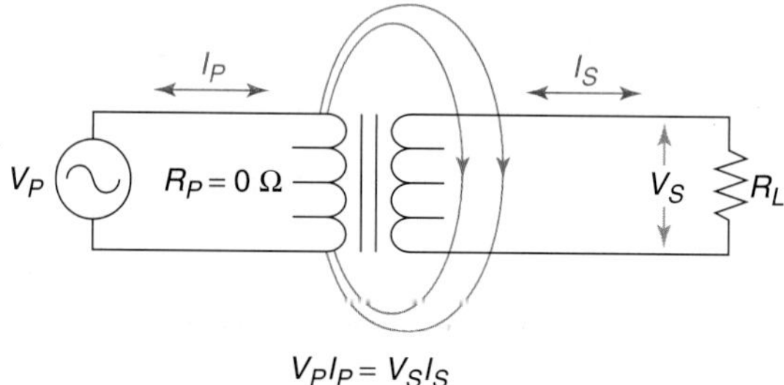

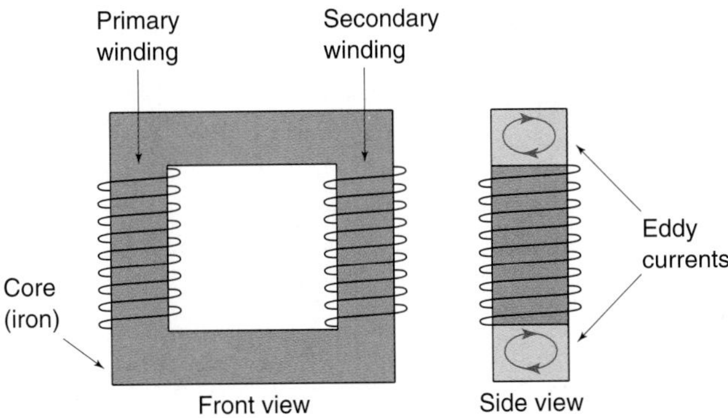


**FIGURE 15.11 Eddy currents in an iron core.**

***Loss Due to Eddy Currents.*** This type of loss is unique to iron-core transformers, like the one shown in Figure 15.11. When magnetic flux is generated in the primary, it passes through the iron core. Since iron is a conductor, the flux passing through the core generates a current within the core, called an **eddy current.** As shown in Figure 15.11, this current travels in a circular motion through the core. As eddy currents pass through an iron core, the resistance of the core causes some amount of power to be dissipated in the form of heat.

**Eddy current**
A circular flow of charge generated in the core of a transformer by a changing magnetic field. [The term *eddy* is normally used to describe anything (air, water, charge, etc.) that flows in a circular motion.]

**Hysteresis loss**
A loss of energy that occurs whenever a magnetic field of one polarity must overcome any residual magnetism from the previous polarity.

***Hysteresis Loss.*** **Hysteresis loss** is a loss of energy that occurs whenever a magnetic field of one polarity must overcome any residual magnetism from the previous polarity. To understand this type of energy loss fully, we need to review briefly the concepts of retentivity and residual flux, and the relationship between the two.

In Chapter 10, we defined retentivity as the ability of a material to retain its magnetic characteristics after a magnetizing force has been removed. For example, if an iron bar is placed in a magnetic field (as shown in Figure 15.12a), the iron becomes magnetized. Even after the magnetic field is removed, the iron bar maintains its magnetic characteristics for a relatively short time. The higher the retentivity of any material, the longer it maintains its magnetic characteristics after any magnetizing force is removed.

The magnetism that remains in a material after a magnetizing force has been removed is referred to as **residual flux.** Residual flux is represented by the $+B_0$ and $-B_0$ points on the curve in Figure 15.12b. As a review, here is a description of the highlighted portion of the hysteresis curve:

**Residual flux**
The magnetism that remains in a material after a magnetizing force has been removed. (This term was first defined in Chapter 10.)

1. The magnetizing force ($H$) reaches some maximum value in the positive direction (designated as $+H_M$). As a result, the flux density within the material also reaches its maximum value (designated as $+B_M$). The value of $+B_M$ depends on the material and the strength of the magnetizing force.
2. When the magnetizing force returns to zero (designated by $H_0$), the flux density within the material does not drop to zero. Rather, it drops to some level lower than $+B_M$ (designated as $+B_0$).

The value of $+B_0$ is determined (to an extent) by the retentivity of the material. For example, Figure 15.12c and d show how the shape of a hysteresis curve is affected by the retentivity of the material. As you can see, low-retentivity materials have much lower values of $B_0$ than high-retentivity materials (all other factors being equal).

When the input to a transformer changes from one polarity to the other, energy must be used to overcome the residual magnetism (hysteresis) in the core. For example, let's say that the input to the transformer in Figure 15.13a is changing from positive-going to negative-going. If $B_{0(+)}$ in the hysteresis curve (Figure 15.13b) represents the residual magnetism in the core, then the input must use energy to cause the change highlighted in the figure. The energy lost in causing this change in flux is hysteresis loss.

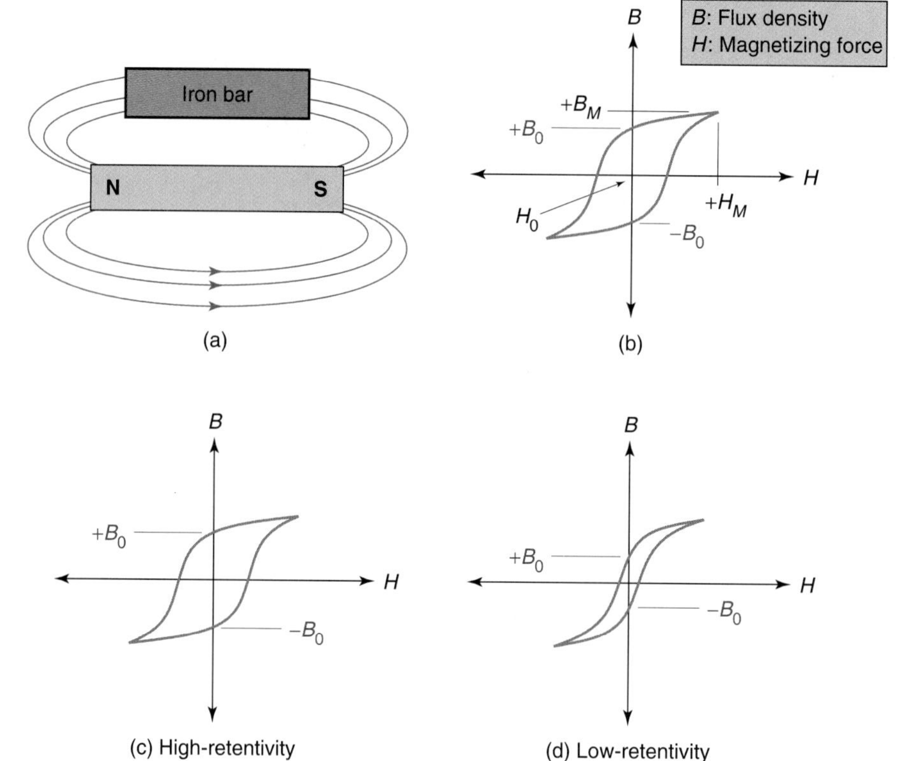


**FIGURE 15.12**

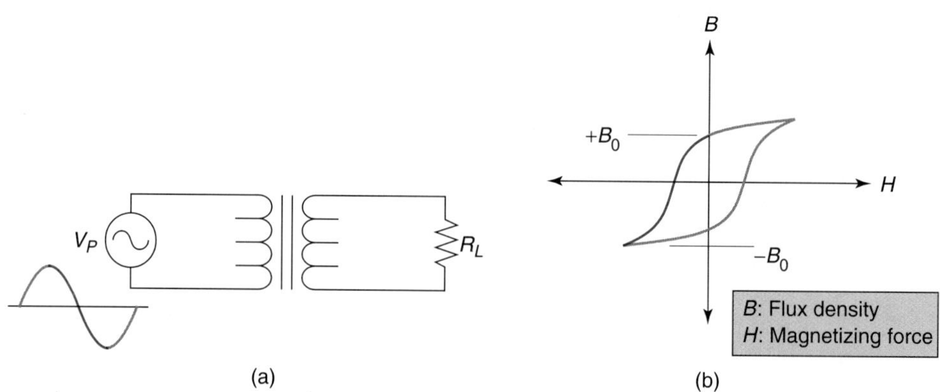


**FIGURE 15.13**

***Reducing Power Losses.*** Each of the losses described in this section are inherent to transformers. None can be eliminated completely, so transformer efficiency never reaches 100%. However, the effects of each type of loss can be reduced using the techniques listed in Table 15.1. Even though the losses described in this section are very real, each results in relatively little power loss. As a result, transformers tend to have high efficiency ratings, typically 90% or higher. For this reason, upcoming discussions are based on the ideal transformer; that is, we will assume that a transformer is 100% efficient (unless indicated otherwise).

**TABLE 15.1 Reducing Transformer Losses**

| *Type of Loss . . .* | *. . . Can Be Reduced by . . .* |
|---|---|
| Copper ($I^2R$) loss | Using a transformer with the largest practical diameter wire (to reduce wire resistance). |
| Eddy current loss | Using a **laminated core.** This type of core (shown in Figure 15.14) is broken up into thin layers, with an oxide coating between each. The insulation provided by the oxide layers prevents eddy currents from being generated (without interfering with magnetic coupling). |
| Hysteresis loss | Using low-retentivity materials in the core, such as air or silicon-steel (an alloy). Hysteresis loss can also be reduced by the use of a laminated core. |

**Laminated core**
A core made up of thin layers with an oxide coating between them. The oxide insulators in a laminated core prevent eddy currents from being established within the core.

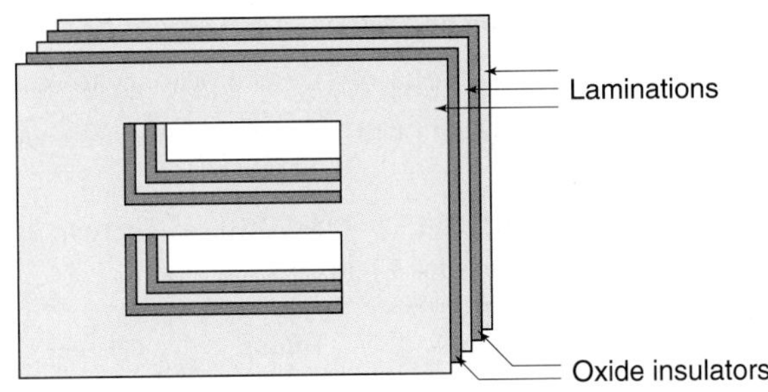


**FIGURE 15.14 A laminated core.**

## Transformer Input and Output Current

◀ *OBJECTIVE 10*

The relationship between transformer input and output power provides a basis for the relationship between transformer input current and output current. As you know, the ideal transformer has equal input and output power values. This relationship was described in equation (15.4) as

$$I_P V_P = I_S V_S$$

This relationship can be rewritten as

$$I_S = I_P \frac{V_P}{V_S} \tag{15.6}$$

Equation (15.6) is used to determine the value of $I_S$ in Example 15.3.

**EXAMPLE 15.3**

Determine the value of secondary current ($I_S$) for the transformer in Figure 15.15a.

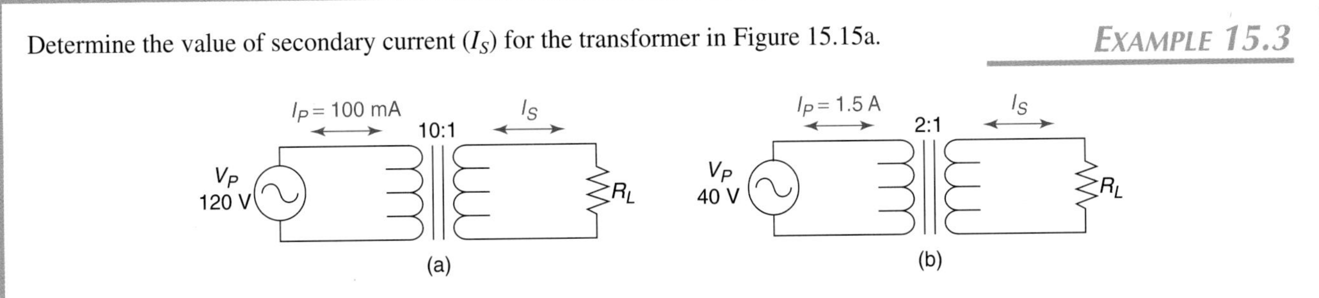


**FIGURE 15.15**

***Solution:*** First, the value of secondary voltage for the transformer is found as

$$V_S = V_P \frac{N_S}{N_P} = (120\ \text{V}_{\text{ac}}) \frac{1}{10} = \mathbf{12\ V_{ac}}$$

Now, using $V_S$ and the values shown in the figure, the value of $I_S$ can be found as

$$I_S = I_P \frac{V_P}{V_S} = (100 \text{ mA}) \frac{120 \text{ V}_{ac}}{12 \text{ V}_{ac}} = \mathbf{1\ A}$$

***PRACTICE PROBLEM 15.3***

Calculate the value of secondary current ($I_S$) for the transformer in Figure 15.15b.

Table 15.2 lists the voltage, current, and power values for the circuit analyzed in Example 15.3. Note that the power values were found as $P = IV$. Using the values shown in the table, we can make several observations about the circuit:

1. The primary and secondary values of power are equal to each other.
2. Voltage decreases by a factor of 10 from primary to secondary.
3. Current increases by a factor of 10 from primary to secondary.

**TABLE 15.2 The Voltage, Current, and Power Values for Figure 15.15a**

| *Circuit* | *Voltage* | *Current* | *Power* |
|---|---|---|---|
| Primary | 120 $V_{ac}$ | 100 mA | 12 W |
| Secondary | 12 $V_{ac}$ | 1 A | 12 W |

The first of these points simply reinforces the ideal power relationship described earlier in this section. The second two points, however, introduce a new and important concept: For any step-up or step-down transformer, *the change in current from input to output is inversely proportional to the change in voltage.* For a step-down transformer:

1. The output voltage is less than the input voltage.
2. The output current is greater than the input current.

For a step-up transformer:

1. The output voltage is greater than the input voltage.
2. The output current is less than the input current.

Note that voltage and current change by the same factor for any transformer. For example, if a step-down transformer decreases voltage by a factor of four, it also increases current by a factor of four. (The same principle applies to any step-up transformer.)

Earlier, you were told that voltage varies directly with the turns ratio of a transformer; that is, the change in voltage from input to output is determined by the turns ratio of the transformer. Since current and voltage vary inversely, current varies inversely with the turns ratio. By formula,

$$\frac{I_S}{I_P} = \frac{N_P}{N_S} \qquad \textbf{(15.7)}$$

If we solve this relationship for secondary current ($I_S$), we get

$$I_S = I_P \frac{N_P}{N_S} \qquad \textbf{(15.8)}$$

This equation can be used to demonstrate the relationship between the values of voltage and current for a step-up transformer, as shown in Example 15.4.

## EXAMPLE 15.4

Calculate the output voltage and current values for the step-up transformer shown in Figure 15.16a.

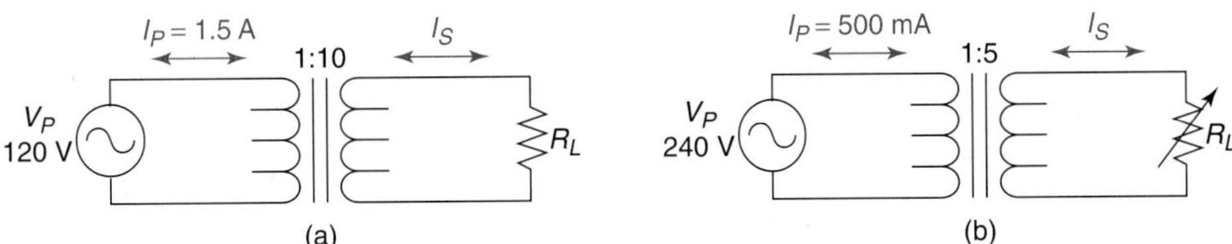


**FIGURE 15.16**

***Solution:*** The secondary voltage is found as

$$V_S = V_P \frac{N_S}{N_P} = (120\text{ V})\frac{10}{1} = \mathbf{1.2\ kV}$$

Using equation (15.8), the value of the secondary current can be found as

$$I_S = I_P \frac{N_P}{N_S} = (1.5\text{ A})\frac{1}{10} = \mathbf{150\ mA}$$

### *PRACTICE PROBLEM 15.4*

Calculate the secondary voltage and current values for the step-up transformer in Figure 15.16b.

If we compare the results from Examples 15.3 and 15.4, we get a clearer picture of the relationships among the turns ratio, voltage values, and current values for step-up and step-down transformers. For convenience, the results from the two examples are listed in Table 15.3. As shown in the table, the step-down transformer: ◄ *OBJECTIVE 11*

1. Has a turns ratio of 10:1.
2. Decreases voltage by a factor of 10.
3. Increases current by a factor of 10.

On the other hand, the step-up transformer:

1. Has a turns ratio of 1:10.
2. Increases voltage by a factor of 10.
3. Decreases current by a factor of 10.

**TABLE 15.3 The Results from Examples 15.3 and 15.4**

| *Transformer* | *Circuit* | *Turns*[a] | *Voltage* | *Current* | *Power* |
|---|---|---|---|---|---|
| Step-down | Primary | 10 | 120 V | 100 mA | 12 W |
| | Secondary | 1 | 12 V | 1 A | 12 W |
| Step-up | Primary | 1 | 120 V | 1.5 A | 180 W |
| | Secondary | 10 | 1.2 kV | 150 mA | 180 W |

[a]These are the numbers given in the turns ratio. They do not equal the actual number of turns in the primary and secondary windings.

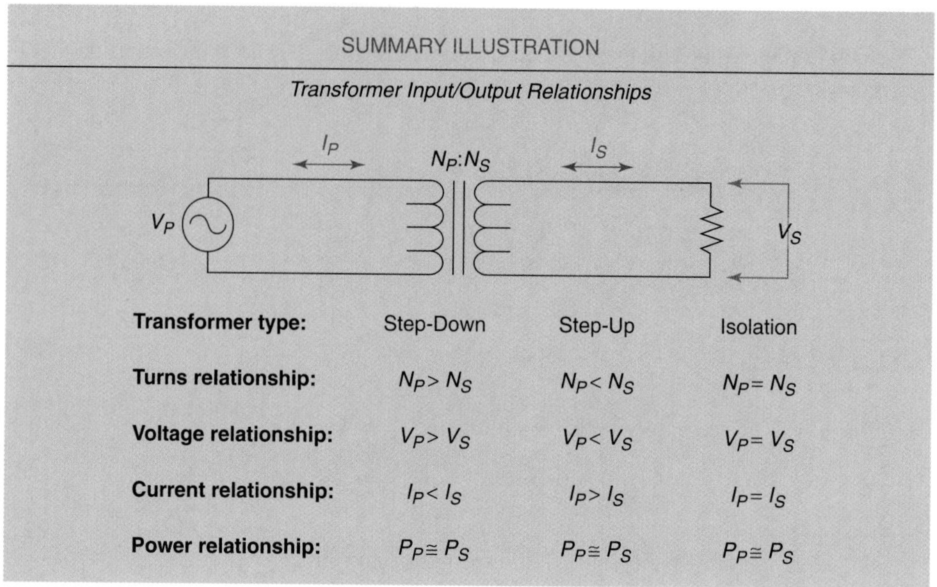

SUMMARY ILLUSTRATION
Transformer Input/Output Relationships
$I_P$
$N_P$:$N_S$
$I_S$
$V_P$
$V_S$
Transformer type: Step-Down Step-Up Isolation
Turns relationship: $N_P > N_S$ $N_P < N_S$ $N_P = N_S$
Voltage relationship: $V_P > V_S$ $V_P < V_S$ $V_P = V_S$
Current relationship: $I_P < I_S$ $I_P > I_S$ $I_P = I_S$
Power relationship: $P_P \cong P_S$ $P_P \cong P_S$ $P_P \cong P_S$


**FIGURE 15.17**

These results agree with the statements made earlier regarding the effect of a transformer's turns ratio on the input and output voltage and current values.

For future reference, the input/output relationships established in this section are summarized in Figure 15.17. Note that the input and output values for the isolation transformer are equal in every category. As a result, the isolation transformer is used strictly to provide dc isolation between two ac circuits without affecting the ac coupling from one to the other.

## Section Review

1. What is the relationship between $N_P$ and $N_S$ for a step-down transformer?
2. What is the relationship between $N_P$ and $N_S$ for a step-up transformer?
3. What is the relationship between $N_P$ and $N_S$ for an isolation transformer?
4. What purpose is served by the polarity dots on the symbol for a transformer?
5. What is assumed when no polarity dots are shown on the symbol for a transformer?
6. What is the ideal relationship between transformer primary power and secondary power?
7. What is the efficiency of a transformer?
8. Describe each of the following: copper loss, eddy current loss, hysteresis loss.
9. How are the losses in Question 8 reduced?
10. Describe the relationship between a transformer's voltage ratio and its current ratio.
11. Describe the relationship between a transformer's turns ratio and its current ratio.

## 15.3 LOAD EFFECTS

*OBJECTIVE 12* ▶ Up to this point, we have assumed a value of primary current ($I_P$) in all our circuits without regard to any value of load resistance. In practice, the resistance of the load determines (in part) the values of the primary and secondary currents. For example, consider the circuit shown in Figure 15.18. The secondary voltage is found using the turns ratio and

**FIGURE 15.18 Using Ohm's Law to calculate secondary current.**

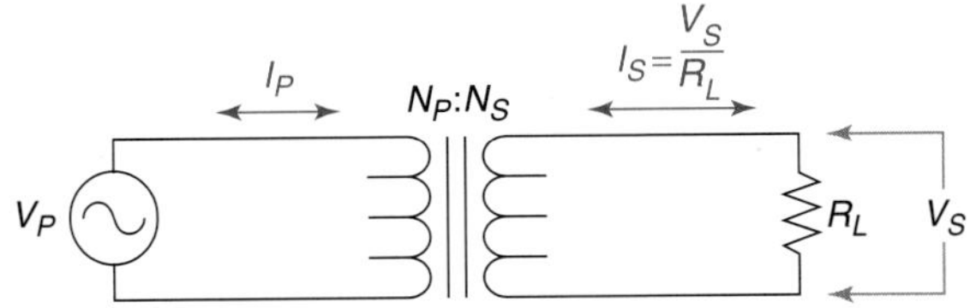


primary voltage (as always). Once the value of $V_S$ is known, the secondary current ($I_S$) is calculated using Ohm's law. Then, the primary current ($I_P$) is found using a transposed form of equation (15.7), as follows:

$$I_P = I_S \frac{N_S}{N_P} \quad \textbf{(15.9)}$$

The process for determining the value of primary current for a given transformer is demonstrated in Example 15.5.

## EXAMPLE 15.5

Determine the value of primary current ($I_P$) for the circuit shown in Figure 15.19a.

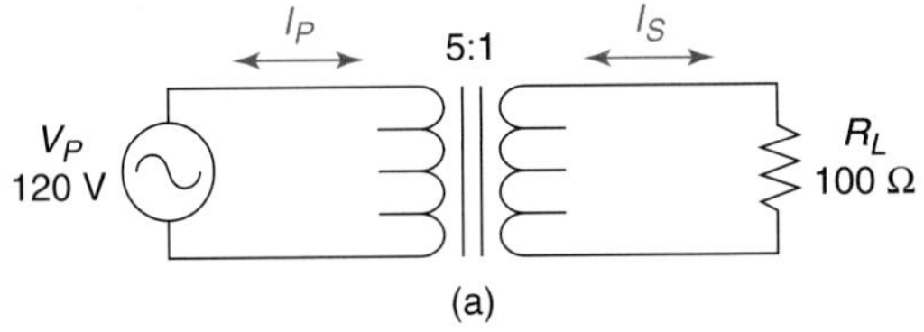


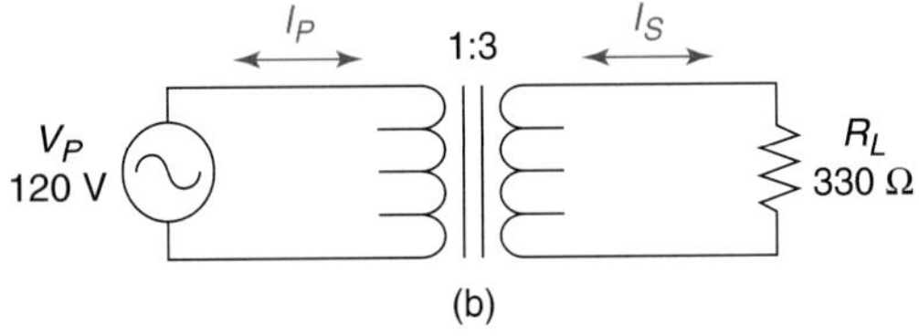


EWB

**FIGURE 15.19**

***Solution:*** First, the secondary voltage is found as

$$\boldsymbol{V_S} = V_P \frac{N_S}{N_P} = (120\text{ V})\frac{1}{5} = \mathbf{24\ V}$$

Now, Ohm's law is used to calculate the secondary current ($I_S$) as follows:

$$\boldsymbol{I_S} = \frac{V_S}{R_L} = \frac{24\text{ V}}{100\ \Omega} = \mathbf{240\ mA}$$

Finally, the value of the primary current is found as

$$\boldsymbol{I_P} = I_S \frac{N_S}{N_P} = (240\text{ mA})\frac{1}{5} = \mathbf{48\ mA}$$

***PRACTICE PROBLEM 15.5***

Calculate the value of primary current for the circuit in Figure 15.19b.

The relationship between primary current and load resistance.

Example 15.5 demonstrates an important transformer characteristic: for any transformer, *the value of primary current is a function of the load demand in the secondary.* Since load demand varies inversely with load resistance, we can say that primary current varies inversely with load resistance. This relationship is illustrated in Example 15.6.

## EXAMPLE 15.6

Calculate the range of primary current values for the circuit in Figure 15.20a.

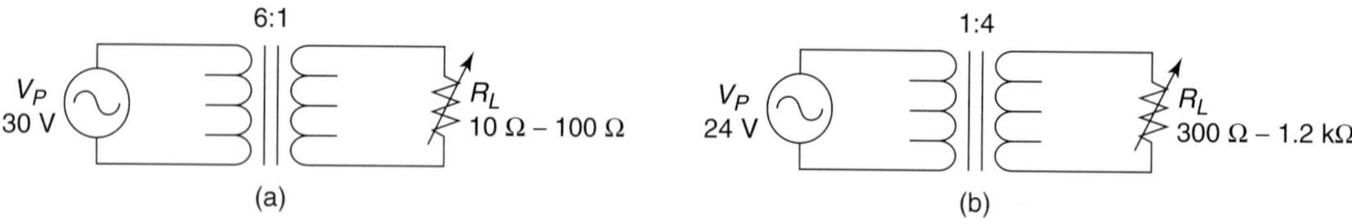


FIGURE 15.20

***Solution:*** Using the input voltage and turns ratio shown, the secondary voltage is found as

$$V_S = V_P\frac{N_S}{N_P} = (30\text{ V})\frac{1}{6} = \mathbf{5\ V}$$

When the load is set to 10 Ω, the secondary current is found as

$$I_S = \frac{V_S}{R_L} = \frac{5\text{ V}}{10\ \Omega} = \mathbf{500\ mA}$$

and the primary current is found as

$$I_P = I_S\frac{N_S}{N_P} = (500\text{ mA})\frac{1}{6} = \mathbf{83.3\ mA}$$

When the load is set to 100 Ω, the secondary current is found as

$$I_S = \frac{V_S}{R_L} = \frac{5\text{ V}}{100\ \Omega} = \mathbf{50\ mA}$$

and the primary current is found as

$$I_P = I_S\frac{N_S}{N_P} = (50\text{ mA})\frac{1}{6} = \mathbf{8.33\ mA}$$

Thus, increasing the load from 10 Ω to 100 Ω causes the primary current to decrease from 83.3 mA to 8.33 mA.

***PRACTICE PROBLEM 15.6***

Calculate the range of primary current values for the circuit in Figure 15.20b.

As you can see, primary current varies inversely with load resistance. Example 15.7 demonstrates a practical situation involving this principle.

## EXAMPLE 15.7

A fuse has been added to the primary circuit in Figure 15.21. The primary current must be limited to 80% of this rating. Determine the minimum allowable setting for $R_L$.

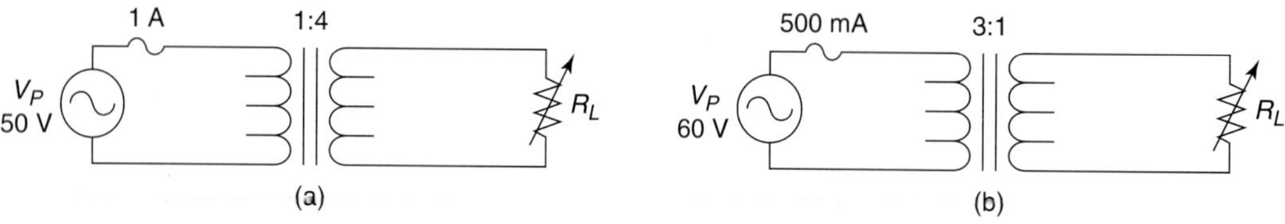


FIGURE 15.21

**Solution:** Using the turns ratio of the transformer, the secondary voltage is found as

*Don't Forget:* We always want to limit current to approximately 80% of the rated value of a fuse to provide a safety margin.

$$V_S = V_P\frac{N_S}{N_P} = (50\text{ V})\frac{4}{1} = \mathbf{200\text{ V}}$$

As stated earlier, the primary current must not be allowed to exceed 80% of the fuse rating. Therefore, maximum primary current is found as

$$I_P = (0.8)(1\text{ A}) = \mathbf{800\text{ mA}} \quad \text{(maximum)}$$

Using this value in equation (15.8), the maximum secondary current is found as

$$I_S = I_P\frac{N_P}{N_S} = (800\text{ mA})\frac{1}{4} = \mathbf{200\text{ mA}}$$

Finally, the minimum allowable setting for the load is found as

$$R_{L(\text{min})} = \frac{V_S}{I_S} = \frac{200\text{ V}}{200\text{ mA}} = \mathbf{1\text{ k}\Omega}$$

If the load is set to a value lower than 1 kΩ, the primary current will exceed the safety limit called for in the design.

***PRACTICE PROBLEM 15.7***

Determine the minimum allowable setting for the load in Figure 15.21b. Assume that the primary current must be limited (again) to 80% of the fuse rating.

## Primary Impedance ($Z_P$)

◄ *OBJECTIVE 13*

In Chapter 14, you learned that impedance ($Z$) is the total opposition to current in an ac circuit, made up of resistance ($R$) and/or reactance ($X$). The total opposition to current provided by the primary of a transformer is generally referred to as **primary impedance ($Z_P$)**.

**Primary impedance ($Z_P$)** The total opposition to current in the primary of a transformer.

The load on a transformer affects the total current in the primary and, therefore, the total opposition to current in the primary. For example, consider the circuit shown in Figure 15.22. The values shown in the circuit are the given (and calculated) values from Example 15.5. The circuit's 100 Ω load determines the value of current in the secondary. This current, in turn, determines the value of the primary current. For the values of primary current and voltage shown, we can use Ohm's law to calculate the value of the primary impedance ($Z_P$) as follows:

$$Z_P = \frac{V_P}{I_P} = \frac{120\text{ V}}{48\text{ mA}} = 2.5\text{ k}\Omega$$

Primary impedance versus primary resistance.

You may be wondering why we have switched to the term *impedance* when we are calculating its value using Ohm's law (as we could any resistance value). The term *impedance* is used to distinguish between the total opposition to current in the primary (a dynamic value) and the dc resistance of the primary winding (a static value). For any transformer, the value of primary impedance depends (in part) on the value of the load. When the load value is changed, the value of primary impedance also changes. On the other hand,

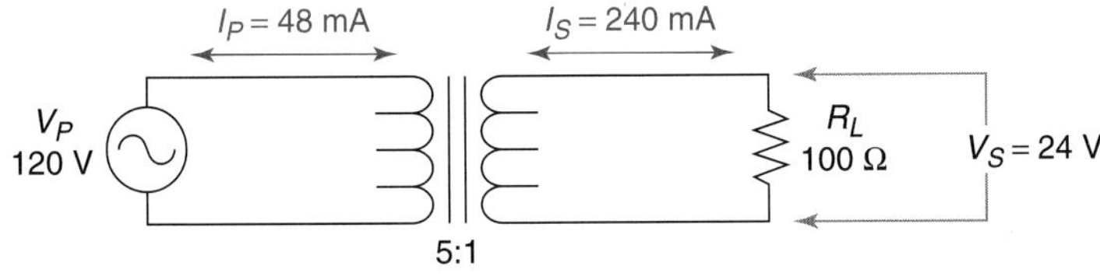


**FIGURE 15.22**

**Primary resistance ($R_P$)** The winding resistance of the primary coil.

the dc resistance of the primary coil, which is referred to as **primary resistance ($R_P$),** is determined by the physical characteristics of the primary winding. Therefore, its value is constant and unaffected by changes in the load. This is why we use the terms *impedance* and *resistance* to distinguish between the two.

*OBJECTIVE 14* ▶

**Secondary impedance ($Z_S$)** The total opposition to current in the secondary circuit of a transformer; generally assumed to equal the value of the load resistance.

The primary impedance of a transformer is directly proportional to the square of the turns ratio and the **secondary impedance ($Z_S$),** as follows:

$$Z_P = Z_S\left(\frac{N_P}{N_S}\right)^2 \qquad \textbf{(15.10)}$$

where $Z_S$ = the total opposition to current in the secondary (generally assumed to equal the opposition provided by the load)

An application of this relationship (which is derived in Appendix E) is provided in Example 15.8.

## EXAMPLE 15.8

Use equation (15.10) to calculate the primary impedance of the transformer in Figure 15.22.

***Solution:*** With a turns ratio of 5 : 1 and a 100 Ω load, the value of the primary impedance can be found as

$$Z_P = Z_S\left(\frac{N_P}{N_S}\right)^2 = (100\ \Omega)\left(\frac{5}{1}\right)^2 = (100\ \Omega)(25) = \mathbf{2.5\ k\Omega}$$

As you can see, this value matches the one we calculated for the circuit using Ohm's law.

***PRACTICE PROBLEM 15.8***

A transformer like the one shown in Figure 15.22 has a 12 : 1 turns ratio and a 50 Ω load. Calculate the value of primary impedance for the component.

It should be noted that equation (15.10) can be transposed to provide an equation for secondary impedance. However, since secondary impedance determines the value of primary impedance (and not the other way around), such an equation serves no practical purpose.

## Putting It All Together

*OBJECTIVE 15* ▶

In this section, you have learned that the voltage, current, and impedance ratios of a transformer can be related to its turns ratio, as follows:

$$\frac{N_P}{N_S} = \frac{V_P}{V_S} \qquad \frac{N_P}{N_S} = \frac{I_S}{I_P} \qquad \frac{N_P}{N_S} = \sqrt{\frac{Z_P}{Z_S}}$$

The impedance relationship shown here was derived using equation (15.10). Transposing this equation,

$$\frac{Z_P}{Z_S} = \left(\frac{N_P}{N_S}\right)^2$$

or

$$\sqrt{\frac{Z_P}{Z_S}} = \frac{N_P}{N_S}$$

Based on these relationships, we can summarize the voltage, current, and impedance characteristics for every step-down transformer as follows:

$$N_P > N_S \qquad V_P > V_S \qquad I_P < I_S \qquad Z_P > Z_S$$

These step-down transformer characteristics are demonstrated in Example 15.9.

Verify the input/output current, voltage, and impedance relationships for the transformer in Figure 15.23.

**EXAMPLE 15.9**

**FIGURE 15.23**

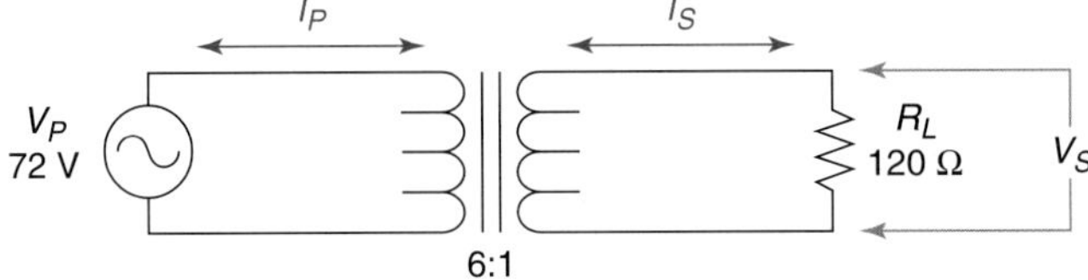


***Solution:*** First, the value of the secondary voltage is found as

$$\boldsymbol{V_S} = V_P \frac{N_S}{N_P} = (72\text{ V})\frac{1}{6} = \mathbf{12\ V}$$

Now, the secondary current can be found as

$$\boldsymbol{I_S} = \frac{V_S}{R_L} = \frac{12\text{ V}}{120\ \Omega} = \mathbf{100\ mA}$$

Using the secondary current and the transformer turns ratio, the primary current can be found as

$$\boldsymbol{I_P} = I_S \frac{N_S}{N_P} = (100\text{ mA})\frac{1}{6} = \mathbf{16.7\ mA}$$

Assuming that the secondary impedance equals the load resistance, the primary impedance can be found as

$$\boldsymbol{Z_P} = Z_S\left(\frac{N_P}{N_S}\right)^2 = (120\ \Omega)\left(\frac{6}{1}\right)^2 = (120\ \Omega)(36) = \mathbf{4.32\ k\Omega}$$

The values given (and calculated) for the circuit are listed below, along with the relationship between each value pair. As you can see, the relationships listed match those for a step-down transformer.

| *Characteristic* | *Primary Value* | *Secondary Value* | *Relationship* |
|---|---|---|---|
| Turns[a] | 6 | 1 | $N_P > N_S$ |
| Voltage | 72 V | 12 V | $V_P > V_S$ |
| Current | 16.7 mA | 100 mA | $I_P < I_S$ |
| Impedance | 4.32 kΩ | 120 Ω | $Z_P > Z_S$ |

[a]Values are listed as given in the turns ratio. They do not equal the actual number of turns for the coils.

***PRACTICE PROBLEM 15.9***

A transformer like the one shown in Figure 15.23 has the following values: $N_P = 12$, $N_S = 1$, $V_P = 36$ V, and $R_L = 10\ \Omega$. Verify the voltage, current, and impedance relationships for the component.

For the step-up transformer, the voltage, current, and impedance characteristics can be summarized as follows:

$$N_P < N_S \qquad V_P < V_S \qquad I_P > I_S \qquad Z_P < Z_S$$

These step-up transformer characteristics are demonstrated in Example 15.10.

## *EXAMPLE 15.10*

Verify the input/output current, voltage, and impedance characteristics for the step-up transformer in Figure 15.24.

**FIGURE 15.24**

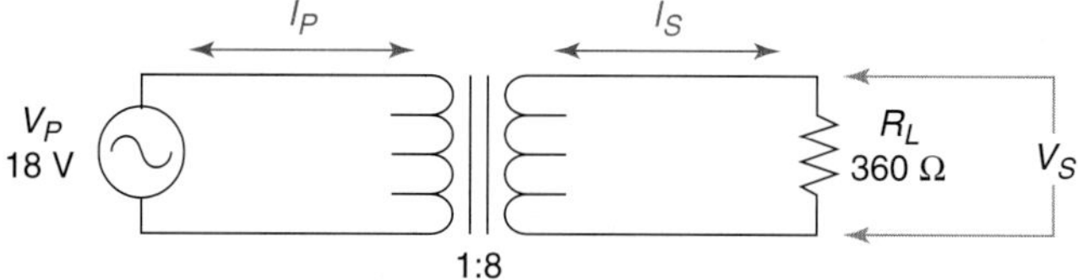


***Solution:*** First, the value of the secondary voltage is found as

$$\boldsymbol{V_S} = V_P\frac{N_S}{N_P} = (18\text{ V})\frac{8}{1} = \textbf{144 V}$$

Now, the secondary current can be found as

$$\boldsymbol{I_S} = \frac{V_S}{R_L} = \frac{144\text{ V}}{360\ \Omega} = \textbf{400 mA}$$

Using the secondary current and the transformer turns ratio, the primary current can be found as

$$\boldsymbol{I_P} = I_S\frac{N_S}{N_P} = (400\text{ mA})\frac{8}{1} = \textbf{3.2 A}$$

Assuming that the secondary impedance equals the resistance of the load, the primary impedance can be found as

$$\boldsymbol{Z_P} = Z_S\left(\frac{N_P}{N_S}\right)^2 = (360\ \Omega)\left(\frac{1}{8}\right)^2 = \textbf{5.63 Ω}$$

The values given (and calculated) for the circuit are listed below, along with the relationship between each value pair. As you can see, the relationships listed match those for a step-up transformer.

| *Characteristic* | *Primary Value* | *Secondary Value* | *Relationship* |
|---|---|---|---|
| Turns[a] | 1 | 8 | $N_P < N_S$ |
| Voltage | 18 V | 144 V | $V_P < V_S$ |
| Current | 3.2 A | 400 mA | $I_P > I_S$ |
| Impedance | 5.76 Ω | 360 Ω | $Z_P < Z_S$ |

[a]Values are listed as given in the turns ratio. They do not equal the actual number of turns for the coils.

### *PRACTICE PROBLEM 15.10*

A transformer like the one shown in Figure 15.24 has the following values: $N_P = 1$, $N_S = 4$, $V_P = 24$ V, and $R_L = 12$ kΩ. Verify the voltage, current, and impedance relationships for the component.

## *Section Review*

1. What is the relationship between a transformer's primary current and the resistance of its load?
2. What is primary impedance ($Z_P$)?

3. Contrast primary impedance and primary resistance.
4. What is the relationship between a transformer's turns ratio and its impedance ratio?

# 15.4 RELATED TOPICS

In this section, we will discuss various topics relating to transformers, their ratings, and their applications. The topics here deal with the more practical aspects of working with transformers.

## Impedance Matching: A Transformer Application

◄ *OBJECTIVE 16*

In Chapter 5, you were introduced to the concept of maximum power transfer. At that time, you were told that:

1. Maximum power transfer from a fixed-resistance source to a variable-resistance load occurs when the source resistance equals the load resistance ($R_S = R_L$).
2. Maximum power transfer from a variable-resistance source to a fixed-resistance load occurs when the source resistance is set to its lowest possible value.

Now, consider the case where both the load and the source resistance are fixed. Such a circuit is shown in Figure 15.25a. As you can see, the circuit has fixed values of $R_S = 100\ \Omega$ and $R_L = 4\ \Omega$.

As constructed, the circuit efficiency is extremely low; that is, very little of the source power is actually transferred to the load. Instead, most of the source power is dissipated by its own internal resistance. This situation is not considered acceptable in any practical situation. (The relationship between $R_S$ and $R_L$ in Figure 15.25a is often described using the term **impedance mismatch.**)

**Impedance mismatch**
A term used to describe a situation where source impedance does not match load impedance.

**Buffer**
An impedance matching circuit.

One method used to correct the power transfer problem shown is to place an impedance-matching circuit, or **buffer,** between the source and the load. A buffer is a circuit with input and output impedance values that closely match the source and load impedances, respectively. For example, look at the buffer that has been added to the circuit in Figure 15.25b. The input impedance of the buffer matches the source resistance, which ensures maximum power transfer from the source to the buffer. At the same time, the output impedance of the buffer matches the load resistance, which ensures maximum power transfer from the buffer to the load. In effect, the buffer has matched the load resistance to the

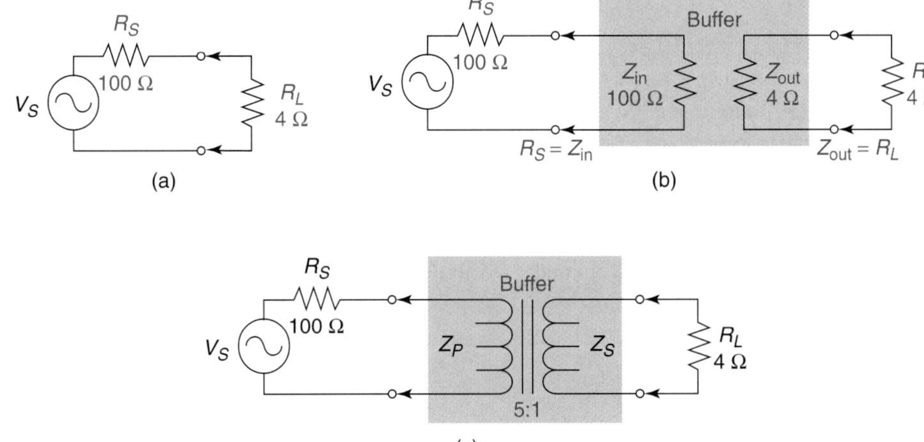


**FIGURE 15.25 A transformer as an impedance-matching circuit (or buffer).**

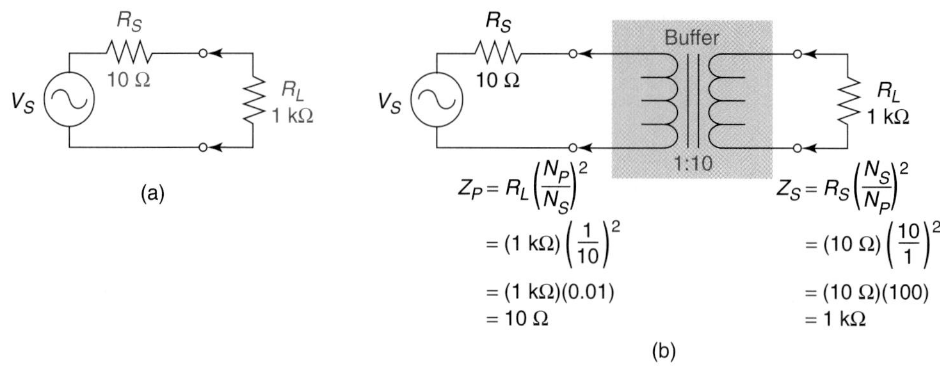


**FIGURE 15.26 Matching a low-resistance source to a high-resistance load.**

source resistance. This impedance match provides for maximum power transfer from the source to the load.

*A Practical Consideration:* A transformer will match a source resistance to a load resistance when

$$\frac{N_P}{N_S} = \sqrt{\frac{R_S}{R_L}}$$

A transformer can be used as a buffer for a circuit with a sinusoidal input, as shown in Figure 15.25c. The primary impedance of the transformer ($Z_P$) acts as the buffer input impedance. The secondary impedance of the transformer ($Z_S$) acts as the buffer output impedance. With the transformer turns ratio shown, the transformer primary impedance can be found as

$$Z_P = R_L\left(\frac{N_P}{N_S}\right)^2 = (4\ \Omega)\left(\frac{5}{1}\right)^2 = (4\ \Omega)(25) = 100\ \Omega$$

As you can see, this value matches the resistance of the source (100 Ω). The secondary impedance of the transformer can be found as

*Note:* The *s* subscripts in the calculation can be a bit confusing. $N_S$ is a *secondary* value. $R_S$ is the *source* resistance (a value from the primary circuit).

$$Z_S = R_S\left(\frac{N_S}{N_P}\right)^2 = (100\ \Omega)\left(\frac{1}{5}\right)^2 = (100\ \Omega)(0.04) = 4\ \Omega$$

which matches the resistance of the load (4 Ω). Since the source and load impedances have been matched, the transformer is providing for maximum power transfer between the two.

It should be noted that a transformer can also be used to match a low-resistance source to a high-resistance load. This is accomplished using a step-up transformer, as shown in Figure 15.26. As the calculations in the figure show, the transformer provides impedance matching between the source and load, and therefore, provides for maximum power transfer between the two.

## Transformer Voltage and Current Ratings

*OBJECTIVE 17* ▶

**Output voltage rating** The rating that indicates the ac output voltage from a transformer when it is provided with a $120\ V_{ac}$ input.

Throughout this chapter, we have assumed that transformers are rated using turns ratios. In fact, many transformers are rated for specific output (secondary) voltages. The **output voltage rating** of a transformer indicates the ac output voltage from the component when it is provided with a $120\ V_{ac}$ input. A transformer rated for a specific output voltage is shown in Figure 15.27. Note that the transformer rating is normally provided directly below (or above) the transformer schematic symbol.

**FIGURE 15.27 A transformer rated at a specific output (secondary) voltage.**

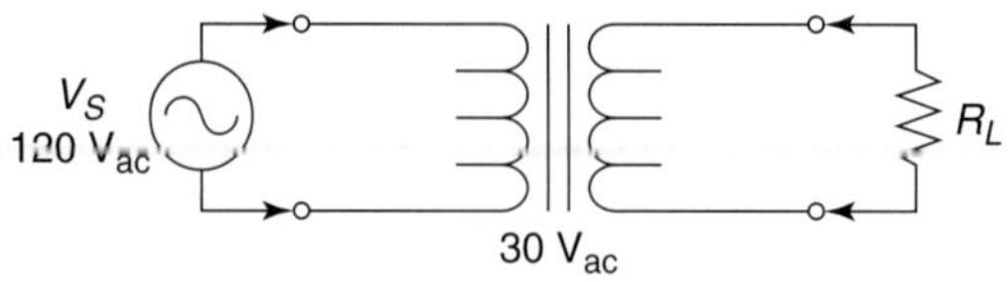

As you can see, the transformer input is 120 $V_{ac}$. As indicated by the transformer rating, the transformer steps this voltage down to 30 $V_{ac}$. With the input and output voltages given, the turns ratio of the transformer can be determined as follows:

$$\frac{N_P}{N_S} = \frac{V_P}{V_S} = \frac{120\ V_{ac}}{30\ V_{ac}} = \frac{4}{1}$$

Once the turns ratio of the transformer is determined, you can calculate the component current and impedance values as shown earlier in this chapter.

## Output Current Ratings

Like any component, a transformer can be damaged (or destroyed) by excessive heat. The power dissipated by the primary and secondary windings is a function of the circuit current and the winding resistance of the coil. By formula,

$$P_d = I^2 R_w \tag{15.11}$$

The **output current rating** of a transformer is a rating that indicates the maximum allowable secondary current for the component. As long as this rating is not exceeded, the current demand on the transformer will not cause excessive heat to be generated in either the primary or secondary circuits.

**Output current rating**
A rating that indicates the maximum allowable value of secondary current.

## Center-Tapped Transformers

◄ *OBJECTIVE 18*

A **center-tapped transformer** is a transformer that has a lead connected to the center of the secondary winding. The schematic symbol for a center-tapped transformer is shown in Figure 15.28.

**Center-tapped transformer**
A transformer that has a lead connected to the center of the secondary winding.

When a third lead is connected to the center of the secondary, the voltage from either end-terminal of the secondary to the center tap is one-half the secondary voltage. This relationship is illustrated in Figure 15.29. The transformer shown is rated for a 12.6 $V_{ac}$ output. The voltage across either end-terminal ($S_1$ or $S_2$) and the center tap is one-half the secondary voltage, 6.3 $V_{ac}$.

It should be noted that the end-terminal voltages, when measured with respect to the center tap, have opposite polarities. This principle is illustrated in Figure 15.30. If we assume that the secondary voltage has the polarity shown, the voltage at $S_1$ is positive with respect to the grounded center tap. At the same time, the voltage at $S_2$ is negative with respect to the grounded center tap. When the polarity of the secondary voltage reverses, so do the polarities of the other voltages.

The polarity relationship shown in Figure 15.30 is the basis for a common center-tapped transformer application: converting one sine wave into two sine waves that are 180° out of phase. This application is illustrated in Figure 15.31. Each waveform in Figure 15.31 is measured from its respective terminal to the grounded center tap. By using a 1:2

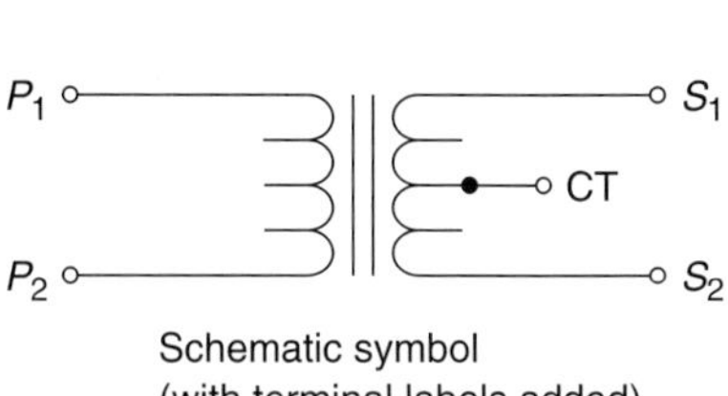


**FIGURE 15.28 A center-tapped transformer schematic symbol.**

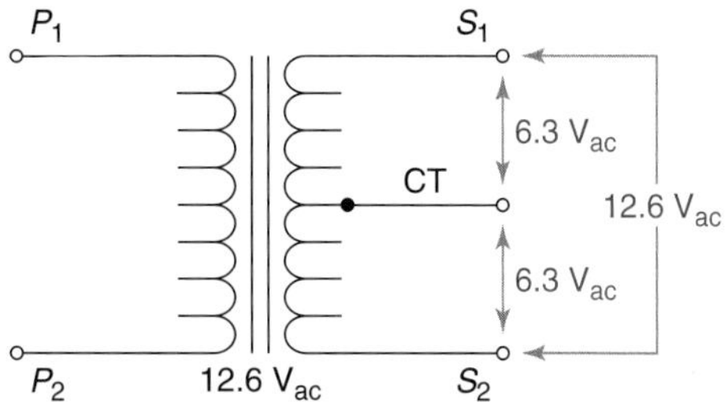


**FIGURE 15.29 A 12.6 $V_{ac}$ center-tapped transformer.**

*A Practical Consideration:* The polarities shown in Figure 15.30 must exist for the two secondary voltages to add up to the secondary voltage, as required by Kirchhoff's voltage law.

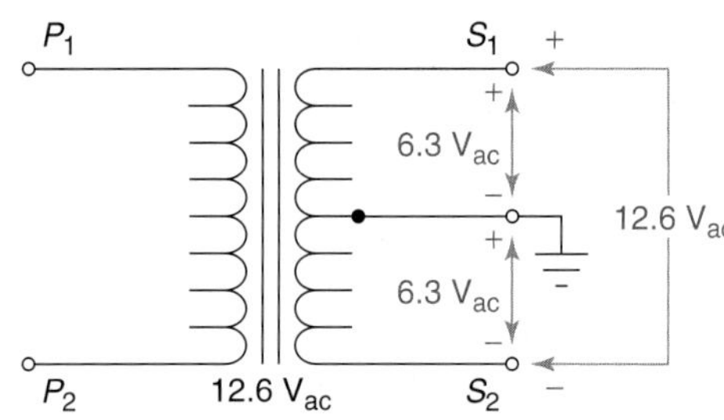


**FIGURE 15.30 Terminal voltage polarities with respect to a grounded center tap.**

*Note:* The waveforms in Figure 15.31 are measured using the center tap as the reference.

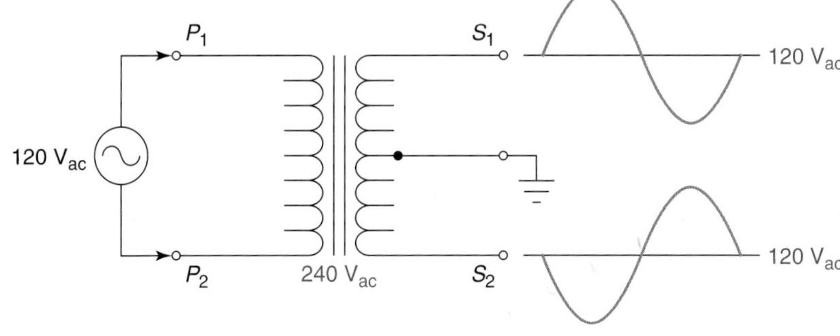


EWB

**FIGURE 15.31 Using a center-tapped transformer to generate sine waves that are 180° out of phase.**

transformer, the output voltages have the same effective (rms) value as the input to the primary. It should be noted, however, that each output current is significantly lower than the primary current.

## Transformers with Multiple Outputs

**dc power supply**
A circuit that converts the line voltage to one or more dc operating voltages for a given electronic system.

Transformers are widely used in dc power supply circuits. In this case, the term **dc power supply** is used to describe a circuit that converts the line voltage to one or more dc operating voltages for an electronic system. Any electronic system that derives power from a power receptacle has a dc power supply. A simplified block diagram of a dc power supply is shown in Figure 15.32.

The large block in the diagram represents three separate circuits, called the rectifier, the filter, and the regulator. These three circuits work together to convert the ac output from the transformer into a steady dc voltage. The conversion of an ac voltage to a steady dc voltage is discussed in a course on solid-state electronics. For now, we are interested simply in establishing the fact that the ac line input is commonly applied to a transformer. Then, the output from the transformer is applied to the conversion circuits.

The transformer in a dc power supply serves several purposes. First, it normally steps the input voltage up or down, depending on the desired dc output from the circuit. Second, it provides isolation between the power supply circuitry and the line input. This

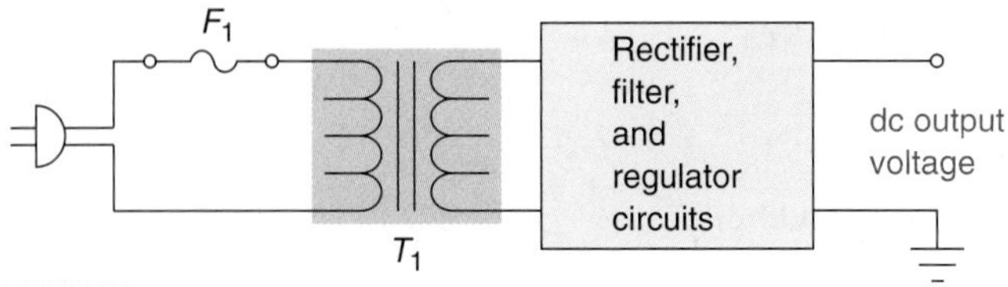


**FIGURE 15.32 A simplified block diagram of a dc power supply.**

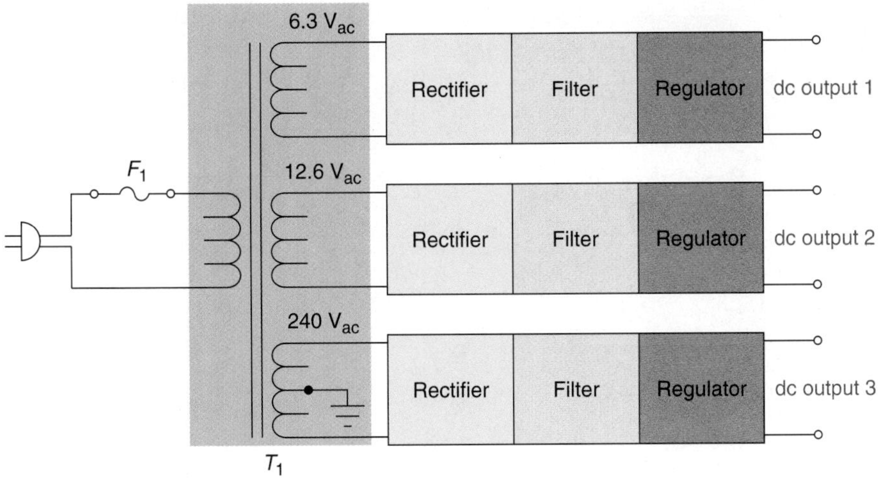


**FIGURE 15.33 A dc power supply with a multiple-secondary transformer.**

prevents any problems in the dc power supply from being coupled back to the electrical circuitry that is providing the line input. For example, let's say that you have a stereo with a transformer-input dc power supply. If a problem were to develop in the stereo, the transformer would prevent the problem from affecting the electrical system that delivers power to the wall receptacle.

In many cases, more than one dc power supply is needed to provide various dc voltages for system operation. For example, a desktop computer may need internal $+5\ V_{dc}$ and $-5\ V_{dc}$ power supplies to drive its integrated circuits, and a $+63\ V_{dc}$ power supply to drive its video display circuitry. When an electronic system needs two or more internal dc supply voltages to operate, they can usually be derived from the ac line input using a **multiple-output transformer**. As the name implies, a multiple-output transformer is one that has two or more secondary outputs. An example of a multiple-output transformer is shown in Figure 15.33.

**Multiple-output transformer**
One with two or more secondary outputs.

When working with a multiple-secondary transformer, keep several things in mind:

Working with multiple-output transformers.

1. The transformer may have a single secondary winding with one or more taps, or it may contain two or more independent secondary windings.
2. Each secondary circuit is analyzed as if it were the only one; that is, each of the secondary circuits is analyzed independently of the other secondary circuits.
3. If a fault develops in one of the secondary circuits, it can affect the rest of the secondary circuits, especially if the transformer has a single secondary winding with multiple taps. Even with multiple secondary windings, a short circuit in one of the secondary circuits can cause the primary fuse to open. In effect, the short causes power to be lost in all the secondary circuits.

We will discuss transformer faults and fault symptoms in more detail later in this section.

## Autotransformers

◄ *OBJECTIVE 19*

An **autotransformer** is a transformer that consists of a single coil with three terminal connections. The basic construction of an autotransformer is shown (along with its schematic symbol) in Figure 15.34a. As you can see, the autotransformer is, for all practical purposes, a tapped inductor.

**Autotransformer**
A special-case transformer that consists of a single coil with three terminal connections; in essence, a tapped inductor.

The turns ratio of an autotransformer depends on:

1. The number of turns between terminals 1 and 3 ($N_{1,3}$).
2. The number of turns between terminals 2 and 3 ($N_{2,3}$).
3. The input/output designations of the terminal pairs.

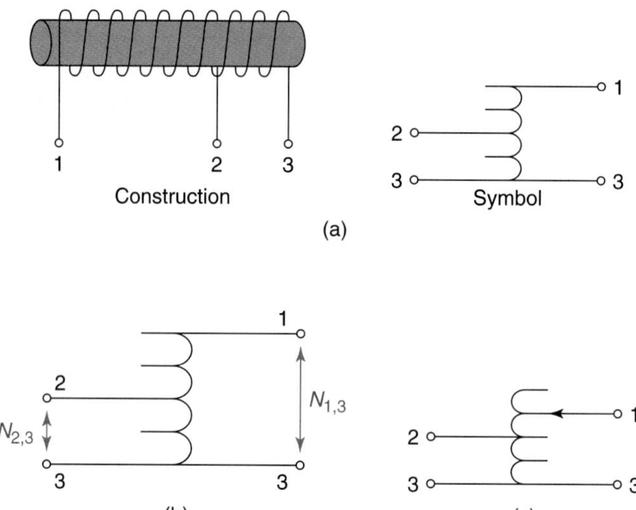


**FIGURE 15.34 Autotransformers.**

The turns values listed are identified in Figure 15.34b. For standard autotransformers, the values of $N_{1,3}$ and $N_{2,3}$ are fixed. For a variable autotransformer, the position of terminal 1 can be varied, meaning that $N_{1,3}$ can be adjusted. $N_{2,3}$, however, remains fixed. (The schematic symbol for a variable autotransformer is shown in Figure 15.34c.)

The input/output designations of the terminal pairs depend on the circuit configuration. Because of its construction, the autotransformer can be wired in either a step-up or a step-down configuration. These two configurations are illustrated in Figure 15.35. If we assume that terminal 2 center taps the autotransformer shown in the figure, then

$$N_{1,3} = 2N_{2,3}$$

This equation indicates that:

1. The circuit in Figure 15.35a has a 1:2 (step-up) turns ratio.
2. The circuit in Figure 15.35b has a 2:1 (step-down) turns ratio.

## Autotransformer Applications

As demonstrated in Figure 15.35, autotransformers can be used as either step-up or step-down components. However, they do not provide dc isolation between the primary and secondary terminal pairs. As such, they are used only in situations where the step-up (or step-down) characteristic is desired and where dc isolation is not needed. For example, an autotransformer could be used in the impedance matching circuit in Figure 15.25.

## Transformer Shielding

*OBJECTIVE 20* ▶

*Reference:* Magnetic shielding is covered on pages 370–371.

In Chapter 10, you were introduced to the concept of magnetic shielding. At that time, you learned that magnetic shielding is used to protect a component or circuit from the magnetic field generated by an inductor.

Like inductors, transformers generate a magnetic field that could interfere with the operation of nearby components. In such cases, transformer shielding is used to prevent any magnetic interference. Transformer shielding is accomplished using a metal casing that surrounds the transformer, as shown in Figure 15.36.

As a reference, an unshielded transformer is shown in Figure 15.36a. Metal covers can be added to this type of transformer to form a magnetic shield, as shown in Figure 15.36b. Figure 15.36c shows a shielded RF (radio-frequency) transformer. In each case,

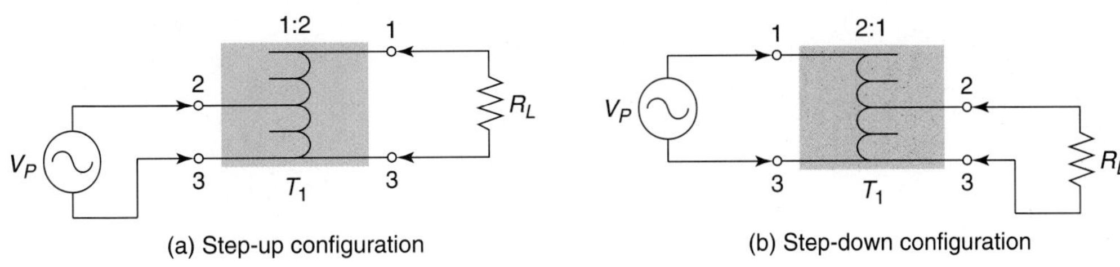

**FIGURE 15.35 Autotransformer configurations.**

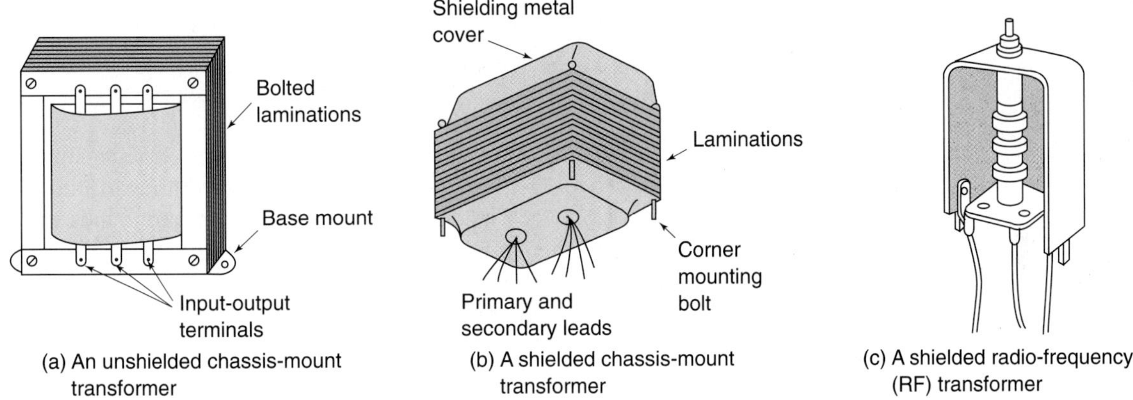


**FIGURE 15.36 Transformer shielding.**

the shield protects any surrounding circuits that would be affected by the presence of a magnetic field.

## Transformer Faults

There are several types of faults that can develop in a transformer:

◀ *OBJECTIVE 21*

1. A shorted primary or secondary winding.
2. A primary-to-secondary short.
3. An open primary or secondary winding.

At this point, we will take a brief look at each of these transformer faults.

***Shorted Windings.*** The effect of a shorted primary winding on a voltage source is illustrated in Figure 15.37a. In this case, the primary coil shorts out the voltage source, and the input fuse to the transformer opens.

A shorted secondary winding can have the same overall effect on the voltage source. The effect of a shorted secondary winding can be explained with the aid of Figure 15.37b. In this case, the impedance of the secondary drops to 0 Ω. This impedance is reflected back to the primary, as follows:

$$Z_P = Z_S\left(\frac{N_P}{N_S}\right)^2 = (0\ \Omega)\left(\frac{N_P}{N_S}\right)^2 = 0\ \Omega$$

This value indicates that the winding resistance of the primary is the only opposition to current provided by the transformer. As a result, a relatively high current demand is placed on the line input to the transformer. This excessive current demand, again, causes the fuse to open.

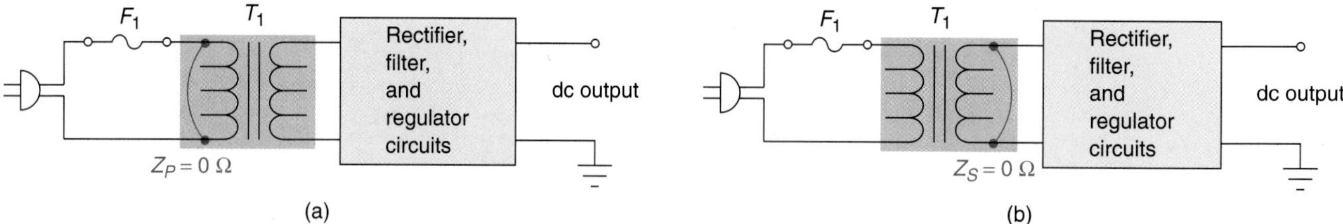


**FIGURE 15.37 Shorted transformer windings.**

When the secondary winding of a transformer is shorted, it may take some time for the fuse in the voltage source to open. The time required depends on the winding resistance of the primary circuit and the internal resistance of the source. When the shorted secondary does not cause the fuse to open immediately, the following symptoms will be present:

1. The transformer will be much hotter than normal.
2. The output from the secondary will be approximately 0 V.

If these symptoms appear, power to the circuit should be disconnected immediately and the transformer should be tested. Testing a transformer involves measuring the dc resistance of the primary and secondary coils. The measured values are compared to the rated values to determine whether or not either coil is shorted.

Sometimes, one of the windings in a transformer can become partially shorted. A partially shorted primary or secondary winding usually won't cause a transformer to fail. However, the partial short will effectively change the turns ratio of the component. Since

$$V_S = V_P \frac{N_S}{N_P}$$

the change in the turns ratio will also cause the output (secondary) voltage to change. Whenever the output voltage from a transformer is out of tolerance (as indicated by the turns ratio or secondary voltage rating) chances are that one of the transformer coils is partially shorted.

***A Primary-to-Secondary Short.*** If a short develops between the windings of a transformer, the primary circuit is connected directly to the secondary circuit. As a result:

1. The step function of the transformer is lost; that is, the turns ratio will change to approximately 1:1.
2. dc isolation between the primary and secondary circuits is lost.

The first symptom can be detected by comparing the measured primary and secondary voltages to the rating of the transformer. The second symptom can be detected by a continuity check between the primary and secondary leads.

***An Open Winding.*** If either winding opens, the current in the transformer secondary drops to 0 A. If an open is located in the primary, secondary voltage will drop to 0 V because there is no magnetic field generated to couple the source voltage from the primary circuit to the secondary circuit. If an open develops in the secondary, there may still be some measurable value of $V_S$. However, the current in the secondary will be 0 A.

## Transformer Color Coding

Transformer primary and secondary wires are color coded to aid in identifying the various leads. One method used to color code transformer wires is introduced in Appendix B.

## Section Review

1. List the two conditions under which maximum power transfer from source to load occurs.
2. What is a buffer?
3. Describe the use of a transformer as a buffer.
4. What input value is assumed for the output voltage rating of a transformer?
5. What is the output current rating of a transformer?
6. What is a center-tapped transformer?
7. What is the phase relationship between the two output waveforms of a center-tapped transformer (relative to the center tap)?
8. What function is performed by the dc power supply of a given electronic system?
9. What is a multiple-output transformer?
10. How is the secondary of a transformer wired to provide multiple outputs?
11. What is an autotransformer?
12. What factors affect the turns ratio of an autotransformer?
13. What advantage does an autotransformer have over a standard transformer? What limitation does it have?
14. Under what circumstances must a shielded transformer be used?
15. List the primary symptoms of each of the following:
    a. A totally shorted primary or secondary winding.
    b. A partially shorted primary or secondary winding.
    c. A primary-to-secondary short.
    d. An open primary or secondary winding.

## Chapter Summary

Here is a summary of the major points made in this chapter:

1. A transformer is a two-coil component that uses electromagnetic induction to pass an ac signal from its input to its output while providing dc isolation between the two.
    a. A transformer provides ac coupling.
    b. A transformer provides dc isolation.
2. A transformer consists of a core and two coils that are called the primary and the secondary.
    a. The core is usually made of air or iron.
    b. The input to a transformer is applied to the primary coil (or winding).
    c. The output from a transformer is taken from the secondary winding.
3. Transformers are classified according to the relationship between their input and output voltages.
    a. The ac output voltage from a step-up transformer is greater than the ac input voltage.
    b. The ac output voltage from a step-down transformer is less than the ac input voltage.
    c. The ac output voltage from an isolation transformer equals the input voltage.
4. Transformer operation is based on the principle of electromagnetic induction, as follows:
    a. An ac input causes the primary winding to generate a magnetic field.
    b. The magnetic field generated by the primary coil cuts through the secondary coil.
    c. When the magnetic field cuts through the secondary coil, an ac voltage is generated across that coil.

5. Since a dc input does not generate a magnetic field, it is not coupled to the secondary circuit. This is how the transformer provides dc isolation.
6. Each loop of wire in a coil is referred to as a turn.
7. The turns ratio of a transformer is the ratio of primary turns to secondary turns.
   a. For a step-down transformer, $N_P > N_S$.
   b. For a step-up transformer, $N_P < N_S$.
   c. For an isolation transformer, $N_P = N_S$.
8. The turns ratio of a transformer is important because it determines the ratio of input voltage to output voltage (see Examples 15.1 and 15.2).
9. Polarity dots are often used to indicate the phase relationship between transformer input and output voltages.
   a. When used, polarity dots identify the sides of the primary and secondary coils that are in phase with each other.
   b. When polarity dots are not used, the input and output voltages are assumed to be in phase.
10. Ideally, all the power provided to the primary of a transformer is transferred to the secondary circuit; that is, $P_P = P_S$.
11. The efficiency rating of a transformer indicates the percentage of primary power transferred to the secondary.
12. Due to transformer losses, no transformer is 100% efficient. The common types of power loss are as follows:
    a. Copper loss is the power dissipated by the winding resistance of the transformer coils.
    b. Eddy currents, which are currents generated in the core by the magnetic field, cause power to be dissipated in the core.
    c. Hysteresis loss is a loss of energy that occurs whenever a magnetic field of one polarity must overcome any residual magnetism from the other polarity. (Hysteresis loss is illustrated in Figure 15.12.)
13. The methods used to reduce the losses listed in item 12 are given in Table 15.1.
14. Transformer efficiency ratings are typically in the range of 90% or higher.
15. The change in current from transformer input to output is inversely proportional to the change in voltage.
    a. Current increases from input to output for a step-down transformer.
    b. Current decreases from input to output for a step-up transformer.
    c. Current and voltage change by the same factor from the input of a transformer to its output.
16. The isolation transformer is used strictly to provide dc isolation between its input and output.
17. The primary and secondary currents for a transformer are determined (in part) by the resistance of the load (see Example 15.5).
18. The total opposition to current provided by the primary of a transformer is generally referred to as primary impedance ($Z_P$).
    a. Primary impedance is a dynamic value that is affected by the resistance of the load.
    b. Primary resistance is a static value. It is the winding resistance of the primary coil.
19. The ratio of primary to secondary impedance equals the square of the turns ratio.

20. The following maximum power transfer relationships were provided in this chapter as a review:

   a. Maximum power transfer from a fixed-resistance source to a variable-resistance load occurs when $R_S = R_L$.

   b. Maximum power transfer from a variable-resistance source to a fixed-resistance load occurs when the source resistance is set to its lowest possible value.

21. The term *impedance mismatch* is used to describe a situation where the values of $R_S$ and $R_L$ in a circuit prevent maximum power transfer between the two.

22. An impedance mismatch can be corrected using an impedance-matching circuit called a buffer.

   a. The buffer is inserted between the source and the load.

   b. The input impedance of the buffer is designed to match the source resistance as closely as possible.

   c. The output impedance of the buffer is designed to match the load resistance as closely as possible.

23. A transformer can be used as a buffer. The transformer will match the source resistance to the load resistance when:

$$\frac{N_P}{N_S} = \frac{R_S}{R_L}$$

   a. A step-down transformer is used to match a high-resistance source to a low-resistance load.

   b. A step-up transformer is used to match a low-resistance source to a high-resistance load.

24. The output voltage rating of a transformer indicates the ac output voltage from the component when it is provided with a 120 $V_{ac}$ input.

25. When the output voltage rating of a transformer is known, the turns ratio can be found as

$$\frac{N_P}{N_S} = \frac{120\ V_{ac}}{V_S}$$

   where $V_S$ is the rated output of the transformer.

26. The output current rating of a transformer indicates the maximum allowable value of secondary current for the component.

27. As long as secondary current is kept below the output current rating of the transformer, both the primary and secondary circuits are protected from power dissipation problems.

28. A center-tapped transformer is one that has an extra lead connected to the center of the secondary winding.

   a. The voltage from either end-terminal of the secondary to the center tap is one-half of $V_S$.

   b. The end-terminal voltages, when measured with respect to the center tap, have opposite polarities (see Figures 15.30 and 15.31).

29. Transformers are widely used in dc power supply circuits.

   a. The dc power supply of an electronic system converts the ac line voltage to the dc voltage(s) required for the system to operate.

   b. A simplified block diagram of a dc power supply circuit is shown in Figure 15.32.

30. The transformer in a dc power supply:

   a. Steps the voltage up or down, depending on the desired values of dc voltage.

   b. Provides isolation between the power supply circuitry and the ac line input.

31. Multiple-output transformers are commonly used in dc power supplies that provide two (or more) dc operating voltages. A multiple-output transformer (as the name implies) provides more than one output (secondary) voltage (see Figure 15.33).
32. An autotransformer is a single-coil transformer with three terminal connections (as shown in Figure 15.34a).
    a. An autotransformer can be used as either a step-up or a step-down component.
    b. Because of its construction, an autotransformer does not provide dc isolation between its input and output terminals.
33. The autotransformer is used primarily in impedance-matching applications.
34. Transformer shielding is used to prevent the magnetic field generated by a transformer from affecting the operation of nearby circuits.
35. The most common faults that can develop in a transformer are:
    a. A shorted primary or secondary winding, either complete or partial.
    b. A primary-to-secondary short.
    c. An open primary or secondary winding.

## Equation Summary

| Equation Number | Equation | Section Number |
|---|---|---|
| **(15.1)** | $\frac{N_P}{N_S} = \frac{V_P}{V_S}$ | 15.1 |
| **(15.2)** | $V_S = V_P\frac{N_S}{N_P}$ | 15.2 |
| **(15.3)** | $P_P = P_S \quad \text{(ideal)}$ | 15.2 |
| **(15.4)** | $I_PV_P = I_SV_S \quad \text{(ideal)}$ | 15.2 |
| **(15.5)** | $\eta = \frac{P_S}{P_P} \times 100 \qquad (\%)$ | 15.2 |
| **(15.6)** | $I_S = I_P\frac{V_P}{V_S}$ | 15.2 |
| **(15.7)** | $\frac{I_S}{I_P} = \frac{N_P}{N_S}$ | 15.2 |
| **(15.8)** | $I_S = I_P\frac{N_P}{N_S}$ | 15.2 |
| **(15.9)** | $I_P = I_S\frac{N_S}{N_P}$ | 15.3 |
| **(15.10)** | $Z_P = Z_S\left(\frac{N_P}{N_S}\right)^2$ | 15.3 |
| **(15.11)** | $P_d = I^2R_w$ | 15.4 |

## Key Terms

The following terms were introduced and defined in this chapter:

| | | |
|---|---|---|
| autotransformer | isolation | primary impedance |
| buffer | isolation transformer | primary resistance |
| center-tapped transformer | laminated core | secondary |
| copper loss | multiple-output transformer | secondary impedance |
| coupling | output current rating | step-down transformer |
| dc power supply | output voltage rating | step-up transformer |
| eddy current | polarity dots | terminals |
| hysteresis loss | primary | transformer |
| impedance mismatch | | turns ratio |

## Practice Problems

1. A transformer has a 30 $V_{ac}$ input and a 10:1 turns ratio. Determine the value of the secondary voltage.
2. A transformer has a 75 $V_{ac}$ input and a 3:1 turns ratio. Determine the value of the secondary voltage.
3. A transformer has a 120 $V_{ac}$ input and a 1:8 turns ratio. Determine the value of the secondary voltage.
4. A transformer has a 250 $mV_{ac}$ input and a 1:100 turns ratio. Determine the value of the secondary voltage.
5. Determine the value of secondary current ($I_S$) for the transformer in Figure 15.38a.
6. Determine the value of secondary current ($I_S$) for the transformer in Figure 15.38b.
7. Calculate the values of secondary voltage and current for the transformer shown in Figure 15.39a.
8. Calculate the values of secondary voltage and current for the transformer shown in Figure 15.39b.
9. Determine the value of primary current ($I_P$) for the circuit shown in Figure 15.40a.
10. Determine the value of primary current ($I_P$) for the circuit shown in Figure 15.40b.

**FIGURE 15.38**

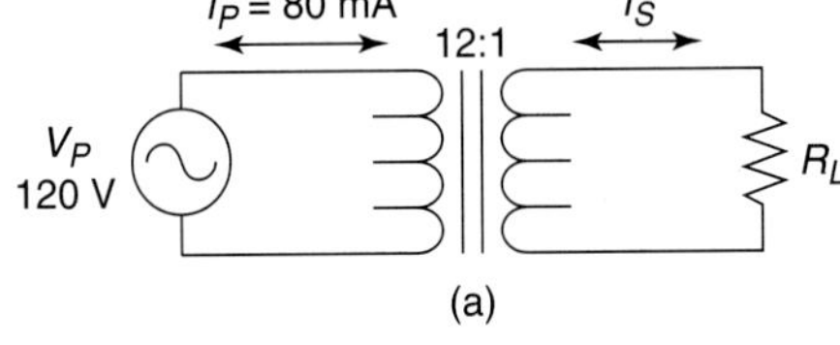


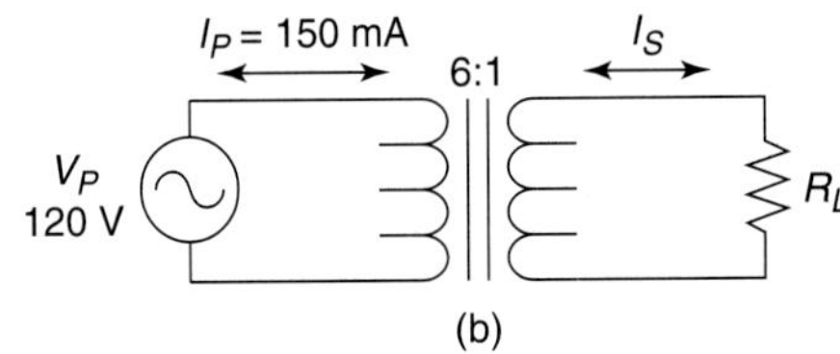


**FIGURE 15.39**

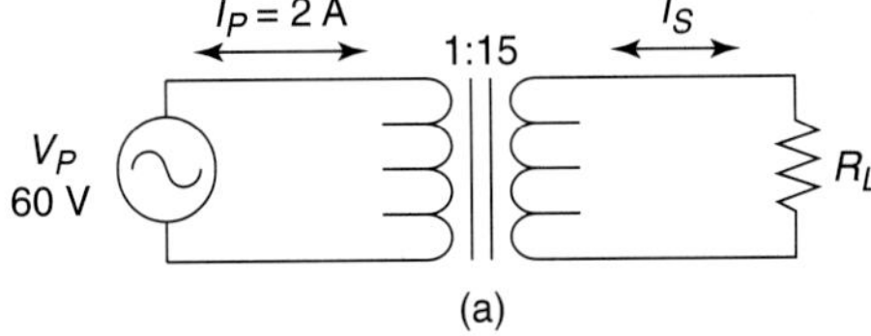


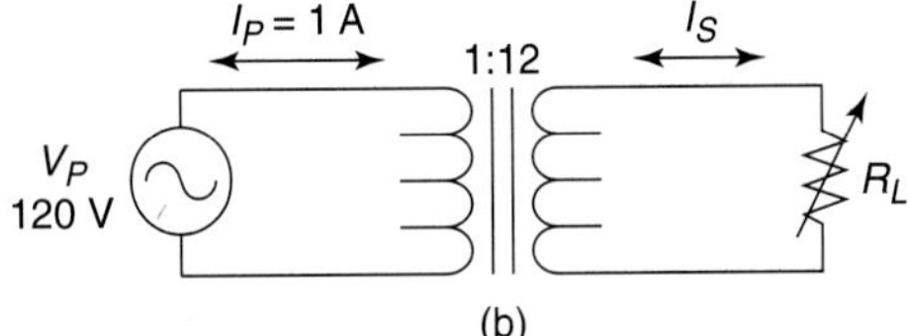


**FIGURE 15.40**

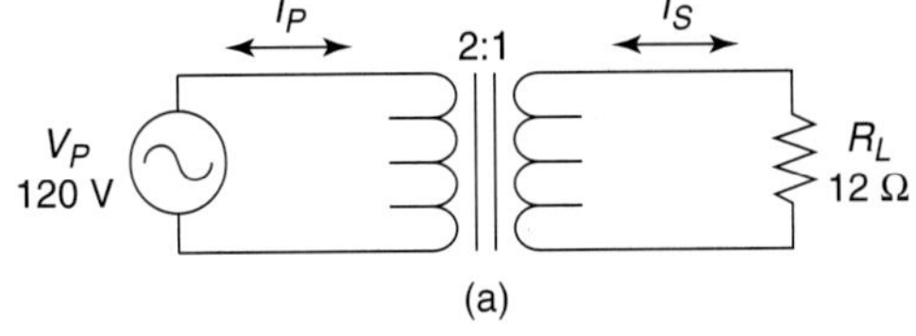


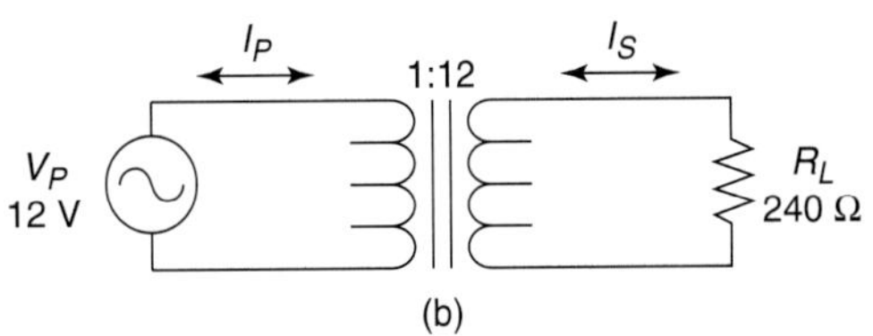

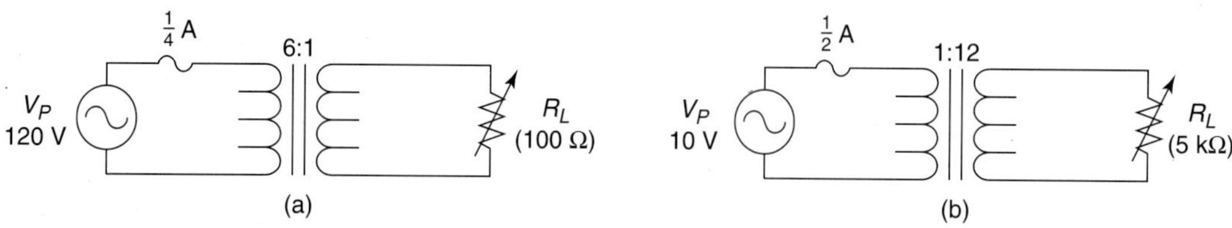


FIGURE 15.41

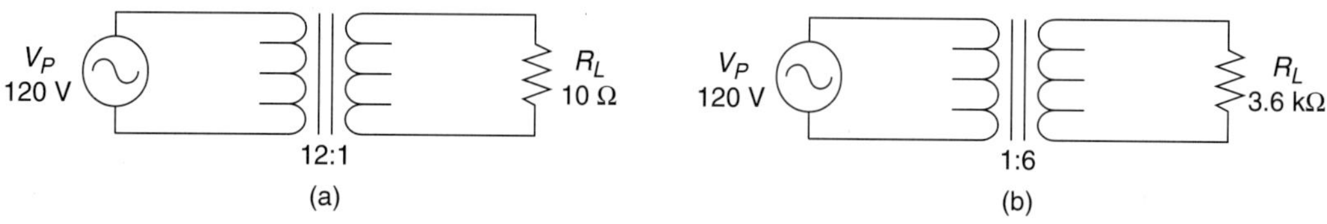


FIGURE 15.42

11. The primary current in Figure 15.41a must be limited to 80% of the primary fuse rating. Determine the minimum allowable setting for $R_L$.
12. The primary current in Figure 15.41b must be limited to 80% of the primary fuse rating. Determine the minimum allowable setting for $R_L$.
13. Refer to Figure 15.40a. Calculate the primary impedance ($Z_P$) for the circuit.
14. Refer to Figure 15.40b. Calculate the primary impedance ($Z_P$) for the circuit.
15. Calculate the values of $V_S$, $I_S$, $I_P$, $Z_P$, and $P_P$ for the circuit in Figure 15.42a.
16. Calculate the values of $V_S$, $I_S$, $I_P$, $Z_P$, and $P_P$ for the circuit in Figure 15.42b.

## *Troubleshooting Practice Problems*

17. Determine the possible fault(s), if any, indicated by the readings in Figure 15.43a.
18. Determine the possible fault(s), if any, indicated by the readings in Figure 15.43b.
19. Determine the possible fault(s), if any, indicated by the readings in Figure 15.44a.
20. Determine the possible fault(s), if any, indicated by the readings in Figure 15.44b.

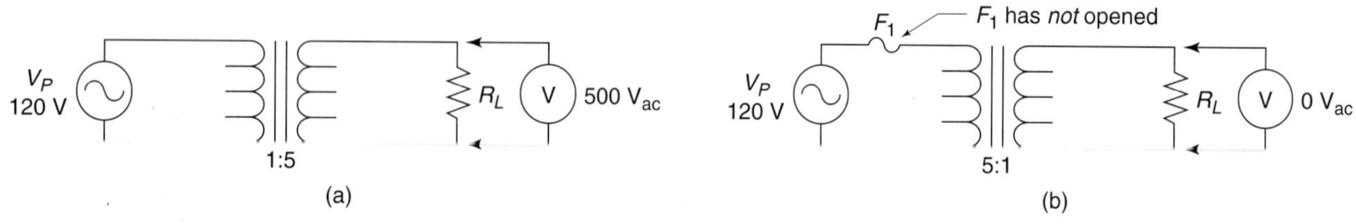


FIGURE 15.43

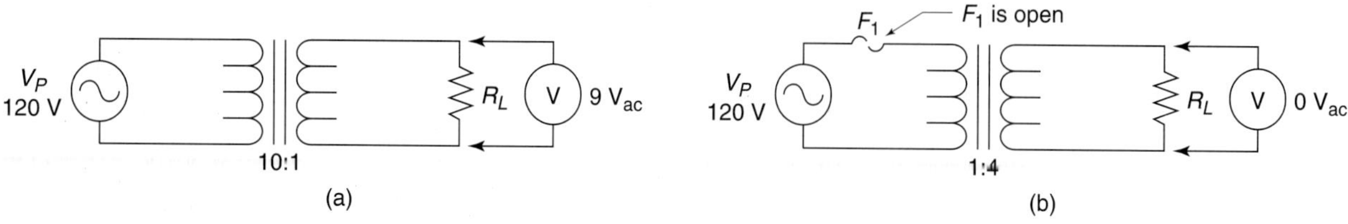


FIGURE 15.44

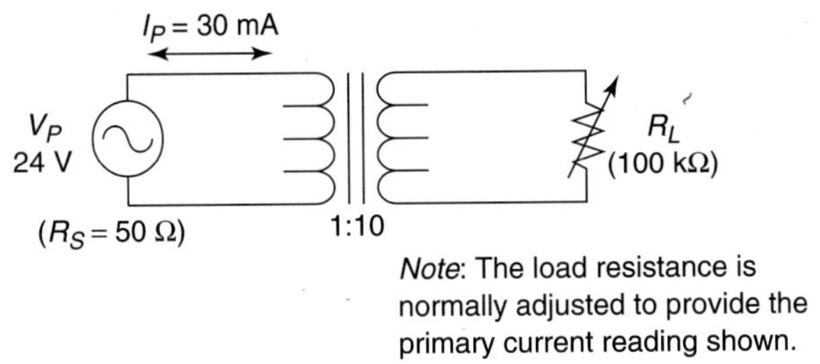


FIGURE 15.45

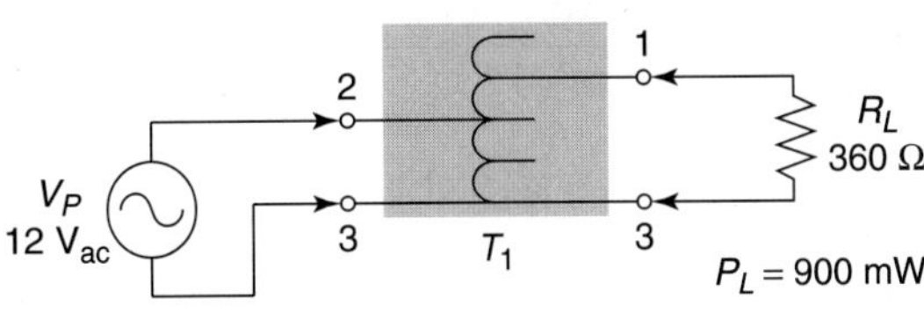


FIGURE 15.46

## The Brain Drain

**21.** For the circuit shown in Figure 15.45, determine:

**a.** The adjusted value of $R_L$.

**b.** The minimum acceptable power rating for the potentiometer, assuming that the power rating must be at least 20% greater than the maximum possible load power.

**22.** Determine the adjusted turns ratio of the autotransformer in Figure 15.46.

## Answers to the Example Practice Problems

**15.1** 10 V
**15.2** 640 mV
**15.3** 3 A
**15.4** 1200 V, 100 mA
**15.5** 3.27 A
**15.6** 320 mA to 1.28 A
**15.7** 16.7 Ω
**15.8** 7.2 kΩ
**15.9** $V_S = 3$ V, $I_S = 300$ mA, $I_P = 25$ mA, $Z_P = 1.44$ kΩ
**15.10** $V_S = 96$ V, $I_S = 8$ mA, $I_P = 32$ mA, $Z_P = 750$ Ω

## EWB Applications Problems

| Figure | EWB File Reference |
|---|---|
| 15.4 | EWBA15_4.ewb |
| 15.19 | EWBA15_19.ewb |
| 15.31 | EWBA15_31.ewb |
| 15.35 | EWBA15_35.ewb |

# 16 CAPACITORS

## OBJECTIVES

*After studying the material in this chapter, you should be able to:*

1. Define capacitance.
2. Describe the physical makeup of a capacitor.
3. Describe the charge and discharge characteristics of capacitors.
4. Discuss capacity and its unit of measure.
5. Describe the relationship between the physical construction of a capacitor and its value.
6. Explain the ac-coupling and dc-isolation characteristics of capacitors.
7. Describe the ac phase relationship between capacitor current and voltage.
8. Calculate the total capacitance of any number of capacitors connected in series or parallel.
9. Compare and contrast the following: dielectric resistance, effective capacitor resistance, and capacitive reactance.
10. Calculate the reactance of a capacitor.
11. Solve basic circuit problems involving capacitive reactance.
12. Discuss the breakdown voltage rating of a capacitor.
13. Describe capacitor leakage current.
14. Explain the quality ($Q$) of a capacitor and calculate its value.
15. Discuss dielectric absorption and its effect on capacitor discharge time.
16. Compare and contrast the commonly used types of capacitors.
17. Discuss the symptoms and testing procedures for leaky capacitors.

The last three chapters have dealt with an electrical property called inductance. At this point, we will turn our attention to another electrical property called *capacitance*. Technically defined, **capacitance** is the ability of a component to store energy in the form of an electrostatic charge. Based on its overall effect, capacitance is often described as the characteristic of a component that opposes any change in voltage. As you will learn, both descriptions describe the property of capacitance accurately.

◄ *OBJECTIVE 1*

**Capacitance**
The ability of a component to store energy in the form of an electrostatic charge. Often described as the characteristic of a component that opposes any change in voltage.

A **capacitor** is a component designed to provide a specific measure of capacitance. Capacitors store energy in the form of an electrostatic charge, and oppose any change in voltage. These characteristics make the capacitor useful in a variety of applications, some of which will be introduced later in this chapter.

**Capacitor**
A component designed to provide a specific measure of capacitance.

## 16.1 CAPACITORS AND CAPACITANCE: AN OVERVIEW

The simplest approach to the study of capacitance begins with a description of the capacitor. In this section, we will look at the construction of a basic capacitor and what happens when it is connected to a dc voltage source. Using this circuit as an example, we will then discuss the property of capacitance.

### Capacitor Construction

A capacitor is made up of two conductive surfaces separated by an insulating layer, as shown in Figure 16.1a. The conductive surfaces are referred to as **plates** and the insulating layer is referred to as the **dielectric.** The materials used as dielectrics are listed and described in Section 16.5.

◄ *OBJECTIVE 2*

**Plates**
The conductive surfaces of a capacitor.

The most commonly used capacitor schematic symbols are shown in Figure 16.1b. Note the similarity between the capacitor symbol and its physical construction.

**Dielectric**
The insulating layer between the plates of a capacitor.

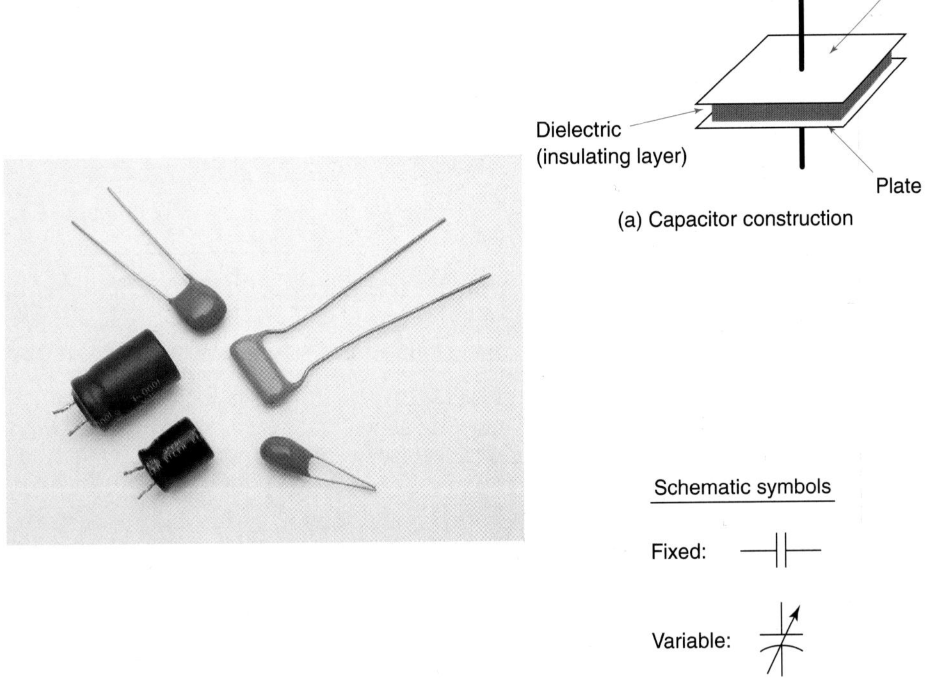


**FIGURE 16.1 Capacitor construction and schematic symbols.**

## Capacitor Charge

*OBJECTIVE 3* ▶ When a capacitor is connected to a dc voltage source, an electrostatic charge is developed across the plates of the component. This concept is illustrated in Figure 16.2. In Figure 16.2a, the capacitor is connected to a dc voltage source via a switch (SW1). Initially, the capacitor has a 0 V difference of potential across its plates. When SW1 is closed (as shown in Figure 16.2b), the source voltage is applied to the component. This connection causes two simultaneous events:

1. The positive terminal of the source draws electrons away from the upper plate of the capacitor. This leaves an excess of positive charges at this plate.
2. The negative terminal of the source forces electrons toward the lower plate of the capacitor. This produces an excess of negative charges at this plate.

The negative and positive charges on the plates of the capacitor exert a force of attraction on each other. This force of attraction permeates the dielectric, keeping the ions at the plates.

The number of charges that a capacitor can accept is determined by its capacity. Once a capacitor reaches its capacity, electron motion shown in Figure 16.2b ceases; that is, the charged capacitor blocks any flow of charge (current) in the circuit. In essence, the charged capacitor can be viewed as a dc open because it blocks any current through the circuit.

How a capacitor stores charge.

If the source is disconnected from the capacitor (as shown in Figure 16.2c), the charges remain at the plates of the component. Since the open switch blocks the flow of charge, the ions remain attracted to each other through the dielectric of the capacitor. Thus, the capacitor retains its "plate-to-plate" charge even after the voltage source has been disconnected. In effect, the capacitor stores energy in the electrostatic field between its plates. This energy remains in the capacitor as long as there is no discharge path for the component. In fact, if the capacitor is removed from the circuit (as shown in the figure), charges are still held at its plates. Note that these charges maintain a difference of potential (voltage) across the component, even though it is no longer connected to a voltage source.

*Note:* By convention, lines of electrical force are always drawn positive to negative. (See Figure 2.7).

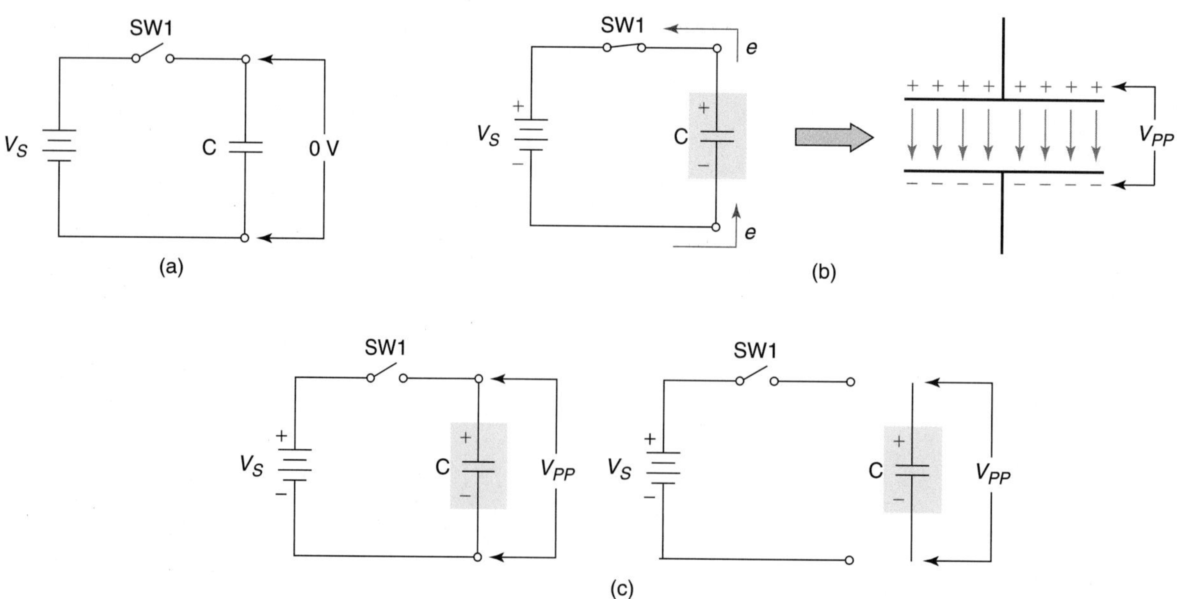


**FIGURE 16.2 Capacitor charging.**

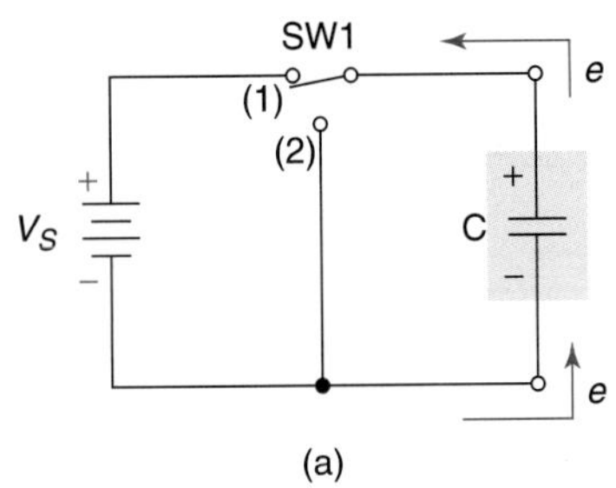


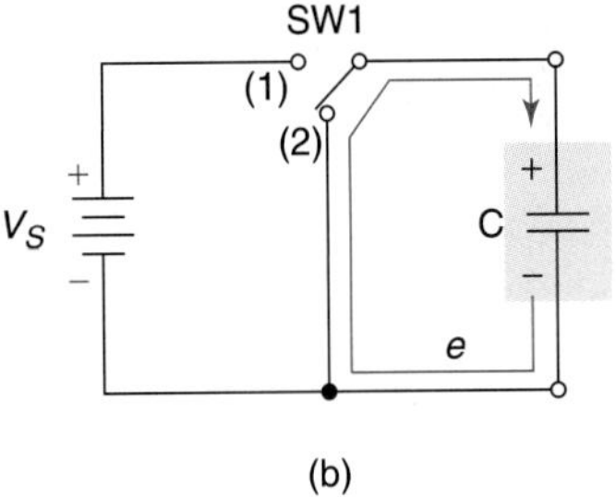


**FIGURE 16.3 Capacitor discharging.**

*A Practical Consideration:* A circuit like the one in Figure 16.3 normally contains a current-limiting resistor in series with the capacitor. It has been omitted here to simplify the principle being illustrated.

## Capacitor Discharge

To discharge a capacitor, an external path must be provided for the flow of charge (current) between the plates of the component. Such a path is shown in Figure 16.3. In Figure 16.3a, the switch is set so that the voltage source is connected to the capacitor. As described earlier, the capacitor charges until it has reached its full storage capacity. At that point, the flow of charge (current) ceases in the circuit.

When the switch is set to position (2), an external current path is provided between the plates of the capacitor. As a result, the capacitor discharges (as shown in the figure). Note that the charge flows from one plate to the other until there is no longer any difference of potential across the plates; that is, the current ceases when

$$V_C = 0 \text{ V}$$

After a capacitor has fully discharged, the difference of potential across its plates remains at 0 V until it is reconnected to a voltage source.

## A Practical Consideration

Since an external current path must be provided to discharge a capacitor, you must ensure that all capacitors have been discharged whenever you are working with capacitive circuits. An open in a capacitive circuit could prevent the capacitor(s) from discharging. For example, refer again to Figure 16.3. An open contact in the switch could prevent the capacitor from discharging when SW1 is set to position (2). In this case, the capacitor could be storing sufficient charge to injure you (or your test equipment) should you come into contact with the component leads. *Never assume that a capacitor is not holding a charge.* To do so is taking a potentially serious risk.

When working with a capacitive circuit, a shorting tool is often used to ensure that all the capacitors are discharged. The shorting tool, which is simply a conductor with an insulated handle, is placed across the terminals of the component. If any residual charge remains in the component, it is shorted out by the conductor.

## Capacity

◄ *OBJECTIVE 4*

Every capacitor can store a specific amount of charge per volt applied. For example, look at the capacitor shown in Figure 16.4. The component is storing two coulombs (C) of charge when its plate-to-plate voltage is 1 V. The amount of charge that a capacitor can store per volt applied is referred to as the **capacity** of the component. (In fact, the term *capacitance* comes from the word *capacity.*) The capacity of the component in Figure 16.4 would be written as

**Capacity**
The amount of charge that a capacitor can store per volt applied.

$$\text{Capacity } (C) = 2 \text{ C/V} \quad \text{(coulombs per volt)}$$

**FIGURE 16.4 Capacitor charge and plate-to-plate voltage.**

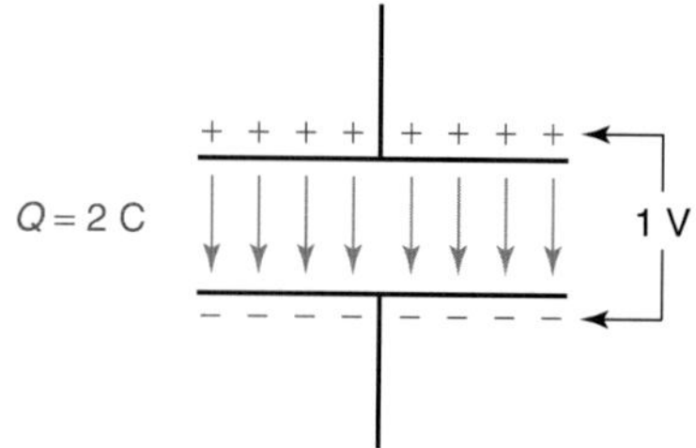


Figure 16.5 shows several capacitors and the capacity rating of each. If you compare capacitors (a) and (b), you'll see that (b) is storing twice as much charge as (a) at the same component voltage. Thus, (b) has twice the capacity of (a). By the same token, (c) has one-fourth the capacity of (b) because it takes four times the voltage to store the same amount of charge.

Based on the capacitors shown in Figure 16.5, we can make the following statement: Capacity is directly proportional to charge and inversely proportional to voltage. By formula,

$$C = \frac{Q}{V} \tag{16.1}$$

where $C$ = the capacity (or capacitance) of the component
$Q$ = the total charge stored by the component
$V$ = the voltage across the capacitor corresponding to the value of $Q$

The use of this equation is demonstrated under each capacitor in Figure 16.5.

Equation (16.1) is somewhat misleading because it implies that the value of a capacitor depends on the values of $Q$ and $V$. In fact, the value of a given capacitor is a function of the component's physical makeup. A more appropriate form of equation (16.1) is

$$Q = CV \tag{16.2}$$

which correctly implies that the charge stored by a capacitor depends on the capacity of the component and the applied capacitor voltage. This relationship will be discussed in greater detail later in this section.

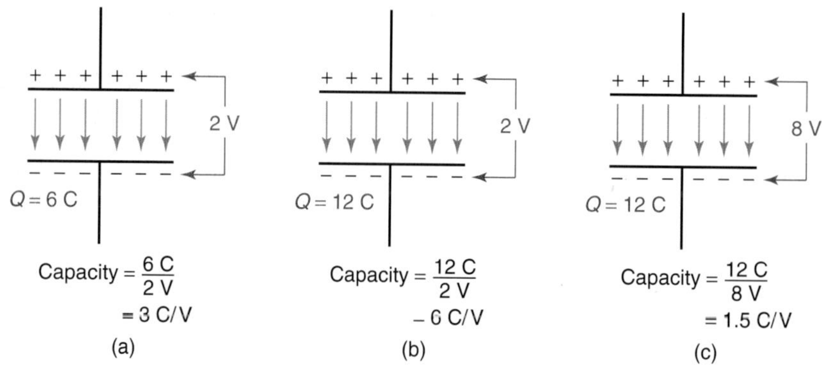


**FIGURE 16.5 Capacity (in coulombs/volt).**

## The Unit of Measure for Capacitance

**Farad (F)**
The unit of measure of capacity. One farad is a capacity of 1 coulomb per volt.

Capacitance is measured in **farads (F).** Using equation (16.1), the farad is defined as a capacity of one coulomb per volt. In other words, it is the amount of capacitance that stores one coulomb of charge for each 1 V difference of potential across its plates. By formula,

$$1\text{ F} = 1\text{ C/V} \tag{16.3}$$

## Capacitor Ratings

Most capacitors are rated in the pico-farad (pF) to micro-farad (μF) range. For some reason, capacitors in the milli-farad (mF) range are commonly rated in thousands of microfarads. For example, one well-known U.S. manufacturing company uses a 68 mF capacitor in one of its products. In their schematics and technical manuals, this component is referred to as a 68,000 μF capacitor.

Most capacitors have fairly poor tolerance ratings; that is, the actual value of a capacitor may, in most cases, vary significantly from its rated value. For this reason, *variable capacitors* are commonly found in circuits where fairly exact values of capacitance are required.

## Physical Characteristics of Capacitors

◀ *OBJECTIVE 5*

Earlier, you were told that the value of a capacitor depends on the physical makeup of the component. The relationship between a capacitor's physical makeup and its capacity is given as

$$C = (8.85 \times 10^{-12})\epsilon_r\frac{A}{d} \tag{16.4}$$

where $C$ = the capacity of the component, in farads
$(8.85 \times 10^{-12})$ = the permittivity of a vacuum, in farads per meter (F/m)
$\epsilon_r$ = the relative permittivity of the dielectric
$A$ = the area of either plate, in square meters
$d$ = the distance between the plates (i.e., the thickness of the dielectric) in meters

Equation (16.4) is useful because it identifies specific capacitor characteristics and helps us to understand their effects on the overall value of the component.

***Permittivity.*** **Permittivity** is a measure of the ease with which lines of electrical force are established within a material. For example, in Figure 16.4, the arrows between the plates represent the force of attraction between the charges. The permittivity of the dielectric is a measure of how easily these lines of force are established within the material. (In a sense, permittivity can be viewed as the electrical equivalent of permeability.) **Relative permittivity** is the ratio of a material's permittivity to that of a vacuum. The product of $\epsilon_r$ and the constant in equation (16.4) equals the actual permittivity of the dielectric material.

**Permittivity**
A measure of the ease with which lines of electrical force are established within a material. In a sense, it is the electrical equivalent of permeability.

**Relative permittivity**
The ratio of a material's permittivity to that of a vacuum ($8.85 \times 10^{-12}$ F/m).

***Plate Area.*** Every ion takes up some amount of physical space. With this in mind, take a look at the capacitors represented in Figure 16.6. The capacitor on the right would be able to store a greater amount of charge than the one on the left because there is more room for

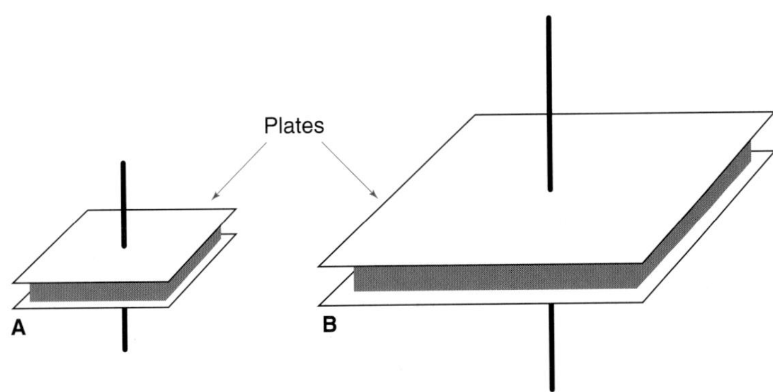
Plates
A
B
Which capacitor provides more plate area for collecting ions?
Which do you think has the greater capacity?

**FIGURE 16.6 Plate area and capacity.**

ions to accumulate on each of the plates. As given in equation (16.4), capacity is directly proportional to the area of either plate.

***Dielectric Thickness.*** In Chapter 2, you were told that the force of attraction between two charged particles is inversely proportional to the square of the distance between them. This relationship, which is described in Coulomb's law, was given in Figure 2.8 as

$$F = k\frac{Q_1Q_2}{r^2}$$

where $F$ = the force of attraction between the charges
$r$ = the distance between the charges

The dielectric thickness of a capacitor determines the distance between its plates. The greater the distance between the plates, the greater the distance between the charges that accumulate on the plates. The greater the distance between the charges, the lower the force of attraction between them. Since it is the force of attraction between the positive and negative ions that keeps them at the plates, capacity varies directly with the value of $F$ (in Coulomb's law). Therefore, capacity varies inversely with the dielectric thickness of a capacitor, as given in equation (16.4).

***A Practical Consideration.*** The relationships given among capacity, plate area, and dielectric thickness may seem purely theoretical at first, given the fact that we cannot physically alter the dimensions of a fixed capacitor. However, you will see that these relationships can be used to explain the relationship between capacitor values and total capacitance in both series and parallel circuits.

## Section Review

1. What is capacitance? What changing value does capacitance oppose?
2. What is a capacitor?
3. Draw a capacitor and label its parts.
4. Describe the means by which a dc voltage source charges a capacitor.
5. What happens to the charge on a capacitor when the component is removed from a circuit?
6. What is needed to discharge a capacitor?
7. What precaution should be taken when working on capacitive circuits?
8. What is capacity?
9. What is the relationship between capacity and charge? Capacity and voltage?
10. Define the unit of measure for capacity.
11. What is permittivity?
12. State and explain the relationship between capacity and plate area.
13. State and explain the relationship between capacity and dielectric thickness.

## 16.2 ALTERNATING VOLTAGE AND CURRENT CHARACTERISTICS

In this section, we will begin our discussion on capacitor response to sinusoidal inputs. Many of the relationships discussed in this section will seem familiar because they are very similar (in form) to the relationships used to describe inductors. At the same time, you'll see that capacitors and inductors are near-opposites when it comes to many of their operating characteristics.

## ac Coupling and dc Isolation: An Overview

Like a transformer, a capacitor is often used to provide both ac coupling and dc isolation between its input and output circuits. The means by which a capacitor provides dc isolation is illustrated in Figure 16.7. Figure 16.7a shows an uncharged capacitor in series with a load. When the switch is closed, the following occurs:

◄ *OBJECTIVE 6*

1. A complete path is provided for the current needed to charge the capacitor.
2. The capacitor charges until it reaches its capacity (in coulombs per volt).
3. Once charged, the capacitor prevents any additional flow of charge (current) through the circuit.

*Note:* The time required for the capacitor to charge is affected by the value of the load resistance. Like an *RL* circuit, a resistive-capacitive (*RC*) circuit has a time constant. *RC* time constants are discussed in detail in Chapter 17.

These statements agree with the principles of capacitor charging that you learned earlier in the chapter. The circuit conditions after the capacitor has charged are shown in Figure 16.7b. Note that:

1. The circuit current has dropped to 0 A.
2. The source voltage is dropped across the capacitor.

Since the capacitor is preventing the flow of charge through the circuit, there is no current through the load. As a result, no voltage is developed across the load. In effect, the capacitor is providing dc isolation between $V_S$ and $R_L$ (once it has charged).

***ac Coupling.*** The ac coupling provided by a capacitor is based on an important component characteristic: *a given capacitor always seeks to maintain its plate-to-plate voltage.* The effect of this principle on capacitor input and output voltages is illustrated in Figure 16.8. In Figure 16.8a, a capacitor is shown to have a 0 V difference of potential across its plates. When the input goes positive (from $t_0$ to $t_1$), the capacitor maintains its 0 V plate-to-plate voltage. As a result, the output makes a transition that is identical to the input. As the input makes its negative-going transition (from $t_1$ to $t_3$), the capacitor continues to maintain its 0 V plate-to-plate voltage. As a result, the output transition is identical (again) to the input transition.

If the capacitor has a plate-to-plate voltage other than 0 V, that voltage shows up as a difference between the peak input and output voltage values. For example, consider the circuit shown in Figure 16.8b. In this circuit, the capacitor is shown to have a 10 V difference of potential between its plates. With the polarity shown, the output peak values are always 10 V more negative than the input peak values. Thus, when the input is at $+20\ \text{V}_{\text{pk}}$, the output reaches its positive peak value of

$$\begin{aligned} V_{S(\text{pk})} - V_C &= +20\ \text{V}_{\text{pk}} - 10\ \text{V} \\ &= +10\ \text{V}_{\text{pk}} \end{aligned}$$

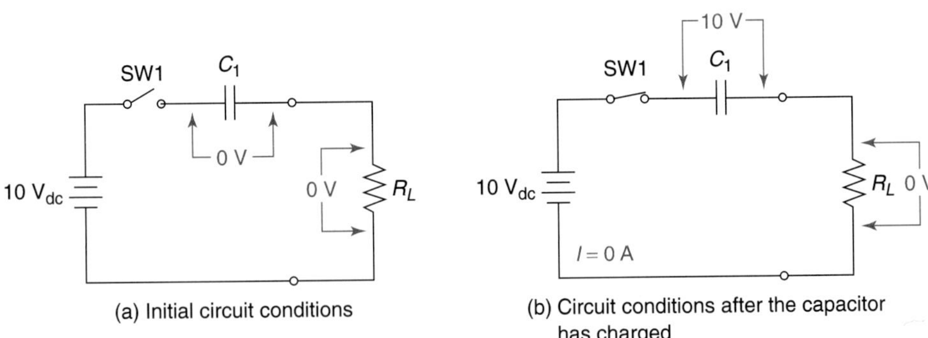


**FIGURE 16.7 Capacitor response to a dc input.**

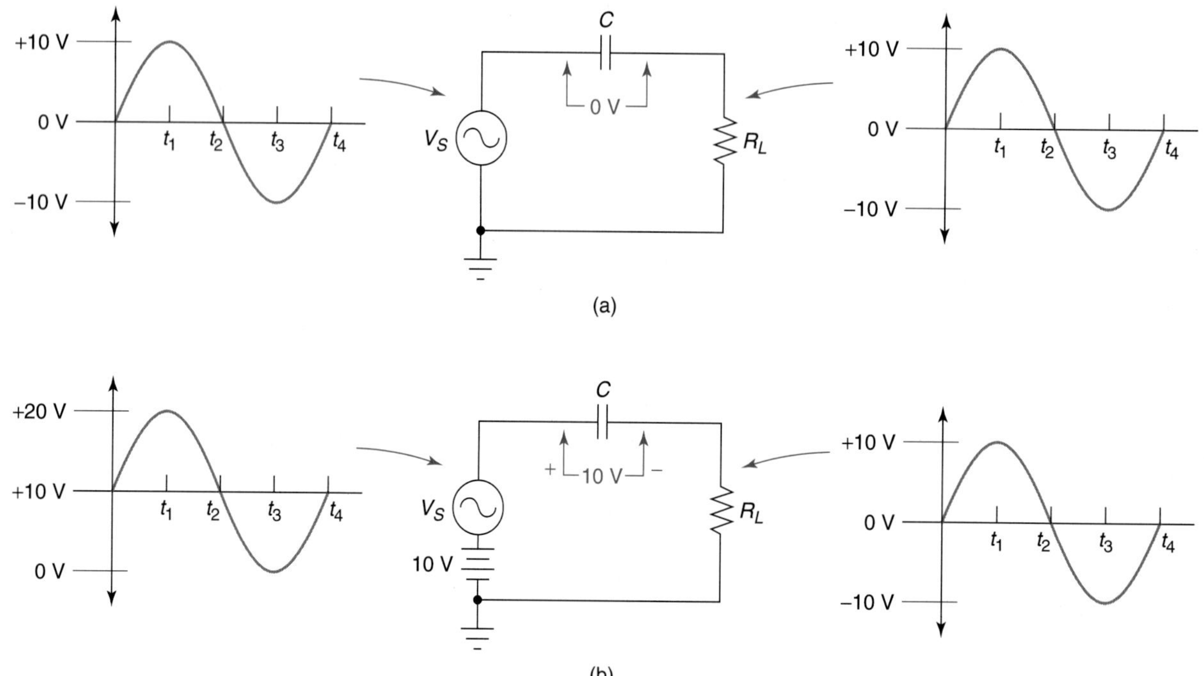


**FIGURE 16.8 ac coupling.**

When the input reaches its negative peak value, the output reaches a negative peak value of

$$V_{S(\text{pk})} - V_C = 0\ \text{V}_{\text{pk}} - 10\ \text{V} = -10\ \text{V}_{\text{pk}}$$

As you can see, the shape of the waveform has not changed from the input to the output. Only its peak (and therefore, average) values have changed. Note that the difference between the average input voltage (+10 V) and average output voltage (0 V) equals the difference of potential across the capacitor.

***Blocking a dc Offset.*** When the input to a capacitor contains a dc offset, the capacitor isolates the dc offset from the output. This concept is illustrated in Figure 16.9. As shown in the figure, the input is a 6 $V_{PP}$ sine wave riding on a +10 $V_{dc}$ offset. The capacitor couples the sine wave to the load while blocking the dc offset. As a result, the output is a 6 $V_{PP}$ sine wave centered around 0 V. In effect, the capacitor has accepted the +10 $V_{dc}$ offset as a plate-to-plate voltage (as shown in the figure).

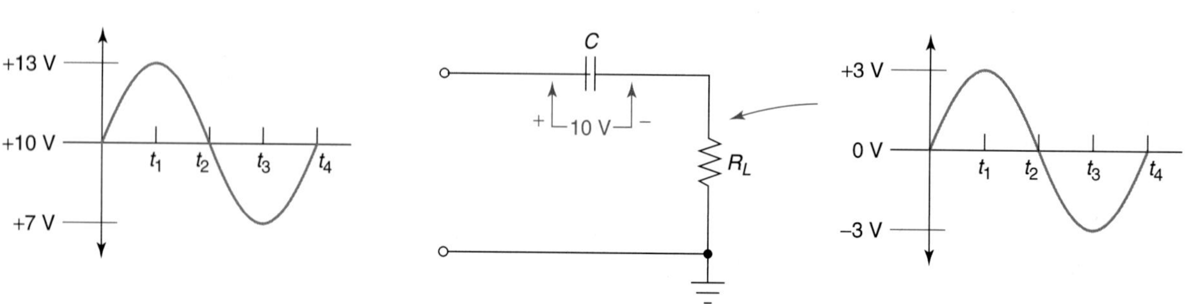


**FIGURE 16.9 A capacitor providing both ac coupling and dc isolation.**

## Capacitor Current

In a purely capacitive ac circuit, the current at any instant is directly proportional to:

1. The capacity of the capacitor(s).
2. The rate of change in capacitor voltage.

These two factors are combined in the following equation:

$$i_C = C\frac{dv}{dt} \qquad \textbf{(16.5)}$$

where $i_C$ = the instantaneous value of capacitor current
$C$ = the capacity of the component(s), in farads
$dv/dt$ = the instantaneous *rate of change* in capacitor voltage

Assuming that the value of $C$ is constant, equation (16.5) indicates that capacitor current:

1. Reaches its maximum value when the rate of change in capacitor voltage ($dv/dt$) reaches its maximum value.
2. Reaches its minimum value when the rate of change in capacitor voltage ($dv/dt$) reaches its minimum value.

As you will see, the relationship between $i_C$ and $dv/dt$ causes a phase angle to exist between capacitor current and voltage.

## Sine Wave Values of dv/dt

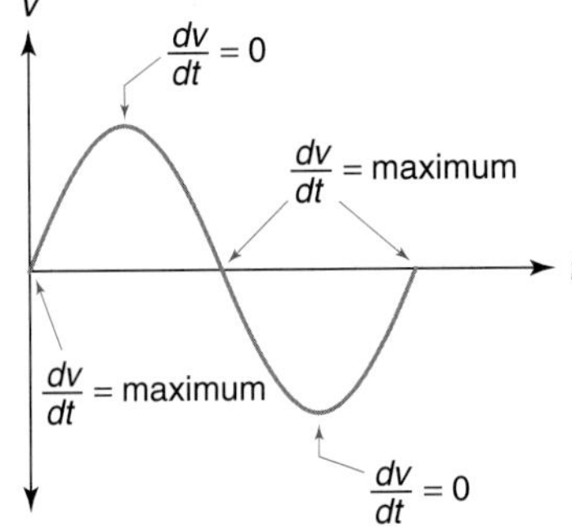


**FIGURE 16.10 Sine wave values of $dv/dt$.**

In Chapter 13, you were shown that the instantaneous rates of change for a sinusoidal current:

1. Reach their maximum values at the zero-crossings.
2. Reach their minimum values at the positive and negative peaks.

The same principles hold true for instantaneous rates of change in voltage, as shown in Figure 16.10.

When the sinusoidal voltage shown in Figure 16.10 reaches either of its peak values, there is an instant when there is no change in voltage. At that instant:

$$\frac{dv}{dt} = 0$$

This is why the peaks in Figure 16.10 are labeled as shown. Since this relationship holds true at both the positive and negative peaks, $dv/dt$ must reach its maximum value at some point between the peaks. For any sinusoidal voltage:

$$\frac{dv}{dt} = \text{(its maximum value)} \qquad \text{when} \qquad v = 0\text{ V}$$

*An Analogy:* In Chapter 13, we used this analogy to explain the relationship between an instantaneous rate of change and the peak value of a sine wave: Picture a ball thrown straight up into the air. At the moment it reaches its maximum height, it must stop for an instant before starting to fall. At that instant, its rate of change in position is zero.

In other words, $dv/dt$ reaches its maximum value at the zero-crossings of the waveform. This is why the zero-crossings in Figure 16.10 are labeled as shown.

## The Phase Relationship Between Capacitor Current and Voltage

◄ *OBJECTIVE 7*

As you know, capacitor current at any instant is directly proportional to the value of $dv/dt$, as given in the relationship

$$i_C = C\frac{dv}{dt}$$

You also know that $dv/dt$ varies inversely with the instantaneous value of capacitor voltage (as shown in Figure 16.10). These relationships are combined in Table 16.1.

The capacitor voltage and current relationships shown in Table 16.1 indicate that capacitor current reaches its maximum value when capacitor voltage reaches its minimum value, and vice versa. This means that each value peaks when the other is at its zero-crossing, as shown in Figure 16.11. The waveforms in Figure 16.11 indicate that there is a 90° phase shift between the capacitor voltage and current waveforms. *If we use the voltage waveform as a reference,* the capacitor current begins its positive half-cycle 90° before the capacitor voltage. This relationship is normally described in one of two ways:

1. Current *leads* voltage by 90°.
2. Voltage *lags* current by 90°.

**TABLE 16.1 Instantaneous Values of Capacitor Voltage and Current**

| *Capacitor Voltage ($v_C$)* | *dv/dt* | *Capacitor Current ($i_C$)* |
|---|---|---|
| Minimum ($v_C = 0$ V) | Maximum | Maximum ($i_C = I_{pk}$) |
| Maximum ($v_C = V_{pk}$) | Minimum | Minimum ($i_C = 0$ A) |

## An Important Point

The phase relationship described in this section applies only to a purely capacitive ac circuit. If we add one or more resistors to a capacitive circuit, we have what is called a resistive-capacitive (RC) circuit. As you will learn in Chapter 17, the phase relationship between circuit current and voltage changes when one or more resistors are added to a capacitive circuit.

*A Memory Aid:* A commonly used aid to remembering inductive and capacitive phase relationships is ELI the ICEman. Here's how it works:

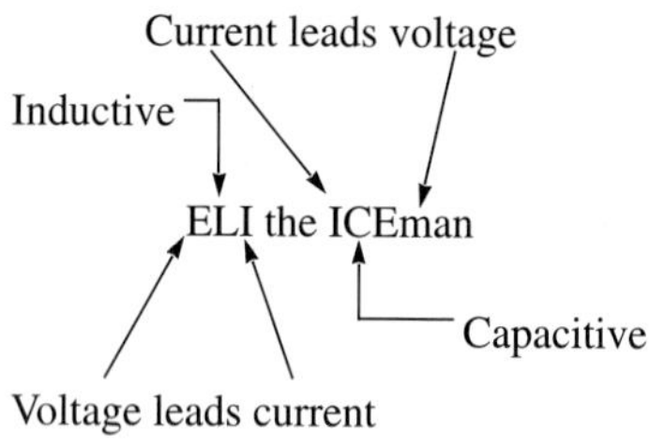


(*Note:* The letter *E* is often used to represent voltage.)

## Capacitive Versus Inductive Phase Relationships

Earlier, you were told that capacitors and inductors have many similar relationships but are near-opposites in terms of their operating characteristics. One of the most striking differences between capacitors and inductors are their current versus voltage phase relationships. These phase relationships are contrasted in Figure 16.12.

The equations in Figure 16.12 are labeled *derivation equations* because we used them to derive the current versus voltage waveforms shown. Despite the similarity between their derivation equations, inductors and capacitors produce opposite phase relationships. In an inductive circuit, current lags voltage. In a capacitive circuit, current leads voltage. In Chapter 18, you will see the effects that these phase relationships have on series and parallel inductive-capacitive (*LC*) circuits.

**FIGURE 16.11 The phase relationship between capacitor voltage and current.**

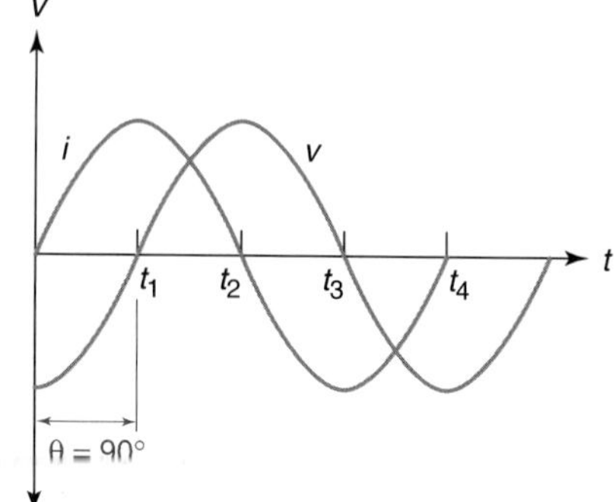

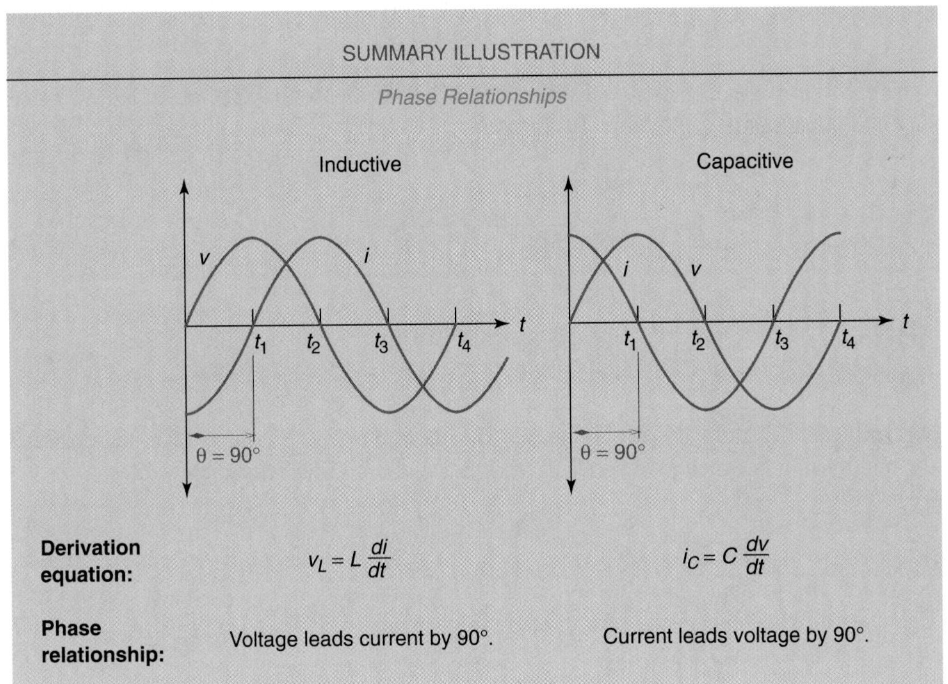


**FIGURE 16.12**

***Section Review***

1. Describe how a capacitor provides dc isolation between a source and its load.
2. On what principle of capacitor operation is the ac coupling characteristic based?
3. Describe the effect of capacitor voltage ($V_C$) on input and output peak (and average) values.
4. What happens when a capacitor receives an ac input that is riding on a dc offset voltage?
5. What is the relationship between capacitor current and the rate of change in capacitor voltage?
6. State and explain the phase relationship between capacitive current and voltage.
7. Contrast inductive and capacitive current versus voltage phase relationships.

## 16.3 SERIES AND PARALLEL CAPACITORS

◄ *OBJECTIVE 8*

In this section, we will establish the methods used to calculate total capacitance in series and parallel circuits. Among other things, we will take an in-depth look at the following relationships:

1. The total capacitance in a series circuit is lower than the lowest value component in the circuit.
2. The total capacitance in a parallel circuit equals the sum of the individual component values.

As these statements imply:

1. Total series capacitance is calculated in the same fashion as total *parallel* inductance or resistance.
2. Total parallel capacitance is calculated in the same fashion as total *series* inductance or resistance.

## Series Capacitors

When two or more capacitors are connected in series, the total capacitance is lower than any individual component value. By formula,

$$C_T = \frac{1}{\frac{1}{C_1} + \frac{1}{C_2} + \ldots + \frac{1}{C_n}} \quad \textbf{(16.6)}$$

where $C_T$ = the total series capacitance
$C_n$ = the highest numbered capacitor in the string

We will establish the basis for equation (16.6) in a moment. Example 16.1 demonstrates the relationship between total series capacitance and individual capacitor values.

### EXAMPLE 16.1

Calculate the total capacitance for the circuit in Figure 16.13.

**FIGURE 16.13**

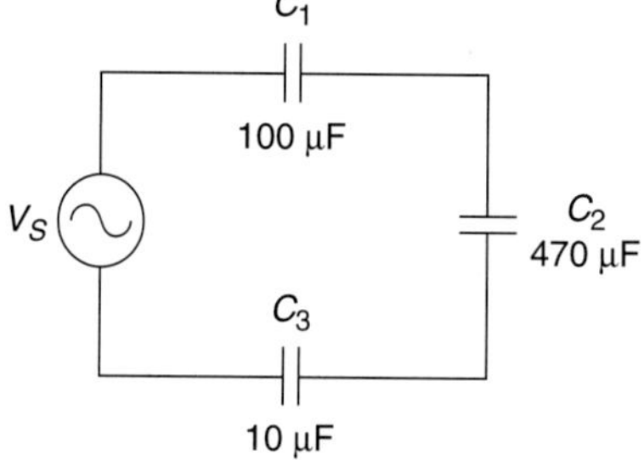


***Solution:*** Using the values shown in the figure, the total series capacitance is found as

$$C_T = \frac{1}{\frac{1}{C_1} + \frac{1}{C_2} + \frac{1}{C_3}}$$

$$= \frac{1}{\frac{1}{100\ \mu F} + \frac{1}{470\ \mu F} + \frac{1}{10\ \mu F}}$$

$$= \mathbf{8.92\ \mu F}$$

When you compare $C_T$ to the individual capacitor values, you can see that the total circuit capacitance is lower than any individual component value. This is always the case when capacitors are connected in series.

***PRACTICE PROBLEM 16.1***

A circuit like the one in Figure 16.13 has the following values: $C_1 = 1\ \mu F$, $C_2 = 2.2\ \mu F$, and $C_3 = 33\ \mu F$. Calculate the total circuit capacitance.

## Alternate Approaches to Calculating Total Series Capacitance

You may have noticed the similarity between equation (16.6) and the relationships introduced earlier for calculating total parallel resistance and inductance. For the sake of comparison, the reciprocal-method equations for two-component circuits are provided as follows:

| **Parallel Resistance** | **Parallel Inductance** | **Series Capacitance** |
|---|---|---|
| $R_T = \dfrac{1}{\dfrac{1}{R_1} + \dfrac{1}{R_2}}$ | $L_T = \dfrac{1}{\dfrac{1}{L_1} + \dfrac{1}{L_2}}$ | $C_T = \dfrac{1}{\dfrac{1}{C_1} + \dfrac{1}{C_2}}$ |

You may recall that there are several special-case approaches to calculating total parallel resistance or inductance. Similar approaches can be used to calculate the total capacitance in a series circuit under specific conditions. For example, the total capacitance of two series capacitors can be found using the *product-over-sum* method, as follows:

$$C_T = \frac{C_1 C_2}{C_1 + C_2} \tag{16.7}$$

For example, let's say we have a series circuit with values of $C_1 = 33\ \mu\text{F}$ and $C_2 = 22\ \mu\text{F}$. The total capacitance for this circuit can be found as

$$C_T = \frac{C_1 C_2}{C_1 + C_2} = \frac{(33\ \mu\text{F})(22\ \mu\text{F})}{33\ \mu\text{F} + 22\ \mu\text{F}} = 13.2\ \mu\text{F}$$

Of course, we also could have used the reciprocal approach given in equation (16.6).

When equal-value capacitors are connected in series, the total circuit capacitance can be found as

$$C_T = \frac{C_n}{n} \tag{16.8}$$

where $C_n$ = the value of any one capacitor in the series circuit
$n$ = the total number of capacitors in the series circuit

For example, the total capacitance of five 33 μF capacitors connected in series can be found as

$$C_T = \frac{C_n}{n} = \frac{33\ \mu\text{F}}{5} = 6.6\ \mu\text{F}$$

As you can see, we can calculate series capacitance values in the same fashion that we calculate parallel resistances and inductances.

## The Basis for Equation (16.6)

In equation (16.2) we established the relationship among capacity, charge, and voltage as follows:

$$C = \frac{Q}{V}$$

If we rewrite this equation to define capacitor voltage, we get

$$V_C = \frac{Q}{C} \tag{16.9}$$

FIGURE 16.14

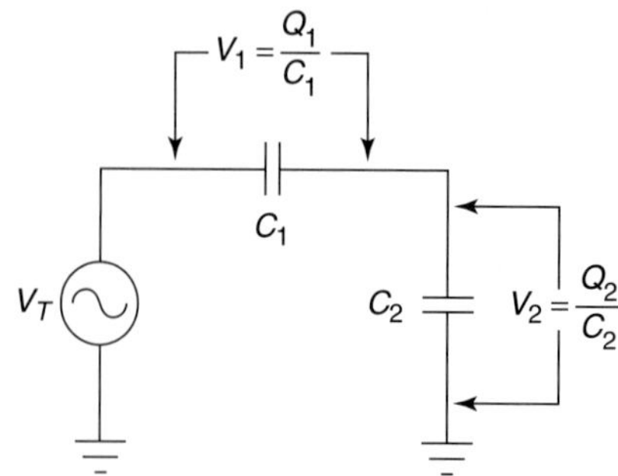


EWB

According to Kirchhoff's voltage law, the sum of the component voltages in a series circuit must equal the applied voltage. For the circuit in Figure 16.14,

$$V_T = V_{C1} + V_{C2}$$

Using equation (16.9), we can rewrite the Kirchhoff's voltage equation as follows:

$$\frac{Q_T}{C_T} = \frac{Q_1}{C_1} + \frac{Q_2}{C_2}$$

Since current is the same throughout a series circuit, so is the value of charge ($Q$). Therefore, $Q_T = Q_1 = Q_2$, and the above equation can be rewritten as

$$\frac{1}{C_T} = \frac{1}{C_1} + \frac{1}{C_2}$$

Taking the reciprocal of each side of the equation gives us

$$C_T = \frac{1}{\dfrac{1}{C_1} + \dfrac{1}{C_2}}$$

which is an abbreviated form of equation (16.6).

## Why Connecting Capacitors in Series Reduces Overall Capacitance

You have seen the derivation of equation (16.6), but that doesn't really help you to understand *why* capacitance is reduced when components are connected in series. The effect of connecting two identical capacitors in series is illustrated in Figure 16.15.

You may recall that the value of a capacitor is inversely proportional to the thickness of the dielectric, as given in the following relationship:

$$C = (8.85 \times 10^{-12})\epsilon_r\frac{A}{d}$$

**FIGURE 16.15 The equivalent of two series capacitors.**

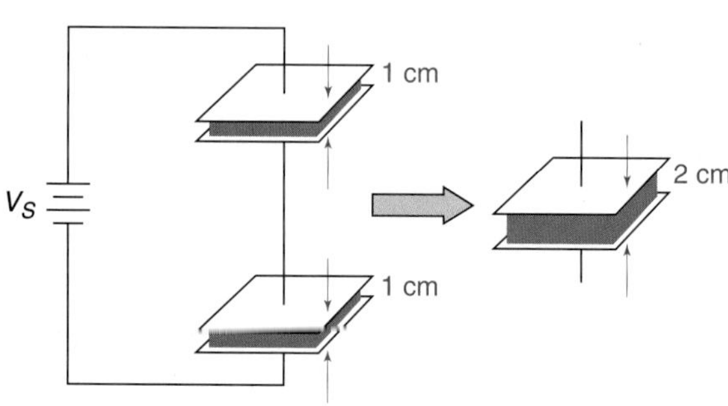

Now, assume that each capacitor in Figure 16.15 has a dielectric thickness of 1 cm. By connecting them in series, the total dielectric thickness between the terminals of the source increases to 2 cm. Since the dielectric thickness has effectively doubled, the total capacitance is one-half the value of either capacitor alone.

## Connecting Capacitors in Parallel

When capacitors are connected in parallel, the total capacitance equals the sum of the individual component values. By formula,

$$C_T = C_1 + C_2 + \ldots + C_n \quad \textbf{(16.10)}$$

where $C_n$ = the highest numbered capacitor in the parallel circuit

The use of equation (16.10) is demonstrated in Example 16.2.

**EXAMPLE 16.2**

Calculate the total capacitance in the parallel circuit shown in Figure 16.16.

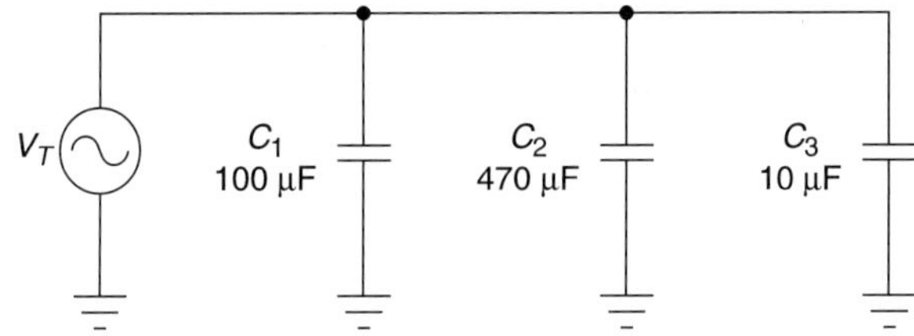


**FIGURE 16.16**

***Solution:*** Using the values shown in the figure, the total capacitance is found as

$$C_T = C_1 + C_2 + C_3 = 100\ \mu\text{F} + 470\ \mu\text{F} + 10\ \mu\text{F} = \mathbf{580\ \mu F}$$

***PRACTICE PROBLEM 16.2***

A circuit like the one in Figure 16.16 has the following values: $C_1 = 1\ \mu\text{F}$, $C_2 = 2.2\ \mu\text{F}$, and $C_3 = 100$ nF. Calculate the total circuit capacitance.

## The Basis for Equation (16.10)

As you know, the total current in a parallel circuit equals the sum of the branch currents. For the circuit in Figure 16.17, the total current is found as

$$I_T = I_1 + I_2$$

**FIGURE 16.17**

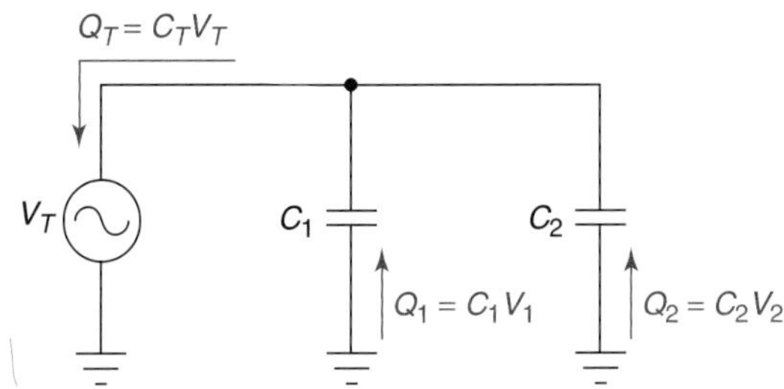

Since current is the flow of charge, it follows that the total circuit charge must equal the sum of the branch charge values; that is,

$$Q_T = Q_1 + Q_2 \tag{16.11}$$

According to equation (16.2), charge can be found as

$$Q = CV$$

Using this relationship, equation (16.11) can be rewritten as

$$C_T V_T = C_1 V_1 + C_2 V_2$$

Since the voltage is equal across parallel branches, $V_T = V_1 = V_2$. Therefore, we can divide both sides of the above equation by *V*, leaving

$$C_T = C_1 + C_2$$

which is an abbreviated form of equation (16.10).

## Why Connecting Capacitors in Parallel Increases Overall Capacitance

You have seen the derivation of equation (16.11), but that doesn't really help you to understand *why* capacitance increases when components are connected in parallel. The effect of connecting two identical capacitors in parallel is illustrated in Figure 16.18.

You were shown earlier that the value of a capacitor is directly proportional to plate area, as given in the following relationship:

$$C = (8.85 \times 10^{-12})\epsilon_r \frac{A}{d}$$

Now, assume that each capacitor in Figure 16.18 has a plate area of 1 $cm^2$. When the capacitors are connected in parallel, the source is providing equal currents to the two components. As such, the source current is encountering a total plate area of 2 $cm^2$. Since the plate area has effectively doubled, so has the total capacitance in the circuit.

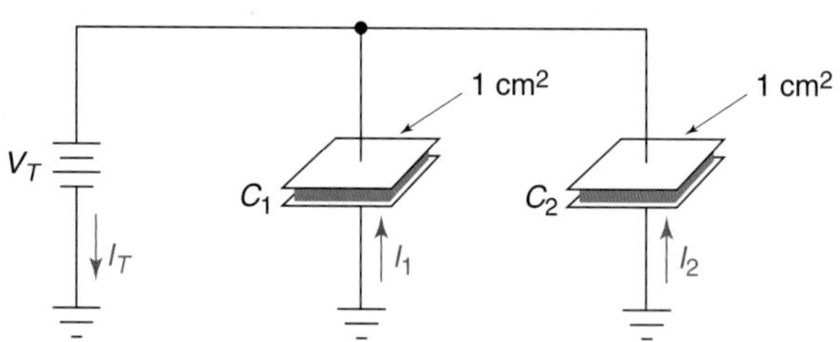


**FIGURE 16.18 Effective plate area increases when capacitors are connected in parallel.**

### Section Review

1. What is the relationship between total capacitance and the individual capacitor values in a series circuit?
2. What is the relationship between total capacitance and the individual capacitor values in a parallel circuit?
3. List the equations commonly used to calculate the total capacitance in a series circuit. Identify the special conditions (if any) of each.
4. Explain the effect that connecting capacitors in series has on the overall value of circuit capacitance.
5. Explain the effect that connecting capacitors in parallel has on the overall value of circuit capacitance.

# 16.4 CAPACITIVE REACTANCE ($X_C$)

In Chapter 13, you were introduced to inductive reactance ($X_L$): the opposition to an alternating current that is provided by an inductor. Capacitors also provide a measurable opposition to alternating current, called capacitive reactance ($X_C$). In this section, we will discuss capacitive reactance, the means by which it is calculated, and its role in ac circuit analysis.

## Capacitor Resistance

◄ *OBJECTIVE 9*

When talking about capacitor resistance, we need to distinguish between the dielectric and effective resistance of the component. Since the terminals of a capacitor are separated by a dielectric (insulator), the true resistance of the component equals the resistance of the dielectric. For all practical purposes, this resistance is generally assumed to be infinite. However, the effective resistance of the component can be quite a bit lower. This concept was illustrated in Figure 16.7, where you were shown that:

1. The effective resistance of an uncharged capacitor is extremely low, allowing a charging current to be generated in the circuit. At the start of a charge cycle, a capacitor effectively acts as a dc short.
2. The effective resistance of a charged capacitor is extremely high, blocking the source voltage from the load. Once charged, a capacitor effectively acts as a dc open.

*Reference:* An ohmmeter can be used in this fashion to test a capacitor for a common fault. Capacitor testing with an ohmmeter is discussed in Section 16.5.

These statements can be verified easily by connecting an ohmmeter to an uncharged capacitor. When first connected to the component, the ohmmeter provides a low resistance reading. However, as the current generated by the ohmmeter charges the capacitor, the resistance reading rises until it reaches the value of the dielectric resistance, for all practical purposes, infinite ohms.

Since the dielectric resistance of a capacitor is extremely high, you would expect the current in a purely capacitive ac circuit to be extremely low. However, this is not necessarily the case. Look at the circuit and waveforms shown in Figure 16.19. The source in Figure 16.19 generates a continual flow of charge (current) to and from the plates of the capacitor. This current is 90° out of phase with the capacitor voltage, as shown in the figure. At the same time, each waveform has a measurable effective (or rms) value. If we were to measure these rms values, we could determine the total opposition to current in the circuit. Using the rms values shown in the figure, this opposition could be found as

$$\text{Opposition} = \frac{V_{\text{rms}}}{I_{\text{rms}}} = \frac{5\ \text{V}_{\text{ac}}}{10\ \text{mA}_{\text{ac}}} = 500\ \Omega$$

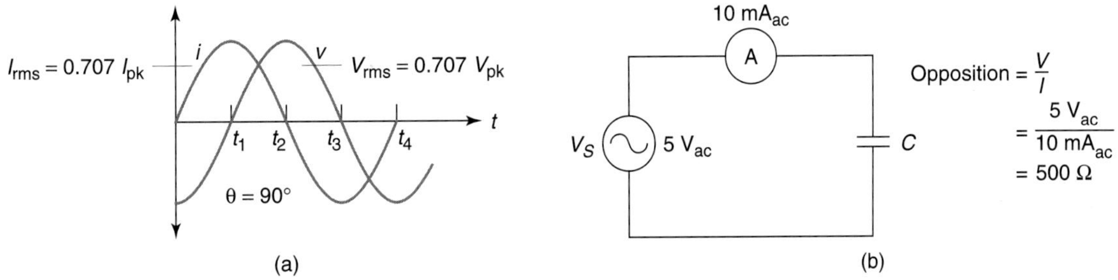

**FIGURE 16.19**

This value is significantly lower than the dielectric resistance of a capacitor. Assuming that the measured values given in the figure are correct, there is only one way to account for the difference between 500 Ω and the dielectric resistance of the capacitor. The capacitor is providing some type of opposition to current other than resistance. This "other" opposition to current is called **capacitive reactance ($X_C$).**

**Capacitive reactance ($X_C$)** The opposition provided by a capacitor to any sinusoidal current.

## Calculating the Value of $X_C$

*OBJECTIVE 10* ▶ You have already been shown one way to calculate the value of $X_C$ for a capacitor. When the rms values of capacitor current and voltage are known, the reactance of the component can be found as

$$X_C = \frac{V_{\text{rms}}}{I_{\text{rms}}} \qquad \textbf{(16.12)}$$

The use of this equation was demonstrated in Figure 16.19.

There is another (and more commonly used) method for calculating the reactance of a capacitor. When the value of a capacitor and the frequency of operation are known, the reactance provided by the component can be found as

$$X_C = \frac{1}{2\pi fC} \qquad \textbf{(16.13)}$$

We will discuss the basis for this equation in a moment. Its use is demonstrated in Example 16.3.

### EXAMPLE 16.3

Calculate the value of $X_C$ for the capacitor in Figure 16.20.

**FIGURE 16.20**

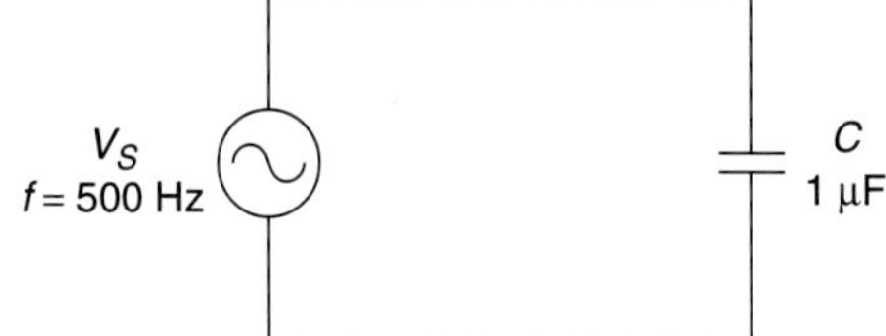


***Solution:*** The circuit is shown to have values of $C = 1\ \mu\text{F}$ and $f = 500$ Hz. Using these values in equation (16.13), we get

$$X_C = \frac{1}{2\pi fC} = \frac{1}{2\pi(500\text{ Hz})(1\ \mu\text{F})} = \mathbf{318\ \Omega}$$

***PRACTICE PROBLEM 16.3***

A 2.2 μF capacitor is operated at a frequency of 120 Hz. Calculate the reactance provided by the component.

In terms of application, equation (16.13) is usually easier to use than equation (16.12); all you need to know is the value of the capacitor and the frequency of operation. At the same time, equation (16.12) requires that you calculate (or measure) the circuit values of voltage and current. In many cases, this is easier said than done.

## *The Basis for Equation (16.13)*

Equation (16.13) is based on several relationships that were established earlier in the chapter; specifically, that the current generated in a capacitive circuit is proportional to:

1. The rate of change in voltage.
2. The total capacity in the circuit.

The instantaneous value of any point on a sine wave is proportional to the sine of the phase ($\theta$). Therefore, the rate of change in voltage is proportional to the rate of change in phase. You may recall that the rate of change in phase for a sine wave equals its angular velocity, found as $\omega = 2\pi f$. Based on these relationships, we can say that capacitor current varies directly with angular velocity. By formula,

$$I_{\text{rms}} \propto 2\pi f$$

As the above equation indicates, the current through a capacitive circuit is directly proportional to the operating frequency. At the same time, it is also directly proportional to the capacity of the component(s). You may recall that:

$$i_C = C\frac{dv}{dt}$$

This relationship indicates that the greater the capacity of a capacitor, the greater the circuit current at a given rate of change in voltage. Since $I_{\text{rms}}$ is a function of all the values of $i_C$ in each cycle, we can say that

$$I_{\text{rms}} \propto C$$

Earlier, we said that

$$X_C = \frac{V_{\text{rms}}}{I_{\text{rms}}}$$

This relationship indicates that, for any given value of $V_{\text{rms}}$, capacitive reactance is inversely proportional to capacitor current. Since capacitor current ($I_{\text{rms}}$) is directly proportional to capacity and operating frequency, $X_C$ must also be inversely proportional to these values.

We have now established the inverse proportionality between:

1. $X_C$ and angular velocity.
2. $X_C$ and capacity.

Combining these relationships into a single equation, we get:

$$X_C = \frac{1}{\omega C} \qquad \text{or} \qquad X_C = \frac{1}{2\pi f C}$$

which is equation (16.13).

## *$X_C$ and Ohm's Law*

In a purely capacitive circuit, capacitive reactance ($X_C$) can be used with Ohm's law to determine the total circuit current, just like resistance. The analysis of a simple capacitive circuit is demonstrated in Example 16.4. ◄ *OBJECTIVE 11*

## EXAMPLE 16.4

Calculate the total current through the circuit in Figure 16.21.

**FIGURE 16.21**

***Solution:*** Using the values shown, the circuit current is found as

$$I_C = \frac{V_S}{X_C} = \frac{18\ V_{ac}}{120\ \Omega} = \mathbf{150\ mA}$$

***PRACTICE PROBLEM 16.4***

A circuit like the one in Figure 16.21 has the following values: $V_S = 8\ V_{ac}$ and $X_C = 320\ \Omega$. Calculate the magnitude of the circuit current.

When the values of capacitance, source voltage, and operating frequency for a circuit are known, the circuit current can be calculated as demonstrated in Example 16.5.

## EXAMPLE 16.5

Calculate the total current for the circuit in Figure 16.22.

**FIGURE 16.22**

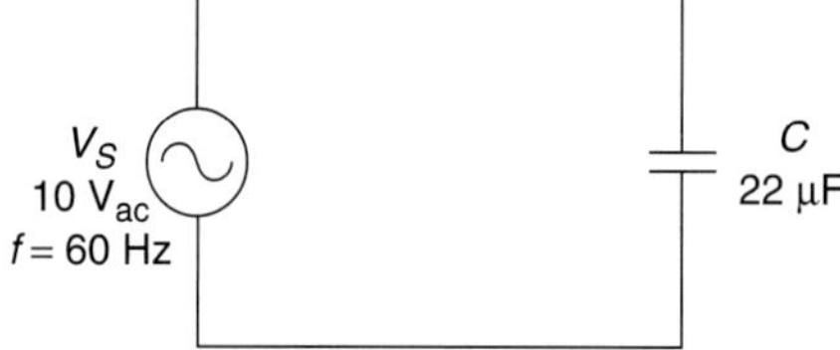


***Solution:*** First, the reactance of the capacitor is found as

$$X_C = \frac{1}{2\pi f C} = \frac{1}{2\pi(60\ \text{Hz})(22\ \mu\text{F})} \cong \mathbf{121\ \Omega}$$

Now, using $X_C$ and the value of the applied voltage, the circuit current is found as

$$I_C = \frac{V_S}{X_C} = \frac{10\ V_{ac}}{121\ \Omega} = \mathbf{82.6\ mA}$$

***PRACTICE PROBLEM 16.5***

A circuit like the one in Figure 16.22 has the following values: $C = 1\ \mu F$, $f = 3$ kHz, and $V_S = 3\ V_{ac}$. Calculate the magnitude of the circuit current.

## Series and Parallel Values of $X_C$

In Chapter 14, you were told that the total reactance in any series or parallel configuration is calculated in the same fashion as total resistance. Figure 16.23 relates this concept to capacitive reactances. (Since you have seen similar equations for both resistance and inductive reactance calculations, we do not need to rework them here.)

**FIGURE 16.23**

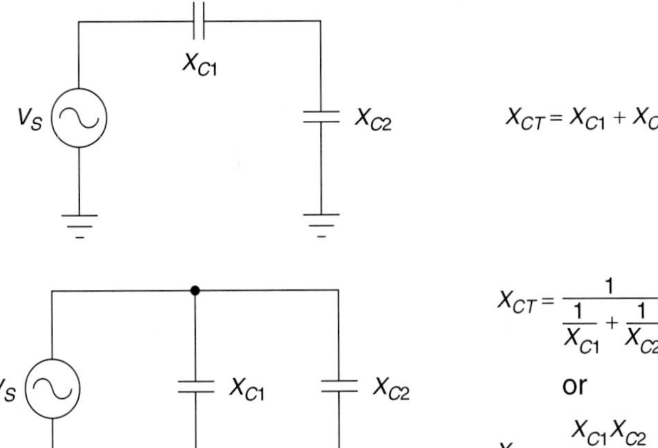


$$X_{CT} = X_{C1} + X_{C2}$$

EWB

$$X_{CT} = \frac{1}{\frac{1}{X_{C1}} + \frac{1}{X_{C2}}}$$

or

$$X_{CT} = \frac{X_{C1}X_{C2}}{X_{C1} + X_{C2}}$$

## Combining Resistance and Capacitive Reactance

In Chapter 17, we will discuss the methods used to combine any resistance and capacitive reactance values into a single impedance value. As you will see, many of the principles covered in that chapter will be very similar to the ones introduced in Chapter 14 for resistance and inductive reactance.

---

***Section Review***

1. Contrast the dielectric and effective resistance of a capacitor.
2. When do capacitors have extremely high resistance values?
3. Explain the resistance readings you get when you connect an ohmmeter to an uncharged capacitor.
4. What is capacitive reactance ($X_C$)?
5. What relationships are commonly used to calculate the value of $X_C$?
6. Which of the relationships listed in number 5 is used most often? Why?
7. In your own words, explain the basis of the following equation:

$$X_C = \frac{1}{2\pi f C}$$

8. How is the magnitude of current in a capacitive circuit calculated?
9. How is the total capacitive reactance calculated in a series circuit? A parallel circuit?

---

# 16.5 RELATED TOPICS

In this section, we will complete the chapter by discussing some miscellaneous topics that relate to capacitors, their applications, and ratings.

## Capacitor Voltage Ratings

◄ *OBJECTIVE 12*

As you know, the plates of a capacitor are separated by a dielectric (insulator). Every insulator, no matter how strong, will break down and conduct if the voltage across the component is sufficient. For example, the capacitor in Figure 16.24a is shown to have a

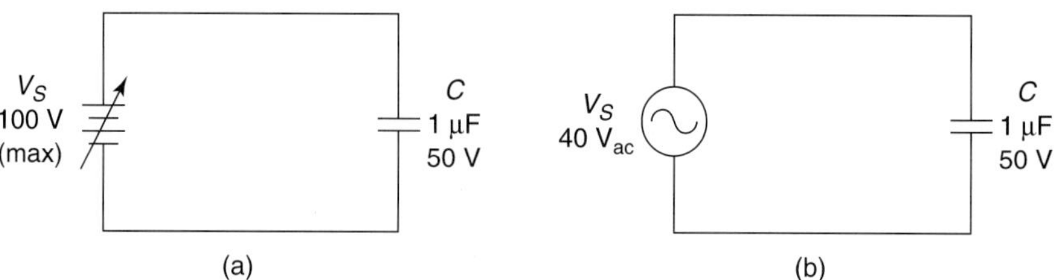


**FIGURE 16.24**

**Breakdown voltage rating**
A capacitor rating that indicates the value of capacitor voltage that may cause the dielectric to break down and conduct.

**breakdown voltage rating** of 50 V. This rating indicates that the dielectric of the component may break down and conduct if the voltage across the component exceeds 50 V.

When using capacitors in ac circuits, care must be taken to ensure that the *peak* capacitor voltage does not exceed the breakdown voltage rating of the capacitor. For example, it would appear that the capacitor in Figure 16.24b could handle the ac input voltage without any problem. However, the source rating is in rms. Converting the rated value of the source to a peak value, we get

$$V_S = \frac{40\ \text{V}_{\text{ac}}}{0.707} = 56.6\ \text{V}_{\text{pk}}$$

While this voltage may not be sufficient to break down the capacitor immediately, it will significantly reduce the life span of the component. For this reason, you should always make sure that the breakdown voltage rating of a given capacitor is greater than the peak capacitor voltage.

## Leakage Current

*OBJECTIVE 13* ▶

**Leakage current**
A low-value current that may be generated through a capacitor. (The presence of any leakage current indicates that the resistance of the dielectric is not infinite.)

Ideally, there is no current through a working capacitor. In practice, this is not the case. Whenever a voltage is applied to a capacitor, a small amount of **leakage current** may be generated through the dielectric. When generated, this current is extremely low in value and rarely affects any circuit operating characteristics or calculations. However, you need to be aware of leakage current because its presence indicates that the true resistance of a capacitor's dielectric is not infinite. Typically, the dielectric resistance of a capacitor is in the mega-ohm range or higher. The fact that capacitors have finite and measurable values of dielectric resistance is important because it affects their quality ($Q$) rating.

## Capacitor Quality (Q)

*OBJECTIVE 14* ▶

In Chapter 13, you were told that the quality ($Q$) rating of an inductor is a figure of merit that indicates how close the component comes to its ideal characteristics. The higher the $Q$ of an inductor, the closer it comes to having the characteristics of the ideal component.

Capacitors also have quality ratings. The higher the $Q$ of a capacitor, the closer it comes to having the characteristics of an ideal component. An ideal capacitor:

1. Has infinite dielectric resistance and, therefore, no leakage current.
2. Does not dissipate any power (because it has no leakage current).
3. Has an infinite breakdown voltage rating (because of its infinite dielectric resistance).

***Calculating the Quality (Q) of a Capacitor.*** In Chapter 13, you learned that $Q$ equals the ratio of reactive power to resistive power. By formula,

*Remember:* A high value for $Q$ means that most of the energy delivered to the component is stored and later returned to the circuit. In the case of a capacitor, the energy is stored in its electrostatic field. Very little energy is used by the component itself.

$$Q = \frac{P_X}{P_R}$$

where $P_X$ = reactive power (VAR)
$P_R$ = resistive power (watts)

For an inductor, the above relationship was simplified to

$$Q = \frac{X_L}{R_w}$$

where $R_w$ is the winding resistance of the coil.

As you can see, the $Q$ of an inductor equals the ratio of reactance to resistance. For a capacitor, the opposite is true. The $Q$ of a capacitor can be found as the ratio of dielectric resistance to capacitive reactance. By formula,

$$Q = \frac{R_d}{X_C} \tag{16.14}$$

where $X_C$ = the reactance of the capacitor at the frequency of operation
$R_d$ = the dielectric resistance of the component

Calculating the $Q$ of a capacitor is demonstrated in Example 16.6.

**EXAMPLE 16.6**

Calculate the $Q$ of the capacitor shown in Figure 16.25.

**FIGURE 16.25**

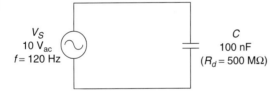


***Solution:*** At the frequency of operation shown, the reactance of the capacitor is found as

$$X_C = \frac{1}{2\pi f C} = \frac{1}{2\pi(120 \text{ Hz})(100 \text{ nF})} = \mathbf{13.26\ k\Omega}$$

With a dielectric resistance of 500 MΩ, the $Q$ of the capacitor is found as

$$Q = \frac{R_d}{X_C} = \frac{500 \text{ M}\Omega}{13.26} \text{k}\Omega = \mathbf{37{,}707}$$

This result indicates that the component stores 37,707 times as much energy as it uses, making it an extremely efficient component.

***PRACTICE PROBLEM 16.6***

A capacitor has values of $C$ = 33 nF and $R_d$ = 250 MΩ. Calculate the $Q$ of the component when it is operated at a frequency of 100 Hz.

***The Basis for Equation (16.14).*** As you have been told, the $Q$ of a reactive component equals the ratio of reactive power to resistive (or true) power. For a capacitor,

$$P_X = \frac{V_C^2}{X_C} \qquad \text{and} \qquad P_R = \frac{V_C^2}{R_d}$$

where $R_d$ is the dielectric resistance of the component. If we substitute these relationships into the basic $Q$ equation, we get

$$\begin{aligned} Q &= \frac{P_X}{P_R} \\ &= \frac{V_C^2}{X_C} \div \frac{V_C^2}{R_d} \\ &= \frac{V_C^2}{X_C} \times \frac{R_d}{V_C^2} \end{aligned}$$

The values of $V_C^2$ in the above equation cancel out, leaving

$$Q = \frac{R_d}{X_C}$$

which is equation (16.14).

***Capacitor Q Versus Frequency.*** The $Q$ of a capacitor is inversely proportional to capacitive reactance. Capacitive reactance, on the other hand, is inversely proportional to frequency. As a result, the $Q$ of a capacitor varies directly with frequency. This concept is illustrated in Example 16.7.

## EXAMPLE 16.7

EWB

The source in Figure 16.26 has the frequency range shown. Calculate the range of $Q$ values for the circuit.

**FIGURE 16.26**

***Solution:*** When $f = 20$ Hz, the reactance of the capacitor is found as

$$X_C = \frac{1}{2\pi f C} = \frac{1}{2\pi(20 \text{ Hz})(0.1\ \mu\text{F})} = \mathbf{79.6\ k\Omega}$$

and the $Q$ of the capacitor is found as

$$Q = \frac{R_d}{X_C} = \frac{150 \text{ M}\Omega}{79.6 \text{ k}\Omega} \cong \mathbf{1.88 \times 10^3}$$

When $f = 20$ kHz, the reactance of the capacitor is found as

$$X_C = \frac{1}{2\pi f C} = \frac{1}{2\pi(20 \text{ kHz})(0.1\ \mu\text{F})} = \mathbf{79.6\ \Omega}$$

and the $Q$ of the capacitor is found as

$$Q = \frac{R_d}{X_C} = \frac{150 \text{ M}\Omega}{79.6\ \Omega} \cong \mathbf{1.88 \times 10^6}$$

As demonstrated here, $Q$ increases when operating frequency increases.

**PRACTICE PROBLEM 16.7**

A circuit like the one in Figure 16.26 has the following values: $f = 1$ kHz to 8 kHz, $C = 2\ \mu F$, and $R_d = 240\ M\Omega$. Calculate the range of $Q$ values for the capacitor.

***One Final Note.*** The values of $Q$ calculated in this section have been in the thousands and higher. In Chapter 13, the values of $Q$ found for inductors were typically less than 100. As these results imply, the typical values of capacitor $Q$ are much greater than those for inductors. This point will be discussed in greater detail in Chapter 18.

## Fixed-Value Capacitors

Most capacitors are named for the material used as the dielectric. Table 16.2 contains a summary of some common types of capacitors and their characteristics. The values and applications given in the table are typical. You may come across applications and component values not listed here.

◄ *OBJECTIVE 15*

**Dielectric absorption** The tendency of a dielectric to absorb charge.

***Dielectric Absorption.*** The polypropylene and teflon listings in Table 16.2 make reference to a property called dielectric absorption. **Dielectric absorption** can be defined as the tendency of a dielectric to absorb charge. Dielectric absorption is illustrated in Figure 16.27.

Ideally, a charged capacitor holds positive and negative ions at its plates, as shown in Figure 16.27a. In practice, some of the charges may actually be absorbed into the dielectric, as shown in Figure 16.27b. During a discharge cycle, it takes longer for the charges in the dielectric to be released than it does for the charges on the plates. As a result, a time delay is added to the discharge cycle of the capacitor.

The time required for a capacitor to release any charge in its dielectric depends on the amount of absorption that has taken place. The more charge a dielectric absorbs, the longer it takes to release that charge. For this reason, capacitors with high dielectric absorption (like electrolytic capacitors) have limited high-frequency capabilities. On the

**TABLE 16.2 Common Capacitors**

| *Type* | *Value Range* | *Voltage Range* | *Tolerance* | *Applications/Comments* | *Frequency Range* |
|---|---|---|---|---|---|
| Ceramic | 10 pF–1 μF | 50 V–30 kV | ±(20+)% | Inexpensive, very common | 500 Hz to 500 MHz |
| Double layer | 0.1–10 F | 1.5–6 V | ± 20% | Memory backup | 1 kHz and below |
| Electrolytic | 0.1 μF–1.6 F | 3–600 V | − 20 to +100% | Power supply filters (usually polarized, very short life) | 1 kHz and below |
| Glass | 10–1000 pF | 100–600 V | ±10% | Long-term stability | 1 kHz to 5 GHz |
| Mica | 1 pF–0.01 μF | 100–600 V | ±10% | Radio-frequency (RF) circuits (excellent overall) | 1 kHz to 5 GHz |
| Mylar | 1 nF–50 μF | 50–600 V | ±20% | Inexpensive, very common | 500 Hz to 500 MHz |
| Oil | 0.1–20 μF | 200 V–10 kV | ±10% | High-voltage filters | 1 kHz and below |
| Polycarbonate[a] | 100 pF–30 μF | 50–800 V | ±10% | High quality, small size | dc to 1000 MHz |
| Polypropylene[a] | 100 pF–50 μF | 100–800 V | ± 5% | High quality, low dielectric absorption | dc to 1000 MHz |
| Polystyrene[a] | 10 pF–2.7 μF | 100–600 V | ±10% | Signal filters High quality, large size | dc to 1000 MHz |
| Porcelain | 100 pF–0.1 μF | 50–400 V | ±10% | Good long-term stability | 500 Hz to 500 MHz |
| Tantalum | 0.1–500 μF | 6–100 V | ±20% | High capacity, polarized | 1 kHz and below |
| Teflon | 1 nF–2 μF | 50–200 V | ± 5% | Highest quality, lowest dielectric absorption | dc to 1000 MHz |
| Vacuum | 1–5000 pF | 2–36 kV | ±10% | Transmitters | 500 Hz to 500 MHz |

Note: All values and ranges are typical.

[a]These are collectively referred to as plastic film capacitors.

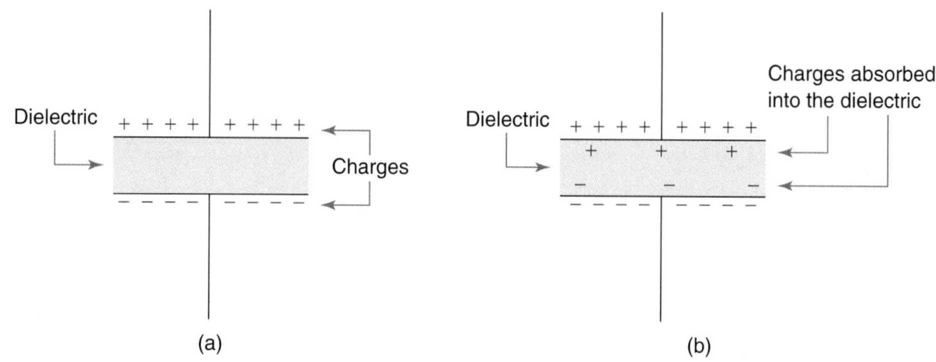


**FIGURE 16.27 Dielectric absorption.**

other hand, capacitors with low dielectric absorption are well suited for high-frequency applications. Polycarbonate, polypropylene, polystyrene, and teflon capacitors all have low dielectric absorption. As a result, they have the highest frequency ranges of the components listed in Table 16.2.

*OBJECTIVE 16* ▶ ***Electrolytic Capacitors.*** Electrolytic capacitors are extremely common components that are used in low-frequency circuits, such as dc power supplies. These capacitors are housed in metal cans that make them relatively easy to identify. Several electrolytic capacitors are shown in Figure 16.28.

Electrolytic capacitors contain an electrolyte that makes it possible to produce high-capacity components that are relatively small. However, the high capacity of the electrolytic capacitor is offset by two distinct drawbacks:

1. Electrolytic capacitors have very low dielectric resistance. As a result, the $Q$ of an electrolytic capacitor is extremely low.
2. The dielectric absorption of an electrolytic capacitor is very high. This, along with the low $Q$ of the component, severely limits its operating frequency range. As shown in Table 16.2, electrolytic capacitors are limited to frequencies below 1 kHz (which prevents them from being used in a majority of electronic circuits).

**FIGURE 16.28 Electrolytic capacitors.**

***Polarized Electrolytic Capacitors.*** Most electrolytic capacitors are polarized. The schematic symbol for a polarized electrolytic capacitor is shown in Figure 16.29a. The positive (or negative) terminal of a polarized electrolytic capacitor is usually identified as shown in Figure 16.29b.

Working with polarized capacitors.

When replacing a polarized capacitor, care must be taken to match the polarity of the capacitor to that of any dc voltage in the circuit. This point is illustrated in Figure 16.30. When inserted correctly (as shown in Figure 16.30a), the polarization of the capacitor matches the polarity of the voltage source. If a polarized electrolytic capacitor is inserted

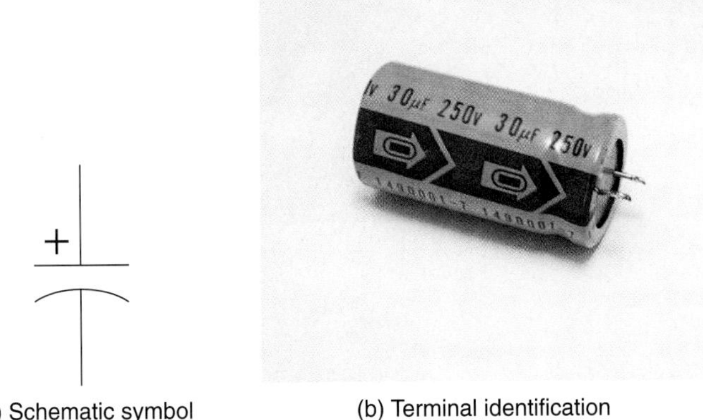


(a) Schematic symbol (b) Terminal identification

**FIGURE 16.29 Polarized electrolytic capacitors.**

incorrectly (as shown in Figure 16.30b), the component will become extremely hot. This condition can be dangerous for two reasons:

1. The component may get hot enough to cause severe burns when touched.
2. As it heats, the electrolyte of the capacitor may break down and burn, producing gases that may be trapped within the metal can. As the electrolyte continues to burn, gas pressure builds within the component and may cause it to explode within a matter of seconds.

As a polarized electrolytic heats up, it produces a strong burning odor. If you smell such an odor, *immediately remove power from the circuit.* Be sure to allow the component to cool for several minutes before attempting to handle it.

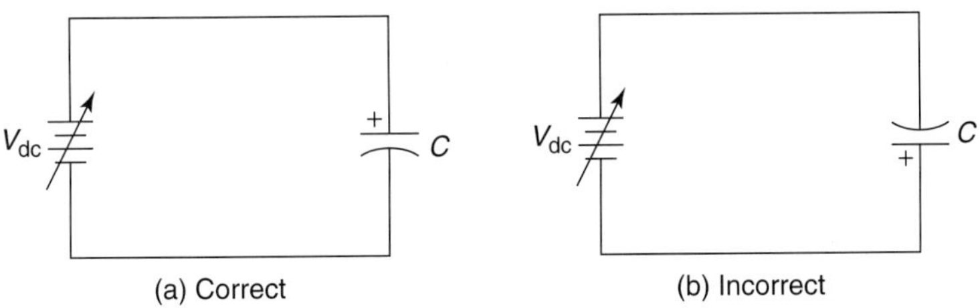


(a) Correct (b) Incorrect

**FIGURE 16.30 Connecting polarized capacitors.**

## Variable Capacitors

Various variable capacitors allow the user to adjust for specific values of capacitance. At this point, we will discuss briefly several types of variable capacitors and their applications.

***Interleaved-Plate Capacitors.*** One common variable capacitor is the interleaved-plate capacitor. Interleaved-plate capacitors are found primarily in high-power systems and older consumer electronics systems, such as stereos and televisions. An interleaved-plate capacitor consists of two groups of plates and an air dielectric. An interleaved-plate capacitor is shown in Figure 16.31a. The capacitor shown is divided into two parts, the stator and the rotor. Each of these parts contains a series of plates separated by air. When the rotor is adjusted, the rotor plates cut into the spaces between the stator plates. When the rotor is adjusted so that its plates are between those of the stator, the plates are said to be *interleaved.*

The effect of interleaving the plates is illustrated in Figure 16.31b. When the rotor shaft is turned in a clockwise direction, the interleaving of the plates increases. As this occurs, the effective plate area of the component increases (as shown in the figure). Since capacity is directly proportional to plate area, turning the rotor shaft clockwise increases the capacity of the component.

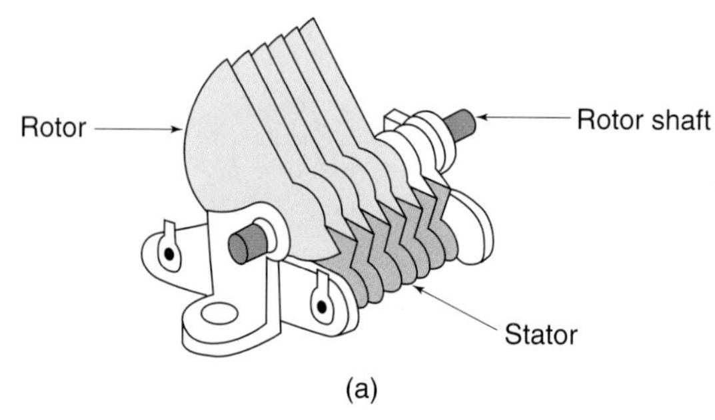


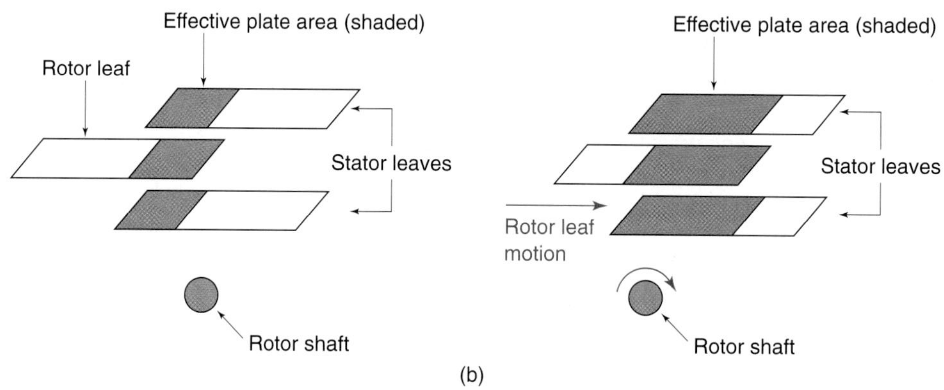


**FIGURE 16.31 Interleaved-plate capacitors.**

***Ganged Capacitors.*** Interleaved-plate capacitors are often ganged, meaning that two or more capacitors share a common rotor shaft. Two ganged interleaved-plate capacitors are illustrated in Figure 16.32a. When the rotor shaft is turned, the two capacitors experience the same change in capacitance. (In a schematic, ganged capacitors are identified as shown in Figure 16.32b.)

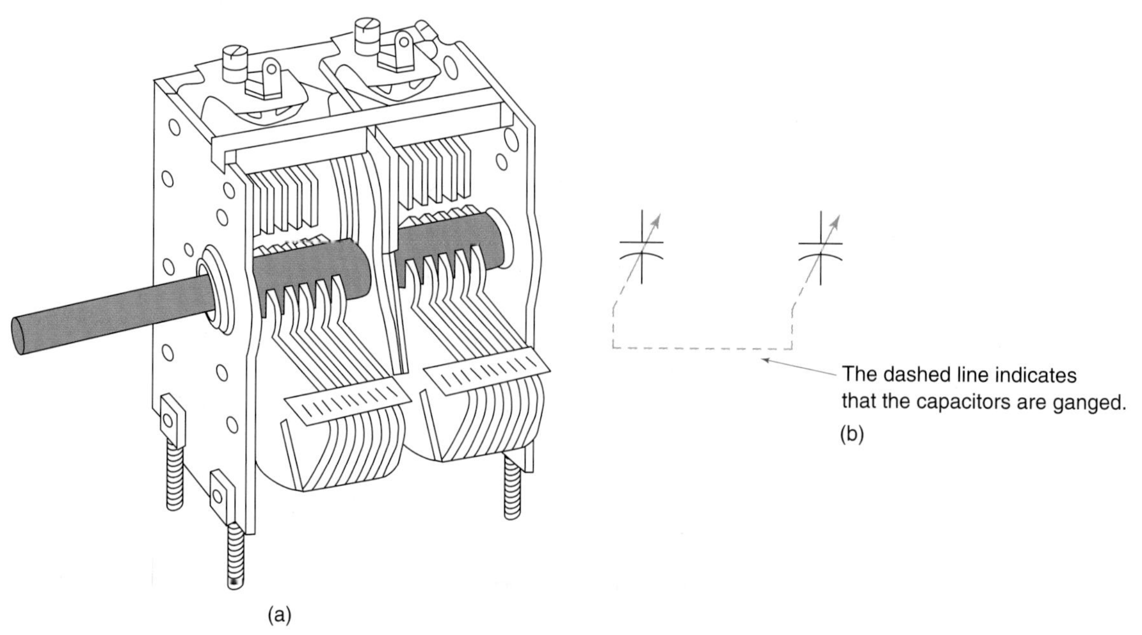


**FIGURE 16.32 Ganged capacitors.**

FIGURE 16.33 **A tuned circuit with ganged capacitors.**

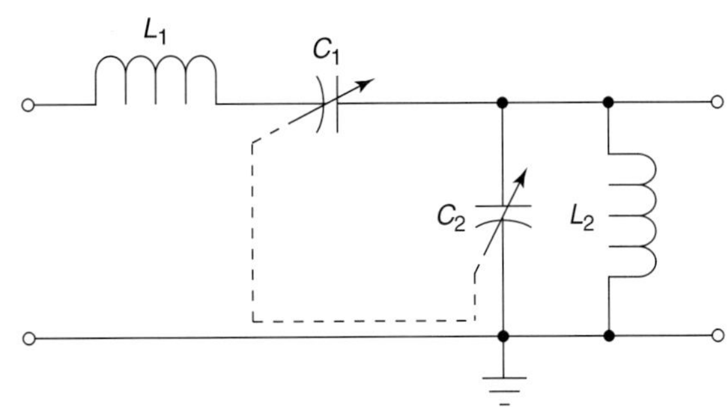


Figure 16.33 shows a typical application for a pair of ganged capacitors. The *LC* (inductive-capacitive) circuit shown is a type of **tuned circuit,** one that provides a specific output over a designated range of frequencies. The designated range of frequencies can be varied using variable inductors or capacitors. For the circuit in Figure 16.33, equal value inductors are used with two ganged capacitors. When the rotor shaft is adjusted, the two capacitors experience the same change in capacitance. Thus, the two *LC* circuits experience the same change in designated frequency range. (The subject of *LC* tuned circuits is discussed in detail in Chapter 19. For now, we simply want to establish the fact that the two circuits can be altered in identical fashion using ganged capacitors.)

**Tuned circuit**
A circuit that provides a specific output over a designated range of frequencies.

***Transmitter-Type Interleaved Capacitors.*** These components are designed specifically for high-power applications. The vanes (metal plates) are spaced further apart than they are for a common interleaved-plate capacitor. Oil dielectrics may be used to provide higher capacity and breakdown voltage ratings.

***Variable Precision Capacitors.*** These capacitors are used primarily as lab standard components and in measurement instruments. Normally, they have a maximum capacity less than or equal to 5000 pF. The capacity ratings have tolerances as low as 0.02%.

***Trimmer Capacitors.*** Trimmer capacitors are low-value components used to make fine adjustments to the total capacitance in a circuit. For example, a trimmer capacitor may be connected as shown in Figure 16.34. When connected as shown, the total capacitance between points A and B equals the sum of the capacitor values. By adjusting the trimmer capacitor, the total capacitance can be varied to any value between 100.5 pF to 110 pF. Trimmer capacitors normally have values in the low pico-farad range. Most often, their dielectrics are air, mica, or ceramic. Typically, trimmer capacitors are used in circuits that operate at frequencies up to (and including) the radio-frequency range.

*A Practical Consideration:* Trimmer capacitors are normally adjusted using a plastic alignment tool in place of a screwdriver, because a metal screwdriver can actually affect the value of a low-capacity (pF) component.

## Leaky Capacitors

◄ *OBJECTIVE 17*

The most common capacitor fault involves a drastic reduction in the resistance of the dielectric. When this occurs, the capacitor begins to leak a significant amount of current; that is, a significant amount of charge flows from plate to plate through the dielectric. An

**FIGURE 16.34 A parallel-connected trimmer capacitor.**

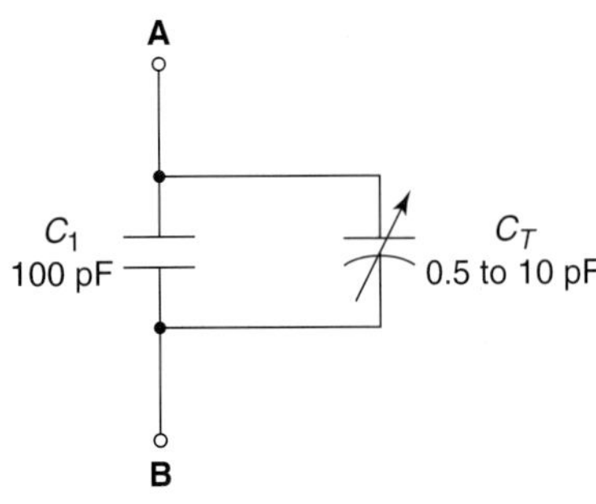

**FIGURE 16.35 The leaky capacitor allows a load voltage to be developed.**

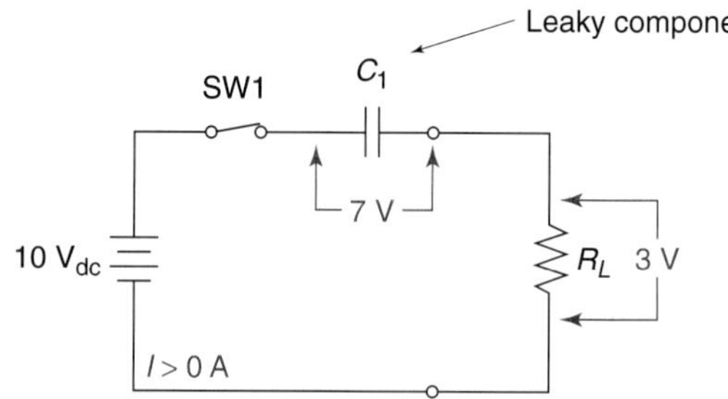


example of a *leaky capacitor* is shown in Figure 16.35. Under normal circumstances, all the source voltage would be dropped across the capacitor, leaving 0 V across the load resistance. However, since the capacitor is leaky, the source voltage can generate a current that develops the 3 V measured across the load.

In some cases, a faulty capacitor does not leak a significant amount of current until after it has warmed up. In such a case, the circuit will operate normally for a moment or two after power is first applied. For this reason, capacitors are always suspect when a circuit fails soon after power up.

Testing a suspect capacitor with an aerosol coolant.

If you suspect that a capacitor is leaky, there are several ways to test the component. In a circuit, an aerosol coolant can be sprayed on the component. If the circuit begins to operate normally, then the capacitor is the source of the fault and must be replaced. Two precautions:

1. *Use only coolants designed for use with electric circuits.* Coolants not approved for use with electric circuits may cause corrosion or may conduct current, presenting a serious fire or injury hazard.
2. *Never use an aerosol coolant on the metal can electrolytic.* The sudden cooling may cause the component to crack and thus leak electrolyte onto the circuit.

Testing a suspect capacitor with an ohmmeter.

When a capacitor cannot be tested in-circuit, it can be tested using an ohmmeter, as shown in Figure 16.36. When the ohmmeter is connected to the component, it should have a low initial resistance reading that climbs to the top of the scale. If a capacitor is leaky, either of the following may occur:

1. The resistance reading will never reach the top of the scale.
2. The resistance reading will reach the top of the scale but will drop off after being left connected for several minutes.

If either of these occurs, the capacitor is faulty and must be replaced. Note that high voltage capacitors may appear normal when tested with an ohmmeter but still fail when subjected to high voltages. In this case, replacement by trial and error is the only practical remedy.

**FIGURE 16.36 Testing a capacitor with an ohmmeter.**

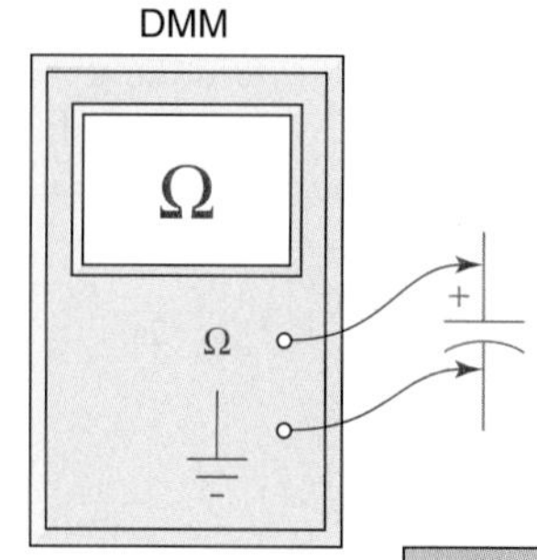


*Note*: The ohms (Ω) jack on the DMM is assumed to be positive with respect to the ground jack. This may not be the case with your DMM. Always check the polarities of the resistance jacks before connecting them to a polarized capacitor like the one shown here.

## Section Review

1. What does the breakdown voltage rating of a capacitor indicate?
2. What is leakage current?
3. List the characteristics of the ideal capacitor.
4. How is the quality ($Q$) of a capacitor calculated?
5. What is the relationship between capacitor $Q$ and operating frequency?
6. What is dielectric absorption?
7. Explain how dielectric absorption affects the discharge time of a capacitor.
8. Why are electrolytic capacitors easy to identify?
9. What is the advantage that electrolytic capacitors have over other types of capacitors?
10. List the drawbacks of using electrolytic capacitors.
11. What precautions must be taken when working with polarized capacitors?
12. What are interleaved-plate capacitors?
13. What are ganged capacitors?
14. How are trimmer capacitors used?

## Chapter Summary

Here is a summary of the major points made in this chapter:

1. Capacitance is the ability of a component to store energy in the form of an electrostatic charge. It is often described as the characteristic of a component that opposes any change in voltage.
2. A capacitor is a component designed to provide a specific measure of capacitance.
3. A capacitor is made up of two conducting surfaces (called plates) separated by an insulating layer (called a dielectric).
4. When a capacitor is connected to a dc voltage source, an electrostatic charge is developed across the plates of the component.
5. When a dc voltage source is connected across a capacitor:
   a. The positive terminal of the source draws electrons away from the positive plate.
   b. The negative terminal of the source forces electrons toward the negative plate.
   c. The positive and negative charges on the plates exert a force of attraction on each other through the dielectric. This force of attraction keeps the charges at the plates.
   d. The number of charges on the plates is determined by the capacity of the component.
6. Once charged:
   a. A capacitor blocks the flow of charge (current) in a dc circuit.
   b. The difference of potential across the capacitor plates is approximately equal to the applied voltage.
7. To discharge a capacitor, an external path must be provided for the flow of charge (current) between the plates of the component.
8. Once a capacitor has fully discharged, there is no longer a difference of potential across its plates.
9. An open in the discharge path for a capacitor can prevent the component from discharging.
   a. Never assume that a capacitor is not holding a charge.
   b. When working with a capacitive circuit, a shorting tool is often used to ensure that all the capacitors have discharged.

10. The capacity of a capacitor is the amount of charge the component can store per volt applied. Capacity is measured in coulombs per volt.
11. The amount of charge stored by a capacitor depends on its capacity and the applied capacitor voltage.
12. Capacitance is measured in farads (F). One farad is a capacity of one coulomb of charge per volt applied.
13. Capacity is determined by the physical characteristics of a capacitor, as follows:
    a. It is directly proportional to the permittivity of the dielectric.
    b. It is directly proportional to the cross-sectional area of the plates.
    c. It is inversely proportional to the distance between the plates (i.e., the thickness of the dielectric).
14. Permittivity is a measure of the ease with which lines of electrical force are established within a material, measured in farads per meter.
15. The higher the permittivity of a material, the easier it is to establish lines of electrical force within that material.
16. Capacitors and inductors are near-opposites when it comes to many of their operating characteristics.
17. A capacitor is often used to provide ac coupling and dc isolation between a source and its load.
18. A capacitor always seeks to maintain its plate-to-plate charge; that is, it opposes any change in its plate-to-plate voltage.
19. When the input to a capacitor contains a dc offset voltage, the capacitor isolates the dc offset from the load.
20. In a purely capacitive circuit, the current at any instant is directly proportional to
    a. The capacity of the capacitor(s).
    b. The rate of change in capacitor voltage.
21. In a circuit with a sinusoidal voltage source, capacitor current reaches its maximum value when capacitor voltage reaches zero, and vice versa.
22. The current-versus-voltage phase relationships of inductors and capacitors are exact opposites.
    a. In a purely inductive circuit, voltage leads current by 90°.
    b. In a purely capacitive circuit, current leads voltage by 90°.
23. When two (or more) capacitors are connected in series, total capacitance is lower than the lowest value component in the circuit.
    a. Total series capacitance is calculated in the same fashion as parallel resistance or inductance.
    b. Connecting capacitors in series effectively increases dielectric thickness (as shown in Figure 16.15), which is why overall capacitance is reduced by the series connection.
24. The total capacitance in a parallel circuit equals the sum of the branch capacitances.
25. Connecting capacitors in parallel effectively increases plate area (as shown in Figure 16.18), which is why overall capacitance is increased by the parallel connection.
26. Capacitors provide a measurable opposition to alternating current, called capacitive reactance ($X_C$).
27. Capacitive reactance is inversely proportional to capacitance and operating frequency.
28. In a purely capacitive circuit, $X_C$ can be used with Ohm's law to calculate circuit current.
29. Series and parallel values of $X_C$ are combined in the same fashion as series and parallel values of resistance (or inductive reactance).

30. The breakdown voltage rating of a capacitor indicates the value of voltage that may cause a capacitor to break down and allow conduction.
31. When using a capacitor in an ac circuit, the peak capacitor voltage must not be allowed to exceed the breakdown voltage rating of the component.
32. Whenever a voltage is applied to a capacitor, a small amount of leakage current may be generated through the dielectric of the component.
33. The presence of capacitor leakage current indicates that the true resistance of the component's dielectric is not infinite.
34. The higher the quality ($Q$) rating of a capacitor, the closer it comes to having the characteristics of the ideal component.
35. The ideal capacitor:
    - **a.** Has infinite dielectric resistance and, therefore, no leakage current.
    - **b.** Does not dissipate any power.
    - **c.** Has an infinite breakdown voltage rating.
36. The $Q$ of a capacitor is directly proportional to dielectric resistance and inversely proportional to $X_C$.
37. The $Q$ of a capacitor varies directly with operating frequency.
38. Typical values of capacitor $Q$ are much greater than those of inductors.
39. Most capacitors are named for their dielectric materials.
40. Dielectric absorption is the tendency of a dielectric to absorb charge.
    - **a.** Ideally, a dielectric is a perfect insulator that blocks positive and negative ions held at the plates.
    - **b.** In practice, some charges may be absorbed into the less-than-perfect dielectric.
41. Dielectric absorption causes a time delay to be added to the discharge time of a capacitor.
42. Capacitors with high dielectric absorption have limited high-frequency capabilities due to the time delay in component discharge.
43. Electrolytic capacitors are extremely common components used in low-frequency applications. Electrolytics:
    - **a.** Are housed in metal cans.
    - **b.** Contain an electrolyte that makes it possible to produce high-capacity components that are relatively small.
    - **c.** Typically have very low dielectric resistance and therefore low $Q$.
    - **d.** Typically have high dielectric absorption and therefore a very limited frequency range.
44. Most electrolytic capacitors are polarized.
45. When replacing a polarized electrolytic capacitor, care must be taken to match the polarity of the capacitor to that of any dc voltage in the circuit.
46. If a polarized electrolytic capacitor is inserted into a circuit backward:
    - **a.** It may become hot enough to cause severe burns when touched.
    - **b.** The burning electrolyte may produce gases, eventually causing the component to explode.
    - **c.** The burning electrolyte generally produces a strong burning odor.
47. If a polarized electrolytic smells like it's burning:
    - **a.** Immediately remove power from the circuit.
    - **b.** Allow several moments for the component to cool before attempting to touch it.
48. The interleaved-plate capacitor is a variable component used primarily in high-power systems and older consumer electronics systems.

49. An interleaved-plate capacitor consists of two groups of plates and an air dielectric, as shown in Figure 16.31.

50. Interleaved capacitors are often ganged, meaning that two (or more) capacitors share a common rotation shaft. When the shaft is rotated, the two capacitors experience the same change in capacitance.

51. The most common capacitor fault involves a drastic reduction in the resistance of the dielectric. As a result, the capacitor begins to leak a significant amount of current.

52. In many cases, a faulty capacitor will not leak a significant amount of current until after it has warmed.

53. A leaky component may be tested in-circuit by spraying it with an aerosol coolant. If the circuit operates normally when cooled, the capacitor must be replaced. A few precautions:

    **a.** Use coolants designed for use with electric circuits.

    **b.** Never spray a metal-can electrolytic capacitor with an aerosol coolant.

54. To test for capacitor leakage out of circuit, connect an ohmmeter to the component. If the capacitor is leaky, either of the following may occur:

    **a.** The resistance reading will never reach the top of the scale.

    **b.** The resistance reading will reach the top of the scale, then decline after the component warms.

55. High voltage capacitors may appear normal when tested with an ohmmeter but still fail in high-voltage circuits. In this case, replacement by trial and error is the only practical remedy.

## Equation Summary

| Equation Number | Equation | Section Number |
|---|---|---|
| **(16.1)** | $C = \frac{Q}{V}$ | 16.1 |
| **(16.2)** | $Q = CV$ | 16.1 |
| **(16.3)** | $1 \text{ F} = 1 \text{ C/V}$ | 16.1 |
| **(16.4)** | $C = (8.85 \times 10^{-12})\epsilon_r \frac{A}{d}$ | 16.1 |
| **(16.5)** | $i_C = C\frac{dv}{dt}$ | 16.2 |
| **(16.6)** | $C_T = \frac{1}{\frac{1}{C_1} + \frac{1}{C_2} + \ldots + \frac{1}{C_n}}$ | 16.3 |
| **(16.7)** | $C_T = \frac{C_1 C_2}{C_1 + C_2}$ | 16.3 |
| **(16.8)** | $C_T = \frac{C_n}{n}$ | 16.3 |
| **(16.9)** | $V_C = \frac{Q}{C}$ | 16.3 |
| **(16.10)** | $C_T = C_1 + C_2 + \ldots + C_n$ | 16.3 |
| **(16.11)** | $Q_T = Q_1 + Q_2$ | 16.3 |

| | | |
|---|---|---|
| **(16.12)** | $X_C = \frac{V_{rms}}{I_{rms}}$ | 16.4 |
| **(16.13)** | $X_C = \frac{1}{2\pi fC}$ | 16.4 |
| **(16.14)** | $Q = \frac{R_d}{X_C}$ | 16.5 |

## Key Terms

The following terms were introduced and defined in this chapter:

capacitance
capacitive reactance ($X_C$)
capacitor
capacity
dielectric
dielectric absorption
electrolytic capacitor
farad (F)
ganged
interleaved-plate capacitor
leakage current
permittivity
plates
trimmer capacitor
tuned circuit

## Practice Problems

1. A capacitor stores 200 mC of charge when the difference of potential across its plates is 25 V. Calculate the value of the component.
2. A capacitor stores 100 μC of charge when the difference of potential across its plates is 40 V. Calculate the value of the component.
3. Determine the charge that a 22 μF capacitor stores when the difference of potential across its plates is 18 V.
4. Determine the charge that a 100 pF capacitor stores when the difference of potential across its plates is 25 V.
5. Calculate the total capacitance in the series circuit shown in Figure 16.37a.
6. Calculate the total capacitance in the series circuit shown in Figure 16.37b.
7. Calculate the total capacitance in the series circuit shown in Figure 16.38a.
8. Calculate the total capacitance in the series circuit shown in Figure 16.38b.

**FIGURE 16.37**

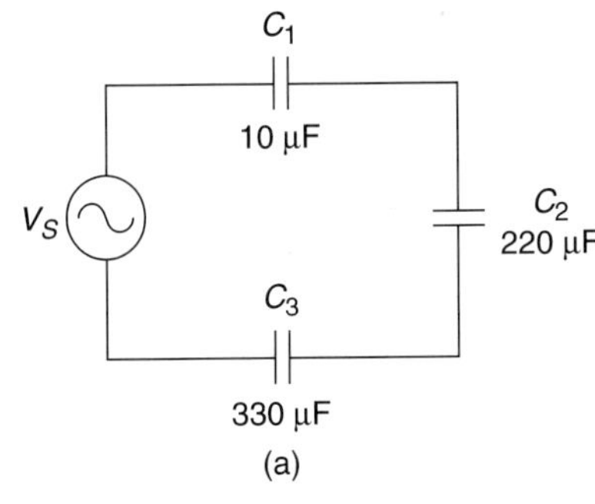

(a)

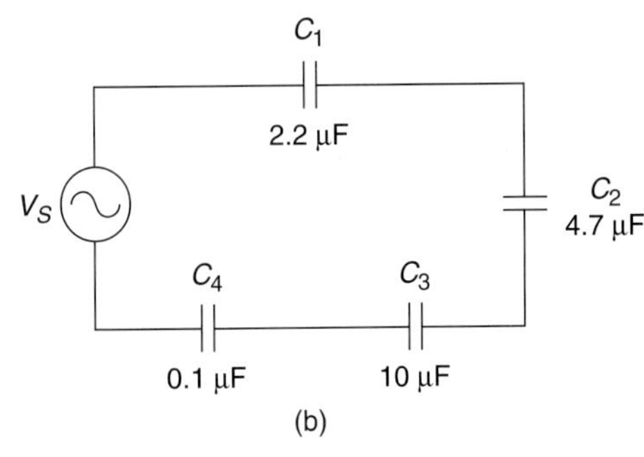

(b)

**FIGURE 16.38**

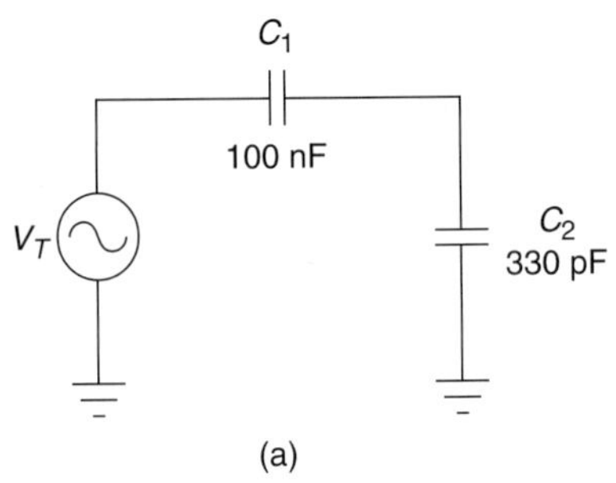

(a)

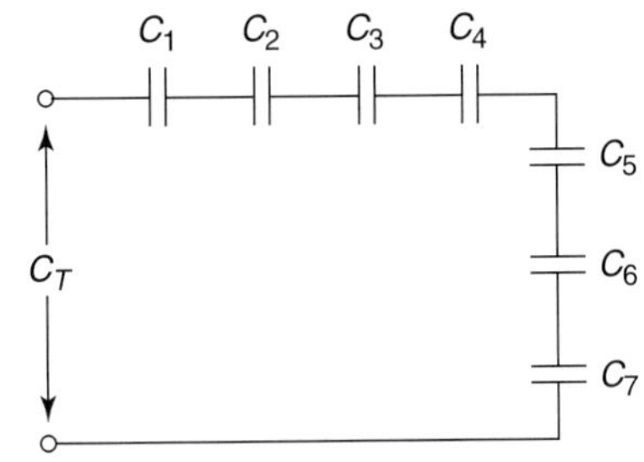

*Note*: Each capacitor has a value of 220 pF.
(b)

EWB

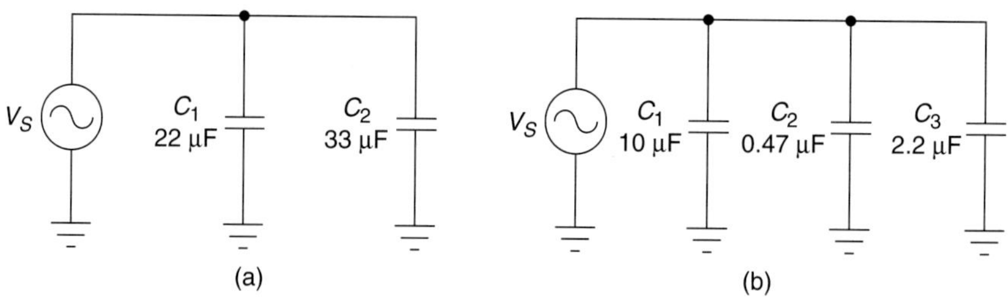


**FIGURE 16.39**

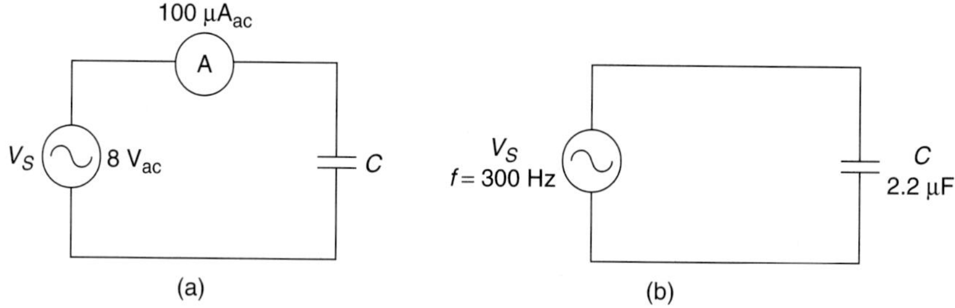


**FIGURE 16.40**

**9.** Calculate the total capacitance in the parallel circuit shown in Figure 16.39a.
**10.** Calculate the total capacitance in the parallel circuit shown in Figure 16.39b.
**11.** Calculate the value of $X_C$ for the circuit shown in Figure 16.40a.
**12.** Calculate the value of $X_C$ for the circuit shown in Figure 16.40b.
**13.** Calculate the value of $X_C$ for the circuit shown in Figure 16.41a.
**14.** Calculate the value of $X_C$ for the circuit shown in Figure 16.41b.
**15.** Calculate the value of total current for the circuit in Figure 16.42a.
**16.** Calculate the value of total current for the circuit in Figure 16.42b.

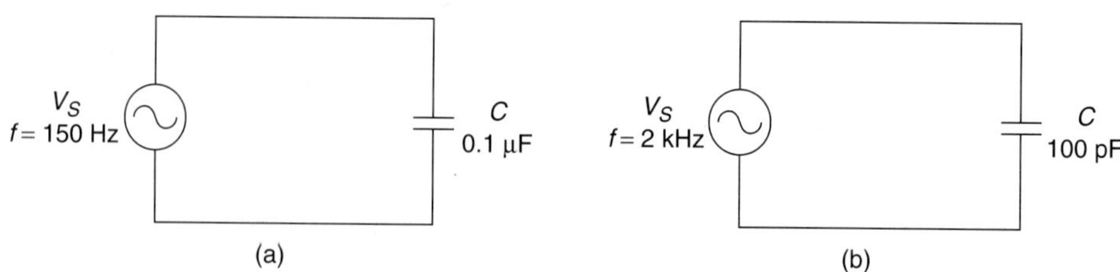


**FIGURE 16.41**

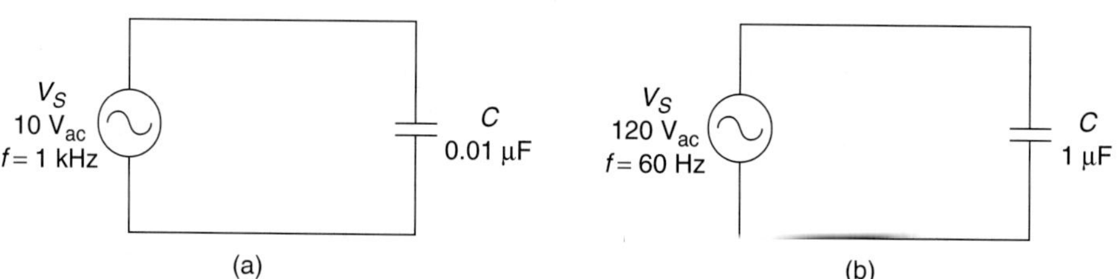


**FIGURE 16.42**

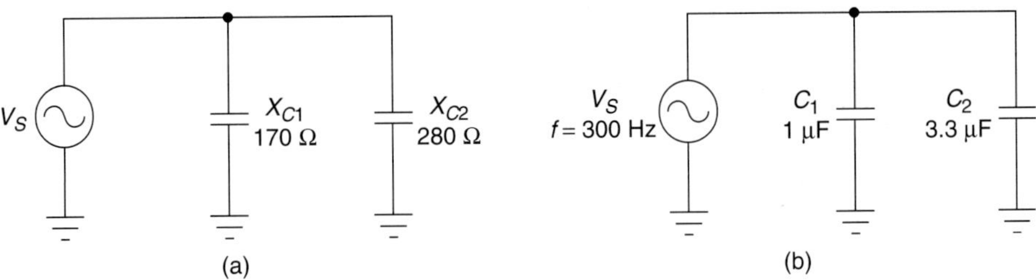


**FIGURE 16.43**

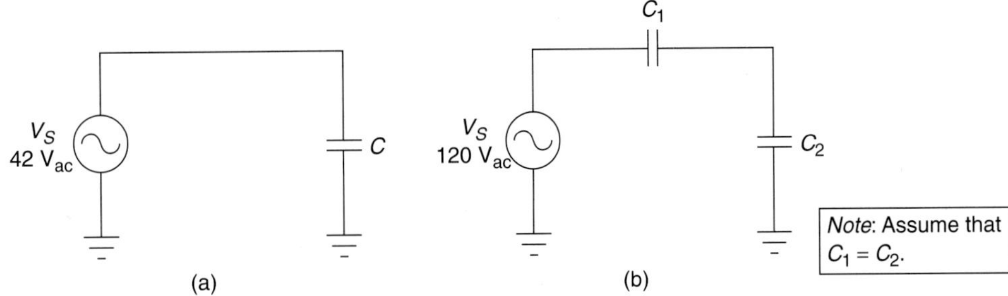


**FIGURE 16.44**

**17.** Calculate the total reactance in the circuit shown in Figure 16.43a.

**18.** Calculate the total reactance in the circuit shown in Figure 16.43b.

**19.** Determine whether or not a 50 V capacitor can be used in the circuit shown in Figure 16.44a.

**20.** Determine whether or not 100 V capacitors can be used in the circuit shown in Figure 16.44b.

**21.** The capacitor in Figure 16.45a has a dielectric resistance of 500 MΩ. Calculate the $Q$ of the component.

**22.** The capacitor in Figure 16.45b has a dielectric resistance of 180 MΩ. Calculate the $Q$ of the component.

**23.** The source in Figure 16.46a has the frequency range shown. Assume that the capacitor has a dielectric resistance of 1000 MΩ and calculate the range of $Q$ for the circuit.

**24.** The source in Figure 16.46b has the frequency range shown. Assume that the capacitor has a dielectric resistance of 200 MΩ and calculate the range of $Q$ for the circuit.

**FIGURE 16.45**

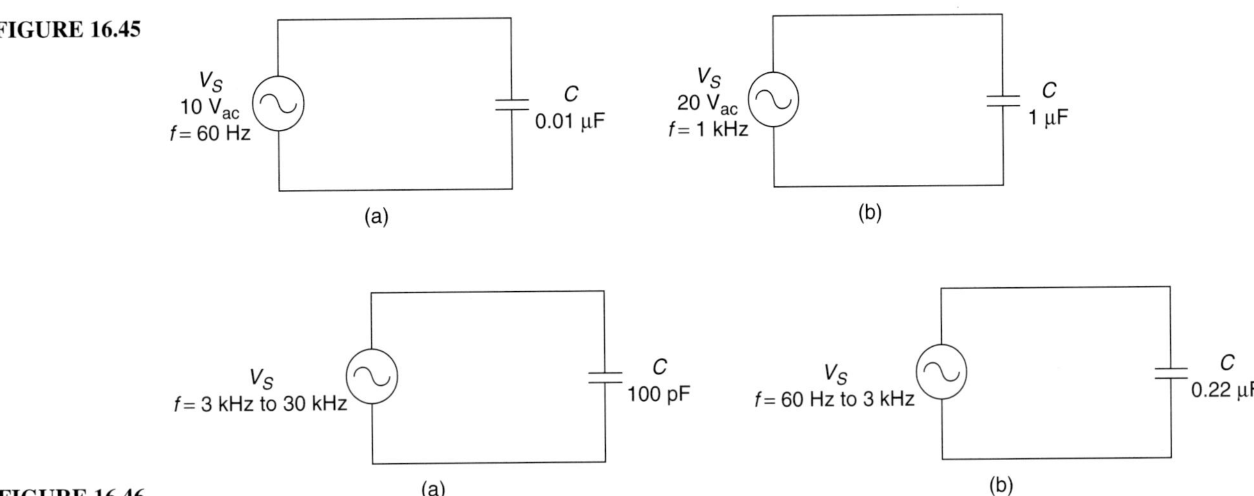


**FIGURE 16.46**

## The Brain Drain

**25.** The relative permittivity of a material is the ratio of its permittivity to that of free space. The relative permittivity ratings of several common dielectrics are as follows:

| Dielectric | Relative Permittivity |
|---|---|
| Air | 1.001 |
| Glass | 7.500 |
| Mica | 5.000 |
| Porcelain | 6.000 |
| Teflon | 2.000 |

Determine which of these dielectrics is used in the capacitor shown in Figure 16.47.

**FIGURE 16.47**

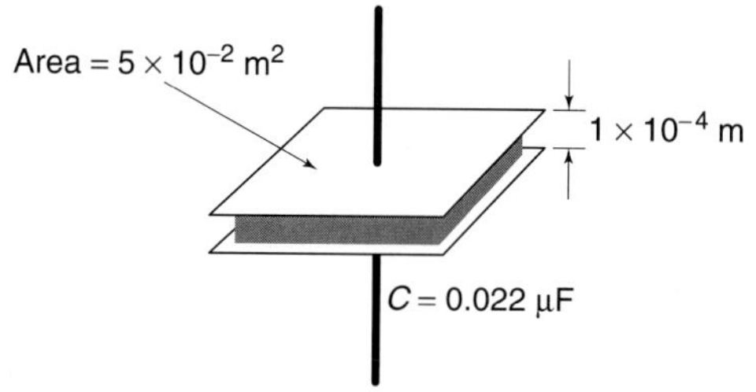


## Answers to the Example Practice Problems

**16.1** $C_T = 673.5$ nF
**16.2** 3.3 μF
**16.3** 603 Ω
**16.4** 25 mA
**16.5** 56.6 mA
**16.6** 5184
**16.7** $3.02 \times 10^6$ to $24.13 \times 10^6$

## EWB Applications Problems

| Figure | EWB File Reference |
|---|---|
| 16.9 | EWBA16_9.ewb |
| 16.11 | EWBA16_11.ewb |
| 16.14 | EWBA16_14.ewb |
| 16.19 | EWBA16_19.ewb |
| 16.23 | EWBA16_23.ewb |
| 16.26 | EWBA16_26.ewb |
| 16.39 | EWBA16_39.ewb |

# 17

# RESISTIVE-CAPACITIVE (RC) CIRCUITS

## OBJECTIVES

*After studying the material in this chapter, you should be able to:*

1. Compare and contrast series resistive, capacitive, and resistive-capacitive (*RC*) circuits.
2. Describe the voltage and phase angle characteristics and measurements for a series *RC* circuit.
3. Describe the impedance and current characteristics of a series *RC* circuit.
4. Perform a complete mathematical analysis of any series *RC* circuit.
5. Describe and analyze the frequency response of a series *RC* circuit.
6. Perform all relevant power calculations for a series *RC* circuit.
7. Compare and contrast parallel resistive, capacitive, and resistive-capacitive (*RC*) circuits.
8. Calculate any current or impedance value for a parallel *RC* circuit.
9. Perform a complete mathematical analysis of any parallel *RC* circuit.
10. Describe and analyze the frequency response of a parallel *RC* circuit.
11. Perform a complete mathematical analysis of a series-parallel *RC* circuit.
12. Describe and explain the current and voltage waveforms found in an *RC* switching circuit.
13. Describe the universal voltage curve for an *RC* switching circuit and use it to approximate the value of the capacitor voltage at any time interval.
14. Calculate the value of capacitor voltage at any point on a rise or decay curve using the universal curve equations.
15. Calculate the time constant and transition time for a series *RC* circuit.
16. State and explain the definition of the term *RC time constant.*
17. Calculate the time required for the capacitor voltage in an *RC* switching circuit to reach a designated value.

Capacitors (like inductors) are often used in conjunction with resistors. When a circuit contains capacitors and resistors, it is referred to as a resistive-capacitive (*RC*) circuit. As you will see, the principles of *RC* circuits are very similar to those of the *RL* circuits we discussed in Chapter 14.

## 17.1 SERIES *RC* CIRCUITS

**Resistive-capacitive (*RC*) circuit** A circuit that contains any combination of resistors and capacitors.

A **resistive-capacitive (*RC*) circuit** is one that contains any combination of resistors and capacitors. Several examples of basic *RC* circuits are shown in Figure 17.1. As we progress through this chapter, we will develop the current, voltage, power, and impedance characteristics of *RC* circuits. First, we will compare series *RC* circuits to purely resistive and purely capacitive series circuits. The goal is to show the similarities and differences among the three.

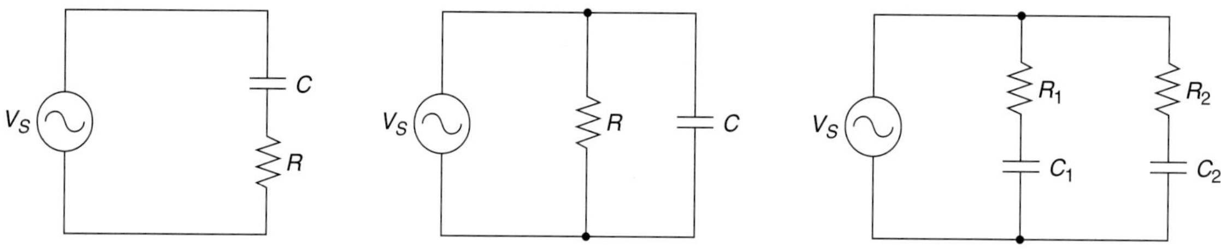


FIGURE 17.1

### *Series Circuits: A Comparison*

*OBJECTIVE 1* ▶ In Chapters 5 and 16, we established the fundamental characteristics of resistive and capacitive series circuits:

1. Current is the same at all points throughout the circuit.
2. The applied voltage equals the sum of the individual component voltages.
3. The total impedance (resistance or reactance) equals the sum of the individual component impedances.

These relationships are provided (as equations) in Figure 17.2.

The relationships listed for the resistive and capacitive circuits should be familiar. As we progress through this section, we will discuss each relationship listed for the *RC* circuit in Figure 17.2. First, let's compare them to those given for the other two circuits. As the $I_T$ equation shows, current is constant throughout the series *RC* circuit, just as it is in the other two circuits. The voltage and impedance relationships, however, are more similar to those of a series resistive-inductive (*RL*) circuit than either of the circuits shown in Figure 17.2.

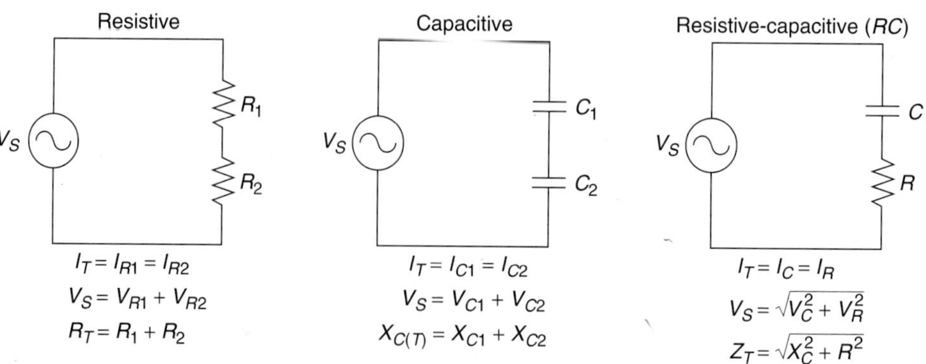


FIGURE 17.2

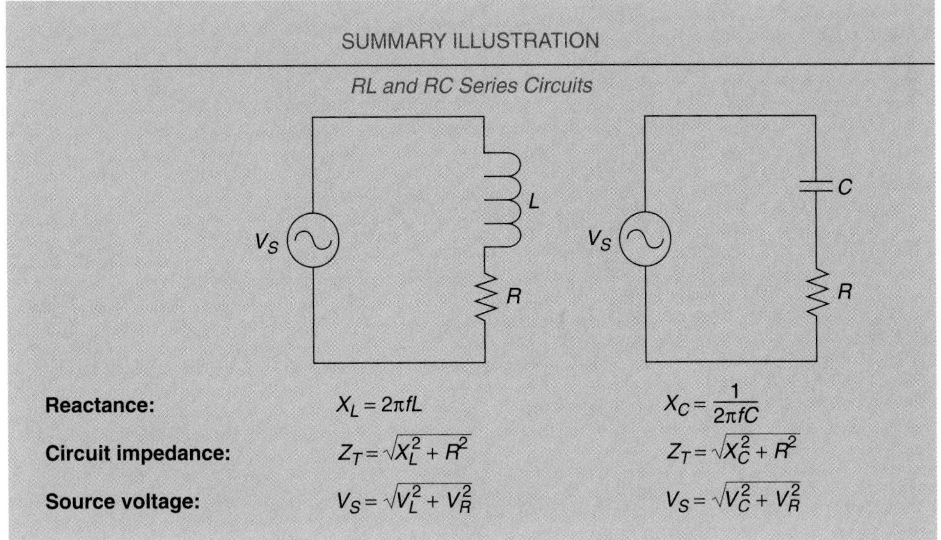

SUMMARY ILLUSTRATION
RL and RC Series Circuits
$V_S$
$L$
$R$
$V_S$
$C$
$R$
Reactance:
$X_L = 2\pi fL$
$X_C = \frac{1}{2\pi fC}$
Circuit impedance:
$Z_T = \sqrt{X_L^2 + R^2}$
$Z_T = \sqrt{X_C^2 + R^2}$
Source voltage:
$V_S = \sqrt{V_L^2 + V_R^2}$
$V_S = \sqrt{V_C^2 + V_R^2}$


**FIGURE 17.3**

The voltage and impedance relationships for series *RC* and series *RL* circuits are compared in Figure 17.3. As you can see, the relationships listed are nearly identical. This should make it relatively easy for you to make the transition from *RL* circuits to *RC* circuits.

## RC Circuit Voltage and Impedance

◄ *OBJECTIVE 2*

In a series *RC* circuit, the geometric sum of the component voltages equals the source voltage. By the same token, the geometric sum of resistance and capacitive reactance equals the total impedance (opposition to current) in the circuit. Even though we are dealing with geometric sums, the *RC* circuit still operates according to the rules we have established for all series circuits:

1. Component voltages add up (geometrically) to equal the source voltage.
2. Resistance and reactance add (geometrically) to equal the total impedance.

Another important characteristic of *RC* circuits is that the values of source voltage and total impedance each have a phase angle that is normally expressed as part of the value. For example, the circuit in Figure 17.4 was analyzed using the principles and techniques you will learn later in this section. As you can see, the source voltage and total impedance each have a phase angle. (The same principle was demonstrated for series *RL* circuits in Section 14.1.)

## Series Voltages

In Chapter 16, you were told that current leads voltage by 90° in a capacitive circuit. In a resistive circuit, voltage and current are always in phase. These two relationships are illustrated in Figure 17.5. If we combine the circuits in Figure 17.5, we get the series circuit and

**FIGURE 17.4**

$V_S$ 12 $V_{ac}$

$X_C$ 150 Ω

$R$ 150 Ω

$I_T = 113$ mA

$V_S = 12\ V\angle{-45°}$

$Z_T = 106\ \Omega\angle{-45°}$

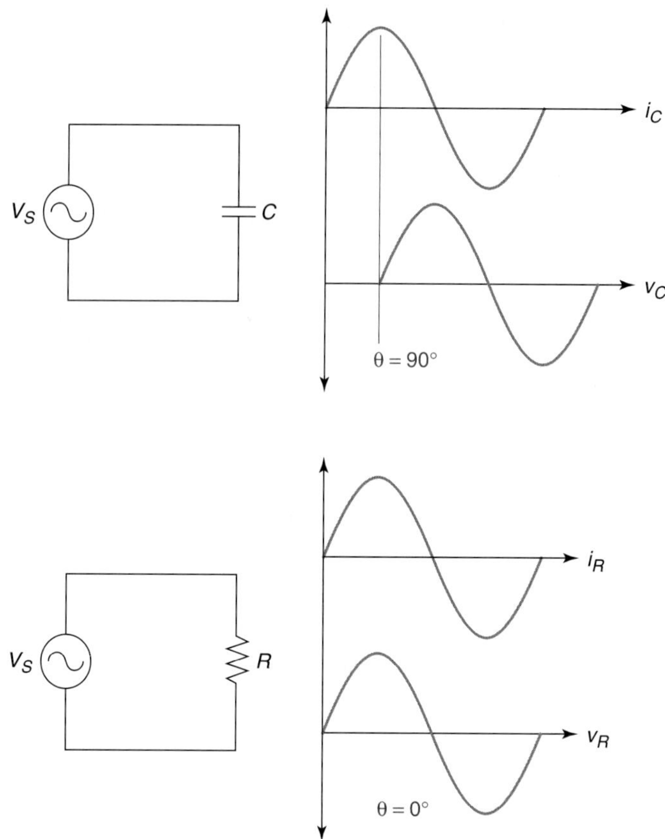


**FIGURE 17.5**

waveforms shown in Figure 17.6. Since current is constant throughout a series circuit, the capacitor and resistor voltage waveforms are referenced to a single current waveform. As you can see:

1. The capacitor voltage lags the circuit current by 90°.
2. The resistor voltage is in phase with the circuit current.

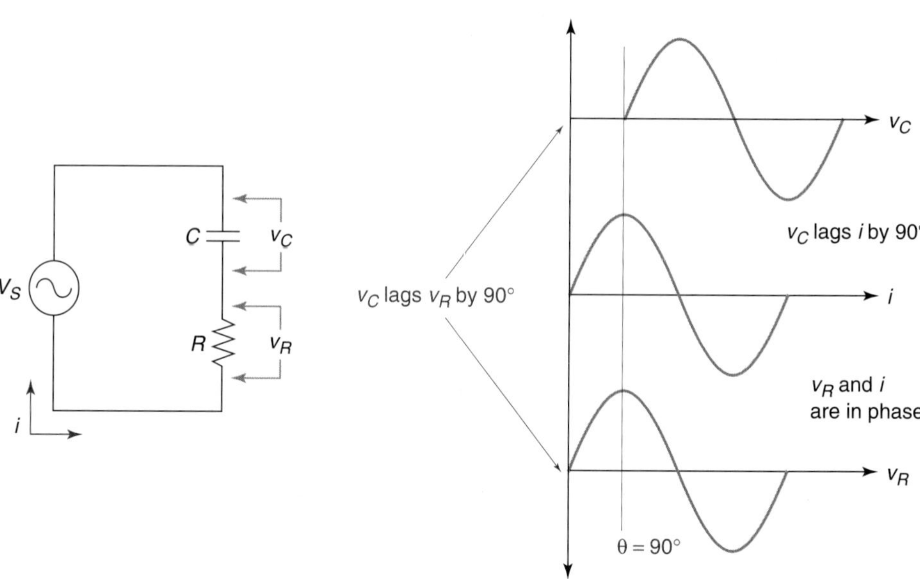


**FIGURE 17.6**

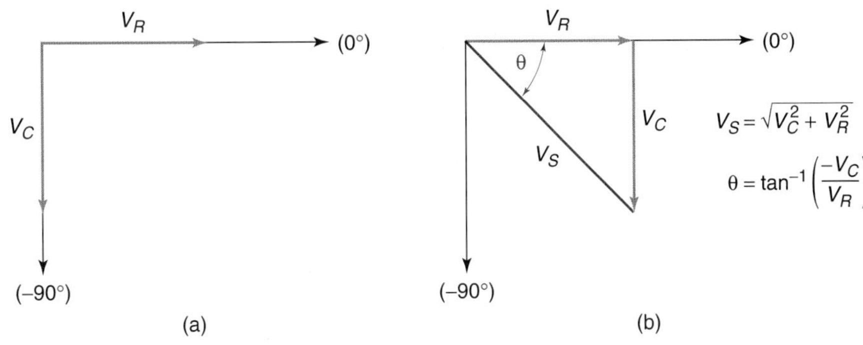


**FIGURE 17.7**

Therefore, capacitor voltage lags resistor voltage by 90°. Another way of expressing this relationship is to say that resistor voltage leads capacitor voltage by 90°. This phase relationship holds true for every series *RC* circuit with an ac source.

Since $V_R$ leads $V_C$ by 90°, the values of the component voltages are represented using phasors, as shown in Figure 17.7a. By plotting the value of $V_R$ on the *x*-axis (0°) and the value of $V_C$ on the negative *y*-axis (−90°), the phasor relationship between the two is represented in the graph.

Like any other series circuit, the source voltage in an *RC* circuit must equal the sum of the component voltages. Since the component voltages in an *RC* circuit are phasor values, they must be added geometrically. Therefore, the sum of the component voltages represented by the phasors in Figure 17.7b can be found as

$$V_S = \sqrt{V_C^2 + V_R^2} \tag{17.1}$$

*The phase angle of $V_S$ indicates the phase difference between the source voltage and the circuit current.* In Chapter 12, you were shown that this phase angle can be found as

$$\theta = \tan^{-1}\frac{y}{x}$$

where $\tan^{-1}$ = the inverse tangent of the fraction
$y$ = the magnitude of the *y*-axis value
$x$ = the magnitude of the *x*-axis value

Relating this equation to Figure 17.7b, the phase angle of the source voltage can be found as

$$\theta = \tan^{-1}\frac{-V_C}{V_R} \tag{17.2}$$

Note that $V_C$ is represented as a negative value in the equation because its phasor falls on the negative *y*-axis (as shown in Figure 17.7). Example 17.1 demonstrates the calculation of source voltage for a series *RC* circuit.

**EXAMPLE 17.1**

The component voltages in Figure 17.8 were measured using an ac voltmeter. Calculate the source voltage for the circuit.

***Solution:*** Using equation (17.1), the magnitude of the source voltage is found as

$$V_S = \sqrt{V_C^2 + V_R^2} = \sqrt{(15\text{ V})^2 + (10\text{ V})^2} = \mathbf{18\text{ V}}$$

The phase angle of the source voltage (relative to the circuit current) can be found as

$$\boldsymbol{\theta} = \tan^{-1}\left(\frac{-V_C}{V_R}\right) = \tan^{-1}\left(\frac{-15\text{ V}}{10\text{ V}}\right) = \tan^{-1}(-1.5) = \mathbf{-56.3°}$$

Therefore, the source voltage for the circuit has a value of $V_S = 18\text{ V}\angle{-56.3°}$.

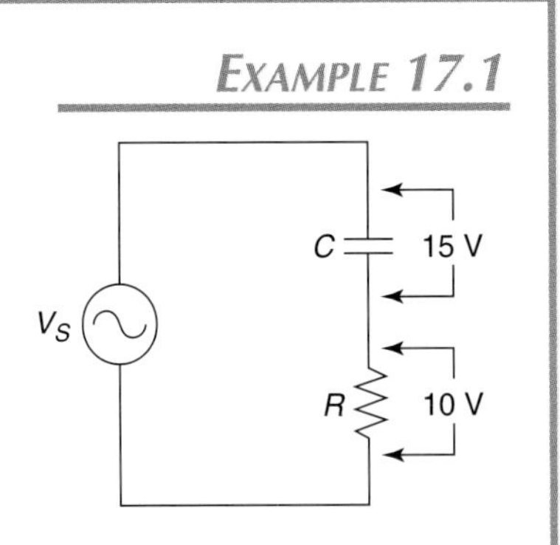


**FIGURE 17.8**

*PRACTICE PROBLEM 17.1*

A circuit like the one in Figure 17.8 has measured values of $V_C = 20\ \text{V}_{\text{ac}}$ and $V_R = 25\ \text{V}_{\text{ac}}$. Calculate the magnitude and phase angle of the source voltage.

The source voltage phase angle.

Note that the phase angle in Example 17.1 indicates that the source voltage lags the circuit current by 56.3°. For any *RC* circuit, the limits on the phase angle between the source voltage and circuit current are given as follows:

*Remember:* In a mathematical relationship involving two negative numbers, the less negative number is considered to be greater than the other number, which is why θ is written as being greater than −90° in equation (17.3).

$$-90^\circ < \theta < 0^\circ \qquad \textbf{(17.3)}$$

This range of values for θ makes sense when you consider that:

1. The phase angle of a purely resistive circuit is 0°.
2. The phase angle of the voltage (relative to the circuit current) in a purely capacitive circuit is −90°.

A circuit containing both resistance and capacitance is neither purely resistive nor purely capacitive. Therefore, it must have a voltage phase angle that is less than 0° and greater than −90°. As you will see, the value of θ for a series *RC* circuit depends on the ratio of capacitive reactance to resistance.

## Measuring the Source Phase Angle (θ)

In Chapter 14, you were shown how to measure the source phase angle for a series *RL* circuit. The same technique can be used to measure the source phase angle for a series *RC* circuit. Since circuit current is in phase with resistor voltage, θ can be measured by measuring the phase angle between the source voltage ($V_S$) and the resistor voltage ($V_R$). Figure 17.9 shows how a dual-trace oscilloscope is used to measure the value of θ for a series *RC* circuit. When the oscilloscope settings are adjusted properly, the display will resemble the one shown in Figure 17.9b. As a review, the procedure for measuring the phase angle with an oscilloscope is demonstrated in Example 17.2.

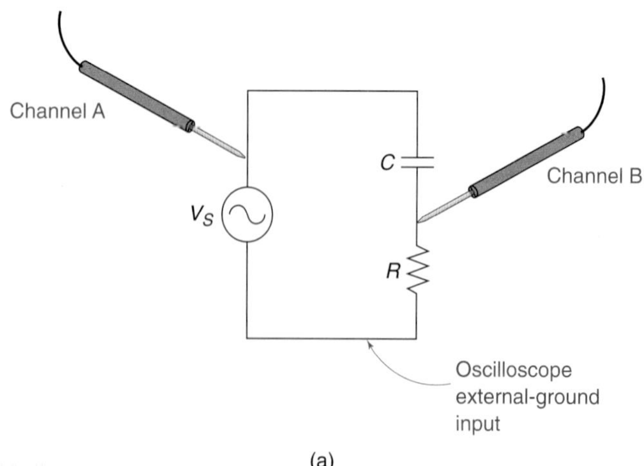


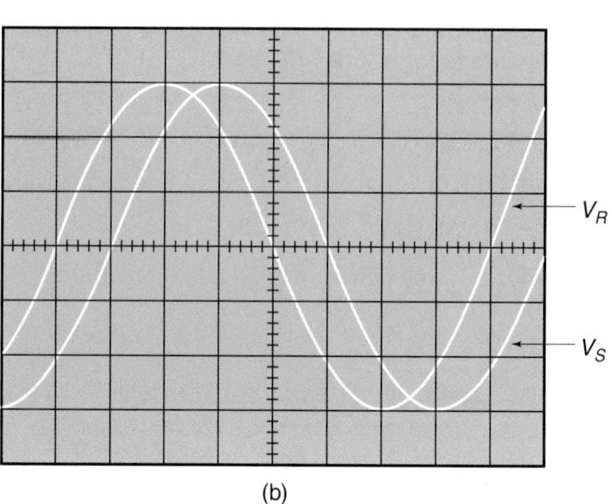


**FIGURE 17.9**

Calculate the phase angle between the waveforms shown in Figure 17.10a.

## EXAMPLE 17.2

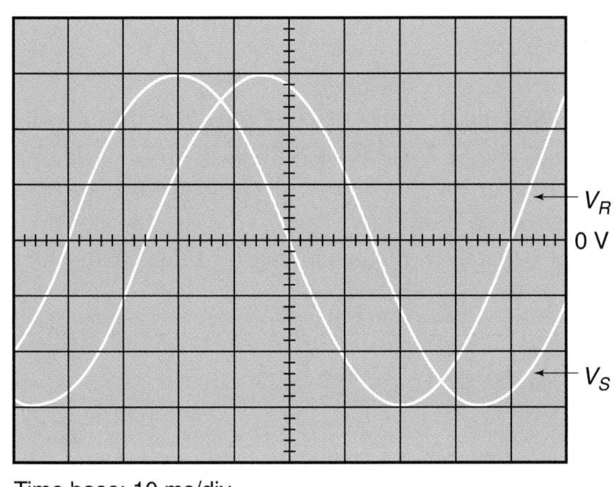

(a)

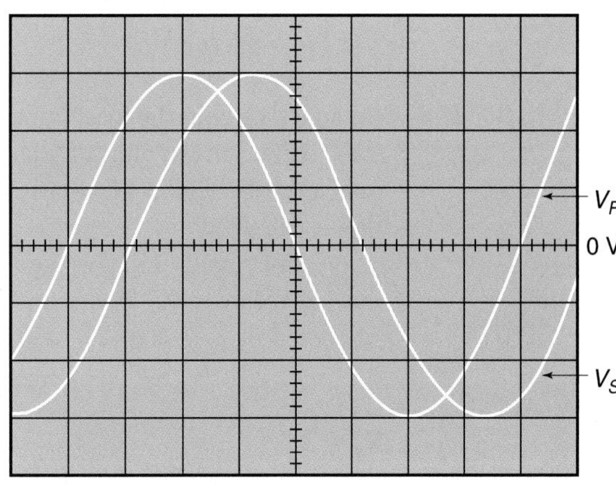

(b)

**FIGURE 17.10**

***Solution:*** As you have been shown, the phase angle between two waveforms can be found as

$$\theta = (360°)\frac{t}{T_C}$$

where $t$ = the time between the zero-crossings of the waveforms
$T_C$ = the cycle time of the waveforms

There are 1.5 major divisions between the time that $V_S$ and $V_R$ cross the 0 V line. With a time base setting of 10 ms/div, the value of $t$ can be found as

$$\boldsymbol{t} = (1.5 \text{ div})\frac{10 \text{ ms}}{\text{div}} = \mathbf{15 \text{ ms}}$$

Each waveform is 8 major divisions in length. Therefore, the cycle time of each waveform can be found as

$$\boldsymbol{T_C} = (8 \text{ div})\frac{10 \text{ ms}}{\text{div}} = \mathbf{80 \text{ ms}}$$

Finally, the phase angle between the waveforms can be found as

$$\boldsymbol{\theta} = (360°)\frac{t}{T_C} = (360°)\frac{15 \text{ ms}}{80 \text{ ms}} = \mathbf{67.5°}$$

The phase angle of the circuit voltage would be written as $\theta = -67.5°$. The negative sign indicates that the source voltage lags the circuit current.

***PRACTICE PROBLEM 17.2***

Calculate the phase angle between the waveforms shown in Figure 17.10b.

Don't forget that we can use the number of divisions for $t$ and $T_C$ in the phase angle equation rather than the actual time values. Had we used the number of divisions for each time period, the phase angle would have been found as

$$\theta = (360°)\frac{1.5 \text{ div}}{8 \text{ div}} = 67.5°$$

This technique, which was introduced in Chapter 14, saves several steps in the calculation process.

## Series Impedance

*OBJECTIVE 3* ▶ In an *RC* circuit, the total impedance is made up of resistance ($R$) and capacitive reactance ($X_C$). Like series voltages, the values of $R$ and $X_C$ for a series circuit are phasor quantities that must be added geometrically. The reason for this can be explained with the help of Figure 17.11.

In Chapter 16, you were told that capacitive reactance can be found using Ohm's law, as follows:

$$X_C = \frac{V_C}{I_C}$$

$X_C$ has a phase angle relative to series circuit current.

where $V_C$ and $I_C$ are both effective (rms) values. Since Chapter 16 dealt with purely capacitive circuits, it was not necessary to consider any phase angle associated with $X_C$. Now that we are dealing with *RC* circuits, we need to establish the fact that there *is* a phase angle associated with capacitive reactance. This phase angle can be determined by including the phase angle of $V_C$ (relative to capacitor current) in the Ohm's law calculation of $X_C$. For the circuit in Figure 17.11:

$$X_C = \frac{V_C\angle -90^\circ}{I_C\angle 0^\circ} = \frac{V_C}{I_C}\angle -90^\circ$$

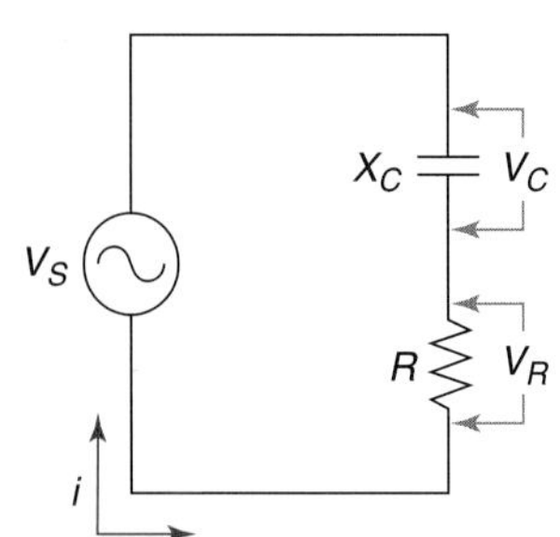

**FIGURE 17.11**

While this result does not include a magnitude, it *does* indicate that capacitive reactance lags circuit current by 90°. Since resistor voltage and current are in phase, resistance ($R$) has a phase angle of 0°. Therefore, capacitive reactance lags resistance by 90° in a series circuit. This means that $X_C$ and $R$ must be treated as phasor values that can be plotted as shown in Figure 17.12. As the figure indicates, the magnitude of $Z$ equals the geometric sum of $X_C$ and $R$. By formula,

$$Z = \sqrt{X_C^2 + R^2} \tag{17.4}$$

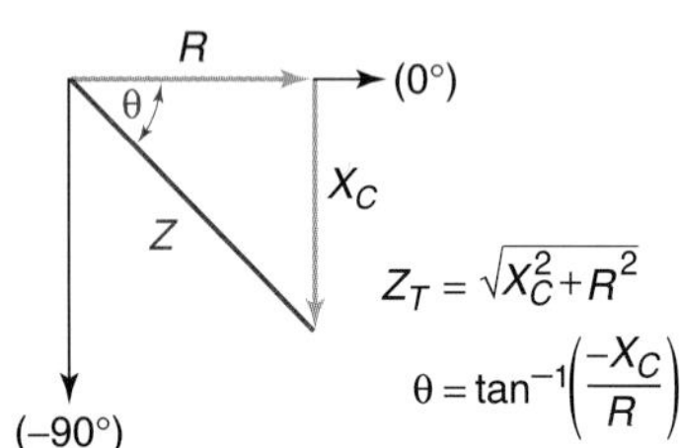

**FIGURE 17.12**

The phase angle of the circuit impedance *(relative to the circuit current)* is found in a similar fashion as the phase angle of the source voltage. By formula,

$$\theta = \tan^{-1}\frac{-X_C}{R} \tag{17.5}$$

Note that $X_C$ is written as a negative value in the equation to indicate its phase relationship with $R$. Example 17.3 demonstrates the procedure for determining the total impedance in a series *RC* circuit.

### EXAMPLE 17.3

Calculate the total impedance for the circuit in Figure 17.13.

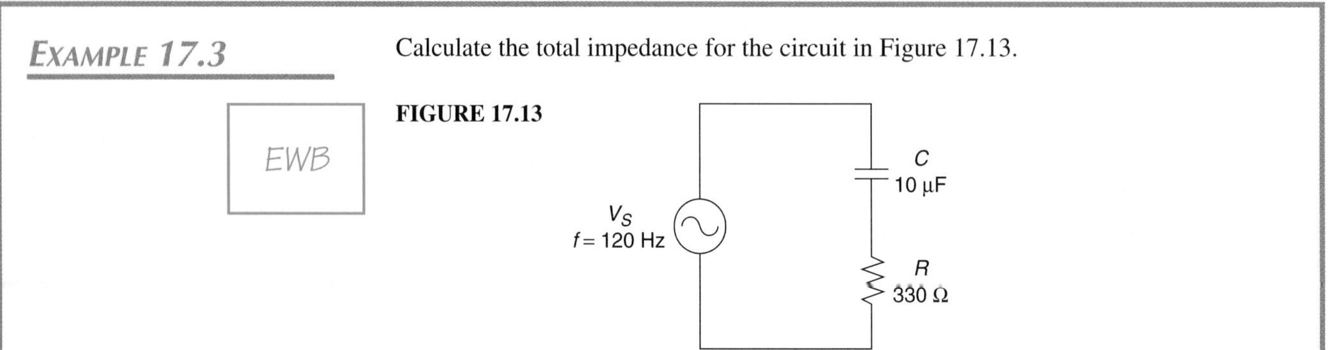

**FIGURE 17.13**

**Solution:** Using the values shown in the figure, the value of $X_C$ is found as

$$X_C = \frac{1}{2\pi f C} = \frac{1}{2\pi(120\ \text{Hz})(10\ \mu\text{F})} \cong \mathbf{133\ \Omega}$$

The magnitude of the circuit impedance can now be found as

$$Z = \sqrt{X_C^2 + R^2} = \sqrt{(133\ \Omega)^2 + (330\ \Omega)^2} \cong \mathbf{356\ \Omega}$$

Finally, the phase angle of the circuit impedance is found as

$$\theta = \tan^{-1}\left(\frac{-X_C}{R}\right) = \tan^{-1}\left(\frac{-133\ \Omega}{330\ \Omega}\right) = \tan^{-1}(-0.403) = \mathbf{-22°}$$

As this result indicates, the circuit impedance (356 Ω) lags the circuit current by 22°.

***PRACTICE PROBLEM 17.3***

A circuit like the one in Figure 17.13 has the following values: $f = 300$ Hz, $C = 0.47\ \mu$F, and $R = 1$ kΩ. Calculate the magnitude and phase angle of the circuit impedance.

## Resolving Series RC Circuit Phase Angles

Any group of phase angles associated with a given circuit must resolve themselves when the circuit analysis is complete. In other words, the phase angles must combine so that all circuit values conform to the basic laws of circuit operation, such as Ohm's law and Kirchhoff's voltage law. With this in mind, look at the circuit in Figure 17.14.

The circuit shown in the figure is basically the same circuit that we analyzed in Example 17.3. For the sake of discussion, a circuit current of 20 mA has been assumed, and Ohm's law has been used to calculate the resulting component voltages. According to equations (17.1) and (17.2),

$$V_S = \sqrt{V_C^2 + V_R^2} = \sqrt{(2.66\ \text{V})^2 + (6.60\ \text{V})^2} = 7.12\ \text{V}$$

and

$$\theta = \tan^{-1}\left(\frac{-V_C}{R}\right) = \tan^{-1}\left(\frac{-2.66\ \text{V}}{6.6\ \text{V}}\right) = \tan^{-1}(-0.403) = -22°$$

Therefore, the source voltage has a value of 7.12 V∠−22°. In Example 17.3, we calculated a total circuit impedance of 356 Ω∠−22°. When we use these two values, Ohm's law provides a circuit current of

$$I = \frac{V_S}{Z_T} = \frac{7.12\ \text{V}\angle -22°}{356\ \Omega\angle -22°} = 20\ \text{mA}\angle 0°$$

which agrees with our assumed value of circuit current. As you can see, all the phase angles for the voltage, current, and impedance values conform to basic circuit principles, which is always the case when a circuit is analyzed properly.

**FIGURE 17.14**

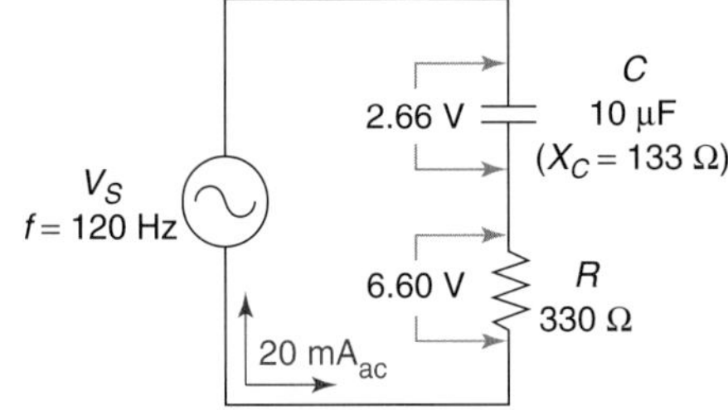

## Calculating Series RC Circuit Current

When you use Ohm's law to calculate the total current in a series $RC$ circuit, you may run into a situation where it appears that the circuit current has a positive phase angle. For example, consider the circuit shown in Figure 17.15. The analysis of this circuit begins by finding the value of $X_C$, as follows:

$$X_C = \frac{1}{2\pi f C} = \frac{1}{2\pi(200\text{ Hz})(1.5\ \mu\text{F})} = 531\ \Omega$$

The next step is to determine the magnitude and phase angle of the circuit impedance, as follows:

$$Z_T = \sqrt{X_C^2 + R^2} = \sqrt{(531\ \Omega)^2 + (150\ \Omega)^2} = 552\ \Omega$$

and

$$\theta = \tan^{-1}\left(\frac{-V_C}{R}\right) = \tan^{-1}\left(\frac{-531\ \Omega}{150\ \Omega}\right) = \tan^{-1}(-3.54) = -74.2°$$

With a circuit impedance of $552\ \Omega\angle -74.2°$, Ohm's law provides us with a circuit current of

$$I = \frac{V_S}{Z_T} = \frac{20\text{ V}\angle 0°}{552\ \Omega\angle -74.2°} = 36.2\text{ mA}\angle 74.2°$$

The meaning of a positive current phase angle.

As you can see, the calculated circuit current contains a positive phase angle. Although this phase angle may seem out of place, it really isn't. The positive phase angle in our result indicates that the circuit current leads the source voltage by 74.2°. As you have been told, this is the same thing as saying that the source voltage lags the circuit current by 74.2°. Therefore, we could express the phase relationship for this circuit in either of two ways:

$$V_S = 20\text{ V}\angle 0° \qquad\qquad V_S = 20\text{ V}\angle -74.2°$$
$$I = 36.2\text{ mA}\angle 74.2° \quad \text{or} \quad I = 36.2\text{ mA}\angle 0°$$

These two sets of numbers say the same thing. The only difference is that the first set of values references the phase angle to the source voltage, and the second references it to the circuit current. *Since current is constant throughout a series circuit, using current as the 0° reference is the preferred method.* Therefore, we rewrite our results as

$$V_S = 20\text{ V}\angle -74.2° \qquad I = 36.2\text{ mA}\angle 0°$$

You will see this adjustment made in several of the upcoming examples.

## Series Circuit Analysis

*OBJECTIVE 4* ▶ You have been shown how the voltage, current, and impedance values are calculated for a basic series $RC$ circuit. At this point, we will work through several circuit analysis problems. The goal is to demonstrate the approaches taken to various series $RC$ circuit analysis problems. We'll start by working through a "straightforward" circuit analysis.

**FIGURE 17.15**

$V_S$ 20 V $f$ = 200 Hz; $C$ 1.5 μF; $R$ 150 Ω

## EXAMPLE 17.4

Determine the voltage, current, and impedance values for the circuit in Figure 17.16.

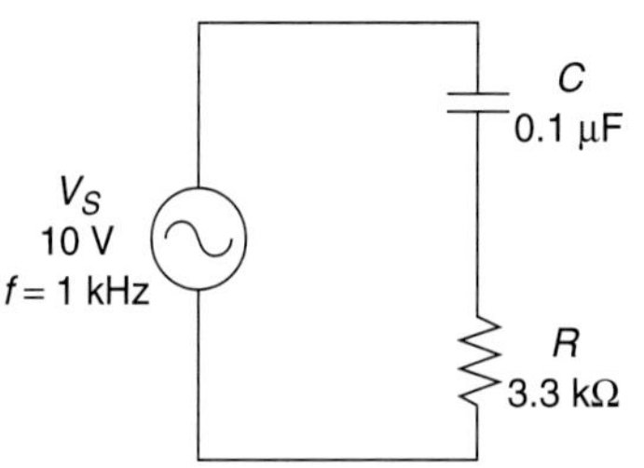


FIGURE 17.16

***Solution:*** First, the value of $X_C$ is found as

$$X_C = \frac{1}{2\pi fC} = \frac{1}{2\pi(1\text{ kHz})(0.1\ \mu\text{F})} = \mathbf{1.59\ k\Omega}$$

The magnitude and phase angle of the circuit impedance can now be found as

$$Z_T = \sqrt{X_C^2 + R^2} = \sqrt{(1.59\text{ k}\Omega)^2 + (3.3\text{ k}\Omega)^2} = \mathbf{3.66\ k\Omega}$$

and

$$\theta = \tan^{-1}\left(\frac{-X_C}{R}\right) = \tan^{-1}\left(\frac{-1.59\text{ k}\Omega}{3.3\text{ k}\Omega}\right) = \mathbf{-25.7^\circ}$$

Using the values of $Z_T$ and $V_S$, the circuit current is found as

$$I = \frac{V_S}{Z_T} = \frac{10\text{ V}\angle 0^\circ}{3.66\text{ k}\Omega\angle -25.7^\circ} = \mathbf{2.73\ mA\angle 25.7^\circ}$$

Remember, the current phase angle is relative to the source voltage. Since current is generally used as the 0° reference in a series circuit, the values of $V_S$ and $I$ are written as

$$V_S = 10\text{ V}\angle -25.7^\circ \qquad I = 2.73\text{ mA}\angle 0^\circ$$

Now, we can calculate the magnitudes of $V_C$ and $V_R$ as follows:

$$V_C = I \cdot X_C = (2.73\text{ mA})(1.59\text{ k}\Omega) = \mathbf{4.34\ V}$$

and

$$V_R = I \cdot R = (2.73\text{ mA})(3.3\text{ k}\Omega) = \mathbf{9.01\ V}$$

At this point, we have calculated all the designated values. To verify our results, we will use the component voltages to calculate the magnitude and phase angle of the voltage source. If everything has been calculated correctly, we will obtain the value of $V_S = 10\text{ V}\angle -27.5^\circ$. Using the calculated values of $V_C$ and $V_R$, the magnitude of the source voltage is found as

$$V_S = \sqrt{V_C^2 + V_R^2} = \sqrt{(4.34\text{ V})^2 + (9.01\text{ V})^2} = \mathbf{10\ V}$$

Now, the phase angle of the source voltage (relative to the circuit current) is found as

$$\theta = \tan^{-1}\left(\frac{-V_C}{V_R}\right) = \tan^{-1}\left(\frac{-4.34\text{ V}}{9.01\text{ V}}\right) = \mathbf{-25.7^\circ}$$

These values match our known source values, so we know we have solved the circuit correctly.

### *PRACTICE PROBLEM 17.4*

A circuit like the one in Figure 17.16 has the following values: $V_S = 12\text{ V}_{ac}$, $f = 1.5\text{ kHz}$, $C = 0.22\ \mu\text{F}$, and $R = 510\ \Omega$. Determine the voltage, current, and impedance values for the circuit. Verify your results using the component voltages (as we did in this example).

Using the voltage-divider equation.

There are always instances when we are interested in calculating only the component voltages in a series *RC* circuit. We can calculate the component voltages using the voltage-divider equation, just as we did with series *RL* circuits. For a series *RC* circuit, the voltage-divider equation is given as

$$V_n = V_S \frac{Z_n}{Z_T} \quad (17.6)$$

where $Z_n$ = the magnitude of $R$ or $X_C$
$Z_T$ = the geometric sum of $R$ and $X_C$

Example 17.5 demonstrates the use of the voltage-divider equation in the analysis of a series *RC* circuit.

## EXAMPLE 17.5

Calculate the component voltages for the circuit in Figure 17.17.

**FIGURE 17.17**

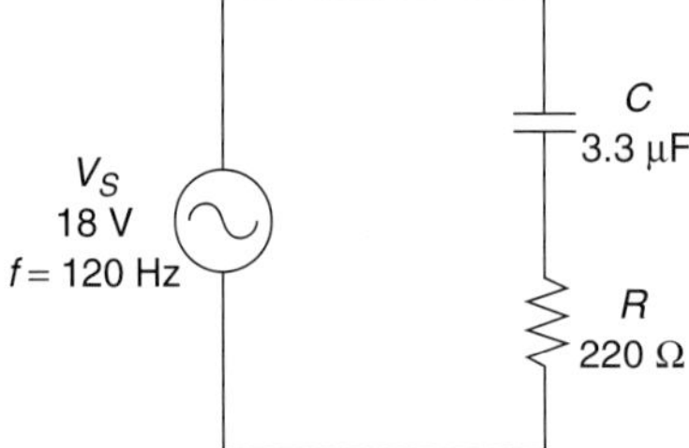


***Solution:*** First, the value of $X_C$ is found as

$$X_C = \frac{1}{2\pi fC} = \frac{1}{2\pi(120\text{ Hz})(3.3\ \mu\text{F})} = \mathbf{402\ \Omega}$$

and the magnitude of the circuit impedance is found as

$$Z_T = \sqrt{X_C^2 + R^2} = \sqrt{(402\ \Omega)^2 + (220\ \Omega)^2} = \mathbf{458\ \Omega}$$

Now, the magnitude of the capacitor voltage can be found as

$$V_C = V_S \frac{X_C}{Z_T} = (18\text{ V})\frac{402\ \Omega}{458\ \Omega} = \mathbf{15.8\ V}$$

and the magnitude of the resistor voltage can be found as

$$V_R = V_S \frac{R}{Z_T} = (18\text{ V})\frac{220\ \Omega}{458\ \Omega} = \mathbf{8.65\ V}$$

***PRACTICE PROBLEM 17.5***

A circuit like the one in Figure 17.17 has the following values: $V_S = 32\ \text{V}_{\text{ac}}$, $f$ = 500 Hz, $C$ = 2.2 μF, and $R$ = 470 Ω. Using the voltage-divider equation, calculate the magnitudes of the component voltages.

The accuracy of the results in Example 17.5 can be verified by comparing the geometric sum of $V_C$ and $V_R$ to the source voltage. (If the component voltages have been calculated correctly, their geometric sum will equal the source voltage.)

When a series *RC* circuit contains more than one capacitor and/or resistor, the analysis of the circuit requires a few more steps. This point is demonstrated in Example 17.6

Calculate the values of $Z_T$, $V_{C1}$, $V_{C2}$, and $V_R$ for the circuit in Figure 17.18.

## EXAMPLE 17.6

**FIGURE 17.18**

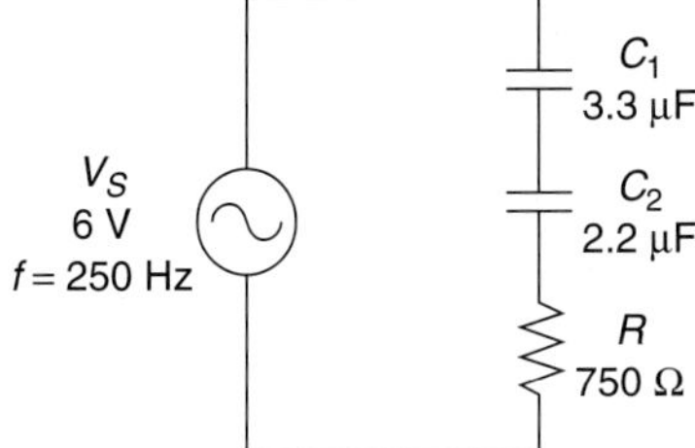


***Solution:*** First, we need to determine the circuit values of $X_C$, as follows:

$$X_{C1} = \frac{1}{2\pi f C_1} = \frac{1}{2\pi(250\text{ Hz})(3.3\ \mu\text{F})} = \mathbf{193\ \Omega}$$

and

$$X_{C2} = \frac{1}{2\pi f C_2} = \frac{1}{2\pi(250\text{ Hz})(2.2\ \mu\text{F})} = \mathbf{289\ \Omega}$$

To calculate the total circuit impedance, we need to determine the total circuit reactance. Since $X_{C1}$ and $X_{C2}$ are in phase, the total reactance equals the algebraic sum of their values, as follows:

$$X_{C(T)} = X_{C1} + X_{C2} = 193\ \Omega + 289\ \Omega = \mathbf{482\ \Omega}$$

*An Alternate Approach:* To calculate the total reactance, we could have solved for total capacitance ($C_T$), and then used $C_T$ in the inductive reactance equation. The result would have been the same.

Using the values of $X_{C(T)}$ and $R$, the magnitude of the circuit impedance can now be found as

$$Z_T = \sqrt{X_{C(T)}^2 + R^2} = \sqrt{(482\ \Omega)^2 + (750\ \Omega)^2} = \mathbf{892\ \Omega}$$

Once we know the magnitude of the circuit impedance, we can use the voltage-divider equation to calculate the magnitudes of the component voltages, as follows:

$$V_{C1} = V_S\frac{X_{C1}}{Z_T} = (6\text{ V})\frac{193\ \Omega}{892\ \Omega} = \mathbf{1.30\ V}$$

$$V_{C2} = V_S\frac{X_{C2}}{Z_T} = (6\text{ V})\frac{289\ \Omega}{892\ \Omega} = \mathbf{1.94\ V}$$

and

$$V_R = V_S\frac{R}{Z_T} = (6\text{ V})\frac{750\ \Omega}{892\ \Omega} = \mathbf{5.04\ V}$$

***PRACTICE PROBLEM 17.6***

A circuit like the one in Figure 17.18 has the following values: $V_S = 25\ \text{V}_{ac}$, $f = 2$ kHz, $R = 750\ \Omega$, $C_1 = 47$ nF, and $C_2 = 0.1\ \mu$F. Calculate the values of $Z_T$, $V_{C1}$, $V_{C2}$, and $V_R$ for the circuit.

Once again, we can verify our results by using the component voltages to calculate the value of the source voltage. Since $V_{C1}$ and $V_{C2}$ are in phase, their values can be added algebraically, as follows:

$$V_{C(T)} = V_{C1} + V_{C2} = 1.30\text{ V} + 1.94\text{ V} = 3.24\text{ V}$$

Then, using the values of $V_{C(T)}$ and $V_R$, the magnitude of the source voltage can be found as

$$V_S = \sqrt{V_{C(T)}^2 + V_R^2} = \sqrt{(3.24\text{ V})^2 + (5.04\text{ V})^2} \cong 6\text{ V}$$

Since our result was approximately 6 V (the given value of the source), we can assume that all our calculations were correct.

In Example 17.6, we used the total circuit reactance, $X_{C(T)}$, to calculate the magnitude of the circuit impedance. By the same token, the value of $X_{C(T)}$ is used to calculate the phase angle of $Z_T$ (relative to the circuit current). For example, the phase angle of $Z_T$ in Figure 17.18 is found as

$$\theta = \tan^{-1}\left(\frac{-X_{C(T)}}{R}\right) = \tan^{-1}\left(\frac{-482\ \Omega}{750\ \Omega}\right) = -32.7°$$

When working with circuits containing multiple capacitors and/or resistors, remember the following:

1. Any two (or more) values that are in phase, such as two or more reactances, can be added algebraically.
2. Any two (or more) values that are not in phase, such as a resistance and a reactance, must be added geometrically.
3. Any overall circuit value, such as $\theta$ or the magnitude of $Z_T$, is calculated using the appropriate total values. For example, $Z_T$ is calculated using $X_{C(T)}$ and $R_T$.

## Series RC Circuit Frequency Response

*OBJECTIVE 5* ▶ In Chapter 14, the term *frequency response* was used to describe any changes that occur in a circuit as a result of a change in operating frequency. For example, consider the circuit shown in Figure 17.19. Table 17.1 shows the circuit response to an increase in operating frequency.

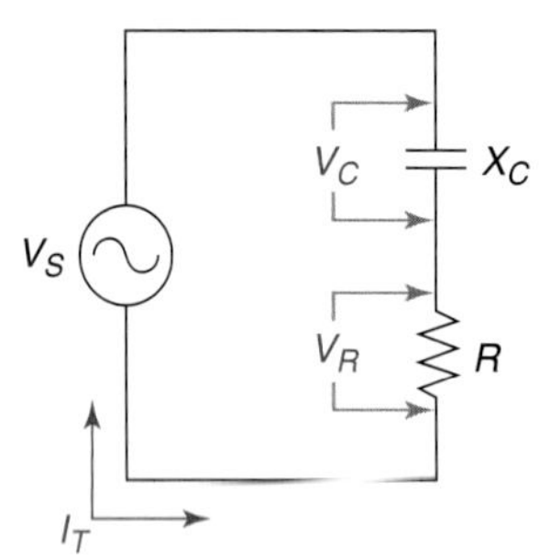


**FIGURE 17.19**

**TABLE 17.1 Series *RC* Circuit Response to an Increase in Operating Frequency**

| *Cause and Effect* | *Relevant Equation* |
|---|---|
| 1. The increase in frequency causes $X_C$ to decrease. | $X_C = \frac{1}{2\pi fC}$ |
| 2. The decrease in $X_C$ causes the magnitude of: | |
| a. $\theta$ to decrease. | $\theta = \tan^{-1}\frac{-X_C}{R}$ |
| b. $Z_T$ to decrease. | $Z_T = \sqrt{X_C^2 + R^2}$ |
| 3. The decrease in $Z_T$ causes $I_T$ to increase. | $I_T = \frac{V_S}{Z_T}$ |
| 4. The decrease in $X_C$ causes $V_C$ to decrease. | $V_C = I_T X_C$ |
| 5. The decrease in $V_C$ causes $V_R$ to increase. | $V_R = \sqrt{V_S^2 - V_C^2}$ |

By the same token, here's what happens if the operating frequency decreases:

1. The decrease in operating frequency causes $X_C$ to increase.
2. The increase in $X_C$ causes the magnitude of $\theta$ to increase.
3. The increase in $X_C$ also causes $Z_T$ to increase.
4. The increase in $Z_T$ causes $I_T$ to decrease.
5. The increase in $X_C$ causes $V_C$ to increase.
6. The increase in $V_C$ causes $V_R$ to decrease.

Example 17.7 demonstrates the effects of a change in operating frequency on series *RC* circuit values.

**EXAMPLE 17.7**

The voltage source in Figure 17.20 has the frequency limits shown. Calculate the circuit voltage, current, and impedance values at the frequency limits of the source.

**FIGURE 17.20**

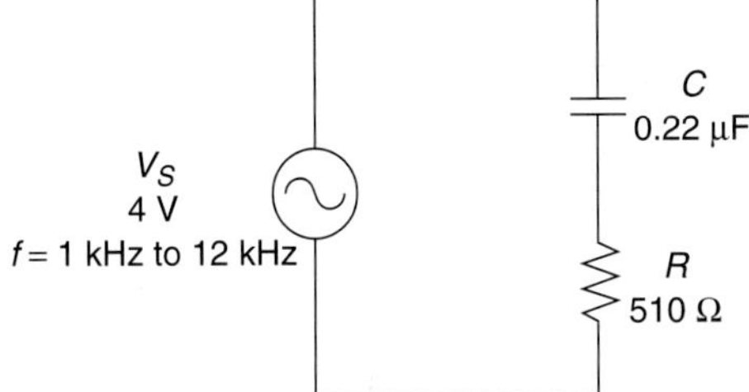

***Solution:*** When the operating frequency of the circuit is 1 kHz, the reactance of the capacitor is found as

$$X_C = \frac{1}{2\pi fC} = \frac{1}{2\pi(1\text{ kHz})(0.22\ \mu\text{F})} = \mathbf{723\ \Omega}$$

The magnitude of the circuit impedance is found as

$$Z_T = \sqrt{X_C^2 + R^2} = \sqrt{(723\ \Omega)^2 + (510\ \Omega)^2} = \mathbf{885\ \Omega}$$

and the value of $\theta$ is found as

$$\theta = \tan^{-1}\left(\frac{-X_C}{R}\right) = \tan^{-1}\left(\frac{-723\ \Omega}{510\ \Omega}\right) = \mathbf{-54.8°}$$

The magnitude of the circuit current is found as

$$I_T = \frac{V_S}{Z_T} = \frac{4\text{ V}}{885\ \Omega} = \mathbf{4.52\ mA}$$

Now, the magnitudes of the component voltages can be found as

$$V_C = I_T X_C = (4.52\text{ mA})(723\ \Omega) = \mathbf{3.27\ V}$$

and

$$V_R = I_T R = (4.52\text{ mA})(510\ \Omega) = \mathbf{2.31\ V}$$

When the operating frequency increases to 12 kHz, the reactance of the capacitor decreases to

$$X_C = \frac{1}{2\pi fC} = \frac{1}{2\pi(12\text{ kHz})(0.22\ \mu\text{F})} = \mathbf{60.3\ \Omega}$$

This decrease in $X_C$ gives us a circuit impedance of

$$Z_T = \sqrt{X_C^2 + R^2} = \sqrt{(60.3\ \Omega)^2 + (510\ \Omega)^2} = \mathbf{514\ \Omega}$$

at a phase angle of

$$\theta = \tan^{-1}\left(\frac{-X_C}{R}\right) = \tan^{-1}\left(\frac{-60.3\ \Omega}{510\ \Omega}\right) = \mathbf{-6.74°}$$

The decreased circuit impedance increases the magnitude of the circuit current to

$$I_T = \frac{V_S}{Z_T} = \frac{4\text{ V}}{514\ \Omega} = \mathbf{7.78\ mA}$$

Finally, the magnitudes of the component voltages at this operating frequency can be found as

$$V_C = I_T X_C = (7.78 \text{ mA})(60.3\ \Omega) = \mathbf{469\ mV}$$

and

$$V_R = I_T R = (7.78 \text{ mA})(510\ \Omega) = \mathbf{3.97\ V}$$

***PRACTICE PROBLEM 17.7***

A circuit like the one in Figure 17.20 has the following values: $V_S = 22\ V_{ac}$, $f =$ 1.5 kHz to 6 kHz, $C = 10$ nF, and $R = 2$ kΩ. Calculate the circuit voltage, current, and impedance values at the frequency limits of the source.

The values obtained in Example 17.7 are listed in Table 17.2. When you compare the results with the cause-and-effect statements in Table 17.1, you'll see that the circuit responded to the change in frequency exactly as described in that table.

**TABLE 17.2 A Comparison of the Two Sets of Results in Example 17.7**

| *Variable* | *Value at f = 1 kHz* | *Value at f = 12 kHz* | *Effect* |
|---|---|---|---|
| $X_C$ | 723 Ω | 60.3 Ω | Decreased |
| $Z_T$ | 880 Ω | 514 Ω | Decreased |
| θ | −54.9° | −6.74° | Decreased[a] |
| $I_T$ | 4.55 mA | 7.78 mA | Increased |
| $V_C$ | 3.29 V | 469 mV | Decreased |
| $V_R$ | 2.32 V | 3.97 V | Increased |

[a]Effect refers to magnitude.

**FIGURE 17.21**

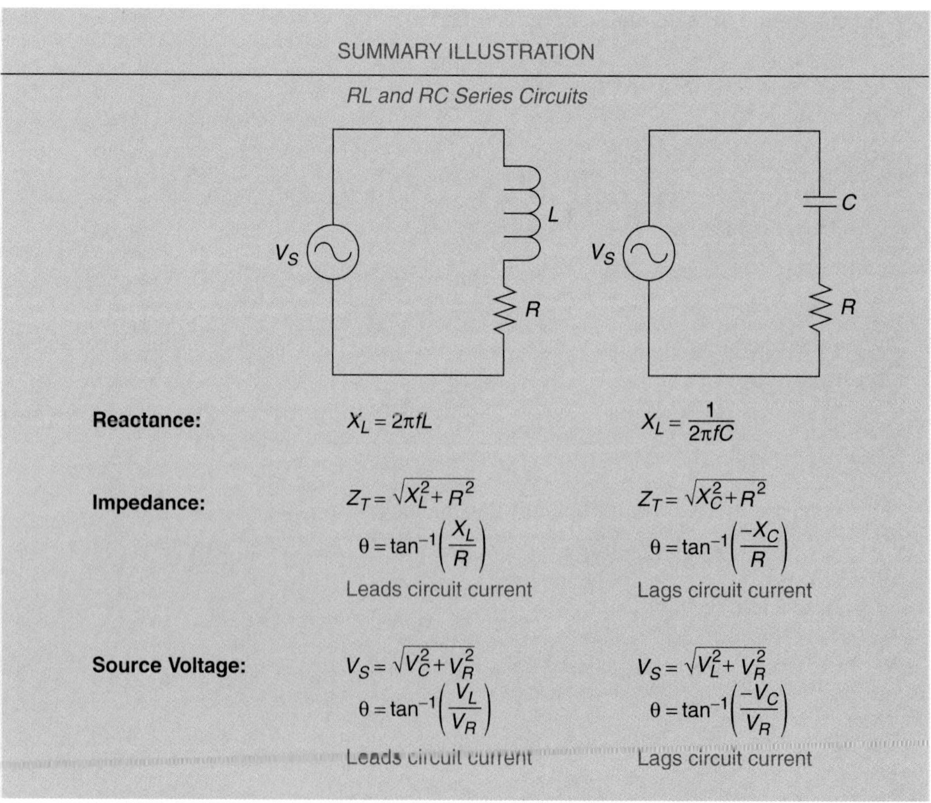

SUMMARY ILLUSTRATION
RL and RC Series Circuits
L
R
C
R
Reactance:
$X_L = 2\pi fL$
$X_L = \frac{1}{2\pi fC}$
Impedance:
$Z_T = \sqrt{X_L^2 + R^2}$
$\theta = \tan^{-1}\left(\frac{X_L}{R}\right)$
Leads circuit current
$Z_T = \sqrt{X_C^2 + R^2}$
$\theta = \tan^{-1}\left(\frac{-X_C}{R}\right)$
Lags circuit current
Source Voltage:
$V_S = \sqrt{V_C^2 + V_R^2}$
$\theta = \tan^{-1}\left(\frac{V_L}{V_R}\right)$
Leads circuit current
$V_S = \sqrt{V_L^2 + V_R^2}$
$\theta = \tan^{-1}\left(\frac{-V_C}{V_R}\right)$
Lags circuit current

## One Final Note

You have been introduced to several relationships in this section. Remember that series *RC* circuits are relatively easy to analyze because they are similar to series *RL* circuits. For future reference, the characteristics of series *RC* circuits are compared to those of series *RL* circuits in Figure 17.21.

### Section Review

1. What are the three fundamental characteristics of every series circuit?
2. In a series *RC* circuit, what is the phase relationship between $V_C$ and $V_R$?
3. In a series *RC* circuit, what does the phase angle of the source voltage represent?
4. Describe the process for measuring the phase angle of the source voltage for a series *RC* circuit.
5. In a series *RC* circuit, what is the phase relationship between capacitive reactance and resistance?
6. In a series *RC* circuit, what does the phase angle of the circuit impedance represent?
7. When using Ohm's law to calculate the circuit current in a series *RC* circuit, what is indicated by a positive phase angle in the result? How is this phase angle normally resolved?
8. How is the voltage-divider equation modified for use in series *RC* circuits?
9. Describe the effect that a change in operating frequency has on the voltage, current, and impedance values in a series *RC* circuit.

# 17.2 POWER CHARACTERISTICS AND CALCULATIONS

Like *RL* circuits, *RC* circuits have significant values of apparent power, reactive power, and true power. In this section, we will look at the fundamental power characteristics and calculations for series *RC* circuits. As you will see, many of the principles covered in this section apply to parallel *RC* circuits as well.

**TABLE 17.3 Power Values in Resistive-Reactive Circuits**

| *Value* | *Definition* |
|---|---|
| **Resistive power ($P_R$)** | The power dissipated by the resistance in an *RC* circuit. Also known as *true power.* |
| **Reactive power ($P_X$)** | The value found using $P = I^2X_C$. The energy (per unit time) found using $I^2X_C$ is actually stored in the capacitor's electrostatic field and is returned to the circuit when the component discharges. $P_X$ is measured in volt-amperes-reactive (VAR) to distinguish it from actual power dissipation values. Also known as *imaginary power.* |
| **Apparent power ($P_{APP}$)** | The combination of resistive (true) power and reactive (imaginary) power. Measured in volt-amperes (VA) because its value does not represent a true power-dissipation value. |
| **Power factor (PF)** | The ratio of resistive power to apparent power, equal to the cosine of the phase angle (cos θ). |

## ac Power Values: A Brief Review

*OBJECTIVE 6* ▶ In Chapters 13 and 14, we discussed apparent power ($P_{APP}$), resistive power ($P_R$), and reactive power ($P_X$) in terms of *RL* circuit operation. Each of these values is described (in terms of *RC* circuit values) in Table 17.3. The values of resistive and reactive power for a series *RC* circuit are calculated as shown in Example 17.8.

### EXAMPLE 17.8

The circuit in Figure 17.22 has the values shown. Calculate the values of resistive and reactive power for the circuit.

***Solution:*** Using the values shown, the reactive power is found as

$$P_X = I^2X_C = (15\text{ mA})^2(500\ \Omega) = \mathbf{112.5\ mVAR}$$

and the value of resistive power is found as

$$P_R = I^2R_1 = (15\text{ mA})^2\ (220\ \Omega) = \mathbf{49.5\ mW}$$

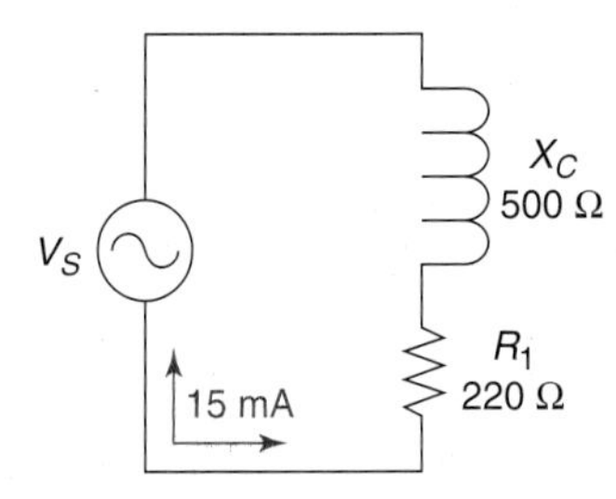


**FIGURE 17.22**

***PRACTICE PROBLEM 17.8***

A circuit like the one in Figure 17.22 has the following values: $I = 22$ mA, $X_C = 200\ \Omega$, and $R = 180\ \Omega$. Calculate the values of $P_X$ and $P_R$ for the circuit.

Dielectric resistance versus effective resistance.

You may be wondering why we didn't consider the dielectric resistance of the capacitor in the resistive power ($P_R$) calculation. The answer lies in the difference between the ideal and practical characteristics of a capacitor. The ideal capacitor has infinite dielectric resistance. This dielectric resistance is represented as shown in Figure 17.23a. Note that the resistance is drawn as a parallel component (just as it is for a current source). A practical representation of a capacitor must include the true dielectric resistance of the component. For example, let's say that the actual dielectric resistance of the component in Figure 17.23 is 100 MΩ. This value is included as a parallel 100 MΩ resistor (as shown in Figure 17.23b). When connected to a voltage source, the current through the 100 MΩ resistor represents the component leakage current.

When driven by an ac source, the reactance of the capacitor is in parallel with the dielectric resistance of the component (as shown in Figure 17.23c). Normally, $X_C << R_d$. Therefore, the dielectric resistance of the capacitor is (for all practical purposes) shorted out by the component's reactance. This is why it is not considered in the resistive power calculation for the circuit.

## Calculating Apparent Power ($P_{APP}$)

Like voltage and impedance, apparent power in a series *RC* circuit equals the geometric sum of resistive power and reactive power. This relationship can be explained with the aid of Figure 17.24. The graph in Figure 17.24a shows the phase relationship between $X_C$ and

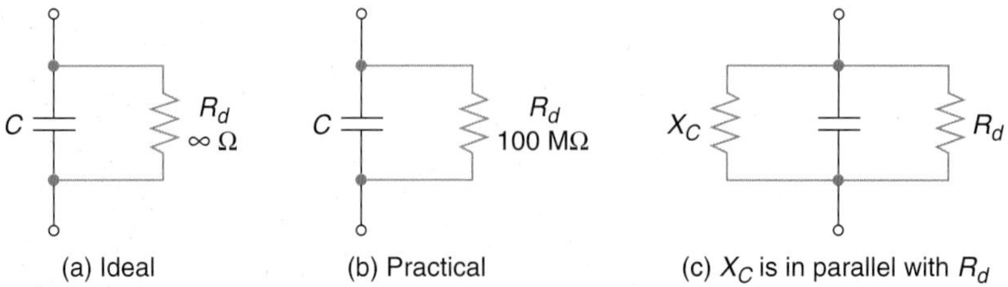


(a) Ideal (b) Practical (c) $X_C$ is in parallel with $R_d$

**FIGURE 17.23 The ideal and practical representations of a capacitor.**

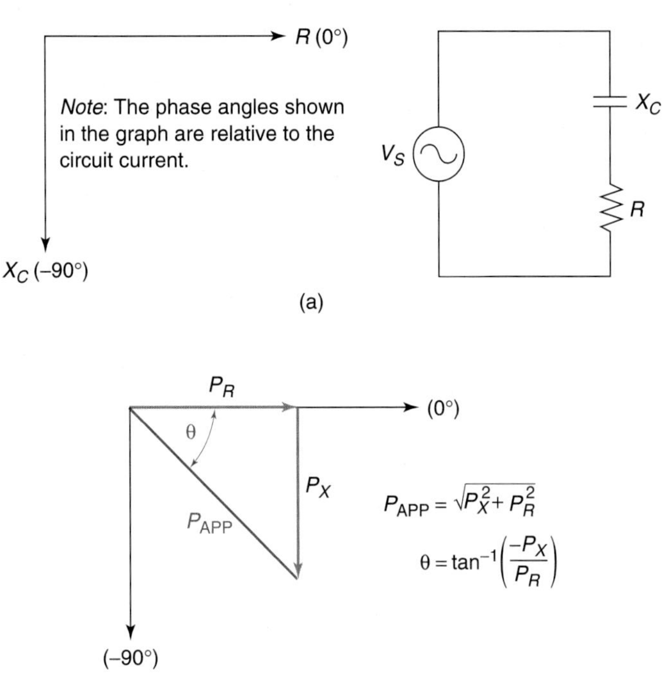


**FIGURE 17.24**

$R$. As indicated, $X_C$ has a $-90°$ phase angle relative to $R$. If we consider this phase angle in our $P_X$ equation, we get

$$P_X = I^2X_C = (I^2\angle 0°)(X_C\angle -90°) = (I^2X_L)\angle -90°$$

While the result does not commit to a magnitude, it does indicate that reactive power has a phase angle of $-90°$ *(relative to the circuit current)*. If we perform the same type of calculation for $P_R$, we get

$$P_R = I^2R = (I^2\angle 0°)(R\angle 0°) = (I^2R)\angle 0°$$

Again, the result does not commit to a magnitude. However, it does indicate that resistive power is in phase with the circuit current. Since reactive power lags circuit current by 90°, it also lags resistive power by the same angle. Therefore, apparent power equals the geometric sum of the magnitudes of $P_X$ and $P_R$ as shown in Figure 17.24b. By formula,

$$P_{APP} = \sqrt{P_X^2 + P_R^2} \tag{17.7}$$

where $P_{APP}$ = the apparent power, measured in volt-amperes (VA)

This is the same relationship that we used to determine the value of apparent power for an *RL* circuit. By the same token, the phase angle equation for apparent power *(relative to the circuit current)* is very similar to the one used for *RL* circuits. By formula,

$$\theta = \tan^{-1}\frac{-P_X}{P_R} \tag{17.8}$$

Note, however, that $(-P_X)$ is used in the equation. Reactive power is written as a negative value because of its phase relationship to $P_R$ (as shown in Figure 17.24b). Example 17.9 demonstrates the use of these equations.

## EXAMPLE 17.9

Calculate the value of apparent power (including its phase angle) for the circuit in Example 17.8.

***Solution:*** In Example 17.8, we calculated the following magnitudes: $P_X =$ 112.5 mVAR and $P_R = 49.5$ mW. Using these values in equation (17.7), the magnitude of apparent power is found as

$$P_{\text{APP}} = \sqrt{P_X^2 + P_R^2} = \sqrt{(112.5 \text{ mVAR})^2 + (49.5 \text{ mW})^2} = \mathbf{122.9 \text{ VA}}$$

and the phase angle is found as

$$\theta = \tan^{-1}\left(\frac{-P_X}{P_R}\right) = \tan^{-1}\left(\frac{-112.5 \text{ mVAR}}{49.5 \text{ mW}}\right) = \tan^{-1}(-2.273) = \mathbf{66.3°}$$

***PRACTICE PROBLEM 17.9***

Calculate the magnitude and phase angle of $P_{\text{APP}}$ for the circuit described in Practice Problem 17.8.

Don't forget: Only a portion of the apparent power in an *RC* circuit is actually dissipated by the resistance. A portion of its value represents the energy stored in the electrostatic field of the capacitor. (This is why apparent power has its own unit of measure.)

## Power Factor

In Chapter 14, you were told that:

1. The power factor for a resistive-reactive circuit is the ratio of resistive power to apparent power. (In other words, it is the ratio of actual power dissipation to apparent power.)
2. For any resistive-reactive circuit, the power factor can be found as:

$$\text{PF} = \cos\theta \qquad \textbf{(17.9)}$$

3. The power factor can be used to calculate resistive power when apparent power and the circuit phase angle are known, as follows:

$$P_R - P_{\text{APP}} \cos 0$$

Example 17.10 shows how this relationship is applied to a series *RC* circuit.

## EXAMPLE 17.10

The waveforms shown in Figure 17.25a were obtained from a series *RC* circuit. The total circuit current was measured at 50 mA. Determine the values of apparent power and resistive power (dissipation) for the circuit.

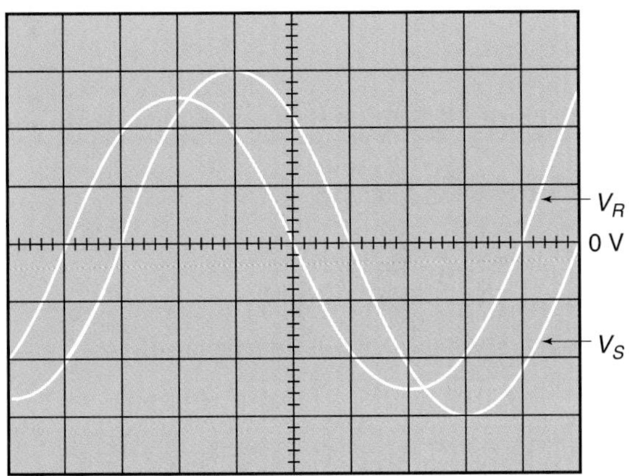


Time base: 20 ms/div
Vertical sensitivity (for $V_S$): 10 V/div
(a)

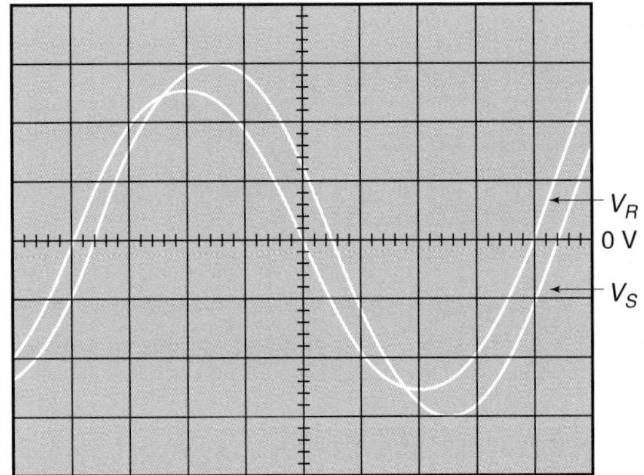


Time base: 100 μs/div
Vertical sensitivity (for $V_S$): 5 V/div
(b)

**FIGURE 17.25**

***Solution:*** First, we need to obtain several values from the oscilloscope display. With a vertical sensitivity of 10 V/div, the source is shown to have a value of 30 $V_{pk}$. Since we are dealing with power calculations, we must convert this value to rms as follows:

$$V_S = 0.707\ V_{pk} = (0.707)(30\ V_{pk}) = \mathbf{21.2\ V_{rms}}$$

Combining this value with the measured circuit current, we get an apparent power of

$$P_{APP} = V_S I_T = (21.2\ V)(50\ mA) = \mathbf{1.06\ VA}$$

Now, we need to determine the phase angle in the circuit. As the display shows, the relevant time periods are represented as follows:

$$t \rightarrow 1\text{ div} \qquad T_C \rightarrow 8\text{ div}$$

and

$$\theta = 360° \frac{t}{T_C} = 360° \frac{1\text{ div}}{8\text{ div}} = \mathbf{45°}$$

Now, we can determine the resistive power (dissipation) in the circuit as follows:

$$P_R = P_{APP} \cos\theta = (1.06\ VA)(\cos 45°) = (1.06\ VA)(0.707) \cong \mathbf{750\ mW}$$

This means that approximately 750 mW of the power drawn from the source is actually dissipated by the resistance in the circuit.

### *PRACTICE PROBLEM 17.10*

The waveforms in Figure 17.25b were measured for a series *RC* circuit. The measured circuit current is 200 mA. Calculate the values of apparent power and resistive power (dissipation) for the circuit.

## Putting It All Together: Series RC Circuit Analysis

Now that we have covered the power characteristics and calculations associated with series *RC* circuits, we will close out the section with a complete circuit analysis problem.

### EXAMPLE 17.11

Perform a complete analysis of the series *RC* circuit shown in Figure 17.26.

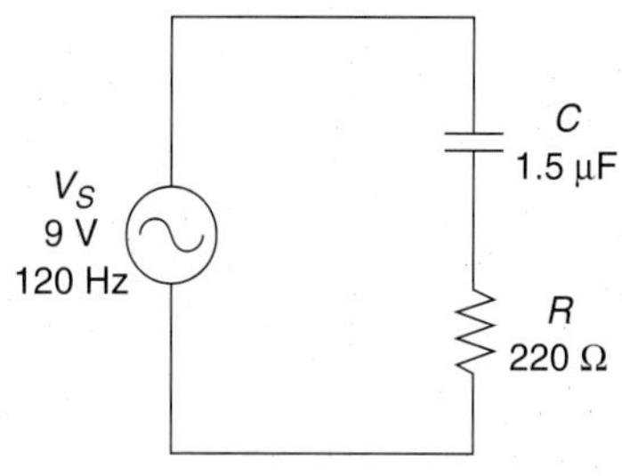


**FIGURE 17.26**

***Solution:*** The reactance of the capacitor is found as

$$X_C = \frac{1}{2\pi f C} = \frac{1}{2\pi(120\text{ Hz})(1.5\ \mu\text{F})} = \mathbf{884\ \Omega}$$

This, combined with the value of $R_1$, gives us a circuit impedance of

$$Z_T = \sqrt{X_C^2 + R^2} = \sqrt{(884\ \Omega)^2 + (220\ \Omega)^2} = \mathbf{911\ \Omega}$$

at a phase angle of:

$$\theta = \tan^{-1}\frac{-X_C}{R} = \tan^{-1}\frac{-884\ \Omega}{220\ \Omega} = \tan^{-1}(-4.018) = \mathbf{-76°}$$

The magnitude of the circuit current can now be found as

$$I_T = \frac{V_S}{Z_T} = \frac{9\text{ V}}{911\ \Omega} = \mathbf{9.88\ mA}$$

So far, we know that the circuit has the following values:

$$Z_T = 911\ \Omega \qquad V_S = 9\text{ V} \qquad I_T = 9.88\text{ mA} \qquad \theta = -76°$$

As you have been told, θ is the phase angle for both impedance and the source voltage, relative to the circuit current. Therefore, the values of $Z_T$ and $V_S$ are written more precisely as

$$Z_T = 911\ \Omega\angle{-76°} \qquad \text{and} \qquad V_S = 9\text{ V}\angle{-76°}$$

Using the voltage-divider equation, the voltage across the resistor can be found as

$$V_R = V_S\frac{R_1}{Z_T} = (9\text{ V}\angle{-76°})\frac{220\ \Omega\angle 0°}{911\ \Omega\angle{-76°}} = \mathbf{2.17\ V\angle 0°}$$

and the voltage across the capacitor can be found as

$$V_C = V_S\frac{X_C}{Z_T} = (9\text{ V}\angle{-76°})\frac{884\ \Omega\angle{-90°}}{911\ \Omega\angle{-76°}} = \mathbf{8.73\ V\angle{-90°}}$$

Finally, the circuit power values can be found as

$$P_R = I_T V_R = (9.88\text{ mA})(2.17\text{ V}) = \mathbf{21.4\ mW}$$
$$P_X = I_T V_C = (9.88\text{ mA})(8.73\text{ V}) = \mathbf{86.3\ mVAR}$$

and

$$P_{APP} = \sqrt{P_X^2 + P_R^2} = \sqrt{(86.3\text{ mVAR})^2 + (21.4\text{ mW})^2} = \mathbf{88.9\ VA}$$

This completes the analysis of the circuit.

***PRACTICE PROBLEM 17.11***

A circuit like the one in Figure 17.26 has the following values: $V_S = 12\ V_{ac}$, $f = 300$ Hz, $C = 1\ \mu$F, and $R = 360\ \Omega$. Perform a complete analysis on the circuit.

1. Compare and contrast the following: apparent power, resistive power, and reactive power.
2. Why is the dielectric resistance of a capacitor normally ignored in the resistive power calculations for an *RC* circuit?
3. State and explain the phase relationship between *RC* circuit reactive power ($P_X$) and resistive power ($P_R$).
4. What is the power factor for an *RC* circuit? What does it tell you?

**Section Review**

# 17.3 PARALLEL *RC* CIRCUITS

A **parallel *RC* circuit** contains one or more resistors in parallel with one or more capacitors. Several parallel *RC* circuits are shown in Figure 17.27. As you can see, none of the circuit branches contain more than one component. (If they did, we'd be dealing with a series-parallel *RC* circuit.)

**Parallel *RC* circuit**
A circuit that contains one or more resistors in parallel with one or more capacitors. Each branch in a parallel *RC* circuit contains only one component (either resistive or capacitive).

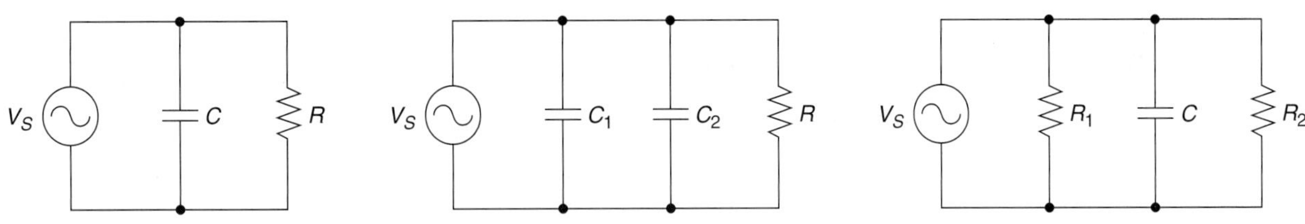


**FIGURE 17.27 Parallel *RC* circuits.**

## *Parallel Circuits: A Comparison*

◀ *OBJECTIVE 7*

As we did with series circuits, we will start by comparing a parallel *RC* circuit to purely resistive and purely capacitive parallel circuits. Comparisons among these circuits are provided in Figure 17.28.

In Chapters 5 and 16, we established the fundamental characteristics of resistive and capacitive parallel circuits:

1. All branch voltages equal the source voltage.
2. The circuit current equals the sum of the branch currents.
3. The total impedance (resistance or reactance) is lower than the lowest branch value and can be found as shown in Figure 17.28.

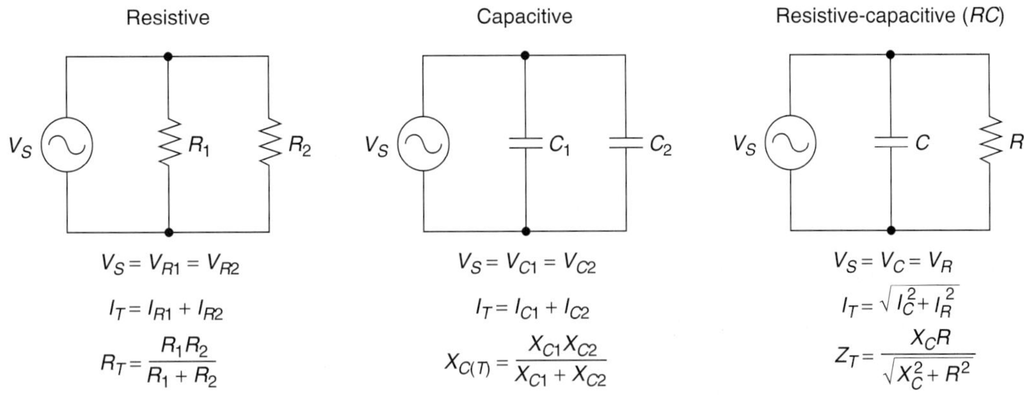


**FIGURE 17.28 Comparisons of resistive, capacitive, and *RC* parallel circuits.**

The relationships listed for the resistive and capacitive circuits in Figure 17.28 should all seem familiar. In each case, all the branch voltages equal the source voltage, and the total circuit current equals the sum of the branch currents. The total impedance for the resistive and capacitive circuits can be found using the product-over-sum methods shown.

*A Practical Consideration:* The product-over sum method of calculating total impedance is emphasized here because the reciprocal method is extremely tedious when applied to parallel *RC* circuits. (This is discussed in detail later in this section.)

The current and impedance relationships shown for the *RC* circuit in Figure 17.28 are explained in detail later in this section. Even so, the relationships are very similar to those for a parallel *RL* circuit:

1. Total current in a parallel *RC* circuit equals the geometric sum of the branch currents.
2. The calculation of total impedance involves the geometric sum of capacitive reactance and resistance (in the denominator of the equation).

These geometric relationships exist because of the presence of phase angles within the circuit.

## Branch Currents

*OBJECTIVE 8* ▶

As you know, current leads voltage by 90° in a capacitive circuit. At the same time, current and voltage in a resistive circuit are always in phase. You saw these two relationships in Figure 17.4. When combined in parallel, a resistor and a capacitor produce the waveforms shown in Figure 17.29.

Since voltage is constant across all branches in a parallel circuit, the capacitor and resistor currents are referenced to a single voltage waveform. As you can see:

1. The capacitor current leads the circuit voltage by 90°.
2. The resistor current is in phase with the circuit voltage.

Therefore, capacitor current leads resistor current by 90°. This phase relationship holds true for every parallel *RC* circuit with an ac voltage source.

Since $I_C$ leads $I_R$ by 90°, the values of the component currents are represented using phasors as shown in Figure 17.30a. By plotting the value of $I_R$ on the *x*-axis (0°) and the value of $I_C$ on the positive *y*-axis (90°), the phase relationship between the two is represented in the graph.

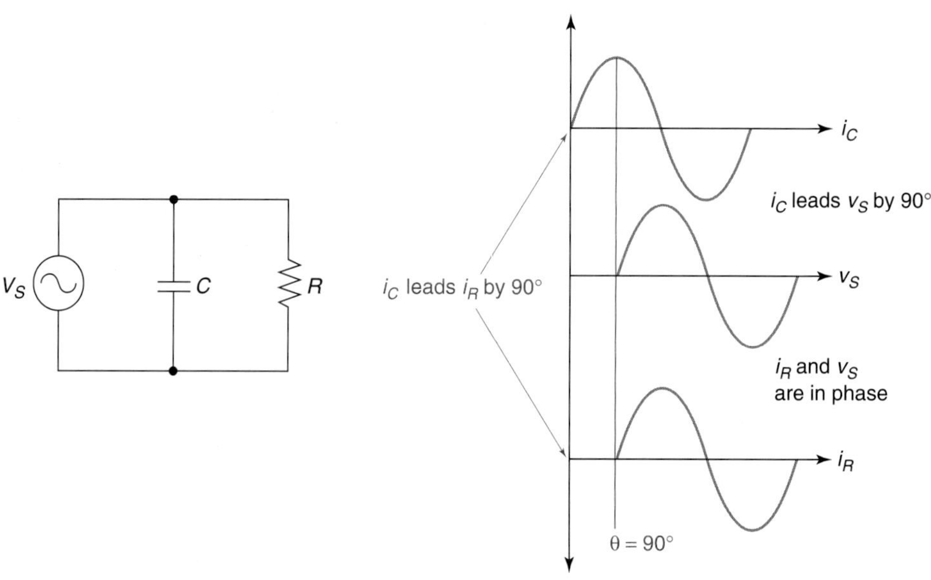


**FIGURE 17.29 Parallel *RC* circuit waveforms.**

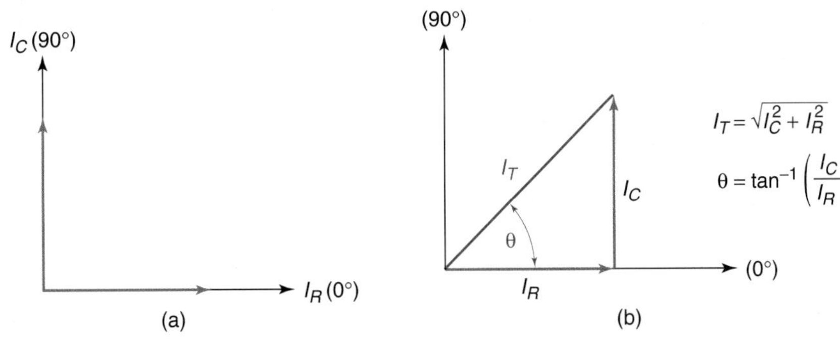


**FIGURE 17.30 Phasor representations of resistor and capacitor current.**

As in any parallel circuit, the sum of $I_C$ and $I_R$ must equal the total circuit current. The phasor relationship between the two values requires that they be added together geometrically. By formula,

$$I_T = \sqrt{I_C^2 + I_R^2} \tag{17.10}$$

The phase angle of $I_T$ in a parallel *RC* circuit is the phase difference between the circuit current and the source voltage. As shown in Figure 17.30b, the value of θ can be found as

$$\theta = \tan^{-1}\frac{I_C}{I_R} \tag{17.11}$$

Since voltage is equal across all branches in a parallel circuit, $V_S$ is used as the reference for all phase angle measurements. Therefore, *the value of* θ *for the circuit current indicates its phase relative to the source voltage.* Example 17.12 demonstrates the calculations of total current and θ for a parallel *RC* circuit.

## EXAMPLE 17.12

The current values in Figure 17.31 were measured as shown. Calculate the total current for the circuit.

**FIGURE 17.31**

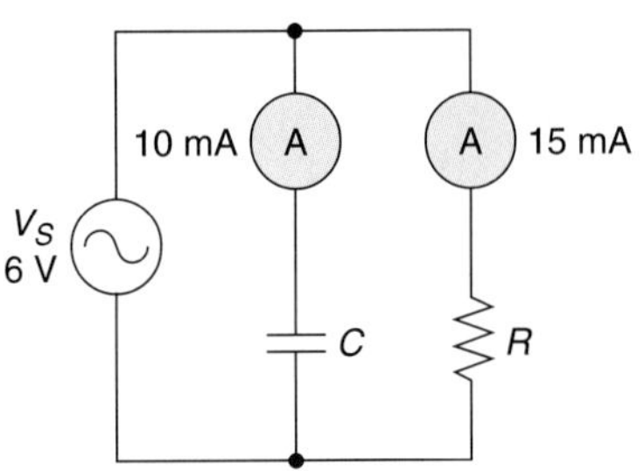


EWB

***Solution:*** Using equation (17.10), the magnitude of the circuit current is found as

$$I_T = \sqrt{I_C^2 + I_R^2} = \sqrt{(10\text{ mA})^2 + (15\text{ mA})^2} \cong \mathbf{18\text{ mA}}$$

The phase angle of the circuit current (relative to the source voltage) can be found as

$$\theta = \tan^{-1}\left(\frac{I_C}{I_R}\right) = \tan^{-1}\left(\frac{10\text{ mA}}{15\text{ mA}}\right) = \tan^{-1}(0.667) = \mathbf{33.7°}$$

The positive value of θ indicates that the circuit current leads the source voltage by 33.7°.

***PRACTICE PROBLEM 17.12***

A circuit like the one in Figure 17.31 has the following values: $I_C$ = 25 mA and $I_R$ = 8 mA. Calculate the magnitude and phase angle of the total circuit current.

The circuit current ($I_T$) phase angle.

For any parallel $RC$ circuit, the limits on the phase angle between circuit current and source voltage are given as follows:

$$0° < \theta < 90° \qquad \textbf{(17.12)}$$

This range of values for $\theta$ makes sense when you consider that:

1. The current in a purely resistive circuit is in phase with the source voltage.
2. The current in a purely capacitive circuit leads the source voltage by 90°.

A parallel circuit containing both capacitance and resistance is neither purely resistive nor purely capacitive. Therefore, it must have a phase angle that is greater than 0° and less than 90°.

## Parallel-Circuit Impedance

The easiest way to calculate the total impedance in a parallel $RC$ circuit begins with calculating the circuit current. Once the circuit current is known, Ohm's law states that the circuit impedance can be found as

$$Z_T = \frac{V_S}{I_T} \qquad \textbf{(17.13)}$$

This approach to calculating the value of $Z_T$ (which was first introduced in Chapter 14) is demonstrated in Example 17.13.

### EXAMPLE 17.13

Calculate the total impedance for the circuit shown in Figure 17.31 (Example 17.12).

***Solution:*** In Example 17.12, we calculated a total circuit current of $I_T$ = 18 mA∠33.7°. With a value of $V_S$ = 6 V, the total circuit impedance can be found as

$$Z_T = \frac{V_S}{I_T} = \frac{6\text{ V}\angle 0°}{18\text{ mA}\angle 33.7°} = \mathbf{333\ \Omega\angle{-33.7°}}$$

***PRACTICE PROBLEM 17.13***

Calculate the total impedance for the circuit described in Practice Problem 17.12. Assume that the source has a value of $V_S = 12\text{ V}_{\text{ac}}$.

## An Observation Based on the Circuit in Example 17.13

As with any parallel circuit, the total impedance in a parallel $RC$ circuit is lower than any branch impedance value. For example, the branch impedance values in Figure 17.31 can be found as

$$X_C = \frac{V_S}{I_C} = \frac{6\text{ V}}{10\text{ mA}} = 600\ \Omega$$

and

$$R = \frac{V_S}{I_R} = \frac{6\text{ V}}{15\text{ mA}} = 400\ \Omega$$

Both of these values are greater than the magnitude of $Z_T$ found in the example (333 Ω).

## The Phase Angle of $Z_T$

In our discussion on series *RC* circuits, it was shown that $Z_T$ and $V_S$ have equal phase angles (relative to the circuit current). In parallel *RL* circuits, $Z_T$ and $I_T$ both have phase angles (relative to the source voltage). However, their values are not equal.

In Example 17.13, the parallel *RL* circuit was found to have the following values:

$$I_T = 18 \text{ mA}\angle 33.7^\circ \qquad Z_T = 333\ \Omega\angle -33.7^\circ$$

As you can see, the current phase angle is *positive* and the impedance phase angle is *negative.* This may seem confusing at first, but it makes sense when you apply Ohm's law to the circuit values. For the overall circuit, the source voltage must equal the product of current and impedance. Using the values in Example 17.13, this calculation works out as follows:

$$\begin{aligned} V_S &= I_T Z_T \\ &= (18 \text{ mA}\angle 33.7^\circ)(333\ \Omega\angle -33.7^\circ) \\ &\cong 6 \text{ V}\angle[33.7^\circ + (-33.7^\circ)] \\ &= 6 \text{ V}\angle(33.7^\circ - 33.7^\circ) \\ &= 6 \text{ V}\angle 0^\circ \end{aligned}$$

Since source voltage is the 0° reference for phase measurements in a parallel *RC* circuit, the result of the calculation makes sense. The only way that the product of current and impedance can have a 0° phase angle is if the two angles are positive and negative equivalents. This holds true for any parallel *RC* (or *RL*) circuit.

## Calculating Parallel-Circuit Impedance

In Chapters 5 and 16, you were shown how to calculate the total impedance (resistance or reactance) of purely resistive and purely capacitive parallel circuits. The commonly used equations for these parallel circuits are listed (as a reference) in Table 17.4.

**TABLE 17.4 Parallel-Circuit Impedance Equations for Two-Component Resistive and Capacitive Circuits**

| *Method* | *Resistive Form* | *Capacitive Form* |
|---|---|---|
| Reciprocal: | $R_T = \dfrac{1}{\dfrac{1}{R_1} + \dfrac{1}{R_2}}$ | $X_{CT} = \dfrac{1}{\dfrac{1}{X_{C1}} + \dfrac{1}{X_{C2}}}$ |
| Product-over-sum: | $R_T = \dfrac{R_1 R_2}{R_1 + R_2}$ | $X_{CT} = \dfrac{X_{C1} X_{C2}}{X_{C1} + X_{C2}}$ |
| Equal-value branches: | $R_T = \dfrac{R}{n}$ | $X_{CT} = \dfrac{X_C}{n}$ |
| | (where $n$ = the number of branches in the circuit) | |

Since $X_C$ and $R$ do not have the same phase angle, the equal-value branches approach to calculating total impedance can never be used on a parallel *RC* circuit. At the same time, the phase angles of $X_C$ and $R$ make the reciprocal method extremely long and tedious. For this reason, the product-over-sum approach to calculating circuit impedance is preferred for parallel *RC* circuits. Using the product-over-sum format, the magnitude of parallel *RC* circuit impedance can be found as

$$Z_T = \frac{X_C R}{\sqrt{X_C^2 + R^2}} \qquad \textbf{(17.14)}$$

Note that the denominator of the fraction contains the geometric sum of the variables. Example 17.14 demonstrates the use of this equation.

## EXAMPLE 17.14

Calculate the magnitude of $Z_T$ for the circuit in Figure 17.32.

***Solution:*** Using the values shown in the figure, the magnitude of $Z_T$ is found as

$$Z_T = \frac{X_C R}{\sqrt{X_C^2 + R^2}} = \frac{(600\ \Omega)(400\ \Omega)}{\sqrt{(600\ \Omega)^2 + (400\ \Omega)^2}} \cong \mathbf{333\ \Omega}$$

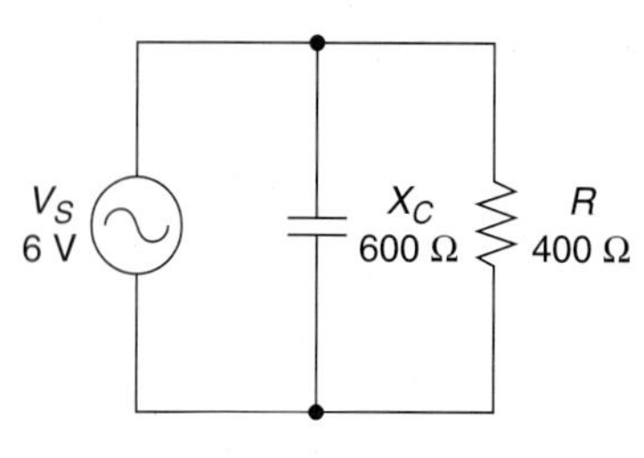


**FIGURE 17.32**

To provide a basis for comparison, the values used in this example match those used as the basis for Example 17.13. If you compare the results of these two examples, you'll see that their magnitudes are the same.

***PRACTICE PROBLEM 17.14***

Refer to Practice Problem 17.13. Calculate the values of $X_C$ and $R$. Then, calculate the value of $Z_T$ using equation (17.14). Compare your result to the magnitude of $Z_T$ found in that practice problem.

***Calculating the Impedance Phase Angle.*** The phase angle of the circuit impedance for a parallel $RC$ circuit can be found as

$$\theta = \tan^{-1} \frac{R}{-X_C} \qquad \textbf{(17.15)}$$

As you can see, the fraction in equation (17.15) is the reciprocal of the one used to calculate the impedance phase angle in a series $RC$ circuit. Example 17.15 applies this equation to the circuit in Figure 17.32.

## EXAMPLE 17.15

Calculate the phase angle of the circuit impedance for the circuit in Figure 17.32 (Example 17.12).

***Solution:*** The circuit has values of $X_C = 600\ \Omega$ and $R = 400\ \Omega$. Using these values in equation (14.19), the value of $\theta$ for the circuit impedance is found as

$$\boldsymbol{\theta} = \tan^{-1}\left(\frac{R}{-X_C}\right) = \tan^{-1}\left(\frac{400\ \Omega}{-600\ \Omega}\right) = \tan^{-1}(-0.667) = \mathbf{-33.7^\circ}$$

***PRACTICE PROBLEM 17.15***

Calculate the phase angle of the circuit impedance for the circuit in Figure 17.31.

In Example 17.13, we calculated the phase angle of the circuit impedance using Ohm's law. When you compare the results of the two examples, you'll see that they are the same.

***The Basis of Equation (17.15).*** The current triangle in Figure 17.30b was used to illustrate the current relationships in a parallel $RC$ circuit. As shown in the figure,

$$\theta = \tan^{-1}\left(\frac{I_C}{I_R}\right)$$

This relationship was given earlier as equation (17.11).

As you have been shown, the phase angle of circuit impedance is the negative-equivalent of the current phase angle. For this reason, we will rewrite equation (17.11) as follows.

$$\theta = \tan^{-1}\left(\frac{-I_C}{I_R}\right)$$

(When written in this form, the equation provides a negative result, which is consistent with the impedance phase angle.) If we rewrite each current value in the equation in the form of $V/R$, we get

$$\theta = \tan^{-1}\left(\frac{\frac{V_S}{-X_C}}{\frac{V_S}{R}}\right) = \tan^{-1}\left(\frac{V_S}{-X_C} \times \frac{R}{V_S}\right)$$

Finally, the values of $V_S$ cancel out, leaving

$$\theta = \tan^{-1}\left(\frac{R}{-X_C}\right)$$

which is equation (17.15).

## Measuring the Current Phase Angle ($\theta$)

In Chapter 14, you were shown how a sensing resistor is added to a parallel circuit so that the current phase angle can be measured with an oscilloscope. The same technique can be used to measure the current phase angle in a parallel *RC* circuit. (The discussion in Chapter 14 can be found in Section 14.3.)

## Parallel Circuit Analysis

In this section, you have been shown how current and impedance values are calculated for a parallel *RC* circuit. Parallel circuit power calculations are identical to those for series circuits, so they have been covered as well. At this point, we will tie all these calculations together by working a circuit analysis problem.

◄ *OBJECTIVE 9*

### EXAMPLE 17.16

Perform a complete analysis of the circuit in Figure 17.33.

**FIGURE 17.33**

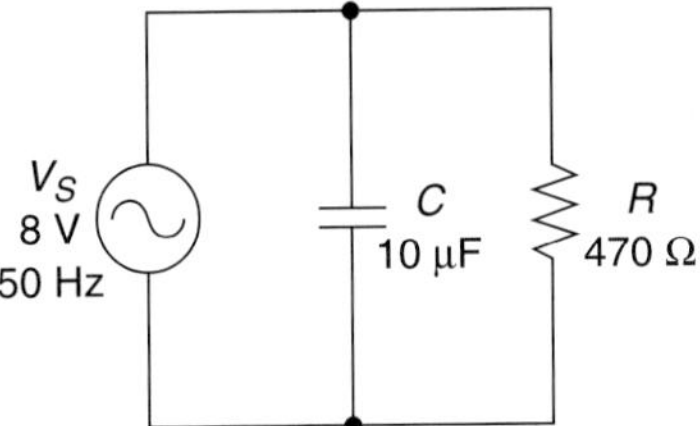


***Solution:*** First, the reactance of the capacitor is found as

$$X_C = \frac{1}{2\pi fC} = \frac{1}{2\pi(50 \text{ Hz})(10 \ \mu\text{F})} = \mathbf{318\ \Omega}$$

Now, the magnitudes of the branch currents can be found as

$$I_C = \frac{V_S}{X_C} = \frac{8 \text{ V}}{318\ \Omega} = \mathbf{25.2 \text{ mA}}$$

and

$$I_R = \frac{V_S}{R} = \frac{8 \text{ V}}{470\ \Omega} = \mathbf{17 \text{ mA}}$$

The magnitude of the circuit current can now be found as

$$I_T = \sqrt{I_C^2 + I_R^2} = \sqrt{(25.2 \text{ mA})^2 + (17 \text{ mA})^2} = \mathbf{30.4 \text{ mA}}$$

and the current phase angle (relative to the source voltage) can be found as

$$\boldsymbol{\theta} = \tan^{-1}\left(\frac{I_C}{I_R}\right) = \tan^{-1}\left(\frac{25.2 \text{ mA}}{17 \text{ mA}}\right) \cong \mathbf{56°}$$

Once the total current is known, the circuit impedance can be found as

$$\boldsymbol{Z_T} = \frac{V_S}{I_T} = \frac{8 \text{ V}\angle 0°}{30.4 \text{ mA}\angle 56°} = \mathbf{263\ \Omega\angle{-56°}}$$

The value of reactive power ($P_X$) for the circuit can be found as

$$\boldsymbol{P_X} = I_C^2 X_C = (25.2 \text{ mA})^2(318\ \Omega) = \mathbf{202\ mVAR}$$

and the value of resistive power dissipation ($P_R$) can be found as

$$\boldsymbol{P_R} = I_R^2 R = (17 \text{ mA})^2(470\ \Omega) = \mathbf{135.8\ mW}$$

Once the values of reactive and resistive power are known, the value of apparent power ($P_{\text{APP}}$) can be found as

$$\boldsymbol{P_{\text{APP}}} = \sqrt{P_X^2 + P_R^2} = \sqrt{(202 \text{ mVAR})^2 + (135.8 \text{ mW})^2} = \mathbf{243.4\ mVA}$$

This completes the analysis of the circuit.

***PRACTICE PROBLEM 17.16***

A circuit like the one in Figure 17.33 has the following values: $V_S = 14\ \text{V}_{\text{ac}}$, $f = 100$ Hz, $C = 4.7\ \mu\text{F}$, and $R = 220\ \Omega$. Perform a complete analysis on the circuit.

There are several approaches we could take for checking the accuracy of the results in Example 17.16. The simplest method is to use the power factor ($\cos\theta$) to calculate the value of resistive power, as follows:

$$P_R = P_{\text{APP}} \cos\theta = (243.4 \text{ mVA})(\cos 56°) = (243.4 \text{ mVA})(0.5592) = 136 \text{ mW}$$

Since this value is approximately equal to the value of $P_R$ found in the example, we can assume that our results are accurate.

## Parallel Circuit Frequency Response

*OBJECTIVE 10* ▶ The frequency response of a parallel *RC* circuit is quite different than that of a series *RC* circuit. For example, consider the circuit shown in Figure 17.34. Table 17.5 shows the circuit response to an increase in operating frequency. By the same token, here's what happens if the operating frequency decreases:

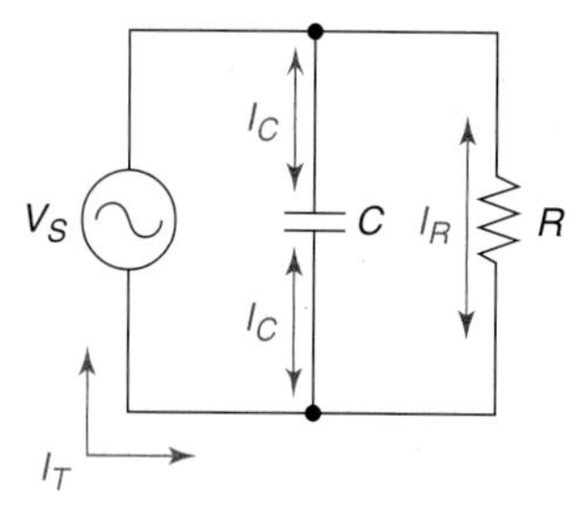


**FIGURE 17.34**

1. The decrease in operating frequency causes $X_C$ to increase.
2. The increase in $X_C$ causes the magnitude of:
   a. $I_C$ to decrease.
   b. $\theta$ to decrease.
   c. $Z_T$ to increase.
3. The decrease in $I_C$ causes $I_T$ to decrease.

Example 17.17 demonstrates the effects of a change in operating frequency on parallel *RC* circuits.

**TABLE 17.5 Parallel *RC* Circuit Response to an *Increase* in Operating Frequency**

| *Cause and Effect* | *Relevant Equation* |
|---|---|
| 1. The increase in frequency causes $X_C$ to decrease. | $X_C = \dfrac{1}{2\pi fC}$ |
| 2. The decrease in $X_C$ causes the magnitude of: | |
| a. $I_C$ to increase. | $I_C = \dfrac{V_S}{X_C}$ |
| b. $\theta$ to increase. | $\theta = \tan^{-1}\dfrac{R}{-X_C}$ |
| c. $Z_T$ to decrease. | $Z_T = \dfrac{X_C R}{\sqrt{X_C^2 + R^2}}$ |
| 3. The increase in $I_C$ causes $I_T$ to increase. | $I_T = \sqrt{I_C^2 + I_R^2}$ |

**EXAMPLE 17.17**

The voltage source in Figure 17.35 has the frequency limits shown. Calculate the circuit current and impedance values at the frequency limits of the source.

**FIGURE 17.35**

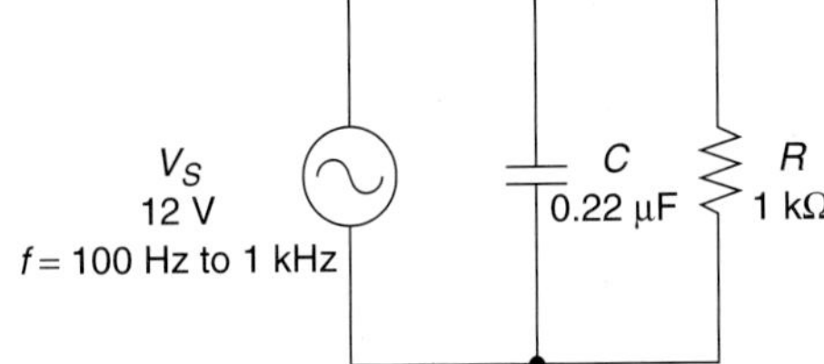


***Solution:*** When the operating frequency is set to 100 Hz, the reactance of the capacitor is found as

$$X_C = \frac{1}{2\pi fC} = \frac{1}{2\pi(100\text{ Hz})(0.22\ \mu\text{F})} = \mathbf{7.23\ k\Omega}$$

The magnitudes of the branch currents are found as

$$I_C = \frac{V_S}{X_C} = \frac{12\text{ V}}{7.23\text{ k}\Omega} = \mathbf{1.66\ mA}$$

and

$$I_R = \frac{V_S}{R} = \frac{12\text{ V}}{1\text{ k}\Omega} = \mathbf{12\ mA}$$

Combining these values, the magnitude of the circuit current is found as

$$I_T = \sqrt{I_C^2 + I_R^2} = \sqrt{(1.66\text{ mA})^2 + (12\text{ mA})^2} = \mathbf{12.1\ mA}$$

and the current phase angle is found as

$$\theta = \tan^{-1}\left(\frac{I_C}{I_R}\right) = \tan^{-1}\left(\frac{1.66\text{ mA}}{12\text{ mA}}\right) = \mathbf{7.88^\circ}$$

Finally, the circuit impedance is found as

$$Z_T = \frac{V_S}{I_T} = \frac{12\text{ V}\angle 0^\circ}{12.1\text{ mA}\angle 7.88^\circ} = \mathbf{992\ \Omega\angle{-7.88^\circ}}$$

When the operating frequency is increased to 1 kHz, the reactance of the capacitor is found as

$$X_C = \frac{1}{2\pi f C} = \frac{1}{2\pi(1\text{ kHz})(0.22\ \mu\text{F})} = \mathbf{723\ \Omega}$$

The magnitudes of the branch currents are found as

$$I_C = \frac{V_S}{X_C} = \frac{12\text{ V}}{723\ \Omega} = \mathbf{16.6\ mA}$$

and

$$I_R = \frac{V_S}{R} = \frac{12\text{ V}}{1\text{ k}\Omega} = \mathbf{12\ mA}$$

Combining these values, the magnitude of the circuit current is found as

$$I_T = \sqrt{I_C^2 + I_R^2} = \sqrt{(16.6\text{ mA})^2 + (12\text{ mA})^2} = \mathbf{20.5\ mA}$$

and the current phase angle is found as

$$\theta = \tan^{-1}\left(\frac{I_C}{I_R}\right) = \tan^{-1}\left(\frac{16.6\text{ mA}}{12\text{ mA}}\right) = \mathbf{54°}$$

Finally, the circuit impedance is found as

$$Z_T = \frac{V_S}{I_T} = \frac{12\text{ V}\angle 0°}{20.5\text{ mA}\angle 54°} = \mathbf{585\ \Omega\angle{-54°}}$$

***PRACTICE PROBLEM 17.17***

A circuit like the one in Figure 17.35 has the following values: $V_S = 9\ V_{ac}$, $f = 120$ Hz to 6 kHz, $C = 1\ \mu$F, and $R = 620\ \Omega$. Calculate the circuit current and impedance values at the frequency limits of the source.

The values obtained in Example 17.17 are listed in Table 17.6. When you compare the results with the cause-and-effect statements in Table 17.5, you'll see that the circuit responded to the change in frequency exactly as described in that table.

**TABLE 17.6 A Comparison of the Two Sets of Results in Example 17.17**

| *Variable* | *Value at f = 100 Hz* | *Value at f = 1 kHz* | *Effect* |
|---|---|---|---|
| $X_C$ | 7.23 kΩ | 723 Ω | Decreased |
| $I_C$ | 1.66 mA | 16.6 mA | Increased |
| $I_T$ | 12.1 mA | 20.5 mA | Increased |
| $\theta$ | 7.88° | 54° | Increased |
| $Z_T$ | 992 Ω | 585 Ω | Decreased |

## One Final Note

In this section, you have been introduced to parallel *RC* circuit relationships and analysis. Once again, we have a situation where all the relationships discussed are similar to those given for *RL* circuits in Chapter 14. For future reference, the basic characteristics of parallel *RC* circuits are compared to those of parallel *RL* circuits in Figure 17.36.

We have taken one approach to analyzing parallel *RC* circuits in this section. Another approach emphasizes the use of *susceptance* and *admittance,* the reciprocals of reactance and impedance, respectively. This approach to analyzing the operation of parallel *RC* circuits is introduced and demonstrated in Appendix F.

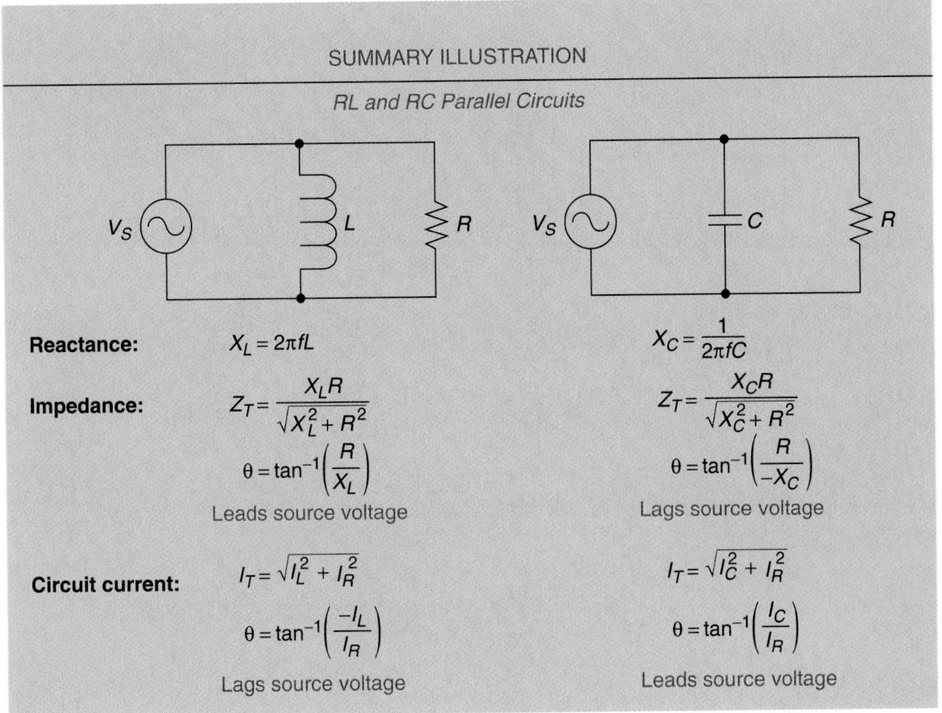


**FIGURE 17.36**

### Section Review

1. What are the three fundamental characteristics of any parallel circuit?
2. In a parallel *RC* circuit, what is the phase relationship between $I_C$ and $I_R$?
3. In a parallel *RC* circuit, what does the current phase angle represent?
4. Why does the current in a parallel *RC* circuit have a positive phase angle?
5. What is the simplest approach to calculating the value of $Z_T$ for a parallel *RC* circuit?
6. When $X_C = R$ in a parallel *RC* circuit, can the equal-value branches approach to calculating $Z_T$ be used? Explain your answer.
7. Describe the response of a parallel *RC* circuit to an increase in operating frequency.
8. Describe the response of a parallel *RC* circuit to a decrease in operating frequency.

## 17.4 SERIES-PARALLEL *RC* CIRCUIT ANALYSIS

◄ *OBJECTIVE 11*

You were first introduced to series-parallel circuit analysis in Chapter 7. At that time, you learned that analyzing a series-parallel circuit is simply a matter of:

1. Combining series components according to the rules of series circuits.
2. Combining parallel components according to the rules of parallel circuits.

The analysis of a series-parallel *RC* circuit gets a bit complicated because of the phase angles involved. For example, consider the circuit shown in Figure 17.37a. This circuit, which was analyzed in Example 17.16, has a total impedance of $Z_T = 263\ \Omega\angle{-56°}$. The circuit impedance is represented as a block ($Z_P$) in the equivalent circuit shown.

If we add a series resistor ($R_S$) to the circuit, we have the series-parallel circuit shown in Figure 17.37b. As the equivalent of this circuit shows, $R_S$ is in series with $Z_P$. The phase difference between $R_S$ and $Z_P$ makes it difficult to add their polar notation values.

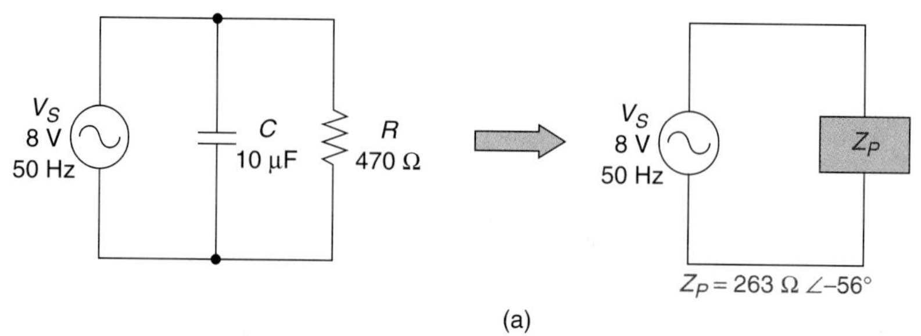


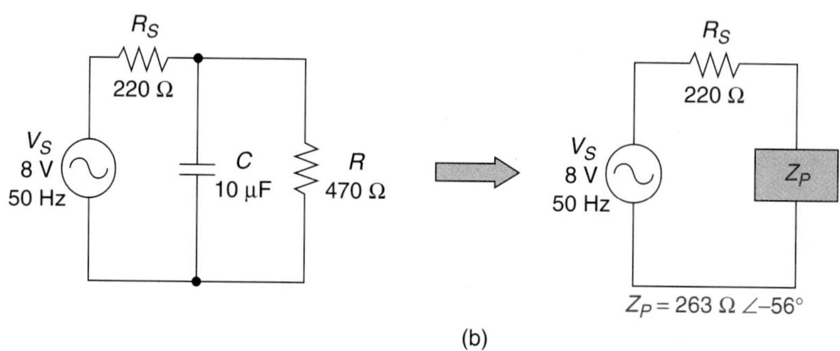


**FIGURE 17.37**

However, they can be added easily when they are converted to rectangular notation (as is the case with *RL* circuits).

## RC Circuit Values: Polar and Rectangular Forms

Earlier in this chapter, you were shown how $X_C$, *R*, and *Z* are plotted as an impedance triangle. An *RC* impedance triangle is shown, with the applicable polar-rectangular conversions, in Figure 17.38. As shown in the figure, the impedance of an *RC* circuit can be converted from polar to rectangular form using

$$R = Z \cos \theta \quad \textbf{(17.16)}$$

and

$$X_C = Z \sin \theta \quad \textbf{(17.17)}$$

**FIGURE 17.38 Polar-to-rectangular conversions.**

*Note:* If any of the relationships shown in Figure 17.38 seem unfamiliar, you may want to review the appropriate discussion in Chapter 12 before continuing.

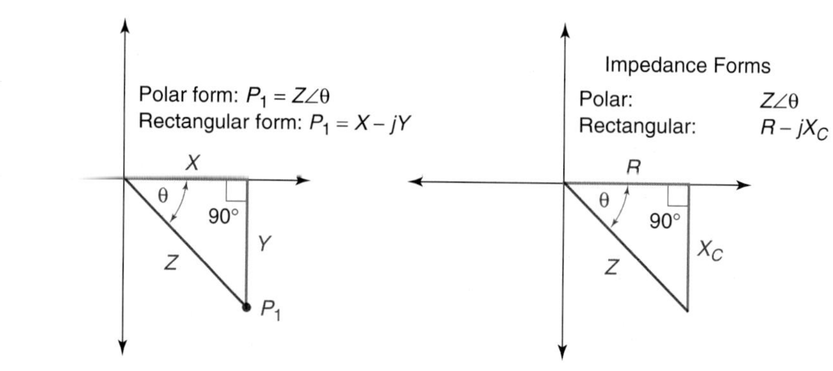


Conversion equations for the impedance triangle:

| Given: $Z\angle\theta$ | Given: $R - jX_C$ |
|---|---|
| $R = Z \cos \theta$ | $Z = \sqrt{X_L^2 + R^2}$ |
| $X_C = Z \sin \theta$ | $\theta = \tan^{-1} \frac{-X_C}{R}$ |

Note that equation (17.17) always provides a negative result, which indicates that $X_C$ lags the circuit resistance by 90°. Therefore, the value of $X_C$ is written as $(-jX_C)$ in the rectangular representation of the value. This point is demonstrated in Example 17.18.

## EXAMPLE 17.18

The parallel combination of $R$ and $C$ in Figure 17.37b is shown to have a value of $Z_P = 263\ \Omega\angle-56°$. Convert this value to rectangular form.

***Solution:*** The value of $R$ is found as

$$\mathbf{R} = Z\cos\theta = (263\ \Omega)(\cos -56°) = \mathbf{147\ \Omega}$$

The value of $X_C$ is found as

$$\mathbf{X_C} = Z\sin\theta = (263\ \Omega)(\sin -56°) = \mathbf{-218\ \Omega}$$

As you can see, equation (17.17) has provided a negative value of $X_C$. Therefore, the rectangular representation of the parallel impedance is written as $Z_P = (147 - j218)\ \Omega$.

*Note:* By now, you probably know that the conversions discussed here can be performed using any standard scientific calculator. However, manual conversions are discussed in this section to reinforce the relationships between the values.

***PRACTICE PROBLEM 17.18***

A parallel $RC$ circuit is found to have a total impedance of $Z_P = 760\ \Omega\angle-40°$. Convert this value to rectangular form.

The result in Example 17.18 indicates that the value of $Z_P$ equals the total impedance of a 147 Ω resistance in series with a 218 Ω capacitive reactance; that is, it provides us with the series equivalent of $Z_P$. This point is illustrated in Figure 17.39.

Using rectangular notation provides us with the series equivalent of any impedance value written in polar notation.

Figure 17.39a shows the circuit used as the basis for Example 17.18. As you can see, the total impedance of the parallel network ($Z_P$) is $263\ \Omega\angle-56°$. The rectangular form of this value (which was calculated in the example) is drawn as a series impedance ($Z_S$) in Figure 17.39b. As you can see, the values in the $Z_S$ equation match those in the figure. If we were to disconnect $R_S$ and measure the impedance of the shaded circuit ($Z_S$), we would measure approximately 263 Ω. In effect, what we have done in the example is to determine the series equivalent of $Z_P$. Once the series equivalent impedance of the parallel $RC$ network has been determined, we can solve easily for the total circuit impedance, as demonstrated in Example 17.19.

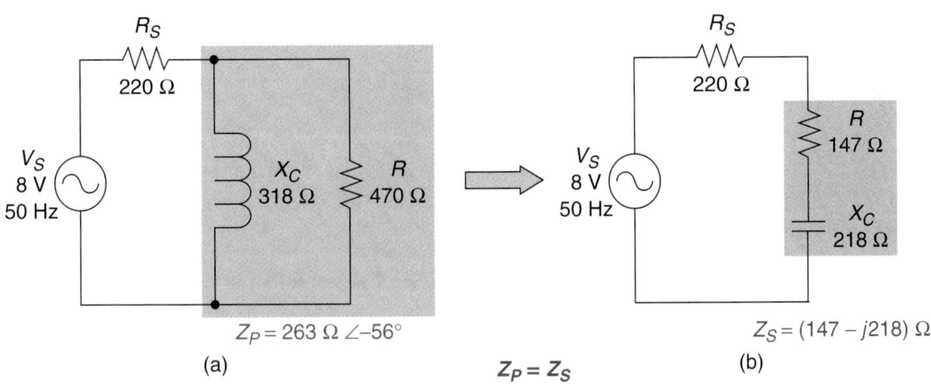


FIGURE 17.39

## EXAMPLE 17.19

Determine the total impedance (in polar form) for the circuit in Figure 17.37b.

***Solution:*** First, the total resistance in the circuit is found as

$$R_T = R_S + R = 220\ \Omega + 147\ \Omega = \mathbf{367\ \Omega}$$

Now, $R_T$ and $X_C$ can be combined to determine the magnitude of the circuit impedance, as follows:

$$Z_T = \sqrt{X_C^2 + R_T^2} = \sqrt{(218\ \Omega)^2 + (367\ \Omega)^2} = \mathbf{427\ \Omega}$$

and the phase angle of the circuit impedance can be found as

$$\theta = \tan^{-1}\left(\frac{-X_C}{R_T}\right) = \tan^{-1}\left(\frac{-218\ \Omega}{367\ \Omega}\right) = \mathbf{-30.7°}$$

Thus, the total impedance of the original series-parallel circuit is $427\ \Omega\angle -30.7°$.

***PRACTICE PROBLEM 17.19***

A circuit like the one in Figure 17.39b has the following values: $R_S = 150\ \Omega$, $R = 348\ \Omega$, and $X_C = 80\ \Omega$. Calculate the total circuit impedance.

In Examples 17.18 and 17.19, we derived the series equivalent for the circuit in Figure 17.37a. The remainder of the circuit analysis would involve solving for the component current, voltage, and power values. Example 17.20 demonstrates the process from start to finish.

## EXAMPLE 17.20

Perform a complete analysis on the circuit shown in Figure 17.40a.

***Solution:*** First, the reactance of the capacitor is found as

$$X_C = \frac{1}{2\pi f C} = \frac{1}{2\pi(250\ \text{Hz})(0.47\ \mu\text{F})} = \mathbf{1.35\ k\Omega}$$

Now, the magnitude of the parallel circuit impedance can be found as

$$Z_P = \frac{X_C R_2}{\sqrt{X_C^2 + R_2^2}} = \frac{(1.35\ \text{k}\Omega)(1.8\ \text{k}\Omega)}{\sqrt{(1.35\ \text{k}\Omega)^2 + (1.8\ \text{k}\Omega)^2}} = \mathbf{1.08\ k\Omega}$$

Since $C$ and $R_2$ are connected in parallel, the phase angle of $Z_P$ is found as

$$\theta = \tan^{-1}\left(\frac{R_2}{-X_C}\right)$$

$$= \tan^{-1}\left(\frac{1.8\ \text{k}\Omega}{-1.35\ \text{k}\Omega}\right)$$

$$= \mathbf{-53.1°}$$

We have determined that the parallel combination of $C$ and $R_2$ has a total impedance of $1.08\ \text{k}\Omega\angle -53.1°$. This impedance is represented as a block ($Z_P$) in Figure 17.40b. The series equivalent of $Z_P$ can now be found using

$$R = Z\cos\theta = (1.08\ \text{k}\Omega)(\cos -53.1°) \cong \mathbf{648\ \Omega}$$

*Note:* The negative value of $X_C$ merely indicates that $X_C$ lags $R$ by 90°.

and

$$X_C = Z\sin\theta = (1.08\ \text{k}\Omega)(\sin -53.1°) = \mathbf{-864\ \Omega}$$

These values are included as $Z_S$ in the series equivalent circuit (Figure 17.40b).

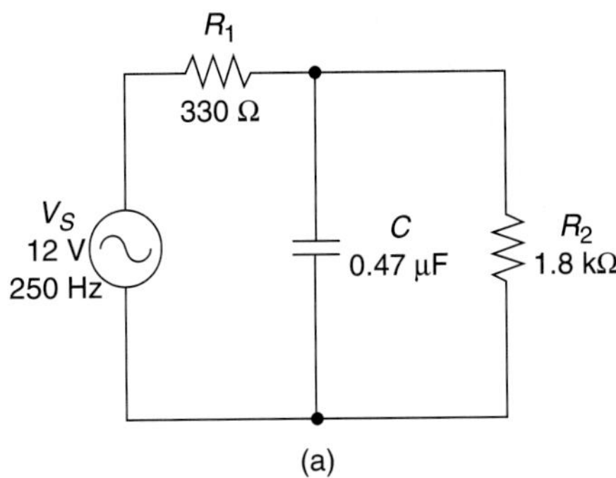


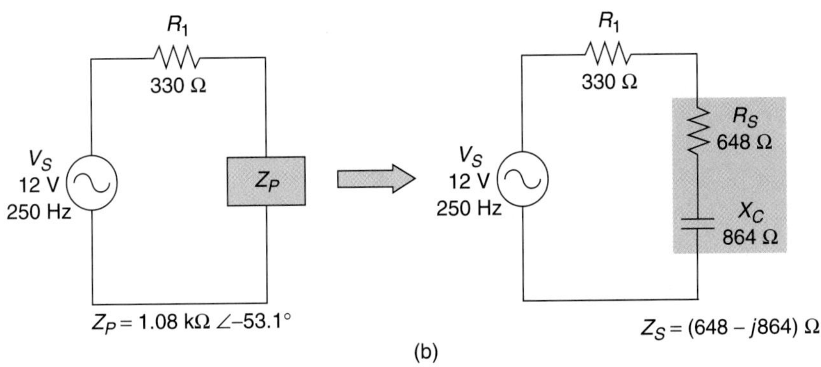


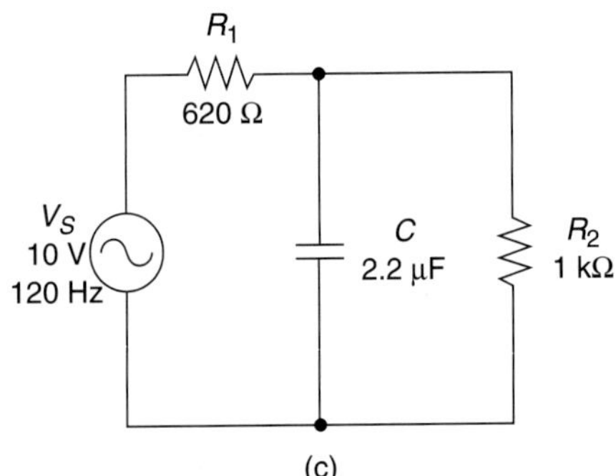


**FIGURE 17.40**

Once the series equivalent circuit has been derived, the resistances in the circuit can be combined as follows:

$$\boldsymbol{R_T} = R_S + R_1 = 648\ \Omega + 330\ \Omega = \mathbf{978\ \Omega}$$

Now, the magnitude of the circuit impedance can be found as

$$\boldsymbol{Z_T} = \sqrt{X_C^2 + R_T^2} = \sqrt{(864\ \Omega)^2 + (978\ \Omega)^2} = \mathbf{1.3\ k\Omega}$$

and its phase angle can be found as

$$\boldsymbol{\theta} = \tan^{-1}\left(\frac{-X_C}{R_T}\right) = \tan^{-1}\left(\frac{-864\ \Omega}{978\ \Omega}\right) = \mathbf{-41.5°}$$

Once the circuit impedance is determined, the circuit current can be found as

$$\boldsymbol{I_T} = \frac{V_S}{Z_T} = \frac{12\ \text{V}}{1.3\ \text{k}\Omega\angle -41.5°} = \mathbf{9.23\ mA\angle 41.5°}$$

This result indicates that circuit current leads the source voltage by 41.5°. Earlier in the chapter, you were told that current is considered to be the zero degree reference in a series circuit. Since we are dealing with a series equivalent circuit, the phase angles are adjusted as follows:

$$V_S = 12 \text{ V}\angle{-41.5^\circ} \qquad \text{and} \qquad I_T = 9.23 \text{ mA}\angle 0^\circ$$

Now, the voltage across $R_1$ can be found as

$$\boldsymbol{V_1} = I_T R_1 = (9.23 \text{ mA})(330\ \Omega) = \mathbf{3.05\ V}$$

Note that the phase angles were left out of the $V_1$ calculation because both are 0°. However, phase angles must be included when calculating the voltage across the parallel circuit impedance ($Z_P$). This voltage ($V_P$) can be found as

$$\boldsymbol{V_P} = I_T Z_P = (9.23 \text{ mA}\angle 0^\circ)(1.08 \text{ k}\Omega\angle{-53.1^\circ}) \cong \mathbf{9.97\ V\angle{-53.1^\circ}}$$

Now that we know the voltage across the parallel circuit, we can calculate the values of $I_C$ and $I_{R2}$. Again, we must include the phase angles in our calculations, as follows:

$$\boldsymbol{I_C} = \frac{V_P}{X_C} = \frac{9.97 \text{ V}\angle{-53.1^\circ}}{1.35 \text{ k}\Omega\angle{-90^\circ}} = 7.39 \text{ mA}\angle[-53.1^\circ - (-90^\circ)] = \mathbf{7.39\ mA\angle 36.9^\circ}$$

and

$$\boldsymbol{I_{R2}} = \frac{V_P}{R_2} = \frac{9.97 \text{ V}\angle{-53.1^\circ}}{1.8 \text{ k}\Omega\angle 0^\circ} = \mathbf{5.54\ mA\angle{-53.1^\circ}}$$

That completes the circuit currents and voltages. The apparent power for the circuit can be found using the values of $I_T$ and $Z_T$, as follows:

$$\boldsymbol{P_{APP}} = I_T^2 Z_T = (9.23 \text{ mA}\angle 0^\circ)^2(1.3 \text{ k}\Omega\angle{-41.5^\circ}) = \mathbf{111\ mVA\angle{-41.5^\circ}}$$

Given the complexity of the circuit, the easiest way to calculate the actual power dissipation (i.e., resistive power) is to use the power factor, as follows:

$$\boldsymbol{P_R} = P_{APP}\cos\theta = (111 \text{ mVA})(\cos -41.5^\circ) = (111 \text{ mVA})(0.749) = \mathbf{83.1\ mW}$$

This completes the analysis of the circuit.

***PRACTICE PROBLEM 17.20***

Perform a complete analysis on the circuit shown in Figure 17.40c.

## *Putting It All Together: Some Observations on Example 17.20*

To make sense of the results in Example 17.20, we need to combine them and see how they fit together. For ease of discussion, Figure 17.41 includes the currents and voltages calculated in the example. Note that the source has been given a phase angle relative to the circuit current, as found in the example.

First, let's look at the parallel circuit ($C$ and $R_2$). Earlier in the chapter, you were given the following relationships for a parallel $RC$ circuit:

1. The voltages across the branches are equal.
2. $I_C$ leads $I_R$ by 90°.
3. The sum of the branch currents equals the total circuit current.

The first of these relationships is shown in the figure. The voltage across each branch equals the value of $V_P$ calculated in the example. The phase relationship between $I_C$ and $I_{R2}$ can be verified by finding the difference between their phases, as follows:

$$\theta = 36.9^\circ - (-53.1^\circ) = 36.9^\circ + 53.1^\circ = 90^\circ$$

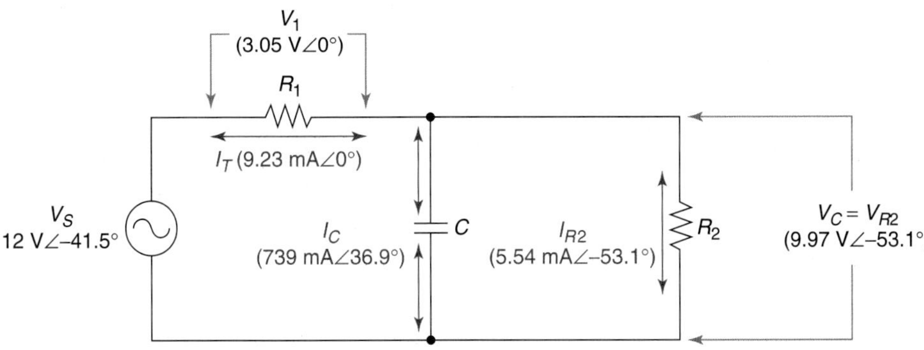


**FIGURE 17.41 The voltages and currents calculated in Example 17.20.**

As you can see, the second relationship given for parallel *RC* circuits also holds true. The third relationship can be verified by converting each of the branch currents to rectangular form. Using the conversion techniques covered in this section, the branch currents in the figure can be written as follows:

$$I_C = (5.91 + j4.44)\text{ mA} \qquad I_{R2} = (3.33 - j4.43)\text{ mA}$$

Adding these currents together, we get

$$\begin{aligned} I_C &= (5.91 + j4.44)\text{ mA} \\ I_{R2} &= \underline{(3.33 - j4.43)\text{ mA}} \\ &\phantom{=} (9.24 + j0.01)\text{ mA} = I_T \end{aligned}$$

This value is approximately equal to 9.24 mA, which closely matches the one shown in the figure. (The difference between the numbers is caused by rounding values throughout the analysis.)

We have shown that the characteristics of the parallel *RC* circuit match those given earlier in the chapter. Now, let's look at the series combination of $R_1$ and $(C \| R_2)$. In any series circuit:

1. The currents through the series components are equal.
2. The source voltage equals the sum of the series component voltages.

We have already shown the first of these relationships to be true. The second can be demonstrated by converting $V_{R1}$ and the voltage across the parallel circuit ($V_P$) to rectangular form, as follows:

$$V_{R1} = (3.05 + j0)\text{ V} \qquad \text{and} \qquad V_P = (5.99 - j7.97)\text{ V}$$

Adding these voltages together, we get

$$\begin{aligned} V_{R1} &= (3.05 + j0.00)\text{ V} \\ V_P &= \underline{(5.99 - j7.97)\text{ V}} \\ &\phantom{=} (9.04 - j7.97)\text{ V} = V_S \end{aligned}$$

We can verify this result by converting it to polar form, as follows:

$$V_S = \sqrt{(9.04\text{ V})^2 + (-7.97\text{ V})^2} = 12.05\text{ V}$$

and calculating the phase angle, as follows:

$$\theta = \tan^{-1}\left(\frac{-7.97\text{ V}}{9.04\text{ V}}\right) = \tan^{-1}(-0.882) = -41.4°$$

These results closely match the source voltage shown in Figure 17.41. As you can see, all the fundamental relationships for series and parallel circuits hold true for the circuit in Example 17.20.

## A Look Ahead

Up to this point, we have concentrated on the operation of *RC* circuits with sinusoidal input signals. In the next section, we will look at the response that series *RC* circuits have to rectangular inputs. More specifically, we will concentrate on the effect that a series *RC* circuit has on a square-wave input. As you will see, we will be dealing once again with circuit time constants.

### Section Review

1. How is the impedance of an *RC* circuit converted from polar form to rectangular form?
2. When a parallel *RC* impedance is converted from polar to rectangular form, what type of circuit conversion has taken place?

# 17.5 SQUARE WAVE RESPONSE: *RC* TIME CONSTANTS

In Chapter 14, you were shown how an *RL* circuit responds to rectangular waveforms. In this section, we will discuss the response of an *RC* circuit to the same type of input.

## *RC* Circuit Waveforms

*OBJECTIVE 12* ▶ When a square wave is applied to a purely resistive circuit, the voltage and current waveforms have identical shapes, as shown in Figure 17.42. When a square wave is applied to a series *RC* circuit, the waveforms are altered as shown in Figure 17.43. The waveforms in Figure 17.43a result from the charging of the capacitor. The effects of capacitor charging on the *RC* circuit waveforms can be explained using the relationships shown in Figure 17.43b. As shown in the figure, the resistor voltage always equals the difference between

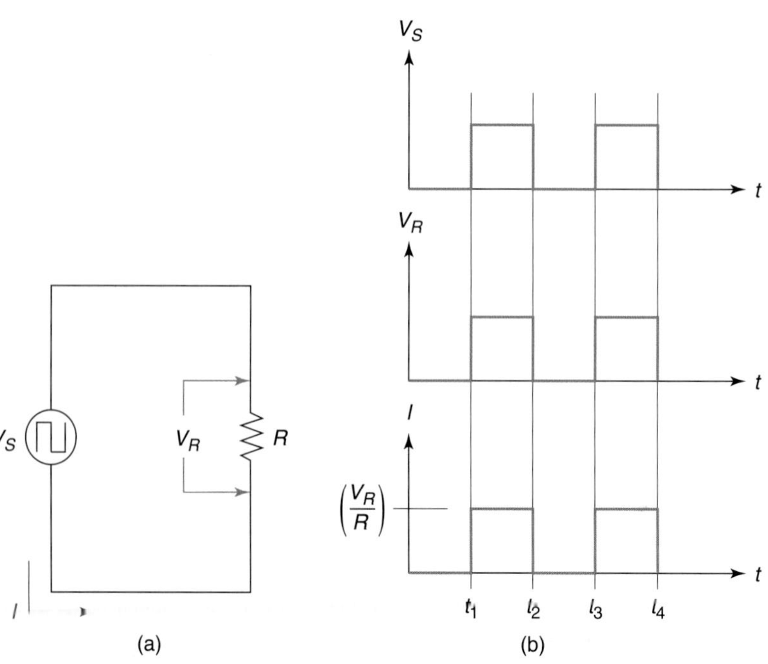


**FIGURE 17.42**

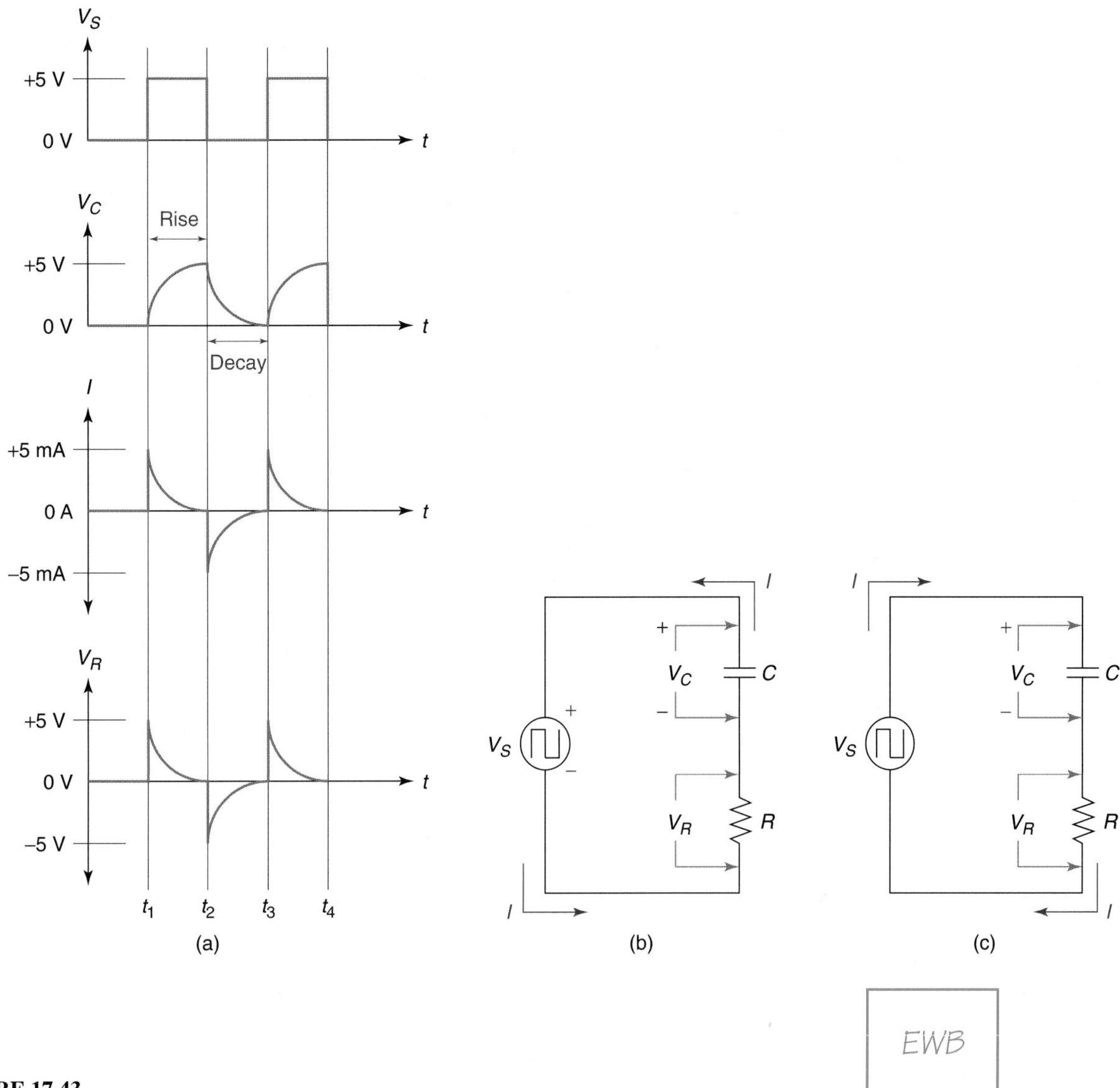


**FIGURE 17.43**

the source voltage and the capacitor voltage. This relationship, which is consistent with Kirchhoff's voltage law, is given as

$$V_R = V_S - V_C$$

When the values of $V_R$ and $R$ are known, Ohm's law allows us to calculate the total circuit current as

$$I_T = \frac{V_R}{R}$$

Combining these equations, we get

$$I_T = \frac{V_S - V_C}{R} \tag{17.18}$$

Using equation (17.18), let's look at the waveforms in Figure 17.43. At $t_1$, the source voltage is shown making a transition from 0 V to +5 V. At the instant of the transition, the capacitor has not begun to charge, and $V_C = 0$ V. Assuming $R = 1$ kΩ, the circuit current at $t_1$ jumps to a value of

$$I_T = \frac{V_S - V_C}{R} = \frac{5\text{ V} - 0\text{ V}}{1\text{ k}\Omega} = 5\text{ mA}$$

This current begins to charge the capacitor (as shown in Figure 17.43b). At some point between $t_1$ and $t_2$, capacitor voltage reaches 2.5 V. At that point,

$$I_T = \frac{V_S - V_C}{R} = \frac{5\text{ V} - 2.5\text{ V}}{1\text{ k}\Omega} = 2.5\text{ mA}$$

As this value indicates, the circuit current decreases as the capacitor charges. Since the charging current decreases, so does the rate at which the capacitor is charging. In other words, the reduced circuit current causes the capacitor to charge at a slower rate (which is why the rate of increase is shown to decrease between $t_1$ and $t_2$).

Eventually, the capacitor charges to the point where its voltage is approximately equal to the source voltage. At that point, there is approximately 0 V across the resistor, and the circuit current is reduced to approximately zero.

At $t_2$, the source voltage makes the transition back to 0 V. When this occurs, the capacitor begins to discharge through $R$ and the source (as shown in Figure 17.43c). Because of the polarity of the capacitor voltage, the circuit current switches direction. The value of the circuit current at the moment the capacitor begins to discharge can be found as

$$I_T = \frac{V_S - V_C}{R} = \frac{0\text{ V} - 5\text{ V}}{1\text{ k}\Omega} = -5\text{ mA}$$

The minus sign in the result indicates that the current has reversed direction, as shown in the figure.

As the capacitor discharges, its plate-to-plate voltage decreases. As $V_C$ decreases, so does the value of the discharge current. For example, at some point between $t_2$ and $t_3$, the capacitor voltage drops to 2.5 V. At that time,

$$I_T = \frac{V_S - V_C}{R} = \frac{0\text{ V} - 2.5\text{ V}}{1\text{ k}\Omega} = -2.5\text{ mA}$$

The fact that capacitor discharge current is decreasing indicates that the component is discharging at a slower rate, which is why the rate of change in the current curve decreases between $t_2$ and $t_3$. Eventually, the capacitor discharges completely. At that time, circuit current is reduced to zero, as is the voltage across the capacitor.

## Resistor Voltage ($V_R$)

As you know, the voltage across a resistor is always proportional to the current through it. Since $V = IR$, the voltage waveform in Figure 17.43 has the same shape as the current waveform. Note that the peak values of the voltage waveform can be found as the product of the peak current values and $R$.

## More on the RC Circuit Waveforms

Several points need to be made regarding the waveforms in Figure 17.43. The first deals with the relationship between the component voltages. Even though we limited our discussion to the times immediately before and after the transitions, $V_R$ and $V_C$ change at the same rate. Therefore, their sum always equals the supply voltage.

The second point we need to make deals with the rate at which $V_C$ makes its transitions. As stated earlier, capacitor voltage does not change at a linear (constant) rate. Rather, the rate of change in $V_C$ decreases as the transition progresses. This concept is illustrated

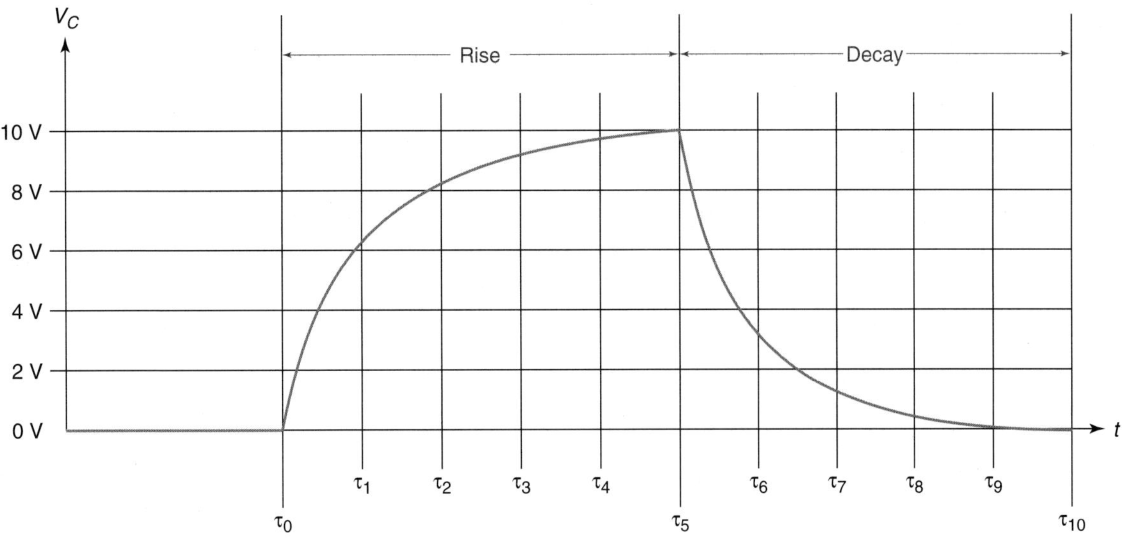


**FIGURE 17.44**

**TABLE 17.7 Estimated Values of $\Delta V_C$ During the Rise Portion of the Curve in Figure 17.44**

| *Time Interval* | *Estimated* $\Delta V_C$ |
|---|---|
| $t_0 \rightarrow t_1$ | 6.2 V |
| $t_1 \rightarrow t_2$ | 2.2 V |
| $t_2 \rightarrow t_3$ | 1.0 V |
| $t_3 \rightarrow t_4$ | 0.2 V |
| $t_4 \rightarrow t_5$ | $<$0.2 V |

**TABLE 17.8 Estimated Values of $\Delta V_C$ During the Decay Portion of the Curve in Figure 17.44**

| *Time Interval* | *Estimated* $\Delta V_C$ |
|---|---|
| $t_5 \rightarrow t_6$ | −6.2 V |
| $t_6 \rightarrow t_7$ | −2.2 V |
| $t_7 \rightarrow t_8$ | −1.0 V |
| $t_8 \rightarrow t_9$ | −0.2 V |
| $t_9 \rightarrow t_{10}$ | $<$−0.2 V |

further in Figure 17.44. For the sake of discussion, the $V_C$ waveform in the figure is assumed to have a peak value of +10 V.

Each half of the waveform in Figure 17.44 is divided into five equal time intervals. Each time interval is identified using the Greek letter tau, $\tau$. The estimated change in voltage ($\Delta V_C$) during each of the first five time intervals is provided in Table 17.7. As you can see, the change in $V_C$ decreases from one time interval to the next during the rising portion of the curve. The same principle holds true for the decay portion of the curve. The estimated values of $\Delta V_C$ for the decay portion of the curve are listed in Table 17.8. When you compare the values of $\Delta V_C$ in these two tables, you can see that the two halves of the curve are symmetrical; that is, the rate of rise in any time interval matches the corresponding rate of decay. This relationship holds true for any *RC* switching circuit.

## The Universal Voltage Curve

◄ *OBJECTIVE 13*

The shape of the curve in Figure 17.44 is a function of the capacitor's opposition to a change in voltage. This means that the wave shape is independent of the peak value of the source voltage. For example, if we doubled the peak source voltage in Figure 17.44, the peak value of $V_C$ would also double. However, the shape of the curve would remain unchanged.

The voltage characteristics of an *RC* switching circuit can be represented using a **universal voltage curve.** (You may recall that a universal curve was used in Chapter 14 to represent the current rise and decay in an *RL* switching circuit.) The universal voltage curve for an *RC* switching circuit is shown in Figure 17.45.

**Universal voltage curve** One that can be used to predict the voltage values at any time in a specified circuit.

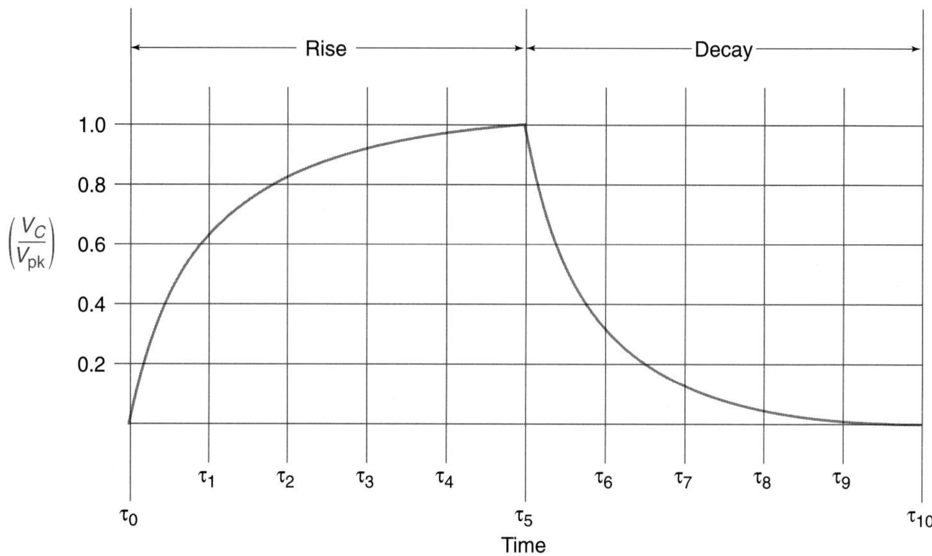


**FIGURE 17.45 The universal voltage curve for a series *RC* switching circuit.**

The $x$-axis of the universal curve is divided into the same time intervals that we saw earlier. The $y$-axis, however, measures the ratio of capacitor voltage ($V_C$) to peak source voltage ($V_{pk}$). For example, at $\tau_1$, the curve indicates that

$$\frac{V_C}{V_{pk}} \cong 0.62$$

At $\tau_2$, the curve indicates that

$$\frac{V_C}{V_{pk}} \cong 0.85$$

and so on.

When the peak source voltage is known, the curve can be used to approximate the value of $V_C$ at any time interval. At any time interval, the value of $V_C$ can be approximated as

$$V_C = \left(\frac{V_C}{V_{pk}}\right) V_{pk} \qquad \textbf{(17.19)}$$

where $\left(\frac{V_C}{V_{pk}}\right)$ = the voltage ratio obtained from the curve

$V_{pk}$ = the known peak source voltage

For example, the voltage ratio at $\tau_4$ in Figure 17.45 is approximately 0.95. If we know that an *RC* circuit has a value of $V_{pk} = 15$ V, then the capacitor voltage at $\tau_4$ can be found as

$$V_C = \left(\frac{V_C}{V_{pk}}\right) V_{pk} = (0.95)(15 \text{ V}) = 14.25 \text{ V}$$

Note that the curve provides only approximate values of capacitor voltage at each time interval. If a more exact value of $V_C$ is desired, it must be calculated using the curve equation.

*OBJECTIVE 14* ►

## The Rise Curve Equation

**Exponential curve**
A curve whose rate of change is a function of a variable exponent.

In Chapter 14, you were told that the universal curve is an **exponential curve,** meaning its rate of change is a function of a variable exponent. For example, the rise portion of the universal voltage curve in Figure 17.46 is actually a plot of the equation

$$\frac{V_C}{V_{pk}} = 1 - e^{-x} \qquad \textbf{(17.20)}$$

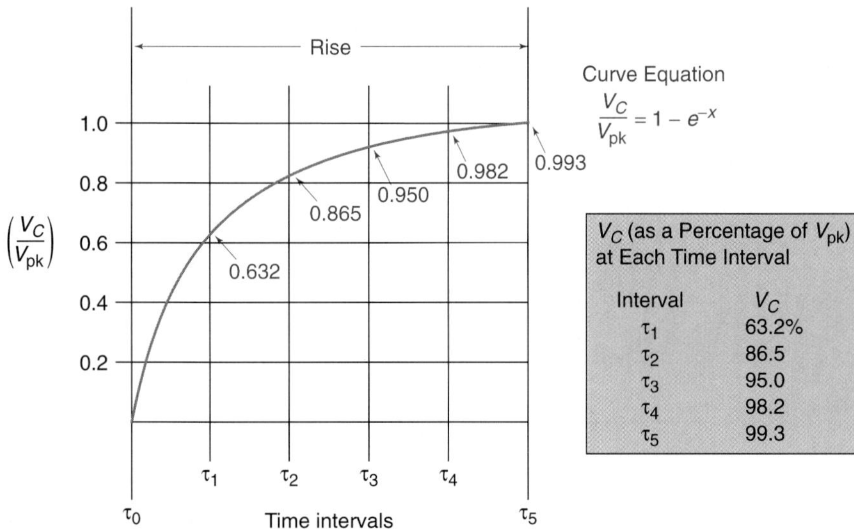


**FIGURE 17.46 The rise portion of the universal voltage curve.**

where $e$ = the base of the natural log system (approximately 2.71828)
$x$ = the variable exponent

The only value on the right-hand side of the equation that is variable is $x$. Therefore, the shape of the curve must be a function of this variable exponent. By definition, that makes the curve an exponential curve.

## Plotting the Universal Curve

If we solve equation (17.20) for whole-number values of $x = 1 \rightarrow 5$, we get the results listed in Table 17.9. If we assume that the values of $x$ in Table 17.9 correlate to the time intervals along the $x$-axis in Figure 17.46, then the curve forms a plot of the results. Using the values shown on the curve, we can express the value of $V_C$ at each time interval as it relates to the value of the peak source voltage ($V_{pk}$). For example, at $\tau_1$:

$$\frac{V_C}{V_{pk}} = 0.632$$

Therefore,

$$V_C = 0.632\ V_{pk}$$

This equation indicates that $V_C$ reaches 63.2% of its peak value at $\tau_1$. The value of $V_C$ (as a percentage of $V_{pk}$) for each time interval is listed in the figure. Based on the percentages shown, we can say that capacitor voltage increases from 0% to 99.3% of its peak value over the course of the five time intervals.

**TABLE 17.9 Results of Equation (17.20) for Increasing Values of $x$**

| *Value of x* | *Equation (17.20) Result* |
|---|---|
| 1 | 0.632 |
| 2 | 0.865 |
| 3 | 0.950 |
| 4 | 0.982 |
| 5 | 0.993 |

## The Decay Curve

The decay curve is simply a mirror image of the rise curve. Like the rise curve, it is an exponential curve. The exponential relationship used to plot the decay curve is

$$\frac{V_C}{V_{pk}} = e^{-x} \qquad \textbf{(17.21)}$$

If we solve this relationship for whole-number values of $x = 1 \rightarrow 5$, we get the results listed in Table 17.10, which are identified on the decay curve in Figure 17.47. Again, each time interval corresponds to a value of $x$ from the table. As was the case with the rise curve,

**TABLE 17.10 Results of Equation (17.21) for Increasing Values of $x$**

| *Value of x* | *Equation (17.21) Result* |
|---|---|
| 1 | 0.368 |
| 2 | 0.135 |
| 3 | 0.050 |
| 4 | 0.018 |
| 5 | 0.007 |

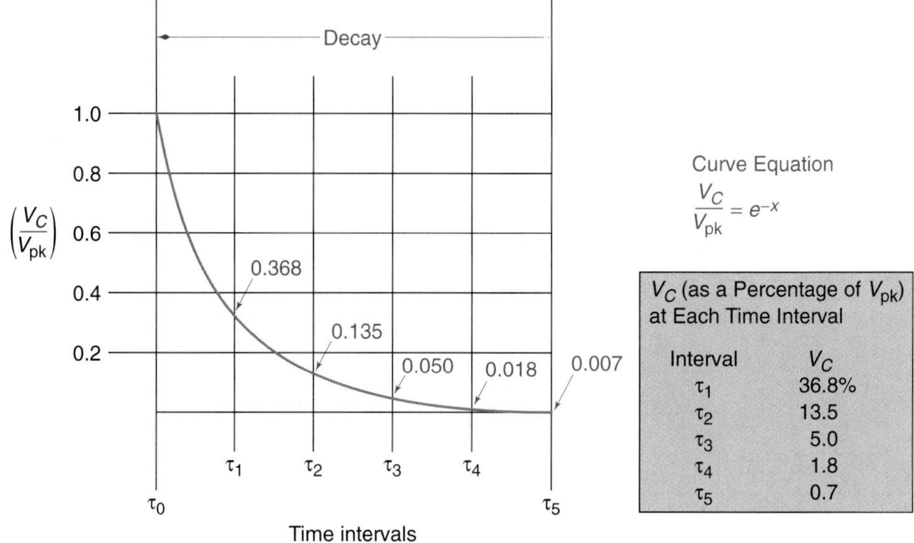


**FIGURE 17.47 The decay portion of the universal voltage curve.**

we can express the value of $V_C$ at each time interval in Figure 17.47 as a percentage of the peak source voltage ($V_{pk}$). The value of $V_C$ (as a percentage of $V_{pk}$) for each time interval is listed in the figure. As you can see, capacitor voltage decreases from approximately 100% to 0.7% of its peak value over the course of the five time intervals.

*OBJECTIVE 15* ▶

## *The RC Time Constant ($\tau$)*

*Note:* A precise definition of *RC time constant* is provided later in this section.

Each time interval in the universal curve represents a real-time value referred to as a time constant ($\tau$). In Chapter 14, you were told that the term *constant* is used because $\tau$ is independent of the circuit current magnitude and operating frequency. For a given series *RC* switching circuit, the value of the *RC* time constant is found as

$$\tau = RC \quad \textbf{(17.22)}$$

where $\tau$ = the duration of each time interval, in seconds
$C$ = the total series capacitance, in farads
$R$ = the total series resistance, in ohms

Example 17.21 demonstrates the use of this equation in determining the time constant for a series *RC* circuit with a square-wave input.

### EXAMPLE 17.21

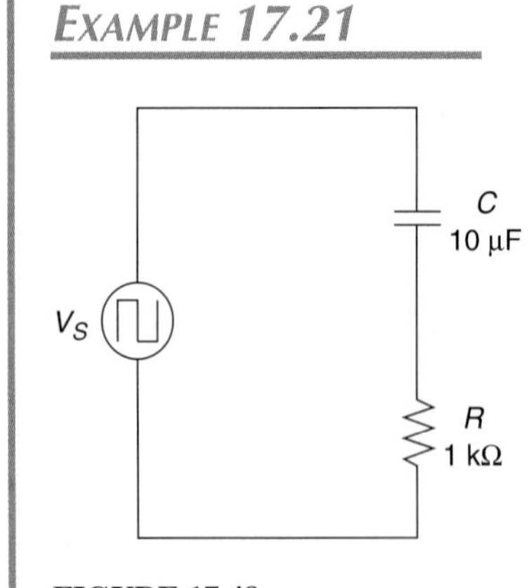


**FIGURE 17.48**

Determine the time constant for the circuit in Figure 17.48.

***Solution:*** Using equation (17.22), the time constant for the circuit is found as

$$\tau = RC = (1\text{ k}\Omega)(10\ \mu\text{F}) = \mathbf{10\ ms}$$

***PRACTICE PROBLEM 17.21***

A circuit like the one in Figure 17.48 has values of $R = 220\ \Omega$ and $C = 0.1\ \mu$F. Calculate the time constant for the circuit.

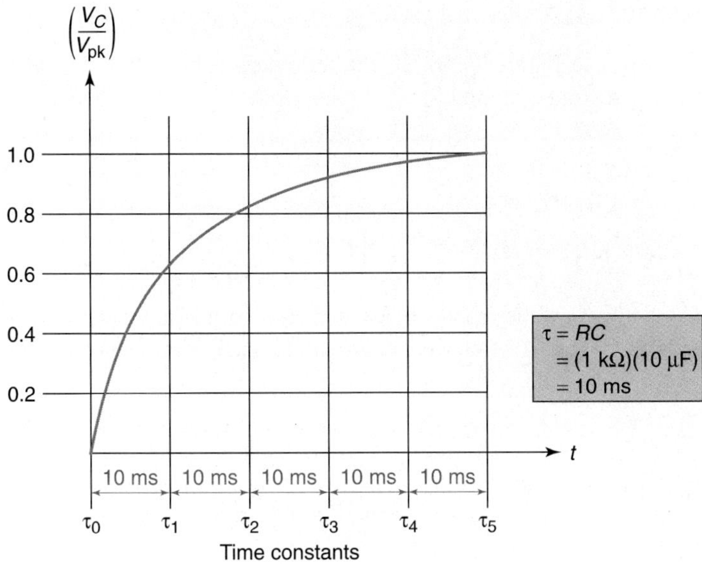


**FIGURE 17.49**

Figure 17.49 shows the rise curve for the circuit in the example. As you can see, each time interval represents the value of the circuit time constant (10 ms). The total time required for $V_C$ to complete its rise is approximately equal to 5 time constants. By formula,

$$T = 5\,\tau \tag{17.23}$$

where $T$ = the time required for $V_C$ to rise to its peak value, in seconds
$\tau$ = the time constant of the circuit, in seconds

Example 17.22 demonstrates the process for determining the total time required for the capacitor voltage in an *RC* circuit to rise from zero to its peak value.

*EXAMPLE 17.22*

Determine the time required for $V_C$ to rise from 0 V to its peak value in the circuit shown in Figure 17.50.

**FIGURE 17.50**

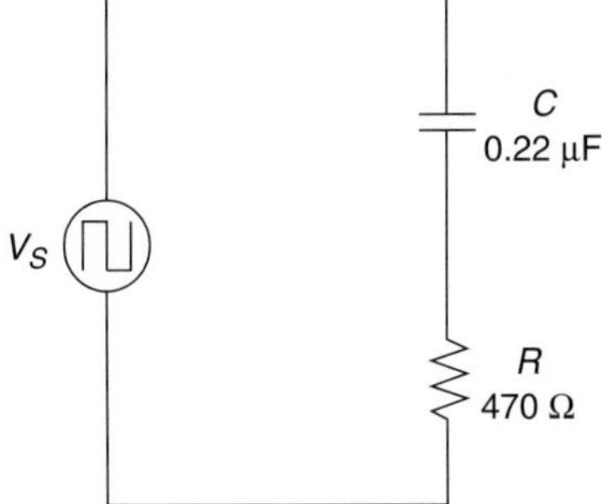


EWB

***Solution:*** First, the value of the circuit time constant is found as

$$\tau = RC = (470\ \Omega)(0.22\ \mu\text{F}) = \mathbf{103.4\ \mu s}$$

It takes 5 time constants for the rise portion of the curve to complete its transition. Therefore, the total time required for the transition is found as

$$T = 5\tau = (5)(103.4\ \mu\text{s}) = \mathbf{517\ \mu s}$$

***PRACTICE PROBLEM 17.22***

A circuit like the one in Figure 17.50 has the following values: $R = 510\ \Omega$ and $C = 4.7\ \mu F$. Determine the time required for $V_C$ to rise from 0 V to its peak value.

## *The Basis for Equation (17.23)*

You may be wondering how the product of resistance and capacitance can yield a value measured in seconds. In Chapter 16, you were shown that capacitor current can be found as

$$i_C = C\frac{dv}{dt}$$

where $dt$ is measured in seconds. During the rise (or decay) portion of the current curve, the source provides the circuit with a dc input. For a dc input, the above equation is rewritten as

$$\Delta I_C = C\frac{\Delta V}{\Delta t}$$

Rewriting the above equation for $C$, we get

$$\frac{(\Delta I)(\Delta t)}{\Delta V} = C \qquad \text{or} \qquad \left(\frac{\Delta I}{\Delta V}\right)(\Delta t) = C$$

Since $\Delta I/\Delta V = 1/R$, the above equation can be rewritten as

$$\left(\frac{1}{R}\right)(\Delta t) = C$$

and multiplying both sides of the equation by $R$, we get

$$\Delta t = RC$$

Using $\tau$ to represent the change in time ($\Delta t$), the above equation can be rewritten as

$$\tau = RC$$

which is equation (17.23).

## *Another Look at the Universal Curve Equations*

Earlier in this section, you learned that the universal curve is actually a plot of the following equations:

| **Rise Curve** | **Decay Curve** |
|---|---|
| $\frac{V_C}{V_{pk}} = 1 - e^{-x}$ | $\frac{V_C}{V_{pk}} = e^{-x}$ |

These equations can be viewed as the voltage equivalents of the universal current curves introduced in Chapter 14 (for *RL* switching circuits). In that chapter, you were told that the variable ($x$) in the curve equations is actually a ratio of two time values. The same holds true for an *RC* switching circuit. By formula,

$$x = \frac{t}{\tau} \qquad \textbf{(17.24)}$$

where
$x$ = the exponent of $e$ in the curve equations
$t$ = the time from the start of the curve
$\tau$ = the duration of one time constant

Refer back to Figure 17.49. As shown in that figure, the duration of one time constant in the curve is 10 ms. The time interval labeled $t_3$ occurs 30 ms after the start of the curve (which is labeled $\tau_0$). Therefore, the power of $e$ at this time interval is found as

$$x = \frac{t}{\tau} = \frac{30 \text{ ms}}{10 \text{ ms}} = 3$$

This result is used to calculate the voltage ratio at $\tau_3$, as follows:

$$\frac{V_C}{V_{pk}} = 1 - e^{-x} = 1 - e^{-3} = 0.95$$

Using equation (17.24) allows us to calculate the voltage ratio at any point on the rise (or decay) curve. This point is demonstrated in Example 17.23.

## EXAMPLE 17.23

The circuit in Figure 17.51 is shown to have a peak source voltage of 50 V. Determine the value of $V_C$ 300 μs after the start of the decay cycle.

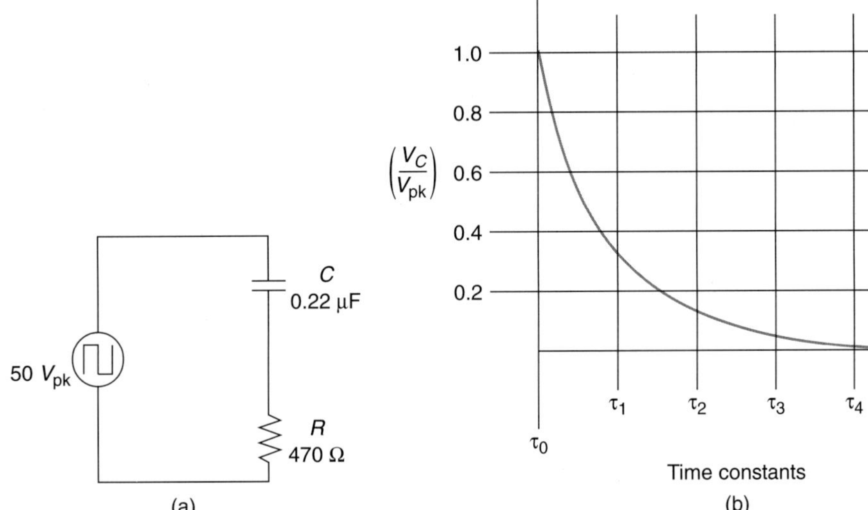


**FIGURE 17.51**

EWB

***Solution:*** In Example 17.22, we calculated a value of $\tau = 103.4$ μs for this circuit. At $t = 300$ μs, the power of $e$ is found as

$$\boldsymbol{x} = \frac{t}{\tau} = \frac{300\ \mu\text{s}}{103.4\ \mu\text{s}} = \mathbf{2.9}$$

Now, the voltage ratio at the designated time is found as

$$\frac{\boldsymbol{V_C}}{\boldsymbol{V_{pk}}} = e^{-x} = e^{-2.9} = \mathbf{0.055}$$

Using this voltage ratio, the value of $V_C$ at $t = 300$ μs is found as

$$\boldsymbol{V_C} = 0.055\ V_{pk} = (0.055)(50\text{ V}) = \mathbf{2.75\ V}$$

When you compare the voltage ratio (0.055) to the curve in Figure 17.51b, you can see that it is very close to the value provided by the curve at the $\tau_3$ interval.

*PRACTICE PROBLEM 17.23*

Refer back to Practice Problem 17.22. Assume that the peak source voltage for the circuit is 12 V and determine the value of $V_C$ when $t = 7.19$ ms.

As you can see, the time ratio in the universal curve equation allows us to determine the circuit current at any point during the decay (or rise) cycle. (We could not make this type of calculation when we assumed that $x$ simply represented a whole-number time interval.)

## Defining the RC Time Constant

*OBJECTIVE 16* ▶

**RC time constant**
The time required for the capacitor voltage in an *RC* switching circuit to increase (or decrease) by 63.2% of its maximum possible $\Delta V$.

In Chapter 14, you were told that an *RL* time constant is the time required for the current in an *RL* switching circuit to increase (or decrease) by 63.2% of its maximum possible $\Delta I$. An ***RC* time constant** is defined as the time required for the capacitor voltage in an *RC* switching circuit to increase (or decrease) by 63.2% of its maximum possible $\Delta V$. This definition can be explained with the aid of Figure 17.52.

The rise curve shown in the figure was derived using the approach demonstrated in Example 17.23; that is, the voltage ratio for each time interval was determined and then used to calculate the value of $V_C$. At any time interval, the maximum possible $\Delta V$ equals the difference between its value at the start of the time interval and its value at the end of the transition. For example, the capacitor voltage at $\tau_0$ in the figure is 0 V. The value of $V_C$ at the end of the transition is shown to be approximately 100 V. Therefore, the maximum possible $\Delta V$ (at $\tau_0$) is found as

$$\Delta V_{max} = 100 \text{ V} - 0 \text{ V} = 100 \text{ V} \quad (\text{at } \tau_0)$$

During the first time constant, the capacitor voltage increases by 63.2% of $\Delta V_{max}$. Therefore, the value of $V_C$ at $\tau_1$ is found as

$$V_C = V_{\tau 0} + 0.632(\Delta V_{max}) = 0 \text{ V} + 0.632(100 \text{ V}) = 0 \text{ V} + 63.2 \text{ V} = 63.2 \text{ V}$$

From $\tau_1$ to $\tau_2$, the capacitor voltage again changes by 63.2% of $\Delta V_{max}$. However, the value of $V_C$ at the start of the time interval is 63.2 V. Therefore, the value of $\Delta V_{max}$ is found as

$$\Delta V_{max} = 100 \text{ V} - 63.2 \text{ V} = 36.8 \text{ V} \quad (\text{at } \tau_1)$$

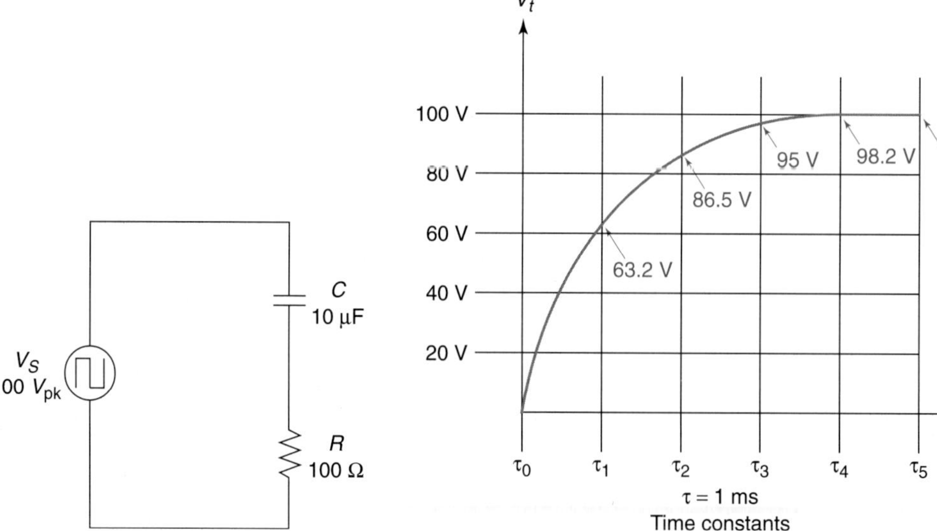


**FIGURE 17.52**

**TABLE 17.11 Voltage Calculations for the Circuit in Figure 17.52**

| *Time Interval* | *Capacitor Voltage* |
|---|---|
| $\tau_2 \rightarrow \tau_3$ | $V_{\tau 2} = 86.5\text{ V}$<br>$\Delta V_{max} = 100\text{ V} - 86.5\text{ V} = 13.5\text{ V}$<br>$V_{\tau 3} = 86.5\text{ V} + 0.632(13.5\text{ V}) = 95\text{ V}$ |
| $\tau_3 \rightarrow \tau_4$ | $V_{\tau 3} = 95\text{ V}$<br>$\Delta V_{max} = 100\text{ V} - 95\text{ V} = 5\text{ V}$<br>$V_{\tau 4} = 86.5\text{ V} + 0.632(5\text{ V}) = 98.2\text{ V}$ |
| $\tau_4 \rightarrow \tau_5$ | $V_{\tau 4} = 98.2\text{ V}$<br>$\Delta V_{max} = 100\text{ V} - 98.2\text{ V} = 1.8\text{ V}$<br>$V_{\tau 5} = 86.5\text{ V} + 0.632(1.8\text{ V}) = 99.3\text{ V}$ |

Using this value of $\Delta V_{max}$, the value of $V_C$ at $\tau_2$ is found as

$$V_C = V_{\tau 1} + 0.632(\Delta V_{max}) = 63.2\text{ V} + 0.632(36.8\text{ V}) = 63.2\text{ V} + 23.3\text{ V} = 86.5\text{ V}$$

As you can see, this value corresponds to the value shown on the curve when $t = \tau_2$. Continuing this series of calculations provides the results given in Table 17.11. In each case, the capacitor voltage at the end of the time interval matches the value shown in the curve (Figure 17.52).

As the calculations have demonstrated, the change in capacitor voltage from one time constant to the next equals 63.2% of the maximum possible transition. This is the basis for the definition of the *RC* time constant.

The voltage relationship described here also applies to the decay curve of an *RC* switching circuit. However, the capacitor voltage during each time interval decreases by 63.2% of the maximum possible transition. Demonstrating this principle through calculations (like those in Table 17.11) is left as a Brain Drain problem at the end of the chapter.

## Time Calculations

◀ *OBJECTIVE 17*

There are situations where it is desirable to know how long it will take for the capacitor voltage in an *RC* switching circuit to reach a designated value. For example, look at the delay circuit shown in Figure 17.53. This circuit uses the capacitor voltage rise of the *RC* circuit to provide a delay between the positive-going transitions of $V_S$ and $V_{out}$. Here's how the circuit works: The switching circuit is designed to switch between two output levels: 0 V and +5 V. The output from the circuit depends on the voltages at its signal inputs (A and B). The only time that the output from the switching circuit is at +5 V is when both of its input voltages ($V_A$ and $V_B$) are greater than or equal to +3 V. With this relationship in mind, let's look at the waveforms in Figure 17.53. At the instant that $V_S$ makes its positive-going transition from 0 V to +5 V, there is no voltage across the capacitor. As the capacitor charges, $V_C$ increases. When $V_C$ reaches +3 V, the switching circuit has the following input voltages:

$$V_A = V_S = +5\text{ V} \qquad V_B = V_C = +3\text{ V}$$

These two voltages meet the requirements for a +5 V output from the switching circuit, so $V_{out}$ makes its positive-going transition. The time difference between the positive-going transitions of $V_S$ and $V_{out}$ is shown as the signal delay in the waveforms.

The analysis of a circuit like the one in Figure 17.53 requires that you know how long it takes for $V_C$ to reach its desired level. The time required for $V_C$ to reach a specified level can be found as

$$t = -\tau \ln\left(1 - \frac{V_C}{V_S}\right) \qquad \textbf{(17.25)}$$

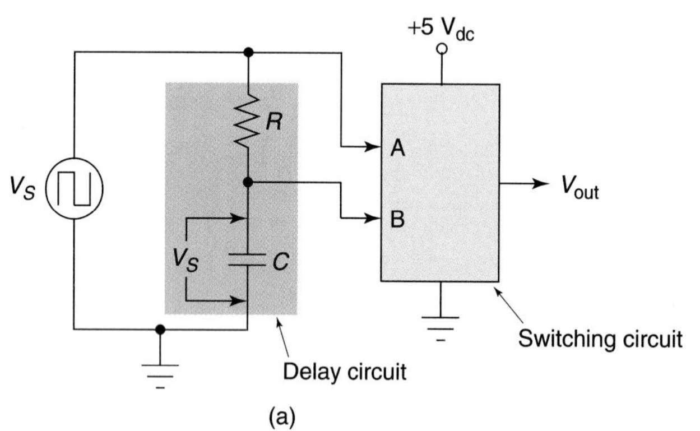


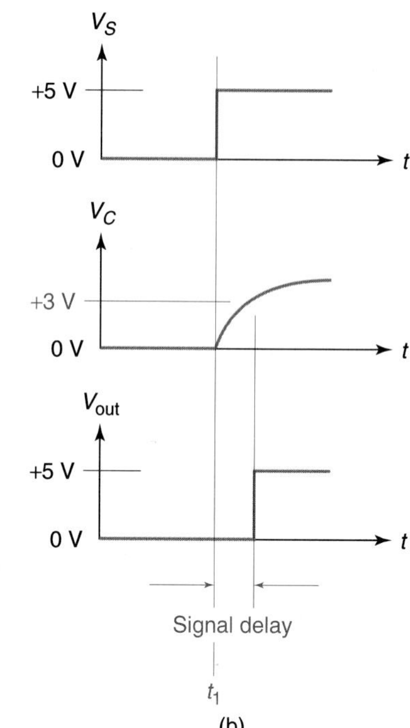


**FIGURE 17.53 An *RC* delay circuit and waveforms.**

where *ln* is the natural log function. Equation (17.25) is derived in Appendix E. Its use is demonstrated in Example 17.24.

## EXAMPLE 17.24

Assume that the circuit in Figure 17.53 has values of $C = 10\ \mu\text{F}$ and $R = 330\ \Omega$. Determine the duration of the signal delay shown in the waveforms.

***Solution:*** First, the time constant of the *RL* circuit is found as

$$\tau = RC = (330\ \Omega)(10\ \mu\text{F}) = \mathbf{3.3\ ms}$$

As the waveforms show, $V_{out}$ makes its transition when $V_C = +3$ V and $V_S = +5$ V. The time required for the $V_C$ transition can be found as

$$t = -\tau \ln\left(1 - \frac{V_C}{V_S}\right) = -(3.3\text{ ms}) \ln\left(1 - \frac{3\text{ V}}{5\text{ V}}\right) = -(3.3\text{ ms}) \ln(0.4) = \mathbf{3.02\ ms}$$

This result indicates that $V_{out}$ will make its positive-going transition 3.02 ms after $V_S$ makes its positive-going transition.

***PRACTICE PROBLEM 17.24***

A circuit like the one in Figure 17.53 has the following values: $C = 2.2\ \mu\text{F}$, $R = 620\ \Omega$, and $V_S = +18\ V_{pk}$. Determine the time required for $V_C$ to reach a value of +10 V.

## One Final Note

The purpose of this section has been to introduce you to some of the basic principles of *RC* switching circuits. As your electronics education continues, you'll learn that the principles covered in this section are part of several applications. For future reference, many of the principles and relationships introduced in this section are summarized in Figure 17.54.

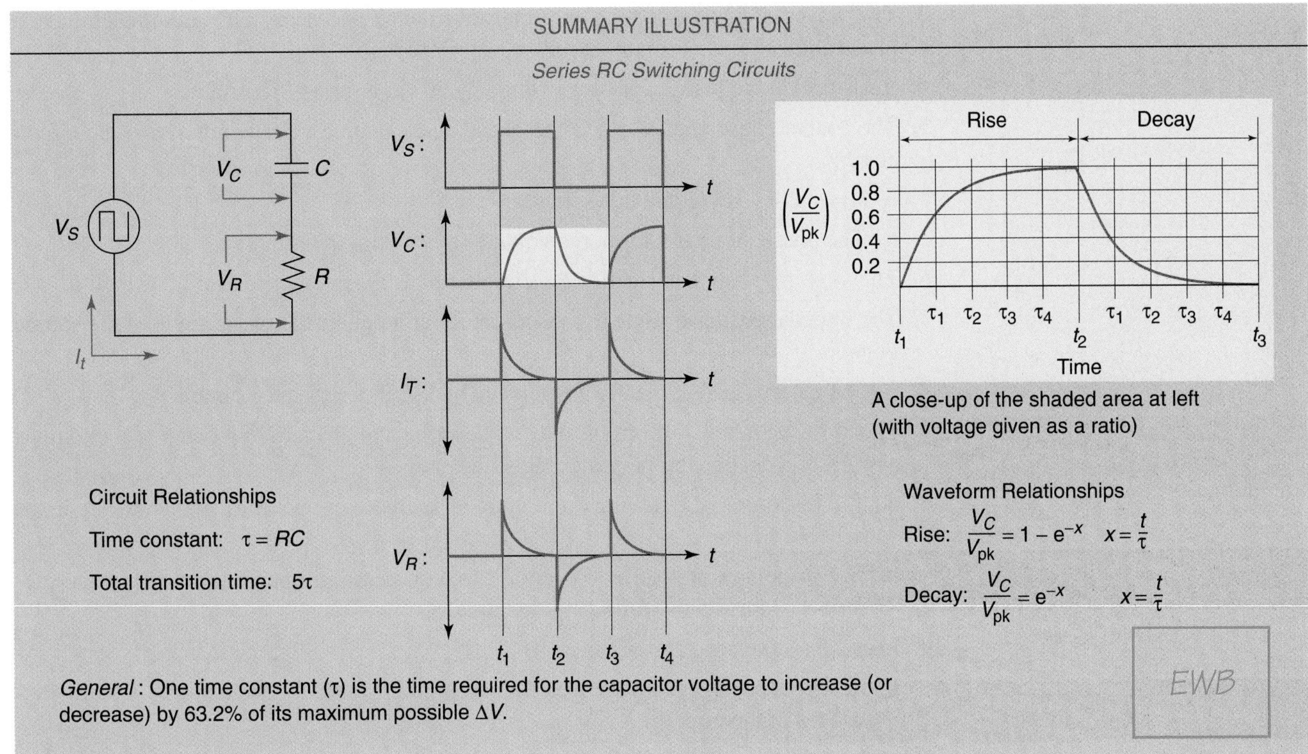


**FIGURE 17.54**

## Section Review

1. Refer to Figure 17.43. What causes the shape of the $V_C$ and $I$ transitions?
2. Explain the shape of the $V_R$ waveform in Figure 17.43.
3. What is the relationship between the shape of a voltage rise (or decay) curve and the peak source voltage?
4. What is a universal voltage curve? What purpose does it serve?
5. What is an exponential curve?
6. What is implied by the word *constant* in the term *time constant*?
7. How do you calculate the value of the time constant in a series *RC* circuit?
8. How many time constants are required to complete a rise (or decay) curve?
9. What is actually represented by the variable exponent ($x$) in the rise and decay curve equations?
10. What is an *RC* time constant? Explain your answer.
11. Describe the operation of the delay circuit in Figure 17.53.

## Chapter Summary

Here is a summary of the major points made in this chapter:

1. A resistive-capacitive (*RC*) circuit is one that contains any combination of resistors and capacitors.
2. In a series *RC* circuit:
   a. The source voltage equals the geometric sum of the resistor and capacitor voltages.
   b. The total impedance equals the geometric sum of resistance and capacitive reactance.

3. The values of source voltage and total impedance in a series *RC* circuit each have a phase angle that is normally expressed as part of the value.
4. Resistor voltage leads capacitor voltage by 90° in a series *RC* circuit.
5. The source phase angle ($\angle\theta$) is the phase of the source voltage relative to the circuit current.
   a. Current leads voltage in a series *RC* circuit.
   b. The phase angle of the source voltage falls within the limits of $-90° < \theta < 0°$.
6. The total impedance in a series *RC* circuit equals the geometric sum of $X_C$ and *R*.
7. The impedance phase angle in a series *RC* circuit equals the phase angle of the source voltage.
8. Circuit current leads capacitive reactance by 90° in a series *RC* circuit.
9. The term *frequency response* is used to describe any changes that occur in a circuit as a result of a change in operating frequency.
   a. The responses of a series *RC* circuit to an increase in operating frequency are listed in Table 17.1.
   b. The responses of a series *RC* circuit to a decrease in operating frequency are described immediately following the table.
10. In any resistive-reactive circuit:
    a. Resistive power (or true power) is measured in watts. $P_R$ is the power actually dissipated in the circuit.
    b. Reactive power (or imaginary power) is measured in volt-amperes-reactive (VAR).
    c. Apparent power (the geometric sum of resistive and reactive power) is measured in volt-amperes (VA).
11. Circuit current leads apparent power in a series *RC* circuit.
12. The power factor for a series *RC* circuit is the ratio of resistive power to apparent power.
    a. The power factor for an *RC* circuit is found as $\text{PF} = \cos\theta$.
    b. When PF and $Z_T$ are known, total circuit resistance can be found as $R_T = Z_T \cos\theta$.
13. The complete mathematical analysis of a series *RC* circuit is demonstrated in Example 17.11.
14. A parallel *RC* circuit is one that contains one or more resistors in parallel with one or more capacitors, each branch containing only one component.
15. In a parallel *RC* circuit:
    a. The total circuit current equals the geometric sum of the currents through the resistive and capacitive branches.
    b. The capacitor current leads the source voltage by 90°.
    c. The resistor current is in phase with the circuit voltage.
16. The current phase angle ($\angle\theta$) in a parallel *RC* circuit is the phase of the circuit current relative to the source voltage.
    a. Since voltage is assumed to have an angle of 0° in a parallel circuit, the current phase angle is always positive.
    b. For a parallel *RC* circuit, the current phase angle (relative to the source voltage) has limits of $0° < \theta < 90°$.
17. The product-over-sum method for calculating total impedance is preferred for parallel *RC* circuits. (The geometric sum of $X_C$ and *R* is used in the denominator.)
18. The impedance phase angle in a parallel *RC* circuit is always the negative equivalent of the current phase angle.

19. The impedance phase angles for series and parallel $RC$ circuits are calculated using reciprocal fractions, as follows:

| **Series** | **Parallel** |
|---|---|
| $\theta = \tan^{-1}\frac{-X_C}{R}$ | $\theta = \tan^{-1}\frac{R}{-X_C}$ |

20. The current phase angle in a parallel $RC$ circuit can be measured using the technique introduced in Chapter 14 for parallel $RL$ circuits (see Section 14.3). Relating that discussion to parallel $RC$ circuits:
    a. A sensing resistor is placed in series with the parallel branches to produce a low-value voltage that is in phase with the circuit current.
    b. The phase angle of the circuit current is measured using an oscilloscope to compare the phase of the voltage across the sensing resistor to that of the voltage source.
    c. To prevent the sensing resistor from seriously affecting the current phase angle, the value of the resistor should be selected according to the following guideline: $R_{sensing} << Z/10$. (Remember that the value of the sensing resistor must be sufficient to generate a measurable voltage at the value of circuit current.)
21. A complete mathematical analysis of a parallel $RC$ circuit is provided in Example 17.16.
22. The responses of a parallel $RC$ circuit to an increase in operating frequency are listed in Table 17.5. The responses to a decrease in operating frequency are provided immediately preceding the table.
23. Analyzing a series-parallel circuit is simply a matter of:
    a. Combining series components according to the rules of series circuits.
    b. Combining parallel components according to the rules of parallel circuits.
24. The analysis of a series-parallel circuit with resistive and reactive components gets a bit complicated because of the phase angles involved.
    a. Geometric addition is used to determine the sum of any two phasors that are at 90° angles.
    b. Rectangular notation is used when adding any two phasors that are at an angle other than 90° (or 0°).
25. To add two phasors that are at angles other than 90° (or 0°):
    a. Convert both phasors to rectangular form.
    b. Add the two rectangular values.
    c. Convert the result back to polar form (if needed).
26. When the total impedance of a parallel $RC$ circuit is converted to rectangular form, the new value is the series equivalent of the parallel impedance network.
27. In an $RC$ switching circuit, the capacitor opposes any change in voltage. As a result, there is a measurable delay between the transition of the source voltage and the complete transition of the capacitor voltage.
28. The voltage transitions in an $RC$ switching circuit are referred to as the rise and decay curves.
29. The $V_R$ curve in an $RC$ switching circuit is shown in Figure 17.43.
30. A linear waveform is one that changes at a constant rate. The rise and decay curves for an $RL$ switching circuit are not linear because the rate-of-change decreases as the transition progresses.
31. A universal curve is one that can be used to predict the operation of a specified type of circuit.
    a. The universal rise curve for an $RC$ switching circuit is an exponential curve, which means that its value is determined by a variable exponent.

b. The equation for the rise portion of the universal curve (Figure 17.44) is $V_C/V_{pk} = 1 - e^{-x}$. In this case, $x$ is the variable exponent that determines the value of the voltage ratio.

32. The decay curve is simply a mirror image of the rise curve. The equation for the decay portion of the universal curve is $V_C/V_{pk} = e^{-x}$.
33. Each time interval in the universal curve represents a real-time value referred to as a time constant ($\tau$). The term *constant* is used because its value is independent of the circuit current magnitude and operating frequency.
34. Capacitor voltage rises (or decays) to its steady state value in 5 time constants.
35. The exponent ($x$) in the curve equations is actually a time ratio. This ratio is given as $x = t/\tau$, where $t$ = the time from the start of the transition.
36. Technically defined, an *RC* time constant is the time required for the capacitor voltage in a switching circuit to increase (or decrease) by 63.2% of its maximum possible transition.

## Equation Summary

| Equation Number | Equation | Section Number |
|---|---|---|
| **(17.1)** | $V_S = \sqrt{V_C^2 + V_R^2}$ | 17.1 |
| **(17.2)** | $\theta = \tan^{-1}\frac{-V_C}{V_R}$ | 17.1 |
| **(17.3)** | $-90° < \theta < 0°$ | 17.1 |
| **(17.4)** | $Z = \sqrt{X_C^2 + R^2}$ | 17.1 |
| **(17.5)** | $\theta = \tan^{-1}\frac{-X_C}{R}$ | 17.1 |
| **(17.6)** | $V_n = V_S\frac{Z_n}{Z_T}$ | 17.1 |
| **(17.7)** | $P_{APP} = \sqrt{P_X^2 + P_R^2}$ | 17.2 |
| **(17.8)** | $\theta = \tan^{-1}\frac{-P_X}{P_R}$ | 17.2 |
| **(17.9)** | $PF = \cos\theta$ | 17.2 |
| **(17.10)** | $I_T = \sqrt{I_C^2 + I_R^2}$ | 17.3 |
| **(17.11)** | $\theta = \tan^{-1}\frac{I_C}{I_R}$ | 17.3 |
| **(17.12)** | $0° < \theta < 90°$ | 17.3 |
| **(17.13)** | $Z_T = \frac{V_S}{I_T}$ | 17.3 |
| **(17.14)** | $Z_T = \frac{X_C R}{\sqrt{X_C^2 + R^2}}$ | 17.3 |
| **(17.15)** | $\theta = \tan^{-1}\frac{R}{-X_C}$ | 17.3 |
| **(17.16)** | $R = Z\cos\theta$ | 17.4 |

| | | |
|---|---|---|
| **(17.17)** | $X_C = Z \sin \theta$ | 17.4 |
| **(17.18)** | $I_T = \frac{V_S - V_C}{R}$ | 17.5 |
| **(17.19)** | $V_C = \left(\frac{V_C}{V_{pk}}\right) V_{pk}$ | 17.5 |
| **(17.20)** | $\frac{V_C}{V_{pk}} = 1 - e^{-x}$ (rise curve) | 17.5 |
| **(17.21)** | $\frac{V_C}{V_{pk}} = e^{-x}$ (decay curve) | 17.5 |
| **(17.22)** | $\tau = RC$ | 17.5 |
| **(17.23)** | $T = 5\tau$ | 17.5 |
| **(17.24)** | $x = \frac{t}{\tau}$ | 17.5 |
| **(17.25)** | $t = -\tau \ln\left(1 - \frac{V_C}{V_S}\right)$ | 17.5 |

## Key Terms

The following terms were introduced and defined in this chapter:

parallel *RC* circuit
*RC* time constant
resistive-capacitive (*RC*) circuit
universal voltage curve

## Practice Problems

1. The component voltages shown in Figure 17.55a were taken using an ac voltmeter. Determine the magnitude and phase angle of $V_S$ for the circuit.
2. The component voltages shown in Figure 17.55b were taken using an ac voltmeter. Determine the magnitude and phase angle of $V_S$ for the circuit.

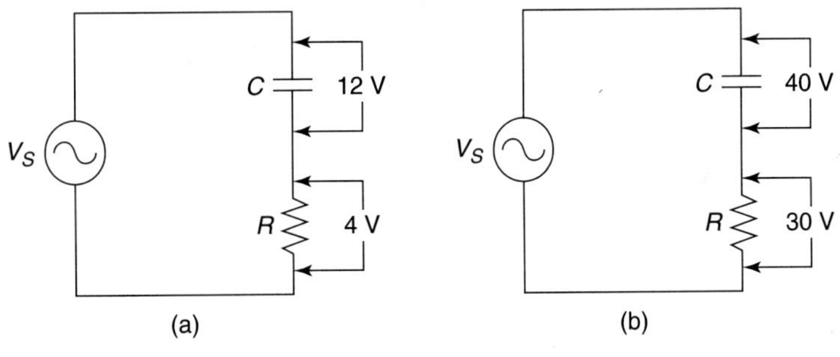


**FIGURE 17.55**

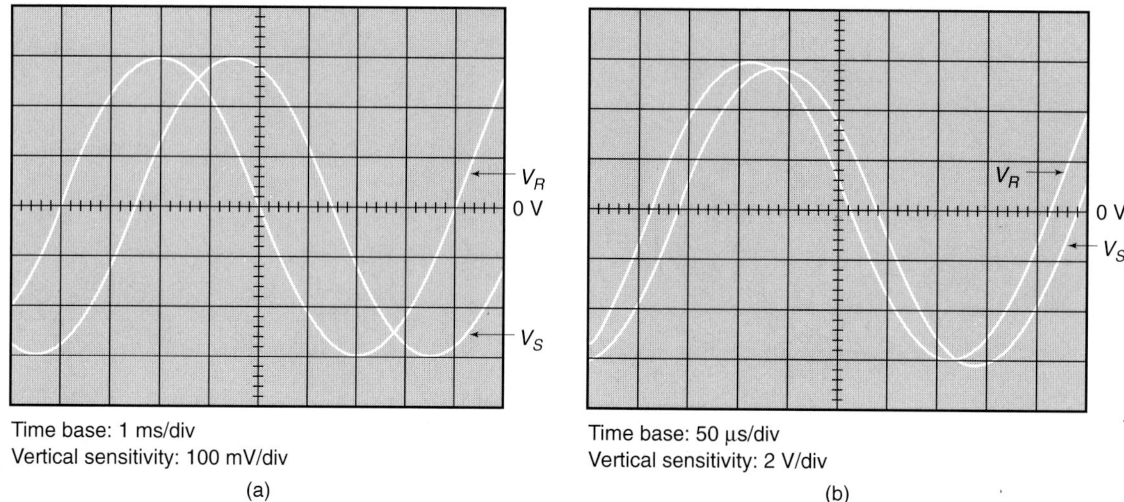


**FIGURE 17.56**

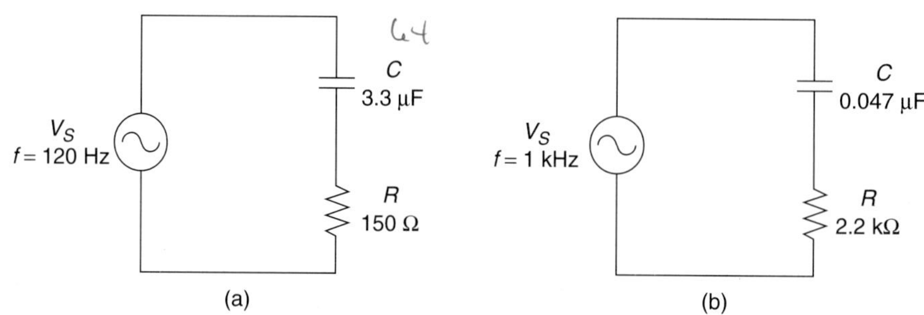


**FIGURE 17.57**

3. Calculate the phase angle between the waveforms in Figure 17.56a.
4. Calculate the phase angle between the waveforms in Figure 17.56b.
5. Calculate the magnitude and phase angle of the circuit impedance for Figure 17.57a.
6. Calculate the magnitude and phase angle of the circuit impedance for Figure 17.57b.
7. Calculate the values of $Z_T$, $I_T$, $V_C$, $V_R$, and $\theta$ for the circuit in Figure 17.58a.
8. Calculate the values of $Z_T$, $I_T$, $X_C$, $V_R$, and $\theta$ for the circuit in Figure 17.58b.

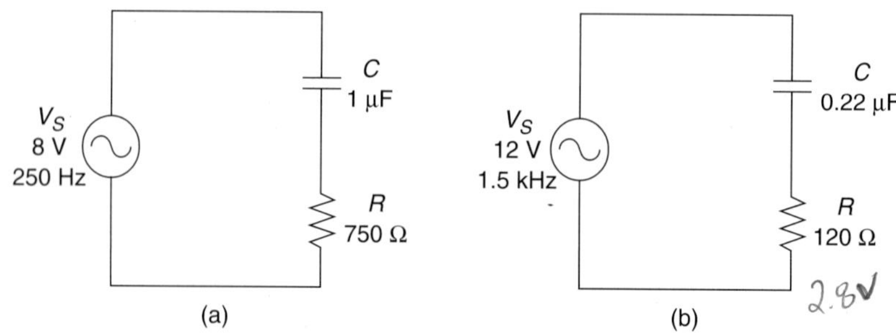


**FIGURE 17.58**

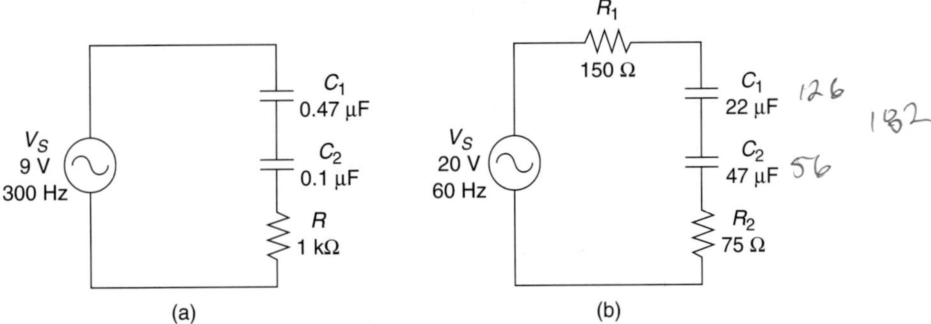


**FIGURE 17.59**

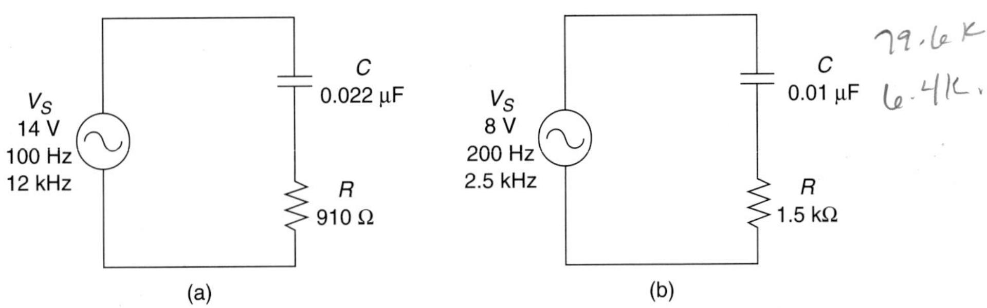


**FIGURE 17.60**

9. Calculate the values of $Z_T$, $I_T$, $V_{C1}$, $V_{C2}$, $V_R$, and $\theta$ for the circuit in Figure 17.59a.
10. Calculate the values of $Z_T$, $I_T$, $V_{C1}$, $V_{C2}$, $V_{R1}$, $V_{R2}$, and $\theta$ for the circuit in Figure 17.59b.
11. The voltage source in Figure 17.60a has the frequency limits shown. Calculate the range of $Z_T$, $V_C$, $V_R$, and $\theta$ values for the circuit.
12. The voltage source in Figure 17.60b has the frequency limits shown. Calculate the range of $Z_T$, $V_C$, $V_R$, and $\theta$ values for the circuit.
13. Refer to Figure 17.58a. Calculate the values of resistive and reactive power for the circuit.
14. Refer to Figure 17.58b. Calculate the values of resistive and reactive power for the circuit.
15. Calculate the magnitude and phase angle of apparent power for the circuit in Problem 13.
16. Calculate the magnitude and phase angle of apparent power for the circuit in Problem 14.
17. Refer to Figure 17.56a. The current for this circuit was measured at 100 μA. Determine the values of $P_{APP}$, cos $\theta$, and $P_R$ for the circuit.
18. Refer to Figure 17.56b. The current for this circuit was measured at 50 mA. Determine the values of $P_{APP}$, cos $\theta$, and $P_R$ for the circuit.
19. Determine the value of $R$ for the circuit in Problem 17.
20. Determine the value of $R$ for the circuit in Problem 18.

FIGURE 17.61

FIGURE 17.62

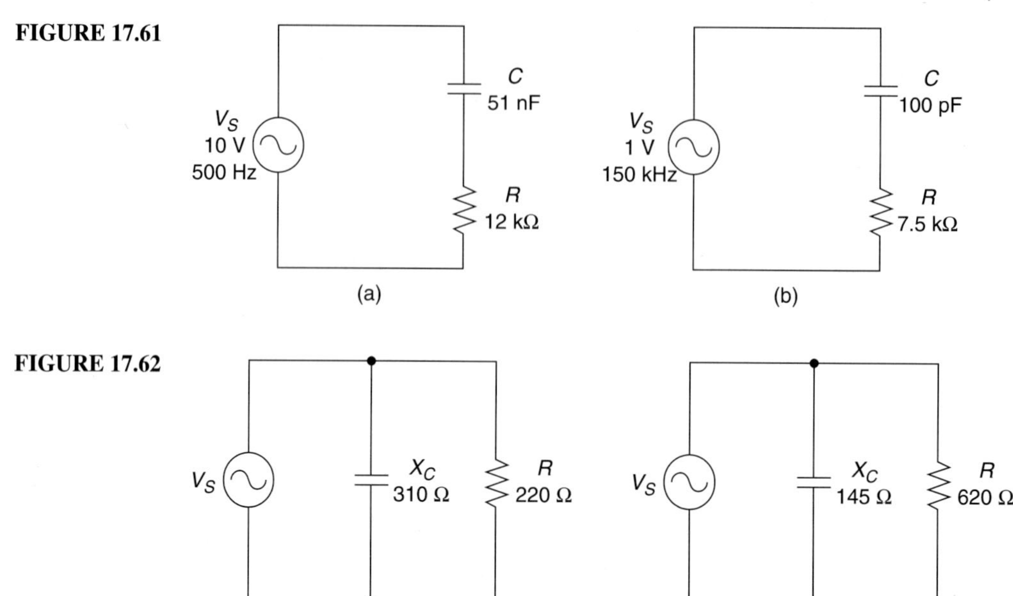


21. Perform a complete ac analysis of the circuit shown in Figure 17.61a.
22. Perform a complete ac analysis of the circuit shown in Figure 17.61b.
23. A parallel *RC* circuit has values of $I_C = 25$ mA and $I_R = 60$ mA. Calculate the magnitude and phase angle of the circuit current.
24. A parallel *RC* circuit has values of $I_C = 320$ mA and $I_R = 140$ mA. Calculate the magnitude and phase angle of the circuit current.
25. Calculate the total impedance for the circuit shown in Figure 17.62a.
26. Calculate the total impedance for the circuit shown in Figure 17.62b.
27. Calculate the magnitude and phase angle of the circuit impedance in Figure 17.63a.
28. Calculate the magnitude and phase angle of the circuit impedance in Figure 17.63b.
29. Perform a complete ac analysis of the circuit in Figure 17.64a.
30. Perform a complete ac analysis of the circuit in Figure 17.64b.

FIGURE 17.63

FIGURE 17.64

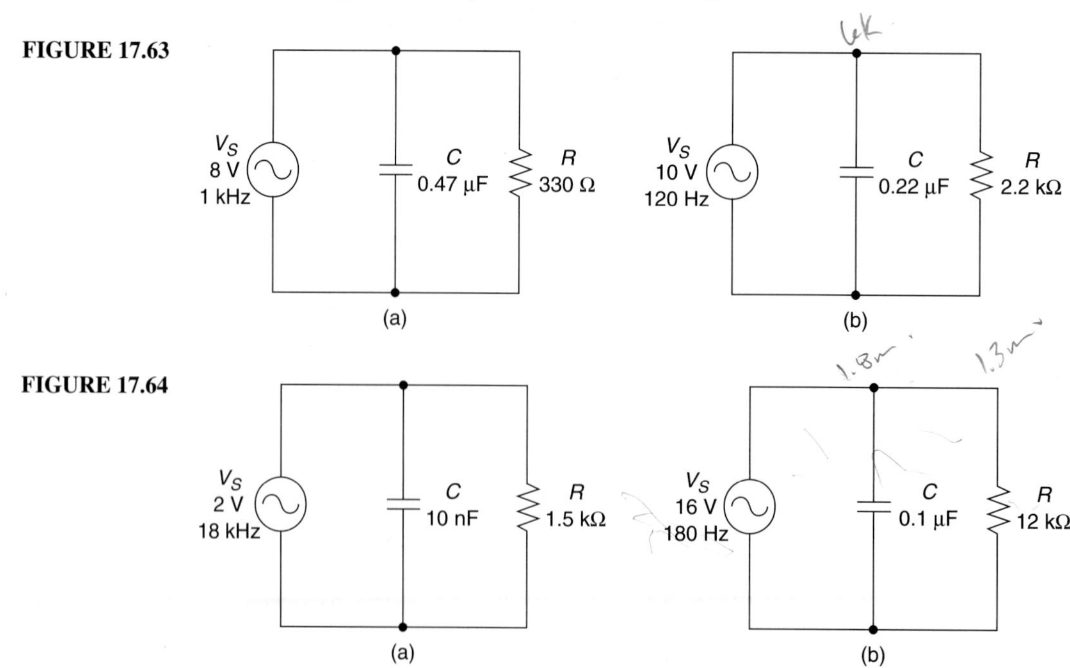

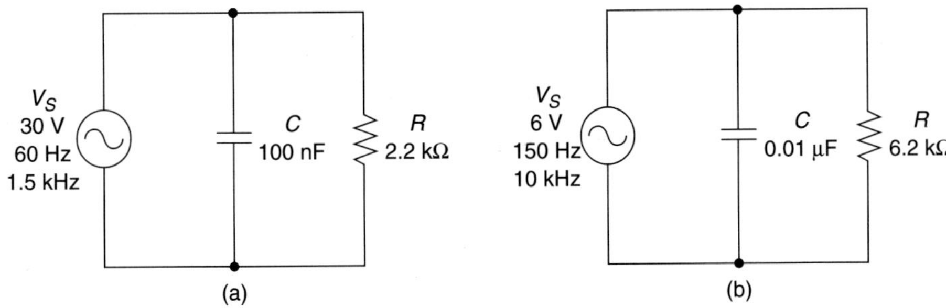


**FIGURE 17.65**

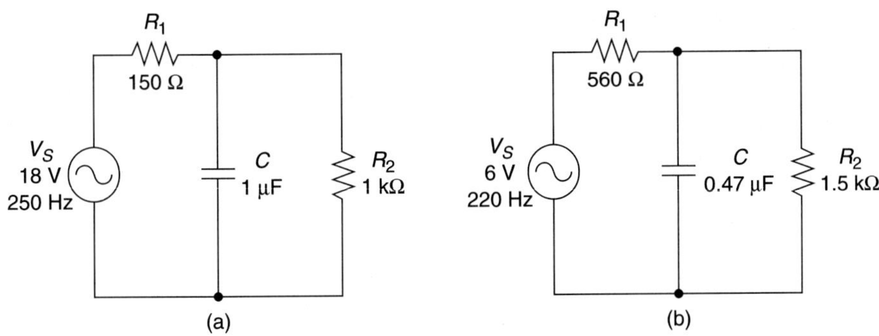


**FIGURE 17.66**

**31.** The voltage source in Figure 17.65a has the frequency limits shown. Calculate the circuit current and impedance values at the frequency limits of the source.

**32.** The voltage source in Figure 17.65b has the frequency limits shown. Calculate the circuit current and impedance values at the frequency limits of the source.

**33.** Refer to Problem 27. Express the circuit impedance in rectangular form.

**34.** Refer to Problem 28. Express the circuit impedance in rectangular form.

**35.** Perform a complete ac analysis on the circuit shown in Figure 17.66a.

**36.** Perform a complete ac analysis on the circuit shown in Figure 17.66b.

**37.** Determine the time constant for the circuit in Figure 17.67a.

**38.** Determine the time constant for the circuit in Figure 17.67b.

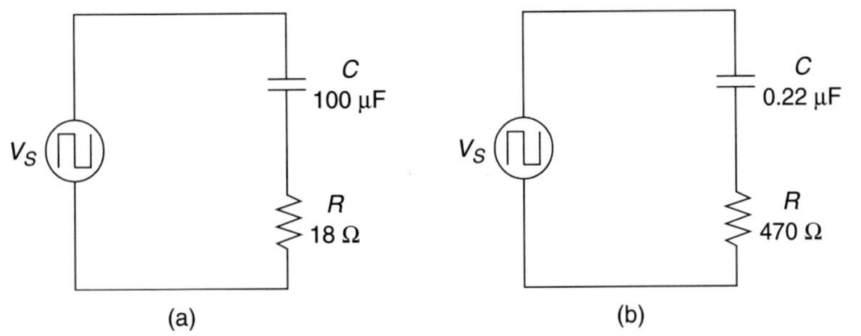


**FIGURE 17.67**

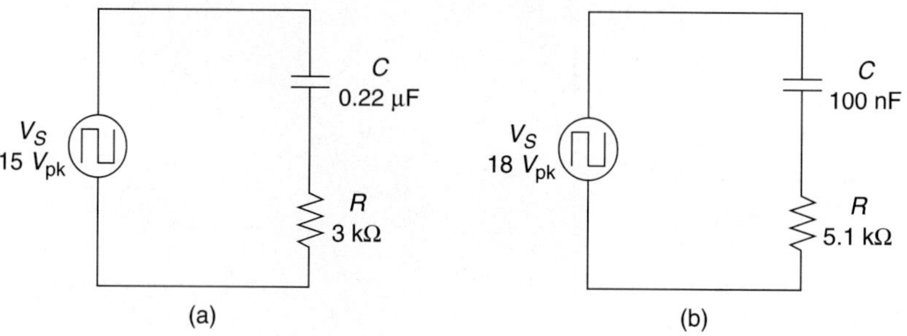


**FIGURE 17.68**

**39.** Determine the time required for $V_C$ in Figure 17.68a to reach its peak value.

**40.** Determine the time required for $V_C$ in Figure 17.68b to reach its peak value.

**41.** An *RC* switching circuit has values of $R = 750\ \Omega$, $C = 220\ \mu F$, and $V_{pk} = 18$ V. Determine the value of $V_C$ at the following times:

**a.** 100 ms after the start of the $V_C$ rise cycle.

**b.** 300 ms after the start of the $V_C$ decay cycle.

**42.** An *RC* switching circuit has values of $R = 120\ \Omega$, $C = 47\ \mu F$, and $V_{pk} = 8$ V. Determine the value of $V_C$ at the following times:

**a.** 8 ms after the start of the $V_C$ rise cycle.

**b.** 4 ms after the start of the $V_C$ decay cycle.

**43.** Refer to Figure 17.68a. Determine the time required for $V_C$ to reach +6 V on its rise curve.

**44.** Refer to Figure 17.68b. Determine the time required for $V_C$ to reach +15 V on its rise curve.

## The Brain Drain

**45.** Using the decay curve and curve equation in Figure 17.47 and a series of calculations like those in Table 17.11, show that $V_C$ decreases by 63.2% of its maximum $\Delta V$ during each time interval.

## Answers to the Example Practice Problems

**17.1** $32\ V\angle -38.7°$

**17.2** $-54°$

**17.3** $1.51\ k\Omega\angle -48.5°$

**17.4** $Z_T = 702\ \Omega\angle -43.4°$, $I = 17.1$ mA, $V_S = 12\ V\angle -43.4°$, $V_C = 8.25$ V, $V_R = 8.72$ V

**17.5** $V_C = 9.42$ V, $V_R = 30.58$ V

**17.6** $Z_T = 2.6\ k\Omega\angle -73.2°$, $V_{C1} = 16.27$ V, $V_{C2} = 7.66$ V, $V_R = 7.22$ V

**17.7** When $f = 1.5$ kHz: $Z_T = 10.8\ k\Omega\angle -79.3°$, $I_T = 2.04$ mA, $V_C = 21.6$ V, $V_R = 4.07$ V. When $f = 6$ kHz: $Z_T = 3.32\ k\Omega\angle -53°$, $I_T = 6.63$ mA, $V_C = 17.6$ V, $V_R = 13.3$ V.

**17.8** 96.8 mVAR, 87.1 mW

**17.9** $130\ mVA\angle -48°$

**17.10** 2.12 VA, 2.09 W

**17.11** $Z_T = 641\ \Omega\angle -55.8°$, $I_T = 18.72$ mA, $V_R = 6.74$ V, $V_C = 9.93$ V, $P_R = 126$ mW, $P_X = 186$ mVAR, $P_{APP} = 225$ mVA

**17.12** $26.3\ mA\angle 72{,}3°$

**17.13** $457\ \Omega\angle -72.3°$

**17.14** $X_C = 480\ \Omega$, $R = 1.5\ k\Omega$, $Z_T = 457\ \Omega$

**17.15** $-33.7°$

**17.16** $X_C = 339\ \Omega$, $I_C = 41.3$ mA, $I_R = 63.6$ mA, $I_T = 75.9$ mA, $\theta = 33°$, $Z_T = 185\ \Omega\angle -33°$, $P_X = 579$ mVAR, $P_R = 891$ mW, $P_{APP} = 1.06$ VA

**17.17** For $f = 120$ Hz: $I_T = 16$ mA$\angle 25.1°$, $Z_T = 562\ \Omega\angle -25.1°$. For $f = 6$ kHz: $I_T =$ 339 mA$\angle 87.6°$, $Z_T = 26.5\ \Omega\angle -87.6°$.

**17.18** $(582 - j489)\ \Omega$

**17.19** $504\ \Omega\angle 9.13°$

**17.20** $X_C = 603\ \Omega$, $Z_P = 516\ \Omega\angle -58.9° = (267 - j442)\ \Omega$, $R_T = 887\ \Omega$, $Z_T = 991\ \Omega\angle -26.5°$, $I_T = 10.1$ mA, $V_S = 10$ V$\angle -26.5°$, $V_{R1} = 6.26$ V, $V_P = 5.21$ V$\angle -58.9°$, $I_C = 8.65$ mA$\angle 31.1°$, $I_{R2} = 5.21$ mA$\angle -58.9°$, $P_{APP} = 101$ mVA$\angle -26.5°$, $P_R = 90.3$ mW

**17.21** 22 μs

**17.22** 12 ms

**17.23** 11.4 V

**17.24** 1.1 ms

---

## EWB Applications Problems

| Figure | EWB File Reference |
|---|---|
| 17.13 | EWB17_13.ewb |
| 17.15 | EWB17_15.ewb |
| 17.20 | EWB17_20.ewb |
| 17.31 | EWB17_31.ewb |
| 17.43 | EWB17_43.ewb |
| 17.50 | EWB17_50.ewb |
| 17.51 | EWB17_51.ewb |
| 17.54 | EWB17_54.ewb |

# 18

# RLC CIRCUITS

## OBJECTIVES

*After studying the material in this chapter, you should be able to:*

1. Discuss the relationship between the component voltages and the operating characteristics of a series *LC* circuit.
2. Discuss the relationship between the reactance values and the operating characteristics of a series *LC* circuit.
3. Calculate the reactance and voltage values for a series *LC* circuit.
4. Discuss the relationship between the branch currents and the operating characteristics of a parallel *LC* circuit.
5. Calculate the magnitude of the current through a parallel *LC* circuit.
6. Calculate the parallel equivalent reactance for any parallel *LC* circuit.
7. Discuss the relationship between the branch reactances and the operating characteristics of a parallel *LC* circuit.
8. Describe resonance and calculate the resonant frequency of an *LC* circuit.
9. Describe the effects of stray capacitance, stray inductance, and oscilloscope input capacitance on resonant frequency.
10. Describe and analyze the operation of series and parallel resonant *LC* circuits.
11. Describe and analyze the operation of any series *RLC* circuit.
12. Describe and analyze the operation of any parallel *RLC* circuit.
13. Determine the impedance, voltage, and current values for a series-parallel *RLC* circuit.

Now that we have discussed *RL* and *RC* circuits, it is time to combine them into *RLC* (resistive-inductive-capacitive) circuits. We will start our discussion of *RLC* circuits with the operating characteristics of *LC* (inductive-capacitive) circuits.

## 18.1 SERIES *LC* CIRCUITS

When an inductor is connected in series with a capacitor (as shown in Figure 18.1), the circuit has the following current and voltage characteristics:

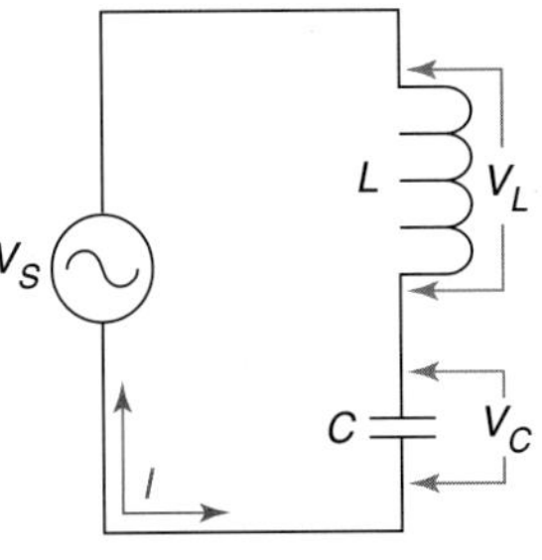


**FIGURE 18.1 A series *LC* circuit.**

1. The inductor and capacitor currents are equal ($I_L = I_C$).
2. The inductor and capacitor voltages are 180° out of phase.

The first principle needs no clarification. Since we are dealing with a series circuit, the current must be equal through all components in the loop. The second principle can be explained with the help of Figure 18.2.

Figure 18.2a shows two circuits, one purely inductive and one purely capacitive. As shown in the figure, the voltage in the inductive circuit leads the current by 90°. In the capacitive circuit, the voltage lags the circuit current by 90°. These phase relationships were discussed in Chapters 13 and 16.

◄ *OBJECTIVE 1*

When an inductor is placed in series with a capacitor, the waveforms in Figure 18.2b are generated. The inductor and capacitor voltages are referenced to circuit current because current is the constant throughout a series circuit. As shown in the figure:

1. The inductor voltage leads the circuit current by 90°.
2. The capacitor voltage lags the circuit current by 90°.

Therefore, there is a 180° difference in phase between the two voltage waveforms. Since they are out of phase by 180°, the total voltage is equal to the difference between $V_L$ and $V_C$. By formula,

$$V_S = V_L - V_C \tag{18.1}$$

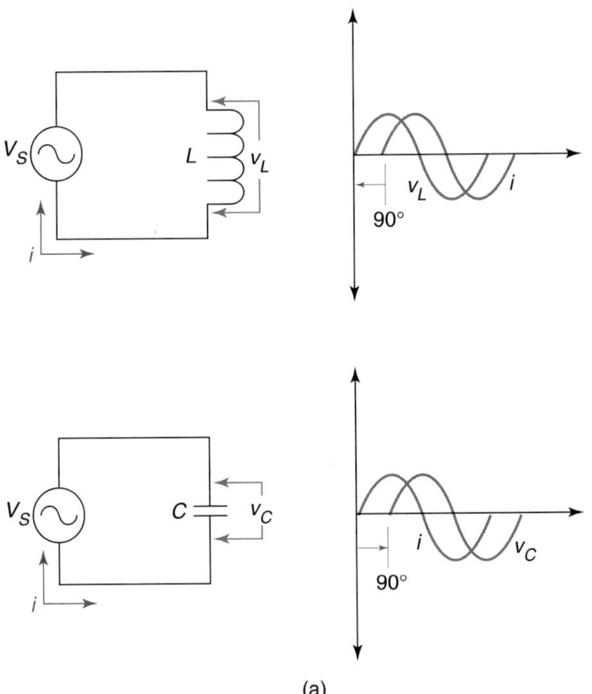


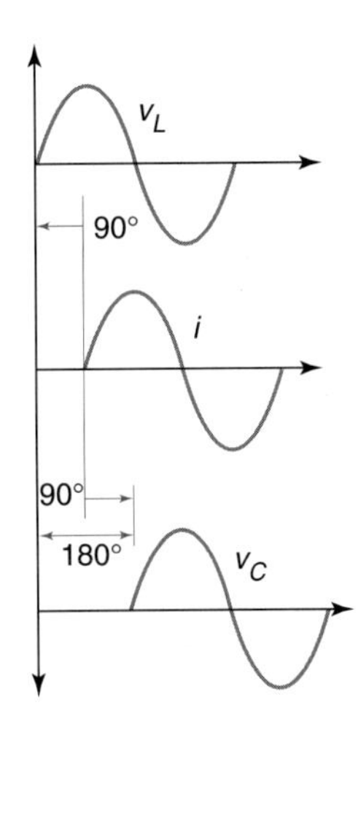


**FIGURE 18.2 Inductor and capacitor phase relationships.**

This equation states that the difference between the inductor and capacitor voltages must equal the source voltage.

## Voltage Relationships and Phase Angles

There are three possible relationships between $V_L$ and $V_C$ in any series *LC* circuit:

1. $V_L > V_C$
2. $V_L < V_C$
3. $V_L = V_C$

Depending on the relationship between $V_L$ and $V_C$, the source sees the circuit as being inductive, or capacitive, or as a short circuit. This principle is illustrated in Figure 18.3. In the graph shown, the positive *x*-axis is used to represent the circuit current. Inductor voltage leads current by 90°, so $V_L$ is represented on the positive *y*-axis. By the same token, capacitor voltage lags current by 90°, so it is represented on the negative *y*-axis. When plotted as shown in the figure, the voltage phasors are 180° out of phase (as are the component voltages).

When $V_L$ and $V_C$ are combined, the direction of the resulting phasor is determined by the relationship between the two values. For example, Figure 18.4a shows the result phasor when $V_L > V_C$. In this case, $V_L = 6$ V and $V_C = 2$ V. The result is a 4 V phasor that lies on the positive *y*-axis. This phasor indicates that:

1. The source sees the circuit as being inductive.
2. The source voltage leads the circuit current by 90°.

When $V_L < V_C$, the result phasor falls on the negative *y*-axis, as shown in Figure 18.4b. In this case, $V_L = 1$ V and $V_C = 4$ V. The result is a 3 V phasor that falls on the negative *y*-axis. This vector indicates that:

1. The source sees the circuit as capacitive.
2. The source voltage lags the circuit current by 90°.

When $V_L = V_C$, the phasors have equal magnitude and are in opposite directions. In this case, the sum of the phasors is 0 V. If we apply Ohm's law to this situation, we get

$$Z = \frac{0\text{ V}}{I} = 0\ \Omega$$

Therefore, when the component voltages are equal, a series *LC* circuit acts as a short circuit. When this is the case, the *LC* circuit shorts out the voltage source.

Two points need to be made:

1. $V_L = V_C$ is a "special case" discussed in detail in Section 18.3. (For now, we will simplify our discussion by limiting ourselves to cases where $V_L \neq V_C$.)

**FIGURE 18.3 Phasor representations of inductor voltage and capacitor voltage.**

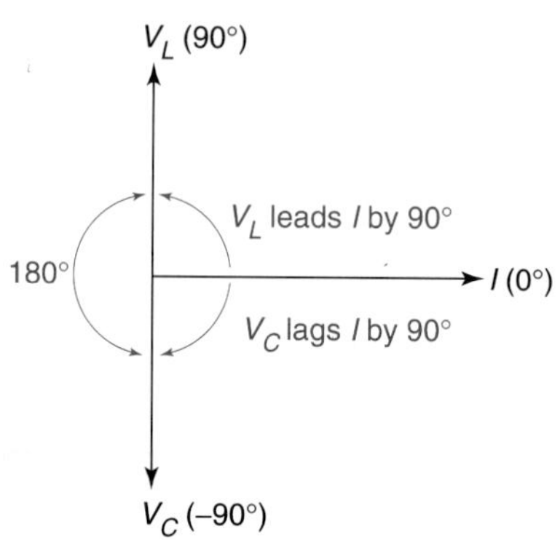

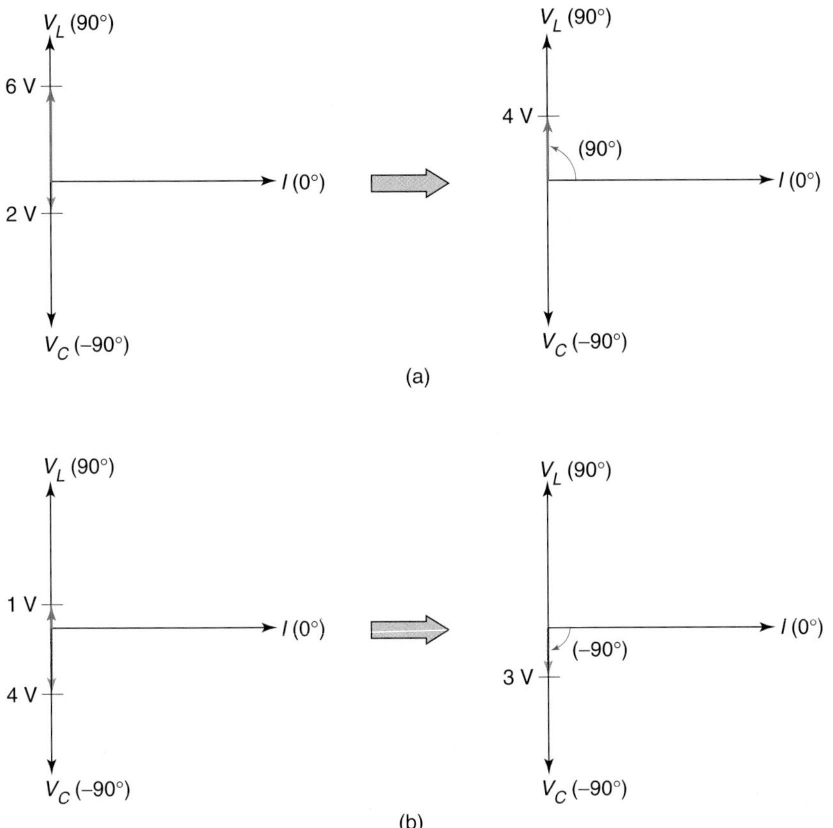


**FIGURE 18.4 Examples of phasor combinations.**

2. The result of equation (18.1) indicates whether the source sees the circuit as capacitive or inductive. When $V_L > V_C$, the result of equation (18.1) is positive, and the source sees the circuit as inductive. When $V_L < V_C$, the result of equation (18.1) is negative, and the source sees the circuit as capacitive.

The second of these points is demonstrated in Examples 18.1 and 18.2.

## EXAMPLE 18.1

Determine the magnitude and phase angle of $V_S$ for the circuit in Figure 18.5.

***Solution:*** Using the values shown in the figure, the value of the source voltage is found as

$$V_S = V_L - V_C = 12 \text{ V} - 18 \text{ V} = \mathbf{-6 \text{ V}}$$

Since we obtained a negative result, we know that the source voltage lags the circuit current by 90°. Therefore, the value of the source is written as

$$V_S = 6 \text{ V}\angle{-90°}$$

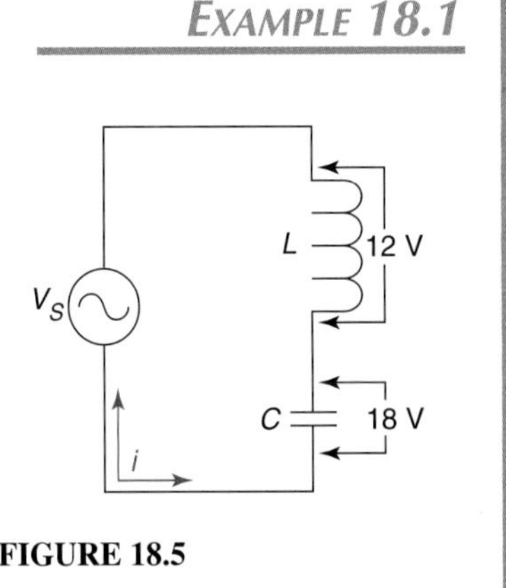


**FIGURE 18.5**

***PRACTICE PROBLEM 18.1***

A circuit like the one in Figure 18.5 has values of $V_L = 1$ V and $V_C = 1.5$ V. Determine the magnitude and phase angle of the source voltage.

The result in Example 18.1 indicates that the source sees the circuit as being capacitive because the value of $V_C$ is greater than the value of $V_L$. Example 18.2 demonstrates the phase relationship between $V_S$ and circuit current when $V_L > V_C$.

## EXAMPLE 18.2

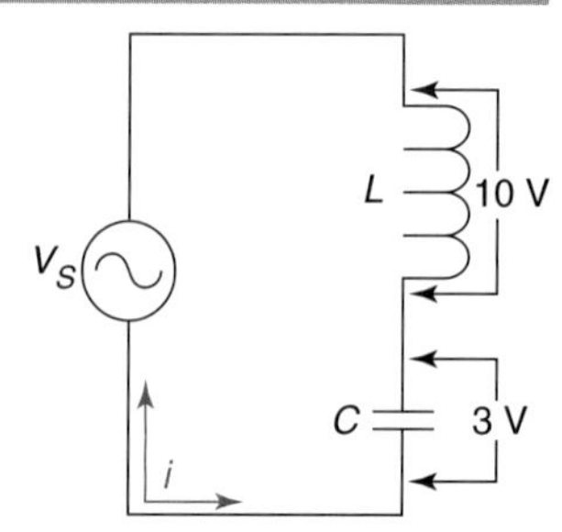


**FIGURE 18.6**

Determine the magnitude and phase angle of $V_S$ for the circuit in Figure 18.6.

***Solution:*** Using the values shown in the figure, the value of the source voltage is found as

$$V_S = V_L - V_C = 10 \text{ V} - 3 \text{ V} = \mathbf{7 \text{ V}}$$

Since we obtained a positive result, we know that the source voltage leads the circuit current by 90°. Therefore, the value of the source voltage is written as

$$V_S = 7 \text{ V}\angle 90°$$

## Series Reactance ($X_S$)

*OBJECTIVE 2* ▶ When an inductor and capacitor are connected in series, the net series reactance ($X_S$) is found as the difference between $X_L$ and $X_C$. By formula,

$$X_S = X_L - X_C \tag{18.2}$$

The basis for this relationship is illustrated in Figure 18.7.

As you know, inductor voltage leads current by 90°. Using Ohm's law, the phase angle of inductive reactance (relative to circuit current) can be found as

$$X_L = \frac{V_L \angle 90°}{I\angle 0°} = \frac{V_L}{I}\angle 90°$$

This calculation does not commit $X_L$ to a numeric value, but it does indicate that inductive reactance leads circuit current by 90°, as represented in Figure 18.7. By the same token, the phase angle of capacitive reactance (relative to circuit current) can be found as

$$X_C = \frac{V_C}{I} = \frac{V_C \angle -90°}{I\angle 0°} = \frac{V_C}{I}\angle -90°$$

This calculation is based on the fact that capacitor voltage lags circuit current by 90°. As you can see, capacitive reactance has a phase angle of −90° relative to the circuit current. With $X_L$ and $X_C$ plotted as shown in Figure 18.7, the phase angle between the reactance phasors is 180°. This is why the net series reactance ($X_S$) is equal to the difference between the two.

As was the case with component voltages, the result of equation (18.2) indicates whether the circuit is seen by the source as inductive, or capacitive, or as a short circuit. If the result of equation (18.2) is positive, the source sees the circuit as inductive, and $X_S$

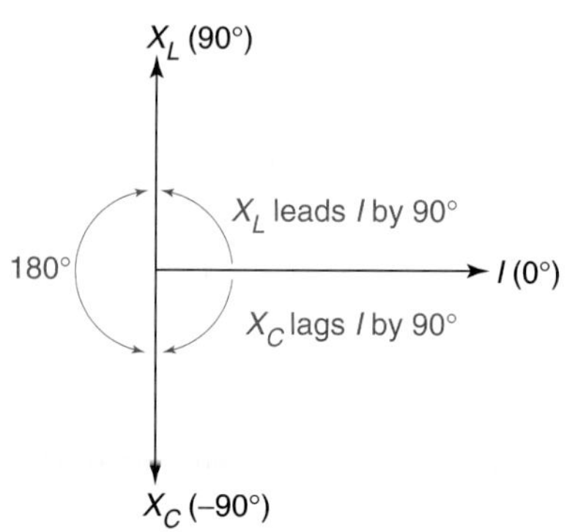


**FIGURE 18.7 The phase relationship between inductive and capacitive reactance.**

leads current by 90°. If the result is negative, the source sees the circuit as capacitive, and $X_S$ lags current by 90°. If $X_L = X_C$, $X_S$ is 0 Ω, and the source sees the combination as a short circuit.

The case of $X_L = X_C$ is discussed in detail in Section 18.3. For now, we will limit our discussion to cases where $X_L \neq X_C$. Example 18.3 demonstrates the phase relationship between $X_S$ and circuit current when $X_L > X_C$.

## EXAMPLE 18.3

◀ *OBJECTIVE 3*

Determine the magnitude and phase angle of $X_S$ for the circuit in Figure 18.8.

**Solution:** Using the values shown in the figure, the net series reactance is found as

$$X_S = X_L - X_C = 400\ \Omega - 100\ \Omega = \mathbf{300\ \Omega}$$

Since we obtained a positive result, we know that $X_S$ leads the circuit current by 90°. Therefore, the value of $X_S$ is written as

$$X_S = 300\ \Omega\angle 90^\circ$$

***PRACTICE PROBLEM 18.3***

A circuit like the one in Figure 18.8 has values of $X_L = 80\ \Omega$ and $X_C = 55\ \Omega$. Determine the magnitude and phase angle of $X_S$.

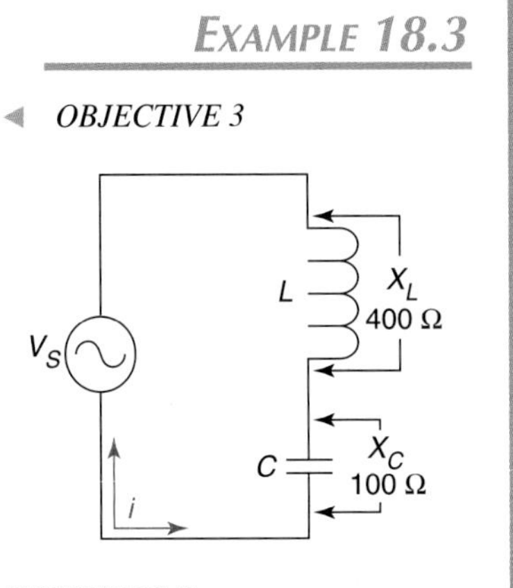


**FIGURE 18.8**

The result in Example 18.3 indicates that the source sees the circuit as being inductive because the value of $X_L$ is greater than the value of $X_C$. Since the circuit is inductive, the source voltage leads the circuit current by 90°. Example 18.4 demonstrates the phase relationship between $X_S$ and circuit current when $X_L < X_C$.

## EXAMPLE 18.4

Determine the magnitude and phase angle of $X_S$ for the circuit in Figure 18.9.

**Solution:** Using the values shown in the figure, the net series reactance is found as

$$X_S = X_L - X_C = 120\ \Omega - 300\ \Omega = \mathbf{-180\ \Omega}$$

Since we obtained a negative result, we know that $X_S$ lags the circuit current by 90°. Therefore, the value of $X_S$ is written as

$$X_S = 180\ \Omega\angle -90^\circ$$

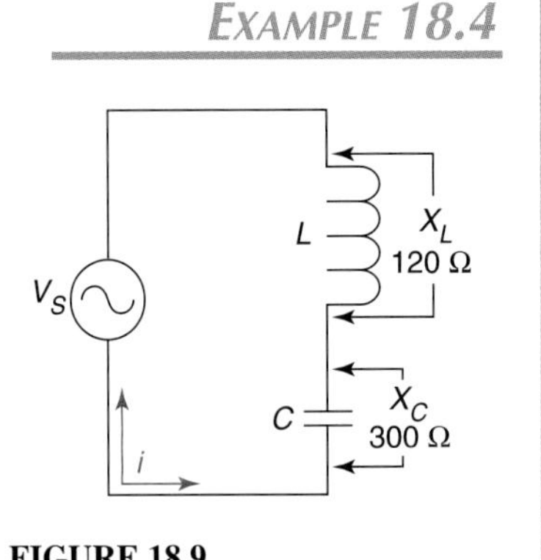


**FIGURE 18.9**

The result in Example 18.4 indicates that the source sees the circuit as being capacitive because the value of $X_C$ is greater than the value of $X_L$. Since the circuit is capacitive, the source voltage lags the circuit current by 90°.

## Putting It All Together: Basic Series LC Circuit Analysis

◀ *OBJECTIVE 3*

In this section, we have established the basic voltage and reactance characteristics of series *LC* circuits. These characteristics are summarized in Table 18.1. The relationships listed for $X_L > X_C$ are verified in Example 18.5.

**TABLE 18.1 Series $LC$ Circuit Conditions When $X_L \neq X_C$**

| *Reactance Relationship* | *Circuit Characteristics* |
|---|---|
| $X_L > X_C$ | $X_S$ leads circuit current by 90°.<br>The source "sees" the circuit as inductive.<br>$V_S$ leads circuit current by 90°. |
| $X_C > X_L$ | $X_S$ lags circuit current by 90°.<br>The source "sees" the circuit as capacitive.<br>$V_S$ lags circuit current by 90°. |

## EXAMPLE 18.5

Determine the magnitudes and phase angles of $V_L$, $V_C$, and $V_S$ for the circuit in Figure 18.10.

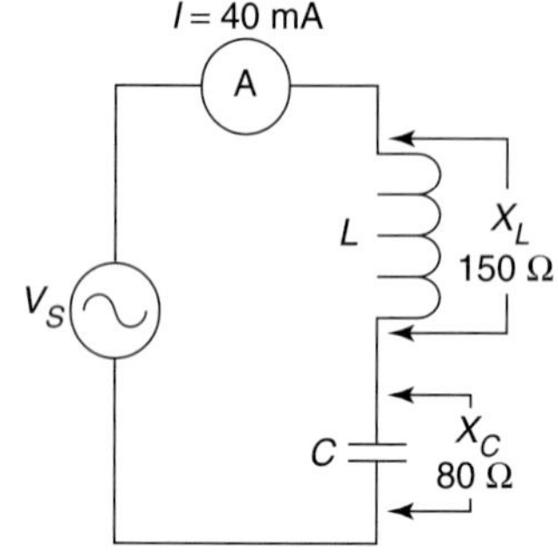


**FIGURE 18.10**

***Solution:*** The value of inductor voltage for the circuit can be found as

$$V_L = I \cdot X_L = (40 \text{ mA}\angle 0°)(150\ \Omega\angle 90°) = \mathbf{6\ V\angle 90°}$$

The value of capacitor voltage for the circuit can be found as

$$V_C = I \cdot X_C = (40 \text{ mA}\angle 0°)(80\ \Omega\angle -90°) = \mathbf{3.2\ V\angle -90°}$$

There are two approaches that can be taken to calculating the magnitude and phase angle of $V_S$. Since we know the values of $V_L$ and $V_C$, the simplest approach is to solve equation (18.1), as follows:

$$V_S = V_L - V_C = 6\text{ V} - 3.2\text{ V} = \mathbf{2.8\ V}$$

The result is positive, so we know that $V_S = 2.8\text{ V}\angle 90°$. The other approach begins with solving for the value of $X_S$ as follows:

$$X_S = X_L - X_C = 150\ \Omega - 80\ \Omega = \mathbf{70\ \Omega}$$

The result of this calculation is positive, so we know that $X_S = 70\ \Omega\angle 90°$. Using this value and the measured circuit current, the value of $V_S$ can be found as

$$V_S = I \cdot X_S = (40 \text{ mA}\angle 0°)(70\ \Omega\angle 90°) = \mathbf{2.8\ V\angle 90°}$$

As you can see, this agrees with the result obtained using the component voltages. Note that the phase angle of $V_S$ indicates the source voltage leads circuit current by 90°. Therefore, the circuit is inductive.

### *PRACTICE PROBLEM 18.5*

A circuit like the one in Figure 18.10 has the following values: $X_L = 2.2\text{ k}\Omega$, $X_C = 800\ \Omega$, and $I = 50$ mA. Calculate the values of $V_L$, $V_C$, and $V_S$ for the circuit. Verify your value of $V_S$ using the values of $X_S$ and circuit current.

The relationships listed for $X_C > X_L$ are verified in Example 18.6.

Determine the magnitudes and phase angles of $V_L$, $V_C$, and $V_S$ for the circuit in Figure 18.11.

## EXAMPLE 18.6

***Solution:*** The value of inductor voltage for the circuit can be found as

$$V_L = I \cdot X_L = (25 \text{ mA}\angle 0°)(250\ \Omega\angle 90°) = \mathbf{6.25\ V\angle 90°}$$

The value of the capacitor voltage for the circuit can be found as

$$V_C = I \cdot X_C = (25 \text{ mA}\angle 0°)(820\ \Omega\angle -90°) = \mathbf{20.5\ V\angle -90°}$$

Using the calculated values of $V_L$ and $V_C$, the value of $V_S$ can be found as

$$V_S = V_L - V_C = 6.25 \text{ V} - 20.5 \text{ V} = \mathbf{-14.25\ V}$$

The result is negative, so we know that $V_S = 14.25 \text{ V}\angle -90°$. Again, we can calculate the value of the source voltage using the total series reactance and circuit current. For the circuit in Figure 18.11, the value of $X_S$ is found as

$$X_S = X_L - X_C = 250\ \Omega - 820\ \Omega = \mathbf{-570\ \Omega}$$

This result indicates that the total series reactance is $570\ \Omega\angle -90°$. Using this value and the measured circuit current, the source voltage can be found as

$$V_S = I \cdot X_S = (25 \text{ mA}\angle 0°)(570\ \Omega\angle -90°) = \mathbf{14.25\ V\angle -90°}$$

As you can see, this agrees with the result obtained using the component voltages. Note that the phase angle of $V_S$ indicates that the source voltage lags the circuit current by 90°. Therefore, the circuit is capacitive.

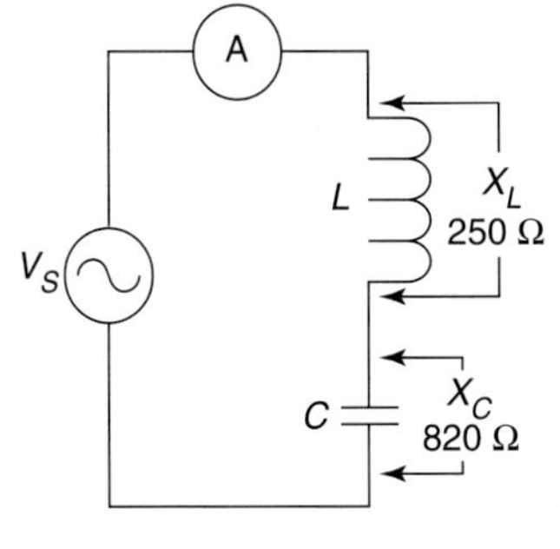

**FIGURE 18.11**

***PRACTICE PROBLEM 18.6***

A circuit like the one in Figure 18.11 has the following values: $X_L = 1.1\ \text{k}\Omega$, $X_C = 3.3\ \text{k}\Omega$, and $I = 4.5$ mA. Calculate the values of $V_L$, $V_C$, and $V_S$ for the circuit. Verify your value of $V_S$ using the values of $X_S$ and circuit current.

## Some Important Points

Because of the unique phase relationships in a series *LC* circuit, the values of $V_S$ and $X_S$ can be lower than one (or both) of the individual component values. For example, look at the following values from Example 18.6:

$$V_L = 6.25 \text{ V} \qquad V_C = 20.5 \text{ V} \qquad V_S = 14.25 \text{ V}$$
$$X_L = 250\ \Omega \qquad X_C = 820\ \Omega \qquad X_S = 570\ \Omega$$

As these values indicate, $V_L < V_S < V_C$ and $X_L < X_S < X_C$. Note that source voltage is never the highest value voltage in the circuit. By the same token, $X_S$ is never the highest value reactance in the circuit.

In any *LC* circuit, the values of $X_L$ and $X_C$ are determined (in part) by the operating frequency of the circuit. In Section 18.3, we will discuss the frequency characteristics of *LC* circuits. First, we will establish the current and reactance characteristics of parallel *LC* circuits.

### Section Review

1. List the current and voltage characteristics of a series *LC* circuit.
2. Explain the phase relationship between series values of $V_L$ and $V_C$.
3. Describe the phase angle of source voltage (relative to circuit current) when:
   a. $V_L > V_C$
   b. $V_C > V_L$

4. State and explain the phase relationship between series values of $X_L$ and $X_C$.
5. Describe the phase angle of net series reactance (relative to circuit current) when:
   a. $X_L > X_C$
   b. $X_C > X_L$
6. Describe the phase relationship between source voltage and circuit current in a series $LC$ circuit when:
   a. $X_L > X_C$
   b. $X_C > X_L$

## 18.2 PARALLEL *LC* CIRCUITS

When an inductor and a capacitor are connected in parallel (as shown in Figure 18.12), the circuit has the following current and voltage characteristics:

1. The inductor and capacitor voltages are equal ($V_L = V_C$).
2. The inductor and capacitor currents are 180° out of phase.

The first principle should be familiar by now. Since we are dealing with a parallel circuit, the voltage must be equal across all components in the circuit. The second principle can be explained with the help of Figure 18.13.

*OBJECTIVE 4* ▶ To provide a reference, Figure 18.13a shows the current and voltage phase relationships for purely inductive and purely capacitive circuits. As you can see:

1. Capacitor current leads capacitor voltage by 90°.
2. Inductor current lags inductor voltage by 90°.

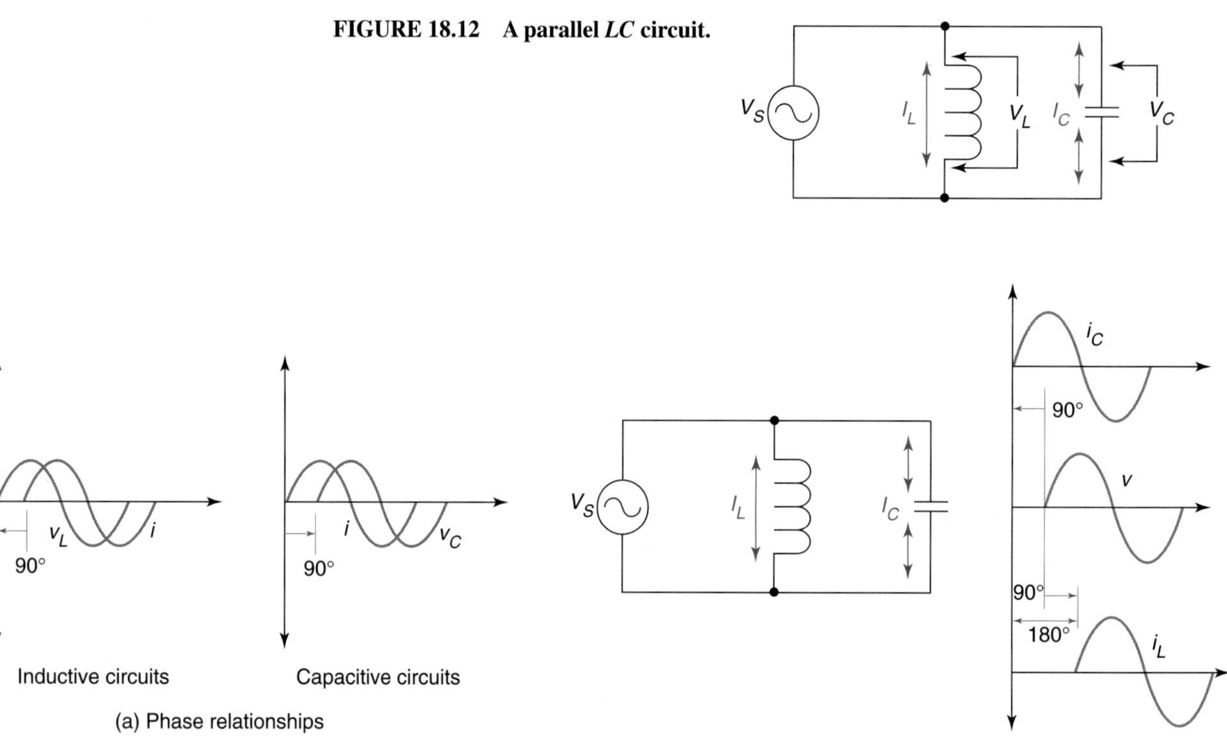


**FIGURE 18.12 A parallel *LC* circuit.**

**FIGURE 18.13 Inductor and capacitor phase relationships.**

When a capacitor is connected in parallel with an inductor, the waveforms in Figure 18.13b are generated. The capacitor and inductor currents are referenced to the source voltage because voltage is equal across parallel branches. As you can see, capacitor current leads the source voltage by 90° while inductor current lags source voltage by 90°. Therefore, there is a 180° difference in phase between the two current waveforms. Since they are out of phase by 180°, the total current is equal to the difference between $I_C$ and $I_L$. By formula,

$$I_T = I_C - I_L \tag{18.3}$$

## Current Relationships and Phase Angles

There are three possible relationships between $I_C$ and $I_L$ in any parallel $LC$ circuit:

1. $I_C > I_L$
2. $I_C < I_L$
3. $I_C = I_L$

Depending on the relationship between $I_C$ and $I_L$, the source sees the circuit as being inductive, or capacitive, or as an open circuit. This principle is illustrated in Figure 18.14. In the graph shown, the positive $x$-axis is used to represent the source voltage. Capacitor current leads source voltage by 90°, so $I_C$ is represented on the positive $y$-axis. By the same token, inductor current lags source voltage by 90°, so it is represented on the negative $y$-axis. When plotted as shown in the figure, the current phasors are 180° out of phase (as are the currents through the components). ◄ *OBJECTIVE 5*

When $I_C$ and $I_L$ are combined, the direction of the resulting phasor is determined by the relationship between the two values. For example, Figure 18.15a shows the result phasor when $I_C > I_L$. In this case, $I_C = 8$ mA and $I_L = 5$ mA. The result is a 3 mA phasor that lies on the positive $y$-axis. This phasor indicates that:

1. The source sees the circuit as capacitive.
2. The circuit current leads the source voltage by 90°.

When $I_C < I_L$, the result phasor falls on the negative $y$-axis, as shown in Figure 18.15b. In this case, $I_C = 2$ mA and $I_L = 6$ mA. The result is a 4 mA phasor that falls on the negative $y$-axis. This phasor indicates that:

1. The source sees the circuit as inductive.
2. The circuit current lags the source voltage by 90°.

When $I_C = I_L$, the phasors have equal magnitude and are in opposite directions. In this case, the sum of the phasors is 0 A. If we apply Ohm's law to this situation, we get

$$Z = \frac{V_S}{I_T} = \frac{V_S}{0\text{ A}}$$

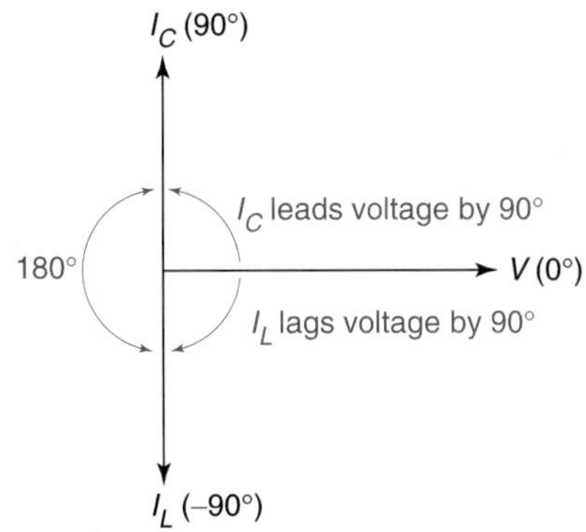


**FIGURE 18.14 Phasor representations of capacitor current and inductor current.**

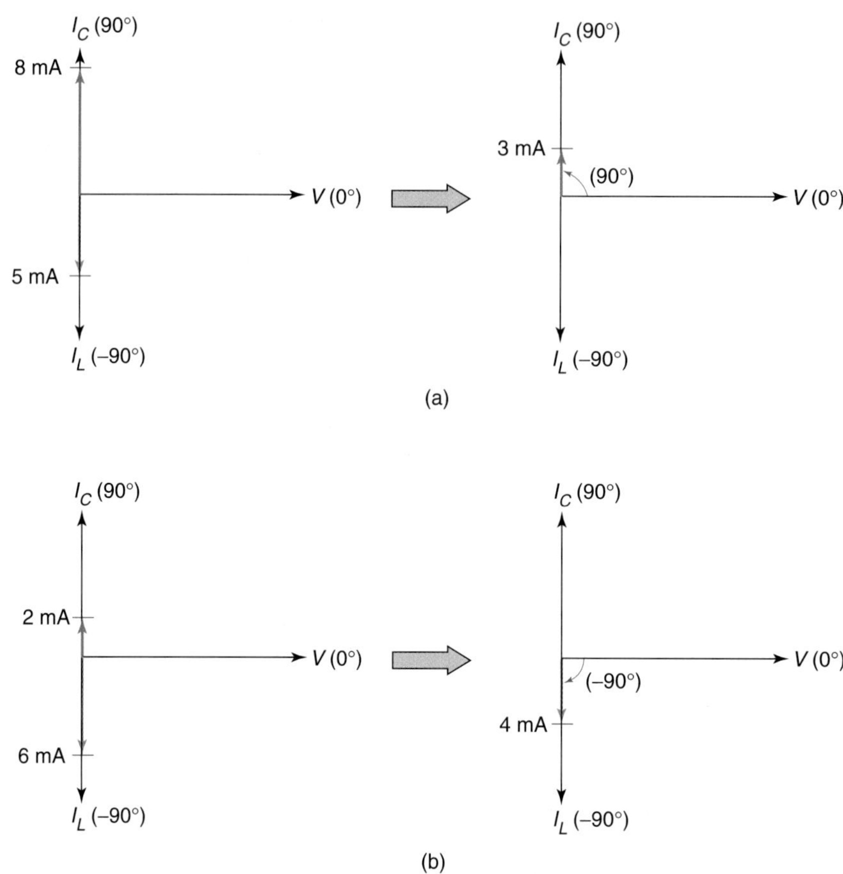


**FIGURE 18.15 Examples of current phasor combinations.**

which approaches infinite ohms. By definition, this is an open circuit. Therefore, when the component currents are equal, a parallel *LC* circuit acts as an open circuit.

Two points need to be made:

1. $I_C = I_L$ is a special case that is discussed in detail in Section 18.3. (For now, we will simplify our discussion by limiting ourselves to cases where $I_C \neq I_L$.)
2. The result of equation (18.3) indicates whether the source sees the circuit as capacitive or inductive. When $I_C > I_L$, the result of equation (18.3) is positive, and the source sees the circuit as capacitive. When $I_C < I_L$, the result of equation (18.3) is negative, and the source sees the circuit as inductive.

The second of these items is demonstrated in Examples 18.7 and 18.8.

## EXAMPLE 18.7

Determine the magnitude and phase angle of the circuit current in Figure 18.16.

**FIGURE 18.16**

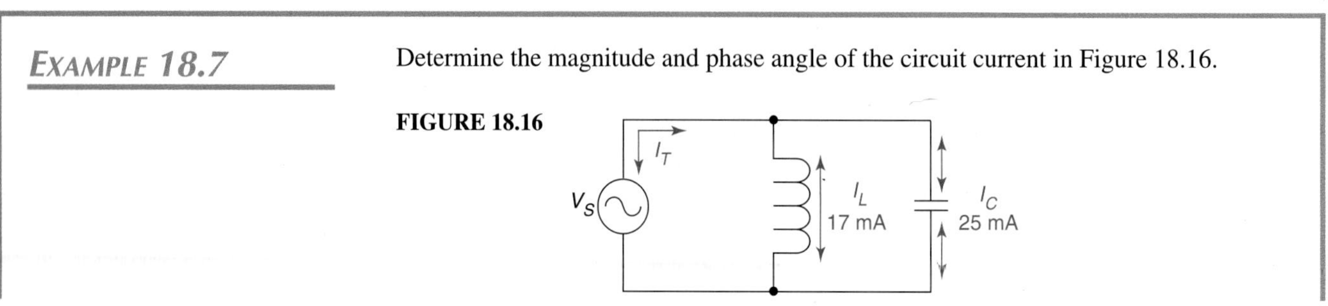

**Solution:** Using the values shown in the figure, the magnitude of the circuit current is found as

$$I_T = I_C - I_L = 25 \text{ mA} - 17 \text{ mA} = \mathbf{8\ mA}$$

Since we obtained a positive result, we know that the circuit current leads the source voltage by 90°. Therefore, the value of the circuit current is written as

$$I_T = 8 \text{ mA}\angle 90°$$

***PRACTICE PROBLEM 18.7***

A circuit like the one in Figure 18.16 has values of $I_C = 120$ mA and $I_L = 75$ mA. Determine the magnitude and phase angle of the circuit current.

The result in Example 18.7 indicates that the source sees the circuit as capacitive because the value of $I_C$ is greater than the value of $I_L$. Example 18.8 demonstrates the phase relationship between circuit current and $V_S$ when $I_C < I_L$.

### EXAMPLE 18.8

Determine the magnitude and phase angle of the circuit current in Figure 18.17.

**FIGURE 18.17**

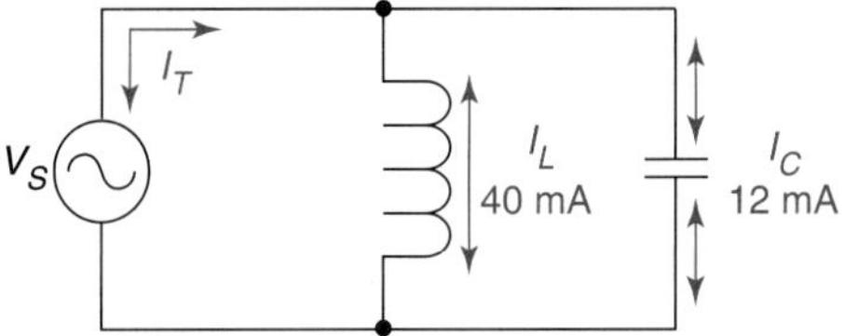


**Solution:** Using the values shown in the figure, the value of the circuit current is found as

$$I_T = I_C - I_L = 12 \text{ mA} - 40 \text{ mA} = \mathbf{-28\ mA}$$

Since we obtained a negative result, we know that the circuit current lags the source voltage by 90°. Therefore, the value of the circuit current is written as

$$I_T = 28 \text{ mA}\angle -90°$$

## Parallel Reactance ($X_P$)

Because of the unique phase relationship between $I_C$ and $I_L$, the total reactance of a parallel *LC* circuit can be greater than one (or both) of the branch reactance values. For example, let's assume that the circuit in Figure 18.17 has a value of $V_S = 6$ V. Using Ohm's law, the reactance values for the circuit can be found as ◄ *OBJECTIVE 6*

$$X_L = \frac{V_S}{I_L} = \frac{6 \text{ V}}{40 \text{ mA}} = 150\ \Omega \qquad \text{and} \qquad X_C = \frac{V_S}{I_C} = \frac{6 \text{ V}}{12 \text{ mA}} = 500\ \Omega$$

Now, if we divide the source voltage by the source current found in Example 18.8, we get a total circuit reactance of

$$X_P = \frac{V_S}{I_T} = \frac{6 \text{ V}}{28 \text{ mA}\angle -90^\circ} = 214.3 \ \Omega\angle 90^\circ$$

As you can see, this value is greater than the value of $X_L = 150 \ \Omega$ that was calculated for the circuit. Don't forget that the total reactance of a parallel *LC* circuit is always greater than the value of one (or both) of the branch reactances.

When a capacitor and an inductor are connected in parallel, the total parallel reactance ($X_P$) can be found as

$$X_P = \frac{1}{\frac{1}{X_L} + \frac{1}{X_C}} \qquad \textbf{(18.4)}$$

or

$$X_P = \frac{X_L X_C}{X_L + X_C} \qquad \textbf{(18.5)}$$

The forms of these equations should be familiar to you by now. However, there is one catch: For the equations to work, capacitive reactance must be entered as a negative value to account for the phase difference between $X_C$ and $X_L$. This point is demonstrated in Example 18.9.

## EXAMPLE 18.9

Calculate the value of $X_P$ for the circuit in Figure 18.18.

**FIGURE 18.18**

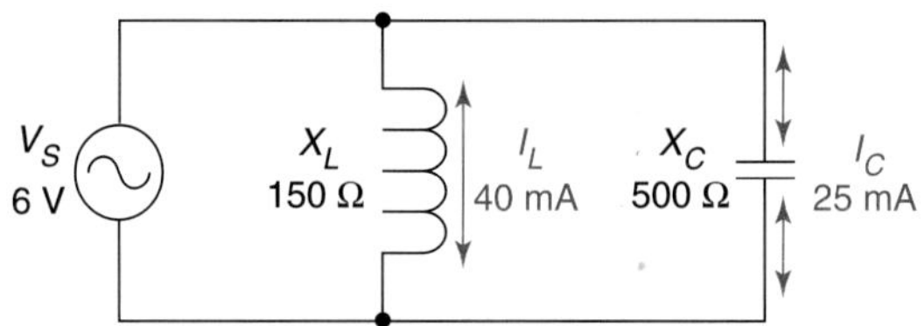


***Solution:*** The circuit shown is the same as Figure 18.17. However, we have added the assumed value of $V_S$ and the calculated values of reactance. For the circuit, the total parallel impedance can be found as

$$X_P = \frac{1}{\frac{1}{X_L} + \frac{1}{X_C}} = \frac{1}{\frac{1}{150 \ \Omega} + \frac{1}{-500 \ \Omega}} = \frac{1}{\frac{1}{150 \ \Omega} - \frac{1}{500 \ \Omega}} = \mathbf{214.3 \ \Omega}$$

or

$$X_P = \frac{X_L X_C}{X_L + X_C} = \frac{(150 \ \Omega)(-500 \ \Omega)}{150 \ \Omega + (-500 \ \Omega)} = \frac{-75{,}000}{-350} \ \Omega = \mathbf{214.3 \ \Omega}$$

As you can see, both equations provide the same result as the Ohm's law calculation we made earlier.

***PRACTICE PROBLEM 18.9***

A circuit like the one in Figure 18.18 has the following values: $V_S = 8$ V, $I_L = 20$ mA, and $I_C = 4$ mA. Calculate the value of $X_P$ using Ohm's law, equation (18.4), and equation (18.5).

When using equation (18.4) or equation (18.5), the sign of the result indicates the phase relationship between source voltage and current. For example, in Figure 18.18, the value of $X_P$ is positive, which indicates that:

◀ *OBJECTIVE 7*

1. The source sees the circuit as inductive.
2. Circuit current lags the source voltage by 90°.

*Note:* The inductive nature of the circuit in Figure 18.18 is also indicated by the fact that $I_L > I_C$.

When equation (18.4) or equation (18.5) yields a negative result:

1. The source sees the circuit as capacitive.
2. The circuit current leads the source voltage by 90°.

This phase relationship is demonstrated in Example 18.10.

**EXAMPLE 18.10**

Calculate the value of $X_P$ for the circuit in Figure 18.19.

**FIGURE 18.19**

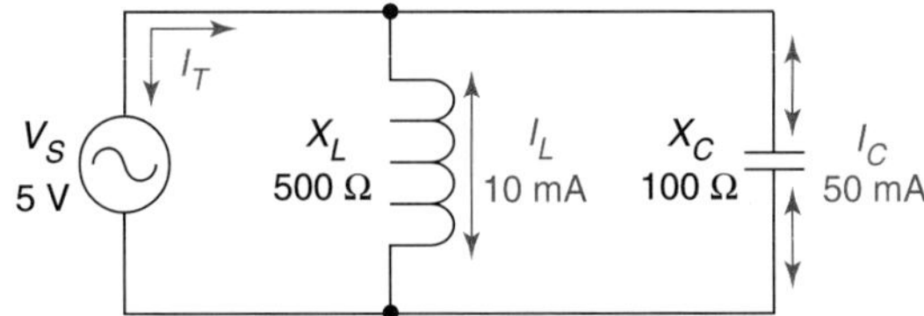


***Solution:*** Using the values shown in Figure 18.19, the value of $X_P$ can be found as

$$X_P = \frac{1}{\frac{1}{X_L} + \frac{1}{X_C}} = \frac{1}{\frac{1}{500\ \Omega} + \frac{1}{-100\ \Omega}} = \frac{1}{\frac{1}{500\ \Omega} - \frac{1}{100\ \Omega}} = \mathbf{-125\ \Omega}$$

or

$$X_P = \frac{X_L X_C}{X_L + X_C} = \frac{(500\ \Omega)(-100\ \Omega)}{500\ \Omega + (-100\ \Omega)} = \frac{-50{,}000}{400}\ \Omega = \mathbf{-125\ \Omega}$$

Since our result is negative, the source sees the circuit as capacitive. For this circuit, $X_P = 125\ \Omega\angle -90°$ and $I_T$ leads $V_S$ by 90°.

## Putting It All Together: Basic Parallel LC Circuit Characteristics

The relationships among the currents, voltages, and reactances in a parallel *LC* circuit can be clarified by comparing the values in Examples 18.9 and 18.10. For convenience, the values of interest are summarized as follows:

| Example | $I_L$ | $I_C$ | $I_T$ | $X_L$ | $X_C$ | $X_P$ |
|---|---|---|---|---|---|---|
| 18.9 | 40 mA | 25 mA | Inductive | 150 Ω | 500 Ω | 214.3 Ω |
| 18.10 | 10 mA | 50 mA | Capacitive | 500 Ω | 100 Ω | 125 Ω |

In Example 18.9, we concluded that the circuit is inductive, and that circuit current lags the source voltage by 90°. This conclusion is supported by the fact that the inductor current is greater than the capacitor current. Therefore, the total current (which is equal to the difference between the two) must be inductive. In Example 18.10, we concluded that the circuit was capacitive. This conclusion is supported by the fact that capacitor current is greater than inductor current. Therefore, the total current must be capacitive.

The point being made is this: The greater branch current determines the overall nature of the circuit. When $I_L > I_C$, the circuit is inductive and the circuit current lags the source voltage by 90°. When $I_C > I_L$, the circuit is capacitive and the circuit current leads the source voltage by 90°.

In terms of reactance relationships, the characteristics of a parallel *LC* circuit can be summarized as shown in Table 18.2. The relationships listed here can be verified by reviewing the values and calculations in Examples 18.9 and 18.10.

**TABLE 18.2 Parallel *LC* Circuit Conditions When $X_L \neq X_C$**

| *Reactance Relationship* | *Circuit Characteristics* |
|---|---|
| $X_L > X_C$ | $I_C > I_L$<br>The calculated value of $X_P$ is negative.<br>The circuit is capacitive.<br>Circuit current leads $V_S$ by 90°. |
| $X_C > X_L$ | $I_L > I_C$<br>The calculated value of $X_P$ is positive.<br>The circuit is inductive.<br>Circuit current lags $V_S$ by 90°. |

## An Important Point

Because of the unique phase relationships in a parallel *LC* circuit:

1. The total circuit current is lower than the highest valued branch current.
2. The value of $X_P$ is greater than one (or both) of the branch reactance values.

These relationships can be verified using the following values from Example 18.10:

$$I_L = 10 \text{ mA} \qquad I_C = 50 \text{ mA} \qquad I_T = 40 \text{ mA}$$
$$X_L = 500\ \Omega \qquad X_C = 100\ \Omega \qquad X_P = 125\ \Omega$$

As you can see, $I_L < I_T < I_C$ and $X_L > X_P > X_C$. Note that total current is never the highest valued current in the circuit. By the same token, $X_P$ is never the lowest valued reactance in the circuit.

### Section Review

1. List the current and voltage characteristics of a parallel *LC* circuit.
2. Explain the phase relationship between parallel values of $I_L$ and $I_C$.
3. Describe the phase angle of circuit current (relative to source voltage) when:
   a. $I_L > I_C$
   b. $I_C > I_L$
4. State and explain the phase relationship between parallel values of $X_L$ and $X_C$.
5. Describe the phase angle of $X_P$ (relative to source voltage) when:
   a. $X_L > X_C$
   b. $X_C > X_L$

6. Describe the phase relationship between circuit current and source voltage in a parallel *LC* circuit when:

   a. $X_L > X_C$

   b. $X_C > X_L$

## 18.3 RESONANCE

◄ *OBJECTIVE 8*

In Chapters 13 and 16, you were shown that reactance varies with frequency. Specifically, you were shown that:

1. Inductive reactance ($X_L$) varies directly with operating frequency.
2. Capacitive reactance ($X_C$) varies inversely with operating frequency.

For every *LC* circuit, there is a specific frequency at which $X_L = X_C$. An *LC* circuit with equal values of $X_L$ and $X_C$ is said to be operating at **resonance.** The frequency at which resonance occurs in an *LC* circuit is referred to as its **resonant frequency** ($f_r$).

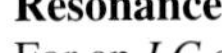

**Resonance**
For an *LC* circuit, the operating condition where $X_L = X_C$.

**Resonant frequency ($f_r$)**
For an *LC* circuit, the frequency at which resonance occurs.

### Resonant Frequency ($f_r$)

As you know, inductive and capacitive reactance can be found as

$$X_L = 2\pi fL \qquad \text{and} \qquad X_C = \frac{1}{2\pi fC}$$

The graph in Figure 18.20 shows the effect that frequency has on inductive and capacitive reactance. Note that $X_L$ and $X_C$ are plotted on the positive and negative axes of the graph to illustrate the 180° phase relationship between them.

The line representing $X_L$ shows that inductive reactance is 0 Ω when $f = 0$ Hz. As frequency increases, the value of $X_L$ increases at a linear rate. Note that $X_L$ approaches infinite ohms ($X_L \rightarrow \infty\ \Omega$) as frequency approaches infinity. The curve representing $X_C$ approaches infinite ohms as frequency approaches 0 Hz. At the other end of the curve, $X_C \rightarrow$ 0 Ω as frequency approaches infinity.

For each *LC* circuit, there is an operating frequency where $X_L = X_C$, as shown in Figure 18.21. This frequency is the resonant frequency of the circuit. The resonant frequency of a given *LC* circuit is found as

$$f_r = \frac{1}{2\pi\sqrt{LC}} \qquad \textbf{(18.6)}$$

where $f_r$ = the frequency at which $X_L = X_C$
$L, C$ = the values of inductance and capacitance, respectively, in the circuit

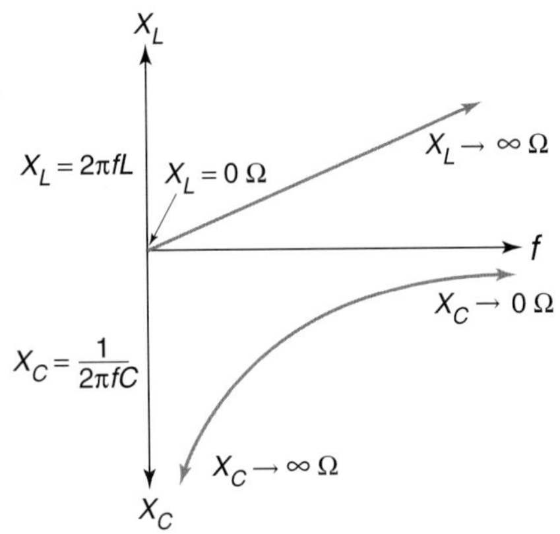


**FIGURE 18.20 Reactance versus frequency curves.**

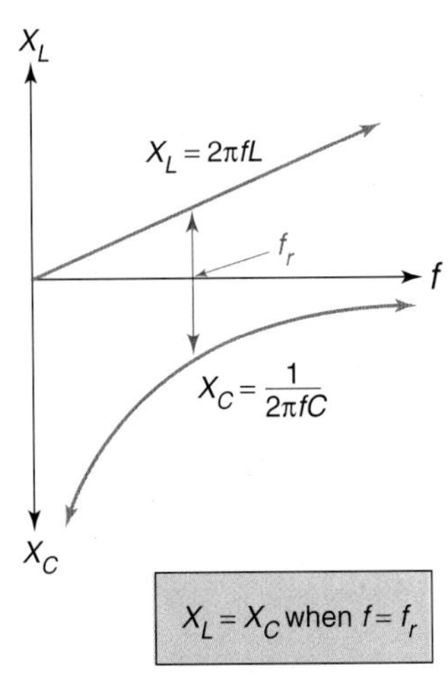


**FIGURE 18.21 Resonant frequency ($f_r$).**

The use of equation (18.6) is demonstrated in Example 18.11.

## EXAMPLE 18.11

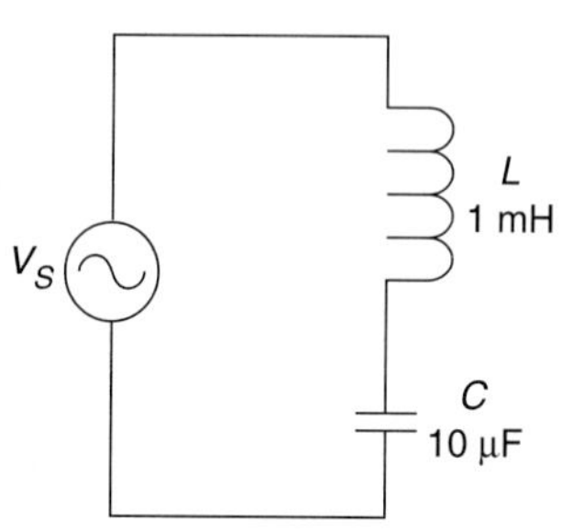


FIGURE 18.22

Determine the resonant frequency for the series $LC$ circuit shown in Figure 18.22.

***Solution:*** Using the values of $L$ and $C$ shown, the resonant frequency of the circuit can be found as

$$f_r = \frac{1}{2\pi\sqrt{LC}} = \frac{1}{2\pi\sqrt{(1\text{ mH})(10\ \mu\text{F})}} = \frac{1}{2\pi\sqrt{10 \times 10^{-9}}} = \mathbf{1.59\ kHz}$$

***PRACTICE PROBLEM 18.11***

A circuit like the one in Figure 18.22 has values of $L = 3.3$ mH and $C = 0.22\ \mu$F. Calculate the resonant frequency of the circuit.

For the circuit in Figure 18.22, we calculated a value of $f_r = 1.59$ kHz. At this frequency, the values of $X_L$ and $X_C$ can be found as

$$X_L = 2\pi fL = 2\pi(1.59\text{ kHz})(1\text{ mH}) \cong 10\ \Omega$$

and

$$X_C = \frac{1}{2\pi fC} = \frac{1}{2\pi(1.59\text{ kHz})(10\ \mu\text{F})} = 10\ \Omega$$

Note that the approximated value of $X_L$ was rounded up from 9.99 Ω. Therefore, we can state that the values of $X_L$ and $X_C$ are (for all practical purposes) equal when the circuit is operated at its resonant frequency.

The principles of resonance also apply to parallel $LC$ circuits, as demonstrated in Example 18.12.

## EXAMPLE 18.12

Determine the value of $f_r$ for the circuit in Figure 18.23. Then calculate the values of $X_L$ and $X_C$ when the circuit is operated at its resonant frequency.

FIGURE 18.23

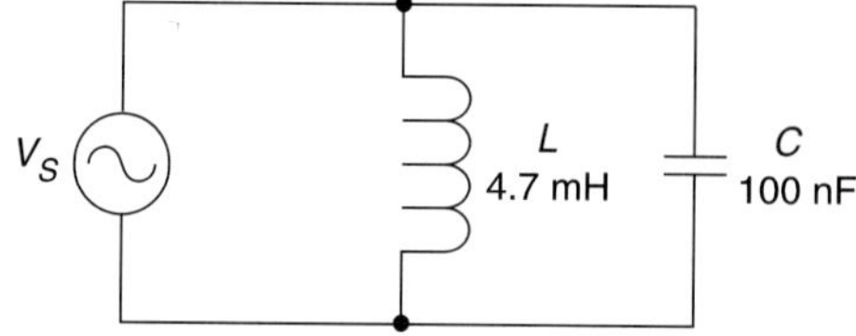


***Solution:*** Using the values of $L$ and $C$ shown, the resonant frequency of the circuit is found as

$$f_r = \frac{1}{2\pi\sqrt{LC}} = \frac{1}{2\pi\sqrt{(4.7\text{ mH})(100\text{ nF})}} = \frac{1}{2\pi\sqrt{4.7 \times 10^{-10}}} = \mathbf{7.34\ kHz}$$

The circuit values of $X_L$ and $X_C$ at this frequency can be found as

$$X_L = 2\pi f L = 2\pi(7.34 \text{ kHz})(4.7 \text{ mH}) \cong 217\ \Omega$$

and

$$X_C = \frac{1}{2\pi f C} = \frac{1}{2\pi(7.34 \text{ kHz})(100 \text{ nF})} \cong 217\ \Omega$$

***PRACTICE PROBLEM 18.12***

A circuit like the one in Figure 18.23 has values of $L = 100\ \mu\text{H}$ and $C = 51$ nF. Determine the value of $f_r$ for the circuit. Then calculate the values of $X_L$ and $X_C$ when the circuit is operated at its resonant frequency.

As demonstrated in Example 18.12, the principles of resonance also apply to parallel *LC* circuits. Specifically:

1. The resonant frequency of a parallel *LC* circuit is found using equation (18.6).
2. When operated at its resonant frequency, the values of $X_L$ and $X_C$ for a parallel *LC* circuit are equal.

Although these principles are the same for series and parallel resonant circuits, you'll soon see that other characteristics of the two circuits vary significantly.

## The Basis for Equation (18.6)

By definition, the resonant frequency of an *LC* circuit is the frequency at which $X_L = X_C$. Using the basic reactance equations, this relationship can be expressed as

$$2\pi f_r L = \frac{1}{2\pi f_r C}$$

Multiplying both sides of the equation by $(2\pi f_r C)$, we get

$$(2\pi f_r L)(2\pi f_r C) = 1 \qquad \text{or} \qquad (2\pi)^2 f_r^2 LC = 1$$

Now, dividing both sides of the equation by $[(2\pi)^2 LC]$ gives us

$$f_r^2 = \frac{1}{(2\pi)^2 LC}$$

Finally, we take the square root of each side of the equation to get

$$f_r = \frac{1}{2\pi\sqrt{LC}}$$

which is equation (18.6).

## Factors Affecting the Value of $f_r$

Equation (18.6) indicates that resonant frequency varies inversely with the values of *L* and *C*; that is, an increase in either *L* or *C* causes a decrease in the value of $f_r$. This relationship is demonstrated in Example 18.13.

## EXAMPLE 18.13

The value of $L$ in Figure 18.23 (Example 18.12) is increased to 10 mH. Calculate the new resonant frequency for the circuit.

***Solution:*** In Example 18.12, we calculated a value of $f_r = 7.34$ kHz for the circuit. Using values of $L = 10$ mH and $C = 100$ nF, the new value of $f_r$ for the circuit is found as

$$f_r = \frac{1}{2\pi\sqrt{LC}} = \frac{1}{2\pi\sqrt{(10\text{ mH})(100\text{ nF})}} = \mathbf{5.03\ kHz}$$

As you can see, increasing the circuit inductance causes a decrease in its resonant frequency. The resonant frequency responds in the same way to an increase in capacitance.

***PRACTICE PROBLEM 18.13***

Refer to Practice Problem 18.12. Assume that the value of $C$ is decreased to 33 nF. Calculate the new value of $f_r$ for the circuit. Compare the new value of $f_r$ to the one you calculated for the original circuit.

There are several other factors that can affect the value of $f_r$ for an $LC$ circuit. We will now take a brief look at several of these factors.

*OBJECTIVE 9* ▶

***Stray Inductance.*** Any time that alternating current passes through a conductor, a magnetic field is generated around that conductor. (Remember: An inductor is simply a length of wire looped to concentrate the magnetic flux produced by the alternating current through the wire.) With this in mind, look at the conductors represented in Figure 18.24.

On the surface, the circuit in Figure 18.24a appears to be purely capacitive. However, the circuit contains lengths of wire that are passing an alternating current. As such, these wires all have some amount of inductance. This inductance is referred to as **stray inductance.**

**Stray inductance**
Inductance generated whenever an alternating current passes through a conductor.

Stray inductance can also be generated by the leads on a component. For example, the leads on the resistor in Figure 18.24b pass an alternating current. Therefore, they generate magnetic flux and have some amount of stray inductance. The amount of stray inductance produced by a component (such as a resistor) can be limited by trimming the component leads to the shortest practical length before inserting the component into a circuit.

Sources of stray inductance (like those shown in Figure 18.24) can affect the resonant frequency of a low-inductance $LC$ circuit. For example, if an $LC$ circuit with a 1 μH inductor contains 0.1 μH of stray inductance, the resonant frequency of the $LC$ circuit can be significantly lower than its calculated value.

It should be noted that stray inductance is rarely a factor in the operation of low-frequency circuits. However, as the operating frequency of a circuit increases, the values of stray inductance within the circuit can have a significant impact on the circuit operation.

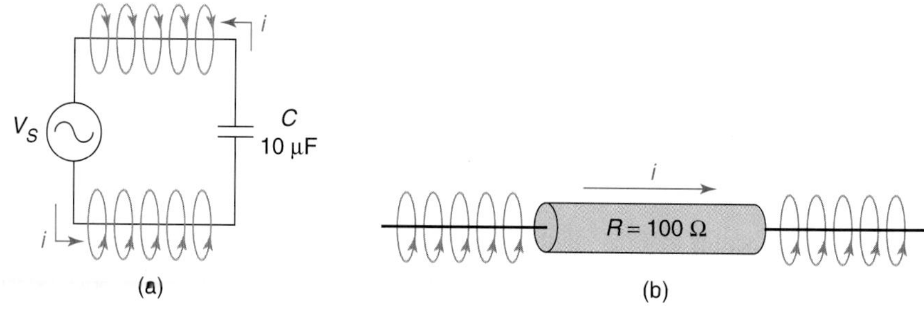


**FIGURE 18.24 Conductors and magnetic fields.**

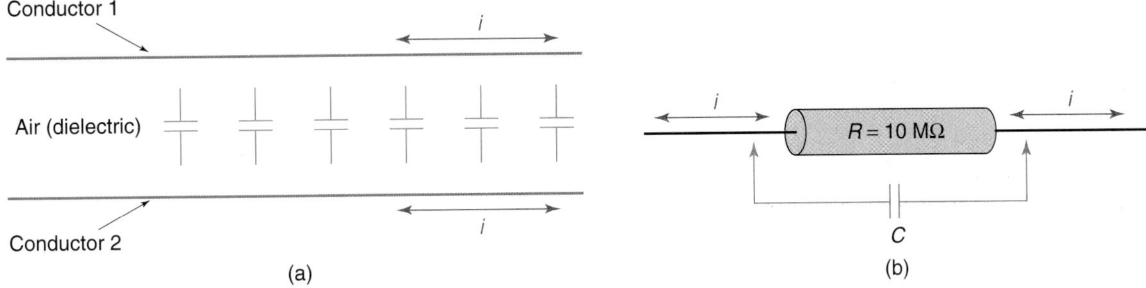


**FIGURE 18.25 Conductors and capacitance.**

***Stray Capacitance.*** Whenever you have two conductors separated by an insulator, you have some amount of capacitance between the conductors. With this in mind, take a look at Figure 18.25.

**Stray capacitance** Capacitance formed between parallel conductors separated by an insulator.

Figure 18.25a shows two conductors, each passing an alternating current. Combined with the air (dielectric) between them, the conductors form a capacitor. The capacitance generated by the conductors is referred to as **stray capacitance.** Note that the capacitance (represented by the capacitor symbols) is distributed along the length of the conductors. For this reason, the stray capacitance between the conductors is sometimes referred to as *distributed capacitance*. The type of stray capacitance shown in Figure 18.25a can be eliminated by placing the conductors at 90° angles (whenever possible). It can also be reduced (or eliminated) by spacing the conductors as far apart as possible.

*A Practical Consideration:* The distributed capacitance between two parallel conductors can cause interference between them. This is why a signal line (like the coaxial cable between a VCR and a television) should not be run parallel to a power cord. Interference between two conductors can be minimized by keeping them perpendicular to each other.

Individual components can also generate stray capacitance. For example, look at the resistor shown in Figure 18.25b. The high resistance of the component can be viewed as a dielectric when compared to the low resistance of the leads. From this viewpoint, the component consists of two conductors separated by a dielectric. Although extremely low in value, the component does generate some stray capacitance. This stray capacitance can affect the resonant frequency of a high-frequency low-capacitance *LC* circuit.

***Oscilloscope Input Capacitance.*** Oscilloscopes have a measurable amount of input capacitance. This input capacitance can affect the resonant frequency of any *LC* circuit under test, as illustrated in Figure 18.26. The capacitor labeled $C_{in}$ represents the input capacitance of the oscilloscope. When connected to the circuit as shown, $C_{in}$ is placed in parallel with $C_1$. As a result, the total capacitance in the circuit changes from $C_1$ to $(C_1 + C_{in})$. This

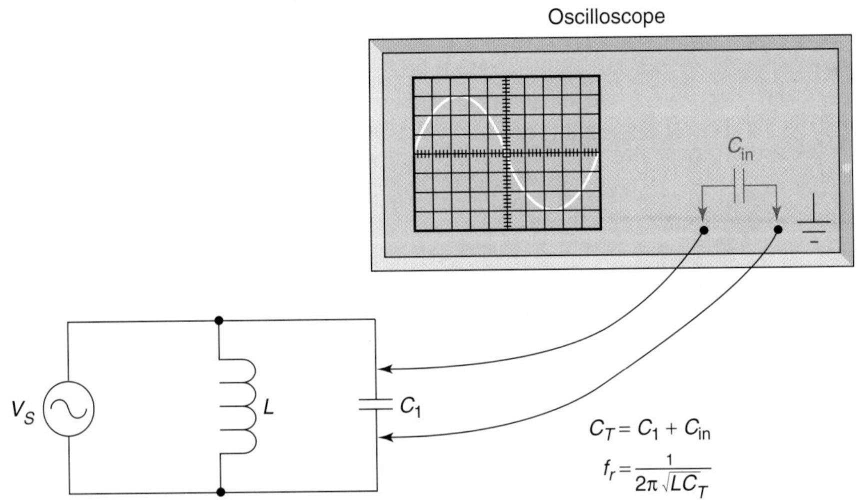


**FIGURE 18.26 Oscilloscope input capacitance.**

*A Practical Consideration:* The amount of capacitance at the input to an oscilloscope is provided on its specification sheet. Oscilloscope input capacitance is typically in the low pF range.

has the effect of lowering the resonant frequency of the circuit. For this reason, the measured value of $f_r$ is typically lower than its calculated value.

## Series Resonant LC Circuits

*OBJECTIVE 10* ► When a series *LC* circuit is operated at resonance, it has the following characteristics:

1. The total reactance of the series resonant circuit is 0 Ω.
2. The voltage across the series *LC* circuit is 0 V.
3. The circuit voltage and current are in phase; that is, the circuit is resistive.

These characteristics can be explained using the circuit and graph in Figure 18.27.

Assuming that the circuit in Figure 18.27 is operating at resonance, the values of $X_L$ and $X_C$ are equal. As shown in the reactance graph, these two values are 180° out of phase. Therefore, the total reactance in the circuit can be found as

$$X_T = X_L - X_C = 0\ \Omega$$

Since the total reactance in the circuit is 0 Ω, the voltage across the series combination of *L* and *C* ($V_{LC}$) can be found as

$$V_{LC} = (I)(X_T) = (I)(0\ \Omega) = 0\ \text{V}$$

Thus, the voltage across a series resonant *LC* circuit is 0 V. For the circuit in Figure 18.27, this means that the source voltage is dropped across its internal resistance. Since $V_S$ is dropped across $R_S$, the total current in the circuit is found as

$$I = \frac{V_S}{R_S}$$

This equation indicates that the circuit is resistive. Since the circuit is resistive, the source voltage and circuit current are in phase.

The fact that $V_{LC} = 0$ V may lead you to believe that $V_L$ and $V_C$ each have a value of 0 V. However, this is not the case. $V_L$ and $V_C$ each have some measurable value that may be much greater than the value of source voltage. However, the sum of the component voltages is 0 V because of their phase relationship. This point is demonstrated in Example 18.14.

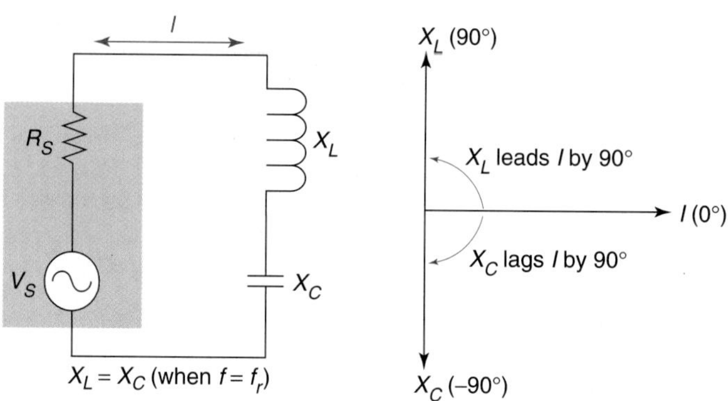


**FIGURE 18.27 Series resonant circuit characteristics.**

Calculate the values of $V_L$, $V_C$, and $V_{LC}$ for the circuit in Figure 18.28.

**EXAMPLE 18.14**

**FIGURE 18.28**

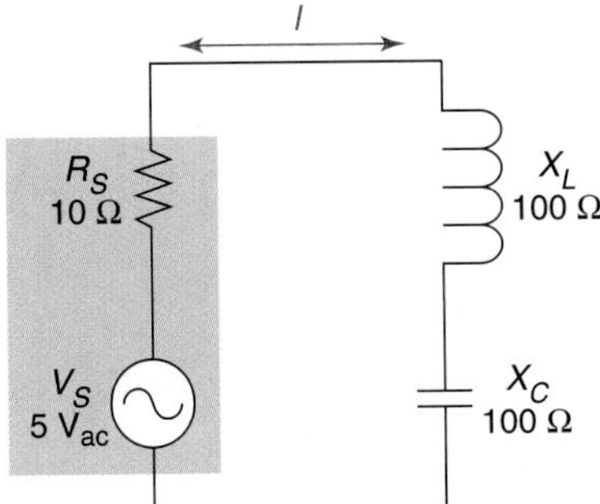


***Solution:*** With equal values of $X_L$ and $X_C$, the total circuit reactance is 0 Ω. Therefore, the source resistance is the only opposition to current in the circuit, and the value of circuit current is found as

$$I = \frac{V_S}{R_S} = \frac{5 \text{ V}_{\text{ac}}}{10 \ \Omega} = \mathbf{500 \ mA}$$

Using $I = 500$ mA, the voltage across the inductor is found as

$$V_L = I \cdot X_L = (500 \text{ mA})(100 \ \Omega\angle 90°) = \mathbf{50 \ V\angle 90°}$$

Using the same current value, the voltage across the capacitor is found as

$$V_C = I \cdot X_C = (500 \text{ mA})(100 \ \Omega\angle -90°) = \mathbf{50 \ V\angle -90°}$$

Since these voltages are equal in magnitude and are 180° out of phase, their sum is 0 V.

***PRACTICE PROBLEM 18.14***

A circuit like the one in Figure 18.28 has the following values: $V_S = 8$ V, $R_S = 40 \ \Omega$, $X_L = 1 \text{ k}\Omega$, and $X_C = 1 \text{ k}\Omega$. Calculate the values of $V_L$, $V_C$, and $(V_L + V_C)$ for the circuit.

As you can see, the circuit has values of $V_L$ and $V_C$ that are each ten times the value of the source voltage. However, their sum is 0 V because of the 180° phase relationship between them.

## *Parallel Resonant LC Circuits*

When a parallel *LC* circuit is operated at resonance, it has the following characteristics:

1. The sum of the currents through the parallel *LC* circuit is 0 A.
2. The circuit has infinite reactance; that is, it acts as an open.

These characteristics can be explained using the circuit and graph shown in Figure 18.29.

Assuming that the circuit in Figure 18.29 is operating at resonance, the values of $X_L$ and $X_C$ are equal. Since the branch reactances and voltages are equal, so are the branch currents. As shown in the graph, $I_L$ and $I_C$ are 180° out of phase. Therefore, the total current in the circuit can be found as

$$I_T = I_C - I_L = 0 \text{ A}$$

Since the net current through the parallel resonant circuit is 0 A, the circuit acts as an open. In effect, the total reactance of the parallel resonant circuit is infinite. These parallel resonant circuit characteristics are illustrated further in Example 18.15.

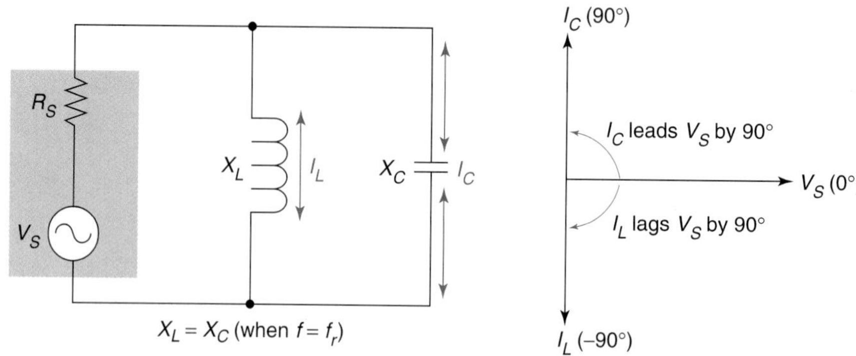


**FIGURE 18.29 Parallel resonant circuit characteristics.**

## EXAMPLE 18.15

Calculate the values of $I_C$, $I_L$, $I_T$, and $X_T$ for the parallel resonant circuit in Figure 18.30.

**FIGURE 18.30**

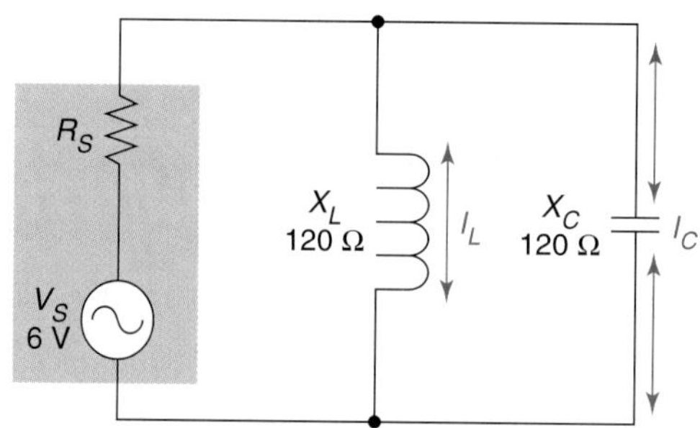


***Solution:*** To demonstrate the phase relationship between the branch currents, we start by calculating the value of capacitor current as follows:

$$\boldsymbol{I_C} = \frac{V_S}{X_C} = \frac{6\text{ V}}{120\ \Omega\angle -90^\circ} = \mathbf{50\ mA\angle 90^\circ}$$

Now, the value of the inductor current is found as

$$\boldsymbol{I_L} = \frac{V_S}{X_L} = \frac{6\text{ V}}{120\ \Omega\angle 90^\circ} = \mathbf{50\ mA\angle -90^\circ}$$

Since the branch currents are equal in magnitude and 180° out of phase, the net current through the circuit is 0 A. Using this value, the total reactance of the circuit approaches infinity, as follows:

$$\boldsymbol{X_T} = \frac{V_S}{I_T} = \frac{6\text{ V}}{0\text{ A}} \rightarrow \boldsymbol{\infty\ \Omega}$$

The arrow at the end of the calculation indicates that the value of $X_T$ *approaches* infinity.

### *PRACTICE PROBLEM 18.15*

A circuit like the one in Figure 18.30 has the following values: $V_S = 10$ V, $X_L = 40\ \Omega$, and $X_C = 40\ \Omega$. Calculate the values of $I_C$, $I_L$, $I_T$, and $X_T$ for the circuit.

FIGURE 18.31

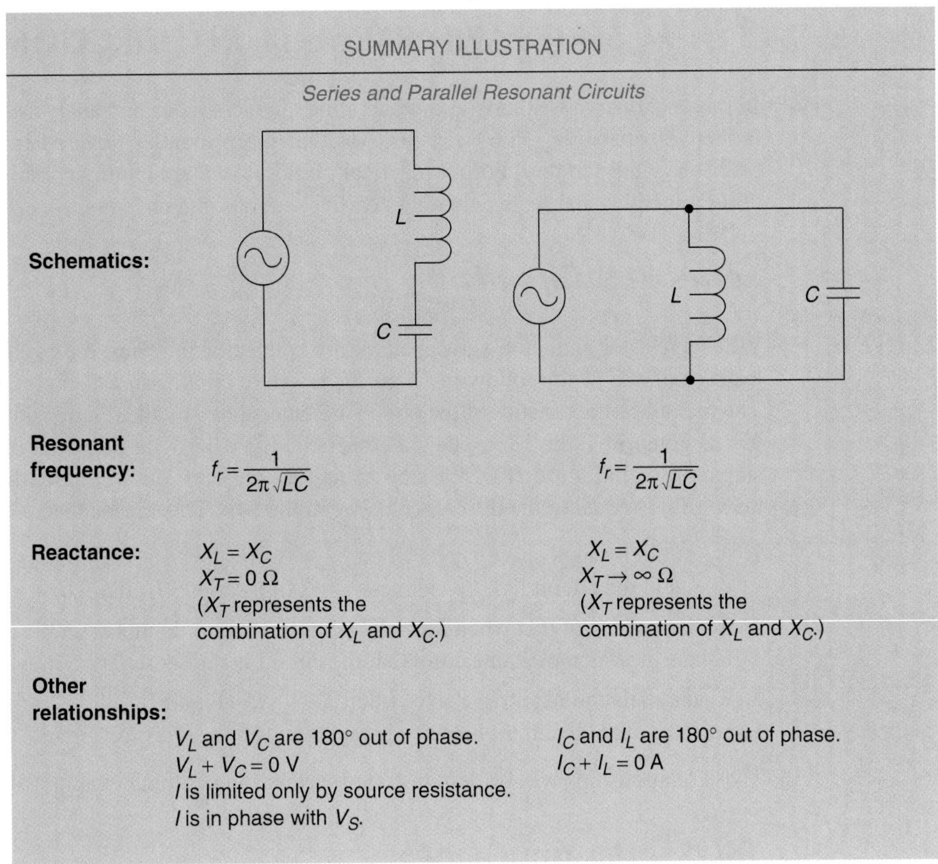

SUMMARY ILLUSTRATION
Series and Parallel Resonant Circuits
Schematics:
L
C
L
C
Resonant frequency:
$f_r = \frac{1}{2\pi\sqrt{LC}}$
$f_r = \frac{1}{2\pi\sqrt{LC}}$
Reactance:
$X_L = X_C$
$X_T = 0\ \Omega$
($X_T$ represents the combination of $X_L$ and $X_C$.)
$X_L = X_C$
$X_T \to \infty\ \Omega$
($X_T$ represents the combination of $X_L$ and $X_C$.)
Other relationships:
$V_L$ and $V_C$ are 180° out of phase.
$V_L + V_C = 0$ V
$I$ is limited only by source resistance.
$I$ is in phase with $V_S$.
$I_C$ and $I_L$ are 180° out of phase.
$I_C + I_L = 0$ A


## Series Versus Parallel Resonance: A Comparison

In this section, you have been introduced to the characteristics of series and parallel resonant *LC* circuits. Figure 18.31 provides a summary comparison of the characteristics of these two types of circuits. When you compare the overall reactance, current, and voltage characteristics of the circuits, you can see that they are nearly opposites.

## One Final Note

The frequency response characteristics of series and parallel *LC* circuits are more complicated than represented in this section. In Chapter 19, we will discuss the frequency response of *LC* (and other) circuits. As you will see, values like inductor *Q*, capacitor *Q*, and circuit resistance all affect the frequency response of *LC* circuits.

### Section Review

1. What is the resonant frequency of an *LC* circuit? How is it calculated?
2. How do changes in component values affect the resonant frequency of an *LC* circuit?
3. What is stray inductance?
4. What is stray capacitance?
5. How does the input capacitance of an oscilloscope affect the resonant frequency of a circuit under test?
6. List the basic characteristics of a series resonant *LC* circuit.
7. List the basic characteristics of a parallel resonant *LC* circuit.

## 18.4 SERIES AND PARALLEL *RLC* CIRCUITS

In this section, we will look at the operating characteristics and analysis of basic resistive-inductive-capacitive (*RLC*) circuits. Most of the principles covered in this section will probably seem familiar. For all practical purposes, we are merely combining circuit operating principles that were covered in earlier sections and chapters.

### Series RLC Circuits

*OBJECTIVE 11* ▶ When a resistor, inductor, and capacitor are connected in series, the circuit has the characteristics of one of the following: a purely resistive circuit, an *RC* circuit, or an *RL* circuit. The overall characteristics of the circuit are determined by the relationship between $X_L$ and $X_C$, as given in Table 18.3. The statements in Table 18.3 are based on the phase relationships shown in Figure 18.32a. Since $X_L$ and $X_C$ are 180° out of phase, the net series reactance ($X_S$) is equal to the difference between the two. This means that:

1. $X_S$ parallels the positive *y*-axis when $X_L > X_C$ (Figure 18.32b). In this case, $X_S$ is inductive, and the circuit has the characteristics of a series *RL* circuit.
2. $X_S$ has a value of 0 Ω when $X_L = X_C$ (Figure 18.32c). In this case, the circuit impedance equals the value of *R*, meaning that the circuit is purely resistive.
3. $X_S$ parallels the negative *y*-axis when $X_L < X_C$ (Figure 18.32d). In this case, $X_S$ is capacitive, and the circuit has the characteristics of a series *RC* circuit.

These relationships will be demonstrated further in upcoming examples.

### Total Series Impedance

Figure 18.32b and d indicates that $X_S$ is at a 90° angle to *R* when its value is inductive or capacitive. When this is the case, the magnitude of the circuit impedance is found as the geometric sum of $X_S$ and *R*. By formula,

$$Z_T = \sqrt{X_S^2 + R^2} \quad \textbf{(18.7)}$$

where $X_S$ is the net series reactance, found as $X_S = X_L - X_C$

The phase angle of the circuit impedance is found as

$$\theta = \tan^{-1}\frac{X_S}{R} \quad \textbf{(18.8)}$$

When $X_L > X_C$, the circuit impedance of a series *RLC* circuit has a positive phase angle, which indicates that the circuit has the characteristics of a series *RL* circuit. This point is demonstrated in Example 18.16.

**TABLE 18.3 Reactance Versus Circuit Characteristics for Series *RLC* Circuits**

| *Reactance Relationship* | *Resulting Circuit Characteristics* |
|---|---|
| $X_L > X_C$ | The net series reactance is inductive, so the circuit has the characteristics of a series *RL* circuit: source voltage and circuit impedance lead the circuit current. |
| $X_L = X_C$ | The resonant *LC* circuit has a net reactance of 0 Ω. Therefore, the circuit is resistive: source voltage and circuit impedance are both in phase with circuit current. |
| $X_L < X_C$ | The net series reactance is capacitive, so the circuit has the characteristics of a series *RC* circuit: source voltage and circuit impedance both lag the circuit current. |

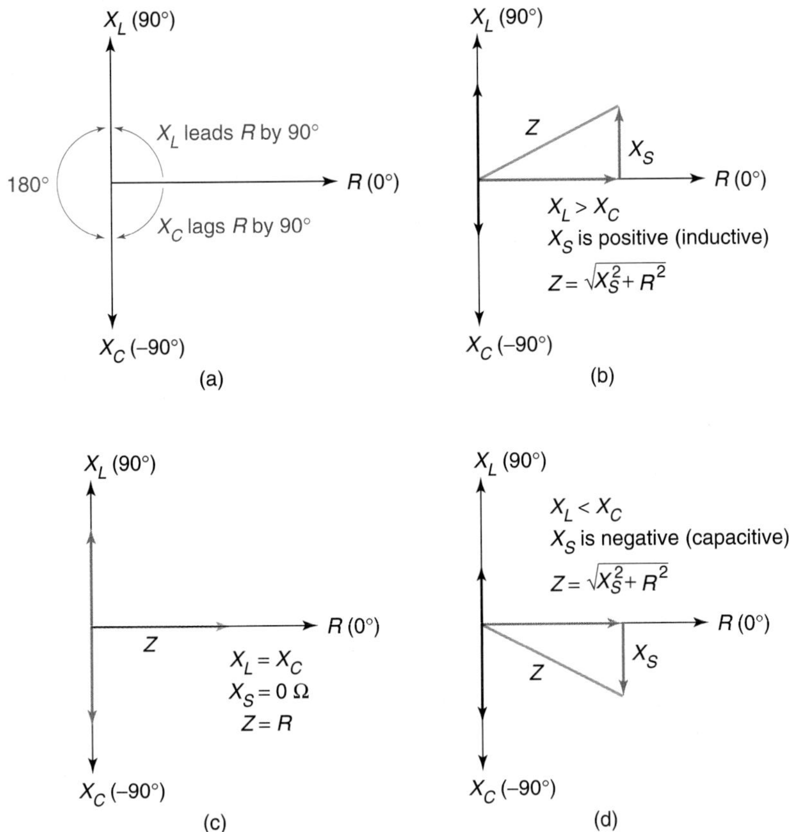


**FIGURE 18.32 The relationship among $X_L$, $X_C$, and $X_S$.**

EXAMPLE 18.16

Determine the magnitude and phase angle of the circuit impedance in Figure 18.33a.

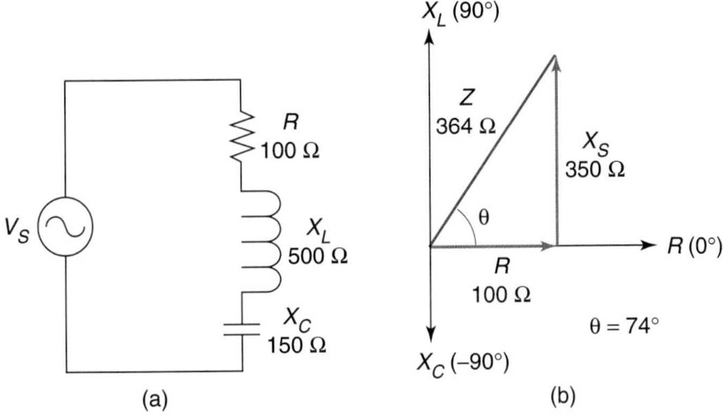


**FIGURE 18.33**

***Solution:*** First, the net series reactance is found as

$$X_S = X_L - X_C = 500\ \Omega - 150\ \Omega = \mathbf{350\ \Omega}$$

Now, the magnitude of the circuit impedance can be found as

$$Z = \sqrt{X_S^2 + R^2} = \sqrt{(350\ \Omega)^2 + (100\ \Omega)^2} = \mathbf{364\ \Omega}$$

and its phase angle can be found as

$$\theta = \tan^{-1}\frac{X_S}{R} = \tan^{-1}\left(\frac{350\ \Omega}{100\ \Omega}\right) = \mathbf{74°}$$

Using the values of $X_S$, $R$, and $Z$, the impedance triangle can be plotted as shown in Figure 18.33b. Comparing this impedance triangle to the one shown in Figure 18.32b demonstrates that the circuit is inductive; that is, it has the phase characteristics of a series *RL* circuit.

***PRACTICE PROBLEM 18.16***

A circuit like the one in Figure 18.33a has the following values: $R = 330\ \Omega$, $X_L = 840\ \Omega$, and $X_C = 600\ \Omega$. Calculate the magnitude and phase angle of the circuit impedance. Then plot the impedance triangle for the circuit.

When $X_L < X_C$, the circuit impedance of a series *RLC* circuit has a negative phase angle, which indicates that the circuit has the characteristics of a series *RC* circuit. This point is demonstrated in Example 18.17.

## EXAMPLE 18.17

Determine the magnitude and phase angle of the circuit impedance in Figure 18.34a.

**FIGURE 18.34**

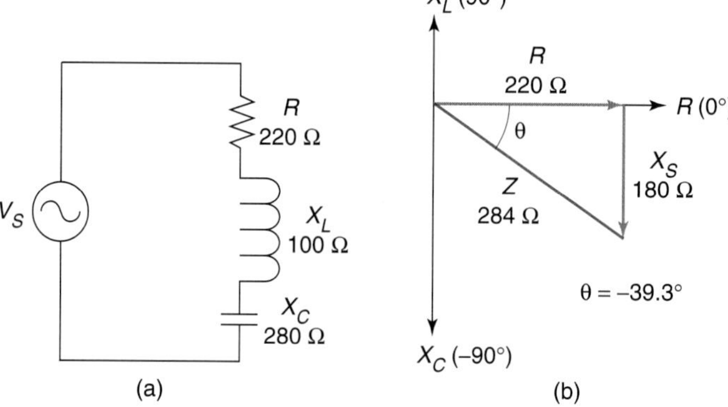


***Solution:*** First, the net series reactance is found as

$$X_S = X_L - X_C = 100\ \Omega - 280\ \Omega = \mathbf{-180\ \Omega}$$

Now, the magnitude of the circuit impedance can be found as

$$Z = \sqrt{X_S^2 + R^2} = \sqrt{(-180\ \Omega)^2 + (220\ \Omega)^2} \cong \mathbf{284\ \Omega}$$

and its phase angle can be found as

$$\theta = \tan^{-1}\frac{X_S}{R} = \tan^{-1}\left(\frac{-180\ \Omega}{220\ \Omega}\right) = \mathbf{-39.3°}$$

Using the values of $X_S$, $R$, and $Z$, the impedance triangle can be plotted as shown in Figure 18.34b. Comparing this impedance triangle to the one shown in Figure 18.32d demonstrates that the circuit is capacitive; that is, it has the phase characteristics of a series *RC* circuit.

***PRACTICE PROBLEM 18.17***

A circuit like the one in Figure 18.34a has the following values: $R = 1\ \text{k}\Omega$, $X_L = 300\ \Omega$, and $X_C = 1.5\ \text{k}\Omega$. Calculate the magnitude and phase angle of the circuit impedance. Then plot the impedance triangle for the circuit.

When a series *RLC* circuit is operating at resonance, the values of inductive and capacitive reactance are equal. When $X_L = X_C$, the net series reactance is 0 Ω (as explained in Section 18.3). With a value of $X_S = 0\ \Omega$, the circuit impedance is found as

$$Z = \sqrt{X_S^2 + R^2} = \sqrt{(0\ \Omega)^2 + R^2} = \sqrt{R^2} = R$$

and the circuit phase angle is found as

$$\theta = \tan^{-1}\frac{X_S}{R} = \tan^{-1}\left(\frac{0\ \Omega}{R}\right) = 0°$$

As these values indicate, a series *RLC* circuit operating at resonance has no phase angles and is purely resistive.

## Series Circuit Frequency Response

You have seen how the values of reactance determine the operating characteristics of a series *RLC* circuit. The effect of reactance on series circuit operation is a primary concern because reactance varies with operating frequency. This means that, in effect, the operating frequency of a series *RLC* circuit determines the circuit characteristics. For example, consider the frequency graphs shown in Figure 18.35.

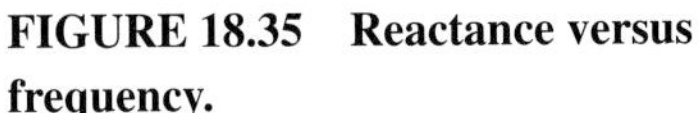

**FIGURE 18.35 Reactance versus frequency.**

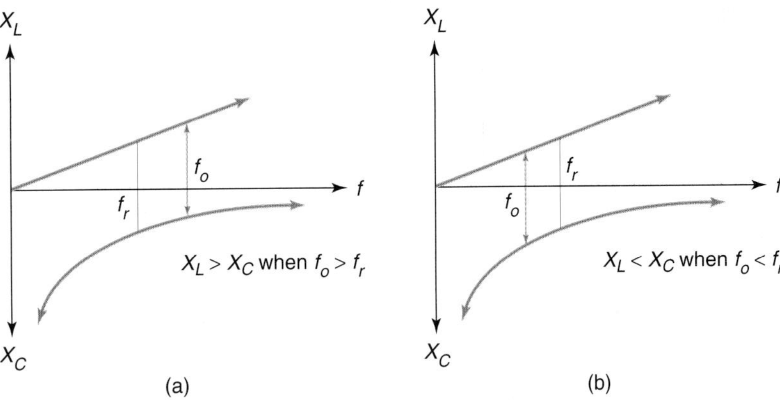


The graph in Figure 18.35a illustrates the relationship between $X_L$ and $X_C$ when the operating frequency ($f_o$) is greater than the circuit's resonant frequency. As shown in the graph, $X_L > X_C$ whenever $f_o > f_r$. This means that a series *RLC* circuit is inductive whenever it is operated above its resonant frequency. By the same token, the graph in Figure 18.35b shows that $X_L < X_C$ whenever $f_o < f_r$. This means that a series *RLC* circuit is capacitive whenever it is operated below its resonant frequency. Of course, the values of $X_L$ and $X_C$ in a series *RLC* circuit cancel each other out when the circuit is operated at resonance. Therefore, a series *RLC* circuit is resistive when $f_o = f_r$. The relationships between operating frequency and the characteristics of a series *RLC* circuit are summarized in Table 18.4.

**TABLE 18.4 The Effects of Operating Frequency on Series *RLC* Circuit Characteristics**

| *Operating Frequency* | *Reactance* | *Other Characteristics* |
|---|---|---|
| $f_o < f_r$ | $X_L < X_C$ | The circuit is capacitive, meaning that it has the operating characteristics of a series *RC* circuit. |
| $f_o = f_r$ | $X_L = X_C$ | The reactances cancel out, meaning that the circuit has the operating characteristics of a purely resistive circuit. |
| $f_o > f_r$ | $X_L > X_C$ | The circuit is inductive, meaning that it has the operating characteristics of a series *RL* circuit. |

We will discuss the effects of operating frequency on circuit operation in greater detail in Chapter 19. For now, we merely want to establish the fact that operating frequency is a determining factor in the operating characteristics of a series *RLC* circuit.

## Series Circuit Analysis

Until now, we have ignored the voltage relationships in series *RLC* circuits. As you were shown in Chapters 14 and 17, current is considered to be the 0° reference in any series circuit. Therefore, the voltage phase angles in a series *RLC* circuit can be determined as follows:

$$\begin{aligned} V_L &= I \cdot X_L \\ &= (I\angle 0°)(X_L\angle 90°) \\ &= (I \cdot X_L)\angle 90° \end{aligned} \qquad \begin{aligned} V_R &= I \cdot R \\ &= (I\angle 0°)(R\angle 0°) \\ &= (I \cdot R)\angle 0° \end{aligned} \qquad \begin{aligned} V_C &= I \cdot X_C \\ &= (I\angle 0°)(X_C\angle -90°) \\ &= (I \cdot X_C)\angle -90° \end{aligned}$$

These calculations indicate that the component voltages in a series *RLC* circuit have the phase relationship illustrated in Figure 18.36.

With the phase relationships shown, the net reactive voltage ($V_{LC}$) equals the difference between $V_L$ and $V_C$. By formula,

$$V_{LC} = V_L - V_C \tag{18.9}$$

As you were shown in Figure 18.4, the net reactive voltage falls on either the positive or negative *y*-axis. In either case, the sum of $V_{LC}$ and $V_R$ is found as the geometric sum of the two values. By formula:

$$V_S = \sqrt{V_{LC}^2 + V_R^2} \tag{18.10}$$

where $V_S$ = the source voltage
$V_{LC}$ = the net reactive voltage, found as $V_{LC} = V_L - V_C$
$V_R$ = the voltage across the resistor

As you can see, this equation is identical in form to equation (18.7), which was used to calculate the magnitude of series *RLC* circuit impedance. By the same token, the voltage phase angle (relative to the circuit current) in a series *RLC* circuit can be found as

$$\theta = \tan^{-1} \frac{V_{LC}}{V_R} \tag{18.11}$$

The process used to determine the magnitude and phase angle of the voltage source in a series *RLC* circuit is demonstrated in Example 18.18.

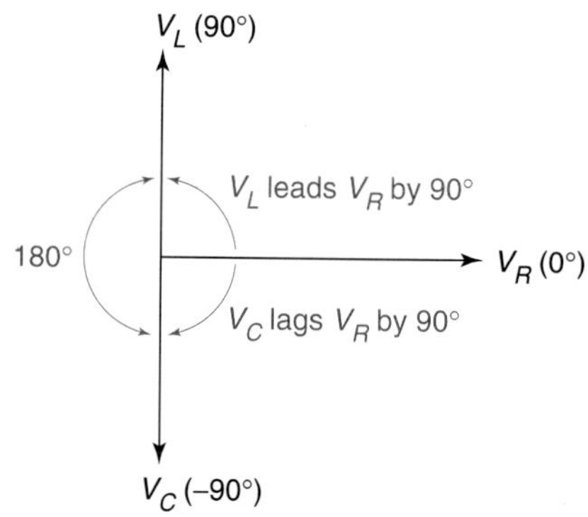


**FIGURE 18.36 Voltage phase relationships in a series *RLC* circuit.**

Determine the magnitude and phase angle of the source voltage in Figure 18.37a.

## EXAMPLE 18.18

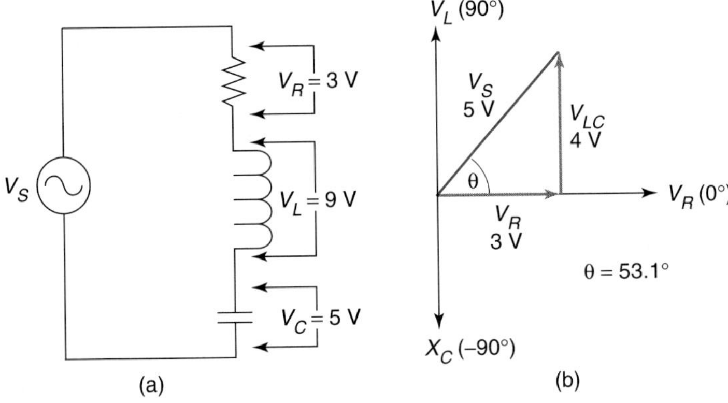


**FIGURE 18.37**

***Solution:*** First, the net reactive voltage is found as

$$V_{LC} = V_L - V_C = 9 \text{ V} - 5 \text{ V} = \mathbf{4 \text{ V}}$$

Now, the magnitude of the source voltage is found as

$$V_S = \sqrt{V_{LC}^2 + V_R^2} = \sqrt{(4 \text{ V})^2 + (3 \text{ V})^2} = \mathbf{5 \text{ V}}$$

Finally, the phase angle of the source voltage (relative to the circuit current) is found as

$$\theta = \tan^{-1}\frac{V_{LC}}{V_R} = \tan^{-1}\left(\frac{4 \text{ V}}{3 \text{ V}}\right) = \mathbf{53.1°}$$

***PRACTICE PROBLEM 18.18***

A circuit like the one in Figure 18.37a has the following values: $V_R = 20$ V, $V_L = 40$ V, and $V_C = 10$ V. Calculate the magnitude and phase angle of the source voltage.

The values given and calculated in Example 18.18 can be plotted as shown in Figure 18.37b. As you can see, the voltage phasors indicate that the circuit is inductive; that is, it has the phase relationships characteristic of a series *RL* circuit.

When $V_L < V_C$, a series *RLC* circuit has the phase relationships characteristic of a series *RC* circuit. This point is illustrated in Example 18.19.

Determine the magnitude and phase angle of the source voltage in Figure 18.38.

## EXAMPLE 18.19

**FIGURE 18.38**

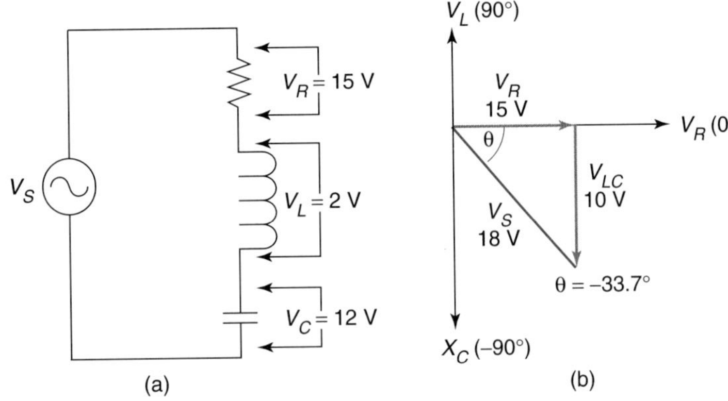

***Solution:*** First, the net reactive voltage is found as

$$V_{LC} = V_L - V_C = 2\text{ V} - 12\text{ V} = \mathbf{-10\text{ V}}$$

Now, the magnitude of the source voltage is found as

$$V_S = \sqrt{V_{LC}^2 + V_R^2} = \sqrt{(-10\text{ V})^2 + (15\text{ V})^2} = \mathbf{18\text{ V}}$$

Finally, the phase angle of the source voltage (relative to the circuit current) is found as

$$\theta = \tan^{-1}\frac{V_{LC}}{V_R} = \tan^{-1}\left(\frac{-10\text{ V}}{15\text{ V}}\right) = \mathbf{-33.7^\circ}$$

***PRACTICE PROBLEM 18.19***

A circuit like the one in Figure 18.38a has the following values: $V_R = 12$ V, $V_L = 4$ V, and $V_C = 22$ V. Calculate the magnitude and phase angle of the voltage source.

The given and calculated values in Example 18.19 can be plotted as shown in Figure 18.38b. As you can see, the voltage phasors indicate that the circuit is capacitive; that is, it has the phase relationships characteristic of a series *RC* circuit.

## Putting It All Together

The complete analysis of a series *RLC* circuit involves determining the values of reactance, impedance, current, and voltage. A complete analysis problem is provided in Example 18.20.

### EXAMPLE 18.20

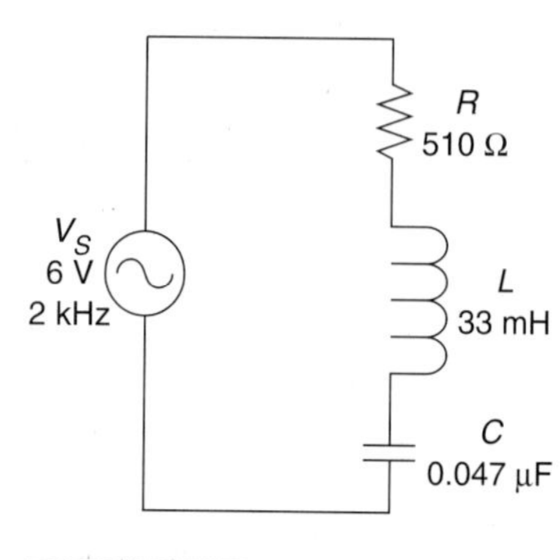

**FIGURE 18.39**

Perform a complete analysis of the circuit shown in Figure 18.39.

***Solution:*** At the operating frequency given, the value of inductive reactance can be found as

$$X_L = 2\pi fL = 2\pi(2\text{ kHz})(33\text{ mH}) = \mathbf{415\ \Omega}$$

and the value of capacitive reactance can be found as

$$X_C = \frac{1}{2\pi fC} = \frac{1}{2\pi(2\text{ kHz})(0.047\ \mu\text{F})} = \mathbf{1.69\text{ k}\Omega}$$

The net series reactance for the circuit can now be found as

$$X_S = X_L - X_C = 415\ \Omega - 1.69\text{ k}\Omega \cong \mathbf{-1.28\text{ k}\Omega}$$

The negative value of $X_S$ indicates that the circuit is capacitive. Therefore, the value of $X_S$ would be expressed more properly as $X_S = 1.28\text{ k}\Omega\angle -90^\circ$. Now, the magnitude of the circuit impedance can be found as

$$Z = \sqrt{X_S^2 + R^2} = \sqrt{(-1.28\text{ k}\Omega)^2 + (510\ \Omega)^2} = \mathbf{1.38\text{ k}\Omega}$$

and its phase angle (relative to the circuit current) can be found as

$$\theta = \tan^{-1}\frac{X_S}{R} = \tan^{-1}\left(\frac{-1.28\text{ k}\Omega}{510\ \Omega}\right) = \mathbf{-68.3^\circ}$$

Once we know the value of the circuit impedance, the magnitude of the circuit current can be found as

$$I = \frac{V_S}{Z} = \frac{6\text{ V}}{1.38\text{ k}\Omega} = \mathbf{4.35\text{ mA}\angle 0^\circ}$$

Note that we have assigned the circuit current a phase angle of 0°. This is always the case with a series *RLC* circuit.

Once the value of circuit current is known, the component voltages can be found as

$$\mathbf{V_R} = I \cdot R = (4.35 \text{ mA}\angle 0°)(510 \,\Omega\angle 0°) = \mathbf{2.22 \text{ V}\angle 0°}$$
$$\mathbf{V_L} = I \cdot X_L = (4.35 \text{ mA}\angle 0°)(415 \,\Omega\angle 90°) = \mathbf{1.81 \text{ V}\angle 90°}$$

and

$$V_C = I \cdot X_C = (4.35 \text{ mA})(1.69 \text{ k}\Omega\angle -90°) = \mathbf{7.35 \text{ V}\angle -90°}$$

To verify our calculated component voltages, we will use them to calculate the value of the source voltage. If our result matches the known source voltage, then our calculations are correct. First, the net reactive voltage is found as

$$V_{LC} = V_L - V_C = 1.81 \text{ V} - 7.35 \text{ V} = \mathbf{-5.54 \text{ V}}$$

This result indicates that the net reactive voltage is capacitive; that is, $V_{LC} = 5.54 \text{ V}\angle -90°$. Combining $V_{LC}$ with $V_R$, we get

$$V_S = \sqrt{V_{LC}^2 + V_R^2} = \sqrt{(-5.54 \text{ V})^2 + (2.22 \text{ V})^2} = \mathbf{5.97 \text{ V}}$$

Finally, the phase angle of the source voltage (relative to the circuit current) is found as

$$\theta = \tan^{-1}\frac{V_{LC}}{V_R} = \tan^{-1}\left(\frac{-5.54 \text{ V}}{2.22 \text{ V}}\right) = \mathbf{-68.2°}$$

The calculated source voltage (5.97 V) is approximately equal to the value given in the figure. Also, the phase angle calculated using the component voltages is approximately equal to the phase angle of the circuit impedance, as it should be for a series resistive-reactive circuit. (The differences are due to the rounding off of values throughout the calculations.) Therefore, we can assume that our calculations were correct.

***PRACTICE PROBLEM 18.20***

A circuit like the one in Figure 18.39 has the following values: $V_S = 9$ V, $f = 10$ kHz, $R = 1.5$ kΩ, $L = 22$ mH, and $C = 33$ nF. Perform a complete analysis of the circuit.

## Parallel RLC Circuits

*OBJECTIVE 12*

When a resistor, capacitor, and inductor are connected in parallel, the circuit has the characteristics of one of the following: a purely resistive circuit, a parallel *RL* circuit, or a parallel *RC* circuit. For example, look at the circuit in Figure 18.40. The overall characteristics of the circuit are determined by the relationship between $I_L$ and $I_C$, as given in Table 18.5.

The statements made in Table 18.5 are based on the phase relationships shown in Figure 18.41a. Since $I_L$ and $I_C$ are 180° out of phase, the net reactive current ($I_{LC}$) is equal to the difference between the two. This means that:

1. $I_{LC}$ parallels the positive *y*-axis when $I_L > I_C$ (Figure 18.41b). In this case, $I_{LC}$ is inductive, and the circuit has the characteristics of a parallel *RL* circuit.

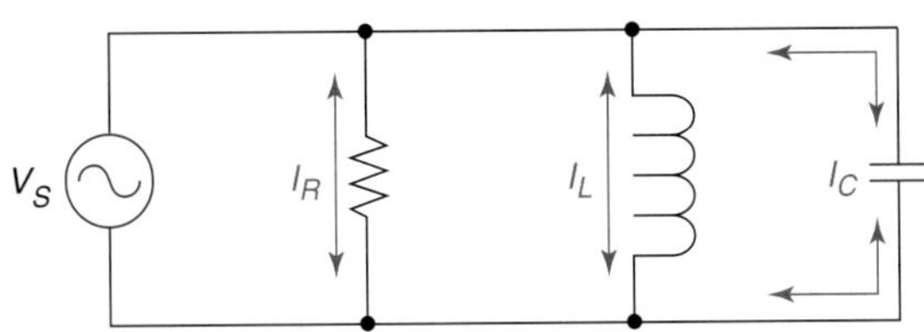


**FIGURE 18.40 Parallel *RLC* circuit currents.**

**TABLE 18.5 Reactive Current Versus Circuit Characteristics**

| *Current Relationship* | *Resulting Circuit Characteristics* |
|---|---|
| $I_L > I_C$ | The net reactive current is inductive, so the circuit has the characteristics of a parallel *RL* circuit: source voltage leads the circuit current and lags the circuit impedance. |
| $I_L = I_C$ | The resonant *LC* circuit has a net current of 0 A. Therefore, the circuit is resistive in nature: source voltage, current, and impedance are all in phase. |
| $I_L < I_C$ | The net reactive current is capacitive, so the circuit has the characteristics of a parallel *RC* circuit: source voltage lags the circuit current and leads the circuit impedance. |

2. $I_{LC}$ has a value of 0 A when $I_L = I_C$ (Figure 18.41c). In this case, the total circuit current equals the current through the resistor, meaning that the circuit is purely resistive.
3. $I_{LC}$ parallels the negative *y*-axis when $I_L < I_C$ (Figure 18.41d). In this case, $I_{LC}$ is capacitive, and the circuit has the characteristics of a parallel *RC* circuit.

These relationships will be demonstrated further in upcoming examples.

## Total Parallel Current

Figure 18.41b and d indicates that $I_{LC}$ is at a 90° angle to $I_R$ when its value is inductive or capacitive. In either case, the magnitude of the circuit current is found as the geometric sum of $I_{LC}$ and $I_R$. By formula,

$$I_T = \sqrt{I_{LC}^2 + I_R^2} \quad \textbf{(18.12)}$$

**FIGURE 18.41 The relationship among $I_L$, $I_C$, and $I_{LC}$.**

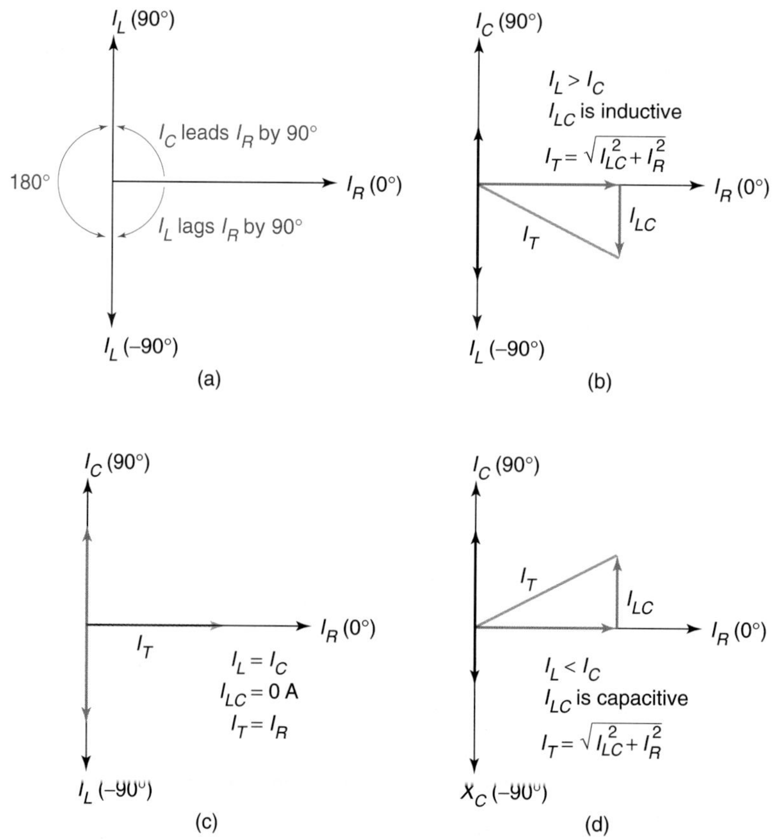

where $I_{LC}$ = the net reactive current, found as $I_{LC} = I_C - I_L$

The phase angle of the circuit current (relative to the source voltage) is found as

$$\theta = \tan^{-1}\frac{I_{LC}}{I_R} \qquad \textbf{(18.13)}$$

When $I_L > I_C$, the total current in a parallel *RLC* circuit has a negative phase angle, which indicates that the circuit has the characteristics of a parallel *RL* circuit. This point is demonstrated in Example 18.21.

*EXAMPLE 18.21*

Determine the magnitude and phase angle of the circuit current in Figure 18.42a.

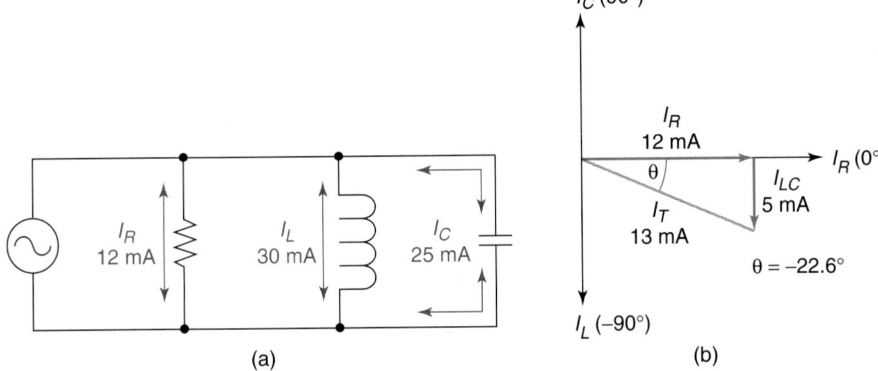


**FIGURE 18.42**

***Solution:*** First, the net reactive current is found as

$$I_{LC} = I_C - I_L = 25 \text{ mA} - 30 \text{ mA} = \mathbf{-5 \text{ mA}}$$

Now, the magnitude of the circuit current can be found as

$$I_T = \sqrt{I_{LC}^2 + I_R^2} = \sqrt{(-5 \text{ mA})^2 + (12 \text{ mA})^2} = \mathbf{13 \text{ mA}}$$

and its phase angle can be found as

$$\theta = \tan^{-1}\frac{I_{LC}}{I_R} = \tan^{-1}\left(\frac{-5 \text{ mA}}{12 \text{ mA}}\right) = \mathbf{-22.6°}$$

Using the values of $I_{LC}$, $I_R$, and $I_T$, the current triangle can be plotted as shown in Figure 18.42b. Comparing this current triangle with the one in Figure 18.41b demonstrates that the circuit is inductive; that is, it has the phase characteristics of a parallel *RL* circuit.

***PRACTICE PROBLEM 18.21***

A circuit like the one in Figure 18.42a has the following values: $I_R$ = 30 mA, $I_L$ = 50 mA, and $I_C$ = 10 mA. Calculate the magnitude and phase angle of the circuit current. Then plot the current triangle for the circuit.

When $I_L < I_C$, the circuit current in a parallel *RLC* circuit has a positive phase angle, which indicates that the circuit has the characteristics of a parallel *RC* circuit. This point is demonstrated in Example 18.22.

## EXAMPLE 18.22

Determine the magnitude and phase angle of the circuit current in Figure 18.43a.

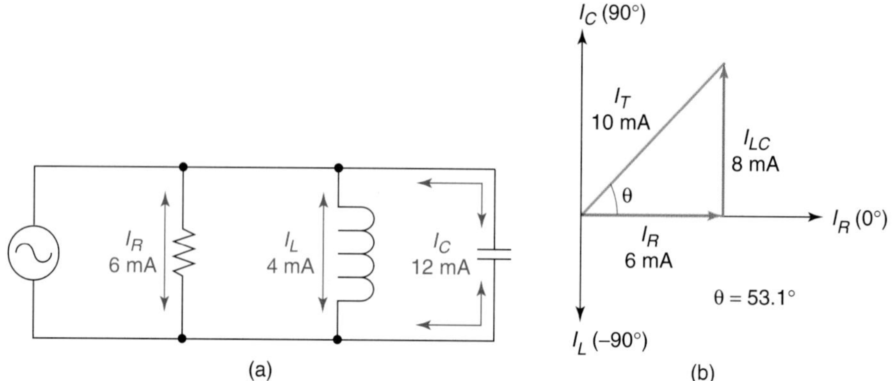


**FIGURE 18.43**

***Solution:*** First, the net reactive current is found as

$$I_{LC} = I_C - I_L = 12 \text{ mA} - 4 \text{ mA} = \mathbf{8\ mA}$$

Now, the magnitude of the circuit current can be found as

$$I_T = \sqrt{I_{LC}^2 + I_R^2} = \sqrt{(8 \text{ mA})^2 + (6 \text{ mA})^2} = \mathbf{10\ mA}$$

and its phase angle can be found as

$$\theta = \tan^{-1}\frac{I_{LC}}{I_R} = \tan^{-1}\left(\frac{8 \text{ mA}}{6 \text{ mA}}\right) = \mathbf{53.1°}$$

Using the values of $I_{LC}$, $I_R$, and $I_T$, the current triangle can be plotted as shown in Figure 18.43b. Comparing this current triangle with the one in Figure 18.41d demonstrates that the circuit is capacitive; that is, it has the phase characteristics of a parallel *RC* circuit.

***PRACTICE PROBLEM 18.22***

A circuit like the one in Figure 18.43a has the following values: $I_R = 12$ mA, $I_L = 40$ mA, and $I_C = 20$ mA. Calculate the magnitude and phase angle of the circuit current. Then plot the current triangle for the circuit.

When a parallel *RLC* circuit operates at resonance, the values of $I_L$ and $I_C$ are equal. Therefore, the net reactive current is 0 A. With a value of $I_{LC} = 0$ A, the circuit current is found as

$$I_T = \sqrt{I_{LC}^2 + I_R^2} = \sqrt{(0 \text{ A})^2 + I_R^2} = \sqrt{I_R^2} = I_R$$

and the circuit phase angle is found as

$$\theta = \tan^{-1}\frac{I_{LC}}{I_R} = \tan^{-1}\left(\frac{0 \text{ A}}{I_R}\right) = 0°$$

As these values indicate, a parallel *RLC* circuit operating at resonance has no phase angles and is purely resistive.

## Parallel Circuit Frequency Response

You have seen how changes in operating frequency affect the operation of a series *RLC* circuit. The effects of operating frequency on parallel *RLC* circuit operation can be described using the graphs shown in Figure 18.44. At first glance, the graphs seem identical to the reactance versus frequency curves discussed earlier in this section. However, if you look at the labels on the positive and negative *y*-axes, you'll see that the curves represent the relationship between component current and operating frequency (assuming that component voltage remains constant).

**FIGURE 18.44 Reactive branch current versus frequency.**

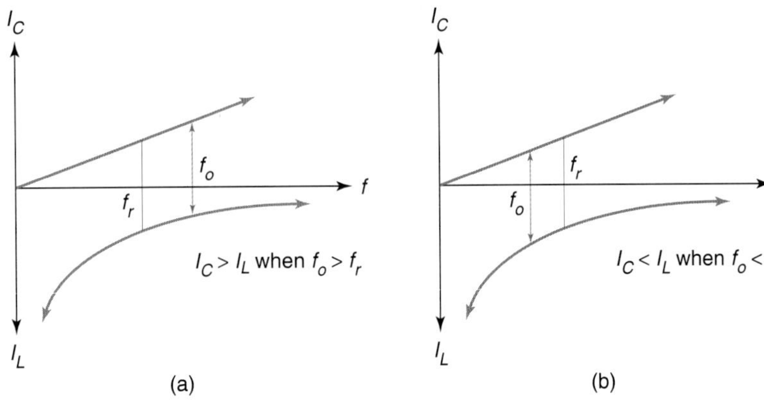


The graph in Figure 18.44a illustrates the relationship between $I_C$ and $I_L$ when the operating frequency ($f_o$) is greater than the circuit's resonant frequency. As shown in the graph, $I_C > I_L$ whenever $f_o > f_r$. This relationship makes sense, given the following:

1. An increase in operating frequency causes a decrease in $X_C$. Assuming that $V_C$ is constant, the decrease in $X_C$ results in an increase in capacitor current ($I_C$); that is, capacitor current varies directly with changes in operating frequency when voltage remains constant.
2. An increase in operating frequency causes an increase in $X_L$. Assuming that $V_L$ is constant, the increase in $X_L$ results in a decrease in inductor current ($I_L$); that is, inductor current varies inversely with changes in operating frequency when voltage remains constant.

Since $I_C = I_L$ at resonance, the above statements indicate that $I_C > I_L$ whenever $f_o > f_r$. This means that a parallel *RLC* circuit is capacitive when it is operated above its resonant frequency. By the same token, the graph in Figure 18.44b indicates that $I_C < I_L$ whenever $f_o < f_r$. This means that a parallel *RLC* circuit is inductive whenever it is operated below its resonant frequency. The relationships between operating frequency and the characteristics of a parallel *RLC* circuit are summarized in Table 18.6. We will discuss the frequency response of resistive-reactive circuits in greater detail in Chapter 19.

**TABLE 18.6 The Effects of Operating Frequency on Parallel *RLC* Circuit Characteristics**

| *Operating Frequency* | *Reactive Current* | *Other Characteristics* |
|---|---|---|
| $f_o > f_r$ | $I_C > I_L$ | The net reactive current is capacitive. As a result, the circuit has the operating characteristics of a parallel *RC* circuit. |
| $f_o = f_r$ | $I_C = I_L$ | The currents cancel, so the net reactive current is 0 A. As a result, $I_T = I_R$, meaning the circuit has the operating characteristics of a purely resistive circuit. |
| $f_o < f_r$ | $I_C < I_L$ | The net reactive current is inductive. As a result, the circuit has the operating characteristics of a parallel *RL* circuit. |

*A Practical Consideration:* If you compare Tables 18.6 and 18.4, you'll see that current *leads* voltage when $f_o < f_r$. For this reason, an *RLC* circuit operating *below* resonance is often referred to as a *lead circuit.* By the same token, current *lags* voltage when $f_o > f_r$. For this reason, an *RLC* circuit operating *above* resonance is often referred to as a *lag circuit.*

## *Parallel Circuit Analysis*

The analysis of a parallel *RLC* circuit normally begins with determining the values of the branch currents and the total circuit current. Then, using the values of $V_S$ and $I_T$, the circuit impedance is calculated. The analysis of a parallel *RLC* circuit is demonstrated in Example 18.23.

### EXAMPLE 18.23

Perform a complete analysis on the circuit shown in Figure 18.45.

EWB

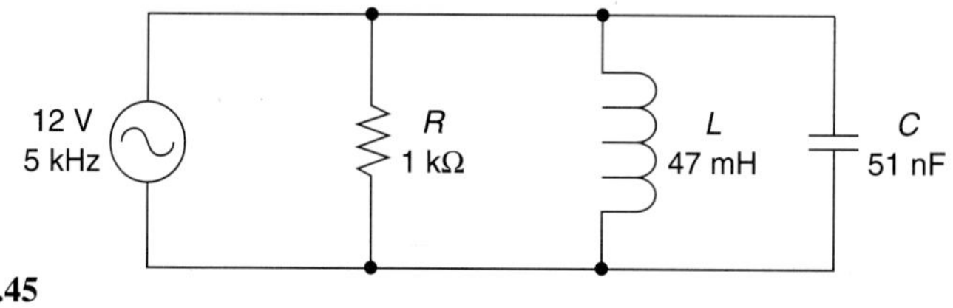

12 V
5 kHz
R
1 kΩ
L
47 mH
C
51 nF

**FIGURE 18.45**

***Solution:*** First, the circuit reactances are found as

$$X_L = 2\pi fL = 2\pi(5\text{ kHz})(47\text{ mH}) = \mathbf{1.48\ k\Omega}$$

and

$$X_C = \frac{1}{2\pi fC} = \frac{1}{2\pi(5\text{ kHz})(51\text{ nF})} = \mathbf{624\ \Omega}$$

The reactive currents can now be found as

$$I_L = \frac{V_S}{X_L} = \frac{12\text{ V}\angle 0^\circ}{1.48\text{ k}\Omega\angle 90^\circ} = \mathbf{8.11\ mA\angle{-90^\circ}}$$

and

$$I_C = \frac{V_S}{X_C} = \frac{12\text{ V}\angle 0^\circ}{624\ \Omega\angle -90^\circ} = \mathbf{19.23\ mA\angle 90^\circ}$$

Note that the phase angles were included in these calculations to demonstrate further the phase relationship between $I_C$ and $I_L$. Since the currents are 180° out of phase, the net reactive current is found as

$$I_{LC} = I_C - I_L = 19.23\text{ mA} - 8.11\text{ mA} = \mathbf{11.12\ mA}$$

The value of $I_R$ is found as

$$I_R = \frac{V_S}{R} = \frac{12\text{ V}}{1\text{ k}\Omega} = \mathbf{12\ mA}$$

Now that we know the values of the branch currents, the magnitude of the total circuit current can be found as

$$I_T = \sqrt{I_{LC}^2 + I_R^2} = \sqrt{(11.12\text{ mA})^2 + (12\text{ mA})^2} = \mathbf{16.36\ mA}$$

and its phase angle can be found as

$$\theta = \tan^{-1}\frac{I_{LC}}{I_R} = \tan^{-1}\left(\frac{11.12\text{ mA}}{12\text{ mA}}\right) = \mathbf{42.8^\circ}$$

Once the value of total circuit current is known, the circuit impedance can be found as

$$Z_T = \frac{V_S}{I_T} = \frac{12\text{ V}\angle 0^\circ}{16.36\text{ mA}\angle 42.8^\circ} = \mathbf{734\ \Omega\angle{-42.8^\circ}}$$

*PRACTICE PROBLEM 18.23*

A circuit like the one in Figure 18.45 has the following values: $V_S = 10$ V, $f = 8$ kHz, $R = 750\ \Omega$, $L = 10$ mH, and $C = 0.1\ \mu$F. Perform a complete analysis on the circuit.

## Power Calculations

We have ignored power calculations throughout this section because they haven't changed. Whether you are dealing with a series or parallel *RLC* circuit, power is calculated as it is for any resistive-reactive circuit. The values of apparent, reactive, and resistive power can be found using the following equations (which were introduced earlier in the text):

$$P_{APP} = \frac{V_S^2}{Z_T} \qquad P_R = P_{APP}\cos\theta \qquad P_X = P_{APP}\sin\theta$$

There are, of course, other relationships that can be used to calculate the power values in an *RLC* circuit. However, the ones listed here are probably the easiest to apply to most circuits.

### Section Review

1. Describe the phase relationships for a series *RLC* circuit under the following conditions:
   a. $X_L > X_C$
   b. $X_L = X_C$
   c. $X_L < X_C$
2. What values are used to plot the impedance triangle for a series *RLC* circuit?
3. Describe the phase relationships for a series *RLC* circuit under the following conditions:
   a. $f_o > f_r$
   b. $f_o = f_r$
   c. $f_o < f_r$
4. Describe the phase relationships for a parallel *RLC* circuit under the following conditions:
   a. $I_C > I_L$
   b. $I_C = I_L$
   c. $I_C < I_L$
5. What values are used to plot the current triangle for a parallel *RLC* circuit?
6. Describe the phase relationships for a parallel *RLC* circuit under the following conditions:
   a. $f_o > f_r$
   b. $f_o = f_r$
   c. $f_o < f_r$

# 18.5 SERIES-PARALLEL *RLC* CIRCUIT ANALYSIS

◄ *OBJECTIVE 13*

Series-parallel *RLC* circuits can be easy or difficult to analyze, depending on whether or not the values of *L* and *C* can be combined directly. For example, consider the circuits shown in Figure 18.46. Both of the series-parallel circuits shown contain an *LC* circuit that can be combined directly. In Figure 18.46a, the parallel *LC* circuit can be combined into a single parallel-equivalent reactance ($X_P$). The series combination of *R* and $X_P$ can be

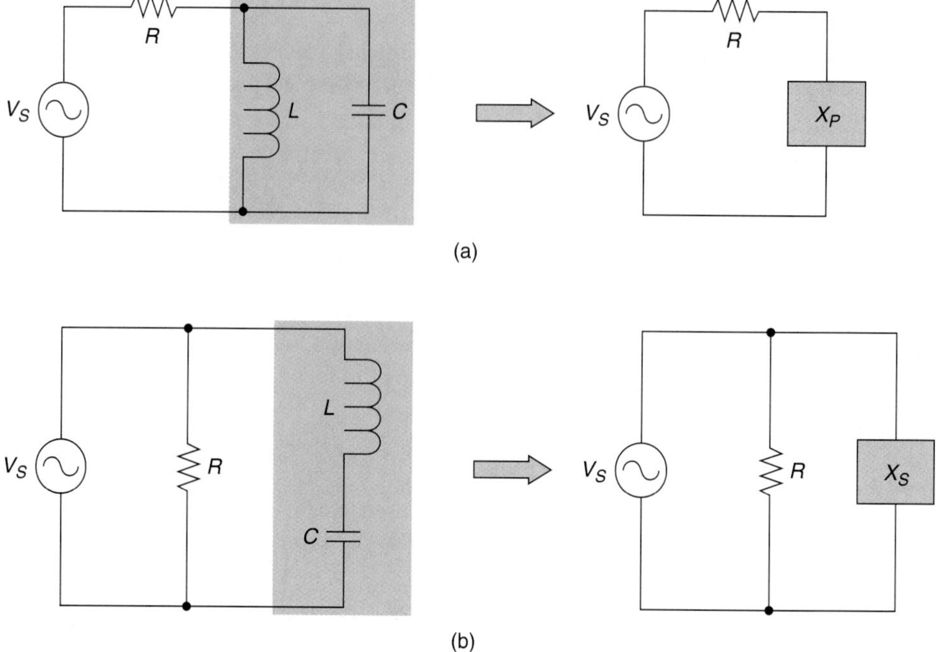


**FIGURE 18.46 Simple series-parallel *RLC* circuits.**

analyzed using the techniques that apply to a series *RC* (or *RL*) circuit. In Figure 18.46b, the series *LC* circuit can be combined into a single series-equivalent reactance ($X_S$). The parallel combination of *R* and $X_S$ can be analyzed using the techniques that apply to a parallel *RC* (or *RL*) circuit.

Series-parallel *RLC* circuit analysis becomes more involved when the reactive components cannot be combined directly. For example, consider the circuits shown in Figure 18.47. Because of the way the components are connected, the reactive components cannot

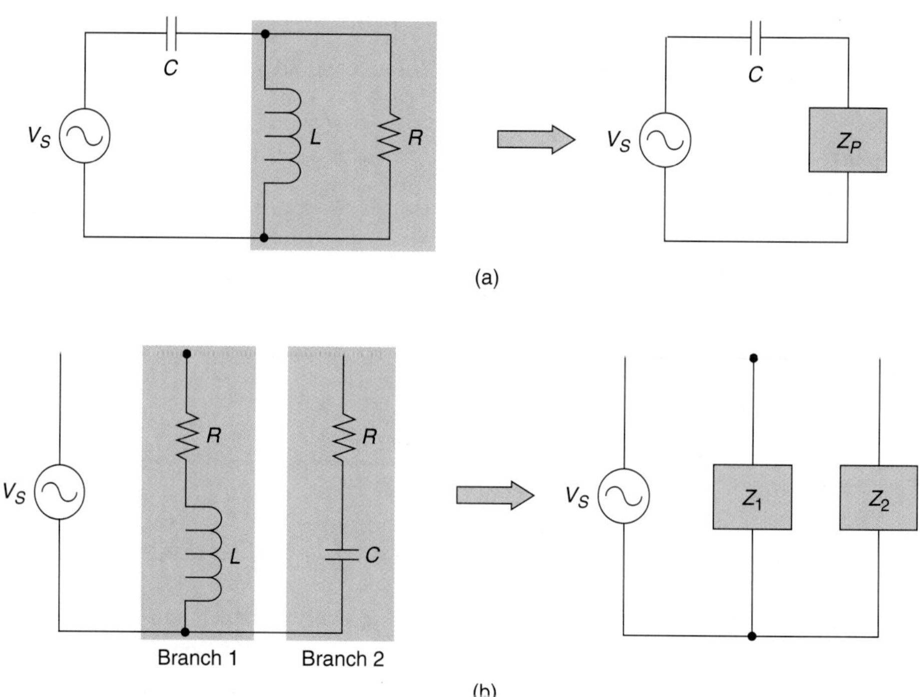


**FIGURE 18.47 Complex series-parallel *RLC* circuits.**

be combined directly. For the circuit in Figure 18.47a, the parallel *RL* circuit must be combined into a parallel equivalent ($Z_P$). Then the circuit can be treated as a series circuit. For the circuit in Figure 18.47b, each branch must be simplified into a single equivalent impedance. Then the circuit can be treated as a parallel circuit.

In this section, we will work through four circuit analysis problems. They will include circuits that are similar to those shown in Figures 18.46 and 18.47. While these circuits do not represent every possible series-parallel *RLC* combination, they provide a basis for analyzing most circuit configurations.

## EXAMPLE 18.24

Determine the impedance, current, and voltage values for the circuit in Figure 18.48a.

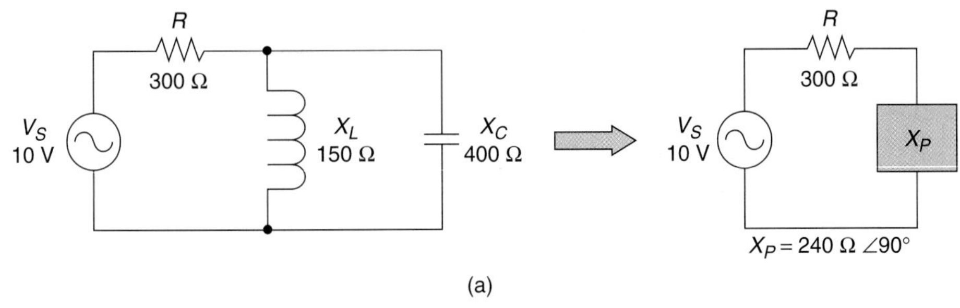


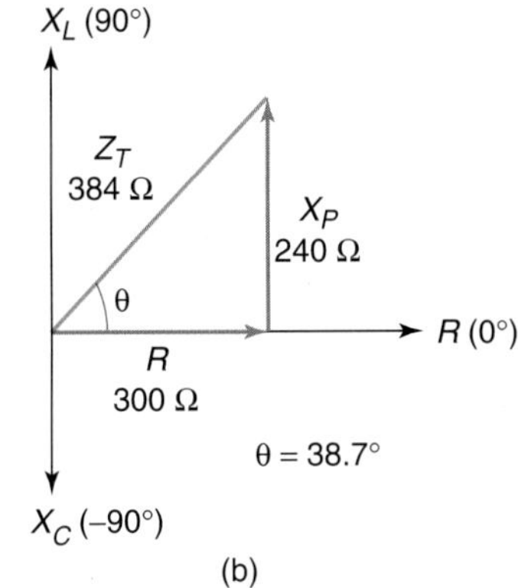


**FIGURE 18.48**

***Solution:*** The first step is to combine the reactances into a single parallel-equivalent value using equation (18.5), as follows:

$$X_P = \frac{X_L X_C}{X_L + X_C} = \frac{(150\ \Omega)(-400\ \Omega)}{150\ \Omega + (-400\ \Omega)} = \mathbf{240\ \Omega}$$

Since the value of $X_P$ is positive, it is inductive. Therefore, its value can be written as $X_P = 240\ \Omega\angle 90°$. This value is included in the series-equivalent circuit shown in Figure 18.48a. Using the values shown in the figure, the impedance triangle for the circuit can be plotted as shown in Figure 18.48b. The magnitude of the circuit impedance is found as

$$Z_T = \sqrt{X_P^2 + R^2} = \sqrt{(240\ \Omega)^2 + (300\ \Omega)^2} = \mathbf{384\ \Omega}$$

and its phase angle is found as

$$\theta = \tan^{-1}\frac{X_P}{R} = \tan^{-1}\left(\frac{240\ \Omega}{300\ \Omega}\right) = \mathbf{38.7°}$$

These values have been included on the impedance triangle in Figure 18.48b. Once the circuit impedance is known, the circuit current can be found as

$$\boldsymbol{I_T} = \frac{V_S}{Z_T} = \frac{10\text{ V}}{384\ \Omega\angle 38.7^\circ} = \mathbf{26\ mA\angle -38.7^\circ}$$

The phase angle in the result indicates that the circuit current lags the source voltage by 38.7°. Since we are dealing with a series-equivalent circuit, current is assumed to have a phase of 0°. Therefore, the values of circuit current and source voltage are written as

$$I_T = 26\text{ mA}\angle 0^\circ \qquad \text{and} \qquad V_S = 10\text{ V}\angle 38.7^\circ$$

Once the value of circuit current is known, the voltage across the parallel *LC* circuit can be found as

$$\boldsymbol{V_{LC}} = I_T X_P = (26\text{ mA}\angle 0^\circ)(240\ \Omega\angle 90^\circ) = \mathbf{6.24\ V\angle 90^\circ}$$

The phase angle of $V_{LC}$ makes sense when you remember the fact that $X_P$ is inductive. Therefore, the voltage across $X_P$ would have to lead current by 90°. Using the calculated value of $V_{LC}$, the values of $I_C$ and $I_L$ can be found as

$$\boldsymbol{I_C} = \frac{V_{LC}}{X_C} = \frac{6.24\text{ V}\angle 90^\circ}{400\ \Omega\angle -90^\circ} = \mathbf{15.6\ mA\angle 180^\circ}$$

and

$$\boldsymbol{I_L} = \frac{V_{LC}}{X_L} = \frac{6.24\text{ V}\angle 90^\circ}{150\ \Omega\angle 90^\circ} = \mathbf{41.6\ mA\angle 0^\circ}$$

The phase angles calculated for $I_C$ and $I_L$ indicate that these two currents are 180° out of phase, which is consistent with our discussions on parallel *LC* circuits. Since the reactive currents are 180° out of phase, the net reactive current is found as

$$\boldsymbol{I_{LC}} = I_C - I_L = 15.6\text{ mA} - 41.6\text{ mA} = \mathbf{-26\ mA}$$

The negative value of $I_{LC}$ indicates that the current lags $V_{LC}$ by 90°. This is consistent with the phase relationship calculated earlier.

Finally, the voltage across the resistor is found as

$$\boldsymbol{V_R} = I_T R = (26\text{ mA}\angle 0^\circ)(300\ \Omega) = \mathbf{7.8\ V\angle 0^\circ}$$

***PRACTICE PROBLEM 18.24***

A circuit like the one in Figure 18.48a has the following values: $V_S = 18$ V, $R = 1.2$ kΩ, $X_L = 3$ kΩ, and $X_C = 1$ kΩ. Determine the impedance, current, and voltage values for the circuit. Then plot its impedance triangle.

Early in Example 18.24, we determined that the source has a value of $V_S = 10\text{ V}\angle 38.7^\circ$. If our calculated values of $V_R$ and $V_{LC}$ are correct, we should be able to use the values to calculate the magnitude and phase angle of the source, as follows:

$$\begin{aligned} V_S &= \sqrt{V_{LC}^2 + V_R^2} \\ &= (6.24\text{ V})^2 + (7.8\text{ V})^2 \\ &\cong 10\text{ V} \end{aligned} \qquad \text{and} \qquad \begin{aligned} \theta &= \tan^{-1}\frac{V_{LC}}{V_R} \\ &= \tan^{-1}\frac{6.24\text{ V}}{7.8\text{ V}} \\ &= 38.7^\circ \end{aligned}$$

These values match those of the source voltage, so we know that the calculations made in the example are correct.

Determine the impedance, current, and voltage values for the circuit shown in Figure 18.49a.

**EXAMPLE 18.25**

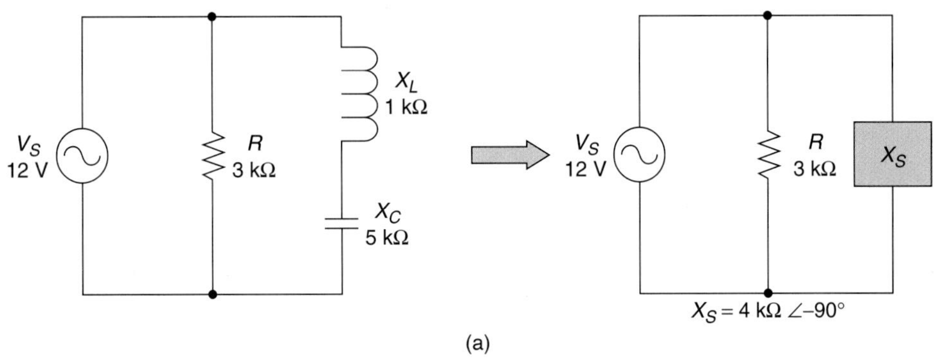

(a)

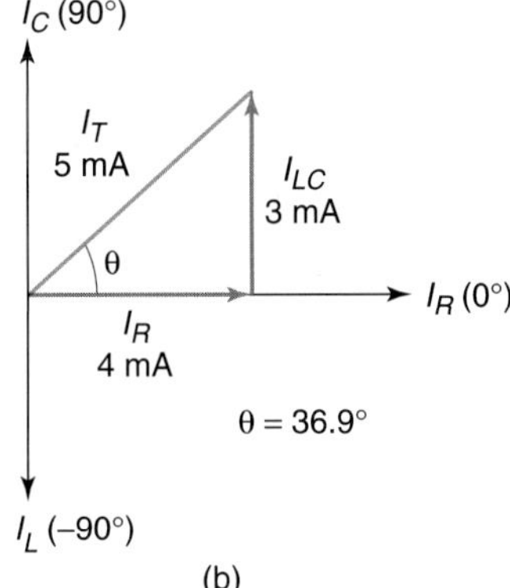

(b)

**FIGURE 18.49**

***Solution:*** First, the net series reactance ($X_S$) is found as

$$X_S = X_L - X_C = 1\text{ k}\Omega - 5\text{ k}\Omega = \mathbf{-4\text{ k}\Omega}$$

The negative value of $X_S$ indicates that it is capacitive. Therefore, the value of $X_S$ is written as $X_S = 4\text{ k}\Omega\angle -90°$. This value is included in the parallel-equivalent circuit in Figure 18.49a.

Now, the branch currents in the circuit can be found as follows:

$$I_R = \frac{V_S}{R} = \frac{12\text{ V}\angle 0°}{3\text{ k}\Omega} = \mathbf{4\text{ mA}\angle 0°}$$

and

$$I_{LC} = \frac{V_S}{X_S} = \frac{12\text{ V}\angle 0°}{4\text{ k}\Omega\angle -90°} = \mathbf{3\text{ mA}\angle 90°}$$

The phase angle of $I_{LC}$ indicates that it leads the source voltage by 90°. Using the calculated current values, the current triangle for the parallel circuit can be plotted as shown in Figure 18.49b. The magnitude of $I_T$ can be found as

$$I_T = \sqrt{I_{LC}^2 + I_R^2} = \sqrt{(3\text{ mA})^2 + (4\text{ mA})^2} = \mathbf{5\text{ mA}}$$

and its phase angle can be found as

$$\theta = \tan^{-1}\frac{I_{LC}}{I_R} = \tan^{-1}\left(\frac{3\text{ mA}}{4\text{ mA}}\right) = \mathbf{36.9^\circ}$$

With a net reactive current of 3 mA, the values of $V_L$ and $V_C$ can be found as

$$\mathbf{V_L} = I_{LC}X_L = (3\text{ mA}\angle 90^\circ)(1\text{ k}\Omega\angle 90^\circ) = \mathbf{3\text{ V}\angle 180^\circ}$$

and

$$\mathbf{V_C} = I_{LC}X_C = (3\text{ mA}\angle 90^\circ)(5\text{ k}\Omega\angle -90^\circ) = \mathbf{15\text{ V}\angle 0^\circ}$$

The phase angles calculated for $V_L$ and $V_C$ indicate that these voltages are 180° out of phase, which is consistent with our discussions on series *LC* circuits. Since the reactive voltages are 180° out of phase, the net reactive voltage ($V_{LC}$) is found as

$$\mathbf{V_{LC}} = V_L - V_C = 3\text{ V} - 15\text{ V} = \mathbf{-12\text{ V}}$$

The negative value of $V_{LC}$ indicates that this voltage lags the branch current by 90°. Note that $V_{LC} = V_S$, which is correct for a parallel circuit.

Finally, the magnitude of the circuit impedance is found as

$$\mathbf{Z_T} = \frac{V_S}{I_T} = \frac{12\text{ V}\angle 0^\circ}{5\text{ mA}\angle 36.9^\circ} = \mathbf{2.4\text{ k}\Omega\angle -36.9^\circ}$$

***PRACTICE PROBLEM 18.25***

A circuit like the one in Figure 18.49a has the following values: $V_S = 6$ V, $R = 560\ \Omega$, $X_L = 2\text{ k}\Omega$, and $X_C = 1.8\text{ k}\Omega$. Determine the impedance, current, and voltage values for the circuit. Then plot its current triangle.

In the last two examples, we were able to combine the *LC* circuit into a single equivalent reactance, and then use that reactance to analyze the circuit. When the values of inductance and capacitance cannot be combined directly, the analysis of the circuit gets more involved. This point is demonstrated in Example 18.26.

## EXAMPLE 18.26

Determine the values of impedance, current, and voltage for the circuit in Figure 18.50a.

***Solution:*** The first step is to combine $X_L$ and $R$ into a single parallel-equivalent impedance ($Z_P$). The magnitude of this impedance can be found using equation (14.18), as follows:

$$\mathbf{Z_P} = \frac{X_L R}{\sqrt{X_L^2 + R^2}} = \frac{(200\ \Omega)(300\ \Omega)}{\sqrt{(200\ \Omega)^2 + (300\ \Omega)^2}} = \mathbf{166\ \Omega}$$

and its phase angle can be found as

$$\theta = \tan^{-1}\frac{R}{X_L} = \tan^{-1}\left(\frac{300\ \Omega}{200\ \Omega}\right) = \mathbf{56.3^\circ}$$

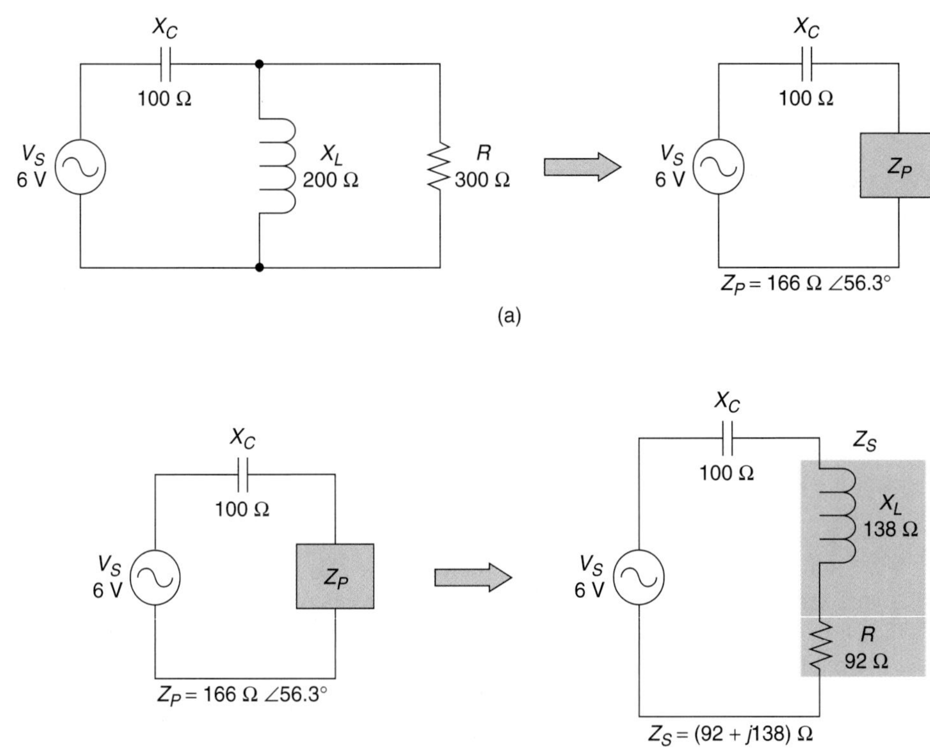


**FIGURE 18.50**

This value has been included in the series-equivalent circuit shown in Figure 18.50a. The next step is to convert $Z_P$ into a series-equivalent impedance. You may recall that this is accomplished by converting $Z_P$ to rectangular form, as follows:

$$X_S = Z_P \sin\theta = (166\ \Omega)(\sin 56.3°) = (166\ \Omega)(0.8320) = \mathbf{138\ \Omega}$$

and

$$R_S = Z_P \cos\theta = (166\ \Omega)(\cos 56.3°) = (166\ \Omega)(0.5548) = \mathbf{92\ \Omega}$$

Combining these two values, we have a series-equivalent impedance of $Z_S = (92 + j138)\ \Omega$. Since $X_S$ is positive, it is represented as an inductive reactance in the series-equivalent circuit (Figure 18.50b).

Once we have reduced the circuit to a simple series *RLC* equivalent, the circuit analysis proceeds using the techniques discussed earlier in this chapter. Using these techniques, the following values are calculated for the series-equivalent circuit:

$$X_{LC} = 38\ \Omega$$
$$Z_T = 100\ \Omega\angle 22.4°$$
$$I_T = 60\ \text{mA}\angle 0°$$
$$V_S = 6\ \text{V}\angle 22.4°$$
$$V_C = 6\ \text{V}\angle -90°$$
$$V_{LR} = 9.96\ \text{V}\angle 56.3°$$

Note that the phase angle of the source voltage ($V_S$) is a result of setting the current phase angle to 0°. The value of $V_{LR}$ represents the voltage across the series *RL* circuit, and therefore, across the original parallel *RL* circuit. Since we now know the voltage across the parallel *RL* circuit, the values of the branch currents can be found as

$$I_L = \frac{V_{LR}}{X_L} = \frac{9.96\ \text{V}\angle 56.3°}{200\ \Omega\angle 90°} = \mathbf{49.8\ mA\angle -33.7°}$$

and

$$I_R = \frac{V_{LR}}{R} = \frac{9.96 \text{ V}\angle 56.3^\circ}{300\ \Omega} = \mathbf{33.2\ mA\angle 56.3^\circ}$$

Since we had already calculated the current through, and voltage across, the capacitor, the analysis of the circuit is complete.

***PRACTICE PROBLEM 18.26***

A circuit like the one in Figure 18.50a has the following values: $V_S = 8$ V, $X_C = 300\ \Omega$, $X_L = 800\ \Omega$, and $R = 150\ \Omega$. Determine the values of impedance, current, and voltage for the circuit.

Before we start the last example problem, let's take a moment to look at the results from Example 18.26. To aid in our discussion, the calculated and given values are all shown in Figure 18.51. Note that $I_L$ lags $I_R$ by 90°, as is always the case for a parallel *RL* circuit. If we have calculated the values of $I_L$ and $I_R$ correctly, the sum of their values should equal the total circuit current. The sum of $I_L$ and $I_R$ can be found as follows:

$$\begin{aligned} I_L &= 49.8\text{ mA}\angle -33.7^\circ = 41.4 - j27.6\text{ mA} \\ I_R &= 33.2\text{ mA}\angle 56.3^\circ \ \ = \underline{18.4 + j27.6\text{ mA}} \\ & \qquad\qquad\qquad\qquad\quad 59.8 + j0 \quad \text{mA} \cong 60\text{ mA}\angle 0^\circ \end{aligned}$$

As you can see, the sum of the branch currents is approximately equal to the total current (as it should be). By the same token, the sum of $V_{LR}$ and $V_C$ should equal the source voltage, as follows:

$$\begin{aligned} V_{LR} &= 9.96\text{ V}\angle 56.3^\circ = 5.53 + j8.29\text{ V} \\ V_C &= 6\text{ V}\angle -90^\circ \quad\ = \underline{0.00 - j6.00\text{ V}} \\ & \qquad\qquad\qquad\qquad\ \ 5.53 + j2.29\text{ V} \cong 6\text{ V}\angle 22.4^\circ \end{aligned}$$

As you can see, the result of the calculation is approximately equal to the source voltage. Since all the values add up correctly, we can assume that all our calculations were correct.

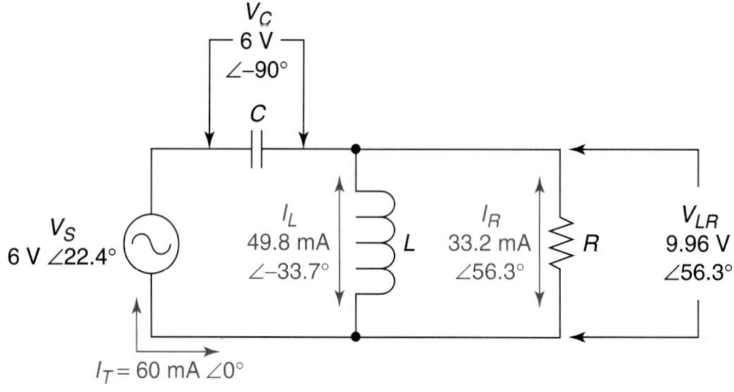


**FIGURE 18.51 The circuit values calculated in Example 18.26.**

Determine the values of impedance, current, and voltage for the circuit in Figure 18.52a.

**EXAMPLE 18.27**

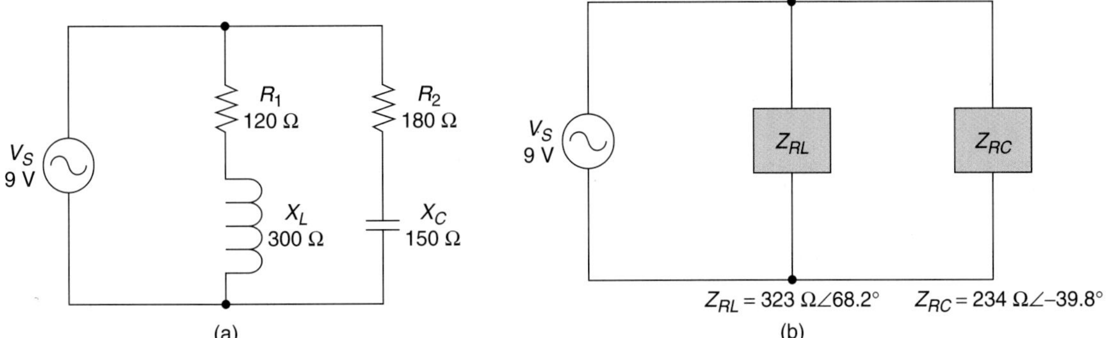


**FIGURE 18.52**

***Solution:*** The first step in solving the circuit is to combine each branch into a single equivalent impedance. The magnitude and phase angle of the *RL* branch impedance can be found as

$$Z_{RL} = \sqrt{X_L^2 + R_1^2} = \sqrt{(300\ \Omega)^2 + (120\ \Omega)^2} = \mathbf{323\ \Omega}$$

and

$$\theta_{RL} = \tan^{-1}\frac{X_L}{R_1} = \tan^{-1}\left(\frac{300\ \Omega}{120\ \Omega}\right) = \mathbf{68.2^\circ}$$

The magnitude and phase angle of the *RC* branch impedance can be found as

$$Z_{RC} = \sqrt{X_C^2 + R_2^2} = \sqrt{(150\ \Omega)^2 + (180\ \Omega)^2} = \mathbf{234\ \Omega}$$

and

$$\theta_{RC} = \tan^{-1}\frac{-X_C}{R_2} = \tan^{-1}\left(\frac{-150\ \Omega}{180\ \Omega}\right) = \mathbf{-39.8^\circ}$$

The branch impedance values are included in the parallel-equivalent circuit shown in Figure 18.52b. Using the values shown in the figure, the values of the branch currents can be found as

$$I_{RL} = \frac{V_S}{Z_{RL}} = \frac{9\ \text{V}\angle 0^\circ}{323\ \Omega\angle 68.2^\circ} = \mathbf{27.9\ mA\angle -68.2^\circ}$$

and

$$I_{RC} = \frac{V_S}{Z_{RC}} = \frac{9\ \text{V}\angle 0^\circ}{234\ \Omega\angle -39.8^\circ} = \mathbf{38.5\ mA\angle 39.8^\circ}$$

Once the values of the branch currents are known, the total current can be found as shown:

$$\begin{aligned} I_{RL} &= 27.9\ \text{mA}\angle -68.2^\circ = 10.4 - j25.9\ \text{mA} \\ I_{RC} &= 38.5\ \text{mA}\angle 39.8^\circ \ \ = \underline{29.6 + j24.6\ \text{mA}} \\ &\qquad\qquad\qquad\qquad\quad 40.0 - j1.3\ \ \text{mA} = 40\text{mA}\angle 1.86^\circ \end{aligned}$$

Using the total circuit current and source voltage, the circuit impedance can be found as

$$Z_T = \frac{V_S}{I_T} = \frac{9\ \text{V}\angle 0^\circ}{40\ \text{mA}\angle 1.86^\circ} = \mathbf{225\ \Omega\angle -1.86^\circ}$$

Now, we'll use the branch currents to solve for the component voltages. For the *RL* branch:

$$\mathbf{V_{R1}} = I_{RL}R_1 = (27.9 \text{ mA}\angle{-68.2°})(120 \ \Omega) = \mathbf{3.35 \ V\angle{-68.2°}}$$

and

$$V_L = I_{RL}X_L = (27.9 \text{ mA}\angle{-68.2°})(300 \ \Omega\angle 90°) = \mathbf{8.37 \ V\angle 21.8°}$$

For the *RC* branch:

$$\mathbf{V_{R2}} = I_{RC}R_2 = (38.5\angle 39.8°)(180 \ \Omega) = \mathbf{6.93 \ V\angle 39.8°}$$

and

$$V_C = I_{RC}X_C = (38.5 \text{ mA}\angle 39.8°)(150 \ \Omega\angle{-90°}) = \mathbf{5.78 \ V\angle{-50.2°}}$$

Now that we have calculated the component voltage values, the analysis of the circuit is complete.

***PRACTICE PROBLEM 18.27***

A circuit like the one in Figure 18.52a has the following values: $V_S = 15$ V, $R_1 = 300 \ \Omega$, $X_L = 200 \ \Omega$, $R_2 = 240 \ \Omega$, and $X_C = 500 \ \Omega$. Determine the values of impedance, current, and voltage for the circuit.

Now, let's look at the voltage values from Example 18.27. To aid in our discussion, the calculated and given voltages are shown in Figure 18.53. First, note that the component voltages in each branch are 90° out of phase, as they should be.

If the component voltages were calculated correctly, then the sum of the voltages in each branch will equal the source voltage. The sum of the *RL* branch voltages can be found as follows:

$$\begin{aligned} V_{R1} &= 3.35 \text{ V}\angle{-68.2°} = 1.24 - j3.11 \text{ V} \\ V_L &= 8.37 \text{ V}\angle 21.8° = \underline{7.77 + j3.11 \text{ V}} \\ & \qquad 9.01 + j0.00 \text{ V} \cong 9 \text{ V}\angle 0° \end{aligned}$$

The sum of the *RC* branch voltages can be found as follows:

$$\begin{aligned} V_{R2} &= 6.93 \text{ V}\angle 39.8° = 5.32 + j4.44 \text{ V} \\ V_C &= 5.78 \text{ V}\angle{-50.2°} = \underline{3.70 - j4.44 \text{ V}} \\ & \qquad 9.02 + j0.00 \text{ V} \cong 9 \text{ V}\angle 0° \end{aligned}$$

As you can see, the component voltages in each branch add up to equal the supply voltage. Therefore, we can assume that all our calculations were correct.

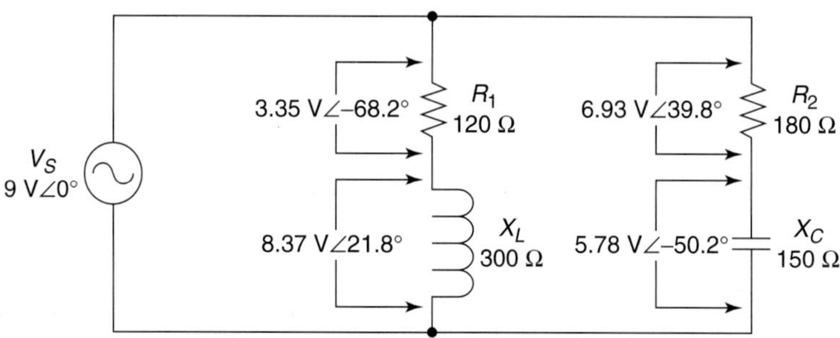


**FIGURE 18.53 The voltage values calculated in Example 18.27.**

## One Final Note

It would be impossible to show examples of every type of series-parallel *RLC* circuit. However, you have seen in this section that analyzing such a circuit is a matter of

1. Simplifying the circuit.
2. Solving for the overall circuit values.
3. Using the overall circuit values to solve for individual component values.

### Section Review

1. In your own words, describe the approach you would take to analyzing a series-parallel *RLC* circuit with values of *L* and *C* that can be combined directly.
2. In your own words, describe the approach you would take to analyzing a series-parallel *RLC* circuit with values of *L* and *C* that cannot be combined directly.

## Chapter Summary

Here is a summary of the major points made in this chapter:

1. When a capacitor and an inductor are connected in series:
   a. The component currents are equal.
   b. The component voltages are 180° out of phase.
2. The net voltage across a series *LC* circuit equals the difference between the component voltages.
3. A series *LC* circuit has the characteristics of a(n)
   a. Inductive circuit when $V_L > V_C$.
   b. Capacitive circuit when $V_C > V_L$.
   c. Short circuit when $V_L = V_C$.
4. When an inductor and capacitor are connected in series, the total series reactance equals the difference between $X_L$ and $X_C$.
5. For any series *LC* circuit, the total series reactance is always less than one (or both) of the component values.
6. A series *LC* circuit has the characteristics of a(n)
   a. Inductive circuit when $X_L > X_C$.
   b. Capacitive circuit when $X_C > X_L$.
   c. Short circuit when $X_L = X_C$.
7. When an inductor and capacitor are connected in parallel:
   a. The component voltages are equal.
   b. The component currents are 180° out of phase.
8. The net current through a parallel *LC* circuit equals the difference between $I_C$ and $I_L$.
9. The net current through a parallel *LC* circuit may be greater than one of the two branch current values.
10. A parallel *LC* circuit has the characteristics of a(n)
   a. Inductive circuit when $I_L > I_C$.
   b. Capacitive circuit when $I_C > I_L$.
   c. Open circuit when $I_L = I_C$.

11. A parallel *LC* circuit has the characteristics of a(n)
    a. Inductive circuit when $X_L < X_C$.
    b. Capacitive circuit when $X_C < X_L$.
    c. Open circuit when $X_L = X_C$.
12. An *LC* circuit with equal values of $X_L$ and $X_C$ is said to be operating at resonance.
13. The frequency at which resonance occurs for a given *LC* circuit is referred to as its resonant frequency ($f_r$).
14. Resonant frequency varies inversely with the values of *L* and *C*; that is, an increase in either *L* or *C* causes $f_r$ to decrease, and vice versa.
15. The resonant frequency of an *LC* circuit can be affected by:
    a. Stray inductance.
    b. Stray capacitance.
    c. Oscilloscope input capacitance when the circuit is under test.
16. When a series *LC* circuit is operating at resonance:
    a. The total series reactance of the *LC* circuit is 0 Ω.
    b. The voltage across the *LC* circuit is 0 V.
    c. The circuit voltage and current are in phase.
17. When a parallel *LC* circuit is operating at resonance:
    a. The sum of the component currents is 0 A.
    b. The circuit has infinite reactance; that is, it acts as an open.
18. Figure 18.31 provides a comparison of series and parallel resonant circuits.
19. A series *RLC* circuit has the characteristics of a(n)
    a. *RL* circuit when $V_L > V_C$.
    b. *RC* circuit when $V_C > V_L$.
    c. Purely resistive circuit when $V_L = V_C$.
20. A series *RLC* circuit has the characteristics of a(n)
    a. *RL* circuit when $X_L > X_C$.
    b. *RC* circuit when $X_C > X_L$.
    c. Purely resistive circuit when $X_L = X_C$.
21. A series *RLC* circuit has the characteristics of a(n)
    a. *RL* circuit when $f_o > f_r$.
    b. *RC* circuit when $f_o < f_r$.
    c. Purely resistive circuit when $f_o = f_r$.
22. A parallel *RLC* circuit has the characteristics of a(n)
    a. *RL* circuit when $I_L > I_C$.
    b. *RC* circuit when $I_C > I_L$.
    c. Purely resistive circuit when $I_L = I_C$.
23. A parallel *RLC* circuit has the characteristics of a(n)
    a. *RL* circuit when $f_o < f_r$.
    b. *RC* circuit when $f_o > f_r$.
    c. Purely resistive circuit when $f_o = f_r$.

## Equation Summary

| Equation Number | Equation | Section Number |
|---|---|---|
| (18.1) | $V_S = V_L - V_C$ | 18.1 |
| (18.2) | $X_S = X_L - X_C$ | 18.1 |
| (18.3) | $I_T = I_C - I_L$ | 18.2 |
| (18.4) | $X_P = \dfrac{1}{\dfrac{1}{X_L} + \dfrac{1}{X_C}}$ | 18.2 |
| (18.5) | $X_P = \dfrac{X_L X_C}{X_L + X_C}$ | 18.2 |
| (18.6) | $f_r = \dfrac{1}{2\pi\sqrt{LC}}$ | 18.3 |
| (18.7) | $Z_T = \sqrt{X_S^2 + R^2}$ | 18.4 |
| (18.8) | $\theta = \tan^{-1}\dfrac{X_S}{R}$ | 18.4 |
| (18.9) | $V_{LC} = V_L - V_C$ | 18.4 |
| (18.10) | $V_S = \sqrt{V_{LC}^2 + V_R^2}$ | 18.4 |
| (18.11) | $\theta = \tan^{-1}\dfrac{V_{LC}}{V_R}$ | 18.4 |
| (18.12) | $I_T = \sqrt{I_{LC}^2 + I_R^2}$ | 18.4 |
| (18.13) | $\theta = \tan^{-1}\dfrac{I_{LC}}{I_R}$ | 18.4 |

## Key Terms

The following terms were introduced and defined in this chapter:

resonance
resonant frequency ($f_r$)
stray capacitance
stray inductance

## Practice Problems

1. A series *LC* circuit has values of $V_L = 10$ V and $V_C = 3$ V. Determine the magnitude and phase angle of the voltage source.
2. A series *LC* circuit has values of $V_L = 6$ V and $V_C = 15$ V. Determine the magnitude and phase angle of the voltage source.
3. Determine the magnitude and phase angle of $X_S$ for the circuit in Figure 18.54a.
4. Determine the magnitude and phase angle of $X_S$ for the circuit in Figure 18.54b.
5. Determine the magnitude and phase angle of $X_S$ for the circuit in Figure 18.55a.
6. Determine the magnitude and phase angle of $X_S$ for the circuit in Figure 18.55b.
7. Determine the values of $V_L$, $V_C$, and $V_S$ for the circuit in Figure 18.56a.
8. Determine the values of $V_L$, $V_C$, and $V_S$ for the circuit in Figure 18.56b.
9. Determine the values of $V_L$, $V_C$, and $V_S$ for the circuit in Figure 18.57a.
10. Determine the values of $V_L$, $V_C$, and $V_S$ for the circuit in Figure 18.57b.

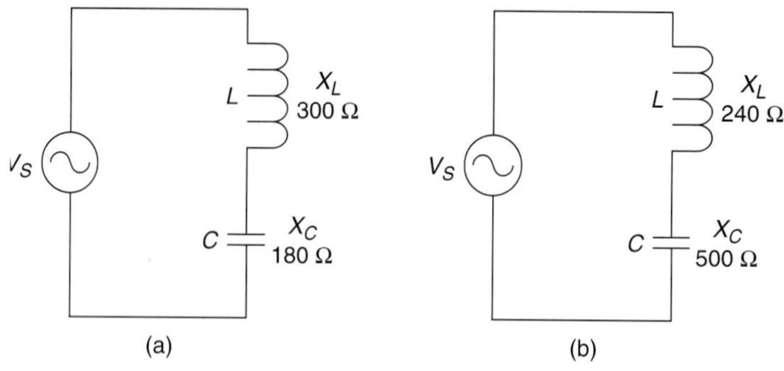


**FIGURE 18.54**

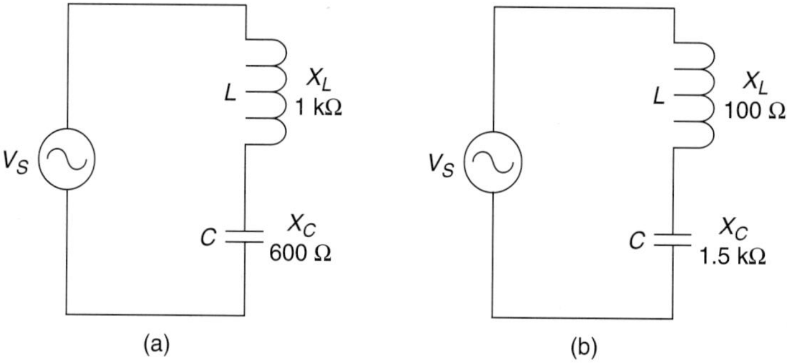


**FIGURE 18.55**

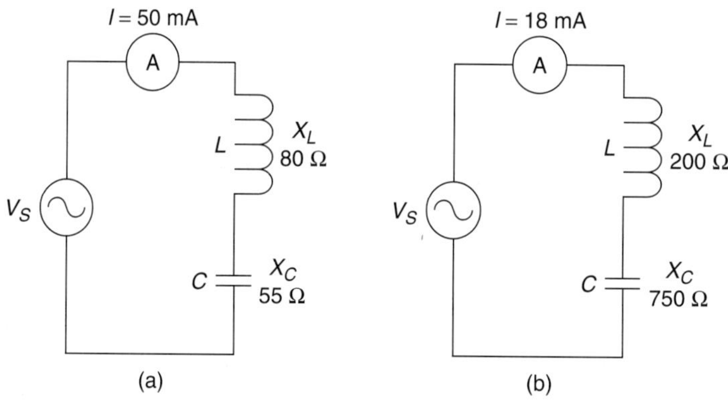


**FIGURE 18.56**

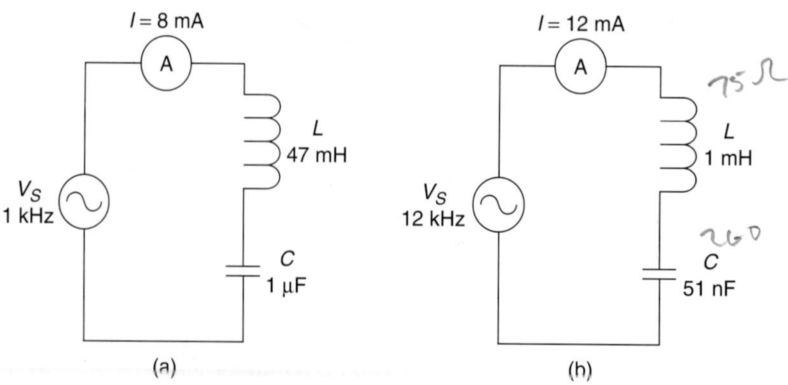


**FIGURE 18.57**

FIGURE 18.58

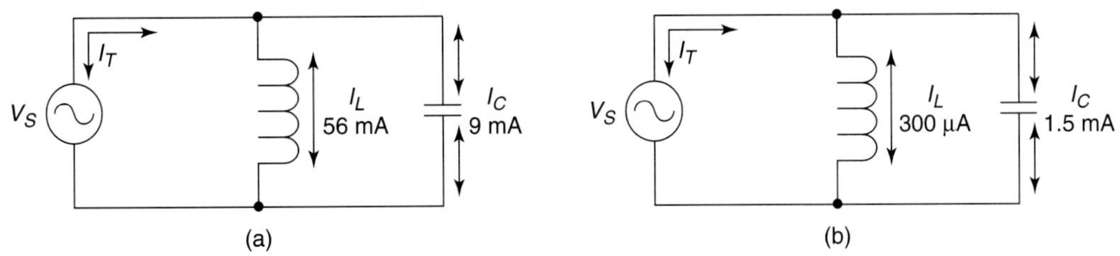

FIGURE 18.59

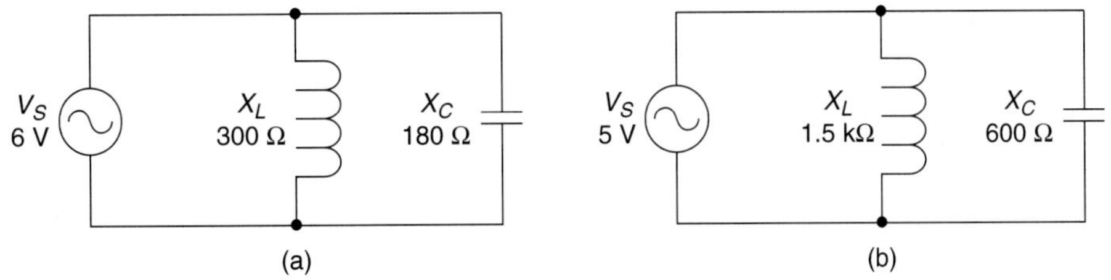

11. Determine the magnitude and phase angle of the circuit current in Figure 18.58a.
12. Determine the magnitude and phase angle of the circuit current in Figure 18.58b.
13. Determine the parallel-equivalent reactance for the circuit in Figure 18.59a.
14. Determine the parallel-equivalent reactance for the circuit in Figure 18.59b.
15. Determine the resonant frequency for the circuit in Figure 18.60a.
16. Determine the resonant frequency for the circuit in Figure 18.60b.
17. Calculate the values of $X_L$ and $X_C$ for the circuit in Figure 18.57a when it is operating at its resonant frequency.
18. Calculate the values of $X_L$ and $X_C$ for the circuit in Figure 18.57b when it is operating at its resonant frequency.
19. Calculate the values of $V_L$, $V_C$, and $V_{LC}$ for the circuit shown in Figure 18.61a.
20. Calculate the values of $V_L$, $V_C$, and $V_{LC}$ for the circuit shown in Figure 18.61b.

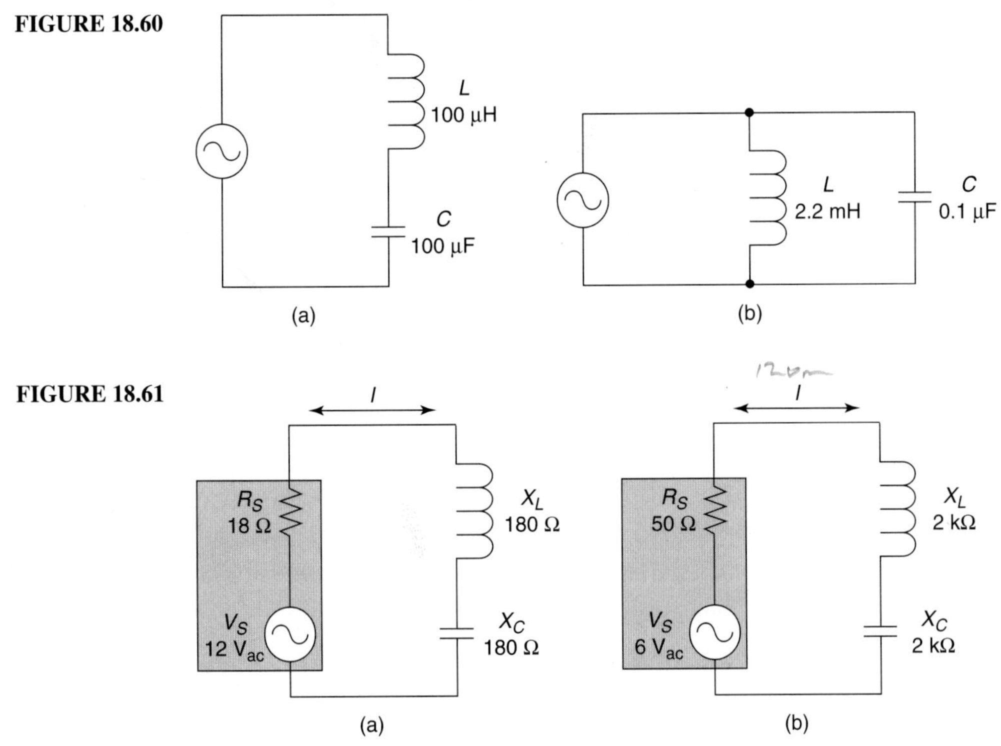


FIGURE 18.60

FIGURE 18.61

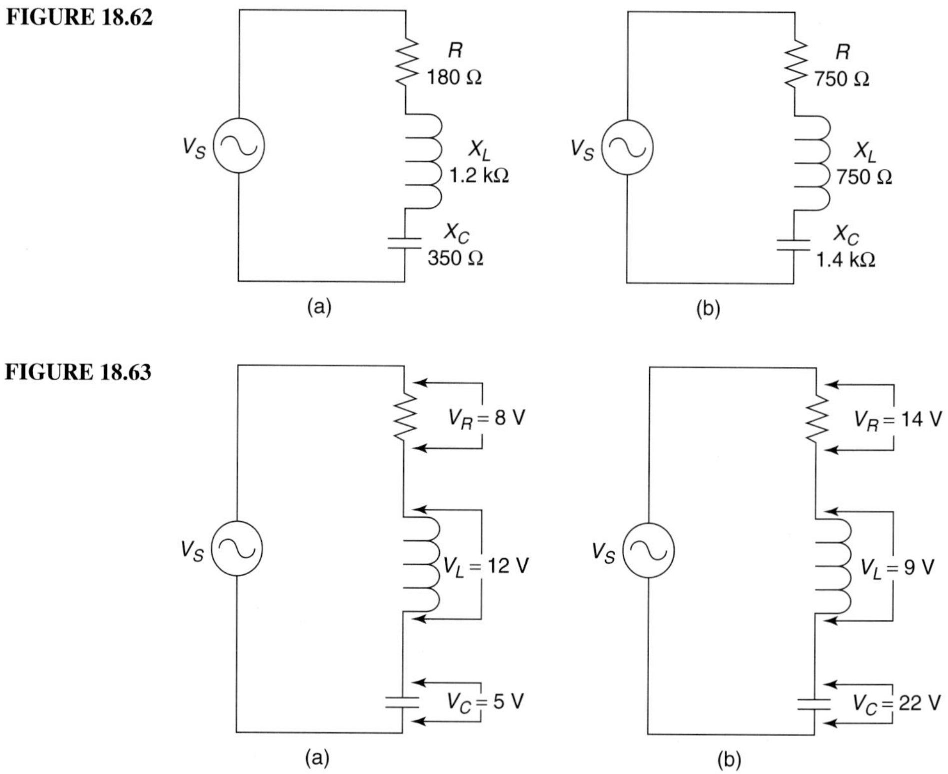


**FIGURE 18.62**

**FIGURE 18.63**

**21.** Determine the magnitude and phase angle of the circuit impedance in Figure 18.62a.

**22.** Determine the magnitude and phase angle of the circuit impedance in Figure 18.62b.

**23.** Determine the magnitude and phase angle of the source voltage in Figure 18.63a.

**24.** Determine the magnitude and phase angle of the source voltage in Figure 18.63b.

**25.** Perform a complete analysis of the circuit shown in Figure 18.64a.

**26.** Perform a complete analysis of the circuit shown in Figure 18.64b.

**27.** Determine the magnitude and phase angle of the circuit current in Figure 18.65a.

**28.** Determine the magnitude and phase angle of the circuit current in Figure 18.65b.

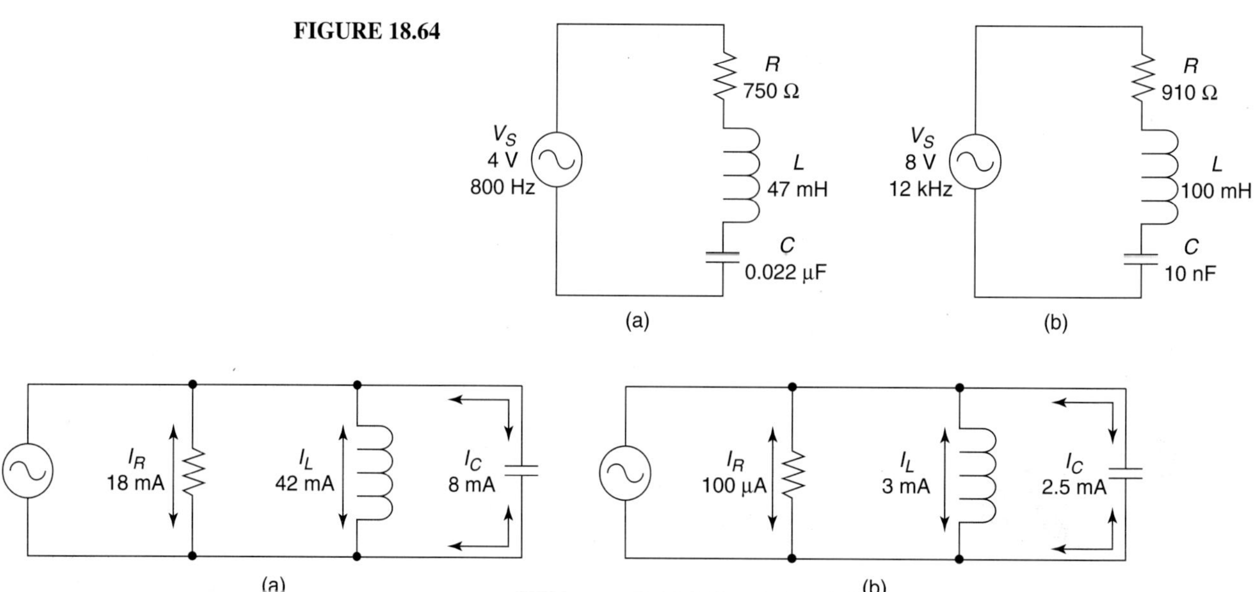


**FIGURE 18.64**

**FIGURE 18.65**

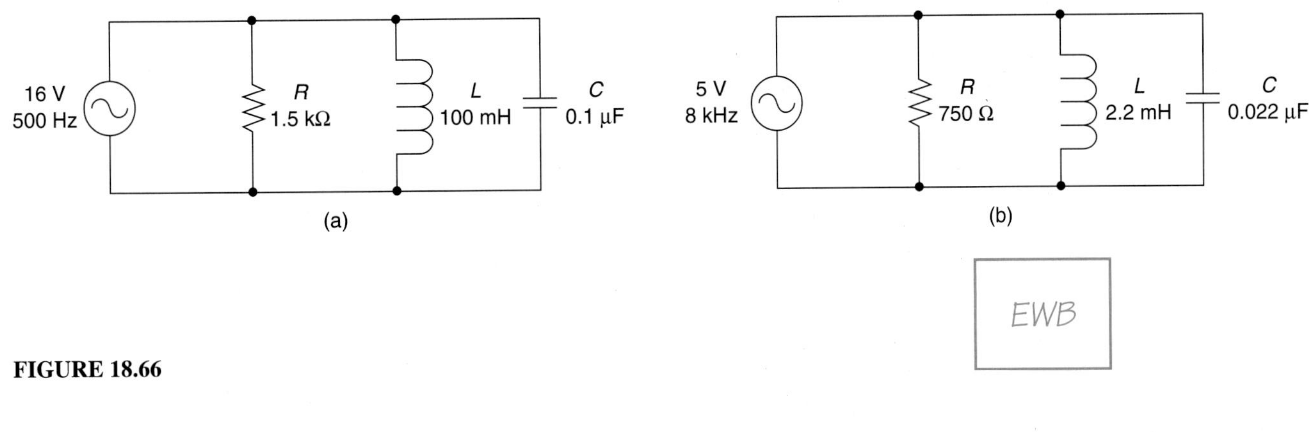


FIGURE 18.66

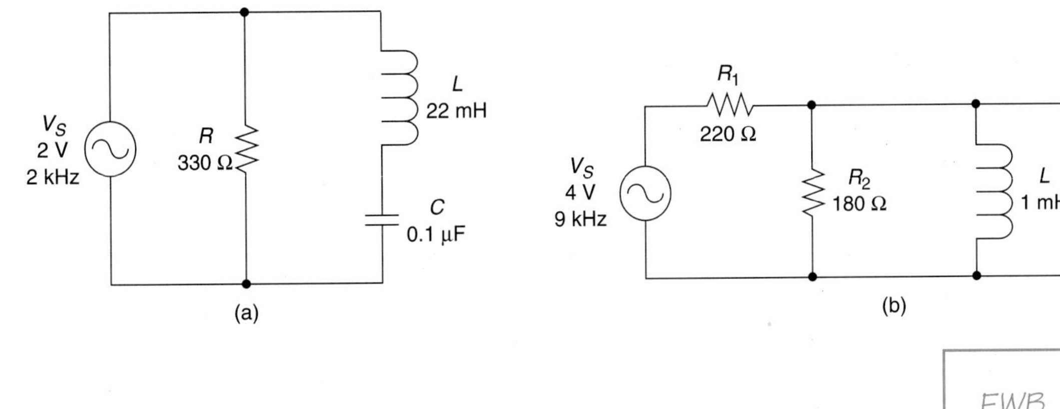


EWB

FIGURE 18.67

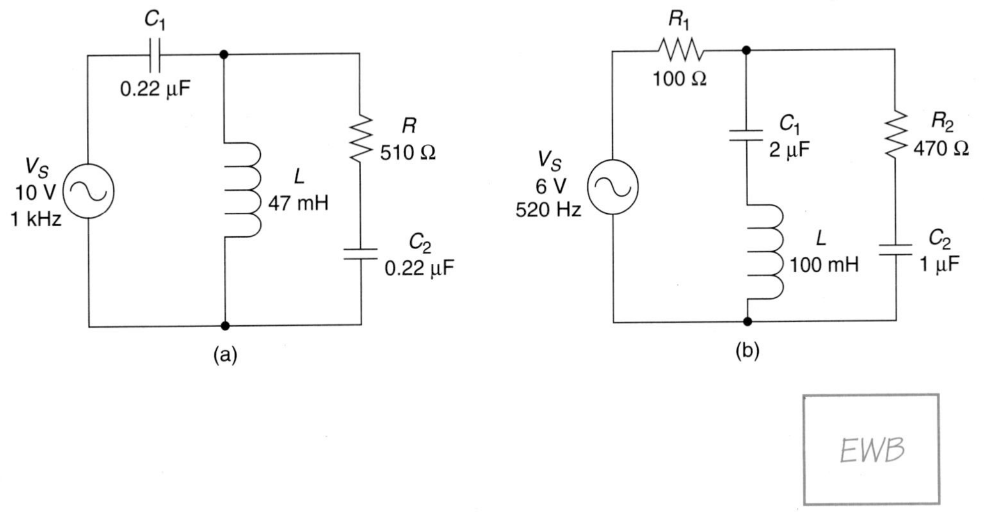


FIGURE 18.68

**29.** Perform a complete analysis of the circuit shown in Figure 18.66a.

**30.** Perform a complete analysis of the circuit shown in Figure 18.66b.

**31.** Determine the impedance, current, and voltage values for the circuit in Figure 18.67a.

**32.** Determine the impedance, current, and voltage values for the circuit in Figure 18.67b.

**33.** Determine the impedance, current, and voltage values for the circuit in Figure 18.68a.

**34.** Determine the impedance, current, and voltage values for the circuit in Figure 18.68b.

## Answers to the Example Practice Problems

**18.1** 500 mV$\angle -90°$
**18.3** 25 $\Omega\angle 90°$
**18.5** $V_L = 110$ V$\angle 90°$, $V_C = 40$ V$\angle -90°$, $V_S = 70$ V$\angle 90°$
**18.6** $V_L = 4.95$ V$\angle 90°$, $V_C = 14.85$ V$\angle -90°$, $V_S = 9.9$ V$\angle -90°$
**18.7** 45 mA$\angle 90°$
**18.9** 500 $\Omega$
**18.11** 5.91 kHz
**18.12** $f_r = 70.5$ kHz, $X_L = 44.3\ \Omega$, $X_C = 44.3\ \Omega$
**18.13** 87.61 kHz; $f_r$ increased as $C$ decreased.
**18.14** $V_L = 200$ V$\angle 90°$, $V_C = 200$ V$\angle -90°$, $(V_L + V_C) = 0$ V
**18.15** $I_C = 250$ mA$\angle 90°$, $I_L = 250$ mA$\angle -90°$, $I_T = 0$ A, $X_T \rightarrow \infty\ \Omega$
**18.16** 408 $\Omega\angle 36°$
**18.17** 1.56 k$\Omega\angle -50.2°$
**18.18** 36.1 V$\angle 56.3°$
**18.19** 21.6 V$\angle 56.3°$
**18.20** $X_L = 1.38$ k$\Omega$, $X_C = 482\ \Omega$, $X_S = 900\ \Omega$, $Z = 1.75$ k$\Omega\angle 31°$, $I_T = 5.15$ mA, $V_R = 7.72$ V, $V_L = 7.11$ V, $V_C = 2.48$ V, $V_{LC} = 4.63$ V, $V_S = 9$ V$\angle 31°$
**18.21** 50 mA$\angle -53.1°$
**18.22** 34.4 mA$\angle 54.5°$
**18.23** $X_L = 503\ \Omega$, $I_L = 19.9$ mA$\angle -90°$, $X_C = 199\ \Omega$, $I_C = 50.2$ mA$\angle 90°$, $I_{LC} = 30.4$ mA$\angle 90°$, $I_R = 13.3$ mA, $I_T = 33.2$ mA$\angle 66.3°$, $Z_T = 302\ \Omega\angle -66.3°$
**18.24** $X_P = 1.5$ k$\Omega\angle -90°$, $Z_T = 1.92$ k$\Omega\angle -51.3°$, $I_T = 9.38$ mA, $V_S = 18$ V$\angle -51.3°$, $V_{LC} = 14.1$ V$\angle -90°$, $I_L = 4.69$ mA$\angle -180°$, $I_C = 14.1$ mA$\angle 0°$, $I_{LC} = 9.37$ mA, $V_R = 11.3$ V
**18.25** $I_{LC} = 30$ mA$\angle -90°$, $I_R = 10.7$ mA$\angle 0°$, $I_T = 31.9$ mA$\angle -70.4°$, $V_L = 60$ V$\angle 0°$, $V_C = 54$ V$\angle -180°$, $V_{LC} = 6$ V, $Z_T = 188\ \Omega\angle 70.4°$
**18.26** $Z_P = 147\ \Omega\angle 79.4°$, $X_S = 145\ \Omega$, $R_S = 27.2\ \Omega$, $X_{LC} = 155\ \Omega\angle -90°$, $Z_T = 157\ \Omega\angle -80.1°$, $I_T = 50.8$ mA, $V_S = 8$ V$\angle -80.1°$, $V_C = 15.2$ V$\angle -90°$, $V_{LR} = 7.94$ V$\angle 79.4°$, $I_L = 9.36$ mA$\angle -10.6°$, $I_R = 49.9$ mA$\angle 79.4°$
**18.27** $Z_{RL} = 361\ \Omega\angle 33.7°$, $Z_{RC} = 555\ \Omega\angle -64.4°$, $I_{RL} = 41.6$ mA$\angle -33.7°$, $I_{RC} = 27.1$ mA$\angle 64.4°$, $I_T = 46.4$ mA$\angle 1.62°$, $Z_T = 323\ \Omega\angle -1.62°$, $V_{R1} =$ 12.5 V$\angle -33.7°$, $V_{R2} = 6.49$ V$\angle 64.4°$, $V_L = 8.32$ V$\angle 56.3°$, $V_C =$ 13.52 V$\angle -25.6°$

## EWB Applications Problems

| Figure | EWB File Reference |
|---|---|
| 18.39 | EWBA18_39.ewb |
| 18.45 | EWBA18_45.ewb |
| 18.66 | EWBA18_66.ewb |
| 18.67 | EWBA18_67.ewb |
| 18.68 | EWBA18_68.ewb |

# 19

# FREQUENCY RESPONSE AND PASSIVE FILTERS

## OBJECTIVES

*After studying the material in this chapter, you should be able to:*

1. Describe attenuation.
2. Describe the relationship between circuit amplitude and cutoff frequency.
3. List the four primary filters and identify their frequency response curves.
4. Identify (and solve for) each of the following for a bandpass (or notch) filter: bandwidth, center frequency, lower cutoff frequency, and upper cutoff frequency.
5. Discuss the relationship among filter *Q*, bandwidth, and center frequency.
6. Describe and calculate the average frequency of a bandpass (or notch) filter.
7. Describe the relationship among filter *Q*, center frequency, and average frequency.
8. List and describe the commonly used logarithmic frequency scales.
9. Represent any power value in bel (B) or decibel (dB) form.
10. Convert dB power gains to standard numeric form.
11. List and discuss the reasons that gain is commonly represented in dB form.
12. Represent any voltage gain value in dB form.
13. Convert dB voltage gains to standard numeric form.
14. Convert current gain from standard numeric form to dB form, and vice versa.
15. Describe and analyze the operation of the *RC* low-pass filter.
16. Compare and contrast Bode plots with frequency response curves.
17. List the standard roll-off rates for *RC* filters.
18. Describe and analyze the operation of the *RL* low-pass filter.
19. Describe and analyze the operation of the *RC* high-pass filter.
20. Describe and analyze the operation of the *RL* high-pass filter.
21. Describe and analyze the operation of series *LC* bandpass filters.
22. Describe and analyze the operation of shunt *LC* bandpass filters.
23. Describe the operation of *LC* notch filters.

We have touched on the topic of frequency response throughout the past five chapters. Now, we will take a more in-depth look at frequency response and the operating characteristics of a group of circuits called *passive filters.* As you will learn, these circuits are designed for specific frequency response characteristics.

## 19.1 FREQUENCY RESPONSE: CURVES AND MEASUREMENTS

The subject of frequency response is complex, involving many new principles and concepts. In this section, we will begin by touching on several of the most general principles. Most of the topics covered in this section are discussed in more detail later in the chapter. For now, we are simply establishing a starting point.

### Attenuation

*OBJECTIVE 1* ► Every circuit responds to a change in operating frequency to some extent. The changes may be too subtle to observe, or they may be dramatic. An example of the latter is illustrated in Figure 19.1.

The circuit represented in Figure 19.1a has input and output waveforms that are nearly identical. When the input frequency is increased by a factor of ten (as shown in Figure 19.1b), the amplitude of the circuit output drops significantly. If frequency doubles again (to a value of $20f$), the amplitude of the output decreases even more. In fact, further increases in operating frequency eventually cause the output to drop out completely.

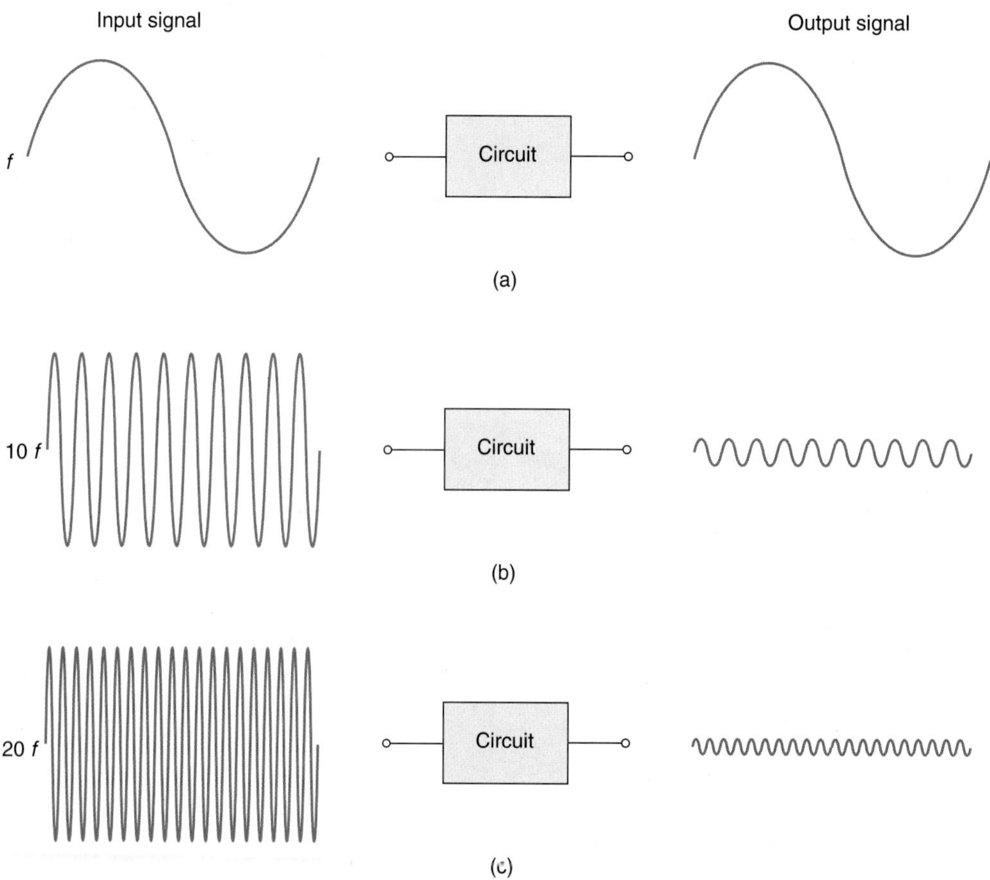


**FIGURE 19.1 Attenuation.**

**FIGURE 19.2 Output amplitude versus frequency.**

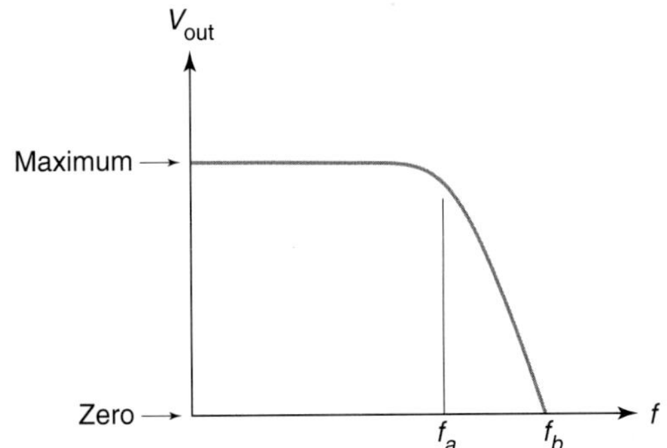


The signal loss caused by the frequency response of a circuit is referred to as **attenuation.** The attenuation caused by the circuit in Figure 19.1 can be graphed as shown in Figure 19.2. As you can see, the amplitude of the circuit output stays relatively constant over the range of frequencies below $f_a$. As operating frequency increases above the value of $f_a$, the attenuation introduced by the circuit causes the amplitude to drop. If the operating frequency reaches the frequency labeled $f_b$, the output amplitude drops to zero. (This means the circuit has no output signal, even though it still has an input signal.)

**Attenuation**
The signal loss caused by the frequency response of a circuit.

## Frequency Response Curves

There are many possible ways to describe the frequency response characteristics of a component or circuit. One of the more common approaches is to describe the effect that a change in operating frequency has on *the ratio of output amplitude to input amplitude.* For example, the curve in Figure 19.3a represents amplitude as a ratio of output voltage to input voltage. As you can see, the voltage ratio has a maximum value of one (1). When $V_{out}/V_{in} = 1$, the input and output voltages are equal. When the circuit is operated at the frequency designated $f_x$, the output voltage is half the input voltage, and so on. While the frequency response curve is not committed to specific input and output amplitudes, it does indicate the relationship between the input and output voltages at any given frequency.

In most cases, frequency response curves use a power ratio as the indicator of amplitude. For example, the curve in Figure 19.3b represents amplitude as the ratio of output power to input power.

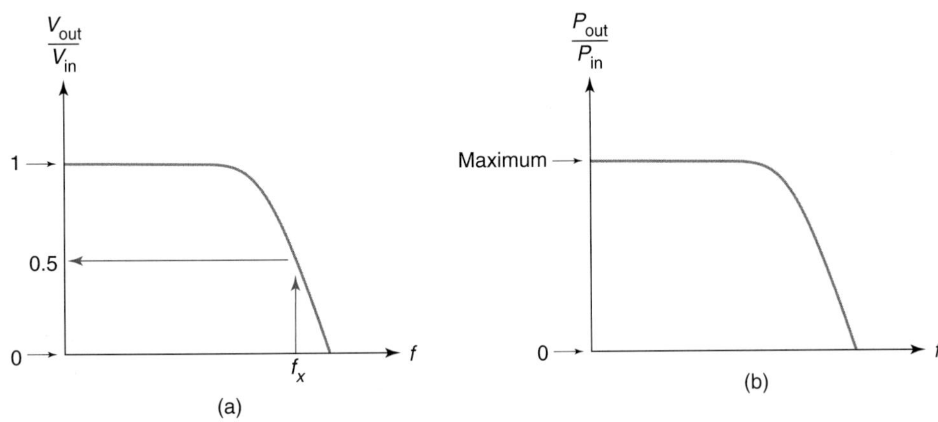


**FIGURE 19.3 Amplitude measurements on frequency response curves.**

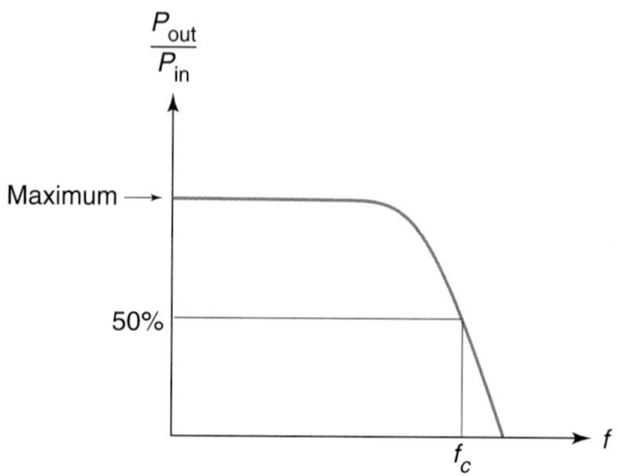


**FIGURE 19.4 Cutoff frequency.**

## Cutoff Frequency ($f_c$)

*OBJECTIVE 2* ▶

**Cutoff frequency ($f_c$)**
The frequency at which the power ratio of a circuit drops to 50% of its maximum value.

In Figure 19.3b, the power ratio begins decreasing once the operating frequency passes a certain value. The frequency at which the power ratio drops to 50% of its maximum value is referred to as the **cutoff frequency ($f_c$).** The cutoff frequency is identified on a frequency response curve in Figure 19.4.

By convention, the cutoff frequency on a curve is the dividing line between acceptable and unacceptable output levels from a component or circuit. In other words, when the power ratio is greater than 50%, the output is considered to be close enough to its maximum value to be acceptable. When the power ratio is less than 50%, the output has been attenuated to the point of being considered unacceptable. It should be noted that the 50% standard for cutoff frequency is an arbitrary value; that is, it is used only because it is agreed upon by professionals in the field. There are no critical changes in circuit or component operation that occur when the power ratio decreases to 50% of its maximum value.

**Filter**
A circuit designed to pass a specific range of frequencies while rejecting others.

The cutoff frequency of a component or circuit is determined by its resistive and reactive values. Later in the chapter, you will learn how its value is calculated and measured.

**Low-pass filter**
A filter designed to pass all frequencies below its cutoff frequency.

## Filters

*OBJECTIVE 3* ▶

**High-pass filter**
A filter designed to pass all frequencies above its cutoff frequency.

**Bandpass filter**
A filter designed to pass all frequencies between two cutoff frequencies.

**Band-stop filter**
A filter designed to reject all frequencies between two cutoff frequencies. Also referred to as a **notch filter.**

Many circuits are designed for specific frequency response characteristics; that is, they are designed to pass a specific range of frequencies while rejecting others. These circuits are generally referred to as **filters.**

There are four basic types of filters, each with its own frequency response curve. The four characteristic filter response curves are shown in Figure 19.5. Figure 19.5a shows the curve for a **low-pass filter.** A low-pass filter is one designed to pass all frequencies below its cutoff frequency. In contrast, a **high-pass filter** is designed to pass all frequencies above its cutoff frequency (as shown in Figure 19.5b).

A **bandpass filter** is one designed to pass the band of frequencies that lies between two cutoff frequencies, as shown in Figure 19.5c. In contrast, a **band-stop filter,** or **notch filter,** is designed to reject the band of frequencies that lies between its cutoff frequencies. As shown in Figure 19.5d, the notch filter passes frequencies below its lower cutoff frequency ($f_{c1}$) and above its upper cutoff frequency ($f_{c2}$).

Since the bandpass and notch filters have two cutoff frequencies, several values are used to describe their operation that are not commonly applied to low-pass and high-pass filters. These values are called *bandwidth* and *center frequency.*

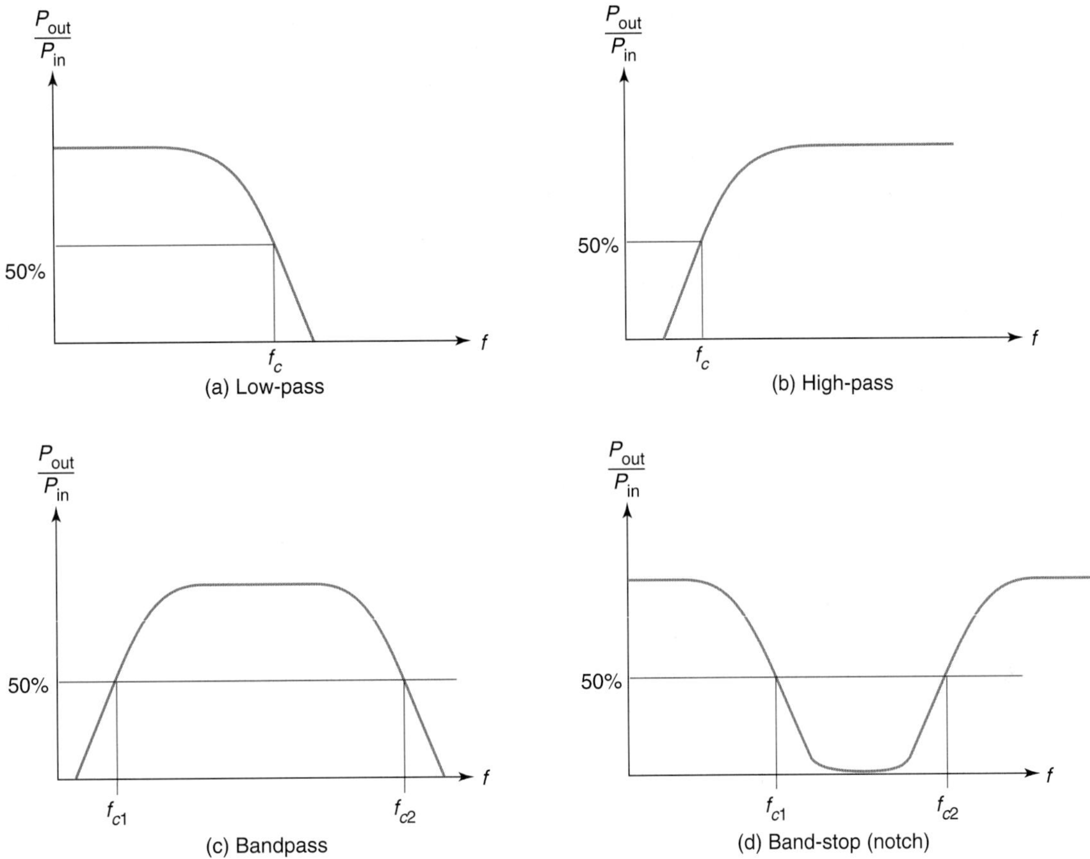


**FIGURE 19.5 Filter frequency response curves.**

## Bandwidth

◄ *OBJECTIVE 4*

The range, or *band,* of frequencies between the cutoff frequencies of a component or circuit is referred to as its **bandwidth.** The concept of bandwidth is illustrated in Figure 19.6. As shown in the figure, bandwidth is equal to the difference between the cutoff frequencies. By formula,

**Bandwidth** The band of frequencies between the cutoff frequencies of a component or circuit.

$$BW = f_{c2} - f_{c1} \quad \textbf{(19.1)}$$

**FIGURE 19.6 Bandwidth.**

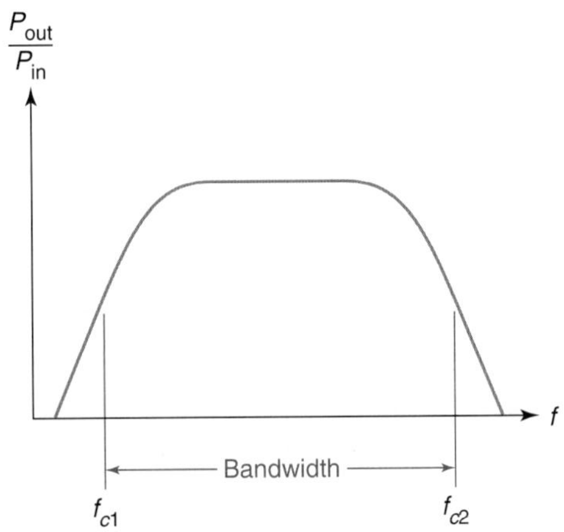

where $BW$ = the bandwidth of the component or circuit, in Hz
$f_{c2}$ = the upper cutoff frequency
$f_{c1}$ = the lower cutoff frequency

The bandwidth of a component or circuit can be found as demonstrated in Example 19.1.

### EXAMPLE 19.1

The bandpass filter has the frequency response curve shown in Figure 19.7. Calculate the bandwidth of the circuit.

FIGURE 19.7

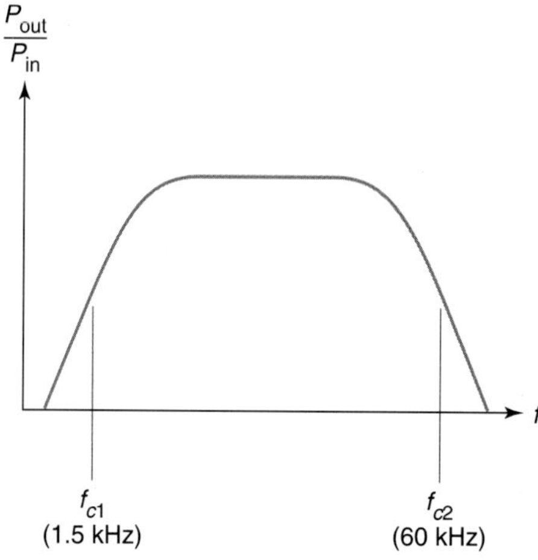


***Solution:*** Using the cutoff frequencies shown on the curve, the bandwidth of the circuit is found as

$$BW = f_{c2} - f_{c1} = 60 \text{ kHz} - 1.5 \text{ kHz} = \mathbf{58.5 \text{ kHz}}$$

As the result indicates, there is a 58.5 kHz spread between the cutoff frequencies of the circuit.

***PRACTICE PROBLEM 19.1***

A curve like the one in Figure 19.7 has the following values: $f_{c2} = 88$ kHz and $f_{c1} =$ 6.2 kHz. Calculate the bandwidth of the circuit represented by the curve.

## Center Frequency ($f_0$)

To describe the frequency response of a bandpass or notch filter accurately, we need to know more than its bandwidth. We also need to know *where* the bandwidth is located on the frequency spectrum. For example, consider the two bandpass curves shown in Figure 19.8. Each of these curves has a bandwidth of 50 kHz, yet clearly they are not the same curve. To distinguish between them, we need to identify the center frequency of each curve.

**Center frequency ($f_0$)** The frequency that equals the geometric average of the cutoff frequencies.

Technically defined, **center frequency ($f_0$)** is the frequency that equals the geometric average of the cutoff frequencies. The geometric average of any two values is equal to the square root of their product. Therefore, center frequency is found as

$$f_0 = \sqrt{f_{c1} f_{c2}} \qquad \textbf{(19.2)}$$

Example 19.2 shows how the value of $f_0$ is found for a component or circuit.

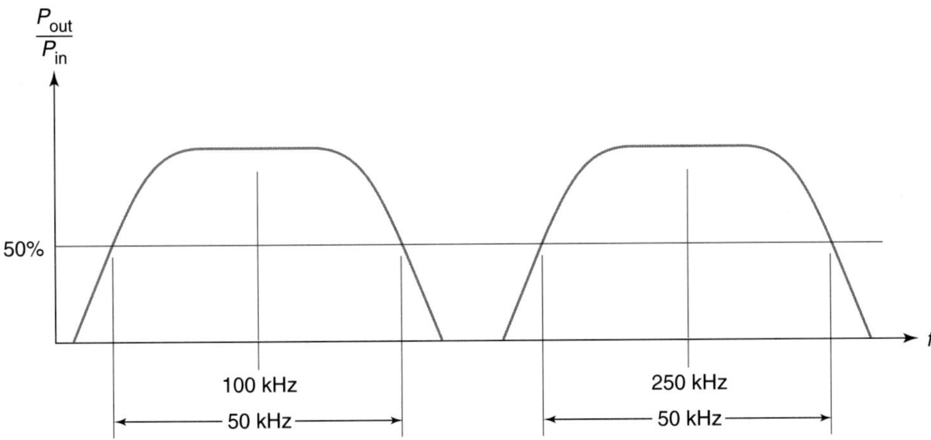


**FIGURE 19.8 Bandpass filters with equal bandwidths.**

*EXAMPLE 19.2*

A bandpass filter has the following values: $f_{c1} = 5$ kHz and $f_{c2} = 20$ kHz. Calculate the value of the circuit's center frequency.

***Solution:*** Using the values given, the value of $f_0$ is found as

$$\mathbf{f_0} = \sqrt{f_{c1}f_{c2}} = \sqrt{(5\text{ kHz})(20\text{ kHz})} = \mathbf{10\text{ kHz}}$$

***PRACTICE PROBLEM 19.2***

A band-stop filter has the following values: $f_{c1} = 300$ Hz and $f_{c2} = 120$ kHz. Calculate the value of the circuit's center frequency.

When we first hear the term *center frequency,* we tend to visualize a frequency that is halfway between the cutoff frequencies. However, this is not necessarily the case. Consider the following results from Example 19.2:

$$f_{c1} = 5\text{ kHz} \qquad f_0 = 10\text{ kHz} \qquad f_{c2} = 20\text{ kHz}$$

Obviously, 10 kHz is not halfway between 5 kHz and 20 kHz. However, it is at the geometric center of the two values, meaning that the ratio of $f_0$ to $f_{c1}$ equals the ratio of $f_{c2}$ to $f_0$. By formula,

$$\frac{f_0}{f_{c1}} = \frac{f_{c2}}{f_0} \tag{19.3}$$

If we apply this relationship to the values obtained in the example, we get

$$\frac{10\text{ kHz}}{5\text{ kHz}} = \frac{20\text{ kHz}}{10\text{ kHz}}$$

which is true. As you can see, the frequency ratios are equal, which is always the case for the geometric average of two values.

## Frequency Calculations

When one cutoff frequency and the center frequency for a filter are known, the other cutoff frequency can be found using a form of equation (19.3). By transposing that equation, we develop these two relationships:

$$f_{c1} = \frac{f_0^2}{f_{c2}} \tag{19.4}$$

and

$$f_{c2} = \frac{f_0^2}{f_{c1}} \tag{19.5}$$

Example 19.3 demonstrates the first of these relationships.

### EXAMPLE 19.3

A notch filter has the following values: $f_0 = 6$ kHz and $f_{c2} = 120$ kHz. Calculate the value of the circuit's lower cutoff frequency.

***Solution:*** Using the values given, the filter's lower cutoff frequency is found as

$$f_{c1} = \frac{f_0^2}{f_{c2}} = \frac{(6\text{ kHz})^2}{120\text{ kHz}} = \mathbf{300\text{ Hz}}$$

***PRACTICE PROBLEM 19.3***

A bandpass filter has the following values: $f_0 = 500$ Hz and $f_{c2} = 5$ kHz. Calculate the value of the filter's lower cutoff frequency.

If you compare the frequencies in Example 19.3 to the values given and calculated in Practice Problem 19.2, you'll see that the same curve was used for the two problems. This fact should help validate the relationship for you. Example 19.4 shows how the upper cutoff frequency of a filter can be determined when the values of $f_0$ and $f_{c1}$ are known.

### EXAMPLE 19.4

A bandpass filter has values of $f_{c1} = 50$ Hz and $f_0 = 500$ Hz. Calculate the value of the filter's upper cutoff frequency.

***Solution:*** Using the values given, the filter's upper cutoff frequency can be found as

$$f_{c2} = \frac{f_0^2}{f_{c1}} = \frac{(500\text{ Hz})^2}{50\text{ Hz}} = \mathbf{5\text{ kHz}}$$

***PRACTICE PROBLEM 19.4***

A notch filter has the following values: $f_{c1} = 400$ Hz and $f_0 = 8$ kHz. Calculate the value of the filter's upper cutoff frequency.

## Frequency Descriptions of Frequency Response Curves

Earlier, you were told that bandwidth and center frequency are commonly used to describe bandpass and notch filters, that they are not normally used to describe low-pass or high-pass filters. Normally, a low-pass or high-pass filter is described using only its cutoff frequency. For example, the curve in Figure 19.9a represents the response characteristics of a low-pass filter with a 20 kHz cutoff frequency. The fact that 20 kHz is an upper cutoff frequency is understood because a low-pass filter is designed to pass frequencies *below* a given value. By the same token, the curve in Figure 19.9b represents the frequency response of a high-pass filter with a 5 kHz cutoff frequency. Again, the fact that 5 kHz is a lower cutoff frequency is understood because of the type of filter being described.

Bandpass and notch filters are normally described in terms of bandwidth and center frequency. For example, the curve in Figure 19.9c represents the operation of a 50 kHz bandpass filter with a 300 kHz center frequency. By the same token, the curve in Figure 19.9d represents the operation of a 15 kHz notch filter with an 80 kHz center frequency.

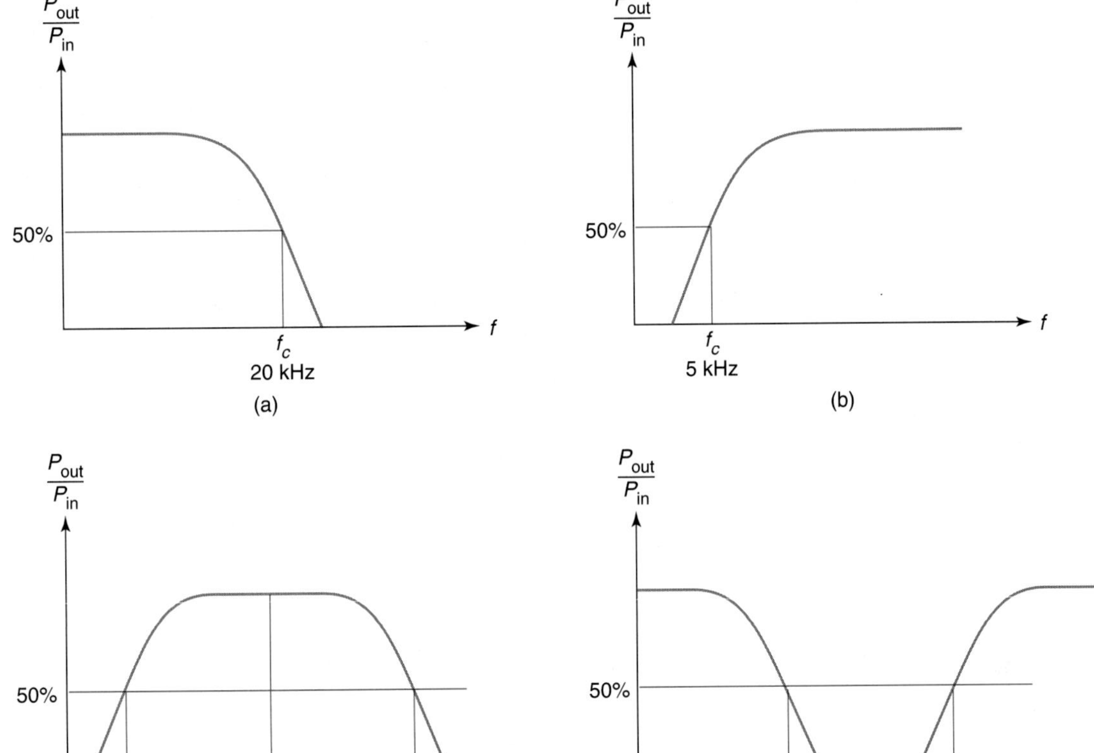


**FIGURE 19.9**

Judging by the curves in Figures 19.9c and 19.9d, you'd think that the cutoff frequencies are equidistant from the center frequency. For example, it should seem that the cutoff frequencies for the circuit represented in Figure 19.9c could be found as

$$f_{c1} = 300 \text{ kHz} - 25 \text{ kHz} \qquad \text{and} \qquad f_{c2} = 300 \text{ kHz} + 25 \text{ kHz}$$

where 25 kHz is half the bandwidth. However, this is not necessarily the case. Determining the cutoff frequencies for a bandpass or notch filter using the center frequency and bandwidth can be more involved than it first appears. The relationship among center frequency, bandwidth, and the values of the cutoff frequencies depends on the quality ($Q$) rating of the filter. We will take a brief look at this rating now, and discuss it again in greater detail later in the chapter.

## Filter Quality ($Q$)

The quality ($Q$) of a bandpass or notch filter equals the ratio of its center frequency to its bandwidth. By formula, ◄ *OBJECTIVE 5*

$$Q = \frac{f_0}{\text{BW}} \qquad \textbf{(19.6)}$$

Example 19.5 demonstrates this relationship.

## EXAMPLE 19.5

Determine the $Q$ of the bandpass filter represented by the curve in Figure 19.9c.

***Solution:*** The curve has values of $f_0 = 300$ kHz and BW $= 50$ kHz. Using these two values, the $Q$ of the filter is found as

$$Q = \frac{f_0}{\text{BW}} = \frac{300 \text{ kHz}}{50 \text{ kHz}} = \mathbf{6}$$

Note that the result has no unit of measure because it is a ratio of one frequency to another.

***PRACTICE PROBLEM 19.5***

Determine the $Q$ of the notch filter represented in Figure 19.9d.

The relationship among $Q$, center frequency, and bandwidth.

An important point needs to be made at this time. Equation (19.6) is somewhat misleading, because it implies that the value of $Q$ depends on the values of center frequency and bandwidth. However, this is not the case. $Q$, as described earlier in the text, is a circuit characteristic that depends on component values. When working with a filter, the $Q$ of the filter and its center frequency are both determined using component values (as will be demonstrated later in this chapter). Then the circuit values of $Q$ and center frequency are used to determine the bandwidth of the circuit, as follows:

$$\text{BW} = \frac{f_0}{Q} \tag{19.7}$$

Example 19.6 provides a more common application of the relationship among $Q$, center frequency, and bandwidth.

## EXAMPLE 19.6

Through a series of calculations, a bandpass filter is determined to have values of $Q = 4.8$ and $f_0 = 96$ kHz. Calculate the circuit bandwidth.

***Solution:*** Using the calculated values of $Q$ and center frequency, the circuit bandwidth is found as

$$\mathbf{BW} = \frac{f_0}{Q} = \frac{96 \text{ kHz}}{4.8} = \mathbf{20 \text{ kHz}}$$

***PRACTICE PROBLEM 19.6***

Using a series of calculations, a notch filter is found to have values of $Q = 3.3$ and $f_0 = 165$ kHz. Determine the circuit bandwidth.

The methods used to calculate the values of $Q$ and $f_0$ vary from one type of filter to another, and are discussed later in this chapter. For now, we simply want to establish the fact that filter bandwidth is determined by the circuit values of $Q$ and $f_0$.

## Average Frequency ($f_{ave}$)

◀ *OBJECTIVE 6*

Every bandpass and notch filter has both a center frequency ($f_0$) and an algebraic average frequency. The **average frequency ($f_{ave}$)** of a filter lies halfway between the cutoff frequencies. By formula,

**Average frequency ($f_{ave}$)** The frequency that lies halfway between the cutoff frequencies.

$$f_{ave} = \frac{f_{c1} + f_{c2}}{2} \qquad \textbf{(19.8)}$$

As demonstrated in Example 19.7, there can be a significant difference between the values of $f_0$ and $f_{ave}$.

**EXAMPLE 19.7**

Determine the values of $f_0$ and $f_{ave}$ for the circuit represented by the curve in Figure 19.10a.

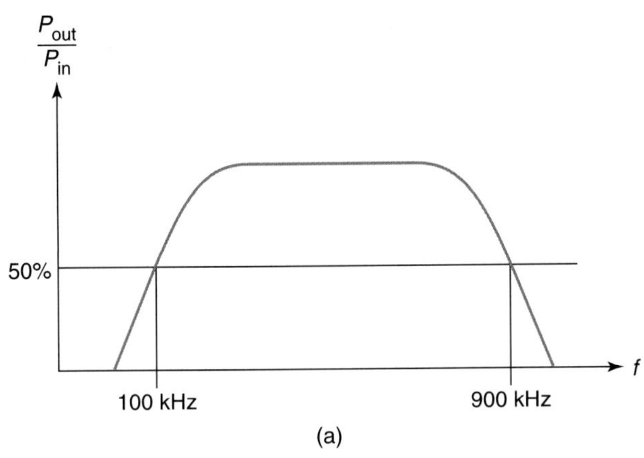


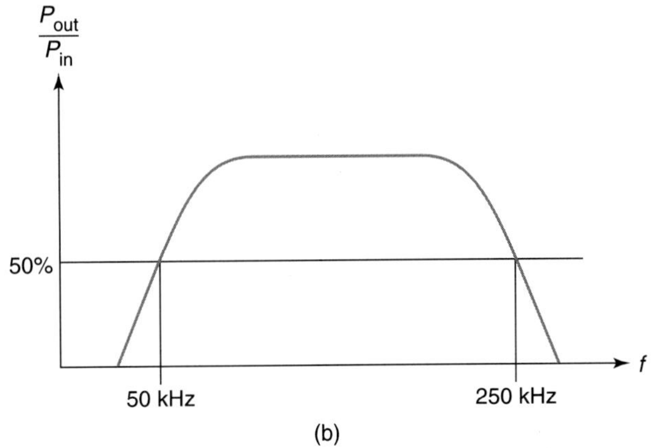


**FIGURE 19.10**

***Solution:*** Using the values shown on the curve, the geometric average of the cutoff frequencies ($f_0$) is found as

$$f_0 = \sqrt{f_{c1} f_{c2}} = \sqrt{(100 \text{ kHz})(900 \text{ kHz})} = \mathbf{300\ kHz}$$

The algebraic average of the cutoff frequencies is found as

$$f_{ave} = \frac{f_{c1} + f_{c2}}{2} = \frac{100 \text{ kHz} + 900 \text{ kHz}}{2} = \mathbf{500\ kHz}$$

***PRACTICE PROBLEM 19.7***

Determine the values of $f_0$ and $f_{ave}$ for the circuit represented by the curve in Figure 19.10b.

As you can see, there can be a significant difference between the two values. That fact is important because the cutoff frequencies are equidistant from the average frequency, as illustrated in Figure 19.11. As you can see, the curve has an 800 kHz bandwidth, and the cutoff frequencies are each 400 kHz away from the average frequency. These relationships can be expressed as follows:

$$f_{c1} = f_{ave} - \frac{\text{BW}}{2} \qquad \textbf{(19.9)}$$

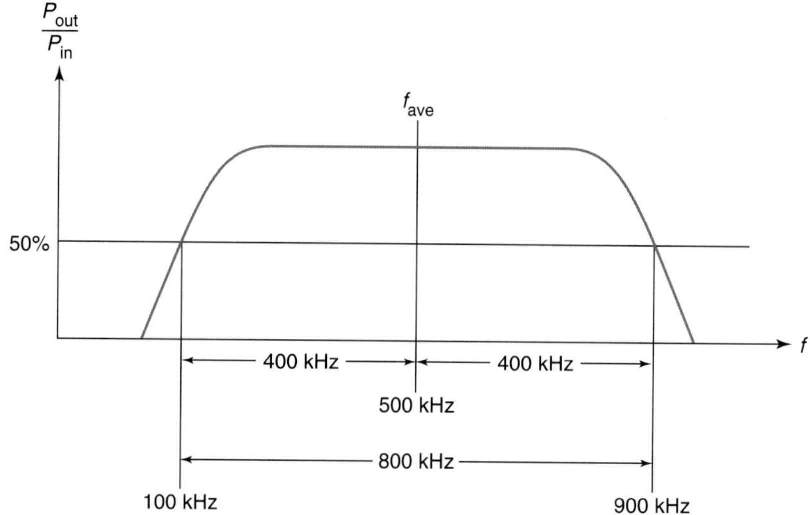


**FIGURE 19.11 Frequency relationships from Example 19.8.**

and

$$f_{c2} = f_{ave} + \frac{BW}{2} \qquad \textbf{(19.10)}$$

Using these equations, we can determine the cutoff frequencies of any bandpass or notch filter when we know its average frequency and bandwidth. There is only one problem: Equation (19.8) assumes that we know the cutoff frequencies of the filter. When the cutoff frequencies of a filter are not known, we need another way of finding $f_{ave}$ so that we can determine the circuit cutoff frequencies.

*OBJECTIVE 7* ▶ ***Determining the Value of $f_{ave}$.*** The values of $f_0$ and $Q$ for a given filter can be used to determine the value of $f_{ave}$, as follows:

$$f_{ave} = f_0\sqrt{1 + \left(\frac{1}{2Q}\right)^2} \qquad \textbf{(19.11)}$$

This equation provides a means of determining the average frequency when the center frequency and $Q$ of a bandpass or notch filter are known. An application of this relationship is demonstrated in Example 19.8.

## EXAMPLE 19.8

A bandpass filter has the following values: $f_0 = 250$ kHz and BW $= 150$ kHz. Determine the values of $f_{ave}$ for the circuit.

***Solution:*** First, the $Q$ of the filter is found as

$$Q = \frac{f_0}{BW} = \frac{250 \text{ kHz}}{150 \text{ kHz}} = \mathbf{1.67}$$

Using equation (19.11), the value of $f_{ave}$ for the circuit can now be found as

$$\begin{aligned} f_{ave} &= f_0\sqrt{1 + \left(\frac{1}{2Q}\right)^2} \\ &= (250 \text{ kHz})\sqrt{1 + \left(\frac{1}{(2)(1.67)}\right)^2} \\ &= (250 \text{ kHz})(1.044) = \mathbf{261 \text{ kHz}} \end{aligned}$$

***PRACTICE PROBLEM 19.8***

A notch filter has the following values: $f_0 = 300$ kHz and BW $= 240$ kHz. Determine the value of $f_{ave}$ for the circuit.

Once the value of $f_{ave}$ is known, the circuit cutoff frequencies can be determined as demonstrated in Example 19.9.

## EXAMPLE 19.9

Determine the cutoff frequencies for the circuit described in Example 19.8.

***Solution:*** From Example 19.8, we know that the circuit has values of $f_{ave} = 261$ kHz and BW $= 150$ kHz. Using these values, the cutoff frequencies of the circuit are found as

$$f_{c1} = f_{ave} - \frac{\text{BW}}{2} = 261 \text{ kHz} - \frac{150 \text{ kHz}}{2} = \mathbf{186\ kHz}$$

and

$$f_{c2} = f_{ave} + \frac{\text{BW}}{2} = 261 \text{ kHz} + \frac{150 \text{ kHz}}{2} = \mathbf{336\ kHz}$$

***PRACTICE PROBLEM 19.9***

Determine the cutoff frequencies for the notch filter described in Practice Problem 19.8.

The values found in Example 19.9 can be verified using equation (19.2), as follows:

$$f_0 = \sqrt{f_{c1}f_{c2}} = \sqrt{(186 \text{ kHz})(336 \text{ kHz})} \cong 250 \text{ kHz}$$

Since this value matches the value of $f_0$ given in Example 19.8, we have verified that our calculations were correct.

***Frequency Approximations.*** When a bandpass or notch filter has a value of $Q \geq 2$, the center and average frequencies can be assumed to be approximately equal in value. By formula,

$$f_{ave} \cong f_0 \qquad \text{when} \qquad Q \geq 2 \tag{19.12}$$

The basis for this relationship can be seen by taking another look at equation (19.11). According to that equation, $f_{ave}$ and $f_0$ differ by a factor of

$$\sqrt{1 + \left(\frac{1}{2Q}\right)^2}$$

If we solve this expression for $Q = 2$, we get

$$\sqrt{1 + \left(\frac{1}{2Q}\right)^2} = \sqrt{1 + \left(\frac{1}{4}\right)^2} = 1.03$$

This result indicates that there is only a 3% difference between the values of $f_{ave}$ and $f_0$ when $Q = 2$. When $Q > 2$, the difference between the values of $f_{ave}$ and $f_0$ becomes even smaller. Therefore, we can assume that the two frequencies are approximately equal to each other as long as $Q \geq 2$. This relationship is demonstrated in Example 19.10.

## EXAMPLE 19.10

A bandpass filter has a center frequency of 400 kHz and a bandwidth of 80 kHz. Calculate the average frequency for the circuit.

***Solution:*** First, the $Q$ of the circuit can be found as

$$Q = \frac{f_0}{\text{BW}} = \frac{400 \text{ kHz}}{80 \text{ kHz}} = \mathbf{5}$$

Now, the value of the average frequency can be found as

$$f_{\text{ave}} = f_0\sqrt{1 + \left(\frac{1}{2Q}\right)^2} = (400 \text{ kHz})\sqrt{1 + \left(\frac{1}{10}\right)^2} = (400 \text{ kHz})(1.005) = \mathbf{402 \text{ kHz}}$$

As you can see, there is only a 2 kHz (0.5%) difference between the values of $f_{\text{ave}}$ and $f_0$ for the circuit. In this case, we could have assumed that the two values were equal without introducing significant errors into any subsequent calculations.

***PRACTICE PROBLEM 19.10***

A notch filter has a center frequency of 300 kHz and a bandwidth of 40 kHz. Calculate the value of $f_{\text{ave}}$ for the circuit. What is the difference percentage between the values of $f_{\text{ave}}$ and $f_0$?

## Frequency Scales

*OBJECTIVE 8* ▶ Up to this point, we have shown only the given and calculated frequencies on our response curves. In practice, frequency response curves are commonly plotted using frequency scales like those shown in Figure 19.12.

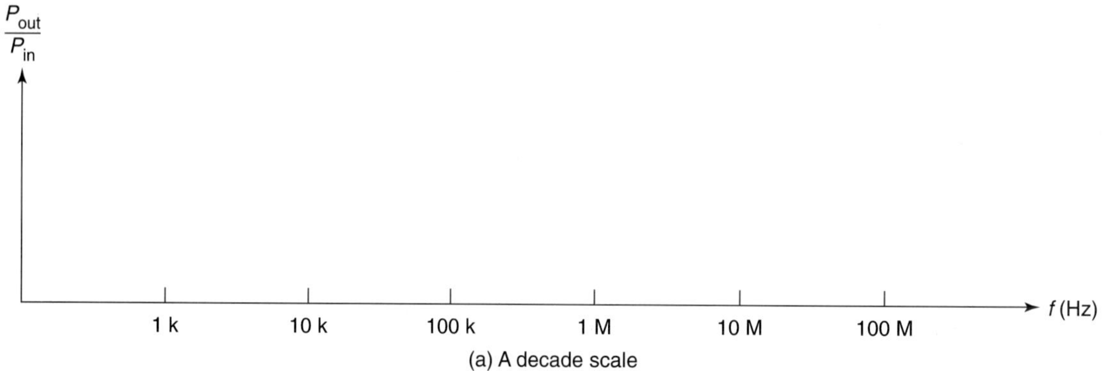


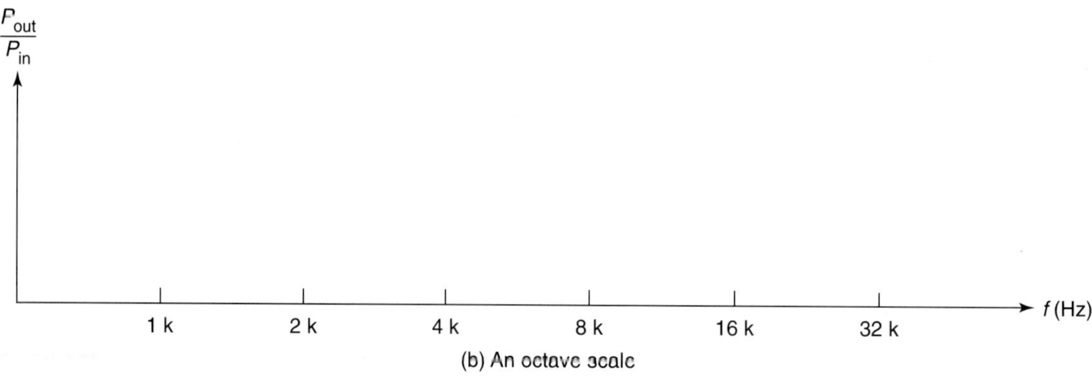


**FIGURE 19.12 Logarithmic frequency scales.**

The scales shown in the figure are referred to as **logarithmic scales,** meaning that the frequency spread from one increment to the next increases at a geometric rate. In a geometric scale, the value of each increment is a whole number multiple of the previous increment. For example, the value of each increment in Figure 19.12a is ten times the value of the previous increment.

**Logarithmic scale**
A scale with frequency spreads that increase from one increment to the next at a geometric rate.

A frequency multiplier of 10 is referred to as a **decade.** For this reason, the scale shown in Figure 19.12a is referred to as a **decade scale.** In contrast, 2 is the multiplier used from one increment to the next in Figure 19.12b. A frequency multiplier of 2 is referred to as an **octave,** so the scale shown is referred to as an **octave scale.** Note that both decade and octave scales are used in practice. When a large frequency range is needed, a decade scale is used. When a more precise representation over a smaller frequency range is needed, an octave scale may be used.

**Decade**
A frequency multiplier of ten.

**Decade scale**
A scale where each increment is ten times the value of the previous increment.

**Octave**
A frequency multiplier of two.

**Octave scale**
A scale where each increment is two times the value of the previous increment.

The difference between decade and octave scales becomes clearer when you compare the two scales shown in Figure 19.12. The decade scale covers a frequency range of 100 MHz. The octave scale represents a frequency range of 32 kHz over the same number of increments. Therefore, more precise frequency approximations are possible using an octave scale.

In general, logarithmic scales are preferred over algebraic frequency scales for two reasons:

Why logarithmic frequency scales are preferred over algebraic scales.

1. An algebraic scale can be extremely long. An algebraic scale uses a constant as the difference between increments; that is, the value of each increment is determined by adding a constant to the value of the previous increment.
2. When a logarithmic scale is used, the center frequency (which is the geometric average of the cutoff frequencies) falls in the physical center of the curve.

The first of these two points is illustrated in Figure 19.13. Figure 19.13a is the same octave scale that you saw in Figure 19.12. Figure 19.13b is an algebraic scale. As you can see, the frequency range between each pair of increments is a constant ($c$), 2 kHz. To represent 32 kHz, the scale in Figure 19.13b would have to extend well off the page.

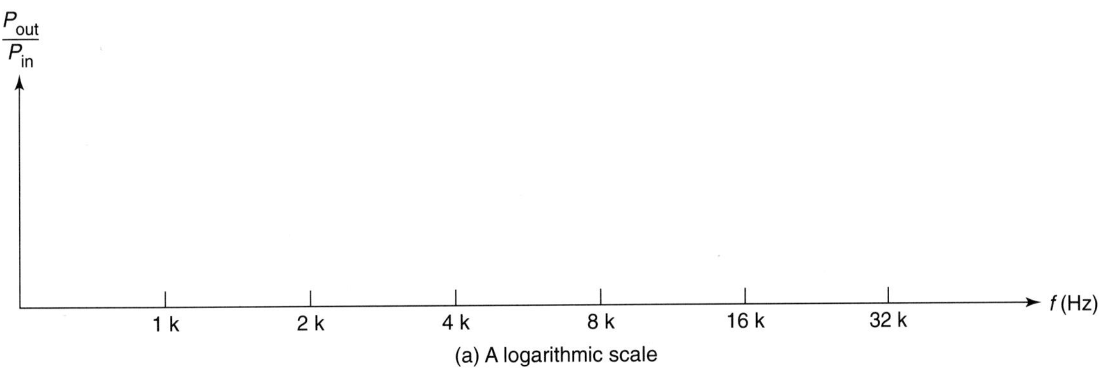

(a) A logarithmic scale

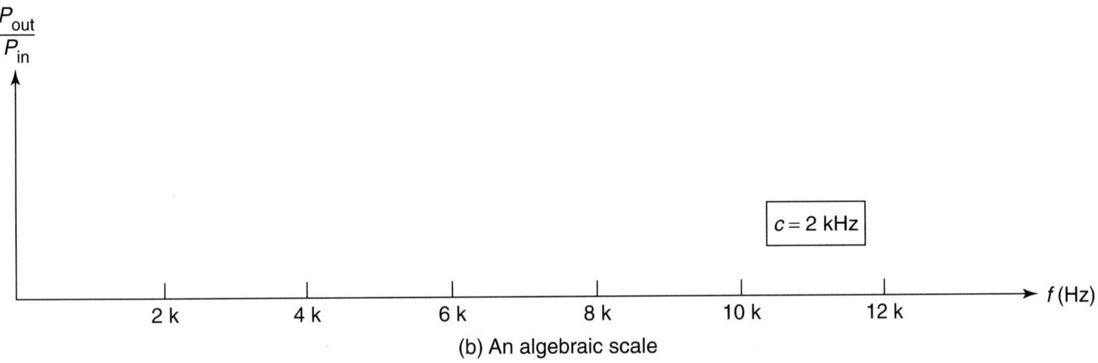

(b) An algebraic scale

**FIGURE 19.13**

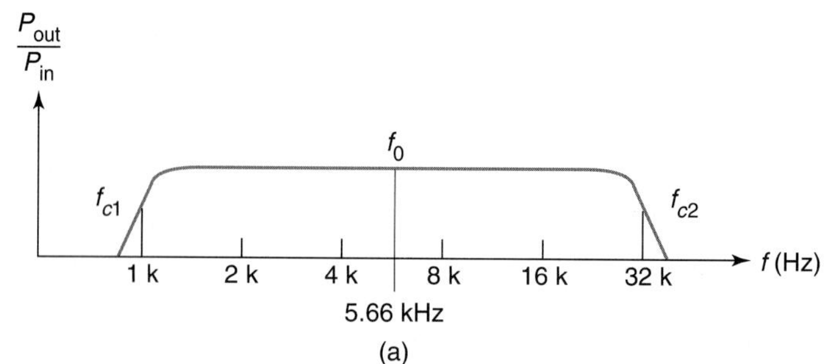


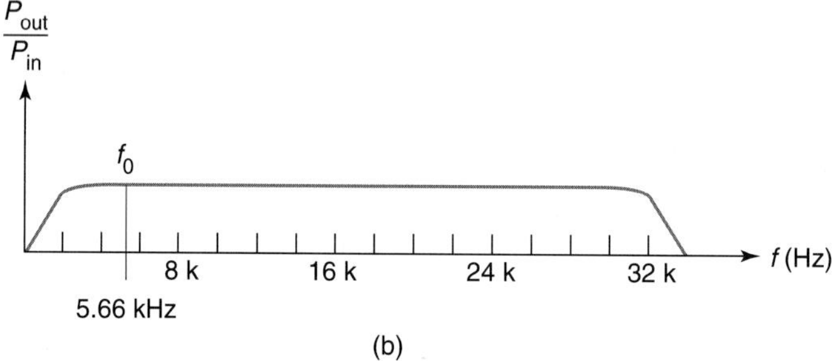


**FIGURE 19.14**

The second point is illustrated in Figure 19.14a. The curve shown was plotted using assumed cutoff frequencies of 1 kHz and 32 kHz. Using equation (19.2), the center frequency of the curve can be found as

$$f_0 = \sqrt{f_{c1}f_{c2}} = \sqrt{(1 \text{ kHz})(32 \text{ kHz})} = 5.66 \text{ kHz}$$

As you can see, the geometric center frequency ($f_0$) falls very close to the physical center of the curve. If an algebraic scale had been used, $f_0$ would have fallen very close to the left end of the scale, as shown in Figure 19.14b.

## Critical Frequencies: Putting It All Together

Low-pass and high-pass filters are identified using a single cutoff frequency, as shown in Figure 19.15. Low-pass filters have an upper cutoff frequency ($f_{c2}$), while high-pass filters have a lower cutoff frequency ($f_{c1}$). In either case, the cutoff frequency is the operating frequency where the ratio of output power to input power drops to 50% of its maximum value. The value of the cutoff frequency for either response curve depends on component values in the circuit itself.

Bandpass and band-stop (or notch) filters each have two cutoff frequencies, as shown in Figure 19.16. A bandpass filter passes the band of frequencies that lies between its cutoff frequencies. A notch filter blocks (attenuates) the band of frequencies that lies between its cutoff frequencies.

**FIGURE 19.15 Low-pass and high-pass frequency response curves.**

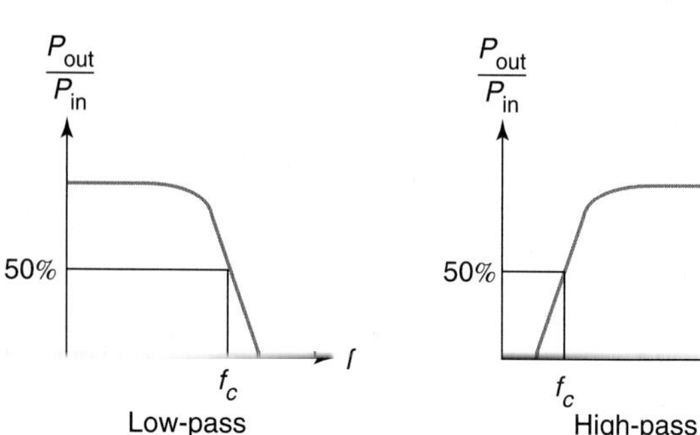

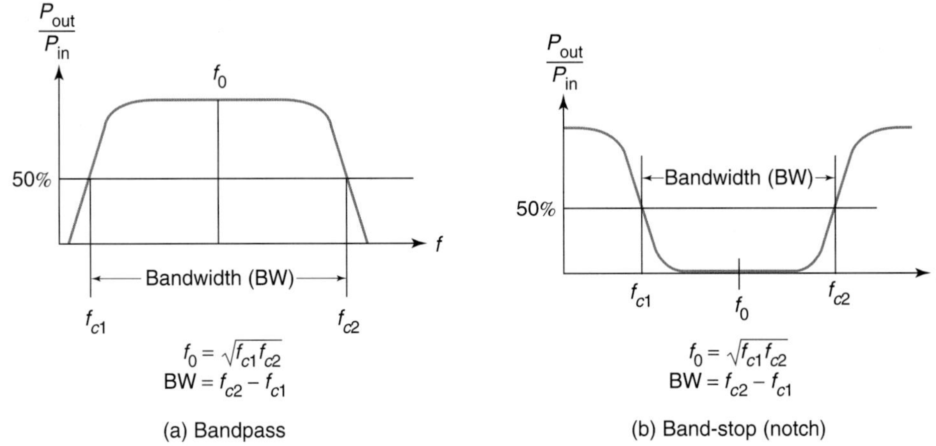


**FIGURE 19.16 Bandpass and band-stop frequency response curves.**

Bandpass and notch filters are normally described in terms of their center frequency ($f_0$) and bandwidth (BW). The center frequency equals the geometric average of the cutoff frequencies, and the bandwidth equals the difference between them. In effect, the center frequency tells us where the curve is located on the frequency spectrum, while the bandwidth tells us the range of frequencies that is passed (or blocked) by the filter.

The average frequency ($f_{ave}$) of a bandpass or notch filter equals the algebraic average of the cutoff frequencies. The value of $f_{ave}$ (not shown in the figure) falls halfway (algebraically) between the cutoff frequencies.

When the $Q$ of a bandpass or notch filter is greater than (or equal to) two, the values of $f_{ave}$ and $f_0$ can be assumed to be equal without introducing any significant error into subsequent calculations. When $Q$ is less than two, the value of $f_{ave}$ must be calculated using equation (19.11). The value of $f_{ave}$ is then used, along with the circuit bandwidth, to determine the values of the cutoff frequencies.

Most frequency response curves are plotted on logarithmic scales. The frequency spread from one increment to the next on a logarithmic scale increases at a geometric rate; that is, the value of each increment is a whole number multiple of the previous increment.

The two most commonly used logarithmic scales are the *decade* and *octave* scales. A decade scale is constructed so that the value of each increment is ten times the value of the previous increment. An octave scale uses *two* as the multiplier between increments. (Decade and octave scales are illustrated in Figure 19.12.)

## Section Review

1. What is attenuation?
2. How is amplitude normally measured on a frequency response curve?
3. What is a cutoff frequency?
4. What is a filter?
5. List the four common types of filters and describe the frequency response of each.
6. What is meant by the term *band*?
7. What is the bandwidth of a bandpass or notch filter?
8. What two values are required to locate and describe accurately the frequency response curve of a bandpass or notch filter?
9. What is the center frequency of a bandpass or notch filter?
10. Using frequency ratios, describe the relationship among $f_{c1}$, $f_{c2}$, and $f_0$.
11. What does the quality ($Q$) rating of a filter equal?
12. What is the average frequency ($f_{ave}$) of a bandpass or notch filter?

13. What determines the relationship between the values of $f_{ave}$ and $f_0$ for a bandpass or notch filter?
14. Under what circumstance is it safe to assume that the values of $f_{ave}$ and $f_0$ for a filter are approximately equal?
15. What is a logarithmic scale?
16. What is a decade scale? What is an octave scale?
17. When would each of the scales mentioned in Question 16 be preferred over the other?
18. Why are logarithmic frequency scales preferred over algebraic frequency scales?

## 19.2 AMPLITUDE MEASUREMENTS: dB POWER GAIN

*OBJECTIVE 9* ▶ Before we begin to look at the frequency response of passive filter circuits, we need to discuss the methods commonly used to represent amplitude on frequency response curves. As stated earlier, amplitude is normally expressed as a ratio of an output value (such as voltage or power) to a corresponding input value. In most cases, these ratios are represented using logarithmic values, called decibels (dB).

**Decibel (dB)**
A logarithmic representation of a number. (A more precise definition follows later in this section.)

A **decibel (dB)** is a logarithmic representation of a number. Decibels are used because they allow us to represent easily very large and very small values. In this section, you will learn how to convert any numeric value to dB form, and vice versa. You will also learn how to perform basic mathematical operations involving dB values.

### Bel Power Gain

**Bel (B)**
A ratio of output power to input power, expressed as a common (base 10) logarithm.

At one time, power ratios were expressed using logarithmic values called bels. The **bel (B)** is a ratio of output power to input power, expressed as a common logarithm. By formula,

$$A_{p(\text{B})} = \log_{10}\frac{P_{\text{out}}}{P_{\text{in}}} \tag{19.13}$$

where $A_{p(\text{B})}$ = the *power gain* of a component or circuit, in bels

$\log_{10}\frac{P_{\text{out}}}{P_{\text{in}}}$ = the common log of the power ratio

**Power gain ($A_p$)**
The ratio of output power to input power for a component or circuit.

**Unity gain**
A gain of one (1).

The ratio of output power to input power for a component or circuit is commonly referred to as **power gain ($A_p$)**. The word *gain* is used in reference to amplifiers, which are capable of increasing the power level of their ac input signals. Therefore, amplifiers can have power ratios greater than one. The circuits we will discuss in this chapter have power ratios less than one, meaning that output power is lower than input power. However, we will still use the term *gain* to describe their output power to input power ratios. Note that the term **unity gain** is commonly used to refer to a value of $A_p = 1$. Using the symbol for power gain, equation (19.13) can be rewritten as

$$A_{p(\text{B})} = \log_{10}A_p \tag{19.14}$$

where

$$A_p = \frac{P_{\text{out}}}{P_{\text{in}}}$$

An application of equation (19.14) is provided in Example 19.11.

**EXAMPLE 19.11**

The ac output power from an amplifier is measured and found to be 100 times the value of the ac input power. Calculate the power gain of the circuit in bels.

***Solution:*** Since ac output power is 100 times ac input power, $A_p = 100$. Using this value in equation (19.14), the power gain of the amplifier (in bels) is found as

$$\mathbf{A_{p(B)}} = \log_{10} A_p = \log_{10}(100) = \mathbf{2\ B}$$

***PRACTICE PROBLEM 19.11***

The ac output power from an amplifier is measured and found to be 8 times the value of the ac input power. Calculate the power gain of the circuit in bels.

## *Bel Characteristics*

Even though amplifiers are not the focus of this chapter, Example 19.11 was included because it provides a basis for a better understanding of the bel. In the example, we determined that $\log_{10}(100) = 2$. This relationship can be expressed as follows: if $\log_{10}(y) = x$, then $10^x = y$. Applying this relationship to the values in Example 19.12, we get:

$$\log_{10}(100) = 2 \qquad \text{therefore} \qquad 10^2 = 100$$

which is true. In other words, *the bel value of a ratio indicates the power of 10 that equals the ratio.* Some other examples of the relationship between a gain value and its corresponding bel value are provided in Table 19.1.

**TABLE 19.1 Bels and Powers of Ten**

| | | |
|---|---|---|
| $\log_{10}(1000) = 3$ | therefore | $10^3 = 1000$ |
| $\log_{10}(500) = 2.699$ | therefore | $10^{2.699} = 500$ |
| $\log_{10}(1) = 0$ | therefore | $10^0 = 1$ |
| $\log_{10}(0.01) = -2$ | therefore | $10^{-2} = 0.01$ |
| $\log_{10}(0.001) = -3$ | therefore | $10^{-3} = 0.001$ |

The values in Table 19.1 demonstrate some important characteristics of bel values. First, note that the table is divided into ratios that are greater than, equal to, and less than one (1). As the table indicates:

1. The bel value is positive when $A_p > 1$.
2. The bel value is zero when $A_p = 1$.
3. The bel value is negative when $A_p < 1$.

Since $A_p$ is the ratio of output power to input power, these statements tell us that:

1. Output power is greater than input power when $A_{p(B)}$ is positive.
2. Output power equals input power when $A_{p(B)} = 0$.
3. Output power is lower than input power when $A_{p(B)}$ is negative.

Another important characteristic of bel values can be seen by comparing the values shown in the first and last lines of Table 19.1. According to the calculations shown,

$$\log_{10}(1000) = 3 \qquad \text{and} \qquad \log_{10}(0.001) = -3$$

As you may have realized, 1000 and 0.001 are reciprocals of each other. The fact that their bel values have equal magnitudes is not a coincidence. For any given value, the following statement applies: if $\log_{10}(y) = x$, then $\log_{10}\frac{1}{y} = -x$. Several examples of this relationship are provided in Table 19.2.

**TABLE 19.2 Examples of Reciprocal Bel Relationships**

| | | |
|---|---|---|
| $\log_{10}(2) = 0.3\text{ B}$ | therefore | $\log_{10}\left(\frac{1}{2}\right) = -0.3\text{ B}$ |
| $\log_{10}(10) = 1\text{ B}$ | therefore | $\log_{10}\left(\frac{1}{10}\right) = -1\text{ B}$ |
| $\log_{10}(10{,}000) = 4\text{ B}$ | therefore | $\log_{10}\left(\frac{1}{10{,}000}\right) = -4\text{ B}$ |

## Decibels (dB)

**Decibel (dB)**
A ratio of output power to input power, expressed as ten times the common logarithm of the ratio.

At some point, the industry decided that the bel was too large a unit of gain to be practical. So the switch was made from bels to tenths of a bel, or **decibels (dB).** Since there are 10 decibels in one bel, the dB power gain of a component or circuit is found as

$$A_{p(\text{dB})} = 10 \log\frac{P_{\text{out}}}{P_{\text{in}}} \qquad \textbf{(19.15)}$$

or

$$A_{p(\text{dB})} = 10 \log A_p \qquad \textbf{(19.16)}$$

Example 19.12 demonstrates the use of these relationships.

### EXAMPLE 19.12

The output power of a certain bandpass filter is half the value of the input power. Calculate the value of dB power gain for the circuit.

***Solution:*** Since output power is half the value of input power,

$$A_p = \frac{P_{\text{out}}}{P_{\text{in}}} = \frac{0.5P_{\text{in}}}{P_{\text{in}}} = \mathbf{0.5}$$

and

$$A_{p(\text{dB})} = 10 \log A_p = 10 \log(0.5) = (10)(-0.3) = \mathbf{-3\ dB}$$

***PRACTICE PROBLEM 19.12***

A bandpass filter is found to have values of $P_{\text{in}} = 1\text{ W}$ and $P_{\text{out}} = 750\text{ mW}$. Calculate the dB power gain of the circuit.

As you can see, the dB power gain calculated in the example was negative. Like bel values, dB power gain is:

1. Negative when $P_{\text{out}} < P_{\text{in}}$.
2. Zero when $P_{\text{out}} = P_{\text{in}}$.
3. Positive when $P_{\text{out}} > P_{\text{in}}$.

The first of these relationships was demonstrated in Example 19.12. When the values of $P_{out}$ and $P_{in}$ for a circuit are equal, $A_p = 1$. Therefore, the second relationship can be demonstrated by converting one (1) to dB form, as follows:

$$A_{p(\text{dB})} = 10 \log A_p = 10 \log(1) = (10)(0) = 0 \text{ dB}$$

The third relationship is demonstrated in Example 19.13.

**EXAMPLE 19.13**

An amplifier is found to have a power gain of $A_p = 2$. Determine the dB power gain of the circuit.

***Solution:*** The value of dB power gain is found as

$$A_{p(\text{dB})} = 10 \log A_p = 10 \log(2) = (10)(0.3) = \mathbf{3\ dB}$$

***PRACTICE PROBLEM 19.13***

An amplifier is found to have a power gain of $A_p = 100$. Determine the dB power gain of the circuit.

As you can see, the value of $A_{p(\text{dB})}$ is positive when $A_p > 1$. By the same token, the reciprocal relationship given for bel values also applies to dB values; that is, reciprocal values have positive and negative dB values with equal magnitude. This can be seen by comparing the results of Examples 19.12 and 19.13. For convenience, the values from these examples are listed as follows:

| Example | Power Gain | Power Gain (in dB) |
|---|---|---|
| 19.12 | 0.5 | −3 dB |
| 19.13 | 2.0 | 3 dB |

Since $0.5 = 1/2$, the power gains are reciprocal values. As you can see, they have positive and negative dB values that are equal in magnitude.

## Converting dB Power Gain to Standard Numeric Form

In many cases, you will need to know how to convert a dB value to standard numeric form. Any dB power gain value can be converted to standard numeric form using the following relationship: ◄ *OBJECTIVE 10*

$$A_p = \log^{-1}\frac{A_{p(\text{dB})}}{10} \tag{19.17}$$

where $\log^{-1}$ represents the *inverse log* function

Power gain values are converted from dB form to standard numeric form as demonstrated in Example 19.14.

**EXAMPLE 19.14**

The gain of a bandpass filter has a maximum value of −2 dB. Determine the maximum ratio of output power to input power for the circuit.

***Solution:*** Using equation (19.17), the power gain of the filter is found as

$$A_p = \log^{-1}\frac{A_{p(\text{dB})}}{10} = \log^{-1}\left(\frac{-2}{10}\right) = \log^{-1}(-0.2) = \mathbf{0.631}$$

This result indicates that the output power has a maximum value of 0.631 $P_{in}$.

### *PRACTICE PROBLEM 19.14*

The gain of a notch filter has a maximum value of $-1.4$ dB. Determine the maximum ratio of output power to input power for the circuit.

Earlier in this section, we defined $A_p$ as the ratio of output power to input power. When the values of power gain and input power for a circuit are known, the output power can be found as

$$P_{out} = A_p P_{in} \tag{19.18}$$

An application of this relationship is demonstrated in Example 19.15.

### EXAMPLE 19.15

A bandpass filter has maximum values of $A_{p(dB)} = -1.8$ dB and $P_{in} = 2.5$ W. Determine the maximum output power for the filter.

***Solution:*** First, we need to determine the value of $A_p$ for the circuit. Using the known value of dB power gain, the value of $A_p$ is found as

$$\mathbf{A_p} = \log^{-1}\frac{A_{p(dB)}}{10} = \log^{-1}\left(\frac{-1.8}{10}\right) = \log^{-1}(-0.18) = \mathbf{0.661}$$

Now, using a value of $A_p = 0.661$, the output power from the circuit can be found as

$$\mathbf{P_{out}} = A_p P_{in} = (0.661)(2.5\text{ W}) = \mathbf{1.65\text{ W}}$$

### *PRACTICE PROBLEM 19.15*

A bandpass filter has maximum values of $A_{p(dB)} = -1.925$ dB and $P_{in} = 900$ mW. Determine the maximum output power for the filter.

## Why dB Values Are Used

*OBJECTIVE 11* ▶ There are two primary advantages to using dB gain values:

1. A very large range of values can be represented using relatively small numbers.
2. The total gain produced by series filters and/or amplifiers can be determined using simple addition.

The first of these advantages can be demonstrated by performing two simple calculations. For a value of $A_p = 1 \times 10^6$:

$$A_{p(dB)} = 10\log A_p = 10\log(1 \times 10^6) = (10)(6) = 60\text{ dB}$$

For a value of $A_p = 1 \times 10^{-6}$:

$$A_{p(dB)} = 10\log A_p = 10\log(1 \times 10^{-6}) = (10)(-6) = -60\text{ dB}$$

As these calculations demonstrate, we can represent any value of $A_p$ within a range of 0.000001 and 1,000,000 using dB values that fall between $\pm 60$ dB. As you will learn, most gain values fall far short of these extremes.

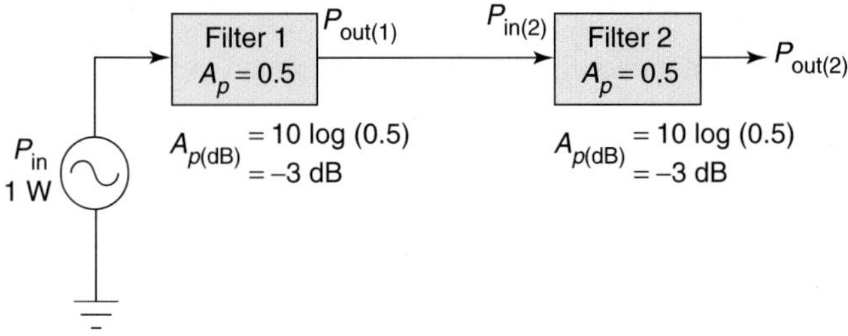


**FIGURE 19.17 Series-connected filters.**

The second advantage can be explained with the aid of Figure 19.17. The blocks represent two filters connected in series, each with a value of $A_p = 0.5$. When two or more filters are connected in series, the total power gain equals the product of the individual gain values. For example, equation (19.18) can be used to calculate the output power from the first filter as follows:

$$P_{out(1)} = A_p P_{in(1)} = (0.5)(1\text{ W}) = 500\text{ mW}$$

With the filters connected as shown, the output from the first filter serves as the input to the second filter. Therefore, $P_{in(2)} = P_{out(1)}$, and

$$P_{out(2)} = A_p P_{in(2)} = (0.5)(500\text{ mW}) = 250\text{ mW}$$

These calculations indicate that the series circuit has overall values of $P_{in} = 1$ W and $P_{out} = 250$ mW. Using these values, the overall power gain of the circuit can be found as

$$A_p = \frac{P_{out}}{P_{in}} = \frac{250\text{ mW}}{1\text{ W}} = 0.25$$

If we convert this overall power gain to dB form, we get

$$A_{p(dB)} = 10 \log A_p = 10 \log(0.25) = (10)(-0.6) = -6\text{ dB}$$

Now, compare this result to the value of $A_{p(dB)} = -3$ dB shown for each filter in the circuit. As you can see, the overall dB gain of the circuit equals the sum of the individual dB gains. This principle, which holds true regardless of the number of series-connected filters, can be expressed as

$$A_{pT(dB)} = A_{p1(dB)} + A_{p2(dB)} + \ldots + A_{pn(dB)} \tag{19.19}$$

## Multistage Filter Gain

When filters are connected in series (as shown in Figure 19.17), they are called a **cascade.** Each filter in the cascade is referred to as a **stage.** The combination of cascaded filters is commonly referred to as an ***n*-stage filter,** where *n* is the number of stages. For example, the cascaded filters in Figure 19.17 would be referred to as a two-stage filter.

When the stage gains of a multistage filter are given in dB, the total output power can be found using the following steps:

1. Determine the overall dB gain of the circuit.
2. Convert the dB gain to standard numeric form.
3. Multiply the input power by the overall gain to determine the total output power.

This procedure is demonstrated in Example 19.16.

**Cascade**
A group of series-connected filters (or other circuits). Each circuit in a cascade is referred to as a **stage.**

***n*-stage filter**
A term commonly used to refer to a cascaded circuit, where *n* is the number of stages.

## EXAMPLE 19.16

Determine the total output power for the circuit represented in Figure 19.18.

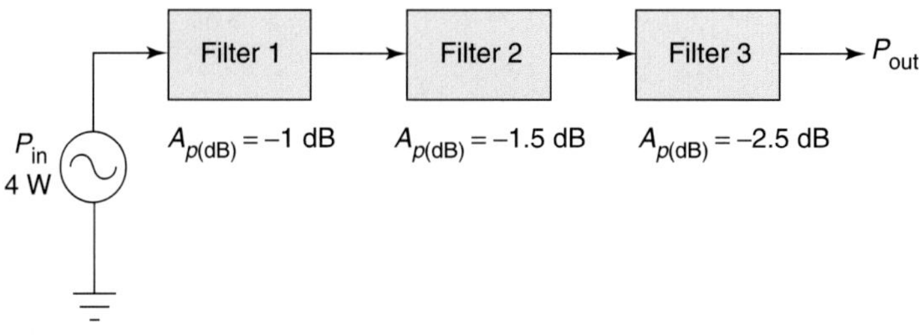


**FIGURE 19.18**

***Solution:*** First, the sum of the stage gains (in dB) is found as

$$\mathbf{A_{pT(dB)}} = (-1 \text{ dB}) + (-1.5 \text{ dB}) + (-2.5 \text{ dB}) = \mathbf{-5 \text{ dB}}$$

Now, the total power gain of the 3-stage filter is converted to standard numeric form as follows:

$$\mathbf{A_p} = \log^{-1}\frac{A_{pT(\text{dB})}}{10} = \log^{-1}\left(\frac{-5 \text{ dB}}{10}\right) = \log^{-1}(-0.5) = \mathbf{0.316}$$

Finally, the output power from the circuit can be found as

$$\mathbf{P_{out}} = A_p P_{in} = (0.316)(4 \text{ W}) = \mathbf{1.26 \text{ W}}$$

***PRACTICE PROBLEM 19.16***

A circuit like the one in Figure 19.18 has the following values: $A_{p1(\text{dB})} = -2$ dB, $A_{p2(\text{dB})} = -3$ dB, $A_{p3(\text{dB})} = -3.5$ dB, and $P_{in} = 2.5$ W. Calculate the output power from the circuit.

## The dBm Reference

Throughout this section, we have referred to the fact that dBs are commonly used to represent the ratio of output power to input power for a filter or amplifier. One limitation on the value of knowing dB power gain is that input power must be known to calculate output power. For example, let's say that a filter has a value of $A_{p(\text{dB})} = -2$ dB. This value cannot be used to determine the output power for the circuit unless we know the exact value of the input power.

**dBm reference**
A dB value that references a power value to 1 mW.

The **dBm reference** differs from $A_{p(\text{dB})}$ because it represents an actual power level. When given in dBm, circuit output power is referenced to 1 mW, as shown in the following relationship:

$$P_{\text{dBm}} = 10 \log\frac{P_{out}}{1 \text{ mW}} \qquad \textbf{(19.20)}$$

Note that any dBm value is an actual power level, rather than a power ratio. This point is illustrated in Example 19.17.

**EXAMPLE 19.17**

An amplifier has an output rated at 2 W. Calculate the value of output power in dBm for the circuit.

***Solution:*** Using equation (19.20), the amplifier output power can be found as

$$P_{\mathbf{dBm}} = 10 \log \frac{P_{\text{out}}}{1\ \text{mW}} = 10 \log \frac{2\ \text{W}}{1\ \text{mW}} = 10 \log(2 \times 10^3) = \mathbf{33\ dBm}$$

As this calculation demonstrates, 2 W has a value of 33 dBm.

***PRACTICE PROBLEM 19.17***

An amplifier has an output rated at 12 W. Calculate the value of output power in dBm for the circuit.

As demonstrated in the example, any value written in dBm corresponds to an actual power level, because 1 mW is always used as the reference. The dBm reference is used most commonly to identify a power limit. For example, a circuit may be rated for a maximum output power of 200 dBm. When this is the case, you need to be able to determine the value of $P_{\text{out}}$ using the dBm rating. This value can be determined using the following relationship:

$$P_{\text{out}} = (1\ \text{mW})\left(\log^{-1} \frac{P_{\text{dBm}}}{10}\right) \qquad \textbf{(19.21)}$$

An application of this relationship is provided in Example 19.18.

**EXAMPLE 19.18**

A circuit is rated for a maximum output of 50 dBm. Calculate the maximum output power that can be provided by the circuit.

***Solution:*** Using equation (19.21), the maximum output power from the circuit can be found as

$$P_{\mathbf{out}} = (1\ \text{mW})\left(\log^{-1} \frac{P_{\text{dBm}}}{10}\right) = (1\ \text{mW})\left(\log^{-1} \frac{50\ \text{dBm}}{10}\right) = (1\ \text{mW})(1 \times 10^5) = \mathbf{100\ W}$$

***PRACTICE PROBLEM 19.18***

A circuit is rated for a maximum output of 25 dBm. Calculate the maximum output power that can be provided by the circuit.

## *Putting It All Together*

Amplitude is normally represented on a frequency response curve using output-to-input ratios. The ratio of output power to input power for a filter or amplifier is referred to as power gain ($A_p$). For an amplifier, the value of $A_p$ is usually greater than one (1). The circuits we will focus on later in this chapter all have values of $A_p$ that are less than one.

When $A_p > 1$, ac output power is greater than ac input power. When $A_p = 1$, ac output power equals ac input power. When $A_p < 1$, ac output power is less than ac input power.

Power ratios are normally represented using logarithmic values called decibels (dB). Decibels allow us to represent a wide range of gain values using relatively small values. The other advantages of using dB gain values are:

1. Reciprocals have positive and negative dB values that are equal in magnitude.
2. The total gain of a multistage filter can be determined by simply adding the dB gains of the individual stages.

At the same time, dB values must be converted to standard numeric form before they can be used in any power calculations. In other words, if you know the values of input power and dB power gain for a filter, you must convert the dB gain to standard numeric form before you can actually calculate the circuit output power.

The dBm reference is an indicator of an actual power value. The dBm value for any power level is determined by referencing that power level to 1 mW. The dBm reference is often used on specification sheets to indicate the maximum possible output power that can be provided by a component or system.

### Section Review

1. What is a bel?
2. What is power gain ($A_p$)?
3. Describe the bel and power relationships corresponding to each of the following conditions:
   a. $A_p > 1$
   b. $A_p = 1$
   c. $A_p < 1$
4. What is the relationship between bel representations of reciprocal values?
5. What is a decibel (dB)?
6. What characteristics do dB values share with bel values?
7. What is the procedure for calculating output power when input power and dB power gain are known?
8. What are the advantages of using dBs to represent gain?
9. What is a cascaded filter?
10. How do you determine the overall dB gain of a multistage filter?
11. What is represented using the dBm reference?
12. What is the most common application for dBm values?

## 19.3 AMPLITUDE MEASUREMENTS: dB VOLTAGE AND CURRENT GAIN

In the last section, the decibel was defined in terms of power gain. In this section, you will see that dBs can also be defined in terms of voltage gain and current gain. As such, the cutoff frequencies of a filter can also be defined in terms of these values.

### Voltage Gain

**Voltage gain ($A_v$)**
The ratio of circuit output voltage to input voltage.

The ratio of circuit output voltage to input voltage is generally referred to as **voltage gain ($A_v$)**. In many cases, voltage gain (rather than power gain) is used as the means of measuring amplitude on frequency response curves. For example, the frequency response curve of a bandpass filter can be plotted as shown in Figure 19.19.

**FIGURE 19.19 Voltage gain measurement on a frequency response curve.**

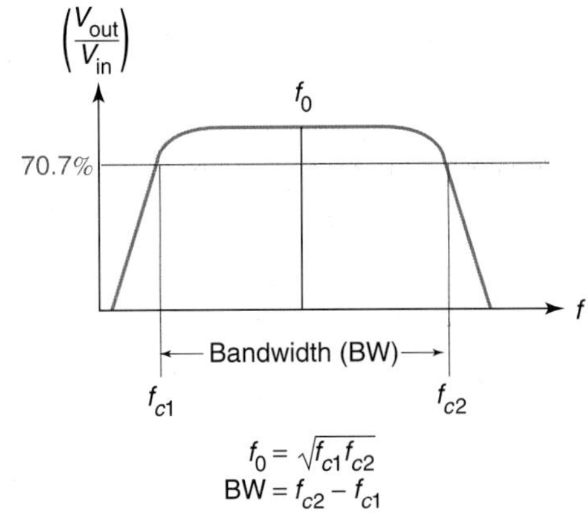


The difference between this curve and those discussed earlier is that the cutoff frequencies are defined in terms of a voltage ratio rather than a power ratio. As shown in the figure, the voltage gain of the curve drops to 70.7% of its maximum value at the cutoff frequencies. By formula,

Voltage gain at the cutoff frequencies.

$$f = f_c \qquad \text{when} \qquad A_v = 0.707\, A_{v(\max)} \tag{19.22}$$

## The Basis for Equation (19.22)

Earlier, the cutoff frequencies of a filter were defined as the frequencies where power gain drops to 50% of its maximum value. This relationship can be written as

$$f = f_c \qquad \text{when} \qquad A_p = 0.5\, A_{p(\max)}$$

Now, let's take a look at the relationship between voltage and power. Assume for a moment that $A_{p(\max)}$ can be defined as

$$A_{p(\max)} = \frac{V_{out}^2}{R}$$

This assumption is valid if we normalize the value of input power, that is, if we assume that the input power has a value of one (1). Here is what happens when $V_{out}$ drops to 70.7% of its peak value:

$$A_p = \frac{(0.707\, V_{out})^2}{R} = \frac{(0.707)^2\, V_{out}^2}{R} = 0.5\frac{V_{out}^2}{R} = 0.5\, A_{p(\max)}$$

As these calculations indicate, power drops to 50% of its maximum value when voltage drops to 70.7% of its maximum value. The same holds true for power gain and voltage gain. $A_p$ drops to 50% of its maximum value when $A_v$ drops to 70.7% of its maximum value. Therefore, $A_v = 0.707\, A_{v(\max)}$ at the cutoff frequencies.

## dB Voltage Gain

◄ *OBJECTIVE 12*

The dB voltage gain of a filter (or amplifier) is found as

$$A_{v(\text{dB})} = 20 \log \frac{V_{out}}{V_{in}} \tag{19.23}$$

or

$$A_{v(\text{dB})} = 20 \log A_v \tag{19.24}$$

where $\quad A_v =$ the ratio of output voltage to input voltage

The calculation of dB voltage gain is demonstrated in Example 19.19.

## EXAMPLE 19.19

A filter has a value of $A_v = 0.707$. Calculate the value of $A_{v(\text{dB})}$ for the circuit.

***Solution:*** Using equation (19.24), the dB voltage gain of the filter can be found as

$$\mathbf{A_{v(dB)}} = 20 \log A_v = 20 \log (0.707) = (20)(-0.15) = \mathbf{-3\ dB}$$

***PRACTICE PROBLEM 19.19***

A filter has a value of $A_v = 0.5$. Calculate the value of $A_{v(\text{dB})}$ for the circuit.

The value of $A_v$ used in Example 19.19 was selected to demonstrate further the relationship between voltage gain and power gain. Here is a comparison of the values from Examples 19.12 and 19.19:

| Example | Gain | Gain (in dB) |
|---|---|---|
| 19.12 | $A_p = 0.5$ | $A_{p(\text{dB})} = -3$ dB |
| 19.19 | $A_v = 0.707$ | $A_{v(\text{dB})} = -3$ dB |

As you can see, the two circuits have the same value of dB gain (−3 dB). This demonstrates, once again, the correlation between a power gain of 0.5 and a voltage gain of 0.707.

## The Basis for Equation (19.24)

As you know, $P = V^2/R$. Therefore:

$$P_{out} = \frac{V_{out}^2}{R_{out}} \quad \text{and} \quad P_{in} = \frac{V_{in}^2}{R_{in}}$$

Using these values in place of $P_{out}$ and $P_{in}$ in equation (19.15), we get

$$A_{(\text{dB})} = 10 \log \frac{P_{out}}{P_{in}} = 10 \log \frac{\dfrac{V_{out}^2}{R_{out}}}{\dfrac{V_{in}^2}{R_{in}}} = 10 \log \left( \frac{V_{out}^2}{V_{in}^2} \times \frac{R_{in}}{R_{out}} \right)$$

If we normalize the value of $R_{in}/R_{out}$, then

$$A_{(\text{dB})} = 10 \log \frac{V_{out}^2}{V_{in}^2} = 10 \log \left( \frac{V_{out}}{V_{in}} \right)^2 = 10 \log (A_v)^2 = 20 \log A_v$$

As this relationship indicates, a conversion factor of 20 is used when calculating $A_{v(\text{dB})}$ in place of the 10 used when calculating $A_{p(\text{dB})}$.

## Converting dB Voltage Gain to Standard Numeric Form

*OBJECTIVE 13* ▶ When you want to convert a dB voltage gain to standard numeric form, the following equation is used:

$$A_v = \log^{-1} \frac{A_{v(\text{dB})}}{20} \tag{19.25}$$

Note that the conversion factor of 20 is used in the denominator of the fraction. Otherwise, the equation is nearly identical to equation (19.17), which we used to convert dB power gain to standard numeric form. Example 19.20 demonstrates the process for converting dB voltage gain to standard numeric form.

**EXAMPLE 19.20**

A filter is rated at $A_{v(\text{dB})} = -2.8$ dB. Calculate the value of $A_v$ for the circuit.

***Solution:*** Using equation (19.25), the value of $A_{v(\text{dB})}$ is converted to standard numeric form as follows:

$$\mathbf{A_v} = \log^{-1}\frac{A_{v(\text{dB})}}{20} = \log^{-1}\left(\frac{-2.8\text{ dB}}{20}\right) = \log^{-1}(-0.14) = \mathbf{0.724}$$

This result tells us that $V_{\text{out}} = 0.724\ V_{\text{in}}$.

***PRACTICE PROBLEM 19.20***

A filter is rated at $A_{v(\text{dB})} = -3.5$ dB. Calculate the value of $A_v$ for the circuit.

A positive dB voltage gain indicates that output signal voltage is greater than input signal voltage. Increasing signal voltage is accomplished using a *voltage amplifier,* a circuit capable of increasing the voltage level of its input signal. The relationship between positive dB voltage gains and the value of $A_v$ is demonstrated in Example 19.21.

**EXAMPLE 19.21**

An amplifier is rated at $A_{v(\text{dB})} = 2.8$ dB. Calculate the value of $A_v$ for the circuit.

***Solution:*** Using equation (19.25) the value of $A_{v(\text{dB})}$ is converted to standard numeric form as follows:

$$\mathbf{A_v} = \log^{-1}\frac{A_{v(\text{dB})}}{20} = \log^{-1}\left(\frac{2.8\text{ dB}}{20}\right) = \log^{-1}(0.14) = \mathbf{1.38}$$

This result tells us that the amplifier provides a value of $V_{\text{out}} = 1.38\ V_{\text{in}}$.

***PRACTICE PROBLEM 19.21***

A filter is rated at $A_{v(\text{dB})} = 3.5$ dB. Calculate the value of $A_v$ for the circuit.

## Reciprocal Relationships

Earlier, you were told that reciprocals have positive and negative dB values that are equal in magnitude. This principle applies to dB voltage gain as well as dB power gain. For example, let's take a look at the values from Examples 19.20 and 19.21. For convenience, these values are summarized as follows:

| Example | Voltage Gain | Voltage Gain (in dB) |
|---|---|---|
| 19.20 | $A_v = 0.724$ | $A_{v(\text{dB})} = -2.8$ dB |
| 19.21 | $A_v = 1.38$ | $A_{v(\text{dB})} = 2.8$ dB |

The voltage gain values listed here are approximate reciprocals; that is, $1/0.724 \cong 1.38$ (and vice versa). As you can see, their dB values have equal magnitudes. The same relationship can be seen by comparing your results from Practice Problems 19.20 and 19.21.

## Changes in dB Gain

Starting in the next section, we will discuss the operation of various filter circuits. Among other things, we will be interested in the rate at which gain decreases when a filter is operated outside its bandwidth.

Even though dB values of voltage and power gain are calculated differently, they change at the same rate. By formula,

$$\Delta A_{p(\text{dB})} = \Delta A_{v(\text{dB})} \tag{19.26}$$

where $\Delta A_{p(\text{dB})}$ = the change in dB power gain
$\Delta A_{v(\text{dB})}$ = the change in dB voltage gain

For example, if the voltage gain of a filter drops by 3 dB, the power gain drops by the same amount. The actual values of $A_v$ and $A_p$ may or may not be equal, but they will change by the same amount. The significance of this relationship will become clear in the next section.

## Current Gain ($A_i$)

OBJECTIVE 14 ▸

**Current gain ($A_i$)**
The ratio of circuit output current to input current.

The ratio of circuit output current to input current is generally referred to as **current gain ($A_i$).** In terms of frequency response, current gain characteristics are identical to those introduced for voltage gain. In fact, current gain relationships are identical to those given earlier for voltage gain. For convenience, these relationships are rewritten for current gain as follows:

$$f = f_c \quad \text{when} \quad A_i = 0.707\, A_{i(\max)} \tag{19.27}$$

$$A_{i(\text{dB})} = 20 \log \frac{i_{\text{out}}}{i_{\text{in}}} \tag{19.28}$$

$$A_{i(\text{dB})} = 20 \log A_i \tag{19.29}$$

$$A_i = \log^{-1} \frac{A_{i(\text{dB})}}{20} \tag{19.30}$$

$$\Delta A_{p(\text{dB})} = \Delta A_{i(\text{dB})} \tag{19.31}$$

Since these relationships are identical to the voltage gain relationships given earlier, we do not need to elaborate on them any further at this point.

### Section Review

1. What is voltage gain ($A_v$)?
2. What is the value of $A_v$ (as a percentage) at the cutoff frequencies of a filter?
3. Compare the calculations of $A_{v(\text{dB})}$ and $A_{p(\text{dB})}$.
4. What is the relationship between $\Delta A_{p(\text{dB})}$ and $\Delta A_{v(\text{dB})}$?
5. What is current gain ($A_i$)?

# 19.4 LOW-PASS FILTERS

Simple *RL* and *RC* circuits can be used as low-pass and high-pass filters. In this section, we will look at *RL* and *RC* low-pass and high-pass circuits, their characteristics, and circuit analysis procedures. We will also discuss an alternate method of graphing frequency response, called a *bode* (pronounced bō-dē) *plot.*

OBJECTIVE 15 ▸

## *RC* Low-Pass Filters

An *RC* circuit acts as a low-pass filter when constructed as shown in Figure 19.20. In the circuit shown, the resistor is positioned directly in the signal path, that is, directly between the source and the load. The capacitor is connected from the signal path to ground, in

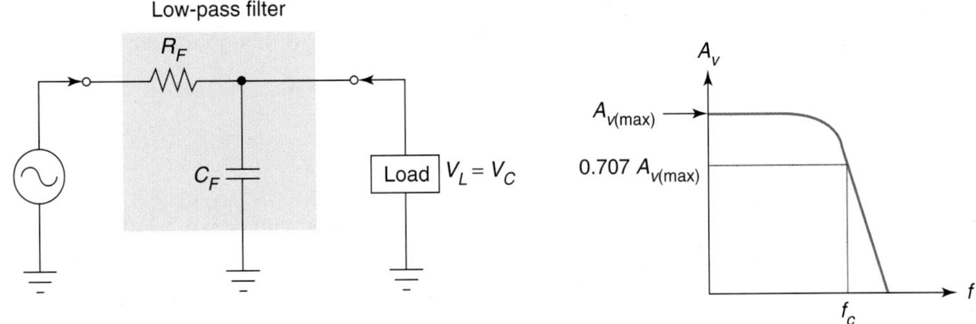


**FIGURE 19.20 An *RC* low-pass filter and its frequency response curve.**

parallel with the load. Therefore, $V_L = V_C$ (as shown in the figure). When a capacitor (or other component) is connected in parallel with the load, it is referred to as a **shunt** component.

**Shunt**
A term used to describe a component or circuit connected in parallel with a load (or other circuit).

The filtering action of the circuit in Figure 19.20 is a result of the capacitor's response to an increase in frequency. This concept is illustrated in Figure 19.21. The reactance curve in Figure 19.21a was introduced in Chapter 18. As the curve indicates, a capacitor has near infinite reactance when its operating frequency is 0 Hz. With this in mind, look at the circuit shown in Figure 19.21b. Assuming that the input frequency is 0 Hz, the capacitor effectively acts as an open circuit, so it has been dropped from the equivalent circuit. In this case, the voltage across the load equals the difference between $V_S$ and $V_R$.

At the opposite end of the reactance curve (Figure 19.21a), the value of $X_C$ approaches 0 Ω. When this is the case, the filter has the equivalent circuit shown in Figure 19.21c. As you can see, the capacitor is represented as a short circuit in parallel with the load. In this case, $V_L = 0$ V.

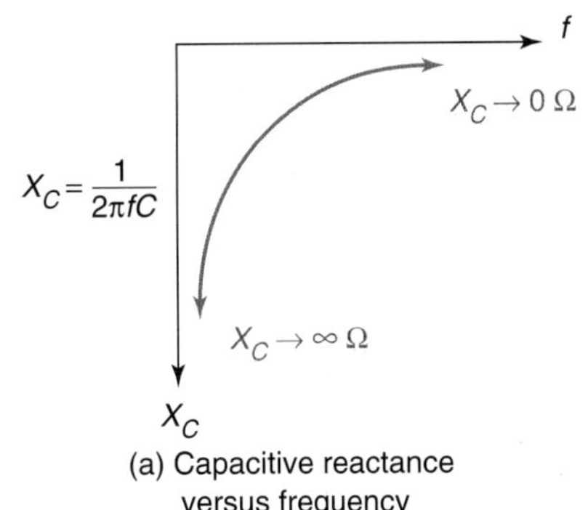


(a) Capacitive reactance versus frequency

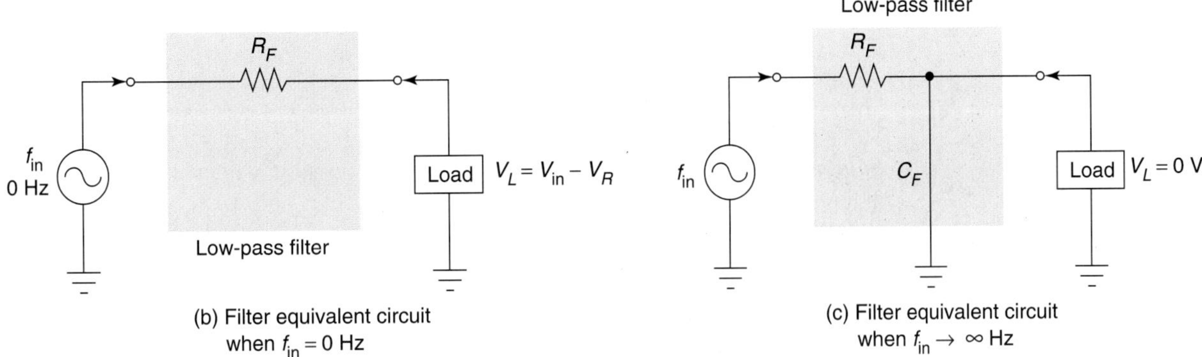


(b) Filter equivalent circuit when $f_{in} = 0$ Hz

(c) Filter equivalent circuit when $f_{in} \rightarrow \infty$ Hz

**FIGURE 19.21 *RC* low-pass filter operating limits.**

Between the extremes represented in Figure 19.21 lies a range of frequencies over which $V_L$ decreases from $(V_S - V_R)$ to 0 V. At this point, our primary interests are establishing methods to determine:

1. The maximum value of $A_v$ for the circuit.
2. The cutoff frequency ($f_c$) for the circuit.

***Maximum Voltage Gain, $A_{v(\text{max})}$.*** The maximum gain of a low-pass filter occurs when the input frequency is 0 Hz. As shown in Figure 19.21b, the load forms a voltage divider with the series resistor ($R$) when $f_{\text{in}} = 0$ Hz. Therefore, the maximum load voltage is found as

$$V_{L(\text{max})} = V_{\text{in}} \frac{R_L}{R_L + R_F} \quad \textbf{(19.32)}$$

As you know, voltage gain is the ratio of output voltage to input voltage. For the $RC$ low-pass filter, the maximum voltage gain can be found as

$$A_{v(\text{max})} = \frac{V_{L(\text{max})}}{V_{\text{in}}} \quad \textbf{(19.33)}$$

Substituting equation (19.32) for the value of $V_{L(\text{max})}$ in equation (19.33), we get

$$A_{v(\text{max})} = \frac{V_{\text{in}}\left(\dfrac{R_L}{R_L + R_F}\right)}{V_{\text{in}}}$$

The values of $V_{\text{in}}$ in the above equation cancel, leaving us with:

$$A_{v(\text{max})} = \frac{R_L}{R_L + R_F} \quad \textbf{(19.34)}$$

Example 19.22 demonstrates the usefulness of this relationship.

## EXAMPLE 19.22

Calculate the maximum voltage gain for the low-pass filter in Figure 19.22a.

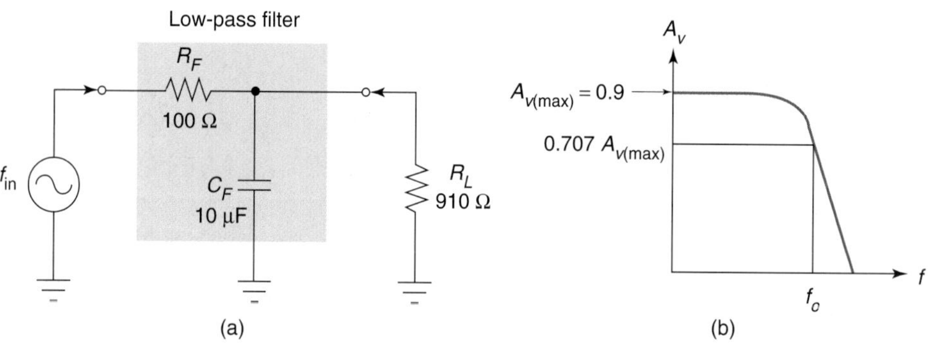


FIGURE 19.22

***Solution:*** Using equation (19.34), the maximum possible voltage gain for the circuit is found as

$$A_{v(\text{max})} = \frac{R_L}{R_L + R_F} = \frac{910\ \Omega}{1010\ \Omega} = \mathbf{0.9}$$

This value is shown in the frequency response curve in Figure 19.22b.

### *PRACTICE PROBLEM 19.22*

A circuit like the one in Figure 19.22a has the following values: $R_F = 51\ \Omega$, $C_F = 47\ \mu F$, and $R_L = 750\ \Omega$. Calculate the value of $A_{v(max)}$ for the circuit.

Equation (19.34) is also important because it defines the limit on the voltage gain of an *RC* low-pass filter. Since $(R_L + R_F)$ must be greater than $R_L$, the value of $A_{v(max)}$ must be less than 1. However, a value of $A_{v(max)} \cong 1$ can be achieved when $R_F << R_L$. This relationship is demonstrated in Example 19.23.

## EXAMPLE 19.23

Calculate the value of $A_{v(max)}$ for the circuit shown in Figure 19.23.

**FIGURE 19.23**

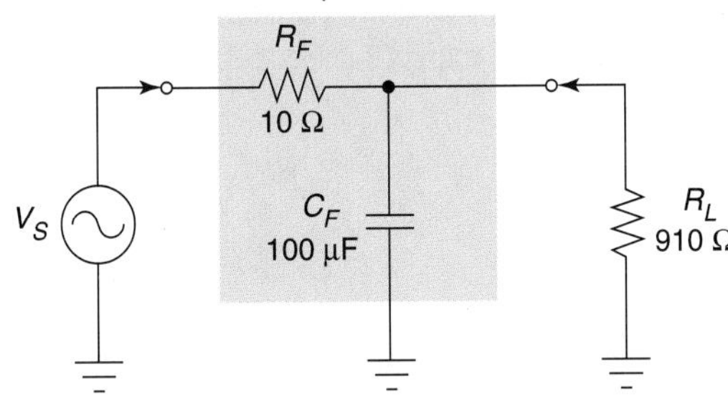


***Solution:*** Using the values shown in the figure, the value of $A_{v(max)}$ can be found as

$$\mathbf{A_{v(max)}} = \frac{R_L}{R_L + R_F} = \frac{910\ \Omega}{920\ \Omega} = \mathbf{0.99}$$

As demonstrated in the example, $A_{v(max)} \cong 1$ when $R_F << R_L$. The greater the difference between the values of $R_F$ and $R_L$, the closer the value of $A_{v(max)}$ comes to one (1). This relationship can be seen by comparing the results from Examples 19.22 and 19.23, where only the value of $R_F$ was changed.

*Remember:* A gain of one (1) is commonly referred to as *unity gain.*

Because of the relationship among $A_{v(max)}$, $R_F$, and $R_L$, *RC* low-pass filters are normally designed using the lowest practical values of series resistance. As you will see, the value of a filter resistor also affects the circuit's cutoff frequency.

***Upper Cutoff Frequency ($f_c$).*** The upper cutoff frequency for a given *RC* low-pass filter is determined by the circuit resistor and capacitor values. By formula,

$$f_c = \frac{1}{2\pi RC} \qquad \textbf{(19.35)}$$

where $f_c$ = the circuit's cutoff frequency
$R$ = the total circuit resistance *as seen by the capacitor*
$C$ = the value of the filter capacitor

We will establish the basis for this equation in a moment. First, we need to take a closer look at the value of $R$ in the equation. As stated above, the value of $R$ used in the equation is the total resistance as seen by the capacitor. This resistance can be identified as shown in Figure 19.24.

Although only two resistors are shown in Figure 19.24a, the circuit has three resistance values. These values are the filter resistor ($R_F$), the load resistance ($R_L$), and the resistance of the source ($R_S$). In Figure 19.24b, the source has been replaced by a resistor

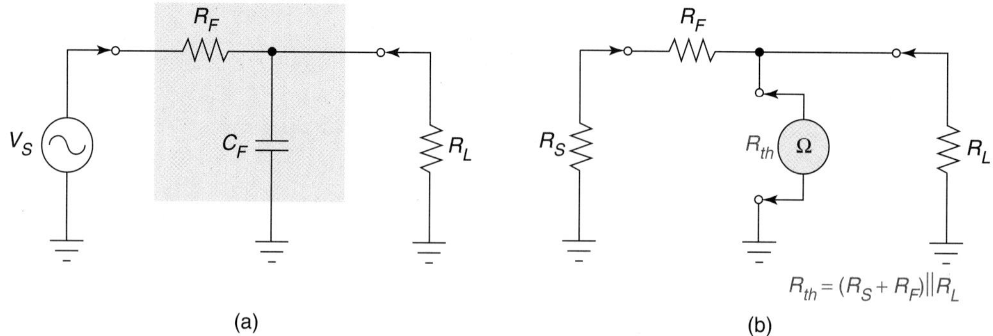


**FIGURE 19.24 Filter resistance as seen by the capacitor.**

*A Practical Consideration:* The term *Thevenin resistance* is traditionally applied to load analysis. However, the term can be used to describe the resistance across any open pair of component terminals.

representing the value of $R_S$, and the capacitor has been removed. The total resistance seen by the capacitor can be measured as a Thevenin resistance ($R_{th}$) across the open capacitor terminals. As shown in the figure, the capacitor sees a parallel combination of ($R_S + R_F$) and $R_L$. Therefore,

$$R_{th} = (R_S + R_F) \| R_L \tag{19.36}$$

and

$$f_c = \frac{1}{2\pi R_{th} C} \tag{19.37}$$

Example 19.24 demonstrates the procedure for calculating the cutoff frequency of an *RC* low-pass filter.

## EXAMPLE 19.24

Determine the cutoff frequency for the circuit shown in Figure 19.25a.

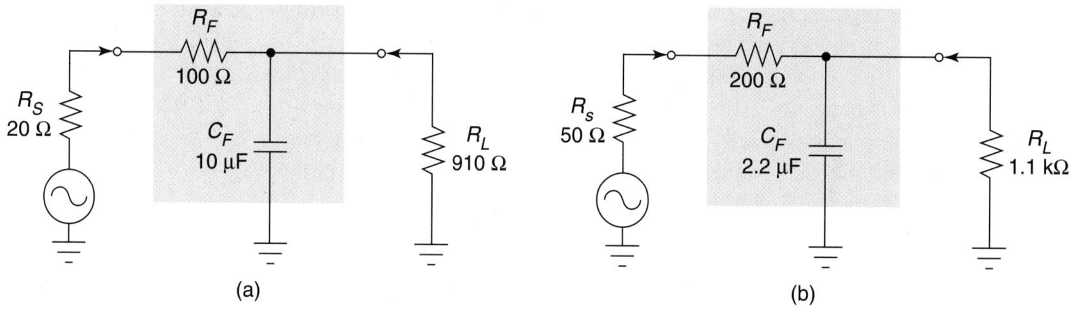


EWB

**FIGURE 19.25**

***Solution:*** Using the resistance values shown in the figure, the value of $R_{th}$ can be found as

$$R_{th} = (R_S + R_F) \| R_L = (100\ \Omega + 20\ \Omega) \| 910\ \Omega = 106\ \Omega$$

This represents the total circuit resistance as seen by the capacitor. Now, the cutoff frequency of the circuit can be found as

$$f_c = \frac{1}{2\pi R_{th} C_F} = \frac{1}{2\pi(106\ \Omega)(10\ \mu\text{F})} = 150\ \text{Hz}$$

This result indicates that the voltage gain of the circuit drops to 70.7% of its maximum value when the operating frequency reaches 150 Hz.

***PRACTICE PROBLEM 19.24***

Determine the cutoff frequency for the circuit shown in Figure 19.25b.

The circuit used in this example is the same one we used in Example 19.22 (with an added value of source resistance). In Example 19.22, we determined that the circuit has a value of $A_{v(\max)} = 0.9$. Now that we know its cutoff frequency, we can plot the frequency response curve for the circuit as shown in Figure 19.26. This curve tells us that the ratio of the circuit output voltage to input voltage has a maximum value of 0.9. As the input frequency is increased to 150 Hz, the voltage ratio drops to a value of 0.636.

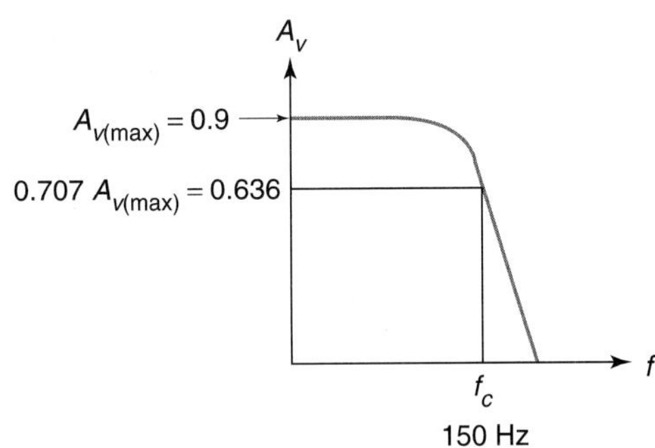


**FIGURE 19.26 The frequency response curve for the circuit in Examples 19.22 and 19.24.**

***The Basis for Equation (19.35).*** The maximum voltage gain of a filter is defined as

$$A_{v(\max)} = \frac{V_{\text{out}(\max)}}{V_{\text{in}}}$$

For the circuit in Figure 19.27, the output voltage is measured across the capacitor. Therefore, the above equation can be rewritten as

$$A_{v(\max)} = \frac{V_{C(\max)}}{V_S}$$

Since the resistor and capacitor form a voltage divider, the value of the output voltage can be found as

$$V_{\text{out}} = V_S \frac{X_C}{\sqrt{X_C^2 + R^2}}$$

**FIGURE 19.27**

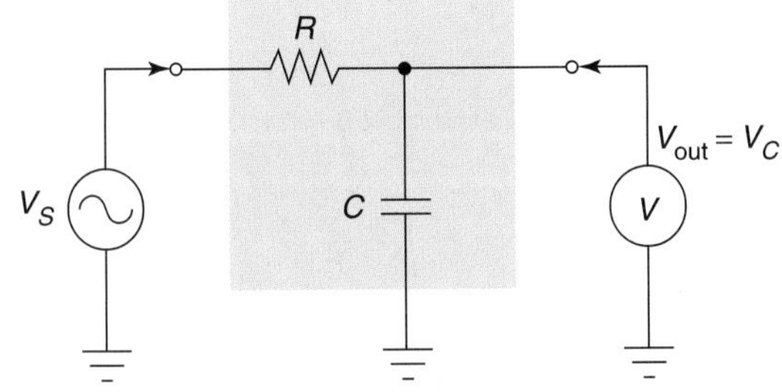

If we divide both sides of the equation by $V_S$, we get

$$\frac{V_{\text{out}}}{V_S} = \frac{X_C}{\sqrt{X_C^2 + R^2}}$$

or

$$A_v = \frac{X_C}{\sqrt{X_C^2 + R^2}}$$

As you know, the value of $X_C$ approaches infinity at 0 Hz. As frequency increases, the value of $X_C$ decreases. Assume for a moment that frequency increases to the point where $X_C = R$. At this frequency,

$$A_v = \frac{X_C}{\sqrt{X_C^2 + R^2}} = \frac{R}{\sqrt{R^2 + R^2}} = \frac{R}{\sqrt{2R^2}} = \frac{R}{1.414R} = 0.707$$

This result indicates that $A_v$ drops from its maximum value to 0.707 $A_{v(\text{max})}$ when $X_C = R$. As you know, the reactance of a capacitor is found as

$$X_C = \frac{1}{2\pi fC}$$

Transposing this equation, we get

$$f = \frac{1}{2\pi X_C C}$$

We know that $X_C = R$ at the cutoff frequency. Therefore,

$$f_c = \frac{1}{2\pi RC}$$

which is equation (19.35).

OBJECTIVE 16 ▶

## Bode Plots

**Bode plot**
A normalized graph that represents frequency response as a change in gain versus operating frequency.

Bode plots are ideal representations of frequency response.

A **bode plot** is a normalized graph that represents frequency response as a change in gain ($\Delta A_v$) versus operating frequency. A frequency response curve and its equivalent bode plot are shown in Figure 19.28. (The frequency response curve was plotted using the results from Examples 19.22 and 19.24. For the sake of comparison, the gain values on the frequency response curve have been converted to dB values.)

A bode plot is an ideal plot of frequency response, because it assumes that gain remains constant until the cutoff frequency is reached. For example, the bode plot in Figure 19.28 shows a value of $\Delta A_v = 0$ dB from 0 Hz to 150 Hz. At frequencies above 150 Hz, the

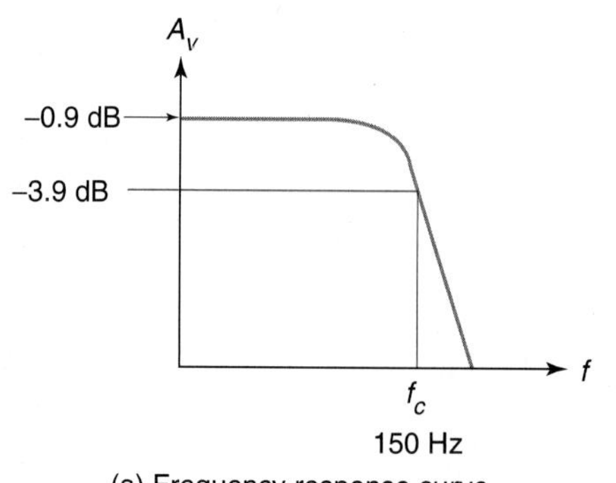

(a) Frequency response curve

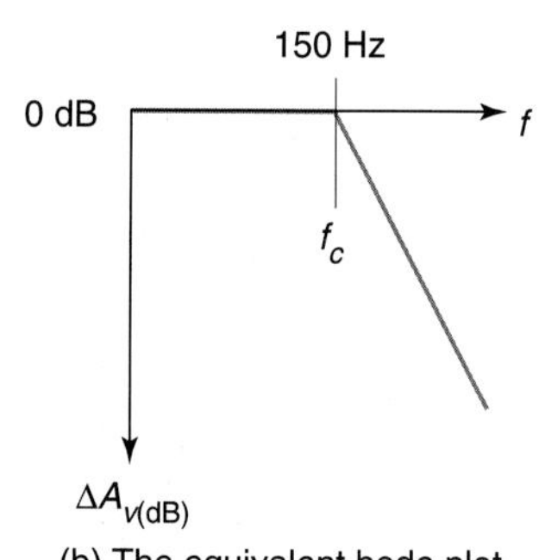

(b) The equivalent bode plot

**FIGURE 19.28 A frequency response curve and its equivalent bode plot.**

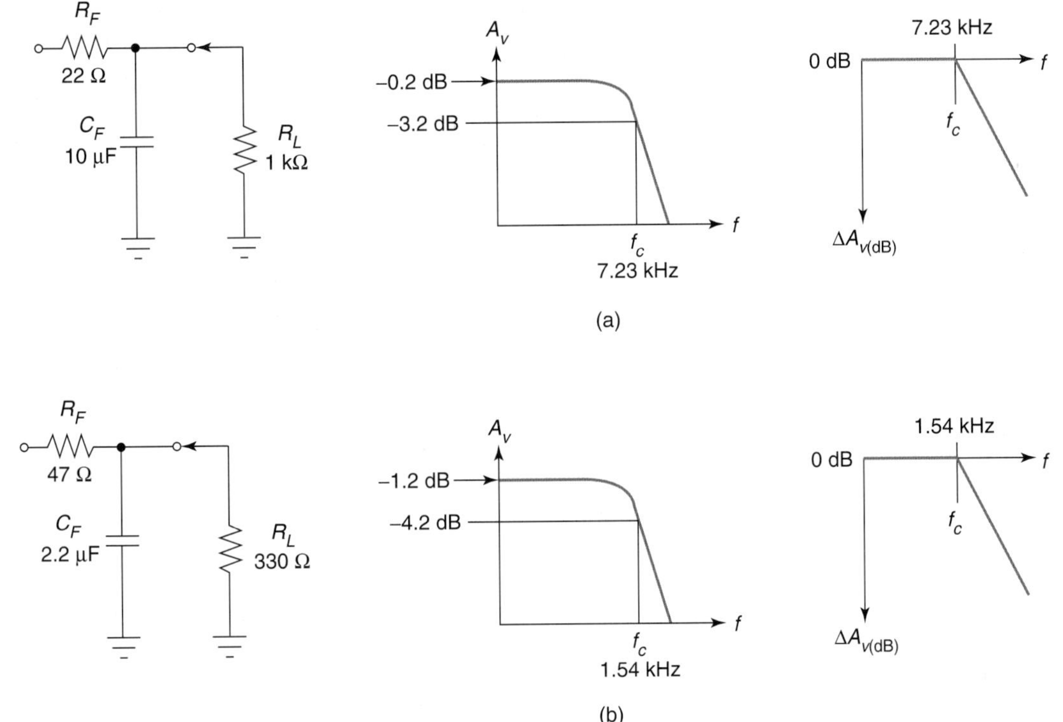


**FIGURE 19.29**

bode plot shows the gain of the circuit as changing at a constant rate. In contrast, the frequency response curve shows voltage gain as beginning to decrease well before the cutoff frequency is reached.

Why bode plots are used.

The value of the bode plot lies in the fact that it uses a change in voltage gain to represent frequency response. As a result, the only difference between one bode plot and another is the value of the cutoff frequency (assuming that the circuits have the same number of stages). For example, compare the circuits and curves shown in Figure 19.29. The circuits have different values of $A_{v(\max)}$, and therefore, different values of $A_v$ at the cutoff frequency. However, the bode plots for the circuits differ only in the cutoff frequency. Thus, a single bode plot can be used to represent the operating characteristics of any low-pass filter without recalculating actual gain values.

**3 dB point or 3 dB frequency** Terms commonly used to describe a given cutoff frequency, based on the fact that gain drops by 3 dB at that frequency.

If you compare the two frequency response curves in Figure 19.29, you'll see that each has a difference of 3 dB between $A_{v(\max)}$ and the value of $A_v$ at the cutoff frequency. In other words, the voltage gain of each circuit at its cutoff frequency can be found as $A_v = A_{v(\max)} - 3$ dB. For this reason, the cutoff frequency of a low-pass filter is often referred to as its **3 dB point,** or **3 dB frequency.**

## Filter Roll-Off Rate

◀ *OBJECTIVE 17*

When the operating frequency of a low-pass filter goes above the circuit cutoff frequency, the gain of the circuit drops at a constant rate. This constant drop in gain is referred to as **roll-off.** The roll-off of a low-pass filter is represented by the highlighted portion of the bode plot in Figure 19.30.

**Roll-off** A term used to describe the decrease in gain that occurs when a filter is operated outside its pass band.

**Roll-off rate** The rate at which gain rolls off, normally given in dB per octave or dB per decade.

The **roll-off rate** of a filter is the rate of change in gain experienced by the circuit when it is operated outside its frequency limit, usually expressed in dB per octave or dB per decade. For example, the bode plot in Figure 19.31a shows a decrease in gain of 6 dB whenever frequency doubles, which is described as a roll-off rate of *6 dB per octave.* The bode plot in Figure 19.31b shows a 20 dB decrease in gain whenever frequency increases by a factor of ten (10), which is described as a roll-off rate of *20 dB per decade.*

FIGURE 19.30 Low-pass filter roll-off.

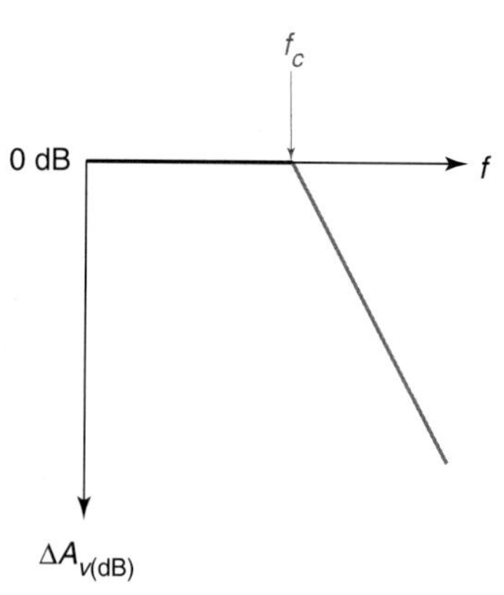


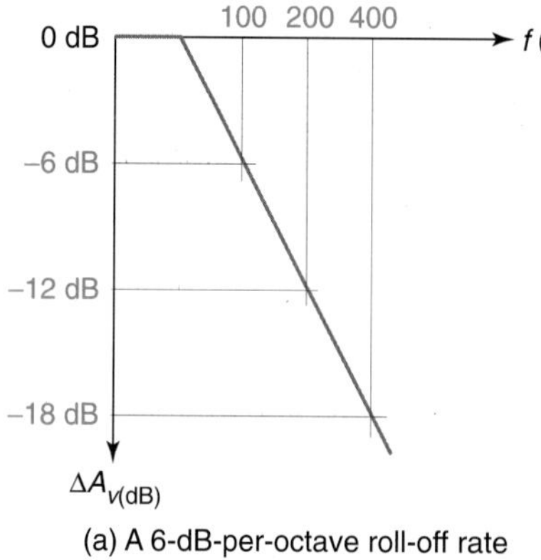


(a) A 6-dB-per-octave roll-off rate

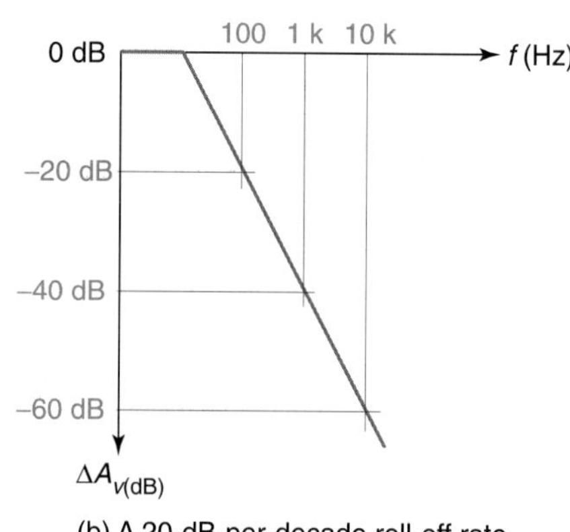


(b) A 20-dB-per-decade roll-off rate

FIGURE 19.31 Filter roll-off rates.

The roll-off rates used in Figure 19.31 were chosen for two reasons that may surprise you:

1. They are equal; that is, a roll-off rate of 6 dB per octave is equal to a roll-off rate of 20 dB per decade.
2. These roll-off rates apply to every single-stage *RC* circuit, regardless of the component values. In other words, every *RC* circuit operated outside its cutoff frequency has a roll-off rate of approximately 6 dB per octave, or 20 dB per decade.

*Note:* Roll-off rates are additive. Therefore, the overall roll-off rate for a multi-stage filter equals the sum of the stage roll-off rates. For example, the overall roll-off rate of a two-stage passive *RC* filter is 12 dB/octave (40 dB/decade).

To make sense of these characteristics, we need to take a look at how the value of $\Delta A_{v(\mathrm{dB})}$ is calculated at specific frequencies.

## Calculating Filter Roll-Off

The change in voltage gain experienced by an *RC* low-pass filter operated outside its pass band can be found as

$$\Delta A_v = \frac{1}{\sqrt{1 + (f_o/f_c)^2}} \tag{19.38}$$

or

$$\Delta A_{v(\mathrm{dB})} = 20 \log \frac{1}{\sqrt{1 + (f_o/f_c)^2}} \tag{19.39}$$

where $f_o$ = the circuit operating frequency
$f_c$ = the circuit cutoff frequency

Equations (19.38) and (19.39) are derived in Appendix E. We can validate the equations by using them to calculate the change in gain experienced by a filter when operated at its cutoff frequency, as shown in Example 19.25.

**EXAMPLE 19.25**

An *RC* low-pass filter is operated at its cutoff frequency. Calculate the change in voltage gain experienced by the circuit.

***Solution:*** The circuit is operating at cutoff, so

$$\left(\frac{f_o}{f_c}\right)^2 = 1$$

and

$$\mathbf{\Delta A_v} = \frac{1}{\sqrt{1 + (f_o/f_c)^2}} = \frac{1}{\sqrt{1+1}} = \frac{1}{1.414} = \mathbf{0.707}$$

In dB form,

$$\mathbf{\Delta A_{v(dB)}} = 20 \log \frac{1}{\sqrt{1 + (f_o/f_c)^2}} = 20 \log (0.707) = \mathbf{-3\ dB}$$

## Low-Pass Filter Roll-Off Rates

Earlier, you were told that the roll-off rate of an *RC* low-pass filter is 6 dB per octave (or 20 dB per decade). The first of these roll-off rates can be verified using the following sequence of calculations:

When $f_o = 2f_c$: $\quad \Delta A_{v(dB)} = 20 \log \frac{1}{\sqrt{1 + 2^2}} = -6.99\ \text{dB} \ \Big|\ \frac{f_o}{f_c} = 2$

When $f_o = 4f_c$: $\quad \Delta A_{v(dB)} = 20 \log \frac{1}{\sqrt{1 + 4^2}} = -12.3\ \text{dB} \ \Big|\ \frac{f_o}{f_c} = 4$

When $f_o = 8f_c$: $\quad \Delta A_{v(dB)} = 20 \log \frac{1}{\sqrt{1 + 8^2}} = -18.1\ \text{dB} \ \Big|\ \frac{f_o}{f_c} = 8$

As you can see, the voltage gain of the circuit changes by approximately 6 dB every time the operating frequency doubles. Therefore, we can say that the circuit has a roll-off rate of approximately 6 dB per octave. Note that the rate comes closer to exactly 6 dB per octave if the sequence of calculations continues to higher multiples of $f_c$.

The second roll-off rate (20 dB per decade) can be verified using the following sequence of calculations:

When $f_o = 10f_c$: $\quad \Delta A_{v(dB)} = 20 \log \frac{1}{\sqrt{1 + 10^2}} = -20\ \text{dB} \ \Big|\ \frac{f_o}{f_c} = 10$

When $f_o = 100f_c$: $\quad \Delta A_{v(dB)} = 20 \log \frac{1}{\sqrt{1 + 100^2}} = -40\ \text{dB} \ \Big|\ \frac{f_o}{f_c} = 100$

When $f_o = 1000f_c$: $\quad \Delta A_{v(dB)} = 20 \log \frac{1}{\sqrt{1 + 1000^2}} = -60\ \text{dB} \ \Big|\ \frac{f_o}{f_c} = 1000$

As you can see, the voltage gain of the circuit decreases by 20 dB every time the operating frequency increases by a factor of 10. Therefore, we can say that the circuit has a roll-off rate of 20 dB per decade.

The same equation was used to verify the 6-dB-per-octave and 20-dB-per-decade roll-off rates, without committing the calculations to specific circuits. Therefore, we can draw the following conclusions:

1. The roll-off rates are equal; i.e., 6 dB per octave equals 20 dB per decade.
2. The roll-off rates are independent of circuit values of $R$ and $C$. (Since $R$ and $C$ are not part of the roll-off equations, they do not weigh into the rate at which gain decreases.)

The values of $R$ and $C$ determine the cutoff frequency of an $RC$ low-pass filter. However, once the cutoff frequency is reached, the circuit gain drops at the same rate as it does for every other $RC$ low-pass filter.

## Putting It All Together: RC Low-Pass Circuit Analysis

The material presented on $RC$ low-pass filters has been extensive, but it is relatively simple to apply. At this point, we will tie all the material together by working through a detailed analysis of an $RC$ low-pass filter. This analysis is presented in Example 19.26.

### EXAMPLE 19.26

Perform a complete gain and frequency response analysis on the $RC$ low-pass filter shown in Figure 19.32. Assume the circuit has a value of $R_S = 50\ \Omega$.

**FIGURE 19.32**

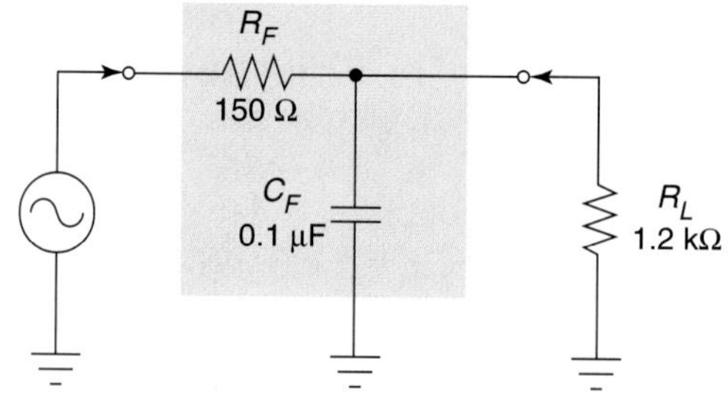


***Solution:*** First, the maximum voltage gain of the filter is found as

$$A_{v(\max)} = \frac{R_L}{R_L + R_F} = \frac{1.2\ \text{k}\Omega}{1.2\ \text{k}\Omega + 150\ \Omega} = \mathbf{0.89}$$

In dB form, the maximum voltage gain is found as

$$A_{v(\text{dB})} = 20 \log A_v = 20 \log(0.89) = \mathbf{-1\ dB}$$

The Thevenin resistance of the circuit can be found as

$$R_{th} = (R_S + R_F) \| R_L = (200\ \Omega) \| (1.2\ \text{k}\Omega) = \mathbf{171\ \Omega}$$

The filter's cutoff frequency is found as

$$f_c = \frac{1}{2\pi R_{th} C_F} = \frac{1}{2\pi(171\ \Omega)(0.1\ \mu\text{F})} = \mathbf{9.3\ kHz}$$

That completes the circuit calculations. The frequency response of the circuit can be represented in one of three ways:

1. Using a frequency response curve.
2. Using a bode plot with octave frequency intervals.
3. Using a bode plot with decade frequency intervals.

All three are shown in Figure 19.33.

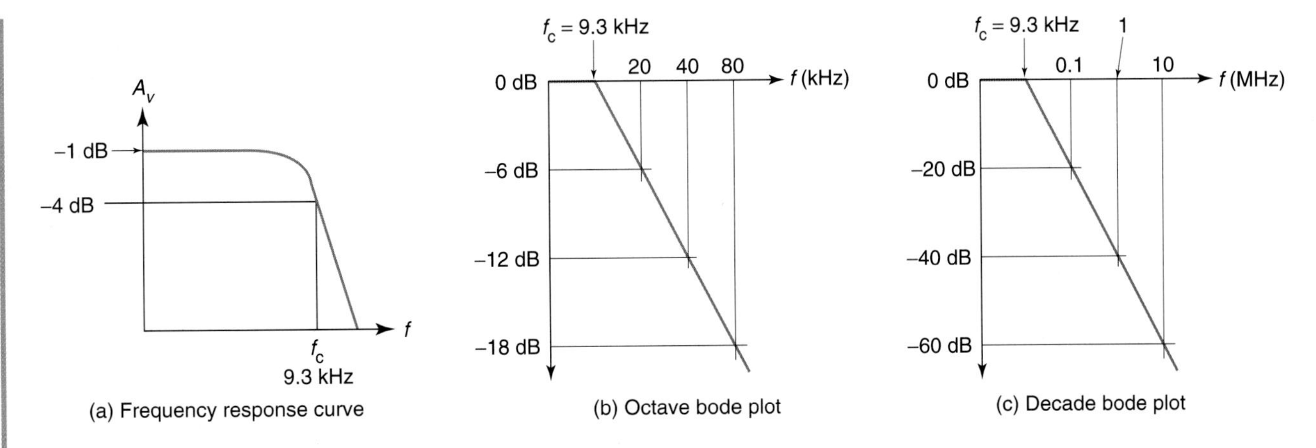


**FIGURE 19.33**

***PRACTICE PROBLEM 19.26***

A circuit like the one in Figure 19.32 has the following values: $R_S = 49\ \Omega$, $R_F = 51\ \Omega$, $C_F = 0.47\ \mu\text{F}$, and $R_L = 1.8\ \text{k}\Omega$. Determine the gain and cutoff frequency values for the circuit.

## RL Low-Pass Filters

◀ *OBJECTIVE 18*

An *RL* circuit acts as a low-pass filter when constructed as shown in Figure 19.34. In the circuit shown, the inductor is the series component and the resistor is the shunt component.

The gain and cutoff frequency calculations for an *RL* low-pass filter are slightly different from those of the *RC* low-pass filter. However, all the concepts relating to response curves, bode plots, and roll-off rates are the same. Therefore, we will focus only on the overall operation of the *RL* low-pass filter, along with its gain and cutoff frequency calculations.

***Circuit Operation.*** The filtering action of the circuit in Figure 19.34 is a result of the inductor's response to an increase in operating frequency. This concept is illustrated in Figure 19.35. The reactance curve in Figure 19.35a was introduced in Chapter 18. As the curve indicates, the reactance of an inductor is 0 Ω when the input frequency is 0 Hz. With this in mind, look at the circuit shown in Figure 19.35b. Assuming that the input frequency is 0 Hz, the inductive reactance is 0 Ω. However, the inductor does have some amount of winding resistance ($R_w$). Therefore, a resistor representing the winding resistance of the coil has been included in the equivalent circuit. In this case, the load voltage equals the difference between the source voltage and the voltage across $R_w$.

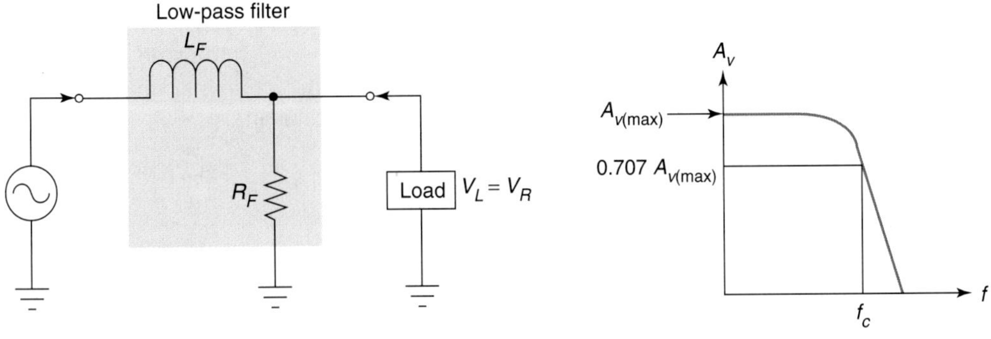


**FIGURE 19.34 An *RL* low-pass filter and its frequency response curve.**

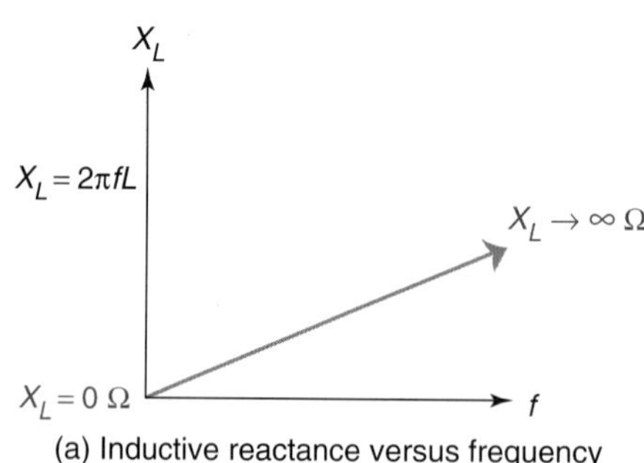


(a) Inductive reactance versus frequency

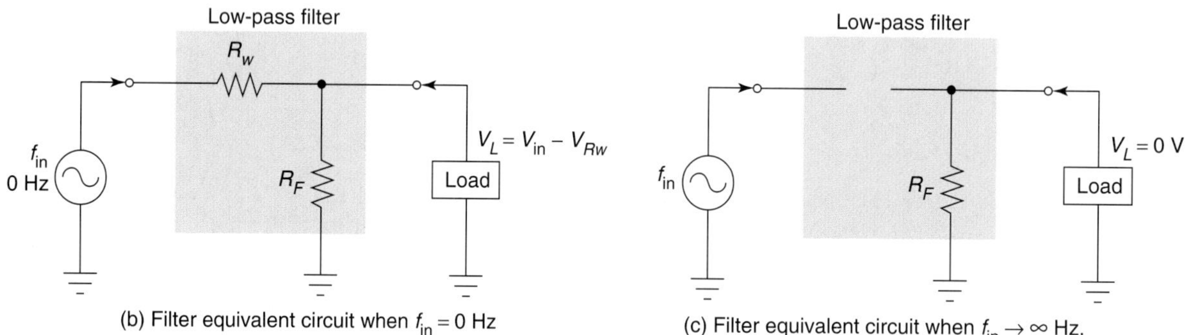


(b) Filter equivalent circuit when $f_{in} = 0$ Hz

(c) Filter equivalent circuit when $f_{in} \to \infty$ Hz. Note that the break in the signal path represents the infinite reactance of the inductor.

**FIGURE 19.35** ***RL*** **low-pass filter operating limits.**

As shown in the reactance curve, increases in operating frequency cause proportional increases in the value of $X_L$. Theoretically, the operating frequency can become high enough for the inductor to act effectively as an open. When this is the case, the filter has the equivalent circuit shown in Figure 19.35c. As you can see, the inductor is represented as a break in the conductor (because of its infinite reactance). In this case, $V_L = 0$ V. Between the extremes represented in Figure 19.35 lies a range of frequencies over which $V_L$ decreases from $(V_S - V_{Rw})$ to 0 V.

***Maximum Voltage Gain, $A_{v(max)}$.*** The maximum gain of any low-pass filter occurs when the input frequency is 0 Hz. As shown in Figure 19.35b, the parallel combination of the load and $R_F$ forms a voltage divider with $R_w$ when $f_{\text{in}} = 0$ Hz. Therefore, the maximum load voltage is found as

$$V_{L(\text{max})} = V_{\text{in}} \frac{R_{EQ}}{R_{EQ} + R_w} \quad \Big| \quad R_{EQ} = R_F \| R_L \tag{19.40}$$

The value of $R_w$ is typically much lower than the parallel combination of $R_F$ and $R_L$. Normally, the value of $R_w$ is assumed to be low enough to ignore when its value is less than one-tenth the value of $(R_F \| R_L)$. In this case, the fraction in equation (19.40) is assumed to equal one (1), leaving the following approximation:

$$V_{L(\text{max})} \cong \text{V}_{\text{in}} \quad \Big| \quad R_w < \frac{R_F \| R_L}{10} \tag{19.41}$$

As you know, voltage gain is the ratio of output voltage to input voltage. For the *RL* low-pass filter, the maximum voltage gain can be found as

$$A_{v(\text{max})} = \frac{V_{L(\text{max})}}{V_{\text{in}}}$$

If we divide both sides of equation (19.40) by $V_{\text{in}}$, maximum voltage gain is found to equal the resistance ratio. By formula,

$$A_{v(\max)} = \frac{R_{EQ}}{R_{EQ} + R_w} \quad \Bigg| \quad R_{EQ} = R_F \| R_L \qquad \textbf{(19.42)}$$

Example 19.27 demonstrates the use of this relationship.

## EXAMPLE 19.27

Calculate the value of $A_{v(\max)}$ for the circuit shown in Figure 19.36a.

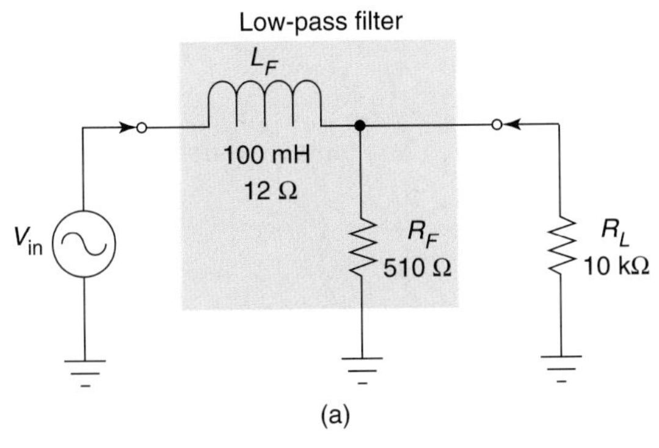


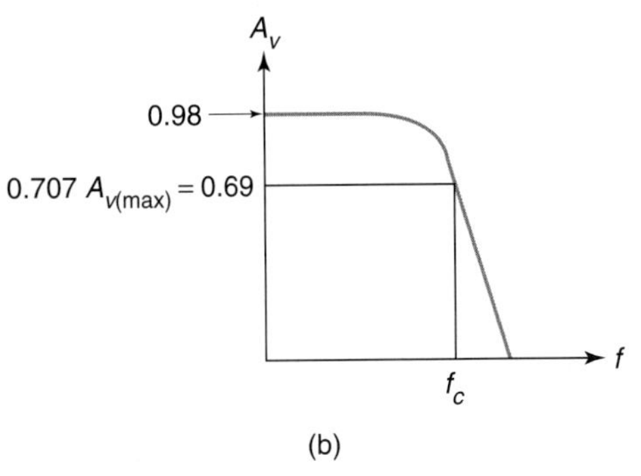


EWB

**FIGURE 19.36**

***Solution:*** First, the values of $R_F$ and $R_L$ are combined into an equivalent parallel resistance as follows:

$$R_{EQ} = R_F \| R_L = (10\text{ k}\Omega)\|(510\ \Omega) = \mathbf{485\ \Omega}$$

Now, the maximum voltage gain of the filter can be found as

$$\mathbf{A_{v(\max)}} = \frac{R_{EQ}}{R_{EQ} + R_w} = \frac{485\ \Omega}{497\ \Omega} = \mathbf{0.98}$$

This value is shown in the frequency response curve in Figure 19.36b.

***PRACTICE PROBLEM 19.27***

A filter like the one in Figure 19.36a has the following values: $R_F = 680\ \Omega$, $R_L = 15\text{ k}\Omega$, $L = 470$ mH, and $R_w = 20\ \Omega$. Calculate the value of $A_{v(\max)}$ for the circuit.

The circuit in Figure 19.36a can be used to demonstrate several important principles. First, you were told that the value of $V_{L(\max)}$ is often assumed to be equal to $V_{\text{in}}$ when $R_W$ is less than one-tenth the value of $(R_F \| R_L)$. If we use this approximation, we get a maximum voltage gain of

$$A_v = \frac{V_{L(\max)}}{V_{\text{in}}} = 1$$

Now, compare this ideal value with the one calculated in Example 19.27. As you can see, there is little practical difference between the two values of $A_{v(\max)}$. In fact, if we convert both values to dB form, we get

$$A_{v(\max)} = 20 \log(1) = 0\text{ dB} \qquad \text{and} \qquad A_{v(\max)} = 20 \log(0.98) = -0.18\text{ dB}$$

In this case, using the approximation causes no significant error in the voltage gain calculation.

Another important point can be seen by comparing the values of the filter resistor ($R_F$) and the load resistance in Figure 19.36. As you can see, the value of $R_F$ is much lower than the value of $R_L$. This relationship is desirable because it makes the frequency response relatively independent of the value of load resistance. This point is discussed in greater detail in a moment.

***Upper Cutoff Frequency ($f_c$).*** The upper cutoff frequency for an *RL* low-pass filter is determined by the inductor and the parallel combination of $R_F$ and $R_L$. By formula,

$$f_c = \frac{R_{EQ}}{2\pi L} \tag{19.43}$$

where

$$R_{EQ} = R_F \| R_L$$

We will establish the basis for this equation in a moment. First, let's apply it to the circuit we have used thus far.

**EXAMPLE 19.28**

Calculate the cutoff frequency for the circuit in Figure 19.36a.

***Solution:*** First, the value of $R_{EQ}$ is found as

$$R_{EQ} = R_F \| R_L = (510\ \Omega) \| (10\ \text{k}\Omega) = \mathbf{485\ \Omega}$$

Now, the filter cutoff frequency is found as

$$f_c = \frac{R_{EQ}}{2\pi L} = \frac{485\ \Omega}{2\pi(100\ \text{mH})} = \mathbf{772\ Hz}$$

***PRACTICE PROBLEM 19.28***

Calculate the cutoff frequency for the filter described in Practice Problem 19.27.

To understand the fact that the load resistance has little effect on the value of $f_c$ when $R_F << R_L$, consider what would happen if the load in Figure 19.36a were to open. With an open load, the value of $R_{EQ}$ would equal the value of $R_F$, which would cause the cutoff frequency of the circuit to change as follows:

$$f_c = \frac{R_{EQ}}{2\pi L} = \frac{510\ \Omega}{2\pi(100\ \text{mH})} = 812\ \text{Hz}$$

As you can see, the absence of the load has little impact on $f_c$ when $R_F << R_L$. Note that an increase in the cutoff frequency of an *RL* low-pass filter is a primary symptom of an open filter resistor. Any time that the cutoff frequency of a low-pass filter increases drastically, check for an open filter resistor.

***The Basis of Equation (19.43).*** Earlier in this section, we established the fact that filter reactance and resistance are equal at the cutoff frequency. For example, look at the circuit in Figure 19.37. A voltage divider is formed by $X_L$ and *R*. The output voltage, which is measured across *R*, can be found as

$$V_{\text{out}} = V_{\text{in}} \frac{R}{\sqrt{X_L^2 + R^2}}$$

FIGURE 19.37

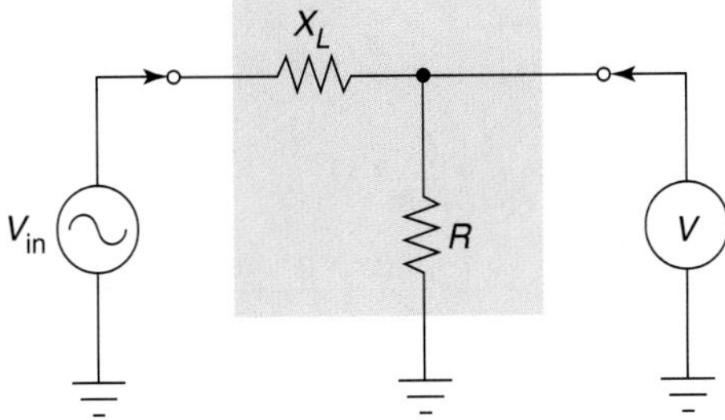


Therefore, the voltage gain of the circuit can be found as

$$A_v = \frac{V_{out}}{V_{in}} = \frac{R}{\sqrt{X_L^2 + R^2}}$$

When $X_L = R$, the above equation gives us a value of

$$A_v = \frac{R}{\sqrt{X_L^2 + R^2}} = \frac{R}{\sqrt{R^2 + R^2}} = \frac{R}{\sqrt{2R^2}} = \frac{R}{1.414R} = 0.707$$

Therefore, the cutoff frequency occurs when $X_L = R$.

As you know, the reactance of an inductor is found as

$$X_L = 2\pi f L$$

Transposing this equation for frequency, we get

$$f = \frac{X_L}{2\pi L}$$

We know that $X_L = R$ when $f = f_c$, therefore

$$f_c = \frac{R}{2\pi L}$$

For an *RL* low-pass filter, the total resistance is the parallel combination of the filter resistor and the load resistance. Therefore, the above equation is rewritten as

$$f_c = \frac{R_{EQ}}{2\pi f} \quad \Big| \quad R_{EQ} = R_F \| R_L$$

which is equation (19.43).

## Roll-Off Rates

*RL* filter roll-off characteristics are identical to those of the *RC* low-pass filter. Specifically:

1. The gain remains relatively constant until the cutoff frequency is reached.
2. When operated above its cutoff frequency, an *RL* low-pass filter has a roll-off rate of 6 dB per octave (or 20 dB per decade).
3. The roll-off rate for an *RL* low-pass filter is independent of the values of *L* and *R*.

In fact, the bode plot for an *RL* low-pass filter is identical to that of an *RC* low-pass filter. This point is illustrated in Example 19.29.

## EXAMPLE 19.29

Derive the bode plot for the circuit shown in Figure 19.38a.

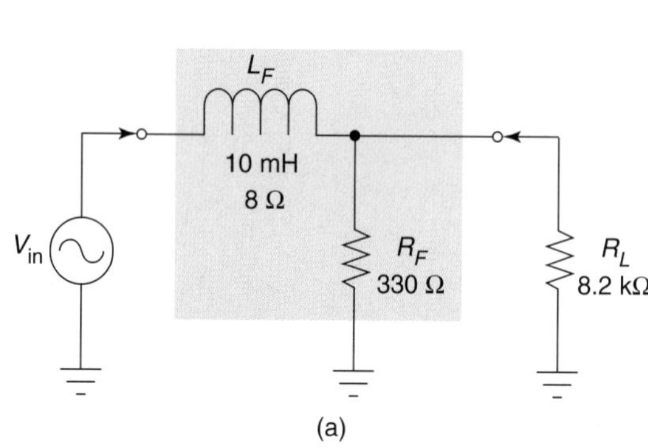


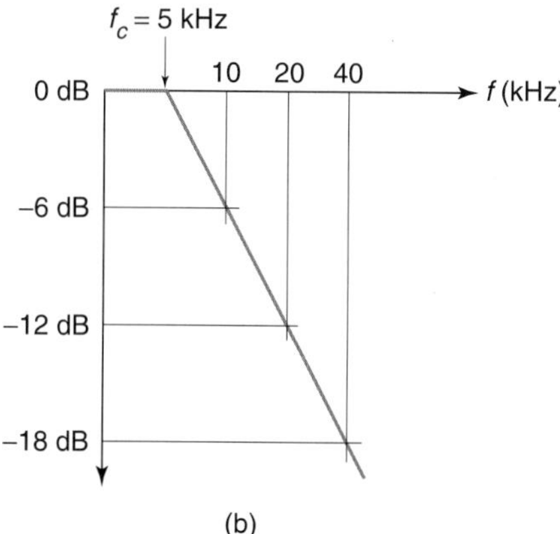


**FIGURE 19.38**

***Solution:*** First, the value of $R_{EQ}$ is found as

$$R_{EQ} = R_F \| R_L = (330\ \Omega) \| (8.2\ \text{k}\Omega) = \mathbf{317\ \Omega}$$

Now, the circuit cutoff frequency is found as

$$f_c = \frac{R_{EQ}}{2\pi L} = \frac{317\ \Omega}{2\pi(10\ \text{mH})} = \mathbf{5\ kHz}$$

Since the bode plot normalizes gain, we do not need to calculate its value. The octave bode plot for the circuit is shown in Figure 19.38b.

***PRACTICE PROBLEM 19.29***

A circuit like the one in Figure 19.38a has the following values: $L = 4.7$ mH, $R_w = 5\ \Omega$, $R_L = 12$ kΩ, and $R_F = 820\ \Omega$. Derive the decade bode plot for the circuit.

## A Look Ahead

We have now discussed the operation and analysis of *RC* and *RL* low-pass filters. In the following section, we will take a similar look at *RC* and *RL* high-pass filters. As you will see, most of the principles discussed in this section apply to those circuits as well.

## Section Review

1. What is a shunt component?
2. Why is the voltage gain of an *RC* circuit always less than one (1)?
3. What is a bode plot?
4. What filter characteristic is normalized on a bode plot?
5. What other names are used to describe the cutoff frequency of a low-pass filter? What is the basis for this name?
6. What is roll-off? What is the roll-off rate of a filter?
7. What is the relationship between the values of *R* and *C* in a low-pass filter and its roll-off rate?

8. What are the standard filter roll-off rates?
9. What characteristics do *RL* and *RC* low-pass filters share?

# 19.5 HIGH-PASS FILTERS

As you know, a high-pass filter is designed to pass all frequencies above its cutoff frequency. High-pass filters can be formed by reversing the positions of the resistive and reactive components in the *RC* and *RL* low-pass filters. In this section, we will discuss the operation of *RC* and *RL* high-pass filters. As you will see, most of the relationships for high-pass filters are nearly identical to those for low-pass filters, because the primary difference between a high-pass filter and its low-pass counterpart is component placement.

## RC High-Pass Filters

◄ *OBJECTIVE 19*

An *RC* circuit acts as a high-pass filter when constructed as shown in Figure 19.39a. For comparison, an *RC* low-pass filter is shown in Figure 19.39b. As you can see, the capacitor and resistor positions are reversed between the two circuits. In the high-pass circuit, the capacitor is in the signal path and the resistor is the shunt component.

The filtering action of the circuit in Figure 19.39a is a result of the capacitor's response to an increase in frequency. This concept is illustrated in Figure 19.40. The reactance curve (which we used to describe low-pass filter operation) shows that capacitive reactance varies inversely with operating frequency. With this in mind, look at the equivalent circuit shown in Figure 19.40b. Assuming that the input frequency is near the high end of the reactance curve, $X_C$ can be assumed to be approximately 0 Ω. In this case, the capacitor is replaced by a direct connection between the source and the load, providing a maximum load voltage equal to the source voltage.

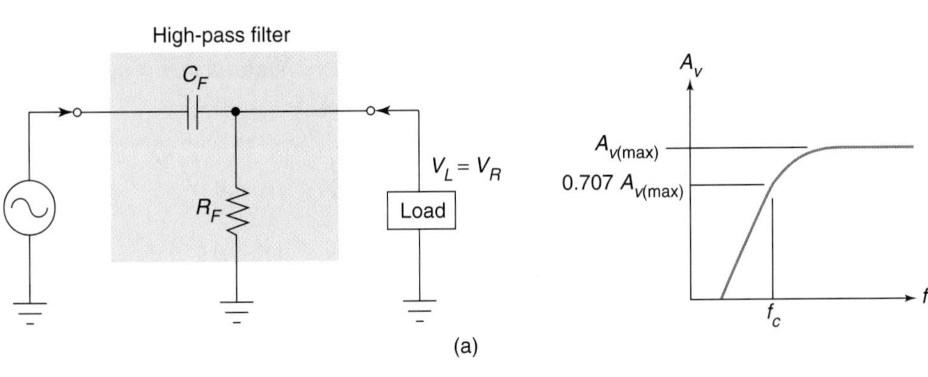


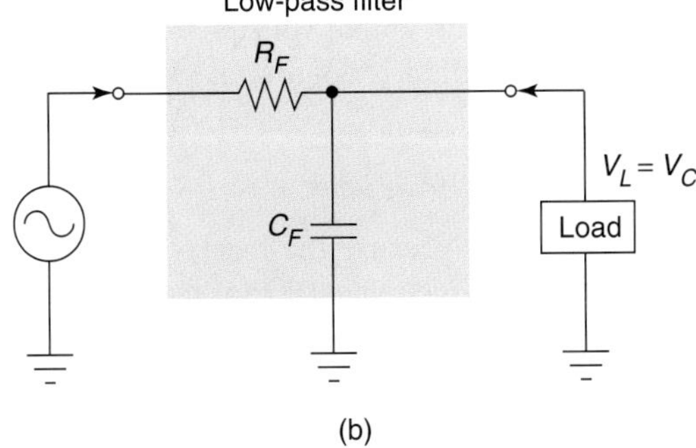


**FIGURE 19.39 An *RC* high-pass filter, its frequency response curve, and low-pass counterpart.**

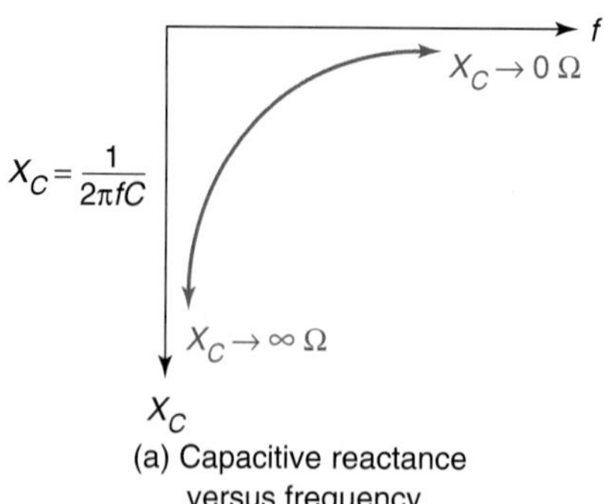

(a) Capacitive reactance versus frequency

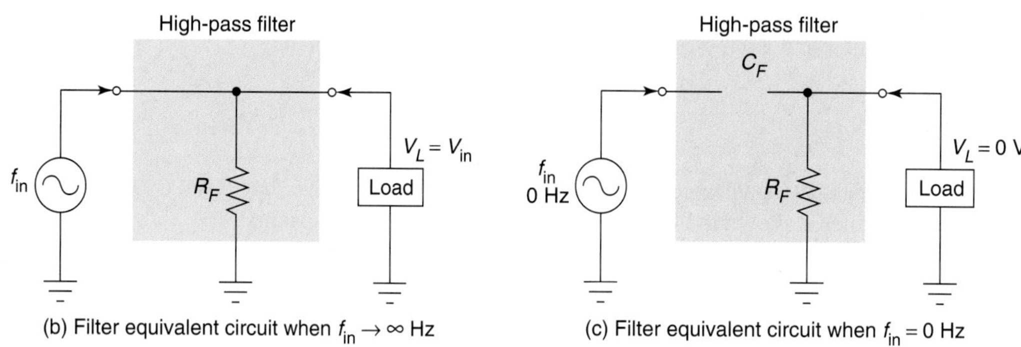

(b) Filter equivalent circuit when $f_{in} \to \infty$ Hz

(c) Filter equivalent circuit when $f_{in} = 0$ Hz

**FIGURE 19.40** ***RC* high-pass filter operating limits.**

If the circuit operating frequency decreases to 0 Hz, the reactance of the capacitor approaches infinite ohms. With this in mind, look at the equivalent circuit shown in Figure 19.40c. As you can see, the infinite reactance of the capacitor is represented as a break in the signal path. Therefore, the source is isolated from the load, and $V_L = 0$ V.

Between the extremes represented in Figure 19.40 lies a range of frequencies over which $V_L$ decreases from $V_L = V_{in}$ to $V_L = 0$ V. Our primary interests at this point are establishing methods to determine:

1. The maximum value of $A_v$ for the circuit.
2. The cutoff frequency ($f_c$) for the circuit.

***Maximum Voltage Gain, $A_{v(max)}$.*** The maximum gain of a high-pass filter occurs when the input frequency increases to the point where $X_C \cong 0\ \Omega$. As shown in Figure 19.40b, the load is coupled directly to the source when $X_C \cong 0\ \Omega$. Therefore, the maximum load voltage is found as

$$V_L \cong V_{in} \tag{19.44}$$

As you know, voltage gain is the ratio of output voltage to input voltage. Because the maximum output from an *RC* high-pass filter is approximately equal to the input voltage, its maximum voltage gain is found as

$$A_{v(\max)} \cong 1 \tag{19.45}$$

Note that an approximation is used, because the reactance of the capacitor never truly reaches 0 Ω. However, $X_C$ gets close enough to 0 Ω for the approximation to be valid.

***Lower Cutoff Frequency ($f_c$).*** The cutoff frequency for an *RC* high-pass filter is determined using the same relationship we established for the low-pass filter. By formula,

$$f_c = \frac{1}{2\pi RC}$$

This relationship was given as equation (19.35) in the last section. We can use this relationship because only the component positions within the filter have changed. In other words, the relationship among $f_c$, $R$, and $C$ doesn't change simply because the component positions have changed. Example 19.30 demonstrates the complete approach to calculating the lower cutoff frequency for an $RC$ high-pass filter.

**EXAMPLE 19.30**

Calculate the cutoff frequency for the circuit shown in Figure 19.41a.

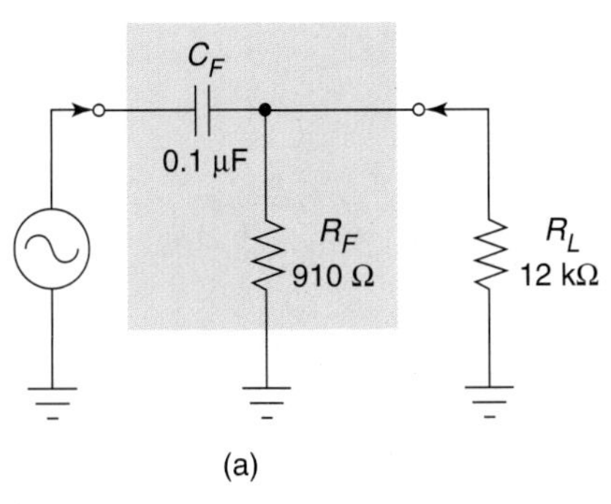


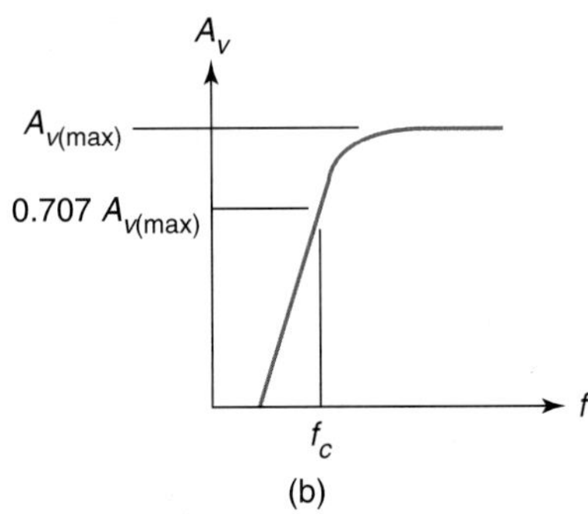


**FIGURE 19.41**

***Solution:*** The total resistance in the circuit equals the parallel combination of the filter resistor and the load resistance. The parallel equivalent of $R_F$ and $R_L$ is found as

$$R_{EQ} = R_F \| R_L = (910\ \Omega) \| (12\ \text{k}\Omega) = \mathbf{846\ \Omega}$$

Using $R_{EQ}$ as the value of $R$ in equation (19.35), we get a cutoff frequency of

$$f_c = \frac{1}{2\pi RC} = \frac{1}{2\pi(846\ \Omega)(0.1\ \mu\text{F})} = \mathbf{1.88\ kHz}$$

This frequency is used to plot the frequency response curve shown in Figure 19.41b.

***PRACTICE PROBLEM 19.30***

A circuit like the one in Figure 19.41a has the following values: $C_F = 0.22\ \mu\text{F}$, $R_L = 15\ \text{k}\Omega$, and $R_F = 750\ \Omega$. Calculate the value of the circuit cutoff frequency.

## High-Pass Filter Bode Plots and Roll-Off Rates

The bode plot for a high-pass filter is simply a mirror image of the plot for a low-pass filter. The bode plot for the high-pass filter in Figure 19.41a is shown in Figure 19.42. As you can see, the bode plot shows how changes in operating frequency affect the circuit dB voltage

**FIGURE 19.42 The bode plot for the circuit in Figure 19.41a.**

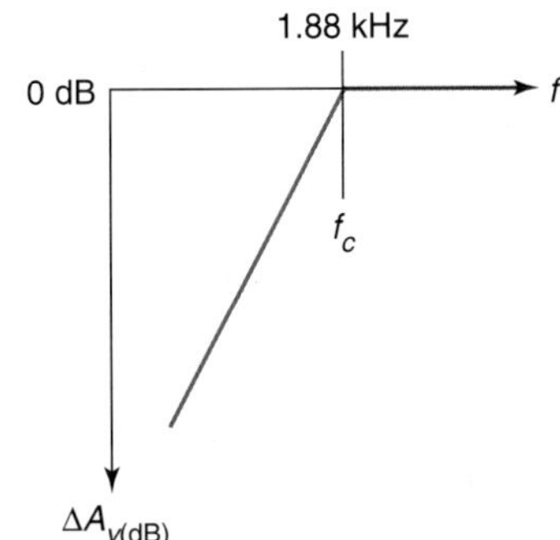

gain. At frequencies above $f_c$, the graph shows that $\Delta A_v = 0$ dB, indicating that the voltage gain of the circuit remains at its maximum value. As the operating frequency decreases below the cutoff frequency, the dB voltage gain is shown to roll off at a constant rate.

As was the case with low-pass filters, all single-stage, high-pass filters have roll-off rates of 6 dB per octave (20 dB per decade). This roll-off rate is based on the following relationships:

$$\Delta A_v = \frac{1}{\sqrt{1 + (f_c/f_o)^2}} \tag{19.46}$$

and

$$\Delta A_{v(\text{dB})} = 20 \log \frac{1}{\sqrt{1 + (f_c/f_o)^2}} \tag{19.47}$$

where $f_c$ = the lower cutoff frequency of the filter
$f_o$ = the circuit operating frequency

These equations are simply the high-pass filter equivalents of equations (19.38) and (19.39). The derivations of those equations are provided in Appendix E.

Using equation (19.47), we can perform a series of calculations to verify the 6 dB per octave roll-off rate for high-pass filters, as follows:

When $f_o = 0.5\,f_c$: $\quad \Delta A_{v(\text{dB})} = 20 \log \frac{1}{\sqrt{1 + 2^2}} = -6.99 \text{ dB} \quad \Big| \; \frac{f_c}{f_o} = 2$

When $f_o = 0.25\,f_c$: $\quad \Delta A_{v(\text{dB})} = 20 \log \frac{1}{\sqrt{1 + 4^2}} = -12.3 \text{ dB} \quad \Big| \; \frac{f_c}{f_o} = 4$

When $f_o = 0.125\,f_c$: $\quad \Delta A_{v(\text{dB})} = 20 \log \frac{1}{\sqrt{1 + 8^2}} = -18.1 \text{ dB} \quad \Big| \; \frac{f_c}{f_o} = 8$

As you can see, the voltage gain of the circuit changes by approximately 6 dB each time the operating frequency drops by 50%. Therefore, we can say that the circuit has a roll-off rate of approximately 6 dB per octave. Note that the rate comes closer to exactly 6 dB per octave if the sequence continues to even lower percentages of $f_c$.

The second roll-off rate (20 dB per decade) can be verified using the following sequence of calculations:

When $f_o = 0.1\,f_c$: $\quad \Delta A_{v(\text{dB})} = 20 \log \frac{1}{\sqrt{1 + 10^2}} = -20 \text{ dB} \quad \Big| \; \frac{f_c}{f_o} = 10$

When $f_o = 0.01\,f_c$: $\quad \Delta A_{v(\text{dB})} = 20 \log \frac{1}{\sqrt{1 + 100^2}} = -40 \text{ dB} \quad \Big| \; \frac{f_c}{f_o} = 100$

When $f_o = 0.001\,f_c$: $\quad \Delta A_{v(\text{dB})} = 20 \log \frac{1}{\sqrt{1 + 1000^2}} = -60 \text{ dB} \quad \Big| \; \frac{f_c}{f_o} = 1000$

As you can see, the voltage gain of the circuit decreases by 20 dB every time the operating frequency decreases by a factor of 10. Therefore, we can say that the circuit has a roll-off rate of 20 dB per decade.

## Putting It All Together: RC High-Pass Filter Analysis

The complete analysis of an *RC* high-pass filter is fairly simple because of its gain characteristics. Such an analysis is presented in Example 19.31.

Perform a complete gain and frequency response analysis of the *RC* high-pass filter shown in Figure 19.43.

## EXAMPLE 19.31

EWB

**FIGURE 19.43**

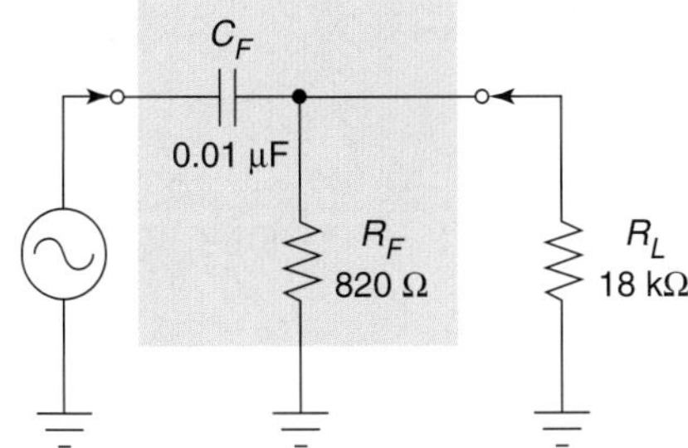


***Solution:*** The reactance of the capacitor is assumed to have a minimum value of 0 Ω when the operating frequency is well above the cutoff frequency. When this is the case, the load is coupled directly to the signal source, $V_{L(\max)} = V_{\text{in}}$, and

$$A_{v(\max)} = 1$$

As you know, this corresponds to a value of $A_{v(\max)} = 0$ dB.

The total resistance in the circuit (as seen by the capacitor) is found as

$$R_{EQ} = R_F \| R_L = (820\ \Omega) \| (18\ \text{k}\Omega) = \mathbf{784\ \Omega}$$

Using 784 Ω as the value of *R*, the cutoff frequency of the filter is found as

$$f_c = \frac{1}{2\pi RC} = \frac{1}{2\pi(784\ \Omega)(0.01\ \mu\text{F})} = \mathbf{20.3\ kHz}$$

That completes the circuit calculations. The frequency response of the circuit can be represented in one of three ways:

1. Using a frequency response curve.
2. Using a bode plot with octave frequency intervals.
3. Using a bode plot with decade frequency intervals.

All three are shown in Figure 19.44. To simplify the illustrations, the circuit cutoff frequency has been rounded off to 20 kHz.

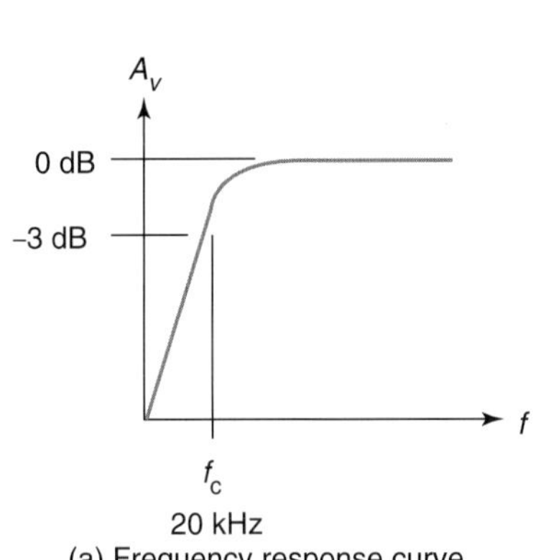

(a) Frequency response curve

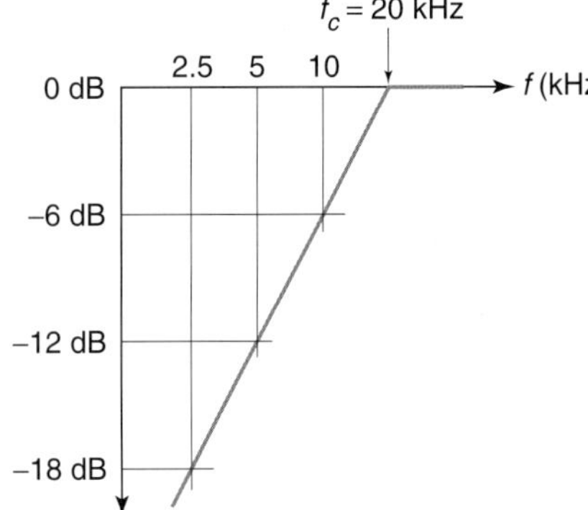

(b) Octave bode plot

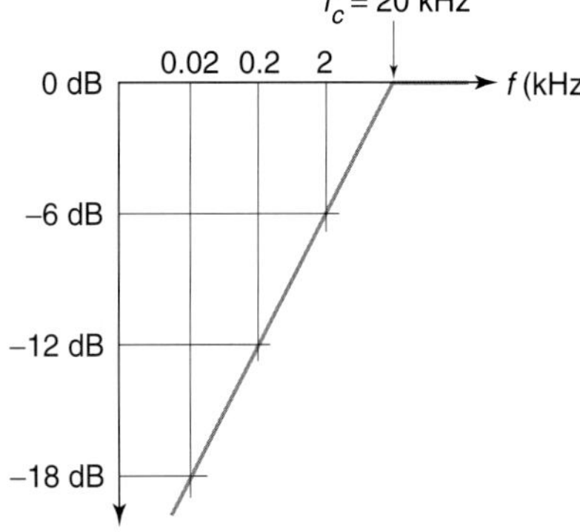

(c) Decade bode plot

**FIGURE 19.44**

***PRACTICE PROBLEM 19.31***

A circuit like the one in Figure 19.43 has the following values: $C_F = 0.47\ \mu\text{F}$, $R_L = 15\ \text{k}\Omega$, and $R_F = 620\ \Omega$. Determine the gain and cutoff frequency values for the circuit.

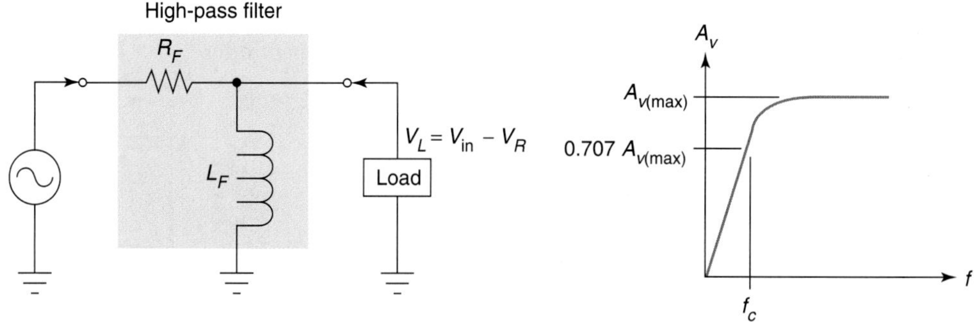


**FIGURE 19.45 An *RL* high-pass filter and its frequency response curve.**

## *RL High-Pass Filters*

*OBJECTIVE 20* ▶ An *RL* circuit acts as a high-pass filter when constructed as shown in Figure 19.45. In the circuit shown, the resistor is the series component and the inductor is the shunt component. As you will see, the gain calculations for this circuit are similar to those for the *RC* low-pass filter.

***Circuit Operation.*** The filtering action of the circuit in Figure 19.45 is a result of the inductor's response to a decrease in operating frequency. This concept is illustrated in Figure 19.46. As the reactance curve indicates, $X_L$ approaches infinity as frequency increases. If the input frequency to the circuit is sufficient, the inductor acts effectively as an open. If we assume that $X_L \cong \infty\ \Omega$, the equivalent circuit in Figure 19.46b applies. As you can see, the inductor is represented as a break in the shunt component path. In this case, $V_L = V_{in} - V_R$.

If the input frequency decreases to 0 Hz, the inductive reactance decreases to 0 Ω. In this case, the equivalent circuit in Figure 19.46c applies. Ignoring the small amount of winding resistance in the coil, the shunt inductor is represented as a shorted path to ground.

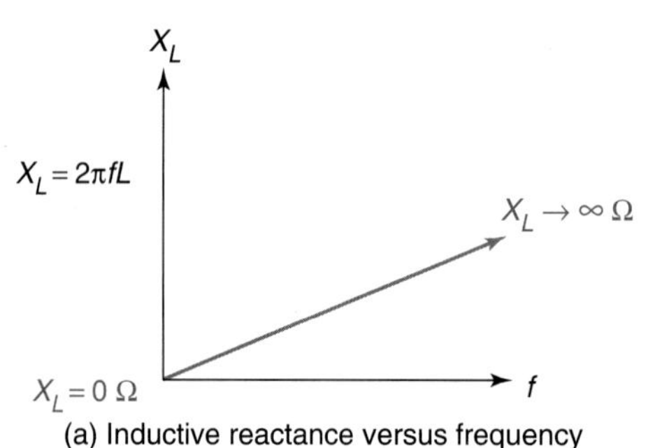


(a) Inductive reactance versus frequency

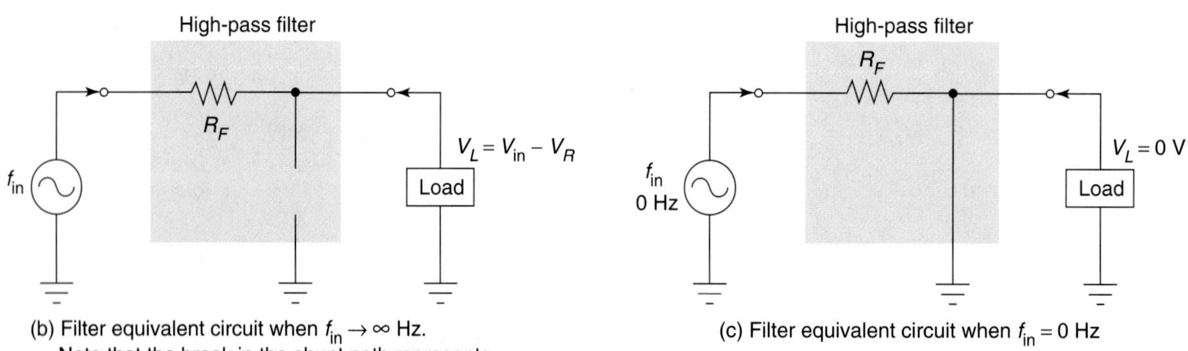


(b) Filter equivalent circuit when $f_{in} \rightarrow \infty$ Hz. Note that the break in the shunt path represents the infinite reactance of the inductor.

(c) Filter equivalent circuit when $f_{in} = 0$ Hz

**FIGURE 19.46 *RL* high-pass filter operating limits.**

In this case, $V_L = 0$ V. Between the extremes represented in Figure 19.46 lies a range of frequencies over which $V_L$ decreases from $(V_{in} - V_R)$ to 0 V.

***Maximum Voltage Gain, $A_{v(max)}$.*** The maximum gain of an *RL* high-pass filter occurs when the operating frequency is high enough to cause the inductor to act effectively as an open. As shown in Figure 19.46b, the load forms a voltage divider with the filter resistor when $X_L \cong \infty\ \Omega$. Therefore, the maximum load voltage is found as

$$V_{L(\max)} = V_{in} \frac{R_L}{R_L + R_F} \tag{19.48}$$

Like any other filter, the voltage gain for the *RL* high-pass filter can be found as

$$A_{v(\max)} = \frac{V_{L(\max)}}{V_{in}}$$

If we divide both sides of equation (19.48) by $V_{in}$, maximum voltage gain is shown to equal the resistance ratio. By formula,

$$A_{v(\max)} = \frac{R_L}{R_L + R_F} \tag{19.49}$$

We have worked through this type of problem numerous times in this chapter, so we won't rework it here. However, two points need to be made:

1. The resistance ratio indicates that the voltage gain of the *RL* high-pass filter is always less than one (1).
2. Maximum voltage gain is achieved by designing an *RL* high-pass filter so that $R_F << R_L$. For this reason, an *RL* high-pass filter is normally designed using the lowest practical value of $R_F$.

***Lower Cutoff Frequency ($f_c$).*** The lower cutoff frequency for an *RL* high-pass filter is determined by the inductor and the parallel combination of $R_F$ and $R_L$. By formula,

$$f_c = \frac{R_{EQ}}{2\pi L}$$

where $R_{EQ} = R_F \| R_L$. This relationship was introduced earlier in the chapter as equation (19.43).

The value of $R_{EQ}$ in equation (19.43) is determined in the same manner we used to measure its value in an *RC* low-pass filter. That measurement for $R_{EQ}$ was shown in Figure 19.24. Since the inductor in the high-pass circuit is in the same position as the capacitor in Figure 19.24, the values of $R_{EQ}$ for the circuits are measured the same way. Example 19.32 demonstrates the cutoff frequency calculations for an *RL* high-pass filter.

Calculate the cutoff frequency for the *RL* high-pass filter in Figure 19.47a.

EXAMPLE 19.32

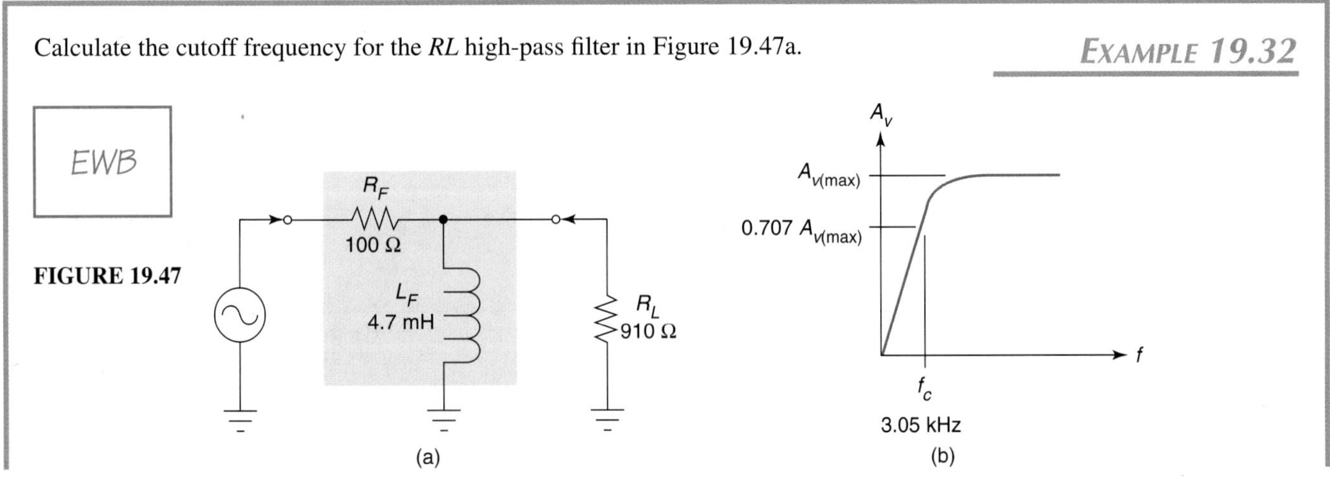


FIGURE 19.47

***Solution:*** First, the value of $R_{EQ}$ is found as

$$\boldsymbol{R_{EQ}} = R_F \| R_L = (100\ \Omega)\|(910\ \Omega) = \mathbf{90\ \Omega}$$

Now, the filter cutoff frequency is found as

$$f_c = \frac{R_{EQ}}{2\pi L} = \frac{90\ \Omega}{2\pi(4.7\ \text{mH})} = \mathbf{3.05\ kHz}$$

The frequency response curve for this circuit is shown in Figure 19.47b.

***PRACTICE PROBLEM 19.32***

A circuit like the one in Figure 19.47a has the following values: $R_F = 200\ \Omega$, $L_F = 1$ mH, and $R_L = 1.8$ k$\Omega$. Calculate the filter cutoff frequency.

As you can see, the analysis of an *RL* high-pass circuit is nearly identical to that of an *RC* low-pass filter. The bode plots for the circuit in Example 19.32 would be similar to those shown in Figure 19.44. The only difference would be the frequencies identified on the graph.

## Section Review

1. In terms of component placement, what is the difference between an *RC* high-pass filter and an *RC* low-pass filter?
2. Explain the filtering action of the *RC* high-pass filter in Figure 19.39a.
3. Why is the maximum voltage gain of an *RC* high-pass filter assumed to be approximately equal to one (1)?
4. Compare high-pass filter roll-off to low-pass filter roll-off.
5. In terms of component placement, what is the difference between an *RL* high-pass filter and an *RL* low-pass filter?
6. Explain the filtering action of the *RL* high-pass filter in Figure 19.45a.

## 19.6 BANDPASS AND NOTCH FILTERS

Earlier in the chapter, you were told that bandpass and notch (band-stop) filters are designed to pass or block a specified range of frequencies. As a review, the primary frequencies are identified on the frequency response curves in Figure 19.48. As you can see, each of these filters has two cutoff frequencies, designated $f_{c1}$ and $f_{c2}$. The difference between

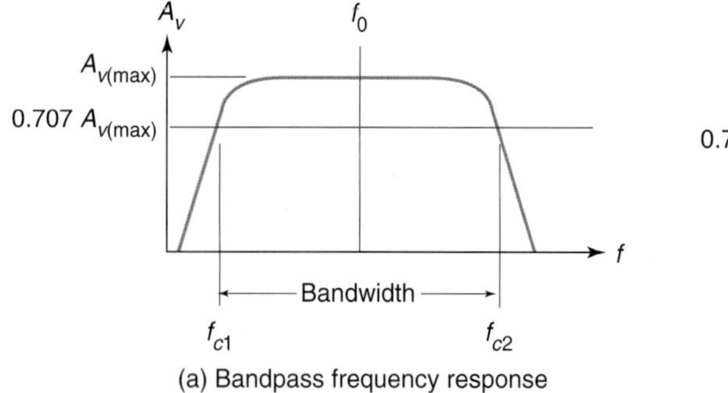

(a) Bandpass frequency response

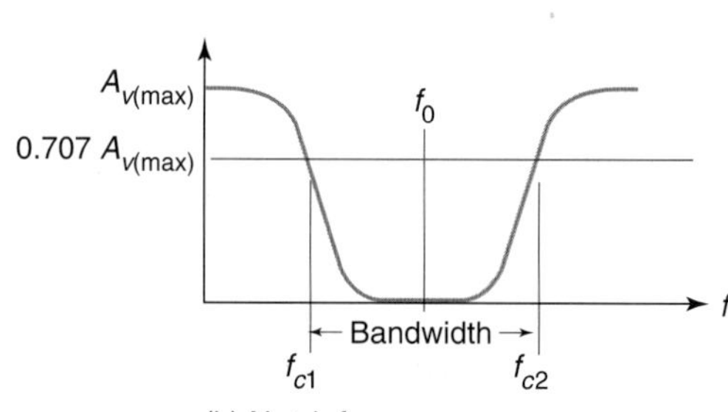

(b) Notch frequency response

**FIGURE 19.48 Bandpass and notch filter frequency response curves.**

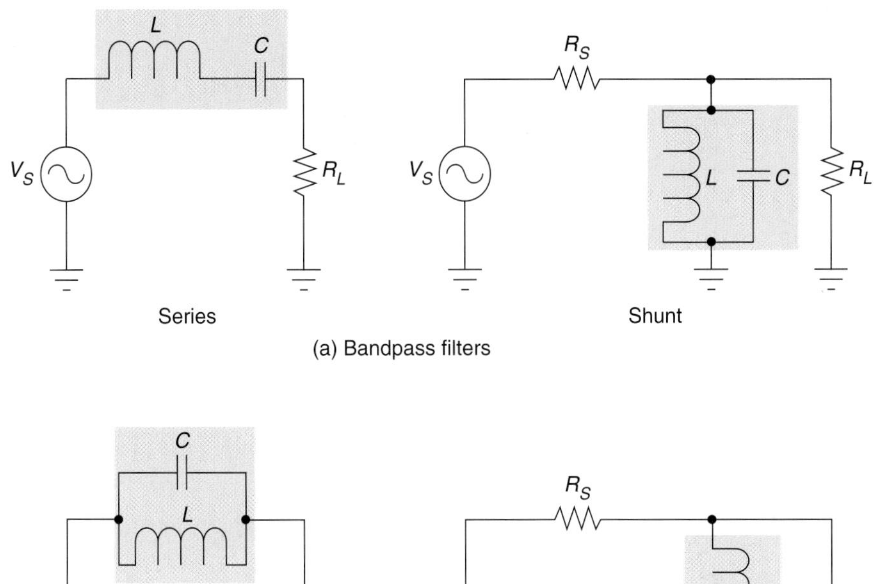


(a) Bandpass filters

C
L
$V_S$
$R_L$
$R_S$
$V_S$
L
C
$R_L$
Series
Shunt

(b) Notch (band-stop) filters

**FIGURE 19.49** ***LC*** **bandpass and notch filters.**

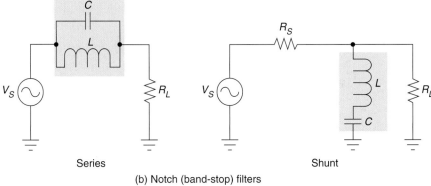

the cutoff frequencies is referred to as the *bandwidth* (BW) of the filter. The geometric average of the cutoff frequencies is referred to as the *center frequency* ($f_0$).

There are many possible approaches to building bandpass and notch filters. The most common passive bandpass and notch filters are *LC* filters, like those shown in Figure 19.49. The circuits in Figure 19.49a are bandpass filters. The circuits in Figure 19.49b are notch filters. In each case, the filtering action is based on the resonant characteristics of the *LC* circuits.

In this section, we will concentrate on the operating characteristics of *LC* bandpass and notch filters. Because their operation is based on the concept of resonance, we will start with a brief review of this principle.

## Resonance: A Review

When an *LC* circuit is operated at resonance, the values of inductive and capacitive reactance are equal in magnitude and 180° out of phase, as illustrated in Figure 19.50. The resonant frequency ($f_r$) of an *LC* circuit is found as

$$f_r = \frac{1}{2\pi\sqrt{LC}}$$

This relationship was introduced in Chapter 18 as equation (18.6).

Figure 19.50 also shows the relationship between $X_L$ and $X_C$ when an *LC* circuit is operated above and below the resonant frequency. When the operating frequency ($f_0$) of an *LC* circuit is lower than $f_r$, $X_L < X_C$. When the operating frequency of an *LC* circuit is higher than $f_r$, $X_L > X_C$. The effects of these relationships on circuit voltages and currents depend on whether the components are connected in series or in parallel.

FIGURE 19.50

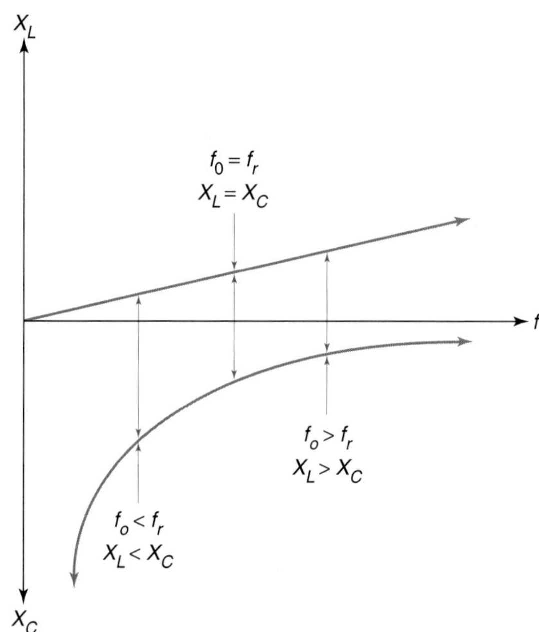


***Series Resonant Circuit Characteristics.*** Figure 19.51 shows a series *LC* circuit, its phase diagrams, and primary circuit calculations. Based on this figure, we can summarize the characteristics of series resonant circuits as shown in Table 19.3.

**TABLE 19.3 Series Resonant Circuit Characteristics**

| *Characteristic* | *Summary* |
|---|---|
| Impedance | 1. $X_C$ and $X_L$ are equal in magnitude and 180° out of phase. Net series reactance ($X_S$) is 0 Ω.<br>2. Since net reactance is 0 Ω, the total circuit impedance equals the source resistance. |
| Current | 1. Determined by the values of $V_S$ and $R_S$.<br>2. Relatively high in magnitude because of the low net opposition. |
| Voltage | 1. Component voltages are relatively high in magnitude because of the high circuit current.<br>2. The component voltages are equal in magnitude and 180° out of phase.<br>3. The net reactive voltage ($V_{LC}$) is approximately 0 V. |

***Parallel Resonant Circuit Characteristics.*** Figure 19.52 shows a parallel *LC* circuit, its phase diagrams, and primary circuit calculations. Based on this figure, we can summarize the characteristics of parallel resonant circuits as shown in Table 19.4.

This review of series and parallel resonant circuits have been very brief. If you feel you need to review these principles further, refer to Section 18.3. For convenience, series and parallel resonant characteristics are summarized in Figure 19.53.

**TABLE 19.4 Parallel Resonant Circuit Characteristics**

| *Characteristic* | *Summary* |
|---|---|
| Voltage | As with any parallel circuit, the branch voltages equal the source voltage. |
| Current | 1. Branch currents are equal in magnitude and 180° out of phase.<br>2. The net circuit current ($I_T$) is 0 A. |
| Impedance | Since the net circuit current is 0 A, the total circuit impedance approaches infinity. |

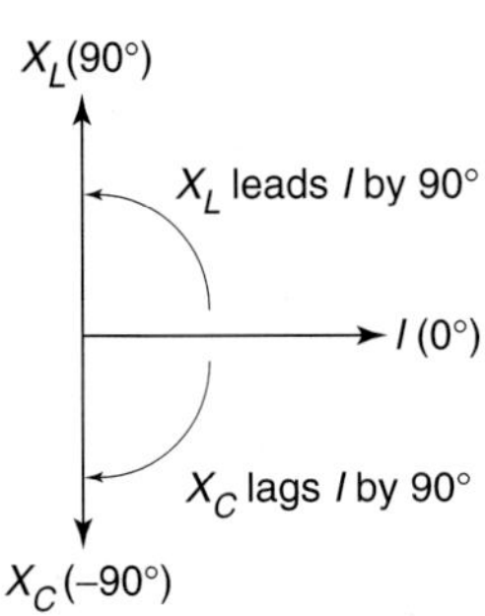


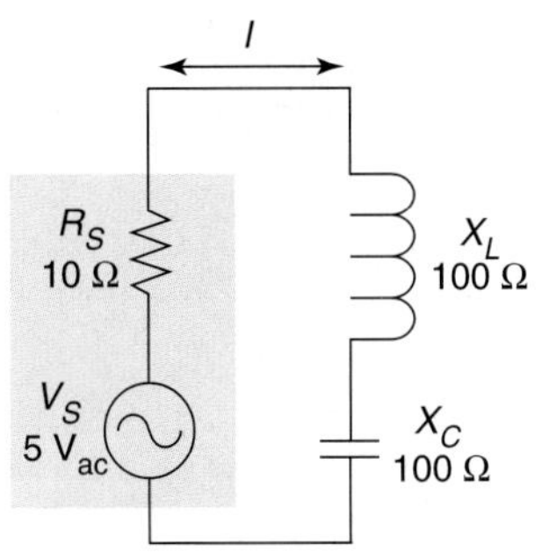


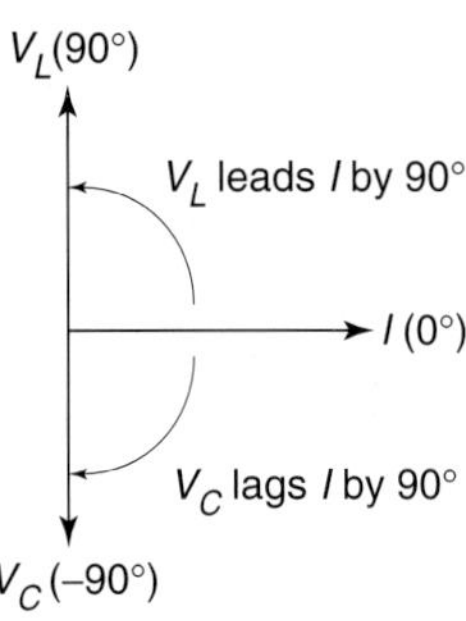


Impedance

$$X_S = X_L - X_C = 0\ \Omega$$

$$Z_T = R_S = 10\ \Omega$$

Current

$$I_T = \frac{V_S}{R_S} = \frac{5\ V_{ac}}{10\ \Omega} = 500\ \text{mA}$$

Voltage

$$V_L = IX_L = (500\ \text{mA})(100\ \Omega\angle 90°) = 50\ V\angle 90°$$

$$V_C = IX_C = (500\ \text{mA})(100\ \Omega\angle -90°) = 50\ V\angle -90°$$

$$V_{LC} = V_L + V_C = 0\ V$$

**FIGURE 19.51**

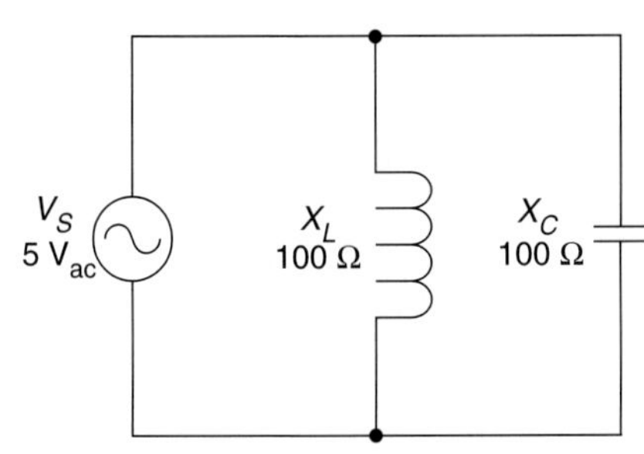


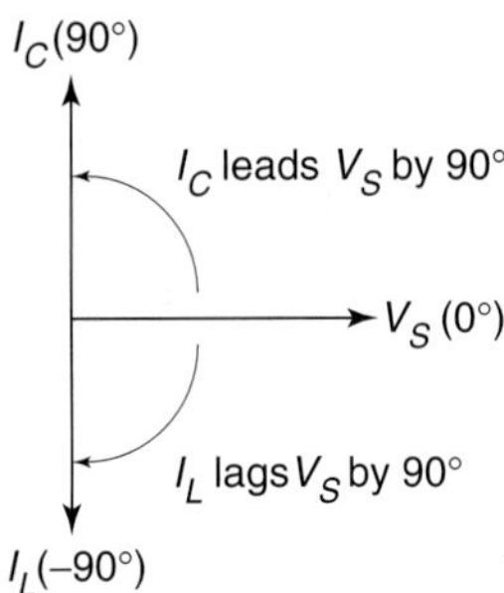


Branch currents

$$I_L = \frac{V_S}{X_L} = \frac{5\ V_{ac}}{100\ \Omega\angle 90°} = 50\ \text{mA}\angle -90°$$

$$I_C = \frac{V_S}{X_C} = \frac{5\ V_{ac}}{100\ \Omega\angle -90°} = 50\ \text{mA}\angle 90°$$

Total current

$$I_T = I_L + I_C = 0\ \text{mA}$$

Total impedance

$$Z_T = \frac{V_S}{I_T} \rightarrow \frac{5\ V_{ac}}{0\ \text{mA}} \rightarrow \infty\ \Omega$$

**FIGURE 19.52**

**SUMMARY ILLUSTRATION**

*Series and Parallel Resonant Circuits*

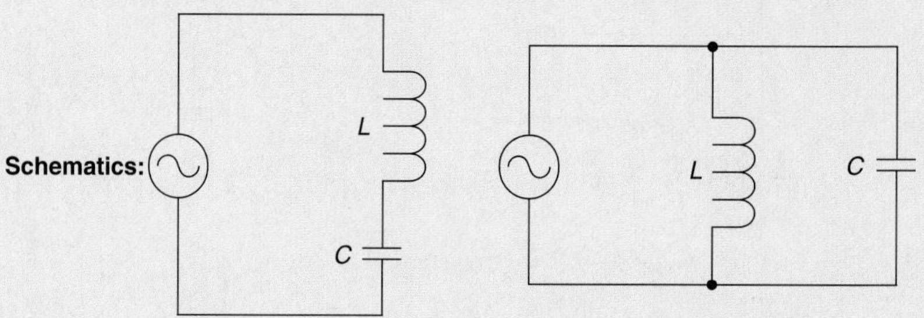


| | Series | Parallel |
|---|---|---|
| **Resonant frequency:** | $f_r = \frac{1}{2\pi\sqrt{LC}}$ | $f_r = \frac{1}{2\pi\sqrt{LC}}$ |
| **Reactance:** | $X_L = X_C$<br>$X_T = 0\ \Omega$ | $X_L = X_C$<br>$X_T \rightarrow \bullet\ \Omega$ |

($X_T$ represents the combination of $X_L$ and $X_C$)

| | Series | Parallel |
|---|---|---|
| **Other relationships:** | $V_L$ and $V_C$ are 180° out of phase.<br>$V_L + V_C = 0$ V<br>$I$ is limited only by source resistance.<br>$I$ is in phase with $V_S$. | $I_C$ and $I_L$ are 180° out of phase.<br>$I_C + I_L = 0$ A |

**FIGURE 19.53**

OBJECTIVE 21 ▶

## *Series LC Bandpass Filters*

The operation of a series *LC* bandpass filter is easiest to understand when the filter is represented as an equivalent circuit like the one in Figure 19.54b. In this circuit, the net series reactance ($X_S$) of the filter is represented as a series component between the source and the load. Specifically, we are interested in the effect of $X_S$ on the circuit output when:

1. $f_o < f_r$
2. $f_o = f_r$
3. $f_o > f_r$

When the circuit in Figure 19.54a is operated below its resonant frequency, the value of $X_C$ is greater than $X_L$. As a result, $X_S$ has some measurable value that is capacitive. For example, consider the effect of operating the circuit at 0 Hz. In this case, $X_L = 0\ \Omega$ and $X_C \rightarrow \infty\ \Omega$. Therefore, we can assume that the net series reactance is

$$X_S = 0\ \Omega - \infty\ \Omega = -\infty\ \Omega$$

If the value of $X_S$ approaches infinity, then the circuit input is isolated completely from the load; that is, $V_L = 0$ V. As the operating frequency increases from 0 Hz toward $f_r$, the value

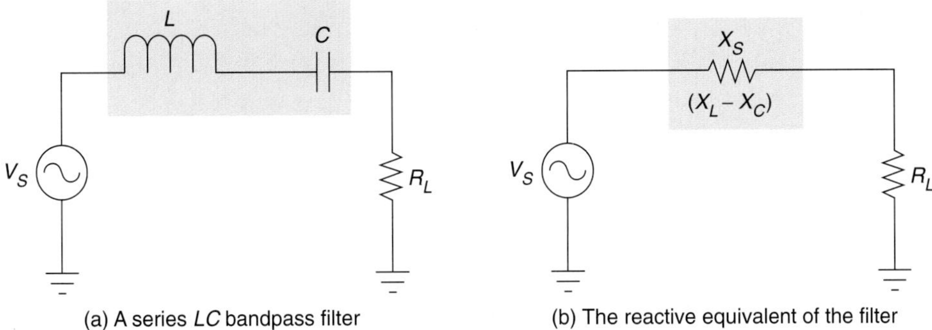


**FIGURE 19.54 A series *LC* bandpass filter and its equivalent circuit.**

of $X_C$ decreases. At the same time, the value of $X_L$ increases. Since $X_S$ equals the difference between $X_L$ and $X_C$, its value decreases rapidly. As the value of $X_S$ decreases, the value of $V_L$ increases, as given in the following relationship:

$$V_L = V_{\text{in}} \frac{R_L}{\sqrt{X_S^2 + R_L^2}} \tag{19.50}$$

where $X_S$ = the net series reactance of the filter, $(X_L - X_C)$

As you can see, this is just a voltage-divider equation for the series combination of the signal source, $X_S$, and the load.

When $f_o = f_r$, the value of $X_S$ is approximately 0 Ω. In this case, the source is coupled directly to the load, so the load voltage is found as

$$V_L = V_{\text{in}}$$

This being the case, the maximum voltage gain of the circuit can be approximated as

$$A_{v(\max)} = \frac{V_L}{V_{\text{in}}} \cong 1$$

We know that this is the maximum voltage gain, because the output voltage cannot be greater than the input voltage.

If the operating frequency of the circuit increases beyond the value of $f_r$, the value of $X_L$ increases and the value of $X_C$ decreases. Since the value of $X_S$ equals the difference between $X_L$ and $X_C$, its value starts increasing rapidly. If the circuit operating frequency approaches infinity, $X_C \rightarrow 0\ \Omega$ and $X_L \rightarrow \infty\ \Omega$. Therefore, we can assume that the net series reactance is

$$X_S = X_L - X_C = \infty\ \Omega - 0\ \Omega = \infty\ \Omega$$

The positive value of $X_S$ indicates that its value is inductive. As $X_S$ approaches infinity again, the circuit input is isolated completely from the load; that is, $V_L = 0$ V.

The circuit operation described here is represented by the frequency response curve in Figure 19.55. As you can see, the circuit has a value of $A_v = 0$ when the operating frequency is 0 Hz. This value is a result of the infinite capacitive reactance, which blocks the input voltage from the load. As the frequency increases toward $f_r$, the value of $X_S$ approaches 0 Ω. Therefore, the output voltage approaches the value of $V_{\text{in}}$, and the circuit voltage gain approaches one (1). As the frequency continues to increase, $X_S$ begins to increase rapidly, causing the load voltage and circuit voltage gain to drop. At some point, the inductive reactance in the filter increases to the point where it effectively isolates the source from the load. At this point, $V_L$ and $A_v$ both return to zero.

In a moment, we will discuss the cutoff frequencies and bandwidth of the curve in Figure 19.55. First, we will take a brief look at the operation of shunt *LC* bandpass filters.

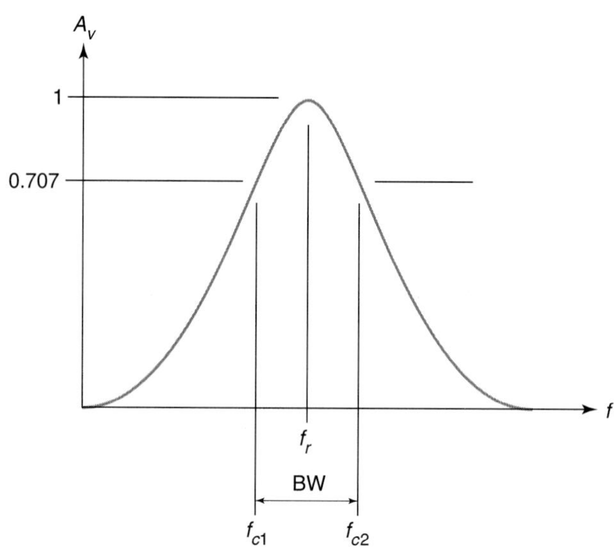


**FIGURE 19.55 The frequency response curve of an *LC* bandpass filter.**

## *Shunt LC Bandpass Filters*

*OBJECTIVE 22* ▶ Once you understand the operation of the series *LC* bandpass filter, the operation of its shunt counterpart is relatively easy to visualize. A shunt *LC* bandpass filter is shown in Figure 19.56, along with its reactive equivalent circuit. Note that $X_P$ in Figure 19.56b represents the parallel combination of $X_L$ and $X_C$, as indicated by the equation.

When the operating frequency of the circuit is 0 Hz, the filter has a value of $X_L = 0\ \Omega$. Therefore, the inductor effectively shorts out the load, and $V_L = 0$ V. Note the presence of the series resistor ($R_S$). This resistor is used to limit the source current when $X_P = 0\ \Omega$.

As the circuit operating frequency increases, the value of $X_L$ begins to increase. At the same time, the value of $X_C$ begins to decrease. As these values change, so does the value of $X_P$. As a result, the voltage across the filter and the load begins to increase. When the operating frequency reaches the value of $f_r$, the parallel *LC* circuit acts as an open; that is, $X_P \to \infty\ \Omega$. When this happens, the filter is effectively removed from the circuit, and the load voltage is found as

$$V_L = V_{in}\frac{R_L}{R_L + R_S}$$

Note that the circuit is normally designed using the minimum possible value of $R_S$ to provide the maximum possible value of load voltage, and therefore, voltage gain.

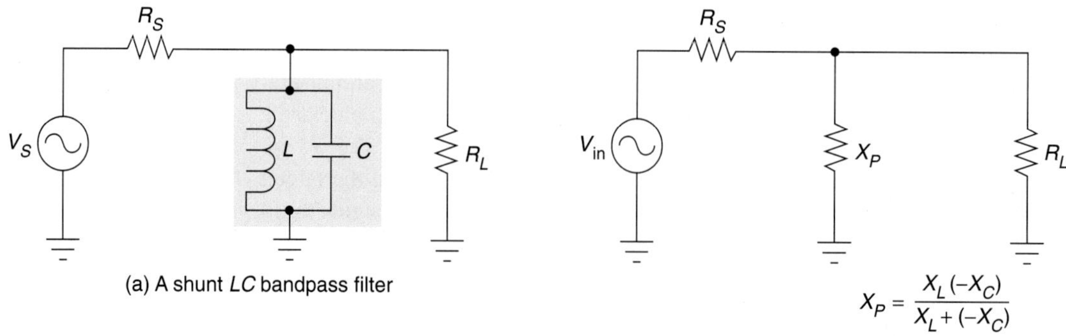


**FIGURE 19.56 A shunt *LC* bandpass filter.**

As the frequency increases above $f_r$, the value of $X_L$ continues to increase. However, the decreasing value of $X_C$ causes $X_P$ to begin decreasing again. As a result, the voltage across the filter starts decreasing, as does the load voltage. If the operating frequency continues to decrease, the reactance of the capacitor approaches 0 Ω. At that point, the capacitor effectively shorts out the load, and $V_L = 0$ V again. Note that the frequency response curve for the circuit in Figure 19.56 would be identical to the one shown in Figure 19.55. Although the two circuits operate differently, they produce the same overall frequency response.

So far, we have ignored the *LC* filter bandwidth and cutoff frequencies. As you will learn, *LC* filter bandwidth is a function of the circuit center frequency and the quality ($Q$) of the filter. (The relationship among $Q$, bandwidth, and center frequency was first discussed in Section 19.1.)

## Filter Quality (Q)

In earlier chapters, we discussed $Q$ as a component characteristic. As a review, the $Q$ of an inductor or capacitor is a measure of how close the component comes to the ideal component. The $Q$ characteristics of inductors and capacitors are summarized in Figure 19.57. The component equivalents shown in the figure represent the reactive and resistive components of the inductor and capacitor.

The ideal inductor would have a winding resistance ($R_w$) of 0 Ω. As a result, the total opposition to current ($Z$) presented by the inductor would equal $X_L$. For a practical inductor, the winding resistance is part of its opposition to current, which is why the value of $R_w$ is included in the actual component impedance calculation. Note that the higher the $Q$ of the component, the closer it comes to the ideal of $Z = X_L$.

The ideal capacitor would have infinite dielectric resistance ($R_d$). As a result, the total opposition to current ($Z$) presented by the capacitor would equal $X_C$. For a practical

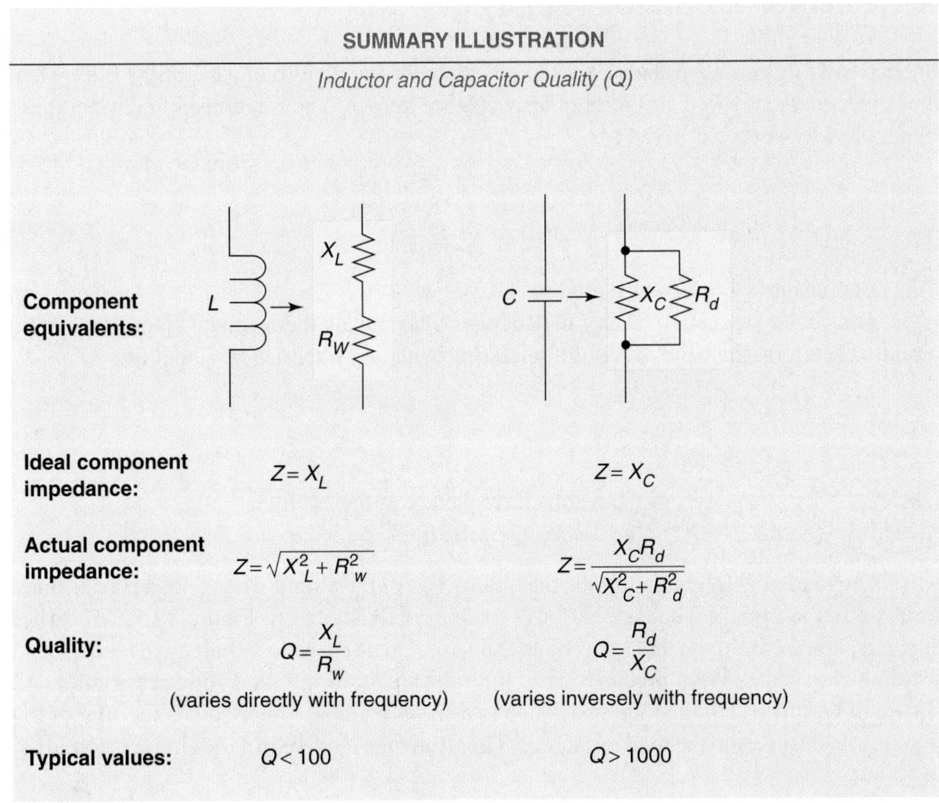

SUMMARY ILLUSTRATION
Inductor and Capacitor Quality (Q)
Component equivalents:
L
X_L
R_W
C
X_C
R_d
Ideal component impedance:
Z = X_L
Z = X_C
Actual component impedance:
Z = √(X_L² + R_w²)
Z = X_C R_d / √(X_C² + R_d²)
Quality:
Q = X_L / R_w
(varies directly with frequency)
Q = R_d / X_C
(varies inversely with frequency)
Typical values:
Q < 100
Q > 1000


**FIGURE 19.57**

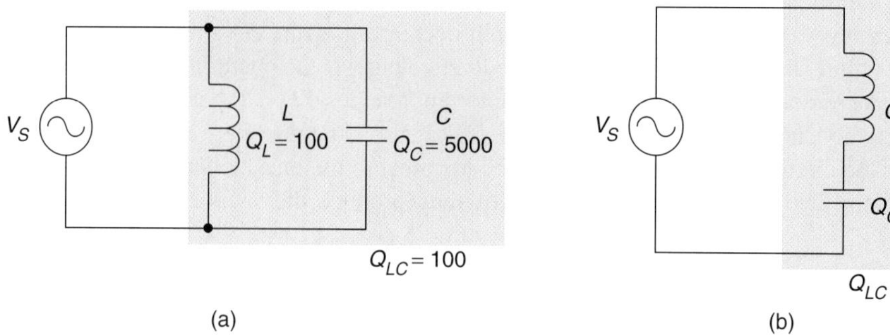


**FIGURE 19.58 Filter values of $Q$.**

capacitor, the dielectric resistance is part of its opposition to current, as demonstrated by the actual component impedance equation. Note that the higher the $Q$ of the component, the closer it comes to the ideal of $Z = X_C$.

Inductors and capacitors generally have the $Q$ ranges shown in the figure. As these ranges indicate, inductor $Q$ is generally much lower than capacitor $Q$. This relationship is significant because the $Q$ of a series or parallel $LC$ circuit equals the $Q$ of the inductor. For example, take a look at the circuits shown in Figure 19.58. In each case, the circuit $Q$ equals that of the inductor. This relationship is easy to accept when you remember that $Q$ represents the quality of the circuit. The quality of any circuit can never be greater than that of its lowest quality component. (This parallels the saying that a chain is only as strong as its weakest link.) It should be noted that the addition of a resistive load to a series or parallel $LC$ filter reduces the $Q$ of the circuit. We will address the effects of loading on filter $Q$ later in this section.

## Series Filter Frequency Analysis

In Section 19.1, you were shown how to calculate the bandwidth of a bandpass filter when the center frequency ($f_0$) and filter quality ($Q$) are known. The relationship between these filter characteristics was given as

$$\text{BW} = \frac{f_0}{Q}$$

This relationship was given as equation (19.2).

The center frequency of an $LC$ bandpass filter equals the resonant frequency of the circuit. Therefore, the bandwidth of a series $LC$ bandpass filter can be found as

$$\text{BW} = \frac{f_r}{Q_L} \tag{19.51}$$

where $f_r$ = the resonant frequency of the $LC$ filter
$Q_L$ = the *loaded* $Q$ of the filter

**Loaded $Q$ ($Q_L$)**
The quality ($Q$) of a filter when a load is connected.

The **loaded $Q$ ($Q_L$)** of a filter is the quality ($Q$) of the circuit when a load is connected to its output terminals. The concept of $Q$ loading is illustrated in Figure 19.59. In earlier chapters, you were shown that the $Q$ of an inductor equals the ratio of inductive reactance to winding resistance. More precisely, it is the ratio of $X_L$ to the total series resistance. As shown in Figure 19.59, the total resistance in a series bandpass filter equals the sum of the winding resistance and the load resistance. Therefore, the loaded $Q$ of the circuit is found as

$$Q_L = \frac{X_L}{R_w + R_L} \tag{19.52}$$

**FIGURE 19.59 Loaded $Q$.**

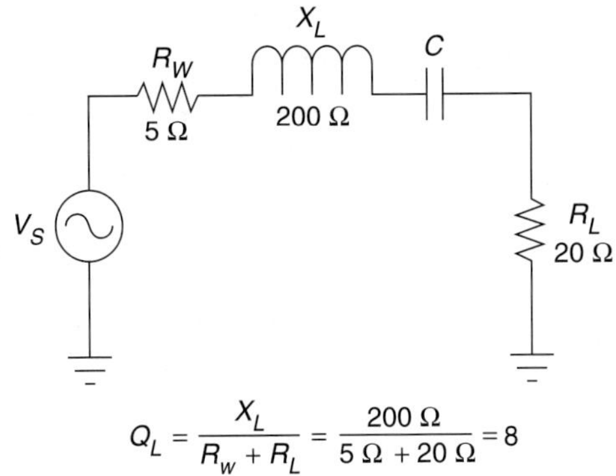


$$Q_L = \frac{X_L}{R_W + R_L} = \frac{200\ \Omega}{5\ \Omega + 20\ \Omega} = 8$$

The calculation of $Q_L$ for the circuit in Figure 19.59 is shown in the figure. As you can see, the loaded $Q$ is significantly lower than the value of $Q$ we would have obtained using only the inductor winding resistance.

To determine the bandwidth of a series $LC$ bandpass filter, you must first calculate its resonant frequency. Then you must calculate the values of $X_L$ and $Q_L$ at the resonant frequency. Finally, the values of $f_r$ and $Q_L$ are used to determine the circuit bandwidth. Example 19.33 demonstrates this series of circuit calculations.

**EXAMPLE 19.33**

Determine the bandwidth of the series $LC$ bandpass filter in Figure 19.60a.

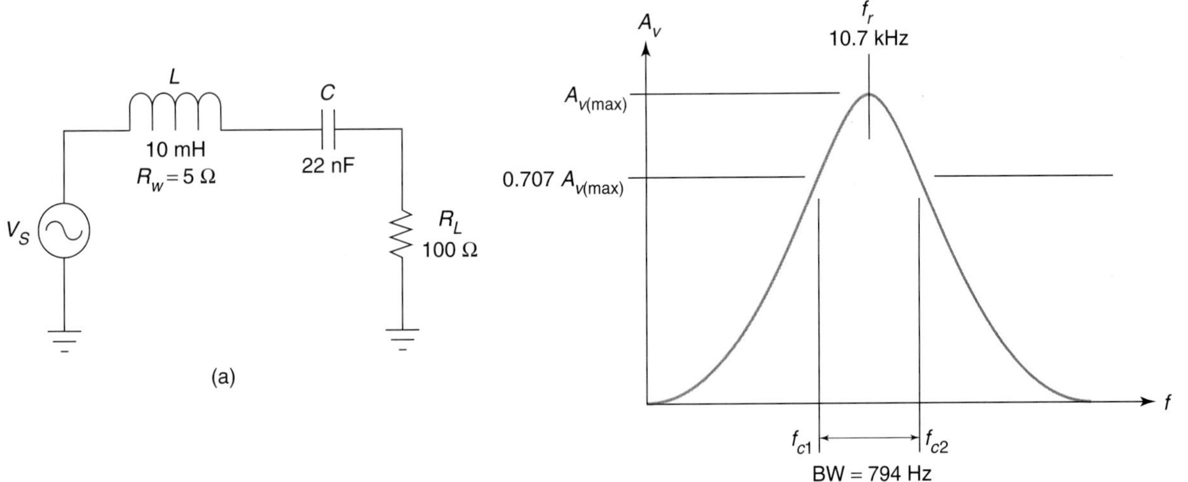

**FIGURE 19.60**

***Solution:*** First, the resonant frequency of the circuit is found as

$$f_r = \frac{1}{2\pi\sqrt{LC}} = \frac{1}{2\pi\sqrt{(10\text{ mH})(22\text{ nF})}} = \mathbf{10.7\ kHz}$$

When the circuit is operated at this frequency, the value of $X_L$ is found as

$$X_L = 2\pi fL = 2\pi(10.7\text{ kHz})(10\text{ mH}) = \mathbf{672\ \Omega}$$

The circuit is shown to have values of $R_w = 5\ \Omega$ and $R_L = 100\ \Omega$. Therefore, the loaded $Q$ of the component (at the resonant frequency) is found as

$$Q_L = \frac{X_L}{R_w + R_L} = \frac{672\ \Omega}{105\ \Omega} = \mathbf{6.4}$$

Finally, the bandwidth of the series bandpass filter can be found as

$$\mathbf{BW} = \frac{f_r}{Q_L} = \frac{10.7\text{ kHz}}{6.4} = \mathbf{1.67\ kHz}$$

The bandwidth and center frequency of the circuit are identified on the frequency response curve in Figure 19.60b.

***PRACTICE PROBLEM 19.33***

A filter like the one in Figure 19.60a has the following values: $L = 22$ mH, $R_w = 8\ \Omega$, $R_L = 330\ \Omega$, and $C = 4.7$ nF. Calculate the bandwidth of the filter.

Once we know the bandwidth and center frequency of a series bandpass filter, we have the information needed to calculate the circuit cutoff frequencies. In Section 19.1, we were shown that the cutoff frequencies of a bandpass filter can be found as

$$f_{c1} = f_{\text{ave}} - \frac{\text{BW}}{2} \qquad \text{and} \qquad f_{c2} = f_{\text{ave}} + \frac{\text{BW}}{2}$$

where $f_{\text{ave}}$ = the frequency that lies halfway between the cutoff frequencies

These relationships were introduced as equations (19.19) and (19.20). When a filter has a value of $Q_L \geq 2$, the values of $f_{\text{ave}}$ and $f_r$ are equal for all practical purposes. Therefore, when $Q_L \geq 2$, the cutoff frequencies for a series bandpass filter can be found as

$$f_{c1} = f_r - \frac{\text{BW}}{2} \qquad \textbf{(19.53)}$$

and

$$f_{c2} = f_r + \frac{\text{BW}}{2} \qquad \textbf{(19.54)}$$

Example 19.34 applies these relationships to the circuit we analyzed in Example 19.33.

## EXAMPLE 19.34

Determine the cutoff frequencies for the circuit in Figure 19.60a.

***Solution:*** In Example 19.33, the circuit was found to have the following values:

$$Q_L = 6.4 \qquad f_r = 10.7\text{ kHz} \qquad \text{BW} = 1.67\text{ kHz}$$

Since $Q_L > 2$, we can calculate the values of the cutoff frequencies as

$$f_{c1} = f_r - \frac{\text{BW}}{2} = 10.7\text{ kHz} - \frac{1.67\text{ kHz}}{2} \cong \mathbf{9.87\ kHz}$$

and

$$f_{c2} = f_r + \frac{\text{BW}}{2} = 10.7\text{ kHz} + \frac{1.67\text{ kHz}}{2} \cong \mathbf{11.54\ kHz}$$

***PRACTICE PROBLEM 19.34***

Calculate the cutoff frequencies for the circuit described in Practice Problem 19.33.

When a filter has a value of $Q_L < 2$, then the value of $f_{ave}$ must be calculated using

$$f_{ave} = f_r\sqrt{1 + \left(\frac{1}{2Q_L}\right)^2}$$

This relationship was introduced earlier (in a generic form) as equation (19.11). Once the value of $f_{ave}$ is determined, then the cutoff frequencies can be determined in the same manner as demonstrated in Example 19.34.

## Shunt Filter Frequency Analysis

The analysis of a shunt *LC* bandpass filter is nearly identical to that of the series filter. The primary difference lies with the calculation of $Q_L$. For example, consider the circuit shown in Figure 19.61a. To determine the value of $Q_L$ for this circuit, we need to combine $R_w$ with the other resistor values. In the configuration shown, calculating the total circuit resistance is extremely difficult. However, the winding resistance of the coil can be represented as a parallel resistance ($R_P$), as shown in Figure 19.61b. This parallel resistance, which represents the effective dc resistance of the filter, has a value that is found as

$$R_P = Q^2 R_w \tag{19.55}$$

where $Q$ = the *unloaded* $Q$ of the filter

This equation is derived in Appendix E.

From the viewpoint of the filter, the three resistors in Figure 19.61b are in parallel. Using the parallel combination of the resistors, the loaded $Q$ of the circuit is found as

$$Q_L = \frac{R_S \| R_P \| R_L}{X_L} \tag{19.56}$$

This equation is also derived in Appendix E. Example 19.35 demonstrates the process used to determine the $Q$ of a loaded shunt bandpass filter.

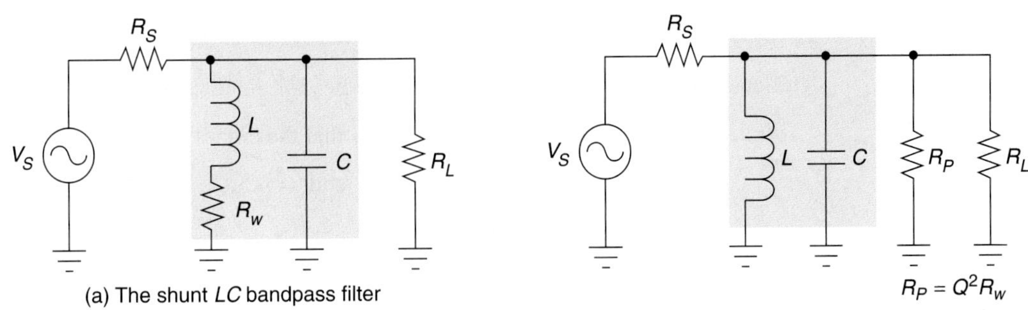

(a) The shunt *LC* bandpass filter

(b) Filter parallel resistance ($R_P$)

**FIGURE 19.61 Loaded *Q* for a shunt *LC* filter.**

**EXAMPLE 19.35**

Determine the value of $Q_L$ for the circuit shown in Figure 19.62.

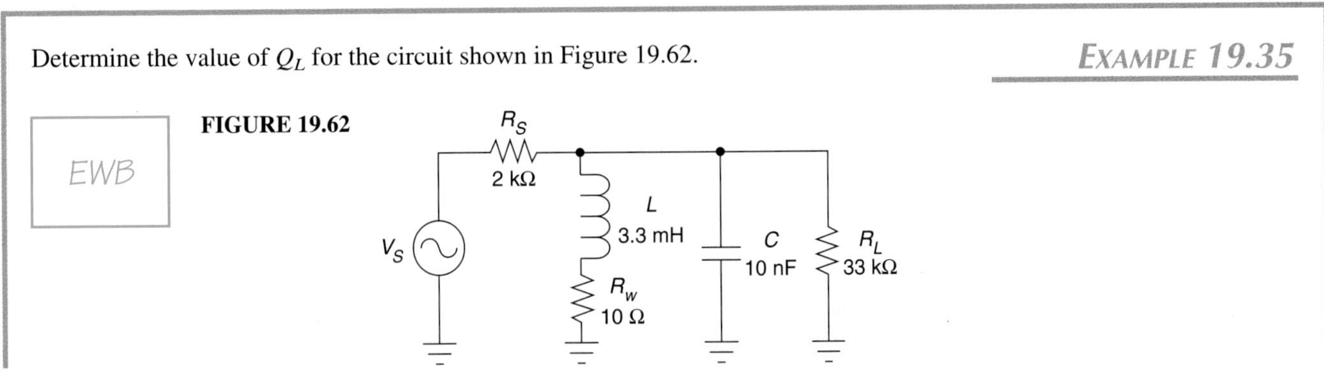

**FIGURE 19.62**

**Solution:** First, the resonant frequency of the circuit is found as

$$f_r = \frac{1}{2\pi\sqrt{LC}} = \frac{1}{2\pi\sqrt{(3.3\text{ mH})(10\text{ nF})}} = \mathbf{27.7\ kHz}$$

Now, the value of $X_L$ at the resonant frequency is found as

$$X_L = 2\pi fL = 2\pi(27.7\text{ kHz})(3.3\text{ mH}) = \mathbf{574\ \Omega}$$

The $Q$ of the inductor can now be found as

$$Q = \frac{X_L}{R_w} = \frac{574\ \Omega}{10\ \Omega} = \mathbf{57.4}$$

Using the inductor $Q$, the value of $R_P$ is found as

$$R_P = Q^2R_w = (57.4)^2(10\ \Omega) = \mathbf{32.9\ k\Omega}$$

Finally, the loaded $Q$ of the filter is found as

$$Q_L = \frac{R_S \| R_P \| R_L}{X_L} = \frac{2\text{ k}\Omega \| 32.9\text{ k}\Omega \| 33\text{ k}\Omega}{574\ \Omega} = \mathbf{3.1}$$

***PRACTICE PROBLEM 19.35***

A circuit like the one in Figure 19.62 has the following values: $R_S = 3.3\text{ k}\Omega$, $L = 10\text{ mH}$, $R_w = 8\ \Omega$, $C = 22\text{ nF}$, and $R_L = 50\text{ k}\Omega$. Calculate the value of $Q_L$ for the filter.

Once the value of $Q_L$ for a shunt bandpass filter has been determined, the rest of the frequency analysis proceeds just as it did for the series bandpass filter. This analysis is demonstrated in Example 19.36.

## EXAMPLE 19.36

Calculate the bandwidth and cutoff frequencies for the filter in Figure 19.62.

**Solution:** From Example 19.35, we know that the circuit has the following values:

$$f_r = 27.7\text{ kHz} \qquad \text{and} \qquad Q_L = 3.1$$

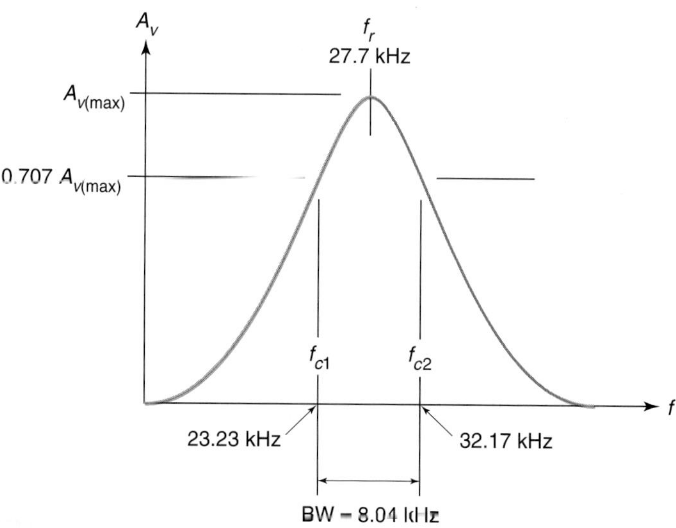


**FIGURE 19.63**

Using these values, the bandwidth of the filter can be found as

$$\mathbf{BW} = \frac{f_r}{Q_L} = \frac{27.7 \text{ kHz}}{3.1} = \mathbf{8.94\ kHz}$$

Since $Q_L > 2$, the filter cutoff frequencies can be found as

$$f_{c1} = f_r - \frac{\text{BW}}{2} = 27.7 \text{ kHz} - \frac{8.94 \text{ kHz}}{2} = \mathbf{23.23\ kHz}$$

and

$$f_{c2} = f_r + \frac{\text{BW}}{2} = 27.7 \text{ kHz} + \frac{8.94 \text{ kHz}}{2} = \mathbf{32.17\ kHz}$$

Using these values and those calculated in Example 19.35, the frequency response curve for the circuit is plotted as shown in Figure 19.63.

***PRACTICE PROBLEM 19.36***

Calculate the cutoff frequencies for the filter described in Practice Problem 19.35.

An interesting point needs to be made: While the ideal parallel *LC* circuit has infinite impedance at resonance, the impedance of the practical circuit is significantly lower. This impedance, which is resistive, is represented by $R_P$. As you will see, $R_P$ has an impact on the maximum voltage gain of the circuit.

## Maximum Voltage Gain

Figure 19.64a shows the circuit we analyzed in the last two examples. Figure 19.64b is the resistive equivalent of the circuit when operated at resonance. The *LC* circuit in the resistive equivalent circuit has been replaced by $R_P$. As you can see, the circuit consists of $R_S$ in series with $(R_P \| R_L)$. As such, the circuit forms a voltage divider, and the load voltage can be found as

$$V_{L(\max)} = V_{\text{in}} \frac{R_{EQ}}{R_{EQ} + R_S} \qquad \mathbf{(19.57)}$$

where $$R_{EQ} = R_P \| R_L$$

If we divide both sides of equation (19.57) by $V_{\text{in}}$, we get

$$\frac{V_{L(\max)}}{V_{\text{in}}} = \frac{R_{EQ}}{R_{EQ} + R_S}$$

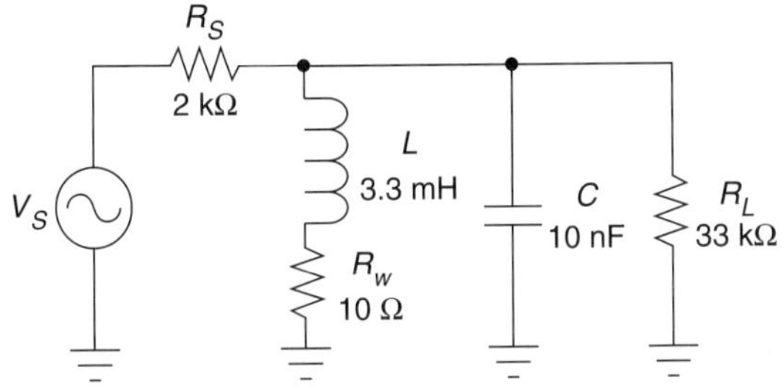


(a) The filter from Examples 19.36 and 19.37

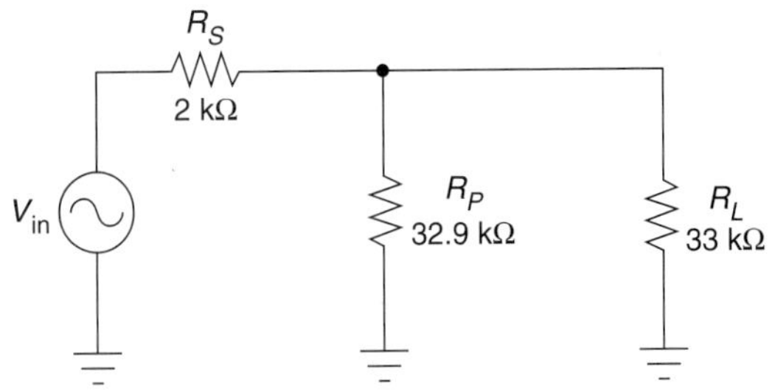


(b) The resistive equivalent of the filter when operated at resonance

**FIGURE 19.64**

or

$$A_{v(\max)} = \frac{R_{EQ}}{R_{EQ} + R_S} \tag{19.58}$$

Using equation (19.58), the maximum voltage gain for the circuit can be found as follows:

$$\boldsymbol{R_{EQ}} = R_P \| R_L = (32.9\ \text{k}\Omega) \| (33\ \text{k}\Omega) = \mathbf{16.5\ k\Omega}$$

and

$$\boldsymbol{A_{v(\max)}} = \frac{R_{EQ}}{R_{EQ} + R_S} = \frac{16.5\ \text{k}\Omega}{18.5\ \text{k}\Omega} = \mathbf{0.892}$$

As you know, voltage gain drops to 70.7% of its maximum value at the cutoff frequencies. For the circuit in Figure 19.64a, the maximum voltage gain was found to be 0.892. Therefore, the voltage gain at each of the cutoff frequencies is found as

$$\boldsymbol{A_v} = 0.707\ A_{v(\max)} = (0.707)(0.892) = \mathbf{0.631}$$

## The Effects of Q on Filter Gain and Roll-Off

The gain and roll-off rate of a bandpass filter vary with the filter quality ($Q$). The effect of filter $Q$ on bandpass frequency response is illustrated in Figure 19.65. As you can see, the two curves are centered on the same resonant frequency ($f_r$). At the same time, the figure shows that a decrease in $Q$ results in a decrease in the value of $A_{v(\max)}$. The decrease in gain is simple to understand when explained using the circuits in Figure 19.64. As you were shown earlier, the value of $A_{v(\max)}$ for the circuit in Figure 19.64a is found as

$$A_{v(\max)} = \frac{R_{EQ}}{R_{EQ} + R_S} \quad \Bigg| \quad R_{EQ} = R_P \| R_L$$

As you have been shown, $R_P = Q^2 R_w$. If the value of $Q$ decreases, the value of $R_P$ decreases. The decrease in $R_P$ results in a decrease in the value of $R_{EQ}$, and thus, in the value of $A_{v(\max)}$.

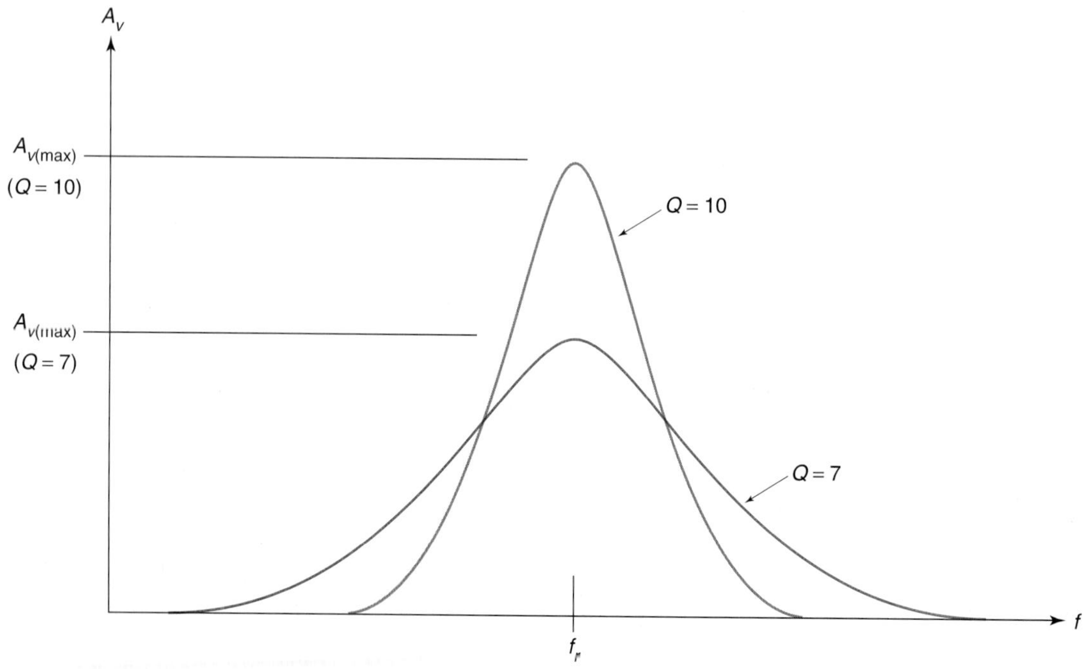


**FIGURE 19.65 The effects of $Q$ on bandpass frequency response.**

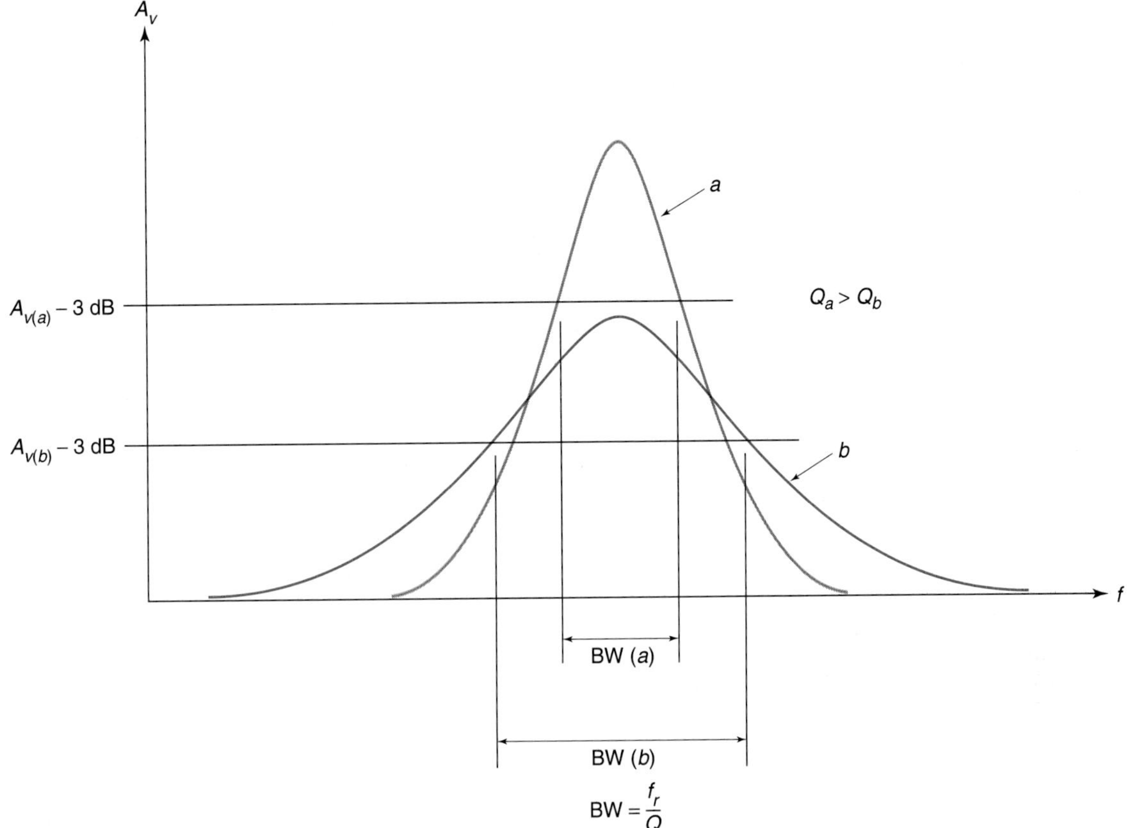


**FIGURE 19.66 The effect of $Q$ on filter bandwidth.**

A decrease in filter $Q$ causes:

1. The circuit bandwidth to increase.
2. The roll-off rates of the filter to decrease.

The relationship between filter $Q$ and bandwidth is illustrated in Figure 19.66. As shown in the figure, the lower $Q$ response curve has a wider bandwidth. The relationship between $Q$ and bandwidth shown in the figure makes sense when you remember that bandwidth is inversely proportional to $Q$ at a given value of $f_r$ (as given by the equation shown in the figure).

The roll-off rate of a filter is a measure of its slope. The greater the slope of the decrease in gain, the greater the roll-off rate. With this in mind, take another look at the curves in Figure 19.66. As you can see, the lower-$Q$ curve ($b$) intersects the higher-$Q$ curve ($a$). This indicates that the lower-$Q$ curve has a more gradual slope and thus a lower roll-off rate.

## Series LC Notch Filters

A series $LC$ notch filter can be constructed by placing a parallel $LC$ circuit in series between the signal source and the load. Such a circuit is shown in Figure 19.67, along with its frequency response curve. ◄ *OBJECTIVE 23*

The operation of the series $LC$ notch filter is easy to understand when you consider the circuit response to each of the following conditions:

$$f_{in} = 0 \text{ Hz} \qquad f_{in} = f_r \qquad f_{in} = \infty \text{ Hz}$$

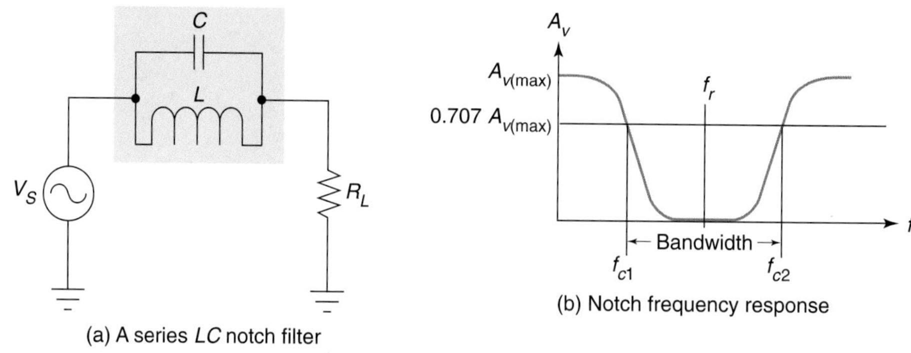


(a) A series *LC* notch filter

(b) Notch frequency response

**FIGURE 19.67 A series *LC* notch filter.**

When $f_{in} = 0$ Hz, the reactance of the filter inductor is 0 Ω. In this case, the signal source is coupled directly to the load via the inductor, and $V_L = V_{in}$. Since voltage gain is the ratio of output to input voltage, the circuit has a value of

$$A_{v(max)} \cong 1 \tag{19.59}$$

The maximum voltage gain of the circuit is approximately 1 because of the winding resistance of the coil. This resistance forms a voltage divider with the load, so load voltage is always less than the input voltage. However, the value of $R_w$ is typically much lower than the load resistance, so the approximation is valid in most cases.

When $f_{in} = f_r$, the impedance of the *LC* filter is infinite for all practical purposes. In this case, the *LC* circuit isolates the source from the load, and $V_L = 0$ V. The increase in the impedance of the *LC* circuit as frequency increases from 0 Hz to the value of $f_r$ accounts for the drop in voltage gain. As the impedance of the *LC* filter increases, the filter drops more and more of the input signal. As a result, the output voltage decreases along with the circuit voltage gain. This decrease in voltage gain accounts for the slope of the frequency response curve as it goes from 0 Hz to $f_r$.

When $f_{in} = \infty$ Hz (a theoretical value), the reactance of the filter capacitor is 0 Ω. In this case, the signal source is coupled directly to the load via the capacitor, and $V_L = V_{in}$. Again, the circuit has a voltage gain approximately equal to one (1). The decrease in filter impedance as frequency increases from $f_r$ to ∞ Hz causes an increase in load voltage. This increase accounts for the gradual increase in voltage gain.

## The Shunt LC Notch Filter

A shunt *LC* notch filter can be constructed by placing a series *LC* circuit in parallel with a load. Such a circuit is shown in Figure 19.68. The frequency response curve of the circuit is identical, for all practical purposes, to the one shown in Figure 19.67b.

The operation of the shunt *LC* notch filter is easy to understand when you consider the circuit response to each of the following conditions:

$$f_{in} = 0 \text{ Hz} \qquad f_{in} = f_r \qquad f_{in} = \infty \text{ Hz}$$

**FIGURE 19.68 A shunt *LC* notch filter.**

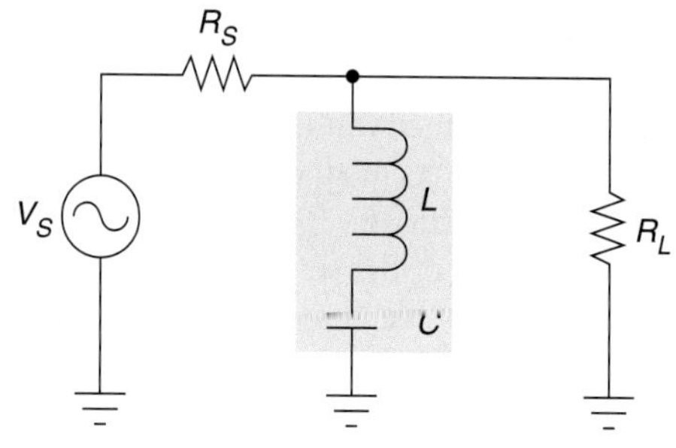

When $f_{\text{in}} = 0$ Hz, the reactance of the capacitor is infinite, so the *LC* circuit acts as an open. As a result, the circuit effectively consists of the signal source, $R_S$, and the load. In similar configurations, the maximum voltage gain has been found to equal the resistance ratio of the series circuit. In the same fashion, the maximum voltage gain of the shunt *LC* notch filter can be found as

$$A_{v(\max)} = \frac{R_L}{R_L + R_S} \tag{19.60}$$

When $f_{\text{in}} = f_r$, the *LC* circuit essentially shorts out the load. In this case, the load voltage is nearly 0 V. The impedance of the *LC* circuit decreases from $\infty$ Ω to 0 Ω as the input frequency increases from 0 Hz to $f_r$. This decrease in impedance causes the voltage across the *LC* circuit to decrease. Since the *LC* circuit is in parallel with the load, load voltage also decreases. This accounts for the drop in $A_v$ (between 0 Hz and $f_r$) shown in the frequency response curve (Figure 19.67b).

When $f_{\text{in}} = \infty$ Hz (a theoretical value), the reactance of the filter inductor is infinite. Therefore, the *LC* circuit acts as an open, just as it did when $f_{\text{in}} = 0$ Hz. Again, the circuit has a voltage gain that is found using equation (19.60). The increase in filter impedance as frequency increases from $f_r$ to $\infty$ Hz causes the voltage across the filter to increase. Since the load is in parallel with the *LC* filter, load voltage also increases. This increase accounts for the rise in $A_v$ (between $f_r$ and $\infty$ Hz) shown in the frequency response curve (Figure 19.67b).

## Section Review

1. List the impedance, current, and voltage characteristics of a series resonant *LC* circuit.
2. List the impedance, current, and voltage characteristics of a parallel resonant *LC* circuit.
3. Describe the frequency response of the series *LC* bandpass filter in Figure 19.54.
4. Describe the frequency response of the shunt *LC* bandpass filter in Figure 19.56.
5. Which component in an *LC* filter has the greatest effect on the value of filter *Q*? Explain your answer.
6. Explain the concept of loaded *Q*.
7. Describe the effects of *Q* on *LC* filter gain and roll-off.
8. Describe the operation of the series *LC* notch filter in Figure 19.67.
9. Describe the operation of the shunt *LC* notch filter in Figure 19.68.

## Chapter Summary

Here is a summary of the major points made in this chapter:

1. Signal loss caused by the frequency response of a circuit is referred to as attenuation.
2. Attenuation is normally described using the ratio of a circuit's output amplitude to its input amplitude. In most cases, a power ratio is used.
3. The frequency at which the power ratio of a circuit drops to 50% of its maximum value is referred to as the cutoff frequency ($f_c$).
4. Filters are circuits designed for specific frequency response characteristics.
5. There are four basic types of filters:
   - **a.** The *low-pass filter* is designed to pass all frequencies below its cutoff frequency.
   - **b.** The *high-pass filter* is designed to pass all frequencies above its cutoff frequency.
   - **c.** The *bandpass filter* is one designed to pass the band of frequencies that lies between two cutoff frequencies.

**d.** The *band-stop (or notch) filter* is designed to block the band of frequencies that lies between two cutoff frequencies.

The basic frequency response curves for these filters are shown in Figure 19.5.

6. The cutoff frequencies of a bandpass (or notch) filter are referred to as the lower cutoff frequency ($f_{c1}$) and the upper cutoff frequency ($f_{c2}$).
7. Since bandpass and notch filters have two cutoff frequencies, two values, bandwidth and center frequency, that are not commonly applied to the low-pass and high-pass filters are used to describe their operation.
8. The bandwidth of a filter is the range (or band) of frequencies between its cutoff frequencies.
9. The bandwidth of a filter equals the difference between its cutoff frequencies.
10. The center frequency ($f_0$) of a filter is the geometric average of its cutoff frequencies.
11. The ratio of $f_{c2}$ to $f_0$ equals the ratio of $f_0$ to $f_{c1}$.
12. Low-pass filters are normally described using their cutoff frequencies. An example is *a low-pass filter with a 10 kHz cutoff frequency.*
13. Bandpass and notch filters are normally described in terms of bandwidth and center frequency. An example is *a 20 kHz bandpass filter with a 150 kHz center frequency.*
14. The quality ($Q$) of a bandpass or notch filter equals the ratio of its center frequency to its bandwidth.
15. The value of filter $Q$ actually depends on circuit component values. The bandwidth of a filter depends on the values of center frequency and $Q$.
16. The average frequency ($f_{\text{ave}}$) of a filter lies halfway between its cutoff frequencies. This frequency is used in conjunction with the circuit bandwidth to determine the values of the cutoff frequencies.
17. When $Q \geq 2$, the values of $f_0$ and $f_{\text{ave}}$ for the filter are approximately equal. When $Q < 2$, the value of $f_{\text{ave}}$ can be found as demonstrated in Example 19.9.
18. Frequency response curves normally use logarithmic frequency scales. The value of each increment on a logarithmic scale is a whole-number multiple of the previous increment.

    **a.** An octave scale uses a frequency multiplier of 2 between increments.

    **b.** A decade scale uses a frequency multiplier of 10 between increments.
19. The ratio of circuit output amplitude to input amplitude is normally expressed using decibels (dB). Decibels are used because they allow us to represent easily very large and very small values.
20. The ratio of output power to input power for a component or circuit is commonly referred to as power gain ($A_p$). The word *gain* is used in reference to amplifiers, which can increase the power level of an ac input signal.
21. Passive filters all have power gains that are less than one (1).
22. The term *unity gain* is commonly used to refer to a value of $A_p = 1$.
23. The bel representation of a value indicates the power of 10 that equals the ratio. To express this representation mathematically: If $\log_{10}(y) = x$, then $10^x = y$. Several examples of this relationship are provided in Table 19.1.
24. The bel representation of a number is:

    **a.** Positive when $A_p > 1$.

    **b.** Zero when $A_p = 1$ (unity gain).

    **c.** Negative when $A_p < 1$.
25. Reciprocals have equal-magnitude positive and negative bel values. To express this relationship mathematically: If $\log_{10}(y) = x$, then $\log_{10} 1/y = -x$. Several examples of this relationship are provided in Table 19.2.

26. Years ago, the industry switched from bels to tenths of a bel, or decibels (dB).
27. There are 10 decibels (dB) in 1 bel (B). Therefore, dB power gain is found as ten times the common log of the power ratio.
28. The relationships provided above in numbers 24 and 25 apply to decibels as well as bels.
29. Power gain (in dB) is converted to standard numeric form, as demonstrated in Example 19.15.
30. Decibel representations of gain are used because:
    a. A very large range of values can be represented using relatively small numbers.
    b. The total gain produced by series filters and/or amplifiers can be determined using simple addition.
31. The primary limitation on dB values is that they cannot be used in circuit input/output calculations without being converted to standard numeric form.
32. When filters (or amplifiers) are connected in series, they are said to be *cascaded.*
33. Each filter (or amplifier) in a cascade is referred to as a stage.
34. The dBm reference represents an actual power value. It references a power level to a constant of 1 mW.
35. The ratio of circuit output voltage to input voltage is generally referred to as voltage gain ($A_v$).
36. The voltage gain of a circuit equals 70.7% of its maximum value at the cutoff frequencies. This relationship is expressed mathematically as $A_v = 0.707\ A_{v(\max)}$ when $f = f_C$. This relationship is based on the fact that power gain equals 50% of its maximum value when voltage gain equals 70.7% of its maximum value.
37. dB voltage gain is found as twenty times the common log of $A_v$.
38. When dB power gain drops to 50% of its maximum value, the change in dB power gain is −3 dB.
39. When dB voltage gain drops to 70.7% of its maximum value, the change in dB voltage gain is −3 dB.
40. The process for converting a dB voltage gain to standard numeric values is demonstrated in Example 19.21.
41. dB current gain principles are identical to those of dB voltage gain.
42. When a capacitor (or other component) is connected in parallel with a load (or some other circuit) it is referred to as a *shunt* component.
43. When analyzing a filter, our primary interest is in determining the maximum value of $A_v$ and the value of the cutoff frequency (or frequencies).
44. The process for determining the maximum voltage gain of an *RC* low-pass filter is demonstrated in Example 19.23.
45. The value of $A_{v(\max)}$ for an *RC* low-pass filter is always less than unity (1).
46. The process for determining the cutoff frequency of an *RC* low-pass filter is demonstrated in Example 19.25.
47. A bode plot is a normalized graph that represents frequency response as a change in voltage gain ($\Delta A_v$) versus operating frequency.
48. A bode plot is an ideal plot of frequency response, because it assumes that gain remains constant until the cutoff frequency is reached (see Figure 19.28).
49. The advantage in using a bode plot is that only the value of the cutoff frequency varies from one filter to another of the same type.
50. The cutoff frequency of a filter is commonly referred to as the *3 dB point* or *3 dB frequency.*

51. The roll-off rate of a filter is the rate of change in gain experienced by the circuit when it is operated outside its frequency limit, normally expressed in dB per octave or dB per decade.

52. A roll-off rate of 6 dB per octave equals a roll-off rate of 20 dB per decade.

53. All single-stage, low-pass *RC* filters experience the same 6-dB-per-octave (20-dB-per-decade) roll-off rates. The rates are independent of the values of *R* and *C*.

54. The complete analysis of an *RC* low-pass filter is demonstrated in Example 19.26.

55. The gain and frequency calculations for an *RL* low-pass filter are slightly different from those of the *RC* low-pass filter. However, all the concepts relating to response curves, bode plots, and roll-off rates are the same.

56. The process used to determine the value of $A_{v(\max)}$ for an *RL* low-pass filter is demonstrated in Example 19.27.

57. The process used to determine the cutoff frequency of an *RL* low-pass filter is demonstrated in Example 19.28.

58. High-pass filters are formed by reversing the positions of the resistive and reactive components in *RC* and *RL* low-pass filters.

59. An *RC* high-pass filter and response curve are shown in Figure 19.39.

60. The value of $A_{v(\max)}$ for an *RC* high-pass filter is approximately equal to one (1).

61. The process used to determine the cutoff frequency of an *RC* high-pass filter is demonstrated in Example 19.30.

62. The bode plot for a high-pass filter is simply a mirror image of the plot for a low-pass filter (see Figure 19.42).

63. The complete analysis of an *RC* high-pass filter is demonstrated in Example 19.32.

64. An *RL* high-pass filter and frequency response curve are shown in Figure 19.45.

65. The process used to determine the cutoff frequency of an *RL* high-pass filter is demonstrated in Example 19.32.

66. The most common passive bandpass and notch filters are *LC* filters.

67. The value of $A_{v(\max)}$ for a series *LC* bandpass filter is approximately equal to one (1).

68. The frequency response curve of an *LC* bandpass filter is shown in Figure 19.55.

69. Inductor and capacitor quality ($Q$) are compared in Figure 19.57.

70. The $Q$ of a filter is approximately equal to (or less than) the $Q$ of its inductor.

71. The loaded $Q$ ($Q_L$) of a filter is the quality of the circuit when a load is connected to its output terminals.

72. The loaded $Q$ of a filter is significantly lower than the unloaded $Q$ of the inductor.

73. The process used to determine the bandwidth of a series *LC* bandpass filter is demonstrated in Example 19.33.

74. The process used to determine the cutoff frequencies of a series *LC* bandpass filter is demonstrated in Example 19.34.

75. The process used to determine the value of $Q_L$ for a shunt *LC* bandpass filter is demonstrated in Example 19.35.

76. The gain and roll-off rates of a bandpass filter vary with the filter quality ($Q$). A decrease in $Q$ causes:

    **a.** An increase in bandwidth.

    **b.** A decrease in the roll-off rates.
    These relationships are illustrated in Figure 19.66.

| Equation Number | Equation | Section Number | Equation Summary |
|---|---|---|---|
| **(19.1)** | $\text{BW} = f_{c2} - f_{c1}$ | 19.1 | |
| **(19.2)** | $f_0 = \sqrt{f_{c1} f_{c2}}$ | 19.1 | |
| **(19.3)** | $\dfrac{f_0}{f_{c1}} = \dfrac{f_{c2}}{f_0}$ | 19.1 | |
| **(19.4)** | $f_{c1} = \dfrac{f_0^2}{f_{c2}}$ | 19.1 | |
| **(19.5)** | $f_{c2} = \dfrac{f_0^2}{f_{c1}}$ | 19.1 | |
| **(19.6)** | $Q = \dfrac{f_0}{\text{BW}}$ | 19.1 | |
| **(19.7)** | $\text{BW} = \dfrac{f_0}{Q}$ | 19.1 | |
| **(19.8)** | $f_{\text{ave}} = \dfrac{f_{c1} + f_{c2}}{2}$ | 19.1 | |
| **(19.9)** | $f_{c1} = f_{\text{ave}} - \dfrac{\text{BW}}{2}$ | 19.1 | |
| **(19.10)** | $f_{c2} = f_{\text{ave}} + \dfrac{\text{BW}}{2}$ | 19.1 | |
| **(19.11)** | $f_{\text{ave}} = f_0\sqrt{1 + \left(\dfrac{1}{2Q}\right)^2}$ | 19.1 | |
| **(19.12)** | $f_{\text{ave}} \cong f_0 \,\Big\vert\, Q \geq 2$ | 19.1 | |
| **(19.13)** | $A_{p(\text{B})} = \log_{10}\dfrac{P_{\text{out}}}{P_{\text{in}}}$ | 19.2 | |
| **(19.14)** | $A_{p(\text{B})} = \log_{10} A_p$ | 19.2 | |
| **(19.15)** | $A_{p(\text{dB})} = 10 \log\dfrac{P_{\text{out}}}{P_{\text{in}}}$ | 19.2 | |
| **(19.16)** | $A_{p(\text{dB})} = 10 \log A_p$ | 19.2 | |
| **(19.17)** | $A_p = \log^{-1}\dfrac{A_{p(\text{dB})}}{10}$ | 19.2 | |
| **(19.18)** | $P_{\text{out}} = A_p P_{\text{in}}$ | 19.2 | |
| **(19.19)** | $A_{pT(\text{dB})} = A_{p1(\text{dB})} + A_{p2(\text{dB})} + \ldots + A_{pn(\text{dB})}$ | 19.2 | |
| **(19.20)** | $P_{\text{dBm}} = 10 \log\dfrac{P_{\text{out}}}{1\text{ mW}}$ | 19.2 | |
| **(19.21)** | $P_{\text{out}} = (1\text{ mW})\left(\log^{-1}\dfrac{P_{\text{dBm}}}{10}\right)$ | 19.2 | |
| **(19.22)** | $f = f_c \quad \text{when} \quad A_v = 0.707\, A_{v(\text{max})}$ | 19.3 | |

| | | |
|---|---|---|
| **(19.23)** | $A_{v(\text{dB})} = 20 \log \dfrac{v_{\text{out}}}{v_{\text{in}}}$ | 19.3 |
| **(19.24)** | $A_{v(\text{dB})} = 20 \log A_v$ | 19.3 |
| **(19.25)** | $A_v = \log^{-1} \dfrac{A_{v(\text{dB})}}{20}$ | 19.3 |
| **(19.26)** | $\Delta A_{p(\text{dB})} = \Delta A_{v(\text{dB})}$ | 19.3 |
| **(19.27)** | $f = f_c \quad \text{when} \quad A_i = 0.707\, A_{i(\max)}$ | 19.3 |
| **(19.28)** | $A_{i(\text{dB})} = 20 \log \dfrac{i_{\text{out}}}{i_{\text{in}}}$ | 19.3 |
| **(19.29)** | $A_{i(\text{dB})} = 20 \log A_i$ | 19.3 |
| **(19.30)** | $A_i = \log^{-1} \dfrac{A_{i(\text{dB})}}{20}$ | 19.3 |
| **(19.31)** | $\Delta A_{p(\text{dB})} = \Delta A_{i(\text{dB})}$ | 19.3 |
| **(19.32)** | $V_{L(\max)} = V_{\text{in}} \dfrac{R_L}{R_L + R_F}$ | 19.4 |
| **(19.33)** | $A_{v(\max)} = \dfrac{V_{L(\max)}}{V_{\text{in}}}$ | 19.4 |
| **(19.34)** | $A_{v(\max)} = \dfrac{R_L}{R_L + R_F}$ | 19.4 |
| **(19.35)** | $f_c = \dfrac{1}{2\pi RC}$ | 19.4 |
| **(19.36)** | $R_{th} = (R_S + R_F) \parallel R_L$ | 19.4 |
| **(19.37)** | $f_c = \dfrac{1}{2\pi R_{th} C}$ | 19.4 |
| **(19.38)** | $\Delta A_v = \dfrac{1}{\sqrt{1 + (f_o/f_c)^2}}$ | 19.4 |
| **(19.39)** | $\Delta A_{v(\text{dB})} = 20 \log \dfrac{1}{\sqrt{1 + (f_o/f_c)^2}}$ | 19.4 |
| **(19.40)** | $V_{L(\max)} = V_{\text{in}} \dfrac{R_{EQ}}{R_{EQ} + R_w} \;\Big\vert\; R_{EQ} = R_F \parallel R_L$ | 19.4 |
| **(19.41)** | $V_{L(\max)} \cong V_{\text{in}} \;\Big\vert\; R_w < \dfrac{R_F \parallel R_L}{10}$ | 19.4 |
| **(19.42)** | $A_{v(\max)} = \dfrac{R_{EQ}}{R_{EQ} + R_w} \;\Big\vert\; R_{EQ} = R_F \parallel R_L$ | 19.4 |
| **(19.43)** | $f_c = \dfrac{R_{EQ}}{2\pi L}$ | 19.4 |
| **(19.44)** | $V_L \cong V_{\text{in}}$ | 19.5 |
| **(19.45)** | $A_{v(\max)} \cong 1$ | 19.5 |

| | | |
|---|---|---|
| **(19.46)** | $\Delta A_v = \dfrac{1}{\sqrt{1 + (f_c/f_o)^2}}$ | 19.5 |
| **(19.47)** | $\Delta A_{v(\text{dB})} = 20 \log \dfrac{1}{\sqrt{1 + (f_c/f_o)^2}}$ | 19.5 |
| **(19.48)** | $V_{L(\max)} = V_{\text{in}} \dfrac{R_L}{R_L + R_F}$ | 19.5 |
| **(19.49)** | $A_{v(\max)} = \dfrac{R_L}{R_L + R_F}$ | 19.5 |
| **(19.50)** | $V_L = V_{\text{in}} \dfrac{R_L}{\sqrt{X_S^2 + R_L^2}}$ | 19.6 |
| **(19.51)** | $\text{BW} = \dfrac{f_r}{Q_L}$ | 19.6 |
| **(19.52)** | $Q_L = \dfrac{X_L}{R_w + R_L}$ | 19.6 |
| **(19.53)** | $f_{c1} = f_r - \dfrac{\text{BW}}{2}$ | 19.6 |
| **(19.54)** | $f_{c2} = f_r + \dfrac{\text{BW}}{2}$ | 19.6 |
| **(19.55)** | $R_P = Q^2 R_w$ | 19.6 |
| **(19.56)** | $Q_L = \dfrac{R_S \| R_P \| R_L}{X_L}$ | 19.6 |
| **(19.57)** | $V_{L(\max)} = V_{\text{in}} \dfrac{R_{EQ}}{R_{EQ} + R_S}$ | 19.6 |
| **(19.58)** | $A_{v(\max)} = \dfrac{R_{EQ}}{R_{EQ} + R_S}$ | 19.6 |
| **(19.59)** | $A_{v(\max)} \cong 1$ | 19.6 |
| **(19.60)** | $A_{v(\max)} = \dfrac{R_L}{R_L + R_S}$ | 19.6 |

## Key Terms

The following terms were introduced and defined in this chapter:

attenuation
average frequency ($f_{\text{ave}}$)
bandpass filter
band-stop filter
bandwidth
bel (B)
bode plot
cascaded
center frequency ($f_0$)
current gain ($A_i$)
cutoff frequency ($f_c$)
dBm reference
decade
decade scale
decibel (dB)
frequency response curve
high-pass filter
loaded $Q$ ($Q_L$)
logarithmic scale
low-pass filter
notch filter
$n$-stage filter
octave
octave scale
roll-off rate
power gain ($A_p$)
stage
3 dB frequency
3 dB point
unity gain
voltage gain ($A_v$)

## Practice Problems

**1.** Complete the table below.

| | $f_{c1}$ | $f_{c2}$ | BW | $f_0$ |
|---|---|---|---|---|
| a. | 5 kHz | 88 kHz | ____ | ____ |
| b. | 640 Hz | 1.8 kHz | ____ | ____ |
| c. | 280 kHz | 1.4 MHz | ____ | ____ |
| d. | 60 Hz | 300 Hz | ____ | ____ |

**2.** Complete the table below.

| | $f_{c1}$ | $f_{c2}$ | BW | $f_0$ |
|---|---|---|---|---|
| a. | 2.2 kHz | 63 kHz | ____ | ____ |
| b. | 120 kHz | 2 MHz | ____ | ____ |
| c. | 400 Hz | 800 Hz | ____ | ____ |
| d. | 24 kHz | 500 kHz | ____ | ____ |

**3.** Complete the table below.

| | $f_{c1}$ | $f_{c2}$ | BW | $f_0$ |
|---|---|---|---|---|
| a. | 800 Hz | ____ | 12 kHz | ____ |
| b. | 5 kHz | ____ | ____ | 10 kHz |
| c. | ____ | 200 kHz | 198 kHz | ____ |

**4.** Complete the table below.

| | $f_{c1}$ | $f_{c2}$ | BW | $f_0$ |
|---|---|---|---|---|
| a. | 400 Hz | ____ | 39.6 kHz | ____ |
| b. | 50 kHz | ____ | ____ | 200 kHz |
| c. | ____ | 400 kHz | 50 kHz | ____ |

**5.** A filter has values of BW = 18 kHz and $f_0$ = 120 kHz. Determine the value of $Q$ for the filter.

**6.** A filter has values of BW = 48 kHz and $f_0$ = 340 kHz. Determine the value of $Q$ for the filter.

**7.** A filter has values of $f_0$ = 52 kHz and $Q$ = 6.8. Determine the circuit bandwidth.

**8.** A filter has values of $f_0$ = 650 Hz and $Q$ = 2.5. Determine the circuit bandwidth.

**9.** Determine the value of $f_{ave}$ for a filter with values of $f_0$ = 88 kHz and $Q$ = 1.2.

**10.** Determine the value of $f_{ave}$ for a filter with values of $f_0$ = 340 kHz and $Q$ = 1.8.

**11.** Determine the value of $f_{ave}$ for the filter described in Problem 7.

**12.** Determine the value of $f_{ave}$ for the filter described in Problem 8.

**13.** A bandpass filter has a 330 kHz center frequency and an 80 kHz bandwidth. Determine the average frequency for the circuit.

**14.** A bandpass filter has a 460 kHz center frequency and a 240 kHz bandwidth. Determine the average frequency for the circuit.

**15.** Determine the cutoff frequencies for the circuit described in Problem 13.

**16.** Determine the cutoff frequencies for the circuit described in Problem 14.

**17.** Complete the table below.

| | $P_{in}$ | $P_{out}$ | $A_p$ | $A_{p(dB)}$ |
|---|---|---|---|---|
| a. | 2 mW | 80 mW | ____ | ____ |
| b. | 48 mW | 300 mW | ____ | ____ |
| c. | 240 mW | 60 mW | ____ | ____ |
| d. | 1 W | 50 mW | ____ | ____ |

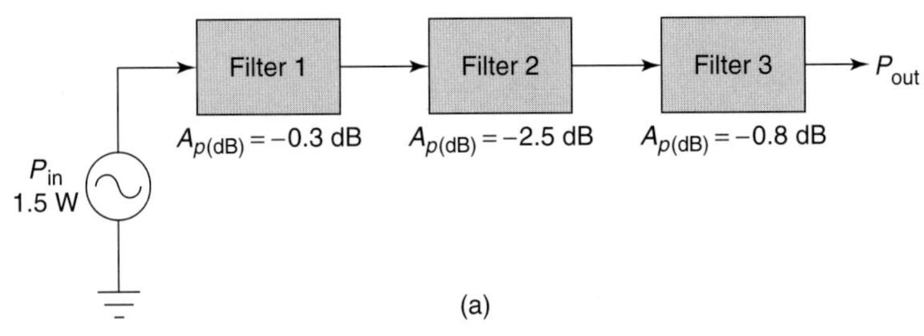


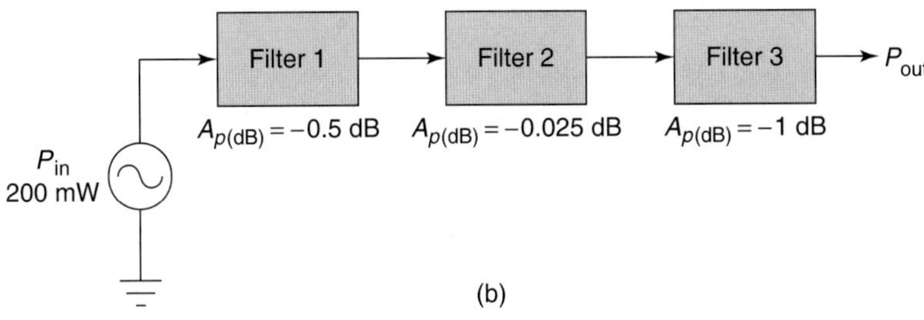


**FIGURE 19.69**

**18.** Complete the table below.

| | $P_{in}$ | $P_{out}$ | $A_p$ | $A_{p(dB)}$ |
|---|---|---|---|---|
| a. | 60 mW | 720 mW | ____ | ____ |
| b. | 50 μW | 30 mW | ____ | ____ |
| c. | 40 mW | 200 μW | ____ | ____ |
| d. | 120 mW | 8 mW | ____ | ____ |

**19.** The gain of a filter is −0.8 dB. Determine the output power when $P_{in} = 800$ mW.

**20.** The gain of a filter is −1.2 dB. Determine the output power when $P_{in} = 250$ mW.

**21.** The gain of a filter is −0.4 dB. Determine the output power when $P_{in} = 1.4$ W.

**22.** The gain of a filter is −1.95 dB. Determine the output power when $P_{in} = 1$ W.

**23.** Determine the output power for the circuit represented in Figure 19.69a.

**24.** Determine the output power for the circuit represented in Figure 19.69b.

**25.** Express each of the following power values in dBm.

**a.** 12 W **c.** 2.5 W

**b.** 180 mW **d.** 300 μW

**26.** Express each of the following power values in dBm.

**a.** 640 mW **c.** 1 mW

**b.** 7 W **d.** 800 nW

**27.** Convert each of the following to a power value.

**a.** 3.5 dBm **c.** −1.2 dBm

**b.** 12 dBm **d.** −3.5 dBm

**28.** Convert each of the following to a power value.

**a.** 2 dBm **c.** 0.9 dBm

**b.** 7.2 dBm **d.** −2.2 dBm

**29.** Convert each voltage gain ($A_v$) value listed to dB form.

**a.** 120 **c.** 0.995

**b.** 40 **d.** 0.978

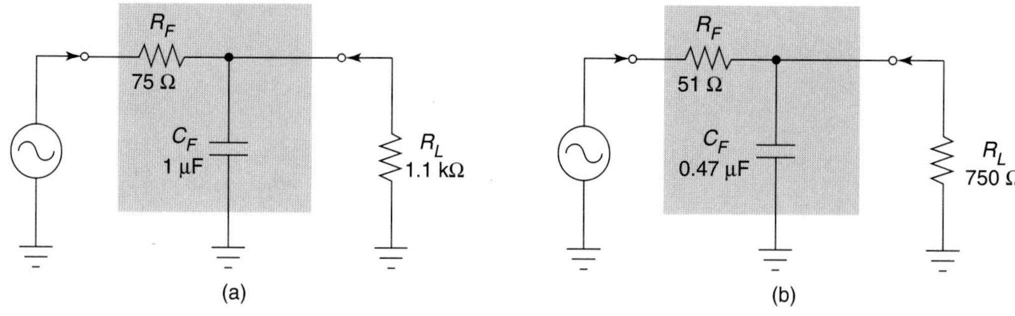


**FIGURE 19.70**

**30.** Convert each voltage gain ($A_v$) value listed to dB form.

**a.** 0.64 **c.** 0.895

**b.** 0.75 **d.** 0.159

**31.** Convert each value of dB voltage gain listed to standard numeric form.

**a.** 1.2 dB **c.** −3.6 dB

**b.** 4 dB **d.** −2.5 dB

**32.** Convert each value of dB voltage gain listed to standard numeric form.

**a.** 0.8 dB **c.** −0.8 dB

**b.** −7.2 dB **d.** −4.8 dB

**33.** Calculate the maximum voltage gain for the low-pass filter in Figure 19.70a.

**34.** Calculate the maximum voltage gain for the low-pass filter in Figure 19.70b.

**35.** Calculate the maximum voltage gain for the low-pass filter in Figure 19.71a.

**36.** Calculate the maximum voltage gain for the low-pass filter in Figure 19.71b.

**37.** Calculate the cutoff frequency for the low-pass filter in Figure 19.70a. Assume the resistance of the source is 25 Ω.

**38.** Calculate the cutoff frequency for the low-pass filter in Figure 19.70b. Assume the resistance of the source is 15 Ω.

**39.** Calculate the cutoff frequency for the low-pass filter in Figure 19.71a. Assume the resistance of the source is 30 Ω.

**40.** Calculate the cutoff frequency for the low-pass filter in Figure 19.71b. Assume the resistance of the source is 32 Ω.

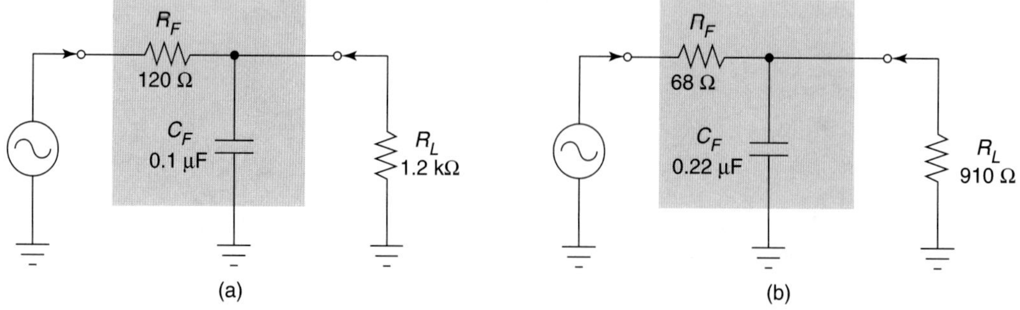


**FIGURE 19.71**

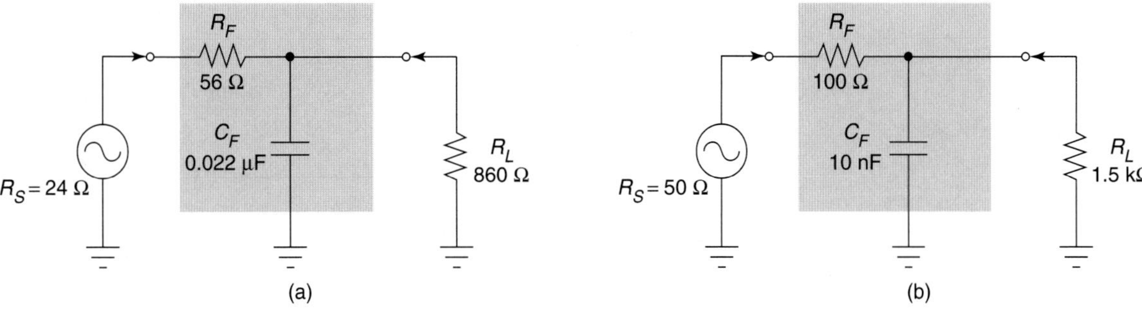


**FIGURE 19.72**

41. Perform a complete gain and frequency analysis on the low-pass filter in Figure 19.72a. Plot the frequency response curve for the circuit.
42. Perform a complete gain and frequency analysis on the low-pass filter in Figure 19.72b. Plot the frequency response curve for the circuit.
43. Calculate the maximum voltage gain for the low-pass filter in Figure 19.73a.
44. Calculate the maximum voltage gain for the low-pass filter in Figure 19.73b.
45. Calculate the maximum voltage gain for the low-pass filter in Figure 19.74a.
46. Calculate the maximum voltage gain for the low-pass filter in Figure 19.74b.
47. Calculate the cutoff frequency for the low-pass filter in Figure 19.73a.
48. Calculate the cutoff frequency for the low-pass filter in Figure 19.73b.

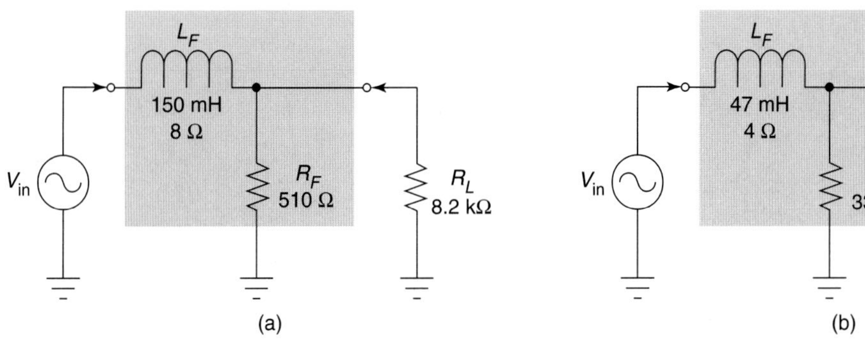


**FIGURE 19.73**

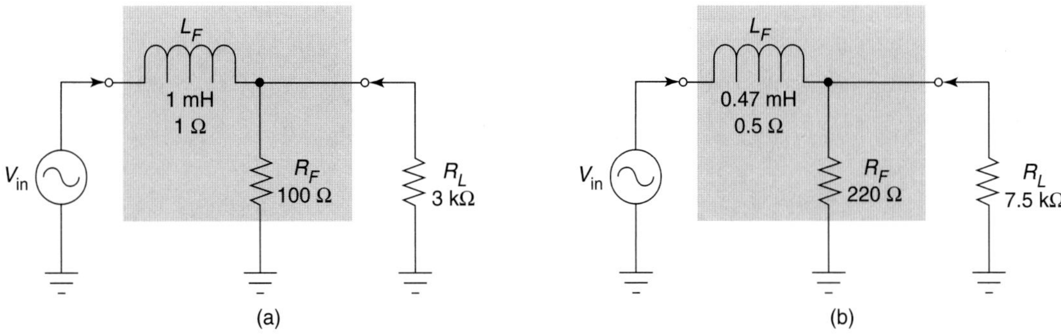


**FIGURE 19.74**

49. Calculate the cutoff frequency for the low-pass filter in Figure 19.74a.

50. Calculate the cutoff frequency for the low-pass filter in Figure 19.74b.

51. Perform a complete gain and frequency analysis on the low-pass filter in Figure 19.75a.

52. Perform a complete gain and frequency analysis on the low-pass filter in Figure 19.75b.

53. Calculate the cutoff frequency for the high-pass filter in Figure 19.76a.

54. Calculate the cutoff frequency for the high-pass filter in Figure 19.76b.

55. Perform a complete gain and frequency analysis on the high-pass filter in Figure 19.77a.

56. Perform a complete gain and frequency analysis on the high-pass filter in Figure 19.77b.

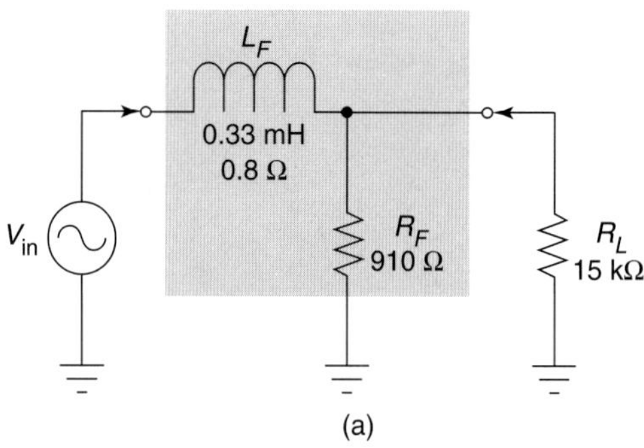


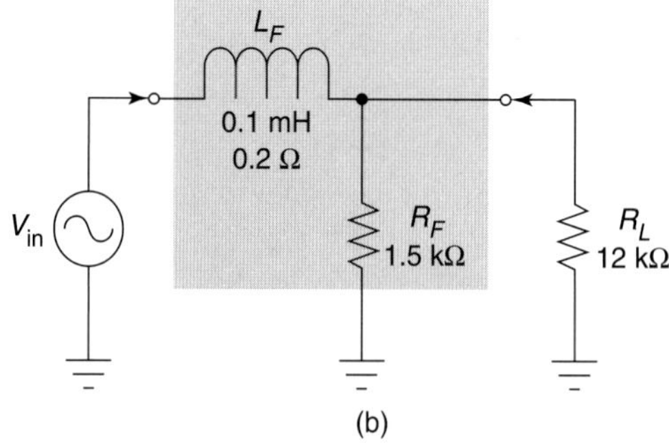


**FIGURE 19.75**

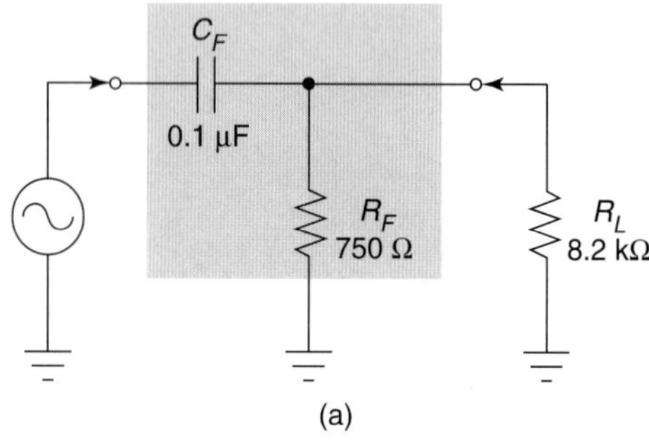


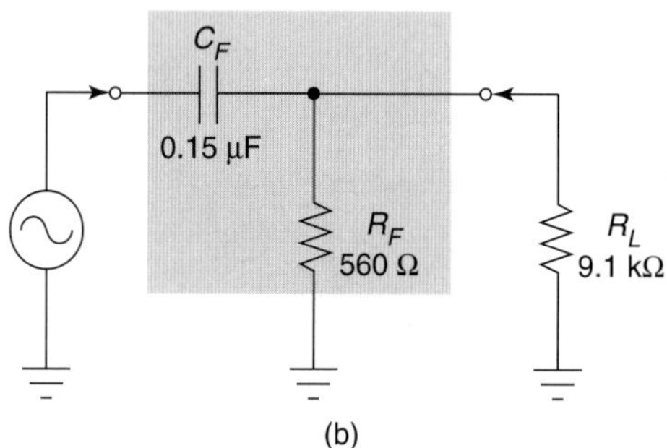


**FIGURE 19.76**

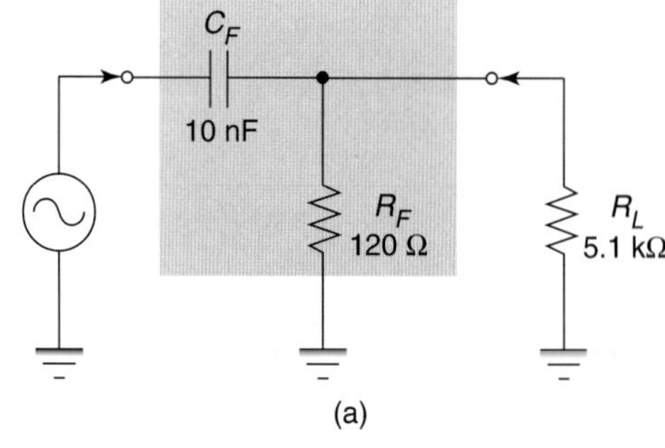


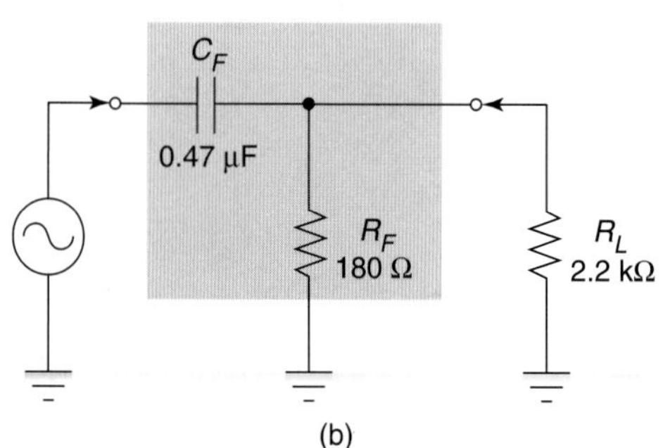


**FIGURE 19.77**

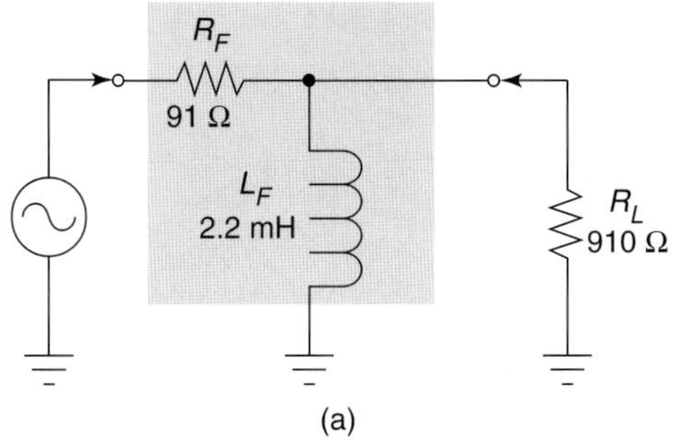


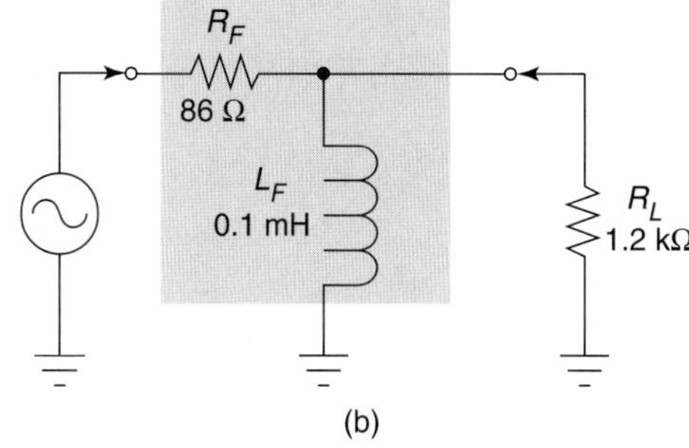


FIGURE 19.78

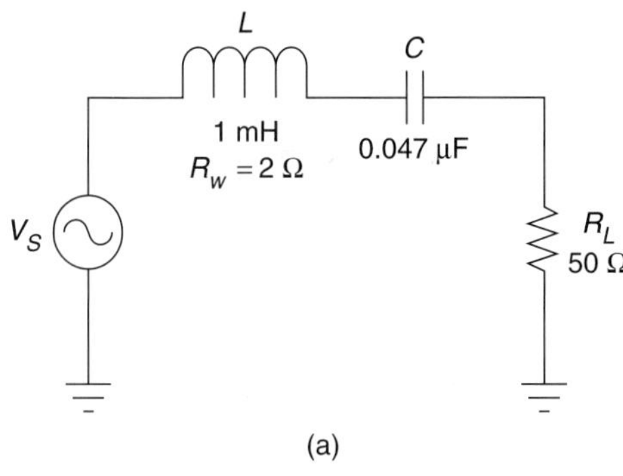


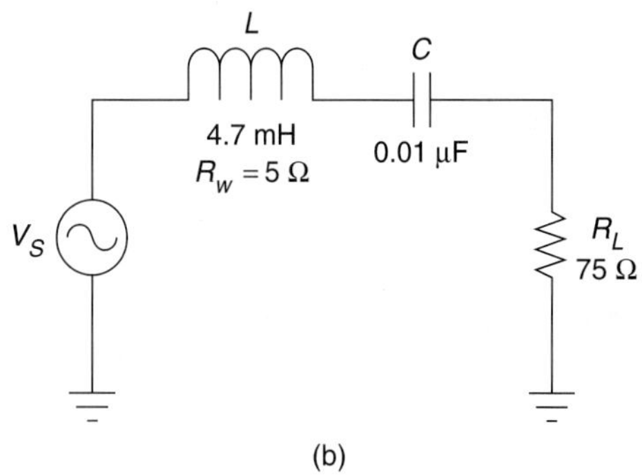


FIGURE 19.79

**57.** Calculate the cutoff frequency for the high-pass filter in Figure 19.78a.

**58.** Calculate the cutoff frequency for the high-pass filter in Figure 19.78b.

**59.** Determine the bandwidth of the bandpass filter in Figure 19.79a.

**60.** Determine the bandwidth of the bandpass filter in Figure 19.79b.

**61.** Determine the cutoff frequencies for the bandpass filter in Figure 19.79a.

**62.** Determine the cutoff frequencies for the bandpass filter in Figure 19.79b.

**63.** Determine the bandwidth and cutoff frequencies for the bandpass filter in Figure 19.80a.

**64.** Determine the bandwidth and cutoff frequencies for the bandpass filter in Figure 19.80b.

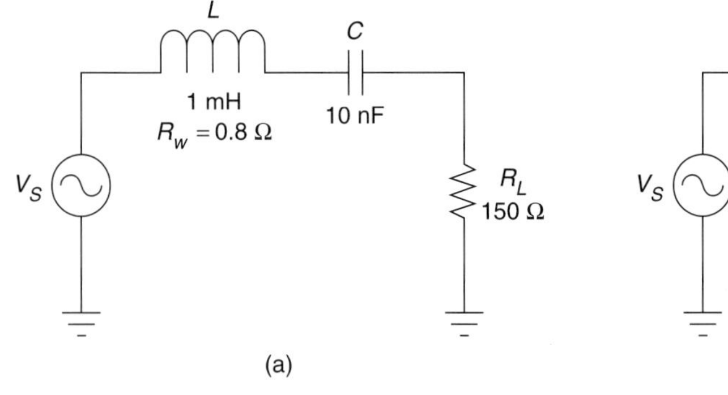


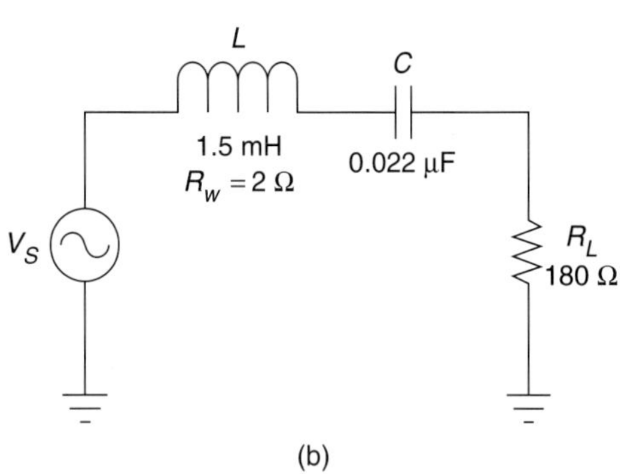


FIGURE 19.80

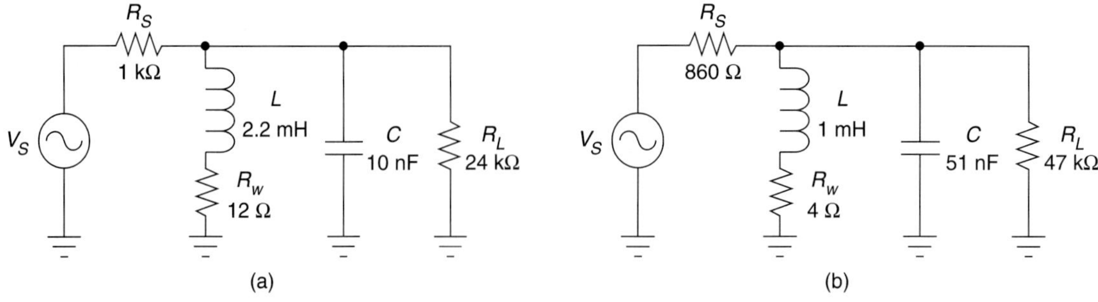


**FIGURE 19.81**

**65.** Determine the value of $Q_L$ for the circuit in Figure 19.81a.

**66.** Determine the value of $Q_L$ for the circuit in Figure 19.81b.

**67.** Calculate the bandwidth and cutoff frequencies for the bandpass filter in Figure 19.81a.

**68.** Calculate the bandwidth and cutoff frequencies for the bandpass filter in Figure 19.81b.

## Answers to the Example Practice Problems

**19.1.** 81.8 kHz
**19.2.** 6 kHz
**19.3.** 50 Hz
**19.4.** 160 kHz
**19.5.** 5.33
**19.6.** 50 kHz
**19.7.** $f_0 = 112$ kHz, $f_{ave} = 150$ kHz
**19.8.** 323 kHz
**19.9.** 203 kHz, 443 kHz
**19.10.** 300.7 kHz, 0.22%
**19.11.** 0.9 B
**19.12.** −1.25 dB
**19.13.** 20 dB
**19.14.** 0.7244
**19.15.** 578 mW
**19.16.** 353 mW
**19.17.** 40.8 dBm
**19.18.** 316 mW
**19.19.** −6 dB
**19.20.** 0.6683
**19.21.** 1.496
**19.22.** 0.9363
**19.24.** 355 Hz
**19.26.** $A_{v(max)} = 0.9724 = -0.24$ dB, $f_c = 3.57$ kHz
**19.27.** 0.9702
**19.28.** 220 Hz
**19.29.** See Figure 19.82.
**19.30.** 1.01 kHz
**19.31.** $A_{v(max)} = 1 = 0$ dB, $f_c = 569$ Hz
**19.32.** 28.7 kHz
**19.33.** 2.45 kHz
**19.34.** 14.4 kHz, 16.9 kHz
**19.35.** 4.35
**19.36.** 9.5 kHz, 12 kHz

**FIGURE 19.82**

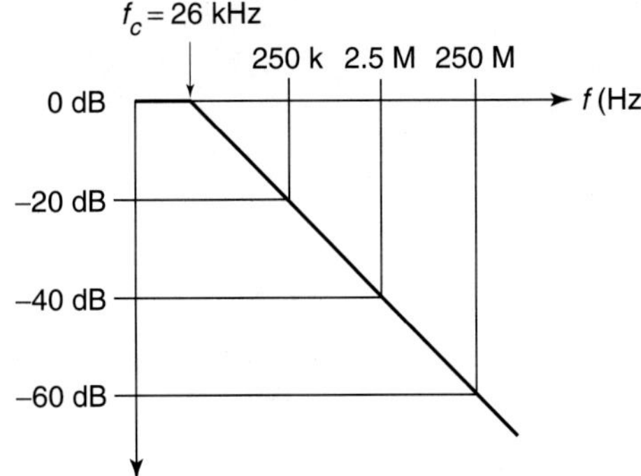


## EWB Applications Problems

| Figure | EWB File Reference |
|---|---|
| 19.25 | EWBA19_25.ewb |
| 19.36 | EWBA19_36.ewb |
| 19.43 | EWBA19_43.ewb |
| 19.47 | EWBA19_47.ewb |
| 19.60 | EWBA19_60.ewb |
| 19.62 | EWBA19_62.ewb |

# APPENDIX A

# Conversions and Units

## *Conversions**

| Multiply | By | To Get |
|---|---|---|
| ampere-turns | 1.257 | gilberts |
| ampere-turns per cm | 2.540 | ampere-turns per inch |
| ampere-turns per inch | 0.3937 | ampere-turns per centimeter |
| ampere-turns per inch | 0.4950 | gilberts per centimeter |
| British thermal units | 1054 | joules |
| British thermal units | $2.928 \times 10^{-4}$ | kilowatt-hours |
| centimeters | 0.3937 | inches |
| centimeters | 0.01 | meters |
| centimeters | 393.7 | mils |
| centimeters | 10 | millimeters |
| circular mils | $5.067 \times 10^{-6}$ | square centimeters |
| circular mils | $7.854 \times 10^{-7}$ | square inches |
| circular mils | 0.7854 | square mils |
| cubic centimeters | $3.531 \times 10^{-5}$ | cubic feet |
| cubic centimeters | $6.102 \times 10^{-2}$ | cubic inches |
| cubic centimeters | $10^{-6}$ | cubic meters |
| cubic centimeters | $2.642 \times 10^{-4}$ | gallons |
| cubic centimeters | $10^{-3}$ | liters |
| cubic feet | $2.832 \times 10^{4}$ | cubic centimeters |
| cubic feet | 1728 | cubic inches |
| cubic feet | 0.02832 | cubic meters |
| cubic feet | 7.481 | gallons |
| cubic feet | 28.32 | liters |
| cubic inches | 16.39 | cubic centimeters |
| cubic inches | $5.787 \times 10^{-4}$ | cubic feet |
| cubic inches | $1.639 \times 10^{-5}$ | cubic meters |
| cubic inches | $2.143 \times 10^{-5}$ | cubic yards |
| cubic inches | $4.329 \times 10^{-3}$ | gallons |
| cubic inches | $1.639 \times 10^{-2}$ | liters |
| cubic meters | $10^{6}$ | cubic centimeters |
| cubic meters | 35.31 | cubic feet |
| cubic meters | $6.1023 \times 10^{4}$ | cubic inches |
| cubic meters | 264.2 | gallons |
| cubic meters | $10^{9}$ | liters |
| degrees (∠) | 60 | minutes |

*Most of the values in this list were obtained from *Modern Electronics,* David Bruce, Appendix J, Reston Publishing Co.: 1984.

| Multiply | By | To Get |
|---|---|---|
| degrees (∠) | $1.745 \times 10^{-2}$ | radians |
| degrees (∠) | 3600 | seconds |
| degrees per second | $1.745 \times 10^{-2}$ | radians per second |
| degrees per second | 0.1667 | revolutions per minute |
| degrees per second | $2.778 \times 10^{-3}$ | revolutions per second |
| dynes | $1.02 \times 10^{-3}$ | grams |
| dynes | $2.248 \times 10^{-6}$ | pounds |
| feet | 30.48 | centimeters |
| feet | 12 | inches |
| feet | 0.3048 | meters |
| feet | ⅓ | yards |
| gauss | 6.452 | lines per square inch |
| gilberts | 0.7958 | ampere-turns |
| gilberts per centimeter | 2.021 | ampere-turns per inch |
| horsepower | 42.44 | BTU per minute |
| horsepower | $3.3 \times 10^{4}$ | foot-pounds per minute |
| horsepower | 550 | foot-pounds per second |
| horsepower | 1.014 | metric horsepower |
| horsepower | 10.7 | kilogram-calories per minute |
| horsepower | 0.7457 | kilowatts |
| horsepower | 745.7 | watts |
| inches | 2.54 | centimeters |
| inches | $10^{3}$ | mils |
| joules | $9.486 \times 10^{-4}$ | British thermal units |
| joules | $10^{7}$ | ergs |
| joules | 0.7376 | foot-pounds |
| joules | $2.39 \times 10^{-4}$ | kilogram-calories |
| joules | 0.102 | kilogram-meters |
| joules | $2.778 \times 10^{-4}$ | watt-hours |
| kilometers | $10^{5}$ | centimeters |
| kilometers | 3281 | feet |
| kilometers | $10^{3}$ | meters |
| kilometers | 0.6214 | miles |
| kilometers | 1093.6 | yards |
| kilowatts | 56.92 | BTU per minute |
| kilowatts | $4.425 \times 10^{4}$ | foot-pounds per minute |
| kilowatts | 737.6 | foot-pounds per second |
| kilowatts | 1.341 | horsepower |
| kilowatts | 14.34 | kilogram-calories per minute |
| kilowatts | $10^{3}$ | watts |
| kilowatt-hours | 3415 | British thermal units |
| kilowatt-hours | 1.341 | horsepower-hours |
| kilowatt-hours | $2.655 \times 10^{6}$ | joules |
| kilowatt-hours | 860.5 | kilogram-calories |
| kilowatt-hours | $3.671 \times 10^{5}$ | kilogram-meters |
| lines per square centimeter | 1 | gauss |
| lines per square inch | 0.155 | gauss |
| ln (N) | 0.4343 | $\log_{10}$ N |
| $\log_{10}$ N | 2.303 | ln (N) |
| lumens per square foot | 1 | foot-candles |
| Maxwells | $10^{-2}$ | micro-webers (μWb) |
| meters | 100 | centimeters |
| meters | 3.2808 | feet |
| meters | 39.37 | inches |

| Multiply | By | To Get |
|---|---|---|
| meters | $10^{-3}$ | kilometers |
| meters | $10^{3}$ | millimeters |
| meters | 1.0936 | yards |
| microfarads | $10^{-6}$ | farads |
| microns | $10^{-6}$ | meters |
| microwebers | 100 | Maxwells |
| miles | $1.609 \times 10^{5}$ | centimeters |
| miles | 5280 | feet |
| miles | 1.6093 | kilometers |
| miles | 1760 | yards |
| millihenries | $10^{-3}$ | henries |
| millimeters | 0.1 | centimeters |
| millimeters | 0.03937 | inches |
| millimeters | 39.37 | mils |
| mils | $2.54 \times 10^{-3}$ | centimeters |
| mils | $10^{-3}$ | inches |
| minutes (∠) | $2.909 \times 10^{-4}$ | radians |
| minutes (∠) | 60 | seconds (∠) |
| quadrants (∠) | 90 | degrees |
| quadrants (∠) | 5400 | minutes |
| quadrants (∠) | 1.571 | radians |
| radians | 57.3 | degrees |
| radians | 3438 | minutes |
| radians | 0.637 | quadrants |
| radians per second | 57.3 | degrees per second |
| radians per second | 0.1592 | revolutions per second |
| radians per second | 9.549 | revolutions per minute |
| revolutions | 360 | degrees |
| revolutions | 4 | quadrants |
| revolutions | 6.283 | radians |
| seconds (∠) | $4.848 \times 10^{-6}$ | radians |
| square centimeters | $1.973 \times 10^{5}$ | circular mils |
| square centimeters | $1.076 \times 10^{-3}$ | square feet |
| square centimeters | 0.155 | square inches |
| square centimeters | $10^{-6}$ | square meters |
| square centimeters | 100 | square millimeters |
| square feet | 929 | square centimeters |
| square feet | 144 | square inches |
| square feet | $9.29 \times 10^{-2}$ | square meters |
| square feet | $3.587 \times 10^{-3}$ | square miles |
| square feet | ⅑ | square yards |
| square inches | $1.273 \times 10^{6}$ | circular mils |
| square inches | 6.452 | square centimeters |
| square inches | $6.944 \times 10^{-3}$ | square feet |
| square inches | 645.2 | square millimeters |
| square kilometers | $10.76 \times 10^{6}$ | square feet |
| square kilometers | $10^{6}$ | square meters |
| square kilometers | 0.3861 | square miles |
| square meters | 10.764 | square feet |
| square meters | $3.861 \times 10^{-7}$ | square miles |
| square miles | $2.788 \times 10^{7}$ | square feet |
| square miles | 2.59 | square kilometers |
| square millimeters | $1.973 \times 10^{3}$ | circular mils |
| square millimeters | 0.01 | square centimeters |

| Multiply | By | To Get |
| --- | --- | --- |
| square millimeters | $1.55 \times 10^{-3}$ | square inches |
| watts | $5.692 \times 10^{-2}$ | BTU per minute |
| watts | $10^{7}$ | ergs per second |
| watts | 44.62 | foot-pounds per minute |
| watts | 0.7376 | foot-pounds per second |
| watts | $1.341 \times 10^{-3}$ | horsepower |
| watts | $1.434 \times 10^{-2}$ | kilogram-calories per minute |
| watt-hours | 3.415 | British thermal units |
| watt-hours | 2655 | foot-pounds |
| watt-hours | 0.8605 | kilogram-calories |
| watt-hours | 367.1 | kilogram-meters |
| watt-hours | $10^{-3}$ | kilowatt-hours |

## Temperature Scales and Conversions

### Scales

| *Scale* | *Unit of Measure* |
| --- | --- |
| Fahrenheit (F) | °F |
| Rankine (R)[a] | °F |
| Celsius (C) | °C |
| Kelvin (K)[b] | K |

[a]Sometimes referred to as *absolute Fahrenheit.*

[b]Sometimes referred to as *absolute temperature.*

### Conversion Equations

$$T_C = \frac{5}{9}(T_F - 32) \qquad T_F = 1.8T_C + 32$$

$$T_C = T_K - 273.15 \qquad T_K = T_C + 273.15$$

$$T_F = T_R - 480 \qquad T_R = T_F + 480$$

## International System of Units (SI)*

| Quantity | Unit of Measure | Symbol for Unit |
| --- | --- | --- |
| acceleration | meters per second squared | $m/s^2$ |
| angular acceleration | radians per second squared | $rad/s^2$ |
| angular velocity | radians per second | rad/s |
| area | square meters | $m^2$ |
| density | kilograms per cubic meter | $kg/m^3$ |
| electric capacitance | farad | F |
| electric charge | coulomb | C |
| electric field strength | volts per meter | V/m |
| electric resistance | ohms | Ω |
| flux of light | lumens | lm |
| force | newtons | N |

*The material in this list was obtained from *Modern Electronics,* David Bruce, Appendix I, Reston Publishing Co.: 1984.

| Quantity | Unit of Measure | Symbol for Unit |
|---|---|---|
| frequency | hertz | Hz |
| illumination | lux | lx |
| inductance | henry | H |
| luminance | candela per square meter | $cd/m^2$ |
| magnetic field strength | ampere per meter | A/m |
| magnetic flux | weber | Wb |
| magnetic flux density | tesla | T |
| magnetomotive force | ampere | A |
| power | watt | W |
| pressure | newtons per meter squared | $N/m^2$ |
| velocity | meters per second | m/s |
| voltage, potential difference, electromotive force | volt | V |
| volume | cubic meters | $m^3$ |
| work, energy, quantity of heat | joule | J |

# APPENDIX B

# Component Standard Values and Color Codes

## *Precision Resistors*

Precision resistors are commonly available in 0.1%, 0.25%, 0.5%, and 1% tolerances. The values listed below are the standard digit combinations for the resistors in this tolerance range. Resistors with 1% tolerance are available only in the magnitudes shown in **bold.**

| | | | | | | | | | | | |
|---|---|---|---|---|---|---|---|---|---|---|---|
| **100** | 101 | **102** | 104 | **105** | 106 | **107** | 109 | **110** | 111 | **113** | 114 |
| **115** | 117 | **118** | 120 | **121** | 126 | **127** | 129 | **130** | 132 | **133** | 135 |
| **137** | 138 | **140** | 142 | **143** | 145 | **147** | 149 | **150** | 152 | **154** | 156 |
| **158** | 160 | **162** | 164 | **165** | 167 | **169** | 172 | **174** | 176 | **178** | 180 |
| **182** | 184 | **187** | 189 | **191** | 193 | **196** | 198 | | | | |
| **200** | 203 | **205** | 208 | **210** | 213 | **215** | 218 | **221** | 223 | **226** | 229 |
| **232** | 234 | **237** | 240 | **243** | 246 | **249** | 252 | **255** | 258 | **261** | 264 |
| **267** | 271 | **274** | 277 | **280** | 284 | **287** | 291 | **294** | 298 | | |
| **301** | 305 | **309** | 312 | **316** | 320 | **324** | 328 | **332** | 336 | **340** | 344 |
| **348** | 352 | **357** | 361 | **365** | 370 | **374** | 379 | **383** | 388 | **392** | 397 |
| **402** | 407 | **412** | 417 | **422** | 427 | **432** | 437 | **442** | 448 | **453** | 459 |
| **464** | 470 | **475** | 481 | **487** | 493 | **499** | | | | | |
| 505 | **511** | 517 | **523** | 530 | **536** | 542 | **549** | 562 | **566** | 569 | **576** |
| 583 | **590** | 599 | | | | | | | | | |
| **604** | 612 | **619** | 626 | **634** | 642 | **649** | 657 | **665** | 673 | **681** | 690 |
| **698** | | | | | | | | | | | |
| 706 | **715** | 723 | **732** | 741 | **750** | 759 | **768** | 777 | **787** | 796 | |
| **806** | 816 | **825** | 835 | **845** | 856 | **866** | 876 | **887** | 898 | | |
| **909** | 920 | **931** | 942 | **953** | 965 | **976** | 988 | | | | |

## *Precision Resistor Color Code*

Precision resistors have *three* significant digits (instead of two), so they require a modified color code. This modified color code is represented in Figure B.1. In many cases, alphanumeric codes are used to indicate the values of lower tolerance resistors. However, you will see resistors that use the color code shown in the figure. Here are some example interpretations of the 5-band code:

Red, Black, Gray, Brown, Blue $= 208 \times 10^1 = 2080\ \Omega = 2.08\ \text{k}\Omega$ (0.25%)

Violet, Brown, Green, Red, Brown $= 715 \times 10^2 = 71{,}500 = 71.5\ \text{k}\Omega$ (2%)

Green, Orange, Blue, Black, Green $= 536 \times 10^0 = 536\ \Omega$ (0.5%)

**FIGURE B.1 The precision resistor color code.**

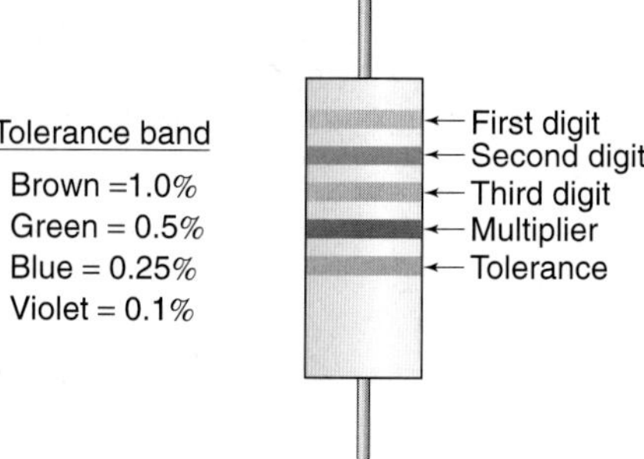


## Resistor Alphanumeric Codes

An *alphanumeric code* is one that uses both letters and numbers to express the value of a component. One of the most common alphanumeric resistor codes works as follows:

**1.** When $R \geq 100\ \Omega$, four digits are used to express the component's value. The first three digits in the code are the digits in the value of the resistor. The fourth digit in the code represents the power of ten multiplier.

*Examples:* $5101 = 510 \times 10^1 = 5100\ \Omega = 5.1\ \text{k}\Omega$

$2703 = 270 \times 10^3 = 270{,}000\ \Omega = 270\ \text{k}\Omega$

The tolerances are indicated by letters following the component value. These letters are identified and coded as follows:

$F = \pm 1\%$ $G = \pm 2\%$ $J = \pm 5\%$ $K = \pm 10\%$ $M = \pm 20\%$

*Examples:* $3302G = 330 \times 10^2 = 33{,}000\ \Omega = 33\ \text{k}\Omega\ (\pm 2\%)$

$1470F = 147 \times 10^0 = 147\ \Omega\ (\pm 1\%)$

**2.** When $R < 100\ \Omega$, two (or three) digits are used, along with the letter "R". The R is used to designate the position of a decimal point within the component value.

*Examples:* $7R32F = 7.32\ \Omega\ (\pm 1\%)$

$R36G = 0.36\ \Omega\ (\pm 2\%)$

$51RJ = 51\ \Omega\ (\pm 5\%)$

# *Transformer Color Coding*

Standard wire colors are commonly used to identify the input and output connections on a given transformer. The colors used and their applications are identified in Figure B.2.

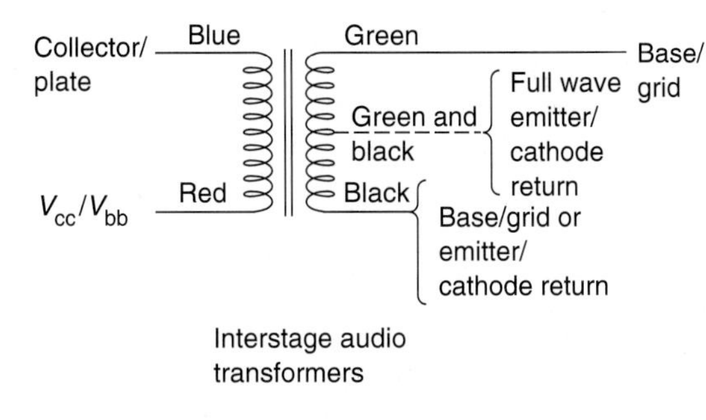


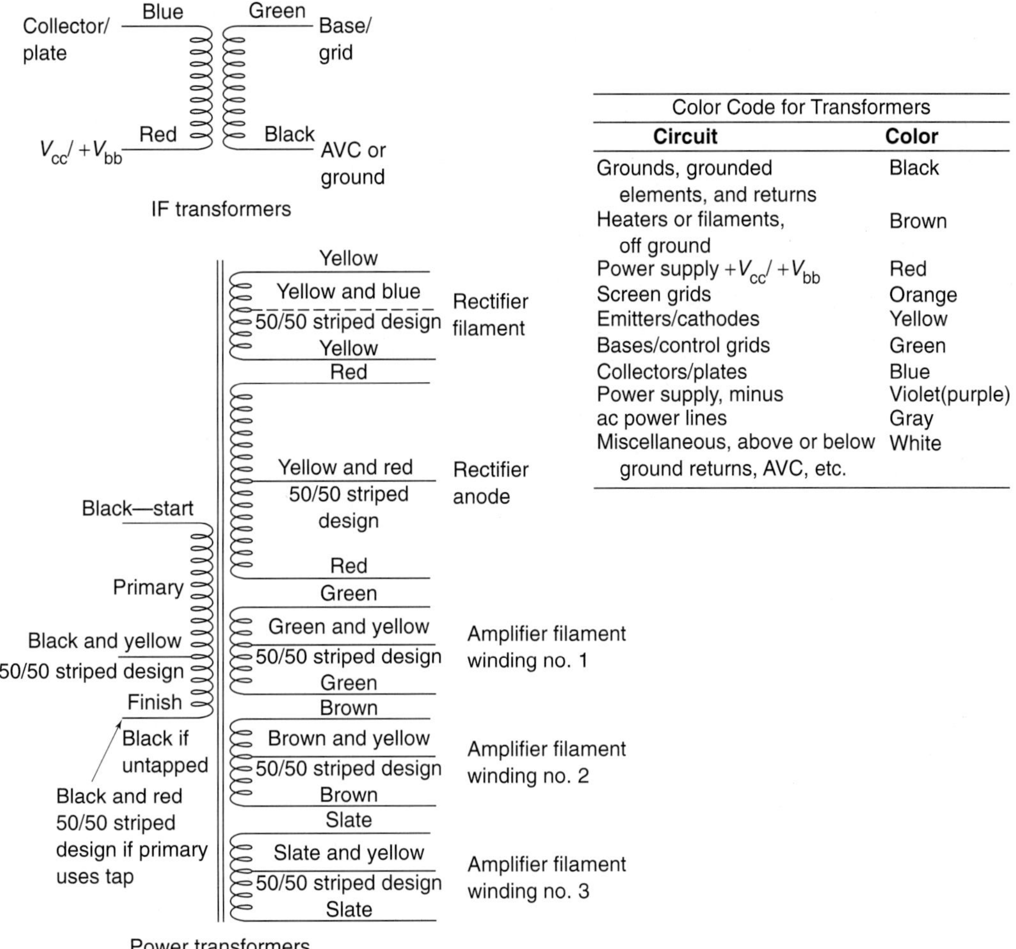


Color Code for Transformers

| Circuit | Color |
|---|---|
| Grounds, grounded elements, and returns | Black |
| Heaters or filaments, off ground | Brown |
| Power supply $+V_{cc}/+V_{bb}$ | Red |
| Screen grids | Orange |
| Emitters/cathodes | Yellow |
| Bases/control grids | Green |
| Collectors/plates | Blue |
| Power supply, minus | Violet(purple) |
| ac power lines | Gray |
| Miscellaneous, above or below ground returns, AVC, etc. | White |

**FIGURE B.2 Transformer color coding.**

# APPENDIX C

# Common Schematic Symbols

Literally hundreds of schematic symbols are used in various applications. Some of the most common schematic symbols are provided (as a reference) in this appendix.

## General Symbols

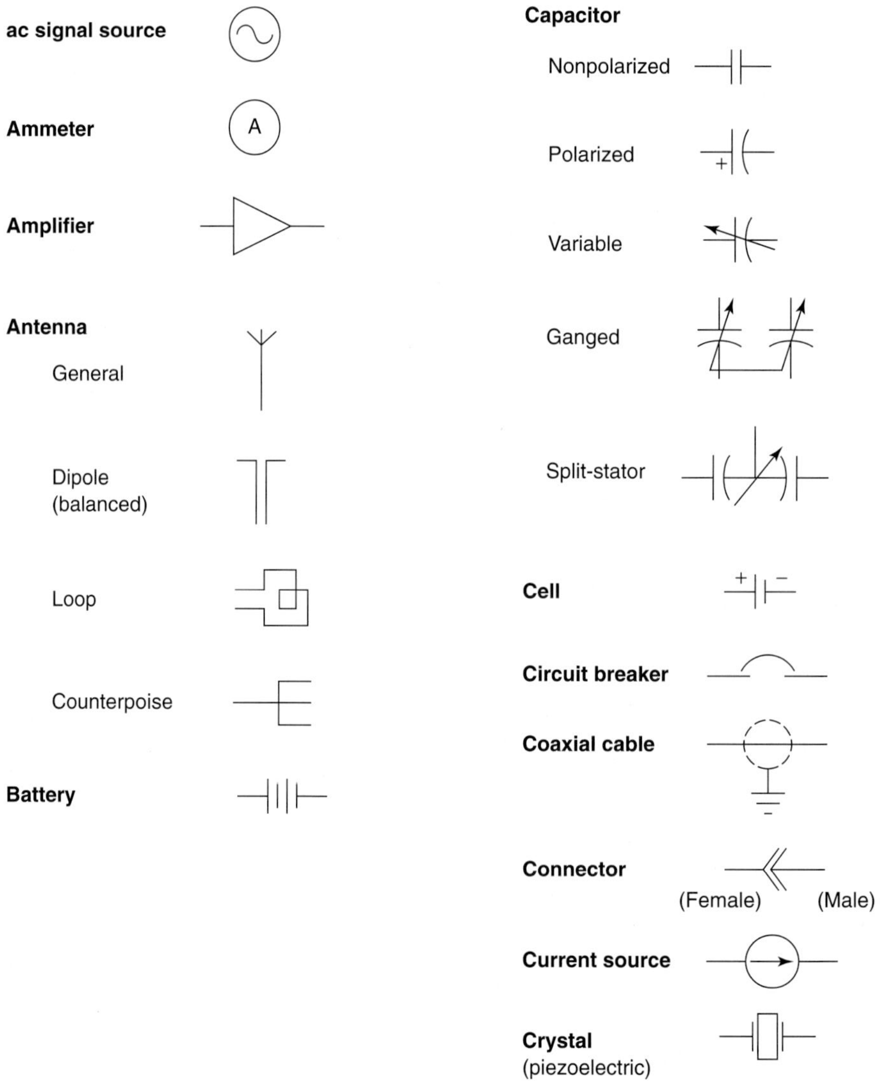

**ac signal source**

**Ammeter**

**Amplifier**

**Antenna**

General

Dipole (balanced)

Loop

Counterpoise

**Battery**

**Capacitor**

Nonpolarized

Polarized

Variable

Ganged

Split-stator

**Cell**

**Circuit breaker**

**Coaxial cable**

**Connector**

(Female) (Male)

**Current source**

**Crystal** (piezoelectric)

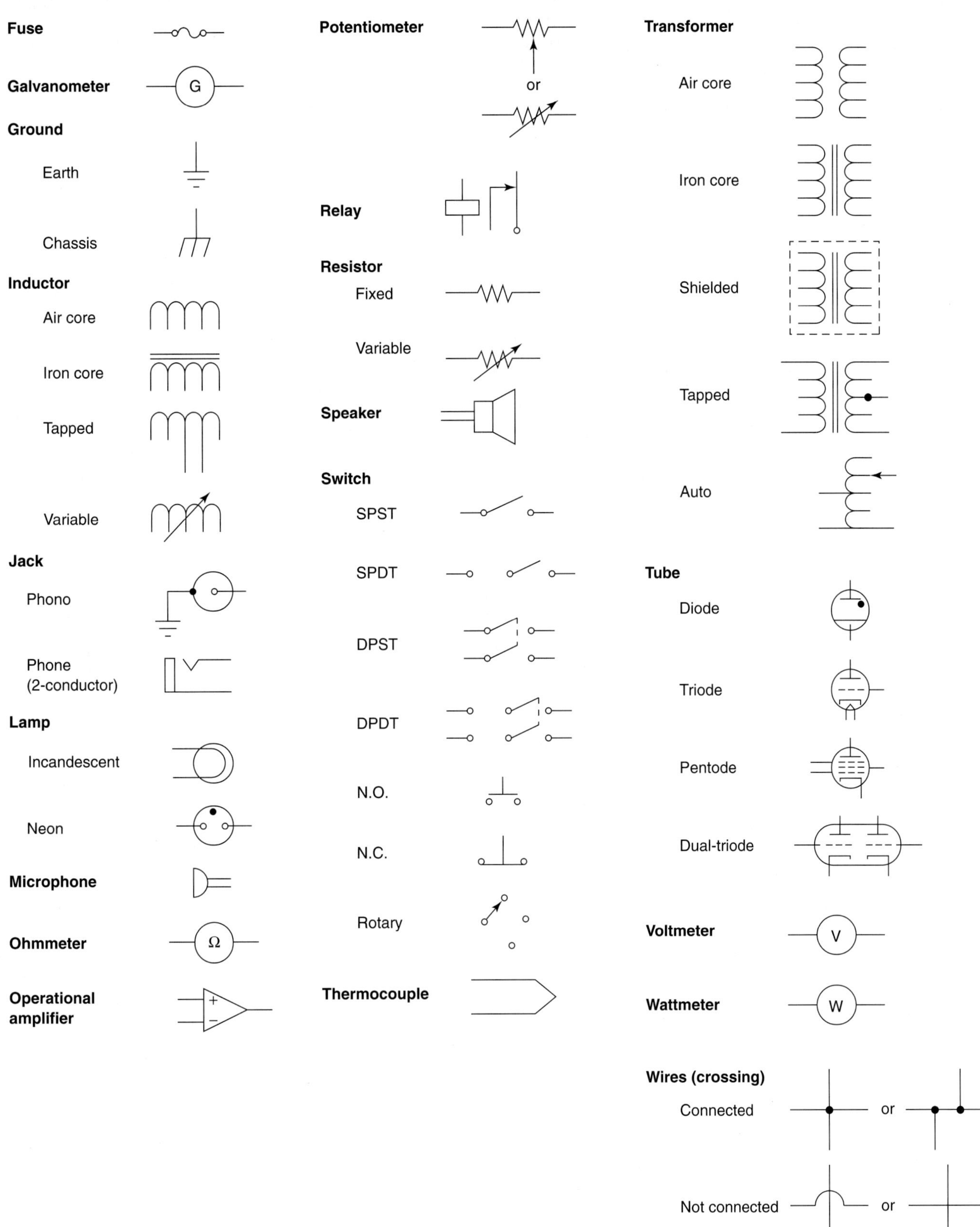
Fuse
Galvanometer
G
Ground
Earth
Chassis
Inductor
Air core
Iron core
Tapped
Variable
Jack
Phono
Phone
(2-conductor)
Lamp
Incandescent
Neon
Microphone
Ohmmeter
Ω
Operational
amplifier
+
−
Potentiometer
or
Relay
Resistor
Fixed
Variable
Speaker
Switch
SPST
SPDT
DPST
DPDT
N.O.
N.C.
Rotary
Thermocouple
Transformer
Air core
Iron core
Shielded
Tapped
Auto
Tube
Diode
Triode
Pentode
Dual-triode
Voltmeter
V
Wattmeter
W
Wires (crossing)
Connected
or
Not connected
or

## Semiconductor Devices

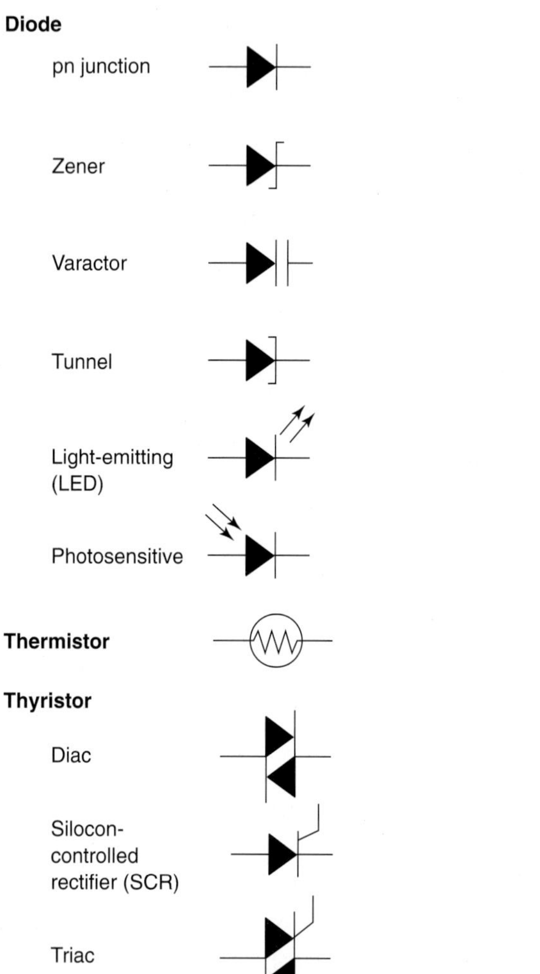


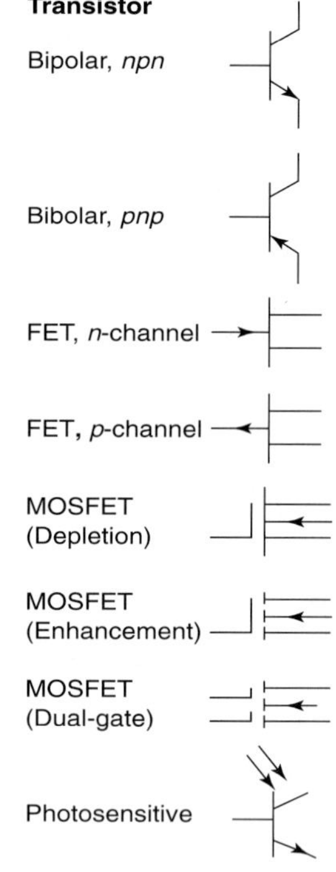


## Logic Symbols

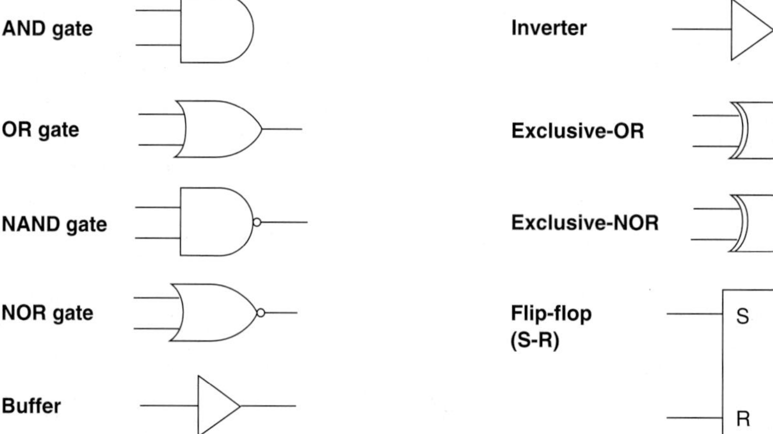

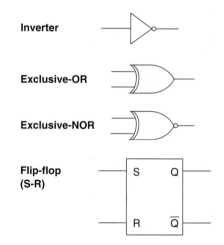

# APPENDIX D

# Glossary

**Active components.** Components with characteristics that are controlled (in part) by an external dc power source.

**Admittance (*Y*).** The reciprocal of impedance, measured in Siemens.

**Air gap.** The area of free space (if any) between two magnetic poles.

**Alter-and-add method.** A method of solving two simultaneous equations in which the equations are added to eliminate one of the variables. This normally involves altering one (or both) of the equations.

**Alternating current (ac).** A term used to describe current that is bidirectional; i.e., the flow of charge changes direction periodically.

**Alternation.** A term used to describe either the positive or negative transition of a waveform; also referred to as a half-cycle.

**American wire gauge (AWG).** A system that uses numbers to identify standard wire sizes (cross-sectional areas).

**Ampacity.** From *amp*ere cap*acity,* the rated limit on current for a wire gauge.

**Ampere (A).** The basic unit of current. A rate of charge flow equal to 1 coulomb per second.

**Ampere-hours (Ah).** A capacity rating for a battery, equal to the product of current (in amperes) and time (in hours). This rating, which is a constant for a battery, provides a means of determining how long a battery will last at a given load demand.

**Ampere-turns (At).** The product of coil current and the number of turns in a coil. Usually represented as $NI$ in equations ($N$ = number of turns, $I$ = coil current).

**Amplifier.** A circuit that can be used to increase the power level of a sine wave (or some other waveform).

**Angular velocity (ω).** The speed of rotation for the rotor in a sine wave generator. The rate at which the phase of a sine wave changes and, therefore, the rate at which its instantaneous values change, equal to $2\pi f$.

**Apparent power ($P_{APP}$).** The product of alternating voltage and current in a reactive circuit, measured in volt-amperes (VA). The term is used because only a portion of the value obtained is actually dissipated by the resistive components.

**Atom.** The smallest particle of matter that retains the physical characteristics of an element. In its simplest model, the atom is made up of protons and neutrons (that form the nucleus) and orbiting electrons.

**Atomic number.** The number of protons in the nucleus of an atom. Elements are commonly identified by their atomic number.

**Attenuation.** The signal loss caused by the frequency response of a component or circuit.

**Atto- (a).** The engineering notation prefix with a value of $10^{-18}$.

**Autotransformer.** A special-case transformer made up of a single coil with three terminal connections (in essence, a tapped inductor).

**Average ac power.** The average power generated over the course of each complete cycle of an ac waveform. On a graph of power, the value that falls halfway between 0 W and peak power. Often referred to as *equivalent dc power.*

**Average breakdown voltage.** The voltage that will force an insulator to conduct; rated in kV/cm.

**Average frequency ($f_{ave}$).** The frequency that lies halfway between the cutoff frequencies of a bandpass or notch filter.

**Balanced.** A term used to describe the bridge operating state where the galvanometer current is 0 A. A bridge is balanced when the voltages on the two sides of the meter are equal.

**Bandpass filter.** A filter designed to pass all frequencies between two cutoff frequencies.

**Band-stop filter.** A filter designed to reject all frequencies between two cutoff frequencies. Also referred to as a *notch filter*.

**Bandwidth (BW).** The frequency spread between the cutoff frequencies of a component or circuit, equal to the difference between those frequencies.

**Battery.** A component that produces a difference of potential (emf). Batteries convert chemical, thermal, or nuclear energy into electrical energy.

**Bel (B).** A ratio expressed as a common (base 10) logarithm. In terms of power gain, the ratio of output power to input power, expressed as the common log of the ratio.

**Binomial.** An expression written as the sum of (or difference between) two terms.

**Biomedical systems.** Systems designed for use in diagnosing, monitoring, and treating medical problems.

**Bode plot.** A normalized graph that represents frequency response as a change in gain versus operating frequency.

**Branch.** A term used to identify each current path in a parallel circuit.

**Branch current.** A term used to describe the current generated through each branch of a parallel circuit, or each loop in a series-parallel circuit.

**Branch current analysis.** A method of analyzing multisource circuits by redefining the circuit voltages as products of resistor values and branch currents.

**Breadboard.** A physical base used for the construction of experimental circuits.

**Breakdown voltage rating.** A capacitor rating that indicates the value of capacitor voltage that may cause the dielectric to break down and conduct.

**Buffer.** An impedance matching circuit.

**Capacitance ($C$).** The ability of a component to store energy in the form of a charge, measured in farads (F). The characteristic of a component that opposes any change in voltage.

**Capacitive reactance ($X_C$).** The opposition provided by a capacitor to any sinusoidal alternating current.

**Capacitor.** A component designed to provide a specific measure of capacitance.

**Capacity.** A measure of how long a battery will last at a given output current, measured in ampere-hours (Ah). The amount of charge that a capacitor can store, measured in coulombs per volt, or farads (F).

**Carbon-composition resistor.** A component that uses carbon to provide a desired value of resistance.

**Cascade.** A group of series-connected filters (or other circuits). Each circuit in a cascade is referred to as a stage.

**Cathode-ray tube (CRT).** A vacuum tube used as the video display for many televisions, video games, etc.

**Cell.** A single unit designed to produce electrical energy through thermal (chemical) or optical (light) means.

**Center frequency ($f_0$).** The frequency that equals the geometric average of the cutoff frequencies of a bandpass or notch filter.

**Center-tapped transformer.** A transformer that has a lead connected to the center of the secondary winding.

**Charge ($Q$).** A force that causes two particles to be attracted to, or repelled from, each other.

**Choke.** A low-resistance inductor designed to provide a high reactance within a designated frequency range.

**Circuit.** A group of components that performs a specific function.

**Circular-mil (CM).** A unit of area found by squaring the diameter (in mils) of a conductor.

**Coefficient of coupling ($k$).** A measure of the degree of coupling that takes place between two (or more) coils.

**Coercive force.** The magnetomotive force required to return the value of flux density within a material to zero.

**Combination circuit.** Another name for a series-parallel circuit.

**Common terminal.** The terminal of a dc power supply that is common to two (or more) voltage sources and thus is used as the reference point.

**Communications systems.** Systems designed to transmit and/or receive information.

**Complex number.** Any number written as the sum of two vectors.

**Computer systems.** Systems designed to store and process information.

**Conductance ($G$).** A measure of the ability of a circuit to conduct (pass current), measured in siemens.

**Conjugate.** For any binomial ($a + b$), the value found by changing the sign between the variables from + to − (or vice versa).

**Coordinate notation.** A method of expressing the value of a point in terms of its $x$-axis and $y$-axis values (or coordinates).

**Copper loss.** Power dissipated by any current-carrying conductor, generally applied to the winding resistance of a coil.

**Coulomb (C).** The basic unit of charge, equal to the total charge of approximately $6.25 \times 10^{18}$ electrons.

**Coulomb's law.** A law stating that the magnitude of force between two charges is directly proportional to the product of the charges and inversely proportional to the square of the distance between them.

**Counter emf.** A term used to describe the voltage induced across an inductor by its own changing magnetic field. (The term serves as a reminder that the induced voltage opposes the change in current that generated the changing magnetic field at the start.)

**Coupling.** When energy is transferred between two or more components (or circuits).

**Current ($I$).** The directed flow of charge through a circuit, measured in amperes.

**Current gain ($A_i$).** The ratio of circuit output current to input current.

**Current source.** A source designed to provide an output current that remains relatively constant over a wide range of load impedance values.

**Cutoff frequency ($f_c$).** Any frequency at which the power ratio of a circuit drops to 50% of its maximum value.

**Cycle.** The complete transition through one positive alternation and one negative alternation of a waveform.

**Cycle time.** The time required to complete one cycle of an ac signal. Also referred to as *period.*

**dc offset.** A term used to describe a dc value (other than zero) that acts as the reference for a sine wave (or some other waveform).

**dc power supply.** A circuit that converts ac line voltage to one or more dc operating voltages for an electronic system. A piece of equipment with dc outputs that can be adjusted to any value within its design limits.

**Decade.** A frequency multiplier equal to ten.

**Decade box.** A device that contains a series of potentiometers, each with different resistance ratings. Adjusting the individual pots allows the user to set the overall (total) resistance to specific values over a wide range.

**Decade scale.** A scale where each increment has ten times the value of the previous increment.

**Decibel (dB).** A logarithmic representation of a number. In terms of power, a ratio of output power to input power, expressed as ten times the common logarithm of the ratio.

**Dielectric.** The insulating layer between the plates of a capacitor.

**Dielectric absorption.** The tendency of a dielectric to absorb charge.

**Digital multimeter (DMM).** A meter that measures voltage, current, and resistance. Many DMMs are capable of making additional measurements, depending on the model.

**Direct current (dc).** A term used to describe current that is unidirectional; i.e., charge flows in one direction only.

**Domain theory.** One possible explanation for the source of magnetism. This theory states that atoms with like magnetic fields join together to form magnetic "domains," each acting like a magnet.

**Dual in-line package (DIP) switches.** SPST switches that are grouped in a single case.

**Dual-trace oscilloscope.** An oscilloscope with two independent traces that can be used to compare two waveforms. Each trace has its own set of vertical controls.

**Duty cycle.** The ratio of pulse width to cycle time, usually given as a percentage.

**Dynamic values.** Values that change as a direct result of normal circuit operation. Dynamic values are measured under specified conditions.

**Eddy current.** A circular flow of charge that is generated in the core of a transformer by a changing magnetic field. (The term *eddy* is normally used to describe anything [air, water, charge, etc.] that flows in a circular motion.)

**Effective value.** Another name for an rms value. It is derived from the fact that an rms value provides the same heating effect as its dc equivalent.

**Efficiency ($\eta$).** The ratio of output power to input power, typically given as a percentage.

**Electrodes.** The terminals of a battery.

**Electrolyte.** The chemical in a cell that interacts with the terminals, producing a difference of potential between them. A chemical used to produce the relatively high capacitance of an electrolytic capacitor.

**Electromagnetic wave.** A waveform that consists of perpendicular electric and magnetic fields.

**Electromotive force (*emf*).** A force that moves electrons. Also called *voltage* and *difference of potential.*

**Electron.** In the simplest model of the atom, the particle orbiting the nucleus that is the source of negative charge.

**Electronics.** The field dealing with the design, manufacturing, installation, maintenance, and repair of electronic equipment.

**Element.** A substance that cannot be broken down into a combination of simpler substances.

**Engineering notation.** A form of scientific notation where standard prefixes are used to designate specific ranges of values.

**Exa- (E).** The engineering notation prefix with a value of $10^{18}$.

**Exponential curve.** A curve whose rate of change is a function of a variable exponent.

**Farad (F).** The unit of measure of capacity. One farad is a capacity of 1 coulomb per volt.

**Fempto- (f).** The engineering notation prefix with a value of $10^{-15}$.

**Filter.** A circuit designed to pass a specific range of frequencies while rejecting (blocking) others.

**Flux density.** The amount of flux per unit area; an indicator of magnetic field strength.

**Frequency ($f$).** A measure of the rate at which the cycles of a waveform repeat themselves, measured in cycles per second. *See* Hertz.

**Frequency counter.** A piece of test equipment that measures the frequency of a waveform.

**Frequency response.** Any changes that occur in a circuit as a result of a change in operating frequency.

**Full-cycle average.** The average of all the instantaneous values of a waveform voltage (or current) throughout one complete cycle. Sometimes referred to as *full-wave average.*

**Full load.** A load that draws maximum current from the source. (A full load occurs when load impedance is at its minimum value.)

**Full-wave average.** Another name for the full-cycle average of a waveform.

**Full-wave rectifier.** A circuit that converts negative alternations to positive alternations (or vice versa) to provide a single-polarity output.

**Fundamental frequency.** The lowest frequency in a harmonic series.

**Fuse.** A device designed to open automatically if its current exceeds a specified value.

**Galvanometer.** A current meter that indicates both the magnitude and direction of a low-value current.

**Giga- (G).** The engineering notation prefix with a value of $10^9$.

**Ground-loop path.** The conducting path between the ground connections throughout a circuit.

**Half-cycle.** A term used to describe the negative and positive alternations of a waveform.

**Half-cycle average.** The average of all the instantaneous values of a waveform voltage (or current) throughout one alternation of a cycle. Sometimes referred to as *half-wave average.*

**Half-wave average.** Another name for the half-cycle average of a waveform.

**Half-wave rectifiers.** A circuit that eliminates the negative (or positive) alternations of a waveform to provide a single-polarity output.

**Harmonic.** A whole-number multiple of a frequency.

**Harmonic series.** A group of harmonic frequencies.

**Harmonic synthesis.** Generating a square wave by adding a fundamental frequency with an infinite number of its odd-order harmonics.

**Henry (H).** The unit of measure of inductance. One henry is the amount of inductance that produces a 1 V difference of potential when current changes at a rate of 1 A/s.

**Hertz (Hz).** The unit of measure for frequency, equal to one cycle per second.

**High-pass filter.** A filter designed to pass all frequencies above its cutoff frequency.

**Hysteresis.** The time lag between the removal of a magnetizing force and the drop in flux density to zero.

**Hysteresis curve.** A curve representing flux density ($B$) as a function of the field intensity (H) of a magnetizing force.

**Hysteresis loss.** A loss of energy that occurs whenever a magnetic field of one polarity must overcome any residual magnetism from the previous polarity.

**Ideal voltage source.** A voltage source that maintains a constant output voltage regardless of the impedance of its load(s).

**Imaginary number.** A term used to describe any number containing the $j$ operator. The name is based on the relationship $j = \sqrt{-1}$.

**Impedance (Z).** The total opposition to current in an ac circuit, consisting of resistance and/or reactance. The geometric sum of resistance and reactance, expressed in ohms.

**Impedance mismatch.** A term used to describe a situation where source impedance does not match load impedance.

**Inductance ($L$).** (1) The ability of a component with a changing current to induce a voltage across itself or a nearby circuit by generating a changing magnetic field, measured in henries (H). (2) The ability of a component to store energy in an electromagnetic field. (3) The property of a component that opposes any change in current.

**Induction.** The process of producing an artificial magnet.

**Inductive reactance ($X_L$).** The opposition that an inductor presents to a changing current, measured in ohms.

**Inductor.** A component designed to provide a specific amount of inductance. A component that opposes a change in current.

**Industrial systems.** Systems designed for use in manufacturing environments.

**Instantaneous value.** The magnitude of a voltage or current at a specified point in time.

**Integrated circuit.** Semiconductor that contains groups of components or circuits housed in a single package.

**Integrated resistor.** Microminiature resistors made using semiconductor materials other than carbon. Because they are so small, many of them can be housed in a single casing.

**Isolation transformer.** A transformer with an equal number of primary and secondary windings; one designed for equal primary and secondary voltages.

***j*-operator.** A mathematical operation resulting in a vector rotation of 90°.

**Kilo- (k).** The engineering notation prefix with a value of $10^3$.

**Kilowatt-hour (kWh).** A practical unit for measuring energy, equal to the amount of energy used by a 1000 W (1 kW) device that is run for one hour.

**Kirchhoff's current law.** A law stating that the algebraic sum of the currents entering and leaving a point must equal zero.

**Kirchhoff's voltage law.** A law stating that the algebraic sum of the voltages around a closed loop must equal zero.

***LCR* bridge.** A piece of test equipment used to measure inductance ($L$), capacitance ($C$), and resistance ($R$).

**Leakage current.** A low-value current that may be generated through a capacitor, indicating that the resistance of the dielectric is not infinite.

**Left-hand rule.** A memory aid that helps you remember the relationship between the direction of electron flow and the direction of the resulting magnetic field.

**Lenz's law.** A law stating that an induced voltage always opposes its source (in keeping with the law of conservation of energy).

**Linear.** A term used to describe any value that changes at a constant rate.

**Load.** The part of a circuit that absorbs (uses) power.

**Load analysis.** A method of predicting the effect that a change in load has on the output from a circuit.

**Loaded *Q*.** The quality ($Q$) of a filter when a load is connected.

**Loaded voltage divider.** A voltage divider designed specifically to provide one (or more) specific load voltages.

**Logarithmic scale.** A scale with frequency intervals that increase from each increment to the next at a geometric rate.

**Low-pass filter.** A filter designed to pass all frequencies below its cutoff frequency.

**Magnetic field.** The area of space surrounding a magnet that contains magnetic flux.

**Magnetic flux ($\phi$).** A term used to describe the total lines of force produced by a magnet.

**Magnetic force.** The force that a magnet exerts on the objects around it.

**Magnetic induction.** Using an external magnetic force to align the magnetic domains within a material, thereby magnetizing the material.

**Magnetic shielding.** Insulating an instrument (or material) from magnetic flux by diverting the lines of force around the instrument (or material).

**Magnetomotive force (mmf).** The force that produces magnetic flux. The magnetic equivalent of electromotive force.

**Magnitude.** Another word for quantity or value.

**Matter.** Anything that has weight and occupies space.

**Maximum power transfer theorem.** A theorem stating that maximum power transfer from a voltage source to a variable load occurs when the load and source resistances are equal.

**Mean.** Another word for average.

**Mega- (M).** The engineering notation prefix with a value of $10^6$.

**Mesh.** A term used to describe a multisource circuit that contains more than one loop.

**Mesh current analysis.** A form of branch current analysis where all loop currents are assumed to be in the same direction, regardless of voltage source polarities.

**Metal-film resistor.** A precision (exact value) resistor with a low temperature coefficient.

**Micro- ($\mu$).** The engineering notation prefix with a value of $10^{-6}$.

**Micro-weber ($\mu$Wb).** A practical unit of measure of magnetic flux, equal to 100 maxwells (lines of force).

**Mil.** A unit of length equal to one-thousandth of an inch.

**Milli- (m).** The engineering notation prefix with a value of $10^{-3}$.

**Millman's theorem.** A theorem stating that any number of voltage sources connected in parallel can be represented as a single voltage source in series with a single resistance.

**Momentary switch.** A switch that makes or breaks the connection between its terminals only while the contact button is pressed.

**Multisource circuit.** A circuit with more than one voltage and/or current sources.

**Mutual inductance ($L_M$).** (1) The process where the magnetic field generated by one inductor induces a voltage across another inductor in close proximity. (2) The inductance generated between two or more inductors in close proximity.

**Nano- (n).** The engineering notation prefix with a value of $10^{-9}$.

**Negative temperature coefficient.** An indicator of the effect of temperature on resistance for a material. A negative temperature coefficient indicates that resistance varies inversely with temperature.

**Neutron.** In the simplest model of the atom, the particle in the nucleus that provides mass but no charge.

**Node.** A word used to describe any point that connects two or more current paths.

**Node voltage analysis.** A circuit analysis technique where the voltage at a node is defined in terms of the other voltages in the circuit.

**No-load output voltage ($V_{NL}$).** The output from a voltage source when its load terminals are open. The maximum possible output from a voltage source.

**Nominal value.** The rated value of a component (or other circuit value).

***n*-order harmonics.** The frequencies (other than the fundamental) in a harmonic series.

**Normally closed (NC) switch.** A pushbutton switch that normally makes a connection between its terminals. Activating the switch breaks the connection.

**Normally open (NO) switch.** A pushbutton switch that normally has a broken path between its terminals. Activating the switch makes the connection.

**Norton's theorem.** A theorem stating that any resistive circuit or network, no matter how complex, can be represented as a current source in parallel with a source resistance.

**Notch filter.** Another name for a band-stop filter.

***n*-stage filter.** A term commonly used to refer to a cascaded circuit, where *n* is the number of stages.

**Nucleus.** The core of an atom. In its simplest model, the nucleus of an atom consists of protons and neutrons. (Hydrogen has no neutrons.)

**Octave.** A frequency multiplier equal to two.

**Octave scale.** A scale where the value of each increment is two times the value of the previous increment.

**Ohm ($\Omega$).** The unit of measure for resistance; the amount of resistance that limits current to 1 ampere when 1 volt is applied.

**Ohm's law.** A law stating that current is directly proportional to voltage and inversely proportional to resistance.

**Open circuit.** A physical break in a conduction path.

**Oscilloscope.** A piece of test equipment that provides a visual representation of a voltage waveform for various amplitude and time measurements.

**Output current rating.** A transformer rating that indicates the maximum allowable value of secondary current.

**Output voltage rating.** A transformer rating that indicates the ac output voltage from a transformer when provided with a 120 $V_{ac}$ input.

**Parallel circuit.** A circuit that provides more than one current path between any two points.

**Parallel-equivalent circuit.** The equivalent of a series-parallel circuit, made up entirely of components connected in parallel.

**Parallel *RC* circuit.** A circuit that contains one or more resistors in parallel with one or more capacitors. Each branch in a parallel *RC* circuit contains only one component (either resistive or capacitive).

**Parallel *RL* circuit.** A circuit that contains one or more resistors in parallel with one or more inductors. Each branch in a parallel *RL* circuit contains only one component (either resistive or inductive).

**Parameter.** A limit.

**Passive components.** A term generally used to describe components with characteristics that are not controlled by an external dc power source.

**Peak-to-peak value.** The difference between the peak values of a waveform.

**Peak value.** The maximum value reached by either alternation of a waveform.

**Permeability ($\mu$).** A measure of the ease with which magnetic lines of force are established within a material.

**Permittivity.** A measure of the ease with which lines of electrical force are established within a material. In a sense, it is the electrical equivalent of permeability.

**Peta- (P).** The engineering notation prefix with a value of $10^{15}$.

**Phase.** The position of a point on a waveform relative to the start of the waveform, usually expressed in degrees.

**Phase angle ($\theta$).** The phase difference between two or more waveforms.

**Phasor.** A vector used to represent a value that constantly changes phase. A vector with constant angular velocity.

**Pico- (p).** The engineering notation prefix with a value of $10^{-12}$.

**Plates.** The conductive surfaces of a capacitor.

**Polarity dots.** Dots placed in the symbol of a transformer to indicate the phase relationship between the transformer input and output.

**Polar notation.** A method of expressing the value of a vector as a magnitude followed by an angle.

**Pole.** The moving contact in a mechanical switch.

**Poles.** The points where magnetic lines of force leave (and return to) a magnet, referred to as the north-seeking pole (N) and the south-seeking pole (S).

**Positive temperature coefficient.** An indicator of the effect of temperature on resistance for a material. A positive temperature coefficient indicates that resistance varies directly with temperature.

**Potentiometer.** A three-terminal resistor whose value can be adjusted (within limits) by the user. A variable resistor.

**Power (*P*).** The amount of energy used per unit time, measured in watts. One watt (W) of power is equal to a rate of one joule per second.

**Power factor (PF).** The ratio of resistive power to apparent power, equal to cos $\theta$.

**Power gain ($A_p$).** The ratio of output power to input power for a component or circuit.

**Power rating.** A measure of a component's ability to dissipate heat, measured in watts (W).

**Primary.** The transformer coil that serves as the component input.

**Primary cell.** A cell that cannot be recharged. Often called dry cells because they contain dry electrolytes.

**Primary impedance.** The total opposition to current in the primary of a transformer.

**Primary resistance.** The winding resistance of a transformer's primary coil.

**Proton.** In the simplest model of the atom, the particle in the nucleus that is the source of positive charge.

**Prototype.** The first working model of a circuit or system.

**Pulse width (PW).** A term generally used to describe the positive alternation of a rectangular waveform.

**Pythagorean theorem.** A theorem stating that, for any right triangle, the square of the hypotenuse is equal to the sum of the squares of the other two sides.

**Quality (*Q*).** (1) A numeric value that indicates how close an inductor (or capacitor) comes to having the power characteristics of the ideal inductor (or capacitor). Sometimes referred to as the figure of merit of the component. (2) The ratio of filter center frequency to bandwidth.

**Radian.** The angle formed at the center of a circle by two radii and an arc of equal length. (*Note: Radii* is the plural form of *radius.*)

**Ramp.** Another name for a sawtooth waveform. The term is also used to describe the actual transition from one peak to the other.

***RC* time constant ($\tau$).** The time required for the capacitor voltage in an *RC* switching circuit to increase (or decrease) by 63.2% of its maximum possible $\Delta V$.

***RL* time constant ($\tau$).** The time required for inductor current in an *RL* switching circuit to increase (or decrease) by 63.2% of its maximum possible $\Delta I$.

**Rectangular notation.** A variation on coordinate notation where the *x*- and *y*-coordinates are written in the form of $x + jy$.

**Rectangular waves.** Waveforms that alternate between two dc levels.

**Rectifier.** A circuit that converts ac to pulsating dc.

**Relative conductivity.** The conductivity of a material as compared to the conductivity of copper, expressed as a percentage.

**Relative permeability ($\mu_r$).** The ratio of a material's permeability to that of free space.

**Relative permittivity.** The ratio of a material's permittivity to that of a vacuum.

**Reluctance ($\Re$).** The opposition that a material presents to magnetic lines of force. The magnetic equivalent of resistance.

**Residual flux.** The magnetism that remains in a material after a magnetizing force has been removed.

**Residual flux density.** The flux density that remains in a material after any mmf is removed.

**Resistance (*R*).** An opposition to current, measured in ohms ($\Omega$).

**Resistive-capacitive (*RC*) circuit.** A circuit that contains any combination of resistors and capacitors.

**Resistive-inductive (*RL*) circuit.** A circuit that contains any combination of resistors and inductors.

**Resistivity.** The resistance of a specified volume of an element or compound.

**Resistor.** A component designed to provide a specific amount of resistance; a component used to limit current in a circuit.

**Resolution.** For a potentiometer, the change in resistance per degree of control shaft rotation. High-resolution pots (those with low resistance per degree ratings) allow more exact control over the adjusted value of the potentiometer.

**Retentivity.** The ability of a material to retain its magnetic characteristics after a magnetizing force is removed.

**Right-hand rule.** A memory aid that helps you to remember the relationship between the direction of conventional current and the direction of the resulting magnetic field.

**Ring magnet.** A magnet that forms a closed loop and therefore has no identifiable poles. There is no magnetic flux in the center of a ring magnet.

**Roll-off.** A term used to describe the decrease in gain that occurs when a circuit is operated outside its pass band.

**Roll-off rate.** The rate at which gain rolls off, normally given in dB per octave or dB per decade.

**Root-mean-square (rms) value.** The value of voltage (or current) that, when used in the appropriate power equation, gives the average ac power (equivalent dc power) of the waveform.

**Rotary switch.** A switch with one or more poles and a series of throws.

**Rotor.** The rotating part in a motor or generator.

**Rowland's law.** A law stating that magnetic flux is directly proportional to magnetomotive force (mmf) and inversely proportional to reluctance.

**Sawtooth.** A term used to describe a waveform that constantly changes at a linear rate.

**Scalar.** A value that has only magnitude.

**Scientific notation.** A system that represents a number as the product of a number and a whole-number power of ten. The number is always written so that the most significant digit falls in the units position.

**Secondary.** The transformer coil that serves as the component output.

**Secondary cell.** A cell that can be recharged. Often called wet cells because they contain liquid electrolytes.

**Secondary impedance.** The total opposition to current in the secondary circuit of a transformer; generally assumed to equal the value of the load impedance.

**Self-inductance.** The ability of an inductor with a changing current to generate a changing magnetic field that, in turn, induces a voltage across the component.

**Semiconductors.** A group of components that began to replace vacuum tubes in the early 1950s. Semiconductors are smaller, cheaper, more efficient, and more rugged than their vacuum tube counterparts.

**Series-aiding inductors.** A two-inductor series connection where the magnetic poles are such that the flux produced by the coils add together, resulting in a total inductance that is greater than the sum of the individual component values.

**Series-aiding voltage sources.** Sources connected so that the current supplied by the individual sources are in the same direction.

**Series circuit.** A circuit that contains only one current path.

**Series-equivalent circuit.** The equivalent of a series-parallel circuit, made up entirely of components connected in series.

**Series-opposing inductors.** A two-inductor series connection where the flux generated by each coil opposes the flux generated by the other, resulting in a total inductance that is less than the sum of the individual component values.

**Series-opposing voltage sources.** Sources connected so that the current supplied by the individual sources are in opposition to each other.

**Series-parallel circuit.** A circuit that contains both series and parallel elements. Also referred to as a combination circuit.

**Shell.** A term used to describe the orbital paths of electrons.

**Short circuit.** An extremely low resistance path between two points that does not normally exist.

**Shunt.** A term used to describe a component or circuit connected in parallel to a load (or some other element).

**Siemens (S).** The unit of measure for conductance. One siemen equals one ampere per volt.

**Signal.** A term commonly used in reference to a waveform.

**Simultaneous equations.** Two or more equations used to describe relationships that occur at the same time. Any group of linear simultaneous equations has one common solution.

**Slope.** The rate at which a value changes, measured as a change per unit time.

**Solder.** A high-conductivity compound that has a relatively low melting point, used to affix components to each other and/or PC boards.

**Source.** The part of a circuit that supplies power.

**Space width (SW).** A term generally used to describe the negative alternation of a rectangular waveform.

**Square wave.** A special-case rectangular wave with equal pulse width and space width values. The duty cycle of any square wave is 50%.

**Stage.** A term used to describe each filter (or circuit) in a cascade.

**Static values.** Values that are constant. Static values do not change as a result of normal circuit operation.

**Step-down transformer.** A transformer with fewer secondary windings than primary windings, resulting in a secondary voltage that is less than the primary voltage.

**Step-up transformer.** A transformer with fewer primary windings than secondary windings, resulting in a secondary voltage that is greater than the primary voltage.

**Substitution method.** A method of solving two simultaneous equations where the equations are combined so that they become one single-variable equation.

**Superposition theorem.** A theorem stating that the response of a circuit to more than one source can be determined by analyzing the circuit's response to each source (alone) and combining the results.

**Susceptance (*B*).** The reciprocal of reactance, measured in Siemens.

**Switch.** A device that allows you to make or break the connection between two or more points in a circuit.

**Switching circuit.** A term generally used to describe any circuit with a square (or rectangular) wave input.

**Symmetrical waveform.** A waveform with cycles made up of identical halves. Waveforms can be symmetrical in time and/or amplitude.

**Taper.** A measure of the rate at which the resistance of a potentiometer changes as the control shaft is rotated between its extremes. Tapers are designated as either linear or nonlinear.

**Telecommunications.** The area that deals with the transmission of data between two or more locations.

**Tera- (T).** The engineering notation prefix with a value of $10^{12}$.

**Terminals.** Another name for the leads on a component.

**Theorem.** A relationship between a group of variables that has never been disproved.

**Thevenin resistance ($R_{th}$).** The resistance measured across the output terminals of a circuit with the load removed. To measure $R_{th}$, the voltage source must be removed and replaced with a wire.

**Thevenin voltage ($V_{th}$).** The voltage measured across the output terminals of a circuit with the load removed.

**3 dB frequency.** A term commonly used to describe a cutoff frequency based on the fact that gain drops by 3 dB at that frequency.

**3 dB point.** Another name for the 3 dB frequency.

**Throw.** The nonmoving contact in a mechanical switch.

**Time base.** The oscilloscope control used to set the period of time represented by the space between adjacent major divisions along the horizontal axis.

**Time constant ($\tau$).** A time interval on any universal rise (or decay) curve that is constant and independent of magnitude and operating frequency.

**Tolerance.** The range of values for a resistor, given as a percentage of its nominal (rated) value.

**Toroid.** An inductor with a doughnut-shaped core.

**Traces.** The conductors that connect components on a printed circuit board.

**Transducer.** Any device that converts energy from one form to another.

**Transformer.** An inductive component that uses electromagnetic energy to pass an ac signal from its input to its output, while (usually) providing dc isolation between the two.

**Transition.** In a switching circuit, the change from one dc level to another.

**Triangular waveform.** A symmetrical sawtooth waveform. In a triangular waveform, the slopes of the transitions are equal.

**Trimmer potentiometer.** A carbon-composition pot designed for low-power applications. Trimmers are produced as stand-up and lay-down components.

**Troubleshooting.** The process of locating faults in a circuit or system.

**True ac waveform.** An ac waveform with peak values that are equal in magnitude.

**Tuned circuit.** A circuit that provides a specific output over a designated range of frequencies.

**Turn.** Another name for a single loop of wire in a coil.

**Turns ratio.** The ratio of primary turns to secondary turns for a given transformer.

**Unity coupling.** A term used to describe a situation where 100% of the flux generated by one coil cuts through a parallel coil.

**Unity gain.** A gain of one (1).

**Universal curve.** A curve that can be used to predict the operation of any specified type of circuit.

**Vacuum tubes.** A group of components whose operation is based on the flow of charge through a vacuum.

**Valence electron(s).** The electron(s) orbiting in the valence shell of an atom. The electron(s) farthest from the nucleus of an atom.

**Valence shell.** For an atom, the outermost orbital shell containing electrons.

**Vector.** Any value that has both magnitude and an angle (or direction).

**Vertical sensitivity.** The oscilloscope control used to set the amount of voltage represented by the space between adjacent major divisions along the vertical axis. Usually referred to as the volts per division (V/div) control.

**Volt (V).** The unit of voltage (difference or potential). One volt is equal to 1 joule per coulomb.

**Voltage (*V*).** The potential that causes the directed flow of charge (current) in a circuit. Voltage is often referred to as *difference of potential.* The force required to move a coulomb of charge, measured in joules per coulomb.

**Voltage divider stability.** The ability of a voltage divider to maintain a stable output voltage despite normal (or anticipated) variations in load demand.

**Voltage gain ($A_v$).** The ratio of circuit output voltage to input voltage.

**Volt-amperes (VA).** The unit of measure for apparent power. This unit is used to distinguish apparent power from resistive and reactive power.

**Volt-amperes-reactive (VAR).** The unit of measure for reactive (imaginary) power. The unit is used to distinguish reactive power from resistive (true) power.

**Volt-ohm-milliammeter (VOM).** The forerunner of the DMM. The VOM uses an analog scale for a readout (rather than a digital display).

**Watt (W).** The unit of measure for power, equal to one joule per coulomb.

**Waveform.** A graph of the relationship between current, voltage, or power and time.

**Wavelength ($\lambda$).** The physical length of a transmitted waveform.

**Wheatstone bridge.** A circuit containing four resistors and a meter bridge that provides resistive measurements.

**Winding resistance ($R_W$).** The dc resistance of the wire used to make a coil.

**Wiper arm.** The potentiometer terminal that is connected to the sliding contact; the control shaft of a potentiometer.

**Wire-wound resistor.** A resistor that uses the resistivity of a length of wire to produce a desired value of resistance.

# APPENDIX E

# Selected Equation Derivations

## Equation 6.6

The total resistance of a two-branch parallel resistive circuit can be found as

$$R_T = \frac{1}{\frac{1}{R_1} + \frac{1}{R_2}}$$

This relationship is a form of equation (6.5). The denominator of the fraction can be rewritten as

$$\frac{1}{R_1} + \frac{1}{R_2} = \left(\frac{1}{R_1} \times \frac{R_2}{R_2}\right) + \left(\frac{1}{R_2} \times \frac{R_1}{R_1}\right) = \frac{R_2}{R_1R_2} + \frac{R_1}{R_1R_2} = \frac{R_1 + R_2}{R_1R_2}$$

Note that each fraction has been multiplied by one (1), in the form of a fraction. If we substitute the final value back into equation (6.5), we get

$$R_T = \frac{1}{\frac{1}{R_1} + \frac{1}{R_2}} = \frac{1}{\frac{R_1 + R_2}{R_1R_2}}$$

or

$$R_T = \frac{R_1R_2}{R_1 + R_2}$$

which is equation (6.6).

## Equation 8.6

Figure E.1 shows three voltage sources connected in parallel with a pair of open load terminals. Since the load terminals are in parallel with the rest of the circuit, the load voltage equals the voltage across all the circuit branches. According to Ohm's law, this voltage can be found as

$$V_T = I_T R_T$$

With the load terminals open, the total current equals the sum of the source currents. By formula,

$$I_T = I_1 + I_2 + I_2$$

Using Ohm's law, the above equation can be rewritten as

$$\frac{V_T}{R_T} = \frac{V_1}{R_1} + \frac{V_2}{R_2} + \frac{V_3}{R_3}$$

**FIGURE E.1**

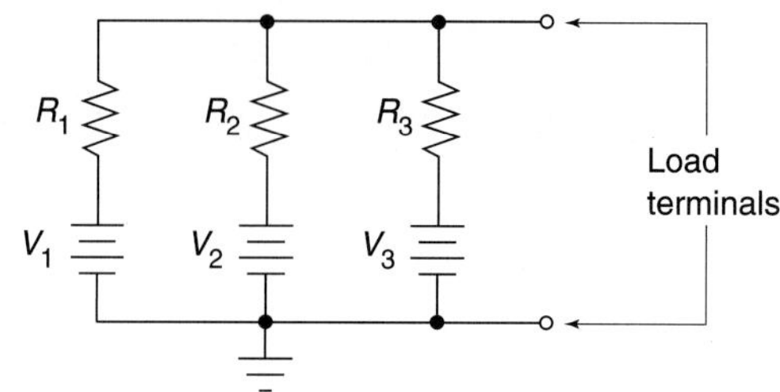


or

$$V_T = R_T\left(\frac{V_1}{R_1} + \frac{V_2}{R_2} + \frac{V_3}{R_3}\right)$$

where $R_T$ is the parallel combination of $R_1$, $R_2$, and $R_3$. Since $V_T$ and $R_T$ can be measured across the open load terminals, they are referred to as the Millman voltage ($V_m$) and Millman resistance ($R_m$). Using these labels, the above equation is rewritten as

$$V_m = R_m\left(\frac{V_1}{R_1} + \frac{V_2}{R_2} + \frac{V_3}{R_3}\right)$$

where $R_m = R_1 \| R_2 \| R_3$

## Equations 8.7, 8.8, and 8.9

For two circuits to be resistive equivalents, they must provide the same resistance readings between any pair of terminals. For example, consider the circuits shown in Figure E.2. Assuming that the $\Delta$ and $Y$ circuits shown are resistive equivalents, then the two must provide the same resistance readings between any pair of terminals. If we were to measure the resistance between the ($x$) and ($y$) terminals for each circuit in Figure E.2, we would obtain the following:

$$R_{(x\rightarrow y)} = R_1 + R_2 \quad \text{(for the } Y \text{ circuit)}$$

and

$$R_{(x\rightarrow y)} = R_A \| (R_B + R_C) \quad \text{(for the } \Delta \text{ circuit)}$$

Since the two are resistive equivalents, the above equations indicate that:

$$R_1 + R_2 = R_A \| (R_B + R_C) = \frac{R_A(R_B + R_C)}{R_A + R_B + R_C} = \frac{R_AR_B + R_AR_C}{R_A + R_B + R_C} \qquad \textbf{(E.1)}$$

The same technique is used to establish the following relationships:

$$R_1 + R_3 = R_C \| (R_A + R_B) = \frac{R_C(R_A + R_B)}{R_A + R_B + R_C} = \frac{R_AR_C + R_BR_C}{R_A + R_B + R_C} \qquad \textbf{(E.2)}$$

**FIGURE E.2**

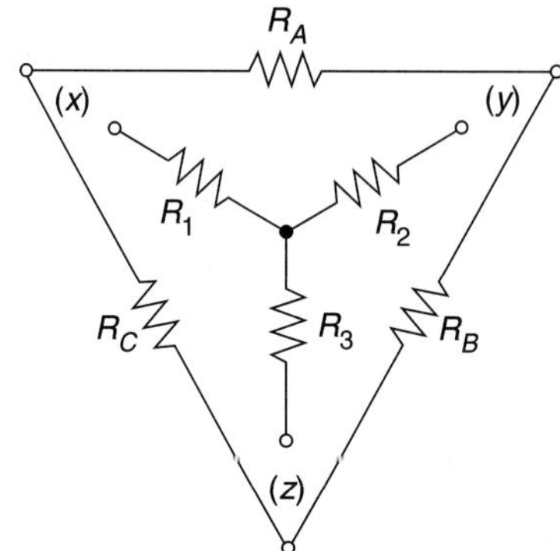

and

$$R_2 + R_3 = R_B \| (R_A + R_C) = \frac{R_B(R_A + R_C)}{R_A + R_B + R_C} = \frac{R_A R_B + R_B R_C}{R_A + R_B + R_C} \tag{E.3}$$

Since these three equations describe the same circuit, they are simultaneous equations, meaning that the three are true and interdependent. When two (or more) equations are true, then any combination of the equations also holds true. For example, if we subtract equation (E.3) from (E.2), we get

$$(R_1 + R_3) - (R_2 + R_3) = \frac{R_A R_C + R_B R_C}{R_A + R_B + R_C} - \frac{R_A R_B + R_B R_C}{R_A + R_B + R_C} = \frac{R_A R_C - R_A R_B}{R_A + R_B + R_C}$$

or

$$R_1 - R_2 = \frac{R_A R_C - R_A R_B}{R_A + R_B + R_C} \tag{E.4}$$

Since equations (E.2) and (E.3) are true, equation (E.4) must also hold true. Now, if we add equation (E.4) to equation (E.1), we get

$$(R_1 + R_2) + (R_1 - R_2) = \frac{R_A R_B + R_A R_C}{R_A + R_B + R_C} + \frac{R_A R_C - R_A R_B}{R_A + R_B + R_C} = \frac{2R_A R_C}{R_A + R_B + R_C}$$

or

$$2R_1 = \frac{2R_A R_C}{R_A + R_B + R_C}$$

Dividing both sides of this equation by 2 gives us

$$R_1 = \frac{R_A R_C}{R_A + R_B + R_C}$$

which is equation (8.7). In a similar fashion:

1. Equation (8.8) is derived by subtracting equation (E.2) from equation (E.1) and adding the result to equation (E.3).
2. Equation (8.9) is derived by subtracting equation (E.1) from equation (E.3) and adding the result to equation (E.2).

## Equations 8.10, 8.11, and 8.12

The $Y$-to-$\Delta$ conversion equations are derived using the $\Delta$-to-$Y$ conversion relationships. For convenience, the $\Delta$-to-$Y$ relationships are listed here:

$$R_1 = \frac{R_A R_C}{R_A + R_B + R_C} \tag{E.5}$$

$$R_2 = \frac{R_A R_B}{R_A + R_B + R_C} \tag{E.6}$$

and

$$R_3 = \frac{R_B R_C}{R_A + R_B + R_C} \tag{E.7}$$

Using equations (E.5) and (E.6), the ratio of $R_1{:}R_2$ can be found as

$$\frac{R_1}{R_2} = R_1\left(\frac{1}{R_2}\right) = \left(\frac{R_A R_C}{R_A + R_B + R_C}\right)\left(\frac{R_A + R_B + R_C}{R_A R_B}\right) = \frac{R_A R_C}{R_A R_B} = \frac{R_C}{R_B}$$

Based on this ratio, we can define $R_C$ as follows:

$$R_C = R_B\frac{R_1}{R_2} \quad \textbf{(E.8)}$$

Now, we will use the same approach to define the value of $R_A$. Using equations (E.7) and (E.5), the ratio of $R_1{:}R_3$ can be found as

$$\frac{R_1}{R_3} = R_1\left(\frac{1}{R_3}\right) = \left(\frac{R_AR_C}{R_A + R_B + R_C}\right)\left(\frac{R_A + R_B + R_C}{R_BR_C}\right) = \frac{R_AR_C}{R_BR_C} = \frac{R_A}{R_B}$$

Based on this ratio, we can define $R_A$ as follows:

$$R_A = R_B\frac{R_1}{R_3} \quad \textbf{(E.9)}$$

Now, we will substitute equations (E.8) and (E.9) for the defined variables in equation (E.5), as follows:

$$R_1 = \frac{R_AR_C}{R_A + R_B + R_C} = \frac{\left(R_B\frac{R_1}{R_3}\right)\left(R_B\frac{R_1}{R_2}\right)}{\left(R_B\frac{R_1}{R_3}\right) + R_B + \left(R_B\frac{R_1}{R_2}\right)}$$

This equation looks a bit intimidating, but it has one very important characteristic. If you look at the compound fraction, you'll see we have eliminated all the lettered variables other than $R_B$. To solve for $R_B$, we must start by simplifying the numerator of the compound fraction, as follows:

$$(R_B\frac{R_1}{R_3})(R_B\frac{R_1}{R_2}) = \frac{(R_BR_1)(R_BR_1)}{R_2R_3} = \frac{(R_BR_1)^2}{R_2R_3}$$

Substituting this result into the numerator of the compound fraction, we get:

$$R_1 = \frac{\frac{(R_BR_1)^2}{R_2R_3}}{\frac{R_BR_1}{R_3} + R_B + \frac{R_BR_1}{R_2}} \quad \textbf{(E.10)}$$

At this point, we have to rewrite the equation so that all the terms have the same denominator. This can be accomplished by rewriting each term in the denominator of equation (E.10) as follows:

$$\frac{R_BR_1}{R_3} \times \frac{R_2}{R_2} = \frac{R_BR_1R_2}{R_2R_3}$$

$$R_B = R_B\frac{R_2R_3}{R_2R_3} = \frac{R_BR_2R_3}{R_2R_3}$$

$$\frac{R_BR_1}{R_2} \times \frac{R_3}{R_3} = \frac{R_BR_1R_3}{R_2R_3}$$

Now, using these values, equation (E.10) can be rewritten as

$$R_1 = \frac{\frac{(R_BR_1)^2}{R_2R_3}}{\frac{R_BR_1R_2}{R_2R_3} + \frac{R_BR_2R_3}{R_2R_3} + \frac{R_BR_1R_3}{R_2R_3}}$$

Now, the common denominator $(R_2R_3)$ can be dropped, leaving

$$R_1 = \frac{(R_BR_1)^2}{R_BR_1R_2 + R_BR_2R_3 + R_BR_1R_3} = \frac{R_BR_1^2}{R_1R_2 + R_2R_3 + R_1R_3}$$

Taking the reciprocal of the above relationship, we get

$$\frac{1}{R_1} = \frac{R_1R_2 + R_2R_3 + R_1R_3}{R_BR_1^2}$$

Finally, multiplying each side of the equation by $(R_BR_1)$ gives us

$$R_B = \frac{R_1R_2 + R_2R_3 + R_1R_3}{R_1}$$

which is equation (8.11). Equations (8.10) and (8.12) are derived in the same fashion: substituting equations (E.8) and (E.9) into equations (E.6) and (E.7).

## Equation 11.5

The average value of any curve equals the area under the curve divided by its length. For example, the average value of the sine-wave alternation shown in Figure E.3 can be found as

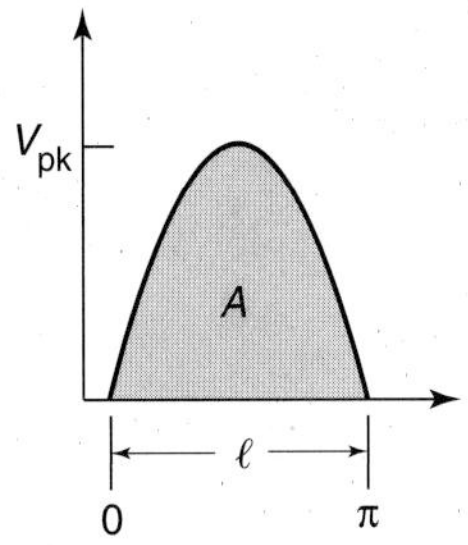

$$V_{\text{ave}} = \frac{A}{\ell}$$

where $A$ is the value of the shaded area under the curve and $\ell$ is the difference between the zero points on the curve.

To find the area under the curve, we start with the curve equation. The equation of the curve in Figure E.3 is given as



FIGURE E.3

$$v = V_{\text{pk}}\sin \omega t \qquad \textbf{(E.11)}$$

where

$\omega t$ = the phase angle, given as a product of angular velocity ($\omega$), in radians, and time

Using calculus, the area under the curve is found as the integral of equation (E.11). By formula,

$$A = \int_0^{\pi} V_{\text{pk}}\sin \omega t \, d(\omega t) \qquad \textbf{(E.12)}$$

Solving equation (E.12), we get

$$\begin{aligned} A &= \int_0^{\pi} V_{\text{pk}}\sin \omega t \, d(\omega t) \\ &= -V_{\text{pk}}[-\cos \omega t]_0^{\pi} \\ &= -V_{\text{pk}}[-\cos \pi - \cos 0] \\ &= -V_{\text{pk}}[-1 - 1] \\ &= 2\text{ V}_{\text{pk}} \end{aligned}$$

Now, we know that the area under the curve equals 2 $V_{pk}$. As shown in Figure E.3, the difference between the zero crossings of the waveform equals $\pi$ (radians). Using these two values in equation (E.11):

$$V_{\text{ave}} = \frac{A}{\ell} = \frac{2\,V_{\text{pk}}}{\pi}$$

which is equation (11.5).

## Equation 14.29

The charge curve is a plot of the equation

$$\frac{I_t}{I_{pk}} = 1 - e^{-\frac{t}{\tau}}$$

Transposing the equation, we get

$$e^{-t/\tau} = 1 - \frac{I_t}{I_{pk}}$$

Taking the natural log (ln) of each side of the equation, we get:

$$\ln\left(e^{-t/\tau}\right) = ln\left(1 - \frac{I_t}{I_{pk}}\right)$$

or

$$-\frac{t}{\tau} = \ln\left(1 - \frac{I_t}{I_{pk}}\right)$$

Finally, multiplying each side of the equation by $(-\tau)$, we get

$$t = -\tau \ln\left(1 - \frac{I_t}{I_{pk}}\right) \quad \textbf{(E.13)}$$

Assuming that the inductor in a series *RL* switching circuit has no winding resistance, the total resistance in the circuit equals *R*. Therefore, the circuit current at any instant is found as

$$I_t = \frac{V_R}{R}$$

and the peak current is found as

$$I_{pk} = \frac{V_S}{R}$$

Using these relationships, the current ratio in equation (E.13) can be rewritten as

$$\frac{I_t}{I_{pk}} = \frac{\frac{V_R}{R}}{\frac{V_S}{R}} = \frac{V_R}{R} \times \frac{R}{V_S} = \frac{V_R}{V_S}$$

Using the ratio of $V_R$ to $V_S$, equation (E.13) can be rewritten as

$$\mathrm{t} = -\tau \ln\left(1 - \frac{V_R}{V_S}\right)$$

which is equation (14.29).

## Equation 15.10

Transformer secondary voltage ($V_S$) is found as

$$V_S = V_P \frac{N_S}{N_P}$$

This relationship was given as equation (15.2). Rewriting this equation for ($V_P$), we get

$$V_P = V_S \frac{N_P}{N_S} \tag{E.14}$$

Transformer secondary current ($I_S$) is found as

$$I_S = I_P \frac{N_P}{N_S}$$

This relationship was given as equation (15.8). Rewriting this equation for ($I_P$), we get

$$I_P = I_S \frac{N_S}{N_P} \tag{E.15}$$

Now, if we divide equation (E.14) by equation (E.15), we get

$$\frac{V_P}{I_P} = \frac{V_S \frac{N_P}{N_S}}{I_S \frac{N_S}{N_P}} = \frac{V_S}{I_S}\left(\frac{N_P}{N_S} \times \frac{N_P}{N_S}\right) = \frac{V_S}{I_S}\left(\frac{N_P}{N_S}\right)^2$$

Since $Z = V/I$, the above equation can be rewritten as

$$Z_P = Z_S \left(\frac{N_P}{N_S}\right)^2$$

which is equation (15.25).

## Equations 19.38 and 19.39

This derivation begins with the circuit shown in Figure E.4. For the series *RC* circuit shown, the ratio of output voltage to input voltage equals the ratio of $X_C$ to total circuit impedance. By formula

$$\frac{v_{\text{out}}}{v_{\text{in}}} = \frac{X_C}{Z_T}$$

or

$$A_v = \frac{X_C}{Z_T} \tag{E.16}$$

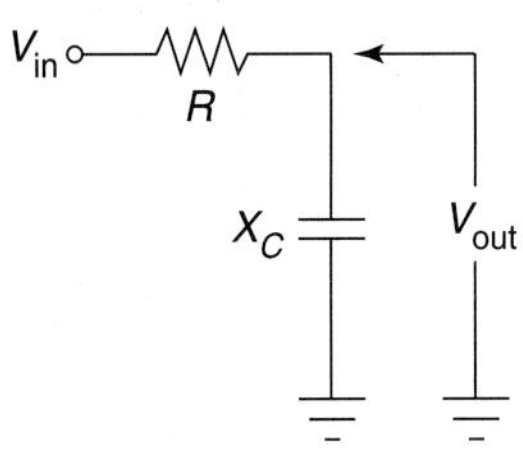

**FIGURE E.4**

In rectangular form, equation (E.16) can be written as

$$A_v = \frac{\frac{1}{j\omega C}}{\frac{1}{j\omega C} + R}$$

Multiplying the numerator and the denominator of the fraction by $j\omega C$, we get

$$A_v = \frac{1}{1 + j\omega RC} \tag{E.17}$$

In Chapter 19, we established the relationship between cutoff frequency and the values of *RC* as follows:

$$f_c = \frac{1}{2\pi RC}$$

or

$$\frac{1}{f_c} = 2\pi RC \qquad \textbf{(E.18)}$$

Since $\omega = 2\pi f$, the value of $j\omega RC$ in the denominator of equation (E.17) can be rewritten as

$$j\omega RC = j(2\pi fRC) = j(2\pi RC)f = j\frac{f}{f_c}$$

Using this value, equation (E.17) is rewritten as

$$\Delta A_v = \frac{1}{1 + j(f/f_c)^2} \qquad \textbf{(E.19)}$$

Note that the equation defines a change in voltage gain. When the ratio of operating frequency to cutoff frequency changes, so does the ratio of output voltage to input voltage. This is why $\Delta A_v$ replaces $A_v$ in equation (E.19).

The value in the denominator of equation (E.19) is in rectangular notation. Converting to polar notation,

$$1 + j(f/f_c)^2 = \sqrt{1^2 + (f/f_c)^2} = \sqrt{1 + (f/f_c)^2}$$

Substituting the geometric sum into equation (E.19), we get

$$\Delta A_v = \frac{1}{\sqrt{1 + (f/f_c)^2}}$$

which is equation (19.38). Since the equation represents a change in voltage gain, a multiplier of 20 is used to convert it to dB form, as follows:

$$\Delta A_{v(\mathrm{dB})} = 20 \log \frac{1}{\sqrt{1 + (f/f_c)^2}}$$

which is equation (19.39).

## Equations 19.55 and 19.56

These derivations start with an assumption: that every series resistive-reactive circuit has a parallel equivalent. These two circuits are represented in Figure E.5.

The quality ($Q$) of the series circuit shown is given as

$$Q = \frac{X_S}{R_S} \qquad \textbf{(E.20)}$$

and the total impedance (in rectangular form) can be expressed as

$$Z_S = R_S \pm jX_S \qquad \textbf{(E.21)}$$

**FIGURE E.5**

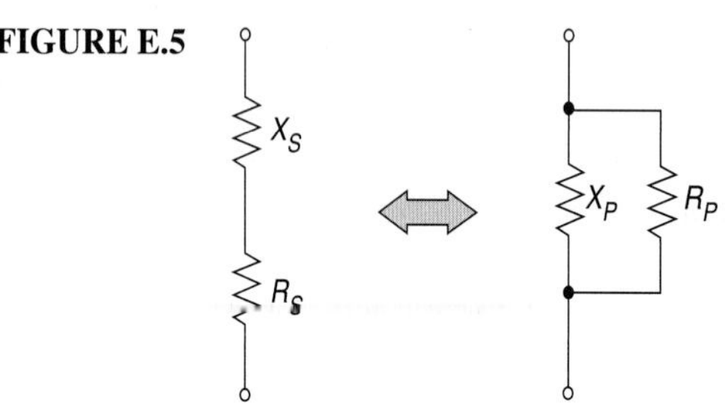

For the parallel circuit in Figure E.5, the total impedance ($Z_P$) can be expressed as

$$Z_P = \frac{(R_P)(\pm jX_P)}{R_P \pm jX_P} \quad \textbf{(E.22)}$$

If we multiply the fraction in equation (E.22) by 1 in the form of

$$\frac{R_P \pm jX_P}{R_P \pm jX_P}$$

we get

$$Z_P = \frac{R_P X_P^2}{R_P^2 + X_P^2} \pm j\frac{R_P^2 X_P}{R_P^2 + X_P^2} \quad \textbf{(E.23)}$$

Now, our original assumption was that the two circuits in Figure E.5 are equivalent circuits. Assuming that $Z_S = Z_P$, equation (E.23) can be rewritten as

$$Z_S = \frac{R_P X_P^2}{R_P^2 + X_P^2} \pm j\frac{R_P^2 X_P}{R_P^2 + X_P^2} \quad \textbf{(E.24)}$$

If we substitute equation (E.21) for the value of $Z_S$ in equation (E.24), we get

$$R_S \pm jX_S = \frac{R_P X_P^2}{R_P^2 + X_P^2} \pm j\frac{R_P^2 X_P}{R_P^2 + X_P^2} \quad \textbf{(E.25)}$$

For the two sides of equation (E.25) to be equal, the real components must equal each other, and the imaginary components must equal each other. Therefore,

$$R_S = \frac{R_P X_P^2}{R_P^2 + X_P^2} \quad \textbf{(E.26)}$$

and

$$X_S = \frac{R_P^2 X_P}{R_P^2 + X_P^2} \quad \textbf{(E.27)}$$

Equations (E.26) and (E.27) can now be substituted into equation (E.20) as follows:

$$Q = \frac{X_S}{R_S} = X_S\frac{1}{R_S} = \frac{R_P^2 X_P}{R_P^2 + X_P^2} \times \frac{R_P^2 + X_P^2}{R_P \times X_P^2} = \frac{R_P^2 X_P}{R_P X_P^2}$$

or

$$Q = \frac{R_P}{X_P} \quad \textbf{(E.28)}$$

Now, using equation (E.28), we can rewrite equation (E.27) as

$$X_S = \frac{X_P}{1 + (1/Q)^2} \quad \textbf{(E.29)}$$

Assuming that $Q >> 2$ (which is normally the case with filters), the value of $(1/Q)^2 << 1$. Based on this assumption, equation (E.29) simplifies to

$$X_S \cong X_P \quad \textbf{(E.30)}$$

Based on equation (E.28), we know that $R_P = QX_P$ or, based on equation (E.30):

$$R_P = QX_S \quad \textbf{(E.31)}$$

Since $X_S = QR_S$ (as implied by equation E.20),

$$R_P = QX_S = Q(QR_S)$$

or

$$R_P = Q^2R_S$$

which is a form of equation (19.55).

In a tuned $LC$ circuit, $X_L$ is the series reactance and $R_W$ is the series resistance. Replacing $R_S$ in the $R_P$ equation, we get

$$R_P = Q^2R_w$$

Finally, replacing $X_P$ with $X_L$ in equation (E.28) gives us

$$Q = \frac{R_P}{X_L}$$

or

$$Q = \frac{R_P \| R_L}{X_L}$$

when a load is connected. This equation is a form of equation (19.56).

# APPENDIX F

# Admittance Analysis of Parallel *RL* and *RC* Circuits

There are many approaches to solving a circuit analysis problem. One approach to solving parallel *RL* and *RC* circuits involves the use of susceptance and admittance. This approach, which was not used in the body of the text, is introduced in this appendix.

## Susceptance and Admittance

In Chapter 2, you were introduced to conductance ($G$), the reciprocal of resistance, measured in Siemens. At that time, you were shown that the conductance of a resistive component is found as

$$G = \frac{1}{R}$$

**Susceptance ($B$)** is the reciprocal of reactance, measured in Siemens. Susceptance can be viewed as the reactive counterpart to conductance. The susceptance of an inductor is found as

$$B_L = \frac{1}{X_L} \qquad \textbf{(F.1)}$$

The susceptance of a capacitor is found as

$$B_C = \frac{1}{X_C} \qquad \textbf{(F.2)}$$

As you know, inductive and capacitive reactance both have a phase angle relative to resistance. These phase angles are represented by the graph in Figure F.1a. As shown, $X_L$ has a positive 90° phase angle, indicating that it leads resistance by 90°. In contrast, $X_C$ has a negative 90° phase angle, indicating that it lags resistance by 90°.

**FIGURE F.1 Phase angles between reactance and resistance.**

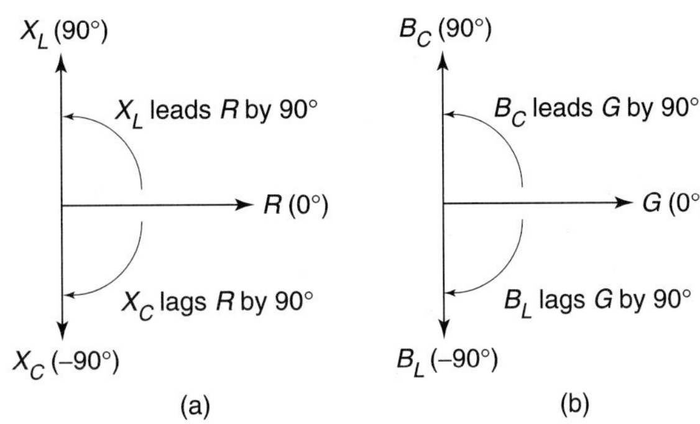

The phase relationships between susceptance and conductance are shown in Figure F.1b. The phase relationships shown are based on the following calculations:

$$G = \frac{1}{R\angle 0^\circ} = \frac{1}{R}\angle 0^\circ$$

$$B_C = \frac{1}{X_C\angle -90^\circ} = \frac{1}{X_C}\angle 90^\circ$$

$$B_L = \frac{1}{X_L\angle 90^\circ} = \frac{1}{X_L}\angle -90^\circ$$

These calculations do not commit $G$, $B_C$, and $B_L$ to specific values, but they do indicate that:

1. Capacitive susceptance ($B_C$) leads circuit conductance by 90°.
2. Inductive susceptance ($B_L$) lags circuit conductance by 90°.

Now, let's compare these phase angles to those for the branch currents in a parallel resistive-reactive circuit. This comparison is provided in Figure F.2. As the graphs indicate:

1. Conductance is in phase with resistor current.
2. Inductive susceptance is in phase with inductor current.
3. Capacitive susceptance is in phase with capacitor current.

As you will see, these three relationships are the basis for the use of conductance and susceptance in parallel *RL* and *RC* circuit analysis.

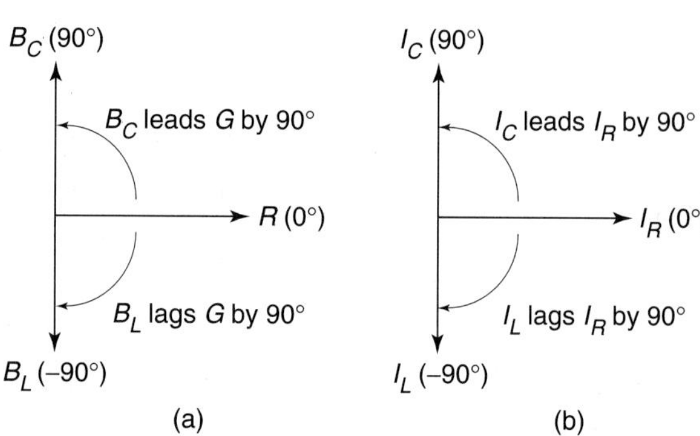


**FIGURE F.2 The phase relationships among susceptance, resistance, and branch currents.**

As you know, impedance is the total opposition to current in a circuit, made up of resistance and reactance. In effect, impedance represents the combined effects of resistance and reactance.

**Admittance (*Y*)** is a measure of the ease with which charge flows in a circuit, made up of conductance and susceptance. In effect, admittance represents the combined effects of conductance and susceptance. The admittance of a component or circuit equals the reciprocal of its impedance, as follows:

$$Y = \frac{1}{Z} \tag{F.3}$$

***Polar and Rectangular Representations of Admittance.*** Like impedance, admittance is normally represented using either polar or rectangular notation. In polar form, the admittance of a circuit is written in the form of

$$Y\angle -\theta$$

where $-\theta$ = the negative equivalent of the impedance phase angle. For example, let's say that a circuit has a value of $Z = 250\ \Omega\angle 60^\circ$. The admittance of the circuit is found as

$$Y = \frac{1}{Z}\angle -\theta = \frac{1}{250\ \Omega}\angle -60^\circ = 4\text{ mS}\angle -60^\circ$$

Using the same calculation, a circuit with a value of $Z = 100\ \Omega\angle -45°$ has an admittance of

$$Y = \frac{1}{Z}\angle -\theta = \frac{1}{100\ \Omega}\angle -(-45°) = 1\ \text{mS}\angle 45°$$

As these calculations indicate, admittance and impedance have equal-magnitude phase angles with opposite signs (+ versus −).

In rectangular notation, admittance is written in the form of

$$Y = G \pm jB$$

For example, let's say that a parallel *RL* circuit has the following values: $G = 10$ mS and $B = 5\ \text{mS}\angle -90°$. The total admittance in this circuit can be expressed as

$$Y = (10 - j5)\ \text{mS}$$

***Notation Conversions.*** Admittance values are converted from one form to another in the same fashion as impedance values. The conversion equations are as follows:

| **Polar to Rectangular** | **Rectangular to Polar** |
|---|---|
| $G = Y\cos\theta$ | $Y = \sqrt{G^2 + B^2}$ |
| $B = Y\sin\theta$ | $\theta = \tan^{-1}\frac{\pm B}{G}$ |

These conversion equations are used in several upcoming examples.

## Parallel RL Circuit Analysis

A parallel circuit can be analyzed using circuit values of conductance, susceptance, and admittance (in place of resistance, reactance, and impedance). This approach is based on the following relationships:

$$I_R = \frac{V_R}{R} = \frac{1}{R}V_R = GV_R \qquad I_L = \frac{V_L}{X_L} = \frac{1}{X_L}V_L = B_LV_L$$

As you can see, branch current equals the product of branch voltage and conductance (or susceptance). The use of these relationships is demonstrated in Example F.1.

**EXAMPLE F.1**

Calculate the value of total current for the circuit in Figure F.3.

**FIGURE F.3**

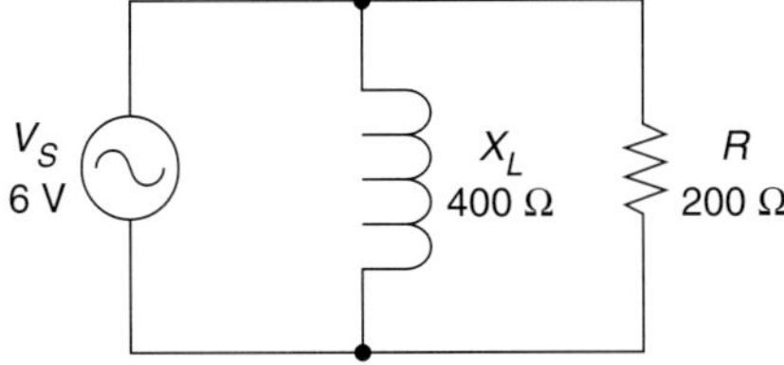


***Solution:*** First, the conductance of the resistor is found as

$$G = \frac{1}{R} = \frac{1}{200\ \Omega\angle 0°} = \mathbf{5\ mS\angle 0°}$$

and the susceptance of the inductor is found as

$$B_L = \frac{1}{X_L} = \frac{1}{400\ \Omega\angle 90°} = \mathbf{2.5\ mS\angle -90°}$$

The values of the branch currents can now be found as

$$\boldsymbol{I_R} = GV_S = (5 \text{ mS}\angle 0°)(6 \text{ V}) = \mathbf{30 \text{ mA}\angle 0°}$$

and

$$\boldsymbol{I_L} = B_L V_S = (2.5 \text{ mS}\angle -90°)(6 \text{ V}) = \mathbf{15 \text{ mA}\angle -90°}$$

As always, the magnitude of the circuit current is found as

$$\boldsymbol{I_T} = \sqrt{I_R^2 + I_L^2} = \sqrt{(30 \text{ mA})^2 + (15 \text{ mA})^2} = \mathbf{33.5 \text{ mA}}$$

and the phase angle of the circuit current is found as

$$\boldsymbol{\theta} = \tan^{-1}\frac{-I_L}{I_R} = \tan^{-1}\frac{-15 \text{ mA}}{30 \text{ mA}} = \tan^{-1}(-0.5) = \mathbf{-26.6°}$$

Several observations should be made about Example F.1. First, note that branch currents have the proper phase angles when calculated using conductance and susceptance values. Second, the phase angle of the circuit current could have been found as

$$\theta = \tan^{-1}\frac{-B_L}{G} = \tan^{-1}\frac{-2.5 \text{ mS}}{5 \text{ mS}} = \tan^{-1}(-0.5) = -26.6°$$

This result indicates that admittance and current in a parallel *RL* circuit have the same phase angle.

When the values of conductance and susceptance for a parallel *RL* circuit are known, they can be used to calculate the circuit impedance. This approach to calculating circuit impedance is demonstrated in Example F.2.

## EXAMPLE F.2

Calculate the impedance magnitude and phase angle for the circuit in Figure F.3.

***Solution:*** In Example F.1, we calculated values of

$$G = 5 \text{ mS}\angle 0° \qquad \text{and} \qquad B_L = 2.5 \text{ mS}\angle -90°$$

These values can be combined into a single admittance. In rectangular notation, the value of this admittance is found as

$$Y = G - jB_L = \mathbf{(5 - j2.5) \text{ mS}}$$

To calculate the circuit impedance, we first convert $Y$ to polar form, as follows:

$$\boldsymbol{Y} = \sqrt{G^2 + B_L^2} = \sqrt{(5 \text{ mS})^2 + (2.5 \text{ mS})^2} = \mathbf{5.59 \text{ mS}}$$

and

$$\boldsymbol{\theta} = \tan^{-1}\frac{-B_L}{G} = \tan^{-1}\frac{-2.5 \text{ mS}}{5 \text{ mS}} = \tan^{-1}(-0.5) = \mathbf{-26.6°}$$

Thus, the total circuit admittance has a value of $Y_T = 5.59 \text{ mS}\angle -26.6°$. This value can now be converted to an impedance, as follows:

$$\boldsymbol{Z_T} = \frac{1}{Y_T} = \frac{1}{5.59 \text{ mS}\angle -26.6°} = \mathbf{179\ \Omega\angle 26.6°}$$

There are several approaches we can take to verifying the result from Example F.2. The simplest approach involves Ohm's law. In Example F.1, we calculated a value of $I_T = 33.5\text{ mA}\angle{-26.6°}$. Using this, along with the source voltage, the circuit impedance can be found as

$$Z_T = \frac{V_S}{I_T} = \frac{6\text{ V}\angle 0°}{33.5\text{ mA}\angle{-26.6°}} = 179\ \Omega\angle 26.6°$$

As you can see, this agrees with the result found in the example. We can also verify the result in the example using the original values of $R$ and $X_L$, as follows:

$$Z_T = \frac{X_L R}{\sqrt{X_L^2 + R^2}} = \frac{(400\ \Omega)(200\ \Omega)}{\sqrt{(400\ \Omega)^2 + (200\ \Omega)^2}} = 179\ \Omega$$

and

$$\theta = \tan^{-1}\frac{R}{X_L} = \tan^{-1}\frac{200\ \Omega}{400\ \Omega} = \tan^{-1}(0.5) = 26.6°$$

Again, the validity of the result in Example F.2 has been verified. Example F.3 demonstrates the entire admittance analysis of a parallel *RL* circuit.

**EXAMPLE F.3**

Determine the current and impedance values for the circuit shown in Figure F.4.

**FIGURE F.4**

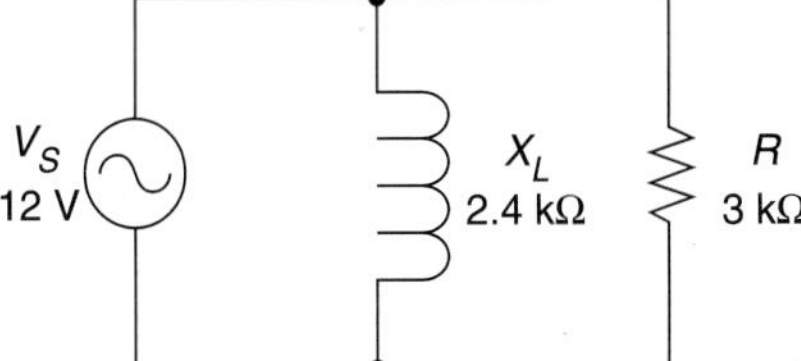


***Solution:*** The susceptance of the inductive branch can be found as

$$\mathbf{B_L} = \frac{1}{X_L} = \frac{1}{2.4\text{ k}\Omega\angle 90°} = \mathbf{417\ \mu S\angle{-90°}}$$

The conductance of the resistive branch can be found as

$$\mathbf{G} = \frac{1}{R} = \frac{1}{3\text{ k}\Omega} = \mathbf{333\ \mu S\angle 0°}$$

The current through the inductive branch can now be found as

$$\mathbf{I_L} = V_S B_L = (12\text{ V})(417\ \mu\text{S}\angle{-90°}) = \mathbf{5\text{ mA}\angle{-90°}}$$

The current through the resistive branch can now be found as

$$\mathbf{I_R} = V_S G = (12\text{ V})(333\ \mu\text{S}\angle 0°) = \mathbf{4\text{ mA}}$$

The circuit current has a value of

$$\mathbf{I_T} = \sqrt{I_L^2 + I_R^2} = \sqrt{(5\text{ mA})^2 + (4\text{ mA})^2} = \mathbf{6.4\text{ mA}}$$

and a phase angle of

$$\boldsymbol{\theta} = \tan^{-1}\frac{-I_L}{I_R} = \tan^{-1}\frac{-5\text{ mA}}{4\text{ mA}} = \tan^{-1}(-1.25) = \mathbf{-51.3°}$$

Finally, the circuit impedance can be found as

$$Z_T = \frac{1}{Y_T} = \frac{1}{G - jB_L} = \frac{1}{(333 - j417)\ \mu\text{S}} = \frac{1}{534\ \mu\text{S}\angle -53.1^\circ} = \mathbf{1.87\ k\Omega\angle 53.1^\circ}$$

## Parallel RC Circuit Analysis

The admittance analysis of a parallel *RC* circuit is nearly identical to that of a parallel *RL* circuit. The only differences are the phase angles associated with the results. This point is demonstrated in Example F.4.

### EXAMPLE F.4

Determine the current and impedance values for the circuit in Figure F.5.

**FIGURE F.5**

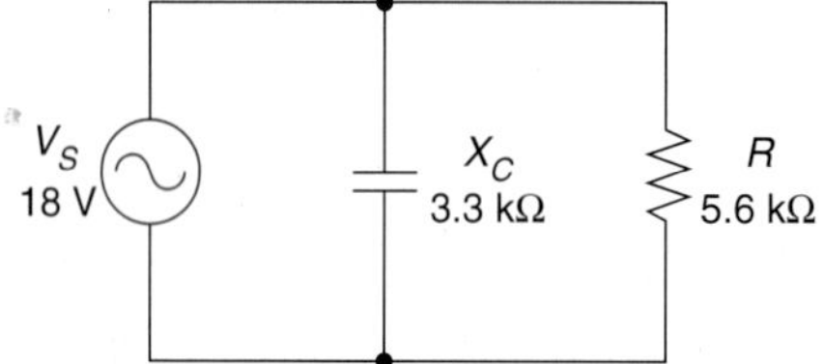


***Solution:*** The susceptance of the capacitive branch can be found as

$$B_C = \frac{1}{X_C} = \frac{1}{3.3\ \text{k}\Omega\angle -90^\circ} = \mathbf{303\ \mu S\angle 90^\circ}$$

The conductance of the resistive branch can be found as

$$G = \frac{1}{R} = \frac{1}{5.6\ \text{k}\Omega} = \mathbf{179\ \mu S\angle 0^\circ}$$

The current through the capacitive branch can now be found as

$$I_C = V_S B_C = (18\ \text{V})(303\ \mu\text{S}\angle 90^\circ) = \mathbf{5.45\ mA\angle 90^\circ}$$

The current through the resistive branch can now be found as

$$I_R = V_S G = (18\ \text{V})(179\ \mu\text{S}\angle 0^\circ) = \mathbf{3.22\ mA}$$

The circuit current has a value of

$$I_T = \sqrt{I_C^2 + I_R^2} = \sqrt{(5.45\ \text{mA})^2 + (3.22\ \text{mA})^2} = \mathbf{6.33\ mA}$$

and a phase angle of

$$\theta = \tan^{-1}\frac{I_C}{I_R} = \tan^{-1}\frac{5.45\ \text{mA}}{3.22\ \text{mA}} = \tan^{-1}(1.69) = \mathbf{59.4^\circ}$$

Finally, the circuit impedance can be found as

$$Z_T = \frac{1}{Y_T} = \frac{1}{G + jB_C} = \frac{1}{(179 + j303)\ \mu\text{S}} = \frac{1}{352\ \mu\text{S}\angle 59.4^\circ} = \mathbf{2.84\ k\Omega\angle -59.4^\circ}$$

As you can see, the approach to the problem was identical to that used for the parallel *RL* circuit analyzed in Example F.3. The only differences were the phase angles in the results.

# APPENDIX G

# Answers to Selected Odd-Numbered Problems

**Chapter 1:**

**1.** a. $3.492 \times 10^3$ b. $9.22 \times 10^2$ c. $2.38 \times 10^7$ d. $-4.46 \times 10^5$ e. $2.29 \times 10^4$ f. $1.22 \times 10^6$ g. $-8.2 \times 10^4$ h. $1 \times 10^0$ i. $3.97 \times 10^9$ j. $-3.8 \times 10^3$ **3.** a. 1.83 EE 23 b. 5.6 ± EE 4 c. 4.2 EE ± 9 d. 9.22 ± EE ± 7 e. 3.3 EE ± 12 f. 1 EE 6 **5.** a. 0.0443 b. −0.000028 c. 0.000762 d. 0.000000000338 e. −0.000000058 f. 0.634 g. 0.00000000111 h. 0.00902 i. −0.0000083 j. 0.000000444 **7.** a. $3.16 \times 10^2$ b. $-2.9273 \times 10^2$ c. $9.94 \times 10^0$ d. $2.7353 \times 10^2$ **9.** a. 38.4 km b. 234 kW c. 44.32 MHz d. 175 kV e. 60 W f. 1.8 kΩ g. 3.2 A h. 2.5 MΩ i. 4.87 GHz j. 22 kΩ **11.** a. 22 MΩ b. 3.3 kW c. 700 GHz d. 510 kΩ e. 85 kW f. 47 μF g. 100 mH h. 660 μs i. 40 ns j. 220 pF

**Chapter 2:**

**1.** 50 mA **3.** 4 μA **5.** 50 V **7.** 100 mV **9.** 212.77 mS **11.** 1.961 μS **13.** 20 kΩ **15.** 4 MΩ **17.** 68.44 mΩ **19.** 200 mA

**Chapter 3:**

**1.** 99 kΩ to 121 kΩ **3.** 342 kΩ to 378 kΩ **5.** a. 160 Ω b. 2.7 MΩ c. 390 kΩ d. 51 Ω e. 4.7 Ω f. 56 kΩ **7.** a. orange, orange, orange b. brown, green, black c. white, brown, yellow d. red, red, red e. brown, red, yellow f. brown, black, blue g. brown, gray, gold h. red, yellow, silver **9.** a. 228 kΩ to 252 kΩ b. 3.528 Ω to 3.672 Ω c. 65.6 Ω to 98.4 Ω d. 0.612 Ω to 0.748 Ω **11.** 440 mΩ

**Chapter 4:**

**1.** a. 5.45 mA b. 44.4 μA c. 17.02 mA d. 48.5 μA e. 17.65 μA **3.** a. 8.2 V b. 6.5 V c. 351 mV d. 26.4 V e. 2.88 V **5.** a. 1.5 Ω b. 4.7 kΩ c. 22 kΩ d. 180 Ω e. 100 kΩ **7.** a. 2.24 W b. 4.8 W c. 11 W d. 14.4 mW e. 12.8 W **9.** a. 2.06 W b. 3.07 W c. 193 mW d. 618.8 mW e. 26.5 mW **11.** a. 47.9 mW b. 1.36 mW c. 1.03 mW d. 28.3 mW e. 6.53 mW **13.** ½ W (rated) **15.** ⅛ W (rated) **17.** a. 1.8% b. 7.42% c. 78.8% d. 34% e. 0.004% **19.** 8.7 kWh **21.** a. 400 V, 40 kΩ b. 500 μA, 64 kΩ c. 27.6 V, 8.37 mA d. 3 kΩ, 675 mW e. 28.8 V, 691 mW **23.** 3.64 mA to 3.83 mA **25.** 2.89 mA to 3.19 mA **27.** 12 Ω

**Chapter 5:**

**1.** a. 2.65 kΩ b. 275 Ω c. 1.24 kΩ d. 106.4 kΩ **3.** a. 470 Ω b. 27 kΩ c. 5.1 kΩ d. 148 Ω **5.** 7 kΩ **7.** 200 mA **9.** 10 mA **11.** 85.6 Ω **13.** 3.63 kΩ **15.** 18 V **17.** 5.5 V **19.** $P_1 = 270$ mW, $P_2 = 330$ mW, $P_T = 600$ mW **23.** $V_{R1} + V_{R2} + V_{R3} - V_S = 7\text{ V} + 2\text{ V} + 6\text{ V} - 26\text{ V} = 0\text{ V}$ **25.** 100 mA **27.** 22.8 mA **29.** 1.64 V **31.** $V_{R1} = 6.82$ V, $V_{R2} = 8.18$ V **33.** $V_{R1} = 2.2$ V, $V_{R2} = 7.8$ V, $V_{R3} = 20$ V **35.** 4.36 V **37.** 12.6 V **39.** $R_T = 770$ kΩ, $I_T = 32.5$ μA, $P_T = 812$ mW, $V_1 = 4.87$ V, $V_2 = 20.1$ V, $P_1 = 158$ μW, $P_2 = 654$ μW **43.** 14.93 V **45.** 14.78 V **47.** 11.88 V, 10.91 V, 6 V **49.** 1.07 W **51.** $R_1$ is shorted **55.** 207.8 μA to 229.7 μA

**Chapter 6:**

**1.** a. 21.4 mA b. 6.9 mA c. 71.3 mA d. 1.81 mA **3.** 8 mA, 40 mA, 48 mA **5.** 9 mA, 6 mA, 3 mA, 18 mA **7.** 167 Ω **9.** 500 Ω **11.** 1.37 kΩ **13.** 2.08 kΩ **15.** 1.2 kΩ **17.** 5.94 V to 2.16 V **19.** 40 mA, 20 mA **21.** 667 μA, 133 μA **23.** $R_T = 761$ Ω, $I_T = 18.4$ mA, $I_1 = 10$ mA, $I_2 = 7$ mA, $I_3 = 1.4$ mA, $P_1 = 140$ mW, $P_2 = 98$ mW, $P_3 = 19.6$ mW, $P_T = 257.6$ mW **27.** $R_3$ is open **29.** $R_3$ is shorted **31.** a. $R_3 = 100$ Ω b. $R_3 = 150$ Ω, $I_S = 120$ mA

**Chapter 7:**

**1.** $I_1 = I_2 = 8$ mA, $I_3 = I_4 = 3$ mA, $V_1 = 9.6$ V, $V_2 = 14.4$ V, $V_3 = 14.1$ V, $V_4 = 9.9$ V, $P_1 = 76.8$ mW, $P_2 = 115.2$ mW, $P_3 = 42.3$ mW, $P_4 = 29.7$ mW, $I_T = 11$ mA, $R_T = 2.18$ kΩ, $P_T = 264$ mW **5.** $V_A = 1.32$ V, $V_B = 4.68$ V, $I_1 = 6$ mA, $I_2 = 13.2$ mA, $I_3 = 9.18$ mA, $I_4 = 9.96$ mA, $I_T = 19.15$ mA, $R_T = 313$ Ω, $P_1 = 7.92$ mW, $P_2 = 17.42$ mW, $P_3 = 43$ mW, $P_4 = 46.6$ mW, $P_T = 115$ mW **9.** The circuit consists of $R_{eq} = 1.16$ kΩ, $R_5$, and $R_L$ in series with the source. **11.** The circuit consists of $(R_2 \| R_L) + R_3 = 2.13$ kΩ and $R_1$ in parallel with the source. **13.** 3.68 V **15.** 4.23 mA **17.** 2.67 W **19.** 5.65 V, 6 V **21.** 5.06 V to 5.62 V **23.** 4.998 V to 5.14 V **25.** $V_2 = 4.5$ V, $V_M = 3.1$ V **27.** 10 kΩ **33.** 727.3 Ω

**Chapter 8:**

**1.** $V_1 = 3.9$ V, $V_2 = 2.1$ V, $V_3 = 9.9$ V **3.** $V_1 = 12.97$ V, $V_2 = 5.03$ V, $V_3 = 4.97$ V **5.** $V_1 = 944$ mV, $V_2 = 944$ mV, $V_3 = 8.623$ V, $V_4 = 2.433$ V **7.** They are equivalents **9.** $I_S = 300$ mA, $R_S = 30$ Ω **11.** $I_S = 66.7$ mA, $R_S = 75$ Ω **13.** $V_S = 6$ V, $R_S = 150$ Ω **15.** $V_{th} = 9$ V, $R_{th} = 2.16$ kΩ **17.** $V_{th} = 5.81$ V, $R_{th} = 419$ Ω **19.** 1.67 V to 6.67 V **21.** 35.45 mW **23.** 3.45 V to 6.33 V **25.** 161 mW **27.** $V_{th} = 2.05$ V, $R_{th} = 114$ Ω **29.** 11.5 mW **31.** $I_n = 4.16$ mA, $R_n = 2.16$ kΩ **33.** $I_n = 13.9$ mA, $R_n = 419$ Ω **35.** 3.64 mA to 13.68 mA **37.** 5.56 V **39.** 4.37 V **41.** 231 mW **43.** $R_1 = 46.2$ Ω, $R_2 = 92.3$ Ω, $R_3 = 69.2$ Ω **45.** $R_A = 95.3$ Ω, $R_B = 52$ Ω, $R_C = 78$ Ω **47.** $V_{L1} = 4.89$ V to 5.66 V, $V_{L2} = 2.43$ V to 2.1 V

**Chapter 9:**

**1.** $x = 4, y = 3$ **3.** a. $x = -6, y = 9$ b. $x = -3, y = -2$ c. $x = 2.5, y = -2.5$ d. $x = -1, y = 3$ **5.** a. −11 b. 41 c. −235 d. −27 **7.** a. $x = -6, y = 9$ b. $x = -3, y = -2$ c. $x = 2.5, y = -2.5$ d. $x = -1, y = 3$ **9.** a. $x = 1, y = 4, z = -3$ b. $x = -2, y = -3, z = 8$ **11.** $I_1 = 40.6$mA, $I_2 = 18.8$mA, $V_1 = 4.88$ V, $V_2 = 1.88$V, $V_3 = 7.13$V **13.** $I_1 = 9.55$mA, $I_2 = 24.5$mA, $V_1 = 1.9$V, $V_2 = 4.9$V, $V_3 = 4.1$V **15.** $I_1 = 21$ mA, $I_2 = 11.5$mA, $V_1 = 3.15$V, $V_2 = 1.15$V, $V_3 = 5.85$V **17.** $I_1 = 2.73$mA, $I_2 = 2.73$mA, $V_1 = 6$V, $V_2 = 9$V, $V_3 = 0$ V **21.** $V_1 = 4.875$ V, $V_2 = 1.875$ V, $V_3 = 7.125$ V **23.** $V_1 = 1.91$ V, $V_2 = 4.91$ V, $V_3 = 4.09$ V

**Chapter 10:**

**1.** 1.8 T **3.** 2.0 T **5.** 600 **7.** $7.96 \times 10^5$ **9.** 268 μT **11.** 53.3 mT

**Chapter 11:**

**1.** 80 ms **3.** 80 μs **5.** 12.5 Hz, 2 kHz **7.** 33.3 kHz **9.** a. 2 ms b. 400 μs c. 1 μs d. 13.3 μs **11.** 7.64 V **13.** 11.5 V **15.** 45.1 mA **17.** 57.9 mA **19.** 48 W **21.** 150 $mV_{pk}$, 300 $mV_{PP}$ **23.** 20 $V_{pk}$, 40 $V_{PP}$ **25.** a. 4.17 μs b. 20.8 μs c. 180° d. 72° **29.** 36° **31.** 120° **33.** 22.5° **35.** 67.5° **37.** a. 12.99 V b. 7.5 V c. −12.99 V d. −2.6 V **39.** −8.82 V **41.** 29.52 mV **43.** 29.52 mV **45.** −5.88 V **47.** 10.61 V, −10.61 V **49.** −3 V **51.** −15 V **53.** 100 cm **55.** 3100 mi **57.** 12.4 mi to 3.4 mi **59.** 16 kHz **61.** 380 kHz **63.** 18.8% **65.** 8.33% **67.** 0 V **69.** −2.86 V **71.** 1.31 kHz, 8.69 kHz **73.** $1.5 \times 10^5$ cm **75.** −2.6 V

**Chapter 12:**

**1.** a. $-10\angle-6°$ b. $39.6\angle189°$ c. $-30\angle25°$ d. $540\angle8°$ **3.** a. $5\angle15°$ b. $-0.93\angle152°$ c. $0.72\angle-30°$ d. $0.25\angle90°$ **5.** a. $-1.05\angle-30°$ b. $0.63\angle-59°$ **7.** a. $4 + j1$ b. $26 + j18$ **9.** a. $6 + j5$ b. $-5 + j11$ **11.** a. $24 + j12$ b. $-4 - j111$ c. $-50 + j70$ d. 34 **13.** 68 **15.** 289 **17.** a. $1.5 + j2$ b. $12 - j3$ c. $0 - j4$ d. $0.29 - j0.55$ **19.** a. $17.68 + j17.68$ b. $25 - j43.3$ c. $9 + j15.6$ d. $0 + j100$ **21.** a. $2.5\angle53.14°$ b. $12.36\angle-14.04°$ c. $4\angle0°$ d. $0.612\angle-61.5$ **23.** $0.13\angle-34.83°$

**Chapter 13:**

**1.** 6.6 mV **3.** 2.5 V **5.** 2.27 μH **7.** 452 μH **9.** 573 mH **11.** 11.55 mH **13.** 0.051 **15.** 87.3 μH **17.** 532 Ω **21.** 15.9 kHz **23.** 150 **25.** 518 **27.** 209 to 1047 **29.** 13.3 kHz **31.** 60 mA, 23.9 kHz

**Chapter 14:**

**1.** $4.12\,V\angle14°$ **3.** 67.5° **5.** $4.83\,k\Omega\angle13.3°$ **7.** $Z_T = 3.95\,k\Omega$, $\theta=46.8°$, $I_T = 760\,\mu A$, $V_L = 2.19\,V$, $V_R = 2.05\,V$ **9.** $Z_T = 1.795\,k\Omega$, $\theta=73.5°$, $I_T = 4.46\,mA$, $V_{L1} = 6.29\,V$, $V_{L2} = 1.39\,V$, $V_R = 2.27\,V$ **11.** $Z_T = 3.72\,k\Omega$ to $7.8\,k\Omega$, $\theta=57.5°$ to $75.1°$, $V_L = 10.1\,V$ to $11.6\,V$, $V_R = 6.46\,V$ to $3.08\,V$ **13.** $P_R = 1.56\,mW$, $P_X = 1.66\,m\,VAR$ **15.** 4.77 mW **17.** $2.28\,mVA\angle46.8°$ **19.** $P_{APP} = 212\,m\,VA$, $\theta = 67.5°$, $P_R = 8.12\,\mu W$ **21.** 812 Ω **23.** $X_L = 377\,\Omega$, $Z = 406\,\Omega$, $\theta=68.3°$, $I = 296\,mA$, $V_L = 112\,V$, $V_R = 44.4\,V$, $P_X = 33\,VAR$, $P_R = 13.1\,W$, $P_{APP} = 35.5\,VA$ **25.** $58.3\,mA\angle-31°$ **27.** $1.6\,k\Omega\angle36.9°$ **29.** $2.91\,k\Omega\angle76°$ **31.** $X_L = 177\,\Omega$, $I_L = 677\,mA$, $I_R = 160\,mA$, $I_T = 696\,mA$, $\theta=-76.7°$, $Z_T = 172\,\Omega\angle76.7°$, $P_X = 81.3\,VAR$, $P_R = 19.2\,W$, $P_{APP} = 83.5\,VA$ **33.** $X_{L1} = 691\,\Omega$, $X_{L2} = 471\,\Omega$, $I_{L1} = 4.34\,mA$, $I_{L2} = 6.37\,mA$, $I_{LT} = 10.7\,mA$, $I_R = 5.88\,mA$, $I_T = 12.2\,mA$, $\theta=-61.2°$, $Z_T = 246\,\Omega\angle61.2°$, $P_{X1} = 13\,mVAR$, $P_{X2} = 19.1\,mVAR$, $P_{XT} = 32.1\,mVAR$, $P_R = 17.6\,mW$, $P_{APP} = 36.7\,mVA$ **35.** When $f = 1$ kHz: $I_T = 50.6\,mA\angle-70.8°$, $Z_T = 59.3\,\Omega\angle70.8°$. When $f = 2.5$ kHz: $I_T = 25.35\,mA\angle-48.9°$, $Z_T = 118\,\Omega\angle48.9°$ **37.** $(706 + j2.82k)\Omega$ **39.** $X_L = 1.89\,k\Omega$, $Z_P = 1.3\,k\Omega$, $\theta=43.6°$, $R_S = 941\,\Omega$, $X_S = 1.3\,k\Omega$, $R_T = 1.27\,k\Omega$, $Z_T = 1.82\,k\Omega\angle45.7°$, $I_T = 5.49\,mA$, $V_S = 10\,V\angle45.7°$, $V_{R1} = 1.81\,V$, $V_P = 7.14\,V\angle43.6°$, $I_L = 3.78\,mA\angle-46.4°$, $I_{R2} = 3.97\,mA\angle43.6°$, $P_{APP} = 54.88\,mVA\angle45.7°$, $P_R = 38.33\,mW$ **41.** 66.7 μs **43.** 667 ns **45.** 17.9 mA, 18.7 mA **47.** 68.1 ns **49.** 182 Ω

**Chapter 15:**

**1.** 3 V **3.** 960 V **5.** 960 mA **7.** 900 V **9.** 60 V **11.** 16.7 Ω **13.** 48 Ω **15.** $V_S = 10\,V$, $I_S = 1\,A$, $I_P = 83.3\,mA$, $Z_P = 1.44\,k\Omega$, $P_P = 10\,W$ **21.** 75 kΩ, 1W

**Chapter 16:**

**1.** 8000 μF **3.** 396 μC **5.** 9.3 μF **7.** 329 pF **9.** 55 μF **11.** 80 kΩ **13.** 10.6 kΩ **15.** 15.9 kΩ **17.** 106 Ω **19.** not acceptable **21.** 1885 **23.** 18,850 **25.** mica

**Chapter 17:**

**1.** $12.65\,V\angle-71.57°$ **3.** −67.5° **5.** $429\,\Omega\angle-69.53°$ **7.** $Z_T = 984\,\Omega$, $I_T = 8.13\,mA$, $\theta=-40.32°$, $V_C = 5.18\,V$, $V_R = 6.1\,V$ **9.** $Z_T = 6.52\,k\Omega$, $I_T = 1.38\,mA$, $\theta=-81.2°$, $V_{C1} = 1.56\,V$, $V_{C2} = 7.33\,V$, $V_R = 1.38\,V$ **11.** For $f = 100$ Hz: $Z_T = 72.34\,k\Omega$, $\theta=-89.3°$, $I_T = 194\,\mu A$, $V_C = 13.998\,V$, $V_R = 176\,mV$. When $f = 12$ kHz: $Z_T = 1.09\,k\Omega$, $\theta=-33.5°$, $I_T = 12.8\,mA$, $V_C = 7.74\,V$, $V_R = 11.68\,V$ **13.** $P_X = 42.1\,mVAR$, $P_R = 49.6\,mW$ **15.** $65.1\,mVA\angle-40.3°$ **17.** $P_{APP} = 212\,mVA$, $\theta=-67.5°$, $P_R = 8.12\,\mu W$ **19.** 812 Ω **21.** $X_C = 6.24\,k\Omega$, $Z_T = 13.5\,k\Omega$, $\theta=-27.5°$, $I_T = 739\,\mu A$, $V_R = 8.87\,V$, $V_C = 4.61\,V$, $P_R = 6.56\,mW$, $P_X = 3.41\,mVAR$, $P_{APP} = 7.39\,mVA$ **23.** $65\,mA\angle22.6°$ **25.** 179 Ω **27.** $236\,\Omega\angle-44.3°$ **29.** $X_C = 884\,\Omega$, $I_C = 2.26\,mA$, $I_R = 1.33\,mA$, $I_T = 2.62\,mA$, $\theta=59.5°$, $Z_T = 763\,\Omega\angle-59.5°$, $P_X = 4.52\,mVAR$, $P_R = 2.65\,mW$, $P_{APP} = 5.24\,mVA$ **33.** $(169 - j165)\Omega$ **35.** $X_C = 637\,\Omega$, $Z_P = 537\,\Omega\angle-57.5°$, For the series equivalent of $Z_P$: $R_S = 288\,\Omega$ and $X_C = 453\,\Omega$, $R_T = 438\,\Omega$, $Z_T = 630\,\Omega\angle-45.9°$, $I_T = 28.6\,mA$, $V_S = 18\,V\angle-45.9°$, $I_C = 24.1\,mA\angle32.48°$, $I_{R2} = 15.3\,mA\angle-57.5°$ **37.** 1.8 ms **39.** 3.3 ms **41.** 8.18 V, 2.92 V **43.** 337 μs

**Chapter 18:**

**1.** $7\,V\angle90°$ **3.** $120\,\Omega\angle90°$ **5.** $400\angle90°$ **7.** $V_L = 4\,V\angle90°$, $V_C = 2.75\,V\angle-90°$, $V_S = 1.25\,V\angle90°$ **9.** $V_L = 2.36\,V\angle90°$, $V_C = 1.27\,V\angle-90°$, $V_S = 1.09\,V\angle90°$ **11.** $47\,mA\angle-90°$ **13.** $450\,\Omega\angle-90°$ **15.** 1.59 kHz **17.** $X_L = X_C = 217\,\Omega$ **19.** $V_L = 120\,V\angle90°$, $V_C = 120\,V\angle-90°$, $V_{LC} = 0\,V$ **21.** $869\,\Omega\angle78°$ **23.** $10.6\,V\angle41.2°$ **25.** $X_L = 236\,\Omega$, $X_C = 9.04\,k\Omega$, $X_S = 8.8\,k\Omega\angle-90°$, $Z_T = 8.83\,k\Omega\angle-85.1°$, $I = 453\,\mu A$, $V_R = 340\,mV$, $V_L = 107\,mV$, $V_C = 4.095\,V$ **27.** $38.5\,mA\angle-62.1°$ **29.** $X_L = 314\,\Omega$, $X_C = 3.18\,k\Omega$, $I_L = 50.96\,mA$, $I_C = 5.03\,mA$, $I_{LC} = 45.9\,mA\angle-90°$, $I_R = 10.7\,mA$, $I_T = 47.1\,mA\angle-76.9°$, $Z_T = 340\,\Omega\angle76.9°$ **31.** $I_{LC} = 3.85\,mA\angle-90°$, $I_R = 6.06\,mA$, $I_T = 7.18\,mA\angle-32.4°$, $V_L = 1.06\,V$, $V_C = 3.06\,V$, $V_R = 2\,V$, $Z_T = 279\,\Omega\angle32.4°$ **33.** $R + X_{C2} = (510 - j723)\Omega = 885\,\Omega\angle-54.8°$, $Z_P = (100 + j380)\Omega$, $= 393\,\Omega\angle75.2°$, $Z_T = (100 - j344)\Omega = 358\,\Omega\angle-73.7°$, $I_T = 27.9\,mA$, $V_S = 10\,V\angle-73.7°$, $V_{C1} = 20.2\,V$, $V_P = 10.96\,V\angle75.2°$, $I_L = 37.1\,mA\angle-14.8°$, $I_R = 12.4\,mA\angle130°$, $V_R = 6.32\,V\angle130°$, $V_{C2} = 8.96\,V\angle40°$

**Chapter19:**

**1.** a. 83 kHz, 20.98 kHz b. 1.16 kHz, 1.07 kHz c. 1.12 MHz, 626 kHz d. 240 Hz, 134 Hz **3.** a. 12.8 kHz, 3.2 kHz b. 20 kHz, 15 kHz c. 2 kHz, 20 kHz **5.** 6.67 **7.** 7.65 kHz **9.** 95.3 kHz **11.** 52.1 kHz **13.** 4.13, 332 kHz **15.** 292.4 kHz, 372.4 kHz **17.** a. 40, 16 dB b. 6.25, 7.96 dB c. 0.25, −6.02 dB d. 0.05, −13 dB **19.** 664 mW **21.** 1.27 W **23.** 655 mW **25.** a. 40.8 dBm b. 22.6 dBm c. 33.98 dBm d. −5.23 dBm **27.** a. 2.24 mW b. 15.9 mW c. 759 μW d. 447 μW **29.** a. 41.6 dB b. 32 dB c. −0.04 dB d. −0.19 dB **31.** a. 1.15 b. 1.59 c. 0.66 d. 0.75 **33.** 0.936 **35.** 0.91 **37.** 1.74 kHz **39.** 11.9 kHz **41.** $A_{v(max)} = 0.939 = -0.55$ dB, $f_c = 98.8$ kHz **43.** 0.984 **45.** 0.99 **47.** 509 Hz **49.** 15.4 kHz **51.** $A_{v(max)} = 0.9991 = -0.008$ dB, $f_c = 414$ kHz **53.** 2.32 kHz **55.** 136 kHz **57.** 5.98 kHz **59.** 8.26 kHz **61.** 19.1 kHz, 27.4 kHz **63.** 38.4 kHz, 62.3 kHz **65.** 1.95

# Index


Absorption, dielectric, 657–58
Ac current and circuits, 379
alternations and cycles in, 381–82
apparent power in, 506–8
with capacitors, 638–43
coupling in, 597, 599, 639–40
cycle time and frequency in, 382–85
vs. dc, 40
dc offsets with, 424–25
full-cycle averages in, 391
half-cycle averages in, 391–92
instantaneous values in, 390–91
magnitude-related measurements in, 397–402
operation of, 379–81
oscilloscope measurements for, 385–88
peak and peak-to-peak values in, 389–90
power in, 393–95, 397
rectangular waves in, 428–33
rms values in, 395–97
sawtooth waves in, 433–34
sine waves in. *See* Sine waves
square waves in. *See* Square waves
static and dynamic values in, 423–24
triangular waves in, 435
Active components, 12
Addition
of complex numbers, 463–64
with polar notation, 458–59
Adjustable capacitors, 659–61
Adjustable resistors. *See* Potentiometers
Admittance, 559, 904–5
in RC circuits, 908
in RL circuits, 905–8
Aerosol coolant, 662
AF (audio-frequency) transformers, 598
Aging components, 152
Air
breakdown voltage rating of, 57
permeability of, 358
Air-core inductors, 511
Air-core transformers, 597–98
Air gaps in magnetism, 371
Alkaline batteries, 77
Alphanumeric resistor codes, 881
Alter-and-add method, 320–23
Alternating current (ac). *See* Ac current and circuits
Alternations
in ac circuits, 381–82
oscilloscope displays, 388
Aluminum
conductivity of, 49
disadvantages of, 55
resistivity rating of, 47
American Wire Gauge (AWG), 55–56
Ammeters, 97
Ampacity, 55–56
Ampere-hours, 76
Ampere-meters, 359
Ampere-turns, 366–68
Amperes, 38
Amperes per weber, 362
Amplification, 156
Amplifiers
audio, 156
dc offsets in, 424–25
Amplitude measurements, 813–14
bel power gain, 806–8
current gain, 818
dB power gain, 808–13
voltage gain, 814–18
Analog meters
magnetic shielding for, 370–71
ring magnets in, 371–72
Angles
phase. *See* Phase angles and relationships
in polar notation, 456–59
of vectors, 454
Angular velocity
in capacitive reactance calculations, 651
in inductive reactance calculations, 503
of rotors, 417–18
Apparent power
in ac circuits, 506–8
in RC circuits
parallel, 700
series, 688–90
series-parallel, 708
in RL circuits, 541–42
in RLC circuits, 771
Atomic number, 31
Atoms, 30–32
in conductors, 45–46
and magnetism, 359–60
in semiconductors, 46
Atoms per unit volume, 45–46
Attenuation, 790–91
Atto- prefix, 24
Attraction
and repulsion, 33–34
of unlike magnetic poles, 353–54
Audio amplifiers, 156
Audio-frequency transformers, 598
Audio-taper potentiometers, 71
Autotransformers, 598, 621–23
Average ac power, 394–95, 397
Average breakdown voltage, 57
Average current, 401–2
Average frequency of filters, 799–802
Average values of rectangular waves, 432–33
Average voltage, 401–2
Averages for center frequency, 794–95
AWG (American Wire Gauge), 55–56

Balanced Wheatstone bridges, 234
Band-stop filters. *See* Notch filters
Bandpass filters, 792–93, 842–43
average frequency of, 799–802
bandwidth of, 796–97, 850–52, 854–55
center frequency for, 796–97
cutoff frequencies for, 842–43, 852–55
LC, 843
series, 846–48, 850–53
shunt, 848–49, 853–57
maximum voltage gain in, 847–49, 855–56
Q effects in, 856–57
quality of, 797–98
Bands on resistors
color codes, 62–67
dark, 152

Bandwidth, 793–94
  of bandpass filters, 796–97, 850–52, 854–55
  and filter quality, 797
  of notch filters, 796–97
Batteries
  capacity of, 76
  cells in, 75–76
  connecting, 79–80
  current in, 41–42, 79
  schematic symbols for, 13
  types of, 77–78
Bel power gain, 806–8
Binomial expressions, 465
Biomedical systems, 4
Black patches in stressed resistors, 152
Black resistor band, 63
Bleeder current, 229–30
Blue resistor band, 63
Boards, PC
  solder bridges on, 125
  traces on, 56–57, 154
Bode plots
  for high-pass filters, 837–38
  for low-pass filters, 824–25, 833–34
Bohr model, 30
Branch analysis. 179–81, 330. *See also* Parallel circuits
  equations for, 330–32
  example problems, 333–36
  observation method in, 332–33
  in RC circuits, 694–96
  in RL circuits, 546–49
Breadboards, 90–97
Breakdown voltage ratings
  for capacitors, 653–54
  for fuses, 86–87
  for insulators, 57
Bridge circuits
  LCR, 496–97
  Thevenin's theorem for, 287–88
  Wheatstone bridge, 233–38
Brown resistor band, 63, 66–67
Buffers, transformers for, 617–18

Calculators
  modes in, 17
  natural logs, 573
  order of operations, 20
  polar and rectangular notation, 473
  radian mode, 419–20
  resistance calculations, 183–84
  scientific notation, 16–17, 19–20
Capacitance, 9, 633, 635–36
  oscilloscope input, 753–54
  of parallel capacitors, 647–48
  of series capacitors, 644–47
  stray, 753
  units for, 636–37
Capacitive reactance
  calculating, 650–51
  and Ohm's law, 651–52
  resistance with, 649–50, 653
  series and parallel values of, 652–53
Capacitors, 9
  ac voltage and current characteristics of, 638–43
  capacity of, 635–36
  charging of, 634
  construction of, 633
  coupling by, 639–40
  discharging of, 635
  electrolytic, 658–59
  fixed-value, 657–59
  dc isolation provided by, 640
  leakage current of, 654
  leaky, 661–62
  in parallel, 647–48
  phase relationships for, 641–43
  physical characteristics of, 637–38
  quality rating of, 654–57
  resistance of, 649–50
  schematic symbols for, 633
  in series, 644–47
  as series-opposing voltage sources, 164
  as shunt components, 819
  susceptance of, 903–4
  tolerance ratings for, 637
  variable, 659–61
  voltage ratings for, 653–54
Capacity
  of batteries, 76
  of capacitors, 635–36
Carbon
  conductivity of, 49
  resistivity rating of, 47
Carbon-composition potentiometers, 71–72
Carbon-composition resistors, 8
  construction of, 58–59
  power ratings of, 67–68
Carbon-zinc batteries, 77
Careers, 2–3
Cascades, gain of, 811–12
Catastrophic failures, 152
Cathode-ray tubes (CRTs), 10–11
Cells in batteries, 75–76
Center frequency, 794–95
  in bandpass and notch filters, 796–97
  and filter quality, 797
  on logarithmic scales, 803–4
Center-tapped transformers, 619–20
Ceramic capacitors, 657
Charge per unit time for current, 38
Charges, 32–33
  for attraction and repulsion, 33–34
  on ions, 33, 35
Charging of capacitors, 634
Chassis grounds, 164–65
Chokes, 512. *See also* Inductors
Circuit analysis
  current sources, 264
  delta and wye circuits, 300–4
  equivalent voltage and current sources, 264–66
  Millman's theorem for, 297–300
  Norton's theorem for. *See* Norton's theorem
  Ohm's law for, 108–10, 120–21
  simultaneous equations for. *See* Simultaneous equations
  source conversions, 266–68
  superposition, 256–63
  Thevenin's theorem for. *See* Thevenin's theorem
  voltage sources, 263–64
Circuit breakers, 89–90
Circuits, 7–8
  calculation problems for, 120–21
  construction of, 90–97
  loads on, 122
  multiload, 285–87
  multisource, 256–63
    mesh current analysis for, 336–43
    Thevenin's theorem for, 282–84
  open. *See* Open circuits
  protecting. *See* Fuses
  short. *See* Short circuits
Circular-mils, 47
Closed loops, 140
Closed switches, 83
Coefficients
  coupling, 493–94, 496–97
  in determinants, 323–27
  temperature, 49, 59
Coercive force, 370
Coils. *See also* Inductors; Transformers
  ampere-turns in, 366–68
  coefficient of coupling for, 493–94, 496–97
  mutual inductance in, 492–93
  parallel-connected, 498–501
  resistance of, 501–2, 506
  series-connected, 494–96
Color codes
  resistor, 62–67, 879–80
  transformer, 624, 880–81
Combination circuits. *See* Series-parallel circuits
Common terminals in dc power supplies, 82
Commons for voltage references, 142
Communications systems, 4
Complete shells, 32
Complex numbers, 462
  adding and subtracting, 463–64
  dividing, 465–67
  imaginary numbers, 467–68
  multiplying, 465
Components. *See* Electronic components
Computer systems, 4

Condensers. *See* Capacitors
Conductance, 43–44
  in parallel circuits, 182–83
    RC, 908
    RL, 905–8
Conductivity, relative, 49
Conductors, 45–46
  breadboard, 91
  loop, 405–6
  PC board traces, 56–57
  resistance of, 47–49
  wire, 54–56
Conjugates, 465–67
Connections
  battery, 79–80
  breadboard, 91
  capacitor, 644–48, 652–53
  coil, 494–96, 498–501
  reactances, 505, 652–53
  voltage source, 160–64
Conservation of energy, 485
Continuous cycles in ac circuits, 381–82
Control shafts, 70, 73
Converting
  delta and wye circuits, 301–4
  to engineering notation, 22–24
  Norton and Thevenin equivalents, 295–97
  to scientific notation, 15–16
  to standard form
    dB power gain, 809–10
    dB voltage gain, 816–17
    scientific notation, 18–19
  temperature, 877
  units, 874–77
  vector notation, 468–74
  voltage and current sources, 266–68
Coolant, 662
Coordinate notation, 460
Copper
  advantages of, 54–55
  atoms of, 31–32
  for battery terminals, 75–76
  conductivity of, 49
  as conductor, 45–46
  free electrons in, 37
  resistivity rating of, 47
Copper losses in transformers, 604, 607
Cores
  coil, 363
  transformer, 597–98, 606
Cosines
  calculating, 469–71
  in power factor calculations, 542–44, 555
Coulombs, 38, 113
Coulomb's law, 33–34
Counter emf, 484–85
Coupled components, 493
Coupling
  by capacitors, 639–40
  by transformers, 597, 599
Coupling coefficient for coils, 493–94, 496–97
Cracks in stressed resistors, 152
Cross-sectional area
  as inductance factor, 487–88, 492
  as reluctance factor, 362
  as resistance factor, 47–48
  of wire, 55–56
CRTs (cathode-ray tubes), 10–11
Current, 8
  ac. *See* Ac current and circuits
  average, 401–2
  in battery circuits, 41–42, 79
  bleeder, 229–30
  with capacitors, 638–43
  in circuit calculation problems, 120–21
  dc. *See* Dc current and circuits
  eddy, 605, 607
  in electromagnetism. *See* Electromagnetism
  *conventional* vs. electron flow, 39–40
  electrons in, 37–39
  induced, 404–5, 484–85
  in inductance, 481–82
  and inductive reactance, 504–5
  in inductors, 489–91
  Kirchhoff's current law, 187–88
  in LC circuits
    parallel, 742–45, 747–48
    phase relationships in, 743–45
    resonant, 844–46
    series, 735, 739–41
  leakage, 654
  in magnetism, 361
  measuring, 98
  mesh, 336–43
  in node voltage analysis, 345
  Norton, 290–91
  in Ohm's law, 14, 106–12, 330–32
  in open circuits, 123
    parallel, 195–96
    series, 147–48
  in parallel circuits, 179–81, 194–96, 212–15
  peak and peak-to-peak, 400–401
  in power calculations, 113–16
  in RC circuits
    parallel, 693–96, 699–700
    phase relationships in, 699–700
    series, 672–73, 680–84
    series-parallel, 707–9
    square wave response to, 710–12
  and resistance, 107
  and resistor tolerance, 121–22
  in resonant circuits, 844–46
  in RL circuits
    parallel, 546–49, 552–56
    phase relationships in, 552–53
    series, 523–24, 531–32, 545
    series-parallel, 565–67
  in RLC circuits
    parallel, 766–68, 770
    series, 764–65
    series-parallel, 773–80
  in series circuits, 134–36, 146–48, 151–52, 208–12
  with series-connected voltage sources, 160–64
  in short circuits, 123–25
    parallel, 196
    series, 151–52
  in transformers, 607–16, 619
  universal (RL switching) curve for, 572–74
  and voltage, 42, 106–7, 110–12
Current capacity
  of PC board traces, 55–56
  of wire, 55–56
Current dividers, 190–93
Current gain, 818
Current-sensing resistors, 552–54
Current sources, 188–90
  analysis of, 264
  converting, 266–68
  equivalent voltage sources, 264–66
  schematic symbols for, 13
Current surges, fuses for, 88
Current transitions in square waves, 568
Cutoff frequencies, 792–94
  for bandpass filters, 842–43
    series, 852–53
    shunt, 854–55
  calculations for, 795–96, 801
  in filter descriptions, 796–97
  for high-pass filters
    RC, 836–37
    RL, 841–42
  on logarithmic scales, 804
  for low-pass filters
    RC, 821–24
    RL, 832
  for notch filters, 842–43
Cycle time
  in ac circuits, 382–85
  on oscilloscope displays, 388
  of rectangular waves, 429
Cycles
  in ac circuits, 381–82
  duty cycle, 429–30
  on oscilloscope displays, 388

dB current gain, 818
dB power gain, 808–9
  benefits of using, 810–11
  converting to standard numeric form, 809–10
  dBm reference for, 812–13
  for multistage filters, 811–12

dB voltage gain, 814–16
changes in, 817–18
converting to standard numeric form, 816–17
reciprocal relationships in, 817
dBm reference, 812–13
Dc current and circuits, 40
isolation of
by capacitors, 640
by transformers, 597, 599, 622
offset voltages
in ac circuits, 424–25
capacitor blocking of, 640
power in, 397
Dc power supplies
for breadboards, 91–92
chokes in, 512
electrolytic capacitors in, 658
outputs in, 80–82
polarity in, 82
transformers in, 620–21
using, 82–83
Decade boxes, 73–74
Decade scales, 802–4
Decades, 428, 803
Decay and decay curves
for RC circuits, 715–16, 718–19
for RL circuits, 571, 574–75, 578–79
Decibel (dB) current gain, 818
Decibel (dB) power gain, 808–9
benefits of using, 810–11
converting to standard numeric form, 809–10
dBm reference for, 812–13
for multistage filters, 811–12
Decibel (dB) voltage gain, 814–16
changes in, 817–18
converting to standard numeric form, 816–17
reciprocal relationships in, 817
Decimal points in scientific notation, 15
Delays
in digital circuits, 581–83
for fuses, 88
in square wave transitions, 568–69
Delta circuits, 300–304
Density, flux, 355–57
ampere-turns in, 367
residual, 370
Derivation equations, 642–43, 893–902
Determinant matrices, 324, 327
Determinants, 323–27
Diamagnetic materials, 362
Dielectric in capacitors, 633
absorption by, 657–58
oil, 661
thickness of, 638, 646–47
Difference of potential, 41–42
Digital circuits, time calculations for, 581–83
Digital multimeters (DMMs), 5, 97–98
for current, 98
internal resistance in, 149
loading by, 231–32
for resistance, 99–100
for voltage, 99
Diodes, 402, 884
DIP (dual in-line package) switches, 85
Direct current (dc). *See* Dc current and circuits; Dc power supplies
Direction
of ac current, 379
of vectors, 454
Discharging of capacitors, 635
Discoloration of stressed resistors, 152
Distributed capacitance, 753
Dividers
current, 190–93
voltage. *See* Voltage dividers
Division
of complex numbers, 465–67
with polar notation, 457–58
DMMs. *See* Digital multimeters (DMMs)
Domain theory, 359–62
Double-pole, double-throw (DPDT) switches, 84
Double-pole, single-throw (DPST) switches, 84
Double-throw switches, 83–84
Dropping magnets, 372
Dry cells, 76
Dual in-line package (DIP) switches, 85
Dual-trace oscilloscopes
measuring phase angles, 412–13
in RC circuits, 676–77
in RL circuits, 527–29, 552–53
for power factor calculations, 691
Duty cycle, 429–30
Dynamic values in ac circuits, 423–24

Earth grounds, 164–65
Eddy currents, 605, 607
EETs (electronics engineering technicians), 3
Effective values, measuring, 402
Efficiency
power, 117–18
transformer, 604, 606
Elasticity of aluminum, 55
Electricity
charges in, 32–35
elements in, 30–37
free electrons in, 35–36
Electrodes in battery cells, 75–76
Electrolytes in battery cells, 75–76
Electrolytic capacitors, 658–59
Electromagnetic induction
laws of, 482–83
in transformer operation, 599–600
Electromagnetic waves, 425–27
Electromagnetism, 363
ampere-turns in, 366–68
and coils, 363, 365
hysteresis in, 368–70
hand rule for, 363, 365
magnetomotive force in, 364, 366
Electromotive force (emf), 41
Electronic components, 7–8, 12–14
aging and stress, 152
capacitors, 9
inductors, 9–10
integrated circuits, 11–12
resistors, 8
semiconductors, 10, 12
stray inductance from, 752
vacuum tubes, 10–11
Electronic systems, 4–7
Electronic test equipment, 5–7
Electronics, 2
Electronics engineering technicians (EETs), 3
Electronics engineers, 2
Electronics technicians, 2–3
Electrons, 30
in current, 37–39
in domain theory, 359–60
flow of, 39–40
free, 35–36
Elements, 30
Emf (electromotive force), 41
Energy. *See also* Power
conservation of, 485
measurements of, 118
Engineering notation
converting to, 22–24
using, 21–22
Engineers, 2
Equal-value branches method
for capacitance, 645
for impedance, 550, 697
for inductance, 499
for resistance, 184–85
Equations
derivation, 642–43, 893–902
simultaneous. *See* Simultaneous equations
Equivalent circuits
Norton's theorem for, 292–93
parallel and serial impedances, 562–65
series-parallel, 217–20, 269–71
Thevenin's theorem for, 274–78
Equivalent voltage and current sources, 264–66
Exa- prefix, 24
Exponential curves
in RC circuits, 714–15
in RL circuits, 573–74

Facsimile (fax) machines, 4
Faraday, Michael, 482
Faraday's laws of induction, 482–84

Farads, 9, 636
Fast-acting fuses, 88
Faults
  open circuits, 123
    in parallel circuits, 195–96
    in series circuits, 147–51
    in series-parallel circuits, 240–41
  short circuits, 123–25
    in parallel circuits, 196
    in series circuits, 151–52
    in series-parallel circuits, 241–42
  in transformers, 623–24
Femto- prefix, 24
Ferromagnetic materials, 362
Field-service representatives (FSRs), 3
Figure of merit for inductors, 508–10
Filter chokes, 512
Filters, 790, 792–93
  average frequency of, 799–802
  bandpass. *See* Bandpass filters
  in dc power supplies, 620–21
  high-pass. *See* High-pass filters
  low-pass. *See* Low-pass filters
  multistage, 811–12
  notch. *See* Notch filters
  quality of, 797–98, 849–51, 853–54
Fire danger in fuse replacement, 87, 196
Fixed-value capacitors, 657–59
Fixed-value resistors, 8, 58
Floating grounds, 164–65
Flux
  and Faraday's laws, 482–84
  and Rowland's law, 364, 366
  units of measure for, 354–55
Flux density, 355–57
  ampere-turns in, 367
  residual, 370
Free electrons, 35–37
Frequency
  in ac circuits, 382–85
  cutoff. *See* Cutoff frequencies
  and harmonics, 427–28
  in inductive reactance calculations, 502–4
  oscilloscope measurements for, 385–88
  resonant. *See* Resonant frequency
  scales for, 802–4
  and wavelength, 427
Frequency counters, 5–6
Frequency response
  amplitude measurements for, 806–18
  attenuation, 790–91
  average frequency, 799–802
  bandwidth, 793–94
  calculations for, 795–96
  and capacitor quality rating, 656–57
  center frequency, 794–97, 803–4
  cutoff frequencies, 792
  of filters, 792–93. *See also* Filters
  Q rating, 510
  of RC circuits
    parallel, 700–702
    series, 684–87
  of RL circuits
    parallel, 557–59
    series, 536–38
  of RLC circuits
    parallel, 769
    series, 761–62
Frequency response curves, 791–92
  frequency descriptions of, 796–97
  voltage gain on, 814–15
FSRs (field-service representatives), 3
Full-cycle average of waveforms, 391
Full loads, 122
Full-wave rectifiers, 401–2
Function generators, 5–6
Fundamental frequencies, 427, 430–31
Fuses, 85–86
  in parallel circuits, 197–98
  ratings for, 86–87
  replacing, 87
  schematic symbols for, 13, 86
  types of, 88–89

Gain
  dB current, 818
  dB power, 808–13
  dB voltage, 814–18
  filter. *See* Voltage gain
Gallium arsenide, 46
Galvanometers, 233–34
Gang-mounted potentiometers, 73–74
Ganged capacitors, 660–61
Generators, hand rule for, 405–6
Geometric averages for center frequency, 794–95
Geometric sums in RL circuits, 524
Germanium, 46
Giga- prefix, 21–22
Gilberts per maxwell, 362, 366
Glass, breakdown voltage rating of, 57
Glossary, 885–92
Gold
  conductivity of, 49
  resistivity rating of, 47
Gold resistor band, 64, 66
Graphite
  conductivity of, 49
  as semiconductor, 45–46
Graphs for alternating current, 379
Gravity, 33
Gray resistor band, 63
Green resistor band, 63
Ground-loop paths, 142, 165
Grounds
  on center-tapped transformers, 619–20
  earth vs. chassis, 164–65
  as voltage references, 142

Half-cycle average of waveforms, 391–93
Half-wave rectifiers, 401–2
Harmonic series, 427
Harmonic synthesis, 430–31
Harmonics, 427–28
Heat
  and aluminum, 55
  free electrons from, 37
  and magnets, 372
  and power, 116–17
  and resistance, 49
  resistor dissipation of, 59, 116–17
Helium atoms, 30–31
Henries, 9, 486, 488
Hertz (Hz), 383
High-pass filters, 792–93
  cutoff frequencies for, 796–97
  RC, 835–39
  RL, 840–42
High-speed instantaneous fuses, 88
High voltage capacitors, 662
Hole flow, 39
Hydrogen atoms, 31
Hysteresis, 368–70
Hysteresis curves, 368–70
Hysteresis loss, 605–7
Hz (hertz), 383

ICs (integrated circuits), 11–12
Ideal current sources, 264
Ideal voltage sources, 156, 263–64
Imaginary numbers, 467–68
Impedance
  vs. admittance, 559
  in LC circuits, 844–46
  matching, transformers for, 617–18, 622–23
  power factor calculations in, 544
  in RC circuits
    parallel, 693–94, 696–700
    series, 672–73, 678–84
  and resistance and reactance, 505
  in RL circuits
    parallel, 546–47, 549–51, 554–56
    series, 523–24, 529–30, 532–36, 545
    series-parallel, 706–7
  in RLC circuits, 770
    parallel, 770
    series, 758–61, 764–65
    series-parallel, 561–67, 773–80
  in transformer circuits, 613–16
In-house technicians, 3
Incomplete shells, 32
Index, here
Inductance, 9, 481
  of coils, 495, 498–99
  current and magnetism in, 481–82
  mutual, 491–93, 500
  in RL circuits, 575–76
  stray, 752

Induction
of current, 404–5, 484–85
Faraday's laws of, 482–84
Lenz's law of, 484–85
and magnetism, 360–61
and transformer operation, 599–600
of voltage, 484–88
Inductive-capacitive circuits. *See* LC circuits
Inductive reactance, 501–2
calculating, 502–4
frequency response of, 536–38, 557
and Ohm's law, 504–5
serial and parallel, 505
Inductors, 9–10, 481
and apparent power, 506–8
chokes, 512
coefficient of coupling for, 493–94, 496–97
induced voltage across, 486–88
iron-core and air-core, 511
parallel-connected, 498–501
phase relationships in, 489–91, 642–43
Q rating of, 508–10
resistance of, 501–2, 506
in RL circuits, 569–70
schematic symbol for, 486
series-connected, 494–96
susceptance of, 903–4
toroids, 511
Industrial systems, 4
Input capacitance, oscilloscope, 753–54
Input current, transformer, 607–13
Input power in efficiency measurements, 117–18
Instantaneous fuses, 88
Instantaneous values
in ac circuits, 390–91
in sine waves, 413–18, 421–22
Insulators
characteristics of, 45–46
ratings for, 57
Integrated circuits (ICs), 11–12
Integrated resistors, 8, 60
Intensity of current, 38
Interference, shielding for, 622–23
Interleaved-plate capacitors, 659–61
Internal resistance
in meters, 148–50
of sources, 156–57
International system of units, 877–78
Internet, 4
Inverse-log functions, 809
Inverse-tangent function, 472–73
Ions, 33, 35
Iron
conductivity of, 49
permeability of, 358
resistivity rating of, 47
retentivity of, 361
Iron-core inductors, 511
Iron-core transformers, 597–98
Isolation
by capacitors, 640
by transformers, 597, 599, 610, 622
Isolation transformers, 599, 610

J-operator
for imaginary numbers, 467–68
for vectors, 461–63
Joules
in power definition, 113
in voltage definition, 42
Junctions, schematic symbols for, 13

Kilo- prefix, 21–22
Kilo-volts per centimeter rating, 57
Kilowatt-hours (kWh), 118
Kirchhoff, Gustav, 140
Kirchhoff's current law, 187–88
Kirchhoff's voltage law, 140–42
for branch current, 330–32
for capacitance, 646
for inductor voltage, 569–70
for multisource circuits, 260–62
for node voltages, 344–45
for phase angles, 531
for RC circuits
series, 679
square wave response to, 711
kWh (kilowatt-hours), 118

Lag
with capacitors, 642
for time delays, 88
Lagging current
in inductor phase relationships, 490–91
in RL circuits, 525–27, 547–49
Lagging voltages with capacitors, 642
Laminated transformer cores, 606
LC bandpass filters, 843
series, 846–48, 850–53
shunt, 848–49, 853–57
LC circuits
parallel, 742–49
resonant frequency in, 749–57, 843–46
series, 735–42, 754–55
LC notch filters, 843
series, 857–58
shunt, 858–59
LCR bridges, 496–97
Lead-acid batteries, 78
Leading current with capacitors, 642
Leading voltage
in inductor phase relationships, 490–91
in RL circuits, 525–27, 547–49
Leads
resistor, 58–59
stray inductance from, 752
Leakage current, 654
Leaky capacitors, 661–62
Left- (right-) hand rule
for electromagnetism, 363, 365
for generators, 405–6
Length
as flux density factor, 367
as inductance factor, 487–88, 492
as reluctance factor, 362
as resistance factor, 47–48
of sine waves, 425–27
Lenz, Heinrich, 484
Lenz's law, 484–85
Lifted traces on PC boards, 154
Light, speed of, 426
Like poles, 353–54
Linear changes, 571
Linear-taper potentiometers, 71
Lithium-iodine batteries, 77–78
Load power in ac circuits, 393–95
Load voltage ranges, Thevenin's theorem for, 279–80
Loaded Q of filters, 850–51, 853–54
Loaded voltage vs. no-load voltage, 225–26
Loads
analysis of. *See* Norton's theorem; Thevenin's theorem
circuit, 122
and maximum power transfer theorem, 158–60
open, 230
in series-parallel circuits, 226–27, 230
transformer, 610–18
voltage divider, 224–25, 227–30
voltmeters as, 230–32
Logarithmic scales, frequency, 802–4
Logarithms
in amplitude measurements, 807–13
for exponential curves, 573
Logic symbols, 884
Loop conductors, 405–6
Low-pass filters, 792–93. *See also* RC low-pass filters; RL low-pass filters
bode plots for, 824–25, 833–34
cutoff frequencies for, 796–97
roll-off rates for, 825–28, 833–34
Lower cutoff frequencies for high-pass filters
RC, 836–37
RL, 841–42

Magnetic fields
in mutual inductance, 491–93
representation of, 355
Magnetic flux
and Faraday's laws, 482–84
and Rowland's law, 364, 366
units of measure for, 354–55
Magnetic flux density, 355–57
ampere-turns, 367
residual, 370

Magnetic force, 353
Magnetic induction of current, 404–5
Magnetic poles, 353–54
Magnetic-type circuit breakers, 90
Magnetism. *See also* Permeability in magnetism
  air gaps in, 371
  domain theory for, 359–62
  electromagnetism. *See* Electromagnetism
  flux density in, 355–57
  and inductance, 481–82, 484–85
  induction of, 360–61
  magnetic flux, 354–55
  magnetic poles, 353–54
  reluctance, 362
  retentivity, 361
  ring magnets, 371–72
  shielding, 370–71
  storing magnets, 372
Magnetomotive force (mmf)
  ampere-turns, 366–68
  and Rowland's law, 364, 366
Magnets
  ring, 371–72
  storing, 372
  temporary, 361
Magnitude
  of alternating current, 379
  in polar notation, 456–59
  of vectors, 454
Magnitude-related values in ac circuits, 397–402
Matrices for simultaneous equations, 323–29
Matrix equations, 324
Matter, 30
Maximum power transfer, 158–60
  Norton's theorem for, 293–95
  Thevenin's theorem for, 280–82
Maximum voltage gain in filters
  bandpass
    series, 847–49
    shunt, 855–56
  high-pass
    RC, 836, 839
    RL, 841
  low-pass
    RC, 820–21, 828–29
    RL, 830–33
  notch, 858–59
Maxwells, 354–55, 362, 366
Mean values in ac circuits, 395–97
Measuring. *See also* Meters
  magnitude-related ac values, 397–402
  phase and time, 408–11, 417–22
Mega- prefix, 21–22
Mercury batteries, 77
Mesh current analysis, 336–38
  equations for, 338–40
  three-loop problem, 340–43
Meshes, 336
Metal casings for transformer shielding, 622–23
Metal-film resistors, 8, 59
Meters, 5, 97–98
  for ac voltage, 401–2
  for current, 98
  galvanometers, 233–34
  internal resistance of, 148–50
  loading by, 230–32
  magnetic shielding for, 370–71
  for resistance, 99–100
  ring magnets in, 371–72
  service, 118
  for voltage, 99
  Wheatstone bridge circuits in, 237
Mhos, 43–44
Mica, breakdown voltage rating of, 57
Micro- prefix, 21–22
Micro-farads, 637
Micro-webers, 355
Milli- prefix, 21–22
Millman's theorem, 297–300
Mils, 47
Mismatches, impedance, 617–18
MMF (magnetomotive force)
  ampere-turns, 366–68
  and Rowland's law, 364, 366
Modes, calculator, 17
Momentary switches, 85
Multi-turn potentiometers, 72–73
Multiload circuits, 285–87
Multimeters, 5, 97–98
  for current, 98
  internal resistance in, 149
  loading by, 231–32
  for resistance, 99–100
  for voltage, 99
Multiple j-operations, 462–63
Multiple opens in series circuits, 150–51
Multiple outputs, transformers with, 620–21
Multiplication
  of complex numbers, 465
  with polar notation, 457–58
Multiplier bands on resistors, 62–67
Multisource circuits, 256–63
  mesh current analysis for, 336–43
  Thevenin's theorem for, 282–84
Multistage filters, 811–12
Mutual inductance, 491–93, 500
Mylar capacitors, 657

N-order harmonics, 428
N-stage filters, 811
Nano- prefix, 21–22
Natural logs, 573
NC (normally closed) switches, 84–85
Negative angles in polar notation, 457
Negative charges, 32–33
Negative-going transitions, 568
Negative ions, 35
Negative j-operations, 461
Negative peak values, 380–81, 389–90
Negative phase angles, 531–32, 549
Negative potential, 41–42
Negative temperature coefficients, 49
Negative voltages in dc power supplies, 82
Neutrons, 31
Nickel-cadmium batteries, 78
No-load output voltage, 156, 225–26
NO (normally open) switches, 84–85
Nodes, 343
  in current paths, 187
  voltage analysis of, 343–46
Nonlinear-taper potentiometers, 71
Nonsymmetrical waveforms, 431–33
Normal instantaneous fuses, 88
Normally closed (NC) switches, 84–85
Normally open (NO) switches, 84–85
North-seeking poles, 353
Norton's theorem, 289–90
  conversion to Thevenin, 295–97
  and equivalent circuits, 292–93
  and maximum power transfer, 293–95
  Norton current, 290–91
  Norton resistance, 291
Notch filters, 792–93, 842–43
  average frequency of, 799–802
  bandwidth and center frequencies for, 796–97
  LC, 843
    series, 857–58
    shunt, 858–59
  quality of, 797
Nucleus, 30
Number of turns as inductance factor, 487–88, 492
Numbers
  complex. *See* Complex numbers
  imaginary, 467–68

Observation method in branch current analysis, 332–33
Octave scales, frequency, 802–4
Octaves, 428, 803
Odd-order harmonics, 430–31
Oersted, Hans Christian, 363
Offset voltages
  in ac circuits, 424–25
  capacitor blocking of, 640
Ohm, Georg Simon, 106
Ohm-centimeter ratings, 47
Ohmmeters
  for capacitor tests, 649, 662
  for resistance, 97
Ohms, 8, 43–44
Ohm's law, 14, 106
  for branch current analysis, 330–32
  and capacitive reactance, 651–52
  for circuit calculations, 108–10, 120–21

Ohm's law, *continued*
and inductive reactance, 504–5
LC circuits, 736, 745–46
node voltage analysis, 344
parallel circuits, 181–82
phase angles, 531
power calculations, 115–16
predictions, 107–8
for RC circuits, 678–79, 711–12
relationships in, 106–7, 110–12
for transformer circuits, 611, 613
and troubleshooting, 111–12
Oil dielectrics, 661
Open circuits, 123
in parallel circuits, 195–96
in series circuits, 147–51
in series-parallel circuits, 240–41
Open loads in series-parallel circuits, 230
Open switches, 83
Open windings in transformers, 624
Operation order in calculators, 20
Orange resistor band, 63, 67
Orbital shells, 31–32
Order of operations in calculators, 20
Order-2 matrices, 323
Order-3 matrices, 323, 327–29
Oscilloscopes, 5, 7
for frequency measurements, 385–88
input capacitance of, 753–54
for peak and peak-to-peak voltages, 397–400
for phase angle measurements, 412–13
in RC circuits, 676–77
in RL circuits, 527–29, 552–53
for power factor calculations, 691
Out of phase waveforms, 411–13
Output current, transformer, 607–13, 619
Output power and efficiency measurements, 117–18
Output voltage ratings for transformers, 618–19
Output voltages
from dc power supplies, 80–82
no-load, 156, 225–26
from transformers, 601–3
Oxide insulators, 606

Paper, breakdown voltage rating of, 57
Parallel circuits
analysis of, 194
conductance of, 182–83
current through, 179–81, 194–96, 212–15
current dividers, 190–93
current sources in, 188–90
Kirchhoff's current law for, 187–88
LC. *See* Parallel LC circuits
open branches in, 195–96
power in, 185–86, 214
RC. *See* Parallel RC circuits
resistance in, 182–85, 212–15
RL. *See* Parallel RL circuits
RLC. *See* Parallel RLC circuits
vs. series circuits, 181–82
shorted branches in, 196
troubleshooting, 197–98
voltage across, 180–81, 212–15
Parallel connections
battery, 79–80
capacitor, 647–48
coil, 498–501
reactances, 505, 652–53
Parallel-equivalent circuits, 217, 219–20
Parallel LC circuits, 742–43
analysis of, 747–48
current through, 742–45, 747–48
phase relationships in, 742–45
reactance in, 745–48
resonant, 755–57, 844–46
Parallel RC circuits, 693–94
admittance and susceptance in, 908
analysis of, 699–700
branch currents in, 693–96
frequency response of, 700–702
impedance of, 693–94, 696–700
vs. parallel RL circuits, 702–3
phase angles in, 694–700
Parallel RL circuits
admittance and susceptance in, 905–8
analysis of, 554–56
characteristics of, 546–47
current through, 546–49, 552–53
frequency response of, 557–59
impedance of, 546–47, 549–51
vs. parallel RC circuits, 702–3
phase angles in, 549–53
Parallel RLC circuits, 765–66
analysis of, 770–71
current through, 766–68, 770
frequency response of, 769
power in, 771
Paramagnetic materials, 362
Parameters for insulators, 57
Passive components, 12
Passive filters, 790, 792–93
average frequency of, 799–802
bandpass. *See* Bandpass filters
in dc power supplies, 620–21
high-pass. *See* High-pass filters
low-pass. *See* Low-pass filters
multistage, 811–12
notch. *See* Notch filters
quality of, 797–98, 849–51, 853–54
PC boards
solder bridges on, 125
traces on, 56–57, 154
Peak-to-peak values in ac circuits and waveforms, 389–90
current, 400–401
full-cycle average, 391
half-cycle average, 391–92
voltage, 399–400
Peak values in ac circuits and waveforms, 380–81, 389–90
current, 400–401
sine waves, 415
voltage, 397–99
Periods in ac waveforms, 380–85
Permanent magnets, 372
Permeability in magnetism, 358–59
as inductance factor, 487–88, 492
of iron-core vs. air-core inductors, 511
relative, 359
as reluctance factor, 362
Permittivity of capacitors, 637
Peta- prefix, 24
PF (power factor). *See* Power factor (PF) calculations
Phase
in center-tapped transformers, 619–20
measuring, 408–11, 417–22
in sine waves, 407–8
and vectors, 454
Phase angles
in apparent power calculations, 541–42
for capacitors, 641–43
for inductors, 489–91, 642–43
in LC circuits
parallel, 742–45
series, 735–41
measuring, 412–13
in RC circuits
parallel, 694–700
series, 673–80
series-parallel, 703–4, 707–8
in RL circuits
parallel, 549–53
series, 524–29, 531–32
series-parallel, 564–65
in RLC circuits
parallel, 765–68, 770
series, 758, 762–64
series-parallel, 774, 776–78
between sine waves, 411–12
between transformer voltages, 603–4
Phasors
in parallel RC circuits, 694–95
in parallel RL circuits, 548–49
in series RC circuits, 675–76
in series RL circuits, 524–25
Pi circuits, 300–304
Pico- prefix, 21–22
Pico-farads, 637
Plastic film capacitors, 657–58
Plates in capacitors, 633, 637–38, 648
Plotting vectors, 454–55
Polar notation, 456–57
addition and subtraction with, 458–59
for admittance, 904–5

converting rectangular to, 472–73
converting to rectangular, 471–72
multiplication and division with, 457–58
for RC circuits, 704–8
rectangular equivalent representations, 468–69
vs. rectangular notation, 467
for RL circuits, 561–66
Polarity
in ac waveforms, 380–81
with center-tapped transformers, 619–20
in dc power supplies, 82
of induced voltage, 484–85
in mesh current equations, 338
of transformer voltages, 603–4
in voltage measurements, 136, 140–42
Polarized electrolytic capacitors, 658–59
Poles
air gaps between, 371
magnetic, 353–54
in switches, 83–84
Polycarbonate capacitors, 657–58
Polypropylene capacitors, 657–58
Polystyrene capacitors, 657–58
Positive angles, 457
Positive charges, 32–33
Positive-going transitions, 568
Positive ions, 35
Positive j-operations, 461
Positive peak values, 380–81, 389–90
Positive potential, 41–42
Positive temperature coefficients, 49
Positive voltages in dc power supplies, 82
Potassium atoms, 32
Potentiometers, 8, 69
carbon-composition vs. wire-wound, 71–72
construction and operation of, 70–71
decade boxes, 73–74
gang-mounted, 73–74
multi-turn, 72–73
ratings for, 71
taper of, 71
trimmer, 72
as voltage dividers, 155–56, 238–40
Power, 113
in ac circuits, 393–95, 397
apparent. *See* Apparent power
calculating, 113–16
efficiency, 117–18
energy measurements, 118
and heat, 116–17
maximum transfer, 158–60, 280–82, 293–95
and Norton's theorem, 293–95
in parallel circuits, 185–86, 194, 214
in RC circuits, 687–91, 700, 708
in RL circuits, 539–45
in RLC circuits, 771
in series circuits, 137–38, 146–47, 210–12
and Thevenin's theorem, 280–82
Power factor (PF) calculations
impedance applications of, 544
in RC circuits, 687–88, 690–91, 700, 708
in RL circuits, 542–43, 545, 555
Power gain
bel, 806–8
dB, 808–13
Power ratings
of potentiometers, 71
of resistors, 67–68
Power supplies. *See* Dc power supplies
Power transfer
Norton's theorem for, 293–95
Thevenin's theorem for, 280–82
in transformers, 603–9
Powers of ten in scientific notation, 15–16
Precision capacitors, 661
Precision resistors
color code for, 879–80
temperature coefficients of, 59
tolerances of, 67
Prefixes, engineering notation, 21–22, 24
Primaries, transformer, 597
impedance of, 613–14
resistance of, 614
shorts, 624
Primary cells, 76
Product-over-sum method, 184
for capacitance, 645
for impedance, 546, 550–51, 697
for inductance, 498
for resistance, 184
Protons, 31, 35
Prototype circuits, 90–91
Pulse width of rectangular waves, 429
Purity in carbon-composition resistors, 58
Pythagorean theorem, 469–71

Q effects on shunt LC bandpass filters, 856–57
Quality (Q) rating
of capacitors, 654–57
of filters, 797–98, 849–51, 853–54
of inductors, 508–11
Quick-acting fuses, 88

Radians, 417–22
Radio frequency (rf) transformers, 598
Ramps in sawtooth waves, 434
Ratings
capacitor, 653–54
fuse, 86–87
insulator, 57
potentiometer, 71
and power, 116–17
resistivity, 47
resistor, 67–68
transformer, 618–19
RC circuits. *See also* Parallel RC circuits; Series RC circuits
admittance in, 908
power characteristics of, 687–93
series-parallel, 703–10
square wave response to. *See* Square waves
RC high-pass filters, 835
analysis of, 838–39
bode plots and roll-off rates for, 837–38
lower cutoff frequencies for, 836–37
maximum voltage gain of, 836, 839
operation of, 835–36
RC low-pass filters
analysis of, 828–29
maximum voltage gain of, 820–21, 828–29
operation of, 818–20
roll-off rates for, 825–28
upper cutoff frequencies of, 821–24
RC time constants, 716–18, 720–23. *See also* Square waves
Reactance
capacitive, 649–53
inductive, 501–5, 536–38, 557
in LC circuits
parallel, 745–48
series, 738–41
in RC circuits, 706
and resistance and impedance, 505
in RLC circuits
parallel, 765–66, 770
series, 758–61, 764–65
series-parallel, 771–73
vs. susceptance, 559
Reactive power, 507
in RC circuits, 687–88
in RL circuits, 539–40
in RLC circuits, 771
Rechargeable batteries, 78
Reciprocal method
for capacitance, 645
for impedance, 550, 697
for inductance, 498–99
for resistance, 183–84
Reciprocal relationships
in bel gain, 807
in dB gain, 817
Rectangular notation, 459–61
for admittance, 904–5
and complex numbers, 463–67
converting polar to, 471–72
converting to polar, 472–73
and imaginary numbers, 467–68
j-operator in, 461–63
polar equivalent representations, 468–69
vs. polar notation, 467
for RC circuits, 704–8
for RL circuits, 561–66

Rectangular waves, 428–29. *See also* Square waves
alternations and cycles in, 382
duty cycle of, 429–30
symmetrical and nonsymmetrical, 431–33
Rectifiers
and ac voltage measurements, 401–2
in dc power supplies, 620–21
Red resistor band, 63, 66–67
References
dBm, 812–13
voltage, 142–43
Regulators in dc power supplies, 620–21
Relative conductivity, 49
Relative permeability, 359
Relative permittivity, 637
Reliability, resistor color bands for, 66–67
Reluctance
and magnetism, 362
and Rowland's law, 364, 366
Replacing
fuses, 87
resistors, 68
Repulsion
and attraction, 33–34
by like magnetic poles, 353–54
Residual flux density, 370
Resistance, 8, 37, 43–44
calculating, 47–49, 183–84
with capacitive reactance, 649–50, 653
of capacitors, 649–50
in circuit calculation problems, 120–21
and current, 107
factors affecting, 47–48
of inductors, 501–2, 506
of leaky capacitors, 661–62
and maximum power transfer theorem, 158–60
measuring, 99–100, 235–36
meter, 148–50
Millman, 298–300
Norton, 291
Ohm's law, 14, 107–12
open circuits, 123
in parallel circuits, 182–85, 194, 212–15
in power calculations, 115–16
and reactance and impedance, 505
in RL circuits, 575–76
in series circuits, 133–34, 146–47, 208–12
in series-parallel circuits, 226–27
short circuits, 123–25
source, 156–57
temperature effects on, 49, 59
Thevenin, 272–74, 822
in transformers and transformer circuits, 605, 610–14, 617–18
variable. *See* Potentiometers
Wheatstone bridges for, 234–36
Resistive-capacitive circuits. *See* RC circuits
Resistive-inductive circuits. *See* RL circuits
Resistive-inductive-capacitive circuits. *See* RLC circuits
Resistive networks in Norton's theorem, 289–90
Resistive power, 507
in RC circuits, 687–88
in RL circuits, 539–41
in RLC circuits, 771
Resistivity, 47–48
Resistors, 8
alphanumeric codes for, 881
carbon-composition, 58–59
color code for, 62–67, 879–80
current-sensing, 552–54
heat dissipation by, 116–17
integrated, 60
metal-film, 59
power ratings of, 67–68
precision, 59, 67, 879–80
replacing, 68
schematic symbol for, 43
signs of excessive heat, 152–53
standard values of, 60–61, 879
tolerance of, 60–62, 66, 121–22
variable. *See* Potentiometers
in voltage dividers, 143–46
wire-wound, 59
Resolution, potentiometer, 73
Resonance, 749
Resonant frequency, 749–52
of LC circuits, 843–44
parallel, 755–57, 844–46
series, 754–55, 757, 844–46
oscilloscope effects on, 753–54
stray capacitance in, 753
stray inductance in, 752
Retentivity
and hysteresis, 368–70
and magnetism, 361
and transformer losses, 605–6
RF chokes, 512
RF transformers, 598
Rheostats, 107
Right triangles, 468–71
Ring magnets, 371–72
Rise and rise curve equations
for RC circuits, 714–15, 717–19
for RL circuits, 571, 573–74, 578–79
RL circuits, 491. *See also* Parallel RL circuits, Series RL circuits
admittance in, 905–8
algebraic and geometric sums in, 524
analysis of, 532–36, 544–45, 554–56, 566–67
characteristics of, 523–24, 546–47
current in, 523–24, 531–32, 546–49, 552–53, 565
frequency response of, 536–38, 557–59
impedance of, 523–24, 529–30, 532–36, 545–47, 549–51, 706–7
phase angles in, 524–29, 531, 549–53, 564–65
polar and rectangular forms for, 561–66
power calculations for, 539–45
series-parallel, 560–67
square wave response of. *See* Square waves
voltages, 525–27, 546–47, 565, 569–70
waveforms, 568–71
RL high-pass filters, 840
lower cutoff frequencies for, 841–42
maximum voltage gain of, 841
operation of, 840–41
RL low-pass filters, 829
maximum voltage gain in, 830–33
operation of, 829–30
roll-off rates for, 833–34
upper cutoff frequency for, 832
RL time constants, 575–78, 580–81. *See also* Square waves
RLC circuits. *See also* Parallel RLC circuits; Series RLC circuits
series-parallel, 771–81
RMS (root-mean-square) values
calculating, 395–97
measuring, 402
Roll-off, 825
in bandpass filters, 856–57
in high-pass filters, 837–38
in low-pass filters, 825–28, 833–34
Q effects on, 856–57
Root-mean-square (rms) values
calculating, 395–97
measuring, 402
Rotary switches, 85
Rotors
angular velocity of, 417–18
in interleaved-plate capacitors, 659–60
in sine wave generation, 405–6
Rowland's law, 364, 366
Rubber
breakdown voltage rating of, 57
as insulator, 45–46

Sawtooth waves, 433–34
Scalar values, 454
Scales, frequency, 802–4
Schematic diagrams, 8
Schematic symbols, 8, 13, 882–84
Scientific calculators
natural logs, 573
polar and rectangular notation, 473
Scientific notation, 14–16
converting to standard form, 18–19
solving problems in, 19–20

Secondaries, transformer, 597
impedance of, 614
voltages across, 601–3, 608–9, 614–16, 618–19
Secondary cells, 76
Self-inductance, 484
Semiconductors, 10, 12, 45
characteristics of, 46
integrated resistors, 60
schematic symbols for, 884
Series-aiding configurations
coils, 494–95
voltage sources, 160–62
Series bandpass filters, 846–48
bandwidth of, 850–52
cutoff frequencies for, 852–53
Series circuits
analyzing, 146–47
current through, 134–36, 208–12
Kirchhoff's voltage law for, 140–42
LC. *See* Series LC circuits
open-resistor fault symptoms, 147–51
vs. parallel circuits, 181–82
power, 137–38, 210–12
RC. *See* Series RC circuits
resistance, 133–34, 208–12
RL. *See* Series RL circuits
RLC. *See* Series RLC circuits
shorted-resistor fault symptoms in, 151–52
troubleshooting, 152–54
voltages, 136–37, 208–12
as voltage dividers, 143–46, 155–56
voltage references for, 142–43
Series connections
battery, 79–80
capacitor, 644–47, 652–53
coil, 494–96
voltage sources, 160–64
Series-equivalent circuits, 217–19
Series LC circuits, 735–36
analysis of, 739–41
as bandpass filters, 846–48, 850–53
as notch filters, 857–58
phase relationships in, 735–41
reactance of, 738–41
resonant, 754–55, 757, 844–46
Series-opposing configurations
coils, 495–96
voltage sources, 162–64
Series-parallel circuits
analyzing, 216–17, 220–24
equivalent, 217–20, 269–71
fault symptoms in, 240–42
load resistance, 226–27
no-load vs. loaded output voltage, 225–26
open loads in, 230
parallel circuits in series, 212–15
RC, 703–4
analysis of, 706–9
polar and rectangular forms for, 704–8
RL, 560–61
analysis of, 566–67
polar and rectangular forms for, 561–66
RLC, 771–81
series circuits in parallel, 208–12
Thevenin equivalent of, 269–71
troubleshooting, 243–44
voltage dividers in
loaded, 224–25
stability of, 227–30
variable, 238–40
voltmeter loading on, 230–32
Wheatstone bridge, 233–38
Series RC circuits, 672
analysis of, 680–84, 692
current in, 672–73, 680
frequency response of, 684–87
impedance of, 672–73, 678–84
phase angles in, 673–80
vs. series RL circuits, 686
voltages in, 672–76
Series RL circuits
algebraic and geometric sums in, 524
analysis of, 532–36, 544–45
characteristics of, 523–24
current through, 523–24, 531–32, 545
frequency response of, 536–38
impedance of, 523–24, 529–30, 532–36, 545
phase relationships in, 524–29, 531–32
power calculations for, 539–45
vs. series RC circuits, 686
voltages in, 523–27
Series RLC circuits, 758
analysis of, 762–65
frequency response of, 761–62
impedance of, 758–61, 764–65
Service meters, 118
Shelf life, battery, 77
Shells, 31–32, 46
Shielding
magnetic, 370–71
transformer, 622–23
Short circuits
in breadboard circuits, 94
characteristics of, 123–25
in parallel circuits, 196
in series circuits, 151–52
in series-parallel circuits, 241–42
Shorted windings in transformers, 621, 623–24
Shunt components, 819
Shunt LC bandpass filters, 848–49
frequency analysis of, 853–55
maximum voltage gain of, 855–56
Q effects on, 856–57
Shunt LC notch filters, 858–59
Siemens, 43
Signal delays in digital circuits, 581–83
Signals, 382
Silicon, 46
Silver
conductivity of, 49
disadvantages of, 54–55
resistivity rating of, 47
Silver resistor band, 64, 66
Simultaneous equations, 318–19
alter-and-add method for, 320–23
determinants for, 323–27
matrices for, 323–29
substitution method for, 319–20
Sine (sin) functions, 469–71
Sine waves
alternations and cycles in, 382
generating, 405–7
harmonics of, 427–28
inductor phase relationships, 489–90
instantaneous values in, 413–18, 421–22
and magnetic induction of current, 404–5
phase and time measurements, 408–11, 417–22
phase angle measurements, 412–13
phase angles, 411–12
phase in, 407–8
wavelength of, 425–27
Single-pole, double-throw (SPDT) switches, 84
Single-pole, single-throw (SPST) switches, 84
Single-throw switches, 83–84
Sinusoidal rate changes, 415
Size of wire, 55–56
Slopes in sawtooth waves, 434
Slow-blow fuses, 88
Solder, 125
Solder bridges, 125
Solid wires, 54
Sources, 122, 188–90
analysis of, 263–64
converting, 266–68
equivalent, 264–66
maximum power transfer theorem, 158–60
Millman's theorem, 297–300
in RC circuits, 676–80
resistance of, 156–57
in RL circuits, 527–29
schematic symbols for, 13
series-connected, 160–64
South-seeking poles, 353
Space width, 429
SPDT (single-pole, double-throw) switches, 84
Speed of light, 426

SPST (single-pole, single-throw) switches, 84
Square waves, 430–31. *See also* Rectangular waves
in RC circuits
decay curve for, 715–16, 718–19
RC time constant, 716–18, 720–23
rise curve for, 714–15, 717–19
universal curves for, 713–15, 718–20
waveforms, 710–13
in RL circuits, 567
decay curve for, 574–75, 578–79
inductor voltage, 569–70
rise curve for, 573–74, 578–79
RL time constant, 575–78, 580–81
time calculations for, 581–83
time intervals for, 575
universal curves for, 572–74, 578–79
voltage and current transitions in, 568
waveforms, 568–71
Stability of voltage dividers, 227–30
Standard form, converting to
dB power gain, 809–10
dB voltage gain, 816–17
scientific notation, 18–19
Standard resistor values, 60–61, 879
Static values in ac circuits, 423–24
Stators in interleaved-plate capacitors, 659–60
Steel
magnetic induction, 360–61
retentivity of, 361
Step-down transformers, 599
autotransformers, 622
characteristics of, 602, 608–10, 614–15
Step-up transformers, 599
autotransformers, 622
characteristics of, 602–3, 608–10, 615–16
Storing magnets, 372
Stranded wires, 54
Stray capacitance, 753
Stray inductance, 752
Stressed components, 152
Substitution method for simultaneous equations, 319–20
Subtraction
of complex numbers, 463–64
in polar notation, 458–59
Superposition, 256
Superposition theorem, 257–63
Surges, fuses for, 88
Susceptance, 559, 903–4
in RC circuits, 908
in RL circuits, 905–8
Switches, 83–84
DIP, 85
normally open and normally closed, 84–85
rotary, 85
schematic symbols for, 13
Switching circuits
square waves in, 567
time calculations for, 581–83
Symmetrical waveforms, 431–33

T circuits, 300–304
Tangent (tan) functions, 469–71
Taper of potentiometers, 71
Technicians, 2–3
Teflon
breakdown voltage rating of, 57
for capacitors, 657–58
Telecommunications, 4
Temperature
and aluminum, 55
converting, 877
and magnets, 372
and resistance, 49, 59
Temporary magnets, 361
Tera- prefix, 21–22
Terminals
battery, 41–42, 75
power supply, 82
transformer, 597
Teslas, 356
Test equipment, 5–7
Thermal energy, free electrons from, 37
Thermal-type circuit breakers, 90
Thermistors, 13, 884
Thermomagnetic-type circuit breakers, 90
Thevenin's theorem, 268–69
bridge circuits, 287–88
conversion to Norton, 295–97
equivalent circuits, 274–78
load voltage ranges, 279–80
maximum power transfer, 280–82
multiload circuits, 285–87
multisource circuits, 282–84
purpose of, 269–71
RC low-pass filters, 822
Thevenin resistance in, 272–74, 822
Thevenin voltage in, 271–72
3-by-3 matrices, 323, 327–29
3 dB points, 825
Throws in switches, 83–84
Thyristors, 884
Time base controls, 386–88
Time constants
in RC circuits, 716–18, 720–23
in RL circuits, 575–78, 580–81
Time delay fuses, 88
Time intervals
in RL time constants, 575
in universal curve equations, 578–79
Time measurements for sine waves, 408–11, 417–22
Time periods in ac waveforms, 380–85
Tolerance ratings
for capacitors, 637
for resistors, 60–62, 66, 121–22
Toroids, 511
Traces, PC board, 56–57, 154
Transducers, 116
Transformers
autotransformers, 621–23
center-tapped, 619–20
color coding for, 624, 880–81
construction of, 597
current in, 607–16, 619
faults in, 623–24
for impedance matching, 617–18, 622–23
load effects on, 610–16
with multiple outputs, 620–21
operation of, 599–600
phase relationships in, 603–4
power transfer in, 603–9
schematic symbols for, 597
shielding, 622–23
turns ratio in, 600–603, 609–10, 621–22
voltage classifications for, 597, 599
voltages in, 601–3, 608–10, 614–16, 618–19
Transistors, schematic symbols for, 884
Transitions, square wave, 568–69
Transmitter-type interleaved-plate capacitors, 661
Triangles, right, 468–71
Triangular waveforms
alternations and cycles, 382
characteristics of, 435
Trig functions, 469–71
Trimmer capacitors, 661
Trimmer potentiometers, 72
Troubleshooting
Ohm's law for, 111–12
parallel circuits, 197–98
series circuits, 152–54
series-parallel circuits, 243–44
True ac waveforms, 389–90
True power vs. apparent power, 506–8
Tubes, 10–11
Tuned circuits, 661
Turns ratio, transformer, 600–601
autotransformers, 621–22
identifying transformers by, 603
and secondary voltages, 601–3, 609–10
2-by-2 matrices, 323
Two-stage filters, 811

Units, 13
for capacitance, 636–37
conversions of, 874–77
international system of, 877–78
Unity coupling, 492–93

Unity gain, 806
Universal curves
  for RC circuits, 713–15, 718–20
  for RL circuits, 572–76, 578–79
Unlike poles, 353–54
Upper cutoff frequencies
  RC, 821–24
  RL, 832
Utility power, 118

VA (volt-amperes) units, 507
Vacuum tubes, 10–11
Valence electrons, 32, 45
Valence shells, 32, 46
VAR (volt-amperes-reactive) units, 507
Variable capacitors, 659–61
Variable resistors. *See* Potentiometers
Variable voltage dividers, 155–56, 238–40
Vectors, 454. *See also* Phasors in RL circuits
  complex numbers for, 463–68
  j-operator for, 461–63
  notation conversions for, 468–74
  plotting, 454–55
  polar notation for, 456–59
  rectangular notation for, 459–68
  representing, 455
  rotation of, 461
Velocity, angular
  in capacitive reactance calculations, 651
  of rotors, 417–18
Vertical sensitivity, oscilloscope, 398–99
Violet resistor band, 63
Volt-amperes-reactive (VAR) units, 507
Volt-amperes (VA) units, 507
Volt-ohm-milliammeters (VOMs)
  characteristics of, 5–6
  loading by, 232
Voltage, 37, 41–42
  in ac circuits, 379–81
  average, 401–2
  battery, 79
  capacitor, 638–43
  in circuit calculation problems, 120–21
  and current, 42, 106–7, 110–12
  dc power supplies for, 80–83
  in delta and wye circuits, 300–301
  Faraday's laws, 482–84
  induced, 486–88
  Kirchhoff's voltage law, 140–42
  in LC circuits
    parallel, 742–43, 747–48
    resonant, 844–46
    series, 735–36, 739–41
  Lenz's law, 484–85
  measuring, 99
  in multisource circuits, 256–63
  node voltage analysis, 343–46
  offset, 424–25, 640
  Ohm's law, 14, 106–12
  across open circuits, 123
    parallel, 195–96
    series, 147–51
  across parallel circuits, 180–81, 195–96, 212–15
  peak and peak-to-peak, 397–400
  polarity of, 136, 140–42
  in power calculations, 113–16
  in RC circuits, 710–12, 720–21
    parallel, 693–94, 699–700
    series, 672–76, 680–84, 692
    series-parallel, 708–9
  rectangular waves, 428–29, 433–34
  references for, 142–43
  in resonant circuits, 844–46
  in RL circuits, 569–70
    parallel, 546–47, 554–56
    series, 523–27, 532–36, 545
    series-parallel, 565–67
  in RLC circuits
    series, 764–65
    series-parallel, 773–80
  sawtooth waves, 433–34
  in series circuits, 136–37, 146–52, 208–12
  short circuits, 124–25
    parallel, 196
    series, 151–52
  square waves, 430–31
  Thevenin, 271–72
  in transformers, 601–3, 608–10, 614–16, 618–19
  voltage dividers for. *See* Voltage dividers
Voltage classifications for transformers, 597, 599
Voltage/current relationships. *See* Phase angles and relationships
Voltage-divider equation, 144
Voltage dividers, 143–46
  vs. current dividers, 190–91
  loaded, 224–25
  potentiometers as, 155–56, 238–40
  in RL circuits, 533–34
  stability of, 227–30
  variable, 238–40
Voltage drops, 140–41
Voltage gain, 811–12
  of bandpass filters
    series, 847–49
    shunt, 855–56
  dB, 814–18
  of high-pass filters
    RC, 836, 839
    RL, 841
  of low-pass filters
    RC, 820–21, 828–29
    RL, 830–33
  of notch filters, 858–59
  Q effects on, 856–57
Voltage ratings
  for capacitors, 653–54
  for fuses, 86–87
  for insulators, 57
Voltage rises, 140–41
Voltage sources
  analysis of, 263–64
  converting, 266–68
  equivalent current sources, 264–66
  Millman's theorem, 297–300
  resistance of, 156–57
  series-connected, 160–64
Voltage transitions in square waves, 568
Voltmeters, 97
  for ac voltage, 401–2
  loading by, 230–32
Volts, 42
Volts/division control, 398–99
Volume controls, 71
VOMs (volt-ohm-milliammeters)
  characteristics of, 5–6
  loading by, 232

Wattage ratings of resistors, 67–68
Watts, 113. *See also* Power
Waveforms, 379
  full-cycle average of, 391
  half-cycle average of, 391–93
  out of phase, 411–13
  rectangular, 428–33. *See also* Square waves
  sawtooth, 433–34
  sine. *See* Sine waves
  symmetrical and nonsymmetrical, 431–33
  true ac, 389–90
Wavelength, 425–27
Webers
  Faraday's laws, 483–84
  magnetic flux, 354–55
Webers per ampere-turn meters, 367
Wet cells, 76
Wheatstone bridges
  construction of, 233
  in meter circuits, 237
  operation of, 233–34
  resistance measurements with, 235–36
  resistance ratios in, 234–35
White resistor band, 63
Winding resistance
  effects of, 540–41
  of inductors, 506, 511
Wiper arms in potentiometers, 70
Wire, 54–55
  for breadboards, 92

Wire, *continued*
current capacity of, 55–56
size of, 55–56
Wire-wound potentiometers, 71–72
Wire-wound resistors, 8, 59
Wye circuits, 300–304

X-axes
on oscilloscope screens, 386
in universal curves, 572
X-coordinates in rectangular notation, 460
X-matrices, 324, 327

Y-axes
on oscilloscope screens, 386
in universal curves, 572
Y-coordinates in rectangular notation, 460
Y-matrices, 324–25, 327
Yellow resistor band, 63, 67

Z-matrices, 327
Zinc for battery terminals, 75–76